Methods in Cell Biology

VOLUME 79

Cellular Electron Microscopy

Series Editors

Leslie Wilson

Department of Molecular, Cellular and Developmental Biology
University of California
Santa Barbara, California

Paul Matsudaira

Whitehead Institute for Biomedical Research
Department of Biology
Division of Biological Engineering
Massachusetts Institute of Technology
Cambridge, Massachusetts

Methods in Cell Biology

VOLUME 79

Cellular Electron Microscopy

Edited by

J. Richard McIntosh

Boulder Laboratory for 3D Electron Microscopy of Cells
Department of MCD Biology
University of Colorado
Boulder, Colorado

ELSEVIER

AMSTERDAM • BOSTON • HEIDELBERG • LONDON
NEW YORK • OXFORD • PARIS • SAN DIEGO
SAN FRANCISCO • SINGAPORE • SYDNEY • TOKYO
Academic Press is an imprint of Elsevier

Front Cover Image: Three-dimensional images of a flagellum from the alga, *Chlamydomonas reinhardtii*, constructed by electron tomography of frozen-hydrated samples and modeled with computer graphics. See Chapter 17 for details and discussion. Reprinted with modifications from Nicastro *et al*., Science 313:5789 (2006) by permission of the authors and the publisher.

Academic Press is an imprint of Elsevier
525 B Street, Suite 1900, San Diego, California 92101-4495, USA
84 Theobald's Road, London WC1X 8RR, UK

This book is printed on acid-free paper. ♾

For information on all Academic Press publications visit our Web site at www.books.elsevier.com

ISBN-13: 978-0-12-370647-8
ISBN-10: 0-12-370647-5

PRINTED IN THE UNITED STATES OF AMERICA
07 08 09 10 9 8 7 6 5 4 3 2 1

CONTENTS

PART II Imaging Fixed Cells in Three Dimensions

PART III Imaging Frozen-Hydrated Cells and Cell Parts

PART V Aspects of Data Collection and Analysis

CONTRIBUTORS

Numbers in parentheses indicate the pages on which the authors' contributions begin.

Ashraf Al-Amoudi* (385), Laboratory for Ultrastructural Analysis, Biophore, University of Lausanne, CH-1015 Lausanne, Switzerland

C. Antony (145), Electron Microscopy Facility, Cell Biology and Biophysics Program, European Molecular Biology Laboratories, 69117 Heidelberg, Germany

Maria A. Aronova (593), National Institute of Biomedical Imaging and Bioengineering, National Institutes of Health, Bethesda, Maryland 20892

Christopher P. Arthur (241), Department of MCD Biology, University of Colorado, Boulder, Colorado 80309

Jotham R. Austin, II (221), Biological Science Division-OSRF, University of Chicago, Chicago, Illinois 60637

Wolfgang Baumeister (615), Max Planck Institute of Biochemistry, Am Klopferspitz 18, D-82152 Martinsried, Germany

A. Beorchia (559), DTI, UMR 6107, UFR de Sciences, BP 1039, 51687 Reims Cédex, France

Christoph Best (615), Max Planck Institute of Biochemistry, Am Klopferspitz 18, D-82152 Martinsried, Germany

Cédric Bouchet-Marquis (385), Laboratory for Ultrastructural Analysis, Biophore, University of Lausanne, CH-1015 Lausanne, Switzerland

James C. Bouwer (643, 721), University of California, San Diego, La Jolla, California 92093

Yifan Cheng (463), Department of Biochemistry and Biophysics, University of California-San Francisco, San Francisco, California 94143

T. Cheutin (559), Institut de Génétique Humaine, CNRS UPR 1142, 34396 Montpellier Cédex 5, France

Wah Chiu (407), National Center for Macromolecular Imaging, Baylor College of Medicine, Houston, Texas 77030; Verna and Marrs McLean Department of Biochemistry and Molecular Biology, Baylor College of Medicine, Houston, Texas 77030

Thomas J. Deerinck (575, 643), National Center for Microscopy and Imaging Research and Center for Research in Biological Systems, University of California, San Diego, La Jolla, California 92093

* Present address: EMBL, Pf. 10.2209, D-69012 Heidelberg.

Yimin Dong (259), Wadsworth Center, New York State Department of Health, Albany, New York 12201; Department of Biomedical Sciences, School of Public Health, State University of New York, Albany, New York 12201

Jacques Dubochet (7, 385), Laboratory for Ultrastructural Analysis, Biophore, University of Lausanne, CH-1015 Lausanne, Switzerland

Susan K. Dutcher (125), Department of Genetics, Washington University School of Medicine, St. Louis, Missouri 63110

Fred Duttweiler (721), University of California, San Diego, La Jolla, California 92093

Mark H. Ellisman (575, 643, 721), National Center for Microscopy and Imaging Research and Center for Research in Biological Systems, University of California, San Diego, La Jolla, California 92093

Mikhail Eltsov (385), Laboratory for Ultrastructural Analysis, Biophore, University of Lausanne, CH-1015 Lausanne, Switzerland

Dudley S. Finch (241), Department of MCD Biology, University of Colorado, Boulder, Colorado 80309

Friedrich Förster (741), Department of Pharmaceutical Sciences, University of California, San Francisco, California 94143

Guido M. Gaietta (575), National Center for Microscopy and Imaging Research and Center for Research in Biological Systems, University of California, San Diego, La Jolla, California 92093

Thomas H. Giddings, Jr. (125), Boulder Laboratory for 3D Electron Microscopy of Cells, Department of Molecular, Cellular, and Developmental Biology, University of Colorado, Boulder, Colorado 80309

Ben N. G. Giepmans (575), National Center for Microscopy and Imaging Research and Center for Research in Biological Systems, University of California, San Diego, La Jolla, California 92093

J. L. Höög (145), Electron Microscopy Facility, Cell Biology and Biophysics Program, European Molecular Biology Laboratories, 69117 Heidelberg, Germany

Reiner Hegerl (741), Max-Planck Institute of Biochemistry, Am Klopferspitz 18, D-82152 Martinsried, Germany

Michael W. Hess (57), Division of Histology and Embryology, Innsbruck Medical University, A-6020 Innsbruck, Austria

Andreas Hoenger (437), Department of Molecular, Cellular, and Developmental Biology, University of Colorado, Boulder, Colorado 80309

Kathryn E. Howell (543), Department of Cellular and Structural Biology, University of Colorado Medical School, Aurora, Colorado 80045

Anthony Hyman (101), Max Planck Institute of Molecular Cell Biology and Genetics (MPI-CBG), 01307 Dresden, Germany

Heather Jiles (799), National Center for Microscopy and Imaging Research, Center for Research in Biological Systems, University of California, San Diego, La Jolla, California 92093

Liang Jin (721), University of California, San Diego, La Jolla, California 92093

Ying Z. Jones (643), University of California, San Diego, La Jolla, California 92093

H. Kaplan (559), IFR53, 51096 Reims Cédex, France

Stuart Kleinfelder (721), University of California, Irvine, Irvine, California 92697

Mark S. Ladinsky (543), Boulder Laboratory for 3D Electron Microscopy of Cells, Department of MCD Biology, University of Colorado, Boulder, Colorado 80309

Albert Lawrence (643), University of California, San Diego, La Jolla, California 92093

Richard D. Leapman (593), National Institute of Biomedical Imaging and Bioengineering, National Institutes of Health, Bethesda, Maryland 20892

Philippe Leblanc (721), University of California, San Diego, La Jolla, California 92093

Shengdong Li (721), University of California, Irvine, Irvine, California 92697

Françoise Livolant (385), Laboratoire de physique des solides, Université Paris sud, F-91 405 Orsay, Cedex, France

Steven J. Ludtke (407), National Center for Macromolecular Imaging, Baylor College of Medicine, Houston, Texas 77030; Verna and Marrs McLean Department of Biochemistry and Molecular Biology, Baylor College of Medicine, Houston, Texas 77030

Mason R. Mackey (643), University of California, San Diego, La Jolla, California 92093

Cem Mangir (799), National Center for Microscopy and Imaging Research, Center for Research in Biological Systems, University of California, San Diego, La Jolla, California 92093

Brad J. Marsh (193), Institute for Molecular Bioscience, Queensland Bioscience Precinct, and Centre for Microscopy and Microanalysis, and School of Molecular and Microbial Sciences, The University of Queensland, Brisbane, Queensland 4072, Australia

Mark Marsh (515), Cell Biology Unit, MRC Laboratory for Molecular Cell Biology and Department of Biochemistry and Molecular Biology, University College London, London WC1E 6BT, United Kingdom

Maryann E. Martone (643, 799), National Center for Microscopy and Imaging Research, Center for Research in Biological Systems, University of California, San Diego, La Jolla, California 92093

Kent McDonald (23, 101), Electron Microscope Laboratory, University of California, Berkeley, California 94720

Bruce F. McEwen (259), Wadsworth Center, New York State Department of Health, Albany, New York 12201; Department of Biomedical Sciences, School of Public Health, State University of New York, Albany, New York 12201

Anna-Clare Milazzo (721), University of California, San Diego, La Jolla, California 92093

Paul Mooney (661), Gatan, Inc., Pleasanton, California 94588

Mary K. Morphew (493), Boulder Laboratory for 3D Electron Microscopy of Cells, Department of Molecular, Cellular and Developmental Biology, University of Colorado, Boulder, Colorado 80309

Thomas Müller-Reichert (101), Max Planck Institute of Molecular Cell Biology and Genetics (MPI-CBG), 01307 Dresden, Germany

Daniela Nicastro (437), Department of Molecular, Cellular, and Developmental Biology, University of Colorado, Boulder, Colorado 80309

Stephan Nickell (615), Max Planck Institute of Biochemistry, Am Klopferspitz 18, D-82152 Martinsried, Germany

M. F. O'Donohue (559), Unité MéDIAN, CNRS UMR 6142, UFR de Pharmacie, 51096 Reims Cédex, France

Eileen T. O'Toole (101,125), Boulder Laboratory for 3D Electron Microscopy of Cells, Department of Molecular, Cellular, and Developmental Biology, University of Colorado, Boulder, Colorado 80309

Marisa S. Otegui (221), Department of Botany, University of Wisconsin, Madison, Wisconsin 53706

Maria Pagratis (241), Department of MCD Biology, University of Colorado, Boulder, Colorado 80309

Annegret Pelchen-Matthews (515), Cell Biology Unit, MRC Laboratory for Molecular Cell Biology and Department of Biochemistry and Molecular Biology, University College London, London WC1E 6BT, United Kingdom

Steven T. Peltier (643, 721), University of California, San Diego, La Jolla, California 92093

Peter J. Peters (373), Division of Tumor Biology, The Netherlands Cancer Institute, Amsterdam, The Netherlands

D. Ploton (559), Unité MéDIAN, CNRS UMR 6142, UFR de Pharmacie, 51096 Reims Cédex, France

David L. Potter (241), Department of MCD Biology, University of Colorado, Boulder, Colorado 80309

Mary C. Reedy (321), Department of Cell Biology, Duke University Medical Center, Durham, North Carolina 27710

Michael K. Reedy (321), Department of Cell Biology, Duke University Medical Center, Durham, North Carolina 27710

Charles E. Robertson (169), Department of Molecular, Cellular and Developmental Biology, University of Colorado, Boulder, Colorado 80309

Kristian Sandberg (769), Boulder Laboratory for 3D Electron Microscopy of Cells, Department of Molecular, Cellular, and Developmental Biology, University of Colorado, Boulder, Colorado 80309

Joy Sargis (799), National Center for Microscopy and Imaging Research, Center for Research in Biological Systems, University of California, San Diego, La Jolla, California 92093

C. Sauvage (559), Unité MéDIAN, CNRS UMR 6142, UFR de Pharmacie, 51096 Reims Cédex, France

David B. Serrell (241), Department of MCD Biology, University of Colorado, Boulder, Colorado 80309

Irina I. Serysheva (407), National Center for Macromolecular Imaging, Baylor College of Medicine, Houston, Texas 77030; Verna and Marrs McLean Department of Biochemistry and Molecular Biology, Baylor College of Medicine, Houston, Texas 77030

Gina E. Sosinsky (575), National Center for Microscopy and Imaging Research and Center for Research in Biological Systems, University of California, San Diego, La Jolla, California 92093

Martin Srayko (101), Max Planck Institute of Molecular Cell Biology and Genetics (MPI-CBG), 01307 Dresden, Germany

Michael H. B. Stowell (241), Department of MCD Biology, University of Colorado, Boulder, Colorado 80309

Sriram Subramaniam (373), Laboratory of Cell Biology, National Cancer Institute, NIH, Bethesda, Maryland 20892

Tatyana Svitkina (295), Department of Biology, University of Pennsylvania, Philadelphia, Pennsylvania 19104

Kenneth A. Taylor (321), Institute of Molecular Biophysics, Florida State University, Tallahassee, Florida 32306

P. Tchélidzé (559), Unité MéDIAN, CNRS UMR 6142, UFR de Pharmacie, 51096 Reims Cédex, France

Masako Terada (643), University of California, San Diego, La Jolla, California 92093

Joshua Tran (799), National Center for Microscopy and Imaging Research, Center for Research in Biological Systems, University of California, San Diego, La Jolla, California 92093

Kristin J. VandenBeldt (259), Wadsworth Center, New York State Department of Health, Albany, New York 12201

Willy Waiho Wong (799), National Center for Microscopy and Imaging Research, Center for Research in Biological Systems, University of California, San Diego, La Jolla, California 92093

Thomas Walz (463), Department of Cell Biology, Harvard Medical School, Boston, Massachusetts 02115

Robert M. Weis (373), Department of Chemistry, University of Massachusetts, Amherst, Massachusetts 01003

Shenping Wu (321), Institute of Molecular Biophysics, Florida State University, Tallahassee, Florida 32306

Nguyen-Huu Xuong (721), University of California, San Diego, La Jolla, California 92093

Peijun Zhang (373), Laboratory of Cell Biology, National Cancer Institute, NIH, Bethesda, Maryland 20892

Benoît Zuber (385), Laboratory for Ultrastructural Analysis, Biophore, University of Lausanne, CH-1015 Lausanne, Switzerland

INTRODUCTION TO ELECTRON MICROSCOPY OF CELLS

Cells have been an important focus of biological research since shortly after their discovery in the seventeenth century. Over the many subsequent years, our understanding of cellular structures has grown as a direct result of the technologies available for their study. During the first century of work, observations with simple microscopes by Leeuwenhoek, Hooke, Malpighi, and others revealed the surprising variety of cell shape and behavior. With the development of achromatic lenses and appropriate condensers for specimen illumination, biologists in the nineteenth century began to see into cells more clearly and to detect the remarkable complexity of their internal organization. Improved fixatives and stains gradually exposed subcellular compartmentalization and the complexity of cytoplasmic fiber systems. These methods also allowed inferences about time-dependent cell behaviors, like Flemming's description of nuclear reorganization at mitosis (summarized in Paweletz, 2001), which were confirmed by Lauterborn's careful studies of living diatoms (reprinted by Pickett-Heaps *et al.*, 1984). From work like this, students of cell structure became convinced that there was a world of important, subcellular organization waiting to be understood, once the tools and methods were up to the task.

Other scientists were, however, skeptical about the microscopic views of cellular organization. These included microbiologists, geneticists, and biochemists who preferred to think of cells as well-stirred compartments. Since cells were too small to interrogate effectively with the tools then available, cytoplasmic structural complexity was conveniently ignored, particularly if one was interested by inheritance, not by cell function or behavior. Likewise, if you began your experiments by homogenizing the cells you were about to study, you could not believe too strongly in the importance of their global organization. Fortunately, however, pioneers in the study of cell structure persisted. Ultimately images obtained with polarized light (Inoue, 1953) and with Zernicke phase contrast microscopy demonstrated the reality of some of the debated intracellular structures, like the fibrous components of the mitotic spindle. This work justified the faith in complex cell organization until electron microscopy (EM) became available and demonstrated the full intricacy of a cell's interior.

The era from the first informative electron micrograph of a cell (Porter *et al.*, 1945) until the mid-1970s was a "Golden Age" for cellular EM. Almost every new cell or tissue examined revealed a novel aspect of cytoplasmic or nuclear organization, and careful work could often relate these structural features to aspects of cell function. The similarity of the finest cellular details to the structures visualized by microscopists working on isolated macromolecules encouraged the belief that the EM was helping to close the gap between optical microscopy of cell architecture and biochemical studies of macromolecules; cell biologists were now seeing macromolecules fixed *in situ*, presumably in the act of doing their biological job.

Despite such progress, the concern persisted that the methods required to get cells ready for EM (e.g., fixation, dehydration, embedding in plastic for sectioning, and staining with heavy metals) were altering cellular structure; not a few biologists thought that the results from EM were unreliable. During this same period, however, alternative preparative methods provided independent evidence for the validity of at least some of the cellular EM that was being done on fixed material. Preparations based on rapid freezing, pioneered in cell biology by Hans Moor (Moor, 1964, 1973), produced images of cells and organelles that resembled samples prepared with more commonly used techniques, suggesting that the growing understanding of cell structure was likely to be valid. Methods for dehydrating and fixing previously frozen samples by freeze-substitution, provided an alternative both to freeze-fracture and conventional chemical fixation (Howard and Aist, 1979; Kellenberger, 1991; Zalokar, 1966). Comparisons between cells studied by freeze-fracture and by freeze-substitution fixation encouraged confidence that EM images could provide accurate insights into the fundamentals of biological organization (Linder and Staehelin, 1979). Likewise, comparisons by low angle X-ray diffraction of live muscle cells and those fixed by rapid freezing and freeze-substitution have confirmed that careful fixation preserves cellular detail to a resolution of about 2 nm (Sosa *et al.*, 1994). Such evidence led many cell biologists to accept the images from high-quality EM as an accurate representation of cell structure.

All this good work was, however, self-terminating. With time, the collection of cell images became so complete that the organelles and intracellular organization they revealed became familiar; images of the Golgi complex or the microtubule cytoskeleton in yet another cell became commonplace. Many biologists came to look down on studies of cellular "fine structure" as repetitive and uninformative. In their view we already knew what the EM could tell us about cell organization, so structural study itself was uninteresting. It only deserved attention if it was linked through experiment to the functions of the organelles observed. Even with the advent of powerful methods for localizing particular gene products in EM images (Tokuyasu, 1980), the focus of cell biologists turned elsewhere, and EM took a back seat.

During the 1980s and 1990s, structural cell biology blossomed as a result of improvements in light microscopy. First, immunofluorescence provided a comparatively quick and reliable way to localize macromolecules of interest within the structure of a reasonably well-preserved cell (Coons and Kaplan, 1950; Lazarides and Weber, 1974). Hybridization of labeled nucleic acid probes with cellular oligonucleotide sequences provided an analogous advance for biologists interested in the cellular organization of gene sequences and RNA molecules (Gall and Pardue, 1969). Then technological advances from outside cell biology began to make a huge difference to what could be seen in the light microscope. Electronic cameras with greatly increased sensitivity enabled live-cell imaging of preparations that contained fluorescent reporters of molecular behavior (Taylor and Wang, 1980). Better lenses that could gather and transmit a greater fraction of the light from faint objects coupled with better devices for digitizing and storing image data made this work both more fruitful and cheaper. Most important, the identification and manipulation of a readily expressed gene whose product was a green fluorescent protein, now widely known as "GFP" (Chalfie *et al.*, 1994), allowed molecular biologists to prepare specimens whose structures they could visualize and analyze in living cells, gaining insight into the function of their favorite gene. The subsequent development of alleles, homologues, and analogues of GFP has led to a tremendous surge in the interest and application of light microscopy to problems in molecular and developmental cell biology (Shaner *et al.*, 2005; Tsien, 1998), particularly for the cytological study of "model organisms," like yeasts, fruit flies, and nematode worms, whose molecular biology has recently seen so much progress. Thanks to all these practical and technological advances, scientists with little specialized training in light microscopy can now obtain high-quality structural data from live cells in real time.

Meanwhile, improved setups for differential interference contrast and polarization optical contrast (Oldenbourg and Mei, 1995) have further extended the power of live-cell imaging. Scanning confocal microscopy has opened new approaches to component localization with fluorescent labels (White *et al.*, 1987) and a continuous stream of innovations [multiphoton imaging (Flesken-Nikitin *et al.*, 2005), spectroscopic analysis of fluoresced light (Koster and Fraser, 2004), second and higher harmonic imaging (Dombeck *et al.*, 2004), and structured illumination (Gustafsson, 2005)] have provided novel and significant ways of getting more information about cell structure. As a result, many biologists are now deeply involved with the study of cell organization. Often, however, the spatial resolution of the light microscope leaves important questions unanswered. EM can in principle solve this problem and provide the needed extension of current work.

Until recently there has been a perception that all cellular structures that could be seen by EM had already been described. "Times are now a-changing," however, thanks to several innovations in EM technologies. The most important of these has been the realization that cells can be frozen so fast that their structure is preserved with something close to perfect fidelity (Dubochet *et al.*, 1983; Moor, 1964). Rapid freezing has been shown to preserve protein structures with atomic detail (Taylor and Glaeser, 1974); these can then be visualized by EM of thin protein crystals (Nogales *et al.*, 1997). Frozen-hydrated organelles and viruses can all be imaged by EM with striking clarity (Dubochet *et al.*, 1988), suggesting that this is a truly reliable way to prepare biological specimens for EM. Indeed, rapid freezing preserves cells so well that on thawing a large fraction of them will grow and divide (McIntosh, 2001; Moor, 1964). These results have encouraged a resurgence of interest in methods for imaging frozen-hydrated cells (Al-Amoudi *et al.*, 2004) and for fixing frozen cells by liquid substitution at low temperatures, the technique of freeze-substitution fixation (Kellenberger, 1991; McDonald and Morphew, 1993).

Thus, there has been fundamental progress in making samples for study by EM that circumvent, or at least ameliorate, the long-standing criticism that EM images of cells are fundamentally limited by fixation artifact.

More recently still, innovations in the hardware and software for generating and analyzing EM images have moved forward briskly. A new generation of microscopes uses electron sources with improved coherence and brightness (field emission guns); these significantly enhance the detail that is visible in low contrast objects, like frozen-hydrated biological samples. Computer-controlled microscope stages have improved the efficiency of imaging, especially for tedious tasks like the collection of many projections from a single sample over a wide range of tilts. Improved electronic cameras have eased the process of image recording and digitization, and filters that remove inelastically scattered electrons have improved the phase contrast imaging of comparatively thick specimens. All these innovations have truly revolutionized EM of large macromolecular complexes, a field that has now become a major methodology for the analysis of macromolecular organization (Frank, 2006; Frank *et al.*, 2002; van Heel *et al.*, 2000). The same advances are also providing new ways of looking at cells (Baumeister, 2002; McIntosh *et al.*, 2005). Many of these innovations are described in the chapters of this book, showing with clarity why they have initiated a resurgence in cellular EM.

One must ask, however, if the light microscope is so good, and particularly since it allows the imaging of live cells, why should one be interested in data from a more cumbersome and expensive technology? The principal answer lies in the space resolutions of these two imaging modalities. The light microscope improves on the eye by ~1000-fold, and the EM improves on the light microscope by a similar amount. Comparison of what one can learn about cells with the naked eye versus what one can see in a good light microscope is a metaphor for the value of EM. That said, there are many factors other than instrument resolution that limit the detail one can see in biological specimens. Many of these limitations are discussed in specific chapters of this book. Nonetheless, it is comparatively easy to obtain electron micrographs of biological material that improve on conventional light imaging by ~40×; this quality allows one to see objects that are only a few nanometers in diameter. Innovations that are reviewed in this book hold the promise of improving the current capabilities of cellular EM by another factor of 2–5, bringing the resolution of this method to the point that one can think rationally about linking electron micrographs directly with the structures of macromolecules that are determined by atomic resolution methods, like crystallography and nuclear magnetic resonance. Thus, modern EM can almost certainly help to forge a link that will ultimately connect light microscopy of live cells with atomic structure determinations, giving us something not far from an "atomic model of a cell."

I put this grand goal in quotation marks, for several reasons. First, the necessary connections will require links between many methodologies, for example, atomic models docked into density maps determined by cryo-EM and cryotomography, fitted into serial-section three-dimensional reconstructions of rapidly frozen, freeze-substitution fixed material imaged by EM tomography, and correlated with light microscopic images of live material. Such an enterprise is fraught with possibilities for mistake, so much care and many controls will always be necessary. Further, it must be stated at the outset that we are not there yet. The chapters of this book will show with considerable authority just where we are in our quest for this kind of structural unity. Some show that for particular macromolecular complexes, like calcium channels and clathrin cages, we are already there, and some sketch pathways that extend into the future. Preparative methods are, in many cases, very good, and imaging is coming along well. A few improvements in both, combined with better ways to record and analyze data, and we may soon have atomic models of complex objects, like the thylakoids of a chloroplast. Accomplishing the same goal for an entire cell from a vertebrate or higher plant, however, will take a good deal more innovation and technological development.

The goal of this book is to provide the interested reader with three important kinds of information: authoritative accounts of how the world's best practitioners of cellular EM currently ply their structural trade, an accurate sense of the limitations imposed by current methods and instruments, and a clear-eyed view of what may plausibly be achieved in the near future. From the material presented we hope that a wide range of cell biologists will learn that they too can profit from taking up the new EM. At the same time, we hope that they will be cautioned about the problems that must still be solved, so they will not be beguiled by false hopes about what can be learned with current methods. We who are already in this field are excited about its productivity and prospects. We hope that a good reading of these chapters will win new adherents and provide the criticism and energy that will take this now burgeoning field over the steps that still lie ahead as we forge the link between the cellular and atomic images of living things.

References

Al-Amoudi, A., Chang, J. J., Leforestier, A., McDowall, A., Salamin, L. M., Norlen, L. P., Richter, K., Blanc, N. S., Studer, D., and Dubochet, J. (2004). Cryo-electron microscopy of vitreous sections. *EMBO J.* **23**(18), 3583–3588.

Baumeister, W. (2002). Electron tomography: Towards visualizing the molecular organization of the cytoplasm. *Curr. Opin. Struct. Biol.* **12,** 679–684.

Chalfie, M., Tu, Y., Euskirchen, G., Ward, W. W., and Prasher, D. C. (1994). Green fluorescent protein as a marker for gene expression. *Science* **263,** 802–805.

Coons, A., and Kaplan, M. (1950). Localization of antigen in tissue cells. II. Improvements in a method for detection of antigen by means of fluorescent antibody. *J. Exp. Med.* **91,** 1–13.

Dombeck, D. A., Blanchard-Desce, M., and Webb, W. W. (2004). Optical recording of action potentials with second-harmonic generation microscopy. *J. Neurosci.* **24,** 999–1003.

Dubochet, J., Adrian, M., Chang, J. J., Homo, J. C., Lepault, J., McDowall, A. W., and Schultz, P. (1988). Cryo-electron microscopy of vitrified specimens. *Q. Rev. Biophys.* **21,** 129–228.

Dubochet, J., McDowall, A. W., Menge, B., Schmid, E. N., and Lickfeld, K. G. (1983). Electron microscopy of frozen-hydrated bacteria. *J. Bacteriol.* **155,** 381–390.

Flesken-Nikitin, A., Williams, R. M., Zipfel, W. R., Webb, W. W., and Nikitin, A. Y. (2005). Use of multiphoton imaging for studying cell migration in the mouse. *Methods Mol. Biol.* **294,** 335–345.

Frank, J. (2006). "Three-Dimensional Electron Microscopy of Macromolecular Assemblies." Oxford University Press, Oxford.

Frank, J., Wagenknecht, T., McEwen, B. F., Marko, M., Hsieh, C. E., and Mannella, C. A. (2002). Three-dimensional imaging of biological complexity. *J. Struct. Biol.* **138,** 85–91.

Gall, J. G., and Pardue, M. L. (1969). Formation and detection of RNA-DNA hybrid molecules in cytological preparations. *Proc. Natl. Acad. Sci. USA* **63,** 378–383.

Gustafsson, M. G. (2005). Nonlinear structured-illumination microscopy: Wide-field fluorescence imaging with theoretically unlimited resolution. *Proc. Natl. Acad. Sci. USA* **102,** 13081–13086.

Howard, R. J., and Aist, J. R. (1979). Hyphal tip cell ultrastructure of the fungus *Fusarium*: Improved preservation by freeze substitution. *J. Ultrastruct. Res.* **66,** 224–234.

Inoue, S. (1953). Polarization optical studies of the mitotic spindle. I. The demonstration of spindle fibers in living cells. *Chromosoma* **5,** 487–500.

Kellenberger, E. (1991). The potential of cryofixation and freeze substitution: Observations and theoretical considerations. *J. Microsc.* **161**(Pt. 2), 183–203.

Koster, R. W., and Fraser, S. E. (2004). Time-lapse microscopy of brain development. *Methods Cell Biol.* **76,** 207–235.

Lazarides, E., and Weber, K. (1974). Actin antibody: The specific visualization of actin filaments in non-muscle cells. *Proc. Natl. Acad. Sci. USA* **71,** 2268–22672.

Linder, J. C., and Staehelin, L. A. (1979). A novel model for fluid secretion by the trypanosomatid contractile vacuole apparatus. *J. Cell Biol.* **83,** 371–382.

McDonald, K., and Morphew, M. K. (1993). Improved preservation of ultrastructure in difficult-to-fix organisms by high pressure freezing and freeze substitution: I. *Drosophila melanogaster* and *Strongylocentrotus purpuratus* embryos. *Microsc. Res. Tech.* **24,** 465–473.

McIntosh, J. R. (2001). Electron microscopy of cells: A new beginning for a new century. *J. Cell Biol.* **153,** F25–F32.

McIntosh, R., Nicastro, D., and Mastronarde, D. (2005). New views of cells in 3D: An introduction to electron tomography. *Trends Cell Biol.* **15,** 43–51.

Moor, H. (1964). Freeze-fixation of living cells and its use in electron microscopy. *Z. Zellforsch. Mikrosk. Anat.* **62,** 546–580.

Moor, H. (1973). "Cryotechnology for the Structural Analysis of Biological Material." Societe Francaise de Microscopie Electronique, Paris.

Nogales, E., Wolf, S. G., and Downing, K. H. (1997). Visualizing the secondary structure of tubulin: Three-dimensional map at 4 Å. *J. Struct. Biol.* **118,** 119–127.

Oldenbourg, R., and Mei, G. (1995). New polarized light microscope with precision universal compensator. *J. Microsc.* **180**(Pt. 2), 140–147.

Paweletz, N. (2001). Walther Flemming: Pioneer of mitosis research. *Nat. Rev. Mol. Cell Biol.* **2,** 72–75.

Pickett-Heaps, J. D., Schmid, A.-M. M., and Tippit, D. H. (1984). Cell division in diatoms. *Protoplasma* **120,** 132–154.

Porter, K. R., Claude, A., and Fullam, E. F. (1945). A study of tissue culture cells by electron microscopy. *J. Exp. Med.* **81,** 233–246.

Shaner, N. C., Steinbach, P. A., and Tsien, R. Y. (2005). A guide to choosing fluorescent proteins. *Nat. Methods* **2,** 905–909.

Sosa, H., Popp, D., Ouyang, G., and Huxley, H. E. (1994). Ultrastructure of skeletal muscle fibers studied by a plunge quick freezing method: Myofilament lengths. *Biophys. J.* **67,** 283–292.

Taylor, D. L., and Wang, Y. L. (1980). Fluorescently labelled molecules as probes of the structure and function of living cells. *Nature* **284,** 405–410.

Taylor, K. A., and Glaeser, R. M. (1974). Electron diffraction of frozen, hydrated protein crystals. *Science* **186,** 1036–1037.

Tokuyasu, K. T. (1980). Immunochemistry on ultrathin frozen sections. *Histochem. J.* **12,** 381–403.

Tsien, R. Y. (1998). The green fluorescent protein. *Annu. Rev. Biochem.* **67,** 509–544.

van Heel, M., Gowen, B., Matadeen, R., Orlova, E. V., Finn, R., Pape, T., Cohen, D., Stark, H., Schmidt, R., Schatz, M., and Patwardhan, A. (2000). Single-particle electron cryo-microscopy: Towards atomic resolution. *Q. Rev. Biophys.* **33,** 307–369.

White, J. G., Amos, W. B., and Fordham, M. (1987). An evaluation of confocal versus conventional imaging of biological structures by fluorescence light microscopy. *J. Cell Biol.* **105,** 41–48.

Zalokar, M. (1966). A simple freeze-substitution method for electron microscopy. *J. Ultrastruct. Res.* **15,** 469–479.

PART I

Preparing Cells for Electron Microscopy

Getting biological samples ready for electron microscopy (EM) is undoubtedly one of the most important steps in the train of tasks that is required for using this kind of microscope well. Electron beams require a vacuum, so water in the sample must be immobilized or removed. Some workers have tried the alternative of introducing a sample into the beam with an environmental chamber that would keep cells hydrated while they were under observation (Turner *et al.*, 1981), but this approach has not provided good images and has been abandoned. Thus, there are basically two ways to get cells ready for EM: drying or freezing. Simple air-drying is disastrously destructive, but even dehydration by sample transfer from water to a water-miscible organic solvent, like acetone or ethanol, is destructive to fragile structural details. Dehydration by liquid transfer must therefore be preceded by a treatment that will toughen the structures of interest. This process, called "fixation," is usually a chemical treatment that stabilizes local interactions. Fixation has a long tradition in cell biology, dating from way before the invention of EM; many scientists have worked to identify protocols that are easy to use and as reliable as possible, where quality is evaluated by the similarity between the structure of a living cell and the structure seen after fixation. Such methods have produced much useful information about cells and many helpful references are available (e.g., Glauert and Lewis, 1998).

As mentioned in the Introduction, however, there are always concerns about the reliability of a given fixation method. The formation of chemical cross-links, for example, by a Schiff base reaction, is quick (a few milliseconds), but the time required for diffusion of the cross-linker into cells can range from seconds (Fitzharris *et al.*, 1972) to many minutes, depending on the relevant diffusion coefficients and the distances that the cross-linkers must travel to reach their site of action. Moreover, not all macromolecules react in the same way with a given chemical fixative, so different cellular materials (protein or nucleic acid polymers, lipids, and polysaccharides) will fix at different rates and with different chemistries. A pair of complementary processes, like the formation and the fusion of vesicles, may become fixed at different speeds, leading to a nonphysiological increase or decrease in the concentration of a visible intermediate, like the vesicles themselves. Fixations can also distort the objects they cross-link, or they can provide stabilization that is insufficient to maintain the structures of interest during dehydration. Thus, many scientists who care deeply about the reliability of structure determination have explored other methods of sample preparation.

Rapid freezing has received much attention as an alternative or complement to fixation, particularly as an initial step for the immobilization of cells. The temperature of small samples can be reduced very rapidly simply by plunging them into a very cold liquid that is near its freezing point, for example, ethane that has been liquefied and almost frozen by liquid nitrogen. In favorable cases, sample freezing by this simple method can drop the sample's temperature at a rate of $\sim 10^5$ °C/sec (Gilkey and Staehelin, 1986). This way, all cellular components are brought to a standstill within milliseconds, encouraging the belief that biological structure is accurately preserved. This desirable situation has the further advantage that cellular water is frozen (solidified) without the formation of ice crystals, resulting in a structureless, glasslike solid that is commonly called "vitreous ice" (also known as amorphous ice). Avoiding crystallization of cellular water is important for preserving the details of cellular structure because ice crystals usually form from pure water, displacing the previously dissolved solutes as they form. Thus, "good" freezing, which vitrifies the water in the sample, is an almost perfect fixative, decreasing the speed of all molecular motions in unison. Bad freezing, on the other hand, leads to crystallization, which alters cellular structures in ways that can be highly problematic. These issues and their relationship to the physics of freezing are discussed in the first chapter of this section, which has been contributed by Dubochet.

A major practical issue in the freezing of biological materials, however, is that many cells are too large to freeze well simply by rapid immersion in a cold liquid. Fortunately, technological advances have led to biologically useful alternatives, like "slam" freezers (Heuser *et al.*, 1979) or a double-jet freezer (Craig and Staehelin, 1988; Müller *et al.*, 1980), either of which can vitrify the surface of a sample that is too large to freeze well in its entirety. More recently, a freezer that uses high hydrostatic pressure to retard ice crystallization has helped biologists to solve this problem (Craig and Staehelin, 1988; McDonald and Morphew, 1993; Moor *et al.*, 1980; Studer *et al.*, 2001). The latter device has made it possible to vitrify large cells and even small organisms, embryos, and pieces of tissue. The issues pertinent to high-pressure freezing of diverse biological samples are discussed in detail by McDonald and by Hess in the chapters that follow.

Once a cell or tissue is well frozen, however, the problem is how to get it into the EM in a form that permits high-resolution imaging. Most cells are too thick for electron imaging in their entirety, although a few exceptions are known (Dubochet *et al.*, 1983; Grimm *et al.*, 1998; Medalia *et al.*, 2002; Nickell *et al.*, 2003; O'Toole *et al.*, 1993). Most biological samples therefore require some form of microtomy by which one cuts slices that are thin enough for fruitful EM. The minimalist approach to this problem is obviously to take the frozen cells and simply cut them at low temperature, making samples that are sufficiently thin for high-quality imaging. Observation of the resulting frozen-hydrated samples must, of course, be done with a microscope that can keep the specimen cold enough that ice crystals do not form (the process of "recrystallization"), but such cold stages have been available for many years. Work from Dubochet's laboratory has pioneered this approach (McDowall *et al.*, 1984), but the process is technically challenging, and it has taken many years to bring cryosectioning and cryomicroscopy of vitrified samples to the point that such samples display the detail one would expect from frozen cells (Al-Amoudi *et al.*, 2004). These issues are mentioned in Chapter 1 by Dubochet, but many factors that are practically important to the cell biologist are described with more detail in the later chapter from Dubochet's group (Chapter 15).

EM of cryosections cut from vitrified cells has great promise, but there are many practical difficulties in applying this method to big samples on a routine basis. Most investigators have therefore sought an alternative methodology. It is now common to use rapid freezing as an initial preparative step to immobilize the sample, then fixation and dehydration at low temperatures, followed by embedding in a plastic resin that permits conventional microtomy and microscopy. This approach is based on water-miscible organic solvents that remain liquid at rather low temperatures (e.g., –90 °C); these can dissolve cellular water by a solid–liquid transition and allow the introduction of cross-linking chemicals while the cell is cold enough to retain its physiological structure, a process called fixation by freeze-substitution. Chapter 2 by McDonald and Chapter 3 by Hess describe several successful methods for achieving this valuable goal.

Students of cell function are often concerned, however, about the physiological state of their sample at the time it is frozen. This is particularly important for the analysis of dynamic biological processes, like the cell cycle or embryonic development. Chapter 4 by Müller-Reichert *et al.* describes the use of commercially available tools to follow cells in life by light microscopy until seconds before they are cryoimmobilized as the initial step in preparation for EM. The scientific value of this method is obvious and great.

A useful frame of mind to take into the reading of the chapters of this section is that recent improvements in methods for rapid freezing, for cryomicrotomy and cryomicroscopy, and for fixation and embedding in plastic by freeze-substitution have provided today's structural cell biologist with a powerful arsenal of techniques by which to prepare materials for scrutiny by EM. The results provide views of cell structure that are highly likely to be accurate representations of the living state. The good news is that such samples can now be imaged with a resolution that surpasses conventional light microscopy by >40-fold. The bad news is that getting there is costly in both time and money.

References

Al-Amoudi, A., Chang, J. J., Leforestier, A., McDowall, A., Salamin, L. M., Norlen, L. P., Richter, K., Blanc, N. S., Studer, D., and Dubochet, J. (2004). Cryo-electron microscopy of vitreous sections. *EMBO J.* **23,** 3583–3588.

Craig, S., and Staehelin, L. A. (1988). High pressure freezing of intact plant tissues. Evaluation and characterization of novel features of the endoplasmic reticulum and associated membrane systems. *Eur. J. Cell Biol.* **46,** 81–93.

Dubochet, J., McDowall, A. W., Menge, B., Schmid, E. N., and Lickfeld, K. G. (1983). Electron microscopy of frozen-hydrated bacteria. *J. Bacteriol.* **155,** 381–390.

Fitzharris, T. P., Bloodgood, R. A., and McIntosh, J. R. (1972). The effect of fixation on the wave propagation of the protozoan axostyle. *Tissue Cell* **4,** 219–225.

Gilkey, J. C., and Staehelin, L. A. (1986). Advances in ultrarapid freezing for the preservation of cellular ultrastructure. *J. Electron Microsc. Tech.* **3,** 177–210.

Glauert, A., and Lewis, P. (1998). "Biological Specimen Preparation for Transmission Electron Microscopy." Princeton University Press, Princeton.

Grimm, R., Singh, H., Rachel, R., Typke, D., Zillig, W., and Baumeister, W. (1998). Electron tomography of ice-embedded prokaryotic cells. *Biophys. J.* **74,** 1031–1042.

Heuser, J. E., Reese, T. S., Dennis, M. J., Jan, Y., Jan, L., and Evans, L. (1979). Synaptic vesicle exocytosis captured by quick freezing and correlated with quantal transmitter release. *J. Cell Biol.* **81,** 275–300.

McDonald, K., and Morphew, M. K. (1993). Improved preservation of ultrastructure in difficult-to-fix organisms by high pressure freezing and freeze substitution: I. *Drosophila melanogaster* and *Strongylocentrotus purpuratus* embryos. *Microsc. Res. Tech.* **24,** 465–473.

McDowall, A. W., Hofmann, W., Lepault, J., Adrian, M., and Dubochet, J. (1984). Cryo-electron microscopy of vitrified insect flight muscle. *J. Mol. Biol.* **178,** 105–111.

Medalia, O., Weber, I., Frangakis, A. S., Nicastro, D., Gerisch, G., and Baumeister, W. (2002). Macromolecular architecture in eukaryotic cells visualized by cryoelectron tomography. *Science* **298,** 1209–1213.

Moor, H., Bellin, G., Sandri, C., and Akert, K. (1980). The influence of high pressure freezing on mammalian nerve tissue. *Cell Tissue Res.* **209,** 201–216.

Müller, M., Meister, N., and Moor, H. (1980). Freezing in a propane jet and its application in freeze-fracturing. *Mikroskopie* **36,** 129–140.

Nickell, S., Hegerl, R., Baumeister, W., and Rachel, R. (2003). *Pyrodictium* cannulae enter the periplasmic space but do not enter the cytoplasm, as revealed by cryo-electron tomography. *J. Struct. Biol.* **141,** 34–42.

O'Toole, E., Wray, G., Kremer, J., and McIntosh, J. R. (1993). High voltage cryomicroscopy of human blood platelets. *J. Struct. Biol.* **110,** 55–66.

Studer, D., Graber, W., Al-Amoudi, A., and Eggli, P. (2001). A new approach for cryofixation by high-pressure freezing. *J. Microsc.* **203,** 285–294.

Turner, J. N., See, C. W., Ratkowski, A. J., Chang, B. B., and Parsons, D. F. (1981). Design and operation of a differentially pumped environmental chamber for the HVEM. *Ultramicroscopy* **6,** 267–279.

CHAPTER 1

The Physics of Rapid Cooling and Its Implications for Cryoimmobilization of Cells

Jacques Dubochet

Laboratory for Ultrastructural Analysis
Biophore, University of Lausanne
CH-1015 Lausanne, Switzerland

The 0°C barrier is important for life because it corresponds to a fundamental transformation of water, the most abundant constituent of biological matter. Freezing takes place in two steps. The first is ice nucleation. It is a rare, entropy-limited process by which the long-range order of the crystal emerges from the disordered liquid. The second step is the rapid, heat-generating growth by which all available water molecules are engulfed in the crystal. Ice formed in the crowded environment of a cell differs from ice grown in pure water because the progression of the crystal surface is impeded by an increasingly concentrated solute. The result is a ramified ice crystal that extends throughout the cell, like the branches of a tree that has grown between aggregated, dehydrated biological material. Water can also be vitrified, namely it can be cooled so rapidly that its molecules are immobilized

0091-679X/07 $35.00
DOI: 10.1016/S0091-679X(06)79001-X

before crystallization can occur. In this case, there is no segregation of solutes from solvent, and the structure of vitrified biological specimens is probably preserved down to atomic resolution. A vitreous sample is, however, metastable; it will crystallize at temperatures above ca. –135 °C. Vitrification of pure water can only be achieved over a limited volume, measuring only micrometers on a side, but this size can be extended by two orders of magnitude in cells and when cooling takes place under high pressure.

Two avenues are possible to obtain specimens for freeze-substitution. One is the artifact-prone crystallization from liquid water. The other is when ice is formed from a vitrified specimen. One can and should make sure to follow the second route.

I. Introduction

Nature has a long experience of crossing the 0 °C barrier and knows how to deal with it. For a few decades now electron microscopists have also been learning to deal with this transition temperature, but they still have much progress to make. The problem is that freezing is quite a dramatic event. When water freezes, the molecules, which in a liquid state communicate exclusively with their nearest neighbor, suddenly adopt cooperative behavior. From being solitary, they become members of a demanding and exclusive group, an ice crystal. Once absorbed into the group, they lose part of their freedom and give up some of their energy. All other molecules become excluded from the resulting solid. The world is now divided into two, ice crystals and . . . all the rest.

Freezing is not good for life. Nature has adopted all kinds of strategies to avoid its deleterious effects. For large animals, a good method is to keep their temperatures well above freezing, but this is expensive; for small organisms it costs too much energy. Plants too have adopted ways of their own. Some have learned how to prevent ice crystal formation. Partial dehydration helps, as does the addition of a cryoprotectant. Indeed, there are proteins that prevent the formation or growth of ice. In other organisms, cryoresistance is achieved by letting ice crystals grow under such conditions that the organism survives. There are amphibians and reptiles that freeze hard in winter and then thaw out happily in the spring. For a long time these multiple facets of freezing have fascinated laymen and specialists. The literature is abundant (Ball, 1999; Luyet and Gehenio, 1940; Réaumur, 1736).

Freezing is also important for human activity. Animal breeding relies on frozen cells (the gametes of prized livestock), a method that also works well with many human cells. The technical problems of preserving viability increase with the size of the sample. However, cryopreservation of tissues and organs is still in its infancy. The case of human embryo preservation may no longer present a limiting technical problem, but it will nevertheless for long remain a hot ethical question. As for the preservation of entire humans in a state of frozen "suspended life," which has apparently become a lucrative business, this obviously relies on deception and ignorance.

Electron microscopists are interested in cryoimmobilization because it offers a way out of the water dilemma: water is the most abundant constituent of living material, but it cannot remain in the vacuum of an electron microscope except at very low temperatures. One avenue is cryospecimen preparation. It consists first of freezing the specimen and then removing the water from the frozen state. Once the specimen is dry, it can be brought to room temperature for observation, and the damage due to dehydration can be less severe than when water is removed from the liquid state.

The other avenue is cryo-electron microscopy in which the water is kept at a temperature low enough to prevent evaporation. As long as cryoimmobilization was synonymous with freezing, and frozen water was ice, the advantage of cryo-electron microscopy was questionable; the method was demanding and it did not eliminate freezing damage. The situation changed with vitrification. This is the process of solidifying water from its liquid state without crystallization. The idea is old, the expectations were high, and the amount of research carried out to achieve it has been considerable. However, a persistent lack of success led to the conclusion that it was fundamentally impossible. The breakthrough came in 1980 when Mayer and Brüggeller (1980) and Dubochet and McDowall (1981) succeeded, against all expectations, in vitrifying small amounts of pure water. From this point, cryo-electron microscopy of biological particles suspended in a thin layer of vitrified water was soon developed into a practical method, which is now widely used (Adrian *et al.*, 1984; Jiang and Ludtke, 2005). It took 20 more years until this approach could be extended to larger samples, such as cells and tissues, by means of *c*ryo-*e*lectron *m*icroscopy *o*f *vi*treous *s*ections (CEMOVIS; Al-Amoudi *et al.*, 2004; Dubochet *et al.*, 2006; Lucic *et al.*, 2005 and Chapter 15, this volume).

In the present chapter, we discuss what happens when liquid water crystallizes on cooling, and how crystal formation rearranges the surrounding biological material. We discuss vitrification, leaving aside the mysterious aspects of the phenomenon. We conclude that cryoimmobilization should always involve vitrification, whether or not the specimen will remain vitreous for subsequent cryo-electron microscopy observation or will be further processed to remove the water before being observed in the dry state. Finally, we insist on the necessity of objectively testing the quality of cryoimmobilization by determining whether or not the sample has been vitrified. Further readings to complement this chapter are found in Angell (2004), Ball (1999), Dubochet (2006), and Dubochet *et al.* (1988, 1991).

II. Freezing Water

In everyday life, we are familiar with three forms of water: vapor, liquid, and ice. At the molecular level we can feel comfortable with some of the simple, pertinent concepts. Vapor is the state in which the molecules are free to fly independently; ice is the state where each molecule has its place in the lattice of a crystal. Liquid water is more complicated. Situated somewhere between the order of crystals and the disorder of a gas, liquid water might be seen either as disordered ice or ordered vapor. For some time a common belief favored the iceberg model in which liquid water was thought to consist of minuscule ice crystals rolling on each other. Bernal (Bernal and Fowler, 1933) demonstrated that this was wrong, mainly by noticing that the iceberg model implied two kinds of interactions—the well-defined bonds in the crystals and the undefined rolling relations between crystals. In fact, only one kind of interaction was actually observed. He proposed the nonintuitive concept that in liquid water the disorder is distributed more or less equally over every molecule. In this view, the difference between liquid and gas is a matter of the degree of attachment at the molecular level. In both cases, one molecule "knows" about another only when they collide. Except for this direct contact, the rest is "everyone on its own." The situation is fundamentally different in ice. Each water molecule "knows" that it belongs to a crystal. It "senses" all its neighbors right up to the last molecule at the most remote corner of the crystal. What happens when an ice crystal is warmed at 0 °C is typical. At this temperature, the liquid state becomes the stable state but, because of the cooperative behavior, no water molecule in the crystal can get warmer until the last little bit of the crystal is melted. Consequently, on heating, a cold ice crystal warms up to 0 °C then stays at this temperature until it has adsorbed all the latent heat of fusion. This is a large amount of energy. It is equal to what is needed to warm the same mass of liquid from 0 to 80 °C (335 kJ/kg). The same amount of energy is released when the liquid freezes into ice. Obviously, changing the principle of order does not go unnoticed.

When one water molecule meets another, it does not make an ice crystal. Does it need three, four, or many to establish the new order? Certainly it requires that some number of molecules, acting together, enter into a very special condition in which the molecules begin to vibrate together. This is the beginning of ice crystal formation. We can grasp the extent to which this nucleation process is extraordinary by the fact that vitrification is possible. It consists in cooling water molecules so rapidly that they come to rest (or close to it) before they have started to form an ice crystal. We can vitrify 1 μm^3 of water—that is, ca. 30×10^9 molecules—in perhaps 10^{-6} sec. Since, on average, each water molecule changes its position every 10^{-12} sec, this means that many billions of molecules, each trying one million positions, have not yet had time to find out how to start a crystal! Even the biggest computer needs a lot of time—and much help—to simulate ice nucleation (Matsumoto *et al.*, 2002). However, classical nucleation theory helps to represent the principle of the phenomena (Franks, 1982). To understand it we shall consider a spherical ice crystal of radius r floating in liquid water at a temperature below 0 °C (Fig. 1). Each water molecule inside the crystal contributes to reducing the free energy, ΔG. This contribution is proportional to the number of molecules in the crystal, namely proportional to r^3. There is another contribution, though; molecules at the surface of the crystal are in an unstable condition, thus contributing to an increase of free energy. Their number is proportional to r^2. Together the two curves always give the same shape; with increasing r, the free

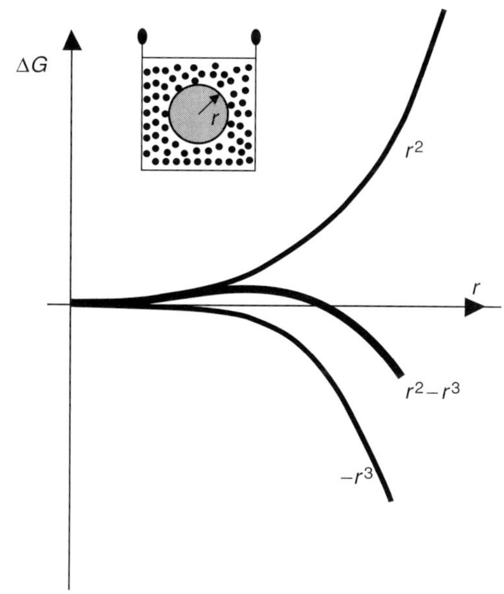

Fig. 1 Classical nucleation theory. Model of an ice crystal in water below 0 °C. The free energy ΔG of the system, represented as a function of the radius r of the crystal, has a favorable—negative—component, due to the molecules in the crystal. This contribution is proportional to the volume of the crystal, namely to r^3. ΔG also has an unfavorable—positive—component due to the molecules at the ice–liquid interface. Their number increases proportionally to r^2. The heavy line results from the sum of these two components.

energy first rises before it falls toward the stable state. In other words, an ice crystal must reach a sufficient size before it can grow by itself; nucleation is a fluctuation-driven process; it is a rare, chance event.

As much as it is difficult to start the growth of an ice crystal, its subsequent growth is rapid and powerful. Below 0 °C, an ice crystal is avid for water. It grows voraciously out of any water molecules that happen into its vicinity. The process is limited, though. Each molecule "falling" on the crystal brings its share of energy—in principle enough to warm this molecule by 80 °C. Thus, a growing ice crystal can be warmed by its growth until it approaches 0 °C, at which temperature it would cease to grow. But not quite! The continuation of crystal growth only takes place at the rate at which heat generation is compensated by external cooling. The spherical model of Fig. 1 does not correspond to optimal growing conditions. Arms extending into still cold water make it possible for more molecules to join the growing ice crystal (Fig. 2). In the end, the ice crystal grows like a tree in the whole available volume. The branching is coarse if the growth is slow, leaving time for temperature equilibration; it is fine in the contrary case. The process is complicated, very complicated indeed, since the interfacial energy of the ramified crystal must also be included (Maddox, 1983). In practice, an ice crystal

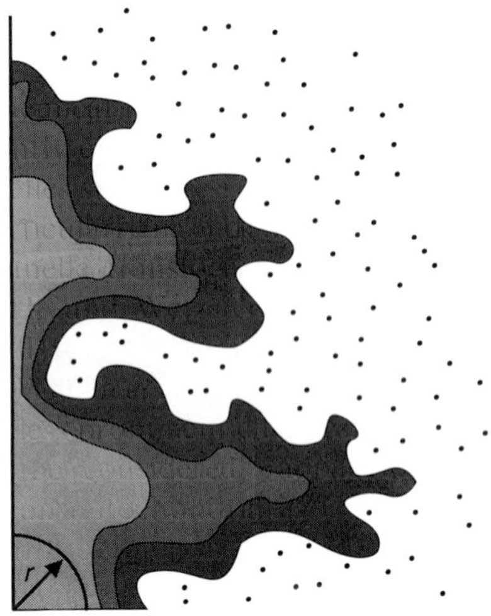

Fig. 2 The ramified growth of an ice crystal in liquid water. The growth of the small, freshly nucleated ice crystal (radius *r*) results from its tendency to adsorb neighboring molecules of the liquid water, limited by the heat generated during the process. Conditions of heat removal, interfacial energy, and ice crystal symmetry together chisel the delicate shape of the resulting crystal.

rapidly grown in supercooled pure water is large and extremely ramified. Sometimes, one may have the chance to observe how a single crystal that extends over the whole volume of a suddenly depressurized bottle of supercooled drinking water is formed within seconds. After forming, it will melt away within a minute because the relative volume of frozen water is small and the branches are very fine. The lucky observer is left with an unforgettable vision of the evanescent lightness of an ice crystal.

III. Biological Material

In biology, we do not deal with pure water but with concentrated solutions of salts and macromolecules. In a cell, biological material is crowded (Goodsell, 1993). No water molecule deals exclusively with other water molecules for long, and none is far away from an ion or from the surface of a macromolecule. For example, we calculate that if the cell's content is modeled exclusively by spherical proteins of 20 kDa, then four layers of water on each protein would account for enough water to comprise the 80% water that is the fraction observed in most real cells. What is the effect of all this nonaqueous material on the properties of water? Two opposed views each have their champions. The dominant one, certainly shared by more than 90% of water specialists and supported by most of the hard-core experimental work, holds that water in the cell is essentially bulk water, namely, the water that run from your tap. Certainly there are water molecules associated with ions and macromolecules but, except for a few, they remain mobile and ready to react as *bona fide* water molecules. The net result is a slight reduction of water activity, corresponding to an apparent reduction of water's concentration. For the rest of the water molecules, all the rules and properties of bulk water hold true (Israelachvili and Wennerström, 1996).

The other view, favored by many biologists, claims that water in the cell is special. Many regard "bound water"—a poorly defined concept—as though it were a different substance (Mentré, 1995). Others speak about the special properties of "vicinal water" (Drost-Hansen, 1969). Not infrequently, the claims have a distinctive tone of vitalism. Among the few experiment-based arguments in favor of the special status of water in living cells, the lifelong work of Ling (Ling, 2001) deserves special consideration. It has been clarified and extended with didactic skill by Pollack (2001).

What shall we do with these conflicting views? I suggest continuing with the classical view that water in a cell is normal because it keeps us on firm ground, but at the same time we should read Pollack in order to keep an agile mind and remain vigilant because surprises are likely to emerge from the many still unsolved questions (Dubochet, 2006).

IV. Freezing Biological Material

What happens to a biological sample on freezing? The scene has been set: there is a physiological solution (ca. 200-mM solutes) bathing the tissue and cells that are crowded with macromolecules. A biological membrane marks the border. In the cells most water molecules are moving 10^{12} times per second; most of the time they are not more than four neighbors away from the surface of the next macromolecule. Ions, soluble molecules, and macromolecules are also moving, although more slowly; they diffuse at speeds that vary roughly as one over the cube root of their molecular weight. Many of the water molecules are partially engaged in interactions with non-water molecules. We can now imagine what happens when the temperature decreases to below 0 °C.

First, nothing happens. The reduced activity of water in the cells causes the ice–water equilibrium temperature to be reduced by some degrees, or much more, depending on the solute concentration (Fig. 3). In extreme cases, for example, in some saturated solutions, it may even happen that no water is left available for crystallization; ice formation is thus prevented under any cooling conditions. This situation can be exploited by cooks to make ice—though saturated sugar is gastronomically not ideal—or by nature to avoid freezing in animals and plants—though the production of a large amount of a water-active solute is expensive. It is also used by electron microscopists. Following Tokoyasu (Tokuyasu and Singer, 1976), a saturated sucrose solution applied to permeabilized cells prevents ice formation, even at relatively slow rates of cooling.

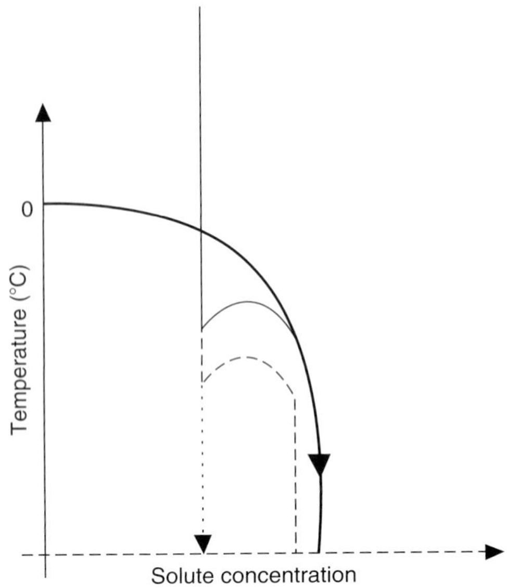

Fig. 3 Ice formation in presence of solute. The equilibrium temperature of water–ice, which is 0 °C for a dilute solution, decreases with solute concentration until there is no water left free for crystallization. On cooling, nothing happens to a solution of defined solute concentration until a nucleation event initiates the crystallization process (black vertical line). From this instant, the heat generated by ice formation opposes further cooling while the concentration of residual solution increases. The process comes to an end when the ice crystal leaves only an unfreezable, concentrated solution. The dashed line represents the case of ice crystal formation under higher cooling speed. The dotted-line represents the case when the cooling speed is so high that vitrification is obtained. In this case, the concentration of the solution remains constant at every point.

A. Homogeneous and Heterogeneous Nucleation

As the temperature falls below 0 °C, ice becomes the preferred state. The sample will crystallize ... eventually, but first it must find its way through the unlikely event of nucleation. The term "homogeneous nucleation" is used when liquid water has to go through the nucleation process by itself, without help from another structure. In fact, pure water is quite bad at homogeneous nucleation and even several degrees below zero are generally not sufficient to initiate crystallization. Without help, supercooling by 10 or 20 °C or even more would be the rule. There is a limit, though, to supercooling. No one has ever observed slowly cooled water approaching −45 °C which is considered to be the limit of supercooling. Minus 45 °C is very cold. Extreme supercooling could be the rule, but in fact it is not; in most cases, water gets help in overcoming the nucleation barrier. The principle of heterogeneous nucleation—namely nucleation favored by an external agent—can be grasped by the fact that in the presence of ice, crystallization of a liquid is easy. A heterogeneous nucleation center can be viewed as a region that mimics an ice crystal, helping water molecules to arrange themselves as ice on its surface. There are plenty of natural heterogeneous nucleation centers. What are they? Small pieces of some material, mimicking in some way the surface of an ice crystal and helping liquid water to find its way toward crystallization. There are good heterogeneous nucleation centers, nearly as good as ice, and there are plenty of mediocre ones. They are plentiful on nearly every natural surface and therefore, in most cases, liquid water freezes very soon below 0 °C. Exceptions are interesting. Here are two examples. (1) Rain water can be very pure. It may happen that the lower zone of the atmosphere is below 0 °C, whereas higher regions are above 0 °C. Rain coming from the higher region is supercooled in the lower region but does not freeze until it comes to the ground where plenty of heterogeneous nucleation centers produce instant ice formation. On a busy road the result may be impressive. (2) Dew remains liquid, even at several degrees below zero. So a plant is frequently damaged when an ice crystal grows out of the dewdrop, leaving the field open for feeding bacteria. Some bacteria have even developed special, surface proteins that are highly efficient heterogeneous nucleation centers. It is amazing that the first commercial application of biotechnology was aimed at preventing the action of ice-nucleating bacteria in strawberry fields in California. It is interesting that structural studies of ice-nucleation proteins have helped us to understand how ice nucleation takes place.

B. Ice Formation in Biological Material

Cell interiors are remarkably free of heterogeneous nucleation centers. Why is this so? Could it be that life, so frequently confronted with the zero-degree boundary, has strictly eliminated all features that favor ice nucleation in the cell? (A. P. MacKenzie, personal communication.) In any case, the result is that ice crystal formation is held up in cells longer than in the surrounding medium because of the high concentration of various materials and the absence of heterogeneous nucleation centers. In general, freezing starts somewhere in the external medium or on the surface of the chamber containing the sample. Heat generated by the crystallization process temporarily prevents further cooling of the freezing zone. It is only when ice formation is completed in the external medium of the freezing zone that the temperature can again go lower. Nucleation becomes possible in the cells and intracellular ice crystals can develop. At the cellular scale, nucleation events are rare. In most cases, a single ice crystal surrounds several cells and, generally, a single ice crystal extends throughout a whole cell.

As compared to ice crystal growth in pure water, the process in a cell differs because of the abundant presence of solute, which is only rarely trapped in the crystal lattice. Crystal growth is not only limited by the heat production/removal equilibrium and surface tension, it has also to find its way in the increasing concentration of solute. Figure 4 illustrates the process. The growing ice crystal adsorbs all the available water it comes into contact with. Around the ice surface the cytoplasm material becomes dehydrated and highly concentrated. If the ice crystal were to stop growing at that instant, diffusion would slowly restore the

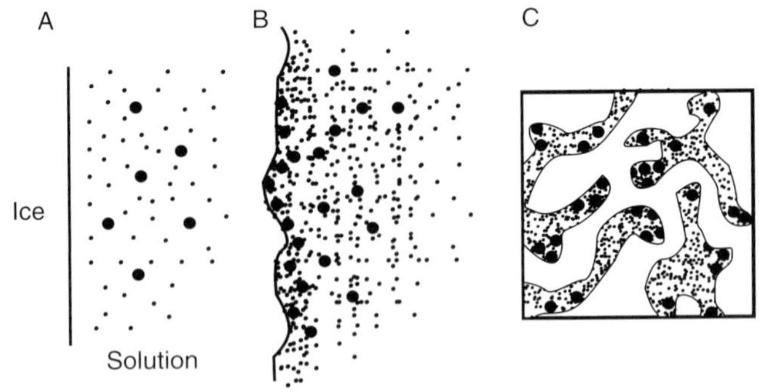

Fig. 4 Ice crystal growth in the concentrated medium of a cell. In the initial state, a hypothetical flat ice crystal is adjacent to an aqueous solution that contains various solutes and macromolecules (A). Depending on their size and diffusion properties, solutes are concentrated at the surface of the progressing ice crystal (B). In the final state, the ice crystal extends over the whole volume of the solution leaving concentrated, dehydrated solution between its tortuous ramifications (C).

cytosolic equilibrium in the remaining space, though at a higher concentration, since part of the water is now in the ice crystal. But ice crystal growth goes on because external cooling continues. It remains just as avid for water as before; it continues to grow wherever water molecules become available. Diffusion of the water molecules toward the crystal and solute molecules away from it become the essential factor. Small molecules can diffuse rapidly; larger ones are slower, thus forming a more and more compact crust at the surface of the crystal. It is probably under these circumstances that irreversible aggregation of the dehydrated material takes place. Of course one should not imagine the growth of an ice crystal as a front progressing uniformly across a rigid, flat surface. As drawn in Fig. 4, the irregularly growing ice crystal becomes more and more ramified because it grows best at the tips of the branches, while remaining irreversibly stuck deep down between the branches, where dehydrated solute has been deprived of water. Slowly grown crystals have large branches without fine ramifications. A rapidly grown crystal has only small branches and complex ramifications.

We understand now how freezing can kill. The image of an ice crystal perforating the cell membrane with its sharp edge is not correct. A membrane is hydrophobic; it is a zone without water; an ice crystal has nothing to do with it directly. Instead, the damage takes place at both a global and a local level. Globally, it originates when an ice crystal devours the available water in one compartment. What the membrane then perceives is the increased solute concentration in this compartment; this results in water stress, osmotic pressure, and an eventually bursting of the membrane. At the local level, dehydration can produce irreversible destruction of biological functions by denaturation and aggregation of the macromolecules. This effect is not always lethal. Most single cell organisms overcome freezing quite well, and the same holds for many plants and some animals. Resistance to freezing is certainly the result of a long evolution, involving subtle adjustments and tightly controlled pathways.

The schema of Fig. 4 is too simple; the reality is in three dimensions. We would not observe the tree in a two-dimensional section, but only a cut through it. In most cases, we would see adjacent cross sections of branches. This is exactly what is observed in electron microscopy sections of frozen samples obtained by freeze-substitution: holes—corresponding to sections cut through the branches of the ice crystal—in a tissue of aggregated, dehydrated material. The holes are large and the mesh of the tissue is coarse when freezing is slow in rapidly diffusing material; the mesh is small and fine when cooling is rapid and when the biological material cannot diffuse away. In no case are the holes vanishing small. For a long time each hole seen in a section of freeze-substituted material was interpreted as a single ice crystal. This is not correct.

How do we know? We know it from direct observation of frozen sections by cryo-electron microscopy. With freeze-substitution, it is not the crystal that is observed but the trace that is left after water has been removed. Conclusions about the nature of the crystal are therefore only indirect. With frozen sections, the atomic orientation of ice crystals is revealed by electron diffraction, and it is found that all the little domains, previously thought to be associated with different crystals, have the same crystal orientation, thus demonstrating that they belong to the same crystal. In most cases, there is only one crystal per cell; sometimes there are a few; never very many.

How do we know that the branches of the ice crystal are never very small? Because this is what is observed. There are also theoretical reasons: an ice crystal is held together by a collective organization of molecules. It cannot be very small in any direction.

At this point, some readers are going to protest. They will say that their preparations are not so bad; they know what is a segregated specimen and they do not publish those bad images. Certainly, they are right and we think there is an explanation (Dubochet *et al.*, 1991). The fact is that crystallization can take another, much less damaging, path. Before coming to this point we must discuss vitrification.

C. Vitrification

At a very high cooling rate, water molecules can be brought to rest, or close to it, before they have time to nucleate a crystal. Water becomes solid while remaining in an amorphous state that can be preserved indefinitely, provided the temperature is kept cold enough. The process is called vitrification. In fact, nobody knows for sure what vitrification is. There are fundamental theoretical questions that are not answered and experimental results that are not understood. A discussion, which should be accessible to nonspecialists, is presented elsewhere (Dubochet, 2006). For lack of certainty as to what vitrification is, we chose the simplest solution, while keeping in mind that our views may need to be revisited.

We will therefore consider that vitrification consists in solidifying liquid water while preserving its basic structure. Vitreous water can thus be seen as liquid water with such a high viscosity that it does not flow by itself during the time of a realistic experiment. Furthermore, we hope that the same principle of conservation holds true for all the biological matter that is immersed or solubilized in the water, so any vitrified sample is structurally identical to the original object at the time of freezing, the only difference being that it is immobilized. Of course, this view is not correct. Temperature is such an important factor in biological equilibria that extreme cooling must have structural consequence. We note also that even if the vitrification time is a small fraction of a second, it is a long time at the scale of molecular processes. However, reports on structural transformations on vitrification are rare, and so far they concern only local changes (Erk *et al.*, 1996).

On the contrary there is plenty of evidence that vitrification is an excellent method for preserving biological samples for electron microscopy. Three lines of argument can be mentioned. (1) Since the pioneering works of Taylor and Glaeser (1974), results obtained by electron microscopy of vitrified specimens demonstrate, in many cases, that biological structures are preserved by freezing down to atomic resolution and, in every case where quantitative comparisons are possible, the level of preservation is found to be excellent. (2) Nowadays, x-ray diffraction is preferentially performed on vitrified specimens. The level of preservation required in these studies is stringent. (3) Last but not least, the level of preservation expected from vitrification should be seen in the context of that which can be obtained by other preparative methods for electron microscopy, those involving either ice crystallization or dehydration. They are not in the same league!

For pure water or dilute aqueous solutions, vitrification can only be achieved by the direct immersion in liquid cryogen of thin layers (<1 μm in thickness) or of droplets whose diameter is less than a few microns (Mayer, 1985). It has never been convincingly demonstrated that vitrification of pure water can be achieved in larger volumes, even if only at the surface. We imagine that the reason lies in retrodiffusion of the heat generated by crystallization deeper in the sample. Vitrification depth can be increased in the presence of a solute. For an animal tissue, the gain is about one order of magnitude. Another order of magnitude can be gained by vitrification under high pressure (Moor, 1987). Whether the sudden application of 2000-bar pressure onto a biological sample can have deleterious effect must certainly be taken into consideration, but available data are globally positive (Cavalier *et al.*, 2006; Studer *et al.*, 1989)—with some exceptions (Leforestier *et al.*, 1996). Combining these two approaches, 100- to 300-μm-thick samples can be routinely vitrified, thus allowing practical manipulation and observations beyond the zone of direct excision damage. Additional cryoprotection must be added in unfavorable cases.

D. Devitrification

A vitrified sample is metastable. It crystallizes on warming. Amazingly, it is not the usual hexagonal ice that is formed but the slightly different cubic ice. Both have the same density, and in both the distances between nearest neighboring molecules are the same. They differ only at the level of the second neighbor (Dubochet, 2006). In comparison with hexagonal ice, cubic ice can be identified by electron diffraction because the latter lacks several reflections of the former. More straightforward is the difference in the direct image; cubic ice crystals are small, typically in the submicromolar range. The devitrification temperature for pure water is around –135 °C, but it depends on the warming rate and on other parameters such as electron beam irradiation, pressure, and, possibly, cooling history (Dubochet *et al.*, 1988). The presence of a cryoprotectant increases the devitrification temperature within a limited range (Lepault *et al.*, 1997). For solute at the concentrations naturally encountered in most cells and tissues, the devitrification temperature remains well below –100 °C. This fact is important for any method making use of freeze-drying (sublimation) because it means that they never deal with vitreous specimens, only with crystalline ice.

Cubic ice also becomes metastable on warming. It transforms into hexagonal ice. The transition probably takes place around –80 °C, but experimental data are scarce since ice cannot be observed in the electron microscope at this temperature. Minus 80 °C also corresponds to the lower range of temperature at which freeze-substitution is performed. Whether cubic ice can exist under these conditions is still an open question, but some evidence suggests that this is not the case in general. We conclude that freeze-substitution deals with hexagonal crystals, sometimes, perhaps, with cubic ice but never with vitreous water.

We discussed above what happens to biological specimens when liquid water crystallizes. The situation is fundamentally different when crystallization takes place by devitrification. In the former case, molecules in solution and biological particles rearrange by diffusing away from the growing ice crystal. They can move over considerable distances. In the second case, they are immobile because their diffusion coefficient is practically zero. The water molecules themselves do not move much. During devitrification they probably rearrange by rotation with only minimal translational displacement. Under these conditions, the devitrification front progresses without significant displacement of macromolecules. The forming ice crystals just embed the biological structures. It could well be that the second phase of the transformation, namely the transformation from cubic to hexagonal ice, also takes place without significant displacement. Then the result is that all the water of a biological specimen is crystallized into large hexagonal crystals without any segregation, and the structural preservation is excellent down to molecular dimensions.

E. Recipe for Optimal Cryoimmobilization

There is only one good way to achieve cryoimmobilization of biological samples: full vitrification. The result is suboptimal—an amiable euphemism—whenever cooling results in ice formation. It is not sufficient that vitrification is achieved in only some regions of the specimens because ice formation causes dehydration and can thereby damage surrounding regions too. As for the verification of successful vitrification, it should not be left to subjective evaluation of nice preservation. If for no other reason, CEMOVIS should be involved here because it provides the only reliable test of vitrification. Any routinely used procedure for cryoimmobilization should be tested in that way. Once vitrification is achieved, two avenues are offered for further processing. CEMOVIS is the method of choice every time optimal resolution is aimed for, or when the observed structure is "liquid" in its native state, that is, it is formed by structures floating in a liquid medium. However, the method is technically demanding, and it is limited by the sensitivity of hydrated specimens to the electron beam. The other avenue is freeze-substitution. In comparison with conventional resin embedding this method is less prone to aggregation artifacts, though it does not eliminate them, since structures that were swimming in their native state do not remain flying after the liquid is removed. The method should be used for observation at intermediate resolutions, for the tomography of thick sections, the reconstruction of large volumes from a series of serial sections, and when immunolabeling or other room temperature operations must be performed on the section.

Acknowledgments

I am thankful to Ms. P. Buhayer and S. Wauters for help with English and drawings.

References

Adrian, M., Dubochet, J., Lepault, J., and McDowall, A. W. (1984). Cryo-electron microscopy of viruses. *Nature* **308,** 32–36.

Al-Amoudi, A., Chang, J. J., Leforestier, A., McDowall, A., Salamin, L.-A., Norlén, L. P. O., Richter, K., Sartori Blanc, N., Studer, D., and Dubochet, J. (2004). Cryo-electron microscopy of vitreous sections. *EMBO J.* **23,** 3583–3588.

Angell, C. A. (2004). Amorphous water. *Ann. Rev. Phys. Chem.* **55,** 559–583.

Ball, P. (1999). "H_2O: A Biography of Water." Weidenfeld and Nicolson, London.

Bernal, J. D., and Fowler, R. H. (1933). A theory of water and ionic solution, with particular reference to hydrogen and hydroxyl ions. *J. Chem. Phys.* **1,** 515–548.

Cavalier, A., Humbel, B. M., and Spehner, D. (eds.) (2006). "Handbook of Cryopreparation Methods for Electron Microscopy." CRC Press (in preparation).

Drost-Hansen, W. (1969). Structure of water near solid interfaces. *Ind. Eng. Chem.* **61,** 10–47.

Dubochet, J. (2006). Vitreous water. *In* "Handbook of Cryopreparation Methods for Electron Microscopy" (A. Cavalier, B. M. Humbel, and A. Spehner, eds.). CRC Press (in preparation).

Dubochet, J., Adrian, M., Chang, J.-J., Homo, J.-C., Lepault, J., McDowall, A. W., and Schultz, P. (1988). Cryo-electron microscopy of vitrified specimens. *Q. Rev. Biophys.* **21,** 129–228.

Dubochet, J., Al-Amoudi, A., Bouchet-Marquis, C., Eltsov, M., and Zuber, B. (2006). How to "read" a vitreous section. *In* "Cellular Electron Microscopy" (J. R. McIntosh, ed.). Academic Press (submitted for publication).

Dubochet, J., and McDowall, A. W. (1981). Vitrification of pure water for electron microscopy. *J. Microsc.* **124,** RP3–RP4.

Dubochet, J., Richter, K., Roy, H.-V., and McDowall, A. W. (1991). Freezing: Facts and hypothesis. *Scanning Microsc. Suppl.* **5,** S11–S16.

Erk, I., Michel, M., and Lepault, J. (1996). Electron cryo-microscopy of vitrified biological specimens: Ideal and real structures of water-lipid phases. *J. Microsc.* **182,** 15–23.

Franks, F. (ed.) (1982). Properties of aqueous solutions at subzero temperatures. *In* "Water: A Comprehensive Treatise. Water and Aqueous Solutions at Subzero Temperatures," Vol. 7, pp. 215–338. Plenum Press, New York, 7 volumes.

Goodsell, D. S. (1993). "The Machinery of Life." Springer, New York.

Israelachvili, J., and Wennerström, H. (1996). Role of hydration and water structure in biological and colloidal interactions. *Nature* **379,** 219–225.

Jiang, W., and Ludtke, S. J. (2005). Electron cryomicroscopy of single particles at subnanometer resolution. *Curr. Opin. Struct. Biol.* **15,** 571–577.

Leforestier, A., Richter, K., Livolant, F., and Dubochet, J. (1996). Comparison of slam-freezing and high pressure freezing effects on the DNA cholesteric liquid crystalline structure. *J. Microsc.* **7,** 4–13.

Lepault, J., Bigot, D., Studer, D., and Erk, I. (1997). Freezing of aqueous specimens: An X-ray diffraction study. *J. Microsc.* **187,** 158–166.

Ling, G. N. (2001). Life at the cell and below—cell level. *In* "The Hidden History of a Fundamental Revolution in Biology." Pacific Press, New York.

Lucic, V., Forster, F., and Baumeister, W. (2005). Structural studies by electron tomography: From cells to molecules. *Ann. Rev. Biochem.* **74,** 833–865.

Luyet, B. J., and Gehenio, P. M. (1940). "Life and Death at Low Temperature." Biodynamica, Normandy, MO.

Maddox, J. (1983). Snowflakes are far from simple. *Nature* **306,** 13.

Matsumoto, M., Saito, S., and Ohmine, I. (2002). Molecular dynamics simulation of the ice nucleation and growth process leading to water freezing. *Nature* **416,** 409–413.

Mayer, E. (1985). Vitrification of pure liquid water. *J. Microsc.* **140,** 3–15.

Mayer, E., and Brüggeller, P. (1980). Complete vitrification in pure liquid water and dilute aqueous solutions. *Nature* **288,** 569–571.

Mentré, P. (1995). "L'eau dans la cellule—Une interface hétérogène et dynamique des macromolécules." Masson, Paris.

Moor, H. (1987). Theory and practice of high pressure freezing. *In* "Cryotechniques in Biological Electron Microscopy" (R. A. Steinbrecht, and K. Zierold, eds.), pp. 175–191. Springer, Heidelberg.

Pollack, G. H. (2001). "Cells, Gels and the Engines of Life." Ebner and Sons Publishers, Seattle.

Réaumur, R. A. F. (1736). Mémoire pour servir à l'histoire des insectes. *In* "Studies on Low Temperatures," Vol. 2, pp. 141–147. Paris.

Studer, D., Michel, M., and Müller, M. (1989). High pressure freezing comes of age. *Scanning Microsc.* **3,** 253–269.

Taylor, K. A., and Glaeser, R. M. (1974). Electron diffraction of frozen, hydrated protein crystals. *Science* **186,** 1036–1037.

Tokuyasu, K. T., and Singer, S. J. (1976). Improved procedures for immunoferritin labeling of ultrathin frozen sections. *J. Cell Biol.* **71,** 894–906.

CHAPTER 2

Cryopreparation Methods for Electron Microscopy of Selected Model Systems

Kent McDonald

Electron Microscope Laboratory
University of California
Berkeley, California 94720

0091-679X/07 $35.00
DOI: 10.1016/S0091-679X(06)79002-1

Specimen preparation for electron microscopy (EM) is a critically important process because distortions or artifacts introduced at this stage will be represented in the final images taken on the microscope. Nowhere is this more important than in electron tomographic three-dimensional (3D) modeling and analysis of cells because the 3D reconstruction efforts can take weeks or even months to complete. Molecular and cell biologists working on model systems are increasingly turning to EM to analyze mutant phenotypes or to immunolocalize molecules at high resolution. In this chapter, we describe in detail EM specimen preparation methods for three of the most powerful and popular model systems in today's biology: *Saccharomyces cerevisiae*, *Caenorhabditis elegans*, and *Drosophila melanogaster*. Each of these organisms is difficult to fix for EM by conventional methods that use buffered fixatives at room temperature because they each have formidable diffusion barriers that hinder the free exchange of reagents. However, when low-temperature fixation methods such as high-pressure freezing and freeze-substitution are used, the preservation of cellular ultrastructure can be excellent. We discuss the basic equipment for high-pressure freezing and how to use it, what to do with frozen material to prepare it for routine EM analysis, cellular tomography, or EM immunolabeling. We suggest some strategies for improving the visualization of particular cytoplasmic components such as membranes, and comment briefly on the problem of artifacts of cryopreparation methods.

I. Introduction

Structural biology today is increasingly turning to 3D methods at the highest possible resolution to visualize biological ultrastructure from molecules to whole cells. Small objects such as individual molecules or molecular assemblies are regularly analyzed in vitreous ice by crystallographic or single particle averaging methods (Nogales and Grigorieff, 2001). Even cell organelles and extremely small whole cells can be studied in ice using cryo-electron tomography (cryo-ET) (Lucic *et al.*, 2005). With larger cells, however, it is difficult to use these methods because the electron microscope beam will not readily penetrate them, and high-quality image formation is impossible (McIntosh *et al.*, 2005). As a result, some kind of sectioning is required. An excellent, but technically difficult choice is to freeze whole cells or tissues and then cut them into sections at very low temperatures; the resulting sections can be imaged in cryo-transmission electron microscopy (cryo-TEM) (Al-Amoudi *et al.*, 2004; Zhang *et al.*, 2004; Chapter 15 by Dubochet *et al.*, this volume). This method, however, is fraught with problems, particularly for serial reconstruction of large volumes, so a more productive approach is to cryoimmobilize cells, freeze-substitute them into resins, and cut resin sections at room temperature for viewing in conventional TEMs. The resulting resolution may be slightly less than with vitreous cryosections, but it is sufficient for the study of a far greater number of questions in cell biology. Perhaps the best available method for analyzing resin sections at high resolution is ET (McIntosh *et al.*, 2005). In this chapter, we will emphasize methods that result in resin-embedded cells that can be viewed by conventional TEM or ET.

Preserving ultrastructure with the greatest fidelity and at the highest possible resolution is essential for structural work that will improve our understanding of the interactions among molecules in a cell. It has been known for years that cryofixation methods for electron microscopy (EM) specimen preparation give the closest approximation to the cell's native state (see historical accounts in: Echlin, 1992; Gilkey and Staehelin, 1986; Robards and Sleytr, 1985; see Murk *et al.*, 2003 for a comparison of conventionally fixed and high-pressure frozen cells). However, only the smallest cells can be fixed without ice crystal damage by plunging them into a cryogen, and even powerful methods like impact freezing can preserve only a few micrometers of cell depth. High-pressure freezing (HPF) is therefore the best available method for cryofixation of most cells and tissues. HPF has been around for about 20 years (Moor, 1987), but it has not yet been widely adopted by the EM community. True, the necessary machines are relatively expensive, but so are many other pieces of EM equipment. EM laboratories typically spend from one-half to several million dollars for microscopes and related instrumentation. It is ironic, then, that the distortions of cellular fine structure introduced by conventional methods of specimen processing can be orders of magnitude greater than the resolving power of the microscopes. We look forward to the day when more laboratories have adopted cryomethods for routine EM analysis because the quality of EM data, and thus our understanding of cell fine structure, will benefit. In this chapter, we will give detailed methods for improving preservation of ultrastructure in some of the most popular and powerful cell systems now being studied. Coincidentally, these are cell types that are more than usually difficult to fix by conventional EM specimen preparation methods.

How does HPF work and why is it a better cryofixation choice than other types of freezing? HPF was developed by Moor and Riehle and first reported at the European Congress of Electron Microscopy in Rome in 1968 (Moor, 1987; Moor and Riehle, 1968). The impetus to develop such a technique was probably related to the freeze-fracture work that was very popular at the time. Because freeze-fracture requires the freezing of a large volume of sample (so it can more easily be fractured), it was then common to get freezing artifacts (ice damage) unless some cryoprotectant was used. HPF is a method that can, in theory, freeze samples that are as much as 600-μm thick without ice damage, even when no cryoprotectant is used. This greatly expands the types of cells and tissues that can be studied by fast-freezing methods, and this remains the reason why HPF is currently so important, although freeze-fracture itself is rarely used these days. In practice, we now know that the 600-μm figure is overly optimistic and that 100–300 μm is a more realistic working figure for most types of sample, especially if some non-penetrating cryoprotectant is added to the sample before freezing (Dubochet, 1995; Sartori *et al.*, 1993). However, as long as the size of any ice crystals that form is smaller than the resolution of the EM method being used, the procedure will usually be considered acceptable.

HPF works by lowering the freezing point and suppressing the rate of ice crystal nucleation and growth (Moor, 1987). This follows from the fact that the volume of water increases as it crystallizes. The rapid application of a high hydrostatic pressure will inhibit this expansion, slowing the crystallization that occurs with cooling and letting the water become immobilized in a vitreous state before crystals can form. The optimal pressure is around 2045 bar and HPF machines are designed to operate in this range. HPF is not as rapid as some other freezing methods. Impact cooling may reach rates of 10^6 K/sec at the surface, where HPF is about 5×10^4 K/sec. At distances away from the surface the rate slows rapidly, so at the center of a 600-μm sample the rate may only be a few hundred Kelvin per second (Moor, 1987). In practice, this means that HPF is not a good method for arresting fast kinetic events such as synaptic vesicle transmission (Heuser and Reese, 1981). There is also the currently unexplored possibility that fast molecular events, like protein polymerization, are affected by HPF in ways that may lead to misleading images.

II. Equipment and Materials

A. High-Pressure Freezers

There are currently three types of commercially available high-pressure freezers: the BAL-TEC HPM 010, the Leica EM PACT2, and the Wohlwend HPF Compact 01. Space does not permit a detailed description of each instrument, but for the BAL-TEC HPM 010 see Moor (1987) and for the EM PACT see Studer *et al.* (2001). There is no published account of the Wohlwend HPF Compact 01, but it is probably similar in design to the BAL-TEC HPM 010 because Wohlwend was an engineer for this machine. In the subsequent discussions, these two machines will be considered together because the loading and specimen carriers are interchangeable.

One significant difference between the EM PACT machine and the others is its portability. It sits on a cart with wheels, is significantly smaller and lighter than the others, and can be moved from place to place, for example, to a sophisticated light microscope where it can be used for correlative light and electron microscope studies. All of these machines work well with a variety of materials, and the choice of which to use is mostly related to the freezing needs of the users and how these match the capabilities of the machines. In the end, it is not the machine that matters most for achieving good freezing. They are like computers in that what comes out is directly related to what goes in. Specimen loading, on the other hand, is far more critical to the successful freezing of samples. In the sections that follow, we will show the variety of specimen loading containers that are available and how best to fill these for freezing.

B. Specimen Carriers

The most significant aspects of specimen carrier design are the heat transfer properties of its materials and its geometry. Among common metals, copper and aluminum have the best heat transfer properties and most carriers are made of one or the other. Gold is another good choice because it is inert and will not react with cells or their growth media solutions, but it is more expensive. Sapphire is a specimen loading accessory that is used for growing cells in culture, and it is an excellent choice because its heat transfer properties are almost twice as good as copper or aluminum and it is transparent.

1. BAL-TEC HPF 010 and Wohlwend HPF Compact 01

a. Cup-Shaped Specimen Carriers

The most common geometry of sample holder for HPF is a simple, round cup. In the BAL-TEC and Wohlwend machines, the cups are 2 mm in diameter, and they range in depth from 100 to 300 μm. Because two cups are fitted together to make a chamber, one can have well depths from 100 to 500 μm in 50-μm increments and a 600-μm-deep well.

b. Jet Freezing Device Carriers

In the Wohlwend HPF Compact 01 system, there is a variant on the simple cup that is a rectangular-shaped depression 200-μm deep in a thin layer of copper (Walther, 2003). These are holders for the BAL-TEC JFD 030 jet freezer device that can be used for both the Wohlwend and BAL-TEC HPF machines by using a special specimen holder tip available from Engineering Office Wohlwend, Sennwald, Switzerland. To cover the well, there is a flat-sided copper piece that fits on the top. A possible advantage of this type of specimen holder is that the walls are thinner than the conventional sample holder for the Wohlwend/BAL-TEC machines. In theory, this should lead to faster cooler rates at the surface of the sample, although if the sample is thicker than 100 μm, it may have no effect on the cooling rates at the center of the sample (Shimoni and Müller, 1998; Studer *et al.*, 1995).

c. Specialty Carriers

In the BAL-TEC Consumables catalog (http://www.bal-tec.com/products/CONSUMABLES.htm), there are a variety of specialized holders available for specialized applications such as freeze-fracture or freezing in gold tubes between two clamp rings (Shimoni and Müller, 1998).

2. Leica EM PACT2

a. Cup-Shaped Carriers

The specimen cup diameters in the EM PACT machines are smaller than those of the other HPF machines. They are 1.2–1.5 mm in diameter and have depths of 100, 200, and 400 μm. These cups are not used in pairs, so the depth combinations are more restricted than with the BAL-TEC styles. In the original cup design, the pressurizing fluid hit the sample directly and caused deformation of the material in line with the hole. Subsequently, the Leica engineers have devised a new type of cup that has a thin, flexible membrane across the cup and this transmits the pressure from the pressurizing fluid to the material in the cup. These new cups come in depths of 100 and 200 μm, are 1.5-mm wide, and are referred to as membrane carriers. They have a further advantage over the previous versions of the Leica cup specimen carriers in that they are much thinner, which reduces the total thermal mass and probably improves the freezing rates.

b. Tube-Shaped Carriers

In the EM PACT, there is a copper tube 16-mm long with an inner diameter of 300 μm that can be used as a capillary tube to draw up cell suspensions for freezing. It will also hold 200-μm diameter cellulose capillary tubing (Hohenberg *et al.*, 1994) filled with cells. This arrangement can give excellent freezing, as long as the air space between the cellulose tubing and copper tube is filled with hexadecene or some other fluid that will effectively transfer heat. This two-step process has the advantage that the distribution of biological material pulled into the cellulose tubing can be checked before freezing. With the copper tubing alone, it is impossible to see if there are trapped air bubbles or whether the material is tightly packed in the tube. The major application of these copper tubes is for vitreous cryosectioning. Once frozen, the tubes can be transferred to a cryoultramicrotome where the copper is trimmed away with a diamond-trimming tool. The resulting block face can be sectioned at $-160\,^{\circ}$C and the sections viewed directly in a cryo-TEM (Al-Amoudi *et al.*, 2004).

c. Live-Cell Carriers

With the addition of an optional rapid transfer system (RTS) to an EM PACT2 machine, users have the ability to do correlative light microscopy (LM) and EM fixation by HPF with a time resolution of about 5 sec. To accomplish this, it is necessary to modify the stage of an inverted light microscope to accept a special stage that will hold the RTS loader. The live-cell carrier is loaded into the tip of the RTS loader and inserted into the stage for observation by LM. At the moment when the operator wants to fix the cells, the RTS loader is removed from the light microscope and inserted into the RTS module, where it is automatically secured into the freezer specimen carrier holder, inserted into the HPF and frozen. Further details of correlative LM/EM by this machine can be found in Chapter 4 by Müller-Reichert *et al.*, this volume.

d. Specialty Carriers

The Leica system also has specialty carriers for sampling by microbiopsy needle (Hohenberg *et al.*, 1996) or for freeze-fracture or cryo-scanning EM.

C. Accessories

Three important accessories to use with HPF specimen carriers are sapphire disks, cellulose microdialysis tubing, and a microbiopsy gun. These are used with certain types of cells and tissues as explained below.

1. *Sapphire disks*. As mentioned earlier, sapphire has an extremely good coefficient of heat transfer, almost twice as good as the metals used for specimen carriers. Sapphire can be machined into small disks that fit into the specimen carriers of both types of HPF machines. These disks have excellent optical properties and can thus be used as substrates for growing tissue culture cells (Hess *et al.*, 2000; Reipert *et al.*, 2004a). The quality of cell growth can be easily checked in a light microscope; in the EM PACT2 with RTS, they can be observed up until 5 sec before freezing.

If the cells on sapphire disks are processed by freeze-substitution (FS) and embedded in resin, the sapphire disk is present in the resin after polymerization. Methods have been worked out for removing the sapphire disk from the cells in the resin and these can be found in McDonald *et al.* (2006).

2. *Microdialysis tubing*. The idea of using kidney dialysis tubing as a carrier for small cells and even small multicellular organisms was first proposed by Hohenberg *et al.* (1994). The tubing has an inner diameter of about 200 μm and a pore exclusion size of about 10-kDa MW. The trick to successfully using the tubing is learning how to crimp the ends so that the cells do not get lost during processing. This is explained in more detail in Chapter 4 by Müller-Reichert *et al.*, this volume.

3. *Microbiopsy device*. Hohenberg *et al.* (1996) were also the first to publish a method using a very small biopsy needle to facilitate rapid sampling of soft tissues. This works much better than cutting up delicate organ tissues with razor blades or scalpels or large punches because it is faster and causes less mechanical damage. The EM PACT2 HPF has a special biopsy loading station that also speeds up the loading.

D. Cryoprotectants/Fillers

Many of the samples loaded into specimen carriers do not completely fill the cavity; something must be used to fill these spaces or else the samples will not freeze well. Fillers should have the following properties:

1. They should transfer heat effectively during cooling. Aqueous solutions are to be avoided because they turn into ice. As this part of the sample

crystallizes, the latent heat of fusion must be removed, greatly slowing the cooling of adjacent regions of the sample. Moreover, once ice forms it becomes a poor conductor of heat.

2. They should be excluded from the cells, so the organization of the biological sample is not altered or compromised by their presence.
3. They should be physiologically compatible with the cells or tissues.
4. They should have cryoprotective properties, that is, bind water molecules such that they will not rearrange into ice.
5. They should separate from the cells/tissues after FS, either by dissolving in the FS solvent or physically separating from the samples.

Cryoprotectants/fillers are classified as either *extracellular*, because they do not penetrate into cells, or *intracellular*, because they are penetrating (Gilkey and Staehelin, 1986). Some common extracellular cryoprotectants include: 1-hexadecene, yeast paste, *Escherichia coli* paste, cold water fish gelatin, sucrose, serum albumin, and high molecular weight carbohydrate compounds such as dextran and ficoll. Some common intracellular cryoprotectants include: glycerol (5–15%), methanol, ethanol, dimethyl sulfoxide, and ethylene glycol.

The issue of cryoprotectants for HPF is one that needs to be considered very carefully. The nonpenetrating cryoprotectants are to be preferred because they should disrupt the internal cell structure less than penetrating cryoprotectants. But these compounds are not always easy to use, or they may have consequences for downstream processing. For example, dextran works well as a cryoprotectant, but after FS, it forms a hard shell around whatever tissue it surrounds, sometimes making sectioning difficult. Unless physically removed or fractured under liquid nitrogen, hexadecene can form a barrier to free exchange of solvents during FS because it does not dissolve at FS temperatures (Hohenberg *et al.*, 1994; Thijssen *et al.*, 1998). For flies, we prefer to make a paste of dry baker's yeast with 10% methanol or buffer. Mix the liquid and dry components in roughly equal volumes, then adjust the final consistency so that the paste is as thick as possible but can still be used to pack around the fly embryos. Some workers like to use the yeast paste that is used on fly egg-laying plates or trays. For worms, we often use the *E. coli* "lawn" that is used as food on worm plates. Worm laboratories will have lots of dishes with *E. coli*, and if you can find plates that are a little older, that is, when the *E. coli* has a thicker consistency, those will work well for filling the spaces around worms. Thus for both flies and worms, we are using materials that normally surround these organisms in their laboratory growth habitat.

Another choice for flies and worms is 20% bovine serum albumin (BSA) made up in phophate-buffered saline (PBS) or M-9 buffer (Sulston and Hodgkin, 1988) for worms. This is used for correlative LM/EM work on single worm embryos (Chapter 4 by Müller-Reichert *et al.*, this volume) and routinely gives good freezing. It has also been shown to be a good cryoprotectant for HPF of tissue culture cells

(Reipert *et al.*, 2004a), murine skin cells (Reipert *et al.*, 2004b), and a variety of other organisms (McDonald *et al.*, 2006 and unpublished results reported to the author).

Intracellular cryoprotectants are less desirable because they have the potential to interfere with cell physiology, and they may change ultrastructure. Space does not permit an extensive discussion of this issue here, but see Gilkey and Staehelin (1986) for more details. Fortunately, for the organisms that are the subject of this chapter, the choice of cryoprotectant is not usually a problem. Yeast, for example, are simply concentrated to a thick pastelike consistency in their normal growth medium. *Drosophila* embryos and *Caenorhabditis elegans* worms are surrounded by impermeable barriers, such as vitelline envelopes or cuticle, so the material surrounding them for the 30 sec or so before freezing appears to have little or no effect on internal ultrastructure.

E. Tools Useful for Most HPF Specimen Loading Operations

1. Fine forceps for picking up specimen cups and other small items. I particularly like the long fine forceps such as Cat. No. 72919-SS from EMS. For the BAL-TEC machine, an additional pair of bent tip forceps such as Cat. No. 72703-D from EMS or other EM vendors are useful for picking up the specimen cups.
2. Paper points (Ted Pella, Redding, CA, Cat. No. 115–18) or some other wicking material.
3. Fine needles.
4. Fine paint brushes. These are useful for transferring delicate samples, or for loading minute amounts of cryoprotectants into the specimen carriers.

III. General Rules for Loading Samples for HPF

Sample loading is the most critical part of the HPF process!!! This point cannot be overemphasized. Hans Moor, the inventor of the technique, said as much in his comprehensive article on the theory and practice of HPF (Moor, 1987). If you want good results from HPF you must be aware of this fact and act accordingly. The machines are robustly designed and work the same way every time, freezing event after freezing event. So if you are not getting good results it is unlikely to be the fault of the machine. To remedy the situation, you have to check what you are doing before the freezing (specimen loading) or after (e.g., FS). In most cases, it is mistakes with specimen loading that are the problem. We will deal with FS issues in a later section. Here, we will give some general rules to consider and to use when you are freezing. If you use them carefully, you will be rewarded with better yields of well-frozen samples.

A. Work Only with Healthy, Unstressed Cells

Cells/tissues must be in optimal physiological condition. Yeast cells in liquid culture, for example, should be in early to mid log phase growth, if at all possible. They should be at the right temperature, in the most physiologically appropriate medium, and in most cases should be on a shaker. Cells should not be concentrated and allowed to sit on ice. Worms are best picked directly from food plates and not washed off with M-9 buffer or otherwise be surrounded by liquid (Section III.D). They should be well fed and be at optimal temperature and humidity for good growth and development. The same goes for flies on egg-laying plates. As discussed in detail by Hess (Chapter 3, this volume), there are other stresses on organisms, including genetic manipulations, that can adversely affect the preservation of ultrastructure.

B. Work Quickly

For all types of preparations, one should go from living to frozen cells in less than a minute. While this is not always possible with some preparations, it is easy for others. If concentration procedures are necessary, they should be done as quickly as possible and be followed by freezing with a minimum lag. One must always consider the kinetics of the process under study and realize that the interval between the time of departure from optimum conditions and the time of freezing is available for the creation of artifacts that can surpass the damage done by chemical fixation. A well-frozen dead or sick cell is not worthy of study.

C. Do Not Let Your Cells/Tissues Dry Out

The volume of the specimen carriers of nearly all shapes and sizes is less than a microliter. This volume will dry out very fast, especially if the humidity around the sample is low. Try working in a moist chamber or otherwise keeping your material moist over the time you are taking samples for freezing. When loading yeast into a cup-shaped specimen carrier, put a top on the cup (BAL-TEC and Wohlwend machines), or insert the sample into the machine (EM PACT2 with RTS) as soon as the cup is filled. As an experiment, load some yeast into a cup then watch under a dissecting microscope to see how long it takes for the surface to change from a shiny (wet) to a matte finish (dry). If your yeast are highly concentrated to begin with, this will take less than a minute.

When working with worms or flies in *E. coli* or yeast paste, have a vial of M-9 buffer, appropriate fly buffer, PBS, or 20% BSA close at hand and add a little with a fine-tipped paint brush if the specimens in the cup begin to dry out. The worms should be moving in the cup (unless they are paralyzed mutants) when you load them into the HPF machine. For fly embryos in yeast paste, use a little 8% methanol solution in water if you need to moisten the paste.

D. Avoid Surrounding Your Cells/Organisms with Aqueous Media

This is a common mistake that can lead to poor results, even when you do everything else correctly. The importance of this issue has to do with the path of heat transfer from the deep interior of your cells or tissues, through the extracellular medium, and into the metal of the specimen holder. The deeper the specimen cup, the more this is a problem. If there is a significant amount of free water around the cells, this will form ice during the freezing process and block or retard the transfer of heat from your cells. This is the rationale for using external cryoprotectants that will bind or replace the free water outside the cells.

Good practical advice regarding these model systems is to: (1) make sure the yeast are more pastelike than liquid when freezing, (2) avoid concentrating worms from a plate by rinsing them into a tube with M-9. If that has to be done for some reason, try to transfer some worms back to an agar plate and let the agar absorb the excess liquid before loading the worms into a specimen cup, and (3) if fly embryos are dechorionated in liquid solutions, transfer them to a Nitex screen or some other filter to let the excess liquid drain off before loading them into the specimen cup.

E. Avoid Mechanical Damage

Sloppy cutting of tissues or the overfilling of specimen cups can damage your cells. Use biopsy needles where feasible, always use the sharpest scalpels and razor blades for bulk cutting. Disposable dermatology biopsy punches in 1.0-, 1.5-, and 2.0-mm diameters are useful for leaves and other hard materials. It is better to have too little material and fill in the spaces with a suitable cryoprotectant than have sample sticking out the top of the carrier that will get crushed when the top piece is put on.

F. Fill the Carrier Correctly

Use a filler that is compatible with the cells being frozen, that has good heat transfer properties, and do not overfill or underfill the carrier. The ideal fill is a *very* slight overfill that will contact the top piece over the cup and ensure that no air is trapped in the cup as the top comes down. Air is not only a barrier to heat transfer but will collapse under pressure and perhaps cause shear that will damage your tissue. In either case the result will be bad freezing. You can tell with the BAL-TEC and Wohlwend systems if there has been air in the cups because they will collapse inward and leave a concave depression on the outer sides. The best way to learn to recognize this phenomenon is to deliberately underfill a cup and observe the subsequent shape of the cup walls.

G. Use the Smallest Volume of Sample You Can

This may be the single most important factor to getting high yields of well-frozen cells. All the HPF machines give you a choice of specimen carriers with different depths or volumes. The important dimension is the depth, for example, of the cup-shaped carriers. If you are freezing a cell suspension such as yeast, always use the shallowest depth available, usually 100 μm. With the BAL-TEC and Wohlwend machines you can even use slot grids to create variable-depth spacers between two flat-sided specimen carriers (McDonald *et al.*, 2006). For worms, the 100-μm-deep carriers also work well because they are deeper than the thickness of adult worms. Fly embryos, on the other hand, are closer to 200 μm in width, so 200-μm-deep carriers must be used. For cells in suspension, dipping a mesh grid into the suspension and sandwiching it between two flat-sided specimen cups or very shallow custom-made cups (Müller *et al.*, 1980; Murk *et al.*, 2003) is another option.

IV. Methods for Specific Organisms

A. Yeasts (*Saccharomyces cerevisiae* and *Schizosaccharomyces pombe*)

1. Background

Yeast has long been associated with developing EM cryomethods, going back to the early days of freeze-fracture studies (Moor, 1967; Moor and Muhlethaler, 1963). Using plunge freezing methods, both *S. pombe* (Tanaka and Kanbe, 1986) and *S. cerevisiae* (Baba and Osumi, 1987) were shown to have vastly improved ultrastructural preservation after FS and thin sectioning. Despite these early observations, studies of yeast ultrastructure have continued to use conventional room temperature methods of specimen preparation, and even in a relatively recent publication (Heiman and Walter, 2000), a modified potassium permanganate method was employed. The idea that freezing followed by FS would improve visualization of fungal cell ultrastructure in thin sections goes back to the pioneering work of Howard and Aist (1979). Mendgen and coworkers began using HPF as the freezing method for a number of different fungi (Knauf and Mendgen, 1988; Knauf *et al.*, 1989; Welter *et al.*, 1988). The work by Ding *et al.* (1993) on *S. pombe* and Winey *et al.* (1995) on *S. cerevisiae* were early efforts using HPF to study yeasts where the focus was on the biology rather than the technique. By now, there are dozens of papers using HPF on yeasts to address a wide variety of biological questions.

2. Materials

a. Vacuum source (pump or house line)
b. Vacuum filtration apparatus, typically a 15-ml Millipore setup
c. 0.45-μm pore size polycarbonate filters, 25-mm diameter
d. Toothpicks or other implement to scrape cells from filter
e. Syringe needle bent at 45° near the tip

f. Yeast cells in liquid culture, growing in early to mid log phase (OD = 0.2–0.4), you will need 5–10 ml for every freezing event, so adjust the total volume to your needs
g. 2% agar made up with YPD (see Difco Manual or search for "Yeast Plates" on a web browser) and poured into 5- to 8-cm-plastic Petri dishes

3. Methods

a. Set up filtration apparatus.
b. Pour 5–15 ml cells into filtration column.
c. Apply suction to pull down cells onto filter, using care not to let them get too dry.
d. Remove column from apparatus, place filter with cells on agar plate.
e. Scrape cells from filter with toothpick tool.
f. Fill well of specimen carrier with cells.
g. Remove excess cells from carrier with syringe needle tool by scraping over the surface. Instead of moving from side to side, put needle across the center of the cup and move to one side. Repeat moving to the opposite side but leave a small ridge of yeast across the middle to ensure that the cup is very slightly overfilled.
h. Freeze.

4. Further Reading

For more detailed discussions of methods for preserving yeast for EM see Giddings *et al.* (2001) and McDonald and Müller-Reichert (2002). In the latter reference, there are also methods for users who do not have access to HPF technology.

B. Nematodes (*C. elegans*)

1. Background

There is a long history of EM associated with *C. elegans* because the organism was developed as a model system with EM in mind (Brenner, 1973). Early studies used serial sections to map out the structure of the anterior sensory anatomy (Ward *et al.*, 1975), of the ventral nerve cord (White *et al.*, 1976), and the pharynx (Albertson and Thomson, 1976). These, and many other *C. elegans* EM references can be found on the *C. elegans* WWW Server (http://elegans.swmed.edu/) using "electron microscopy" as a search term in the Literature Search link. While the quality of specimen preservation was sufficient for tracing cell lineages as well as other descriptive work, it would not be good enough to meet today's standards for preservation at the molecular level.

2. Materials

a. A tool for picking worms off a plate of *E. coli*
b. *E. coli* for use with the picking tool, older plates with thicker, "stickier" bacteria work best
c. Alcohol lamp for flame sterilizing the worm-picking tool
d. Specimen carriers and accessories for the appropriate HPF machine:
   1. EM PACT machines. Typically we use the100-μm-deep membrane carriers
   2. BAL-TEC machines. For this instrument you will need:
      –Type B specimen carriers (flat on one side)
      –TEM slot grids (EMS Maxtaform Cu-Rh 1 $\times$ 2 mm^2, Cat. No. M2010-CR)
      –1-Hexadecene
      –60-mm plastic Petri dish fitted with filter paper. Saturate the filter paper with the hexadecene
      –A number 0 red sable paint brush
      –A number 11 scalpel blade
      –Fine needles
e. M-9 worm buffer solution
f. 20% BSA in M-9 buffer

3. Methods

a. EM PACT2 HPF machine:
   1. Place a 100-μm membrane carrier in the loading device.
   2. Slightly overfill the cup of the membrane carrier with 20% BSA solution in M-9.
   3. Using the worm-picking tool loaded with *E. coli*, pick 15–20 worms off the worm plate. Touch these to the BSA in the cup and allow the worms to swim off into the specimen carrier or push them off the pick with a fine needle.
   4. Use the paper points or other wicking material to lower the level of liquid so that there is only a slight excess of material in the cup.
   5. Load and freeze in the EM PACT machine.
b. BAL-TEC HPF machine:
   1. Place a number of Type B specimen cups (flat side down) and slot grids on the 1-hexadecene saturated filter paper.
   2. Blot off excess 1-hexadecene on the specimen cup on dry filter paper and put into the tip of the specimen cup holder flat side up.
   3. Put the slot grid (1-hexadecene side down) on top of the specimen cup. Do not blot excess off the slot grid.

4. Using the paint brush, put a small amount of M-9 buffer in the slot. It will bead up and should occupy only about half the slot area.
5. Pick 20–25 worms with the worm pick or *E. coli* paste.
6. Use the tip of the scalpel blade to scrape the worms and *E. coli* into the M-9 droplet in the slot grid.
7. Use fine needles to mix the worms and *E. coli*. Allow the solution to thicken and the volume to reduce. Do not let the paste dry out completely. The worms should continue to move throughout this process. If the paste becomes too dry, you can reconstitute it with M-9 from the paint brush. This procedure takes a little practice to get right, and practicing ahead of time will reduce frustration and promote the necessary skills.
8. When the area of the slot grid is only slightly overfilled with worms or *E. coli* solution, take another specimen carrier from the 1-hexadecene pad and place it flat side down on top of the worms.
9. Close the tip of the specimen loader and freeze in the BAL-TEC or Wohlwend HPF machine.

4. Further Reading

For more detailed information about cryomethods for preserving nematodes for EM see Müller-Reichert *et al.* (2003a).

C. *Drosophila* Embryos

1. Background

EM studies of *Drosophila* go back to the early studies of Lowman (1958), Mahowald (1962, 1963), and King (1960; King *et al.*, 1966; Koch and King, 1966) among others. More than yeast or worms, EM studies of *Drosophila* have been published regularly until the present day, although few researchers have chosen to use rapid freezing and FS methods. *Drosophila* was among the first organisms to be studied by HPF (Müller and Moor, 1984) and has been the object of HPF study in some more recent publications (Bardsley *et al.*, 1993; McDonald and Morphew, 1993; Schulte *et al.*, 2003).

2. Materials

1. Embryos on food plates or harvested and dechorionated. Do not store embryos in liquid solutions. Keeping them moist on agar or food plates is the best solution.
2. Yeast paste. Many fly embryo plates are seeded with yeast paste as food for the flies. You can use this as a filler or make up a separate paste. Making up a separate paste using 8% methanol as a wetting agent for dry yeast will give a better yield of well-frozen embryos and will not affect them adversely.
3. Number 0 red sable paint brush.
4. 8% methanol solution in an Eppendorf tube.

3. Methods

1. Working on an agar or food plate, use the paint brush to bring together 20–25 embryos in a small clump. If the embryos are covered with a liquid layer, wick that off, or let them sit a while until the surface appears dry.
2. Take a small amount (practice will tell you how much) of yeast paste on the tip of the paint brush and mix it with the clump of embryos until all are thoroughly coated with yeast.
3. Place a 200-μm-deep cup-shaped specimen carrier (either BAL-TEC or Leica) on the surface next to the coated embryos and transfer the yeast–embryo mixture into the cup so that it is full but not overfull. The ideal fill is just to the level of the top edge of the cup. If it is too full, the cells will be compressed when the top is put on. If it is not full enough there will be an air space and heat transfer will be seriously impeded.
4. Transfer the specimen carrier to the HPF specimen loader and freeze.

4. Non-embryo *Drosophila* Tissues

Processing adult or larval fly tissues is difficult because of the cuticle or other hard materials that cover the cells. These materials probably do not interfere much with the freezing process itself, but rather the subsequent FS steps where organic solvents and fixatives still have to cross the diffusion barrier and replace the cell water inside the tissues. Another problem is that the adult and larval tissues are generally too large to fit into even the largest HPF specimen carriers. Therefore, it is necessary to dissect the desired parts away from the intact organism before freezing and this may also have negative consequences for cell fine structure.

5. Further Reading

For more information and details about HPF and subsequent EM processing of *Drosphila* tissues see McDonald (1994) and McDonald *et al.* (2000).

V. Postfreezing Processing

Freezing specimens is only the first of several steps between living tissue and the electron microscope. Care must be taken at each step to ensure that high-resolution structural information is not lost. Common options after HPF include: freeze-fracturing and metal coating, vitreous cryosectioning, and FS and embedding in resin. In this chapter, we will only concern ourselves with FS and embedding because this is by far the most common way to prepare samples after HPF.

A. Freeze-Substitution

A literature search for "freeze-substitution" will turn up a large number of examples, some going back 40 years or more. The idea of using low-temperature dehydration and fixation methods for EM is usually credited to Fernandez-Moran (1960), although the idea of FS was well known in light microscope cytology (reviewed in Feder and Sidman, 1958). Freeze-drying for EM is a much older technique and was used in some very early EM observations of biological material (Richards *et al.*, 1943). Van Harreveld and Crowell (1964) are often cited for using osmium-acetone as an EM FS fixative. For the history-minded, the book by Echlin (1992) has references to the earlier literature of cryomethods. Other papers to consult for general principles and applications of FS are by Steinbrecht and Müller (1987), Hippe-Sanwald (1993), and Nicolas and Bassot (1993).

FS means to substitute an organic solvent, usually acetone, for the cell water at a low temperature, typically −78 to −90 °C. Other solvents such as methanol or ethanol are sometimes used, but the convention for most people today is to use acetone. Unless one is doing low-temperature embedding with Lowicryls, it is usually necessary to add a fixative to the FS solvent. Osmium tetroxide is the most common fixative, but paraformaldehyde and/or glutaraldehyde can also be used. Unlike conventional processing, the fixation takes place after or perhaps during the dehydration steps. At −90 °C, acetone will dissolve the cell water over a period of hours to days, but the fixatives are not thought to be very reactive at this temperature (Humbel and Müller, 1986). However, the fixative becomes distributed throughout the entire cell or tissue, and when the temperature is permissive for fixation, the fixative is "in place" and does not need to diffuse through relatively great distances, as it does for conventional processing based on diffusion of reagents. The temperature where fixation begins is not well known, although the studies of Humbel and Müller (1986) suggest that glutaraldehyde is active at about −50 °C and osmium tetroxide at −30 °C or so. Another important reference that deals specifically with fixation of chromatin at low temperatures is Horowitz *et al.* (1990). In this study, cross-linking of chromatin with 3% glutaraldehyde in acetone begins at −45 °C.

1. Which FS Method to Choose for Your Cells?

FS is a process that is poorly understood, and there has been little systematic research to probe basic mechanisms. Consequently, there are many different methods published in the literature. The majority of these methods appear to work, so it seems to be a process with a large margin of error. The practical problem for the new user, however, is which method to choose. One suggestion is to search the literature to see if there are good results on the cell and tissue type that you are working on. This seems like a good idea, but be aware that differences in the reagent chemistry from one laboratory to another or subtle procedures not published in the original paper can mean that the method will not give the same good result. Another approach is to select a general protocol that seems to work for a wide variety of tissues and use that as your starting point. This is the approach we have taken in our laboratory. By always starting with the same method, we see the variations that are due to the chemistry of the cells themselves. If we do not like the results, we can begin to change some of the variables in a systematic way to see how we can improve them. Getting a good FS result may require a lot of problem solving and experimentation with the steps in the process. The steps include not only those of time, temperature, and chemistry of the FS process itself, but there are a number of variables during HPF that must also be taken into account.

2. Variables During Freezing

Most of these points such as working with optimally healthy cells, choosing the right filler, and so on have already been discussed in Section III. In addition, there are some issues that we can mention here that might help illustrate the subtle nature of the factors that can affect freezing.

a. The Genotype of the Cells Being Frozen

It is our experience that some mutants do not look as good as wild type even when they are processed in exactly the same way. As discussed in detail by Hess (Chapter 3, this volume), this is only one of several ways that organisms become altered in ways that will adversely affect the preservation of ultrastructure.

b. The Way the Cells Are Grown Can Make a Big Difference

For example, we have heard anecdotally from a yeast researcher that differences in the carbon source in a yeast medium (e.g., acetate vs glucose) can make a big difference in the visualization of mitochondrial membranes. Another plant researcher claims that using mannitol instead of sorbitol in the buffer will make the difference in the final appearance of plant protoplast cells in the microscope (Lonsdale *et al.*, 1999).

3. Variables During the FS Run

a. How the FS Cocktail Is Made and Stored

We make batches of fixative in cryotubes and freeze them for use later. They are stored in liquid nitrogen. This has the advantage that when you are done with HPF, you can put your sample directly into the fixative and start an FS or you can store the sample in the cryovial/fixative until you are ready to do the FS. We have published a detailed recipe for preparing FS fixatives in several earlier papers (McDonald, 1999; McDonald and Müller-Reichert, 2002; McDonald *et al.*, 2000); this method works well for us. The important thing is not to leave the fixative mixture at room temperature any longer than necessary. Also, when freezing the fixative in cryovials, make sure it is upright when frozen so the liquid does not get into the threads of the cryovial cap. If this does happen, it makes it difficult to unscrew the cap when loading samples under liquid nitrogen.

We always make up our FS solutions from small bottles of just-opened acetone, as opposed to acetone over molecular sieves. We believe, as has been mentioned in the literature (Steinbrecht and Müller, 1987), that acetone extracts something from molecular sieves that can react with osmium tetroxide. Because uranyl acetate is not readily soluble in acetone, we make up a 5% stock solution in methanol and dilute that with acetone to get a final concentration of 0.1%.

b. The Chemistry of the FS Medium

This includes choice of solvent(s), choice of fixative and concentration, and whether or not additives are added to the fixative–solvent cocktail. For example, we have chosen 1% osmium tetroxide plus 0.1% uranyl acetate in acetone as our standard FS cocktail for starting morphological studies. Acetone substitutes the water at a much slower rate than methanol, and this turns out to be a desirable property. If the substitution goes too rapidly, as it can do with methanol, the ultrastructure may suffer (McDonald, 1994; Steinbrecht, 1993). It is hard to generalize about this because every tissue is different and studies using methanol have produced good preservation of cell fine structure (Müller *et al.*, 1980). Increasing the concentration of osmium tetroxide may add contrast to some samples, but we find that 1% works well for the model organisms covered in this chapter. Additives to the solvent–fixative cocktail can make a difference to the final appearance of cells. We always include about 0.1–0.2% uranyl acetate with both our osmium and glutaraldehyde FS cocktails. We find that it adds contrast to membranes and seems to prevent destruction of ultrastructure that can occur if the cells are in osmium/acetone alone at room temperature for more than an hour. With the uranyl acetate added, one can leave the cells at room temperature for longer periods and the solution will not turn black as quickly, another bit of evidence suggesting that uranium could be affecting the osmium interactions with biological material. There are other ways to add membrane contrast, and one of the more intriguing is to add water to the FS cocktail (Walther and Ziegler, 2002). This approach is covered in more detail in Chapter 3 by Hess, this volume.

c. Time and Temperature

Most FS protocols start with the samples held at a temperature of −78 to −90 °C for some amount of time, then warming of the sample over another prescribed time period. The initial period varies from a few hours to a week, and the warm-up time can be similarly varied. There are few systematic studies of these variables, but Steinbrecht did some experimentation on FS times for moth antennae some years ago (Steinbrecht, 1982; Steinbrecht and Müller, 1987). These showed that if the time at −80 °C were long enough (7 days), the sample could be warmed up in 1 min without significant ice damage. Likewise, time at −80 °C could be reduced to 5 min if the warm-up time were extended over 6 h. The problem with this experiment, and all similar experiments, is that they may only be true for the specimens under study. Given the extreme heterogeneity of biological tissues, generalizations are not really possible. The approach taken by most investigators is to be conservative, leaving tissues at −80 to −90 °C for 8–72 h, and using warm-up times from 6 to 24 h.

In our laboratory, we tend to use a protocol that varies the period at −90 °C but leaves the warm-up period constant. For example, if we are doing an FS at the beginning of the week, we hold at −90 °C for as little as 5 h. If we are freeze-substituting over a weekend, it might hold for as long as 36 h. We chose a warm-up rate of 5 °C/h, and we halt the warm-up at −25 °C for 12 h, because we found that we get more membrane contrast that way (Section V.D). Then we continue warm-up to 0 °C where we hold the samples until we are ready for further processing. At that point, samples are warmed rapidly to room temperature, rinsed three times over 15 min in pure acetone and infiltrated with resin for embedding.

B. Embedding

The choice of embedding resin is another variable that can influence the final appearance of your high-pressure frozen, freeze-substituted sample. We generally use an Epon-Araldite formulation of medium hardness (Mollenhauer, 1964) because of its excellent sectioning and beam stability properties. For yeast, we find that a mixture of Epon and Spurr's resins (McDonald and Müller-Reichert, 2002) works very well for revealing cytoplasmic details (Fig. 1). Generally, we do not use Spurr's resin for anything because it will extract membranes and other cytoplasmic components (Hess, 2003 and Chapter 3, this volume). However, because the cytoplasm of yeast is so extremely dense compared to most cells, perhaps a little extraction is useful in this particular case.

C. FS and Embedding for Tomography

For samples of resin-embedded material analyzed by ET, the contrast comes mostly from electron-dense molecules used in FS (osmium and uranyl acetate) and/or poststaining (uranyl acetate and lead citrate). In some cells like yeast and in some structures, like the spindle pole body, the electron density can become so high after FS and poststaining that it is difficult to visualize organelle substructure. A strategy for making medium-density images of well-fixed samples is to freeze-substitute them in relatively high concentrations of glutaraldehyde instead of osmium tetroxide, and use only the poststaining solutions to generate contrast (McDonald and Müller-Reichert, 2002; O'Toole *et al.*, 2002). Embedding in a methacrylate resin such as Lowicryl HM20 also seems to help. For worm tissues, we find that our standard FS protocols work well for tomographic studies (O'Toole *et al.*, 2003). So far, we have not done any tomographic studies of *Drosophila*, but based on our experience with routine sectioning studies, we expect that standard FS methods will work well for most tomographic analyses.

D. FS and Embedding for Immunocytochemistry

For immunocytochemical studies our standard fixative is 0.2% glutaraldehyde plus 0.1% uranyl acetate in acetone. Again, we find that the addition of uranyl acetate has a beneficial effect on membrane contrast, but does not seem to interfere with most antigen–antibody reactions, at least not at this concentration. For yeast, we sometimes use another fixative combination made up of 0.1%

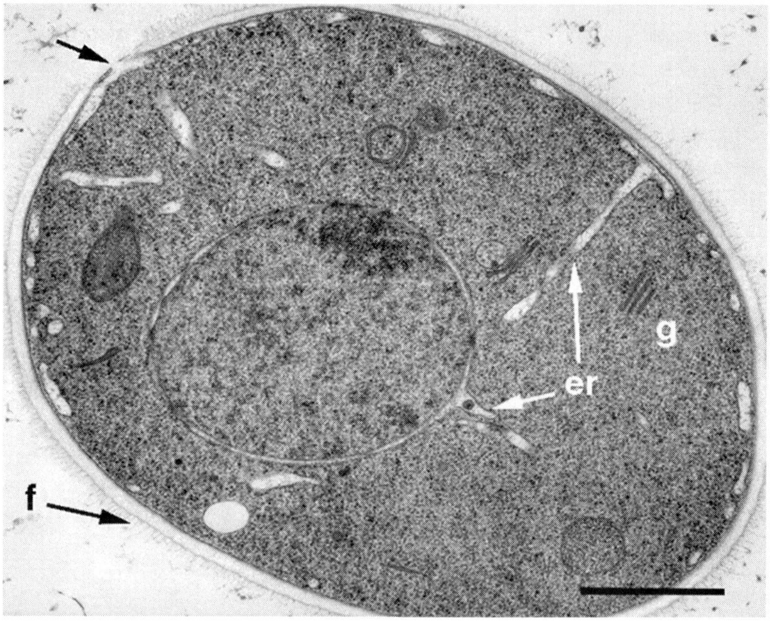

Fig. 1 *S. cerevisiae* visualized in a 60-nm section after HPF-FS and embedment in Epon-Spurr's resin. Endoplasmic reticulum (er) extends from the nuclear envelope toward the cell cortex. Three cisternae of a Golgi apparatus (g) are visible as are the fine projections called fimbriae (f) at the surface of the cell wall. The cell wall is broken (arrow, upper left), as sometimes happens with high-pressure frozen yeast cells. Scale bar = 1.0 μm.

glutaraldehyde, 0.5% uranyl acetate, and 0.01% osmium tetroxide in acetone. This small amount of osmium is sufficient to add membrane contrast, but it still will permit immunolabeling, at least with some antibodies (McDonald and Müller-Reichert, 2002; Müller-Reichert *et al.*, 2003b).

The embedding medium that we prefer for immunolabeling work is LR White resin, the hard formulation. We order our LR White from Ted Pella, Inc. (Redding, CA) because it is shipped with the accelerator separate from the resin. Accelerator is only added when the resin is made up. Some other vendors have the accelerator in the resin at shipping, and you do not know how long it has been on their shelves before that time. We find that LR White is adequate for worms (Fig. 2; Kosinski *et al.*, 2005) as well as flies (Bardsley *et al.*, 1993; McDonald *et al.*, 2000) and yeast (Cid *et al.*, 2001). Lowicryl is probably a superior resin for preserving cell morphology, but when you are working with small, colorless samples in a −50 °C environment, it is sometimes difficult to see what you are doing. The exception to that problem is yeast, which tend to form a firm disk after FS that usually releases from the HPF specimen carrier. These disks can be transferred to gelatin capsules in the FS device and polymerized at low temperatures. We recommend using Lowicryl HM20 resin and polymerizing with UV irradiation at −50 °C.

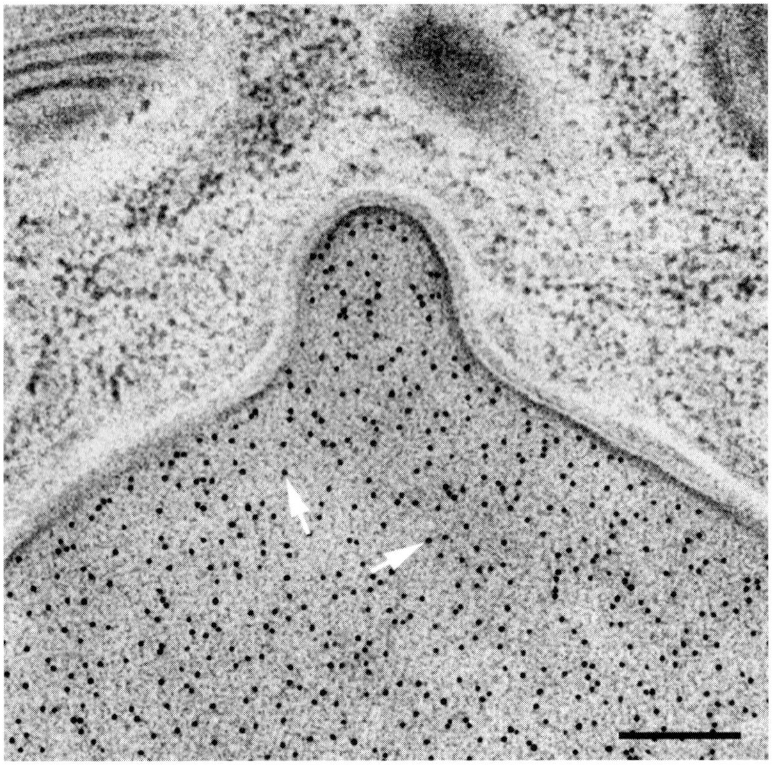

Fig. 2 Part of an amoeboid *C. elegans* sperm following HPF, FS in 0.2% glutaraldehye plus 0.1% uranyl acetate in acetone, and embedding in LR White. 10-nm gold particles (arrows) show the distribution of major sperm protein (MSP). Antibody against MSP courtesy of David Greenstein, Vanderbilt University Medical School. Scale bar = 200 nm.

For worms and fly embryos we prefer to embed in thin layers of resin, so we can screen the organisms in the light microscope prior to sectioning. For this approach, it is necessary to use LR White resin and work at room temperature. Although this method was originally developed in conjunction with microwave polymerization (Lonsdale *et al.*, 2001), we now use oven polymerization more routinely. We construct the thin layer wells as shown in Lonsdale *et al.* (2001) but put them in a sealed Rubbermaid storage container that has been flooded with nitrogen gas to exclude oxygen. This is then put into the oven overnight, and the next day the chamber is flooded with N_2 gas again and put back into the oven for one more day to ensure complete polymerization. After screening with the light microscope, individual worms/fly embryos are cut out, remounted, and precisely oriented for sectioning. For yeast, we almost always embed small pieces of yeast pellets in LR White in flat-bottomed capsules (Ted Pella Cat. No. 133P) polymerized in the microwave (Cid *et al.*, 2001).

E. A Case Study of Adjusting FS Variables to Improve Visualization

When you look at a sample in the microscope and find that you do not like the way certain features look, you should consider how to change the FS methods to improve the situation. As an example, we present here the changes that evolved in our work to get improved visualization of the yeast nuclear envelope (Fig. 3). In our earliest attempts at FS with yeast, we used 2% osmium tetroxide in acetone

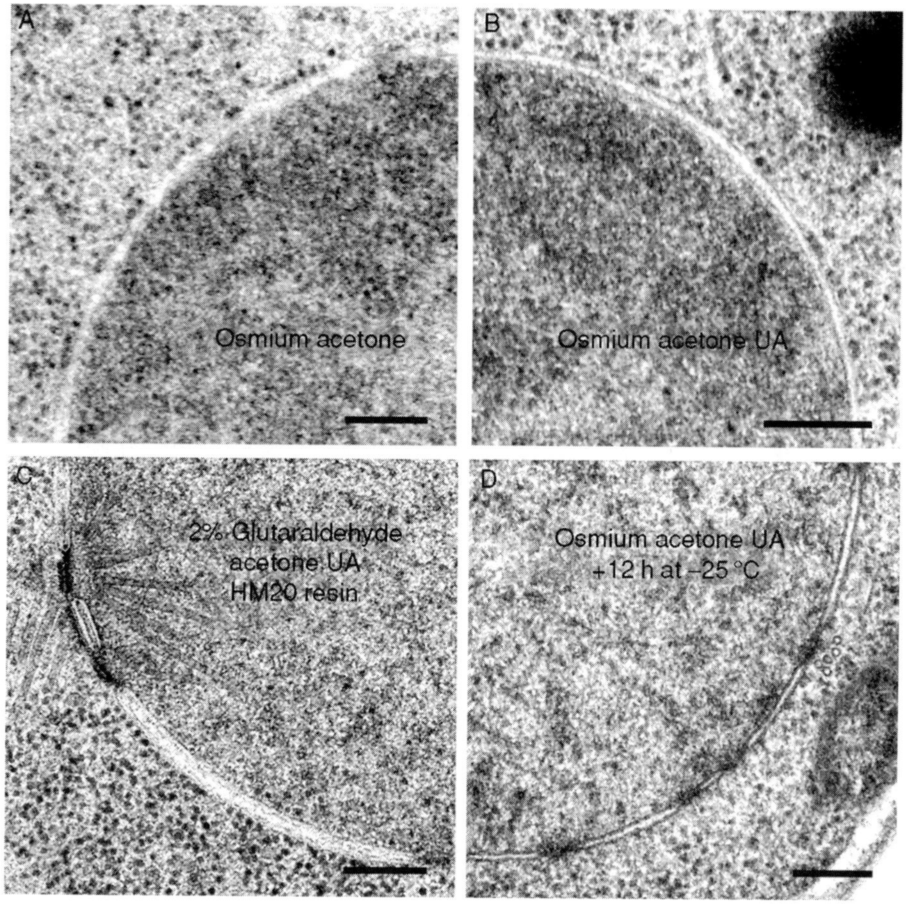

Fig. 3 Changes in the appearance of the nuclear envelope of *S. cerevisiae* cells as a function of changes in the FS protocol. Beginning with 2% osmium tetroxide in acetone (A) there is no clear density in the nuclear envelope; it is detectable mostly because of the absence of staining in the lumen between the double membranes. (B) The addition of 0.1% uranyl acetate to the osmium acetone mixture creates faint dark lines in the areas where we expect to see membrane. By using the same FS cocktail as in (B), but holding the FS temperature at −25 °C for 12 h during the warm-up period, the membrane contrast is increased even further (D). If the cells are fixed "hard" with 2% glutaraldehyde during FS, then some membrane contrast can be generated from uranium and lead poststains alone (C). Scale bars = 200 nm.

as the FS fixative cocktail, because this was considered standard at the time. The results were okay, but a number of people were concerned about the lack of contrast in the membranes (Fig. 3A). Sometime later, Mary Morphew of the Boulder Laboratory added uranyl acetate to the FS mix, and this seemed to present a more familiar membrane image (Fig. 3B). Then, as a result of experiments on other organisms with low membrane contrast, we came up with the idea of holding the FS warm up at −25 °C overnight to add even more membrane contrast (Fig. 3D). This is now our current standard starting FS protocol for all new organisms. In general, if you want more contrast, you can incubate samples in this FS cocktail for variable periods, and the longer you hold it the more contrast you will get, up to the point when the reaction saturates. However, it is also possible to visualize membranes without using osmium at all. In Fig. 3C, the cells have been prepared for tomographic analysis using 2% glutaraldehyde plus 0.1% uranyl acetate as the FS fixative. The contrast here comes mostly from the poststaining of sections in uranyl acetate and lead citrate. These cells were embedded in Lowicryl HM20, which also seems to help the contrast.

We must mention that the desire to see membranes rendered as dark lines is an interesting paradox. This view does not match what one would predict from the distribution of lipid components in membranes, but it is one that has evolved as a consistent artifact of conventional EM preparations; most people are more comfortable when they see membranes this way (see also Chapter 3 by Hess, this volume). We should also say that membrane contrast in freeze-substituted samples is highly variable from organism to organism and even within cells. It probably reflects the differences in the membrane chemistry as well as the FS protocol. In general, membranes in well-frozen samples will look less contrasty than the same membranes prepared by conventional methods. For the organisms that are the focus of this chapter, however, the yeast are the most problematic. Worms (Figs. 4 and 5) and fly tissues generally give reasonable to good membrane contrast with the methods we recommend here. For an alternative approach to the FS of yeast cell membranes, see Walther and Ziegler (2002). We have tried this method in our laboratory, and the images of yeast membranes are excellent.

F. Sectioning

Material prepared by rapid freezing and FS is more dense than that prepared by conventional methods because there is less extraction of the cytoplasm. For yeast cells, especially, it is a good idea to cut sections thinner than you might normally do. We typically section at 40–60 nm when studying yeast cells, and these sections look sharper than those cut thicker (see Fig. 4 in McDonald and Müller-Reichert, 2002). For worms and flies, we vary the thickness between 60 and 100 nm, depending on the cytoplasmic features we are interested in.

The time and type of poststaining on the sections can also be important to how the cells look in the microscope. With the procedures for FS and embedding that we normally use, we find that poststaining in 2% uranyl acetate made in 70%

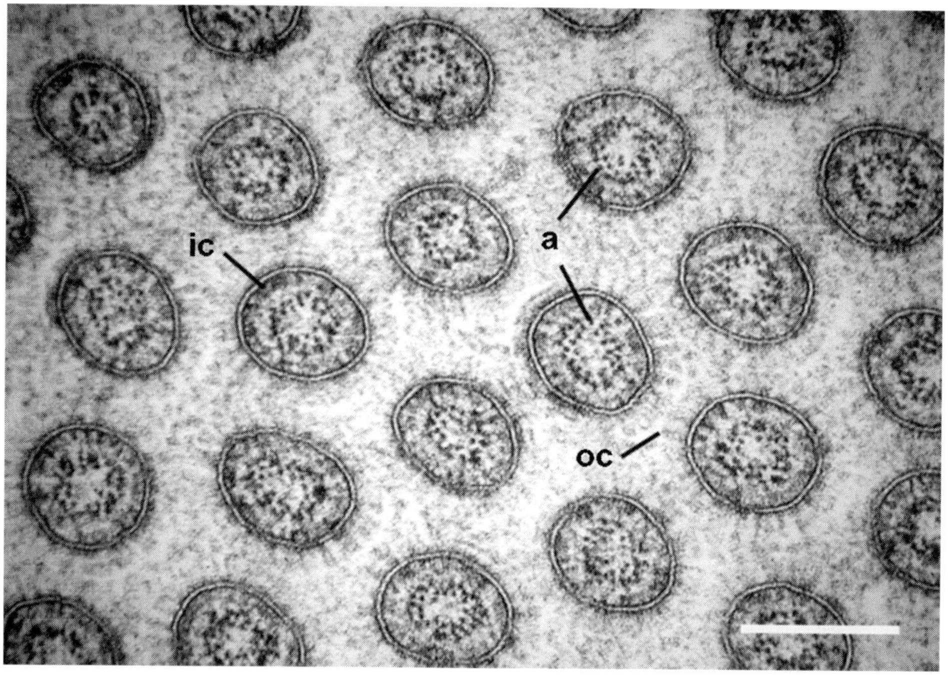

Fig. 4 A cross section through microvilli on the gut epithelial cells of *C. elegans*. Individual actin filaments (a) are visible in the cores of the microvilli, and molecules coating both the inner (ic) and outer (oc) sides of the microvillar membrane are evident. Scale bar = 200 nm.

methanol for 5 min, followed by 3 min in Reynolds (1963) lead citrate gives good results. Longer staining times can actually reduce the contrast on yeast sections because while the overall staining density is greater, the contrast goes down because the range of tones is less. For sections of fly or worm material, we may go as long as 10 min in methanolic uranyl acetate and 5 min in lead citrate. Staining sections with 1% aqueous tannic acid for 3 min (then rinse well) prior to staining them in uranyl acetate and lead citrate can also be very useful if you are interested in cell walls, or certain classes of membranes and endocytic vesicles (Fig. 6).

G. Microscopy and Evaluation of Results

Looking at sections or micrographs of material prepared by rapid freezing and freeze-substitution fixation (RF-FSF) techniques will present some unfamiliar images to even experienced EM users if they are only familiar with conventional fixation (CF) methods. The cytoplasm of RF-FSF material will be denser and have less contrast than CF material (Figs. 7 and 8). This is because the cytoplasm is less extracted and there is less collapse and concentration of proteins and other structures (Kellenberger, 1987). One of the features of a cell that is easiest

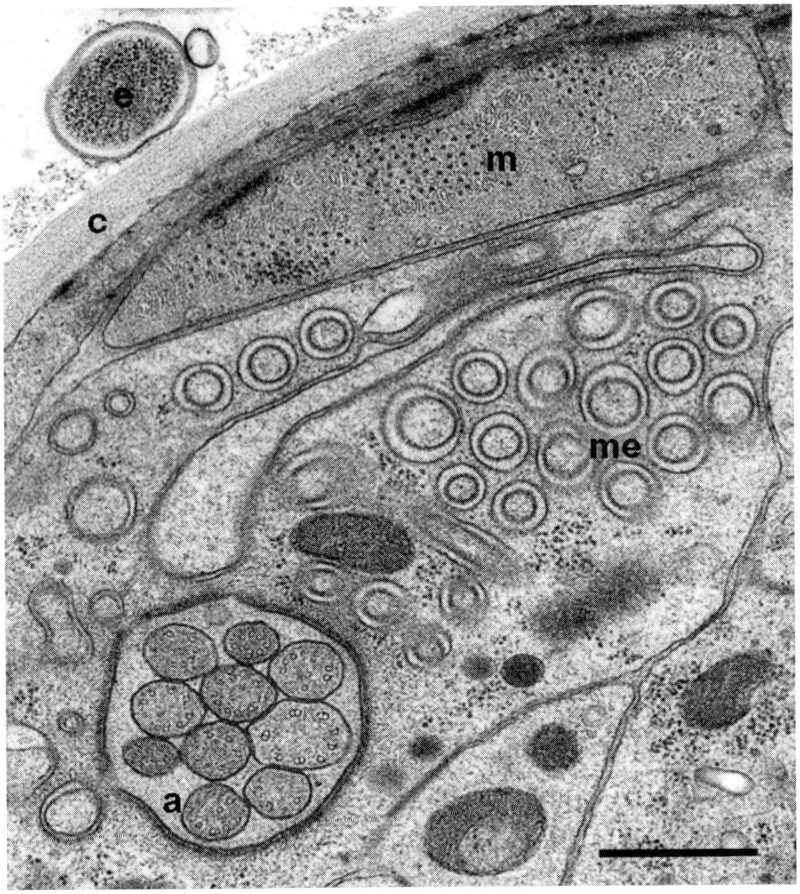

Fig. 5 Details in a cross section of the head region of *C. elegans*. An *E. coli* cell (e) can be seen on the outside of the cuticle (c). A cross section through the body muscle (m) is just inside the cuticle, and concentric membranes (me) and amphid cilia (a) are seen further in. Scale bar = 0.5 μm.

to assess regarding fixation quality is the shape of the nucleus. We know from LM of living cells that most nuclei are round or ovoid; therefore, a section through the nucleus should look like a circle or an ellipse (Fig. 7A). If only part of the nucleus is visible, the nuclear envelope should form a smooth curve. If the nucleus has an irregular outline (Figs. 7B and 8B), then it has most likely suffered from distortions during specimen preparation. Of course, some nuclei, such as those occurring in certain classes of leucocytes, are lobed naturally but these are usually characterized by deep invaginations in the nuclear envelope, as opposed to the kinds of patterns shown in Figs. 7B and 8B. Other organelle membranes can show similar distortions due to collapse during specimen preparation. Space

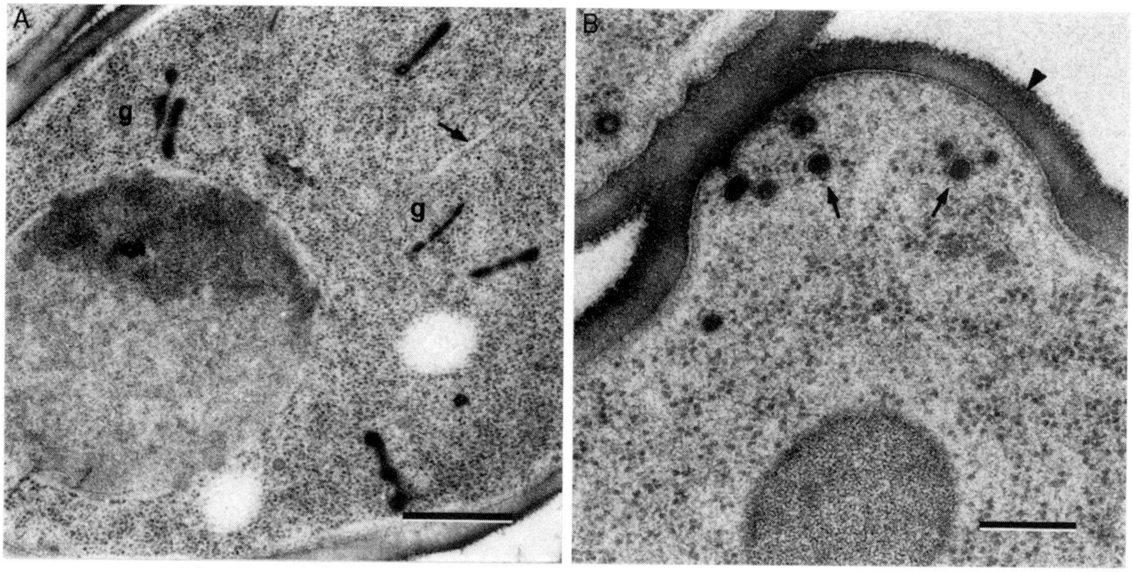

Fig. 6 Use of special poststain procedures on sections to enhance certain cytoplasmic details. A method developed by Susan Hamamoto at the University of California, Berkeley uses a 3-min incubation in 1% aqueous tannic acid, followed by extensive rinsing in dH_2O before conventional poststaining with uranyl acetate and lead citrate. Golgi membrane cisternae (g in A) as well as certain vesicles located in the emerging bud (arrows in B) react with the tannic acid plus the other poststains and become darkly stained. Reproduced from McDonald and Müller-Reichert (2002), courtesy of Academic Press, San Diego. Scale bar in (A) = 0.5 μm; Scale bar in (B) = 0.25 μm.

does not permit showing all the differences that one encounters between cells prepared by cryomethods and CF, but a study of the literature should help make the differences clear. References to hundreds of papers using HPF and FS methods can be found on the University of California Electron Microscope Laboratory Web site (http://em-lab.berkeley.edu/EML/protocols/hpflit.php), or one can search bibliographic databases for similar items. It should be pointed out, however, that not all papers claiming to get improved results by cryomethods actually do so. Freezing brings its own set of specimen preparation artifacts, mostly those that result from ice crystal damage.

H. Artifacts of HPF

Alterations of cell ultrastructure due to poor freezing are well known and not restricted to HPF work. In fact, for large pieces of tissue, they are less common in HPF than with other cryoimmobilization methods. These artifacts are usually referred to as "ice damage" and occur when the rate of freezing is too slow to prevent rearrangement of cellular water molecules into ice crystals. Because the rate of heat transfer is highest at the surface that is in contact with the cooling

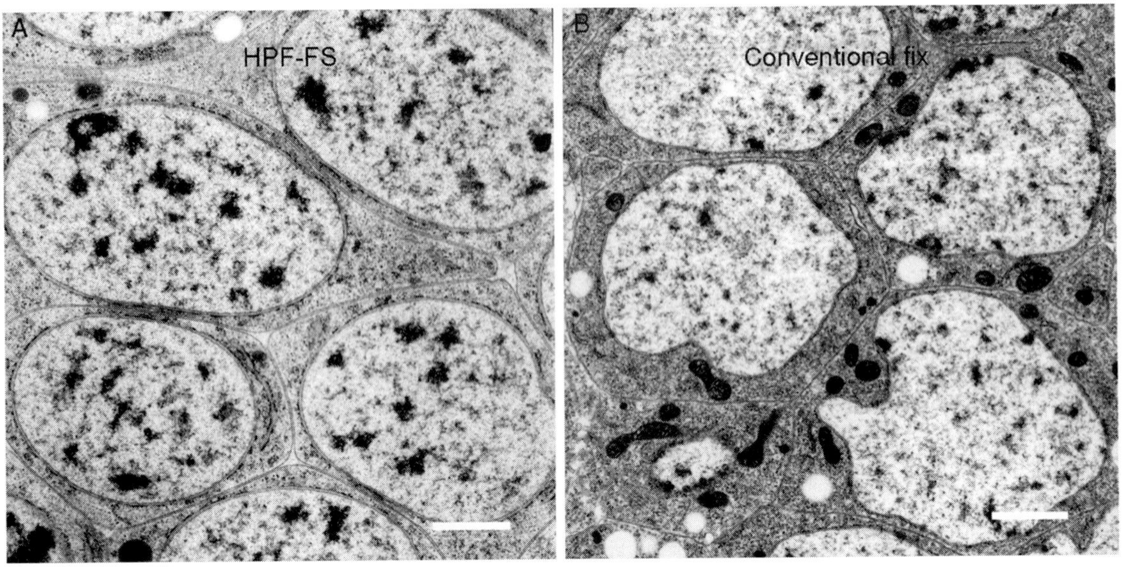

Fig. 7 Comparison of *Drosophila* embryo cells processed by high-pressure freezing and freeze-substitution (HPF-FS) as seen in (A), with the same tissue prepared by CF (Fig. 7B). Note the overall lower contrast in (A) compared to (B), and how the membranes in the HPF-FS material show smooth contours compared to those prepared conventionally. Scale bars = 1.0 μm.

source and slows down rapidly away from the surface, it is not uncommon to find a gradient of ice damage from the surface toward the interior of cells and tissues (Moor, 1987; Shimoni and Müller, 1998). The difference between HPF and other methods is the depth of good freezing before ice crystals appear (Robards and Sleytr, 1985). Nevertheless, poor freezing technique with any instrument will result in ice damage and HPF is no exception. Students of HPF need to recognize the various forms of ice damage before they unwittingly publish poor results while claiming superior preservation. Unfortunately, the literature of freezing does not include many papers that specifically address the issue of ice damage and what it looks like. Steinbrecht (1982, 1985, 1993) has been one of the few to systematically study this problem. The beginner with freezing techniques would benefit from reading these papers. The problem is not with severe ice damage, because that is easy to recognize; the subtle distortions are more problematic. Another valuable source of information for beginners about what ice crystal damage looks like is the careful examination of their own material. It is unlikely in the extreme that all freezing efforts will be successful, even for the most experienced HPF researcher. When one looks at sections of freeze-substituted and embedded samples, it is not uncommon to find a gradient from excellent preservation to subtle or even gross ice damage within a single section (Fig. 1 in Keene and McDonald, 1993). Those new to cryotechniques can benefit greatly from taking lots of images of poorly

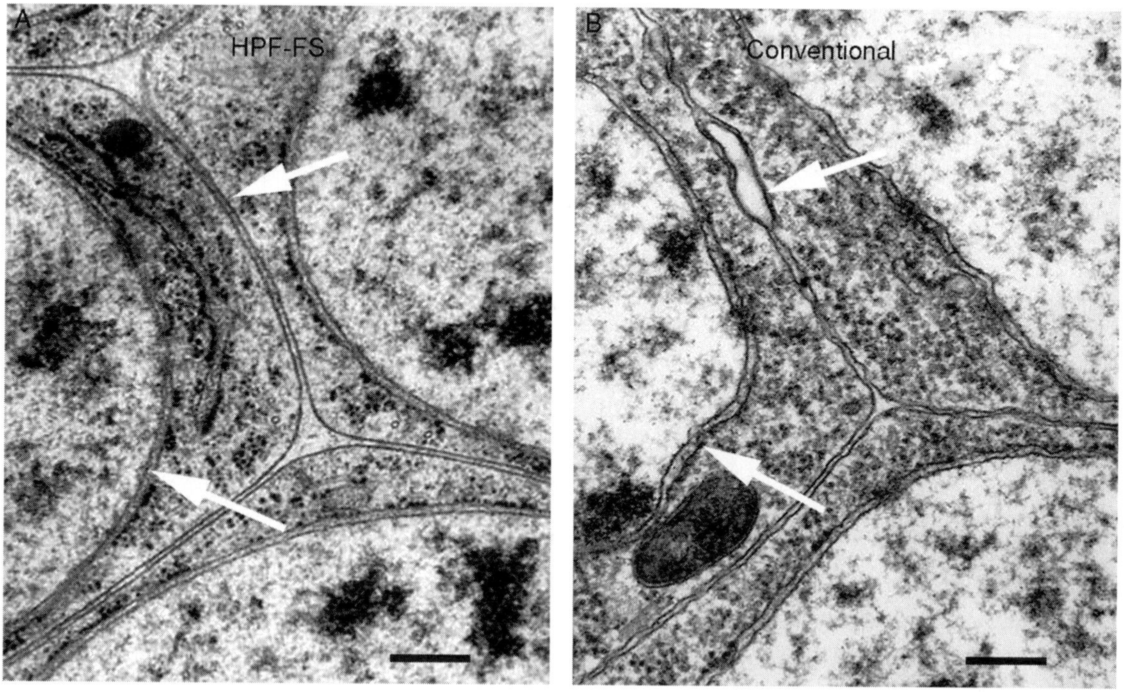

Fig. 8 Regions taken from Fig. 7 displayed at higher magnification, comparing the details of membrane morphology (arrows) in cells prepared by HPF-FS (A) and by conventional methods (B). Note how the membranes in the conventionally prepared samples are distorted compared to the smooth membrane profiles in the HPF-FS sample. Scale bars = 200 nm.

frozen material to help them learn how to recognize even subtle manifestations of ice damage.

Different cell components also vary in their susceptibility to ice damage. Those that are highly charged, and thus bind a lot of water, are most likely to show damage first. Chromatin is a good example; it is one of the first places to check when evaluating the quality of freezing. Mitochondria are another essentially universal cell component that show ice damage by having a clear "halo" around their outer membrane. Ice damage to both chromatin and mitochondria in a worm cell are illustrated in Fig. 9.

Besides ice damage, there are some structural aberrations that may be due to the high pressure. One of these is rupture of the yeast cell wall as shown in Fig. 1. These "cracks" often appear where two cells are touching, although they are not restricted to that arrangement. Another artifact that is probably pressure-related is the "eruption" of dense protein storage vacuoles that are often found in embryonic tissues (see Fig. 6 in McDonald, 1999). As more data come in from high-resolution tomographic studies, we may also find new categories of pressure artifacts that have not previously been noticed.

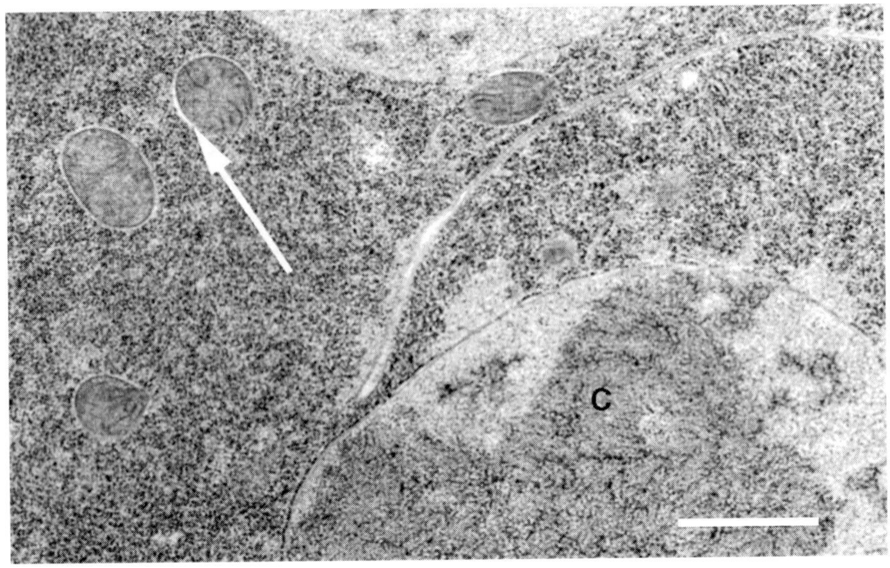

Fig. 9 Ice-damaged *C. elegans* cells. Even at low magnification it is possible to recognize the pattern of ice crystal damage in the chromatin (c) and the clear halos that surround some mitochondria (arrow). Scale bar = 1.0 μm.

VI. Summary

Following an era when many people believed LM would make EM mostly unnecessary, EM is returning as a premier investigative tool in molecular cell biology. In fact, LM used in conjunction with EM is emerging as one of the most powerful approaches to learning the structural basis of cell function. Both LM and EM are evolving technically to provide 3D visualizations of cells with increased resolution. On the EM side, one of the most exciting developments is the emergence of cellular electron tomography (cET) as an imaging method. Because cET can image molecules at 5- to 6-nm resolution or better in the context of the intact cell, specimen preservation techniques are of critical importance. In this chapter, we have given detailed methods for preserving ultrastructure in several model systems that are among the most widely used in cell biological research. When cells of *S. cerevisiae*, *C. elegans*, and *D. melanogaster* are processed by HPF and fixed at low temperature for subsequent resin embedding, the quality of the ultrastructure and preservation of antigenicity are superior to more conventional EM methods. We anticipate that EM and cET of these systems will play an increasingly important role in understanding the molecular basis of cell function.

References

Al-Amoudi, A., Chang, J. J., Leforestier, A., McDowall, A., Salamin, L. M., Norlen, L. P., Richter, K., Blanc, D., Studer, D., and Dubochet, J. (2004). Cryoelectron microscopy of vitreous sections. *EMBO J.* **23,** 3583–3588.

Albertson, D. G., and Thomson, J. N. (1976). The pharynx of *C. elegans*. *Philos. Trans. R. Soc. Lond. B Biol. Sci.* **275,** 299–325.

Baba, M., and Osumi, M. (1987). Transmission and scanning electron microscopy examination of intracellular organelles in freeze-substituted *Kloeckera* and *Saccharomyces cerevisiae* yeast cells. *J. Electron Microsc. Technol.* **5,** 249–261.

Bardsley, A., McDonald, K., and Boswell, R. E. (1993). Distribution of tudor protein in the *Drosophila* embryo suggests separate functions based on site of localization. *Development* **119,** 207–219.

Brenner, S. (1973). The genetics of behaviour. *Br. Med. Bull.* **29,** 269–271.

Cid, V. J., Shulewitz, M. J., McDonald, K. L., and Thorner, J. (2001). Dynamic localization of the Swe1 regulator, Hsl7 during the *Saccharomyces cervisiae* cell cycle. *Mol. Biol. Cell* **12**(6), 1654–1669.

Ding, R., McDonald, K., and McIntosh, J. R. (1993). Three-dimensional reconstruction and analysis of mitotic spindles from the yeast, *Schizosaccharomyces pombe*. *J. Cell Biol.* **120,** 141–152.

Dubochet, J. (1995). High-pressure freezing for cryoelectron microscopy. *Trends Cell Biol.* **5,** 366–368.

Echlin, P. (1992). "Low-Temperature Microscopy and Analysis." Plenum Press, New York.

Feder, N., and Sidman, R. L. (1958). Methods and principles of fixation by freeze-substitution. *J. Biophys. Biochem. Cytol.* **4,** 593–601.

Fernandez-Moran, H. (1960). Low-temperature preparation techniques for electron microscopy of biological specimens based on rapid freezing with liquid Helium II. *Ann. NY Acad. Sci.* **85,** 689–713.

Giddings, T. H., Jr., O'Toole, E. T., Morphew, M., Mastronarde, D. N., McIntosh, J. R., and Winey, M. (2001). Using rapid freeze and freeze-substitution for the preparation of yeast cells for electron microscopy and three-dimensional analysis. *Methods Cell Biol.* **67,** 27–42.

Gilkey, J. C., and Staehelin, L. A. (1986). Advances in ultrarapid freezing for the preservation of cellular ultrastructure. *J. Electron Microsc. Tech.* **3,** 177–210.

Heiman, M. G., and Walter, P. (2000). Prm1p, a pheromone-regulated multispanning membrane protein, facilitates plasma membrane fusion during yeast mating. *J. Cell Biol.* **151,** 719–730.

Hess, M. W. (2003). On plants and other pets: Practical aspects of freeze substitution and resin embedding. *J. Microsc.* **212,** 44–52.

Hess, M. W., Müller, M., Debbage, P. L., Vetterlein, M., and Pavelka, M. (2000). Cryopreparation provides new insight into the effects of Brefeldin A on the structure of the HepG2 golgi apparatus. *J. Struct. Biol.* **130,** 63–72.

Heuser, J. E., and Reese, T. S. (1981). Structural changes after transmitter release at the frog *Rana pipiens* neuro muscular junction. *J. Cell Biol.* **88,** 564–580.

Hippe-Sanwald, S. (1993). Impact of freeze substitution on biological electron. *Microsc. Res. Tech.* **24,** 400–422.

Hohenberg, H., Mannweiler, K., and Müller, M. (1994). High-pressure freezing of cell suspensions in cellulose capillary tubes. *J. Microsc. (Oxford)* **175,** 34–43.

Hohenberg, H., Tobler, M., and Müller, M. (1996). High pressure freezing of tissue obtained by fine needle biopsy. *J. Microsc.* **183,** 133–139.

Horowitz, R. A., Giannasca, P. J., and Woodcock, C. L. (1990). Low-temperature preparation of chromatin and nuclei. *J. Microsc. (Oxford)* **157,** 205–224.

Howard, R., and Aist, J. (1979). Hyphal tip cell ultrastructure of the fungus *Fusarium*: Improved preservation by freeze-substitution. *J. Ultrastruct. Res.* **66,** 224–234.

Humbel, B., and Müller, M. (1986). Freeze substitution and low temperature embedding. *In* "The Science of Biological Specimen Preparation 1985" (M. Müller, R. P. Becker, A. Boyde, and J. Wolosewick, eds.), pp. 175–183. SEM, Inc., AMF O'Hare, IL.

Keene, D. R., and McDonald, K. (1993). The ultrastructure of the connective tissue matrix of skin and cartilage after high pressure freezing and freeze-substitution. *J. Histochem. Cytochem.* **41,** 1141–1153.

Kellenberger, E. (1987). The response of biological macromolecules and supramolecular structures to the physics of specimen cryopreparation. *In* "Cryotechniques in Biological Electron Microscopy" (R. A. Steinbrecht, and K. Zierold, eds.), pp. 35–63. Springer-Verlag, Berlin.

King, R. C. (1960). Oogenesis in adult *Drosophila melanogaster*. IX. Studies on the cytochemistry and ultrastructure of developing oocytes. *Growth* **24,** 265–323.

King, R. C., Aggarwal, S. K., and Bodenstein, D. (1966). The comparative submicroscopic morphology of the ring gland of *Drosophila melanogaster* during the second and third larval instars. *Z. Zellforsch. Mikrosk. Anat.* **73,** 272–285.

Knauf, G. M., and Mendgen, K. (1988). Secretion systems and membrane-associated structures in rust fungi after high pressure freezing and freeze-fracturing. *Biol. Cell* **64,** 363–370.

Knauf, G. M., Welter, K. M., Müller, M., and Mendgen, K. (1989). The haustorial host-parasite interface in rust-infected bean leaves after high-pressure freezing. *Physiol. Mol. Plant Pathol.* **34,** 519–530.

Koch, E. A., and King, R. C. (1966). The origin and early differentiation of the egg chamber of *Drosophila melanogaster*. *J. Morphol.* **119,** 283–303.

Kosinski, M., McDonald, K., Schwartz, J., Yamamoto, I., and Greenstein, D. (2005). *C. elegans* sperm bud vesicles to deliver a meiotic maturation signal to distant oocytes. *Development* **132,** 3357–3369.

Lonsdale, J. E., McDonald, K. L., and Jones, R. L. (1999). High pressure freezing and freeze substitution reveal new aspects of fine structure and maintain protein antigenicity in barley aleurone cells. *Plant J.* **17,** 221–229.

Lonsdale, J. E., McDonald, K. L., and Jones, R. L. (2001). Microwave polymerization in thin layers of LR White allows selection of specimens for immunogold labeling. *In* "Microwave Techniques and Protocols" (R. T. Giberson, and R. S. Demeree, eds.), pp. 139–153. Humana Press, Inc., Totowa, NJ.

Lowman, F. G. (1958). Electron-microscope studies of *Drosophila* salivary-gland chromosomes. *Chromosoma* **8,** 30–52.

Lucic, V., Forster, F., and Baumeister, W. (2005). Structural studies by electron tomography: From cells to molecules. *Annu. Rev. Biochem.* **74,** 833–865.

Mahowald, A. P. (1962). Fine structure of pole cells and polar granules in *Drosophila melanogaster*. *J. Exp. Zool.* **3,** 201–215.

Mahowald, A. P. (1963). Ultrastructural differentiations during formation of the blastoderm in the *Drosophila melanogaster* embryo. *Dev. Biol.* **1963,** 186–204.

McDonald, K. (1994). Electron microscopy and EM immunocytochemistry. *In* "*Drosophila melanogaster*: Practical Uses in Cell and Molecular Biology" (L. S. B. Goldstein, and E. Fyrberg, eds.), Vol. 44, pp. 411–444. Methods in Cell Biology. Academic Press, San Diego.

McDonald, K., and Morphew, M. K. (1993). Improved preservation of ultrastructure in difficult-to-fix organisms by high pressure freezing and freeze substitution: I. *Drosophila melanogaster* and *Strongylocentrotus purpuratus* embryos. *Microsc. Res. Tech.* **24,** 465–473.

McDonald, K., and Müller-Reichert, T. (2002). Cryomethods for thin section electron microscopy. *In* "Guide to Yeast Genetics and Molecular and Cell Biology, Parts B and C" (C. Guthrie, and G. Fink, eds.), Vol. 351, pp. 96–123. Methods in Enzymology. Academic Press, San Diego.

McDonald, K. L. (1999). High pressure freezing for preservation of high resolution fine structure and antigenicity for immunolabeling. *In* "Electron Microscopy Methods and Protocols" (M. A. N. Hajibagheri, ed.), Vol. 117, pp. 77–97. Methods in Molecular Biology. Humana Press, Totowa, NJ.

McDonald, K. L., Morphew, M., Verkade, P., and Müller-Reichert, T. (2007). Recent advances in high pressure freezing: Equipment and specimen loading methods. *In* "Electron Microscopy Methods and Protocols" (J. Kuo, ed.), 2nd edn., Vol. 369, pp. 143–173. Methods in Molecular Biology. Humana Press, Totowa, NJ.

McDonald, K. L., Sharp, D. J., and Rickoll, W. (2000). Preparing thin sections of *Drosophila* for examination in the transmission electron microscope. *In* "*Drosophila*: A Laboratory Manual" (W. Sullivan, M. Ashburner, and S. Hawley, eds.), 2nd edn., pp. 245–271. CSHL Press, Cold Spring Harbor, NY.

McIntosh, J. R., Nicastro, D., and Mastronarde, D. (2005). New views of cells in 3-D: An introduction to electron tomography. *Trends Cell Biol.* **15,** 43–51.

Mollenhauer, H. H. (1964). Plastic embedding mixtures for use in electron microscopy. *Stain Technol.* **39,** 111–114.

Moor, H. (1967). Endoplasmic reticulum as the initiator of bud formation in yeast. *Arch. Mikrobiol.* **57,** 135–146.

Moor, H. (1987). Theory and practice of high pressure freezing. *In* "Cryotechniques in Biological Electron Microscopy" (R. A. Steinbrecht, and K. Zierold, eds.), pp. 175–191. Springer-Verlag, Berlin.

Moor, H., and Muhlethaler, K. (1963). Fine structure in frozen-etched yeast cells. *J. Cell Biol.* **17,** 609–628.

Moor, H., and Riehle, U. (1968). Snap-freezing under high pressure: A new fixation technique for freeze-etching. Proceedings of the Fourth European Regional Conference of Electron Microscopy, Vol. 2, pp. 33–34.

Müller, M., Marti, T., and Kriz, S. (1980). Improved structural preservation by freeze substitution. *In* "Electron Microscopy 1980" (P. Brederoo, and W. de Priester, eds.), Vol. 2, pp. 720–721. Proceedings of the Seventh European Congress on Electron Microscopy, Foundation, Leiden, The Netherlands.

Müller, M., and Moor, H. (1984). Cryofixation of thick specimens by high-pressure freezing. *In* "The Science of Biological Specimen Preparation" (J. P. Revel, T. Barnard, and G. H. Haggis, eds.), pp. 131–138. Proceedings of the Second Pfefferkorn Conference, April 23–28. SEM, Inc., AMF, O'Hare, IL.

Müller-Reichert, T., O'Toole, E. T., Hohenberg, H., and McDonald, K. L. (2003a). Cryoimmobilization and three-dimensional visualization of *C. elegans* ultrastructure. *J. Microsc. (Oxford)* **212,** 71–80.

Müller-Reichert, T., Sassoon, I., O'Toole, E., Romao, M., Ashford, A. J., Hyman, A. A., and Antony, C. (2003b). Analysis of the distribution of the kinetochore protein Ndc10p in Saccharomyces cerevisiae using 3-D modeling of mitotic spindles. *Chromosoma* **111,** 417–428.

Murk, J. L., Posthuma, G., Koster, A. J., Geuze, H. J., Verkleij, A. J., Kleijmeer, M. J., and Humbel, B. M. (2003). Influence of aldehyde fixation on the morphology of endosomes and lysosomes: Quantitative analysis and electron tomography. *J. Microsc.* **212,** 81–90.

Nicolas, M. T., and Bassot, J.-M. (1993). Freeze sustitution after fast-freeze fixation in preparation for immunocytochemistry. *Microsc. Res. Tech.* **24,** 474–487.

Nogales, E., and Grigorieff, N. (2001). Molecular machines: Putting the pieces together. *J. Cell Biol.* **152,** F1–F10.

O'Toole, E., Winey, M., McIntosh, J. R., and Mastronarde, D. N. (2002). Electron tomography of yeast cells. *In* "Guide to Yeast Genetics and Molecular and Cell Biology, Parts B and C" (C. Guthrie, and G. Fink, eds.), Vol. 351, pp. 81–95. Methods in Enzymology. Academic Press, San Diego.

O'Toole, E. T., McDonald, K. L., Mäntler, J., McIntosh, J. R., Hyman, A. A., and Müller-Reichert, T. (2003). Morphologically distinct microtubule ends in the mitotic centrosome of *Caenorhabditis elegans*. *J. Cell Biol.* **163,** 451–456.

Reipert, S., Fischer, I., and Wiche, G. (2004a). High-pressure freezing of epithelial cells on sapphire coverslips. *J. Microsc. (Oxford)* **213,** 81–85.

Reipert, S., Fischer, I., and Wiche, G. (2004b). High pressure immobilization of murine skin reveals novel structural features and prevents extraction artifacts. *Exp. Dermatol.* **13,** 419–425.

Reynolds, E. S. (1963). The use of lead citrate at high pH as an electron-opaque stain in electron microscopy. *J. Cell Biol.* **17,** 208–212.

Richards, A. G., Steinbach, H. B., and Anderson, T. F. (1943). Electron microscope studies of squid giant nerve axoplasm. *J. Cell. Comp. Physiol.* **21,** 129–143.

Robards, A. W., and Sleytr, U. B. (1985). Low temperature methods in biological electron microscopy. *In* "Practical Methods in Electron Microscopy" (A. M. Glauert, ed.), Vol. 10. Elsevier, Amsterdam.

Sartori, N., Richter, K., and Dubochet, J. (1993). Vitrification depth can be increased more than 10-fold by high-pressure freezing. *J. Microsc. (Oxford)* **172,** 55–61.

Schulte, J., Tepass, U., and Auld, V. J. (2003). Gliotactin, a novel marker of tricellular junctions, is necessary for septate junction development in *Drosophila*. *J. Cell Biol.* **161,** 991–1000.

Shimoni, E., and Müller, M. (1998). On optimizing high-pressure freezing: From heat transfer theory to a new microbiopsy device. *J. Microsc. (Oxford)* **192,** 236–247.

Steinbrecht, R. A. (1982). Experiments on freezing damage with freeze substitution using moth antennae as test objects. *J. Microsc. (Oxford)* **125,** 187–192.

Steinbrecht, R. A. (1985). Recrystallization and ice-crystal growth in a biological specimen, as shown by a simple freeze substitution method. *J. Microsc. (Oxford)* **140,** 41–46.

Steinbrecht, R. A. (1993). Freeze substitution for morphological and immunocytochemical studies in insects. *Microsc. Res. Tech.* **24,** 488–504.

Steinbrecht, R. A., and Müller, M. (1987). Freeze substitution and freeze drying. *In* "Cryotechniques in Biological Electron Microscopy" (R. A. Steinbrecht, and K. Zierold, eds.), pp. 149–172. Springer-Verlag, Berlin.

Studer, D., Graber, W., Al-Amoudi, A., and Eggli, P. (2001). A new approach for cryofixation by high-pressure freezing. *J. Microsc. (Oxford)* **203,** 285–294.

Studer, D., Michel, M., Wohlwend, M., Hunziker, E. B., and Buschmann, M. D. (1995). Vitrification of articular cartilage by high-pressure freezing. *J. Microsc. (Oxford)* **179,** 321–332.

Sulston, J., and Hodgkin, J. (1988). Methods. *In* "The Nematode *Caenorhabditis elegans*" (W. B. Wood, ed.), pp. 587–606. Cold Spring Harbor Laboratory Press, Cold Spring Harbor, NY.

Tanaka, K., and Kanbe, T. (1986). Mitosis in the fission yeast *Schizosaccharomyces pombe* as revealed by freeze substitution electron microscopy. *J. Cell Sci.* **80,** 253–268.

Thijssen, M. H., Van Went, J. L., and Van Aelst, A. C. (1998). Heptane and isooctane as embedding fluids for high-pressure freezing of *Petunia* ovules followed by freeze-substitution. *J. Microsc. (Oxford)* **192,** 228–235.

Van Harreveld, A., and Crowell, J. (1964). Electron microscopy after rapid freezing on a metal surface and substitution fixation. *Anat. Rec.* **149,** 381–385.

Walther, P. (2003). Recent progress in freeze-fracturing of high-pressure frozen samples. *J. Microsc. (Oxford)* **212,** 34–43.

Walther, P., and Ziegler, A. (2002). Freeze substitution of high-pressure frozen samples: The visibility of biological membranes is improved when the substitution medium contains water. *J. Microsc. (Oxford)* **208,** 3–10.

Ward, S., Thomson, N., White, J. G., and Brenner, S. (1975). Electron microscopical reconstruction of the anterior sensory anatomy of the nematode *Caenorhabditis elegans*. *J. Comp. Neurol.* **160,** 313–337.

Welter, K., Müller, M., and Mendgen, K. (1988). The hyphae of *Uromyces appendiculatus* within the leaf tissue after high pressure freezing and freeze substitution. *Protoplasma* **147,** 91–99.

White, J. G., Southgate, E., Thomson, J. N., and Brenner, S. (1976). The structure of the ventral nerve cord of *C. elegans*. *Philos. Trans. R. Soc. Lond. B Biol. Sci.* **275,** 327–348.

Winey, M., Mamay, C., O'Toole, E. T., Mastronarde, D. N., Giddings, T. H., Jr., McDonald, K. L., and McIntosh, J. R. (1995). Three-dimensional ultrastructural analysis of the *Saccharomyces cerevisiae* mitotic spindle. *J. Cell Biol.* **129,** 1601–1615.

Zhang, P., Bos, E., Heymann, J., Gnaegi, H., Kessel, M., Peters, P. J., and Subramaniam, S. (2004). Direct visualization of receptor arrays in frozen-hydrated sections and plunge-frozen specimens of *E. coli* engineered to overproduce the chemotaxis receptor Tsr. *J. Microsc. (Oxford)* **216,** 76–83.

CHAPTER 3

Cryopreparation Methodology for Plant Cell Biology★

Michael W. Hess
Division of Histology and Embryology
Innsbruck Medical University
A-6020 Innsbruck, Austria

★ Dedicated to Dr. Martin Müller in appreciation of his invaluable contributions to cryo-electron microscopy.

0091-679X/07 $35.00
DOI: 10.1016/S0091-679X(06)79003-3

This chapter summarizes the published literature on, and reflects our practical experience with cryoimmobilization, freeze-substitution, and resin embedding for ultrastructural and immunocytochemical transmission electron microscopy (TEM) studies of plants and fungi. For cell biological applications, this combination of state-of-the-art methods is considered as probably the most powerful approach to the understanding of cell organization at approximately nanometer resolution. Based on examples from organisms from all kingdoms it is postulated that cells suffering from any kind of stress (be it experimentally induced or naturally occurring) are particularly prone to fixation artifacts when subjected to conventional chemical fixation. Thus, morphological features reflecting cellular stress are difficult to define without ambiguity, since these ultrastructural alterations are always overlaid by more or less conspicuous fixation artifacts. Cryoimmobilization appears, therefore, to be an indispensable prerequisite, not just a complementary approach, for reliable electron microscopy (EM) analysis of such "challenged cells" [challenges include, e.g., infections, heat-shock, senescence, (programmed) cell death, drug administration, transfection, gene silencing]. Cryoimmobilization of plant and fungal bulk specimens by means of high-pressure freezing is described step-by-step, as well as the necessary follow-up procedures. The following aspects are discussed in detail: (1) sampling for high-pressure freezing; (2) specimen mounting and space fillers; (3) high-pressure freezing, specimen storage, and transport; (4) freeze-substitution and embedding with epoxy or (meth)acrylate resins for morphology, immunolabeling, microanalysis, and autoradiography. In addition, other cryo-based EM-preparation techniques are briefly commented as well [freeze-drying plus resin embedding, sodium dodecyl sulfate (SDS)-digested freeze-fracture replica labeling according to Fujimoto, TEM of frozen hydrated sections, cryo-scanning electron microscopy (SEM) of frozen block faces].

I. Introduction

A. Historical Context

The idea that low-temperature methods of specimen preparation are superior to conventional chemical methods for cytological studies was proposed over 100 years ago (freeze-drying: Altmann, 1890). The use of organic solvents at low temperature to replace cell water [freeze-substitution (FS)] was first proposed by Simpson for light microscopy (LM) (Simpson, 1941). Likewise, the theoretical foundations of key-methodologies for modern cryo-electron microscopy (cryo-EM) were published a while ago, including proofs-of-principle for several cell biological applications [e.g., freeze-fracture: Steere, 1957; FS for EM: Van Harreveld and Crowell, 1964; high-pressure freezing (HPF): Moor and Riehle, 1968; cryoultramicrotomy: Fernandez-Moran and Dahl, 1952]. With the exception of the quick-freeze/freeze-fracture approach (Heuser *et al.*, 1979), however, it was not until the past decade that continuous refinements of commercial cryopreparation equipment, together with the development of proper instrumentation for automated electron tomography, have made cryo-based EM a feasible approach for solving real cell biological questions (McIntosh, 2001).

B. Techniques and Organisms Discussed Here[1]

HPF, FS, and resin embedding are currently the most feasible approaches for state-of-the-art morphology and (immuno-)cytochemistry in the EM; thus, these are the subject of the present tutorial. A selection of other methods will be briefly discussed as well: Freeze-drying plus resin embedding, SDS-digested freeze-fracture replica labeling (SDS-FRL), transmission electron microscopy (TEM) of frozen hydrated sections, and cryo-SEM of fractured or sectioned block faces.

Flowering plants and higher fungi are the primary organisms considered here. Although not all systematic groups of higher fungi (Neomycota, true fungi) are closely related to plants,[2] they all pose similar technical problems for the electron microscopist. Shared morphological and functional features (elaborate cell walls and vacuoles) strongly determine and limit the quality of ultrastructure preservation of these organisms in a similar manner. Thus, similar methodological approaches are required. Ferns, mosses, lichens, and algae are not separately discussed in detail; for further specific information on yeasts, see Chapter 2 by McDonald, this volume.

II. Rationale

A. Cryoimmobilization Versus Traditional Chemical Fixation at Ambient Temperatures: Considerations Based on First Principles

Rapid freezing undoubtedly arrests subcellular dynamics much more quickly and reliably than conventional chemical fixation at ambient temperatures. Whether *cryoimmobilization* is an *indispensable prerequisite* or just a *complementary approach* for cellular EM is worth discussing in some detail—particularly when considering the technical and financial effort needed for proper cryo-based EM.

"Healthy" cells from any kind of eukaryotic organisms should be discussed first. Clearly, any attempt properly to define "healthy" cells/tissues/organisms is problematic because biological variability and physiological tolerance do interfere

[1] The literature list is intended to be comprehensive but not exhaustive; it contains, to the best of my knowledge, representative examples that provide further reading. Thus, I apologize to all whose work—though related to the topic—has not been included. For ease of access to the primary literature original papers and books are usually preferred over Meeting Abstracts and academic dissertations and thesis that were written and/or published earlier by the same authors. The recommendations given here reflect my personal experience and preferences and are, therefore, subjective.

[2] A close natural relationship to plants is only recognized for the Basidiomycota, for example for rusts and mycorrhiza-forming "mushrooms." This is no longer the case for the Ascomycota, for example yeasts and moulds.

with such a definition; nevertheless, this term may serve an appropriate starting point for the following reflections. The slow penetration and quite selective stabilizing capacity of chemical fixatives, such as aqueous aldehydes and osmium tetroxide (OsO_4), frequently cause artifactual morphological alterations, particularly in dynamic organelles/structures. Severe, partly unpredictable distortions have been described, particularly in endomembrane compartments and the cytoskeleton (including organelle translocations) (Ashford *et al.*, 1999; Dailey and Bridgman, 1991, 1993; He and Wetzstein, 1995; Hoch, 1991; Kaminskyj *et al.*, 1992; Mersey and McCully, 1978; Murk *et al.*, 2003; Orlovich and Ashford, 1993; Palmieri and Kiss, 2005; Wilson *et al.*, 1990). Other, apparently more stable and/or static components suffer less prominent changes. However, it is far from clear that any organelle can still be considered "apparently stable/static" in the light of (recent) observations on mitochondrial fusion and fission (Bereiter-Hahn and Voth, 1994), which occur also in plants (Arimura *et al.*, 2004), transient plastidial stromules (Bourett *et al.*, 1999; Gunning, 2005) or dark-induced thylakoid dynamics (Pfeiffer and Krupinska, 2005b). In any case, the issue of differential susceptibility to fixation-induced damage pertains not only at the level of organelles but also at the levels of cells, tissues, and organisms: certain structures and/or developmental stages are more fragile and vulnerable than others. Nevertheless, the experienced cytologist should be able—at least theoretically—to detect aforementioned artifacts in conventional preparations from healthy cells, thus, avoiding erroneous interpretations. If severe artifacts are encountered, cryo-based preparation provides a valuable complement or alternative.

By contrast, one is confronted with a completely *new class of interpretative problems*, when "challenged cells" are investigated with the high resolution provided by the EM. As used in this context the term "challenged cells" includes apparent (1) pathophysiological features and (2) degenerative processes, but also (3) latent stress and/or (4) extreme metabolic states. Such challenges occur as a natural part of the organisms' life cycle, for example infections, heat-shock, senescence, (programmed) cell death, or they may be experimentally induced, for example by drugs or transfection. Logical reflection leads to a very simple—not to say trivial—but widely neglected conclusion: the by-definition toxic nature of chemical fixatives makes it impossible to distinguish unequivocally the ultrastructural changes caused by the treatment under study, for example the drug administration or induced cell death, from those effects and artifacts that result from the action of the fixatives themselves. With cryoimmobilization, however, such cumulative and/or synergistic effects need not be considered; the segregation artifacts that result from ice crystal growth, a characteristic problem with less-than-perfect cryo-based preparation, will usually yield highly characteristic patterns that can readily be recognized for what they are.

Several examples from different types of challenged cells/tissues/organisms should substantiate this notion. In all these cases, the cryo-based EM data do differ from those obtained with conventional chemical fixation. The first example deals with the genuine subcellular pathology of fungal, plant, and mammalian specimens as induced by treatment with certain drugs (Driouich *et al.*, 1993; Hess *et al.*, 2000; Howard and Aist, 1980; Noguchi and Watanabe, 1999; Rupes *et al.*, 1995; Zhang *et al.*, 1993), as well as with light-induced distortions in the retinas of albino rats (Szczesny *et al.*, 1996). Certain cases of programmed cell death serve a second class of examples: for example, the differentiation of vascular tissue (Samuels *et al.*, 2002), self-incompatibility reactions (Geitmann *et al.*, 2004), the eve of life of the pollen-nourishing anther tapetum (Hess and Hesse, 1994), and finally, the fate of persisting and degrading synergids in plant ovules (Thijssen *et al.*, 1997). A third example is the severe ultrastructural alterations that are caused by latent stress or disorder; these changes, too, are hard to interpret properly when obscured by fixation artifacts. Suboptimal, but physiologically tolerable cultivation conditions may influence the quality of chemical fixation. Marine flatworms (*Macrostomum* sp.—Turbellaria) fixed immediately with glutaraldehyde/OsO_4 after collection from their natural environment were regularly much better preserved than their siblings that were living and even reproducing under optimized laboratory conditions[3] (R. Rieger, personal communication). Genetically modified organisms provide additional examples for latent stress and, in consequence, increased susceptibility to fixation artifacts. When preparing knockout mice (fetal stages and newborn animals) by conventional chemical fixation, we have already encountered two cases where the transgenic animals displayed certain prominent, ultrastructural changes in the relevant tissues. These changes could not be directly correlated with the genetic modifications. The littermate animals, however, were preserved well by the fixation applied. Eventually, the aforementioned morphology turned out to be artifactual—and not the hoped-for phenotypes; these were unraveled only later (Teis, Huber, Hess: unpublished data), both by chemical fixation and HPF. The persistent occurrence of these fixation artifacts, therefore, seems to reflect merely latent stress or disorder of the knockout tissues. Finally, cells with extremely high or low metabolic activity sometimes behave similarly to challenged cells. They touch the limits of physiological tolerance and are therefore poorly—or at the best suboptimally—preserved by conventional techniques. Hypertrophic, mineralizing cartilage is one classical example (Hunziker *et al.*, 1984), and recently, similar findings were reported for the metabolically highly active keratinocytes of the upper epidermis (Pfeiffer *et al.*, 2000). Their highly disordered ultrastructure observed after chemical fixation was previously misinterpreted as direct morphological evidence of "degenerative" processes (though it is clear that these ageing cells are near the brink of the grave). The opposite extreme,

[3] The observations on stressed, and in consequence poorly fixed little worms should remind us again of a more general dilemma in cell biological research; namely, of the various limitations inherent to laboratory cultures, especially of metazoa—be these conventional, static 2D monolayer cell cultures or physiologically more relevant, organotypic 3D culture models (Felder *et al.*, 2002; Kim *et al.*, 2004; Schmeichel and Bissell, 2003; Zahir and Weaver, 2004), or organisms in culture. Said in another way, whatever the topic of the investigations, one must always consider that the morphological, physiological, and molecular features of cells in culture do fundamentally differ from those *in vivo* (i.e., within the living animal), and from those of freshly taken tissues biopsies.

that is, cells with highly reduced metabolic activity, such as quiescent and dormant cells, should serve a last example. Cold-acclimated vascular cambium during rest in winter reacts with severe fixation artifacts to conventional chemical fixation (Rensing and Samuels, 2004), as do partially or fully desiccated organisms, such as the resurrection plant *Selaginella lepidophylla* (Platt *et al.*, 1997; Thomson and Platt, 1997) or mature pollen (Tiwari *et al.*, 1990).

Taken all these observations into account, it seems necessary to postulate that challenged cells are more susceptible to artifactual deformations that can result from conventional chemical fixation. Similar notions were published some time ago, but obviously they have not strongly influenced the practices in EM laboratories (Trump and Ericsson, 1965) "... increased fragility of organelles which would render them more susceptible to damage during ... preparation"; (Bereiter-Hahn and Voth, 1979) "... results indicate a higher sensitivity against fixation artifacts in cells under pathological conditions than normal cells." In consequence, the use of cryoimmobilization methods is, in my opinion, *mandatory for* proper EM of *challenged cells*, and is *highly recommended for* the study of *healthy cells*, too.

B. Why Plants and Fungi Are Usually Difficult to Fix, but Rather Easy to Freeze?

Elaborate cell walls and vacuoles act in concert to make even healthy plants and fungi difficult to fix with conventional techniques. The penetration of chemical fixatives into live plant and fungal cells is severely impeded by cell walls that form a separate barrier around each individual cell—similar to vertebrate cartilage and bone (but unlike arthropods that become fixed adequately once the cuticle is locally pierced). Intercellular air, as well as waxy cuticles may further slow down fixation so that it takes minutes until the subcellular dynamics of plants/fungi are at least stopped (Hoch, 1991; Mersey and McCully, 1978); unfortunately, this does not mean the underlying structures are already stabilized. Thus, after the cessation of cytoplasmic streaming or vesicular transport severe distortions like the collapse and rearrangement of endomembrane compartments can still occur (McCully and Canny, 1985; Orlovich and Ashford, 1993; Wilson *et al.*, 1990). Concomitant vacuolar leakage is suspected of dislocating the acidic vacuolar contents into the cytoplasm, thereby further contributing to poor ultrastructure preservation (Hayat, 2000; Kellenberger, 1991); notably, however, the vacuolar membrane (tonoplast) of certain cells proved surprisingly stable during fixation with ice-cold glutaraldehyde (Dong *et al.*, 1994).

In the following sections, favorable features of plants and fungi are to be mentioned, which make cryoimmobilization of these organisms quite easy. Most importantly, sampling is in general not difficult. Dissecting plants and fungi in order to fit them into the specimen carriers of most freezing devices available does generally not traumatize the cells (with the possible exception of vascular tissue) nor is temperature control or sample dehydration problematic (i.e., sampling speed).

This strongly contrasts with the various, partly unsolved sampling problems encountered with metazoa (multicellular animals) that are too large to fit as a whole into the carriers of freezing devices [for a detailed discussion, see the contribution by Hohenberg *et al.* (1996)]. Moreover, plants and fungi frequently synthesize and accumulate natural cryoprotectants, for example low-molecular carbohydrates, dehydrins. These may be another important factor contributing to high yields of well-frozen plant and fungal samples.

III. Methods

This paragraph details step-by-step the methods we have consistently employed for various plant and fungal specimens. Further technical aspects and methods are mentioned in Section V and as footnotes in this chapter.

A. Cryoimmobilization

1. Sampling for HPF

HPF permits adequate cryoimmobilization of bulk specimens (Section V). For several years we have been successfully employing several, continuously maturing generations of the HPM 010 HPF apparatus (BAL-TEC, Balzers, Liechtenstein) (Hess, 1993, 2003). In general, we use the protocol developed by Martin Müller's group in Zürich (Studer *et al.*, 1989), with slight modifications.

The specimens' dimensions must not exceed a diameter or length of 2 mm, and a nominal thickness of 0.5 mm for fitting into the cup-shaped specimen carriers (planchettes) of the HPM 010 HPF machine. [Note that this does not mean such voluminous bulk specimens being always well frozen; thus a sample thickness of 0.2 (to 0.4) mm is recommended as detailed in Section V.] The current instruments from Leica have slightly smaller carriers (inner diameter of the Leica flat specimen holders; 1.4 mm), whereas the machines from Wohlwend Engineering (Sennwald, Switzerland) use the same carriers as BAL-TEC.

Root tips or stamens are simply excised with a scalpel, usually immersed in the liquid that subsequently serves to fill the interior volume of the freezing planchette (see below). Tissue disks (2 mm in diameter) are punched with an ophthalmological or dermatological punching device. Suspensions of larger cell aggregates (e.g., barley callus cultures) or fungal hyphae (e.g., *Aspergillus*) are pipetted or smeared directly into the planchettes and covered with culture medium. Suspensions of small organisms, such as yeasts and small algae (be they unicellular or not), can be processed for HPF with the aid of semipermeable cellulose tubes, which can serve as microcontainers.[4] These 200-μm-wide capillary tubes are filled with a

[4] Note that these cellulose microcontainers are also suitable for cultivating plant and animal cells (Tiedemann *et al.*, 1998).

suspension of cells by capillary action and then cut into 2-mm-long pieces. Subsequently, they can be conveniently handled like rigid tissue (Hohenberg *et al.*, 1994).

2. Specimen Mounting and Space Fillers

The replacement of any air inside the specimen carriers is a prerequisite for successful HPF (Craig and Staehelin, 1988; Müller and Moor, 1984). Otherwise, remaining air acts as insulator, thus impeding efficient cold and heat transfer. Furthermore, air bubbles collapse as the pressure rises, and the resulting local flow of fluid can damage the specimen mechanically. Thus, many researchers (including myself) try to replace also the intercellular air of plant and fungal tissues prior to HPF. As space filler, we have usually used 1-hexadecene (from Sigma, St. Louis, MO or Fluka, Buchs, Switzerland). Liquids other than hexadecene have also proved to be suitable space fillers in our hands: that is, cell culture medium [minimal essential medium (MEM), but without antibiotics] or yeast growth medium YPD [consisting of 10% yeast extract (H1702 from Laboratorios Conda, Spain) + 20% bacto peptone (L37, from Oxoid Ltd., Basingstoke, UK) + 2% glucose]; for a detailed discussion on space fillers, see the relevant paragraph of Section V.

Intercellular air *within* the tissue is replaced by immersing the specimens into the space-filling liquids and by exposing them subsequently to a gentle vacuum for 20–60 sec. The vacuum is produced either by a water-jet pump or simply by sucking a vacuum by hand with a 10-ml-plastic syringe.[5]

3. HPF, Specimen Storage, and Transport

Working principle and operation of HPF machines[6] is not described here [for information, see the work by Moor (1987), Müller and Moor (1984), Studer *et al.* (2001); and McDonald (1999)].

For sample storage in liquid nitrogen (LN_2) the planchettes with the frozen specimens are transferred into commercial or homemade metal containers (note that any handling of the planchettes with the frozen specimens is done *under* LN_2 using precooled instruments and containers). Transport of frozen samples from our

[5] For evacuating plant/fungal tissue by hand, 2 ml of liquid and the specimens are loaded into a 10-ml syringe, excess air is pushed out, and the syringe tip is closed with a thumb; a moderate, but sufficient vacuum is then produced by pulling the piston of the still closed syringe as far as possible and holding this position for 30–60 sec. If necessary, this procedure is repeated two to three times.

[6] One hint on the operation of instruments from BAL-TEC, Wohlwend, or the currently no longer sold "big, old" Leica EM-HPF instrument: it is of outstanding importance to remove the specimen holder from the machine as fast as possible and immerse it in LN_2 (though the holder is quite massive and sometimes sticks firmly in the machine). Otherwise, one risks warming of the specimens- and well-frozen samples may deteriorate due to water recrystallization. Current Leica models transfer the frozen specimen automatically to LN_2.

collaborators' laboratories to our own has been conveniently performed with the aid of special travel Dewar flasks[7] that maintain LN_2 temperature for a week or longer.

B. Freeze-Substitution

FS stabilizes frozen specimens through dehydration at temperatures (far) below 0 °C, using organic solvents that are optionally supplemented with other fixatives (for details, see Section V). We have had success using either commercially available, programmable[8] instruments (i.e., FSU 010 from BAL-TEC, or AFS from Leica) or homemade equipment (made from an insulated, passively warming-up brass construction: Hess and Glaser, 1993).

Out of the various solvents that have proven effective for FS, we have regularly employed dry acetone and ethanol (analytical grade) from freshly opened bottles or dried with calcium chloride over at least 24 h. The solvents are used either pure or containing a selection of fixatives/stains. The following fixatives added to the solvents yield good results in our hands:

i. OsO_4 (0.25–2%), dissolved from crystals[9] in acetone;
ii. formaldehyde (0.7–3.7%), diluted in ethanol from standard 37% formalin stock[10];

[7] Travel Dewars to maintain LN_2 temperatures are designed and approved for transport in aircrafts as well (e.g., Taylor-Wharton CX100 dry shipper). Nowadays, however, especially international transport of such Dewars is heavily restricted by governmental security regulations and the airlines' and courier services' own safety rules. If they are transported at all when containing liquid or gaseous N_2, longer travel and delivery times must be taken into account.

[8] Notably, FS devices that heat the chamber for temperature maintenance do not always work correctly (i.e., the set and displayed temperatures may differ considerably from those really obtained in the chamber: P. Wild, personal communication and our own observations). Thus, measuring and/or monitoring the chamber's temperature with an additional, external thermocouple (e.g., Pt 100), is recommended.

[9] Concerning the preparation of OsO_4–acetone mixtures, two safety concerns and possible sources of danger must be emphasized here. (1) Colleagues of mine from chemistry departments worried about adding OsO_4 crystals to acetone at RT because such mixtures are highly reactive, thus, potentially explosive. However, we have never encountered problems with this practice nor have we heard about accidents from this procedure carried out in other laboratories. A plausible explanation may be that we chill the OsO_4 crystals to LN_2 temperatures prior to use. We do so for two reasons. First, because deep-frozen OsO_4 crystals easily detach from the glass ampoules they are contained in; therefore, the ampoules can be emptied completely without submerging them in the solvent. Second, because deep-frozen OsO_4 is not as reactive or volatile as compared to ambient temperatures. (2) For chilling the sealed OsO_4 ampoules to LN_2 temperatures, we proceed in two steps: first, we place them on the dry surface of a LN_2-cooled metal block for several minutes, and then we immerse the prechilled vials into LN_2. This two-step procedure does not rule out, but minimizes the risk of, accidental glass breakage due to fast temperature changes. In any case, it is self-evident that the use of a fully functional fume hood, efficient face and eye protection (plastic glasses, safety shield), and laboratory gloves are mandatory.

[10] Ordinary formalin stock instead of paraformaldehyde can readily be used for FS (in contrast to conventional aqueous specimen fixation for EM). The results are, according to our experience, indistinguishable, and dissolving paraformaldehyde powder in heated methanol is not trivial; for a protocol, see the work by McDonald (1994).

iii. uranyl acetate (UO_2^{2+}) (0.02–0.5%), added either to osmium-acetone (Os-Ac) or to formaldehyde-ethanol (owing to the poor solubility of UO_2^{2+}in acetone, it is recommended to prepare a 15% UO_2^{2+}-stock in methanol, which can be stored at −20 °C and diluted *ad libidum* with acetone or ethanol).

The fixatives are carefully added to the solvents in a beaker at room temperature[9] (RT) followed by brief, gentle mixing (e.g., by pipetting the cocktail once up and down). Subsequently, cryovials are quickly filled with the FS cocktails, closed and immediately frozen in upright position with LN_2 to prevent premature degradation. Cryovials with screw-caps which permit one-hand operation are preferred, for example 2-ml Nalgene vials (from Nalge Nunc International, Rochester, NY) plus the appropriate Nalgene cryoracks with locking bottoms that match the cryovials (note that these vials are not LN_2-certified, but we have never encountered problems).

For FS the cryoimmobilized samples are transferred under LN_2 into the cryovials containing the frozen cocktails [if hexadecane is used for space filling, the frozen hexadecene should be mechanically removed from the specimens prior to FS (Hohenberg *et al.*, 1994); for details, see Section V]. Subsequently, the lids are loosely screwed onto the vials to permit safe evaporation of excess N_2 because the volume ratio of gaseous to LN_2 is 700:1. Finally, the vials are placed into the precooled FS device and after about 1 h the lids are tightened.

FS is carried out for at least 8–12 h at approximately −90 °C (our standard schedule). Subsequently, the samples are allowed to warm slowly to 0 °C over a period of ~24 h (for low-temperature embedding the end temperature is lower, see Section V). When automated FS devices are used, the temperature rise is 10–30 °C/h and includes steps of 8 h at −60 and −30 °C. This should allow chemical fixatives/stains included in the FS media to stabilize the specimens at low temperatures. After warming, the samples are optionally left at +4 °C or at RT for another 30–60 min to increase membrane staining (extended incubation at this temperature range, however, should be avoided because of increased protein degradation) (Matsko and Müeller, 2005) and/or lipid extraction. Finally, the samples are rinsed three times with dry acetone and/or ethanol (RT, 5 min each).

C. Resin Embedding with Epoxides or LR White

Infiltration with a graded resin series is carried out at RT and/or +4 °C. Two aspects proved essential for obtaining homogeneous blocks that are easy to section. First, one should start with as low as 5% resin in solvent to prevent detachment of the cytoplasm from the cell wall (McCully and Canny, 1985); second, continue infiltration for up to 3 days with continual agitation (resin concentration is increased stepwise: 5%, 10%, 30%, 70%, 100%; the latter changed twice). The resins are polymerized 18–24 h by heat or by UV light at +4 °C.

The epoxy resins of our choice are Epon (hard grade) or Epon-Araldite [mixture 2 (Mollenhauer, 1964), for the exact recipes see also the work by Glauert and Lewis (1998), Hayat (2000), and Matsko and Müeller (2005)] both cured at +60 °C, and containing the accelerator already from the beginning of the infiltration process.

Our preferred acrylic resin is LR White (London Resin Co., Woking, Surrey, UK). However, London Resins (like the Lowicryl resins) do not harden in the presence of oxygen. Therefore, the specimens are placed either in closed gelatine capsules (size 00) or in 0.5-ml Eppendorf vials (made of high-density polyethylene), but not in polypropylene embedding capsules (e.g., BEEM capsules). Polymerization is performed at +55 °C or by UV light at +4 °C. The polymerization of acrylic resins is generally exothermic (Glauert and Young, 1989). For curing acrylics in the cold we, therefore, use a heat sink. (An aluminum block with indentations that fit the embedding capsules and that are filled with pure ethanol to improve thermal contact serves this purpose well.)

D. Microtomy, Staining, and TEM

For trimming, we use a motor-driven trimming device (Reichert Ultratrim, from Leica Microsystems, Vienna, Austria) that is equipped with a diamond miller (from Diatome, Biel, Switzerland). Alternatively, the specimen blocks are roughly trimmed by hand with a razor blade. Two aspects seem worth mentioning in this context. First, Lowicryl- or Unicryl-methacrylate embeddings should not be trimmed by milling devices because of the still highly reactive dust particles from the polymerized resin, which are hazardous (H. Sitte, personal communication). Second, we have observed that freeze-substituted plant and fungal cells connect less firmly to the surrounding resin than does chemically fixed material [this is especially true for Epon and/or Araldite because they "glue" less than Spurr's (Mollenhauer, 1988); furthermore, for acrylic resins since they do not bind covalently to the specimens]. Thus, when trimming by hand, it is recommended not to touch the specimen itself with the razor blade. Excess resin on top of the pyramid should be carefully removed by a milling device or by semithin sectioning; otherwise the specimens locally separate from, or even break out of, the resin. To complete the trimming process it is, furthermore, advisable to cut ~10 semithin sections (300 nm, but not thicker), followed immediately by another 10 ultrathin sections (100 nm) that will be disposed of. Thereby, the block surface becomes perfectly polished so really smooth, coherent ultrathin sections can subsequently be obtained. (Such "polishing" is again recommended for the aforementioned Epon and/or Araldite embeddings from FS plants and fungi, whereas embeddings with Spurr's are less delicate in this respect.)

Semithin and ultrathin sections (300 nm and 80–100 nm, respectively) are cut at speeds of 0.4–1 mm/sec by using a Reichert Ultracut S (from Leica) equipped with a diamond knife (from Diatome, Biel, Switzerland). Ultrathin sections are collected onto uncoated grids or on formvar- or parlodion-carbon-coated grids. Semithin sections are stained for LM with aqueous toluidine blue (0.5%). Ultrathin sections are sequentially poststained with aqueous UO_2^{2+} (0.5%, 5–40 min/RT) and lead citrate (Pb: 3 min/RT). Alternatively, we stain with hot ethanolic phosphotungstic acid [EPTA (Locke and Krishnan, 1971): 1% PTA in 95% ethanol, 5 min/+60 °C] to enhance the contrast of certain, poorly visible membranes (e.g., plastid envelope, endoplasmic reticulum) and the cytoskeleton, including single microfilaments (Hess, 2003). The sections are viewed at 50–80 kV with a TEM in brightfield mode. EM tomography of (semi-)thin sections is not dealt with here; for further information on this topic, see the relevant contributions in this volume.

IV. Materials

The majority of the author's results discussed here were obtained using anthers from *Ledebouria socialis* (Hyacinthaceae) and other monocotyledonous angiosperms. Sections through differentiating anthers provide quite a representative, though not complete, picture of the various tissues in higher plants.

Other specimens successfully prepared by the methods described here were: callus cultures from *Hordeum* (barley); leaves from *Arabidopsis*, *Lepidium sativum* (garden cress), and *Allium cepa* (onion); root tips from *Lepidium*, gymnosperm roots with their ectomycorrhizal fungi (Scots pine *Pinus sylvestris* plus *Suillus bovinus*), hyphae of the filamentous mold *Aspergillus fumigatus*.

V. Discussion

A. Cryoimmobilization: Choice of Freezing Method

The method of choice for cryoimmobilizing bulk specimens from higher plants and the voluminous, tissuelike fungal fruiting bodies (mushrooms) is rapid freezing under a pressure of 2100 bar (HPF: Moor, 1987; Moor and Riehle, 1968). HPF lowers the freezing point of water, and both crystal growth and nucleation rates are markedly reduced (Müller and Moor, 1984). The thickness of plant and fungal samples largely devoid of disturbing ice crystal damage ranges from 0.2 to 0.4 mm, on average. This is primarily depending on the physicochemical and physiological properties of the samples, for example, their contents of water and of natural cryoprotectants within the cytoplasm. Water and cryoprotectants are not evenly distributed throughout the subcellular compartments, nor throughout the different cell types that bulk specimens usually contain. In consequence, it may happen that strongly cryoprotected cells (e.g., pollen grains) are well frozen, even when located in the very center of a 0.6-mm-thick bulk specimen, while the surrounding tissues are obviously freeze-damaged. Moreover, a few morphological artifacts that appear to result from the high pressure have been described (Ding *et al.*, 1992; Hyde *et al.*, 1991), but in general they do not pose serious problems. Finally, it must be emphasized that the temporal resolution achieved with HPF is generally not as good as that with rapid freezing at ambient pressure (Moor, 1987; Ryan and Knoll, 1994; Studer *et al.*, 1995); furthermore, there exists a certain time difference between freezing at the specimen's surface and at its very center (M. Müller and J. Listemann, personal communication)—these features of HPF should be considered when studying ultrafast processes.

Rapid freezing at ambient pressure is technically less demanding, but adequate freezing is usually confined to the outermost 10–15 μm of a specimen [for a review on slam, plunge, jet, and spray freezing, see the work by Sitte *et al.* (1987)]. Some of these freezing techniques prove to be as useful for thin fungal hyphae (Ashford *et al.*, 1999; Howard and Aist, 1980; Muller *et al.*, 1980; Rupes *et al.*, 1995; Zalokar, 1966) whereas higher plants are less often well preserved by these methods [for good examples, see the work by Ding *et al.* (1991a,b, 1992) and Huang *et al.* (1993); for further details on rapid freezing and especially HPF, see Chapter 1 by Dubochet, this volume].

B. Space Fillers for HPF

Air replacement *around* the specimens is generally accepted to be a prerequisite for successful HPF (Craig and Staehelin, 1988; Craig *et al.*, 1987; Müller and Moor, 1984). This is achieved with (inert) liquids (see below) or moist yeast paste containing 10% methanol (McDonald, 1994, 1999).

With respect to the intercellular air *inside* plant tissue and fungal plectenchyma, however, laboratory habits do vary. Some researchers customarily evacuate prophylactically any kind of plant or fungal sample (Hess, 2003); others apparently do not evacuate [as judged from Materials and Methods in their publications (Kiss *et al.*, 1990; Mims *et al.*, 2003)]; finally, some evacuate only when the samples do not sink in the space-filling liquid (McDonald, 1999). This third approach is possibly the most pragmatic one, considering that root tips contain much less intercellular air than mature leaves, for example. For replacing intercellular air, only liquids with relative low viscosity and/or surface tension can be recommended [e.g., 1-hexadecene: Studer *et al.*, 1989; or cell culture media]. By contrast, for filling the cavities of the specimen carriers (planchettes or metal tubes) around the sample, other, more viscous solutions have also been employed [e.g., polyvinylpyrollidone or polyvinyl alcohol (Müller and Moor, 1984), low-melt agarose (Lancelle and Hepler, 1989), gelatine, serum albumin, dextran, sucrose, mannitol].

Hexadecene is a paraffin oil with low surface tension; it easily wets the sometimes hydrophobic specimen surface and quickly replaces intercellular air (Studer *et al.*, 1989). Thick-walled plant and fungal samples appear not to suffer from physiological impairment when exposed to hexadecene for some minutes (Studer *et al.*, 1989), which agrees with our observations. Other organisms such as nematodes (Hohenberg *et al.*, 1994), flatworms, or *Hydra* (our unpublished data) stop moving soon after contact with hexadecene. Furthermore, this reagent is an irritant and is potentially allergenic. Thus, protective gloves are advised. Finally, hexadecane is not soluble in acetone, ethanol, or methanol at FS temperatures (Hohenberg *et al.*, 1994).

Therefore, frozen hexadecene must be removed mechanically from the specimen surface, at least locally. This is done under a dissecting microscope with the aid of a scalpel or a needle at low temperature (e.g., below −100 °C inside the cryochamber of an ultramicrotome or even around LN_2 temperature at the almost dry surface of a LN_2-chilled metal block). Otherwise, FS media cannot readily infiltrate and dehydrate the specimens at the desired low temperatures. Intercellular hexadecane is not a problem for proper FS of tissues or plectenchyma according to our experience, but it may become one, when hyphae of filamentous fungi are buried by hexadecane. Taken all together, for plant and fungal bulk specimens, hexadecane is generally convenient to work with [it optimizes the transfer of pressure and temperature to the specimen (Schwarz *et al.*, 1993; Studer *et al.*, 1989) and produces negligible amounts of heat during freezing, unlike water or buffer; for further details on space filling liquids].

Cell culture media or yeast growth medium are, in our hands, feasible alternatives, both giving results—at least—comparable to hexadecane (Hess, 2003). An aqueous solution of dextran (e.g., MW 38,000) is another liquid that is regularly used for space filling around the specimen. It is supposed to have little osmotic activity (Kiss *et al.*, 1990; Staehelin *et al.*, 1990). However, it is not fully compatible with UO_2^{2+} and tannic acid in the FS media, thus leading to sectioning problems (Giddings, 2003; Kiss *et al.*, 1990; Mims *et al.*, 2001); furthermore, some researchers found it to be sensitive to radiolysis in the electron beam, causing problems with EM tomography (R. McIntosh and H. Hohenberg, personal communication).

Other liquids are less frequently employed. When choosing sucrose or mannitol, one must explore their osmotic effect, for example, by cultivating the plants being studied with increasing solute concentrations for accommodation (Lonsdale *et al.*, 1999; Murata *et al.*, 2002; Segui-Simarro and Staehelin, 2006) or to make sure that the conditions used are not deleterious to the sample's health. Heptane and isooctane have been used for increasing the yield of usable embryo sacs, but the outer two to three layers of the surrounding ovule tissues became seriously damaged by these alkanes (Thijssen *et al.*, 1998). Finally, supplementing physiological buffers with 3–8% methanol has recently become popular again for replacing intercellular air in plant tissue (Hippe *et al.*, 1989; Lee *et al.*, 2000; Pfeiffer and Krupinska, 2005a; Pfeiffer *et al.*, 2003). It was originally introduced for leaves with fungal infections (Welter *et al.*, 1988), and these authors pointed to the time- and dose-dependent toxicity and reversible ultrastructural alterations one must be aware of.

C. Freeze-Substitution

FS is a hybrid cryopreparation technique that meets many of the requirements for state-of-the-art cell biology. It combines the benefits of fast, physical sample immobilization with feasible chemical sample stabilization at very low temperatures and the convenience of resin microtomy at RT [for reviews on FS, see the work by Harvey (1982); Humbel and Müller (1986); Humbel and Schwarz (1989); Schwarz *et al.* (1993); and Steinbrecht and Müller (1987)].

1. General Comments on FS Media

Acetone, methanol, and ethanol are the solvents most commonly used for FS, usually complemented with fixing and/or staining reagents (e.g., OsO_4, UO_2^{2+}, aldehydes). When aiming at morphological analyses, most researchers nowadays prefer acetone. It produces samples that yield pictures with the most aesthetic appeal, and it is, therefore, supposed to minimize preparative distortions.

Anhydrous methanol and acetone will dehydrate a sample of average size (<1.5 mm^3) within 2–4 h at temperatures as low as −90 °C (Humbel and Müller, 1986; Humbel and Schwarz, 1989; Steinbrecht and Müller, 1987), exchanging with all but the structurally bound water, which is not removed until the temperature rises to about −30 °C [Kellenberger, 1991; MacKenzie, 1972; see also Chapter 1 by Dubochet, this volume, for further information on cellular water in the context of cryo-EM]. In addition to acetone and methanol, ethanol has been shown to be suitable for FS (Bourett and Howard, 1991; Eppenberger-Eberhardt *et al.*, 1997), but as yet there are no measurements of its FS properties. FS at temperatures lower than −90 °C has so far not been successfully performed: acetone and methanol freeze already at about −94 °C; other solvents so far tested for FS at lower temperatures have not yielded results within a reasonable period of time (Humbel and Müller, 1986). FS times between 8 h (Humbel and Müller, 1986) and 2–3 days are common practice with these three solvents. Prolonged FS may be necessary for desiccated tissues, for voluminous non-HPF samples, or if the solvents are not dry.

The temperatures at which recrystallization of water becomes a problem in cryoimmobilized biological specimens are not exactly known. In the case of pure water the transition from amorphous (i.e., vitreous) to cubic ice occurs at about −130 °C (Dubochet *et al.*, 1991; Moor, 1987). For cellular systems with a high amount of associated and structurally bound water approximately −85 °C seem to be critical (Müller and Moor, 1984; Robards and Sleytr, 1985); sometimes even higher recrystallization temperatures are plausible (e.g., <−70 °C for erythrocytes: Nei, 1973; <−43 °C for silkmoth sensory hairs: Steinbrecht, 1993). This likely explains the good results achieved with dry-ice/acetone mixtures for maintaining FS temperatures—given properly cryoimmobilized samples.

Most researchers take great care to use water-free acetone for FS. Acetone that contains more than 2.5% water dissolves ice properly and quickly only at temperatures higher than −80 °C (Van Harreveld *et al.*, 1965). At −90 °C the highest water content of acetone, to be still tolerable for FS, is 2%. In practice, however, one must consider that FS at −90 °C lasts four times longer with acetone containing 1% water than with dry acetone (Humbel and Müller, 1986); by contrast, methanol and ethanol at −90 °C tolerate water contents of 29% and 13%, respectively (Humbel and Müller, 1986; Steinbrecht and Müller, 1987) without delaying

FS [note that water-containing FS cocktails (Walther and Ziegler, 2002) are discussed below].

During FS the cryoimmomobilized, frozen specimens become dehydrated and proteins and nucleic acids become—at least partly—stabilized by the organic solvents (Bohrmann and Kellenberger, 2001; Kellenberger, 1991); at the same time the specimens become infiltrated with the fixatives and/or stains contained in the FS cocktail. The exact temperatures at which these fixatives can interact at appreciable rates with the various cellular components are largely unknown, but they are usually not as low as −90 °C. OsO_4 may react with macromolecules at a minimum temperature of about −70 °C (White *et al.*, 1976), but this may not hold universally true (M. Müller, personal communication). UO_2^{2+} is supposed to act at even lower temperatures (Humbel and Müller, 1986). Glutaraldehyde starts to act at −50 °C, but it works readily only at or above −30 °C (Humbel and Müller, 1986). Finally, tannic acid likely binds to macromolecules already at −80 °C, or else the protocols of Parthasarathy, Ding, and Hepler could not work (Ding *et al.*, 1991a; Meindl *et al.*, 1992); for details on these protocols see Section V.C.2). To summarize, fixing and staining activities of these additives largely occur during the subsequent warming-up phase. Therefore, slow warming is recommended to give fixatives time to stabilize the specimens at the lowest possible temperature, thereby limiting extraction artifacts. A period of 18–24 h to reach 0 °C is widely used, for example in the protocol with 8 h at −90 °C, 8 h at −60 °C, and 8 h at −30 °C [temperature increase 60 °C/h between the steps (Humbel and Müller, 1986)].

Just as we are unsure about the temperatures at which macromolecules become fixed, we are also poorly informed about the detailed mechanisms for how they are fixed (Griffiths, 1993). Depending on the kind of liquid the fixatives are dissolved in (water-free organic solvents vs aqueous buffers), fixatives presumably interact with cellular components in a quite different manner (Kellenberger, 1991).

2. Morphological Applications

Anhydrous acetone containing OsO_4, usually at a concentration of 1–4%, is one of the most simple, though highly efficient recipes for FS (Os-Ac: Van Harreveld and Crowell, 1964). One major advantage of Os-Ac (combined with epoxide embedding and section poststaining with uranyl and lead salts) is the positive contrasting of membranes, ribosomes, chromatin, and the cytoskeleton of plants (Figs. 1–3 and 8), higher fungi and various metazoa.[11] The appearance of membranes as "railroad tracks" of heavy metal does not reflect the distribution of membrane lipids and the configuration of proteins within membranes at the molecular level (Edelmann, 2002; Fleischer *et al.*, 1967; Matsko and Mueller, 2005; Sjöstrand, 1990). However, it compares nicely with half a century's EM literature from chemically fixed samples and meets—in a pragmatic way—many

[11] Certain organisms not belonging to aforementioned systematic groups, however, fail to show consistently positive membrane contrast after FS with acetone containing 1–2% OsO_4 (e.g., yeasts and molds).

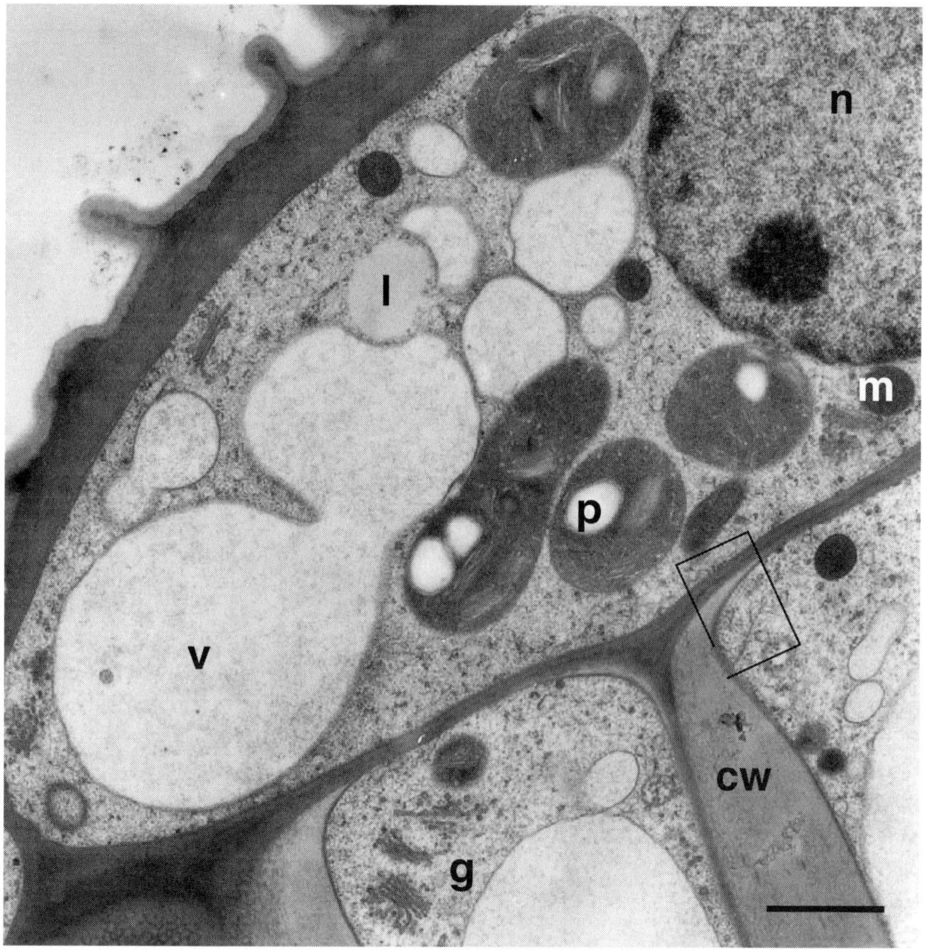

Fig. 1 *Tillandsia* anther: epidermis and endothecium [with thickened cell walls (cw)]. HPF, FS with OsO_4-Ac, Epon-Araldite embedding; section poststaining with aqueous uranyl acetate (UO_2^{2+}) and lead citrate; scale bar = 1 μm. Adequately preserved and contrasted organelles are shown: nucleus (n), mitochondria (m), Golgi stacks (g), vacuoles (v), plastids (p); endoplasmic reticulum is inconspicuous and/or poorly stained, storage lipid droplets (l) are partly preserved and not completely extracted; microtubules located in the framed area are shown enlarged in Fig. 2.

requests of contemporary cell biology (Griffiths, 1993). Positive contrast can readily be achieved with OsO_4 concentrations as low as 0.25%, in both plant and mammalian cells (our unpubublished data). Lower OsO_4 concentrations such as 0.1% (Pfeiffer and Krupinska, 2005a) or 0.01% (McDonald and Müller-Reichert, 2002) frequently yield mixed contrast, with positively contrasted plasma membranes, mitochondria, and endosomes/vacuoles, but with poorly or even negatively stained Golgi membranes.

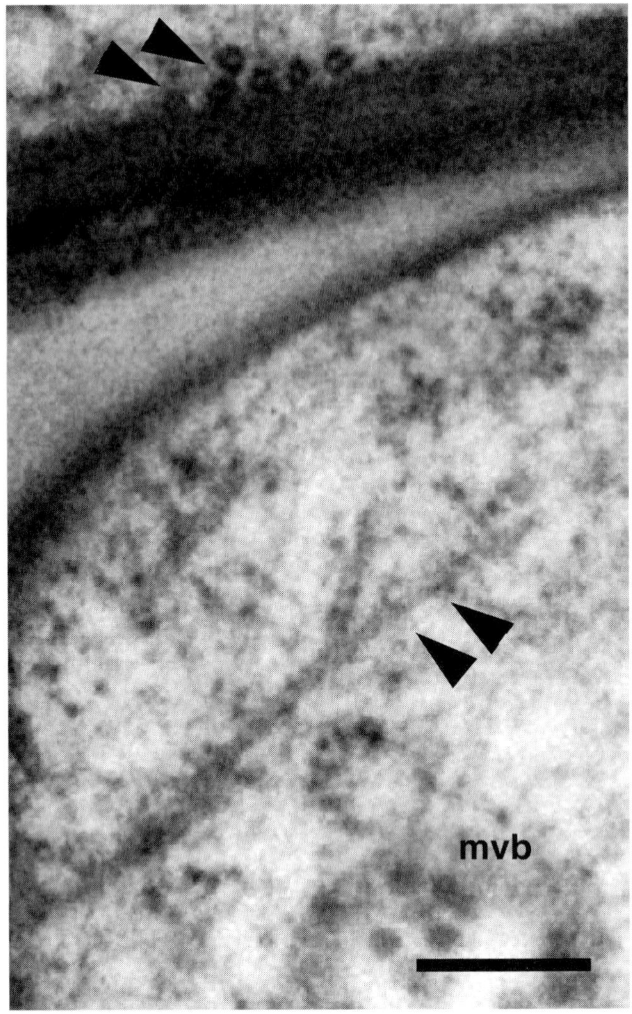

Fig. 2 *Tillandsia* anther epidermis: detail from Fig. 1. Microtubules (double arrowheads) in cross and transversal section, multivesicular body (mvb); HPF, FS with OsO_4-Ac, Epon-Araldite embedding; UO_2^{2+} and lead section staining; scale bar = 200 nm.

UO_2^{2+} combined with Os-Ac further improves the preservation and/or staining of membranes and the cytoskeleton (e.g., Figs. 4 and 5). UO_2^{2+}can be added directly to Os-Ac or used at +4 °C or RT for *en bloc* staining after completion of Os-Ac-FS and postfixation. UO_2^{2+} is not readily soluble in cold acetone (at −85 °C acetone is already saturated with 0.2% UO_2^{2+}), but concentrations as low

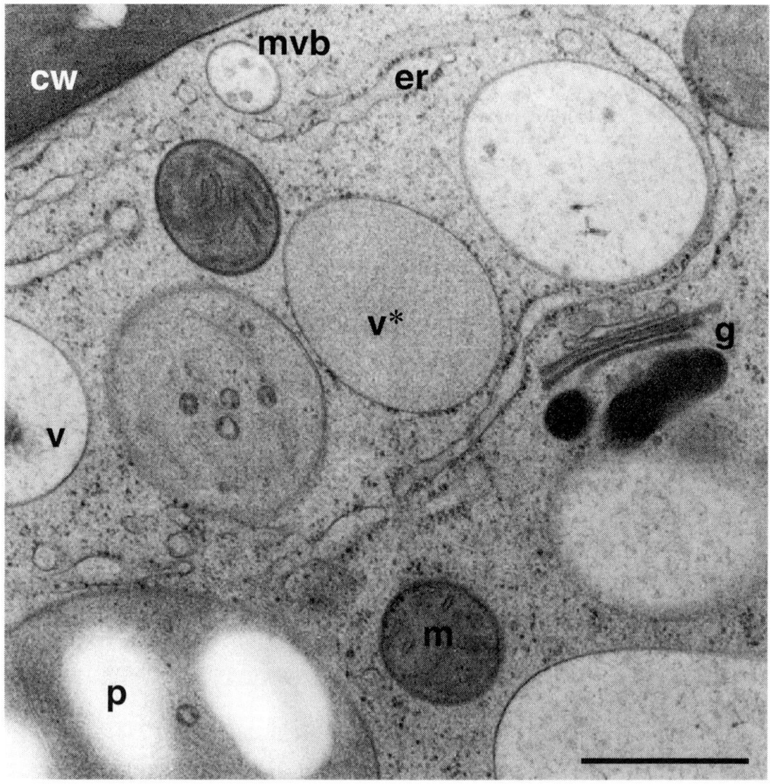

Fig. 3 Epidermis (rhizodermis) of *Lepidium* root tip. HPF, FS with OsO_4-Ac, Epon-embedding; UO_2^{2+}/lead staining; scale bar = 500 nm. Adequately preserved and stained organelles can be seen: for example, different kinds of vacuoles (v), some of which (v*) being possibly derived from the endoplasmic reticulum; endoplasmic reticulum (er) is sufficiently, though not strongly contrasted.

as 0.01% have proven effective in adding contrast (Hoch, 1991). Together, Os-Ac and UO_2^{2+} (Os-UO_2^{2+}-Ac), generally yield well-preserved and stained samples from organisms of all kingdoms, e.g., plants and fungi (Hess, 1995; Lichtscheidl *et al.*, 1990; Lonsdale *et al.*, 1999; Mims *et al.*, 2001; Noguchi, 1994; Segui-Simarro and Staehelin, 2006; Welter *et al.*, 1988), animals (Hess *et al.*, 2000; Kirschning *et al.*, 1998; Murk *et al.*, 2003; Studer *et al.*, 1995), bacteria (Graham and Beveridge, 1990b; Hobot *et al.*, 1985); see also Figs. 1–5 and 8 (this Chapter).

Certain membranes in these specimens, however, are sometimes not well set off after FS with Os-Ac (with/without UO_2^{2+} added) and standard poststaining with uranyl and lead salts [i.e., endoplasmic reticulum, plastid envelope (Hess and Hesse, 1994; Meindl *et al.*, 1992)]. This phenomenon depends on cell type, species, and the tilt angle at which the thin sections are viewed at in the EM. Moreover,

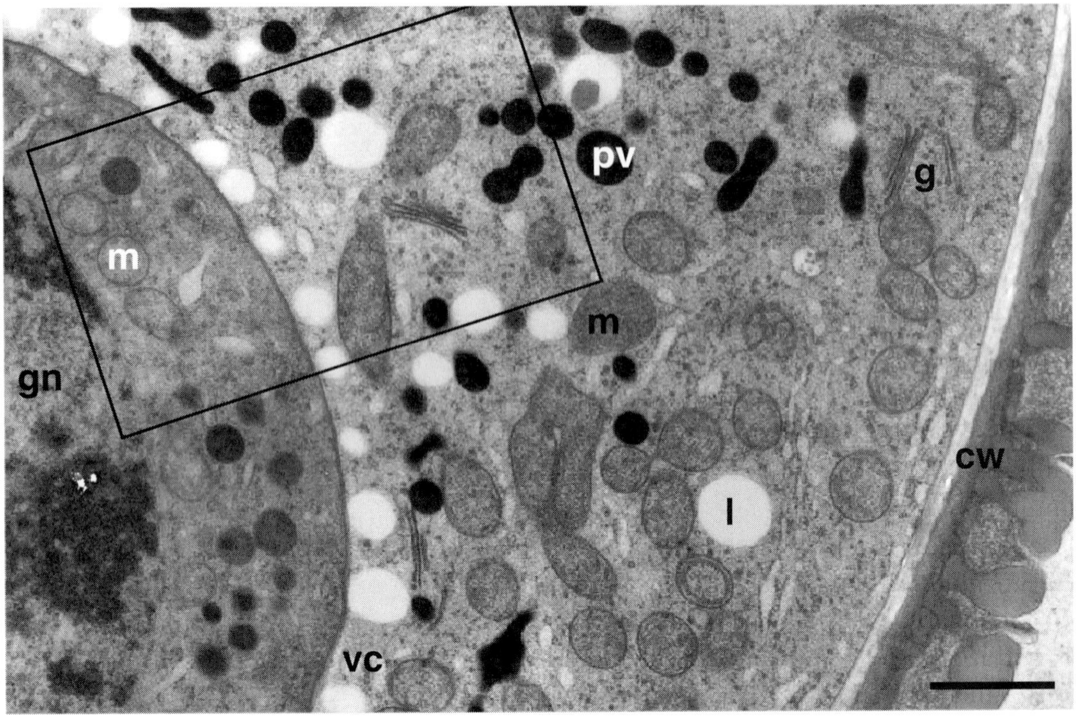

Fig. 4 Young *Ledebouria* pollen grain being composed of a vegetative cell (vc) and a smaller generative cell (gc) embedded within (gn marks the nucleus of the gc). HPF, FS with OsO_4-UO_2^{2+}-Ac, Epon-Araldite embedding; UO_2^{2+}/lead staining; scale bar = 1 μm. Both kinds of cells are adequately preserved and stained. Storage lipid droplets (l) appear completely extracted; note the pleiomorphic protein-storing vacuoles [pv; for cytological details, see the work by Hess (1995)]; framed area enlarged in Fig. 5.

it depends on the freezing and FS quality: the better the freezing and the FS, the lower the contrast. To overcome this problem, Walther and Ziegler (2002) suggested adding up to 5% water to Os-UO_2^{2+}-Ac media, based on investigations of pancreas tissue, mammalian cell cultures, and yeast. Since there were no comparable data on plants, we performed pilot experiments with this recipe on roots and young leaves from garden cress. We obtained some generally usable, strongly contrasted specimens (especially from roots: Figs. 6 and 7); notably, however, storage lipid droplets (i.e., triglycerides) and the neighboring cytoplasm suffered from more or less clearly visible structural damage (Fig. 6). As yet, we have no explanation for these (not tissue specific) artifacts. It is in general still enigmatic whether and how water in these acetone-based substitution cocktails might influence membrane stabilization and/or stainability. Presumably, the 5% water impedes FS by acetone until higher temperatures are reached (Walther and

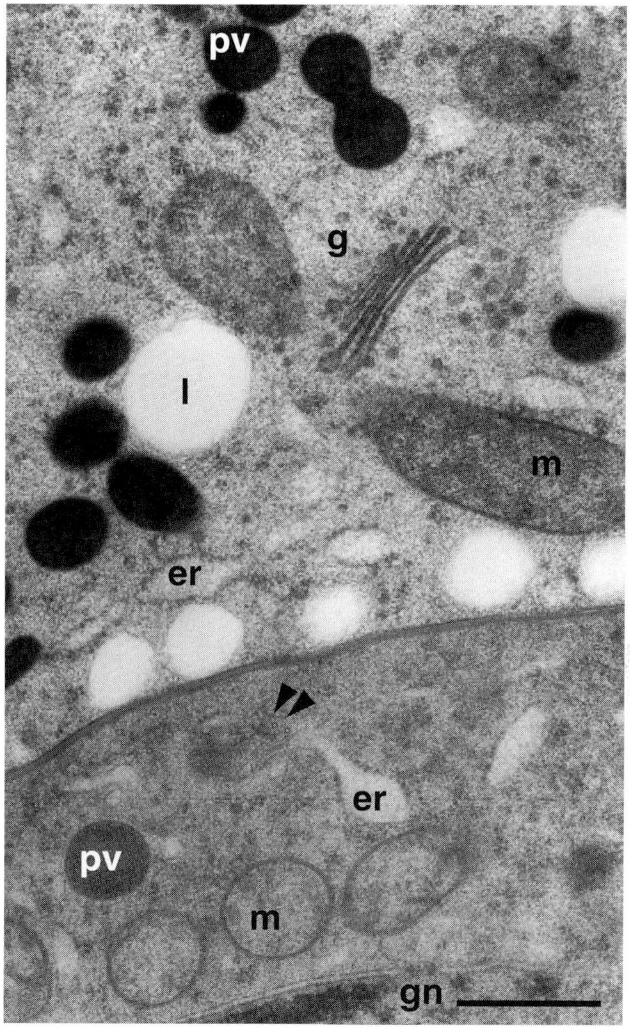

Fig. 5 *Ledebouria* pollen: detail from Fig. 4. HPF, FS with OsO_4-UO_2^{2+}-Ac, Epon-Araldite embedding; UO_2^{2+}/lead staining; scale bar = 500 nm. Note the differential staining of cytoplasm and organelles of generative and vegetative cell, which are at this developmental stage separated by two closed apposed plasma membranes only.

Ziegler, 2002): at about −65 °C acetone readily dissolves ice, even in the presence of 5% water (Van Harreveld *et al.*, 1965), and under these conditions, UO_2^{2+} and possibly also OsO_4 can fix the specimen. Said another way, one may speculate that membrane stabilization by $UO_2^{2+}$and OsO_4 under these conditions might take

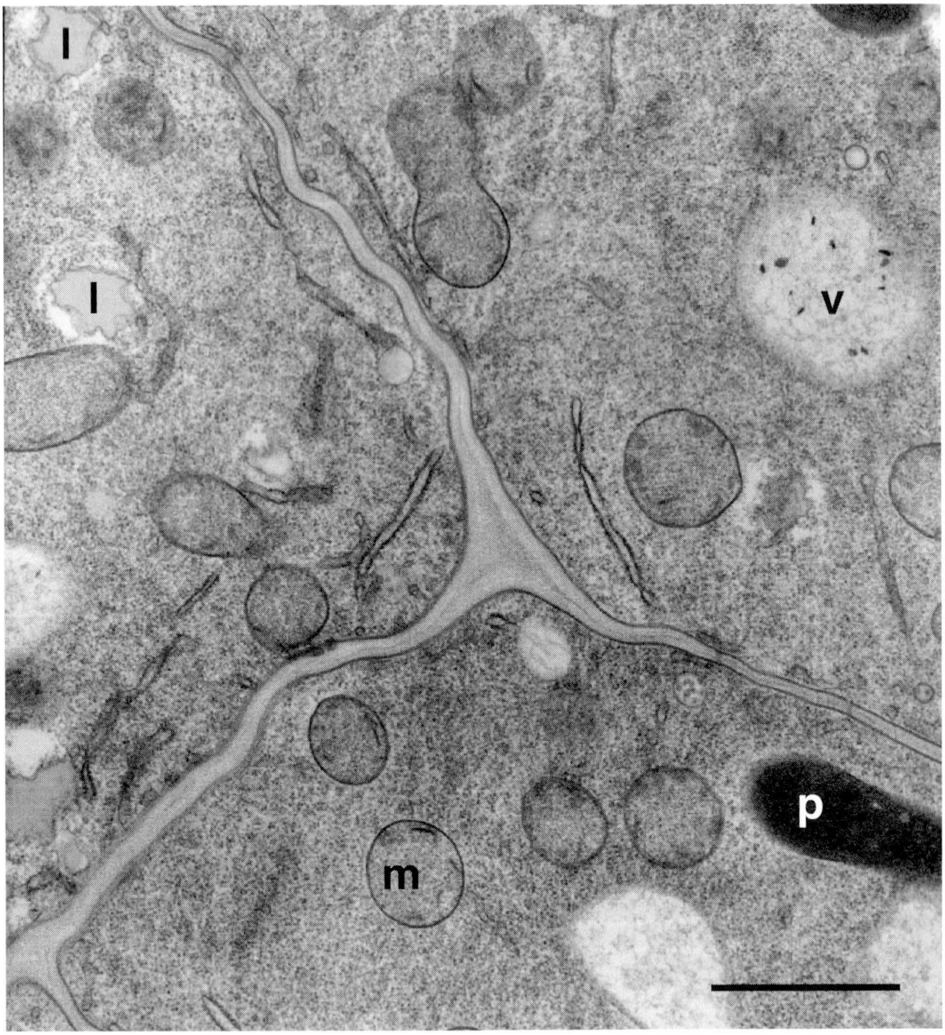

Fig. 6 *Lepidium* root tip after HPF and FS with water-containing OsO_4-UO_2^{2+}-Ac [i.e., 5% H_2O + 2% OsO_4 + 0.1% UO_2^{2+} in acetone, according to Walther and Ziegler (2002)]. Epon embedding; UO_2^{2+}/lead section poststaining; scale bar = 1 μm. Adequate ultrastructure preservation and strong contrast of membranes is seen. Note, however, that storage lipid droplets and the adjacent cytoplasm suffer from clearly visible artifactual distortion; interestingly, the contents of the lipid droplets (l) are not completely extracted (to leave a clearly defined spherical hole), as it is frequently the case with cryoimmobilized cells subjected to FS with water-free solvents (see, e.g., Fig. 5).

place already during FS, and not during the warming-up phase; furthermore, the chemical interactions between membrane constituents and both UO_2^{2+} and OsO_4 might be modified by the presence of 5% water, although there is as yet no

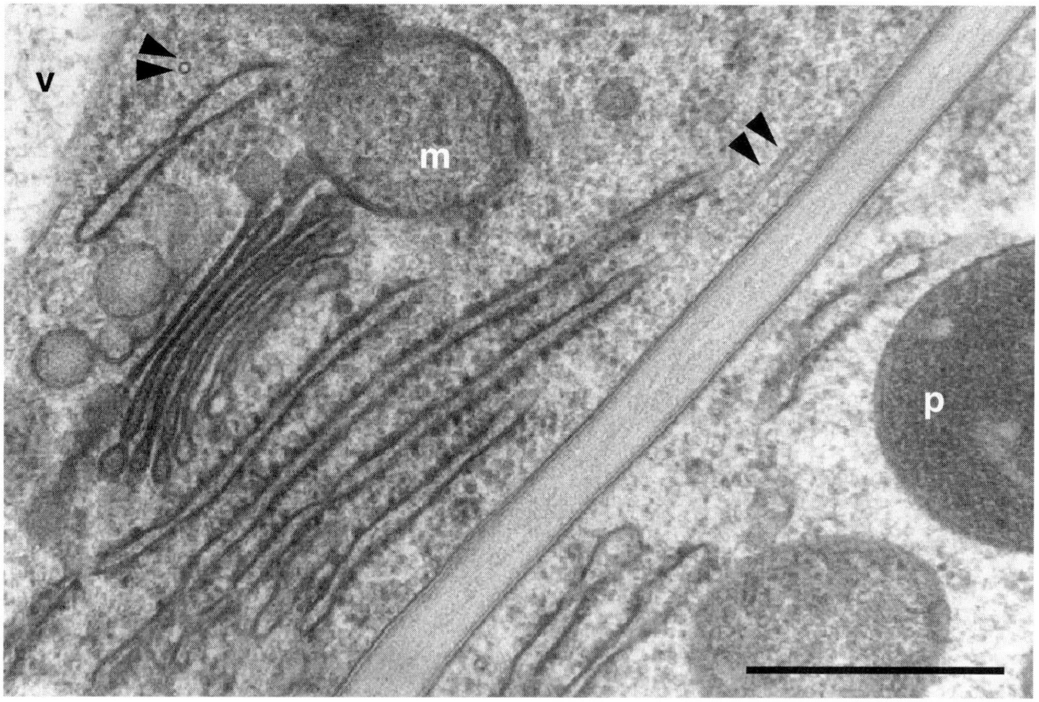

Fig. 7 *Lepidium* root tip after HPF and FS with water-containing OsO_4-UO_2^{2+}-Ac, Epon embedding; UO_2^{2+}/lead staining; scale bar = 500 nm. Adequate ultrastructure preservation and strong contrast of microtubules (double arrowheads) and membranes is observed: plasma membrane, Golgi stack, ER, mitochondria, plastid (p).

experimental evidence for this scenario. In any case, one can expect that FS at −65 °C instead of −90 °C will lead to increased recrystallization of water, but it is hard to determine the extent to which this may reduce the yield of usable samples (i.e., samples showing no visible or only minute segregation patterns). It is noteworthy that a similar but methanol-based cocktail does not appear to show comparable staining and/or preservation patterns ["Müller's complex medium": methanol + 3% water, 1% OsO_4, 0.5% UO_2^{2+}, 3% glutaraldehyde (Humbel and Müller, 1986)]. To summarize, the approach by Walther and Ziegler (2002) might be useful for certain applications, but further detailed evaluation is needed.

Tannic acid is the last fixing and staining additive that seems, in my opinion, worth mentioning in the context of morphological applications. "Tannic acid" is a complex mixture of natural polyphenolic substances of different molecular weights [i.e., (poly)glycosides of gallic acid]. It is mainly obtained from Aleppo nutgalls and has long been known as a mordant for stains. This reagent was used for optimizing the contrast of the tobacco root cytoskeleton (Ding *et al.*, 1991a), and for staining the coat of Golgi-derived vesicles in the alga *Micrasterias* (Meindl *et al.*, 1992). Both groups performed FS at −90/−80 °C with dry acetone plus 0.1% tannic acid, followed by acetone rinses and postfixation at the same temperature with acetone containing OsO_4 (2%) + UO_2^{2+} (at 2% or 0.05%). Because of the heterogeneous composition of different batches of "tannic acid" and the slow, sometimes uneven penetration of this reagent into plant and fungal tissue, reproducible results are apparently not easy to achieve. [I was told by several colleagues that only low molecular weight tannic acid from Mallinckrodt, Inc. (Cat. No. 1764; St. Louis, MO) was suitable for FS, but this product seems no longer to be available.] FS with tannic acid or pure gallic acid in acetone (both without any OsO_4 or UO_2^{2+} treatment) is not recommended for morphology (our unpublished data).

Methanol plus OsO_4 is a rarely used combination that is worth a trial if completely or even partially dehydrated plant tissue should be prepared by FS. In the case of the poikilohydric fern *S. lepidophylla*, only this particular cocktail (applied for 7 days) yielded usable specimens out of a variety of protocols tested (Platt *et al.*, 1997; Thomson and Platt, 1997); by contrast, other researchers obtained convincing data on dry pollen grains with Os-Ac [also applied for 7 days (Tiwari *et al.*, 1990)].

Other additives and solvents are, in my opinion, not the best choice if the focus of interest is morphology [for respective literature on plants/fungi, see the work by Giddings (2003), Hess (2003), and Hoch (1991): $KMnO_4$; Ashford *et al.* (1999): acrolein; Zechmann *et al.* (2005): glutaraldehyde; Pfeiffer and Krupinska (2005a), Zechmann *et al.* (2005): safranin O; Kaeser (1989): water-free, modified "Müller's complex medium," enriched with 50% DMP].

3. FS Media for Immuno-Electron Microscopy

As with any immuno-EM work, there is no straightforward, universally suitable recipe to recommend for FS and embedding—in contrast to morphological applications, where Os-Ac/Epon always gives some usable result. Generally useful rules can, however, be summarized as follows. The solvent of choice is acetone, but ethanol will also work. Aldehydes and UO_2^{2+} are the most suitable additives (for examples see Table I). Acrylic resins are usually preferred over epoxides.

Regarding the solvents it seems worth mentioning that FS with pure acetone, but without any other stabilizers preserves not only antigenicity but also the ultrastructure of certain cell types surprisingly well [for further reading on the possible mechanisms of this low-temperature solvent fixation, see the work by Bohrmann and Kellenberger (2001); Kellenberger (1991)]. The best results with FS in pure solvents have been achieved by embedding in Lowicryls (Monaghan and Robertson, 1990; Monaghan *et al.*, 1998; Usuda *et al.*, 1990; von Schack *et al.*, 1993), but adequate preservation of subcellular gross morphology was also

Table I
Methacrylate and Acrylate Resin Embedding of Freeze-Substituted Plants and Fungi for Immuno-EM (Selected Protocols)

Resin (infiltration temperature/curing temperature)	FS media	References
Lowicryl HM20 (−20/−50 °C)	Acetone + UO_2^{2+} glutaraldehyde	Otegui *et al.*, 2002
Lowicryl HM20 (−35/−35 °C ≫ RT)	Methanol + UO_2^{2+}	Lee *et al.*, 2000
Lowicryl HM20 (−35/−35 °C)	Acetone + UO_2^{2+}	Hippe *et al.*, 1989
Lowicryl HM20 (−50/−50 °C ≫ RT)	Acetone + safranin O	Pfeiffer *et al.*, 2003
Lowicryl HM20 (−50/−50 °C ≫ RT)	Acetone + UO_2^{2+}	Pfeiffer and Krupinska, 2005b
LR Gold (−25/−25°C)	Acetone + UO_2^{2+} glutaraldehyde + methanol	Staiger *et al.*, 1994
LR White (−10 °C/RT)	Acetone	Lancelle and Hepler, 1989
LR White (+4/+4 °C)	Acetone	Hess, 1995; Hess *et al.*, 1995; Puhakainen *et al.*, 2004
LR White (+4/+4 °C)	Ethanol + acrolein	Bourett and Howard, 1991
LR White (+45/+80 °C): microwave assisted	Acetone + UO_2^{2+} glutaraldehyde	Lonsdale *et al.*, 1999

obtained with Epon and/or Araldite (Hess, 2003; Nicolas, 1991; Roch *et al.*, 1997; Usuda *et al.*, 1990; von Schack *et al.*, 1993) or LR White (Hess *et al.*, 1995; von Schack *et al.*, 1993; see also Figs. 9 and 10 of this chapter). Notably, certain antigen-antibody combinations fail in postembedding immunolabeling when the FS media contain any additional fixatives but do work with pure solvents (Hess *et al.*, 1995; Hess and Valenta, 1997). Of course, such specimens suffer from more or less visible extraction, especially when the resin is polymerized at elevated temperatures (i.e., above approximately −30 °C, where also structurally bound water is removed), and their overall contrast is often low.

Despite the possibility of using pure solvents as the one and only stabilizing reagent for FS and postfixation, better structural preservation is usually obtained by adding UO_2^{2+} (0.02–0.5%), and/or glutaraldehyde (0.1–0.5%), and/or formaldehyde [1–4%; see also Table I and work by Humbel and Schwarz (1989)]. UO_2^{2+} binds to proteins and phospholipids and considerably reduces lipid extraction during FS(Diakou *et al.*, 2002; Hayat, 2000; Humbel and Müller, 1986; Humbel and Schwarz, 1989; Weibull *et al.*, 1984), aldehydes react with terminal amino groups. Sometimes, an appropriately balanced mixture of all these three fixatives had been found necessary for the proper retention and detection of easily translocating antigens (Eppenberger-Eberhardt *et al.*, 1997). Occasionally, acrolein has also been employed (Bourett and Howard, 1991), but, to my knowledge, there is no clear evidence about its advantages over other, less harmful reagents. Depending on the concentration of the aldehyde stock solution used for dilution and the final aldehyde concentration desired, a certain amount of water is usually introduced into the FS cocktail (except when using anhydrous glutaraldehyde in acetone, as distributed by EMS Electron Microscopy Sciences, Hatfield, PA). Thus, ethanol is sometimes more suitable than acetone because of its capacity to dissolve ice at −90 °C even in the presence of 13% water. Some investigators have found that UO_2^{2+} and formaldehyde as well as low concentrations of glutaraldehyde (0.1–0.2%) in the FS media interfere only weakly, if at all, with immunolabeling (Bittermann *et al.*, 1992; Schwarz and Humbel, 1989). This agrees to some extent with the observations made with Tokuyasu's method for immunolabeling (Griffiths, 1993; Liou *et al.*, 1996). Sometimes, however, traces of UO_2^{2+} in the cocktail may be sufficient to prevent immunolabeling completely (Hess and Valenta, 1997; Singh *et al.*, 1993; Taylor *et al.*, 1994).

Other additives are not frequently used, nor are they recommended without reservation if the main goal is immuno-EM of protein antigens. Namely, OsO_4 in the FS cocktails only occasionally preserved protein antigenicity [see the work by Bourett *et al.* (1999), Favre *et al.* (1998), Nicolas (1991), Samuels *et al.* (1995)]. By contrast, immunolabeling of carbohydrates and lipids is apparently more compatible with OsO_4-Ac (Hess and Frosch, 1994; Kirschning *et al.*, 1998; Samuels *et al.*, 1995; Zhang and Staehelin, 1992) and even with epoxide embedding [Hess and Frosch, 1994; Kirschning *et al.*, 1998; Samuels *et al.*, 1995; see also Fig. 8 of this chapter: pollen wall immunolabeling with JIM7 (antibody kindly provided by J. P. Knox: Knox *et al.*, 1990]. Similarly, it was shown that certain lectins do adequately stain such Os-Ac/Epon specimens (Tonning *et al.*, 2006). Tannic acid, or its major component gallic acid, have not often proven to be suitable for immuno-EM (T. Beveridge, personal communication), nor has $KMnO_4$. Safranin O has been introduced for correlative fluorescence and immuno-EM (Pfeiffer *et al.*, 2003), but this protocol needs further evaluation. Most recently, Mueller and Matsko demonstrated the suitability of epoxy-resin monomers as stabilizing additives (Matsko and Mueller, 2005). In particular, they mixed 20% Epon-Araldite with acetone for FS and postfixation of *Caenorhabditis elegans* and lung fibroblasts [conditions for FS and warming-up according to work by Humbel and Müller (1986)]. Well-preserved and sufficiently stained ultrastructural details were shown, as well as first results from tubulin immunolabeling. Our pilot experiments with this recipe on cress roots and leaves were successful with respect to ultrastructure preservation but remain still to be evaluated in detail.

D. Embedding

1. Epoxy Resins

Epoxy resins are the first choice for morphological studies, but they are also suitable for immunolabeling of certain antigens (especially carbohydrates and lipids) and for lectin cytochemistry (see above). Propylene oxide is frequently included as an intermediate in standard protocols for conventionally fixed

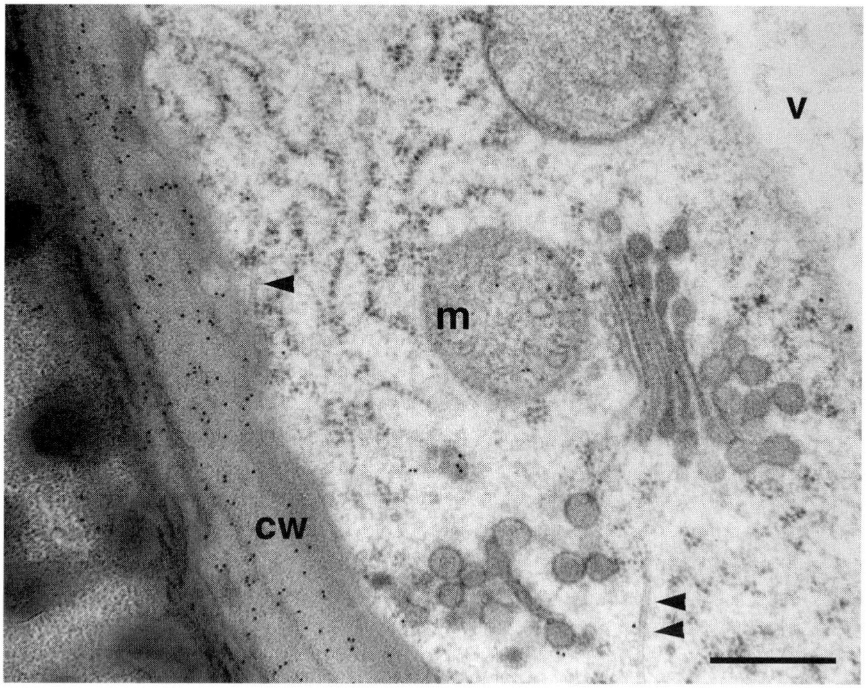

Fig. 8 *Ledebouria* vacuolate microspore [during the formation of the inner cell wall layer, the intine; for cytological details, see the work by Hess (1993)]. HPF, FS with OsO_4-Ac, Epon-Araldite embedding; on-section immunolabeling with JIM7 monoclonal antibodies (courtesy J. P. Knox) against methyl-esterified epitopes of pectin visualized with goat anti-rat 10-nm colloidal gold conjugates (from British Biocell, Cardiff, U.K.); section poststaining with aqueous uranyl acetate and lead citrate; scale bar = 500 nm.

material (i.e., aqueous glutaraldehyde-OsO_4) for facilitating resin infiltration of particularly impermeable tissues. Published FS/resin embedding protocols, however, only rarely mention this highly volatile and carcinogenic reagent (Glauert and Lewis, 1998); it is, indeed, dispensable, according to our experience with a variety of specimens that are hard to penetrate. Whether its use has any negative effect on the quality of ultrastructure preservation of freeze-substituted samples is not clear. Graham and Beveridge (1990b) explicitly recommend not to use propylene oxide for freeze-substituted eubacteria because of its deleterious effects. By contrast, there exist good micrographs from other freeze-substituted, propylene oxide-treated samples (Ikeda *et al.*, 1984; Meindl *et al.*, 1992; Van Harreveld *et al.*, 1965) and this compound was even used as an FS media for microautoradiography (Fisher and Housley, 1972).

Epon, Araldite (including their substitutes), and Epon-Araldite mixtures do not show fundamental differences as to the quality of ultrastructure preservation [for recipes, see the work by Glauert and Lewis (1998), Hayat (2000), Mollenhauer (1964)]. Though viscous, they readily infiltrate any kind of samples (given enough time for, and continual agitation during, infiltration). Their virtues include ease of sectioning, good preservation of structure, their poststaining properties, and their beam stability (Figs. 1–8). Spurr's low-viscosity resin (Spurr, 1969) is very popular among plant scientists, even though it is, like propylene oxide, highly toxic and carcinogenic. However, specimens that have been freeze-substituted with Os-Ac and embedded with Spurr's are sometimes not satisfactorily preserved and/or stained-unlike conventionally fixed samples. More or less severe extraction artifacts have been observed, depending on the specimen and the FS protocol: namely, solubilization of membranes, resulting in indistinct contrast (Hastie, 1990), but also extraction of cytoplasm and organelle matrix constituents (Hess, 1990, 2003). Specimen contrast can be increased in this case by supplementing Os-Ac with UO_2^{2+} for FS (Noguchi, 1990), or by postfixation with $HfCl_4$ (Hoch, 1991; Lancelle *et al.*, 1986), or with +40 °C-hot Os-Ac postfixation followed by +4 °C-cold UO_2^{2+}-methanol (Murata *et al.*, 2002). Thus, there do exist good micrographs from freeze-substituted plants and fungi embedded in Spurr's (Samuels *et al.*, 1995; Tiedemann *et al.*, 1998; Zheng and Staehelin, 2001) or Spurr's-Epon mixtures (Mims *et al.*, 2001, 2003). Furthermore, it is possible that extremely impermeable specimens, such as fully desiccated tissues, do require Spurr's for adequate infiltration and sectioning (Thomson and Platt, 1997; Wesley-Smith, 2001) since it really acts as a "glue" (Mollenhauer, 1988). In this context, it must be emphasized that Epon too, has recently been found to cause artifactual extraction of certain specialized, lipid-enriched endosomal/lysosomal compartments in mammalian cells, but these problems were solved with modified FS protocols (Hess, 2003).

Together, Epon and/or Araldite can definitely be recommended as the best resins available for preserving the morphology of freeze-substituted plants and fungi. If their viscosity causes real problems one may consider mixing one's own cocktail of Epon and/or Araldite substitutes instead of using kits (Mascorro and Kirby, 1992) [for recipes, see also the work by Glauert and Lewis (1998): there exist remarkable differences in the final viscosity of such mixtures, depending merely on the choice of resin monomers from different suppliers].

2. Acrylic Resins

Two different kinds of (meth)acrylate resins (hereafter termed acrylic resins) are available to permit cytochemical localization of substances that are more sensitive to their environment [for review, see the work by Glauert and Lewis (1998); Newman and Hobot (2001)]: (1) Lowicryls and Unicryl and (2) the London Resins LR White and LR Gold. Acrylic resins do not bind covalently to the specimen and the section surface is much rougher than with epoxides; consequently, the efficiency of on-section immunolabeling or lectin labeling is usually higher (Kellenberger *et al.*, 1987). Apart from these obvious advantages acrylic resins are characterized by less optimal ultrastructure preservation as compared to epoxides

(Kellenberger, 1998; Nicolas, 1991). This feature results in part from their high potential to extract not only lipids but also other molecules (Glauert and Lewis, 1998) from polymerization damages (Zalokar, 1966) and from the rough (corrugated) surface of the sections (Kellenberger, 1998).[12]

Lowicryls (Chemische Werke Lowi, Waldkraiburg, Germany) and Unicryl (British Biocell, Cardiff, UK) are closely related in their composition to glycol methacrylate embedding media (GMA: Rosenberg *et al.*, 1960). They are highly toxic and/or carcinogenic, and, most importantly, potentially highly skin sensitizing.[13] Owing to these allergenic properties Lowicryls have not been used in our laboratory; thus, no data of our own are presented here. Fortunately, details on the chemical composition, health hazards, and safe handling of these methacrylate resins can be found in several publications (Glauert and Lewis, 1998; Newman and Hobot, 2001; Tobler and Freiburghaus, 1990). The various types of Lowicryls allow for specimen infiltration and UV-polymerization at temperatures of −35 °C and below (Acetarin *et al.*, 1986; Glauert and Lewis, 1998). This is certainly a great advantage for freeze-substituted material, since lipid extraction may be limited and the structurally bound water is still retained at these temperatures [for further details, see the work by Kellenberger (1991)]. Unicryl, however, is polymerized at a minimum temperature of +4 °C (Newman and Hobot, 2001), thus definitely lacking this advantage of Lowicryls [moreover, the ultrastructural preservation of our freeze-substituted plant and mammalian cells embedded in Unicryl was usually inferior to any other embedding (Hess, 2003)]. Lowicryl resins have been shown as to be quite suitable for animal tissues (Monaghan *et al.*, 1998) but are not that widely used for freeze-substituted plant material. This may be a result of the limited infiltration of bulk plant specimens at low or very low temperatures (Staiger *et al.*, 1994). Nevertheless, convincing examples for Lowicryl embeddings from freeze-substituted plant specimens are shown in Table I.

London Resins (LR White, LR Gold; from London Resin Co.) are supposed to be less toxic and allergenic than Lowicryls and Unicryl. Unfortunately, however,

[12] Another feature that is limited to acrylic resins should be mentioned here as well. Sections subjected to on-section immunolabeling prior to poststaining with uranyl and lead salts frequently show quite poorly contrasted subcellular details, irrespective of the FS- or fixation-protocol applied. Sections from the same block, directly poststained and viewed in the EM, however, show much more detailed ultrastructure. This phenomenon likely reflects some artifactual extraction of the sample. A similar phenomenon (though from chemically fixed LR White embeddings) was reported by others (Berryman and Rodewald, 1990). However, I am not aware of any suggestion of how to improve substantially the visibility of subcellular details in this case, except for shortening the immunolabeling procedure as much as possible (Berryman and Rodewald, 1990).

[13] The risk of developing severe allergic dermatitis as a result of Lowicryl resin monomers seems to vary considerably between individuals. As much as 25% of an average population appears highly susceptible, whereas another 25% seem not responding at all to exposure to Lowicryl vapors; the remaining 50% are in between these two extremes (M. Dürrenberger, personal communication).

their exact chemical composition has not been published (the only component with proven high toxicity is the accelerator, a tertiary amine, needed only for chemically catalyzed resin polymerization). Especially LR White has become quite popular for preparing freeze-substituted plant and fungal material for (immuno-) cytochemistry (Table I). It easily infiltrates any kinds of samples and readily polymerizes with only moderate heat (+50 to +52 °C) or by UV light at cold temperatures [i.e., +4 °C down to −15 °C (Glauert and Lewis, 1998) or even −20 °C (Newman and Hobot, 2001)]. LR Gold, which was originally designed for infiltration and polymerization at −25 °C, is less widely used (Staff *et al.*, 1999; Staiger *et al.*, 1994; Taylor *et al.*, 1994). Lancelle and Hepler (1989) demonstrated the suitability of FS with pure acetone as the one and only stabilizing reagent, followed by LR white embedding for actin immunolabeling in plants. We succeeded with this recipe in localizing various aldehyde-fixation-sensitive and/or water-soluble molecules in pollen (Hess and Valenta, 1997; Hess *et al.*, 1995), leaves (Puhakainen *et al.*, 2004), and mammalian cell cultures (Valenta *et al.*, 1998). Despite the quite week fixation resulting from FS with pure acetone and temperatures of at least +4 °C during resin infiltration and polymerization, gross ultrastructural features were satisfactorily preserved with this protocol (Figs. 9 and 10). Organelles/endomembrane compartments, microtubules and actin microfilament bundles could be identified unequivocally. Subtle details, such as single microfilaments or vesicle coats (COP-coats), were not detectable, however, and membrane lipids were lost. Sometimes, indeed, eye-striking extraction artifacts were observed—but easily recognized as such (e.g., in some cases complete solubilization of the generative cell in *Ledebouria* pollen: Hess, 2003; Hess *et al.*, 1995). This agrees with notions from others, concerning the strong extractive effects of LR White on "mildly" fixed samples (Coetzee and van der Merwe, 1989; McDonald, 1994).

An interesting possibility to reduce the infiltration and polymerization times for acrylic and epoxy resins may be the use of microwave irridation with the aid of specially designed laboratory microwave ovens (Demaree and Giberson, 2001; Giberson *et al.*, 1997). With this method one can expect not only to save time but possibly also limit extraction artifacts. This technique has been successfully applied to LR White embeddings of freeze-substituted aleurone cells (Lonsdale *et al.*, 1999). Recently, the technique and respective instrumentation were modified by constructing a vacuum chamber for carrying the specimens within the oven (http://www.tedpella.com/Micro_html/vacpro.htm); however, when using vacuum infiltration and polymerization one should consider the risk of partial evaporation of volatile resin constituents, thus, the alteration of the resins' properties (Mollenhauer, 1993).

E. FS and Embedding Media for Microanalysis and Autoradiography

For microanalysis, too, FS is generally better suitable than conventional chemical fixation. As to diffusible ions, however, certain limitations of this approach must be considered. In this case, more reliable results are likely obtained

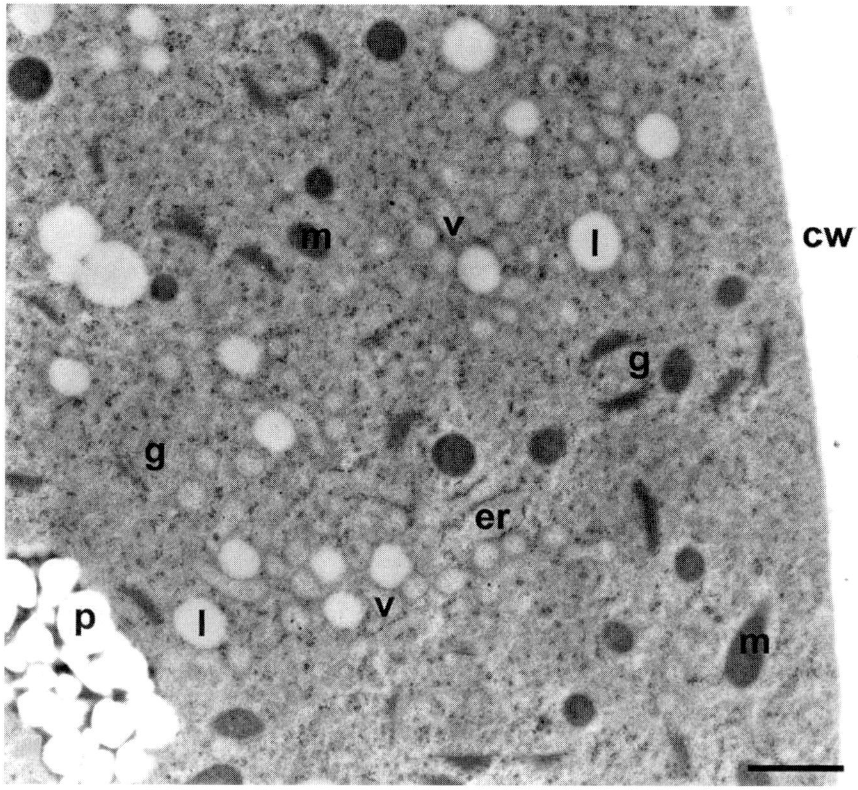

Fig. 9 Mature *Ledebouria* pollen. HPF and FS with pure acetone followed by LR White embedding (polymerized by UV light at +4 °C) preserved the subcellular gross morphology and permit immunolabeling of actin epitopes with commercial monoclonal antibodies against chicken gizzard actin (from Amersham Biosciences, Buckinghamshire, U.K.) visualized with goat anti-mouse 5-nm-colloidal gold [from Sigma; for cytological details see the work by Hess (1995) and Hess *et al.* (1995)]. Plastids (p), mitochondria (m), Golgi stacks (g), rough ER (er), storage lipid droplets (l), and vacuoles (v) can be identified. UO_2^{2+}/lead staining; scale bar = 1 μm.

by freeze-drying followed by resin embedding (Edelmann, 1994, 2002; Steinbrecht and Müller, 1987), or even better by cryoultramicrotomy, optionally combined with freeze-drying of the sections (Frey *et al.*, 1997; Zierold and Steinbrecht, 1987); however, these methods are technically more demanding. Thus, FS, resin embedding and dry sectioning seem, according to the literature, to be an acceptable compromise for many, though not all applications (Harvey, 1980). Tetrahydrofuran and diethylether appear more suitable than methanol, ethanol, or acetone as FS media, and epoxides are preferred over acrylics (Table II). Autoradiographic analyses, though currently not widely used for plant and fungal specimens, can be successfully combined with FS and resin embedding both at

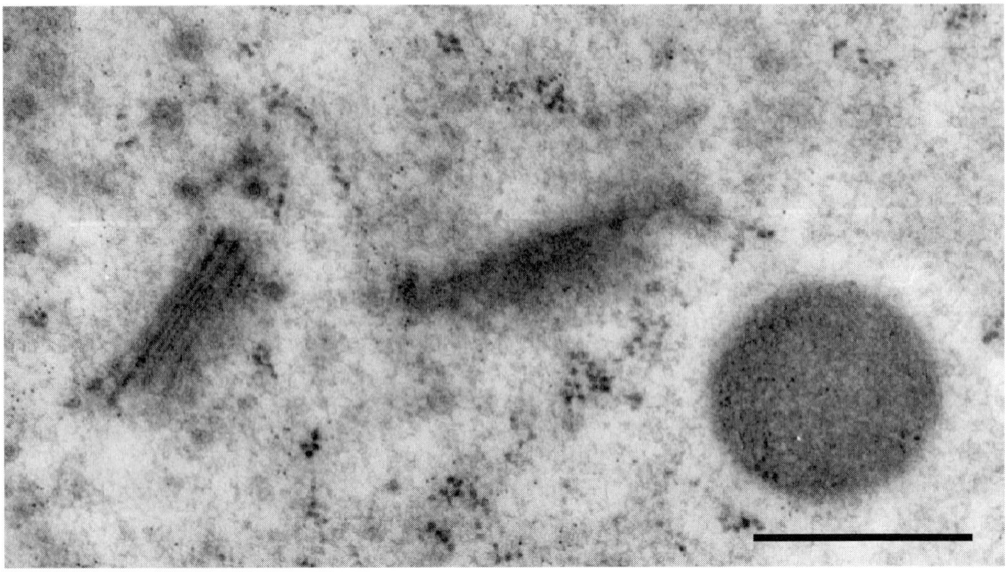

Fig. 10 Mature *Ledebouria* pollen. HPF and FS with pure acetone followed by LR White embedding (polymerized by UV light at +4 °C), same section as Fig. 9. Immunolabeling of actin in the vicinity and in association of Golgi stacks and mitochondria (with mouse anti-chicken gizzard actin from Amersham, visualized with goat anti-mouse 5-nm-colloidal gold, from Sigma). UO_2^{2+}/lead staining; scale bar = 500 nm.

Table II
FS and Embedding Procedures for Microanalysis and Autoradiography

Solvent	Fixative	Resin	References	Application
Diethylether		Spurr's	Harvey, 1980	EDX
Diethylether		Araldite	Palsgard *et al.*, 1994	EDX
Tetrahydrofuran		Araldite	Palsgard *et al.*, 1994	EDX
Diethylether	Acrolein	Spurr's	Ashford *et al.*, 1999	EDX
Tetrahydrofuran		Spurr's	Ashford *et al.*, 1999	EDX
Diethylether	Acrolein	Araldite	Bidwell *et al.*, 2004	EDX
Acetone	Gallic acid	Epon	Graham and Beveridge, 1990a	EDX of bacteria[a]
Acetone	Glutaraldehyde	Epon	Otegui *et al.*, 2002	EDX
Diethylether	Acrolein	Spurr's	Altus and Canny, 1985	Microautoradiography for LM
Acetone, or methanol, or propylene oxide		Epon	Fisher and Housley, 1972	Microautoradiography for LM

[a]This recipe also yields acceptable ultrastructure preservation of plant samples (our unpubublished data).

EDX = energy dispersive X ray analysis.

the LM and EM level (Ikeda, 1985; Nagata, 1998); for protocols and further references see the work by Maunsbach and Afzelius (1999) and Table II.

F. Other Cryopreparation Techniques

Freeze-drying followed by resin embedding could be a valuable complement for cell biological applications, particularly for microanalysis and immunocytochemistry [(Edelmann, 1994, 2002); note that these papers deal with mammalian cells only]. According to the literature and personal discussions with L. Edelmann I have understood that this approach is technically quite demanding, especially with plant and fungal samples. Thus, morphological data that could compete somehow with the quality obtained by FS might be difficult to obtain. First, very small samples are required; the shortest side of the specimen may not exceed 0.3 mm. Second, the thick walls of plant and fungal specimens often prevent effective freeze-drying. For the investigation of somewhat comparable samples, that is, mammalian cell suspensions contained within 200-μm-wide capillary tubes [made of cellulose with a wall thickness of 20 μm (Hohenberg *et al.*, 1994)], the frozen capillaries had to be sliced open; otherwise, freeze-drying was impaired (Edelmann, 2002). The further development of this approach toward a feasible protocol for plant science cannot be foreseen, especially because the Leica CFD cryosorption freeze-drying machine is no longer available. This instrument not only allowed for automatically controlled freeze-drying (that can be achieved with freeze-fracturing or coating devices as well), but also for appropriate resin infiltration and polymerization within the same specimen chamber (thus, no risky specimen transfer was needed). Finally, L. Edelmann—a real pioneer with huge experience in this challenging technique—has already retired, and the research focus of his department has changed (this department was formerly directed by the meanwhile also retired H. Sitte, another pioneer in cryo-based EM).

Cryosectioning and TEM of non-cryo-protected, frozen-hydrated plants were, to my knowledge, for the first time convincingly demonstrated by M. Michel, M. Müller, and colleagues (Michel *et al.*, 1991, 1992). Despite the progress that has subsequently been achieved (Al-Amoudi *et al.*, 2003), this approach has so far not been applied to plants and fungi for solving real cell biological questions. Further detailed information on the status of cryo-electron microscopy of vitrified sections (CEMOVIS) can be found in Chapter 15 by Dubochet and Marquis, this volume.

Freeze-fracturing and shadowing (Steere, 1957) was considerably improved in the 1960s by plant cell biologists (Moor and Mühlethaler, 1963) and later also successfully applied to the morphological analysis of high-pressure frozen root cells (Craig and Staehelin, 1988); for recent developments on HPF combined with freeze-fracturing, see the work by Walther (2003b). Postshadow immunolabeling of replicas was already demonstrated in the 1980s (Rash *et al.*, 1982). However, it was only Fujimoto's technique of SDS-FRL that has initiated a revival of the freeze-fracturing approach (Fujimoto, 1995, 1997; Hagiwara *et al.*, 2005).

Recently, SDS-FRL has also been applied to yeasts (Konomi *et al.*, 2003) and modified for plants (Itoh and Kimura, 2001; Kimura *et al.*, 1999).

Cryo-scanning EM of cryoimmobilized bulk samples is the last technique to be briefly mentioned here. Walther and Müller cut cryosections from high-pressure frozen animal and plant specimens and studied the resulting block face after platinum-carbon coating, using high-resolution cryo-field-emission SEM (Walther and Muller, 1999). Further refinements of this complementary approach were recently published [(Walther, 2003a,b); of course, it is equally possible to view the replicas from the block face (after cleaning) by conventional TEM].

VI. Concluding Remarks

Especially in the case of obviously stressed cells or organisms, EM of conventionally, chemically fixed specimens does not allow the genuine morphological effects of the challenges under investigation to be defined correctly. Thus, primary, "challenge-induced" morphological alterations cannot be properly distinguished from secondary alterations that comprise "challenge-induced" features mixed up with, and/or blurred by, "challenge-amplified" fixation artifacts.

Cryoimmobilization of native specimens by means of HPF provides the best basis currently available for trapping subcellular dynamics of plant and fungal bulk specimens with very high spatial and temporal resolution for ultrastructural analyses.

The old principle of designing experiments to be as simple as possible holds true for HPF, FS, and resin embedding as well. It is recommended that one tries the easy way first, using well-established protocols and avoiding complicated approaches. Especially for the beginner it is better to produce a generous number of similar specimens using just one or two acceptably working, "universal recipes" for cryoimmobilization and FS than to run several protocols in parallel. There are several reasons for doing so. First, it is clear that skill matures with experience; so it can happen that two slightly different, but generally simple, straightforward protocols applied by a "technical novice" might produce a wide variety of (inconsistent) results whereas definitely different (and even complicated) HPF/FS/embedding protocols in the hands of a trained researcher still yield consistent results. Second, it is very likely that the quality of the results depends much more on failures and/or variations in sampling prior to freezing, on the freezing itself, on postfreeze sample transfer (Moor, 1987), and on the interpretation of the morphological patterns observed, than on modifications of the time-temperature regimen during FS or the choice of fixing/staining additives. [Moreover, it seems that some (over-) sophisticated protocols have been recommended in the literature, but have never been *properly* compared with more simple recipes.] Third, it must be emphasized that many of the morphological details and the biological variation of objects under study have not been described so far on the basis of rapidly frozen material. Thus, one should first get familiar with these, partly novel ultrastructural features by using acceptably working procedures before starting with trial-and-error-based modifications of the protocols for freezing, FS, and embedding.

To summarize, it is essential to take meticulous care with the pre- and post-freezing manipulations and to aim at a thorough understanding of the whole chain of procedures used for preparation and analysis. One should not allow oneself to be lulled into a false sense of security simply because one is using the gold standard HPF, possibly combined with another gold standard such as electron tomography.

Acknowledgments

I am very grateful to Martin Müller (Laboratory EM-1, ETH-Zürich) and his collaborators and pupils (René Hermann, Martin Michel, Daniel Studer, Paul Walther, and Ernst Wehrli) for their long-lasting hospitality and friendship. They introduced me into cryopreparation methods and have continuously supported my work until now. Alan Schulman, Marko Jääskelainen (Institute of Biotechnology, Helsinki), Robin Sen (Macaulay Institute, Aberdeen), and Hubertus Haas (Biocenter, Innsbruck Medical University) are acknowledged for providing plant and fungal samples, J. Paul Knox (John Innes Institute, Norwich) for providing JIM 5 and 7 antibodies. Furthermore, I wish to thank Heinz Schwarz (MPI-Tübingen), Bruno Humbel (University of Utrecht), Richard McIntosh (University of Colorado, Boulder), Paul Debbage (Innsbruck Medical University), and Martin Müller for critical reading, discussion, and editing of the chapter; Paul Walther (University of Ulm) for discussions on "watery freeze-substitution"; Karin Gutleben for skillful technical assistance; and Peter Wild (University of Zürich), Reinhard Rieger (University of Innsbruck), Terry Beveridge (University of Guelph), Hellmut Sitte (Seefeld-Austria and Homburg/Saar), Markus Dürrenberger (University of Basel), Richard McIntosh, Paul Walther, and Martin Müller for communicating unpublished observations. The paragraph on challenged cells (Section II.A) is a modified version of a congress talk and is, therefore, reproduced with permission from the Extended Abstracts, see (Hess, 2005). The work in Dr. Hess's laboratory is supported by grants from Österreichische Nationalbank-Jubiläumsfonds (P-11050), Tiroler Wissenschaftsfonds (P-UNI-0404/100), and by TILAK—Tiroler Landeskrankenhaus—Universitätskliniken.

References

Acetarin, J. D., Carlemalm, E., and Villiger, W. (1986). Developments of new Lowicryl resins for embedding biological specimens at even lower temperatures. *J. Microsc.* **143**(Pt. 1), 81–88.

Al-Amoudi, A., Dubochet, J., Gnaegi, H., Luthi, W., and Studer, D. (2003). An oscillating cryo-knife reduces cutting-induced deformation of vitreous ultrathin sections. *J. Microsc.* **212,** 26–33.

Altmann, R. (1890). "Die Elementarorganismen und ihre Beziehungen zu den Zellen." Veit, Leipzig.

Altus, D. P., and Canny, M. J. (1985). Loading of assimilates in wheat leaves. II. The path from chloroplast to vein. *Plant Cell Environ.* **8,** 275–285.

Arimura, S., Yamamoto, J., Aida, G. P., Nakazono, M., and Tsutsumi, N. (2004). Frequent fusion and fission of plant mitochondria with unequal nucleoid distribution. *Proc. Natl. Acad. Sci. USA* **101,** 7805–7808.

Ashford, A. E., Vesk, P. A., Orlovich, D. A., Markovina, A. L., and Allaway, W. G. (1999). Dispersed polyphosphate in fungal vacuoles in *Eucalyptus pilularis/Pisolithus tinctorius ectomycorrhizas. Fungal Genet. Biol.* **28,** 21–33.

Bereiter-Hahn, J., and Voth, M. (1979). Metabolic state dependent preservation of cells by fixatives for electron microscopy. *Microsc. Acta* **82,** 239–250.

Bereiter-Hahn, J., and Voth, M. (1994). Dynamics of mitochondria in living cells: Shape changes, dislocations, fusion, and fission of mitochondria. *Microsc. Res. Tech.* **27,** 198–219.

Berryman, M. A., and Rodewald, R. D. (1990). An enhanced method for post-embedding immunocytochemical staining which preserves cell membranes. *J. Histochem. Cytochem.* **38,** 159–170.

Bidwell, S. D., Crawford, S. A., Woodrow, I. E., Sommer-Knudsen, J., and Marshall, A. T. (2004). Sub-cellular localization of Ni in the hyperaccumulator, *Hybanthus floribundus* (Lindley) F. Muell. *Plant Cell Environ.* **27,** 705–716.

Bittermann, A. G., Knoll, G., Nemeth, A., and Plattner, H. (1992). Quantitative immuno-gold labelling and ultrastructural preservation after cryofixation (combined with different freeze-substitution and embedding protocols) and after chemical fixation and cryosectioning. Analysis of the secretory organelle matrix of *Paramecium trichocysts. Histochemistry* **97,** 421–429.

Bohrmann, B., and Kellenberger, E. (2001). Cryosubstitution of frozen biological specimens in electron microscopy: Use and application as an alternative to chemical fixation. *Micron* **32,** 11–19.

Bourett, T. M., Czymmek, K. J., and Howard, R. J. (1999). Ultrastructure of chloroplast protuberances in rice leaves preserved by high-pressure freezing. *Planta* **208,** 472–479.

Bourett, T. M., and Howard, R. J. (1991). Ultrastructural immunolocalization of actin in a fungus. *Protoplasma* **163,** 199–202.

Coetzee, J., and van der Merwe, C. F. (1989). Extraction of carbon 14-labeled compounds from plant tissue during processing for electron microscopy. *J. Electron Microsc. Tech.* **11,** 155–160.

Craig, S., Gilkey, J. C., and Staehelin, L. A. (1987). Improved specimen support cups and auxiliary devices for the Balzers high pressure freezing apparatus. *J. Microsc.* **148**(Pt. 1), 103–106.

Craig, S., and Staehelin, L. A. (1988). High pressure freezing of intact plant tissues. Evaluation and characterization of novel features of the endoplasmic reticulum and associated membrane systems. *Eur. J. Cell Biol.* **46,** 81–93.

Dailey, M. E., and Bridgman, P. C. (1991). Structure and organization of membrane organelles along distal microtubule segments in growth cones. *J. Neurosci. Res.* **30,** 242–258.

Dailey, M. E., and Bridgman, P. C. (1993). Vacuole dynamics in growth cones: Correlated EM and video observations. *J. Neurosci.* **13,** 3375–3393.

Demaree, R. S., Jr., and Giberson, R. T. (2001). Overview of microwave-assisted tissue processing for transmission electron microsopy. *In* "Microwave Techniques and Protocols" (R. S. Demaree, Jr., ed.), pp. 1–12. Humana Press, Totowa, NJ.

Diakou, P., Fedou, S., Carde, J. P., Cassagne, C., Moreau, P., and Maneta-Peyret, L. (2002). Immunolocalization of long-chain acyl-CoAs in plant cells. *Biochim. Biophys. Acta* **1583,** 85–90.

Ding, B., Turgeon, R., and Parthasarathy, M. V. (1991a). Microfilament organization and distribution in freeze substituted tobacco plant tissues. *Protoplasma* **165,** 96–105.

Ding, B., Turgeon, R., and Parthasarathy, M. V. (1991b). Routine cryofixation of plant tissue by propane jet freezing for freeze substitution. *J. Electron Microsc. Tech.* **19,** 107–117.

Ding, B., Turgeon, R., and Parthasarathy, M. V. (1992). Effect of high-pressure freezing on plant microfilament bundles. *J. Microsc.* **165,** 367–376.

Dong, Z., McCully, M. E., and Canny, M. J. (1994). Retention of vacuoles contents of plant cells during fixation. *J. Microsc.* **175,** 222–228.

Driouich, A., Zhang, G. F., and Staehelin, L. A. (1993). Effect of brefeldin A on the structure of the Golgi apparatus and on the synthesis and secretion of proteins and polysaccharides in sycamore maple (Acer pseudoplatanus) suspension-cultured cells. *Plant Physiol.* **101,** 1363–1373.

Dubochet, J., Richter, K., Roy, H. V., and McDowall, A. W. (1991). Freezing: Facts and hypothesis. *Scanning Microsc. Suppl.* **5,** S11–S15; discussion S15–S16.

Edelmann, L. (1994). Optimal freeze-drying of cryosections and bulk specimens for X-ray microanalysis. *Scanning Microsc. Suppl.* **8,** 67–76; discussion 76–81.

Edelmann, L. (2002). Freeze-dried and resin-embedded biological material is well suited for ultrastructure research. *J. Microsc.* **207,** 5–26.

Eppenberger-Eberhardt, M., Aigner, S., Donath, M. Y., Kurer, V., Walther, P., Zuppinger, C., Schaub, M. C., and Eppenberger, H. M. (1997). IGF-I and bFGF differentially influence atrial natriuretic factor and alpha-smooth muscle actin expression in cultured atrial compared to ventricular adult rat cardiomyocytes. *J. Mol. Cell. Cardiol.* **29,** 2027–2039.

Favre, R., Cermola, M., Nunes, C. P., Hermann, R., Muller, M., and Bazzicalupo, P. (1998). Immuno-cross-reactivity of CUT-1 and cuticlin epitopes between *Ascaris lumbricoides*, *Caenorhabditis elegans*, and *Heterorhabditis*. *J. Struct. Biol.* **123,** 1–7.

Felder, E., Jennings, P., Seppi, T., and Pfaller, W. (2002). LLC-PK(1) cells maintained in a new perfusion cell culture system exhibit an improved oxidative metabolism. *Cell Physiol. Biochem.* **12,** 153–162.

Fernandez-Moran, H., and Dahl, A. O. (1952). Electron microscopy of ultrathin frozen sections of pollen grains. *Science* **116,** 465–467.

Fisher, D. B., and Housley, T. L. (1972). The retention of water-soluble compounds during freeze-substitution and microautoradiography. *Plant Physiol.* **49,** 166–171.

Fleischer, S., Fleischer, B., and Stoeckenius, W. (1967). Fine structure of lipid-depleted mitochondria. *J. Cell Biol.* **32,** 193–208.

Frey, B., Brunner, I., Walther, P., Scheidegger, C., and Zierold, K. (1997). Element localization in ultrathin cryosections of high-pressure frozen ectomycorrhizal spruce roots. *Plant Cell Environ.* **20,** 929–937.

Fujimoto, K. (1995). Freeze-fracture replica electron microscopy combined with SDS digestion for cytochemical labeling of integral membrane proteins. Application to the immunogold labeling of intercellular junctional complexes. *J. Cell Sci.* **108**(Pt. 11), 3443–3449.

Fujimoto, K. (1997). SDS-digested freeze-fracture replica labeling electron microscopy to study the two-dimensional distribution of integral membrane proteins and phospholipids in biomembranes: Practical procedure, interpretation and application. *Histochem. Cell Biol.* **107,** 87–96.

Geitmann, A., Franklin-Tong, V. E., and Emons, A. C. (2004). The self-incompatibility response in Papaver rhoeas pollen causes early and striking alterations to organelles. *Cell Death Differ.* **11,** 812–822.

Giberson, R. T., Demaree, R. S., Jr., and Nordhausen, R. W. (1997). Four-hour processing of clinical/diagnostic specimens for electron microscopy using microwave technique. *J. Vet. Diagn. Invest.* **9,** 61–67.

Giddings, T. H. (2003). Freeze-substitution protocols for improved visualization of membranes in high-pressure frozen samples. *J. Microsc.* **212,** 53–61.

Glauert, A. M., and Lewis, P. R. (1998). "Biological Specimen Preparation for Transmission Electron Microscopy." Portland Press, London.

Glauert, A. M., and Young, R. D. (1989). The control of temperature during polymerization of Lowicryl K4M: There is a low temperature embedding method. *J. Microsc.* **154,** 101–113.

Graham, L. L., and Beveridge, T. J. (1990a). Effect of chemical fixatives on accurate preservation of *Escherichia coli* and *Bacillus subtilis* structure in cells prepared by freeze-substitution. *J. Bacteriol.* **172,** 2150–2159.

Graham, L. L., and Beveridge, T. J. (1990b). Evaluation of freeze-substitution and conventional embedding protocols for routine electron microscopic processing of eubacteria. *J. Bacteriol.* **172,** 2141–2149.

Griffiths, G. (1993). "Fine Structure Immunocytochemistry." Springer, Berlin.

Gunning, B. E. (2005). Plastid stromules: Video microscopy of their outgrowth, retraction, tensioning, anchoring, branching, bridging, and tip-shedding. *Protoplasma* **225,** 33–42.

Hagiwara, A., Fukazawa, Y., Deguchi-Tawarada, M., Ohtsuka, T., and Shigemoto, R. (2005). Differential distribution of release-related proteins in the hippocampal CA3 area as revealed by freeze-fracture replica labeling. *J. Comp. Neurol.* **489,** 195–216.

Harvey, D. M. (1982). Freeze-substitution. *J. Microsc.* **127**(Pt. 2), 209–221.

Harvey, D. M. R. (1980). The preparation of botanical samples for ion localization at the subcellular level. *Scanning Electron Microsc.* **1980**(II), 409–419.

Hastie, R. (1990). Ultrastructure of human basophil leukocytes studied after spray freezing and freeze-substitution. *Lab. Invest.* **62,** 119–130.

Hayat, M. A. (2000). "Principles and Techniques of Electron Microscopy. Biological Applications." Cambridge University Press, Cambridge.

He, Y., and Wetzstein, H. Y. (1995). Fixation induces differential tip morphology and immunolocalization of the cytoskeleton in pollen tubes. *Physiol. Plant.* **93,** 757–763.

Hess, M. W. (1990). Spurr-resin affects the ultrastructure of freeze-fixed/freeze-substituted pollen grains. *Ultramicroscopy* **34,** 219–221.

Hess, M. W. (1993). Cell wall development in freeze-fixed pollen: Intine formation of *Ledebouria socialis* (Hyacinthaceae). *Planta* **189,** 139–149.

Hess, M. W. (1995). High-pressure freeze fixation reveals novel features during ontogenesis of the vegetative cell in Ledebouria pollen: An ultrastructural and cytochemical study. *Biochem. Cell Biol.* **73,** 1–10.

Hess, M. W. (2003). Of plants and other pets: Practical aspects of freeze-substitution and resin embedding. *J. Microsc.* **212,** 44–52.

Hess, M. W. (2005). Challenged cells: A challenge for the electron microscopist. *In* "Proceedings of the Seventh Multinational Congress on Microscopy, June 26–30, 2005, Portoroz, Slovenia" (S. Fidler, ed.), pp. 107–110. Slovene Society for Microscopy, Department for Nanostructured Materials, Jozef Stefan Institute, Ljubljana.

Hess, M. W., and Frosch, A. (1994). Subunits of forming pollen exine and Ubisch bodies in freeze substituted *Ledebouria socialis* Roth. *Protoplasma* **182,** 10–14.

Hess, M. W., and Glaser, A. (1993). A simple and inexpensive device for freeze substitution at 183 K/−90 °C. *Biotech. Histochem.* **68,** 211–214.

Hess, M. W., and Hesse, M. (1994). Ultrastructural observations on anther tapetum development of freeze-fixed *Ledebouria socialis* Roth (Hyacinthaceae). *Planta* **192,** 421–430.

Hess, M. W., Mittermann, I., Luschnig, C., and Valenta, R. (1995). Immunocytochemical localisation of actin and profilin in the generative cell of angiosperm pollen: TEM studies on high-pressure frozen and freeze-substituted *Ledebouria socialis* Roth (Hyacinthaceae). *Histochem. Cell Biol.* **104,** 443–451.

Hess, M. W., Muller, M., Debbage, P. L., Vetterlein, M., and Pavelka, M. (2000). Cryopreparation provides new insight into the effects of brefeldin A on the structure of the HepG2 Golgi apparatus. *J. Struct. Biol.* **130,** 63–72.

Hess, M. W., and Valenta, R. (1997). Profilin revealed in pollen nuclei: Immuno-electron microscopy of high-pressure frozen *Ledebouria socialis* Roth (Hyacinthaceae). *Sex Plant Reprod.* **10,** 283–287.

Heuser, J. E., Reese, T. S., Dennis, M. J., Jan, Y., Jan, L., and Evans, L. (1979). Synaptic vesicle exocytosis captured by quick freezing and correlated with quantal transmitter release. *J. Cell Biol.* **81,** 275–300.

Hippe, S., Düring, K., and Kreuzaler, F. (1989). *In situ* localization of a foreign protein in transgenic plants by immunoelectron microscopy following high pressure freezing. Freeze substitution and low temperature embedding. *Eur. J. Cell Biol.* **50,** 230–234.

Hobot, J. A., Villiger, W., Escaig, J., Maeder, M., Ryter, A., and Kellenberger, E. (1985). Shape and fine structure of nucleoids observed on sections of ultrarapidly frozen and cryosubstituted bacteria. *J. Bacteriol.* **162,** 960–971.

Hoch, H. C. (1991). Preservation of cell ultrastructure by freeze-substitution. *In* "Electron Microscopy of Plant Pathogens" (D.-E. Lesemann, ed.), pp. 1–16. Springer, Berlin, Heidelberg, New York, London, Paris, Tokyo, HongKong, Barcelona.

Hohenberg, H., Mannweiler, K., and Muller, M. (1994). High-pressure freezing of cell suspensions in cellulose capillary tubes. *J. Microsc.* **175**(Pt. 1), 34–43.

Hohenberg, H., Tobler, M., and Muller, M. (1996). High-pressure freezing of tissue obtained by fine-needle biopsy. *J. Microsc.* **183**(Pt. 2), 133–139.

Howard, R. J., and Aist, J. R. (1980). Cytoplasmic microtubules and fungal morphogenesis: Ultrastructural effects of methyl benzimidazole-2-ylcarbamate determined by freeze-substitution of hyphal tip cells. *J. Cell Biol.* **87,** 55–64.

Huang, B. Q., Strout, G. W., and Russell, S. D. (1993). Fertilization in *Nicotiana* tabacum: Ultrastructural organization of propane-jet-frozen embryo sacs *in vivo*. *Planta* **191,** 256–264.

Humbel, B., and Müller, M. (1986). Freeze substitution and low temperature embedding. *In* "The Science of Biological Specimen Preparation 1985" (J. J. Wolosewick, ed.), pp. 175–183. SEM, Inc., AMF O'Hare, Chicago, IL.

Humbel, B. M., and Schwarz, H. (1989). Freeze-substitution for immunochemistry. *In* "Immuno-Gold Labeling in Cell Biology" (J. L. M. Leunissen, ed.), pp. 115–134. CRC Press, Boca Raton, FL.

Hunziker, E. B., Herrmann, W., Schenk, R. K., Müeller, M., and Moor, H. (1984). Cartilage ultrastructure after high pressure freezing, freeze substitution, and low temperature embedding. I. Chondrocyte ultrastructure–implications for the theories of mineralization and vascular invasion. *J. Cell Biol.* **98,** 267–276.

Hyde, G. J., Lancelle, S., Hepler, P. K., and Hardham, A. R. (1991). Sporangial structure in *Phytophthora* is disrupted after high pressure freezing. *Protoplasma* **165,** 203–208.

Ikeda, H. (1985). An ultrastructural analysis of the inclusion body in the type II pneumocyte processed by rapid freezing followed by freeze-substitution–an autoradiographic study. *J. Electron Microsc. (Tokyo)* **34,** 398–410.

Ikeda, H., Ichikawa, A., and Ichikawa, M. (1984). The effects of freeze-substitution media on the ultrastructure of inclusion bodies in type II pneumocytes of mouse lung processed by the cryofixation method. *J. Electron Microsc. (Tokyo)* **33,** 242–247.

Itoh, T., and Kimura, S. (2001). Immunogold labeling of terminal cellulose-synthesizing complexes. *J. Plant Res.* **114,** 483–489.

Kaeser, W. (1989). Freeze-substitution of plant tissues with a new medium containing dimethoxypropane. *J. Microsc.* **154,** 273–278.

Kaminskyj, S. G. W., Jackson, S. L., and Heath, I. B. (1992). Fixation induces differential polarized translocations of organelles in hyphae of *Saprolegnia ferax*. *J. Microsc.* **167,** 153–168.

Kellenberger, E. (1991). The potential of cryofixation and freeze substitution: Observations and theoretical considerations. *J. Microsc.* **161**(Pt. 2), 183–203.

Kellenberger, E. (1998). Learning about truth and biases through experience: Section surface corrugation, protein denaturation, and staining. *Microsc. Res. Tech.* **42,** 33–42.

Kellenberger, E., Durrenberger, M., Villiger, W., Carlemalm, E., and Wurtz, M. (1987). The efficiency of immunolabel on Lowicryl sections compared to theoretical predictions. *J. Histochem. Cytochem.* **35,** 959–969.

Kim, J. B., Stein, R., and O'Hare, M. J. (2004). Three-dimensional *in vitro* tissue culture models of breast cancer–a review. *Breast Cancer Res. Treat.* **85,** 281–291.

Kimura, S., Laosinchai, W., Itoh, T., Cui, X., Linder, C. R., and Brown, R. M., Jr. (1999). Immunogold labeling of rosette terminal cellulose-synthesizing complexes in the vascular plant vigna angularis. *Plant Cell* **11,** 2075–2086.

Kirschning, E., Rutter, G., and Hohenberg, H. (1998). High-pressure freezing and freeze-substitution of native rat brain: Suitability for preservation and immunoelectron microscopic localization of myelin glycolipids. *J. Neurosci. Res.* **53,** 465–474.

Kiss, J. Z., Giddings, T. H., Staehelin, L. A., and Sack, F. D. (1990). Comparison of the ultrastructure of conventionally fixed and high pressure frozen/freeze substituted root tips of *Nicotiana* and *Arabidopsis*. *Protoplasma* **157,** 64–74.

Knox, J. P., Linstead, P. J., King, J., Cooper, C., and Roberts, K. (1990). Pectin esterification is spatially regulated both within cell walls and between developing tissues of root apices. *Planta* **181,** 512–521.

Konomi, M., Fujimoto, K., Toda, T., and Osumi, M. (2003). Characterization and behaviour of alpha-glucan synthase in *Schizosaccharomyces pombe* as revealed by electron microscopy. *Yeast* **20,** 427–438.

Lancelle, S. A., Callaham, D. A., and Hepler, P. K. (1986). A method for rapid freeze fixation of plant cells. *Protoplasma* **131,** 153–165.

Lancelle, S. A., and Hepler, P. K. (1989). Immunogold labelling of actin on sections of freeze-substituted plant cells. *Protoplasma* **150,** 72–74.

Lee, Y. K., Hippe-Sanwald, S., Lee, S. C., Hohenberg, H., and Hwang, B. K. (2000). *In situ* localization of PR-1 mRNA and PR-1 protein in compatible and incompatible interactions of pepper stems with *Phytophthora capsici*. *Protoplasma* **211,** 64–75.

Lichtscheidl, I. K., Lancelle, S. A., and Hepler, P. K. (1990). Actin-endoplasmic reticulum complexes in *Drosera*. Their structural relationship with the plasmalemma, the nucleus, and organelles in cells prepared by high pressure freezing. *Protoplasma* **155,** 116–126.

Liou, W., Geuze, H. J., and Slot, J. W. (1996). Improving structural integrity of cryosections for immunogold labeling. *Histochem. Cell Biol.* **106,** 41–58.

Locke, M., and Krishnan, N. (1971). Hot alcoholic phosphotungstic acid and uranyl acetate as routine stains for thick and thin sections. *J. Cell Biol.* **50,** 550–557.

Lonsdale, J. E., McDonald, K. L., and Jones, R. L. (1999). High pressure freezing and freeze substitution reveal new aspects of fine structure and maintain protein antigenicity in barley aleurone cells. *Plant J.* **17,** 221–229.

MacKenzie, A. P. (1972). Freezing: Freeze-drying and freeze-substitution. *Scanning Electron Microsc.* **2,** 273–280.

Mascorro, J. A., and Kirby, G. S. (1992). Designing multi-viscosity embedding media utilizing novel resin/anhydride/catalyst combinations. *Microsc. Res. Tech.* **20,** 105–106.

Matsko, N., and Mueller, M. (2005). Epoxy resin as fixative during freeze-substitution. *J. Struct. Biol.* **152,** 92–103.

Maunsbach, A. B., and Afzelius, B. A. (1999). "Biomedical Electron Microscopy. Illustrated Methods and Interpretations." Academic Press, San Diego, London, Boston, New York, Sydney, Tokyo, Toronto.

McCully, M. E., and Canny, M. J. (1985). The stabilization of labile configurations of plant cytoplasm by freeze-substitution. *J. Microsc.* **139,** 27–33.

McDonald, K. (1999). High-pressure freezing for preservation of high resolution fine structure and antigenicity for immunolabeling. *Methods Mol. Biol.* **117,** 77–97.

McDonald, K., and Müller-Reichert, T. (2002). Cryomethods for thin section electron microscopy. *Meth. Enzymol.* **351,** 96–123.

McDonald, K. L. (1994). Electron microscopy and EM immunocytochemistry. *Methods Cell Biol.* **44,** 411–444.

McIntosh, J. R. (2001). Electron microscopy of cells: A new beginning for a new century. *J. Cell Biol.* **153,** F25–F32.

Meindl, U., Lancelle, S. A., and Hepler, P. K. (1992). Vesicle production and fusion during lobe formation in *Micrasterias* visualized by high-pressure freeze fixation. *Protoplasma* **170,** 104–114.

Mersey, B., and McCully, M. (1978). Monitoring the course of fixation of plant cells. *J. Microsc.* **114,** 49–76.

Michel, M., Gnägi, H., and Müller, M. (1992). Diamonds are a cryosectioner's best friend. *J. Microsc.* **166,** 43–56.

Michel, M., Hillmann, T., and Müller, M. (1991). Cryosectioning of plant material frozen at high pressure. *J. Microsc.* **163,** 3–18.

Mims, C. W., Celio, G. J., and Richardson, E. A. (2003). The use of high pressure freezing and freeze substitution to study host-pathogen interactions in fungal diseases of plants. *Microsc. Microanal.* **9,** 522–531.

Mims, C. W., Rodriguez-Lother, C., and Richardson, E. A. (2001). Ultrastructure of the host-parasite interaction in leaves of *Duchesnea indica* infected by the rust fungus *Frommeëla mexicana* var. *indicae* as revealed by high pressure freezing. *Can. J. Bot.* **79,** 49–57.

Mollenhauer, H. H. (1964). Plastic embedding mixtures for use in electron microscopy. *Stain Technol.* **39,** 111–114.

Mollenhauer, H. H. (1988). Artifacts caused by dehydration and epoxy embedding in transmission electron microscopy. *In* "Artifacts in Biological Electron Microscopy" (K. L. Klomparens, ed.), pp. 43–64. Plenum Press, New York, London.

Mollenhauer, H. H. (1993). Artifacts caused by dehydration and epoxy embedding in transmission electron microscopy. *Microsc. Res. Tech.* **26,** 496–512.

Monaghan, P., and Robertson, D. (1990). Freeze-substitution without aldehyde or osmium fixatives: Ultrastructure and implication for immunocytochemistry. *J. Microsc.* **158,** 355–363.

Monaghan, P., Perusinghe, N., and Muller, M. (1998). High-pressure freezing for immunocytochemistry. *J. Microsc.* **192**(Pt. 3), 248–258.

Moor, H. (1987). Theory and practice of high pressure freezing. *In* "Cryotechniques in Biological Electron Microscopy" (K. Zierold, ed.), pp. 175–191. Springer, Berlin, Heidelberg.

Moor, H., and Mühlethaler, K. (1963). Fine structure in frozen-etched yeast cells. *J. Cell Biol.* **17,** 609–628.

Moor, H., and Riehle, U. (1968). Snap-freezing under high pressure: A new fixation technique for freeze etching. *In* "Electron Microscopy 1968. Pre-Congress Abstracts of Papers Presented at the Fourth European Regional Conference, September 1–7" (D. A. Bocciarelli, ed.), pp. 33–34. Tipografia Poliglotta Vaticana, Rome.

Muller, M., Meister, N., and Moor, H. (1980). Freezing in a propane jet and its application in freeze-fracturing. *Mikroskopie* **36,** 129–140.

Müller, M., and Moor, H. (1984). Cryofixation of thick specimens by high-pressure freezing. *In* "The Science of Biological Specimen Preparation" (G. H. Haggins, ed.), pp. 131–138. SEM, Inc., AMF O'Hare, Chicago, IL.

Murata, T., Karahara, I., Kozuka, T., Thomas, H. G., Jr., Staehelin, L. A., and Mineyuki, Y. (2002). Improved method for visualizing coated pits, microfilaments, and microtubules in cryofixed and freeze-substituted plant cells. *J. Electron Microsc. (Tokyo)* **51,** 133–136.

Murk, J. L., Posthuma, G., Koster, A. J., Geuze, H. J., Verkleij, A. J., Kleijmeer, M. J., and Humbel, B. M. (2003). Influence of aldehyde fixation on the morphology of endosomes and lysosomes: Quantitative analysis and electron tomography. *J. Microsc.* **212,** 81–90.

Nagata, T. (1998). Techniques of radioautography for medical and biological research. *Braz. J. Med. Biol. Res.* **31,** 185–195.

Nei, T. (1973). Growth of ice crystals in frozen specimens. *J. Microsc.* **99,** 227–233.

Newman, G. R., and Hobot, J. A. (2001). "Resin Microscopy and On-Section Immunocytochemistry." Springer, Berlin, Heidelberg.

Nicolas, G. (1991). Advantages of fast-freeze fixation followed by freeze-substitution for the preservation of cell integrity. *J. Electron Microsc. Tech.* **18,** 395–405.

Noguchi, T. (1990). Consumption of lipid granules and formation of vacuoles in the pollen tube of *Tradescantia reflexa*. *Protoplasma* **156,** 19–28.

Noguchi, T. (1994). Formation and decomposition of vacuoles in *Botryococcus* in relation to the trans-Golgi network. *Protoplasma* **180,** 29–38.

Noguchi, T., and Watanabe, H. (1999). Brefeldin A effects on the trans-Golgi network and Golgi bodies in *Botryococcus braunii* are not uniform during the cell cycle. *Protoplasma* **209,** 193–206.

Orlovich, D. A., and Ashford, A. E. (1993). Polyphosphate granules are an artefact of specimen preparation in the ectomycorrhizal fungus *Pisolithus tinctorius*. *Protoplasma* **173,** 91–102.

Otegui, M. S., Capp, R., and Staehelin, L. A. (2002). Developing seeds of Arabidopsis store different minerals in two types of vacuoles and in the endoplasmic reticulum. *Plant Cell* **14,** 1311–1327.

Palmieri, M., and Kiss, J. Z. (2005). A novel technique for flat-embedding cryofixed plant specimens in LR white resin. *Microsc. Res. Tech.* **68,** 80–84.

Palsgard, E., Lindh, U., and Roomans, G. M. (1994). Comparative study of freeze-substitution techniques for X-ray microanalysis of biological tissue. *Microsc. Res. Tech.* **28,** 254–258.

Pfeiffer, S., Beese, M., Boettcher, M., Kawaschinski, K., and Krupinska, K. (2003). Combined use of confocal laser scanning microscopy and transmission electron microscopy for visualisation of identical cells processed by cryotechniques. *Protoplasma* **222,** 129–137.

Pfeiffer, S., and Krupinska, K. (2005a). Chloroplast ultrastructure in leaves of *Urtica dioica* L. analyzed after high-pressure freezing and freeze-substitution and compared with conventional fixation followed by room temperature dehydration. *Microsc. Res. Tech.* **68,** 368–376.

Pfeiffer, S., and Krupinska, K. (2005b). New insights in thylakoid membrane organization. *Plant Cell Physiol.* **46,** 1443–1451.

Pfeiffer, S., Vielhaber, G., Vietzke, J. P., Wittern, K. P., Hintze, U., and Wepf, R. (2000). High-pressure freezing provides new information on human epidermis: Simultaneous protein antigen and lamellar lipid structure preservation. Study on human epidermis by cryoimmobilization. *J. Invest. Dermatol.* **114,** 1030–1038.

Platt, K. A., Oliver, M. J., and Thomson, W. W. (1997). Importance of the fixative for reliable ultrastructure preservation of poikilohydric plant tissues. Observations on dry, partially, and fully hydrated tissues of *Selaginella lepidophylla*. *Ann. Bot.* **80,** 599–610.

Puhakainen, T., Hess, M. W., Makela, P., Svensson, J., Heino, P., and Palva, E. T. (2004). Overexpression of multiple dehydrin genes enhances tolerance to freezing stress in Arabidopsis. *Plant Mol. Biol.* **54,** 743–753.

Rash, J. E., Johnson, T. J., Hudson, C. S., Giddings, F. D., Graham, W. F., and Eldefrawi, M. E. (1982). Labelled-replica techniques: Post-shadow labelling of intramembrane particles in freeze-fracture replicas. *J. Microsc.* **128**(Pt. 2), 121–138.

Rensing, K. H., and Samuels, A. L. (2004). Cellular changes associated with rest and quiescence in winter-dormant vascular cambium of *Pinus contorta*. *Trees* **18,** 373–380.

Robards, A. W., and Sleytr, U. B. (1985). Low temperature methods in biological electron microscopy. *In* "Practical Methods in Electron Microscopy" (A. M. Glauert, ed.). Elesevier, Amsterdam.

Roch, A. M., Nicolas, M. T., and Quash, G. (1997). Ultrastructural immunolocalization of polyamines in HeLa cells subjected to fast-freezing fixation and freeze substitution. *Histochem. Cell Biol.* **107,** 303–312.

Rosenberg, M., Bartl, P., and Lesko, J. (1960). Water-soluble methacrylate as an embedding medium for the preparation of ultrathin sections. *J. Ultrastruct. Res.* **4,** 298–303.

Rupes, I., Mao, W.-Z., Astrom, H., and Raudaskoski, M. (1995). Effects of nocadazole and brefeldin A on microtubule cytoskeleton and membrane organization in the homobasiodiomycete *Schizophyllum commune*. *Protoplasma* **185,** 212–221.

Ryan, K. P., and Knoll, G. (1994). Time-resolved cryofixation methods for the study of dynamic cellular events by electron microscopy. *Scanning Microsc.* **8,** 259–288.

Samuels, A. L., Giddings, T. H., Jr., and Staehelin, L. A. (1995). Cytokinesis in tobacco BY-2 and root tip cells: A new model of cell plate formation in higher plants. *J. Cell Biol.* **130,** 1345–1357.

Samuels, A. L., Rensing, K. H., Douglas, C. J., Mansfield, S. D., Dharmawardhana, D. P., and Ellis, B. E. (2002). Cellular machinery of wood production: Differentiation of secondary xylem in *Pinus contorta* var. latifolia. *Planta* **216,** 72–82.

Schmeichel, K. L., and Bissell, M. J. (2003). Modeling tissue-specific signaling and organ function in three dimensions. *J. Cell Sci.* **116,** 2377–2388.

Schwarz, H., Hohenberg, H., and Humbel, B. M. (1993). Freeze substitution in virus research: A preview. *In* "Immuno-Gold Electron Microscopy in Virus Diagnosis and Research" (B. T. Eaton, ed.), pp. 349–376. CRC Press, Boca Raton, FL.

Schwarz, H., and Humbel, B. M. (1989). Influence of fixatives and embedding media on immunolabelling of freeze-substituted cells. *Scanning Microsc. Suppl.* **3,** 57–64.

Segui-Simarro, J. M., and Staehelin, L. A. (2006). Cell cycle-dependent changes in Golgi stacks, vacuoles, clathrin-coated vesicles and multivesicular bodies in meristematic cells of Arabidopsis thaliana: A quantitative and spatial analysis. *Planta* **223**(2), 223–236.

Simpson, W. L. (1941). An experimental analysis of the Altmann technique of freeze-drying. *Anat. Rec.* **80,** 173–189.

Singh, M. B., Taylor, P. E., and Knox, R. B. (1993). Special preparation methods for immunocytochemistry of plant cells. *In* "Immunocytochemistry. A Practical Approach" (J. E. Beesley, ed.), pp. 77–102. IRL Press, Oxford.

Sitte, H., Edelmann, L., and Neumann, K. (1987). Cryofixation without pretreatment at ambient pressure. *In* "Cryotechniques in Biological Electron Microscopy" (K. Zierold, ed.), pp. 87–113. Springer, Berlin, Heidelberg.

Sjöstrand, F. S. (1990). Common sense in electron microscopy: About cryofixation, freeze-substitution, low temperature embedding and low denaturation embedding. *J. Struct. Biol.* **103,** 135–139.

Spurr, A. R. (1969). A low-viscosity epoxy resin embedding medium for electron microscopy. *J. Ultrastruct. Res.* **26,** 31–43.

Staehelin, L. A., Giddings, T. H., Kiss, J. Z., and Sack, F. D. (1990). Macromolecular differentiation of Golgi stacks in root tips of Arabidosis and Nicotiana seedlings as visualized in high pressure frozen and freeze-substituted samples. *Protoplasma* **157,** 75–91.

Staff, I. A., Schäppi, G., and Taylor, P. E. (1999). Localisation of antigens in ryegrass pollen and in airborne micronic particles. *Protoplasma* **208,** 47–57.

Staiger, D., Kappeler, S., Muller, M., and Apel, K. (1994). The proteins encoded by two tapetum-specific transcripts, Sa tap35 and Sa tap44, from *Sinapis alba* L. are localized in the exine cell wall layer of developing microspores. *Planta* **192,** 221–231.

Steere, R. L. (1957). Electron microscopy of structural detail in frozen biological specimens. *J. Biophys. Biochem. Cytol.* **3,** 45–60.

Steinbrecht, R. A. (1993). Freeze-substitution for morphological and immunocytochemical studies in insects. *Microsc. Res. Tech.* **24,** 488–504.

Steinbrecht, R. A., and Müller, M. (1987). Freeze-substitution and freeze-drying. *In* "Cryotechniques in Biological Electron Microscopy" (K. Zierold, ed.), pp. 149–172. Springer, Berlin, Heidelberg.

Studer, D., Michel, M., and Muller, M. (1989). High pressure freezing comes of age. *Scanning Microsc. Suppl.* **3,** 253–268; discussion 268–269.

Studer, D., Michel, M., Wohlwend, M., Hunziker, E. B., and Buschmann, M. D. (1995). Vitrification of articular cartilage by high-pressure freezing. *J. Microsc.* **179**(Pt. 3), 321–332.

Studer, D., Graber, W., Al-Amoudi, A., and Eggli, P. (2001). A new approach for cryofixation by high-pressure freezing. *J. Microsc.* **203,** 285–294.

Szczesny, P. J., Walther, P., and Muller, M. (1996). Light damage in rod outer segments: The effects of fixation on ultrastructural alterations. *Curr. Eye Res.* **15,** 807–814.

Taylor, P. E., Staff, I. A., Singh, M. B., and Knox, R. B. (1994). Localization of the two major allergens in rye-grass pollen using specific monoclonal antibodies and quantitative analysis of immunogold labelling. *Histochem. J.* **26,** 392–401.

Thijssen, M. H., Mittempergher, F., van Aelst, A. C., and van Vent, J. L. (1997). Improved ultra-structural preservation of *Petunia* and *Brassica* ovules and embryo sacs by high pressure freezing and freeze substitution. *Protoplasma* **197,** 199–209.

Thijssen, M. H., van Went, J. L., and van Aelst, A. C. (1998). Heptane and isooctane as embedding fluids for high-pressure freezing of *Petunia* ovules followed by freeze-substitution. *J. Microsc.* **192,** 228–235.

Thomson, W. W., and Platt, K. A. (1997). Conservation of cell order in desiccated mesophyll of *Selaginella lepidophylla* ([Hook and Grev.] Spring). *Ann. Bot.* **79,** 439–447.

Tiedemann, J., Hohenberg, H., and Kollmann, R. (1998). High-pressure freezing of plant cells cultured in cellulose microcapillaries. *J. Microsc.* **189,** 163–171.

Tiwari, S. C., Polito, V. S., and Webster, B. D. (1990). In dry pear (*Pyrus communis* L.) pollen, membranes assume a tightly packed multilamellate aspect that disappears rapidly upon hydration. *Protoplasma* **153,** 157–168.

Tobler, M., and Freiburghaus, A. U. (1990). Occupational risks of (meth)acrylate compounds in embedding media for electron microscopy. *J. Microsc.* **160**(Pt. 3), 291–298.

Tobler, M., and Freiburghaus, A. U. (1991). Exceptional protective power of the 4H glove defeats occupational risks in electron microscopy. *J. Microsc.* **163**(Pt. 1), RP1–RP2.

Tonning, A., Helms, S., Schwarz, H., Uv, A. E., and Moussian, B. (2006). Hormonal regulation of mummy is needed for apical extracellular matrix formation and epithelial morphogenesis in *Drosophila. Development* **133,** 331–341.

Trump, B. F., and Ericsson, J. L. (1965). The effect of the fixative solution on the ultrastructure of cells and tissues. A comparative analysis with particular attention to the proximal convoluted tubule of the rat kidney. *Lab. Invest.* **14,** 1245–1323.

Usuda, N., Ma, H. J., Hanai, T., Yokota, S., Hashimoto, T., and Nagata, T. (1990). Immunoelectron microscopy of tissues processed by rapid freezing and freeze-substitution fixation without chemical fixatives: Application to catalase in rat liver hepatocytes. *J. Histochem. Cytochem.* **38,** 617–623.

Valenta, R., Natter, S., Seiberler, S., Wichlas, S., Maurer, D., Hess, M., Pavelka, M., Grote, M., Ferreira, F., Szepfalusi, Z., Valent, P., and Stingl, G. (1998). Molecular characterization of an autoallergen, Hom s 1, identified by serum IgE from atopic dermatitis patients. *J. Invest. Dermatol.* **111,** 1178–1183.

Van Harreveld, A., and Crowell, J. (1964). Electron microscopy after rapid freezing on a metal surface and substitution fixation. *Anat. Rec.* **149,** 381–385.

Van Harreveld, A., Crowell, J., and Malhotra, S. K. (1965). A study of extracellular space in central nervous tissue by freeze-substitution. *J. Cell Biol.* **25,** 117–137.

von Schack, M. L., Fakan, S., Villiger, W., and Muller, M. (1993). Cryofixation and cryosubstitution: A useful alternative in the analyses of cellular fine structure. *Eur. J. Histochem.* **37,** 5–18.

Walther, P. (2003a). Cryo-fracturing and cryo-planing for in-lens cryo-SEM, using a newly designed diamond knife. *Microsc. Microanal.* **9,** 279–285.

Walther, P. (2003b). Recent progress in freeze-fracturing of high-pressure frozen samples. *J. Microsc.* **212,** 34–43.

Walther, P., and Muller, M. (1999). Biological ultrastructure as revealed by high resolution cryo-SEM of block faces after cryo-sectioning. *J. Microsc.* **196**(Pt. 3), 279–287.

Walther, P., and Ziegler, A. (2002). Freeze substitution of high-pressure frozen samples: The visibility of biological membranes is improved when the substitution medium contains water. *J. Microsc.* **208,** 3–10.

Weibull, C., Villiger, W., and Carlemalm, E. (1984). Extraction of lipids during freeze-substitution of *Acholeplasma laidlawii*-cells for electron microscopy. *J. Microsc.* **134,** 213–216.

Welter, K., Müller, M., and Mendgen, K. (1988). The hyphae of *Uromyces appendiculatus* within the leaf tissue after high pressure freezing and freeze substitution. *Protoplasma* **147,** 91–99.

Wesley-Smith, J. (2001). Freeze-substitution of dehydrated plant tissues: Artefacts of aqueous fixation revisited. *Protoplasma* **218,** 154–167.

White, D. L., Andrews, S. B., Faller, J. W., and Barrnett, R. J. (1976). The chemical nature of osmium tetroxide fixation and staining of membranes by X-ray photoelectron spectroscopy. *Biochim. Biophys. Acta* **436,** 577–592.

Wilson, T. P., Canny, M. J., McCully, M. E., and Lefkovitch, L. P. (1990). Breakdown of cytoplasmic vacuoles. A model of endoplasmic membrane rearrangement. *Protoplasma* **155,** 144–152.

Zahir, N., and Weaver, V. M. (2004). Death in the third dimension: Apoptosis regulation and tissue architecture. *Curr. Opin. Genet. Dev.* **14,** 71–80.

Zalokar, M. (1966). A simple freeze-substitution method for electron microscopy. *J. Ultrastruct. Res.* **15,** 469–479.

Zechmann, B., Muller, M., and Zellnig, G. (2005). Effects of different fixation and freeze substitution methods on the ultrastructural preservation of ZYMV-infected *Cucurbita pepo* (L.) leaves. *J. Electron Microsc. (Tokyo)* **54,** 393–402.

Zhang, G. F., Driouich, A., and Staehelin, L. A. (1993). Effect of monensin on plant Golgi: Re-examination of the monensin-induced changes in cisternal architecture and functional activities of the Golgi apparatus of sycamore suspension-cultured cells. *J. Cell Sci.* **104**(Pt. 3), 819–831.

Zhang, G. F., and Staehelin, L. A. (1992). Functional compartmentation of the Golgi apparatus of plant cells. *Plant Physiol.* **99,** 1070–1083.

Zheng, H. Q., and Staehelin, L. A. (2001). Nodal endoplasmic reticulum, a specialized form of endoplasmic reticulum found in gravity-sensing root tip columella cells. *Plant Physiol.* **125,** 252–265.

Zierold, K., and Steinbrecht, R. A. (1987). Cryofixation of diffusible elements in cells and tissues for electron probe microanalysis. *In* "Cryotechniques in Biological Electron Microscopy" (K. Zierold, ed.), pp. 272–282. Springer, Berlin, Heidelberg.

CHAPTER 4

Correlative Light and Electron Microscopy of Early *Caenorhabditis elegans* Embryos in Mitosis

Thomas Müller-Reichert,* Martin Srayko,* Anthony Hyman,* Eileen T. O'Toole,† and Kent McDonald‡

*Max Planck Institute of Molecular Cell Biology and Genetics (MPI-CBG)
01307 Dresden, Germany

†Boulder Laboratory for 3D Electron Microscopy of Cells
University of Colorado
Boulder, Colorado 80309

‡Electron Microscope Laboratory
University of California
Berkeley, California 94720

0091-679X/07 $35.00
DOI: 10.1016/S0091-679X(06)79004-5

The early embryo of *Caenorhabditis elegans* is one of the most powerful model systems in which to study cell division. Here we have developed a correlative light and electron microscopic approach to stage early *C. elegans* embryos prior to high-pressure freezing and electron tomography. We use cellulose microcapillaries to contain the embryos in short transparent tubes. These permit the viewing of early mitotic events under the light microscope. Using the newly developed rapid transfer system (RTS) of the Leica EMPACT2 high-pressure freezer, we have reduced the time between light microscopic observation and cryoimmobilization to about 5 sec. This correlative approach allows systematic structure–function studies in staged early wild-type embryos and those treated with RNA-mediated interference (RNAi) to reduce the level of specific gene products.

I. Introduction

The *Caenorhabditis elegans* early embryo is one of the most powerful model systems in which to study various subprocesses of mitosis, such as nuclear envelope assembly/disassembly, centrosome dynamics, centriole assembly, formation of the mitotic spindle, kinetochore assembly, chromosome segregation, and cytokinesis (for a review see Oegema and Hyman, 2005). For analysis of these processes by correlative microscopy, *C. elegans* offers several advantages. (1) The stereotypic early development of the embryo has been studied and characterized in detail (for reviews see Cowan and Hyman, 2004; Pelletier *et al.*, 2004a; Schneider and Bowerman, 2003). (2) The appearance, migration, rotation, and disappearance of the pronuclei around prometaphase can be easily recognized by noninvasive microscopy, allowing precise staging of embryonic development. (3) Mitotic spindle components, such as microtubules, kinetochores, or centrosomes, can be visualized using GFP-tagged strains. Moreover, strains are available which express multiple-tagged proteins (Oegema *et al.*, 2001). (4) RNA-mediated interference (RNAi) is used routinely to generate oocytes whose cytoplasm is reproducibly depleted of targeted essential gene products. Following fertilization, these oocytes are then analyzed as they attempt to go through the first round of mitosis (reviewed in Oegema and Hyman, 2005). Application of RNAi allows the comparison of wild-type and mutant embryos. (5) The reliability of RNAi technology has it made possible to perform genome-wide screens to identify all the components involved in the early mitoses of *C. elegans* (Sönnichsen *et al.*, 2005). Taken together, these features allow one to study the dynamics and ultrastructure of cell division in a systematic manner that was not possible until recently.

Correlative approaches in microscopy have been used to obtain complementary information about a given system (Kapoor *et al.*, 2006; Rieder and Cassels, 1999). For example, light microscopy (LM) in combination with GFP-tagging, on the one hand, offers the advantage of studying centrosome dynamics in living systems, but with obvious limitations in the obtained resolution. Electron microscopy (EM), on the other hand, delivers high-resolution "snapshot" images of centrosomes, either purified (Chrétien *et al.*, 1997) *in vitro* (Schnackenberg *et al.*, 1998) or *in vivo* (O'Toole *et al.*, 2003b), but the yield of information about the dynamics of the process under study is low. On the basis of only "snapshot" images, it is not always possible to unambiguously reconstruct a complex process, like the assembly of a cellular structure. For analysis of an organelle like the centrosome in the early *C. elegans* embryo, one would like to be able to observe spindle dynamics by live-cell imaging and stop the process immediately at a specific stage by rapid cryofixation; this would permit subsequent, high-resolution analysis *in situ*, preferably by electron tomography. The prerequisites for such a correlative approach are: (1) to optimize the time resolution of the technique and (2) to use a method that gives optimal preservation of the ultrastructure.

One approach to achieving the first goal (i.e., to stop the developing *C. elegans* embryo at specific stages of early mitosis) has been laser-assisted chemical fixation (Dammermann *et al.*, 2004; Kirkham *et al.*, 2003; Pelletier *et al.*, 2004b; Priess and Hirsh, 1986). Taking advantage of the fact that early embryos are protected by coverings that allow them to continue to divide in the presence of fixative, selected embryos were attached to an Aclar plastic surface and "bathed" in M-9 buffer containing 2% glutaraldehyde. A laser beam was used to perforate the embryo's eggshell, allowing the fixative to diffuse into the embryo through a small (1.5 μm) hole and to stop development. Since this method achieved a rapid cessation of the embryonic development and permitted a staging of the embryo prior to conventional EM, it has been applied to understand the role of specific genes in centriole duplication (Dammermann *et al.*, 2004; Kirkham *et al.*, 2003; Pelletier *et al.*, 2004b). The major disadvantage of laser-assisted fixation, however, is its poor rate of fixation and the resulting insufficient structural preservation of the early embryo (for a general discussion on the disadvantages of chemical fixation on cellular fine structure, the reader is referred to Murk *et al.*, 2003). In order to achieve the kind of resolution necessary for electron tomographic studies, we have previously used high-pressure freezing as the means of primary fixation (O'Toole *et al.*, 2003b). Whole worms were high-pressure frozen and embedded in a thin layer of resin, so individual worms could be remounted and sectioned longitudinally. Embryos at the desired stages can be identified in serial sections of these longitudinally oriented worms. Using electron tomography of such high-pressure frozen material, we have described the three-dimensional (3D) structure of the mitotic *C. elegans* centrosome and identified both open and closed morphologies of centrosome-associated microtubule ends (O'Toole *et al.*, 2003b). However, a disadvantage of this procedure is that the whole worms contain populations of embryos at many phases of development, so the stage of mitosis had to be determined after sectioning. Here, we describe a method that combines the advantages of: (1) a correlative LM approach that allows us to stage development and fixation times with high time resolution, (2) high-pressure freezing for the best possible preservation of ultrastructure, and (3) electron tomography for visualization and analysis of high-resolution structure in 3D.

II. Rationale

Rationale of the described technique is to combine the observation of live mitotic embryos with high-pressure freezing of clearly defined stages, allowing us to obtain tomographic reconstructions of mitotic spindle components in well-preserved embryos of *C. elegans* as they develop. The methods presented here include procedures to transfer the isolated embryos into capillary tubes, to stage selected embryos by LM, to immobilize these staged embryos by high-pressure freezing, and to fix them by freeze-substitution. Finally, we describe thin-layer embedding and 3D reconstruction of spindle components by electron tomography (Fig. 1). In the following paragraphs, we give step-by-step descriptions of these methods.

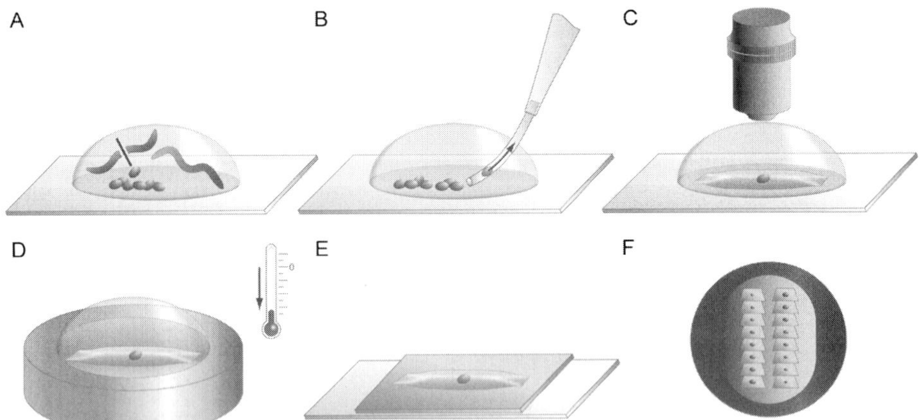

Fig. 1 Diagram illustrating the processing of isolated *C. elegans* embryos for correlative LM/EM. (A) Release of early embryos. (B) Collection of a selected embryo into capillary tubing. (C) Staging of the capillary-contained embryo by LM. (D) Stage-defined fixation by high-pressure freezing. (E) Thin-layer embedding of the freeze-substituted embryo. (F) Serial sectioning for subsequent electron tomography.

III. Methods

A. Staging of Isolated Embryos by LM

This section describes the preparation of isolated *C. elegans* embryos for live-cell imaging under conditions suitable for subsequent high-pressure freezing. The idea is to make use of capillary tubing to contain the embryos in short transparent tubes so that one can observe early developmental events under the light microscope, prior to time-resolved high-pressure freezing (McDonald *et al.*, 2006).

As a first step, we prepare a "loading device" to collect isolated early embryos into capillary tubes and a "crimping" tool to cut pieces of tubes of appropriate sizes. Loading devices were prepared by cutting 2-cm-long pieces of capillary dialysis tubing (Hohenberg *et al.*, 1994), mounting them into gel loader tips, and using nail polish for sealing. Finally, the loading device was mounted on a Pipetman. The crimping tool was made from a scalpel tip by cutting or filing off the final millimeter of the blade, and then shaping it with a whetstone to the shape of a chisel (Fig. 2). The crimping tool was important to enable pieces of tubing of appropriate sizes to be cut and simultaneously closed during the cutting procedure.

Using a dissecting scope, worms were cut open in small Petri dishes using M-9 buffer containing 20% bovine serum albumin (BSA). Embryos were released by cutting the worm in the gonad region with two injection needles. In order to get a high yield of released early embryos, it is preferred to take young adults for cutting. It is also advantageous to choose embryos in early stages of development, for example, when the pronuclei first appear (Fig. 3B). In the wild-type embryo development, from pronuclear appearance to the time of their first meeting takes about 13 min. This time is sufficient to load an embryo into the capillary tube and get ready for live-cell imaging. Thus, an early embryo was selected and sucked into the capillary tube, using the loading device mounted on a Pipetman.

Fig. 2 The "crimping tool" as made from a scalpel tip. The final millimeter of the blade is cut off and then shaped with a whetstone. (A) Side view. (B) Top view.

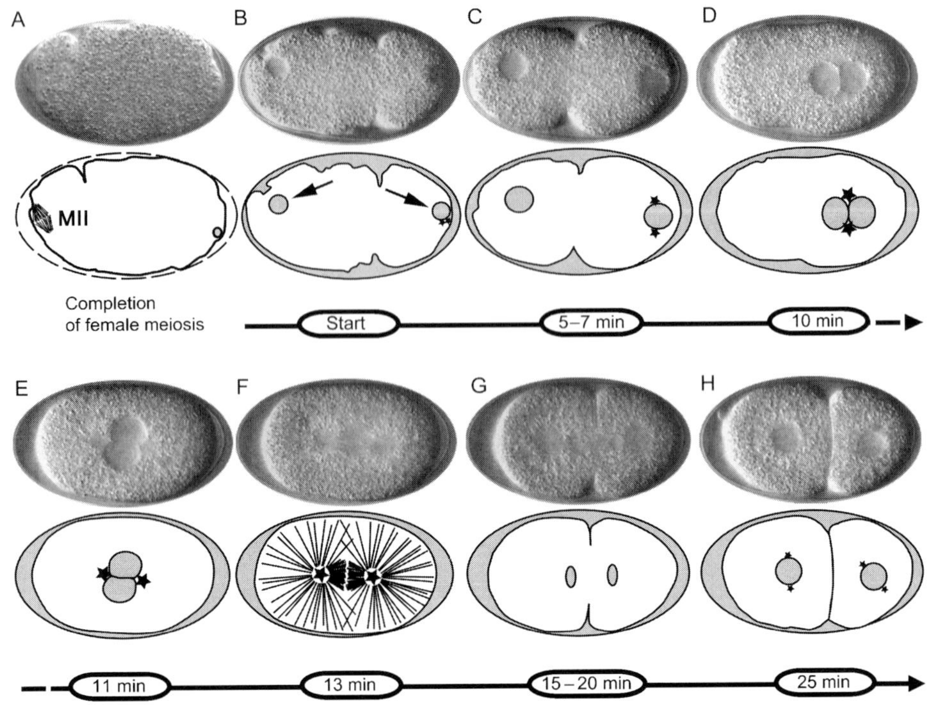

Fig. 3 Early development of the *C. elegans* embryo. Stages are illustrated by differential interference contrast (DIC) images and corresponding schematic drawings. (A) Completion of female meiosis II. (B) Pronuclear appearance. (C) Pronuclear migration. (D) Pronuclear meeting. (E) Pronuclear rotation. (F) Metaphase of the first mitotic division. (G) Telophase. (H) Two-celled embryo. The position and the disappearance of the two pronuclei (arrows) are indicative of specific developmental stages. Asterisks indicate the position of the centrosomes.

The tubing was submerged into the BSA-containing M-9 buffer in the Petri dish, then the "crimping tool" was used to cut the region of the tubing containing the early embryo to a final length of about 1 mm. It is important that this piece of tubing is about this length because it has to fit into the specimen carrier used later for high-pressure freezing. The ends of the tube should be closed during the crimping in order to avoid leaking out of the early embryo. Therefore, the tube should be cut and simultaneously pressed down with the flattened side of the crimping tool.

The closed tube was then transferred from the Petri dish to a droplet of BSA-containing M-9 buffer on a glass slide to observe the further development of the embryo in real time using a light microscope equipped with phase contrast, differential interference contrast (DIC) or fluorescence optics (Fig. 4). In the latter case, a line expressing either GFP::β-tubulin GFP::histone or GFP::γ-tubulin and GFP::histone can be used to observe the stages of mitotic spindle assembly. It is

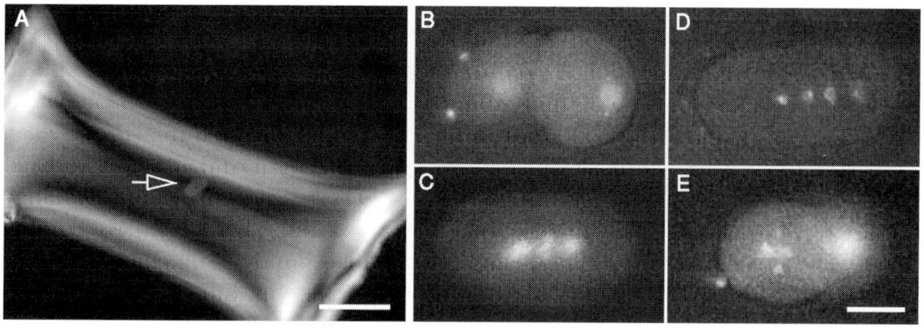

Fig. 4 LM of the developing *C. elegans* embryos. (A) Early *C. elegans* embryo (arrow) contained in a closed capillary tube. (B–E) Staging of contained embryos by fluorescence microscopy. (B) Pronuclear appearance. (C) Metaphase. (D) Anaphase. (E) Two-celled embryo. (C) Strain expressing GFP::β-tubulin and GFP::histone and (B, D, and E) strain expressing GFP::γ-tubulin and GFP::histone were used to visualize spindle components. Scale bar in (A) = 100 μm and in (E) = 10 μm.

also possible to observe the developing embryo with phase contrast because the appearance, position, and migration of the two pronuclei are good indicators of the developmental events during the first mitotic division. The developmental history of every selected embryo should be documented. A time-lapse movie, however, is not absolutely required for documentation.

B. High-Pressure Freezing and Freeze-Substitution

Next, the staged embryo must be cryoimmobilized to "freeze" the observed developmental process in time. The key here is to minimize the time window between observation and freezing. For this, we applied the newly developed EMPACT2+RTS (Leica Microsystems, Vienna, Austria) high-pressure freezer, a portable machine that can be easily moved to the site where staging of the embryo is performed. In addition, the rapid transfer system (RTS) allows a fast loading of the specimen into a preloaded high-pressure freezer under standardized conditions (McDonald *et al.*, 2006). Following high-pressure freezing, the isolated embryos have to be processed for freeze-substitution.

In preparation for freezing, the RTS has to be preloaded with a freezing "pod" (Fig. 5). In addition, a specimen carrier in which the sample will be placed at an appropriate time point must be premounted in the so-called rapid loader (for details see McDonald *et al.*, 2006). Routinely, we used 100-μm-deep "membrane carriers," designed to separate the compartment holding the capillary tube from the pressure by a thin "membrane" of copper. Shortly before freezing, the membrane carrier was filled with M-9 buffer containing 20% BSA (see the preceding section). To avoid "overfilling" of the holder, excess buffer can be sucked away with strips of filter paper. At an appropriate developmental stage, the

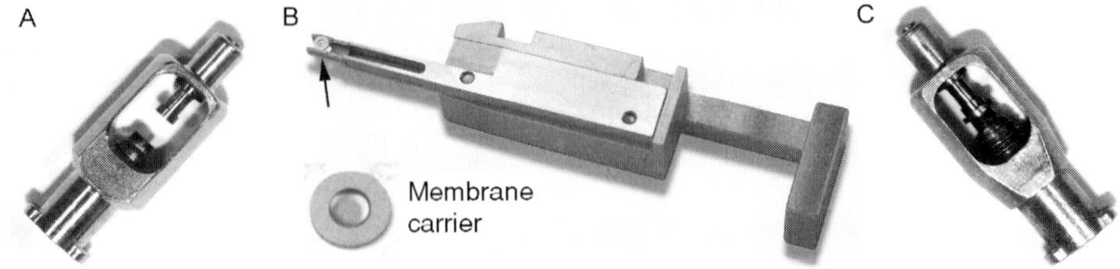

Fig. 5 Loading devices for the RTS of the EMPACT2 high-pressure freezer. (A) The freezing "pod." (B) "Rapid loader" with premounted specimen carrier (arrow). A "membrane carrier" is shown at higher magnification. (C) Closed freezing "pod" with specimen carrier (McDonald *et al.*, 2006).

embryo-containing capillary tube was quickly transferred from the microscope slide to the cryoprotectant-filled specimen carrier using tweezers. As soon as the capillary tube was loaded in the membrane carrier, the rapid loader was transferred to the RTS. The rapid loader fits exactly into the pod. Sliding the rapid loader into the RTS triggers the pod to close automatically and shoot into the EMPACT2, ultrarapidly freezing the sample. Applying this approach, the isolated embryo was transferred from the light microscope to the high-pressure freezer within about 5 sec. After freezing, the membrane carriers were released from the pod and stored in liquid nitrogen until further use.

The membrane carriers containing the tubes were then transferred to precooled cryovials filled with the freeze-substitution "cocktail." Freeze-substitution was carried out in anhydrous acetone containing 1% osmium tetroxide and 0.1% uranyl acetate (for detailed information on the composition and preparation of the freeze-substitution "cocktail", the reader is referred to McDonald and Müller-Reichert, 2002). Using a Leica EM AFS, we maintained the samples at −90 °C for a minimum of 8 h and a maximum of 24 h. Freeze-substituted samples were then allowed to warm to room temperature at a rate of 5 °C/h. Specimens were washed three to five times in fresh anhydrous acetone. After these acetone washes, samples were gradually infiltrated with Epon/Araldite resin (1 part resin:3 parts acetone for 2 h, 1:1 for 3 h, 3:1 for 4 h, and 100% resin for 4 h, then overnight, then for 4 h). Importantly, throughout all these steps, the capillary tubes remained in the BSA-filled membrane carriers and were not removed prior to the final flat-embedding step to avoid the loss of the small capillary tubes during rinsing and resin infiltration.

C. Thin-Layer Embedding, Serial Sectioning, and Prescreening of Samples

After resin infiltration, the embryo-containing capillary tubes are further processed for flat embedding. The objective here is to embed the tubes in thin, optically clean layers of resin on microscope slides, allowing an easy and routine handling of the embryo for serial sectioning. Thin-layer embedding and serial sectioning of whole worms have been described previously (Müller-Reichert *et al.*, 2003). Here, the protocol is adapted to the needs of processing embryo-contained cellulose capillary tubes.

Routinely, labeled microscope slides were wiped clean with a soft cloth, submerged in a Teflon® solution, and allowed to dry. The slides were then polished to get a clean transparent surface, and two layers of Parafilm® were placed on the margins of the glass slide to serve as spacers. About 200–300 μl of Epon/Araldite were placed on the microscope slide and allowed to disperse on the surface of the slide. Subsequently, the membrane carrier containing the capillary tube was placed in the resin droplet and held tightly with appropriate tweezers. The capillary tube was then removed from the specimen carrier with a sharpened tungsten needle. This needle was also used to remove excess BSA from the tubing. Importantly, the tubing tends to be brittle after the freeze-substitution steps, and care must be taken to avoid any mechanical damage to the embryo during these procedures. We embedded two capillary tubes on a single microscope slide with proper labeling indicating the specimen type. After removal of the empty membrane carrier, the embryo in the tube has to be checked so that the orientation is the same as that seen in the light microscope. Orientation of the embryo in the tube can best be checked by simultaneously rotating the tube while looking at the position of the embryo within the tube. This step is important later when the polymerized sample must be remounted and trimmed for serial sectioning. Tubes in their final orientation were then softly pressed down to the surface of the glass slide, and a second, coated and cleaned glass slide was put on top of the resin samples. This "sandwich" was placed on an appropriate support in the oven (about 48 h at 60 °C) to polymerize the resin. After resin polymerization, one of the microscope slides was removed by pressing a razor blade between the resin and one of the glass slides. At this stage, the tubes were on a microscope slide suitable for view in a light microscope and embedded in a thin layer of resin, ready to be remounted on "dummy" blocks for microtomy.

Thereafter, polymerized samples were observed with a light microscope to relocate the capillary tubes on the microscope slide. An objective marker with a diamond tip was used to mark a circle around the tube. The purpose of this circling is twofold: first, tubes are clearly marked during subsequent steps (i.e., during remounting) and second, the circle is used to determine the position of the embryo within the resin. When both embryo and circle are in about the same focus level then the embryo is positioned at the upper surface of the resin. In contrast, when the focus level of the embryo is not identical with the circle then the embryo is positioned close to the lower surface of the resin layer. In the latter case, the sample has to be remounted upside down on the "dummy" block. The tubes with surrounding BSA are usually not stained homogenously during the freeze-substitution step and the resulting "landscape" can be used to sketch the position of the embryo in the tube before remounting. Such a sketch is helpful later during trimming of the specimen. Using a scalpel, small quadrates of 1 × 1 mm^2 resin containing the capillary tube were cut out under a stereo microscope. A rapidly setting Epoxy glue was used to remount the samples on "dummy" blocks with the early embryos facing up. Polymerization of the glue at 60 °C was completed within about 30 min. At this step, the sketch (see above) was helpful to relocate the embryo in the remounted piece of resin.

After polymerization, the remounted sample was trimmed for serial sectioning. The trapezoid block face should have a height slightly larger than the embryo (~0.1 mm). The width of the block face has to be large enough to facilitate serial sectioning (~0.3 mm). Semithick (300–400 nm) sections were then collected in ribbons (with about 10–13 sections per ribbon) and 2–3 ribbons were collected on Formvar-coated copper slot grids. At this stage, it is important to keep the ribbons and the EM grids in sequence. Sections were routinely poststained with 2% uranyl acetate in 70% methanol followed by Reynold's lead citrate.

Stained specimens were then viewed by low-voltage EM (100 kV) to identify features of interest in the serial sections. At this stage, it is advantageous to start screening at low magnification to store the position of each section. By recalling the stored positions when at higher magnification, the sections that show features of interest can be identified. It was important here to document the position of the selected sections (e.g., left ribbon, third section from the top) for further electron tomography.

D. Electron Tomography

Next, we record electron micrographs of a series of tilted views of selected features on the prescreened grids. This tilt series is used to compute the tomogram from which we can generate 3D models (Müller-Reichert *et al.*, 2003; O'Toole *et al.*, 2003b). For details on this method, the reader is referred to Frank (1992).

In preparation for electron tomography, 15-nm-colloidal gold particles were affixed to both surfaces of the semithick sections. Samples were placed in a high-tilt, tomography stage and imaged with an intermediate-voltage electron microscope operated at 300 kV. Serial, tilted views were collected over a ±60° range at 1° increments about two orthogonal axes. Images were acquired using a 2K × 2K CCD camera at a pixel size of 1.01 nm using software for automated image capture (SerialEM). Multiple frames, or montages, were collected to reconstruct an area larger than could be seen as a single CCD frame (Marsh *et al.*, 2001; O'Toole *et al.*, 2003b). Image processing was routinely carried out using the IMOD software package (Kremer *et al.*, 1996). The serial, tilted views were aligned using the positions of the fiducial markers (i.e., the gold particles), and tomograms were computed using the R-weighted back-projection algorithm. Tomograms from each axis were then aligned to each other and combined (Mastronarde, 1997). The IMOD modeling program was used for image display and 3D modeling of mitotic features such as centrioles, microtubules, microtubule ends, and chromatin (Fig. 6).

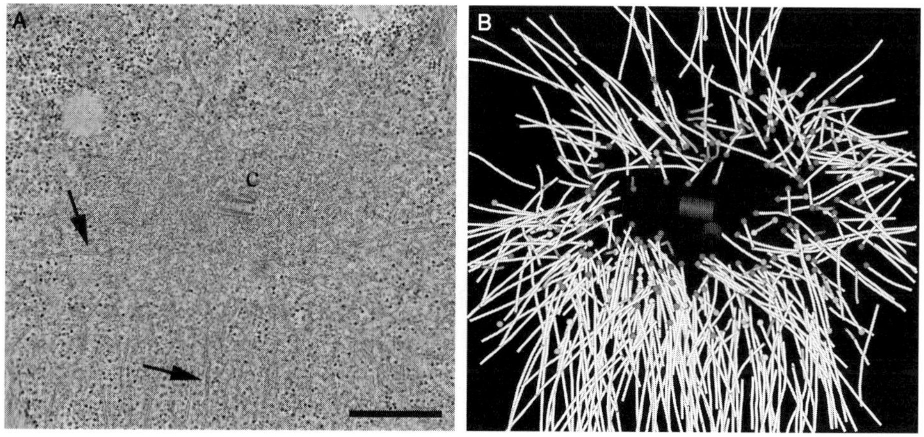

Fig. 6 Electron tomography and 3D modeling of mitotic spindle components in an isolated, eight-celled *C. elegans* embryo. (A) Tomographic slice showing the organization of microtubules (arrows) surrounding the centrioles (c). (B) Corresponding 3D model illustrating microtubules (white lines), centrioles (blue cylinders), and microtubule minus ends (closed, red spheres; open, yellow spheres) and microtubule plus ends (blue spheres). Scale bar = 1 μm. (See Color Insert.)

IV. Instrumentation and Materials

A. Staging of Early Embryos

Instrumentation: Stereo dissecting microscope with light source, light microscope equipped with either DIC or epifluorescence (for example, a Zeiss Axioplan 2 with a 20 × 0.5 NA Plan-Neofluar Apochromat dry objective, Hamamatsu Orca 12 bit digital camera controlled by MetaView Software, time-lapse intervals of 10–15 sec).

Materials: C. elegans strain XA3501 expressing GFP::β-tubulin and GFP::histone to visualize microtubules and chromatin, strain TH31 expressing GFP::γ-tubulin and GFP::histone to visualize centrosomes and chromatin, dialysis tubing with an inner diameter of 200 μm (Leica Microsystems), a micropippetor (0.5- to 10-μl size), two syringe needles, small plastic Petri dishes, number 11 scalpel tip, whetstone, gel loader tips, nail polish, glass slides, fine-tipped tweezers.

Reagents: Worm buffer M-9 [22-mM potassium phosphate monobasic (KH_2PO_4), 19-mM NH_4Cl, 48-mM sodium phosphate dibasic (Na_2HPO_4), 9-mM NaCl], BSA (Sigma-Aldrich, Steinheim, Germany).

B. High-Pressure Freezing and Cryoprocessing of Staged Embryos

Instrumentation: Stereomicroscope with light source, high-pressure freezer (Leica EMPACT2+RTS), freeze-substitution machine (Leica EM AFS, Leica Microsystems).

Materials: Membrane carriers (100-μm deep).

Reagents: Anhydrous acetone (EM grade), osmium tetroxide, uranyl acetate.

C. Thin-Layer Embedding, Serial Sectioning, and Prescreening of Samples

Instrumentation: Objective marker with a diamond tip (Leica), ultramicrotome (Leica UCT, Leica Microsystems), electron microscope operated at 100 kV (TECNAI 12, FEI, Eindhoven, The Netherlands) equipped with a compustage.

Materials: Glass slides, Parafilm®, "dummy" blocks.

Reagents: Teflon® solution (MS-143V TFE Release Agent, Miller-Stephenson Chemical Co., Inc., Danbury, CT), Epon/Araldite, fast glue, Formvar, copper slot grids, methanol, uranyl acetate, lead citrate.

D. Electron Tomography

Instrumentation: Intermediate-voltage electron microscope operated at 300 kV (we use a TECNAI F30 FEG), high-tilt rotating stage (Gatan model 650, Pleasanton, CA), 2K $\times$ 2K CCD camera (e.g., Gatan), image capture software package (SerialEM), and 3D reconstruction software (IMOD). Details can be found at http://bio3d.colorado.edu

Materials: Parafilm®, fine-tipped tweezers.

Reagents: 15-nm-colloidal gold (Ted Pella, Redding, CA).

V. Discussion

We are interested in the 3D analysis of spindle components in early *C. elegans* embryos that have a known developmental history. Our method for correlative LM/EM has enabled us for the first time to select and stage mitotic *C. elegans* embryos prior to high-pressure freezing and subsequent electron tomography. In the next sections, we discuss various methodological aspects of this technique and, briefly, its potential for the analysis of other systems.

A. Live-Cell Imaging

Live-cell imaging prior to EM provides a short developmental history of the specimen that can help with accurate interpretation of ultrastructural information (Kapoor *et al.*, 2006; Rieder and Cassels, 1999). This is especially relevant for studies of mitosis, where it is important to determine the precise cell-cycle stage of a developing embryo and/or compare wild-type and mutant structures at identical mitotic stages. In previous studies, whole worms were cryoimmobilized and the stage of embryonic cell was determined later (O'Toole *et al.*, 2003b). Here, we have staged single embryos prior to high-pressure freezing. For the purpose of stage determination, live-cell imaging, using either conventional phase contrast or DIC, can be applied. In the case of the early *C. elegans* embryos, phase contrast appears to be sufficient for most applications, as the early development of the embryo offers morphological criteria indicative of specific stages of the first mitotic division (Fig. 3). For instance, the appearance, migration, rotation, and disappearance of the pronuclei have routinely been followed by this method for embryonic staging. However, metaphase and early anaphase are more difficult to determine, with either phase or DIC optics. Although the application of video microscopy is not a prerequisite for documenting mitotic stages, it might be important when a specific mitotic event (e.g., onset of mitotic spindle assembly) warrants ultrastructural analysis. Imaging live embryos expressing GFP-tagged proteins that mark specific cellular components, such as centrosomes, kinetochores, histones, or spindle microtubules, provides a clear advantage for the documentation of early development, although the quality of the obtained images is not optimal (see below).

In general, the samples must be recovered during the imaging process; given that embryonic cell division in *C. elegans* is rapid, one must work quickly. For this reason, the embryos (encapsulated in dialysis tubing) are submerged in small volumes of BSA-containing M-9 buffer on the slide, but are imaged with "high-dry" objectives; this facilitates rapid recovery of the embryos. The BSA serves as a cryoprotectant but, unfortunately, it further reduces image quality. Despite the poor quality of the images, we found that most major cellular events can be distinguished (Fig. 4), and freezing of the sample can proceed within seconds after the end of imaging by LM.

B. Capillary Tubes

Nematodes have previously been studied in capillary tubes, which allow them to be incubated, gently concentrated, and continually observed in their cultivating medium until their cryoimmobilization in this state by high-pressure freezing (Hohenberg *et al.*, 1994). Here, short pieces of capillary tubes were used to "handle" isolated single *C. elegans* embryos during preparation for both light and electron microscopic observations. The advantages of cellulose capillary tubes for the described approach can be summarized as follows (Müller-Reichert *et al.*, 2003). (1) Capillary tubes can easily be cut into short pieces. (2) Development of the embryo can be continually monitored by LM because the capillary tubes are transparent. (3) Freezing in capillary tubes gives a high yield of adequately and reproducibly high-pressure frozen samples when combined with the use of an adequate cryoprotectant (see below). (4) The high porosity of the cellulose tubes means that freeze-substitution and embedding procedures are not prolonged. The tubes have pores with various molecular weight cutoffs, ranging from 5 to 20 kDa; these allow free passage of all substitution media and resin monomers. (5) The resin-embedded tubing is suitable for serial thin and semithin sectioning.

A disadvantage of the method, however, is that the tubing cannot always be sealed permanently, so the embryos "leak" out during the light microscopic observation. To prevent this, it is important to press the crimping tool down carefully during cutting.

C. Use of BSA as a "Filler"

The most critical part of high-pressure freezing is probably the loading of samples into the sample holder of the high-pressure freezer (McDonald *et al.*, 2006). It is absolutely necessary to fill the entire volume of the capillary tube, as well as the inner cavity of the specimen holder (e.g., the membrane carrier), with an appropriate medium. If there are air spaces, these will either interfere with effective heat transfer or collapse as the pressure rises, leading to the possibility that the specimen is crushed. In addition, the water content in the surrounding medium should be as low as possible. Usually, addition of an extracellular cryoprotectant serves to prevent the formation of ice crystals during the freezing process. To serve both purposes, people have previously used 1-hexadecene, a hydrophobic and chemically inert paraffin oil of rather low viscosity; it works both as "filler" and cryoprotectant (Dahl and Staehelin, 1989; Studer *et al.*, 1989). For the previous preparation of whole worms, nematodes were enclosed within buffer-containing cellulose capillary tubes and submerged in 1-hexadecene-filled sample holders, thereby avoiding direct contact between the worms and the paraffin oil (Hohenberg *et al.*, 1994). But because 1-hexadecene remains solid at freeze-substitution temperatures (Thijssen *et al.*, 1998), it is important to manually remove frozen 1-hexadecene from around the tubing under liquid nitrogen prior to freeze-substitution (Hohenberg *et al.*, 1994). Here, we have used BSA-containing M-9 buffer for routine specimen preparation and loading.

Our strategy has been to dissect whole worms in the future freezing medium, load the resulting embryos into capillary tubes, image mitotic divisions, and rapidly freeze the living embryos, all in the same medium. Importantly, when transferring the embryo-containing tube from the microscope slide to the prefilled membrane carrier, dilution effects, due to the presence of different media, can be excluded. Recent results have shown that 20% BSA gives good freezing for a variety of specimens, such as marine sponges, choanoflagellates, *Drosophila*, and wasp embryos, *C. elegans* hermaphrodites (McDonald *et al.*, 2006), murine skin cells (Reipert *et al.*, 2004a), PtK2 cells (Reipert *et al.*, 2004b), and MDCK II cells grown on filters (Manninen *et al.*, 2005). Importantly, BSA caused no problems during later steps of preparation (i.e., during serial sectioning).

D. High-Pressure Freezing

The EMPACT2 offers a variety of specimen holders (Chapter 2 by McDonald, this volume; McDonald *et al.*, 2006). The general holder is 1.2- or 1.5-mm wide and 200-μm deep (with a central hole for the pressure tube). We routinely use the newly developed 1.5-mm-wide and 100-μm-deep "membrane" carrier (without a central hole). The copper "membrane" prevents direct contact of the specimen with the pressurizing fluid. In general, this specimen holder gives excellent and reproducible freezing results with isolated early embryos.

The described approach was developed for the tomographic analysis of staged *C. elegans* embryos. A prerequisite for this correlative LM/EM technique, however, is that the time between staging by LM and fixation for EM is as short as possible. The "standard" loading procedure for the EMPACT2 (i.e., when using the high-pressure freezer without the RTS) involves loading of the capillary tube into the sample holder, manually closing the sample holder after insertion into the "pod," insertion of the sample holder into the high-pressure freezer, and starting the freezing process. To minimize the time required for specimen loading, the use of the newly developed RTS associated with the EMPACT2 high-pressure freezer (Leica) is advantageous (McDonald *et al.*, 2006). When the rapid loader is used for automatic insertion of the sample holder into the preloaded pod (Fig. 5), the RTS reduces the time between observation and freezing to about 5 sec. Moreover, the automatic closure of the pod allows standardization of the loading/freezing procedure, thereby contributing to the high reproducibility of our method. It should be noted here that freezing of the isolated embryos is by no means restricted to the EMPACT2+RTS. The question, however, is whether very fast loading times can be achieved with the other commercially available high-pressure freezers. In summary, the freezing method described here gives excellent and reproducible results for ultrastructural studies.

E. Serial Sectioning

Obtaining superior preservation of cellular fine structure is a significant step in undertaking a 3D structure study (McIntosh *et al.*, 2005). It is also critical, however, to be able to localize and/or identify specific structures in a sample both quickly and repeatably (Müller-Reichert *et al.*, 2003). As an example, when one is interested in the structure and duplication of the centriole, one has to be able to routinely find centrosomes in different mitotic stages of either wild-type or RNAi embryos (Dammermann *et al.*, 2004; Kirkham *et al.*, 2003; Pelletier *et al.*, 2004b). Currently, the method of choice for this identification procedure is serial sectioning of plastic embedded samples. With good ribbons of serial sections on slot grids, one can sequentially image slices of the entire *C. elegans* embryo and identify the particular sections that contain the centrosome. Critically, for the identification of the embryonic centrosome, it is absolutely necessary to collect all the serial sections of a series because the small *C. elegans* centrioles are contained within only two to three thin sections of about 70 nm.

F. Electron Tomography and 3D Modeling

In principle, a series of thin sections can be used for 3D reconstruction by computationally "stacking" images of all successive sections and then modeling features of interest in 3D. However, this method has the limitation that the resolution perpendicular to the plane of the section is no better than twice the average section thickness (McDonald *et al.*, 1996). Electron tomography is therefore the method of choice for visualizing the fine structure of cellular organelles with high *z*-resolution (reviewed in McEwen and Marko, 2001; Frank, 2002). Conceptually similar to CT scans in medical imaging, this method is based on the use of serial, tilted views of a semithick section (300–400 nm) to create a computer-generated volume that can be sliced and imaged in any direction. As examples, the spindle pole body (Bullitt *et al.*,

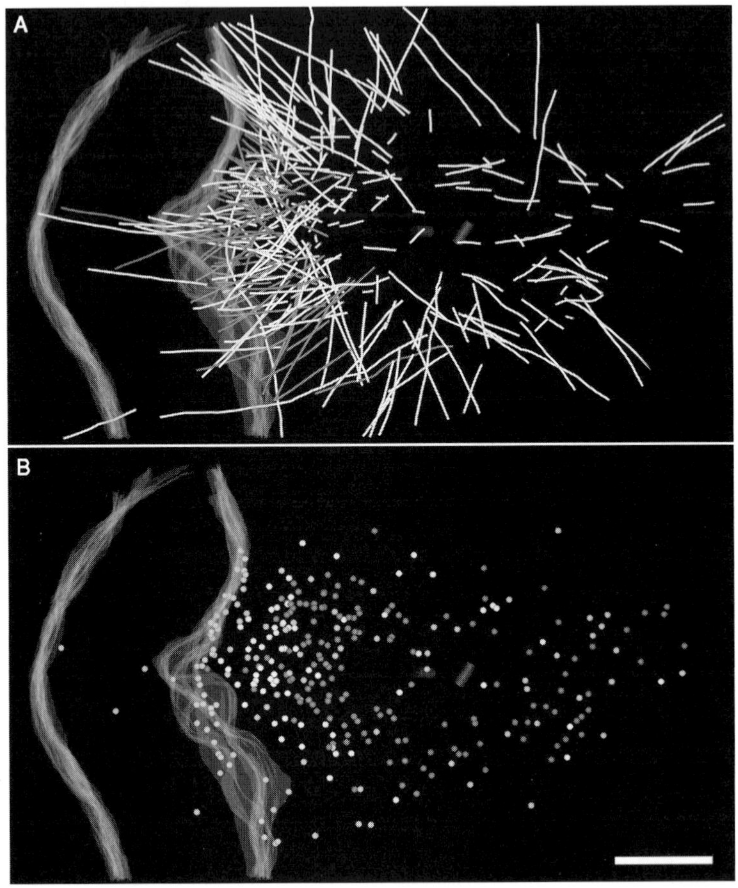

Fig. 7 Reconstruction of a spindle collapse in an RNAi embryo depleted of the worm orthologue of TPX-2. (A) 3D model computed from a 2 × 1 montage, showing boundary of a chromosome (blue), the position of spindle microtubules (red, white lines), and centrioles (blue cylinders). Microtubules that ended on the chromosome were defined as kinetochore microtubules (red). (B) Position of closed (red spheres) and open (yellow spheres) microtubule minus ends in the vicinity of the centrosome. Microtubule plus ends are marked with blue spheres. A high percentage of open microtubule minus ends were associated with kinetochore microtubules, and these ends were situated in a cluster at the outer edge of the centrosome (Özlü *et al.*, 2005). Scale bar = 1 μm. (See Color Insert.)

1997; O'Toole *et al.*, 1999) and the mitotic spindle organization (O'Toole *et al.*, 1999; Yoder *et al.*, 2005) in *Saccharomyces cerevisiae*, the basal body structure in *Chlamydomonas* (O'Toole *et al.*, 2003a), the centrosome in *C. elegans* (O'Toole *et al.*, 2003b), *Drosophila* (Moritz *et al.*, 1995) and *Spisula* (Schnackenberg *et al.*, 1998), and the mammalian kinetochore (McEwen *et al.*, 1998) have been reconstructed in 3D by electron tomography. Illustrating the potential of the technique, we have successfully applied electron tomography to analyze the ultrastructure of an aster in a blastomere of an isolated eight-celled embryo (Fig. 6). Recently, we have also performed electron tomography on *C. elegans* embryos found in whole worms treated with RNAi; these show a spindle that collapse in metaphase due to depletion of the worm orthologue of TPX-2 (Fig. 7; Özlü *et al.*, 2005). 3D models of the depleted embryos revealed an obvious decrease in kinetochore microtubule length and an interesting distribution of open microtubule minus ends (i.e., a cluster of open minus ends situated on the chromosome-facing side of the centrosome). Tomography has currently been used to analyze the role of katanin in *C. elegans* meiosis (Srayko *et al.*, 2006) and to investigate centriole formation in wild-type and RNAi-treated *C. elegans* embryos in mitosis (Pelletier *et al.*, 2006).

G. Concluding Remarks

The correlative LM/EM approach described here is certainly not limited to the study of early embryos from *Caenorhabditis*. The early *Drosophila* embryo offers similar possibilities for structure–function studies (Raff, 2004). Staging early *Drosophila* embryos can be achieved by using GFP-tagged strains, and depletion of targeted gene products is routinely achieved by RNAi. Capillary tubes and the EMPACT2+RTS have been used successfully to freeze *Drosophila* embryos in the presence of 20% BSA (McDonald *et al.*, 2006). It will be challenging, however, to adapt our correlative approach to the analysis of *Drosophila* Schneider S2 and mammalian cells in culture.

We are certainly at the beginning of exploiting the use of model systems such as *C. elegans* and *Drosophila* for systematic structure–function studies. There is no doubt that it will be a challenge to take advantage of the data obtained from genome-wide screens in order to understand the mechanics of mitosis by correlative LM/EM.

Acknowledgments

The authors would like to thank Paul Verkade for sharing expertise on the newly developed EMPACT2+RTS and Jana Mäntler and Susanne Kretschmar for excellent technical assistance.

References

Bullitt, E., Rout, M. P., Kilmartin, J. V., and Akey, C. W. (1997). The yeast spindle pole body is assembled around a central crystal of Spc42p. *Cell* **89,** 1077–1086.

Chrétien, D., Buendia, B., Fuller, S. D., and Karsenti, E. (1997). Reconstruction of the centrosome cycle from cryoelectron micrographs. *J. Struct. Biol.* **120,** 117–133.

Cowan, C. R., and Hyman, A. A. (2004). Asymmetric cell division in *C. elegans*: Cortical polarity and spindle positioning. *Annu. Rev. Cell Dev. Biol.* **20,** 427–453.

Dammermann, A., Müller-Reichert, T., Pelletier, L., Habermann, B., Desai, A., and Oegema, K. (2004). Centriole assembly requires both centriolar and pericentriolar material proteins. *Dev. Cell* **7,** 815–829.

Dahl, R., and Staehelin, L. A. (1989). High pressure freezing for the preservation of biological structure: Theory and practice. *J. Electron Microsc. Tech.* **13,** 165–174.

Frank, J. (1992). "Electron Tomography." Plenum Press, New York.

Hohenberg, H., Mannweiler, K., and Müller, M. (1994). High-pressure freezing of cell suspensions in cellulose capillary tubes. *J. Microsc. (Oxford)* **175,** 34–43.

Kapoor, T. M., Lampson, M. A., Hergert, P., Cameron, L., Cimini, D., Salmon, E. D., McEwen, B. F., and Khodjakov, A. (2006). Chromosomes can congress to the metaphase plate before biorientation. *Science* **311,** 388–391.

Kirkham, M., Müller-Reichert, T., Oegema, K., Grill, S., and Hyman, A. A. (2003). SAS-4 is a *C. elegans* centriolar protein that controls centrosome size. *Cell* **112,** 575–587.

Kremer, J. R., Mastronarde, D. N., and McIntosh, J. R. (1996). Computer visualization of three-dimensional image data using IMOD. *J. Struct. Biol.* **116,** 71–76.

Manninen, A., Verkade, P., Le Lay, S., Torkko, J., Kasper, M., Füllekrug, J., and Simons, K. (2005). Caveolin-1 is not essential for biosynthetic apical membrane transport. *Mol. Cell. Biol.* **25,** 10087–10096.

Marsh, B. J., Mastronarde, D. N., Buttle, K. F., Howell, K. E., and McIntosh, J. R. (2001). Organellar relationships in the Golgi region of the pancreatic beta cell line, HIT-T15, visualized by high resolution electron tomography. *Proc. Natl. Acad. Sci. USA* **98,** 2399–2406.

Mastronarde, D. N. (1997). Dual-axis tomography: An approach with alignment methods that preserve resolution. *J. Microsc. (Oxford)* **120,** 343–352.

McDonald, K., and Müller-Reichert, T. (2002). Cryomethods for thin section electron microscopy. *Methods Enzymol.* **351,** 96–123.

McDonald, K., O'Toole, E. T., Mastronarde, D. N., Winey, M., and McIntosh, J. R. (1996). Mapping the three-dimensional organization of microtubules in mitotic spindles of yeast. *Trends Cell Biol.* **6,** 235–239.

McDonald, K. L., Morphew, M., Verkade, P., and Müller-Reichert, T. (2006). Recent advances in high-pressure freezing. *Methods Mol. Biol.* **369,** 143–174.

McEwen, B. F., Hsieh, C., Matteyses, A. L., and Rieder, C. L. (1998). A new look at kinetochore structure in vertebrate somatic cells using high-pressure freezing and freeze substitution. *Chromosoma* **107,** 366–375.

McEwen, B. F., and Marko, M. (2001). The emergence of electron tomography as an important tool for investigating cellular ultrastructure. *J. Histochem. Cytochem.* **49,** 553–563.

McIntosh, J. R., Nicastro, D., and Mastronarde, D. (2005). New views of cells in 3-D: An introduction to electron tomography. *Trends Cell Biol.* **15,** 43–51.

Moritz, M., Braunfeld, M. B., Fung, J. C., Sedat, J. W., Alberts, B. M., and Agard, D. A. (1995). Three-dimensional structural characterization of centrosomes from early *Drosophila* embryos. *J. Cell Biol.* **130,** 1149–1159.

Müller-Reichert, T., O'Toole, E. T., Hohenberg, H., and McDonald, K. L. (2003). Cryoimmobiliztion and three-dimensional visualization of *C. elegans* ultrastructure. *J. Microsc. (Oxford)* **212,** 71–80.

Murk, J. L., Posthuma, G., Koster, A. J., Geuze, H. J., Verkleij, A. J. K. leijmeer, M. J., and Humbel, B. M. (2003). Influence of aldehyde fixation on the morphology of endosomes and lysosomes: Quantitative analysis and electron tomography. *J. Microsc. (Oxford)* **212,** 81–90.

Oegema, K., Desai, A., Rybina, S., Kirkham, M., and Hyman, A. A. (2001). Functional analysis of kinetochore assembly in *Caenorhabditis elegans*. *J. Cell Biol.* **153,** 1209–1226.

Oegema, K., and Hyman, T. (2005). Cell division. *In* "WormBook" (J. M. Kramer, and D. G. Moerman, eds.). The *C. elegans* Research Community, WormBook, doi/10.1895/wormbook. 1.72.1, http://www.wormbook.org

O'Toole, E. T., Giddings, T. H., McIntosh, J. R., and Dutcher, S. K. (2003a). Three-dimensional organization of basal bodies from wild-type and δ-tubulin deletion strains of *Chlamydomonas reinhardtii*. *Mol. Biol. Cell* **14,** 2999–3012.

O'Toole, E. T., McDonald, K. L., Mäntler, J., McIntosh, J. R., Hyman, A. A., and Müller-Reichert, T. (2003b). Morphologically distinct microtubule ends in the mitotic centrosome of *Caenorhabditis elegans*. *J. Cell Biol.* **163,** 451–456.

O'Toole, E. T., Winey, M., and McIntosh, J. R. (1999). High-voltage electron tomography of spindle pole bodies and early mitotic spindles in the yeast *Saccharomyces cerevisiae*. *Mol. Biol. Cell* **10,** 2017–2031.

Özlü, N., Srayko, M., Kinoshita, K., Habermann, B., O'Toole, E. T., Müller-Reichert, T., Schmalz, N., Desai, A., and Hyman, A. A. (2005). An essential function of the *C. elegans* orthologue of TPX2 is to localize activated Aurora A kinase to mitotic spindles. *Dev. Cell* **9,** 237–248.

Pelletier, L., Müller-Reichert, T., Srayko, M., Özlü, N., Schlaitz, A.-L., and Hyman, A. A. (2004a). The *C. elegans* centrosome during early embryonic development. *In* "Centrosomes in Development and Disease" (E. Nigg, ed.), pp. 227–250. Wiley-VCH, Weinheim.

Pelletier, L., Özlü, N., Hannak, E., Cowan, C., Habermann, B., Ruer, M., Müller-Reichert, T., and Hyman, A. A. (2004b). The *Caenorhabditis elegans* centrosomal protein SPD-2 is required for both pericentriolar material recruitment and centriole duplication. *Curr. Biol.* **14,** 863–873.

Pelletier, L., O'Toole, E., Schwager, A., Hyman, A. A., and Müller-Reichert, T. (2006). Centriole assembly in *Caenorhabditis elegans*. *Nature* (in press).

Priess, J. R., and Hirsh, D. I. (1986). *Caenorhabditis elegans* morphogenesis: The role of the cytoskeleton in elongation of the embryo. *Dev. Biol.* **117,** 156–173.

Raff, J. W. (2004). Centrosomes in a developing organism: Lessons from *Drosophila*. *In* "Centrosomes in Development and Disease" (E. Nigg, ed.), pp. 251–278. Wiley-VCH, Weinheim.

Reipert, S., Fischer, I., and Wiche, G. (2004a). High-pressure cryoimmobilization of murine skin reveals novel structural features and prevents extraction artifacts. *Exp. Dermatol.* **13,** 419–425.

Reipert, S., Fischer, I., and Wiche, G. (2004b). High-pressure freezing of epithelial cells on sapphire coverslips. *J. Microsc. (Oxford)* **213,** 81–85.

Rieder, C. L., and Cassels, G. (1999). Correlative light and electron microscopy of mitotic cells in monolayer cultures. *Methods Cell Biol.* **61,** 297–315.

Schnackenberg, B. J., Khodjakov, A., Rieder, C., and Palazzo, R. E. (1998). The disassembly and reassembly of functional centrosomes *in vitro*. *Proc. Natl. Acad. Sci. USA* **95,** 9295–9300.

Schneider, S. Q., and Bowerman, B. (2003). Cell polarity and the cytoskeleton in the *Caenorhabditis elegans* zygote. *Annu. Rev. Genet.* **37,** 221–249.

Sönnichsen, B., Koski, L. B., Walsh, A., Marshall, P., Neumann, B., Brehm, M., Alleaume, A. M., Artelt, J., Bettencourt, P., Cassin, E., Hewitson, M., Holz, C., *et al.* (2005). Full-genome RNAi profiling of early embryogenesis in *Caenorhabditis elegans*. *Nature* **434,** 462–469.

Srayko, M., O'Toole, E. T., Hyman, A. A., and Müller-Reichert, T. (2006). Katanin disrupts the microtubule lattice and increases polymer number in *C. elegans* meiosis. *Curr. Biol.* **16,** 1944–1949.

Studer, D., Michel, M., and Müller, M. (1989). High-pressure freezing comes of age. *Scanning Microsc. Suppl.* **3,** 253–268; discussion 268–269.

Thijssen, M. H., Van Went, J. L., and Van Aelst, A. C. (1998). Heptane and isooctane as embedding fluids for high-pressure freezing of *Petunia* ovules followed by freeze-substitution. *J. Microsc. (Oxford)* **192,** 228–235.

Yoder, T. J., McElwain, M. A., Francis, S. E., Bagley, J., Muller, E. G., Pak, B., O'Toole, E. T., Winey, M., and Davis, T. N. (2005). Analysis of a spindle pole body mutant reveals a defect in biorientation and illuminates spindle forces. *Mol. Biol. Cell* **16,** 141–152.

PART II

Imaging Fixed Cells in Three Dimensions

Once cells have been prepared for electron microscopy (EM), there are many ways to image them, each with its own strengths and limitations. The chapters in this section of the book provide examples of both specimens and techniques that have produced important new results in structural cell biology, but these authors are also candid about the limitations of their approaches. The focus of this section is on biological material that has been fixed, dehydrated, and embedded, as opposed to simply frozen. Thus, a wide range of biology is within reach, including big cells and parts of whole organisms. On the other hand, each of these investigations must confront the issue of fixation and what it may have done to the structures under study. Moreover, when biological material is embedded in plastic to facilitate microtomy, some stain must be used to give the well-preserved macromolecules a useful level of contrast relative to the surrounding material. Thus, the images under study are really those of stain, not of protein and nucleic acids, and so on. This feature of EM applied to embedded samples adds yet another level of uncertainty as one reaches for every greater reliability and resolution. These are caveats that any critical reader must keep in mind when evaluating images from this kind of microscopy.

That said, the chapters of this section show the rather amazing power of modern EM to determine cellular structures at impressive resolution (4–6 nm) and over significant volumes (up to $\sim$120 μm^3, with bigger reconstructions on their way). The order of the chapters that follow is somewhat arbitrary, but I have pursued a logic. Chapter 5, by O'Toole *et al.*, describes the use of rapid freezing, freeze-substitution fixation and embedding, microtomy, and EM to compare wild-type and mutant basal bodies in three dimensions (3D), employing EM tomography (ET) to visualize phenotypes with informative resolution and clarity. These authors lay out the basics of ET and provide access for the interested reader to Web sites from which one can download tutorials that explain the relevant procedures at a level of detail that would be impossible in a book. Thus, this chapter makes two key points: how to do tomography, and what tomography can do for the cell biologist who has mutant cells to study and phenotypes to compare.

Höög and Antony, Chapter 6, have employed these methods to study the microtubule cytoskeleton of the fission yeast *Schizosaccharomyces pombe*, an organism whose mechanisms for shape determination have already been well studied with genetic and molecular methods at the light microscopic level. This chapter demonstrates the power of having both good resolution for examining structural details (e.g., the shape of microtubule tips) and large volume overviews (to follow the trajectories of microtubules as they course through the cytoplasm). The unique ability of ET to reveal both detail at $\sim$5 nm and an overview that extends to several micrometers, both in 3D, is beautifully described in this chapter.

Robertson, Chapter 7, applies related methods to the study of a small Archeon, *Sulfolobus solfataricus*. This microorganism lacks a nucleus, like all of its kind, but it possesses bacterial flagella, a complex cell wall, a plasma membrane, numerous ribosomes, subregions that are rich in DNA, and several previously unobserved structural features that might be organelles. Again, ET allows the reconstructions of an entire small organism, this time facilitating the exploration of a cell that is not yet well understood. Structural information may provide ways to elucidate this cell's unusual biology (a chemoorganotrophic extremophile that lives happily at pH 3 and 75°C) and fascinating phylogeny (the absence of a nuclear envelop that is reminiscent of eubacteria and a molecular organization of DNA and transcriptional machinery that is distinctly eukaryotic).

Marsh and colleagues (Chapter 8) apply ET to the exploration of a medically important mammalian cell, the "β-cell" from the Islets of Langerhans, which is responsible for insulin biosynthesis and blood sugar regulation. The emphasis of this group is on cytomembrane systems, including the endoplasmic reticulum, the Golgi complex, and components of both the secretory and the endocytic pathways. Here the technological thrust is to reconstruct entire mammalian cells, a feat that remains to be accomplished but appears to be not very far off.

Otegui and Austin, Chapter 9, show how ET can be used to examine the relationships between cytomembranes and cytoskeletal elements in plant cells, giving special emphasis to the structures that form at the end of mitosis to achieve cytokinesis in *Arabadopsis*. Their work again relies on rapid freezing and freeze-substitution fixation and the ability of ET to reveal both local detail and long-range order. They are studying a significant fraction of the cell's volume, but their emphasis is not on getting more volume; it is on using the volume under study to explore the function of a large and wide-ranging organelle whose structural details are important for its function.

Arthur *et al.* (Chapter 10) take on the issue of how one might use ET in combination with both light microscopy and controlled electrical stimulation to unravel the responses of neurons to patterns of stimulation that induce altered behaviors at synapses. These issues are at the heart of our effort to understand learning and memory, major problems in neuroscience. One can hope that the approaches described here will soon bring significant light to this important but currently enigmatic subject.

McEwen *et al.* (Chapter 11) present an authoritative review of the use of EM to pursue functional studies of the mitotic spindle. They describe a wealth of methods that have been used over the years and make the important point that a single technological approach (even one that is currently fashionable, like ET) is not a panacea; one must suit the method to the scientific questions being asked, a timeless point that is relevant to all good structural and experimental biology.

Svetkina (Chapter 12) gives an excellent example of this point of view by showing how a combination of carefully chosen methods can produce highly informative images about a complex cellular event: gliding cell motility. Her work demonstrates the power of appropriate controls and of comparisons with other methods, like light microscopy, for ferreting out artifact and getting ever closer to EM samples that accurately reflect a biologically important process. The correlations she describes give confidence that the results achieved are very likely to be valid, but they also force us to keep in mind that EM of anything other than a frozen-hydrated cell involves the risk that a given preparative method may be misleading.

Chapter 13 by Taylor *et al.* is a *tour de force* that describes a union of specimen choice, sample preparation, EM itself, and image analysis that is helping to solve a long-standing biological problem: the structural cycle of a myosin crossbridge that accompanies filament sliding in striated muscle. This is really two chapters in one, but the text has been left as a single unit because the details of specimen preparation are inextricable from the reliability of the samples and thence from the images and image analyses that provide information about molecular behavior. This work uses carefully timed rapid freezing, freeze-substitution fixation, embedding and staining to obtain samples for EM from an organ that is far too large for whole-cell microscopy, so embedding, meticulous microtomy, and staining are required. The work has been done with scrupulous attention to detail and in the context of atomic structural information about some of the most prevalent molecular components of the myofibril. As a result, one can easily believe that reliable molecular information is coming from these heavy metal representations of biological macromolecules.

The summary view of this section is that expert practitioners of EM are able to use a combination of careful sample preparation and 3D imaging at good resolution to study samples whose physiology is subject to well-controlled experimental conditions (through mutation, selection of stage, definition of environment, or stimulation and temporal control of rapid freezing). From such work, important and novel information can emerge.

CHAPTER 5

Understanding Microtubule Organizing Centers by Comparing Mutant and Wild-Type Structures with Electron Tomography

Eileen T. O'Toole,* Thomas H. Giddings, Jr.,* and Susan K. Dutcher†

*Boulder Laboratory for 3D Electron Microscopy of Cells
Department of Molecular, Cellular, and Developmental Biology
University of Colorado, Boulder, Colorado 80309

†Department of Genetics, Washington University School of Medicine
St. Louis, Missouri 63110

Microtubule organizing centers guide the formation of complex cytoskeletal arrangements in eukaryotic cells by controlling the initiation of microtubules. Thus, a careful study of their structure and function can benefit from three-dimensional (3D)

0091-679X/07 $35.00
DOI: 10.1016/S0091-679X(06)79005-7

imaging methods such as electron tomography (ET). We have used improved methods of specimen preparation and dual-axis ET to study the structure and organization of one type of microtubule organizing center, the basal body complex in the unicellular alga, *Chlamydomonas reinhardtii* (O'Toole *et al.*, 2003a). In this chapter, we describe our methods for rapid freezing and freeze-substitution of whole *Chlamydomonas* cells and the use of ET to study the 3D organization of the basal body complex in intact cells. Using these methods, we compare this complex in wild-type cells with the one from a mutant strain that carries a basal body assembly defect. This use of ET to compare mutants with wild-type cells permits the identification of defects in specific structural components and furthers our understanding of the assembly and function of the basal body complex in *Chlamydomonas*.

I. Introduction

Microtubule organizing centers (MTOCs) are responsible for the initiation of cytoplasmic and spindle microtubules (MTs), and they play essential roles in many critical biological processes (as reviewed in Doxsey *et al.*, 2005). Examples of MTOCs include the centrosome that nucleates the growth of most spindle MTs in animal cells, the spindle pole body in yeast, and the basal body that initiates the MTs of cilia and flagella. MTOCs can form complex, three-dimensional (3D) arrangements in cells. This complexity is perfectly illustrated in the basal body complex from the biflagellate green alga, *Chlamydomonas reinhardtii*. In this organism, the basal body complex consists of two mature basal bodies, two immature or probasal bodies, four sets of rootlet MT bundles, and several fiber systems that connect the individual components (Cavalier-Smith, 1974; Johnson and Porter, 1968; Ringo, 1967). The basal body complex can be isolated (Dutcher, 1995). These preparations, which are enriched for basal bodies, contain 150–250 polypeptides. However, despite intensive research on MTOCs, little is known about that factors that regulate the duplication, assembly, and function of this complex organelle.

C. reinhardtii is an excellent model organism for the study of MTOCs because a combination of genetic and molecular analyses can be used to identify the genes that are important for basal body assembly and function (reviewed in Dutcher, 2003; Preble *et al.*, 2000). For example, it is known that the *UNI3* gene, which encodes δ-tubulin, and the *BLD2* gene, which encodes ε-tubulin, are needed for the proper assembly of triplet MTs in basal bodies (Dutcher and Trabuco, 1998; Dutcher *et al.*, 2002; Goodenough and St. Clair, 1975). In addition, mutations at three loci called *VFL* (*variable flagellar number*) result in defects in basal body segregation (Adams *et al.*, 1985; Salisbury *et al.*, 1988; Silflow *et al.*, 2001; Taillon *et al.*, 1992; Wright *et al.*, 1983, 1985). Mutations in the centrin gene, *VFL2*, results in defects in the assembly of fibers that connect mature basal bodies to the nucleus and to each other, resulting in segregation defects (Salisbury *et al.*, 1988; Wright *et al.*, 1985). Thus, the component structures of the basal bodies themselves as well as the connecting fibers in the basal complex are important for the proper assembly, segregation, and function of this organelle.

The basal body complex, like other MTOCs, is a 3D structure, and a careful examination of its structure can therefore benefit from 3D imaging methods such as electron tomography (ET). ET has been shown to be a powerful method with which to study the 3D structure of complex biological structures (reviewed in Baumeister, 2002; Frank, 1992; Lucic *et al.*, 2005; McIntosh *et al.*, 2005). It results in computer-generated reconstructions that can be sliced and imaged in any orientation at ~6-nm resolution in 3D. ET has been used to study the 3D geometry of MTOCs such as centrosomes isolated from *Drosophila* and *Spisula* (Moritz *et al.*, 1995a,b; Schnackenberg *et al.*, 1998; Vogel *et al.*, 1997), the yeast spindle pole body (Bullitt *et al.*, 1997; O'Toole *et al.*, 1999), and the mitotic centrosome in *Caenorhabditis elegans* (O'Toole *et al.*, 2003b). In addition, several chapters in this volume illustrate the utility of ET to study the structure and organization of organelles in intact cells, and the reader is referred to these chapters for specific examples in other systems.

Recently, genetic analyses have been combined with ET to study MTOC structure and function. The spindle pole body of *Saccharomyces cerevisiae* has been studied in wild-type and mutant strains to analyze defects in spindle pole body assembly and spindle organization (Bullitt *et al.*, 1997; Yoder *et al.*, 2005). ET has also been used to assess the effect of TPXL-1 depletion on the organization of spindle MTs in the mitotic centrosome in *C. elegans* (Ozlu *et al.*, 2005). Specific to this chapter, ET has been used to determine the roles of particular tubulin isoforms in the structure and function of basal bodies from *Chlamydomonas* (O'Toole *et al.*, 2003a). The combination of mutant analysis and ET should help us to understand the assembly and function of this MTOC.

Here we describe the steps necessary to prepare *Chlamydomonas* for ET and provide examples that illustrate how these methods can be used to compare the 3D morphology of both basal bodies and basal body-associated structures in wild-type and mutant strains.

II. Rationale

The rationale for this chapter is to describe a detailed method of specimen preparation involving rapid freezing, followed by freeze-substitution fixation to optimally preserve the structure of the basal body complex. Because the basal body complex is a 3D organelle, we also outline the steps required for tomographic analysis to image/examine these structures in 3D. These methods are then applied to strains with defects in basal body assembly to characterize the phenotypes of mutant strains relative to wild type. The figures we present are selected, single frames, extracted from complete tomographic volumes or models. Several movies of complete tomographic reconstructions and models have been posted as video supplements on our laboratory's Web site (http://bio3d.colorado.edu); these will illustrate the data with more scope and detail for the interested reader.

III. Methods

A. High-Pressure Freezing, Freeze-Substitution, and Embedding

For 3D structure studies of organelles within their cellular context, optimum cell preservation is the critical first step. Conventional chemical fixation involves diffusion of chemical fixatives into the cell, followed by room temperature or 0 °C dehydration and embedding. This method can lead to artifacts, particularly in organisms that have natural barriers to diffusion, such as the cell wall in *Chlamydomonas* (see also Chapter 1 by Dubochet, Chapter 2 by McDonald, and Chapter 3 by Hess, this volume). We have therefore developed an improved fixation protocol that uses high-pressure freezing followed by freeze-substitution fixation (O'Toole *et al.*, 2003a; Preble *et al.*, 2001). A comparison of *Chlamydomonas* cells fixed with a standard chemical fixation (A) and by high-pressure freezing and freeze-substitution (B) is shown in Fig. 1. The cell fixed by conventional chemical fixation shows dramatic artifacts, such as extraction of the nucleoplasm and collapse of the membrane-bounded compartments, like the contractile vacuole and chloroplasts.

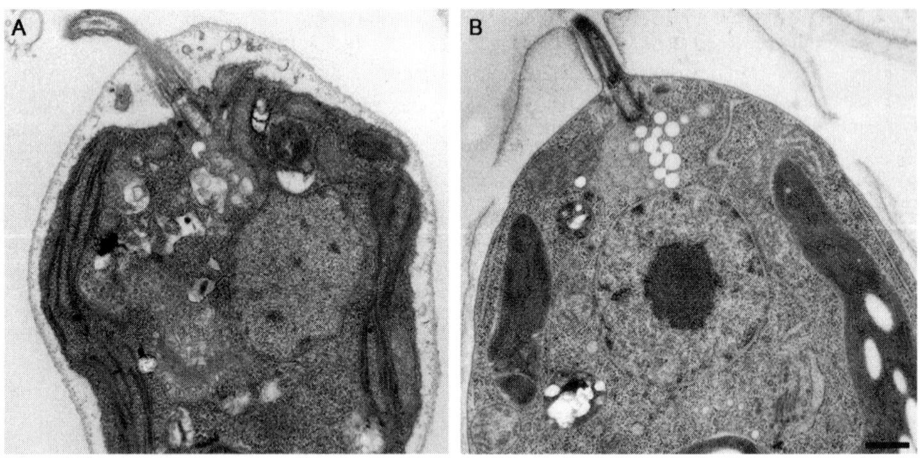

Fig. 1 High-pressure freezing followed by freeze-substitution results in excellent preservation of cellular fine structure. (A) *Chlamydomonas* cell prepared by routine room temperature chemical fixation. Organelles such as the nucleus and contractile vacuole appear extracted and collapsed, thus compromising the 3D organization of the cell. (B) Cells prepared by high-pressure freezing followed by freeze-substitution and embedding are well preserved, as indicated by the round nucleus and intact membranes of the contractile vacuole. The 3D integrity of the region containing the basal body complex is preserved. Scale bar = 500 nm.

Such effects distort the cell and compromise the 3D organization of the region where the basal body complex resides. The cell fixed by high-pressure freezing and freeze-substitution, on the other hand, appears well preserved, particularly in the apical region of the cell where the basal body complex is found (Fig. 1B).

To begin, whole cells grown in liquid cultures are used for high-pressure freezing, so the basal body complex is preserved within its cellular context. Cells are inoculated into 150-ml sterile flasks containing 50 ml of Sagar and Granick medium I (1953) supplemented with 8-mM sodium acetate (Preble *et al.*, 2001). The liquid cultures are then grown at 25 °C under constant illumination on top of a light box, generally for 2–3 days or until the cultures are in log phase growth.

In preparation for high-pressure freezing, aliquots of the liquid cultures are transferred to 50-ml conical centrifuge tubes and spun at 500 × *g* for 5 min. The resulting pellets are gently resuspended in a medium containing 150-mM mannitol, which acts as a cryoprotectant, minimizing the growth of ice crystals during freezing. Mannitol was chosen as it has very minor effects on cell viability compared to a variety of other cryoprotectants tested at similar molarities. The cells are then transferred to 1.5-ml Eppendorf tubes at room temperature for 1 h. The samples are spun again at 500 × *g* for 5 min and the supernatant decanted. The resulting loose cell pellet is then transferred to the bottom half of an aluminum sample planchette (Technotrade, Intl., Manchester, NH) and the top half is placed flat-side down on top of the sample. The sample is then transferred to the freezer transfer tool and rapidly frozen using a high-pressure freezer. The work described in this chapter has been performed with a BAL-TEC HPM-010 high-pressure freezer (Boeckeler Instruments, Inc., Tucson, AZ), although other models of high-pressure freezers are available (Section IV). After freezing, the two halves of the specimen-containing planchette are separated, and the samples are quickly transferred to liquid nitrogen and stored in those conditions until further use.

In preparation for electron microscopy (EM), the frozen cells are freeze-substituted at low temperature in a solvent that contains fixative, and then embedded in resin (Chapter 2 by McDonald, this volume). For freeze-substitution, the frozen samples are transferred under liquid nitrogen to cryovials (Nalge Nunc, Intl., Rochester, NY) containing 1 ml of a frozen, freeze-substitution cocktail. We have obtained good structural preservation of *Chlamydomonas* by freeze-substituting into 2% OsO_4 and 0.1% uranyl acetate in anhydrous acetone for 3 days at −90 °C, then gradually warming the sample to room temperature. Another method that works well to contrast membranes and MTs is to freeze-substitute for 3 days at −90 °C in 0.5% glutaraldehyde and 0.1% tannic acid in acetone, rinse in acetone, followed by freeze-substitution in 2% OsO_4 in acetone at −20 °C for 1 day, then warm to 4 °C. Before embedding, the samples are removed from the high-pressure freezer hats using the pointed end of a number 5 forceps and rinsed three times in acetone.

The freeze-substituted cells are then infiltrated with resin, using several graded dilutions of resin and acetone. We generally use Epon/Araldite but Spurr's resin can also be used. The samples are infiltrated for 60 min in each resin concentration at room temperature on a rocking platform, beginning at 3:1 (acetone:resin), then going to 1:1, 1:3, and 100% resin. Samples are then kept overnight in 100% resin. The final step involves transfer of the samples into BEEM capsules containing 100% resin with an accelerator (DMP30 for Epon/Araldite). The samples are then transferred to a 60 °C oven for polymerization.

B. Preparation of Sections for EM

Because the samples contain pellets of cells in random orientations, it is useful to collect serial sections. We typically trim a block face that is a wide trapezoid of small height to maximize the number of serial sections that will fit on each EM grid. Individual cell profiles containing the basal body complex can then be identified and tracked quickly through the serial sections. It is useful first to cut serial thin (50–70 nm) sections to examine the quality of preservation of a particular sample. We also use thin sections to collect serial images that help to document the phenotype of a mutant strain and can aid in the interpretation of tomographic data. For tomography, semithick (200–300 nm) sections are used, and the serial section ribbons are collected onto Formvar-coated [0.7% (w/v)] copper slot grids.

Prior to imaging in the EM, the sections must first be additionally stained with heavy metals. For thin sections, we typically use 2% uranyl acetate in 70% methanol for 5 min followed by Reynold's lead citrate for 3 min. Sections for tomography require longer staining times due to their increased section thickness. These semithick sections are stained for 8 min in 2% uranyl acetate in 70% methanol followed by Reynold's lead citrate for 5 min.

To prepare grids for tomography, alignment markers are placed on the top and bottom of the sections. Grids are floated on ~20 μl of a commercial solution of 15-nm colloidal gold particles (BBI, Intl., Cardiff, UK) for 5 min and then rinsed by dipping the grid in a drop of dH_2O. The grid is then inverted, and the process is repeated for the opposite side. The gold particles, affixed to each surface, will serve as fiducial markers for subsequent alignment of the tilted views necessary for ET (Ladinsky *et al.*, 1994; O'Toole *et al.*, 1999).

C. Image Acquisition and Tomographic Reconstruction

1. EM and Tilt-Series Acquisition

It is useful to examine thick sections first in a standard TEM operating at 100 kV to identify the profiles of cells that contain the basal body complex. These cells can be mapped through the serial sections by imaging at low magnification, and the map can then be used to locate the same cells in a higher voltage instrument where contrast is greatly reduced, but the resolution of detail in the thick section is improved.

For cellular ET, it is convenient to image a tilt series of thick (>200 nm) sections, and this requires an intermediate (200–300 kV) or high-voltage (750–1000 kV) EM equipped with a eucentric tilting stage. The increased acceleration voltage is important for allowing the beam electrons to penetrate the thick section without plural or inelastic scattering, especially at high tilts (60°–70°) where the section thickness doubles or nearly triples. In our laboratory, we currently record images using a Tecnai F30 EM operating at 300 kV. It is useful to have automatic image acquisition software to facilitate recording of the serial images. We use the SerialEM software package developed in our laboratory (Mastronarde, 2005) although other software packages are available (Section IV). The SerialEM software uses Digital Micrograph (Gatan, Inc., Pleasanton, CA) to capture images on a 2048 × 2048 pixel CCD camera (Gatan, Inc., Pleasanton, CA) at a pixel size of ~1 nm. To begin, the slot grid containing the serial sections is placed in a high-tilt holder, the region of interest is located, and the grid is tilted to 60° or 70°. Serial, tilted views of the section are then collected every degree over a ±60° or 70° range. After the first tilt series has been acquired, the grid is rotated 90° and a second tilt series is imaged. With the automated image acquisition software, it is possible to collect one or more dual-axis tilt series in <2 h.

2. Tomographic Reconstruction and Image Analysis

We carry out dual-axis tomographic reconstruction using the IMOD software package, as described in Kremer *et al.* (1996) and Mastronarde (1997) (see Section IV for a list of other tomography software packages). This package contains all of the programs needed to align serial tilt images, compute tomograms, combine dual-axis tomograms, and join serial section tomograms. The IMOD suite of programs is managed by a graphical user interface, eTomo, which allows the user to step through a series of process control buttons to start a particular program or set of programs and create a tomogram from the original tilt series (Fig. 2). The eTomo control window is much like a flow chart, with the user proceeding through each step from top to bottom. To begin, a *Pre-processing* step runs a program to eliminate pixels with extreme intensity values, which are created by x-rays during data collection. Next is a *Coarse Alignment* step, which runs a program to find a set of translational alignment transforms by cross correlation; these transforms are then applied to create the coarse-aligned stack. The next step in the series is *Fiducial Model Generation.* In this step, modeling software is used to mark the positions of 20–40 gold particles on an image taken at low tilt (Fig. 3A, open circles). The program will then track and model the positions of the same gold particles automatically in all other views (see Video Supplement 1). An iterative, *Fine Alignment* step is then carried out that uses the positions of the gold particles to create a set of transforms to align the serial, tilted views. In this step, the alignment program uses the positions of the gold particles and a least-squares approach to solve for all the tilt angles, magnification changes, shifts and rotations, and even image distortions that will best align the fiducials on the images from all the tilted views (Luther *et al.*, 1988). The final transforms are then applied to the images of the tilted views to create an aligned image stack.

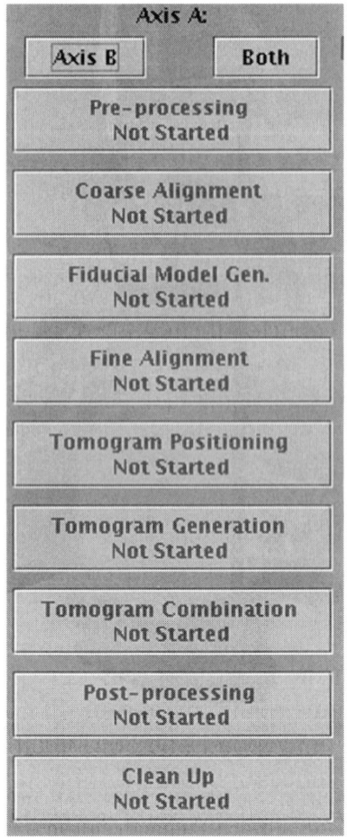

Fig. 2 Process control buttons in the eTomo main window allow the user to step through the procedures necessary to create a tomographic reconstruction. The process control buttons are arranged much like a flow chart that suggests the order of processing from top to bottom. When one of the buttons is selected, the right side of the window will fill in with information and fields associated with a specific process, allowing the user to modify the necessary parameters and execute specific programs required by that processing step.

A tomographic reconstruction is then computed using the density information in the aligned, tilted views. Because the tomograms can be quite large (>1 GB) and the reconstructions are not always flat or centrally located within the volume, it is useful to run the *Tomogram Positioning*. This step uses a 10-pixel subset of the data file to find the adjustments needed to compute tomograms that are as flat as possible and fit in the smallest volume possible. The final reconstruction is then computed in the *Tomogram Generation* step using an R-weighted back-projection algorithm (Gilbert, 1972). The entire procedure is then repeated after rotating the grid in its holder. This creates a tomogram of the same area constructed from a series of tilts about the orthogonal axis (axis B).

A limitation of ET derives from the fact that it is not possible to collect a full range of tilted views from a thick section (±90°). This constraint means that a wedge of information is missing from the reconstruction. This lack can in turn

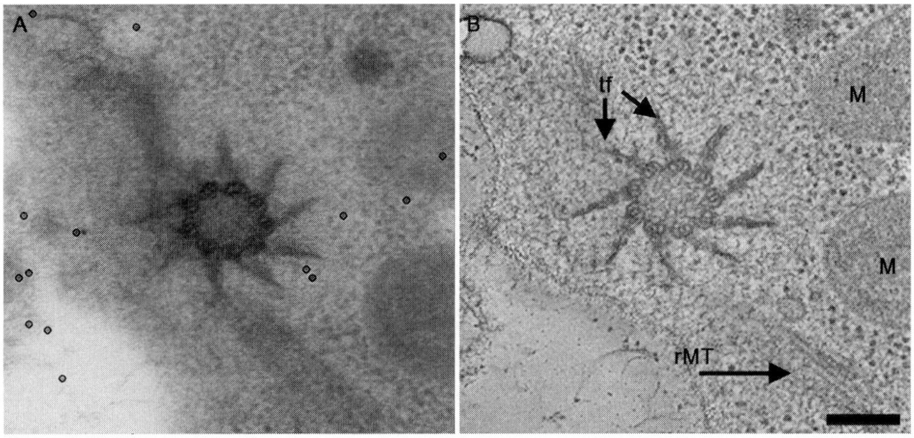

Fig. 3 (A) Zero-degree tilt projection of a 200-nm section that contains a portion of a basal body complex. Colloidal gold particles, used as alignment markers, are indicated with open circles. Not much detail is seen in this comparatively thick section because so much cellular material is superimposed in the projected image. (B) A selected 1.6-nm slice through the dual-axis tomographic reconstruction. Here, a tremendous amount of cellular fine structure can be seen, including details of the triplet MT blades that form the basal body, transitional fibers that radiate out from the triplet blades (tf), a rootlet MT bundle (rMT), and mitochondria (M). Scale bar = 200 nm. A movie of the aligned tilt series corresponding to Fig. 3A can be viewed in Video Supplement 1 and that of the complete tomogram can be viewed in Video Supplement 2 (http://bio3d.colorado.edu).

lead to artifacts that lower the resolution of the reconstruction, particularly along the direction perpendicular to the plane of the section. Our laboratory has developed a dual-axis ET approach to minimize the impact of the missing wedge (Mastronarde, 1997). After the two tomograms have been computed from tilt series around the "A and B axes," which are orthogonal, the *Tomogram Combination* step aligns them and combines their information into a single reconstruction. Figure 3B shows a selected 1.4-nm-thick tomographic slice computed from the combined reconstruction. A tremendous amount of cellular detail can now be seen, including details of the triplet MTs in the basal body and the transitional fibers attached to them, individual ribosomes, rootlet MTs, and mitochondria. A movie of serial, tomographic slices through the complete volume can be seen in Video Supplement 2. A *Post-processing* step is then run to trim the combined tomogram, apply a density scaling, and convert the file from integers to bytes, ultimately reducing the size of the reconstruction file. Finally, a *Clean Up* step is done to archive original image stacks and remove the intermediate files, which are large and no longer useful.

Because the basal body complex spans ~600 nm in its smallest dimension, it is not possible to reconstruct the entire complex from one semithick section; serial tomography is needed. Dual-axis tomograms from serial sections can be aligned and combined (Ladinsky *et al.*, 1994, 1999; Marsh *et al.*, 2001). Figure 4 shows

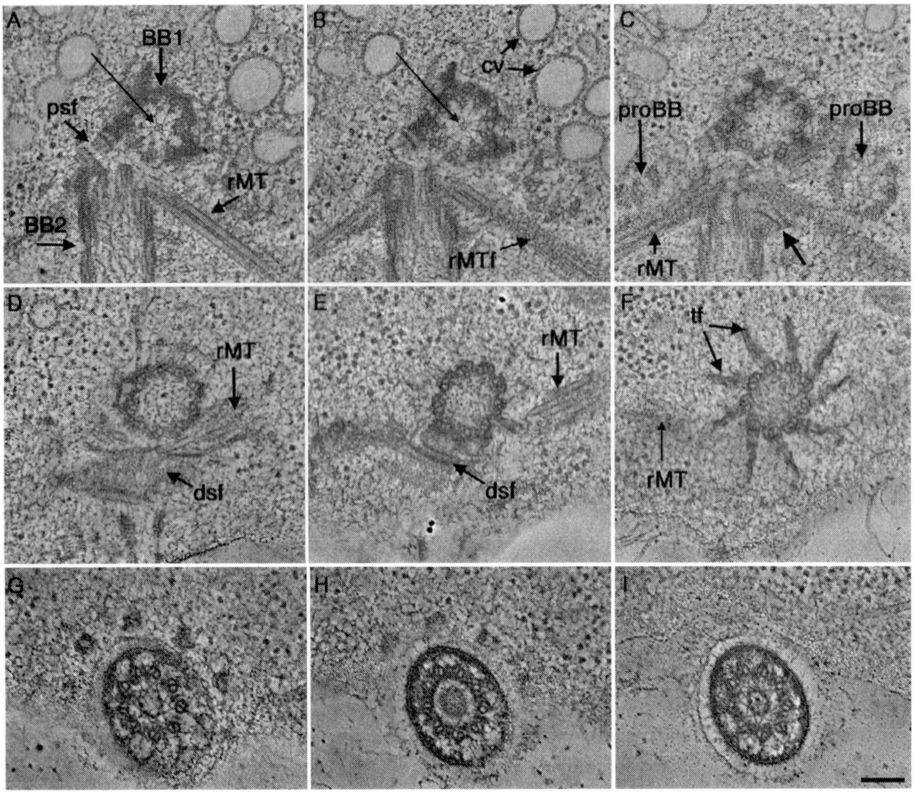

Fig. 4 Reconstructing a larger volume using serial section tomography. Selected 1.6-nm tomographic slices ranging from the proximal region of a basal body complex (A–C), through the distal region (D–F), and on to the transition zone (G–I). Features unique to the proximal basal body complex include a pinwheel structure in the basal body (A; arrow), proximal striated fibers (A; psf), probasal bodies (C; proBB). The distal region of the basal body includes a distal striated fiber (D, E; dsf) and transitional fibers (F; tf). Bundles of rootlet MTs can be seen traversing through the reconstructed volume (A–F; rMT). Distal to the basal body is the transition zone consisting of doublet MTs and assembly of 9-pointed stellar fibers (G–H). Further details of the structures found in the wild-type basal complex can be found in O'Toole *et al.* (2003a). Scale bar = 100 nm. A movie showing serial, tomographic slices through the complete reconstruction can be viewed in Video Supplement 3 (http://bio3d.colorado.edu).

selected tomographic slices from such a reconstruction that extends from the proximal base of the basal body through the "transition zone," which is a specialized region distal to the basal body (O'Toole *et al.*, 2003a). One mature basal body is seen in cross section (BB1) and the other in longitudinal view (BB2). The proximal pinwheel within the basal body (Fig. 4A, arrow), as well details of transitional fiber structure on the distal portion of the basal body, can be seen radiating from the triplet MT blades (Fig. 4F, tf). Other components of the basal body complex such as bundles of rootlet MTs (rMT), probasal bodies (proBB), and the proximal and distal striated fibers (psf, dsf) are evident. Distal to the basal body, the transition zone contains the transition from triplet to doublet MTs and an elaborate set of stellate fibers (Fig. 4G–I). A movie through the complete serial section tomogram can be viewed in Video Supplement 3. In this movie, breaks in the series indicate the transition from one tomographic reconstruction to the next. The 3D relationships among the many components of the basal body complex become evident when moving through the serial, tomographic slices (Video Supplement 3).

D. 3D Modeling of the Basal Body Complex

Modeling and analysis of tomographic data can be accomplished with the program 3dmod (Kremer *et al.*, 1996), although other software packages that accomplish these tasks are available (Section IV). MTs of the basal body and rootlet bundles can be tracked and modeled, and a projection of the model can be displayed to study the relationships of these organelles in 3D. Features of interest are modeled in the serial slices extracted from the tomogram. Because a tomographic slice is a sample through a 3D volume, its orientation can be rotated to obtain the best view of a particular organelle. To facilitate this process, 3dmod includes an image slicer window that displays a slice extracted from the 3D volume in any position or orientation (Fig. 5). With this feature, analysis of the morphology of MT ends in the basal body complex can be done by extracting a slice of image data one-voxel thick and adjusting its orientation to contain the axis of the MT in a single view. An example of this image feature can be seen in Fig. 5, and a simulation of the associated rotations can be seen in the Video Supplement 4.

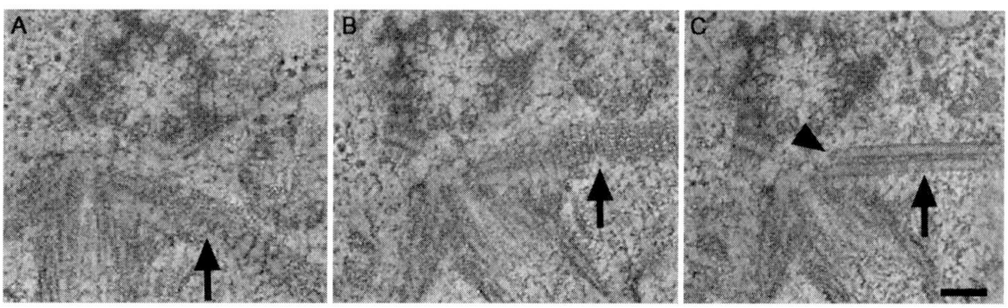

Fig. 5 Use of an image slicer tool to model features in MTOCs. (A) Single, tomographic slice showing a rootlet MT bundle that is running oblique to the plane of the section (arrow). (B) The tomographic slice was rotated so that the axis of the rootlet MT bundle is now horizontal (arrow). (C) The tomographic slice was then tilted about a y-axis to view the MTs in a single plane (arrow); image details, like the capped MT ends, can now be seen (arrowhead). Scale bar = 100 nm. A movie that simulates the use of the slicer tool can be viewed in Video Supplement 4 (http://bio3d.colorado.edu).

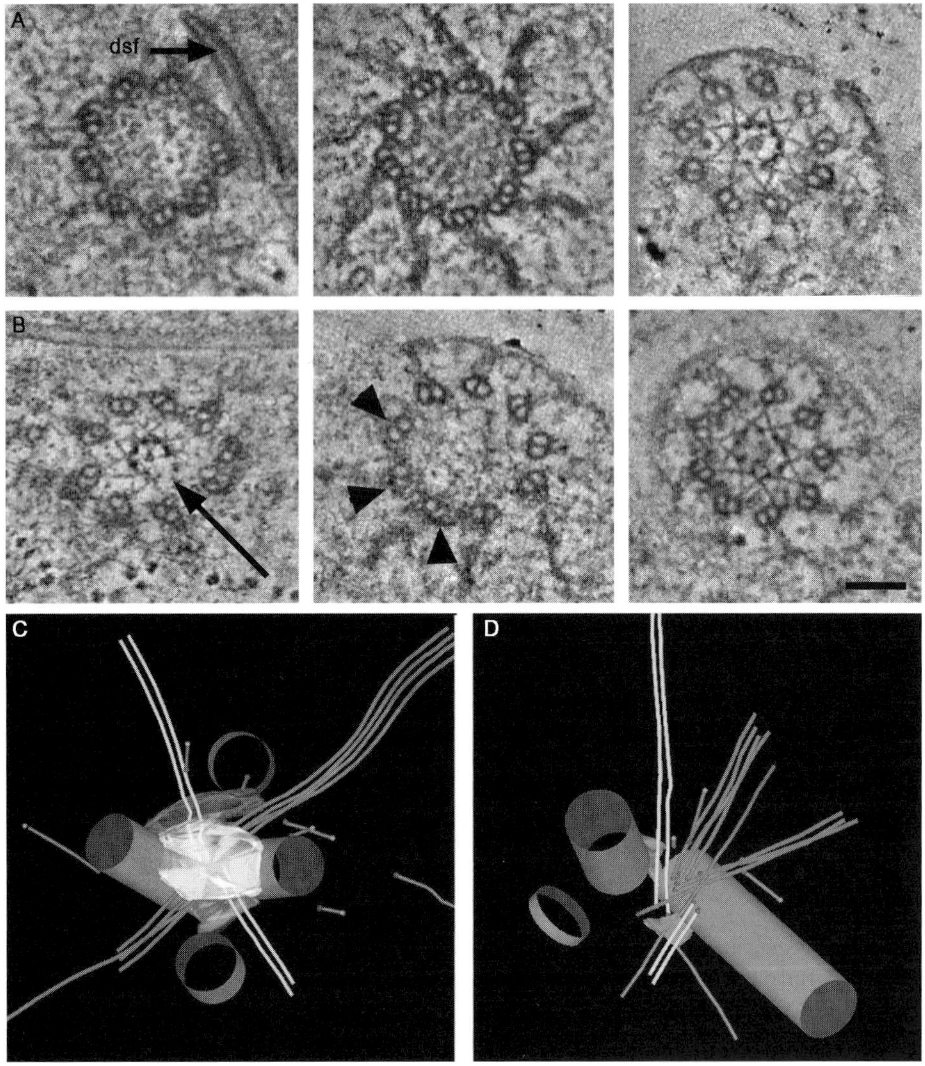

Fig. 6 Comparing the structure of the basal body complex in wild-type and mutant strains. (A) Wild-type basal bodies are formed from triplet MT blades. A distal striated fiber (dsf) connects to a subset of triplet MTs and transitional fibers can be detected on the distal portion (middle). The transition zone is formed from doublet MTs, and stellate fibers assemble a 9-pointed star (right). (B) The basal body from a δ-tubulin deletion strain assembles ectopic transition zone material within the basal body proper (left; arrow). The basal body is formed primarily from doublet MTs (left), although a subset of triplet MTs can be detected on the distal region of the basal body (middle; arrowheads). The mutant strain assembles a normal transition zone (right). (C) A 3D projection of features modeled from a serial section tomogram containing a complete wild-type basal body complex. Rootlet MTs (green, blue lines) form a cross-like arrangement with mature basal bodies (long red cylinders) and probasal bodies (short red cylinders).

1. Using Tomography to Compare Mutant and Wild-Type Structures

Figure 6 shows how serial tomography was used to step through the complete basal body complex from wild-type *Chlamydomonas* and a strain that contains a deletion of the δ-tubulin gene (Fig. 6; O'Toole *et al.*, 2003a). Stepping through serial, 1.4-nm tomographic slices allow the user to detect subtle, structural differences between mutant and wild-type strains. Selected tomographic slices from the wild-type cell show the presence of a distal striated fiber (dsf) and transitional fibers (tf) attached to the triplet MT blades of the basal body. The transition zone contains doublet MTs and a 9-pointed star. Basal bodies from the mutant strain, on the other hand, contain mostly doublet MTs, although a small set of triplets could be detected in the extreme distal region (Fig. 6B, middle; arrowheads). This mutant also displays ectopic transition zone material in the proximal region of the mutant basal body (Fig. 6B, left; arrow) over and above the more or less normal material seen in its transition zone (Fig. 6B, right). The latter resembles the transition zone in wild type (Fig. 6A, right).

MTs of the basal bodies, rootlet MTs, and contours marking the position of the fiber systems can also be modeled, and projections of the model can be displayed and rotated to study its 3D geometry. Structural differences between the two strains become obvious in the projected models shown in Fig. 6C and D. The basal body complex from the wild-type cell (Fig. 6C) is organized with the rootlet MTs (green and blue lines) arranged in a cross-like array with mature basal bodies and the probasal bodies (red) centered between the rootlet MT bundles. The proximal striated fibers (yellow) and the distal striated fiber (pink) connect the mature basal bodies. Several cytoplasmic MTs (magenta lines) can also be detected in the vicinity of the basal body complex.

In contrast, the model of the basal body complex from the δ-tubulin deletion strain shows dramatic organizational differences from wild type (Fig. 6D). Rootlet MT bundles (green and blue lines) are not organized in the characteristic cross-like arrangement and the probasal bodies are misplaced (short red cylinders). Two proximal striated fibers (yellow) connect the mature basal bodies at their proximal base, but the distal striated fiber is markedly absent. By modeling features through 600 nm of the serial tomograms, the 3D relationships of the basal body complex can readily be appreciated, and dramatic differences between mutant and wild-type strains become evident.

Proximal striated fibers (yellow) and a distal striated fiber (pink) connect the mature bodies at their proximal and distal ends, respectively. Several cytoplasmic MTs (magenta lines) can also be detected. (D) Projected 3D model of a basal body complex from a δ-tubulin deletion mutant shows a disorganization of rootlet MT bundles and the absence of a distal striated fiber. Probasal bodies are misplaced. Scale bar = 100 nm. (See Color Insert.)

IV. Instrumentation and Materials

A. High-Pressure Freezing and Freeze-Substitution

1. Instrumentation

Due to the thickness of a *Chlamydomonas* cell, a high-pressure freezer is needed to freeze the samples without visible ice crystal damage. Models available are the BAL-TEC HPM-010, the Leica EM PACT2 (Leica Microsystems, Inc., Bannockburn, IL), and the HPF Compact 01 (Technotrade, Intl., Manchester, NH).

Automated freeze-substitution devices are available including the Leica AFS, but homemade devices can also be used; see Chapter 2 by McDonald, this volume, for specific details.

2. Materials

Sagar and Granick medium I (1953) supplemented with 8-mM sodium acetate (Rich media) for growing cells.

Aluminum high-pressure freezer planchettes
Clinical centrifuge
50-ml conical centrifuge tubes
1.5-ml Eppendorf tubes
Pipettes
Medium with 150-mM mannitol prior to freezing
Nunc cryovials
2% OsO_4, 0.1% uranyl acetate in acetone
EM grade acetone
Epon/Araldite resin
BEEM capsules

B. Electron Tomography

1. Instrumentation

ET of semithick (200–400 nm) sections requires imaging in a microscope operating at intermediate voltage (200 or 300 kV) or high voltage (750–1000 kV). There are currently three NIH-supported National Resource Research Facilities in the United States that are set up with the equipment and staff to help users do tomography (Boulder Laboratory for 3D Electron Microscopy of Cells, University of Colorado, Boulder, http://bio3d.colorado.edu; National Center for Microscopy and Imaging Research, University of California, San Diego, www-ncmir.uscd.edu; and the Wadsworth Center, Albany, NY, http://www.wadsworth.org).

Automated image acquisition programs are available from commercial vendors [FEI Company, Eindhoven, The Netherlands; Tietz Video and Image Processing Systems (TVIPS), Gauting, Germany; and Gatan, Inc., Pleasanton, CA]. Several software packages are freely available and include TOM (http://www.biochem.mpg.de/tom), UCSF Tomography (http://msg.ucsf.edu/tomography), and Serial EM (http://bio3d.colorado.edu/SerialEM).

2. Materials

Copper/rhodium slot grids

0.7% (w/v) Formvar in ethylene dichloride

2% uranyl acetate in 70% methanol; rinse solutions of 70%, 50%, 20% methanol

Reynold's lead citrate

10- or 15-nm colloidal gold

C. Tomographic Reconstruction and Modeling

Our laboratory has developed the IMOD software package, which contains all of the programs needed for calculating tomograms and for the display and modeling of features within the reconstruction. This software package is freely available on our Web site (http://bio3d.colorado.edu), along with useful tutorials and guides for its use. The IMOD software package runs on multiple platforms, including Linux, Mac OSX, and Windows. The programs used for tomographic reconstruction are managed by a graphical user interface eTomo. The eTomo interface facilitates the ease with which users step through the various steps of the process, much like a flow chart. A tutorial on how to calculate a sample data set containing a *Chlamydomonas* basal body is available on our Web site (http://bio3d.colorado.edu). Image display and modeling is carried out with the 3dmod viewing program from the IMOD software package. This program can be run by command lines; it contains windows for image display, the slicer tool for rotating slices of image data and for modeling features of interest in the reconstruction.

Other software packages developed elsewhere are also freely available. These include EM3D (http://em3d.stanford.edu), TOM (http://www.biochem.mpg.de/tom), SPIDER (http://www.wadsworth.org), and XMIPP (http://xmipp.cnb.csic.es/).

V. Discussion

ET is a useful method for addressing important structure–function relationships in cell biology. Its strengths include a significant increase in 3D resolution compared with light microscopy and a sensitivity to detect fine-scale structures that are often obscured in standard thin sections for EM. By modeling cellular organelles in 3D, differences that were not anticipated can be revealed. However, tomography is not without limitation. The costs in both the time of the operator and money for the hardware can be substantial. The time-intensive nature of image acquisition, computation, and analysis makes the sample sizes relatively small, thus limiting extensive quantitative analyses. ET is therefore best suited for questions that would benefit from fine-scale resolution over relatively small 3D volumes.

The methods described in this chapter offer a unique approach for studying MTOC structure and function. The unicellular alga, *Chlamydomonas*, was used as a model system to study the basal body complex. The cell wall of *Chlamydomonas* is a natural barrier to diffusion, so the cells are prone to fixation artifacts when conventional chemical fixations are used. Details of our improved method of specimen preparation involving rapid freezing followed by freeze-substitution are outlined; they illustrate how to achieve excellent preservation of cellular fine structure, including the 3D architecture of the basal body complex. Use of this method has confirmed and extended the original observations on the basal body complex put forth in the landmark papers by others (Cavalier-Smith, 1974; Johnson and Porter, 1968; Ringo, 1967).

The basal body complex, like other MTOCs, is a 3D organelle, so tomography is a useful method to study its complex structure. In addition, serial section tomography enables fine-scale resolution over a larger cellular volume, to reconstruct the entire basal body complex within an ~600-nm volume. The slicer tool was used to unambiguously model and track features of interest that change over small distances. This is especially useful when tracking complex arrays of MTs that originate from the MTOC.

Mutations that cause defects in the basal body proper may also lead to additional 3D defects that are likely downstream from the original mutation. Strains of *Chlamydomonas* that carry a deletion or a reduction in the amount of δ-tubulin have been shown to contain primarily doublet MTs, rather than triplets, in their basal bodies (Dutcher and Trabuco, 1998; Garreau de Loubresse *et al.*, 2001). However, by modeling the whole basal body complex in 3D, additional defects can be observed (O'Toole *et al.*, 2003a). For example, ectopic assemblies of transition zone structures were identified because the serial tomographic slices sectioned through the entire basal body complex identified the material within the proximal region of the basal body as well as in the transition zone. The 1.4-nm tomographic slices contained the resolution necessary to detect a small subset of triplet MT blades at the distal tip of the mutant basal body. Reconstructing the complete volume of the basal body complex and modeling its components made it clear that in the mutant, the distal striated fiber was not assembled onto the basal body. The rootlet MT bundles are normally anchored under this fiber (Ringo, 1967) and its absence results in a disorganization of the rootlet bundles and a mispositioning of the probasal bodies. The use of ET is therefore of value for identifying 3D structural defects in mutant strains and establishes another phenotypic dimension in MTOC structure–function studies.

VI. Summary

The unicellular alga, *Chlamydomonas*, is a useful model system in which to study MTOCs because of the powerful genetic and molecular techniques available for this organism. Recent advances in specimen preparation that involve rapid freezing followed by freeze-substitution result in close-to-optimal preservation of the basal body complex. Coupling genetics with 3D structure studies of optimally preserved cells further enhances our understanding of basal body assembly and function. Finally, these methods can be applied to a wide range of biological systems to address structure–function relationships in other organisms.

Acknowledgments

We thank David Mastronarde for developing the automated image capture software and for helpful advice on tomographic reconstruction. We thank Andrew Staehelin for use of the high-pressure freezer and Mary Morphew for useful suggestions for specimen preparation. This work was supported by Grant RR-000592 from the National Center for Research Resources of the National Institutes of Health to J. R. McIntosh.

References

Adams, G. M. W., Wright, R. L., and Jarvik, J. W. (1985). Defective temporal and spatial control of flagellar assembly in a mutant of *Chlamydomonas reinhardtii* with variable flagellar number. *J. Cell Biol.* **100,** 955–964.

Baumeister, W. (2002). Electron tomography: Towards visualizing the molecular organization of the cytoplasm. *Curr. Opin. Struct. Biol.* **12,** 679–684.

Bullitt, E., Rout, M. P., Kilmartin, J. V., and Akey, C. W. (1997). The yeast spindle pole body is assembled around a central crystal of Spc42p. *Cell* **89,** 1077–1086.

Cavalier-Smith, T. (1974). Basal bodies and flagellar development during the vegetative cell cycle and the sexual cycle of *Chlamydomonas reinhardtii*. *J. Cell Sci.* **16,** 529–556.

Doxsey, S., McCollum, D., and Theurkauf, W. (2005). Centrosomes in cellular regulation. *Annu. Rev. Cell Dev. Biol.* **21,** 411–434.

Dutcher, S. K. (1995). Purification of basal bodies and basal body complexes from *Chlamydomonas reinhardtii*. *Methods Cell Biol.* **47,** 323–324.

Dutcher, S. K. (2003). Elucidation of basal body structure and centriole function in *Chlamydomonas reinhardtii*. *Traffic* **4,** 443–451.

Dutcher, S. K., Morrissette, N. S., Preble, A. M., Rackley, C., and Stanga, J. (2002). Epsilon tubulin is an essential component of the centriole. *Mol. Biol. Cell* **13,** 3859–3869.

Dutcher, S. K., and Trabuco, E. C. (1998). The *UNI3* gene is required for assembly of basal bodies of *Chlamydomonas* and encodes delta-tubulin, a new member of the tubulin superfamily. *Mol. Biol. Cell* **9,** 1293–1308.

Frank, J. (1992). Introduction: Principles of electron tomography. *In* "Electron Tomography: Three Dimensional Imaging with the Transmission Electron Microscope" (J. Frank, ed.), pp. 1–13. Plenum Press, New York.

Garreau de Loubresse, N., Ruiz, F., Beisson, J., and Klotz, C. (2001). Role of delta-tubulin and the C-tubule in assembly of *Paramecium* basal bodies. *BMC Cell Biol.* **2,** 4–10.

Gilbert, P. F. C. (1972). The reconstruction of a three-dimensional structure from projections and its application to electron microscopy. II. Direct Methods. *Proc. R. Soc. Lond. Bio. Sci.* **182,** 89–102.

Goodenough, U. W., and St. Clair, H. S. (1975). *BALD-2*: A mutation affecting the formation of doublet and triplet sets of MTs in *Chlamydomonas reinhardtii. J. Cell Biol.* **66,** 480–491.

Johnson, U. G., and Porter, K. R. (1968). Fine structure of cell division in *Chlamydomonas reinhardtii.* Basal bodies and MTs. *J. Cell Biol.* **38,** 403–425.

Kremer, J. R., Mastronarde, D. N., and McIntosh, J. R. (1996). Computer visualization of three-dimensional image data using IMOD. *J. Struct. Biol.* **116,** 71–76.

Ladinsky, M. S., Kremer, J. R., Furcinitti, P. S., McIntosh, J. R., and Howell, K. E. (1994). HVEM tomography of the trans-Golgi network: Structural insights and identification of a lace-like vesicle coat. *J. Cell Biol.* **127,** 29–38.

Ladinsky, M. S., Mastronarde, D. N., McIntosh, J. R., Howell, K. E., and Staehelin, L. A. (1999). Golgi structure in three dimensions: Functional insights from the normal rat kidney cell. *J. Cell Biol.* **144,** 1135–1149.

Lucic, V., Forster, F., and Baumeister, W. (2005). Structural studies by electron tomography: From cells to molecules. *Annu. Rev. Biochem.* **74,** 833–865.

Luther, P. K., Lawrence, M. C., and Crowther, R. A. (1988). A method for monitoring the collapse of plastic sections as a function of electron dose. *Ultramicroscopy* **24,** 7–18.

Marsh, B. J., Mastronarde, D. N., Buttle, K. F., Howell, K. E., and McIntosh, J. R. (2001). Organellar relationships in the Golgi region of the pancreatic beta cell line, HIT-T15, visualized by high resolution electron tomography. *Proc. Natl. Acad. Sci. USA* **98,** 2399–2406.

Mastronarde, D. N. (1997). Dual-axis tomography: An approach with alignment methods that preserve resolution. *J. Struct. Biol.* **120,** 343–352.

Mastronarde, D. N. (2005). Automated electron microscope tomography using robust prediction of specimen movements. *J. Struct. Biol.* **152,** 36–51.

McIntosh, J. R., Nicastro, D., and Mastronarde, D. (2005). New views of cells in 3D: An introduction to electron tomography. *Trends Cell Biol.* **15,** 43–51.

Moritz, M., Braunfeld, M. B., Fung, J. C., Sedat, J. W., Alberts, B. M., and Agard, D. A. (1995a). Three-dimensional structural characterization of centrosomes from early *Drosophila* embryos. *J. Cell Biol.* **130,** 1149–1159.

Moritz, M., Braunfeld, M. B., Sedat, J. W., Alberts, B., and Agard, D. A. (1995b). Microtubule nucleation by γ-tubulin-containing rings in the centrosome. *Nature* **378,** 638–640.

O'Toole, E. T., Giddings, T. H., McIntosh, J. R., and Dutcher, S. K. (2003a). Three-dimensional organization of basal bodies from wild-type and d-tubulin deletion stains of *Chlamydomonas reinhardtii. Mol. Biol. Cell* **14,** 2999–3012.

O'Toole, E. T., McDonald, K. L., Mantler, J., McIntosh, J. R., Hyman, A. A., and Mueller-Reichert, T. (2003b). Morphologically distinct microtubule ends in the mitotic centrosome of *Caenorhabditis elegans. J. Cell Biol.* **163,** 451–456.

O'Toole, E. T., Winey, M., and McIntosh, J. R. (1999). High voltage electron tomography of spindle pole bodies and early mitotic spindles in the yeast *Saccharomyces cerevisiae. Mol. Biol. Cell* **10,** 2017–2031.

Ozlu, N., Srayko, M., Kinoshita, K., Habermann, B., O'Toole, E. T., Muller-Reichert, T., Schmaltz, N., Desai, A., and Hyman, A. A. (2005). An essential function of the *C. elegans* ortholog of TPX2 is to localize activated aurora A kinase to mitotic spindles. *Dev. Cell* **9,** 237–248.

Preble, A. M., Giddings, T. H., and Dutcher, S. K. (2000). Basal bodies and centrioles: Their function and structure. *Curr. Top Dev. Biol.* **49,** 207–233.

Preble, A. M., Giddings, T. H., and Dutcher, S. K. (2001). Extragenic bypass supressors of mutations in the essential gene *BLD2* promote assembly of basal bodies with abnormal microtubules in *Chlamydomonas reinhardtii. Genetics* **157,** 163–181.

Ringo, D. L. (1967). Flagellar motion and fine structure of the flagellar apparatus in *Chlamydomonas. J. Cell Biol.* **33,** 543–571.

Sagar, R., and Granick, S. (1953). Nutritional studies with *Chlamydomonas reinhardtii*. *Ann. NY Acad. Sci.* **56,** 831–838.

Salisbury, J. L., Baron, A. T., and Sanders, M. A. (1988). The centrin-based cytoskeleton of *Chlamydomonas reinhardtii*: Distribution in interphase and mitotic cells. *J. Cell Biol.* **107,** 635–641.

Schnackenberg, B. J., Khodjakov, A., Rieder, C. L., and Palazzo, R. E. (1998). The disassembly and reassembly of functional centrosomes *in vitro*. *Proc. Natl. Acad. Sci. USA* **16,** 9295–9300.

Silflow, C. D., LaVoie, M., Tam, L.-W., Tousey, S., Sanders, M., Wu, W., Borodovsky, M., and Lefebvre, P. A. (2001). The Vfl1 protein in Chlamydomonas localizes in a rotationally asymmetric pattern at the distal ends of the basal bodies. *J. Cell Biol.* **153,** 63–74.

Taillon, B. E., Adler, S. A., Suhan, J. P., and Jarvik, J. W. (1992). Mutational analysis of centrin: An EF-hand protein associated with three distinct contractile fibers in the basal body apparatus of *Chlamydomonas*. *J. Cell Biol.* **119,** 1613–1624.

Vogel, J. M., Stearns, T., Reider, C. L., and Palazzo, R. E. (1997). Centrosomes isolated from *Spisula solidissima* oocytes contain rings and an unusual stoichiometric ratio of alpha/beta tubulin. *J. Cell Biol.* **137,** 193–202.

Wright, R. L., Chojnacki, B., and Jarvik, J. W. (1983). Abnormal basal-body number, location, and orientation in a striated fiber-defective mutant of *Chlamydomonas reinhardtii*. *J. Cell Biol.* **96,** 1697–1707.

Wright, R. L., Salisbury, J., and Jarvik, J. W. (1985). A nucleus-basal body connector in *Chlamydomonas reinhardtii* that may function in basal body localization or segregation. *J. Cell Biol.* **101,** 1903–1912.

Yoder, T. J., McElwain, M. A., Francis, S. E., Bagley, J., Muller, E. G., Pak, B., O'Toole, E. T., Winey, M., and Davis, T. N. (2005). Analysis of a spindle pole body mutant reveals a defect in biorientation and illuminates spindle forces. *Mol. Biol. Cell* **16,** 141–152.

CHAPTER 6

Whole-Cell Investigation of Microtubule Cytoskeleton Architecture by Electron Tomography

J. L. Höög and C. Antony

Electron Microscopy Facility, Cell Biology and Biophysics Program
European Molecular Biology Laboratories, 69117 Heidelberg, Germany

Understanding the three-dimensional (3D) architecture of whole cells by electron microscopy (EM) is a significant challenge in contemporary cell biology. Electron tomography (ET) and computing sciences have been evolving rapidly to now make possible large-scale 3D reconstructions at EM resolution. The fission yeast *Schizosaccharomyces pombe* is our cellular model for this work because of its smaller size and relatively simple organization. These features provide an ideal situation in

0091-679X/07 $35.00
DOI: 10.1016/S0091-679X(06)79006-9

which to develop an ET approach to the study of its entire microtubule (MT) array in interphase. Here, we report methods and materials that can be combined in order to perform cellular ET on comparatively large volumes. In particular, we focus on specific adaptations, like montage imaging in the plane of each section and the acquisition of serial tomograms that can be joined to encompass the full cell volume, including its MT bundles. A new screening method is introduced that helps in finding organelles of interest, even when they are not visible in single images of the thick sections. Further, instructions for constructing precise 3D models with the IMOD software package are given, with particular emphasis on representation of microtubular elements. Finally, we describe how IMOD can be used efficiently to perform quantitative analysis of the subcellular elements depicted in the 3D models. This chapter should provide a guiding frame for the reader who intends to investigate particular cell features that should be understood as 3D structures viewed with good resolution (~5 nm) within large volumes in cells (~100 μm^3).

I. Introduction

Electron microscopy (EM) studies of cells and of macromolecular complexes within cells are currently undergoing a huge transformation. The improvements of both the methods and the instrumentation for EM are now allowing subcellular structures and organelles to be characterized in three dimension (3D) with unprecedented detail and reliability (McIntosh *et al.*, 2005; O'Toole *et al.*, 1999). This is in part the contribution of cellular electron tomography (ET), a method which can produce 3D reconstructions of fixed and plastic embedded cells or frozen hydrated samples (Baumeister *et al.*, 1999; Frank, 1992). Here, we will focus only on plastic embedded cells with the aim of studying cytoplasmic microtubule (MT) networks. Because such networks often extend throughout a cell, we have had to develop an approach that can reconstruct a large fraction of a cell's volume and that enables us to track MTs through the entirety of the volume under study. The method we have developed is a combination of serial thick sections and systematic montaging during the acquisition of images from each section. This has allowed us to collect information from a large fraction of, if not the full cell volume, these comparatively small cells. The ability to align and stitch together tomograms from semithick serial sections was crucial for this method. Finally, the success of this large-scale approach was greatly helped by the fact that the relevant software packages have become increasingly user friendly.

The fission yeast (*Schizosaccharomyces pombe*) has been our organism because it is suitable as a model system for the study of many aspects of MT-dependent cell processes. The MT dynamics of these cells have been described in depth by light microscopy (LM) (Asakawa *et al.*, 2006; Brunner and Nurse, 2000; Drummond and Cross, 2000; Janson *et al.*, 2005; Sawin, 2004). However, the detailed internal structure of interphase MT bundles cannot be resolved by LM. Here, we describe how cryofixed samples reconstructed by ET provide important details, such as the positions of MT ends and features of their structure, which reveal the polarity of the individual components of an MT bundle. Even finer details, like molecular complexes that crossbridge and bundle MTs, were also resolved. The strength of ET combined with a biological model of small size, like the fission yeast, is to provide reliable and precise reconstructions of subcellular structures and organelles within a large fraction, or even the entirety, of a cell's volume. Nevertheless, the usefulness of the method is limited by the sampling size (*n*), which, given the labor required for serial section ET, will always be less than the large number of cells allowed in LM studies. We describe here how one can make good use of ET to investigate complete cell structure at a comparatively large scale. We predict that the use of ET will spread in the near future and that its contributions to cell biology, as well as in developmental biology, will strengthen our understanding of subcellular architecture.

II. Rationale

MT bundles have traditionally been studied by thin serial section reconstructions of pictures taken from MTs imaged in cross sections. However, this method provided us with an unreliable picture of interphase MTs in *S. pombe* since these cytoplasmic fibers, in contrast to mitotic spindles, are not arranged in parallel bundles (Ding *et al.*, 1993). The interphase MT arrays (IMAs) of fission yeast bend and diverge; consequentially they are hard to track by the serial section reconstruction method. We have found that ET gives a more accurate picture of the cytoskeletal arrangement when one is working with such irregular structures. Additionally, ET provides more detailed information than serial thin sections, since its 3D resolution is essentially isotropic, while that of serial thin sections is limited along the dimension perpendicular to the image by the section thickness (usually >40 nm). The good, isotropic resolution of the tomograms allows one to determine MT polarity and visualize electron-dense protein complexes that are associated with the MTs. Further, interactions between MTs and other organelles, or indeed any ultrastructural features, can be studied in the resulting tomograms. Thus, the labor spent, although significant, yields much valuable information about the entire cell architecture.

III. Methods

A. Sample Preparation

1. Cell Culture

Fission yeast strains are best maintained by well-established methods that have already been described (Moreno *et al.*, 1991). Briefly, yeast cells can be stored in a glycerol-containing medium at –80 °C for years, then awakened by cultivation on a plate of rich medium and incubated at 25–32 °C for 1–4 days. A liquid preculture (10 ml) of a standard rich medium (YE5S) is then inoculated with a small amount of cells from the initial plate. This preculture is placed at 25–32 °C in an incubator with constant agitation. The preculture is then used to inoculate a larger culture, taking into account the generation time of the strain that will be studied (usually about 3 h). The optical density (OD at 595 nm) is checked and should be between 0.25 and 0.75 for cells to be in log-phase growth. Cell cryoimmobilization is best performed when the culture has an OD between 0.5 and 0.75, since this higher density gives better performance during the harvesting of cells for rapid freezing, thus optimizing freezing quality (see below). Cells can also be cultivated in the defined Edinburgh minimal medium (EMM). However, cells usually grow more slowly in defined media, and they often show some morphological alterations, for example, more vacuolar structures within the cells.

2. High-Pressure Freezing

As ET of subcellular organization is a multistep procedure (i.e., expensive and time consuming, see Table I), it is of great importance that specimens are initially preserved as well as possible. Thus, cryoimmobilization by high-pressure freezing (HPF) is highly recommended when possible. HPF freezes biological specimens up to several hundred microns thick under a pressure of over 2000 bar. Its virtues and limitations are discussed in Chapter 2 by McDonald and Chapter 3 by Hess, this volume. Here, we will say simply that the instrument works well for the preservation of fission yeast when employed as described below, and for us it was the method of choice. At the time of freezing, 15 ml of cell culture at OD ~ 0.5 (at 595 nm) is filtered to a paste using a Millipore 15-ml filtration setup with a 25-mm polycarbonate circular filter (0.4-μm pore size). Cells are then gently scraped from the filter using a toothpick, and the paste is loaded into a specimen carrier for whatever HPF is available (BAL-TEC 200-μm-deep carriers or Leica slot carriers). A syringe needle bent at 45° is used to flatten the cell paste contained within the carrier. The carriers are then closed and mounted on the specific specimen holder of the HPF machine and frozen.

3. Freeze-Substitution

Freeze-substitution (FS) is a valuable and common way to process biological samples following HPF cryoimmobilization (Steinbrecht and Müller, 1987). During FS, the water is slowly substituted for an organic solvent at low temperature (around –90 °C) for a long time (2–3 days). This method performs cell dehydration with minimal cellular damage (Giddings *et al.*, 2001; Steinbrecht, 1993; Steinbrecht and Müller, 1987). Shrinkage or collapse of cell structures would inevitably occur at room temperature. Yeast cells are best freeze-substituted in acetone supplemented with some fixatives (McDonald and Morphew, 1993; Chapter 2 by McDonald,

Table I
Steps in Producing a Large Tomogram Reconstruction (White Boxes) and the Software Needed (Black Boxes)

Sequence of experiments	Associated software
Cell culture	
Cryofixation and embedding	
Serial sectioning	
Contrasting and fiducial gold application	
Acquisition	SerialEM
Tomogram reconstruction	eTOMO
Joining of serial tomograms	eTOMO
3D model construction	3dmod, 3dmodv, mtsmooth, smoothsurf
Data analysis	mtk, nda, imodinfo

this volume). The water content of the cells is extracted in the FS solution at –90 °C for 48–72 h in a Leica AFS device (Section IV).

Following cryoimmobilization, the carriers containing the frozen fission yeast are placed into a dehydrated acetone FS solution supplemented with 0.1% dehydrated glutaraldehyde (GA), 0.25% uranyl acetate (UA), and 0.01% osmium tetroxide (OsO_4). This mixture gave good results for visualizing MT arrays in yeast cells (Müller-Reichert *et al.*, 2003); however, the composition of the FS solution can be varied to some extent, as we also obtained very good SPBs and MTs preservation in fission yeast using an FS solution made of 0.5% dehydrated GA, 0.5% UA, and 1% OsO_4. Surprisingly, membrane contrast can be well improved by adding 1% water, or more, to the FS solution (Walther and Ziegler, 2002).

After the dehydration step, the temperature is progressively raised to –45 °C with an increment of 5–10 °C/h. The carriers are then rinsed at this temperature with pure dehydrated acetone and placed together with their container into the chamber of a cryoultramicrotome set at –45 °C in order to separate the pellets from the carriers with a syringe needle. Using a cryoultramicrotome for this operation makes this step more comfortable, as one can benefit from the microscope on the cryo-attachment. The free pellets are then placed back into the plastic tubes, which are transferred back into the AFS machine and further processed by progressive infiltration with resins. In the case of Lowicryl (HM20) embedding, several steps are required to start the infiltration at –45 °C: acetone/Lowicryl 3:1 for 60 min, 1:1 for 2–3 h, 1:3 for 2–3 h, pure Lowicryl 2 times for 60 min, then overnight. Finally, a last resin change is made before starting polymerization by the action of UV light. For the preparation of Epon blocks, the resin is infiltrated in the same way after rinsing with pure acetone, but this is done around –30 °C or even at room temperature.

Lowicryl polymerization is started at –45 °C for 48 h and is continued during the rise of temperature (10 °C/h) up to room temperature, where the blocks are left for another 12 h under the UV lamp. For Epon blocks, the polymerization is made overnight in an oven set at 60 °C.

4. Choice of FS Resin

Advantages of acrylic resins, like Lowicryl (HM20), include their low freezing point as well as their waterlike viscosity even at –35 °C (Griffiths, 1993). Therefore, embedding can be started at very low temperatures when the proteins are best stabilized and the samples are optimally preserved (see Chapter 2 by McDonald and Chapter 3 by Hess, this volume). The resulting blocks have very good sectioning properties when one uses a diamond knife with a 35° included angle. In our study, we found the shrinkage of Lowicryl to be lesser in general as compared with Epon which is known to shrink around 30% when exposed to the electron beam (Braunfeld *et al.*, 1994; Carlemalm *et al.*, 1982). Further, we have found Lowicryl to be slightly less electron dense than Epon. As such, it gives better contrast and permits thicker sections for ET.

5. Serial Sectioning

Serial sections 200- to 250-nm thick are ideal for tomography when using a microscope that operates at 200 or 300 kV. A plastic block is trimmed to make the block face a trapezoid with a low height (base = ~300 μm and h = ~100 μm). Serial sections are cut from the plastic block on an ultramicrotome using a 35° diamond knife. In case the consecutive sections do not attach to each other to form a consistent ribbon, a solution made from a droplet of rubber cement dissolved in xylene (ca. 1:20 dilution) may be applied to the block face. After this adhesive is dried, newly made sections should attach to each other better. Chloroform vapors (from a small ball of cotton moved over the sections) help the sections to flatten if they have wrinkles.

As the large volume approach requires serial sections, slot grids with a large central hole (2 × 1 mm^2) are preferable. These grids will be coated with a rather thick film of Formvar (90–100 nm, which shows a golden interference color); this provides adequate section stability over the entire hole in the grid. Fresh Formvar-coated grids should be made and should not be carbon-coated, as the films become more fragile when old and/or carbon coated.

To align the sections to the middle of the grid, we recommend attaching a long hair diagonally across the knife boat. One end of the hair is attached with tape to the knife's external wall; the other end is attached at the middle of the back edge of the pool. As ribbons of sections come off the knife's edge, they can be parked along the hair, gently aligned with the axis of the slot grid, and therefore easily picked up from underneath. This allows the microtomist to place the ribbon in a precise and controlled way on the grid.

6. Contrasting and Application of Fiducial Markers for Tilt Alignment

Lowicryl sections are well contrasted by staining grids with 2% UA in 70% methanol for 4 min, followed by a lead citrate stain (Reynolds solution) for 1 min. Longer staining darkens the cells within the sections without increasing their contrast. For Epon sections methanolic or aqueous UA contrasting solutions are suitable, and these can be applied to the sections for a longer time (up to 15 min). Lead citrate contrasting is again required, as with Lowicryl sections.

To make a tomogram, one needs a series of images of the same area of the specimen, taken over a wide range of tilts, typically $\pm 60°$ in increments of $1°$. These images must then be accurately aligned. To facilitate this process, we apply gold particles to both sides of the grid after contrasting; these can act as fiducial points to bring the multiple tilted views into alignment. Depending on which magnification is used for acquiring the images (14,500× to 25,000×), the colloidal gold particles should be 10–15 nm in diameter. The issue is to have the image of the gold particle large enough on the CCD camera used for image recording that there are ~10 pixels across the diameter of each particle. To put the particles onto the sample, grids are floated on 5-μl droplets of a solution of colloidal gold (15 nm in diameter, OD = 2) for 3 min on each side. The grids are then quickly rinsed on water droplets and dried by blotting.

B. Tomogram Acquisition

1. Large-Scale Adaptations

In yeast cells, the cytoplasm is filled with electron-dense ribosomes that obscure the MTs in a semithick section, like the ones used for tomography (Fig. 1A). Thus, the selection of cell sections to image must occur "blind." The importance of choosing the right cell must be stressed, since the time spent acquiring and aligning the images and in calculating the tomogram is so considerable that the number of cells studied is usually very low. We developed two methods to tackle this problem. First, a quick screen to examine the content of a section was designed; second, we developed a large volume approach that would include MTs, regardless of their

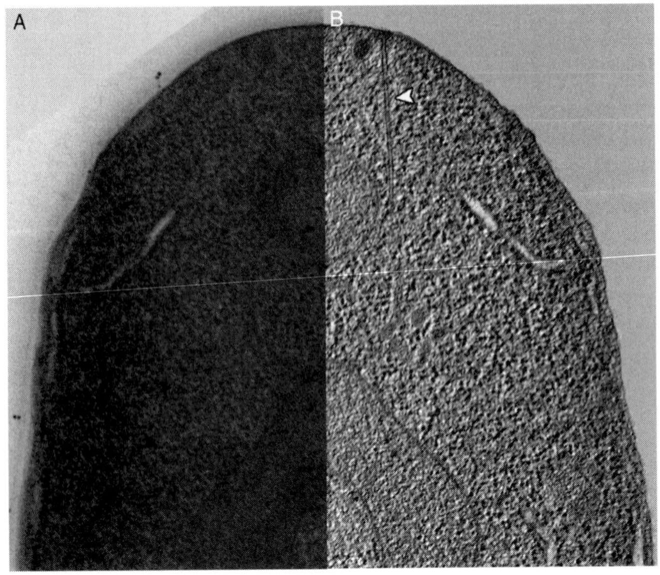

Fig. 1 MTs are not visible in a semithick section of the dense cytoplasm of yeast. (A) A zero angle projection of a 250-nm section, (B) a slice cut from the tomographic reconstruction of the same cell with an MT clearly visible (white arrowhead; picture has been rotated to create a mirror image). This section actually had three MTs.

position in the cell. In both cases image resolution was compromised to achieve the largest possible volume, but our compromises leave MTs still clearly visible (Fig. 1B).

The alignment of interphase MTs almost parallel to the axis of the fission yeast cell (Fig. 2A) made it possible to choose cells that would be favorable for whole-cell MT reconstruction using images from only a single axis of tilt. We chose to image cells that were oriented with their long axis approximately parallel to the tilt axis, with the results that we had good resolution and contrast of MT walls, even with single-axis tomography (Fig. 2B). Though dual-axis tomograms provide more isotropic resolution and contrast (Mastronarde, 1997), our strategy provided reconstructions with adequate quality, even though they were based on only half the data, a serious issue when large volume reconstructions are considered.

Our first approach to picking favorable cells was based on producing a preliminary or "screening" tomogram. For this we acquired pictures in which the pixels were binned to ~2 nm and images were collected at 4° increments over ±60°. These data were used to build a tomogram using cross correlation to align the tilted views without fiducial alignments. The resulting tomogram provides a low-resolution 3D image of the contents in the section studied. When these reconstructions revealed MTs lying flat within the section, we collected the data necessary for a serious reconstruction (1.5° increments over ±60–65° tilt, 1.5-nm pixel). The MT bundles could then be followed with confidence through the serial sections.

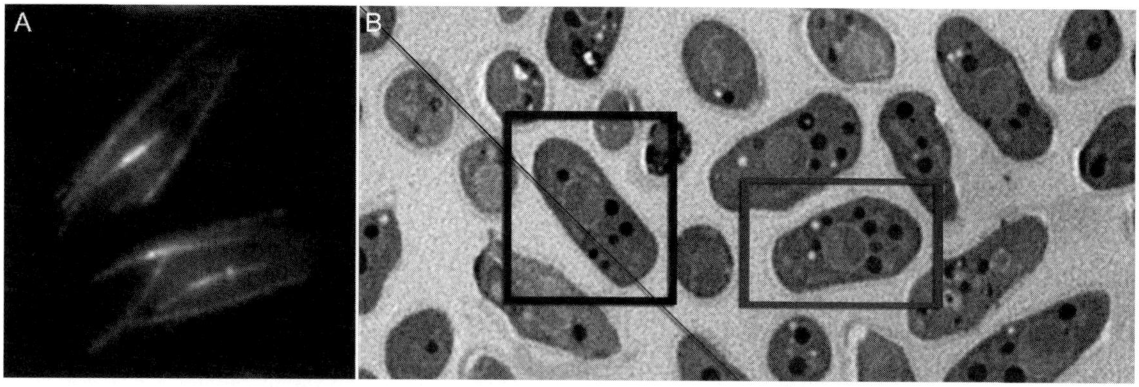

Fig. 2 Linear arrangement of MT interphase arrays in fission yeast enables single-axis tomograms to give the resolution required (A) fluorescence picture with GFP-tubulin (courtesy Dr. Damian Brunner). (B) A suitable, well-embedded, short cell (black square) aligned along tilt axis (black line) in low magnification. In this orientation, IMAs should be well resolved in single-axis tomograms. Cell not suitable for a single-axis tilt series is shown in gray square.

The second approach was made possible by the ease with which one can acquire montaged images, using the SerialEM acquisition software (for details see http://bio3d.colorado.edu). Short cells (7–9 μm) in the right orientation were found and tracked through serial sections (Fig. 3A). This process was simplified by the newly developed Navigator software, which works within SerialEM. Navigator makes a low-magnification montage of the whole EM grid. The resulting map can be used to identify promising areas and to zoom up in magnification, seeking a cell that looks well preserved. One can then zoom out and make sure that the cell can be followed through the serial sections. Each cell that is suitable can then be marked with a colored point (Fig. 3B) and can be relocalized automatically by the computer-controlled stage, even after the grid has been removed from the microscope, thanks to the use of four registration points. Only when a cell was traceable through many sections (5–16) did we go on to acquire montaged image from each section, covering the whole-cell length at a resolution appropriate for reconstruction (Fig. 3C and D; 1×3 montage, 1.5° increment, ± 60–65°, 1.5-nm pixel). Serial sections were then joined in z to reconstruct an *S. pombe* cell that is around 4 μm in diameter. Thus, reconstruction of large parts or even whole cells can be achieved, and IMAs can be partially or even completely reconstructed (Fig. 3E and Höög *et al.*, submitted for publication).

2. Image Acquisition

When the electron beam hits the plastic-embedded sample, shrinkage is induced (Luther *et al.*, 1988). This shrinkage causes problems during the subsequent image alignment, so it is important to minimize the shrinkage during acquisition of a tilt

Fig. 3 Finding a cell that can be viewed in serial sections is easier using Navigator. (A) A low-magnification Navigator "map" showing serial sections of one cell marked in yellow and another in pink. (B) Zoomed up picture of a marked cell in the navigator map. (C) Model showing how a cell should ideally be sectioned to improve the possibility that its MTs are lying flat in the sections. (D) Montaged pictures of nine serial sections of the same cell. (E) A 3D model of a large cell volume reconstructed from nine serial tomograms.

series. If a preillumination step is added, most shrinkage has occurred before the first picture is taken. During this "pre-burn," the electron beam is spread to include a large area around the sample, to have a uniformity of the shrinkage in the whole area, and is left there for around 5 min.

Several programs are available to facilitate the collection of data for tomography; the choice of which to use is determined by the application and which hardware is available. Some examples are TOM, TVIPS, and SerialEM. SerialEM is specially adapted to the FEI Tecnai series and has proven very useful for MT studies, especially when large volumes and montaging are necessary (Mastronarde, 2005).

Tilt increment and angular range desired are entered into the software which then undertakes a repeated macro of tilting the specimen holder, tracking the sample, focusing, final exposure, and saving the image to file. For work with stained, plastic-embedded samples that seeks about 5-nm resolution, pictures are usually taken with about 0.2-μm defocus for optimal result (for details see Chapter 5 by O'Toole, this volume).

C. Tomogram Calculation and 3D Model Reconstruction

1. eTomo and Tomogram Calculation

After a series of tilted two-dimensional (2D) projections of the specimen has been taken, it must undergo a set of processes to generate the final tomogram. Also here there are several software packages available, depending on the application and nature of your data. IMOD is the only software, to our knowledge, with integrated ability to calculate montaged tomograms (Mastronarde, 1997; Sandberg *et al.*, 2003). Further, it has a graphical user interface, eTomo, and a large set of programs designed specially for dealing with MTs once the tomogram has been generated. For an in-depth introduction and a tutorial to this software see the Web page http://bio3d.colorado.edu/ and Chapter 5 by O'Toole, this volume.

The issues special to our own work relate to data acquisition for montaged, serial tomograms. For a montaged tomogram around 40 fiducials per frame are chosen and local alignment is used. Local alignments separate the image stack into smaller regions by choosing areas that include a certain number of fiducials on each side of the section (default is eight on one side and three on the other, but this is flexible). The distortions in these regions are then considered separately by the algorithms that estimate the position of the gold fiducial; this greatly improves the estimates. The larger the area imaged, the greater the distortions, so local alignments matter even more with large areas like those in image montages.

2. Joining of Serial Tomograms

To create reconstructions of large volumes, it is essential to join tomograms of serial sections. While one might imagine that a greater volume could be reconstructed simply by imaging a thicker section, in our hands this leads to poorer image resolution and a reduction in the visibility of cellular detail. In tomograms of thicker section objects like yeast MTs, which are surrounded by a dense distribution of darkly staining ribosomes, become invisible (Fig. 1).

Recently, the procedure for joining serial tomograms with IMOD has been greatly simplified by a new function in the eTomo graphical user interface. The joining consists of three steps: set up, alignment, and join. Set up begins with a definition of the order in which the tomograms should be joined. If they are properly labeled from a known set of serial sections, this is unambiguous. If section order is unknown, it may take a while to compare the tops and bottoms of each section and see who links with whom. Next, one must define the slices from each tomographic reconstruction that should be used for the visual alignment of the serial sections. These slices should be chosen so that organelles and the cell's plasma membrane are clearly visible, but they should still be as close to the top or bottom of the reconstructed volume as possible (Fig. 4B and C). Furthermore, the slices that will be used as the first and last in the actual joining of adjacent tomograms must be chosen. These should normally be the first and last section of the tomographic stack, so as not to lose any of the biological information available (Fig. 4A and D).

The alignment step is crucial for the generation of a large serial tomogram from which it will be possible to generate a precise 3D model. The principle is simple: a slice from the top of one tomogram should be similar to and be aligned with a slice from the bottom of its neighbor. In practice, this is not always easy because of beam-induced distortion of the sections (Luther *et al.*, 1988 and below). By displaying images from the top and bottom of adjacent sections simultaneously, however, one can see the translations, etc. that must be done to align the two tomograms with each other as well as possible (Fig. 4F–G). eTomo provides an initial, automatic alignment, which then can be manually improved, if necessary, with the "manual image distortion and alignment" program, Midas. The transforms that will bring about proper alignment are saved in a text file that is then utilized in the final join step. Here, the serial section tomogram reconstructions are stitched together into one large image file (Fig. 4H). To ensure the best possible result, a test join option is available in the software. This option creates an image stack with every *n*th picture (so the file is not too big for rapid calculation and display); this can easily be opened and scrutinized before the final join is submitted. It is often necessary to do iterative steps of trial joins and alignments to find the best final joined tomogram possible.

3. Taking a Snapshot from the Tomogram

It is often useful to prepare 2D images that are slices from the tomogram. A snapshot from a tomogram is often the image data from a single tomographic slice, which is usually only 1- to 2-nm thick. Sometimes an image from a tomogram can be made more useful and/or pleasing by averaging the 2D information in several adjacent tomographic slices. In IMOD, this can be accomplished with

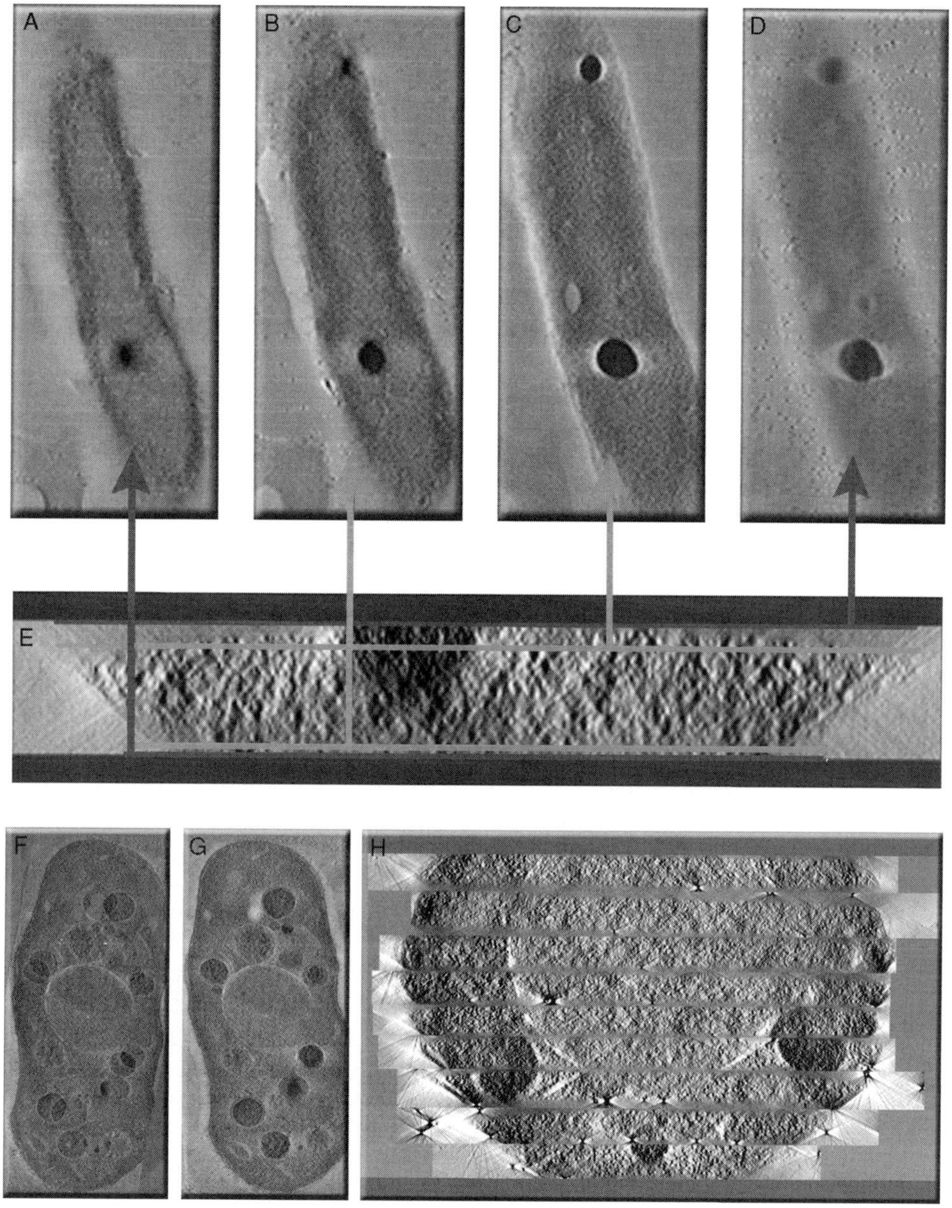

Fig. 4 Principle of choosing the slices for joining adjacent tomograms. (A) and (D) are sections that are used for the actual join. Ideally, they are the top and bottom sections of the tomogram (red lines in E), whereas (B) and (C) are used for alignment since intracellular detail is more clear (green lines in E). (F) Unaligned sections from top (green) and bottom (magenta) of adjacent serial sections, (G) sections in (F) have now been aligned, (H) side view of an image file containing nine joined serial tomographic reconstructions. (See Color Insert.)

the slicer tool. Slicer can also be used to change the orientation of the sample that is cut from the tomogram, so it will best display a feature of interest, like a cytoskeletal fiber (see Chapter by O'Toole, this volume).

4. Construction of a 3D Model

1. Tracking MTs and other filaments using *3dmod*. Segmentation of the image data in 3D reconstructions is sometimes necessary to clearly visualize and display a feature of interest, like the IMA. Segmentation can create graphic objects that represent such features without the visual complexity that is characteristic of tomographic imaging of well-fixed cytoplasm. Model can be built from such graphic objects; these often simplify structure interpretation and can be used for measuring distances between organelles. Still, one must remember that making a model is an interpretation of the data. Thus, a model should not be used instead of snapshots of the real tomogram unless such simplification is necessary. The program suite called IMOD includes the program *3dmod*, which allows a choice of closed or open contours to represent objects on each plane in the tomogram. These contours can subsequently be used to define a graphic object that accurately represents the cellular structure of interest.

A major advantage of having a computerized 3D reconstruction to study is that modeling is greatly facilitated. For example, if one is studying an MT in the IMA, one can cut the reconstructed volume at any angle. In *3dmod*, one can use the slicer option to extract from the tomogram a slice that contains an MT's axis. This greatly facilitates tracking the MT because one can follow the polymer through the reconstructed volume by placing the cursor in its lumen (Fig. 5A) and adjusting the orientation of the slice until the MT is lying flat in the field of view. One can step through tomographic slices to find the middle of the MT (when it looks the widest) and then draw a line (an open contour) down the middle of the fiber (Fig. 5B). Since MTs are flexible, the line that represents their axis will often curve, and realignment along the way is almost always necessary. Once the end of the MT is reached, one can represent the end structure with a simple label, such as a point that is color coded to represent a feature of interest, such as whether the end is open/closed or skirted/blunt/sheet/closed (O'Toole *et al.*, 1999, 2003).

When an MT bundle passes through serial plastic sections, it is sometimes not immediately clear how the contours on one section connect with contours on the next. The process of making these connections accurately can be simplified by modeling each of the MT fragments individually (separate contours), filling them in *z*, and then studying both the model and the model superimposed on the tomogram to decide which two contours to connect. Once these determinations have been made, one can join the two contours together. The contours are shown in the model as lines, which are not very similar to real MTs. To provide a more realistic 3D image, the contour can be meshed to form a tube whose diameter represents that of an actual MT. It is also sometimes necessary to smooth the model MT as it passes from one tomogram to the next in a serial set.

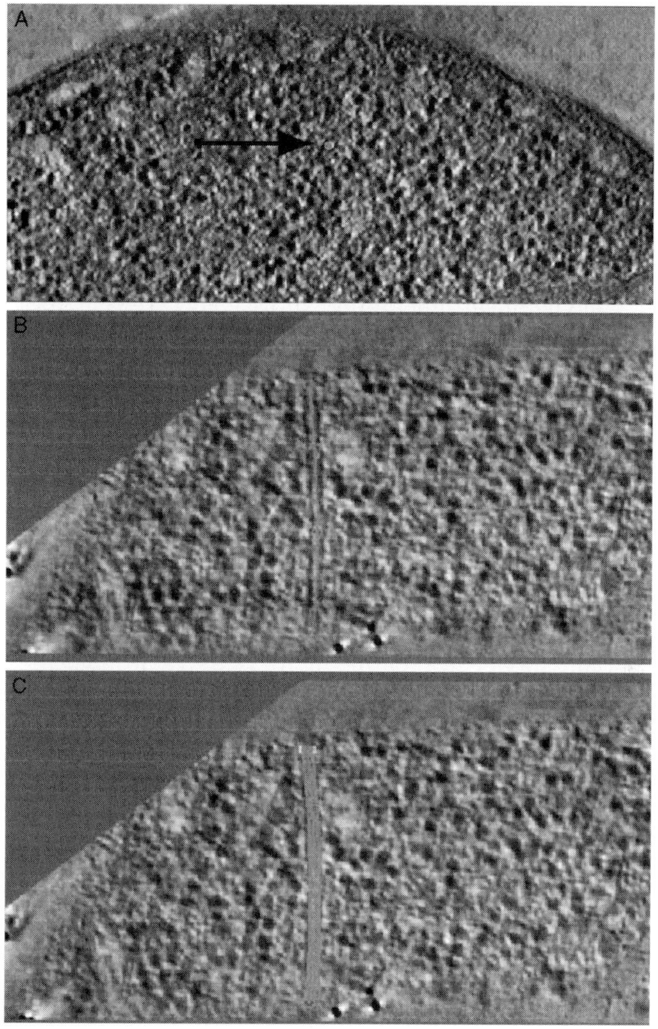

Fig. 5 Modeling of MTs (A) Zap window with a green modeling point in the lumen of an MT, (B) same MT properly oriented and modeled in the slicer window, (C) the MT has been "filled in Z" which simplifies tracking, modeling, and joining of MTs stretching over several serial sections.

2. After the features of special interest have been modeled, it is often valuable to provide a context for the object, for example, by modeling the cell's nucleus, plasma membrane, or other membrane-bound organelles. The goal is to represent membrane bound organelles as surfaces that display the position and shape of the membrane in space. The program used should allow the use of closed contours,

which will always connect the last point digitized with the first. This is ideal for representing membrane-bounded structures where in general there are no free edges. To outline membrane-bound organelles in the tomogram, for example, to show the cell's shape, it is not necessary to place a line around the structure of interest in every tomographic slice. Every third, fifth, or even tenth section is usually sufficient, so long as there is not a significant change in the shape of the organelle over that distance. Even after careful drawing of individual contours, however, some smoothing may be desirable. Most modeling programs include functions to help with this problem. For a more detailed description of how to model membrane-bounded organelles, see Chapter 8 by Marsh, this volume.

D. Analysis of 3D Data

1. Measuring Within a Model

When a plastic-embedded sample is in the electron microscope, the beam vaporizes some of the plastic, causing the sample to shrink anisotropically (Luther *et al.*, 1988). This shrinkage is mostly in the axis perpendicular to the section plane, so the resulting distortions must be accounted for when making measurements in the volume. In 3dmod, this may be done by changing the *z*-scale in the edit model header menu. It is easy, however, to overestimate the amount of shrinkage. If the image file consists of many serial tomograms, an apparently appropriate correction will lead the organelles to look stretched along the *z*-axis. When one has several serial tomograms in one set, it is sometimes easiest to find the *z*-scale factor by trial and error, seeking a value that make the nucleus or appropriate small vesicles appear round. Once the tomogram has been made isotropic, all volumes, lengths, and areas can be measured readily. In IMOD, this is accomplished with the *imodinfo* program. It is important to remember, though, that the issues of beam-induced distortions of the sample may make some of these measurements more relative than absolutely quantitative.

2. Quantifying Feature Proximity in IMOD

There are programs within the IMOD suite that ease the quantitative analysis of 3D models of cellular structures. Our study of MT bundles in interphase of *S. pombe* has been greatly facilitated by these programs. For example, the average distance of MTs from one another can be determined using *nda* (neighbor density analysis). This distance can then be used in *mtpairing* to see which MTs are placed relative to one another by this characteristic distance or something close to it. This can help in deciding when an MT is in a bundle and how long the actual bundle is. Clearly, such decisions are to some extent matters of definition, but with proper quantification, one can be explicit about how such decisions were made.

Preferred distances between 3D objects (i.e., membrane-bound organelles) and MTs can be determined using *mtk* (MT "kissing" program) (Marsh *et al.*, 2001). It defines the distances of closest approach between MTs and any chosen surface. The program displays the frequencies of such distances as the number of objects per cubic micrometer at the given distance. Thus, one can examine whether a given organelle is positioned relative to MTs in a way that it departs from randomness. The departure from random can, in turn, be taken as evidence for some level of interaction between the two objects. For more MT analysis and processing programs see Section IV.

IV. Materials

A. Cells and Media

S. pombe heterothallic minus (h−) mating type of wild type strain (PN 972) was grown in suspension culture at 32 °C on yeast extract with five amino acid supplements (YE5S; i.e., adenine, histidine, leucine, uracyl, and lysine, for details, see Moreno *et al.*, 1991) until the culture reached midlog-phase growth. Alternatively, cells were grown in the defined EMM.

B. High-Pressure Freezer, Freeze-Substitution Resins

Both the BAL-TEC HPM 010 high-pressure freezer (BAL-TEC, Balzers, Lichtenstein) and the Leica EM PACT1 (Leica Microsystems, Vienna, Austria) were used to freeze yeast cells (McDonald *et al.*, 2006). The advantage of the BAL-TEC machine is that a rather large volume can be loaded in the carrier (2 mm in diameter and 200 μm in depth). Therefore, one can process a large pellet of cells, which, once embedded in plastic, can be cut into several smaller blocks using a fine saw. With the Leica EM PACT1 device (Studer *et al.*, 2001) we mostly used home-made carriers (2 mm of diameter and 1-mm thick containing a sample cavity 200 μm in depth and 1.2 mm in diameter).

The slot carriers (Leica) were advantageous as they could freeze a larger pellet, which was easier to handle than the flat carriers (Leica). With the latter, pellets do not easily come off as a single object on rinsing the carriers with acetone at the end of the FS run. Moreover, we had variable quality of freezing using these carriers.

FS was always performed in a Leica EM AFS device (Leica Microsystems, Vienna, Austria) using the special perforated plastic tubes, which greatly facilitate the solvent and resin exchange.

Lowicryl HM20 (Polysciences, Warrington, PA) was mixed in dehydrated acetone during the progressive infiltration steps.

C. Microtome

Plastic semithin sections were cut at a thickness between 210 and 250 nm using a Leica-Reichert Ultramicrotome (Leica Microsystems, Vienna, Austria). We used diamond knives with a 35° included angle (Diatome, Biel, Switzerland). The sections were collected on copper–palladium slot grids (2×1 mm^2) freshly coated with Formvar (1% polyvinyl formal in chloroform). We used cationic gold particles (15 nm) (British BioCell International, UK) as fiducial markers.

D. Microscope Hardware

Tilt series were taken on an FEI Tecnai TF20 transmission electron microscope (TEM), operating at 200 kV. The images were collected on a Gatan Ultrascan 890 digital camera or a Gatan Ultrascan 895 digital camera (4 port readout, 15 μm/pixel). The latter camera was about three times faster in acquiring a tilt series; it was therefore particularly useful for collecting montages of many serial sections.

An FEI Tecnai TF30, operating at 300 keV was used when thicker sections were employed. A Gatan Megascan 795 camera was mounted on this TEM.

The tilt series were acquired over ±65°, using a Gatan 670 high-tilt holder, a Fischione 2020 high-tilt holder, or a Gatan 650 high-tilt-rotate holder. The preferred holder was the Fischione holder which allowed high-tilt imaging over a larger area of the grid. Thus, it was easier to image a larger fraction of the serial sections on the grid. The Gatan 650 high-tilt-rotate holder had a complex grid attachment system, which made the slot grids more prone to break during specimen changes. Further, the area on the grid where high tilts were possible was comparatively small. However, this holder had the advantage that a well-preserved cell could be turned into the right orientation for optimal resolution with single-axis tomography without the removal of the grid.

E. Computers

Sun (Opteron CPUs) computers running the Fedora operating system and the IMOD software package were used to calculate tomograms. The IMOD software works only with upper-end Nvidia graphic cards. The large file sizes produced demand 4–16 GB of RAM. All data were stored on a 2Tb network server, therefore only standard size local hard drives were necessary. Further, a three-buttoned mouse is essential.

F. Software Package IMOD©

IMOD is a software package containing around 140 different programs for image processing, tomographic reconstruction, modeling, and display (Kremer *et al.*, 1996). As described above, it includes features that facilitate 3D reconstruction of EM serial sections, and it can be used for optical sections as well.

In addition to all the programs required for tomogram reconstruction, there are several programs developed specially for research on MTs. Here is a short list of the MT related programs that are the most important when working with tomograms:

mtoverlap: This program displays and measures the overlaps of MTs coming from two distinct foci, like the poles of a mitotic spindle.

mtpairing: This program determines when MTs are paired and the length of this region of pairing. Further, this program can recolor MTs depending on their polarity (based on the positions of their end points in *z*).

mtk: MT kissing analyzes the distance in 3D between MTs and other objects, such as mitochondria or the nuclear envelope (Marsh *et al.*, 2001).

nda: Neighbor density analysis measures the frequency with which MTs are spaced by a given distance and identifies the preferred distance between MTs in a bundle. To use this program, a model must first be converted using *resamplemod* (Mastronarde *et al.*, 1993).

mtsmooth: This program smooths the trajectory of open contours, such as MTs, to improve their presentation. It should not be applied to models that will subsequently be analyzed by *nda* or *mtk*.

resamplemod: Some of these programs were developed for serial section 3D reconstruction. This program was developed to convert a 3D model built from a tomogram into a model it can handle by turning the model 90° about the *y*-axis and resectioning it.

For more information on these and other programs, visit the Web site http://bio3d.colorado.edu/. Here, there is a listing of all programs available in the IMOD software package, which is downloadable at no cost.

V. Discussion

The method described here has been shown to provide information that would not have been available from serial thin section 3D reconstructions. Reconstructions built from serial thin sections are limited in 3D resolution by the thickness of the sections and cannot give isotropic 3D information. In EM tomography, a semithick section is reconstructed in 3D, and models can be built by slicing the tomogram in any direction, depending on the question to be answered. With this method, MTs that diverge from a bundle, or bundles that change direction, are not lost from the reconstruction. Further, tomography provides an image resolution of 6–12 nm in slices, 1- to 2-nm thick that can be extracted from the reconstruction by the computer. This resolution is comparable with the resolution in images of thin sections, and almost the same resolution is obtained in the dimension perpendicular to the section's plane. Thus, a model created from a tomogram is of much higher accuracy than one made from serial sections. Furthermore, bundles of fibers are difficult and time consuming to track through serial thin section, even when using an EM equipped with a sample holder that is capable of both tilt and rotation, so MTs of a given bundle can always be made parallel to the electron beam axis. In a tomogram, on the other hand, all bundles in the volume can be tracked and eventually modeled. Thus, not only information about the bundles themselves but also their interactions can be described.

However, when a tilt series is being acquired in the EM, the plastic that embeds the sample is being evaporated by the electrons. The resulting shrinkage occurs mostly along the axis perpendicular to the section, and it can be as dramatic as 30–50%. The process of fiducial alignment helps to correct for shrinkage in the other axes, and after calculation, an approximate correction for the major shrinkage can be applied. However, the sections do not shrink uniformly. The higher the density of material that binds heavy metal stain, the less the shrinkage. Thus, joining tomograms of two adjacent sections so, cellular structures fit perfectly between the sections is very difficult. This means that MTs that extend over serial sections may appear kinked in the model. Consequentially, measurements made from a model include errors that may be considerable and are very hard to estimate.

This problem can be minimized by choosing a cell whose MTs lie as flat as possible, using the prescreen method described in this chapter. MTs that lie approximately parallel to the plane of section should not extend over many serial sections. When using the large volume approach, small selections of the total amount of MTs with no or little kinks are used for the length statistics. Further, the lesser shrinkage of the Lowicryl resin helps to minimize the problem.

Even larger errors might sneak in to a model when it is not possible to determine which MTs should be joined at a boundary between sections. For this, there is no cure except for careful examination of the model, using tools like slicer in *3dmod*.

For this study, it was crucial to reconstruct large fractions of a cell's volume. To accomplish this in a reasonable period of time, we compromised on image resolution. Dual-axis tomograms were avoided by choosing cells whose orientation provided sufficient resolution of the MTs under study, that is, both the long axis of the cells and the MTs were parallel to the tilt axis. Further, we used a slightly lower magnification and a higher tilt increment than are now common in cellular tomography. Nonetheless, the MTs were still clearly visible, and sometimes, beautiful MT end structures were seen. These enabled a determination of MT polarity. Of course, such a compromise in resolution may lead to mistakes, such as incorrect classification of the ends or even missed MTs. On the basis of the quality of the images we have obtained, however, we believe that the likelihood of the latter problem is minute. Moreover, errors made as a consequence of lower-than-optimal resolution should be counteracted by the larger sample size we have achieved by studying large volumes and looking at quite a few cells. Further, we used complementary dual-axis tomograms of small areas for comparison with the single-axis tomograms. For example, fine details visible in our single-axis reconstructions were seen in dual-axis tomograms as well (Höög *et al.*, submitted for publication). This suggests that the resolution in the single-axis tomograms maintained an acceptable quality for our purposes.

VI. Summary

The various steps of the complete procedure to reconstruct and model large cell volumes by ET can be divided into three main parts (Table I):

1. Sample preparation: The first step involves handling cell cultures and performing cryoimmobilization of cells using a high-pressure freezer. This is followed by FS in acetone and plastic infiltration to generate blocks of embedded samples. From these, thick sections are cut, collected in an ordered series, contrasted, and fiducial gold particles are added to facilitate precise alignment of the projections in the tilt series.
2. Electron microscopy: The second step requires a TEM (200–300 kV) equipped with a CCD camera and controlled by appropriate acquisition software. This is used to acquire a series of 2D projections at various orientations (tilt series). Montaged images in the x–y plane can be made in order to cover the full area of interest at each angle during acquisition.
3. Computation: The final reconstruction requires a precise alignment of all images from the tilt series, which is helped by fiducial markers. Lastly, all tomograms in the series are calculated and joined in the z-axis to create a large volume. A 3D model is then created from contours of features of interest. These can be painted in the tomogram, and quantitative measurements can be performed on these models, using various, readily available programs.

Acknowledgments

We thank the Boulder Laboratory for 3D Electron Microscopy of Cells, in particular J. R, McIntosh, C. Schwartz, and D. Mastronarde for use of microscopes, invaluable help in developing software, and methods to fit our purposes as well as important discussions. We also thank D. Brunner, E. Busch, F. Nedelec, D. Foethke, I. Hagan, and K. McDonald for helpful discussions and constant support; A. Noon and C. Bleck for calculation of tomograms; Daniel Studer for his help with cryoimmobilization using EM PACT1 Leica device and C. Roome for expertise in informatics. J.L.H. is supported by a Marie Curie Early Stage Research Training Fellowship of the European Community's Sixth Framework Programme (FP6) under contract number MEST-CT-2004–504640.

References

Asakawa, K., Kume, K., Kanai, M., Goshima, T., Miyahara, K., Dhut, S., Tee, W. W., Hirata, D., and Toda, T. (2006). The V260I mutation in fission yeast α-tubulin Atb2 affects microtubule dynamics and EB1-Mal3 localization and activates the Bub1 branch of the spindle checkpoint. *Mol. Biol. Cell* **17,** 1421–1435.

Baumeister, W., Grimm, R., and Walz, J. (1999). Electron tomography of molecules and cells. *Trends Cell Biol.* **9,** 81–85.

Braunfeld, M. B., Koster, A. J., Sedat, J. W., and Agard, D. A. (1994). Cryo automated electron tomography: Towards high-resolution reconstructions of plastic-embedded structures. *J. Microsc.* **174**(Pt. 2), 75–84.

Brunner, D., and Nurse, P. (2000). New concepts in fission yeast morphogenesis. *Philos. Trans. R. Soc. Lond. B Biol. Sci.* **355,** 873–877.

Carlemalm, E., Garavito, R. M., and Villiger, W. (1982). Resin development for electron-microscopy and an analysis of embedding at low-temperature. *J. Microsc. (Oxford)* **126,** 123–143.

Ding, R., McDonald, K. L., and McIntosh, J. R. (1993). Three-dimensional reconstruction and analysis of mitotic spindles from the yeast, *Schizosaccharomyces pombe. J. Cell Biol.* **120,** 141–151.

Drummond, D. R., and Cross, R. A. (2000). Dynamics of interphase microtubules in *Schizosaccharomyces pombe. Curr. Biol.* **10,** 766–775.

Frank, J. (1992). "Electron Tomography: Three Dimensional Imaging with Transmission Electron Microscope." Plenum Press, New York.

Giddings, T. H., Jr., O'Toole, E. T., Morphew, M., Mastronarde, D. N., McIntosh, J. R., and Winey, M. (2001). Using rapid freeze and freeze-substitution for the preparation of yeast cells for electron microscopy and three-dimensional analysis. *Methods Cell Biol.* **67,** 27–42.

Griffiths, G. (1993). "Fine Structure Immunocytochemistry." Springer-Verlag, Berlin.

Janson, M. E., Setty, T. G., Paoletti, A., and Tran, P. T. (2005). Efficient formation of bipolar microtubule bundles requires microtubule-bound gamma-tubulin complexes. *J. Cell Biol.* **169,** 297–308.

Kremer, J. R., Mastronarde, D. N., and McIntosh, J. R. (1996). Computer visualization of three-dimensional image data using IMOD. *J. Struct. Biol.* **116,** 71–76.

Luther, P. K., Lawrence, M. C., and Crowther, R. A. (1988). A method for monitoring the collapse of plastic sections as a function of electron dose. *Ultramicroscopy* **24,** 7–18.

Marsh, B. J., Mastronarde, D. N., Buttle, K. F., Howell, K. E., and McIntosh, J. R. (2001). Organellar relationships in the Golgi region of the pancreatic beta cell line, HIT-T15, visualized by high resolution electron tomography. *Proc. Natl. Acad. Sci. USA* **98,** 2399–2406.

Mastronarde, D. N. (1997). Dual-axis tomography: An approach with alignment methods that preserve resolution. *J. Struct. Biol.* **120,** 343–352.

Mastronarde, D. N. (2005). Automated electron microscope tomography using robust prediction of specimen movements. *J. Struct. Biol.* **152,** 36–51.

Mastronarde, D. N., McDonald, K. L., Ding, R., and McIntosh, J. R. (1993). Interpolar spindle microtubules in PTK cells. *J. Cell Biol.* **123,** 1475–1489.

McDonald, K., and Morphew, M. K. (1993). Improved preservation of ultrastructure in difficult-to-fix organisms by high pressure freezing and freeze substitution: I. *Drosophila melanogaster* and *Strongylocentrotus purpuratus* embryos. *Microsc. Res. Tech.* **24,** 465–473.

McDonald, K. L., Morphew, M., Verkade, P., and Mueller-Reichert, T. (2006). "Recent Advances in High Pressure Freezing: Equipment and Specimen Loading Methods." Humana Press, Totowa, NJ.

McIntosh, R., Nicastro, D., and Mastronarde, D. (2005). New views of cells in 3D: An introduction to electron tomography. *Trends Cell Biol.* **15,** 43–51.

Moreno, S., Klar, A., and Nurse, P. (1991). Molecular genetic analysis of fission yeast *Schizosaccharomyces pombe. Methods Enzymol.* **194,** 795–823.

Müller-Reichert, T., Sassoon, I., O'Toole, E., Romao, M., Ashford, A. J., Hyman, A. A., and Antony, C. (2003). Analysis of the distribution of the kinetochore protein Ndc10p in *Saccharomyces cerevisiae* using 3-D modeling of mitotic spindles. *Chromosoma* **111,** 417–428.

O'Toole, E. T., Winey, M., and McIntosh, J. R. (1999). High-voltage electron tomography of spindle pole bodies and early mitotic spindles in the yeast *Saccharomyces cerevisiae. Mol. Biol. Cell* **10,** 2017–2031.

O'Toole, E. T., McDonald, K. L., Mantler, J., McIntosh, J. R., Hyman, A. A., and Muller-Reichert, T. (2003). Morphologically distinct microtubule ends in the mitotic centrosome of *Caenorhabditis elegans. J. Cell. Biol.* **163,** 451–456.

Sandberg, K., Mastronarde, D. N., and Beylkin, G. (2003). A fast reconstruction algorithm for electron microscope tomography. *J. Struct. Biol.* **144,** 61–72.

Sawin, K. E. (2004). Microtubule dynamics: Faint speckle, hidden dragon. *Curr. Biol.* **14,** R702–R704.

Steinbrecht, R. A. (1993). Freeze-substitution for morphological and immunocytochemical studies in insects. *Microsc. Res. Tech.* **24,** 488–504.

Steinbrecht, R. A., and Müller, M. (1987). Freeze substitution and freeze-drying. *In* "Cryotechniques in Biological Electron Microscopy" (R. A. Steinbrecht, and K. Zierold, eds.), pp. 149–172. Springer-Verlag, Berlin.

Studer, D., Graber, W., Al-Amoudi, A., and Eggli, P. (2001). A new approach for cryofixation by high-pressure freezing. *J. Microsc.* **203,** 285–294.

Walther, P., and Ziegler, A. (2002). Freeze substitution of high-pressure frozen samples: The visibility of biological membranes is improved when the substitution medium contains water. *J. Microsc.* **208,** 3–10.

CHAPTER 7

Electron Microscopy of Archaea

Charles E. Robertson

Department of Molecular, Cellular and Developmental Biology
University of Colorado, Boulder
Colorado 80309

Archaea are the recently discovered third "domain" of life. Previously, the two known domains were Eucarya and Bacteria. When Archaea were first discovered, they were grouped with Bacteria because both kinds of organisms lacked nuclei. With the advent of efficient DNA sequencing, however, the genomes of Eucarya, Archaea, and Bacteria have shown that the information-processing systems (molecules involved in DNA replication, transcription, and translation) of Archaea are surprisingly similar to, although much simpler than, the information-processing systems in Eucarya. Therefore, Archaea are an interesting model system in which

0091-679X/07 $35.00
DOI: 10.1016/S0091-679X(06)79007-0

to study the basic elements of information-processing in two of the three existent domains of life. The small size of Archaea (2–5 μm) precludes high-resolution structural studies by light microscopy. Electron tomography allows the creation of full-cell models of Archaea with a resolution of ~10 nm. While electron tomography of frozen, hydrated specimens offers great potential for higher resolution models, high-pressure-frozen, freeze-substitution-fixed, plastic-embedded, and heavy metal stained specimens are at present the most reliable samples for informative studies of whole archaeal cell structure. Construction of such full-cell models has been difficult in the past due to the labor-intensive nature of the image segmentation process. Software-assisted segmentation with application-specific scripts, implemented with software packages such as MATLAB, substantially reduces the manual labor required to segment full-cell models, opening the door to routine creation of models that represent whole Archaeons in a variety of metabolic and genetic states.

I. Introduction

From the advent of light microscopy until near the end of the twentieth century it was taken for granted that life was of two basic kinds: eukaryotes, whose cells contained nuclei, and prokaryotes that lacked them. Carl Woese was the first to realize that the sequence of nucleotides in ribosomes offered a molecular method to investigate the relationships between all forms of life. Comparison of ribosomal nucleotide sequences made it possible to infer the evolutionary relationships among organisms. The greater the similarity between their sequences, the more closely related the organisms. Using the modest nucleotide sequencing technology of the 1970s, Woese worked out the sequences of the small ribosomal RNA (rRNA) from several different cultured microbes and was surprised to find that there were three sequence signatures rather than the expected two. This led Woese to propose that all forms of life are in one of the three domains: Eucarya, Bacteria, or the new domain that was eventually named Archaea (Woese and Fox, 1977). The discovery of Archaea was a milestone in biology as it marked the point when molecular methods, rather than microscopy, were required to discriminate among the fundamental kinds of life on Earth. The systematic comparison of nucleotide sequences to infer evolutionary relationships has developed into the field of research known as molecular phylogenetics. The molecular phylogenetic relationships between sequences, and thus organisms, are graphically represented by a tree. The general form of the small subunit rRNA phylogenetic tree of life is shown in Fig. 1.

Woese's early work, based on small subunit rRNA sequence, has been substantiated by comparisons of the sequences of many different proteins. The protein comparisons show that Archaea possess characteristics in common with both Bacteria and Eucarya. Most of the archaeal proteins involved in bioenergetics are similar to those found in Bacteria, while most archaeal proteins that process information are similar to those in Eucarya (Schleper *et al.*, 2005).

In the years immediately subsequent to their discovery, Archaea became stereotyped as extremeophiles. Culturable Archaea were found living at high temperatures and low pH in geothermal areas such as Yellowstone National Park, as well as extremely halophilic environments such as the Dead Sea. Some Archaea were found to possess the ability to obtain energy by reducing carbon dioxide to methane (methanogenisis). Examples of representative Archaea are summarized in Table I.

With the advent of polymerase chain reaction (PCR) technology, it became possible in the mid-1980s to amplify selected genes directly from environmental samples (DeLong, 2005). PCR bypassed the problematic requirement of culturing an organism in order to detect it. Now it is routine to assay environmental samples for rRNA genes and deposit their sequences in publicly available sequence databases.

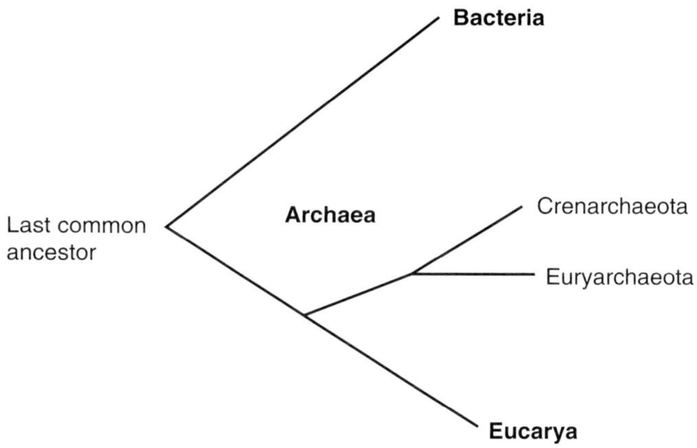

Fig. 1 A phylogenetic tree based on small subunit rRNA sequence showing that the Archaea branch with the Eucarya (Robertson *et al.*, 2005). The actual root of the tree is somewhat uncertain, but best estimates place the root on the bacterial branch.

Table I
Characteristics of Representative Cultured Archaea

Name	Temperature (°C)	pH	Proteins	Shape	Energy?	Oxygen?	Comment
Sulfolobus solfataricus	75	3	2977	Lobed coccus	Chemorganotroph	Aerobic	
Halobacterium	37	7.4	2622	Rod	Chemorganotroph	Aerobic	Will grow in saturated solutions of NaCl (32%)
Methanocaldococcus jannaschii	85	6.5	1758	Cocci	Chemolithoautotroph	Strict anaerobe	Methanogen

From (Madigan and Martinko, 2006) and (NCBI, 2006).

Two insights have resulted from this environmental sampling. First, it appears that as much as 99% of all known microbes do not culture or are very difficult to culture. Second, Archaea are found every where, not just in extreme environments (Robertson *et al.*, 2005). They may in fact dominate some of the largest biomes on the planet such as 100-m below the surface of the ocean (Pace *et al.*, 1985).

II. Rationale

Archaea are little studied compared to bacterial and eucaryal microbes, partly because of their recent discovery but also because Archaea are not known to cause human disease (Eckburg *et al.*, 2003); this restricts archaeal research funding. However, Archaea possess several characteristics that may give insight into the fundamentals of the eucaryal nucleus. At present the entire genomes of 25 Archaea have been fully sequenced. The genomic data show that the information-processing systems of Archaea and Eucarya are surprisingly similar to each other, while both are different from Bacteria. For example, both Eucarya and Archaea use TATA binding proteins to initiate transcription, while Bacteria use sigma factors. In most ways, archaeal DNA and RNA polymerases are like miniature versions of their eucaryal counterparts. A feature shared by Eucarya and Archaea, but not by Bacteria, is that many Archaea package their DNA by wrapping it around histones (Cubonova *et al.*, 2005). The similarity of the information-processing systems of Eucarya and Archaea raises intriguing questions. Might it be possible to learn something of the origin of the eucaryal nucleus through structural studies of Archaea? How similar are the structures of the nuclear bodies of the two non-nuclear domains of life? Precisely what is the nature of nucleoplasm? Insight into these questions and many others may be obtained by applying the best currently available imaging technology to Archaea and Bacteria. This chapter will discuss the issues associated with the construction of three-dimensional (3D), full-cell models of a specific Archaeon at better than 10-nm resolution.

A. Selection of an Archaeon

The organism that has been most used for archaeal research is *Sulfolobus solfataricus*, which makes it the obvious choice for structural investigations. The one drawback to *Sulfolobus solfataricus* as a model organism is, however, that there is not a readily available technology with which to manipulate its genetics. This was true of all Archaea until relatively recently (Rother and Metcalf, 2005).

B. Selection of Imaging Technology

The ability of optical microscopes to discern the separation of objects is limited by the wavelength of visible light to about 200 nm. This means that, at most, five points of light can be resolved across an organism 1 μm in diameter. Even with a good camera, the images of micron-sized organisms appear "blocky" as the individual

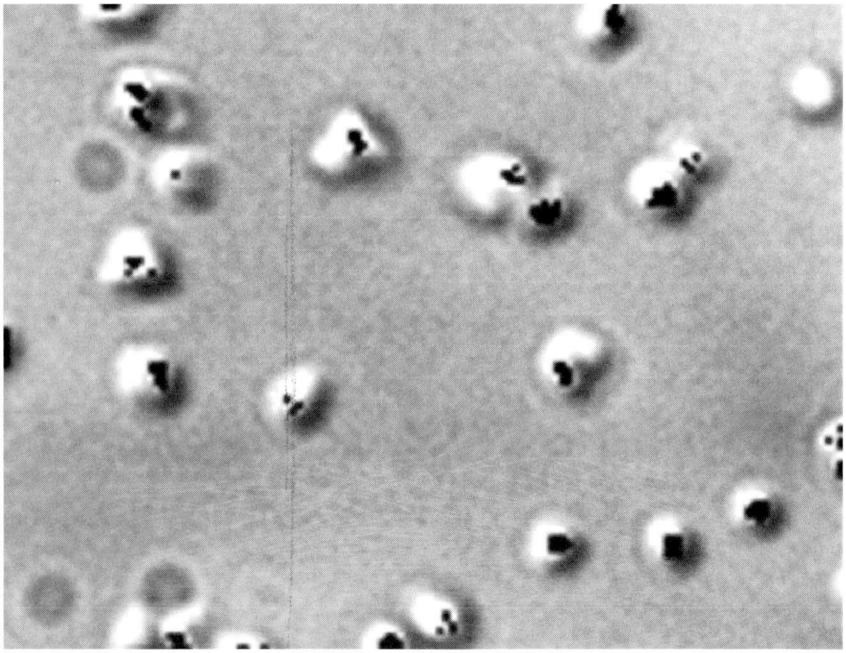

Fig. 2 An image of log phase, DAPI treated *Sulfolobus solfataricus* captured with a fluorescence/phase contrast microscope. The image is formed by the superposition of the fluorescent DAPI (black) image over the same cells imaged via differential interference phase contrast. As DAPI is a DNA intercalator, it indicates the position of the nuclear bodies of the cells. The cells are ~2 μm in diameter.

pixels become evident (Fig. 2). The resolution limitation means that the fine structural details of the smallest microbes are inaccessible with light microscopy.

Transmission electron microscopy (TEM) provides resolution better than a nanometer. TEM of thin sections (~50-nm thick) readily allows discrimination of cell membranes and ribosomes, which means that the practical resolution is ~10 nm or better. The largest drawback to TEM of thin sections for the structural study of small microbes is object superposition. That is, multiple cellular structures, such as proteins, ribosomes, and DNA, and so on, are stacked on top of each other in the projection of a thin section, so the image on a camera is the sum of all of the intervening structures. This superposition makes it impossible to discern the 3D structure of the cell at good resolution. Electron tomography (ET) provides an effective and practical solution to this problem.

C. Preparing *Sulfolobus solfataricus* for Electron Tomography

The ideal electron microscopy (EM) sample preparation protocol for *Sulfolobus solfataricus* would minimize sample handling and chemical treatment. Cryo-EM and cryo-ET are good candidates for achieving these goals (see Chapter 1 by Dubochet, Chapter 16 by Serysheva, Chapter 14 by Zhang *et al.*, and Chapter 13 by Taylor *et al.*, this volume). Figure 3 shows an image of a plunge-frozen

Sulfolobus solfataricus cell. The image shows both the flagella and the interlocked subunits of the proteins in the "S-layer" that covers the surface of *Sulfolobus solfataricus* and many other Archaea. These structures are not as visible with the traditional EM techniques that use plastic embedding and heavy metal staining. However, the central portion of the frozen-hydrated cell of Fig. 3 is featureless because the organism is too thick (~2.5 μm) for 300-keV electrons to penetrate without multiple scattering. An obvious next step would be to cut serial cryosections that can be imaged (250 nm or less) from the frozen-hydrated specimens. Unfortunately, sectioning frozen-hydrated specimens is difficult (see Chapter 1 by Dubochet, this volume). Unlike plastic, which cuts smoothly, ice tends to fracture at irregular intervals leaving fissures in the section's surface. The fissures cause enough distortion that it is probably impossible to track structural features between serial cryosections. Many alternatives are being actively explored to overcome this fissuring problem.

Two other issues must be accommodated in cryo-EM and cryo-ET of cells: the lack of contrast and the dose sensitivity of the frozen-hydrated specimens lead to low signal-to-noise ratios. Studies of specimens composed of repeated subunits, for example eucaryal flagella, can overcome low signal-to-noise ratios by using image averaging (see Chapter 29 by Förster and Hegerl, this volume). Image averaging is currently less useful for full-cell studies because most of the key structures are not composed of repeated subunits. Low signal-to-noise ratios make it very difficult to use automatic segmentation techniques, such as MATLAB-assisted segmentation (Section III.J.1), because such algorithms depend on consistent,

Fig. 3 An image of frozen-hydrated *Sulfolobus solfataricus*. Note that the flagella seen here are not seen in plastic-embedded samples. The very dark clusters of spheres are made up of 10-nm gold fiducials.

high-contrast imaging for structure discrimination. Moreover, cryo-EM, like plastic embedding, suffers from a lack of a molecule-specific staining technology, and immuno-electron tomography (Section II.C.4) protocols are not possible with frozen-hydrated sections.

Until the current limitations of cryo-ET are overcome, more traditional EM sample preparation approaches must be used. The following series of techniques has proven to be successful for *Sulfolobus solfataricus*:

1. High-pressure freezing (HPF): HPF prevents the formation of crystalline ice that can cause significant distortion of structure.
2. Freeze-substitution (FS): FS replaces the water molecules in a specimen with a nonpolar hydrocarbon at a temperature that sustains the amorphous ice created by HPF.
3. Plastic embedding: The nonpolar hydrocarbon is replaced by a plastic resin that is subsequently hardened by exposure to ultraviolet (UV) light.
4. Serial sectioning: A microtome with a diamond knife is used to cut the plastic-embedded specimen into relatively thick sections (250 nm). The sequence of the sections is rigorously preserved by placing them in order on a grid.
5. Heavy metal staining: Heavy metals such as osmium, lead, and uranium nonspecifically label all biomolecules in a specimen, assuring high-contrast EM images.
6. Tilt-series acquisition: Two tilt series, at right angles, are taken of each serial section of a single cell.
7. Reconstruction of each section by tomography: Back projection is used to reconstruct the 3D volume represented in each tilt series. The dual-axis tomograms for each serial section are then combined.
8. Concatenation of serial tomograms: The combined tomograms from each serial section are concatenated to form one whole-cell volume reconstruction.
9. MATLAB-assisted segmentation: Image-processing software routines are used to identify and create 3D models of selected cellular structures.

This collection of techniques will subsequently be referred to as embedded full-cell electron tomography (EFCET). At present EFCET is the only known technique that can yield whole-cell models of microbes at 10-nm resolution. Well-known cellular structures such as the cell membrane, the membrane bilayer, ribosomes, and polysomes, as well as a variety of structures of currently unknown function can be mapped, counted, measured, and compared between cells in a variety of different physiological states.

1. The Rationale of EFCET

Attention to the detail of specimen preparation protocols is crucial to the ultimate success of EFCET. Each cell or tissue type must be empirically approached anew. Fixation of microbes with a traditional preservative, such as 2.5%

gluteraldehyde, yields cells that often appear distorted. The most common aldehyde preservation artifact is that the edges of the cell membranes appear ruffled. This type of preservation artifact is common in the published TEM images of Archaea and Bacteria (Madigan and Martinko, 2006). Empirically, two methods have been found to yield good preservation of *Sulfolobus solfataricus*: plunge freezing (PF) into liquid ethane and HPF. Both of these techniques attempt to minimize the formation of ice crystals that distort morphology. After either type of freezing, the specimens can be freeze-substituted, which replaces water molecules with organic solvent molecules. The organic solvent is subsequently replaced with a plastic resin that facilitates cutting the cells into sections thin enough for ET.

2. Freezing *Sulfolobus solfataricus*

PF is a simple technique designed to freeze small samples so quickly that ice crystals are unlikely to form (see Chapter 1 by Dubochet, this volume). PF is accomplished by applying a few microliters of *Sulfolobus solfataricus* suspended in its growth medium to a small grid that is placed into an apparatus that plunges the grid into ethane, liquefied with liquid nitrogen. Cells from PF grids can subsequently be imaged in the frozen-hydrated state (Fig. 3) or freeze-substituted. The central idea behind HPF is to use a machine in which the water in a specimen is cooled quickly and under high pressure, so the resulting ice is vitreous rather than crystalline (see Chapter 2 by McDonald and Chapter 3 by Hess, this volume). PF has two drawbacks compared to HPF. First, cell preservation in PF is not as consistent as HPF. Second, there are few cells in the PF sections, which means that very large areas must be searched to locate a well-preserved cell. For these reasons, the HPF method is preferred for EFCET of *Sulfolobus solfataricus*.

3. Cryoprotectants

Freezing *Sulfolobus solfataricus* in its culture medium often results in cells that show signs of freeze damage. Freeze damage takes many forms, but a "chicken wire" pattern in the EM images is common and generally indicates that the biomolecules in the specimen were displaced by the formation of crystalline ice, (Fig. 4A). A variety of means have been developed for minimizing freeze damage through the addition of a "cryoprotectant." Cryoprotectants are of many kinds, but they fall into two categories: extracellular, or nonpenetrating, and intracellular (see Chapter 2 by McDonald, this volume). Nonpenetrating cryoprotectants are generally thought of as benign, because they alter only the extracellular milieu, but Archaea are sensitive to their environment, so even these reagents must be used with caution and their effect on a particular Archaeon must be determined empirically. For example, the typical response of *Sulfolobus solfataricus* to an osmotically active cryoprotectant is shown in Fig. 5. The implosion seen in Fig. 5 is most likely due to the flow of water across the cell's membrane. Many cryoprotectant strategies were evaluated for *Sulfolobus solfataricus*. A high molecular weight,

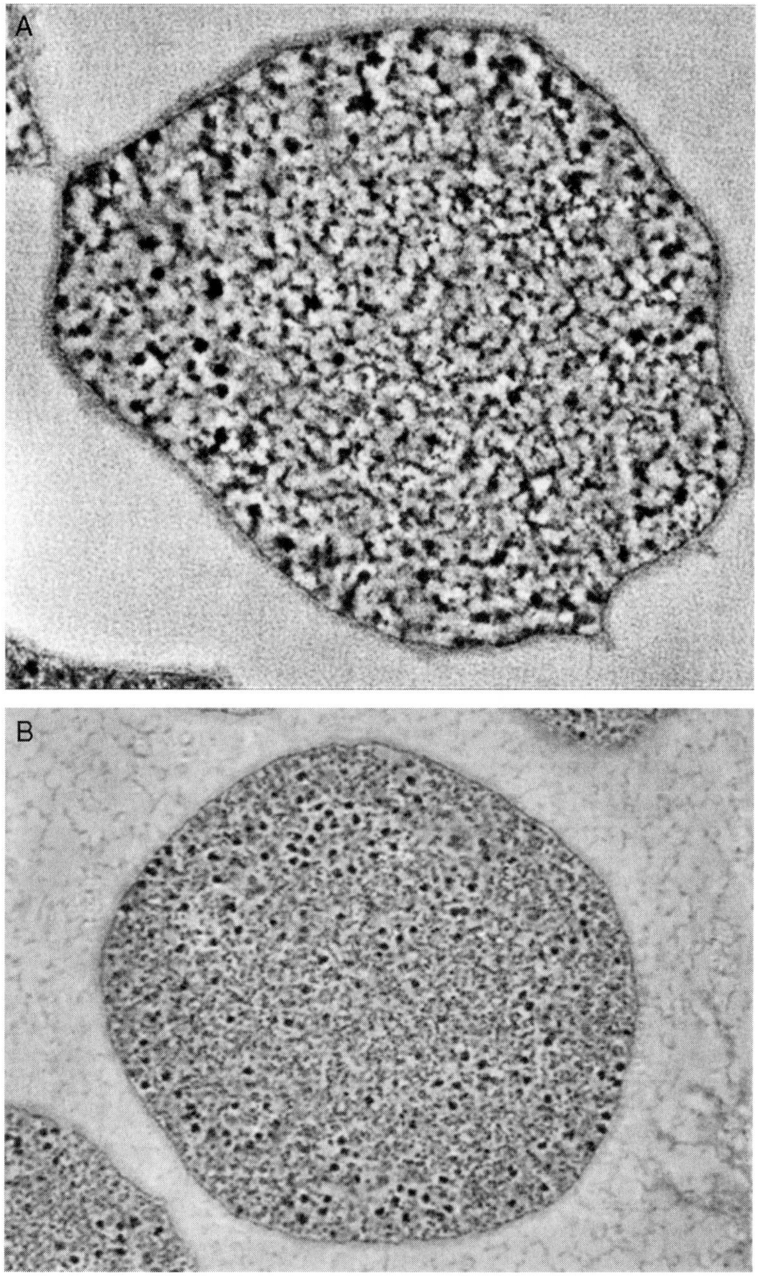

Fig. 4 An example of high pressure freezing of *Sulfolobus solfataricus* that shows evidence of freeze damage (A) compared with a well-frozen cell (B). The cell in (A) was frozen in growth medium ATCC 1723. The cell in (B) was frozen in ATCC 1723 with 8% Ficoll 70. Cells in (A) and (B) are ~2 μm in diameter.

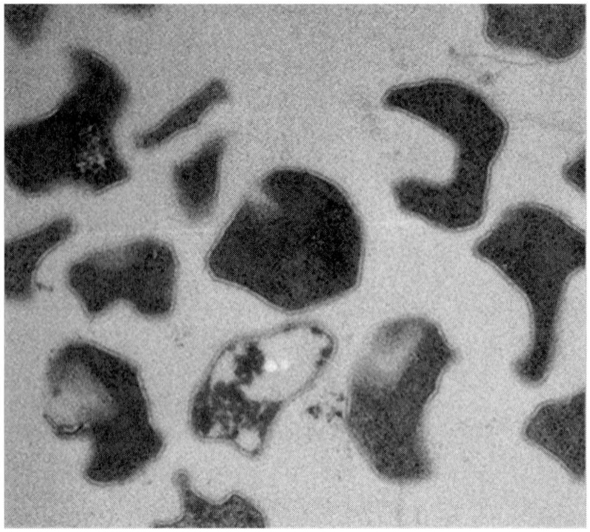

Fig. 5 High-pressure frozen *Sulfolobus solfataricus* showing osmotic damage. The cells were frozen in growth medium ATCC 1723 containing 150-mM mannitol. The large cell in the center is ~2 μm in diameter.

hydrophilic polymer of sucrose, Ficoll, proved to be effective in reproducibly eliminating freeze damage in high-pressure-frozen *Sulfolobus solfataricus*, Fig. 4B.

4. Locating Specific Biomolecules with Electron Tomography: Immuno-Electron Tomography

Although they provide excellent contrast and resolution, heavy metal staining cannot identify specific biomolecules. Heavy metals bind to biomolecules nonspecifically, so structures must be identified by context (e.g., cell membrane) or frequency (ribosomes) or via other (often indirect) information provided by other techniques. Direct identification of specific molecules by EM and ET can be done via immunocytochemistry. In one effective method for immuno-electron tomography a specimen is embedded in a resin, such as Lowicryl, that preserves the antigenic qualities of biomolecules. The sample is then sectioned, and an antibody to the biomolecule to be detected is then applied to the sections surface (see Chapter 19 by Morphew, this volume). Other methods of immuno-electron tomography apply the antibody after fixation and a permeablization step, but before embedding (see Chapter 22 by Cheutin *et al.*, this volume). The relative merits of different approaches to immuno-electron tomography are presented in Section IV of this book and will not be discussed here. We have found that antibody labeling of sections cut from Lowicryl-embedded Archaea is a useful companion to ET and have employed it to advantage.

III. Methods and Materials

A. Cultivation of *Sulfolobus solfataricus*

The morphology of *Sulfolobus solfataricus*, as observed via light microscopy, changes throughout the cell cycle (Lundgren and Bernander, 2005). To elucidate its morphology at higher resolution, it is important to have a means to reproducibly grow many cells that are at a chosen point of interest in the *Sulfolobus solfataricus* cell cycle. It has been found that dilution of a stationary phase *Sulfolobus solfataricus* culture provides a simple means of producing predominantly synchronized cell populations (Hjort and Bernander, 1999).

Starter cultures of *Sulfolobus solfataricus* may be obtained from either the American Type Culture Collection (ATCC, 2006) or the Deutsche Sammlung von Mikroorganismen und Zellkulturen (DSMZ, 2006) microbe collections. *Sulfolobus solfataricus* is a chemoorganotroph that grows well in or on a variety of media given the correct range of temperature and pH. A convenient way to maintain small cultures of *Sulfolobus solfataricus* is to use 100-ml glass bottles containing 50 ml of the *Sulfolobus* growth medium, ATCC 1723, in a commercial drying oven set to the desired temperature. The lids of the bottles should be loose as *Sulfolobus solfataricus* is aerobic, but to prevent excessive evaporation of the medium, the lid and neck of the bottle should be draped with a small square of aluminum foil. A fresh bottle should be started every 7–10 days, as the organisms exhaust the nutrients in the medium. Two milliliters of a robustly growing culture are adequate to seed 50 ml of ATCC 1723. However, when starting a fresh culture from frozen starter stock, or when dramatically increasing the volume of a culture, it is best to keep the starter volume at least 20% of the final volume. *Sulfolobus solfataricus* may be successfully stored at −80 °C indefinitely in a 20% ethylene glycol/ATCC 1723 solution. Given the medium, pH, and temperature that *Sulfolobus solfataricus* prefers, environmental contamination in *Sulfolobus solfataricus* cultures is unusual. However, standard microbiologic handling protocols should be followed: autoclaved bottles and autoclaved or sterile-filtered medium.

1. Growing Large *Sulfolobus solfataricus* Cultures

HPF requires large volumes of cells (discussed in Section III.C). Fortunately, *Sulfolobus solfataricus* is easy to cultivate in large quantities. The cultivator apparatus shown in Fig. 6 has proven to be quite effective. This cultivator keeps the thermophilic *Sulfolobus solfataricus* at the desired temperature, allows easy monitoring of cell density, and provides the vigorous sparging required for high-density *Sulfolobus solfataricus* cultures. A 250-ml starter culture is created from five individual static cultures grown in a drying oven. The starter culture is added to 1750 ml of ATCC 1723 after the cultivator and medium have reached the target temperature. The measured doubling time for *Sulfolobus solfataricus* in the apparatus of Fig. 6 at 75 °C is 4 h. This yields 2 liter of log phase *Sulfolobus solfataricus*

Fig. 6 A cultivation apparatus for *Sulfolobus solfataricus*. Growth occurs in the 2-liter-jacketed spinner flask at center (Bellco part 1965–51100, Vineland, NJ). The flask is maintained at 75 °C by heated water from the sealed water bath to the left of the spinner flask. The water is circulated through the jacket of the spinner flask by a centrifugal pump (the white box atop the water bath). The growth medium is sparged with air supplied from an aquarium pump through a sparging stone attached to the right-hand port of the spinner flask (covered with aluminum foil.) The left-hand port of the spinner flask is used to remove samples of the culture. Due to the elevated temperature, sparging, and the altitude of the laboratory, 1560 m, the growth medium evaporates in a matter of hours. To prevent evaporation, a condenser is attached to the lid of the spinner flask. This apparatus maintains the temperature to ±1 °C without significant evaporation (<10 ml) for periods as long as 10 days.

available for experimentation ~14 h after the culture is started. Culture density is measured via OD at 600 nm and by direct cell count with a Petroff-Hauser counter.

B. Cryoprotectants

An effective cryoprotectant for HPF of *Sulfolobus solfataricus* is a 8% solution of the sucrose polymer Ficoll 70 in ATCC 1723. A difficulty with Ficoll in this application is that the low pH of ATCC 1723 medium (pH 3) hydrolyzes the Ficoll, which releases sucrose that is metabolized by the *Sulfolobus solfataricus*, changing its metabolic state and morphology. Therefore, it is important to minimize the duration of contact between the Ficoll and the medium prior to freezing. This is accomplished by making a stock solution of concentrated Ficoll (40%) in

ATCC 1723 at pH 5.8 (this is the approximate pH of ATCC 1723 before it is titrated to pH 3 with sulfuric acid). Just before use, the 40% Ficoll/ATCC 1723 solution is titrated to pH 3. The *Sulfolobus solfataricus* should be kept at its optimum growth temperature, 75 °C, as long as possible prior to freezing. Mix the *Sulfolobus solfataricus* culture with the concentrated, pH-adjusted Ficoll solution in an Erlenmeyer flask placed in a water bath mounted on a wheeled cart that can transport the water bath and specimen to the high-pressure freezer at the nominal growth temperature of *Sulfolobus solfataricus*.

C. Concentrating Cells for High-Pressure Freezing

Even high-density cultures ($>10^9$ cell/ml^3) yield sections with too few cells per unit area for EM, so some form of cell concentration is required. Centrifugation and filtration are equally effective, but centrifugation is somewhat easier. The cells in the Erlenmeyer flask are decanted into 50-ml conical tubes and centrifuged at 5000 rpm for 8 m at room temperature (the conical tubes retain enough heat to keep the cells healthy for this short time).

D. High-Pressure Freezing

After concentration, the *Sulfolobus solfataricus* cells are transferred to a planchette suitable for HPF (see Chapter 2 by McDonald and Chapter 3 by Hess, this volume). A simple procedure by which to accomplish this is as follows. The fluid above the pelleted cells is decanted and discarded. Empty planchettes are pressed onto a square of parafilm placed on the stage of a dissecting microscope. The cell pellet is aspirated into a very narrow glass pipette via a rubber bulb. The tip of the pipette is guided into the central well of the lower section of the planchette using the dissecting microscope. The cells in the pipette are expressed into the planchette, slightly overfilling the cavity. Forceps are used to place the upper section of the planchette over the lower section. The proper placement of the cells in the planchette is crucial to HPF success because the entire volume of the planchette must be filled with cells. Voids of any size will result in improperly frozen cells. Some amount of fluid should be expressed when closing the freezer hat to assure that no voids are present. The area immediately surrounding the dissecting microscope should be humidified to assure that the cells in the planchette (which should still be substantially above room temperature) do not dry before they are frozen. A cold mist humidifier immediately adjacent to the microscope helps to alleviate the changes in cell morphology that are associated with pellet dehydration. The planchette is then placed in the high-pressure freezer and frozen as quickly as possible. Due to the somewhat unpredictable nature of the high-pressure-freezing process, it is recommended that at least three to five planchettes be frozen at a time to assure the availability of at least one good block.

E. Freeze Substitution, Embedding, and Sectioning

Many process variants exist for FS, embedding, and sectioning. The protocols described by Morphew (2006) are recommended for preparation of *Sulfolobus solfataricus* cells for ET. Good results have been obtained by FS in 1% OsO_4 in acetone followed by infusion with Epon and UV polymerization. Thin sections (70–80 nm) should be cut from each block and examined at low voltage (80 kV) before committing the time and energy needed to prepare high-quality serial thick section grids. After the thick section grids are prepared they should be stained with 2% uranyl acetate for 8 m, rinsed in distilled water, stained for 3 m in Reynolds lead citrate, and rinsed in distilled water again.

F. Fiducial Application

The best quality tomograms of plastic-embedded specimens require gold fiducials on both sides of a grid. However, application of the gold can be problematic (too few or too many fiducials) and can lead to grid failure. Three to four microliters of colloidal gold (British Biocell International Cardiff, UK) are applied with an Eppendorf pipette to each side of a grid held vertically in self-closing tweezers. Fifteen-nanometer colloidal gold is recommended. This size of gold is larger than optimum for good tomogram alignment, but smaller gold is often difficult to track through the densely stained cytoplasm of *Sulfolobus solfataricus*. The gold should remain on the grid for ~5 m, then be removed by carefully wicking the gold solution off of the grid with the edge of a slightly dampened, lint-free tissue. Care should be taken to avoid contact between the tissue and the Formvar. Rinse each side of the grid (using the same technique) with distilled water. Note that allowing the colloidal gold solution to dry onto the surfaces of the grid will leave a crystalline residue that will appear in the EM images. Once deposited, this residue is very difficult to remove by rinsing. The only way to determine if the gold application was successful is to examine the grid with the EM. The application of fiducials is unpredictable and often requires several iterations. As colloidal gold ages, it has a tendency to clump and may have to be discarded long before all of the gold in a container has been used up.

G. Imaging and Modeling Overview

The steps that lead to a 3D model of an entire *Sulfolobus solfataricus* cell are as follows:

1. Preshrink the resin and Formvar by exposure to the electron beam.
2. Create a grid map.
3. Select a cell to image.
4. Acquire two tilt series of each serial section of the cell (the reasons for two tilt series rather than one will be examined in Section III.H).

The operations performed after the images are acquired are as follows:

1. Construct dual-axis tomograms from the tilt series.
2. Create a single image stack from the merged tomograms of each serial section.
3. Segment the final image stack to create the 3D model.

A typical exponential growth phase *Sulfolobus solfataricus* cell is 1–2 μm in diameter, which means that dual-axis tomograms of six to eight 250-nm sections are required to image the volume of an entire cell. The tilt series for each axis is composed of 161 images that, with setup, take about an hour to acquire. Note that this duration may vary widely depending on the available microscope and camera. This means that 12–16 h of microscope time are needed for each cell. Processing the images takes ~6 h (user interaction with the computer) for each dual-axis tomogram. This means that the total time for tomogram creation/cell is 36–48 h. In total, a minimum of 48–64 h of operator time is needed to get the data for a single cell to the point where 3D model creation can begin. It should be clear from these calculations that careful planning is essential to maximize the efficiency of this process.

1. Preparing for Microscopy

Once a block of embedded cells has passed low-voltage thin-section screening, several serial thick (~250 nm) section grids can be prepared. To ensure that the entire grid is readily accessible a small sample rod that is capable of high tilt angles is recommended. Resin and Formvar shrink (collapse) along the *z*-axis when exposed to the beam of the EM (Luther *et al.*, 1988). The initial shrinkage is rapid, dramatic, and nonlinear, but the section stabilizes with increasing accumulated electron dose. The first exposure to the beam is a critical time for grids. In some cases the rapid collapse of the Formvar results in a hole near an edge of the slot. While some holes are repairable by applying another layer of Formvar, in most cases a hole makes a grid unusable; it must be discarded. The tendency for grids to break appears to be much worse for Archaea and Bacteria than for Eucarya, which may be due to the highly localized, intense staining of the tiny cells in contrast to the relative transparency of the resin and Formvar. However, grid breakage is reduced by initially exposing fresh grids to a strong but widely spread beam (objective aperture out) at low magnification for a few minutes.

2. Cell Selection

Careful selection of the correct cell to image is essential. All the serial sections of the cell must be well preserved and readily accessible at high tilt angles with the grid slot both parallel and perpendicular to the tilt axis of the sample rod. Meeting these requirements is not easy because grid sections often fold over, bend, or crack, which can make it problematic to image all of the serial sections that make up an entire cell. There are many possible strategies for identifying and selecting a cell. One approach is to take snapshots along the long axis of the grid at a medium magnification. The snapshots can be examined (even printed) at leisure and a candidate cell selected. A disadvantage of this approach is that it is wasted effort if the grid breaks. As the operator gains more experience with the procedures involved, especially the skill of recognizing a pattern of cells that contains the cell of interest, it is possible to recognize and search out sequential sections of the same cell "on-the-fly."

3. Serial Section Tracking

Impediments to recognizing serial sections on-the-fly by inspection though the microscope eyepiece are twofold. First, the cells are so small that a magnification on the order of 25,000× (~1-nm pixels) is required to see them clearly. However, at this magnification it is often impossible to recognize the features that distinguish different fields of view. Conversely, working at a lower magnification may allow easy identification of corresponding regions on the serial sections, but the cell of interest is now too small to recognize. The second obstacle to on-the-fly section tracking is that the outlines of 1-μm diameter cells can change remarkably from one 250-nm section to the next, especially near the ends of the cells. This often causes the tracking of serial sections of a cell to be a protracted, iterative process. A middle course between creating a printed map of a grid and on-the-fly cell recognition is to create an image stack that consists of two or three images per section, each image at a magnification that makes it easy to identify the features of interest. The images in this stack are especially useful to track the outlines of cells as they near their ends. The image stack can be displayed on a second monitor, allowing the entire pattern of cells from the previous section to be compared with an image of the subsequent section to be sure that the same cell is selected in each.

H. Acquiring Tilt Series

Commonly used strategies for the collection of tilt-series data are presented in several other chapters of this book. For Archaea we have found that dual-axis tomography is preferable to single-axis work. We have come to prefer smaller than conventional angles of tilt increment. A 0.75° tilt increment eases the task of tracking fiducials through the densely stained archaeal cytoplasm when converting tilt series to tomograms.

1. Keeping the Grid Intact

Keeping a grid intact becomes more urgent as tilt series accumulate over the 12–16 h needed to image an entire cell. Grids are most likely to fail during handling steps such as moving the grid from the grid storage container into the sample rod, taking the rod in or out of the vacuum of the microscope, rotating the grid 90° for the second tilt axis, and so on. This suggests a simple, minimalist strategy: acquire the tilt series for all serial sections for the first axis (slot parallel with the sample rod) then rotate the grid (slot at 90° to the sample rod) and acquire all of the tilt series for the second axis. This simple strategy has proven to be the most successful approach to capturing an entire *Sulfolobus solfataricus* cell; all of the necessary data are captured in one, potentially lengthy, microscope session.

I. Converting Tilt Series to Full-Cell Tomographic Reconstructions

IMOD is an effective set of software tools that converts raw tilt series into finished dual-axis tomograms and facilitates the concatenation of serial tomograms into a single image stack ready for segmentation. Comprehensive IMOD documentation and tutorials are available on the internet (Mastronarde, 2006) and are discussed in this volume (see Chapter 5 by O'Toole). No changes to the standard IMOD methodologies are required to process tomographic reconstructions of Archaea.

J. Segmentation: Delineating Objects of Interest

Segmentation is the term used to describe the demarcation of objects of interest within an image or series of images (see Chapter 8 by Marsh, Chapter 9 by Otegui and Austin, and Chapter 30 by Sandberg). In the context of this chapter, the end goal is to create 3D models of *Sulfolobus solfataricus* features such as the cell membrane or ribosomes. Manually tracing the outlines of such features is the classic approach to segmentation. In many ways, manual modeling is the most accurate and sophisticated modeling technique, as technology does not yet match the human brain in pattern recognition. Unfortunately, manual modeling can be quite time consuming. For example, a tomographic image stack for a single *Sulfolobus solfataricus* cell contains ~900 tomographic slices. A manual model of the cell membrane at this resolution requires that an operator trace the membrane on each tomographic section (layer) of the cell. Thus, a model of the cell membrane can be made up of 900 traces or "contours." Each contour of the cell membrane may take the operator anywhere from 30 sec to 1.5 min, depending on operator dexterity, computer speed, and so on. Such a model of the cell membrane for a single *Sulfolobus solfataricus* cell takes 17.5–24 h. The cell membrane is, however, a simple image. Ribosomes are more difficult to model because they are so numerous. An individual ribosome may be present on only 15–20 layers of a tomogram, but there are 5000–10,000 ribosomes in a log phase *Sulfolobus solfataricus* cell. Manually tracing 75,000–200,000 contours is considered by most to be out of the question, so objects such as ribosomes are generally manually modeled by an alternative approach: representation of the feature of interest with a geometric shape, like a sphere. Now, the operator only needs to mark the center of a sphere and an appropriate radius. While this approach definitely saves time, the use of predefined shapes to model features of interest has drawbacks. First, there is uncertainty of the location of the precise center of a ribosome. Finding a good approximation for this point requires that the operator move up and down the *z*-dimension of the image stack, which takes 20–30 sec per ribosome, roughly the time required to create a cell membrane contour.

Another disadvantage of the geometric shape modeling approach is that in almost all cases a geometric shape does not accurately represent the biological feature of interest, especially when the geometric shape has fixed parameters, such as a constant radius. Ribosomes in *Sulfolobus solfataricus* are not spherical, although a sphere can bound them. The bounding sphere overestimates the volume of the *Sulfolobus solfataricus* ribosomes by ~30%. If the precise shape or volume of ribosomes is important to the analyses being performed, a different approach to modeling is required.

The obvious alternative to manual modeling is to use software to attempt to emulate some aspects of the process the human mind goes through to recognize features in images.

1. MATLAB-Assisted Segmentation

Creating purpose-built software to segment biological images has been an expensive, time-consuming task that is beyond the means of most investigators because it requires one or more professional programmers to deal with the computer technology, and one or more biologists who know what to look for in the images. The medical community has done well in marshaling the resources required to build good segmentation software, as exemplified by the 3D images of CAT or MRI data that are routinely generated by minimally trained technicians at the local hospital's imaging center. Computer technology (hardware and software) has advanced significantly in the last few years. Commercial software products, such as MATLAB or Mathematica, now allow nonprogrammers to create sophisticated image-processing tools with inexpensive computers that were originally designed for today's sophisticated consumer of video games. The image-processing capabilities of Mathematica and MATLAB are similar. The choice between these software packages is usually based on the availability of local expertise to assist the novice user along the initial learning curve. MATLAB and Mathematica are approachable software development environments for noncomputer professionals because they allow problems to be described and solved at a higher level of abstraction compared to conventional programming languages such as C++. This high level of abstraction allows scientists to focus on science rather than computer science.

MATLAB should, however, be seen as an adjunct to the segmentation process; it is not a panacea. With MATLAB, an investigator will spend a substantial amount of time very accurately and iteratively, defining the nature of the biological feature of interest in terms of parameters such as shape, size, pattern, and transparency. The time spent in defining the parameters of the feature of interest is returned many times over during the processing of multiple image data sets. The segmentation of archaeal ribosomes is a good example. While it took a few days to write and debug MATLAB code that will segment ribosomes, the resulting code will segment the ribosomes of an entire *Sulfolobus solfataricus* cell in a few hours. Because of the highly interactive nature of the resulting segmentation process, it deserves a name of its own: MAS, which is an acronym for MATLAB-assisted segmentation. The accuracy of MAS is better than that of a human user because the computer does not become fatigued or distracted, no matter how many contours it creates.

2. One Structure at a Time

An effective strategy for MAS is to create a distinct MATLAB script for each biological feature of interest in the electron tomograms. MAS scripts are based on the image-processing functions provided in MATLAB's image-processing toolbox. Additionally, as a user becomes more familiar with the nature of the MAS process, the user will create his or her collection of reusable scripts that perform often-used operations. For this *Sulfolobus solfataricus* research, MAS scripts were created to segment the *Sulfolobus solfataricus* cell membrane, polysaccharide complexes, ribosomes, and ribosome exclusion zones. Additionally, software was written to allow the MAS-generated contours to be imported as models into the IMOD software package, allowing manual models and MAS models to freely coexist.

3. Modeling Ribosomes with MATLAB

Ribosomes are densely stained, relatively large objects that are profusely distributed throughout the *Sulfolobus solfataricus* cell. The attributes used to segment ribosomes are therefore: density, size, shape, extent over multiple tomographic slices, and frequency of occurrence. These criteria are relatively easy to specify in MATLAB. A bit of experimentation with an actual *Sulfolobus solfataricus* image stack showed that most of the objects that were likely to be ribosomes had a central core that extended over at least 10 tomographic slices and occupied an area of at least 50 pixels on each slice. The functions in the MATLAB image-processing tool kit made it easy to specify the amount of overlap that had to be present as a ribosome went from layer to layer in the tomographic reconstruction. The image-processing toolbox also made it easy to translate vague shape and image density requirements with a single statement similar to the following:

```
idx = find(([stats.Eccentricity] < .85) &
     ([stats.Solidity] > .75));
```

The Eccentricity parameter handles the shape requirement while the Solidity parameter deals with the image density requirement. An example of a single slice of a *Sulfolobus solfataricus* cell with segmented ribosomes is shown in Fig. 7.

Using a recent model PC with Microsoft Windows XP operating system and the academic version of MATLAB, the cell membrane and all of the ribosomes in an *Sulfolobus solfataricus* cell were segmented and converted into an IMOD model in ~7 h (Fig. 8).

The potential user is referred to The Mathworks Web site, www.mathworks.com, as the starting point for learning how to do MAS. Another author in this volume, Sandberg (2006) provides an online tutorial specific to using MATLAB to process EM images.

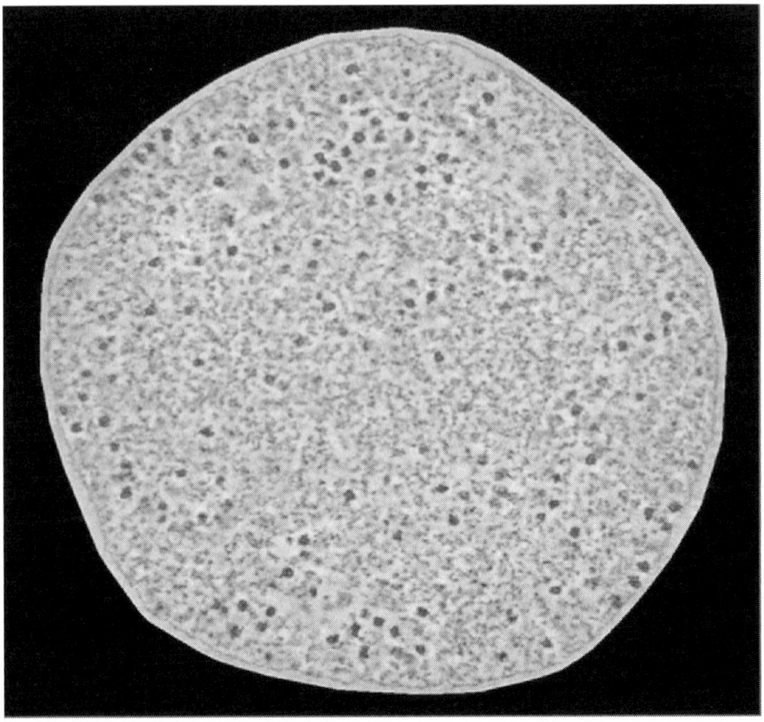

Fig. 7 A tomographic section of a log phase *Sulfolobus solfataricus* cell showing ribosomes (red outlines) segmented with a MATLAB script. (See Color Insert.)

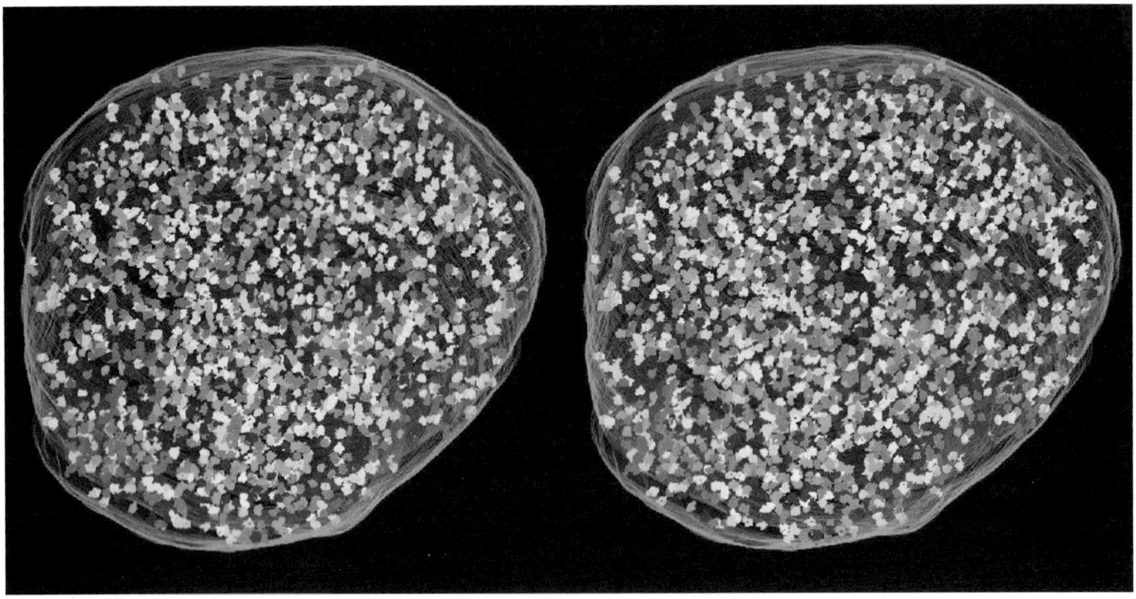

Fig. 8 A stereo pair image of a 3D, MATLAB segmented model of *Sulfolobus solfataricus* showing the cell membrane (green) and ribosomes (various). Multiple colors are used to facilitate visual discrimination of individual ribosomes. The model is based on tomographic reconstructions from six serial sections of a log phase cell. (See Color Insert.)

IV. Discussion

This chapter has discussed the use of a series of techniques, called EFCET (Section II.C) that allows routine creation of 3D high-resolution models of the Archaeon *Sulfolobus solfataricus*. While offering the best current resolution, EFCET has a variety of drawbacks that should be considered.

The largest single molecule in an Archaeon, the DNA, cannot be directly detected with EFCET due to the absence of a DNA-specific label that can be detected against a background of heavy metal stain. The lack of a DNA label is an example of the greatest opportunity for improvement for EFCET: the need for a molecule-specific, electron-dense stain. The lack of a method to label specific molecules is the greatest disadvantage of EFCET when compared to light-based techniques such as laser confocal microscopy.

An issue that adds some uncertainty to EFCET models is that there are gaps of uncertain size between the tomograms that represent the reconstructed serial sections. The reason for the gaps is not certain, but it is probably linked to mass loss from electron beam ablation or some form of evaporation due to the high vacuum of the electron microscope. Estimates of the size of the gaps have been made based on the amount of structure missing from serial sections of eucaryal structures of known size. The eucaryal studies estimate the amount of missing material is 15- to 40-nm per section (McIntosh, personal communication).

The greatest criticism of plastic-embedded, heavy metal stained specimens is that they have been so extensively modified by the preparation techniques that the specimens may not accurately represent the structural biology of interest. As a result, for many years now, much effort has been focused on perfecting cryo-EM to allow high-resolution studies of frozen, hydrated specimens as discussed in Section II.C. Despite the current limitations of cryo-ET, the technology is being pursued aggressively and holds the promise of revealing biological structure without the distortion that may be introduced by cross-linking, dehydration, and heavy metal staining used in the EFCET methodology.

V. Summary

The information-processing systems of Archaea are surprisingly similar to, but much simpler than, the analogous systems in Eucarya. These similarities, discovered via genomics, indicate that structural studies of Archaea may yield insights into the fundamental structures and mechanisms of the human branch in the tree of life. The small size of Archaea is an impediment to their detailed structural study and generally precludes light microscopy. A relatively recent enhancement of EM, ET, allows the 3D reconstruction of a specimen at resolutions on the order of 10 nm, making it the technology of choice for structural studies of Archaea. Archaea must be prepared in some manner before imaging to withstand the rigors of the vacuum of the electron microscope. The two preparation technologies are either freezing alone (cryo-EM), or freezing followed by FS fixation, plastic embedding, and heavy metal staining. Cryo-EM holds the promise of structure elucidation without the artifacts that might be introduced by plastic embedding and heavy metal staining. However, at present, plastic embedding is the only technique that reproducibly allows the creation of full-cell 3D models by software-assisted segmentation. The key advantages of plastic embedding over cryo-EM at this time are: plastic-embedded specimens are relatively easy to section, they have a high-electron-dose tolerance, making them good candidates for dual-axis ET, whereas cryo-EM specimens are sensitive to electron dose, making it difficult to acquire even a single-axis tomogram. Plastic-embedded specimens provide high-contrast, high signal-to-noise ratio images, due to their staining with heavy metals, while cryo-EM images have relatively little contrast and low signal-to-noise ratios. Future technology advances will surely overcome the present day limitations of cryo-EM, but in the mean time, useful work can be accomplished with ET of plastic-embedded specimens. The remaining great technological barrier for ET of any kind is the need to develop electron-dense, molecule-specific, labeling techniques to localize specific macromolecules in the structures we see with ET.

References

ATCC (2006). "American Type Culture Collection Web site" from www.atcc.org

Cubonova, L., Sandman, K., Hallam, S. J., DeLong, E. F., and Reeve, J. N. (2005). Histones in crenarchaea. *J. Bacteriol.* **187**(15), 5482–5485.

DeLong, E. F. (2005). Microbial community genomics in the ocean. *Nat. Rev. Microbiol.* **3**(6), 459–469.

DSMZ (2006). "Deutsche Sammlung von Mikroorganismen und Zellkulturen Web site" from www.dsmz.de

Eckburg, P. B., Lepp, P. W., and Relman, D. A. (2003). Archaea and their potential role in human disease. *Infect. Immun.* **71**(2), 591–596.

Hjort, K., and Bernander, R. (1999). Changes in cell size and DNA content in Sulfolobus cultures during dilution and temperature shift experiments. *J. Bacteriol.* **181**(18), 5669–5675.

Lundgren, M., and Bernander, R. (2005). Archaeal cell cycle progress. *Curr. Opin. Microbiol.* **8**(6), 662–668.

Luther, P. K., Lawrence, M. C., and Crowther, R. A. (1988). A method for monitoring the collapse of plastic sections as a function of electron dose. *Ultramicroscopy* **24**(1), 7–18.

Madigan, M. T., and Martinko, J. M. (2006). "Brock Biology of Microorganisms." Pearson Prentice Hall, Upper Saddle River, NJ.

Mastronarde, D. N. (2006). "IMOD User Documentation." Retrieved January 8, 2006, from http://bio3d.colorado.edu

Morphew, M. K. (2006). "Practical Mehods in High-Pressure Freezing, Freeze-Substitution, Embedding, and Immunocytochemistry for Electron Microscopy." Retrieved January 7, 2006, from http://bio3d.colorado.edu

NCBI (2006). NCBI Microbial Genome Database, National Center for Biotechnology Information.

Pace, N. R., Stahl, D. R., Lane, D. J., and Olsen, G. J. (1985). Analyzing natural microbial populations by rRNA sequences. *ASM News* **51,** 4–12.

Robertson, C. E., Harris, J. K., Spear, J. R., and Pace, N. R. (2005). Phylogenetic diversity and ecology of environmental Archaea. *Curr. Opin. Microbiol.* **8**(6), 638–642.

Rother, M., and Metcalf, W. W. (2005). Genetic technologies for Archaea. *Curr. Opin. Microbiol.* **8**(6), 745–751.

Sandberg, K. (2006). "Introduction to Image Processing in MATLAB." Retrieved Janary 21, 2006, from http://amath.colorado.edu/courses/4720/2000Spr/Labs/Worksheets/Matlab–tutorial/matlabimpr.html

Schleper, C., Jurgens, G., and Jonuscheit, M. (2005). Genomic studies of uncultivated archaea. *Nat. Rev. Microbiol.* **3**(6), 479–488.

Woese, C. R., and Fox, G. E. (1977). Phylogenetic structure of the prokaryotic domain: The primary kingdoms. *Proc. Natl. Acad. Sci. USA* **74**(11), 5088–5090.

CHAPTER 8

Reconstructing Mammalian Membrane Architecture by Large Area Cellular Tomography

Brad J. Marsh

Institute for Molecular Bioscience, Queensland Bioscience Precinct
The University of Queensland, Brisbane
Queensland 4072, Australia

Centre for Microscopy and Microanalysis
The University of Queensland, Brisbane
Queensland 4072, Australia

School of Molecular and Microbial Sciences
The University of Queensland, Brisbane
Queensland 4072, Australia

Microscopy has provided crucial insights into the fundamental features and architecture of mammalian cells and organelles for now over a century. These glimpses of cellular fine structure have thus guided our impressions—as molecular

METHODS IN CELL BIOLOGY, VOL. 79

0091-679X/07 $35.00
DOI: 10.1016/S0091-679X(06)79008-2

cell biologists—of how mammalian cells and their numerous membrane-bound internal compartments are organized within three dimensions (3D), although for the most part these extrapolations have come from static two-dimensional (2D) images taken from the light or electron microscope. Recently, however, we have finally been afforded the chance to dissect subcellular membrane architecture and dynamics at the nanoanatomical level in 3D through development of the technique referred to as electron microscope (EM) tomography, also frequently termed "cellular tomography" (ET). With ET now in hand as a tool that effectively allows one to study molecular membrane traffic "in context" (*in situ*), it has become increasingly important to push the continued advancement of this method so that large cellular volumes can be reconstructed without sacrificing the ability to clearly resolve structures of interest. Likewise, it has become critical to do this as quickly and efficiently as possible in order to generate statistically significant sample sizes that offer reliable insights into cell organization under different physiological or experimental conditions. In this chapter, some of the technical developments, as well as key biological questions that have driven the development of large area ET, will be presented and discussed.

I. Introduction and Rationale

The mammalian Golgi complex was first observed by light microscopy over 100 years ago by Camillo Golgi (Golgi, 1898) and visualized by electron microscopy by Ernest Fullam, Albert Claude, and Keith Porter in 1944 (Porter *et al.*, 1945). Despite this, the limitations of current and/or conventional cell biological and biochemical techniques for elucidating the structure–function relationships of such a complicated organelle can be evidenced by the controversies which remain today with respect to even the most fundamental mechanisms of transport to, through and from this organelle (Elsner *et al.*, 2003; Gu *et al.*, 2001; Lippincott-Schwartz *et al.*, 2000; Marsh and Howell, 2002; Storrie and Nilsson, 2002). A myriad of approaches ranging from conventional two-dimensional (2D) EM of thin-sectioned material, *in vitro* functional (cell free) assays of reconstituted Golgi membranes, live-cell imaging studies to organellar proteomics analyses have so far failed to satisfactorily answer basic questions related to molecular membrane traffic through the Golgi (Breuza *et al.*, 2004; Farquhar and Palade, 1998; Mellman and Simons, 1992; Orci *et al.*, 1989a; Palade, 1988; Rothman and Orci, 1992; Wu *et al.*, 2004; Yates *et al.*, 2005).

Part of the problem lies in the fact that during the 1980s, as progress was accelerating in molecular biological and live-cell imaging studies of the Golgi and intracellular membrane traffic, conventional fine structure studies of intracellular membranes in mammalian cells were for the most part considered passé due both to their static nature and the limitations of 2D analysis, and resources for EM in general were significantly downgraded (Griffiths, 2004; McIntosh, 2001). At the same time, high-resolution EM studies that focused on structure were for the most part dismissed in favor of alternative EM approaches that for the first time allowed the immunolocalization of an antigen or antigens of interest on standard thin sections (typically 60–100 nm), and thus provided crucial information regarding the subcellular distributions of proteins (Griffiths *et al.*, 1984; Orci *et al.*, 1984; Slot and Geuze, 1983). However, as with any other method, techniques for immunolabeling suffer from their own limitations. As noted above, Golgi membranes in particular are often highly convoluted, and so it frequently becomes difficult to unambiguously follow connectivity from one section to the next. Other limitations stem from the fact that only antigens exposed at the section surface can be labeled, and that steric hindrance due to the immunogold itself can limit spatial resolution to between 10 and 20 nm in some cases; labeling efficiency is also compromised. In addition, immunolabeling studies of mammalian membrane traffic have relied primarily on classical chemical fixation techniques to stabilize membranes for visualization in the EM. As this process takes anywhere from seconds to minutes, labile structures and highly dynamic events, which we now know from live-cell imaging studies to be typical for many of the steps in intracellular membrane traffic, are unlikely to be reliably or reproducibly captured (Gilkey and Staehelin, 1986; Lippincott-Schwartz *et al.*, 2000, 2001). It is now more apparent than ever that conventional chemical fixation often captures the cell's (and/or organelle's) response to the fixative as opposed to immobilizing cellular structures in a "close-to-native" state (Biel *et al.*, 2003; Dubochet, 1995). Thus, it follows that reliable descriptions of morphology for the membranes of a rapidly changing organelle such as the Golgi can only be achieved through the use of methods for preservation that immobilize all cellular activity within milliseconds, such as plunge freezing or high-pressure freezing. To be supremely useful, however, such methods for improved ultrastructural preservation of biological specimens will eventually have to incorporate a capacity for the precise localization and identification of proteins, lipids, and inorganic ions of interest within fast/high-pressure frozen cells or tissues (see Chapters in Part IV, this volume).

Three-dimensional (3D) studies of Golgi organization by stereo-scanning EM (Ho *et al.*, 1999; Rambourg *et al.*, 1974), together with "pseudo" 3D studies by analysis of stereo pairs of 2D images of thick specimens tilted in the EM—carried out from the early 1970s through the late 1990s—contributed enormously to the body of knowledge about the 3D organization of Golgi and endoplasmic reticulum (ER) membranes (Hama *et al.*, 1994; Lindsey and Ellisman, 1985; Rambourg and Clermont, 1997). However, all of these studies were ultimately frustrated by the fact that the superimposition of 3D information in the 2D images restricted resolution along the *z*-axis to the thickness of the section at best (Marsh, 2005).

The application of high-resolution ET to the study of Golgi membranes and associated tubules and vesicles in mammalian cells, first in conventionally prepared (Ladinsky *et al.*, 1994), then in fast-frozen/freeze-substituted normal rat kidney (NRK) cells (Ladinsky *et al.*, 1999), convincingly demonstrated the utility of this approach for providing novel insight into dynamic trafficking events at the Golgi complex and, in particular, at the exit face of the Golgi—the *trans*-Golgi network (TGN) (Griffiths and Simons, 1986). This ability to truly dissect the resultant 3D information voxel-by-voxel and/or in any desired orientation (see Fig. 2 in Ladinsky *et al.*, 1994) revealed the underlying power of this technique for starting to address long-standing questions related to Golgi structure–function relationships.

At the inception of our own work, and as noted above, 3D EM studies of the Golgi and associated subcellular compartments had been performed in a number of cell types (Katsumoto *et al.*, 1991; Ladinsky *et al.*, 1994; Rambourg and Clermont, 1997; Tanaka *et al.*, 1986), but not for the insulin-secreting beta cells of the pancreas. Most of what is currently accepted as "conventional wisdom" regarding the mechanisms for insulin granule formation/maturation, trafficking and release emerged from conventional 2D EM surveys of thin sections cut from chemically fixed beta cells in the 1970s and 1980s (Howell and Bird, 1989; Howell and Tyhurst, 1984; Orci, 1986; Orci *et al.*, 1988). However, disparities had started to appear between some of the key concepts regarding structure–function relationships in the beta cell that came out of those early structural studies, and more recent advances in understanding the molecular mechanisms underlying insulin granule biogenesis and exocytosis (Arvan and Castle, 1998; Goodge and Hutton, 2000; Guest *et al.*, 1997; Kowluru and Morgan, 2002; Kuliawat and Arvan, 1994; Molinete *et al.*, 2000; Rutter, 1999; Tsuboi and Rutter, 2003).

Thus, in light of the dramatic improvements that had taken place since the late 1980s in terms of methods for high-quality/reliable sample preparation (Gilkey and Staehelin, 1986; McDonald and Morphew, 1993), instrumentation for EM in general and ET in particular, and hardware/software for complex 3D analysis of biological data (Kremer *et al.*, 1996; Mastronarde, 1997), we felt the time was right to revisit questions regarding precisely where and how are insulin secretory granules formed, and how changes in the organization of the Golgi and other subcellular membranes under key physiological conditions relate to structure–function relationships in the beta cells of the endocrine pancreas.

II. Methods and Materials

For our own studies of the insulin biosynthetic pathway, it was clear from the outset that to attain the most reliable insights into the membrane traffic itinerary of insulin in the pancreatic beta cell, our best approach would be to closely follow the example of Ladinsky *et al.* (1999). Ladinsky and colleagues employed a combination of the then disparate techniques of fast-freezing/freeze-substitution (followed by plastic embedment) with dual-axis ET of four serial semithick (250 nm) sections. This modification of the standard single-axis tomographic approach improves the resolution and symmetry of cellular structures in all 3D (Mastronarde, 1997; Penczek *et al.*, 1995; Taylor *et al.*, 1984). The end result was a comparatively large ($\sim 1 \times 1 \times 4\ \mu m^3$) volume of the Golgi ribbon in an NRK cell reconstructed at high ($\sim$7 nm) resolution, which demonstrated near optimal structural preservation and clarity throughout. From this, Ladinsky *et al.* were able to computationally dissect Golgi membranes and associated vesicle and tubule budding profiles with maximal confidence, and were afforded a reasonably large portion of the NRK Golgi ribbon within a single cell for analysis.

A. Mammalian Cell and Tissue Culture

Since ET of large cellular regions using the techniques that will be described below still remains a relatively painstaking and labor-intensive process in comparison to conventional ET, it has been crucial for us to ensure that cells and/or tissue to be studied in detail are as healthy and as physiologically viable as possible. We initiated our studies with the Syrian hamster-derived beta cell line, HIT-T15, because at early passages it retains the capacity to synthesize and secrete insulin in response to extracellular glucose (Fig. 1) (Marsh *et al.*, 2001a; Santerre *et al.*, 1981).

The fact that these cells are modestly granulated compared with beta cells *in vivo* provided an advantage; it allowed us to conduct a detailed analysis of interactions between the microtubule cytoskeleton and the key organelles involved in membrane traffic/secretion of insulin (i.e., the ER, Golgi, and insulin granules) (Novikoff *et al.*, 1975; Rios and Bornens, 2003; Yorde and Kalkhoff, 1987; see Fig. 2). We initially tried various techniques for rapid freezing, since immortalized beta cell lines are usually cultured as a crude 2D monolayer in the absence of supplementation with growth factors and/or extracellular matrix (Ohgawara *et al.*, 1995), and thus are theoretically thin enough to be amenable to plunge freezing. However, we consistently found that a higher fraction of beta cells were well preserved by high-pressure freezing. This technique is one of a number of different fast-freezing methods that can be used to vitrify cells/tissue, so as to capture cellular events in a native, frozen-hydrated state (see Chapter 1 by Dubochet, Chapter 2 by McDonald, and Chapter 3 by Hess, this volume).

Although immortalized beta cell lines are easily obtained and cultured, and continue to serve as useful models for the study of the insulin secretory pathway and diabetes (Gleason *et al.*, 2000; Poitout *et al.*, 1995), there are obvious and inherent limitations to their usefulness. In general, they contain relatively low levels of insulin (and low numbers of insulin granules) compared to normal adult pancreas (Breant *et al.*, 1992), and they respond poorly (in terms of granule release) to physiologically relevant levels of glucose (Breant *et al.*, 1992; Poitout *et al.*, 1996), suggesting that their mechanisms for sorting/processing insulin might differ substantially from beta cells *in situ* in intact islets. Additionally, islet beta cells are polarized *in vivo* (Bonner-Weir, 1988; Orci *et al.*, 1989b). Although transformed beta cell lines' cells can be induced to polarize experimentally (Cortizo *et al.*, 1990), they are typically not polarized when grown in culture. Such a loss of cell polarity undoubtedly affects cellular trafficking pathways (Lombardi *et al.*, 1986).

Consequently, we modified our cryopreparative techniques for use with intact islets of Langerhans obtained from the pancreata of adult, female Balb/c mice. Care was taken to ensure that islets were as healthy as possible at the point of

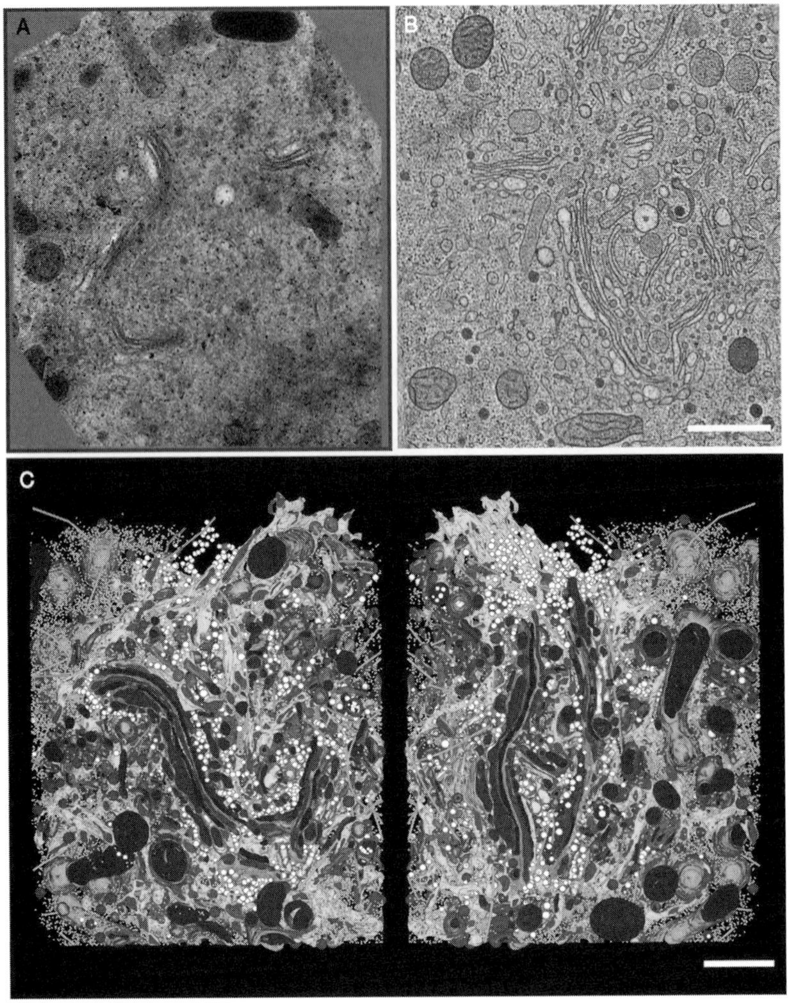

Fig. 1 Organization of the Golgi region at high (~6 nm) resolution and in 3D in a mammalian cell that makes and secretes insulin in a regulated manner (Marsh *et al.*, 2001a) (Copyright 2001, National Academy of Sciences, USA). (A) A single 2D image of a 400-nm plastic section cut from an immortalized rodent beta cell line (HIT-T15) prepared by high-pressure freezing and freeze-substitution, followed by plastic infiltration/embedding. This image is one of 80 such images that together comprise a "tilt series," collected for each axis by imaging the specimen every 1.5° as it was tilted over a range of ±60°. These images were aligned using 10-nm gold fiducials, as described in the chapter by O'Toole, this volume. (B) The tomograms calculated from each set of aligned tilts are then brought into register and combined in 3D to produce a single, dual-axis reconstruction. To enable the study of a larger portion of the Golgi ribbon this process was repeated for multiple 400-nm sections, and the serial reconstructed volumes were aligned with each other to create a single, large ($\sim 3.1 \times 3.2 \times 1.2\ \mu m^3$) volume reconstruction. Once all of the visible structures in the Golgi region had been modeled using the IMOD software package (Kremer *et al.*, 1996), any given (modeled) object could be studied closely in 3D either alone or in context with any other object(s). (C) The Golgi complex with seven cisternae

cryopreservation by first culturing them under normal conditions either for several hours or overnight after isolation. This step allows the isolated tissue to recover following the isolation procedure itself (Sandler and Andersson, 1984), and promotes both the viability (e.g., glucose responsiveness, protein synthesis, and insulin secretory capacity) and integrity of isolated islets by allowing repair/reformation of the connective tissue/collagen capsule at the periphery of rodent islets (Bonner-Weir, 1989; Wang and Rosenberg, 1999). Islets were not cultured or used for experiments more than 72 h following isolation.

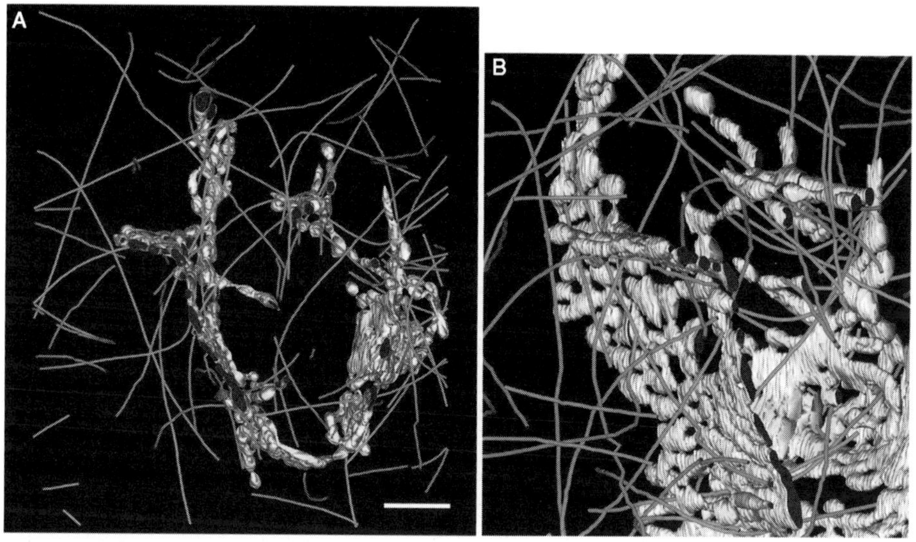

Fig. 2 3D display of part of the Golgi ribbon shown in Fig. 1 revealing the *in situ* physical relationships between the *cis*-most cisterna and the microtubule cytoskeleton. (A) Microtubules closely follow and occasionally form contacts with the membranes of the *cis*-most cisterna. Note that in the modeled region, microtubules do not exhibit a radial organization. Scale bar = 500 nm. (B) A higher-magnification view oriented to show that the paths of some microtubules closely follow the membranes over considerable distances. Microtubules traversing the Golgi stack can also be observed. Original images are from Marsh *et al.* (2001a) and are reproduced with permission from Proceedings of the National Academy of Sciences USA.

(*cis–trans*: C1–C7) is at the center. The color coding is as follows: C1, light blue; C2, pink; C3, cherry red; C4, green; C5, dark blue; C6, gold; C7, bright red. The Golgi is displayed in the context of all surrounding organelles, vesicles, ribosomes, and microtubules: ER, yellow; membrane-bound ribosomes, blue; free ribosomes, orange; microtubules, bright green; dense core vesicles, bright blue; clathrin-negative vesicles, white; clathrin-positive compartments and vesicles, bright red; clathrin-negative compartments and vesicles, purple; mitochondria, dark green. Two views of the modeled (segmented) Golgi region are provided, rotated 180° with respect to each other around the vertical axis. This figure has been reproduced with permission from: "Lessons from tomographic studies of the mammalian Golgi, *Biochim. Biophys. Acta* (2005), **1744,** 273–292." Scale bars = 500 nm. (See Color Insert.)

B. Fast-Freezing and Freeze-Substitution

As noted earlier, to preserve membrane structure and organization in beta cells with high fidelity, we initially employed two independent techniques for cryoimmobilization: plunge freezing (Ladinsky *et al.*, 1999) and high-pressure freezing (McDonald and Morphew, 1993). While plunge freezing yielded some improvement in the preservation of cellular ultrastructure over conventional chemical fixation methods, more reproducible results were generally obtained by high-pressure freezing. This is likely due to the fact that this latter technique inhibits ice crystal growth in thicker samples, even though the rate of cooling is lower than with plunge freezing (Gilkey and Staehelin, 1986; McDonald and Morphew, 1993). Importantly, no major differences could be discerned in secretory granule, vesicle or Golgi morphologies between plunge and high-pressure frozen cells, assuring us that high pressure per se did not appreciably perturb cellular membranes or architecture.

For immortalized beta cell lines such as HIT-T15 that essentially grow as monolayers, cells were seeded onto small plastic chips cut from Thermanox™ coverslips (Nalge Nunc International, Rochester, NY) and cultured for 2–3 days prior to experiments (Marsh *et al.*, 2001a). Immediately prior to freezing, each chip was transferred into a small brass, sandwich device comprised of two, interlocking halves referred to as a "planchette" (Swiss Precision, Inc., Millbrae, CA), also prewarmed to 37 °C. The half of the planchette holding the plastic chip on which the cells had grown was filled with regular medium warmed to 37 °C and buffered with HEPES (10 mM). The second half of the planchette was filled with medium containing 10% dialyzed Ficoll (Sigma, St. Louis, MO) as an extracellular cryoprotectant, and was placed on top of the half of the planchette containing the cells. After brief (<10 sec) manipulation into the tip of the planchette holder, cells were frozen within 10–20 msec under high pressure (~2100 atm) using a HPM 010 high-pressure freezer (BAL-TEC AG, Balzers, Liechtenstein). Frozen specimens were stored under liquid nitrogen. Water was removed from these specimens by freeze-substitution with anhydrous acetone (Ernest F. Fullam, Inc., NY) containing 0.5% glutaraldehyde/0.1% uranyl acetate (UA) at −90 °C for 1–2 days, followed by substitution with acetone containing 1% OsO_4/0.1% UA (Electron Microscopy Sciences, PA) at −70 °C. Specimens were allowed to warm to 0 °C over 2–3 days, whereupon they were rinsed five times with anhydrous acetone. Samples were then warmed to room temperature, infiltrated with increasing concentrations of Embed812-Araldite resin (Electron Microscopy Sciences, PA), and flat-embedded between two Teflon-coated glass microscope slides (Miller-Stephenson Chemical Co., Inc., Sylmar, CA; VWR Scientific, Inc., West Chester, PA). Resin was polymerized at 60 °C in a dry oven over 1–2 days.

Isolated islets cultured overnight as a suspension of free-floating cell clusters were similarly maintained at 37 °C on a humidified heating block in culture medium buffered by the addition of 10-mM HEPES immediately prior to freezing. Typically, 10–30 islets (depending on size) were gently manipulated into one-half of the prewarmed holder prefilled with HEPES-buffered medium containing fetal bovine serum (FBS) 10% (v/v) (Sigma, St. Louis, MO). All manipulations were carried out on Parafilm® placed on top of an inverted heating block warmed to 37°C under a dissecting microscope. The second half of the planchette, filled with RPMI containing 10% dialyzed Ficoll (MW = 70 kDa) and 0.5% Type IX ultra-low temperature gelling agarose (Sigma, St. Louis, MO) as extracellular cryoprotectants, was then placed on top of the half of the sample holder containing the islets. The pair of interlocking hats was then secured into place at the tip of the specimen holder and the islets frozen under high pressure, freeze-substituted, and plastic embedded essentially as described previously and above (Marsh *et al.*, 2001a, 2004).

C. Microtomy and Grid Preparation

Well-preserved beta cells (in the case of beta cell lines) or islets were individually identified by phase-contrast light microscopy, excised from the resin, and remounted onto plastic stubs. HIT-T15 cells on plastic chips were mounted in an orientation that permitted *en face* sectioning; orientation is irrelevant for intact islets. Ribbons of thin (40–60 nm) or thick (300–400 nm) sections were cut on a microtome (Leica Microsystems, Vienna, Austria) for conventional 2D survey at 80–120 kilo electron volt (keV) to assess the quality of cell/islet preservation and to select regions for subsequent study or for high-tilt ET on instruments operating at higher voltages (≥300 keV) (see below), respectively (Marsh *et al.*, 2001a, 2004). Ribbons of serial thick sections were collected onto Formvar-coated copper (2 × 1 mm) slot grids (Electron Microscopy Sciences, Hatfield, PA) and poststained with 2% aqueous UA or 3% UA in 70% methanol (15 min) and Reynold's lead citrate (3 min). These samples often require an additional carbon-coating step to minimize charging/movement in the electron beam, particularly when tilted for tomography (Marsh, 2005). Colloidal gold particles (10 nm) were then deposited on both surfaces of these sections for use as fiducial markers during subsequent image alignment.

D. Basic Instrumentation Requirements for ET of Thick Plastic Sections

At various points in our studies, we have successfully viewed and collected tilt series images from ribbons of equivalently stained 300- to 400-nm-thick sections using (in decreasing order of operating voltage) a JEM-1000 high-voltage EM (HVEM) operated at either 500 keV, 750 keV, or 1 million electron volt (MeV) (JEOL USA, Inc., Peabody, MA), a JEM-4000FX intermediate-voltage EM (IVEM) (JEOL USA, Inc.) operated at 400 keV, and with a Tecnai F30 field emission gun (FEG) IVEM (FEI) operated at 300 keV. Regardless of operating voltage, the grid holding the specimen was tilted in a eucentric goniometer typically at either 1° or 1.5° intervals over a range of 120°–140° about two orthogonal axes (Mastronarde, 1997), as summarized in the legend that accompanies Fig. 1. In earlier tomographic studies of the Golgi region in mammalian cells (Ladinsky *et al.*, 1994, 1999; Marsh *et al.*, 2001a), single images taken at each tilt were collected on film (SO163, Kodak Eastman Company, or 23D56, Agfa-Gevaert NV), with the negatives subsequently digitized on a motorized light table using a high-resolution STAR-1 cooled charge-coupled device (CCD) camera (Photometrics, Ltd., Tucson, AZ), yielding a final pixel size of ~2.3 nm. Although the processing of film and digitizing of EM negatives is labor intensive and time consuming, this procedure inherently afforded those investigating extended membrane structures, like the Golgi, a much larger field of view for a "single snapshot" at each tilted view than could be accomplished otherwise without sacrificing resolution by imaging at a lower magnification to overcome the smaller field of view of a typical CCD (Ladinsky *et al.*, 1999; Marsh *et al.*, 2001a). Even with film, however, to take advantage of as much of the image area on the EM negative as possible required the user to digitally tile a larger image from "montaged" image arrays in *x* and *y* with defined areas of overlap (typically 10% of the pixel dimensions of each piece of the montage), using software to precisely control the digitizing camera as well as movement of the motorized light table (Fig. 1).

E. Acquisition of Digital Tilt Series from Large Cellular Areas by Automated IVEM/HVEM Montaging Tomography

Thick sections of 300–400 nm were typically imaged at 12,000× using a JEM-1000 HVEM operating at 750 keV (JEOL USA) or at 15,500×, 20,000×, or 23,000× using a Tecnai F30 IVEM (FEI). Motorized, tilt-rotate specimen holders (Models 650 and CT3500TR; Gatan, Pleasanton, CA) were available on both kinds of microscopes. Tilt series data were digitally recorded using semiautomated methods for CCD data acquisition, image focus, and alignment as the sections were serially tilted through 1° or 1.5° increments over a range of 120°–140° about two orthogonal axes. The camera and microscope were controlled by the program, SerialEM (Mastronarde, 2005).

To best understand complex spatial and structural relationships for extended cytomembranes, such as the Golgi or ER, it is important to examine a large cellular area. We overcame the limited field of view typically afforded by CCD imaging on the EM by using a variation of the methods for digital montaging of tiled image arrays, discussed above. This required the automation of image montaging using image shift directly in the EM, which is now implemented in the SerialEM data collection package (Mastronarde, 2005). Briefly, a set of images was collected at different coordinates in *x* and *y* at each tilt, using the software to automatically reposition and capture each individual image piece, yielding a montage of overlapping frames at each tilt (Fig. 3A). For our studies of the Golgi region in insulin-secreting cells, we have typically used montages of either 2×2 or 3×3 panels in *x* and *y* (depending on the magnification and thus the final CCD pixel size) to accommodate a total cellular area $\geq 4\ \mu m^2$ per reconstruction (Fig. 3).

Although there is a concomitant increase in the time required to digitally collect a montaged tilt series, overall the yield of data for the time invested is higher with this procedure (Mastronarde, 2005). For example, a 3×3 montage with images collected every 1.5° over an angular tilt range of 120° (±60°) means an increase in the number of individual image frames from 81 to 729, but the steps of tilting, focusing, and tracking are performed more or less as they would be for a single panel tilt series rather than for every frame. The largest additional temporal cost related to data collection/image acquisition—aside from the CCD acquisition time—is usually due to the additional brief (anywhere from 0.5 to 4 sec) preirradiation required just prior to image capture for each piece of the montage, to combat specimen charging that is often inherent to tomography of thick, stained, plastic sections.

F. 3D Reconstruction of Large Cellular Areas

The computer software package IMOD comprises a set of image processing, modeling and display programs used for tomographic reconstruction from EM tilt series, as well as for 3D reconstruction of EM serial sections and optical sections (Kremer *et al.*, 1996). Briefly, the package contains tools for assembling and aligning data within multiple types and sizes of image stacks, viewing 3D data from any orientation, modeling and display of the image files, and to obtain accurate quantitative information in either 2D or 3D. IMOD was developed primarily by David Mastronarde, Rick Gaudette, Sue Held, and Jim Kremer at the Boulder Laboratory for 3D Electron Microscopy of Cells.

Using the IMOD package that incorporates the eTomo graphical user interface for the alignment and tomographic reconstruction of digitally tiled images, individual pieces of the montaged tilt series images acquired for each of the two orthogonal axes were first brought into register with one another by cross-correlation of overlapping image data to determine the relative displacements in x and y between adjacent frames for a given tilted view (Fig. 3A, B, and B′). Occasionally it was necessary to manually correct errors in correlation-based frame registration using the computer tool MIDAS to ensure precise positioning of individual pieces of the montage, and/or to account for any shifts in x and/or y that were too large to be handled automatically. Likewise, cross-correlation was used to calculate the set of translations in x and y that needed to be applied from one tilted view to the next through the stack of tilt series images to generate a crudely prealigned stack. Subsequently, individual frames or pieces of the montage were "blended" across the zone of overlap, using redundant image data in the adjacent image pieces and fused to yield a single image for each tilted view (Fig. 3C and C′).

The images of the now-blended tilt series were then treated essentially like a typical image stack and more accurately aligned by tracking the positions of ~200 of the 10-nm gold fiducial markers on the top and bottom surface(s) of the sections, preferentially with an equal number of fiducials on both surfaces. Ideally, most of the fiducial points selected and tracked for the first axis are also tracked for the second axis (Fig. 3C, C′, D, and D′). This step is accomplished using the program **transferfid**, which automatically determines the most reliable estimate of how fiducial markers correspond between the tilt series collected around the

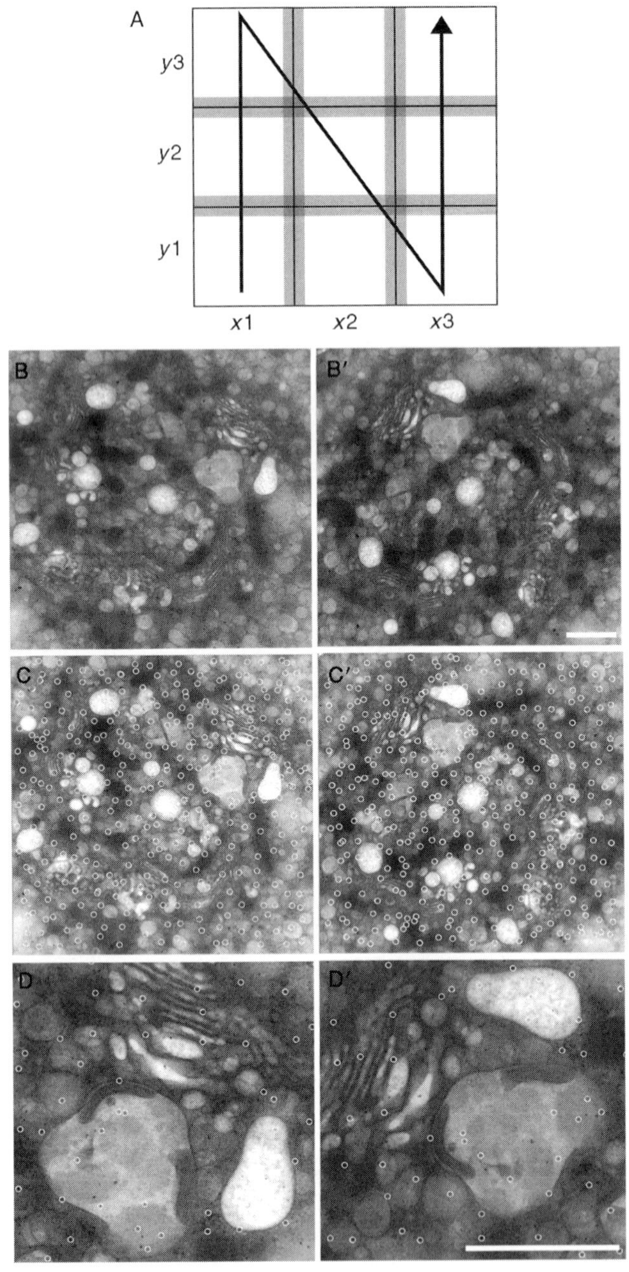

Fig. 3 Digital montaging and image alignment of large areas for ET begins with data collected on the IVEM as set of overlapping pieces or frames in *x* and *y*, collected over a large number of tilts over a range of 120°–140°, and around two orthogonal axes (Mastronarde, 2005). (A) This diagram illustrates how image shift is used to move from one position to the next in a regular array in *x* and *y* (e.g., *x*1,*y*1,*z*1;

two axes. **transferfid** rotates the "seed" fiducial model generated for a single tilt view (typically close to, or at, 0° tilt) from the first axis by 90° (trying first clockwise then counterclockwise to find the best fit) and searches for the pair of views from each tilt series that has the best fiducial correspondence. The program then transfers the fiducial model from the first series to generate a seed model for the second axis. The program also indicates which fiducials failed to transfer and which corresponding fiducials should be specified when setting up to combine the tomograms from each axis (Fig. 4), a particularly crucial step for dual-axis reconstruction, since the coordinates of corresponding fiducials are used for the initial calculation of 3D rotation, distortion, and shift between the volumes.

A large number of fiducials is required for accurately tracking and aligning large area tilt series data because of the heterogeneous distortions that result from specimen exposure to the electron beam. In our experience, the selection of ~200 fiducials for a cellular area $\geq 4\ \mu m^2$ provides a sufficiently high ratio of known versus unknown parameters to accurately calculate image alignment while accounting for differential section thinning, specimen shrinkage, and minor bending. While such distortions may be negligible when handling tilt series data collected from a relatively small area, they are critically important to solve for when attempting to accurately reconstruct large cellular regions (i.e., $>3\ \mu m$ in x and y) at high (5–7 nm) resolution (Marsh *et al.*, 2001a). To deal with these issues, we modified the tilt alignment procedure so that subsets of fiducial markers were used to solve for local alignments in a limited area. **tiltalign** first calculates a single global distortion solution by employing all of the fiducial markers to solve for tilt angle, rotation, overall magnification, and linear distortion within the plane of the section, including distortion due to specimen thinning in the electron beam. The program next finds a series of solutions in a regular array of overlapping local areas

---

*x*1,*y*2,*z*1; *x*1,*y*3,*z*1; *x*2,*y*1,*z*1; *x*2,*y*2,*z*1; and so on), acquiring an image at each position, for a given tilt. For a montage with an odd number of pieces such as a 3 × 3 array, an initial image of the central piece in the array (i.e., at position *x*2,*y*2) is taken and used as a reference for the accurate alignment/positioning of each piece of the montage. Subsequent precise registration of individual pieces in IMOD is achieved by the correlation of image data in zones of overlap (shaded, typically set at around 10% of the total pixel area of each frame) between adjacent frames (see below). (B) Raw montaged images shown for 0° tilt. (B′) A comparable image from the second tilt series which is taken after the specimen has been rotated 90° counterclockwise. When viewed in **3dmod**, the crude registration of elements in the montage can be seen because of the minor differences in contrast between adjacent pieces and/or minor displacement by up to several pixels (compare B and B′). Once individual frames have been blended across the regions of overlap and fused to generate a single image for each tilted view, such borders can be no longer distinguished. A large number (~200) of colloidal gold markers are selected from images in the first tilt series (C and D), then transferred to the corresponding view from the second (C′ and D′). These provide the information necessary to solve for distortion (including rotation, translation, and magnification) and accurately transform and align individual tilted views in the series. They also help in matching and combining the volumes from each axis to make a dual-axis tomogram (see also Fig. 4). Scale bars = 1 μm.

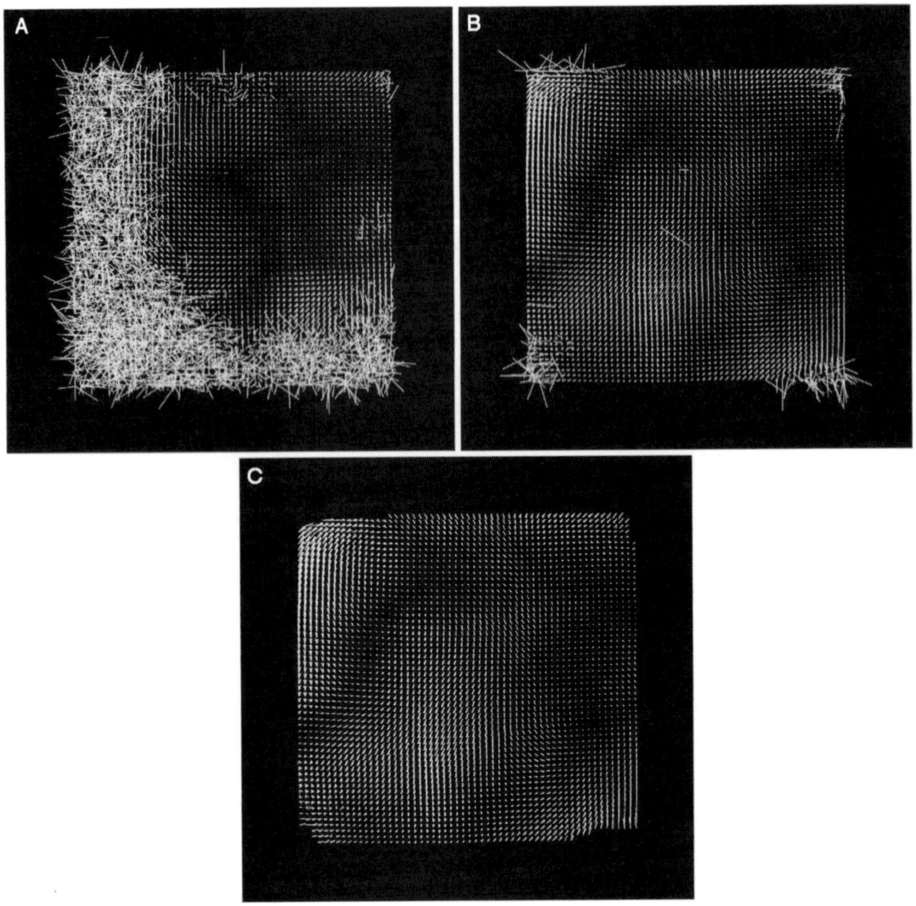

Fig. 4 Matching and combining the tomograms generated from data collected around two orthogonal tilt axes depends on knowing the displacement between corresponding subvolumes in the two tomograms. The coordinates of the fiducials that correspond between the two tomograms allow an initial determination of rotation, distortion, and translation between the two tomograms. However, to combine two volumes precisely one must generate a set of nonlinear transformations by cross-correlating "patches" of a defined size between the first volume and the transformed second volume at an array of positions in 3D. The displacements between the two volumes at each position, which are the output from this process, can be visually examined in the form of a "patch vector model."(A) Such models are often highly informative, since each displacement vector is visualized as an exaggerated line whose length is 10 times the actual length of the original displacement vector. Careful examination of the patch vector model usually provides the user with visual cues as to the nature of the transformations necessary to bring the volumes into register. When combining large cellular areas from montaged tilt series, the large displacements seen in this figure typically result from the use of patches that are too small to provide sufficient information for a good correlation between the corresponding subvolumes. Although standard patches/subvolumes of $80 \times 80 \times 40$ pixels would normally provide a good measure of agreement between tomograms, for the large area generated from a 3×3 montage used in this example, an increase in patch size in z (i.e., to $80 \times 80 \times 70$ pixels) dramatically improved the resulting displacement model (B). However, as noted in Section II, it is essential to exclude from the combine

or "patches" for each 2D projection. These local solutions are incremental to the global solution, so all of the alignment parameters are constrained to vary slowly through the tilt series. The positions of the fiducial markers in 3D are fixed at the values found in the global solution, as the dual-axis tomographic approach requires that the two reconstructed volumes must be accurately aligned with one another (Mastronarde, 1997) (Fig. 4). These constraints dramatically reduce the number of parameters that have to be solved for and increase the ratio of measurements to unknowns, thus allowing accurate local solutions to be obtained from as few as six gold fiducial particles. **tiltalign** then produces a set of linear transformations based on the global solution; these are used to produce globally aligned images. For each local area, it also produces a set of refining transformations based on accurately aligning the images for a reconstruction of that area alone. An R-weighted back-projection program then uses the refining transformations to determine which pixel in a projected image back projects to a given voxel in the reconstructed volume (Wilson *et al.*, 1992). Tomographic volumes calculated by R-weighted back projection from each set of aligned tilts are then matched to each other in 3D by nonlinear transformation using the program **patchcrawl3d**. Basically, **patchcrawl3d** determines the 3D displacement between the two volumes at a regular array of positions using sets of patches of defined sizes in x, y, and z, and then runs the program **corrsearch3d** to determine the displacement of each patch in one tomogram relative to the corresponding patch in the reference tomogram (typically the tomogram from the first axis is used as the reference volume) (Fig. 4). As for fiducial tracking and image alignment for montaged tilt series data, we have had to modify the standard approaches to determine the optimal displacement solutions for matching and combining tomograms of large cellular areas.

Although relatively small patches of overlap will provide sufficient information for accurate combine solutions for small image areas, larger patch sizes are generally required to accurately match one volume to another for the large image areas that result from digitally montaged tilt series. For an example of the dramatic effect of modest changes in patch size specification during tomogram matching and combination, compare the "patch vector models" presented in Fig. 4A and B. It is also important to identify if there are regions within the areas to be combined that should not be used for correlating the tomograms, either because they do not contain sufficient material (e.g., the lumen of a vacuole) or because the reconstruction quality is poor (e.g., typically at the corners of large tomograms) (Fig. 4B).

---

solution erroneous vectors that result from areas in the tomogram that do not contain sufficient dense material for correlation, or from areas at the edges or corners that demonstrate poor reconstruction quality. (B) Such wayward vectors are evident in regions of the correlation model that correspond to the lumens of vacuoles (e.g., Figs. 3B and C and 5A) and at the corners of the model. (C) After manually editing out bad vectors from the model, only the remaining vectors will contribute to the correlated combine solution for matching the two volumes accurately in 3D.

The patch vector model can then be edited so these areas are excluded from the combine solution (see Fig. 4C and accompanying legend). Finally, volumetric image data from each of the two tomograms is combined to produce a single, high-resolution, dual-axis 3D reconstruction (Mastronarde, 1997) (Fig. 5A).

To further increase the cellular volume available for analysis, 3D reconstructions calculated in this manner from serial plastic sections can then be aligned to one another in *z* as previously described (Ladinsky *et al.*, 1999; Marsh *et al.*, 2001a), to produce a final single large volume for analysis and segmentation. The interactive program MIDAS was used to generate a general linear transform that accounted for rotation, distortion, and stretch from one section to its nearest neighbor based on the best-fit alignment of subsets of slices extracted from the bottom of one tomogram and the top of adjacent tomogram.

G. 3D Segmentation and Quantitative Analysis

Membranes of the Golgi complex, ER, compartments of the endosomal–lysosomal pathway, and associated tubules/vesicles within the tomographic volumes were segmented, extracted, and viewed with the IMOD package using **3dmod** (Kremer *et al.*, 1996). Typically, cellular tomograms are stored and viewed as stacks of pixel-thick slices, oriented parallel to the plane of section. However, one is also able to use the **slicer** tool that is part of the program **3dmod** to arbitrarily rotate the data in *x, y*, and *z* to visualize membranous connections between compartments or between membrane-bound compartments and the microtubule cytoskeleton with minimal ambiguity (O'Toole *et al.*, 1999; Marsh *et al.*, 2004).

Although the pixel size—and hence the tomographic slice thickness—are determined by the magnification at which the data were collected (and whether the image data were binned or interpolated), the actual depth of biological material to which these data correspond is somewhat larger, since sections cut from plastic-embedded specimens collapse on initial exposure to the electron beam (Kremer *et al.*, 1990; Luther *et al.*, 1988; Mastronarde, 1997). Despite the fact that each tomographic slice resembles a conventional EM image or "micrograph," it corresponds to slices of material far thinner than can be cut by ultramicrotomy. For example, for a tomographic data set with a nominal pixel size of 2.3 nm, a single pixel-thick slice parallel to the plane of section actually corresponds to ~3.8 nm of material prior to the specimen collapse (Marsh *et al.*, 2001a). Consequently, we typically rescale 3D surface-rendered models generated from segmented cellular tomograms by an appropriate amount in the *z* dimension to account for this phenomenon, and to more accurately represent and quantify the topology of the structures in the specimen prior to data collection in the EM (Ladinsky *et al.*, 1999; Marsh, 2005).

Features of interest to us, such as membrane bilayers and microtubules contrasted with heavy metals, are usually modeled by an expert user who manually segments/traces the membrane contours of an organelle or compartment through

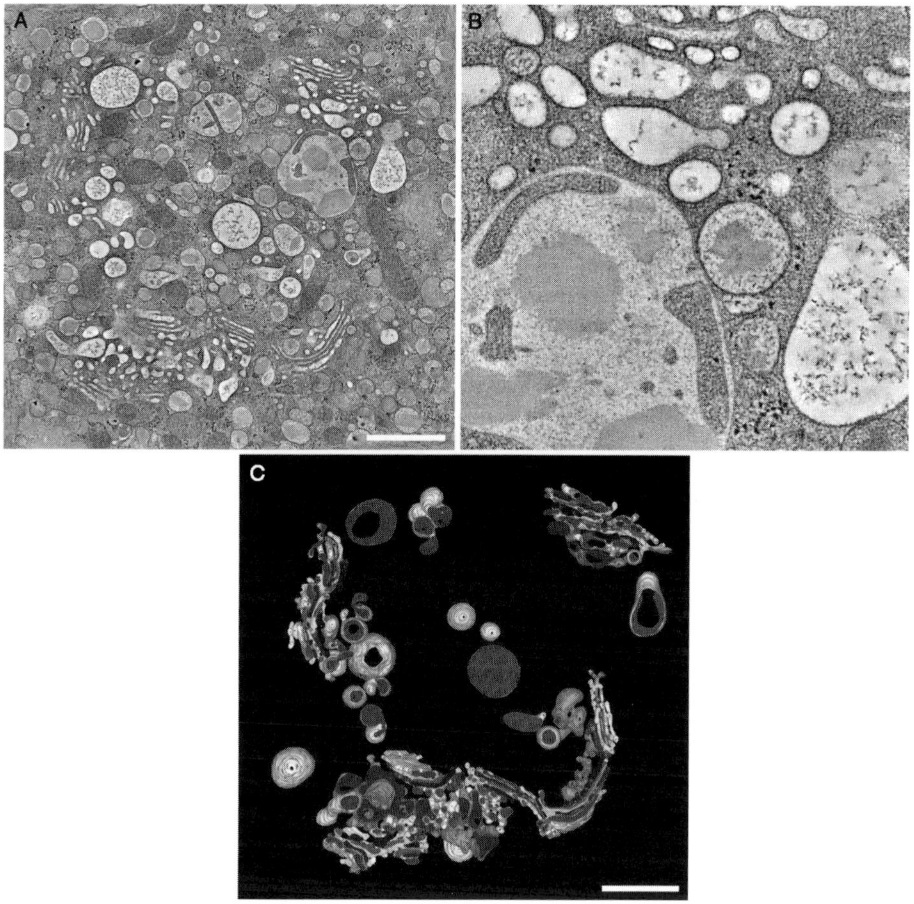

Fig. 5 (A) Part of the extended Golgi ribbon of a glucose-stimulated islet beta cell is evident as the "stacks" of Golgi cisternae that wind through the cytoplasm in the dual-axis tomogram computed from the data represented in Fig. 3, which yielded a reconstructed cellular volume of $5.92 \times 5.92 \times 0.35\ \mu m^3$. Tilt series images were acquired as a 3×3 digital montage at 1° increments over approximately $\pm 63^\circ$ on a Tecnai F30 IVEM operated at 300 keV using a motorized tilt-rotate holder (Gatan). Individual images at each tilt were brought into register by cross-correlation of image data in zones of overlap between adjacent frames, and blended to generate a single image at each tilt. Tilt series images were first aligned with one another by cross-correlation to yield a prealigned stack, and were then accurately aligned by tracking the positions of 10-nm-gold fiduciary markers as discussed in Section II. The tomograms calculated from each set of aligned tilts were then brought into register and combined in 3D to produce the single, dual-axis reconstruction presented here. (B) Membranes of the Golgi, insulin granules, and other compartments were modeled by manual segmentation of the membrane bilayer through sequential tomographic slices (each 1.03-nm thick) in z using the program **3dmod** that is part of the IMOD software package. (C) Polygonal meshes fitted between adjacent contours provided surfaces both for visualization and quantification in 3D. In addition to the stacked cisternal membranes of the Golgi that are color coded as described in the legend for Fig. 1, a number of immature granules and/or condensing vacuoles are visible (colored in light blue). Scale bars = 1 μm. (See Color Insert.)

successive slices in *z*, or along an arbitrary plane in **slicer**, using computer tools in the **3dmod** program. Since each tomographic data set is stored arbitrarily as a set of 2D slices in *x* and *y* (1 pixel thick in *z*), the user physically traces a given membrane-bound compartment through successive slices in *z* (Fig. 5B). Triangular or polygonal meshes fitted between adjacent contours then provide data for surface meshing and visualization (Kremer *et al.*, 1996; Ladinsky *et al.*, 1999). Typically, different colors are used to define membrane contours that belong either to a given compartment or compartment type. For example, in the case of a typical Golgi region in one of our beta cell tomograms, a particular color might be used to identify the membranes of a single continuous cisterna that occupies a certain position or hierarchical level in the stack. The same color would also be used to identify cisternae that can be determined as either functionally equivalent or occupying an equivalent position in stacks that are spatially distinct but remain part of the same Golgi ribbon for a given cell (Ladinsky *et al.*, 1999; Marsh *et al.*, 2001a) (Fig. 5C).

In addition to providing surface data that can be used to generate a "skin" that can simply be visualized in 3D, triangular meshes fitted between contours allow one to compute quantitative parameters, such as the surface area and volume for any segmented compartment, including accurate measurement of fenestrations or holes in Golgi membranes (Ladinsky *et al.*, 1999; Marsh *et al.*, 2001a,b; Otegui *et al.*, 2001). In our case, since contours were drawn down the middle of the membrane bilayer, the interior compartment volumes were computed by subtracting the surface area × 3.5 nm (half of the membrane's thickness) from the volumes inside the contours. To ensure that quantitative data generated from 3D models are as accurate as possible, the ends of closed compartments and vesicles are "capped"—so that they are closed off completely with a surface mesh—as well as "smoothed." Smoothing routines work by fitting local polynomials to the surfaces and replacing each point with a corresponding point from the fitted surface (Ladinsky *et al.*, 1999). This operation helps to compensate for minor irregularities in hand-drawn contours, due to user error, and results in surfaces that shift smoothly from the plane of one contour to the next.

As noted above, and presented in Figs. 1 and 2, individual microtubules could also be tracked in 3D as a set of points with a curvilinear trajectory to characterize and quantify *in situ* associations between the microtubule cytoskeleton and the membrane surface of different organelles and compartments. This process was frequently aided by use of the **slicer** tool in **3dmod**, which allows the user to reorient the data until the microtubule can be viewed clearly *en face* (for more detailed explanation of the use of **slicer** for tracking and visualizing microtubules, please see Chapter 5 by O'Toole, Chapter 6 by Höög and Antony, and Chapter 9 by Otegui and Austin, this volume).

Once all objects of interest have been segmented in the cellular reconstructions, spatial relationships among the modeled objects in 3D can be quantified by measuring distances between objects within a plane and computing an average density of neighboring items as a function of distance between objects, using the programs **nda** and **mtk** (Marsh *et al.*, 2001a). This is similar to previous quantitative morphometric approaches to study the distribution of insulin granules and spatial relationships with the beta cell microtubule cytoskeleton (Yorde and Kalkhoff, 1987). This technique, referred to as "neighbor density analysis," requires that each object of a given geometrical type (i.e., sphere, line, or complex/convoluted surface defined by a mesh of triangles) serve in turn as a reference from which the distances to all nearby neighboring objects of a particular kind are measured (McDonald *et al.*, 1992). Quantification of neighbor density was obtained by dividing the number of items at a given range of distances by the approximate total volume at those distances from all such reference objects, where volume was estimated from the size of a spherical shell at that distance from the object (Marsh *et al.*, 2001a).

III. Discussion

The development and continued advancement of methods for reconstructing large cellular areas in 3D at comparatively high resolution (5–7 nm) have been a prerequisite for EM tomographic studies of the mammalian Golgi ribbon, which frequently extends over distances greater than 10 μm (Cooper *et al.*, 1990; Rambourg and Clermont, 1997). As discussed above, these methods evolved from earlier tomographic studies of large regions of the mammalian Golgi that were originally acquired on film and subsequently digitized to provide a large cellular area for 3D reconstruction and analysis. Key to the success of this method is the use of protocols that retain sufficient resolution to readily identify protein coats such as clathrin (Ladinsky *et al.*, 1999; Marsh *et al.*, 2001a). Importantly, these studies forced the development of new mathematical approaches for handling large area tilt series data and attempting to deal with the heterogeneous changes in specimen geometry that occur over large areas; these had previously made it difficult to accurately align the image data without significantly compromising tomogram quality and resolution (Marsh, 2005; Marsh *et al.*, 2001a). These methods have led directly to tomographic studies of other extended cytomembranes such as somatic and syncytial-type cell plate formation in plants (Otegui *et al.*, 2001; Segui-Simarro *et al.*, 2004) (also, see Chapter 9 by Otegui and Austin, this volume). With large format (>2K $\times$ 2K pixels) CCD cameras now readily available—and in fact often the default for ET—tomographic studies of large cellular areas (even without image montaging) are now more commonplace. Consequently, the application of methods for dealing with distortions in image data over large areas has become a requisite for the generation of high-quality cellular tomograms (Marsh *et al.*, 2001a).

The specific application of these techniques to elucidate structure–function relationships among the key organelles involved in the biosynthesis and trafficking of insulin, first in an immortalized cell culture model (Marsh *et al.*, 2001a,b) and then in glucose-stimulated beta cells preserved within intact pancreatic islets isolated from mice (Marsh *et al.*, 2004), has provided a number of new insights into mammalian cell biology more generally, and the insulin biosynthetic pathway in particular. In some instances, these high-resolution 3D data have provided the first unequivocal evidence for concepts that had been previously received with skepticism by the molecular membrane traffic community. One such example is the idea that multiple, distinct *trans*-cisternae are consumed in the process of sorting and packaging cargo for exit, originally a cornerstone of Novikoff's Golgi-ER-lysosome (GERL) hypothesis (Novikoff, 1964). Associated data that have also provided convincing and clear evidence for other key aspects of the GERL hypothesis, such as the extent of intimacy between specialized regions of the ER and *trans*-Golgi cisternae (Ladinsky *et al.*, 1999; Marsh *et al.*, 2001a,b), have thus provoked serious reconsideration of the complexity of events at and near the *trans*-Golgi. Moreover, these studies have provided compelling evidence that all three major mechanisms of intra-Golgi transport (cisternal progression-maturation, vesicle- and tubule-mediated trafficking) play important roles in intra-Golgi trafficking, demonstrating that they can act in concert in the same region of the Golgi ribbon, and revealing that one mechanism may appear dominant over another depending on the level of protein/lipid traffic and the physiological state of the cell/tissue (Marsh *et al.*, 2001b, 2004; Mironov *et al.*, 1997). Tomographic studies remain underway to elucidate whether this kind of structure–function variation occurs along the Golgi ribbon within a single cell.

In some cases, data from our studies have raised important questions about fundamental concepts regarding maintenance of the sequential processing hierarchy that is a considered a hallmark of Golgi organization. Using physiologically relevant concentrations of glucose to stimulate proinsulin biosynthesis and insulin secretion from beta cells *in situ* in primary cultured mouse islets, we have unequivocally demonstrated the existence of direct connections between nonequivalent cisternae in islet beta cells stimulated to traffic large amounts of secretory protein cargo (Marsh *et al.*, 2004). With short-term stimulation by extracellular glucose (1 h), proinsulin biosynthesis occurs exclusively at the translational level (Wicksteed *et al.*, 2003), resulting in a rapid wave (within ~15–30 min) of proinsulin moving into and through the Golgi (Alarcón *et al.*, 1993). Although tubular connections between Golgi cisternae at their periphery or at branches in the Golgi ribbon had been described by others previously (Rambourg and Clermont, 1997; Weidman, 1995), we observed membrane tubules that connected nonadjacent cisternae by reaching directly through holes or fenestrae in the neighboring cisterna to effectively "bypass" interceding cisternae (Marsh *et al.*, 2004). Such direct intra-Golgi connections could provide a continuous lumen for the expedited forward transit of large amounts of newly synthesized proinsulin between nonequivalent cisternae that are normally distinct from one another, and may also facilitate the rapid retrieval of retrograde traffic (Marsh *et al.*, 2004). These data have provided some of the strongest evidence to date for the concept of lateral transfer in the Golgi complex, in contrast to sequential linear processing.

A. Future Directions

Despite the fact that our "large" reconstruction of a Golgi region (Fig. 1) only covered ~1% of the volume of a beta cell, it provided a rare opportunity to visualize the dense packing of subcellular organelles, ribosomes, vesicles, and microtubules in the cytoplasm of a mammalian cell. This view cannot yet be afforded by any other method (Marsh, 2005; Marsh *et al.*, 2001a). The continued development of extended methods for the accurate tomographic reconstruction of large cellular areas, to the point where it becomes possible to generate a complete set of 3D spatial coordinates for a mammalian cell in toto at high resolution, is likewise expected to provide important and unique insights into cellular organization that cannot come from other methods or even from lower resolution ET of whole cells. Such a "Visible Cell" project would lead to the generation of a cellular atlas with sufficient resolution to distinguish all subcellular compartments and filaments of interest at once. Such a Visible Cell atlas will likely play a crucial role in advancing our understanding of (mammalian) cellular systems biology (Bork and Serrano, 2005; Lehner *et al.*, 2005; Nickell *et al.*, 2006), and should be seen as a fundamental prerequisite to realistically simulating/predicting the spatiotemporal coordinates of complex molecular membrane trafficking events *in silico*.

Meanwhile, comparatively large area cellular reconstructions carried out at relatively high throughput/high resolution through the use of large format/fast readout CCD cameras will provide important new insights into other key aspects of insulin trafficking in the beta cell that occur after secretory cargo packaging and insulin granule biogenesis at the *trans*-Golgi. Specifically, current and future studies are aimed at elucidating the specifics of granule recruitment to the cell surface for release into the bloodstream after glucose levels in the blood rise above normal. Quantitatively determining how interactions between granules and the microtubule and actin cytoskeletons change in islet beta cells exposed to low, normal, and stimulatory concentrations of glucose for different times will be essential to develop a complete understanding of these events (Fig. 6). Microtubules play an important role in potentiating the second, sustained phase of insulin release by transporting insulin granules from the Golgi region (where they are formed) to the plasma membrane, to replenish the readily releasable granule pool following a rise in intracellular [Ca^{2+}] (Boyd *et al.*, 1982; Donelan *et al.*, 2002; Malaisse *et al.*, 1975). This process is believed to be dependent on the recruitment of the cytoplasmic motor protein, kinesin, to the granule membrane (Balczon *et al.*, 1992; Donelan *et al.*, 2002; Varadi *et al.*, 2002). Evidence also suggests that the binding of small GTP-binding proteins such as Rab3A and Rab27A to yet uncharacterized "effector" proteins plays a crucial role in regulating these events in islet beta cells (Coppola *et al.*, 2002; Kajio *et al.*, 2001; Park *et al.*, 2002; Yaekura *et al.*, 2003; Yi *et al.*, 2002). By tracking the frequency of such interactions along the length of any given microtubule over a relatively large distance, combined with a capacity to scrutinize interactions between the microtubule and granule membrane at high resolution, we expect to precisely establish the sequence of steps and molecular

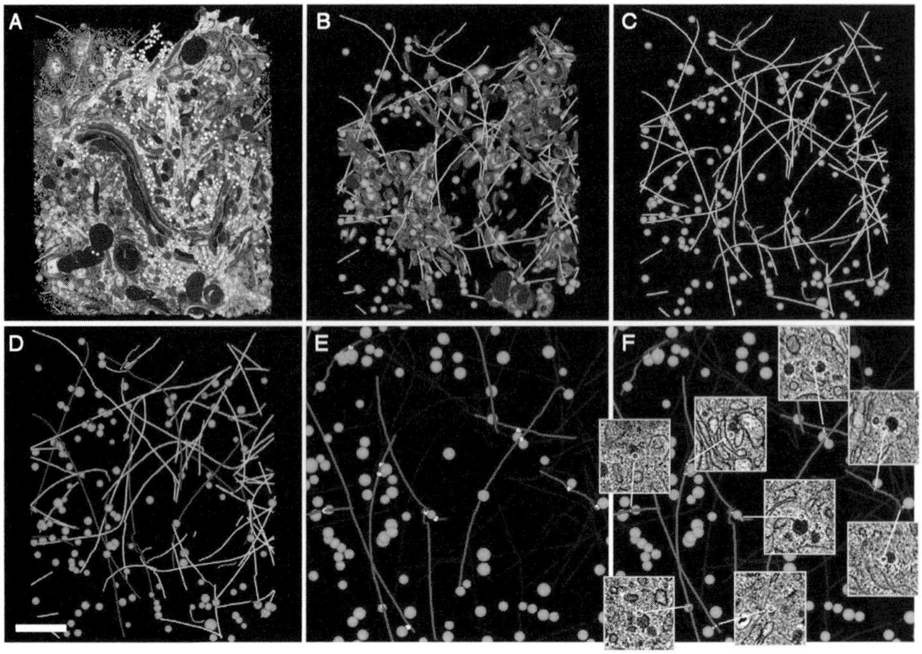

Fig. 6 Our early study of the insulinoma cell line, HIT-T15, demonstrated that we could dissect the beta cell modeled in 3D at ~6-nm resolution as required, that is, each modeled object could be extracted and viewed in any chosen orientation and in the context of any other object(s) to analyze structural associations with confidence (Marsh *et al.*, 2001a). (A) Shown are the membranes that comprise the Golgi, together with the ER, ribosomes, microtubules, secretory granules, vesicles, endosomal–lysosomal compartment, and mitochondria (for color key see legend for Fig. 1C). In that study, microtubules with a total combined length of over 80 μm were segmented to accurately quantify associations between the microtubule cytoskeleton and different organelles and compartments *in situ*. In that analysis each microtubule was divided into regularly spaced (10 nm) subsegments, and the distances from each segment to all neighboring objects of a particular kind were measured. These data revealed the distances of closest approaches from each subsegment to other objects, which in turn gave indications of which objects interact with microtubules and which do not. (B) The paths of the microtubules (bright green) can be followed more readily, as they are displayed only in the context of insulin secretory granules (bright blue) and endosomal–lysosomal compartments (purple and red), which are also likely to play an important role in proinsulin processing/insulin trafficking (Turner and Arvan, 2000). Here, we demonstrate the ability to pinpoint sites of close approach between microtubules and insulin containing compartments in tomograms of beta cells. Using the program **mtk**, microtubules and insulin granules shown in (C) were analyzed in 3D to determine all distances of approach between them. After examining the histogram of distances of close approach between all microtubules and granules, we limited our search of the modeled data to approaches on the order of $\leq$20 nm. This identified the subset of microtubules highlighted in red in (D). (E) Using visible "markers" placed by the software between the granules and the nearby microtubules, we were able to identify the structures at sites of close approach ($\leq$20 nm) in the modeled data (F). Such sites represent the best chance of directly identifying microtubule-associated proteins and cytoplasmic motors involved in microtubule-dependent insulin transport *in situ*. Scale bar = 500 nm. (See Color Insert.)

machinery involved in moving insulin from the peri-Golgi region to the cell surface, and to better understand the nature of the subcellular defects that accompany states of impaired insulin intracellular trafficking and exocytosis/secretion.

IV. Summary

In cooperation with David Mastronarde at the Boulder Laboratory for 3D Electron Microscopy of Cells, we have developed methods for high resolution, large area ET of thick, stained plastic sections for the 3D reconstruction and analysis of mammalian membrane architecture at high fidelity in the insulin-secreting beta cells of the endocrine pancreas. This approach, when employed in conjunction with high-pressure freezing and freeze-substitution, has allowed us to generate tomograms of comparatively large cellular volumes (each typically measuring at least $4 \times 4 \times 0.4\ \mu m^3$) with which to study the key organelles and compartments involved in the synthesis, processing, and trafficking of insulin with unprecedented reliability. These data have afforded us new insights into beta cell biology and function that include: (1) the complexity of structural relationships among the Golgi, ER, and compartments of the endosomal–lysosomal system (Marsh *et al.*, 2001a); (2) evidence that multiple transport mechanisms act in concert in the same region of the Golgi ribbon (Marsh *et al.*, 2001b); (3) a role for the ER in regulating membrane traffic/sorting at the *trans*-Golgi, the presumptive site where proinsulin is sorted and packaged into nascent secretory granules (Marsh *et al.*, 2001a,b); (4) evidence that multiple, distinct *trans*-cisternae, frequently referred to as the TGN, detach and fragment as membrane is consumed in the process of packaging secretory cargo for exit (Marsh *et al.*, 2001a); and (5) the unequivocal demonstration of direct intra-Golgi connections that facilitate "cisternal bypass" to provide a continuous lumen for the expeditious transit of Golgi traffic (Marsh *et al.*, 2004).

Acknowledgments

We thank the entire staff of The Boulder Laboratory for 3D Electron Microscopy of Cells for their personal support and professional assistance from the very onset of these studies, and Kathryn Howell and John Hutton of the University of Colorado Health Sciences Center for additional tutelage and intellectual support. This work was supported by P41-RR00592 to J.R.M., GM42629 and P01-GM61306 to K. E. H., and a Juvenile Diabetes Research Foundation International (JDRF) Postdoctoral Fellowship (3-1999-538) to B.J.M. B.J.M. is currently supported by JDRF (2-2004-275) and NIH/NIDDK (DK-71236) funding and is a senior research affiliate of the ARC Special Research Centre for Functional and Applied Genomics. I would like to especially thank Adam Costin, Janette Galea, Garry Morgan, and Peter van der Heide for generating and segmenting data that are presented here.

The Advanced Cryo-Electron Microscopy Laboratory housed at the Institute for Molecular Bioscience is a major node of the federally funded Nanostructural Analysis Network Organisation's Major National Research Facility (NANO-MNRF), and is jointly supported by the Queensland State government's "Smart State Strategy" initiative. We thank Dr. Jamie Riches of the NANO-MNRF for instrument calibrations and upkeep, and nanoTechnology Systems (Greensborough, VIC, Australia) for critical maintenance and upgrades of the 300 keV Tecnai F30 EM within the laboratory.

References

Alarcón, C., Lincoln, B., and Rhodes, C. J. (1993). The biosynthesis of the subtilisin-related proprotein convertase PC3, but no that of the PC2 convertase, is regulated by glucose in parallel to proinsulin biosynthesis in rat pancreatic islets. *J. Biol. Chem.* **268,** 4276–4280.

Arvan, P., and Castle, D. (1998). Sorting and storage during secretory granule biogenesis: Looking backward and looking forward. *Biochem. J.* **332,** 593–610.

Balczon, R., Overstreet, K. A., Zinkowski, R. P., Haynes, A., and Appel, M. (1992). The identification, purification, and characterization of a pancreatic beta-cell form of the microtubule adenosine triphosphatase kinesin. *Endocrinology* **131,** 331–336.

Biel, S. S., Kawaschinski, K., Wittern, K. P., Hintze, U., and Wepf, R. (2003). From tissue to cellular ultrastructure: Closing the gap between micro- and nanostructural imaging. *J. Microsc.* **212,** 91–99.

Bonner-Weir, S. (1988). Morphological evidence for pancreatic polarity of beta-cell within islets of Langerhans. *Diabetes* **37,** 616–621.

Bonner-Weir, S. (1989). Pancreatic islets: Morphology, organization, and physiological implications. *In* "Molecular and Cellular Biology of Diabetes Mellitus. Vol. I. Insulin Secretion" (B. Draznin, S. Melmed, and D. LeRoith, eds.), pp. 1–11. Alan R. Liss, Inc., New York.

Bork, P., and Serrano, L. (2005). Towards cellular systems in 4D. *Cell* **121,** 507–509.

Boyd, A. E., III, Bolton, W. E., and Brinkley, B. R. (1982). Microtubules and beta cell function: Effect of colchicine on microtubules and insulin secretion *in vitro* by mouse beta cells. *J. Cell Biol.* **92,** 425–434.

Breant, B., Lavergne, C., Astesano, A., Ferrand, N., Asfari, M., Boissard, C., Anteunis, A., and Rosselin, G. (1992). Development of the beta cells. *Mt. Sinai J. Med.* **59,** 175–185.

Breuza, L., Halbeisen, R., Jeno, P., Otte, S., Barlowe, C., Hong, W., and Hauri, H. P. (2004). Proteomics of endoplasmic reticulum-Golgi intermediate compartment (ERGIC) membranes from brefeldin A-treated HepG2 cells identifies ERGIC-32, a new cycling protein that interacts with human Erv46. *J. Biol. Chem.* **279,** 47242–47253.

Cooper, M. S., Cornell-Bell, A. H., Chernjavsky, A., Dani, J. W., and Smith, S. J. (1990). Tubulovesicular processes emerge from trans-Golgi cisternae, extend along microtubules, and interlink adjacent trans-golgi elements into a reticulum. *Cell* **61,** 135–145.

Coppola, T., Frantz, C., Perret-Menoud, V., Gattesco, S., Hirling, H., and Regazzi, R. (2002). Pancreatic beta-cell protein granuphilin binds Rab3 and Munc-18 and controls exocytosis. *Mol. Biol. Cell* **13,** 1906–1915.

Cortizo, A., Espinal, J., and Hammonds, P. (1990). Vectorial insulin secretion by pancreatic beta-cells. *FEBS Lett.* **272,** 137–140.

Donelan, M. J., Morfini, G., Julyan, R., Sommers, S., Hays, L., Kajio, H., Briaud, I., Easom, R. A., Molkentin, J. D., Brady, S. T., and Rhodes, C. J. (2002). Ca2+-dependent dephosphorylation of kinesin heavy chain on beta-granules in pancreatic beta-cells. Implications for regulated beta-granule transport and insulin exocytosis. *J. Biol. Chem.* **277,** 24232–24242.

Dubochet, J. (1995). High-pressure freezing for cryoelectron microscopy. *Trends Cell Biol.* **5,** 366–368.

Elsner, M., Hashimoto, H., and Nilsson, T. (2003). Cisternal maturation and vesicle transport: Join the band wagon! (Review). *Mol. Membr. Biol.* **20,** 221–229.

Farquhar, M. G., and Palade, G. E. (1998). The Golgi apparatus: 100 years of progress and controversy. *Trends Cell Biol.* **8,** 2–10.

Gilkey, J. C., and Staehelin, L. A. (1986). Advances in ultrarapid freezing for the preservation of cellular ultrastructure. *J. Electron Microsc. Tech.* **3,** 177–210.

Gleason, C. E., Gonzalez, M., Harmon, J. S., and Robertson, R. P. (2000). Determinants of glucose toxicity and its reversibility in the pancreatic islet beta-cell line, HIT-T15. *Am. J. Physiol. Endocrinol. Metab.* **279,** E997–E1002.

Golgi, C. (1898). Sur la structure des cellules nerveuses des ganglions spinaux. *Arch. Ital. Biol.* **30,** 278–286.

Goodge, K. A., and Hutton, J. C. (2000). Translational regulation of proinsulin biosynthesis and proinsulin conversion in the pancreatic beta-cell. *Semin. Cell Dev. Biol.* **11,** 235–242.

Griffiths, G. (2004). Ultrastructure in cell biology: Do we still need it? *Eur. J. Cell Biol.* **83,** 245–251.

Griffiths, G., McDowall, A., Back, R., and Dubochet, J. (1984). On the preparation of cryosections for immunocytochemistry. *J. Ultrastruct. Res.* **89,** 65–78.

Griffiths, G., and Simons, K. (1986). The trans Golgi network: Sorting at the exit site of the Golgi complex. *Science* **234,** 438–443.

Gu, F., Crump, C. M., and Thomas, G. (2001). Trans-Golgi network sorting. *Cell. Mol. Life Sci.* **58,** 1067–1084.

Guest, P. C., Bailyes, E. M., and Hutton, J. C. (1997). Endoplasmic reticulum Ca2+ is important for the proteolytic processing and intracellular transport of proinsulin in the pancreatic beta-cell. *Biochem. J.* **323,** 445–450.

Hama, K., Arii, T., and Kosaka, T. (1994). Three-dimensional organization of neuronal and glial processes: High voltage microscopy. *Microsc. Res. Tech.* **29,** 357–367.

Ho, H. C., Tang, C. Y., and Suarez, S. S. (1999). Three-dimensional structure of the Golgi apparatus in mouse spermatids: A scanning electron microscopic study. *Anat. Rec.* **256,** 189–194.

Howell, S. L., and Bird, G. S. (1989). Biosynthesis and secretion of insulin. *Br. Med. Bull.* **45,** 19–36.

Howell, S. L., and Tyhurst, M. (1984). Insulin secretion: The effector system. *Experientia* **40,** 1098–1105.

Kajio, H., Olszewski, S., Rosner, P. J., Donelan, M. J., Geoghegan, K. F., and Rhodes, C. J. (2001). A low-affinity Ca2+-dependent association of calmodulin with the Rab3A effector domain inversely correlates with insulin exocytosis. *Diabetes* **50,** 2029–2039.

Katsumoto, T., Inoue, M., Naguro, T., and Kurimura, T. (1991). Association of cytoskeletons with the Golgi apparatus: Three-dimensional observation and computer-graphic reconstruction. *J. Electron Microsc. (Tokyo)* **40,** 24–28.

Kowluru, A., and Morgan, N. G. (2002). GTP-binding proteins in cell survival and demise: The emerging picture in the pancreatic beta-cell. *Biochem. Pharmacol.* **63,** 1027–1035.

Kremer, J. R., Mastronarde, D. N., and McIntosh, J. R. (1996). Computer visualization of three-dimensional image data using IMOD. *J. Struct. Biol.* **116,** 71–76.

Kremer, J. R., O'Toole, E. T., Wray, G. P., Mastronarde, D. N., Mitchell, S. J., and McIntosh, J. R. (1990). Characterization of beam-induced thinning and shrinkage of semi-thick sections in the HVEM. *In* "Proc. XII Int. Congr. Electr. Microsc.," **3,** 752–753.

Kuliawat, R., and Arvan, P. (1994). Distinct molecular mechanisms for protein sorting within immature secretory granules of pancreatic beta-cells. *J. Cell Biol.* **126,** 77–86.

Ladinsky, M. S., Kremer, J. R., Furcinitti, P. S., McIntosh, J. R., and Howell, K. E. (1994). HVEM tomography of the trans-Golgi network: Structural insights and identification of a lace-like vesicle coat. *J. Cell Biol.* **127,** 29–38.

Ladinsky, M. S., Mastronarde, D. N., McIntosh, J. R., Howell, K. E., and Staehelin, L. A. (1999). Golgi structure in three dimensions: Functional insights from the normal rat kidney cell. *J. Cell Biol.* **144,** 1135–1149.

Lehner, B., Tischler, J., and Fraser, A. G. (2005). Systems biology: Where it's at in 2005. *Genome Biol.* **6,** 338.

Lindsey, J. D., and Ellisman, M. H. (1985). The neuronal endomembrane system. I. Direct links between rough endoplasmic reticulum and the cis element of the Golgi apparatus. *J. Neurosci.* **5,** 3111–3123.

Lippincott-Schwartz, J., Roberts, T. H., and Hirschberg, K. (2000). Secretory protein trafficking and organelle dynamics in living cells. *Annu. Rev. Cell Dev. Biol.* **16,** 557–589.

Lippincott-Schwartz, J., Snapp, E., and Kenworthy, A. (2001). Studying protein dynamics in living cells. *Nat. Rev. Mol. Cell. Biol.* **2,** 444–456.

Lombardi, T., Montesano, R., and Orci, L. (1986). Loss of polarization of plasma membrane domains in transformed pancreatic endocrine cell lines. *Endocrinology* **119,** 502–507.

Luther, P. K., Lawrence, M. C., and Crowther, R. A. (1988). A method for monitoring the collapse of plastic sections as a function of electron dose. *Ultramicroscopy* **24,** 7–18.

Malaisse, W. J., Malaisse-Lagae, F., Van Obberghen, E., Somers, G., Devis, G., Ravazzola, M., and Orci, L. (1975). Role of microtubules in the phasic pattern of insulin release. *Ann. NY Acad. Sci.* **253,** 630–652.

Marsh, B. J. (2005). Lessons from tomographic studies of the mammalian Golgi. *Biochim. Biophys. Acta* **1744,** 273–292.

Marsh, B. J., and Howell, K. E. (2002). The mammalian Golgi-complex debates. *Nat. Rev. Mol. Cell. Biol.* **3,** 789–795.

Marsh, B. J., Mastronarde, D. N., Buttle, K. F., Howell, K. E., and McIntosh, J. R. (2001a). Organellar relationships in the Golgi region of the pancreatic beta cell line, HIT-T15, visualized by high resolution electron tomography. *Proc. Natl. Acad. Sci. USA* **98,** 2399–2406.

Marsh, B. J., Mastronarde, D. N., McIntosh, J. R., and Howell, K. E. (2001b). Structural evidence for multiple transport mechanisms through the Golgi in the pancreatic beta-cell line, HIT-T15. *Biochem. Soc. Trans.* **29,** 461–467.

Marsh, B. J., Volkmann, N., McIntosh, J. R., and Howell, K. E. (2004). Direct continuities between cisternae at different levels of the Golgi complex in glucose-stimulated mouse islet beta cells. *Proc. Natl. Acad. Sci. USA* **101,** 5565–5570.

Mastronarde, D. N. (1997). Dual-axis tomography: An approach with alignment methods that preserve resolution. *J. Struct. Biol.* **120,** 343–352.

Mastronarde, D. N. (2005). Automated electron microscope tomography using robust prediction of specimen movements. *J. Struct. Biol.* **152,** 36–51.

McDonald, K., and Morphew, M. K. (1993). Improved preservation of ultrastructure in difficult-to-fix organisms by high pressure freezing and freeze substitution: I. Drosophila melanogaster and Strongylocentrotus purpuratus embryos. *Microsc. Res. Tech.* **24,** 465–473.

McDonald, K. L., O'Toole, E. T., Mastronarde, D. N., and McIntosh, J. R. (1992). Kinetochore microtubules in PTK cells. *J. Cell Biol.* **118,** 369–383.

McIntosh, J. R. (2001). Electron microscopy of cells. A new beginning for a new century. *J. Cell Biol.* **153,** F25–F32.

Mellman, I., and Simons, K. (1992). The Golgi complex: *In vitro* veritas? *Cell* **68,** 829–840.

Mironov, A. A., Weidman, P., and Luini, A. (1997). Variations on the intracellular transport theme: Maturing cisternae and trafficking tubules. *J. Cell Biol.* **138,** 481–484.

Molinete, M., Irminger, J. C., Tooze, S. A., and Halban, P. A. (2000). Trafficking/sorting and granule biogenesis in the beta-cell. *Semin. Cell Dev. Biol.* **11,** 243–251.

Nickell, S., Kofler, C., Leis, A. P., and Baumeister, W. (2006). A visual approach to proteomics. *Nat. Rev. Mol. Cell. Biol.* **7,** 225–230.

Novikoff, A. B. (1964). GERL, its form and function in neurons of rat spinal ganglia. *Biol. Bull.* **127,** 358.

Novikoff, A. B., Yam, A., and Novikoff, P. M. (1975). Cytochemical study of secretory process in transplantable insulinoma of Syrian golden hamster. *Proc. Natl. Acad. Sci. USA* **72,** 4501–4505.

O'Toole, E. T., Winey, M., and McIntosh, J. R. (1999). High-voltage electron tomography of spindle pole bodies and early mitotic spindles in the yeast *Saccharomyces cerevisiae*. *Mol. Biol. Cell* **10,** 2017–2031.

Ohgawara, H., Miyazaki, J., Karibe, S., Tashiro, F., Akaike, T., and Hashimoto, Y. (1995). Embedded-culture of pancreatic beta-cells derived from transgenic mouse insulinoma as a potential source for xenotransplantation using a diffusion chamber. *Cell Transplant.* **4,** 307–313.

Orci, L. (1986). The insulin cell: Its cellular environment and how it processes (pro)insulin. *Diabetes Metab. Rev.* **2,** 71–106.

Orci, L., Malhotra, V., Amherdt, M., Serafini, T., and Rothman, J. E. (1989a). Dissection of a single round of vesicular transport: Sequential intermediates for intercisternal movement in the Golgi stack. *Cell* **56,** 357–368.

Orci, L., Ravazzola, M., and Perrelet, A. (1984). (Pro)insulin associates with Golgi membranes of pancreatic B cells. *Proc. Natl. Acad. Sci. USA* **81,** 6743–6746.

Orci, L., Thorens, B., Ravazzola, M., and Lodish, H. F. (1989b). Localization of the pancreatic beta cell glucose transporter to specific plasma membrane domains. *Science* **245,** 295–297.

Orci, L., Vassalli, J. D., and Perrelet, A. (1988). The insulin factory. *Sci. Am.* **259,** 85–94.

Otegui, M. S., Mastronarde, D. N., Kang, B. H., Bednarek, S. Y., and Staehelin, L. A. (2001). Three-dimensional analysis of syncytial-type cell plates during endosperm cellularization visualized by high resolution electron tomography. *Plant Cell* **13,** 2033–2051.

Palade, G. E. (1988). Cell fractionation: Importance to cell-free systems development. *Prog. Clin. Biol. Res.* **270,** xix–xx.

Park, J. B., Kim, J. S., Lee, J. Y., Kim, J., Seo, J. Y., and Kim, A. R. (2002). GTP binds to Rab3A in a complex with Ca2+/calmodulin. *Biochem. J.* **362,** 651–657.

Penczek, P., Marko, M., Buttle, K., and Frank, J. (1995). Double-tilt electron tomography. *Ultramicroscopy* **60,** 393–410.

Poitout, V., Olson, L. K., and Robertson, R. P. (1996). Insulin-secreting cell lines: Classification, characteristics and potential applications. *Diabetes Metab.* **22,** 7–14.

Poitout, V., Stout, L. E., Armstrong, M. B., Walseth, T. F., Sorenson, R. L., and Robertson, R. P. (1995). Morphological and functional characterization of beta TC-6 cells—an insulin-secreting cell line derived from transgenic mice. *Diabetes* **44,** 306–313.

Porter, K. R., Claude, A., and Fullam, E. F. (1945). A study of tissue culture cells by electron microscopy: Methods and preliminary observations. *J. Exp. Med.* **81,** 233–246.

Rambourg, A., and Clermont, Y. (1997). Three-dimensional structure of the Golgi apparatus in mammalian cells. *In* "The Golgi apparatus" (E. G. Berger and J. Roth, eds.), pp. 37–61. Birkhauser-Verlag, Basel.

Rambourg, A., Marraud, A., and Thiery, G. (1974). Scanning electron microscopy by transmission: Its value for tridimensional study of cell organelles. *CR Acad. Sci. Hebd. Seances. Acad. Sci. D* **279,** 283–284.

Rios, R. M., and Bornens, M. (2003). The Golgi apparatus at the cell centre. *Curr. Opin. Cell Biol.* **15,** 60–66.

Rothman, J. E., and Orci, L. (1992). Molecular dissection of the secretory pathway. *Nature* **355,** 409–415.

Rutter, G. A. (1999). Insulin secretion: Feed-forward control of insulin biosynthesis? *Curr. Biol.* **9,** R443–R445.

Sandler, S., and Andersson, A. (1984). The significance of culture for successful cryopreservation of isolated pancreatic islets of Langerhans. *Cryobiology* **21,** 503–510.

Santerre, R. F., Cook, R. A., Crisel, R. M., Sharp, J. D., Schmidt, R. J., Williams, D. C., and Wilson, C. P. (1981). Insulin synthesis in a clonal cell line of simian virus 40-transformed hamster pancreatic beta cells. *Proc. Natl. Acad. Sci. USA* **78,** 4339–4343.

Segui-Simarro, J. M., Austin, J. R., II, White, E. A., and Staehelin, L. A. (2004). Electron tomographic analysis of somatic cell plate formation in meristematic cells of Arabidopsis preserved by high-pressure freezing. *Plant Cell* **16,** 836–856.

Slot, J. W., and Geuze, H. J. (1983). Immunoelectron microscopic exploration of the Golgi complex. *J. Histochem. Cytochem.* **31,** 1049–1056.

Storrie, B., and Nilsson, T. (2002). The Golgi apparatus: Balancing new with old. *Traffic* **3,** 521–529.

Tanaka, K., Mitsushima, A., Fukudome, H., and Kashima, Y. (1986). Three-dimensional architecture of the Golgi complex observed by high resolution scanning electron microscopy. *J. Submicrosc. Cytol.* **18,** 1–9.

Taylor, K. A., Reedy, M. C., Cordova, L., and Reedy, M. K. (1984). Three-dimensional reconstruction of rigor insect flight muscle from tilted thin sections. *Nature* **310,** 285–291.

Tsuboi, T., and Rutter, G. A. (2003). Insulin secretion by 'kiss-and-run' exocytosis in clonal pancreatic islet beta-cells. *Biochem. Soc. Trans.* **31,** 833–836.

Turner, M. D., and Arvan, P. (2000). Protein traffic from the secretory pathway to the endosomal system in pancreatic beta-cells. *J. Biol. Chem.* **275,** 14025–14030.

Varadi, A., Ainscow, E. K., Allan, V. J., and Rutter, G. A. (2002). Involvement of conventional kinesin in glucose-stimulated secretory granule movements and exocytosis in clonal pancreatic beta-cells. *J. Cell Sci.* **115,** 4177–4189.

Wang, R. N., and Rosenberg, L. (1999). Maintenance of beta-cell function and survival following islet isolation requires re-establishment of the islet-matrix relationship. *J. Endocrinol.* **163,** 181–190.

Weidman, P. J. (1995). Anterograde transport through the Golgi complex: Do Golgi tubules hold the key? *Trends Cell Biol.* **5,** 302–307.

Wicksteed, B., Alarcon, C., Briaud, I., Lingohr, M. K., and Rhodes, C. J. (2003). Glucose-induced translational control of proinsulin biosynthesis is proportional to preproinsulin mRNA levels in islet beta-cells but not regulated via a positive feedback of secreted insulin. *J. Biol. Chem.* **278,** 42080–42090.

Wilson, C. J., Mastronarde, D. N., McEwen, B., and Frank, J. (1992). Measurement of neuronal surface area using high-voltage electron microscope tomography. *Neuroimage* **1,** 11–22.

Wu, C. C., MacCoss, M. J., Mardones, G., Finnigan, C., Mogelsvang, S., Yates, J. R., III, and Howell, K. E. (2004). Organellar proteomics reveals Golgi arginine dimethylation. *Mol. Biol. Cell* **15,** 2907–2919.

Yaekura, K., Julyan, R., Wicksteed, B. L., Hays, L. B., Alarcon, C., Sommers, S., Poitout, V., Baskin, D. G., Wang, Y., Philipson, L. H., and Rhodes, C. J. (2003). Insulin secretory deficiency and glucose intolerance in Rab3A null mice. *J. Biol. Chem.* **278,** 9715–9721.

Yates, J. R., III, Gilchrist, A., Howell, K. E., and Bergeron, J. J. (2005). Proteomics of organelles and large cellular structures. *Nat. Rev. Mol. Cell. Biol.* **6,** 702–714.

Yi, Z., Yokota, H., Torii, S., Aoki, T., Hosaka, M., Zhao, S., Takata, K., Takeuchi, T., and Izumi, T. (2002). The Rab27a/granuphilin complex regulates the exocytosis of insulin-containing dense-core granules. *Mol. Cell. Biol.* **22,** 1858–1867.

Yorde, D. E., and Kalkhoff, R. K. (1987). Morphometric studies of secretory granule distribution and association with microtubules in beta-cells of rat islets during glucose stimulation. *Diabetes* **36,** 905–913.

CHAPTER 9

Visualization of Membrane–Cytoskeletal Interactions During Plant Cytokinesis

Marisa S. Otegui* and Jotham R. Austin, II†

*Department of Botany, University of Wisconsin
Madison, Wisconsin 53706

†Biological Science Division-OSRF, University of Chicago
Chicago, Illinois 60637

Plant cytokinesis is a complex process that requires coordinated interactions between the cytoskeleton and the membrane-trafficking machinery. The combination of cryofixation/freeze-substitution methods with electron tomography (ET) has revealed amazing structural details of this unique cellular process. In this chapter, we focus on the two main variables associated with these imaging techniques: sample preparation and ET of plastic sections. We discuss which tissue and culture cells are more suitable for plant cytokinesis studies and how these samples can be high-pressure frozen and freeze-substituted to preserve the cytokinetic elements. We present techniques for imaging large samples for ET such as montaged images and the assembly of serial electron tomograms. We also discuss different hand-segmentation (modeling) approaches for membranes and microtubules and strategies for the identification of macromolecules contained in reconstructed tomographic models. Finally, we present information on available computer programs that can

0091-679X/07 $35.00
DOI: 10.1016/S0091-679X(06)79009-4

be used to obtain quantitative information from tomographic models. We specifically focus on the calculation of density, surface area, and volume of cellular structures as well as spatial relationships between organelles.

I. Introduction

Cytokinesis is the process by which the cytoplasm is partitioned between daughter cells after nuclear division is complete. In plants, cytokinesis involves the coordinated redistribution of organelles and the building of the cell plate, a membranous structure that grows centrifugally through the fusion of Golgi-derived vesicles, starting at the center of the division plane. The plate then matures into the new cell wall. The formation of the cell plate is aided by the phragmoplast, a complex array of cytoskeleton elements, cell plate-forming vesicles, and their associated proteins. The phragmoplast microtubule array arises from remnants of the anaphase spindle and consists of two antiparallel sets of microtubules, oriented with their plus ends facing the division plane. A filamentous cell plate assembly matrix (CPAM) surrounds the growing cell plate and is thought to coordinate both cell plate and microtubule dynamics. There is a close relationship between the cytoskeletal elements and the membrane components of the cytokinetic machinery. In fact, the phragmoplast architecture changes over time as the cell plate expands and matures.

In some of the cells involved in sexual reproduction, unconventional types of cytokinesis have been reported. During the development of spores, embryo sacs, and the endosperm of some species, cytokinesis does not immediately follow nuclear division, so multinucleate cells or syncytia are formed. When cytokinesis does take place, small clusters of microtubules called mini-phragmoplasts and cell plates assemble between sister and non-sister nuclei. Overall, these nonconventional cytokinetic mechanisms are quite similar to the conventional ones, exhibiting some differences in the organization of the phragmoplast microtubules and the morphology of some cell plate intermediates (Jurgens, 2005; Otegui and Staehelin, 2000b, 2004).

The combination of high-pressure freezing/freeze-substitution and electron tomography (ET) has revealed amazing details about the complex three-dimensional (3D) organization of membranes and cytoskeletal elements in the cytokinetic apparatus of both somatic and reproductive cells. Four stages have been recognized in the conventional cytokinesis of somatic cells:

1. *Phragmoplast initiation at late anaphase*: Golgi-derived vesicles then travel along microtubules toward the assembly sites. The CPAM assembles in the division plane and encloses fusing vesicles, early cell plate intermediates, and the plus ends of most phragmoplast microtubules (Austin *et al.*, 2005; Segui-Simarro *et al.*, 2004).

2. *Formation of a solid phragmoplast*: It is defined by the appearance of a continuous CPAM that extends across the phragmoplast.

3. *Transitional phragmoplast*: The microtubules and the CPAM located in the central part of the cell plate, which is the more mature cell plate domain, disassemble and reassemble at the peripheral growing edges.

4. *Ring-shaped phragmoplast*: The microtubules and CPAM domains located at the edges of the expanding cell plate mediate the centrifugal expansion of the cell plate until it fuses with the parental cell wall (Segui-Simarro *et al.*, 2004). During this time, the cell plate itself also undergoes changes in conformation, from single fusing vesicles to membranous tubulovesicular and tubular networks and finally to a fenestrated sheet that matures into a young cell wall. Cell plate maturation involves changes in polysaccharide composition, removal of excess membrane by endocytosis, and the disassembly of the CPAM (Samuels *et al.*, 1995; Segui-Simarro *et al.*, 2004).

Although the presence of a dense matrix surrounding the cell plate was detected by conventional electron microscopy (EM) (Gunning, 1982; Otegui and Staehelin, 2000a; Samuels *et al.*, 1995), its distribution and assembly pattern during cytokinesis was understood only when electron tomographic studies were performed (Austin *et al.*, 2005; Otegui and Staehelin, 2004; Otegui *et al.*, 2001; Segui-Simarro *et al.*, 2004). The composition of the CPAM is not known, but it appears to mediate membrane fusion events and to stabilize microtubule ends during cytokinesis.

Electron tomographic analyses on conventional (somatic) and nonconventional cytokinetic mechanisms in plants have also allowed us to visualize elements of the molecular machinery responsible for vesicle transport and membrane fusion/trafficking such as kinesin-like bridges that connect phragmoplast microtubules and vesicles in endosperm cell plates, dynamin rings on cell plate intermediates, and exocyst-like tethering structures (Otegui and Staehelin, 2004; Otegui *et al.*, 2001; Segui-Simarro *et al.*, 2004). In addition, a recent electron tomographic study on somatic phragmoplasts has provided a complete map of the distribution and geometry of microtubule plus ends inside and outside the CPAM, indicating that the CPAM plays a role in organizing and stabilizing the phragmoplast microtubule array (Austin *et al.*, 2005).

II. Rationale

Cytokinesis is a highly regulated process that requires coordinated interactions between the cytoskeleton and the membrane-trafficking pathways. The combination of cryofixation/freeze-substitution methods with ET has revealed amazing details in this complex process. In some cases, ET has allowed us to visualize proteins and protein complexes expected to be found in the phragmoplast and cell plates such as kinesin-like molecules and dynamin rings, respectively (Otegui *et al.*, 2001; Segui-Simarro *et al.*, 2004). In other cases, ET stands out as a method to predict the involvement of certain proteins and protein complexes, whose role in plant cytokinesis has yet to be characterized such as exocyst-like complexes and CPAM-associated molecules (Austin *et al.*, 2005; Otegui and Staehelin, 2004; Otegui *et al.*, 2001; Segui-Simarro *et al.*, 2004). In addition, ET has provided novel information on quantitative changes in cell plate surface area and volume that have been essential in estimating membrane dynamics during cytokinesis.

It has also revealed the 3D organization and distribution of the phragmoplast microtubule arrays during the various stages of cytokinesis. No other method has achieved comparable 3D resolution to help us understand plant cytokinesis at the molecular level in the cellular context. This chapter focuses on the two main variables associated with these imaging techniques: sample preservation and ET of plastic sections.

III. Materials and Methods

To take full advantage of 3D electron tomographic data, (1) the sample must be optimally processed (Gilkey and Staehelin, 1986) and (2) the tomographic data must be assembled correctly and modeled accurately.

A. Processing Dividing Cells for ET

1. Finding Dividing Cells in Plants

Plant stem cells with high mitotic rates can be found in meristems such as those at the apexes of both shoots and roots. Since flowering plants have the unique ability to continuously produce new organs, stem cell populations in meristems are maintained throughout the plant life cycle. In *Arabidopsis thaliana*, both shoot and root apical meristems are routinely used to study cell division. In 8-day-old *Arabidopsis* seedlings, the apical root meristem region consists of ~52 cells each of which divides every 18 h (Beemster and Baskin, 1998). Therefore, the chance of finding at least one cell undergoing cytokinesis in a given root tip is relatively high. The use of reversible mitosis inhibitors, such as the G1/S phase arrest-inducing reagent aphidicolin, can provide a certain degree of synchrony in intact meristems (Planchais *et al.*, 2000; Sala *et al.*, 1986), facilitating the study of cells at a specific stage of the cell cycle.

Plant cell culture lines are also commonly used to study plant cytokinesis (Demidov *et al.*, 2005; Gu and Verma, 1997; Marcus *et al.*, 2005; Nebenfuhr *et al.*, 2000; Samuels *et al.*, 1995; Sano *et al.*, 2005). The tobacco Bright Yellow-2 (BY-2) cell line developed by Nagata and coworkers (Nagata and Kumagai, 1999; Nagata *et al.*, 1992) responds well to synchronization protocols and can provide mitotic indexes of ~39–77%. For studying cytokinesis, BY-2 cells are grown in a medium containing 3–5 μg/ml aphidicolin for 24 h. After washing out the aphidicolin-containing medium, cells are allowed to grow in fresh medium for 3 h. Then 6-μM propyzamide, a microtubule assembly inhibitor, is added to the medium for 6 h (Kakimoto and Shibaoka, 1988). After 90–180 min of washing out the propyzamide, most dividing cells undergo cytokinesis (Samuels *et al.*, 1995).

Developing embryos are also an excellent source of dividing cells. In *Arabidopsis*, the analysis of developing seeds allows for the simultaneous study of somatic cytokinesis in embryo cells and an unconventional cytokinesis that occurs during endosperm cellularization (Otegui and Staehelin, 2000b). The endosperm in

Arabidopsis starts to cellularize at the micropylar region, when the embryo has reached the late globular stage (~5–6 days after pollination).

Other types of nonconventional cytokinesis, such as postmeiotic cytokinesis in male sporocytes, can be studied in reproductive tissues. In *Arabidopsis*, flower primordia arise from the base toward the apex of the inflorescence. Starting to count from the most recently opened flower (position 1) to the youngest, still closed flower bud, anthers containing microsporocytes undergoing meiosis are usually found in the flower buds in position 6 or 7.

2. Processing of Plant Samples by High-Pressure Freezing/Freeze-Substitution

Cryofixation techniques (at high or atmospheric pressures) combined with freeze-substitution allows for the fast immobilization of cellular components, avoiding the artifacts created by the slow action of fixatives during conventional chemical fixation (Gilkey and Staehelin, 1986). In practice, the freezing of biological samples with no visible ice crystal damage at atmospheric pressure can only be accomplished in samples up to 10-μm thick (Gilkey and Staehelin, 1986). Therefore, if one wants to study larger samples, such as root tips or developing seeds, high-pressure freezing is the method of choice (see also Chapter 2 by McDonald and Chapter 3 by Hess, this volume).

Small pieces of samples are loaded in freezing planchettes containing cryoprotectants such as 0.1-M sucrose or 1-hexadecene. If working with *Arabidopsis*, 1-mm-long root tips, intact developing seeds, ovules, and anthers can be loaded directly into the planchettes, commonly known as "freezing hats." If working with cultured cells grown in a medium with sucrose, a soft pellet of cultured cells can likewise be loaded directly into the freezing hats.

Freeze-substitution is performed by placing the cryofixed samples in solvents with fixatives for 3–5 days at −80 °C. This can be achieved by placing the cryovials in dry ice, in a −80 °C freezer, or in an automatic substitution device. The freeze-substitution medium should be chosen according to the cellular structures one wants to analyze. To achieve good preservation and staining of microtubules and membranes in general, 2% OsO_4 in acetone is recommended (Hess, 2003). There are also several membrane contrast-enhancing approaches that can be used during freeze-substitution, such as (1) initial freeze-substitution with 0.1% tannic acid and 0.5% glutaraldehyde or 0.2% uranyl acetate and 0.5% glutaraldehyde for several days at −80 °C, followed by OsO_4 (Giddings, 2003); (2) freeze-substitution with 0.1% potassium permanganate (Giddings, 2003); and (3) addition of 1–5% water in the substitution medium (Walther and Ziegler, 2002).

In addition, better contrast of the membranes and cytoskeleton elements is achieved by staining the plastic sections with heavy metal-containing reagents such as lead citrate and uranyl acetate. We use a methanolic solution of uranyl acetate (2% uranyl acetate in 70% methanol) because it increases the contrast of membrane and cytoskeletal elements more than aqueous solutions do. It is important to note that only one surface of the section is exposed to the heavy metal solutions. As a result, the staining of the section across its thickness may not be homogeneous, especially when using certain resins such as Spurr's. Although this is not a problem in thin sections (~60–70 nm in thickness) used for conventional EM studies, it can cause heterogeneous staining in the semithin sections (250–500 nm in thickness) used in ET. If the staining of the section is not homogeneous due to the limited penetration of the heavy metal solutions, the contrast of the cellular structures becomes poorer in the tomographic slices that correspond to the section surface not exposed to the heavy metals. To minimize this problem, we do not use sections thicker than 300 nm.

After poststaining the sections with heavy metals, 5 μl of 15-nm colloidal gold solution is applied to each side of the sections for 5 min and the excess solution is removed by touching the grid with filter paper. The gold particles are used as fiducials, aiding in the fine alignment of the tilt images during the tomographic reconstruction.

B. Visualizing Plant Cytokinesis by ET

1. Imaging Cell Plates and Phragmoplasts

Once a dividing cell has been selected, the images are collected by tilting the specimen in the electron beam. The resulting two-dimensional (2D) projections are aligned and 3D density maps or electron tomograms are calculated using a weighted back-projection algorithm (see Chapter 5 by O'Toole, this volume). The resolution and overall quality of the tomographic reconstruction depends on many factors. Given that the quality of the tilt series (image focus and alignment) is optimal, the main factors affecting resolution are: (1) the magnification at which the images are collected, (2) the angular interval, (3) the angular range, and (4) section thickness (Marsh, 2005; McEwen and Frank, 2001). The tomographic analysis of phragmoplasts and cell plates requires collecting images across large areas of the cell. Imaging at low magnification is not recommended because of the resulting loss in resolution. If a charge-coupled device (CCD) camera is used to collect the images, the magnification should be high enough so that each pixel projects onto the specimen with a size of ~1 nm. When the structure of interest covers a large area, a single camera field may be insufficient to record the structure under study, but montaged images can be used to image large areas without compromising resolution (Fig. 1). With the accurate tilting stages and high-tilt sample holders currently available, most data collections for ET are performed at 1° angular intervals and over an angular range of ±60° to 70°. To maximize the specimen area that can be imaged at high-tilt angles, single slot specimen grids are recommended.

To determine the best *z* dimension (thickness) of the tomograms several factor should be considered. As explained above, semithick sections over 300 nm may not be uniformly poststained with heavy metals. In addition, if an intermediate-voltage (300 kV) electron microscope is used, sections thicker than this will result in poor resolution images, particularly at high-tilt angles. However, since cell plates and phragmoplast are large and heterogeneous structures, for example, different

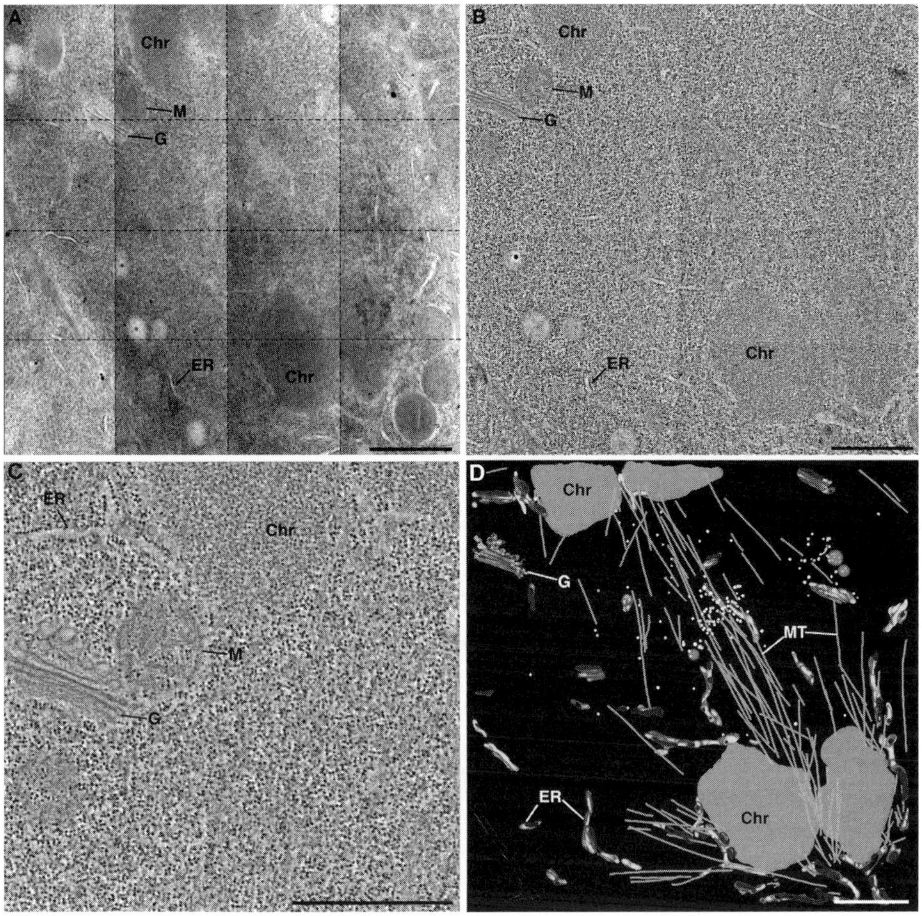

Fig. 1 Tomographic analysis of montaged images collected with a JEM-1000 high-voltage electron microscope (Jeol, Akishima, Tokyo, Japan) operating at 750 kV. (A) Montaged image of an untilted, semithick (300 nm) section of an *Arabidopsis* meristem anaphase spindle. Sixteen individual frames, each covering an area of 1.4 × 1.4 μm^2, were collected with a Gatan digital camera (Gatan, Pleasanton, CA) at 12,000× and at a resolution of 1024 × 1024 pixels (pixel size = 1.42 nm). (B) Image from a dual-axis tomographic reconstruction (view of five combined 2.2-nm-thick tomographic slices) derived from the montaged images shown in (A) using the IMOD package. (C) A higher magnification view of a subregion from B showing the detail that is visible in a tomographic reconstruction. (D) Tomographic model derived from the dual-axis tomogram depicted in (B). Chr, chromosomes (teal); G, Golgi stack (orange); M, mitochondrion; MT, microtubule (green); ER, endoplasmic reticulum (dark blue). Scale bars = 1 μm.

developmental intermediates coexist in the same cell plate, it is important to sample large volumes. To achieve this without exceeding the recommended specimen thickness, serial tomograms can be obtained from serial sections (Otegui and Staehelin, 2004; Segui-Simarro *et al.*, 2004). Ideally, many serial sections are placed on the same grid, usually by cutting and transferring a ribbon of sections.

A dividing cell can then be located and imaged on each relevant section in the ribbon. However, it is important to note that this approach suffers from a 15- to 25-nm gap of missing material between serial tomograms (Ladinsky *et al.*, 1999). For highly complex cellular structures, this gap in information may cause difficulties in the alignment of serial tomograms.

Contrary to medical computed tomography in which 2D projections of the patient can be collected over a full 360°, the angular range allowed by the conventional specimen holders used for ET is more restricted, resulting in a wedge of missing information between the maximal tilt angle collected and 90° (see Chapter 5 by O'Toole, this volume). This results in distorted tomograms with anisotropic resolution (McEwen and Frank, 2001). To improve the isotropy in resolution, we collect images from two orthogonal axes and combine the two resulting tomograms into a dual-axis tomogram; this reduces the missing wedge of information to a missing pyramid (Mastronarde, 1997). This approach greatly improves the visualization of linear elements such as microtubules and actin filaments that in single-axis tomograms tend to fade out when they are oriented in certain directions relative to the axis of tilt (Mastronarde, 1997).

For calculating, displaying, and analyzing electron tomograms, we use the IMOD package developed primarily by David Mastronarde, Rick Gaudette, Sue Held, and Jim Kremer at the Boulder Laboratory for 3D Electron Microscopy of Cells. Information on how to download and use this free software package can be found at http://bio3d.colorado.edu/imod/.

In conclusion, the combination of montaged images and dual-axis, serial tomography allows for the analysis of large volumes: several square micrometers in the x–y plane, and depending on the number of serial tomograms, 0.5–1 μm in z, or even more. The resulting images of cell plates and phragmoplasts having 4- to 7-nm resolution in 3D are shown in Fig. 1. The principal limitation on the size of a tomogram is the capability of the computer. If the final tomogram after combining several serial tomograms is too big to be loaded for analysis, the 3dmod program of the IMOD package allows for defining coordinates in x, y, and z to load just a subvolume of the tomogram, but work becomes slower and more cumbersome.

For some specific studies, for example, where one needs a general map of the distribution of the cell plate, phragmoplast elements, and associated organelles, reconstructions of serial sections using conventional EM are recommended (Segui-Simarro and Staehelin, 2006). However, the resolution in z for serial reconstructions is limited to about twice the section thickness (McEwen and Marko, 1999). Therefore, if thin serial sections of ~50 nm are used, the resolution in z of the 3D reconstruction will be 100 nm (compared to 4–7 nm in ET).

2. Modeling of Cytoskeletal Elements and Membranes

Although the automated collection of tilt series for ET is a fast and well-controlled process, thanks to high-quality CCD cameras and improved computer-controlled microscopes, the hand segmentation (modeling) of the cellular structures is still a very time consuming and somehow subjective task. When working with cell plate tomograms, in particular, the daunting task of modeling thousands of microtubules and vesicles, the complex cell plate membranous intermediates, and macromolecular complexes can be overwhelming. In this section, we will discuss how to model microtubules and membranes and other cytokinesis-related structures using the 3dmod program, so that 3D quantitative information can be obtained and compared between tomograms.

Modeling is the creation of graphic objects that accurately represent the 3D positions of features of interest in a tomogram. The term segmentation is often used because in a sense one is cutting out one kind of data from all the rest of the information in an image, simplifying the issues that will be considered (see also Chapter 30 by Sandberg, Chapter 25 by Best *et al.*, and Chapter 29 by Förster and Hegerl, this volume). The IMOD package contains a program, 3dmod, that allows an operator to draw on the image data, placing points, chosen shapes (circles, and so on), or sets of point (lines or curves that match the structure of interest) as "overlays" on the image data. Each such representation is called a contour, and these are generally drawn as 2D figures on a single plane of image data extracted from the tomogram. Contours can be either closed (if they represent something like the membrane that surrounds a vesicle or cisterna) or open (if they represent a fiber, like a microtubule). Closed contours will always connect the last point entered with the first, so the set of point has no detectable end. Open contours will always have a beginning (first point entered) and an end (last point), and these can be distinguished and marked if so desired.

Contours that represent parts of a larger object, like the successive slices through a vesicle, cisterna, or tubule that one sees on many tomographic slices, can all be grouped as parts of a single object. It is usually convenient to group open contours into objects too, since each contour may represent a member of a set. For example, all the phragmoplast microtubules that lie on one side of the cell's mid-plane are of one polarity (orientation), while the microtubules on the other side are antiparallel. Thus, it is sensible and convenient for subsequent analysis to make all of the contours that represent microtubules in one of these clusters a single object. Likewise, many vesicles that share a significant biological property, like having a clathrin coat or a particular size, can be organized as contours within a single object. Some objects can be modeled as "scattered points," where a sphere is computed from a single model point placed in the center of the object, for example, a vesicle, and its size can be adjusted to match the diameter of the vesicle. A general guide that provides a comprehensive description of the 3dmod modeling program can be found at http://bio3d.colorado.edu/IMOD.

As a general rule, it is easier to combine two separate objects into one than to create two objects out of one. It is also important to think about how best to organize objects, with the thought of later processing the objects quantitatively (lengths, volumes, surface areas, 3D spatial relationships), and presenting the objects as figures in a manuscript or lecture.

a. Microtubules

When modeling microtubules, it is best to start out with two objects, one for microtubules that are contained entirely within the tomographic volume, and another object for those that are only partially within the volume. It may also be advantageous to create separate objects for groups of microtubules, as described above for the opposing sets of microtubules in phragmoplasts. This is a useful modeling approach when quantitative information is required from different sets of fibers.

Microtubules can be modeled at two levels of resolution: (1) as hollow cylinders that represent the whole microtubule (Fig. 2A–C and E) or (2) as groups of protofilaments (Fig. 2F and G). The protofilament modeling process is more time consuming and is used to analyze microtubule end geometries. In both cases, the contours can be drawn in two different ways: (1) directly in the "Zap" window, which displays tomographic slices cut parallel to the surface of the physical section that was reconstructed (Fig. 2B) or (2) in the "slicer" window of the 3dmod program (Fig. 2C) (see also Chapter 5 by O'Toole and Chapter 6 by Höög and Antony, this volume). Both of these modeling strategies use open contours and work by placing points along the length of the microtubules (Fig. 2D). These modeling techniques can, of course, be used for any kind of filament, including other components of the cytoskeleton and material of the extracellular matrix.

In the Zap window, one must first identify one end of the microtubule and place the first point, as close to the center of the microtubule as possible. Every fifth tomographic slice along the microtubule length, one then places another point until reaching the other end of the microtubule. Alternatively, in the slicer window one can manipulate the x, y, z sliders to find a tomographic slice that contains as long a segment as possible of the microtubule axis (Fig. 2C). Now one can quickly mark points *ad lib.* along the microtubule axis and mark both ends of the microtubule in any useful way.

A 3D representation of the microtubule can be obtained by using the command "imodmesh" with the options -t for tube, -d (the pixel diameter of the tube, 25 nm for a microtubule), and -E (this will cap the end of the tube). Complete instructions for use of the imodmesh command can be found at http://bio3d.colorado.edu/imod/doc/man/imodmesh.html. The formation of a mesh on digitized contours can be an important part of model generation; it is useful both for display and for quantification. A mesh generated around the contour that represents a microtubule allows this polymer to be represented by a graphic object that is an accurate model for microtubule's size and shape. The closed contours that represent a membrane-bounded compartment can also be meshed in ways that are valuable for analysis (see below).

As shown in Austin *et al.* (2005), a wealth of information can be obtained from the careful modeling of microtubule end geometries (Fig. 2F and G). It must be understood that since the thickness of a tomographic slice is ~1–3 nm and since ET

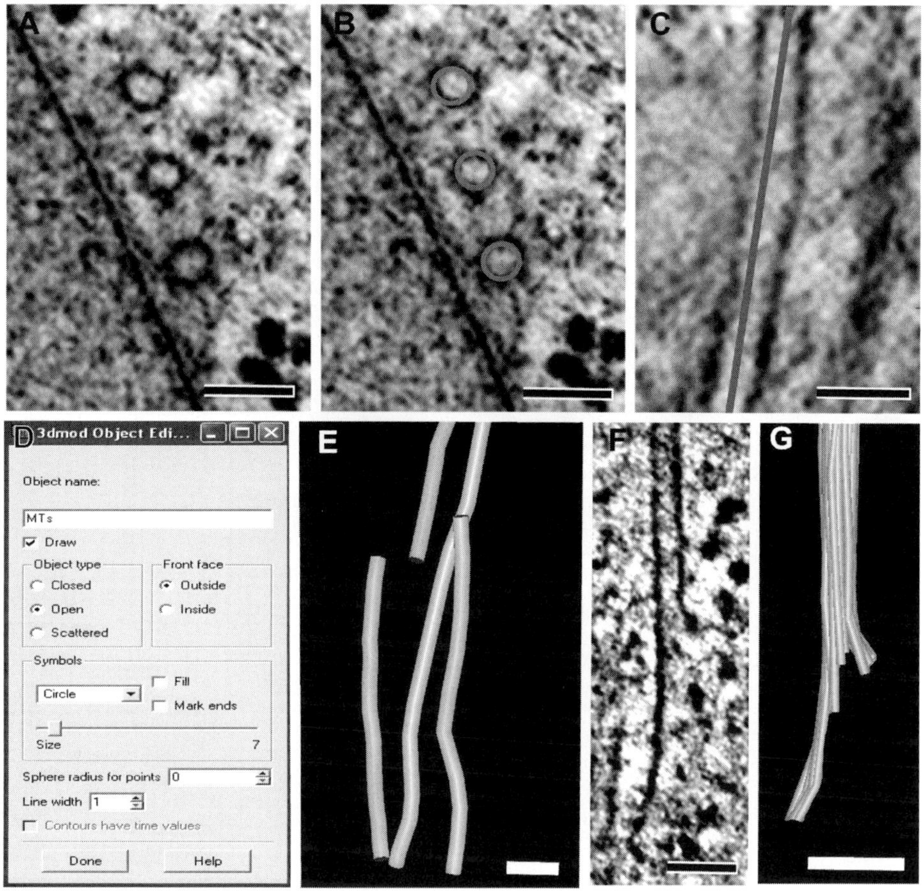

Fig. 2 Modeling of phragmoplast microtubules using the Zap window (A and B) or the slicer window (C). (D) Object specifications for modeling microtubules using 3dmod. (E) 3D model of microtubules depicted in (A–C). Microtubules are rendered as hollow cylinders 25 nm in diameter. (F–G) Example of the "relative mapping" of protofilaments. Protofilaments are rendered as individual cylinders ~4 nm in diameter. Scale bars = 50 nm.

of thick sections always suffers from the data that are missing (due to the limitations on tilt angle during data collection), it is not possible to model every ~4-nm protofilament in those microtubules that are oriented parallel to the section surface. Therefore, the microtubule architecture is "relatively mapped" and the 13 protofilaments may not be identifiable in all of the microtubules in a tomographic volume. Hence "ghost" protofilaments can be added to make the microtubule look cylindrical. After the modeling is completed imodmesh should be used as explained above, but specifying a smaller diameter (-d = 4 nm for protofilaments versus 25 nm for the microtubule diameter).

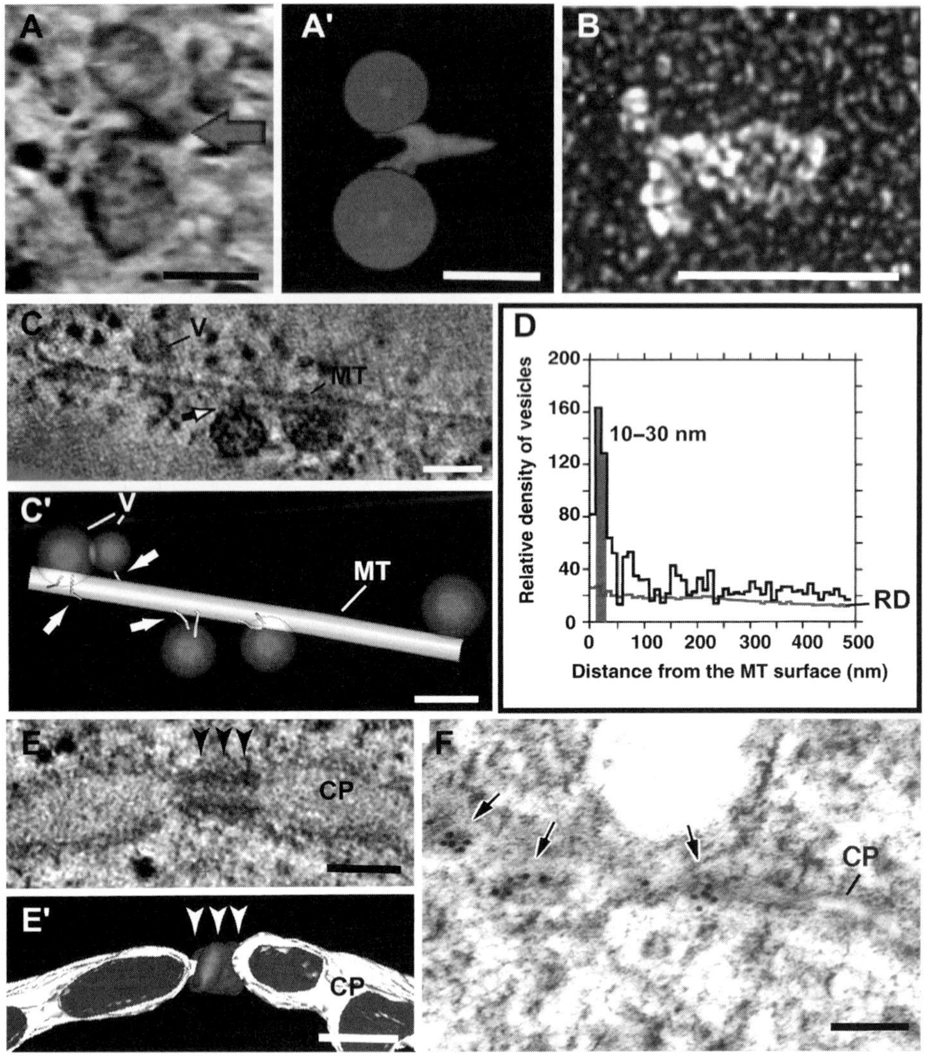

Fig. 3 Examples of identification of protein complexes in ET of plastic sections. (A–B) Exocyst complexes bridging cell plate-forming vesicles in somatic cell plates. (A) Tomographic slice and (A′) tomographic model of two vesicles connected by a structure (arrow), whose shape and size correlate with those of purified mammalian brain exocyst complexes (B). (C–D) Molecules connecting phragmoplast microtubules and vesicles. (C) Tomographic slice through a microtubule (MT) and associated vesicles (V) in the cellularizing endosperm of *Arabidopsis*. (C′) Tomographic model of the tomogram depicted in (C). Note the rodlike linkers (arrows) connecting microtubules and vesicles. These bridges resemble kinesin-like proteins. (D) Quantitative analysis of distances between microtubules and vesicles. The analysis was performed with the MTK program of the IMOD package. Note that a high number of vesicles are located between 10 and 30 nm from the microtubule surfaces. This distance is consistent with the presence of kinesin-like molecules connecting microtubules and vesicles (Otegui *et al.*, 2001). RD, random distribution generated by the MTK program. (E–F) Membrane-surrounding rings in

b. Membranes

During plant cytokinesis there are several membranous compartments that must be considered (cell plate membrane, endoplasmic reticulum, the plasma membrane of the parental cell, vesicles, and various organelles) and model as separate objects. Nonconnected portions of the cell plate or different domains of the endoplasmic reticulum can be modeled as different objects as well.

Quantitative information is calculated based on models that have been meshed, a process that generates graphic objects that are more similar to the structure of interest than the contours themselves. Meshes are, however, derived from contours, so it is extremely important to draw the contours accurately. When modeling membranous organelles, we recommend placing contours in the middle of the lipid bilayer (Ladinsky *et al.*, 1999; Otegui *et al.*, 2001). Imodmesh will then generate a family of triangles that connect neighboring points within a contour and nearby points on adjacent contours. The resulting triangles represent a surface in space that is an excellent approximation to all the contour information, so they are used for all subsequent quantifications of area, volume, and distance. They can also be used to generate a shaded surface that provides a good visual representation of the modeled object for view and qualitative study.

c. Other Macromolecular Elements

One of the most powerful uses of 3D ET is the 4- to 7-nm resolution that allows the visualization of macromolecules in their cellular context. During cytokinesis, hundreds of different types of macromolecules are known to participate, but not until the recent studies using 3D tomography have some of these molecules been visualized at sufficient resolution to elucidate their precise localization.

Molecular complexes can be identified in electron tomograms by comparing and correlating their electron-density maps with existing structural information derived from X-ray crystallography, EM, and/or nuclear magnetic resonance. As an example, tethering structures that connect cell plate-forming vesicles in somatic cells have been tentatively identified as exocyst complexes, based on their resemblance to exocyst-tethering complexes purified from mammalian brain tissue (Hsu *et al.*, 1998; Segui-Simarro *et al.*, 2004) (Fig. 3A and B).

However, it is important to note that sample processing, embedding, and poststaining with heavy metals can often limit the faithful identification of molecular structures. With this in mind, it is highly recommended to confirm the identity

---

endosperm cell plates. (E) Tomographic slice and (E′) corresponding tomographic model of a tubular domain of an endosperm cell plate. Note the presence of an electron-dense collar (arrowheads) associated with a constricted membrane tubule. The identity of the rings as dynamin AtADL1A polymers was confirmed by immunolabeling with specific antibodies (F). Scale bars = 50 nm. Panels A and A′ reprinted from Segui-Simarro *et al.* (2004) (Copyright © 2004 American Association of Plant Biologists). Panel B reprinted (with permission from Elsevier) from Hsu *et al.* (1998) and from Segui-Simarro *et al.* (2004) (Copyright 2004 American Association of Plant Biologists). Panels C, C′, D, and F from Otegui *et al.* (2001) (Copyright 2001 American Association of Plant Biologists).

of the molecules with complementary identification approaches. For example, structures connecting microtubules and cell plate-forming vesicles were visualized in electron tomograms of the *Arabidopsis* endosperm, but more specific details of their shape and sizes were defined by performing a quantitative analysis of the spatial relationship between microtubules and vesicles (Otegui *et al.*, 2001) (Fig. 3C and D). Most vesicles were located between 10 and 30 nm from the microtubule surface, which is consistent with the presence of kinesin-like motor proteins connecting vesicles and microtubules.

Another approach to identifying macromolecules in tomographic reconstructions is to perform correlative immunolabeling experiments. As an example, our immunolabeling experiments with specific antibodies suggest that the electron-dense rings constricting cell plate membranes in endosperm cell plates are dynamin ADL1A polymers (Otegui *et al.*, 2001) (Fig. 3E and F) (for additional evidence on this approach, see Chapter 21 by Ladinsky and Howell, this volume).

The CPAM is a particularly difficult structure to model, since it is neither enclosed in membrane boundaries nor differentially stained. However, a clear feature that defines this area is the exclusion of ribosomes. To model the CPAM, we first establish the density of ribosomes at different distances from the cell plate surface. This observation defines the width of a ribosome exclusion zone, and we interpret the absence of ribosomes as a result of the presence of the CPAM (Fig. 4A). On the basis of these calculations, we have estimated that the width of the ribosome excluding zone varies between 50 and 150 nm in different developmental stages in syncytial (Otegui and Staehelin, 2004; Otegui *et al.*, 2001) and somatic cell plates (Segui-Simarro *et al.*, 2004). Once the width of the CPAM is established, contours are drawn at the calculated distances from the cell plate (Fig. 4B–D).

C. Quantitative Analysis

One of the powerful advantages of ET is the possibility of obtaining quantitative information from tomographic models. In the plant cytokinesis field, quantitative tomographic data have been essential for studying the dynamics of the CPAM during cell plate formation, for revealing the spatial relationship between vesicles and microtubules, and for estimating the amount of membrane that traffics through the cell plate, as well as the frequency and distribution of microtubule plus ends in the phragmoplast. In this section, we will focus on how these types of calculations can be performed.

1. Volume and Surface Area

By calculating the surface area and volume of cell plate intermediates and vesicles, one can estimate the dynamics of membrane transport to and from the cell plate. As an example, we have estimated that 500,000 Golgi-derived vesicles

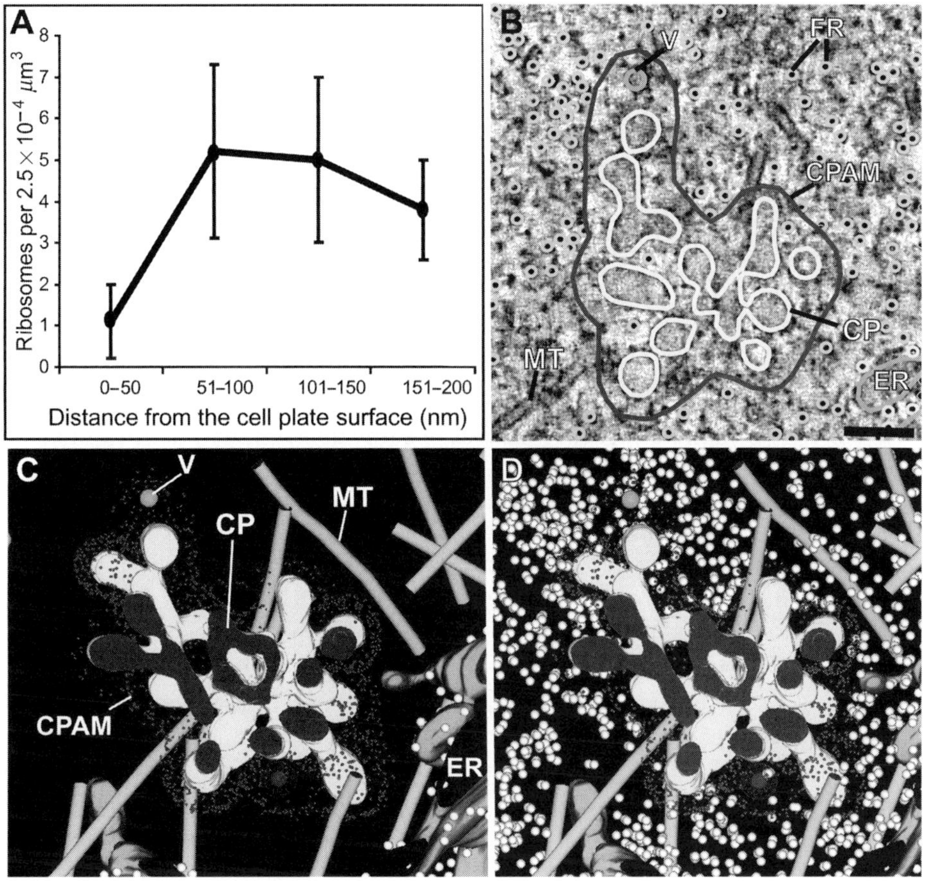

Fig. 4 Modeling of the CPAM. (A) Quantitative analysis of ribosome density at different distances from an endosperm cell plate at the tubular stage. (A) Drawing of contours delimiting the CPAM. (C) 3D tomographic model of CPAM and cell plate (CP) derived from tomogram depicted in (A). (D) Same tomographic model including free ribosomes. ER, endoplasmic reticulum; FR, free ribosomes; MT, microtubule; V, vesicle. Scale bar = 100 nm. (See Color Insert.)

must fuse to form a new cell wall in the *Arabidopsis* endosperm. We have also learned that 75% of the cell plate membrane is removed by endocytosis by the end of cytokinesis (Otegui *et al.*, 2001). To make these calculations, we extracted randomly located boxes of a defined size along the cell plate. We meshed the portions of cell plates enclosed in these boxes and calculate total surface and volume of different modeled structures using the imodinfo program. Because the contours are placed in the middle of the cell plate membrane, the interior volume was calculated by subtracting the surface area multiplied by 3.5 nm (half of the membrane width) from the volume inside the contours (Marsh *et al.*, 2001).

Imodinfo provides information about IMOD models such as lists of objects, contours and point data, lengths and centroids of contours; and surface areas and volumes of objects or surfaces. The measurements are generally expressed in terms of the pixel size, which is described in the model header as a number in nanometers. The extent to which the section thinned in the electron beam (thinning factor) is also described in the model header and is applied to the calculations.

To calculate surface area and volume of vesicles, we considered these structures as spheres and used the "scattered point" type of objects to model them. Surface and volume were calculated based on the radius (r) of the scattered points:

$$\text{Volume} = (4/3)\pi r^3 \quad \text{Surface} = 4\pi r^2$$

From these calculations, and under the assumption that biological membrane cannot expand by stretching (Wolfe and Steponkus, 1983), one can calculate how many Golgi-derived vesicle equivalents are required to build a define area of the cell plate. One can also calculate how many endocytic vesicles should be removed from a cell plate surface unit during its maturation. A complete description of the imodinfo program and its options can be found in http://bio3d.colorado.edu/imod/doc/man/imodinfo.html.

2. Density

For calculating densities of ribosomes, vesicles, or other structures modeled as spheres, we recommend having all the individual items as points in the same contour of the same object. For example, the object containing all the ribosomes in Fig. 4A has one contour and 24,580 points. To calculate density of ribosomes and other structures per volume unit, the number of ribosomes in defined size boxes extracted from the model can be calculated using the imodinfo program.

3. Spatial Relationship: MTK Analysis

To measure the distances between the objects in 3D and to compute an average density of neighboring items as a function of distance between objects, one can use the MTK program (Kremer *et al.*, 1996; Marsh *et al.*, 2001). This program considers structures of three kinds: open contours, like microtubules; scattered point objects, like vesicles that were modeled by spheres of defined radii; and meshed, closed contour objects, like Golgi cisternae and cell plate domains. For microtubules, each contour is considered as a separate line. It is possible to calculate the spatial relationship between any object and a whole microtubule, or to break the microtubule into multiple segments and measure the closest approaches between a chosen object and each of the segments. Distances can be measured from the central axis or surface of one object to the central axis or surface of another. A complete explanation of MTK and its commands can be found in http://bio3d.colorado.edu/imod/doc/man/mtk.html.

To determine the significance of the spatial relationship between cellular structures, MTK can generate a random distribution of the structures under analysis and compare it to the actual distribution (Fig. 3D). These random distributions can be obtained by shifting each object of interest to a new position within the tomographic volume, rejecting those positions that placed objects outside the modeled area or that resulted in collision with other objects (Marsh *et al.*, 2001; Otegui *et al.*, 2001). By computing the average distribution of spacings from many sets of randomized positions, one obtains a robust "null hypothesis" with which to compare the actual distribution data. As explained above, these calculations have been essential in understanding the spatial relationship between phragmoplast microtubules and cell plate-forming vesicles in the *Arabidopsis* endosperm. They have also illuminated the involvement of structures that resemble kinesin-like molecules in the transport of vesicles to the division plane (Otegui *et al.*, 2001).

IV. Discussion

Cytokinesis is one of the most complex processes that plant cells execute. In recent years, numerous studies analyzing mutants and fluorescently tagged molecules have shed light on this intricate process, which requires the coordination of the cytoskeleton, the secretory and the endocytic pathways, and the membrane fusion machinery (see Jurgens, 2005 and Otegui *et al.*, 2005, for recent reviews). The electron tomographic analysis of somatic and nonconventional types of cytokinesis has provided a unique approach to the field: the possibility of analyzing cytokinesis-related structures at the resolution that approaches the molecular level within a cellular context (Austin *et al.*, 2005; Otegui and Staehelin, 2004; Otegui *et al.*, 2001; Segui-Simarro *et al.*, 2004).

One of the most exciting and challenging tasks in ET is the identification of molecules and protein complexes in their cellular context. As mentioned above, sample processing and poststaining with heavy metals used in the ET of plastic sections can limit the identification of molecular structures. By contrast, cryo-ET uses vitrified biological samples and therefore, the resolution is not limited by structural preservation of the sample or the staining with heavy metal salts (Nickell *et al.*, 2006). However, few eukaryotic cells are thin enough to be imaged directly as a whole and therefore, most samples must be sectioned. This entails a number of technical difficulties, since the ice-embedded frozen-hydrated samples have to be sectioned and maintained below devitrification temperature (approximately $-137\,^{\circ}$C). Although cryotomographic studies of eukaryotic cells, with or without sectioning, are becoming more and more common (Hsieh *et al.*, 2006; Medalia *et al.*, 2002; Nicastro *et al.*, 2005), no cryotomographic data on higher plant cells has yet been published.

Since the identification of molecules based on the recognition of their structural patterns is limited in ET of plastic sections, there is an urgent need to develop techniques to label molecules in the semithick plastic sections. Although conventional postsectioning immunogold techniques can be used in combination with ET (Zeuschner *et al.*, 2006), this approach is limited to the identification of molecules at the surface of the sections; postembedding methods cannot get inside the section (but see Chapter 21 by Ladinsky and Howell, this volume). On the other hand, immunolabeling prior to dehydration and resin infiltration seriously compromises cellular preservation. An attractive alternative is the design of genetically encoded, electron-dense protein tags that can be visualized in the electron microscope. Proteins with electron-dense tags could be expressed in transgenic plants and identified in the tomographic reconstructions. These tags should be small proteins or peptides with high affinity for heavy metals. Metallothionein, a small protein of ~60 amino acids that can bind 14–17 gold atoms, appears to be a promising genetically encoded tag for EM studies (Mercogliano and DeRosier, 2006).

In addition, a very exciting approach is to develop genetically encoded tags that allow for correlative analyses by light and electron microscopy. One of these is the green fluorescent protein, whose fluorescence can induce photooxidation of diaminobenzidine in the right chemical environment, causing the localized formation of an insoluble osmiophilic precipitate visible by EM. Another kind of tag consists of a tetracysteine peptide that binds biarsenical, membrane-permeant fluorophores, such as FlAsH (a fluorescein derivative) and ReAsH (a resorufin derivative). Direct correlation of live cell images with high resolution EM detection is possible because bound ReAsH not only becomes fluorescent but can also be used to initiate the photoconversion of diaminobenzidine (Gaietta *et al.*, 2002; Chapter 23 by Sosinsky *et al.*, this volume). This technique has been successfully used to study trafficking of connexin-43 in gap junctions of mammalian cells (Gaietta *et al.*, 2002), but it has not been applied to plants yet.

Although a wealth of information on plant cytokinesis has been obtained from electron tomographic studies, there are still key questions that ET in combination with emerging imaging techniques can help to answer in the future, such as: what is the molecular composition of the CPAM? How is cell plate formation affected in plant cells that carry mutations in CPAM components? How does the CPAM control microtubule stability? How does the CPAM associate with the cell plates and phragmoplast microtubules? How membrane-fusion-related molecules reach the cell plate?

V. Summary

By combining cryofixation/freeze-substitution, which provides near *in vivo* preservation of cellular structures, with 3D ET that has a resolution of 4–7 nm, we can now begin to understand where macromolecules precisely function within cellular processes. This new level of resolution, as demonstrated in the plant cell cytokinesis studies, has led to a wealth of information on cell plate membrane dynamics, microtubule organization, CPAM dynamics, and distribution pattern of a number of macromolecules.

References

Austin, J. R., II, Segui-Simarro, J. M., and Staehelin, L. A. (2005). Quantitative analysis of changes in spatial distribution and plus-end geometry of microtubules involved in plant-cell cytokinesis. *J. Cell Sci.* **118,** 3895–3903.

Beemster, G. T. S., and Baskin, T. I. (1998). Analysis of cell division and elongation underlying the developmental acceleration of root growth in *Arabidopsis thaliana*. *Plant Physiol.* **116,** 1515–1526.

Demidov, D., Van Damme, D., Geelen, D., Blattner, F. R., and Houben, A. (2005). Identification and dynamics of two classes of Aurora-like kinases in Arabidopsis and other plants. *Plant Cell* **17,** 836–848.

Gaietta, G., Deerinck, T. J., Adams, S. R., Bouwer, J., Tour, O., Laird, D. W., Sosinsky, G. E., Tsien, R. Y., and Ellisman, M. H. (2002). Multicolor and electron microscopic imaging of connexin trafficking. *Science* **296,** 503–507.

Giddings, T. H. (2003). Freeze-substitution protocols for improved visualization of membranes in high-pressure frozen samples. *J. Microsc.* **212,** 53–61.

Gilkey, J., and Staehelin, L. A. (1986). Advances in ultrarapid freezing for the preservation of cellular structure. *J. Electron Microsc. Tech.* **3,** 177–210.

Gu, X., and Verma, D. P. S. (1997). Dynamics of phragmoplastin in living cells during cell plate formation and uncoupling of cell elongation from the plane of cell division. *Plant Cell* **9,** 157–169.

Gunning, B. E. S. (1982). The cytokinetic apparatus: Its developmental and spatial regulation. *In* "The Cytoskeleton in Plant Growth and Development" (C. Lloyd, ed.), pp. 229–292. Academic Press, London.

Hess, M. W. (2003). Of plants and other pets: Practical aspects of freeze-substitution and resin embedding. *J. Microsc.* **212,** 44–52.

Hsieh, C.-E., Leith, A., Mannella, C. A., Frank, J., and Marko, M. (2006). Towards high-resolution three-dimensional imaging of native mammalian tissue: Electron tomography of frozen-hydrated rat liver sections. *J. Struct. Biol.* **153,** 1–13.

Hsu, S.-C., Hazuka, C. D., Roth, R., Foletti, D. L., Heuser, J., and Scheller, R. H. (1998). Subunit composition, protein interactions, and structures of the mammalian brain sec6/8 complex and septin filaments. *Neuron* **20,** 1111–1122.

Jurgens, G. (2005). Cytokinesis in higher plants. *Annu. Rev. Plant Biol.* **56,** 281–299.

Kakimoto, T., and Shibaoka, H. (1988). Cytoskeletal ultrastructure of phragmoplast-nuclei complexes isolated from cultured tobacco cells. *Protoplasma* **2**(Suppl.), 95–103.

Kremer, J. R., Mastronarde, D. N., and McIntosh, J. R. (1996). Computer visualization of three-dimensional image data using IMOD. *J. Struct. Biol.* **116,** 71–76.

Ladinsky, M. S., Mastronarde, D. N., McIntosh, J. R., Howell, K. E., and Staehelin, L. A. (1999). Golgi structure in three dimensions: Functional insights from the normal rat kidney cell. *J. Cell Biol.* **144,** 1135–1149.

Marcus, A. I., Dixit, R., and Cyr, R. J. (2005). Narrowing of the preprophase microtubule band is not required for cell division plane determination in cultured plant cells. *Protoplasma* **226,** 169–174.

Marsh, B. J. (2005). Lessons from tomographic studies of the mammalian Golgi. *Biochim. Biophys. Acta–Mol. Cell Res.* **1744,** 273–292.

Marsh, B. J., Mastronarde, D. N., Buttle, K. F., Howell, K. E., and McIntosh, J. R. (2001). Organellar relationship in the Golgi region of the pancreatic beta cell line, HIT-T15, visualized by high-resolution electron tomography. *Proc. Natl. Acad. Sci. USA* **98,** 2399–2406.

Mastronarde, D. (1997). Dual-axis tomography: An approach with alignment methods that preserve resolution. *J. Struct. Biol.* **120,** 343–352.

McEwen, B. F., and Frank, J. (2001). Electron tomographic and other approaches for imaging molecular machines. *Curr. Opin. Neurobiol.* **11,** 594–600.

McEwen, B. F., and Marko, M. (1999). Three-dimensional transmission electron microscopy and its application to mitosis research. *Methods Cell Biol.* **61,** 81–111.

Medalia, O., Weber, I., Frangakis, A. S., Nicastro, D., Gerisch, G., and Baumeister, W. (2002). Macromolecular architecture in eukaryotic cells visualized by cryoelectron tomography. *Science* **298,** 1209–1213.

Mercogliano, C. P., and DeRosier, D. J. (2006). Gold nanocluster formation using metallothionein: Mass spectrometry and electron microscopy. *J. Mol. Biol.* **355,** 211–223.

Nagata, T., and Kumagai, F. (1999). Plant cell biology through the window of the highly synchronized tobacco BY-2 cell line. *Methods Cell Sci.* **21,** 123–127.

Nagata, T., Nemoto, Y., and Hasezawa, S. (1992). Tobacco BY-2 cell line as the "HeLa" cell in the cell biology of higher plants. *Int. Rev. Cytol.* **132,** 1–30.

Nebenfuhr, A., Frohlick, J., and Staehelin, L. (2000). Redistribution of Golgi stacks and other organelles during mitosis and cytokinesis in plant cells. *Plant Physiol.* **124,** 135–151.

Nicastro, D., McIntosh, J. R., and Baumeister, W. (2005). 3-D structure of eukaryotic flagella in a quiescent state revealed by cryo-electron tomography. *Proc. Natl. Acad. Sci. USA* **102,** 15889–15894.

Nickell, S., Kofler, C., Leis, A. P., and Baumeister, W. (2006). A visual approach to proteomics. *Nat. Rev. Mol. Cell Biol.* **7,** 225–230.

Otegui, M. S., and Staehelin, L. A. (2000a). Syncytial-type cell plates: A novel kind of cell plate involved in endosperm cellularization of Arabidopsis. *Plant Cell* **12,** 933–947.

Otegui, M. S., and Staehelin, L. A. (2000b). Cytokinesis in flowering plants: More than one way to divide a cell. *Curr. Opin. Plant Biol.* **3,** 493–502.

Otegui, M. S., Mastronarde, D. N., Kang, B. H., Bednarek, S. Y., and Staehelin, L. A. (2001). Three-dimensional analysis of syncytial-type cell plates during endosperm cellularization visualized by high resolution electron tomography. *Plant Cell* **13,** 2033–2051.

Otegui, M. S., and Staehelin, L. A. (2004). Electron tomographic analysis of post-meiotic cytokinesis during pollen development in Arabidopsis thaliana. *Planta* **218,** 501–515.

Otegui, M. S., Verbrugghe, K. J., and Skop, A. R. (2005). Midbodies and phragmoplasts: Analogous structures involved in cytokinesis. *Trends Cell Biol.* **15,** 404–413.

Planchais, S., Glab, N., Inze, D., and Bergounioux, C. (2000). Chemical inhibitors: A tool for plant cell cycle studies. *FEBS Lett.* **476,** 78–83.

Sala, F., Galli, M. G., Pedrali-Noy, G., and Spadari, S. (1986). Synchronization of plant cells in culture and in meristems by aphidilcolin. *Methods Enzymol.* **118,** 87–96.

Samuels, L. A., Giddings, T. H., and Staehelin, L. A. (1995). Cytokinesis in tobacco BY-2 and root tip cells: A new model of cell plate formation in higher plants. *J. Cell Biol.* **130,** 1345–1357.

Sano, T., Higaki, T., Oda, Y., Hayashi, T., and Hasezawa, S. (2005). Appearance of actin microfilament 'twin peaks' in mitosis and their function in cell plate formation, as visualized in tobacco BY-2 cells expressing GFP-fimbrin. *Plant J.* **44,** 595–605.

Segui-Simarro, J. M., Austin, J. R., II, White, E. A., and Staehelin, L. A. (2004). Electron tomographic analysis of somatic cell plate formation in meristematic cells of Arabidopsis preserved by high-pressure freezing. *Plant Cell* **16,** 836–856.

Segui-Simarro, J. M., and Staehelin, L. A. (2006). Cell cycle-dependent changes in Golgi stacks, vacuoles, clathrin-coated vesicles and multivesicular bodies in meristematic cells of *Arabidopsis thaliana*: A quantitative and spatial analysis. *Planta* **223,** 223–236.

Walther, P., and Ziegler, A. (2002). Freeze substitution of high-pressure frozen samples: The visibility of biological membranes is improved when the substitution medium contains water. *J. Microsc.* **208,** 3–10.

Wolfe, J., and Steponkus, P. L. (1983). Mechanical properties of the plasma membrane of isolated protoplast. *Plant Physiol.* **71,** 276–285.

Zeuschner, D., Geerts, W. J. C., van Donselaar, E., Humbel, B. M., Slot, J. W., Koster, A. J., and Klumperman, J. (2006). Immuno-electron tomography of ER exit sites reveals the existence of free COPII-coated transport carriers. *Nat. Cell Biol.* **8,** 377–383.

CHAPTER 10

Electron Tomographic Methods for Studying the Chemical Synapse

Christopher P. Arthur, David B. Serrell, Maria Pagratis, David L. Potter, Dudley S. Finch, and Michael H. B. Stowell

Department of MCD Biology
University of Colorado
Boulder, Colorado 80309

This chapter focuses on the use of electron tomography (ET) to study the chemical synapse. It discusses the different types of chemical synapses and their properties as well as the biological questions that are addressable using ET. We present

METHODS IN CELL BIOLOGY, VOL. 79

0091-679X/07 $35.00
DOI: 10.1016/S0091-679X(06)79010-0

methods for culturing neurons on a variety of substrates suitable for ET and discuss the development of novel tools that combine electrophysiological methods with ET methods for studying active synapses.

I. Introduction and Rationale

Our current understanding of synapse formation and organization comes from a variety of studies that used immunofluorescence (or GFP-tagged proteins), fluorescent dyes, and/or electron microscopy (EM) of fixed embedded samples (Garner *et al.*, 2002). All these methods are limited in their resolution and their inability to distinguish a functioning synapse from one that is inactive or undergoing formation or decay. In addition, both detailed structural information about the molecular and supramolecular organization of the synapse and knowledge about the dynamic events associated with synaptic transmission are lacking. Such information is important for understanding phenomena such as synaptogenesis, synaptic transmission and plasticity, and synaptic regeneration following neurological damage. During the last few years, neuroscientists have begun to dig more deeply into the structures and process of synaptic behavior by using electron tomography (ET). In this chapter, we will review these methods, recent studies of synapses, their limitations, and consider the ways in which future developments may allow for the systematic structural investigation of synaptic architecture under controlled conditions.

II. Neurons, Neural Networks, and the Synapse

Living neural networks (LNN) are one of Nature's greatest technological achievements. They can be massively parallel and dynamic computational devices with built-in plasticity and repair mechanisms. The most complex LNN, a human brain, comprises over 10^{14} connections, more connections than all the computing power manufactured between 1981 and 2001 (Stowell, 2003). The chemical synapse is at the heart of a functioning LNN. During brain development, chemical synapses are constructed following the establishment of contact between a growth-cone at the end of a growing axon and the cell that will receive its signal. Synaptogenesis is the process by which this complex, intercellular connection is formed. Once formed, a functional synapse has a dynamic life. Of course it transmits signals between neurons, but it is also subject to slower modulations that affect its behavior and function. Individual synaptic signals undergo modulations, such as long-term potentiation (LTP) and long-term depression (LTD) (Bliss *et al.*, 2003; Gilbert, 1998; Lisman, 2003), and in neurodegenerative diseases they can disintegrate. These adaptive changes in synapse function, which can occur on timescales ranging from seconds to minutes, are the foundation for more permanent changes in LNN wiring that may involve the formation of new synapses (Knott *et al.*, 2002;

Trachtenberg *et al.*, 2002), giving rise to biologically indelible memory. All these phenomena are central to our understanding of memory and learning, and they also serve as a foundation for understanding biological computation in general and the development of neuromorphic technologies in the future. Naturally, a great deal of effort has been spent on trying to understand the genetic and molecular bases of neuronal function and in particular memory and learning.

III. ET and the Synapse

The last decade has seen a strong resurgence of interest in biological EM. While this trend has been led by dramatic progress in EM of macromolecules, the analysis of cellular ultrastructure has not lagged far behind (McIntosh, 2001). Several technological advances have facilitated this renaissance. On the instrumentation front, field emission guns (FEGs), which provide electron beams with good coherence and brightness, have improved the imaging of frozen-hydrated samples. Computer-controlled specimen stages have improved the regulation of sample position and automated data collection. Greatly improved CCD cameras have significantly increased the speed of image acquisition and analysis. Finally, as in almost every area of research, the dramatic decrease in price for computational speed and data storage has also played an important role. However, as in most areas of research, perhaps the most dramatic improvements have been the result of greatly improved methods for sample preparation; in particular, rapid freezing of biological specimens has opened the door for higher resolution studies of macromolecules and cells in their native states (Dubochet *et al.*, 1983; Lepault *et al.*, 1983). As a result, the use of ET for ultrastructural analysis has been applied extensively to a variety of cells and organisms such as *Caenorhabditis elegans* (Müller-Reichert *et al.*, 2003), *Saccharomyces cerevisiae* kineticore distribution (O'Toole *et al.*, 2002), and cytoplasmic membrane structures (Duman *et al.*, 2002). The emergence and interest in cryo-ET has produced some exceptional results for small cells (Grimm *et al.*, 1998; Nickell *et al.*, 2003) and isolated organelles or cell components (Medalia *et al.*, 2002; Nicastro *et al.*, 2000). EM tomographic studies of neurons have focused primarily on the neuromuscular junction (Harlow *et al.*, 2001; Rizzoli and Betz, 2004) and the ribbon synapse of frog saccular hair cells (Lenzi *et al.*, 1999); for a review of ultrastructural studies of the synapse see Torrealba and Carrasco (2004). In addition, ultrastructural studies of stimulation-dependent changes in the synapse have been carried out for the ribbon synapse of saccular hair cells (Lenzi *et al.*, 2002) as well as the frog neuromuscular junction (Rizzoli and Betz, 2004). These studies have all used chemical fixation, a method that greatly reduces the quality and reliability of the biological representation derived from ET. A major effort of current studies is to develop tools that will allow the study of stimulation-dependent phenomena by ET, using techniques such as rapid-freezing and freeze-substitution (RF-FS), high-pressure freezing and freeze-substitution (HPF-FS), and potentially, cryo-ET. At present, such studies are not possible because there is no technology that can combine neuronal patterning, axon stimulation, and optical observation with ET for a single platform. This chapter focuses on the use of ET to study the chemical synapse and in particular the small synapse of the brain. The best characterized of these are the glutamatergic synapses between primary cultured hippocampus neurons. Although mention is made of other neuronal systems and types of synapses, our focus will be on the hippocampus synapses, which have functionally well-characterized plasticity events and can be regarded as the "gold standard" sample for studying memory and learning.

IV. Types of Neurons and Synapses

While all metazoans (save for sponges) are a potential source of neurons for structural investigation, a few model organisms have been primarily utilized. These range from the invertebrate neurons of *Aplysia* to vertebrate hippocampus neurons. While each type of neuron and neuron source may have particular experimental advantages or disadvantages, the biological question is the main driver in neuron choice. Here we will give a brief description of three different types of neurons that have been used to investigate synaptic function with structural methods. At this stage, it is important not only to distinguish between chemical and electrical synapses but also between inhibitory and excitatory synapses, as well as the type of neuronal connection that is made. Chemical synapses involve the release of neurotransmitters and require the packaging of neurotransmitter in "quanta," the term used to refer to the amount of neurotransmitter in a single synaptic vesicle. Electrical or fast synapses are mediated by connexons or gap junctions; they do not involve transmitter release and are not known to exhibit plasticity in the traditional sense; however, recent studies of Mauthner cells suggest that such connections can undergo a form of plasticity (Pereda *et al.*, 2004). In the following sections, we discuss the three main synapses that have been studied using EM and ET: neuromuscular junction synapses, ribbon synapses, and hippocampal neuron synapses. While all of these synapses have common features, they differ in a number of important ways. For example, the number and arrangement of vesicles at the synapse show considerable variation. Neuromuscular synapses contain hundreds to thousands of vesicles; they are thought to be stable and not to undergo plasticity events, such as LTP and LTD, but they do undergo more subtle remodeling during postnatal growth (Wilson and Deschenes, 2005). Hippocampal neurons and synapses are undoubtedly the most studied because they display plasticity phenomena with great clarity. Hippocampus neurons form synapses that can undergo a variety of plasticity events, and it is for this reason that much of our knowledge about synaptic plasticity derives from data obtained on hippocampus neurons.

A. Neuromuscular Junction Synapse

The neuromuscular junction is an archetypal synapse, and many of the early structural studies of synapses used the frog neuromuscular junction (Fig. 1A). Seminal studies by Heuser and colleagues (Heuser and Reese, 1981; Heuser *et al.*, 1971) provided much of our fundamental structural understanding of the neuromuscular chemical synapse in both the resting and active states. Recent EM tomographic studies of this system have described a series of structural features that link synaptic vesicles to the "active zone," from which neurotransmitter will subsequently be secreted (Harlow *et al.*, 2001). More recent studies by Betz and colleagues have investigated vesicle recycling at the frog neuromuscular junction using fixed samples and ET (Rizzoli and Betz, 2004).

B. Ribbon Synapses

Structural studies on the ribbon synapse found in auditory, vestibular, and bipolar photoreceptor cells of vertebrates (Sterling and Matthews, 2005) have demonstrated that these synapses have a remarkably regular arrangement of synaptic vesicles (Fig. 1B). Although the precise role of this regular arrangement is not fully understood, it has been conjecturally viewed as a "conveyer belt" for vesicle delivery; plasticity phenomena in these synapses appear to revolve around variations in the size and number of vesicles on the conveyer belt (Vollrath and Spiwoks-Becker, 1996).

C. CNS Synapses

The synapses of the brain can occur at a wide range of places on the neurons. The interactions can be axo-axonic, somato-axonic, somato-dendritic, dendro-axonic, and even dendro-dendritic. The types of activity at these various sites can

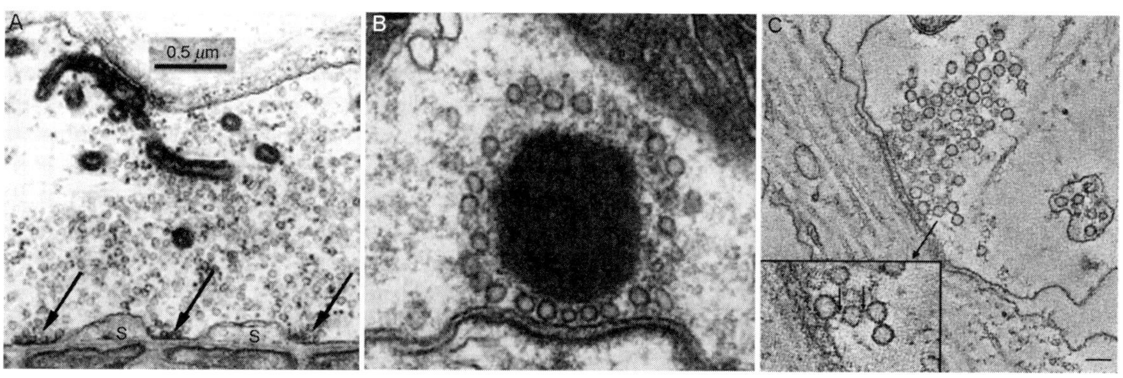

Fig. 1 Electron micrographs of the three main types of synapse being studied by ET. (A) Neuromuscular junction, (B) ribbon synapse, from Sterling and Matthews (2005), and (C) a small synapse of the brain, Stowell *et al.* (unpublished data). The main structural differences between these types of synapses are the number and organization of synaptic vesicles.

be either excitatory (glutamatergic) or inhibitory (GABAergic). The earliest ultrastructural investigation of the brain described two types of synapses, type 1 or asymmetric and type 2 or symmetric (Colonnier, 1968; Gray, 1959). The vast majority of synapses of the brain are asymmetric (Fig. 1C). Ellisman and colleagues have investigated structural changes that occur in the rat hippocampus following ischemia (Martone *et al.*, 2000) and found an intriguing vesicle rearrangement following ischemia.

V. Sources of Neurons for Structural Studies

Many structural investigations of neurons have utilized intact tissue. The obvious advantages of such methods are the nativelike state of the neurons and their associated synapses, as well as of the surrounding cells, primarily glial. One of the disadvantages is the processing time required between a given treatment and sample preservation by either rapid freezing or chemical fixation; there is also difficulty from the thickness of the samples and poor reproducibility of the sample treatments. For these reasons, we and others have focused on the use of primary cultured hippocampus neurons for studying the architecture of synaptic plasticity. The ability to reliably culture neurons *in vitro* (Banker and Goslin, 1991) has opened up a number of areas in neuronal studies. It has permitted the observation of neural patterning (Fromherz and Schaden, 1994), growth-cone targeting (Ming *et al.*, 2002), and synaptogenesis (Ahmari *et al.*, 2000), to name just a few of the areas that have benefited from this methodology. One of the most important aspects of culture methods for hippocampus neurons is that the process can be done at relatively low cell density. Such culture conditions open the possibility of collecting cryo-electron tomographic data on individual synapses that have been grown under defined conditions and subject to defined stimulation protocols. For structural studies, neurons can be cultured on a variety of substrates, including glass, carbon-coated Formvar or collodion (nitrocellulose), as well as filter disks. Our laboratory has also investigated a number of substrates that are compatible with lithographic manufacturing procedures and that allow thin electron-transparent windows to be fabricated. These include silicon, silicon nitride, and anodized alumina oxide (AAO) (Fig. 2).

While various substrates will support primary neuronal culture, several hurdles must typically be overcome; these vary with the substrate. In the case of silicon and silicon nitride substrates, pretreatment of the substrates with oxygen plasma greatly improves the reproducibility of the cultures. In addition, a variety of more standard surface treatments are needed to help the substrate to support neurons. These include the application of Matrigel™, polylysine, and/or polylysine/laminin. Our most reproducible and successful results have been achieved with polylysine/laminin. Our typical procedure starts with brains from P0 to P2 neonatal mice, dissected in calcium-magnesium-free Hank's balanced salt solution (HBSS). After the brain was separated into the two hemispheres, the C-shaped structure in the

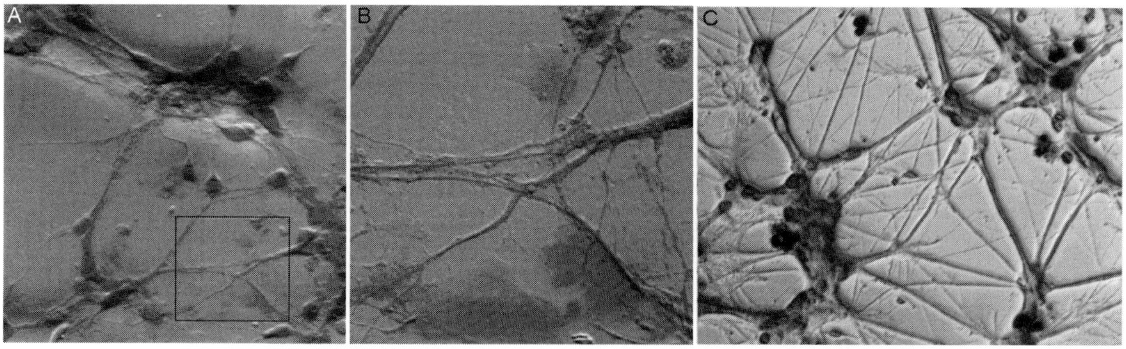

Fig. 2 Light micrographs of low-density, primary-cultured murine hippocampus neurons grown on silicon oxide/nitride windows showing a high degree of differentiation and synapse formation. (A) Field view, (B) Close-up of the boxed area in (A) ($500 \times 500\ \mu m^2$), showing the degree of branching and connectivity. (C) Primary neurons grown on AAO substrates.

posterior half of both hemispheres, hippocampi, were removed under a dissection microscope at 10× to 15× magnification. The meninges and choroid plexus were next removed, and the tissue treated with papain for 1½ h. The tissue was then gently triturated with a fire-polished Pasteur pipet. This material was layered on top of a Percoll gradient (30–60%) and centrifuged for 30 min at 3000 rpm without brake. The first band from the bottom of the tube was collected and concentrated by centrifugation. The resulting pellet was washed with HBSS medium several times and resuspended in neurobasal medium without serum, supplemented with B27, glutaMAX (Invitrogen) streptomycin/penicillin, and glutamate. Generally, 5.0×10^6 cells/ml were plated on polylysine/laminin-treated substrates immobilized on glass coverslips and allowed to stand for 40 min. Then, 2.5 ml of medium was added to the plate and the neurons were incubated at 37°C for 2 days. Half of the medium was then replaced with fresh medium without glutamate. Subsequently, the medium was replaced every week. The number of plated cells varies slightly depending on the substrate, as not all substrates provide good initial cell adhesion. This simple procedure, modified from that of Banker, provides reproducible cultures on a wide range of substrates treated with laminin/polylysine.

VI. Methods for Assessing Neuron and Synapse Integrity

Although primary cultured hippocampus neurons are regarded as the "gold standard" of neuronal systems, it is important to use various quantitative techniques to assess the proper formation of functional synapses prior to structural studies. This can be achieved in several ways. One method is to employ a battery of known synaptic markers for an analysis by fluorescence microscopy. A second method is to use quantitative polymerase chain reaction (QPCR) to measure the

Table I
Synaptic Markers for QPCR

Presynaptic markers	Postsynaptic markers
Bassoon	Gephyrin
Piccolo	Grip
Munc 13	Neuroligin
Cask	NMDA receptor
RIM 1	PSD95
RIM 2	SAP 102
Snap 25	AMPA receptor
Syntaxin	CaM kinase
Synaptobrevin	Syntenin

level of expression of known synaptic marker proteins during synaptogenesis. Table I lists a series of presynaptic and postsynaptic markers that can be used to quantitatively assess the formation of synapses from neurons cultured on various substrates. The advantage of QPCR is that one can assay a much larger number of markers in a single experiment. Previous work has utilized a battery of 500 gene markers, of which approximately half are involved in synaptogenesis (Saito *et al.*, 2005). Thus, QPCR allows one to develop a quantitative although general measure of the biological relevance of neurons cultured on various substrates relative to traditionally cultured neurons; with the resulting information one can rapidly home in on optimal conditions. The important fundamental difference between these methods is that QPCR provides a global readout of all the neurons whereas fluorescence methods allow one to examine individual neurons and synapses within a culture. One obvious question to address using such methods is the difference between the amounts of various synaptic markers in cells grown on various substrates. To date, we have not observed any substrate-dependent differences. The main application of synaptic marker quantification is to assess the differentiation of neurons, which in turn enables a study of synaptogenesis, and to monitor the viability and reproducibility of our cultures.

VII. Rapid Freezing and HPF of Neurons

Neurons grown on various substrates can either be plunge frozen or high-pressure frozen to maintain their structural integrity. Both of these methods are adequately covered in other sections of this book. Here it suffices to say that HPF has the advantage that larger sample masses can be well preserved. We have cultured neurons on silicon substrates patterned by photolithography to display 50–100 μm SU8 epoxy-based polymer surface features and obtained good sample

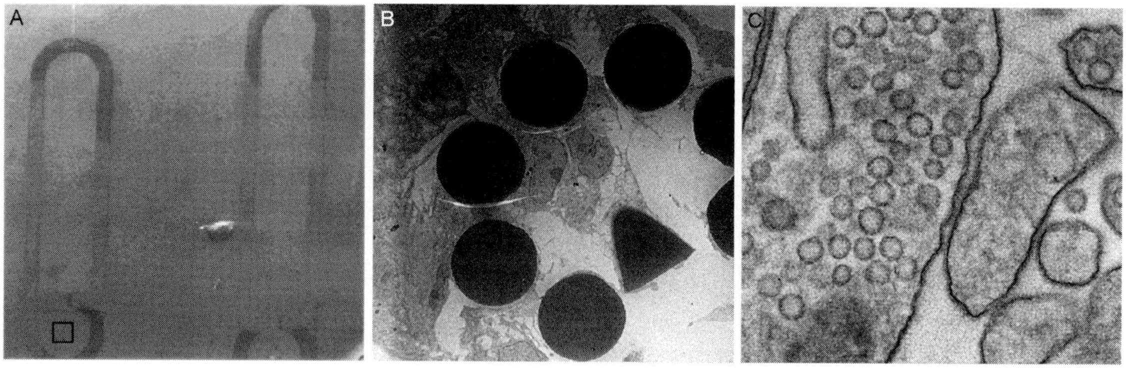

Fig. 3 Thin sections cut from samples of 10-day-old hippocampus neurons grown on a silicon substrate patterned with SU8 and plunge frozen in liquid ethane. (A) Overview (2×2 mm^2) of two neuron wells with an SU8 picket fence and outgrowth track. (B) More detailed view [boxed area in (A) 200×200 μm^2] showing neuronal cell bodies within the "picket fences" (dark circles). (C) A single slice (1×1 μm^2) from a tomogram taken from the sections in (B), showing well-preserved synaptic vesicles.

preservation by either plunge freezing in liquid ethane or HPF, followed by freeze-substituting with standard procedures. The results of such studies are shown in Fig. 3. Following plunge freezing and freeze-substitution, we cut thin sections that demonstrated good overall preservation of the sample and localization of neurons within the designed surface structures. The importance of this result is that it demonstrates good sample preservation for structural studies on substrates that are suitable for functional analysis using field potential electrophysiological methods as discussed below.

VIII. Modeling and Analysis of Synaptic Structures

Once suitable tomograms have been collected, data processing and analysis proceeds in a similar fashion to other samples. Other chapters of this book deal extensively with modeling, so here we have only described the aspects that are important for understanding synaptic structure and function.

A. Size and Organization of the Vesicle Pool

The size and organization of the pool of synaptic vesicles is one of the first elements to be investigated. As discussed above, the size of this pool can vary greatly, depending on the type of neuron. In addition, stimulation-dependent changes can occur, and dramatic changes can occur when important trafficking protein are either removed or inhibited. A classic example of this phenomenon is

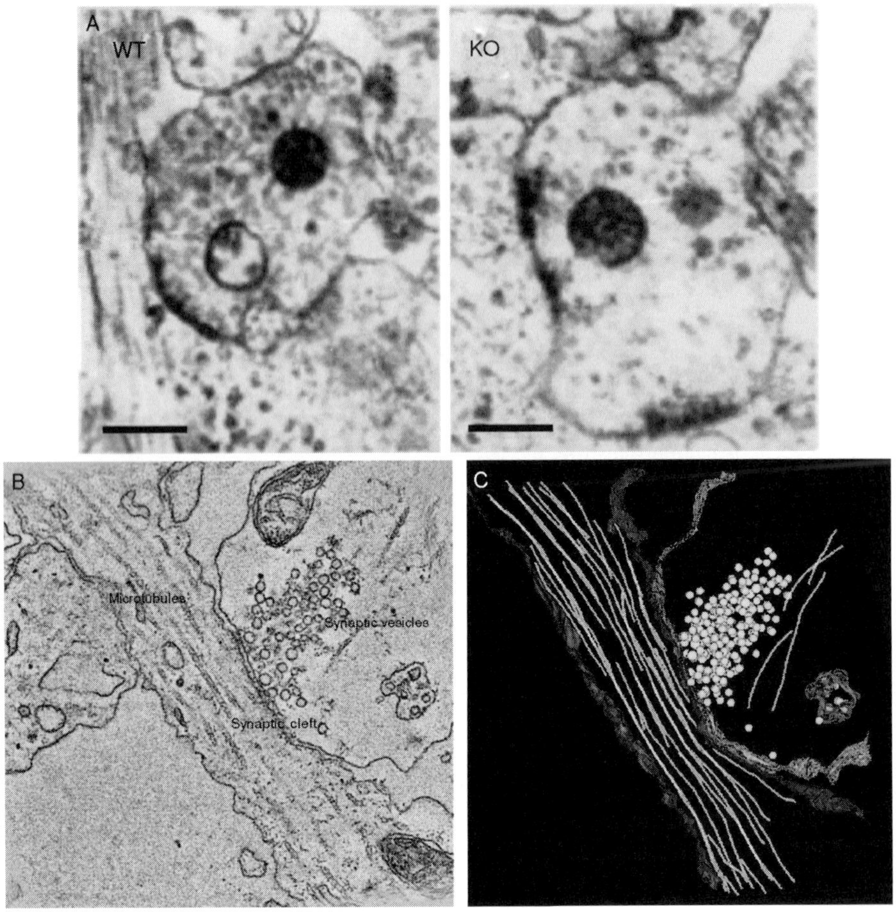

Fig. 4 (A) Images from neonate mice of wild-type (WT) and *kif1a*-knockout (KO) genotypes. The knockout mice show a distinct loss of the clustered synaptic vesicles seen in the WT. (B) Twenty-one-day-old primary cell culture of *murine hippocampus* pyramidal neurons grown on a treated silicon substrate with clearly observed synaptic vesicles and microtubules. (C) Three-dimensional (3D) model of the tomographic data in (B) created using the IMOD suite of programs. (See Color Insert.)

the loss of typical asymmetric synapses in *kif1a*-knockout mice. Kif1A is a kinesin-like motor protein responsible for trafficking to the presynaptic active zone. In the *kif1a*-knockout mice there is a pronounced accumulation of synaptic-like vesicle at the Golgi and the loss of a well-organized synaptic pool (Fig. 4A). Another important study was that by Betz and colleagues (Rizzoli and Betz, 2004) in which they combined ET with photoconversion of dye-labeled vesicles to study the structural organization of the vesicle pool at the neuromuscular junction.

B. Organization of the Postsynaptic Density Proteins

The postsynaptic density (PSD) and the organization of the various channels and channels associated complexes are of particular interest to those studying synaptic plasticity. Dramatic PSD reorganization is known to occur following stimulations such as siezure (Martone *et al.*, 2000). Recently, an intriguing study that combined ET, electrophysiology, and modeling, investigated the role of acetylcholine isoforms in the chick ciliary ganglion synapse (Coggan *et al.*, 2005). This study provided compelling evidence that ectopic neurotransmission is functionally relevant, thereby expanding the modes of biologically important neurotransmission. This study also highlights the need to determine the localization of neurotransmitter receptors in order to fully understand synaptic transmission and its modulation.

C. Synaptic Cleft Complexes

The organization of the cell adhesion proteins that interact across the synaptic cleft is important not only for understanding synaptogenesis but for its important consequences with regard to receptor localization and arrangement. A recent cryo-ET study has given some insight into this organization and discovered that there are substantial lateral connections within the synaptic cleft as well as the expected intercellular connections. (Lucic *et al.*, 2005).

D. Endosomal Compartments

The vesicle cycle of the small synapse of the brain is an intriguing and controversial area of study. At least two modes of vesicle fusion and recycling have been observed: full fusion and "kiss and run" (Fernandez-Peruchena *et al.*, 2005). Understanding how these different types of fusion, and subsequent endocytosis, are regulated is critical to understanding presynaptic modes of plasticity. While studies at the neuromuscular junction (Rizzoli and Betz, 2004) have implicated a random distribution of "kiss and run" vesicles, such studies have not been performed on the small synapse of the CNS. It will be important to understand how these different mechanisms of vesicle fusion are regulated and how this regulation affects the activity of a given synapse.

E. Microtubule and Neurofilament Arrangements

The organization of the microtubule network associated with both the axon and the dendrite is of particular interest when trying to understand both synaptogenesis and long-term plasticity. Although little has been done to investigate the detailed organization of microtubules, it is well established that they are critical for cargo transport to both the pre- and postsynaptic zones. Disruption of microtubules prevents the formation of new synaptic boutons (Ruiz-Canada *et al.*, 2004), and as discussed above, microtubule motor proteins are essential for proper formation of a synaptic pool. The proposed roles of neurofilaments have included a "girder"-like function for the organization of acetylcholine receptors as well as a role in defining microtubule organization for material transport (Paysan *et al.*, 2000).

IX. Stimulation-Dependent Changes in Synaptic Structure

One of the most intriguing aspects of synaptic structure is the structural change that occurs during specific stimulations. While very few studies of this nature have yet been performed, this is certainly an area that will be moving forward with great speed and anticipation. We mentioned above the studies of the frog neuromuscular junction vesicle cycle, as well as PSD reorganization following seizure in hippocampal neurons. Another important study was the investigation of the auditory ribbon synapse of the *Xenopus* saccular hair cells, before and after stimulation (Lenzi *et al.*, 2002). Figure 5 shows the results from this study, which demonstrated a dramatic reorganization of the ribbon synapse vesicle pool following stimulation. The ability to perform defined stimulation protocols and then to collect structural data provides the possibility of important insights into the organization and mechanistic functioning of synapses. Such data cannot be obtained with any other techniques.

X. The Future of EM Tomography and Synapses

One of the most fascinating areas for synaptic structure studies by ET is the ultrastructural analysis of synapses that have recently undergone electrical stimulation with recording. While this kind of study can in principle be performed using traditional electrophysiological methods, as were used for studies on the frog neuromuscular junction, such methods are impractical when studying early stages of plasticity. Only one or a few neurons per culture could be properly stimulated and high-pressure frozen for subsequent ET analysis. The solution to such a problem is to utilize multielectrode arrays. Current technologies for studying neurons in arrays have been optimized for signal generation and detection, using either cultured neurons or brain slices. Several commercial systems are presently available that employ large-format arrays and are optimized mainly for slice work and perfusion. However, these devices are not suitable for EM-tomographic analysis because the samples cannot be rapidly frozen or even high-pressure frozen. Both the size and the thermal mass of these samples are too great to allow rapid cooling, a requirement for adequate sample preservation. To achieve cooling rates suitable for HPF, while retaining the capability to stimulate and pattern the neurons, we have designed and fabricated new devices specifically for this purpose (Fig. 6). With such tools, researchers will be able to recapitulate a variety

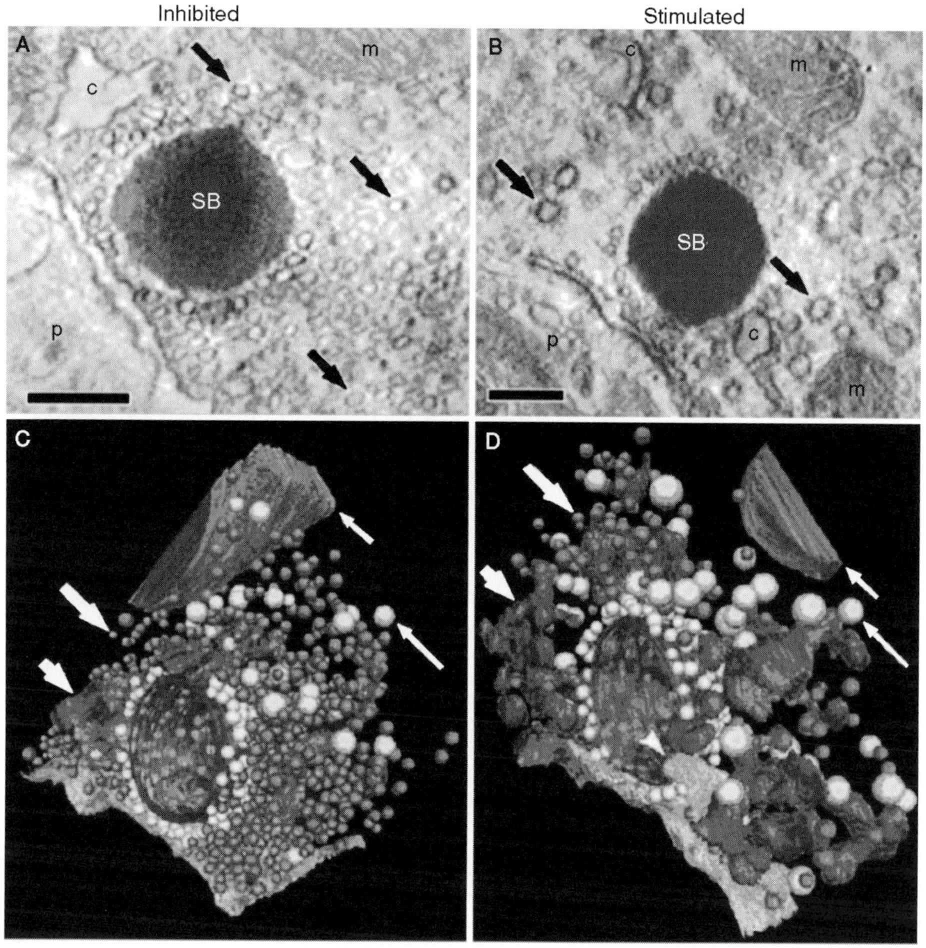

Fig. 5 Stimulation-dependent changes in the saccular ribbon synapse, reproduced from Lenzi *et al.* (2002). Electron tomographic reconstructions of hair cell synapses exposed to 0-Ca^{2+} saline (A and C) or high-K^+ saline (B and D), which stimulates the synapse. (A and B) Single *x–y* planes through the reconstructed volumes are shown. The hair cell cytoplasm contains the synaptic body (SB) surrounded by synaptic vesicles, outlying cytoplasmic vesicles [arrows in (A)], coated vesicles [arrows in (B)], membrane-bounded cistern (c), mitochondria (m), and postsynaptic cell (p). (C and D) 3D structure of presynaptic organelles from the same reconstructions shown in (A) and (B). The SB (blue) was rendered as hollow and semitransparent and lies adjacent to the hair cell's plasma membrane (red). SB-associated vesicles (yellow) surround the SB; outlying vesicles (green, thick long arrows) lie further out in the cytoplasm. Also visible are coated vesicles (gold, thin long arrows), cisternae (purple, thick short arrows), and mitochondria (blue, thin short arrows). (D) Regions of presynaptic density (pink) lie beneath the SB on the plasma membrane, which formed a tubular invagination (arrowhead). The flat surfaces of the mitochondria and the open face of the SBs denote the edges of the reconstruction. Synapses 1 and 11; scale bars = 200 nm. (See Color Insert.)

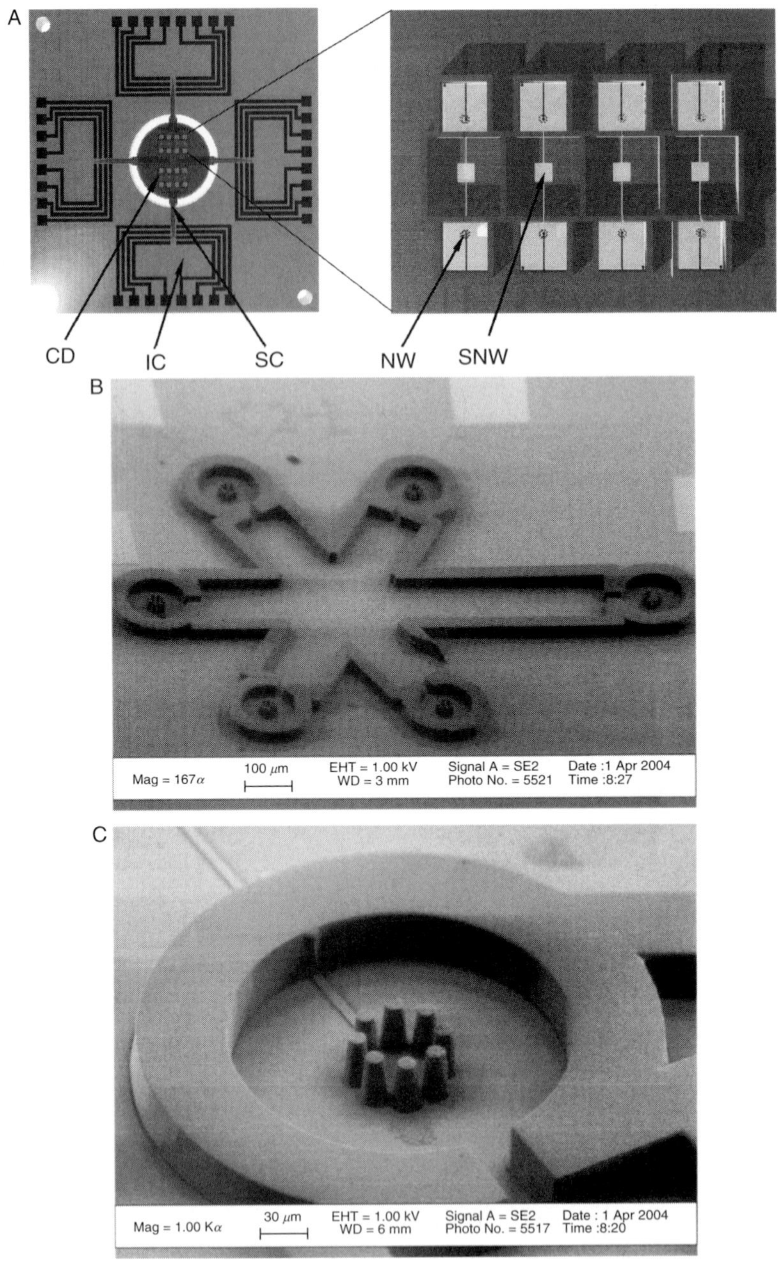

Fig. 6 (A) Silicon chip for use in a high-pressure freezer. The silicon chip is designed so that the inner core of electrodes can be snapped off and introduced directly into the cap of an HPF device. Left: overview of the chip designed for the HPF; CD, central disk for culturing neurons; IC, interconnect for direct hookup to both a multichannel systems or Panasonic MED-64 stimulation and recording amplifier system; SC, snap-off connections between the CD and the main chip body. Right: detail of overview; NW, neuron well; SNW, silicon nitride window for synaptic observation. (B) SEM image of a silicon EM chip with SU8 features for culturing neurons. (C) Detail of (B) showing the picket fence cell body containment posts.

of synaptic phenomena that have previously been characterized by fluorescent light microscopy and electrophysiology, and then go on to observe the identical synapse using ET.

XI. Conclusions

The use of ET to study synaptic structure and function holds tremendous promise, yet there are substantial hurdles that need to be overcome. The ultimate goal of such studies is to identify and ascertain the organization of membranes and supramolecular complexes in a resting synapse or in a synapse after treatment with well-defined stimulation protocols. Many of the relevant molecular complexes are of sufficient mass (>200 kDa) that their appearance will cause clear contrast in conventional EM. These include such important assemblies as the SNARE complex, NSF, and NMDA-PSD-95, the bassoon and piccolo complexes that order synaptic vesicles, polysomes, Ca^{2+}-release channels, microtubules, and associated transport vesicles. While the study of several systems, such as the neuromuscular junction and ribbon synapses following stimulation, has been achieved and important insights gained, such studies cannot currently be performed with primary cultured hippocampus neurons. This limitation can be overcome by employing microfabrication methods and multielectrode stimulation technologies, so that well-defined stimulation protocols can be achieved and their effects rapidly analyzed by EM and ET. In the end, a knowledge of the distribution and organization of relevant macromolecular complexes, including membranes, will give us important new insights into the workings of the synapse.

Acknowledgments

We would like to thank our colleagues at the University of Colorado for helpful comments and advice, J. Richard McIntosh for patients and insight, and the Beckman Foundation for financial support.

References

Ahmari, S. E., Buchanan, J., and Smith, S. J. (2000). Assembly of presynaptic active zones from cytoplasmic transport packets. *Nat. Neurosci.* **3,** 445–451.

Banker, G., and Goslin, K. (1991). "Culturing Nerve Cells." MIT Press, Cambridge, MA.

Bliss, T. V., Collingridge, G. L., and Morris, R. G. (2003). Introduction. Long-term potentiation and structure of the issue. *Philos. Trans. R. Soc. Lond. B Biol. Sci.* **358,** 607–611.

Coggan, J. S., Bartol, T. M., Esquenazi, E., Stiles, J. R., Lamont, S., Martone, M. E., Berg, D. K., Ellisman, M. H., and Sejnowski, T. J. (2005). Evidence for ectopic neurotransmission at a neuronal synapse. *Science* **309,** 446–451.

Colonnier, M. (1968). Synaptic patterns on different cell types in the different laminae of the cat visual cortex. An electron microscope study. *Brain Res.* **9,** 268–287.

Dubochet, J., McDowall, A. W., Menge, B., Schmid, E. N., and Lickfeld, K. G. (1983). Electron microscopy of frozen-hydrated bacteria. *J. Bacteriol.* **155,** 381–390.

Duman, J. G., Pathak, N. J., Ladinsky, M. S., McDonald, K. L., and Forte, J. G. (2002). Three-dimensional reconstruction of cytoplasmic membrane networks in parietal cells. *J. Cell Sci.* **115,** 1251–1258.

Fernandez-Peruchena, C., Navas, S., Montes, M. A., and Alvarez de Toledo, G. (2005). Fusion pore regulation of transmitter release. *Brain Res. Brain Res. Rev.* **49,** 406–415.

Fromherz, P., and Schaden, H. (1994). Defined neuronal arborizations by guided outgrowth of leech neurons in culture. *Eur. J. Neurosci.* **6,** 1500–1504.

Garner, C. C., Zhai, R. G., Gundelfinger, E. D., and Ziv, N. E. (2002). Molecular mechanisms of CNS synaptogenesis. *Trends Neurosci.* **25,** 243–251.

Gilbert, C. D. (1998). Adult cortical dynamics. *Physiol. Rev.* **78,** 467–485.

Gray, E. G. (1959). Axo-somatic and axo-dendritic synapses of the cerebral cortex: An electron microscope study. *J. Anat.* **93,** 420–433.

Grimm, R., Singh, H., Rachel, R., Typke, D., Zillig, W., and Baumeister, W. (1998). Electron tomography of ice-embedded prokaryotic cells. *Biophys. J.* **74,** 1031–1042.

Harlow, M. L., Ress, D., Stoschek, A., Marshall, R. M., and McMahan, U. J. (2001). The architecture of active zone material at the frog's neuromuscular junction. *Nature* **409,** 479–484.

Heuser, J., Katz, B., and Miledi, R. (1971). Structural and functional changes of frog neuromuscular junctions in high calcium solutions. *Proc. R. Soc. Lond. B Biol. Sci.* **178,** 407–415.

Heuser, J. E., and Reese, T. S. (1981). Structural changes after transmitter release at the frog neuromuscular junction. *J. Cell Biol.* **88,** 564–580.

Knott, G. W., Quairiaux, C., Genoud, C., and Welker, E. (2002). Formation of dendritic spines with GABAergic synapses induced by whisker stimulation in adult mice. *Neuron* **34,** 265–273.

Lenzi, D., Crum, J., Ellisman, M. H., and Roberts, W. M. (2002). Depolarization redistributes synaptic membrane and creates a gradient of vesicles on the synaptic body at a ribbon synapse. *Neuron* **36,** 649–659.

Lenzi, D., Runyeon, J. W., Crum, J., Ellisman, M. H., and Roberts, W. M. (1999). Synaptic vesicle populations in saccular hair cells reconstructed by electron tomography. *J. Neurosci.* **19,** 119–132.

Lepault, J., Booy, F. P., and Dubochet, J. (1983). Electron microscopy of frozen biological suspensions. *J. Microsc.* **129**(Pt. 1), 89–102.

Lisman, J. (2003). Long-term potentiation: Outstanding questions and attempted synthesis. *Philos. Trans. R. Soc. Lond. B Biol. Sci.* **358,** 829–842.

Lucic, V., Yang, T., Schweikert, G., Forster, F., and Baumeister, W. (2005). Morphological characterization of molecular complexes present in the synaptic cleft. *Structure* **13,** 423–434.

Martone, M. E., Hu, B. R., and Ellisman, M. H. (2000). Alterations of hippocampal postsynaptic densities following transient ischemia. *Hippocampus* **10,** 610–616.

McIntosh, J. R. (2001). Electron microscopy of cells: A new beginning for a new century. *J. Cell Biol.* **153,** F25–F32.

Medalia, O., Weber, I., Frangakis, A. S., Nicastro, D., Gerisch, G., and Baumeister, W. (2002). Macromolecular architecture in eukaryotic cells visualized by cryoelectron tomography. *Science* **298,** 1209–1213.

Ming, G. L., Wong, S. T., Henley, J., Yuan, X. B., Song, H. J., Spitzer, N. C., and Poo, M. M. (2002). Adaptation in the chemotactic guidance of nerve growth cones. *Nature* **417,** 411–418.

Müller-Reichert, T., Hohenberg, H., O'Toole, E. T., and Mcdonald, K. (2003). Cryoimmobilization and three-dimensional visualization of *C-elegans* ultrastructure. *J. Microsc. (Oxford)* **212,** 71–80.

Nicastro, D., Frangakis, A. S., Typke, D., and Baumeister, W. (2000). Cryo-electron tomography of neurospora mitochondria. *J. Struct. Biol.* **129,** 48–56.

Nickell, S., Hegerl, R., Baumeister, W., and Rachel, R. (2003). Pyrodictium cannulae enter the periplasmic space but do not enter the cytoplasm, as revealed by cryo-electron tomography. *J. Struct. Biol.* **141,** 34–42.

O'Toole, E. T., Winey, M., McIntosh, J. R., and Mastronarde, D. N. (2002). Electron tomography of yeast cells. *Methods Enzymol.* **351,** 81–95.

Paysan, J., Conroy, W. G., Coggan, J. S., and Berg, D. K. (2000). The neurofilament infrastructure of a developing presynaptic calyx. *J. Comp. Neurol.* **425,** 284–294.

Pereda, A. E., Rash, J. E., Nagy, J. I., and Bennett, M. V. (2004). Dynamics of electrical transmission at club endings on the Mauthner cells. *Brain Res. Brain Res. Rev.* **47,** 227–244.

Rizzoli, S. O., and Betz, W. J. (2004). The structural organization of the readily releasable pool of synaptic vesicles. *Science* **303,** 2037–2039.

Ruiz-Canada, C., Ashley, J., Moeckel-Cole, S., Drier, E., Yin, J., and Budnik, V. (2004). New synaptic bouton formation is disrupted by misregulation of microtubule stability in aPKC mutants. *Neuron* **42,** 567–580.

Saito, S., Honma, K., Kita-Matsuo, H., Ochiya, T., and Kato, K. (2005). Gene expression profiling of cerebellar development with high-throughput functional analysis. *Physiol. Genomics* **22,** 8–13.

Sterling, P., and Matthews, G. (2005). Structure and function of ribbon synapses. *Trends Neurosci.* **28,** 20–29.

Stowell, M. H. (2003). Calculated from IDC data which lists ~1 billion computers manufactured between 1981 and 2001. (http://www.idc.com/.)

Torrealba, F., and Carrasco, M. A. (2004). A review on electron microscopy and neurotransmitter systems. *Brain Res. Brain Res. Rev.* **47,** 5–17.

Trachtenberg, J. T., Chen, B. E., Knott, G. W., Feng, G., Sanes, J. R., Welker, E., and Svoboda, K. (2002). Long-term *in vivo* imaging of experience-dependent synaptic plasticity in adult cortex. *Nature* **420,** 788–794.

Vollrath, L., and Spiwoks-Becker, I. (1996). Plasticity of retinal ribbon synapses. *Microsc. Res. Tech.* **35,** 472–487.

Wilson, M. H., and Deschenes, M. R. (2005). The neuromuscular junction: Anatomical features and adaptations to various forms of increased, or decreased neuromuscular activity. *Int. J. Neurosci.* **115,** 803–828.

CHAPTER 11

Using Electron Microscopy to Understand Functional Mechanisms of Chromosome Alignment on the Mitotic Spindle

Bruce F. McEwen,[*,†] Yimin Dong,[*,†] and Kristin J. VandenBeldt[*]

[*]Wadsworth Center, New York State Department of Health
Albany, New York 12201

[†]Department of Biomedical Sciences, School of Public Health
State University of New York, Albany, New York 12201

METHODS IN CELL BIOLOGY, VOL. 79

0091-679X/07 $35.00
DOI: 10.1016/S0091-679X(06)79011-2

For more than 40 years, transmission electron microscopy has provided a detailed description of the mitotic spindle ultrastructure that cannot be obtained by any other approach. However, simply refining ultrastructural knowledge through the use of modern methods of cellular electron microscopy does not ensure new insights into functional mechanisms. Rather, new insights are achieved by using the most appropriate cellular electron microscopy method to answer specific mechanistic questions. In some cases, classical methods are more effective than modern techniques for answering a particular question. In this Chapter, 12 studies are used to illustrate how cellular electron microscopy has been effectively employed to elucidate mechanisms of mitotic function. The studies span an interval from 1971 to the present, and they have used a range of techniques including serial-section analysis, correlative light and electron microscopy, immuno-electron microscopy, electron tomography, high-pressure freezing followed by freeze-substitution, and imaging of isolated molecular components. The emphasis is on (1) the rationale for use of a particular technique and (2) the reason why the question addressed by the study is important for understanding mitosis. We hoped that this chapter will provide insight into the ways in which cellular electron microscopy can be effectively employed to study the functional mechanisms of mitosis and other cellular processes.

I. Introduction

The application of transmission electron microscopy (TEM) to cellular structures as large and complex as the mitotic spindle requires an interdisciplinary approach that brings together structural biology, cell biology, biochemistry, and genetics. Structural biology is often a discovery-based approach, with the underlying assumption that simply solving the structure will provide useful information that leads to more insightful hypotheses. However, the techniques used in cellular electron microscopy, including electron tomography, do not provide the high-resolution structural data afforded by TEM methods such as electron crystallography, helical reconstruction, or single particle methods. Since the overall structures are already known for the basic components of the mitotic apparatus, the use of modern technology to solve spindle structure at modestly higher resolution will not necessarily provide useful information. Therefore, it is imperative that the electron microscopist be aware of the driving biological questions behind the structural problem and that he or she design the study appropriately. The approach of this chapter is to review examples of studies in which TEM methods have been successfully used to address questions concerning the structure and function of the mitotic apparatus in vertebrate cells. We also describe one study of kinetochore microtubules (kMTs) in the nematode *Caenorhabditis elegans*.

A. Brief Review of Mitosis

Because genomic stability is critical for the survival of all eukaryotic organisms, the mitotic apparatus has evolved to carry out genomic segregation during cell division with a high degree of accuracy (Mitchison and Salmon, 2001). The process begins in prophase with the condensation of replicated chromosomes. During late prophase, the microtubule (MT) cytoskeleton breaks down and begins to re-form as a bipolar spindle. Entry into prometaphase is marked by breakdown of the nuclear envelope. This releases chromosomes into the cytoplasm where they become attached to the bipolar spindle (Fig. 1). During this initial stage of prometaphase, many of the

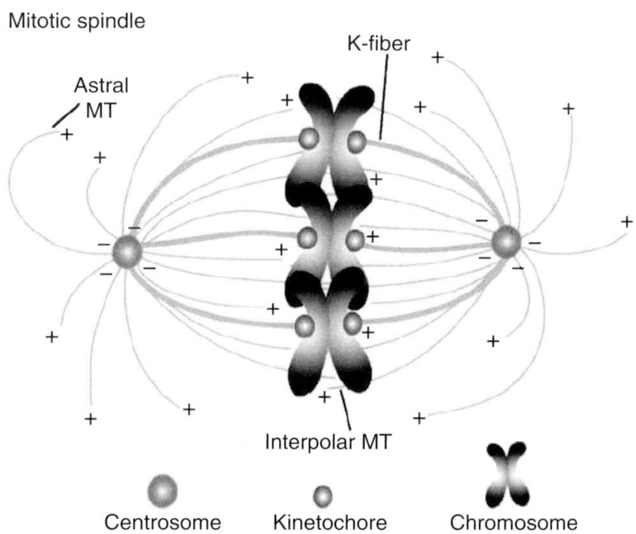

Fig. 1 Illustration of the major spindle components in vertebrate cells. The gray curved lines are MTs with the plus and minus ends indicated. The K-fibers are attached to the chromosomes at the kinetochore and they link the chromosomes with centrosomes. The figure also illustrates the interpolar MTs that overlap at the metaphase plate and the astral MTs that radiate away from the spindle. (Adapted with permission from Kline-Smith and Walczak, 2004.)

chromosomes are attached to only one of the spindle poles. For proper genomic segregation, all chromosomes must become bi-oriented, with duplicate chromatids attached to opposite spindle poles. Once all of the chromosomes are bi-oriented and aligned at the spindle equator, the cell is in metaphase. The time spent in metaphase varies by species and tissue type and can last as long as 20–30 min (Rieder *et al.*, 1994). Metaphase ends with the onset of anaphase, during which the replicate chromatids separate, migrate toward their attached spindle poles, and thereby achieve the requisite genomic segregation. Anaphase is followed by telophase, during which the nuclear envelopes re-form, so the nuclei can reenter interphase, and the cytoplasm is partitioned by the process of cytokinesis.

The major components of the mitotic apparatus are the chromosomes, centrosomes, kinetochores, and spindle MTs (Fig. 1). This chapter focuses on kinetochores and spindle MTs, while centrosomes are discussed in Chapter 5 by O'Toole *et al.*, this volume. Chromosomes are not discussed because structural studies on the compaction of DNA into chromosomes are beyond the scope of the chapter and because current models postulate a limited functional role for chromosomes during mitosis.

B. Challenges in Studying the Mitotic Apparatus by Electron Microscopy

The mitotic spindle in vertebrates is a challenging structure to study because it is large and complex and is only present in one copy per cell during a single, relatively short period of the cell cycle (mitosis). The complexity of the mitotic spindle makes it necessary to obtain relatively high-resolution information in order to understand the functional mechanisms. However, the mitotic spindle is too large to be effectively imaged as a whole-mount preparation (McIntosh *et al.*, 1979), so it must be cut into relatively thin sections for imaging (generally 50–500 nm, depending on the application). Consequently, only a small portion of the spindle will be present in a single section. Thus, even in synchronized cell cultures, a TEM study of mitosis involves a lot of searching, first to find a cell in the appropriate stage of mitosis and then to find the structure of interest within a serial-section series of the selected cell. These issues present a major obstacle, particularly for high-resolution cryo-TEM.

Another obstacle is the scarcity of *in vitro* studies of the mitotic apparatus. *In vitro* spindles can be formed from *Xenopus* extracts, and this system has become popular for biochemical and immunodepletion studies (Heald *et al.*, 1996; Maddox *et al.*, 2003; Sawin and Mitchison, 1991; Walczak *et al.*, 2002). Spindles have also been isolated from mammalian cells, but these spindles are primarily intended for use in proteomic studies and their functionality remains to be tested (Sillje and Nigg, 2006). Although the *Xenopus* system has been used effectively in light microscopy (LM) studies, there have only been a limited number of TEM images published. This is in part because the isolated spindles are too large for whole-mount imaging, yet the spindles are difficult to isolate as compacted pellets for embedding and sectioning. Similarly, few functional assays have been developed for isolated components of the mitotic apparatus, and even fewer of these isolated components have been imaged by TEM.

Another challenge in the study of the ultrastructure of the mitotic apparatus is that structural and live-cell imaging studies have been developed on model systems other than the traditional molecular and genetic model organisms. For example, while yeast is the most extensively developed and well-behaved eukaryotic model for genetic and molecular studies, the cytology of yeast chromosomes and kinetochores is primitive, with little definable ultrastructure (Ding *et al.*, 1993; O'Toole *et al.*, 1999). Furthermore, some key components of higher eukaryotic mitosis, such as CENP-E, are not present in yeast cells. Similarly, in the popular genetic models *C. elegans* and *Drosophila* kinetochores show significant molecular and structural differences from the mammalian kinetochores (Howe *et al.*, 2001; Maiato *et al.*, 2006; O'Toole *et al.*, 2003). This situation necessitates that homologies be found, both on the molecular and structural levels, between several model organisms, including *Saccharomyces cerevisiae*, *C. elegans*, *Drosophila*, vertebrates, and plants.

II. Basic Methodology

A. Conventional Fixation and Embedding

Conventional specimen preparation for electron microscopy (EM) has been widely used in studies of the mitotic spindle since the 1960s (Brinkley and Stubblefield, 1966). Conventional specimen preparation includes fixation with glutaraldehyde, postfixation with osmium tetroxide, dehydration through a graded alcohol series, and flat embedding in Epon. Often, specimens are also stained *en bloc* by incubating with tannic acid followed by uranyl acetate (McEwen *et al.*, 1997). *En bloc* staining improves the contrast of MTs relative to the background. Epon blocks containing the specimen are serially sectioned and collected on Formvar-coated grids. The sections are then poststained with uranyl acetate and lead citrate to enhance contrast, and serial sections are viewed by EM.

The standard protocol used in our laboratory for conventional TEM preparations of cell culture monolayers is fixation with 2.5% glutaraldehyde in PHEM buffer (an MT stabilizing buffer: 60-mM PIPES, 25-mM HEPES, 10-mM EGTA, and 2-mM $MgCl_2$, pH = 6.9) for 30 min at room temperature (McEwen *et al.*, 1997). Cells are then washed three times in PHEM buffer for 5 min each, incubated with 0.15% tannic acid in distilled water for 1 min at room temperature, washed with distilled water for 5 min, and postfixed with 2% osmium tetroxide for 1 h at 4 °C. Cells are then washed three times in distilled water for 5 min each, stained *en bloc* with 1% uranyl acetate for 1 h at 4 °C, and washed twice in distilled water for 5 min each. These steps are followed by dehydration in a graded ethanol dehydration series including 5-min washes in 20, 30, 40, 60, 80, and 95% ethanol, and three 10-min washes in 100% ethanol. Cells are then transferred into 50/50 ethanol/propylene oxide for 5 min, 100% propylene oxide three times for 10 min each, and infiltrated with 50/50 propylene oxide/Epon overnight. Finally, cells are infiltrated in 100% uncatalyzed Epon for 1 h and changed into 100% catalyzed Epon twice for 1 h each. The Epon is cured by incubation in a 60 °C oven for 2 days.

B. Cell Monolayers/Flat Embedding

Monolayer cell cultures offer important advantages for TEM imaging of mitotic cells. Cell monolayers are necessary for correlative video LM/EM in which the same specimen is examined by the researcher at both the light and ultrastructural levels with knowledge of cell cycle history (Section IV). Prior to fixation, monolayer cultures facilitate monitoring the mitotic index of the culture. After fixation and flat embedding, monolayers enable screening at the LM level for cells in a particular stage of mitosis and control of the orientation of cells during sectioning.

Cells must be grown on an optically clear substrate, such as a glass coverslip, in order that they can be filmed or screened by LM. A diamond objective scribe (objective marker with diamond tip, available from LM venders including Carl Zeiss, Inc., Thornwood, NJ) can be used to mark specific cells. Alternatively, the locations of specific cells can be determined using coverslips that are etched with a finder grid (Electron Microscopy Sciences, Fort Washington, PA), but it can be difficult to detect the locations after embedding. The cells are then conventionally processed and flat-embedded. Following embedment, cells are scanned under a phase-contrast LM, scribe marks are transferred to the Epon, and the glass coverslips are removed by hydrofluoric etching. The resulting blocks are trimmed down to the area of the scribe mark and sectioned. In monolayer cultures, the spindle is generally oriented in the longitudinal direction, with the spindle equator and poles clearly visible by phase LM (McDonald *et al.*, 1992; McIntosh and Landis, 1971; Rieder, 1981). This enables the selected cells to be excised from the flat Epon wafer and mounted on a new block to be sectioned in any chosen direction—generally, either longitudinally or along the spindle axis.

C. High-Pressure Freezing and Freeze-Substitution

High-pressure freezing followed by freeze-substitution (HPF/FS) is a specimen preparation method that preserves the native structure of specimens to a much greater extent than do conventional specimen preparations (Al-Amoudi *et al.*, 2004a; McDonald, 1999; see Chapter 2 by McDonald and Chapter 3 by Hess, this volume). HPF involves rapid freezing of samples under high pressure in order to retard ice-crystal formation. During the subsequent FS step, the specimen is dehydrated (i.e., an organic solvent is substituted for water) and chemically fixed at –90 °C. Since the dehydration and fixation steps of EM specimen preparation are performed at low temperature on samples that have been immobilized by the rapid-freezing step, specimens prepared by HPF/FS show less extraction of the cytoplasm, and fewer distortions of structural elements such as MTs and membranes, than do specimens prepared by conventional methods (see Chapter 2 by McDonald, this volume, and Section V.C).

The standard protocol used in our laboratory for HPF/FS preparations is to grow the cells as monolayers on 3-mm diameter sapphire disks (0.050-nm thick; Rudolf Breugger, Tessin, Switzerland). Sapphire disks are used because of their ability to conduct heat at a high rate for the freezing step and because of their transparent properties, which enables the scanning of cells by phase LM both before and after HPF/FS (Reipert *et al.*, 2004). Prior to seeding of the cells, the sapphire disks are acid-washed, coated with carbon, autoclaved, coated with fibronectin, and sterilized by UV light. After seeding with cells, the cultures are grown until the disk reaches 60–90% confluence. Individual disks are then screened for the presence of mitotic cells, and disks that have strong cell growth and one or more mitotic cells are dipped briefly into medium plus 10% Ficoll (to protect the specimen from cryo damage), mounted in HPF planchettes, and rapidly frozen in a HPM 010 (BAL-TEC, Balzers, Liechtenstein) high-pressure freezer. The high-pressure freezer raises the pressure in the specimen chamber to 2100 atm and freezes the specimen within 12 msec, using two jets of liquid nitrogen (see Chapter 2 by McDonald and Chapter 3 by Hess, this volume).

FS is performed in a FSU 010 (BAL-TEC, Balzers) freeze-substitution machine, using a protocol designed by Mary Morphew (University of Colorado, Boulder, CO, personal communication), as described by VandenBeldt *et al.* (2006). Briefly, the FS is a two-step process beginning with a 24-h incubation at –90 °C in dry acetone plus 0.5% glutaraldehyde and 0.1% tannic acid. The specimens are then rinsed with acetone at –90 °C and incubated for another 24 h at –90 °C in acetone plus 1.0% osmium tetroxide and 0.1% uranyl acetate. Specimens are then slowly brought to room temperature and infiltrated with Epon. Before the Epon is cured, the specimen disks are scanned for cells and placed cell-side up on glass coverslips. Epon is added to the coverslips, and in this way the disks are "flat embedded" on the coverslips. It is critical to have the cell-side up at this stage because there is not sufficient Epon on the bottom side of the disks to form a workable specimen bloc. The side of the disks containing the cells is determined under phase LM by alternately bringing both surfaces of the disk into focus. As with conventional fixation, glass coverslips can be removed by etching with hydrofluoric acid. The sapphire disks, however, are resistant to acid etching and must be removed by freeze-shock. Removal of the disks is facilitated by scraping away the thin layer of Epon that forms on the bottom side of the disks, and then cutting the Epon away from the circumference of the disks.

Cell suspensions can be collected as pellets and prepared by HPF/FS. Some cell cultures, such as HeLa cells, can be synchronized and/or enriched for mitotic cells by mitotic shake off, and then centrifuged to form a compact pellet. Pellets are resuspended in a 1:1 ratio of pellet to medium containing 40% Ficoll (MW = 70,000), dextran (MW = 70,000), or bovine serum albumin as a cryoprotectant. The slurry is then placed in a high-pressure freezer specimen carrier and rapidly frozen.

Frozen specimen pellets can be directly sectioned by cryoultramicrotomy. Alternatively, frozen pellets can be processed for FS or simply stored under liquid nitrogen until use. The Leica HPF and FS machines are alternatives to the Balzers models (see Chapter 2 by McDonald and Chapter 4 by Müller-Reichert *et al.*, this volume).

D. Electron Tomography

Electron tomography is a three-dimensional (3D) reconstruction method in which a 3D volume is computed from a series of projection images taken over a wide range of tilt angles (Lucic *et al.*, 2005; McEwen and Marko, 2001; McIntosh *et al.*, 2005; Chapter 5 by O'Toole, Chapter 6 by Höög and Antony, Chapter 8 by Marsh, Chapter 10 by Arthur *et al.*, Chapter 14 by Zhang *et al.*, and Chapter 25 by Best *et al.*, this volume). Electron tomography provides much greater depth resolution than does serial thin-section 3D reconstruction because it avoids the mechanical limitation of how thin a specimen can be cut (McEwen and Marko, 1999). Compared to single-particle, helical, and electron crystallographic 3D reconstruction methods, electron tomography offers the advantage of not requiring the specimen to occur as multiple copies of an identical unit motif. However, the trade-off for the ability to reconstruct pleiomorphic structures is lower resolution, because the signal-to-noise ratio of polymorphic structures cannot be improved by computational averaging methods, due to the lack of a unit motif. Hence, electron tomography offers an intermediate level between the high resolution afforded by the averaging methods mentioned above and the lower resolution afforded by serial thin-section reconstruction. Electron tomography is particularly useful for studying irregular cellular structures, such as the kinetochore and the mitotic spindle, at moderately high resolution.

When electron tomography is applied to mitosis-related questions, it is necessary first to locate a mitotic cell and then to find the specific structural component of the mitotic apparatus that is under study. For vertebrate cells grown as monolayers and flat-embedded, mitotic cells can be identified by phase-contrast LM. We generally cut 150- to 200-nm-thick serial sections of selected cells because this thickness represents a good compromise between having sufficient depth to include a reasonable portion of the structure, while being sufficiently thin enough to allow high-quality images to be recorded at higher tilt angles (60°–70°). The latter consideration is important to maximizing resolution and minimizing distortion from missing angular information. Before collecting tomographic tilt series, projection images of the serial-section series are recorded on film at a magnification low enough to see most of the spindle. The film negatives of the serial-section series are then examined on a large light table to select Epon sections containing structures of interest. Colloidal gold is applied to the surface of the selected sections, and electron tomography is performed by recording dual-axis tilt series with a tilt range of ±70°, and with an angular tilt interval being based on a cosine scheme (McEwen *et al.*, 2002; Taylor *et al.*, 1997). The tilt angle interval at the untilted image is 1°–2°. Image alignment and 3D reconstruction are computed using a combination of the SPIDER and eTomo/IMOD software packages (Frank *et al.*, 1996; Kremer *et al.*, 1996). Serial slices obtained from reconstructions are analyzed to localize structures of interest, and the interactive segmentation utility in Amira software (TGS, Inc., San Diego, CA) is used to further model the structures in the reconstructions.

III. Organization of MTs in the Mitotic Spindle

A. Use of Serial Thin-Section Analysis to Determine the Spatial Organization of Spindle MTs

Knowledge of the basic MT organization of the spindle is critical for an understanding of functional mechanisms of mitosis such as: (1) spindle elongation by the sliding of overlapping MTs; (2) chromosome motion; (3) the continuous flux of MT subunits from the kinetochore through the K-fiber to the spindle pole; and (4) the role of continuous MT depolymerization at spindle poles in maintaining tension across sister kinetochores. Determination of the organization of spindle MTs is challenging because of the large numbers and high density of spindle MTs, and because spindle MTs are long and thin. The small diameter of MTs in general and the dense packing of spindle MTs in particular preclude acquisition of an accurate description of MT numbers by LM analysis, even with modern deconvolution fluorescence methods. It is also challenging to track spindle MTs by TEM. Although longitudinal sections provide the clearest overview of the spindle and components such as the kinetochore, serial longitudinal sections have proven unreliable for tracking spindle MTs because the length and variable orientation of spindle MTs ensure that few of them will be fully contained in a single longitudinal section. Therefore, it is necessary to track MTs across serial thin sections. However, in practice, tracking of spindle MTs between serial thin sections cut in the longitudinal direction is inaccurate due to the large numbers and relatively low contrast of MTs in the mitotic spindle. For this reason, most tracking of spindle MTs has been accomplished on sections that were cut transverse to the spindle axis (transverse sections).

In an early effort, McIntosh and Landis (1971) collected serial sections for three cells each, for metaphase, mid-anaphase, late anaphase, and telophase WI-3 cells, and the interzone of anaphase HeLa cells. The investigators then counted MT cross-sections in every fourth transverse section (see also Brinkley and Cartwright, 1971). In general, MTs cut in cross-section were readily recognized, but those in an oblique orientation were more problematic. To maximize the number of cross-sectional views of MTs, the investigators tilted each serial section by up to 45° in the TEM in order to find the optimal viewing angle. Bias in counting was minimized by having two or more observers make the counts, including personnel not connected with the project. This same strategy was also used by McEwen *et al.* (1997) to quantify MTs bound to kinetochores, and more recently by VandenBeldt *et al.* (2006), to classify the plus-end conformations of kMTs (Sections IV and VI). Although the McIntosh and Landis study was able to quantify numbers of MTs along the spindle, the limited graphic capabilities of the time did not enable the investigators to track individual MTs.

Rieder (1981) circumvented the problem of high MT density in the spindle by exploiting the selective stability of the kMTs to cold. By gently cooling PtK1 cells to 6°–8° for 4–6 h, he found approximately the same number of kMTs as had been found at room temperature, but very few non-kMTs. Selective depolymerization of the non-kMTs greatly simplified the tracking of individual kMTs in transverse serial sections (Fig. 2). He also reduced the labor involved by cutting serial sections 250-nm thick and viewing them at 1000 MV on a high-voltage electron microscope. Individual MTs in the K-fiber bundle were tracked by the marking of MT cross-sections on an acetate sheet placed over a print from one section, with the sheet then placed over a print of the next serial section to identify the same MTs (McDonald *et al.*, 1977). Rieder found that about half of the MTs in the K-fiber are continuous from kinetochore to pole.

More recently, computer imaging has enabled more precise and efficient tracking of spindle MTs (Mastronarde *et al.*, 1993; McDonald *et al.*, 1992). In this

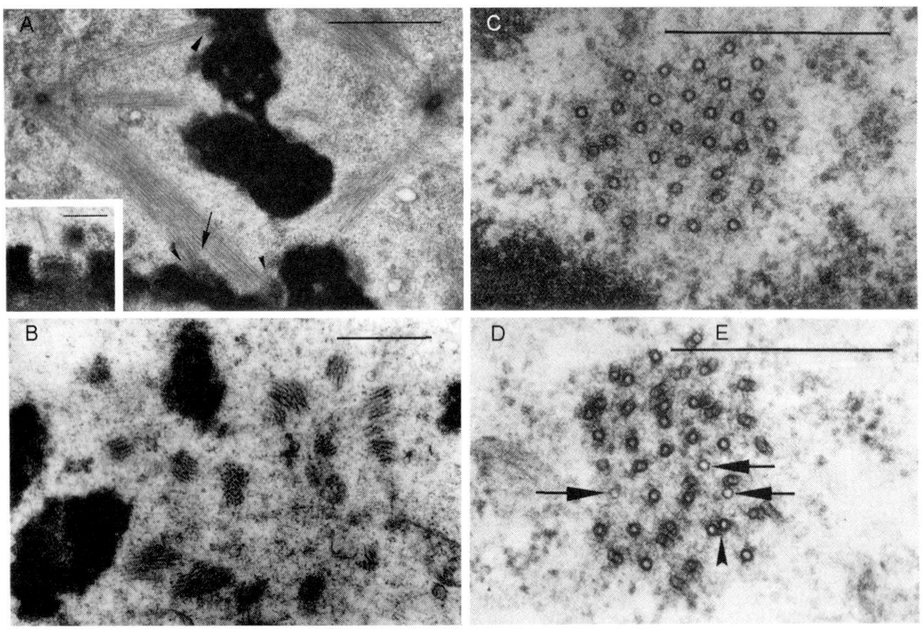

Fig. 2 Tracing of cold-stable K-fibers using serial-section TEM. (A) Image of a longitudinal section of the spindle. Arrow indicates a K-fiber that is fused with the K-fiber of the adjacent kinetochore. Arrowheads indicate the kinetochores. One of the kinetochores is shown at higher magnification in the inset. Scale bar = 2.5 μm. Inset scale bar = 0.5 μm. (B) Image of a transverse section of the spindle near the metaphase plate in a PtK1 cell. Scale bar = 2.0 μm. (C, D) Higher-magnification images of a single K-fiber from transverse sections close to the level of the kinetochore. Arrows in D indicate the kMTs that are slightly dense and terminate in this section. The arrowhead in D indicates two kMTs that are close to one another. Scale bars in C and D = 0.5 μm. (Adapted with permission from Rieder, 1981.)

approach, digitized images of transverse thin sections are put in register by visual alignment of successive sections using a computer graphics utility that allows the user to overlay one image on another and interactively move the image being aligned. The visual alignment is then refined by using the positions of 6–8 distinct MTs to correct for distortions that arise when cutting plastic section. Well-aligned serial sections are then cross-correlated with an average image of MT cross-sections to automatically identify MTs. A map of MT positions is overlaid on the image of the section and edited by visual inspection and comparison of neighboring sections. An interactive cursor provides an efficient implementation of editing choices. This approach enabled McDonald *et al.* (1992) to track K-fiber MTs in untreated metaphase PtK1 cells, and to thereby describe interactions with non-kMTs and avoid possible distortions from cold treatment. The digital format is also well suited for effective 3D graphic representations of spindle MTs (Fig. 3). The latter study

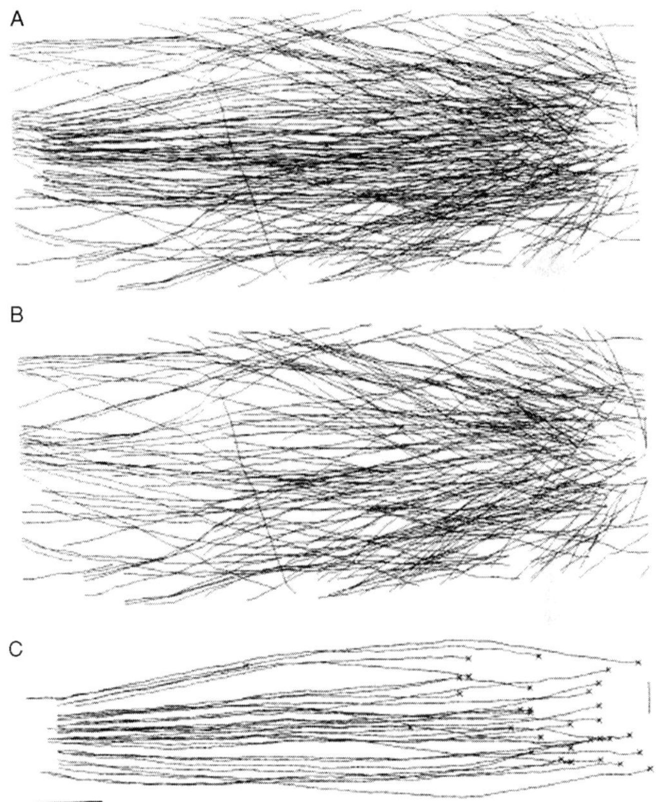

Fig. 3 Computer tracings of a K-fiber and surrounding MTs from a metaphase PtK1 cell. The centrosome is located on the right and the kinetochore is on the left side of each figure. The vertical bar indicates the position of the centrosome. (A) A tracing of all the MTs in and around the K-fiber. (B) A tracing of the non-kMTs. (C) A tracing of the kMTs. The Xs indicate locations of the minus ends of the kMTs. Scale bar = 0.5 μm. (Adapted with permission from McDonald *et al.*, 1992.)

found that most kMTs end near the spindle pole and that the K-fiber maintains a tight bundle along its entire length. Non-kMTs invade the K-fiber and can have a weak interaction with kMTs.

B. Use of Electron Tomography to Track MTs and Determine MT Dynamics

Electron tomography was used to track MTs in a K-fiber from serial longitudinal sections of spindles in *S. cerevisiae* and *C. elegans* (Fig. 4; O'Toole *et al.*, 1999, 2003).

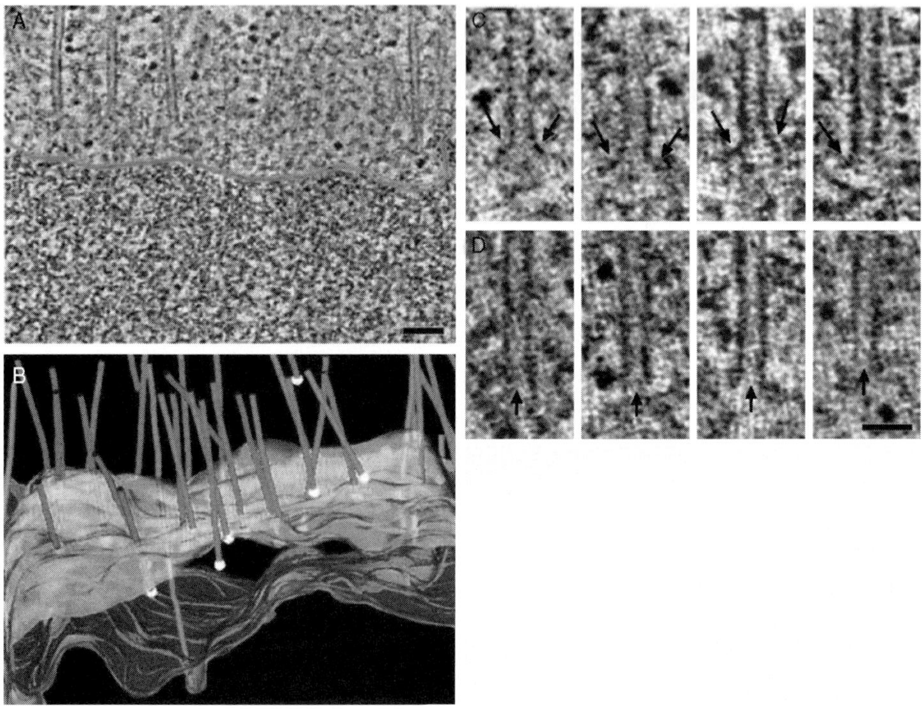

Fig. 4 Use of electron tomography to trace and characterize spindle MTs. (A) A 2.7-nm-thick slice from the electron tomographic 3D reconstruction of a kinetochore in a *C. elegans* cell. Note that despite the lack of a clear plate structure, the kMT plus ends are embedded in a ribosome-exclusion zone; the blue line indicates the outer edge of this zone. The green line indicates the inner edge of the ribosome exclusion zone that is in contact with the heterochromatin. Scale bar = 100 nm. (B) Volume-rendered view of a graphic model of the kinetochore area. The red lines indicate the kMTs with their plus ends (yellow dots) embedded in the 194-nm-thick kinetochore region. The blue contour indicates the outer surface of the ribosome-exclusion zone. The green contour indicates the inner surface of the kinetochore that is adjacent to the heterochromatin. (C, D) Windows each containing a single kMT plus end from the 2.7-nm-thick slices from the tomographic reconstruction shown in A. (C) Open and slightly flared conformations of kMT plus ends, with arrows indicating the flares. (D) Open and blunt conformations of kMT plus ends, with arrows indicating the open ends. Scale bar = 50 nm. (Adapted with permission from O'Toole *et al.*, 2003.) (See Color Insert.)

The combination of tomographic reconstructions from serial sections 250- to 400-nm thick, and modern computer-based methods for tracking and alignment, enabled O'Toole and coworkers to avoid the uncertainties of tracking spindle MTs in serial longitudinal sections. Use of longitudinal sections had the advantage of greater accuracy in determining the location of MT ends relative to the spindle pole and the kinetochore regions.

Tomographic 3D reconstructions of longitudinal sections also revealed the conformations of MT ends (Fig. 4). *In vitro* studies have indicated that the conformation of an MT end is correlated with the polymerization state at that end, with a straight conformation being associated with growth, and a curved conformation being associated with shrinkage (Sections IV.B and VI). Tomography revealed a capping structure at minus ends of many spindle MTs (O'Toole *et al.*, 1999, 2003). Presumably, this capping structure is the γ-tubulin ring complex, which nucleates spindle MTs at the poles (see Chapter 5 by O'Toole *et al.*, this volume). In their 2003 study, O'Toole and colleagues found that 78% of the kMTs in a metaphase *C. elegans* cell had flared plus ends, indicating that a large percentage of kMTs are depolymerizing at the plus end. About 50% of kMTs had capped minus ends. In contrast, 80% of non-kMTs had capped minus ends. These results have important implications for the possible selective severing of kMTs at their minus ends.

C. Determining the Polarity of Spindle MTs

MTs are polar polymers, as evidenced by the differential growth rate of the plus and minus ends (Allen and Borisy, 1974; Desai and Mitchison, 1997; Howard and Hyman, 2003). Polarity is important because MT motors are unidirectional, and the orientation of the MT polarity will therefore determine the directionality of MT-dependent transport, including chromosome motion. The orientation of MT polarity is also an indicator of where an MT is initiated. Determination of polarity in the spindle is challenging because polarity is not readily discernible in the MT structure, except in frozen-hydrated preparations of MTs imaged at comparatively high resolution (2–3 nm) (Chretien *et al.*, 1996).

The difficulty in determining the polarity of spindle MTs was overcome by the use of a high-molarity PIPES buffer (0.5-M PIPES, pH 6.9; 1-mM $MgCl_2$, 1-mM EDTA, and 1-mM GTP) that promotes tubulin assembly onto the walls of existing MTs (Heidemann and McIntosh, 1980; see also Haimo *et al.*, 1979). These *in vitro*-grown appendages are apparently related to naturally occurring appendages such as the MT doublets and triplets found in axonemes and centrosomes (Chapter 5 by O'Toole *et al.*, this volume). The critical feature of the assay is that the tubulin appendages grow asymmetrically and are hook-shaped when the MT is viewed in cross-section (Fig. 5). Using basal bodies and isolated asters for calibration, Heidemann and McIntosh determined that 90% of the hooks were right-handed (arrayed clockwise) when MT cross-sections were viewed from the plus-end direction. Euteneuer and McIntosh used this system to determine that kMTs in the K-fiber were oriented with their plus ends at the kinetochore (Fig. 5; Euteneuer and McIntosh, 1981; McIntosh

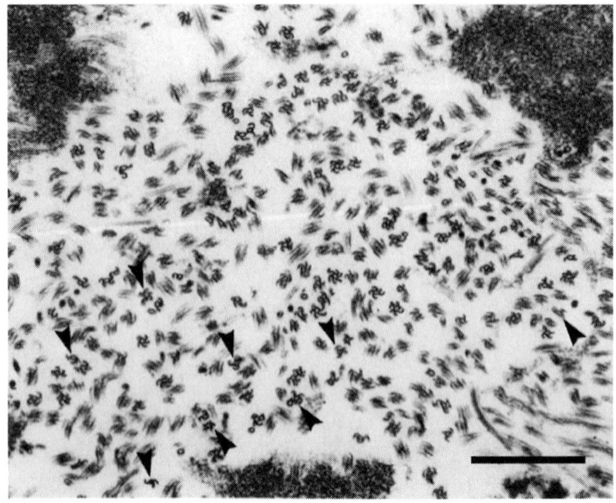

Fig. 5 Image of a transverse section, illustrating the use of tubulin assembly to determine the polarity of spindle MTs in an early anaphase PtK1 cell. Tubulin appendages were grown on the walls of spindle MTs by incubation of extracted mitotic cells with tubulin in a high-molarity Pipes buffer (see text). Transverse serial sections were cut through the cell and viewed from the direction of the centrosome toward the metaphase plate. The majority of appendages seen in the MT cross-sections are counter-clockwise pointing hooks, indicating that the MTs are being viewed from the direction of the minus end toward the plus end. Hence, it is the kMT plus ends that are embedded in the kinetochore outer plate. Arrowheads point to the few MTs that have clockwise-oriented hooks. Scale bar = 0.5 μm. (Adapted with permission from Eutenuer and McIntosh, 1981.)

and Euteneuer, 1984). To accomplish this, they extracted PtK1 cells with detergent and incubated them with the assembly buffer and exogenous tubulin. After fixation and plastic embedding, serial transverse sections were cut and examined as discussed above. This determination of the polarity of kMTs still plays a critical role in evaluating models of K-fiber function, including chromosome alignment.

IV. Deciphering Mechanisms of Chromosome Alignment and K-Fiber Formation Using Correlative Microscopy

A. Correlative LM/EM

The strength of EM is that it provides high-resolution information on cellular architecture. However, the images are static snapshots because EM alone does not allow direct temporal tracing of cellular organelle movements. On the other hand, there has been tremendous progress in video time-lapse recording of LM images produced by differential interference contrast (DIC), phase contrast, or fluorescence imaging, but here the resolution is limited (Khodjakov and Rieder, 2006). Correlative video LM/EM combines these two technologies to enable the following of cells by

LM up to a critical time point, after which they are rapidly fixed and examined at high resolution by EM (Rieder and Bowser, 1985; Rieder and Cassels, 1999). This combination allows for cellular ultrastructure to be linked to critical cellular functions.

In order for cells to be examined by both LM and EM, they must be grown as monolayers on glass coverslips, or sapphire disks, so that cellular events can be followed. For DIC LM, a coverslip is mounted in either a Rose chamber or a perfusion chamber filled with culture medium (Rieder and Cassels, 1999). Cells are filmed and image frames are recorded up to a desired point in the cell cycle. The cells are then perfused with glutaraldehyde that has been shown to fix cells within 1–2 sec. After marking the cell with an objective scribe or the recording the location on the finder grid (Section II.B), the coverslip is removed from the filming chamber and processed for TEM by conventional flat-embedding methods. Recently, technology has been developed for rapid HPF of cells within 5 sec of recording the last recorded image (Chapter 4 by Müller-Reichert *et al.*, this volume).

B. Establishment of the Search and Capture Model for Chromosome Attachment

The search and rescue model for attachment of kinetochores to the mitotic spindle is based on the model for MT dynamics called dynamic instability (Kirschner and Mitchison, 1986; Mitchison and Kirschner, 1984). The term dynamic instability describes the growth and shrinkage from MT ends as two distinct biochemical states related to the hydrolysis of GTP (Desai and Mitchison, 1997; Mitchison and Kirschner, 1984). This concept of MT dynamics explains how growing and shrinking MTs can exist side-by-side in the same solution or cellular environment. MT turnover due to dynamic instability is an order of magnitude more rapid during mitosis than during interphase (Gelfand and Bershadsky, 1991). The idea of the search and capture model is that MTs continuously grow out from the centrosome but then shrink back, unless they are stabilized by binding to a kinetochore (Brinkley and Cartwright, 1971; Rieder, 1981; Rieder *et al.*, 1990; Zhai *et al.*, 1995). Rapid turnover of non-kMTs ensures an efficient exploration of space within the cell because MTs growing in an unproductive search direction are short-lived.

The first direct evidence for the search and capture model came from a classic correlative LM/EM study (Fig. 6; Rieder and Alexander, 1990). Newt lung cells were filmed by video DIC imaging, and the cells were rapidly fixed when it became apparent that a chromosome had attached to the spindle. Newt lung cells were used because they are large cells that lie flat during mitosis and have very large chromosomes. These characteristics facilitated the filming of mitosis and the recognition of critical times at which to fix the cells. Video LM analysis documented that newly attached chromosomes typically move rapidly toward one of the spindle poles. Serial-section EM analysis of a subset of the cells that had been filmed demonstrated that the initial poleward movement occurs along the lateral surface of one or more MTs (Fig. 6). This was the first direct demonstration that

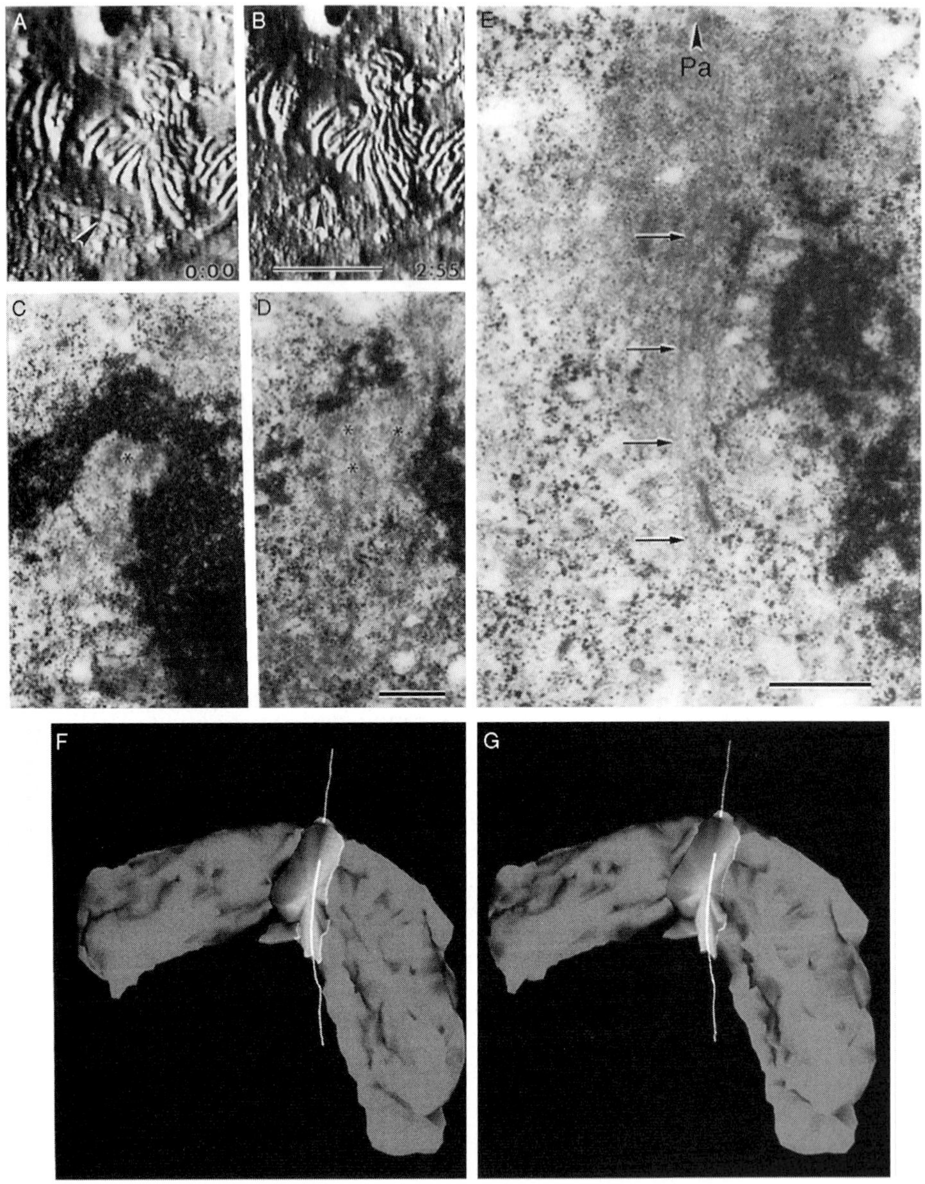

Fig. 6 Use of correlative LM/EM to demonstrate initial chromosome attachment through a lateral interaction with spindle MTs. (A, B) Two frames from a DIC video recording of initial chromosome attachment in newt lung cells. Arrowheads denote a chromosome shortly before (A) and during (B) the process of initial attachment. Scale bar = 25 μm. (C–E) Selected TEM images from serial sections through the same chromosome that is indicated by the arrowheads in A and B. Asterisks in C and D indicate the sister kinetochores in different sections. Arrows in E point to the only MT found in the

chromosomes could become mono-oriented (attached to one spindle pole) through the interaction with a single MT growing from the centrosome.

C. Elucidation of Mechanisms for Chromosome Congression

Congression is the process by which a chromosome moves away from the nearer spindle pole and becomes aligned at the spindle equator (Rieder and Salmon, 1994; Skibbens *et al.*, 1993). Correlative LM/EM analysis was used to show that the direction of chromosome movement does not depend on the ratio of kMTs bound to each sister kinetochore (McEwen *et al.*, 1997). Previously, it was assumed that congression was always initiated when the sister kinetochore facing away from the attached spindle pole bound MTs emanating from the opposite spindle pole. One difficulty with this older model for congression is that the newly attached kinetochore is predicted to exert a more powerful poleward pulling force than its sister kinetochore, even though the newly attached kinetochore is also predicted to have fewer kMTs attached. The number of kMTs bound is an important parameter because the two leading candidates for generating a poleward pulling force at the kinetochore, MT motors and MT depolymerization, both depend on the number of kMTs bound to each kinetochore. To resolve this conflict, Hays and Salmon (1990) postulated that the poleward force generated at each kinetochore is proportional to the number of kMTs times their lengths (i.e., the force is applied all along the kMTs). To test Hay and Salmon's prediction, McEwen *et al.* (1997) used correlative LM/EM to determine the number of kMTs attached to the leading and trailing kinetochores of chromosomes that were undergoing congression at the time of fixation. Contrary to the earlier predictions, chromosomes could congress to the equator with fewer MTs on the leading kinetochore than could be accounted for by differences in MT length alone. The study by McEwen *et al.* also found no correlation between the numbers of MTs bound on each sister kinetochore and the direction of motion for chromosomes that were oscillating about the spindle equator. Therefore, force generation at the kinetochore does not depend solely on the number of kMTs bound.

A second problem for the traditional model of chromosome congression is that some mono-oriented chromosomes oscillate relatively close to the attached spindle pole, which means that the unbound sister kinetochore is correspondingly far from the opposite spindle pole and has a low probability of attaching to MTs emanating from that pole. Hence, if bi-oriented attachment was the only means by which a cell

---

vicinity of either sister kinetochore on this chromosome. This MT forms a lateral interaction with the leading kinetochore. The label Pa and the arrowhead in panel E indicate the direction of the polar area. Scale bars = 0.5 μm. (F, G) Stereo images of the volume reconstructed from the serial sections of the chromosome indicated in A and B. Note that the kinetochore attaches to the lateral surface of the MT. (Adapted with permission from Rieder and Alexander, 1990.)

could achieve equatorial alignment, it would take an unreasonably long time for the cell to align all of its chromosomes at the spindle equator (Wollman *et al.*, 2005). Recently, Kapoor *et al.* (2006) used correlative LM/EM, in conjunction with 3D video fluorescence imaging, to demonstrate that the unbound sister kinetochore of a mono-oriented chromosome can "hitch a ride" along the lateral surface of a neighboring K-fiber bound to a kinetochore on a chromosome that was already aligned at the spindle equator (Fig. 7). This completely novel mechanism for chromosome congression in the absence of bi-orientation is dependent on the centromere motor protein CENP-E, and it appears to operate in addition to congression after bi-orientation. Thus, correlative LM/EM also revealed a major functional role for CENP-E and explained why depletion of CENP-E results in some chromosomes stuck at the spindle poles.

V. Toward a Molecular Understanding of Kinetochore Architecture

A. Kinetochore Function and Composition

The kinetochore is a proteinaceous complex that is located adjacent to the heterochromatin at the primary constriction of the mitotic (and meiotic) chromosome (Chan *et al.*, 2005; Cleveland *et al.*, 2003; Maiato *et al.*, 2004). The kinetochore structure does not fully mature until after nuclear envelope breakdown, when molecular components of the kinetochore located in the cytoplasm are able to bind (Maiato *et al.*, 2004; Rieder and Salmon, 1998; Roos, 1973). In addition to binding spindle MTs and thereby attaching the chromosome to the mitotic spindle, the kinetochore functions to facilitate spindle formation, regulate chromosome alignment, segregate sister chromatids, and control the mitotic cell cycle checkpoint. All of these functions appear to be mediated through kinetochore interaction with the dynamic plus ends of kMTs.

The kinetochore in *S. cerevisiae* is composed of more than 60 proteins that are combined into at least 17 subcomplexes (Biggins and Walczak, 2003; Cheeseman *et al.*, 2002; De Wulf *et al.*, 2003; Westermann *et al.*, 2003). The subcomplexes are in turn arranged into three domains: inner, middle, and outer. The inner domain includes proteins such as CENP-A, CENP-C, and the Mis12 complex. This domain interacts directly with the heterochromatin and forms the constitutive core of the kinetochore (Earnshaw and Migeon, 1985; Meluh and Koshland, 1995; Obuse *et al.*, 2004). The middle domain, which includes the Ndc80 complex, functions as a structural skeleton that bridges between the inner and outer domains (Cheeseman *et al.*, 2004; DeLuca *et al.*, 2005; Emanuele *et al.*, 2005; Wigge and Kilmartin, 2001). The middle-domain proteins are required for MT attachment. The outer domain includes CENP-E, CLIP-170 in vertebrates, and DASH/Dam1 complex in *S. cerevisiae*. This domain is essential for the spindle checkpoint activation (Cheeseman *et al.*, 2001; De Wulf *et al.*, 2003; Li *et al.*, 2002).

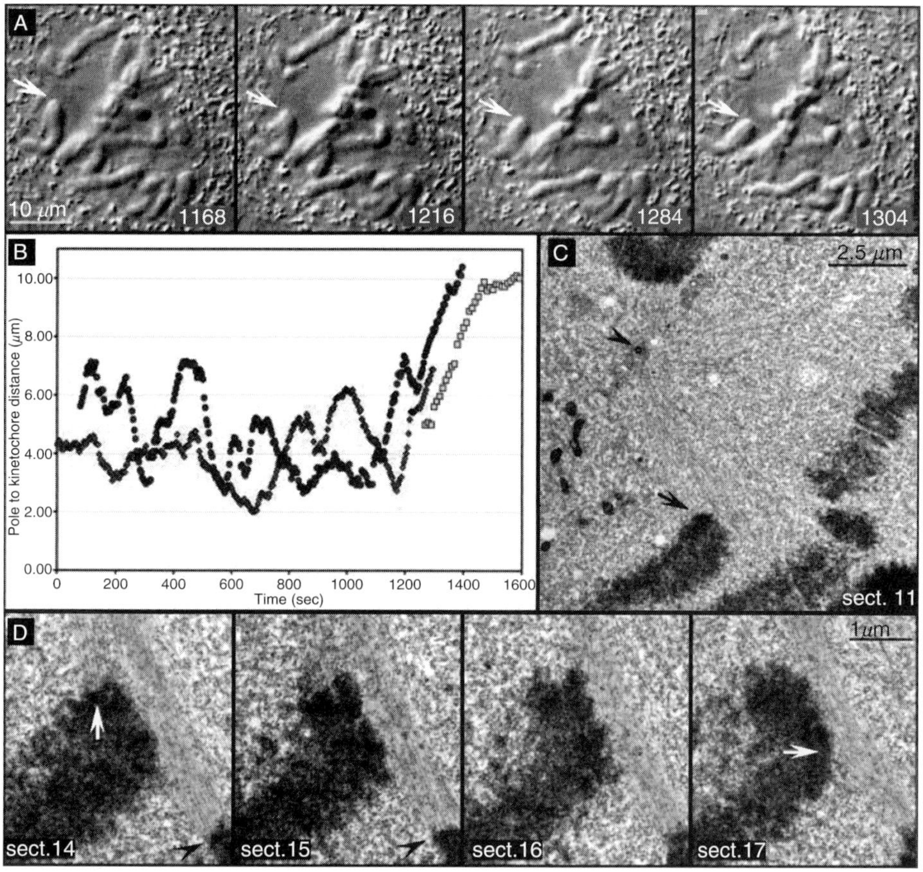

Fig. 7 Use of correlative LM/EM to demonstrate chromosome congression in the absence of bi-oriented attachment. (A) Selected frames from the DIC video recording of a congressing chromosome. The time marker in seconds is indicated in the lower right-hand corner of each frame. The congressing chromosome is indicated by the white arrows. The cell was fixed when the chromosome had almost reached the metaphase plate (just after frame 1304). (B) Tracking records of three congressing chromosomes. Chromosome movement is plotted as distance of the attached kinetochore from the spindle pole versus time. The red curve is the tracing record of the chromosome shown in A, C, and D. (C) Low-magnification TEM view of the cell shown in A. The black arrow indicates the congressing chromosome tracked by the red curve in B. Black arrowhead indicates the centriole. (D) Serial sections of the chromosome indicated in A and C and corresponding to the red tracking curve in B. The white arrow in sect. 14 indicates the trailing kinetochore. The K-fiber attached end-on to the trailing kinetochore is highlighted in red shading in sects. 14 and 15. A neighboring K-fiber that is highlighted in yellow-green shading in sects. 14–17 is attached end-on to a kinetochore on a fully congressed chromosome (black arrowhead in sects. 14 and 15). The leading kinetochore of the congressing chromosome laterally interacts with this same K-fiber (white arrow in sect. 17). (Adapted with permission from Kapoor *et al.*, 2006.) (See Color Insert.)

Homologues of several of the *S. cerevisiae* kinetochore proteins have been found in vertebrates and other model organisms such as *C. elegans* and *Drosophila* (Biggins and Walczak, 2003; Cheeseman *et al.*, 2002, 2004; Emanuele *et al.*, 2005). Importantly, the vertebrate kinetochore also has three molecular domains, with the middle domain required for MT binding.

B. The Classical Trilaminar Model of the Kinetochore Ultrastructure

The three-domain model for the molecular composition of the kinetochore reflects the classic trilaminar model of vertebrate kinetochores that was derived from early TEM studies on model cell lines such as PtK1, PtK2, Chinese hamster, mouse A9, and fetal rat (Brinkley and Stubblefield, 1966; Jokelainen, 1967; Roos, 1973). These studies indicated that the unbound vertebrate kinetochore is composed of three layers: a 30- to 40-nm electron-dense outer plate, in which the MT plus ends terminate; an electron-translucent middle layer; and an electron-dense inner plate, aligned adjacent to the heterochromatin (Fig. 8A). When MTs are not bound, the kinetochore also has a fibrous corona extending from the distal side of the outer plate. The corona disappears on attachment of MT plus ends to the kinetochore and reappears when the cells are treated with colcemid (Cassimeris *et al.*, 1990).

C. Effects of Specimen Preparation on Kinetochore Ultrastructure

Dramatic differences were found in the ultrastructure of kinetochores prepared by conventional and by HPF/FS (Fig. 8; McEwen *et al.*, 1998). In HPF/FS specimens, the outer plate appeared as a fibrous mat extending to 50–75 nm, a width that is at least 25% greater than that observed for conventionally fixed samples (Fig. 8B). The corona was shown to be a ribosome-exclusion zone 100- to 150-nm thick, and the translucent middle zone was not seen. These differences were attributed to the specimen condensation that occurs in conventional preparations because the fixation and dehydration steps take place at room temperature. Other signs of superior structural preservation in the HPF/FS preparations include reduced extraction of the cytoplasm and the straightness of spindle MTs (Figs. 4 and 8; Section VI; and Chapter 1 by Dubochet, Chapter 2 by McDonald, and Chapter 3 by Hess, this volume).

D. Immunolocalization of Kinetochore Components

Immuno-EM is a technique for precisely localizing proteins with an antibody conjugated to an electron-dense tag that is visible in the TEM (Chapter 19 by Morphew, Chapter 23 by Sosinsky *et al.*, Chapter 22 by Cheutin *et al.*, and Chapter 21 by Landinsky and Howell, this volume). Thus far, only preembed methods have been published for immuno-EM localization of kinetochore proteins. In the preembed method, cells are grown on coverslips, permeabilized with a detergent, lightly fixed with a low concentration of glutaraldehyde or paraformaldehyde, incubated with

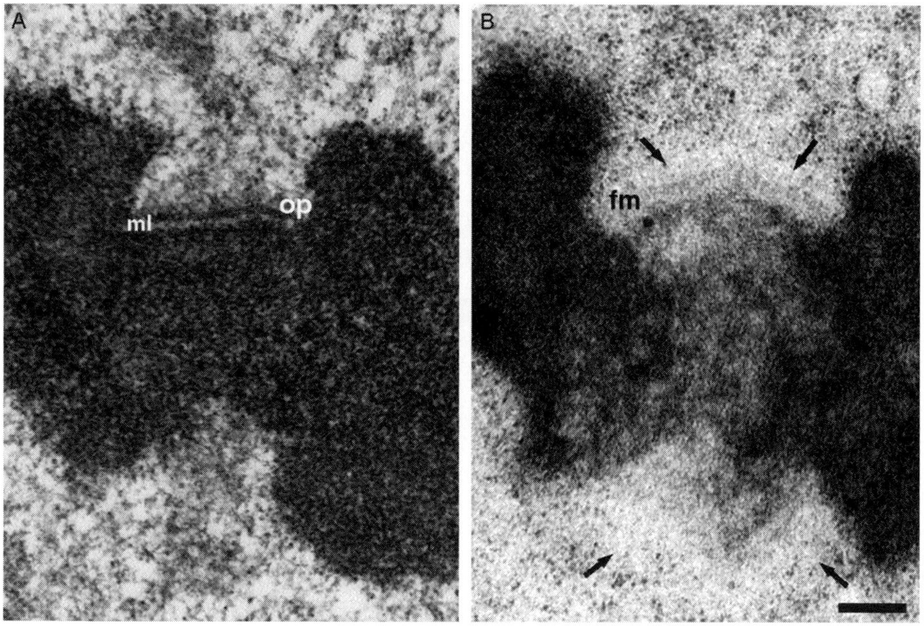

Fig. 8 Ultrastructure of the unbound kinetochore prepared by conventional fixation and by HPF/FS. (A) An unbound kinetochore from a colcemid-treated cell fixed using a TEM conventional protocol. The outer plate (op) appears as a compact, electron-dense band separated from the surface of heterochromatin by an electron-translucent middle layer (ml). A fibrous corona extends radially from the distal surface of the outer plate. (B) A pair of sister unbound kinetochores from a nocodazole-treated cell prepared using HPF/FS. The outer plate appears as a fibrous mat (fm) that is more extended and less dense than the outer plate in A. The translucent middle layer has disappeared. Black arrows indicate the ribosome-exclusion zone, which is composed of delicate fibers. Scale bar = 250 nm. (Adapted with permission from McEwen *et al.*, 1998.)

a primary antibody against the protein of interest, and finally incubated with a secondary antibody that is generally conjugated with colloidal gold. The specimens are then fixed with 2.5% glutaraldehyde and processed by conventional methods. To improve antibody penetration, the secondary antibody is often conjugated to a small gold particle or gold complex (0.8 or 1.4 nm in diameter; Baschong and Stierhof, 1998; Hainfeld, 1996). These small gold particles are not visible in sections of cells by TEM. Therefore, silver enhancement is applied after antibody labeling and second glutaraldehyde fixation, in order to increase the size of the gold particles (Baschong and Stierhof, 1998; Burry *et al.*, 1992; Vandre and Burry, 1992). Preembed immuno-EM generally produces good penetration of the antibodies into the cell and high contrast, but the native specimen structure is severely compromised, due to extraction of the cytoplasm. However, the trilaminar kinetochore structure can still be recognized after preembed labeling (Fig. 9).

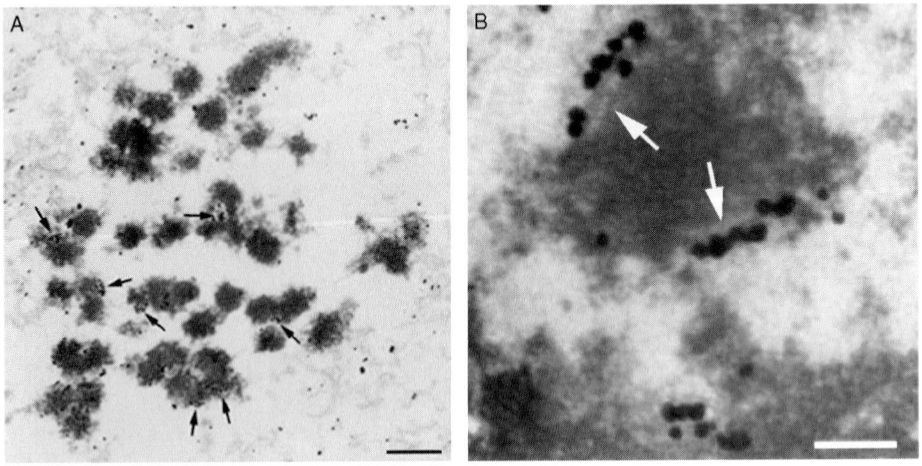

Fig. 9 Localization of Hec1 (component of the mammalian Ndc80 complex) in HeLa cells by immuno-EM. (A) Low-magnification image. The arrows indicate the aggregation of the gold particles near the kinetochore area. Scale bar = 1 nm. (B) High-magnification view of a chromosome in cross-section and sister kinetochores. Despite the detergent extraction used during specimen preparation, the kinetochore outer plate is visible. White arrows point to gold labeling that is evident on both sides of the outer plate. Scale bar = 200 nm. (Adapted with permission from DeLuca *et al.*, 2005.)

Despite such compromised structural preservation, preembed immuno-EM has been useful for establishing a correspondence between the ultrastructure and molecular composition of the kinetochore. The first applications of immuno-EM to the kinetochore established the location of the inner-domain proteins CENP-A, CENP-B, and CENP-C, and demonstrated that DNA was not located in the outer kinetochore plate (Cooke *et al.*, 1990, 1993; Saitoh *et al.*, 1992; Warburton *et al.*, 1997). This was followed by demonstration that CENP-F was located on the distal side of the outer plate (Rattner *et al.*, 1993). The motor protein CENP-E was found predominantly in the corona, with some labeling on the outer plate, during prometaphase and metaphase (Cooke *et al.*, 1997; Yao *et al.*, 1997). Recently, the Ndc80 complex was localized on the outer plate, indicating that the middle domain of the molecular model probably corresponds to the outer plate of the structural model (Fig. 9; Deluca *et al.*, 2005). Although preembed labeling has been important as a first step for establishing the correspondence between the molecular and structural domains of the vertebrate kinetochore, higher-resolution localization of kinetochore proteins will require methods that better preserve the cellular ultrastructure.

E. Imaging of Isolated Protein Complexes from the Kinetochore

An alternative strategy for locating proteins and protein complexes without destroying ultrastructure is to search electron tomographic reconstructions for molecular components whose structures are known (Bohm *et al.*, 2000;

Frangakis *et al.*, 2002; Rath *et al.*, 2003; Chapter 25 by Best *et al.*, this volume). This strategy requires that 3D structures be available for both the organelle of interest and the predominant components. These structural determinations must have sufficient resolution that components can be distinguished from one another in the organelle. Although this approach of looking for "molecular signatures" has not yet been applied to the kinetochore, it has been successfully used to fit different conformations of myosin head molecules into electron tomographic 3D reconstructions of insect flight muscle (Taylor *et al.*, 1999; Chapter 13 by Taylor *et al.*, this volume) and to fit the structure of cadherin molecules into electron tomographic 3D reconstructions of adherens junctions (He *et al.*, 2003). These studies were carried out in plastic sections of specimens prepared by HPF/FS, demonstrating that frozen-hydrated imaging is not always required for detection of molecular components.

The molecular complexity of the kinetochore (Section V.A) presents a major challenge for identification of molecular components within the kinetochore. The general disorder and rapid turnover of the outer-domain proteins, and the location of inner-domain components in the dense heterochromatin region, present additional obstacles to the detection of components within these domains (Howell *et al.*, 2000; Shah *et al.*, 2004). However, the major components of the middle domain appear to be a limited set of proteins and protein complexes that are relatively stable in their location at the structurally distinct outer plate (Cheeseman *et al.*, 2004; DeLuca *et al.*, 2002; Wigge and Kilmartin, 2001). Hence, localization of molecular subcomplexes within the kinetochore outer plate is a feasible starting point for reconstruction of the molecular architecture of the kinetochore.

A major step toward identifying molecular components in the kinetochore outer plate was achieved during the past year, with the first images of the isolated Dam1 and Ndc80 complexes (Ciferri *et al.*, 2005; Miranda *et al.*, 2005; Wei *et al.*, 2005; Westermann *et al.*, 2005, 2006). The 10-member Dam1 complex of *S. cerevisiae* was expressed in bacteria, isolated, and then imaged in both negatively stained and frozen-hydrated specimen preparations (Fig. 10; Miranda *et al.*, 2005; Westermann *et al.*, 2005). Negative staining is fast and provides high contrast, but MTs and similar structures collapse during the drying stage of the negative staining procedure, and the accumulation of stain can obscure structural details (Izard *et al.*, 2004). Therefore, images of negatively stained specimens were used to establish the general features of the Dam1 complex, including the fact that it forms a ring that encircles the plus ends of kMTs (Miranda *et al.*, 2005; Westermann *et al.*, 2005, 2006). Images of frozen-hydrated specimens were used to obtain more precise measurements of the ring's dimensions and to elucidate the interaction of the ring with the MT lattice (Fig. 10; Westermann *et al.*, 2005).

The isolated Ndc80 complex from *S. cerevisiae* was visualized by rotary shadowing. Rotary shadowing is carried out by coating specimens with heavy metal atoms that are evaporated onto the surface of the specimen under a vacuum. This method produces a 3D effect when viewed in the TEM. After rotary shadowing, the Ndc80 complex isolated from *S. cerevisiae* appeared as a rod 56 nm in length

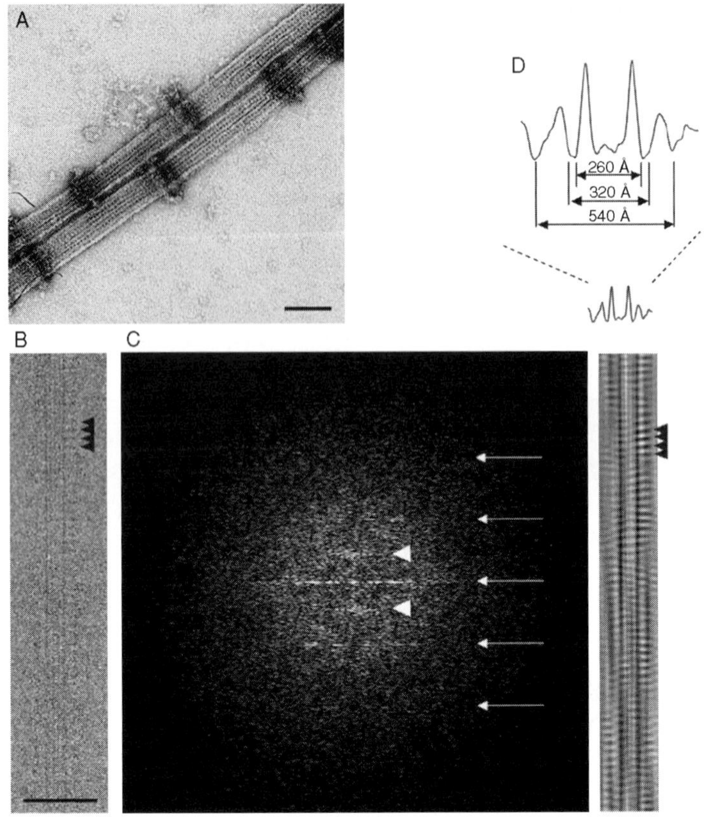

Fig. 10 *In vitro* characterization of the isolated Dam1 complex and its interaction with MTs. (A) TEM image of a negatively stained preparation of bacterially expressed Dam1 complex binding to MTs. The complex forms rings encircling the MTs. Scale bar = 50 nm. (B) TEM image of a frozen-hydrated preparation of the Dam1 complex binding to an MT. Dam1 complex rings are indicated by arrowheads. Scale bar = 50 nm. (C) Power spectrum of the image in B. White arrows indicate the layer lines corresponding to the MT lattice. White arrowheads indicate the layer lines corresponding to the Dam1 ring complex. The spacing between Dam1 rings is 10 nm. (D) A Fourier-filtered image of the MT shown in B and C. Arrowheads indicate Dam1 rings around the MT. The inset above the filtered image is the density profile from which the diameters of the MT and the Dam1 ring structure are measured. (Adapted with permission from Westermann *et al.*, 2005.)

(Wei *et al.*, 2005). Rotary shadowing combined with antibody labeling demonstrated that the subunit proteins of the Ndc80 complex are arranged as two heterodimers connected tail-to-tail. The vertebrate form of the Ndc80 complex was visualized by atomic force microscopy rather than rotary shadowing (Ciferri *et al.*, 2005). Although it is also rod-shaped, it is only 40 nm in length.

Given the rapid progress in identifying and isolating molecular components of the kinetochore, searching for "molecular signatures" is likely to be a productive approach for the near future. The current images of the Dam1 and Ndc80 complexes have significantly lower resolution than the molecular models that have been used previously for "molecular signature" searches. Nevertheless, the complexes are sufficiently large, and have sufficiently distinctive shapes, that it may be possible to locate the complexes in tomographic reconstructions of the kinetochore outer plate. However, a general caveat to the approach of locating structures of individual components within the full kinetochore is that the structure of isolated components might not be the same as is found *in vivo*. For example, although the isolated 50-nm Dam1 ring is large and distinct, it has not been detected in tomographic reconstructions of the budding yeast kinetochore (McIntosh, 2005). Therefore, while the search for macromolecular components has been successful in some studies, differences between the isolated and *in vivo* forms could cause problems for other components.

VI. Beyond Morphology: Use of Electron Tomography to Model Kinetochore Control of MT Dynamics

A. Nature of the Problem

Chromosome alignment during mitosis is frequently accompanied by a dynamic switching between elongation and shortening of K-fibers from their plus ends (Gorbsky *et al.*, 1988; Mitchison *et al.*, 1986; Skibbens *et al.*, 1993; Wadsworth *et al.*, 1989). Since the first observation of this switching behavior, it has been assumed that MT dynamics (Section IV.B) are coordinated with the direction in which a kinetochore is moving. However, K-fibers (Fig. 1) in vertebrates typically contain 15–40 kMTs, and kinetochores can change the direction of their motion within a second (McEwen *et al.*, 1997; Skibbens *et al.*, 1993). Hence, a long-standing question has been how the dynamics of individual kMTs within the same K-fiber are coordinated.

B. Using High-Throughput Electron Tomography as an Assay for kMT Dynamics

Two developments have been critical to answering the question of how MT dynamics are coordinated with chromosome motion. First, a series of cryo-TEM studies on *in vitro*-assembled MTs demonstrated that MT ends have a straight or open sheet conformation during the polymerizing phase, and a curved or splayed end conformation during the depolymerizing phase (Arnal *et al.*, 2000; Chretien *et al.*, 1995; Janosi *et al.*, 2002; Mandelkow *et al.*, 1991; Wang and Nogales, 2005). Both the straight and curved conformations have been detected on the plus ends of kMTs (Figs. 4 and 11; O'Toole *et al.*, 2003; VandenBeldt *et al.*, 2006). Thus, in principle, it is possible to use electron tomography to determine the dynamic state of individual kMTs in the K-fiber at the moment of fixation or rapid freezing.

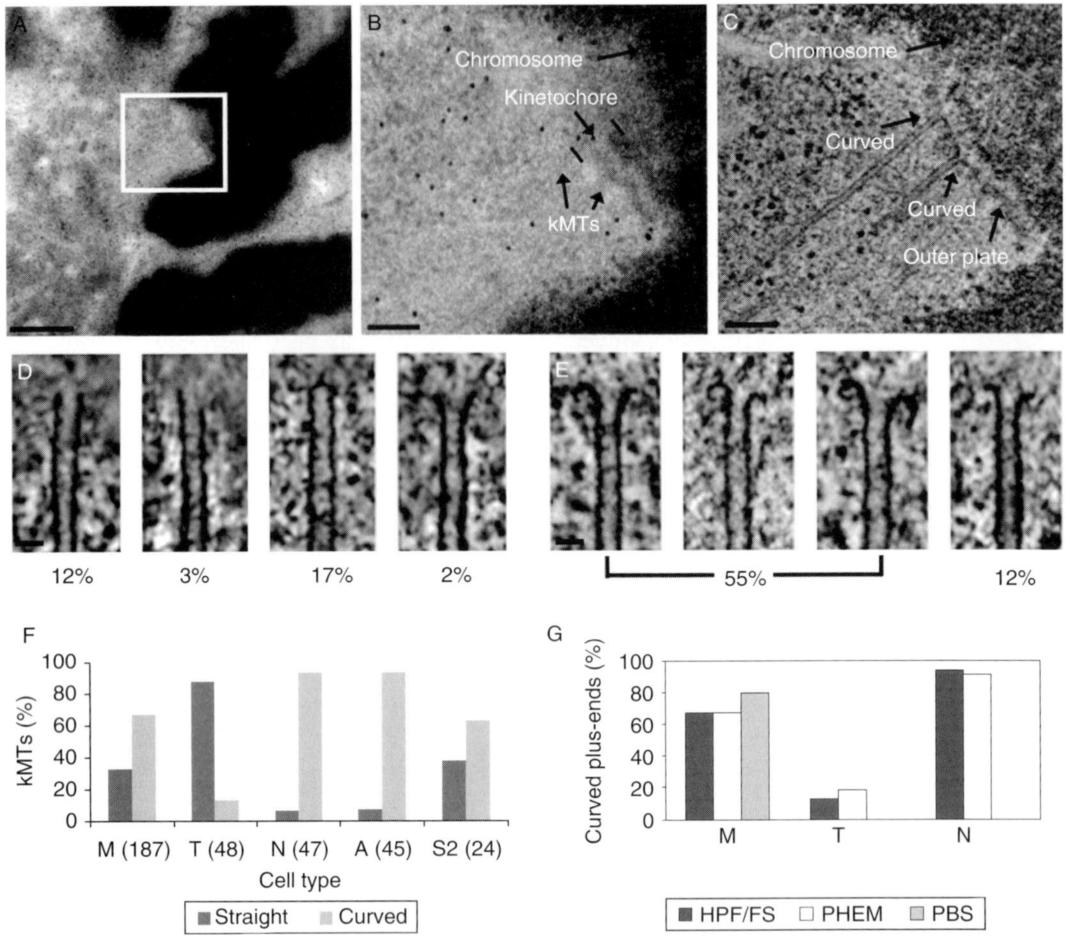

Fig. 11 Conformations of kMT plus ends observed by electron tomography. (A) Low-magnification TEM image of a PtK1 chromosome during metaphase. The cell was prepared by HPF/FS. Scale bar = 1 μm. (B) A higher magnification TEM image of the boxed area in A. The chromosome, kinetochore, and kMTs are labeled. Scale bar = 200 nm. (C) 1.6-nm-thick slice from the tomographic reconstruction of the same area in B. Note that there are two MTs ending in the outer plate in this slice. The plus-end conformations of both MTs are curved. Scale bar = 200 nm. (D, E) Examples of MTs with straight (D) and curved (E) conformations at the plus ends. The percentages of kMT plus ends with similar conformations detected in metaphase cells are presented underneath the corresponding images. Scale bar = 25 nm. (F) Bar graph showing the percentages of kMT plus ends classified with straight and curved conformations under differing conditions, and in different species. M, metaphase; T, taxol-treated metaphase; N, nocodazole-treated metaphase; and A, anaphase PtK1 cells; and S2, metaphase *Drosophila* S2 cells. The numbers in the parentheses after each letter indicate the number of MTs scored. (G) Bar graph showing the percentages of straight plus-end conformations detected in PtK1 cells after various fixation procedures. Metaphase cells and cells treated with taxol and nocodazole are compared. Bar graphs for cells fixed using HPF/FS, a conventional protocol with PHEM buffer, and a conventional protocol with PBS buffer are indicated by the key at the bottom of the graph. In each group, the cells fixed by HPF/FS and those fixed by the conventional protocol in PHEM buffer had similar percentages of straight MT plus ends. (Adapted with permission from VandenBeldt *et al.*, 2006; also, VandenBeldt and McEwen, unpublished data.)

The second critical development has been the implementation of automated and semiautomated methods to rapidly collect, align, and process tilt-series data sets (Baumeister *et al.*, 1999; Grimm *et al.*, 1997, 1998; Koster *et al.*, 1997; Mastronarde, 1997; Niedermann *et al.*, 1997; Penczek *et al.*, 1995; Walz *et al.*, 1997; Ziese *et al.*, 2002). Although automated data collection was originally developed for the purpose of reducing the electron dose during the collection of tilt series from frozen-hydrated specimens, it has had the additional benefit of significantly improving the efficiency of data collection for plastic-embedded specimens. Some of the key methodological and hardware developments have been high-quality digital imaging, highly eucentric and stable tilt stages, and software for automatic tracking and focusing of the specimen (Ziese *et al.*, 2002). Computation of tomographic reconstructions from tilt-series data sets has also become significantly more efficient, in large part due to the availability of ever-faster computers with vast memory and storage spaces. The development of alignment algorithms and user-friendly interfaces has also improved the efficiency and quality of tomographic reconstructions (Frank *et al.*, 1996; Kremer *et al.*, 1996; Mastronarde, 1997; Penczek *et al.*, 1995). As a result of these developments, it is now possible to collect a dual-axis tilt series in 80–90 min, and to compute the tomographic reconstruction in less than an hour. Hence, electron tomography of plastic-embedded specimens has become a relatively high-throughput method of obtaining 3D ultrastructure (Chapter 8 by Marsh, this volume).

These improvements in throughput enabled VandenBeldt *et al.* (2006) to use electron tomography as an assay for the plus-end conformations of over 500 individual kMTs in the K-fibers of PtK1 and *Drosophila* S2, taken at various times during mitosis, after various drug treatments, and using different procedures for specimen preparation (Fig. 11). The tomographic reconstructions were computed from relatively thin serial sections, 140- to 250-nm thick, in order to achieve better quality in the high-tilt images, and thereby greater clarity in the reconstructions. For analysis, the x, y, and z coordinates of two points along each kMT in the reconstructions were used to extract a small subvolume containing the plus end with the same orientation as kMT plus ends in all of the other extracted subvolumes (Fig. 11D and E). The plus ends were then manually classified by three investigators, independently inspecting the subvolumes without knowledge of the kMT origin.

The study produced a classification of the plus-end conformations of kMTs in metaphase PtK1 and *Drosophila* S2 cells and a way to distinguish them by statistically significant discriminators (VandenBeldt *et al.*, 2006). The study also assessed changes to the resulting "dynamic profile" in PtK1 cells on application of the MT stabilizing and destabilizing drugs taxol and nocodazole and on entry into anaphase (Fig. 11F). Surprisingly, two-third of the kMTs in metaphase cells displayed the curved conformation associated with depolymerization (Fig. 11F), even though K-fibers are known to have a net tubulin incorporation during metaphase, to compensate for subunit loss at the minus end (Maddox *et al.*, 2003; Mitchison and Salmon, 1992). Another surprise was the lack of coordination among kMTs within the same K-fiber (VandenBeldt *et al.*, 2006).

C. A New Model for Kinetochore Control of MT Dynamics

The results of the plus-end assay described above suggested a new model in which kMT attachment to the kinetochore outer plate prevents dissociation and shrinkage of kMTs when they attempt to depolymerize. Through a dynamic cycling of individual kMTs between the polymerization and depolymerization phases, K-fibers as a whole can have a net gain of tubulin at their plus ends, even though the majority of the individual kMTs within the K-fibers are in the depolymerizing state. Thus, the kinetochore outer plate mediates the effects of control mechanisms (and drug treatments) that bias MT dynamics and change the probability that kMTs are in a given dynamic state. However, the kinetochore does not directly control the plus-end dynamics of individual kMTs. Hence, the answer to the question of how the kinetochore coordinates the dynamics of 30 or more MT plus ends appears to be that it does not.

D. Conventional Fixation Versus HPF/FS

Since high pressure is known to depolymerize MTs, a possible caveat to the results presented in Fig. 11 is that a significant percentage of the kMT plus ends could be switching to the depolymerizing phase during the few milliseconds the specimens are exposed to high pressure before the rapid-freezing step (Salmon, 1975). Such a transition would skew the results and could explain why two-thirds of the plus ends on kMTs have the curved conformation. To control for this possibility, VandenBeldt *et al.* (2006) compared the conformations obtained from conventionally fixed specimens with those of specimens prepared by HPF/FS. Conventional fixation of PTK1 cells did not increase the percentage of kMT plus ends that had the straight conformation, regardless of whether the fixation was carried out in phosphate-buffered saline (PBS) or the MT-stabilizing buffer, PHEM (Section II.A). The percentage of kMT plus ends with the straight conformation actually decreased slightly when the PBS buffer was used (Fig. 11G).

These results indicate that our customary fixation in PHEM buffer is superior to fixation in PBS, and comparable to fixation using HPF/FS, for preserving the dynamic state of MT ends. Fixation in PHEM also gave results similar to fixation using HPF/FS, for PTK1 cells treated with nocodazole and taxol, and anaphase cells fixed in PHEM had a high percentage of kMTs with curved plus ends (Fig. 11F and G). The control experiments demonstrate that fixation in PHEM does not artificially bias kMT plus ends to adopt the straight conformation. Thus, the results in Fig. 11G indicated that conventional fixation in PHEM buffer can be used as a viable alternative to fixation using HPF/FS, for studies of MT dynamics in monolayer tissue-culture cells.

VII. Conclusions and Future Directions

This chapter has endeavored to provide the reader with concrete examples of how TEM has been used to answer questions regarding the structure and function of the mitotic spindle. The examples presented have highlighted a variety of techniques ranging from conventional fixation followed by serial-section analysis to use of HPF/FS followed by electron tomography.

This chapter has devoted very little attention to the use of frozen-hydrated specimens. This may seem a surprising omission, given the importance of vitreous specimens in modern TEM. However, the use of frozen-hydrated specimens has thus far had minimal impact on studies of the mitotic spindle and its interaction with kinetochores. The reason for this is that vertebrate tissue-culture cells are too large to be viewed as whole-mount specimens and must therefore be cut by cryoultramicrotomy (Al-Amoudi *et al.*, 2004b; Hsieh *et al.*, 2006). This is a technically challenging approach, and at present there are only a few laboratories capable of it. However, recent progress in obtaining images and electron tomographic reconstructions from frozen-hydrated sections suggests that in the future a significant portion of TEM studies on the mitotic spindle and kinetochores will be done on frozen-hydrated specimens.

Nevertheless, the recently published study on the congression of chromosomes in the absence of bi-orientation (Fig. 7; Kapoor *et al.*, 2006) demonstrates that conventional fixation and serial thin-section analysis can still provide important mechanistic insights. Therefore, this chapter concludes by reemphasizing the point made at the onset that a structural study of the mitotic apparatus should be focused on a biological question and should use the techniques most appropriate for answering that question. If the question requires detailed knowledge of cellular ultrastructure, then specimens should be prepared by HPF/FS, or imaged in frozen-hydrated preparation, and 3D reconstructions should be computed by electron tomography. On the other hand, if the question merely requires a general knowledge of cellular architecture and how it changes with protein depletions or different functional stages, then the use of conventional fixation and serial-section reconstruction is often sufficient, and is always less time consuming.

Acknowledgments

The authors wish to thank Dr. Christian Renken and Adriana Verschoor for critical comments on the chapter. We also acknowledge support from NIH grant R01 GM06627 and NSF grant MCB 0110821 to B.F.M. and NIH grant RR41 01219 to Joachim Frank for support of the Wadsworth Center's Resource for Visualization of Biological Complexity. The authors are also grateful for technical support from the Wadsworth Center's Core facility for EM.

References

Allen, C., and Borisy, G. G. (1974). Structural polarity and directional growth of microtubules of *Chlamydomonas* flagella. *J. Mol. Biol.* **90,** 381–386.

Al-Amoudi, A., Chang, J. J., Leforestier, A., McDowall, A., Salamin, L. M., Norlen, L. P., Richter, K., Blanc, N. S., Studer, D., and Dubochet, J. (2004a). Cryo-electron microscopy of vitreous sections. *EMBO J.* **23,** 3583–3588.

Al-Amoudi, A., Norlen, L. P., and Dubochet, J. (2004b). Cryo-electron microscopy of vitreous sections of native biological cells and tissues. *J. Struct. Biol.* **148,** 131–135.

Arnal, I., Karsenti, E., and Hyman, A. A. (2000). Structural transitions at microtubule ends correlate with their dynamic properties in *Xenopus* egg extracts. *J. Cell Biol.* **149,** 767–774.

Baschong, W., and Stierhof, Y. D. (1998). Preparation, use, and enlargement of ultrasmall gold particles in immunoelectron microscopy. *Microsc. Res. Tech.* **42,** 66–79.

Baumeister, W., Grimm, R., and Walz, J. (1999). Electron tomography of molecules and cells. *Trends Cell Biol.* **9,** 81–85.

Biggins, S., and Walczak, C. E. (2003). Captivating capture: How microtubules attach to kinetochores. *Curr. Biol.* **13,** R449–R460.

Bohm, J., Frangakis, A. S., Hegerl, R., Nickell, S., Typke, D., and Baumeister, W. (2000). Toward detecting and identifying macromolecules in a cellular context: Template matching applied to electron tomograms. *Proc. Natl. Acad. Sci. USA* **97,** 14245–14250.

Brinkley, B. R., and Cartwright, J., Jr. (1971). Ultrastructural analysis of mitotic spindle elongation in mammalian cells *in vitro*. Direct microtubule counts. *J. Cell Biol.* **50,** 416–431.

Brinkley, B. R., and Stubblefield, E. (1966). The fine structure of the kinetochore of a mammalian cell *in vitro*. *Chromosoma* **19,** 28–43.

Burry, R. W., Vandre, D. D., and Hayes, D. M. (1992). Silver enhancement of gold antibody probes in pre-embedding electron microscopic immunocytochemistry. *J. Histochem. Cytochem.* **40,** 1849–1856.

Cassimeris, L., Rieder, C. L., Rupp, G., and Salmon, E. D. (1990). Stability of microtubule attachment to metaphase kinetochores in PtK1 cells. *J. Cell Sci.* **96**(Pt. 1), 9–15.

Chan, G. K., Liu, S. T., and Yen, T. J. (2005). Kinetochore structure and function. *Trends Cell Biol.* **15,** 589–598.

Cheeseman, I. M., Drubin, D. G., and Barnes, G. (2002). Simple centromere, complex kinetochore: Linking spindle microtubules and centromeric DNA in budding yeast. *J. Cell Biol.* **157,** 199–203.

Cheeseman, I. M., Enquist-Newman, M., Müller-Reichert, T., Drubin, D. G., and Barnes, G. (2001). Mitotic spindle integrity and kinetochore function linked by the Duo1p/Dam1p complex. *J. Cell Biol.* **152,** 197–212.

Cheeseman, I. M., Niessen, S., Anderson, S., Hyndman, F., Yates, J. R., III, Oegema, K., and Desai, A. (2004). A conserved protein network controls assembly of the outer kinetochore and its ability to sustain tension. *Genes Dev.* **18,** 2255–2268.

Chretien, D., Fuller, S. D., and Karsenti, E. (1995). Structure of growing microtubule ends: Two-dimensional sheets close into tubes at variable rates. *J. Cell Biol.* **129,** 1311–1328.

Chretien, D., Kenney, J. M., Fuller, S. D., and Wade, R. H. (1996). Determination of microtubule polarity by cryo-electron microscopy. *Structure* **4,** 1031–1040.

Ciferri, C., De Luca, J., Monzani, S., Ferrari, K. J., Ristic, D., Wyman, C., Stark, H., Kilmartin, J., Salmon, E. D., and Musacchio, A. (2005). Architecture of the human ndc80-hec1 complex, a critical constituent of the outer kinetochore. *J. Biol. Chem.* **280,** 29088–29095.

Cleveland, D. W., Mao, Y., and Sullivan, K. F. (2003). Centromeres and kinetochores: From epigenetics to mitotic checkpoint signaling. *Cell* **112,** 407–421.

Cooke, C. A., Bazett-Jones, D. P., Earnshaw, W. C., and Rattner, J. B. (1993). Mapping DNA within the mammalian kinetochore. *J. Cell Biol.* **120,** 1083–1091.

Cooke, C. A., Bernat, R. L., and Earnshaw, W. C. (1990). CENP-B: A major human centromere protein located beneath the kinetochore. *J. Cell Biol.* **110,** 1475–1488.

Cooke, C. A., Schaar, B., Yen, T. J., and Earnshaw, W. C. (1997). Localization of CENP-E in the fibrous corona and outer plate of mammalian kinetochores from prometaphase through anaphase. *Chromosoma* **106,** 446–455.

De Wulf, P., McAinsh, A. D., and Sorger, P. K. (2003). Hierarchical assembly of the budding yeast kinetochore from multiple subcomplexes. *Genes Dev.* **17,** 2902–2921.

DeLuca, J. G., Dong, Y., Hergert, P., Strauss, J., Hickey, J. M., Salmon, E. D., and McEwen, B. F. (2005). Hec1 and nuf2 are core components of the kinetochore outer plate essential for organizing microtubule attachment sites. *Mol. Biol. Cell.* **16,** 519–531.

DeLuca, J. G., Moree, B., Hickey, J. M., Kilmartin, J. V., and Salmon, E. D. (2002). hNuf2 inhibition blocks stable kinetochore-microtubule attachment and induces mitotic cell death in HeLa cells. *J. Cell Biol.* **159,** 549–555.

Desai, A., and Mitchison, T. J. (1997). Microtubule polymerization dynamics. *Annu. Rev. Cell Dev. Biol.* **13,** 83–117.

Ding, R., McDonald, K. L., and McIntosh, J. R. (1993). Three-dimensional reconstruction and analysis of mitotic spindles from the yeast, *Schizosaccharomyces pombe*. *J. Cell Biol.* **120,** 141–151.

Earnshaw, W. C., and Migeon, B. R. (1985). Three related centromere proteins are absent from the inactive centromere of a stable isodicentric chromosome. *Chromosoma* **92,** 290–296.

Emanuele, M. J., McCleland, M. L., Satinover, D. L., and Stukenberg, P. T. (2005). Measuring the stoichiometry and physical interactions between components elucidates the architecture of the vertebrate kinetochore. *Mol. Biol. Cell* **16,** 4882–4892.

Euteneuer, U., and McIntosh, J. R. (1981). Structural polarity of kinetochore microtubules in PtK1 cells. *J. Cell Biol.* **89,** 338–345.

Frangakis, A. S., Bohm, J., Forster, F., Nickell, S., Nicastro, D., Typke, D., Hegerl, R., and Baumeister, W. (2002). Identification of macromolecular complexes in cryoelectron tomograms of phantom cells. *Proc. Natl. Acad. Sci. USA* **99,** 14153–14158.

Frank, J., Radermacher, M., Penczek, P., Zhu, J., Li, Y., Ladjadj, M., and Leith, A. (1996). SPIDER and WEB: Processing and visualization of images in 3D electron microscopy and related fields. *J. Struct. Biol.* **116,** 190–199.

Gelfand, V. I., and Bershadsky, A. D. (1991). Microtubule dynamics: Mechanism, regulation, and function. *Annu. Rev. Cell Biol.* **7,** 93–116.

Gorbsky, G. J., Sammak, P. J., and Borisy, G. G. (1988). Microtubule dynamics and chromosome motion visualized in living anaphase cells. *J. Cell Biol.* **106,** 1185–1192.

Grimm, R., Barmann, M., Hackl, W., Typke, D., Sackmann, E., and Baumeister, W. (1997). Energy filtered electron tomography of ice-embedded actin and vesicles. *Biophys. J.* **72,** 482–489.

Grimm, R., Singh, H., Rachel, R., Typke, D., Zillig, W., and Baumeister, W. (1998). Electron tomography of ice-embedded prokaryotic cells. *Biophys. J.* **74,** 1031–1042.

Haimo, L. T., Telzer, B. R., and Rosenbaum, J. L. (1979). Dynein binds to and crossbridges cytoplasmic microtubules. *Proc. Natl. Acad. Sci. USA* **76,** 5759–5763.

Hainfeld, J. F. (1996). Labeling with nanogold and undecagold: Techniques and results. *Scanning Microsc. Suppl.* **10,** 309–322; discussion 322–325.

Hays, T. S., and Salmon, E. D. (1990). Poleward force at the kinetochore in metaphase depends on the number of kinetochore microtubules. *J. Cell Biol.* **110,** 391–404.

He, W., Cowin, P., and Stokes, D. L. (2003). Untangling desmosomal knots with electron tomography. *Science* **302,** 109–113.

Heald, R., Tournebize, R., Blank, T., Sandaltzopoulos, R., Becker, P., Hyman, A., and Karsenti, E. (1996). Self-organization of microtubules into bipolar spindles around artificial chromosomes in *Xenopus* egg extracts. *Nature* **382,** 420–425.

Heidemann, S. R., and McIntosh, J. R. (1980). Visualization of the structural polarity of microtubules. *Nature* **286,** 517–519.

Howard, J., and Hyman, A. A. (2003). Dynamics and mechanics of the microtubule plus-end. *Nature* **422,** 753–758.

Howe, M., McDonald, K. L., Albertson, D. G., and Meyer, B. J. (2001). HIM-10 is required for kinetochore structure and function on *Caenorhabditis elegans* holocentric chromosomes. *J. Cell Biol.* **153,** 1227–1238.

Howell, B. J., Hoffman, D. B., Fang, G., Murray, A. W., and Salmon, E. D. (2000). Visualization of Mad2 dynamics at kinetochores, along spindle fibers, and at spindle poles in living cells. *J. Cell Biol.* **150,** 1233–1250.

Hsieh, C. E., Leith, A., Mannella, C. A., Frank, J., and Marko, M. (2006). Towards high-resolution three-dimensional imaging of native mammalian tissue: Electron tomography of frozen-hydrated rat liver sections. *J. Struct. Biol.* **153,** 1–13.

Izard, J., McEwen, B. F., Barnard, R. M., Portuese, T., Samsonoff, W. A., and Limberger, R. J. (2004). Tomographic reconstruction of treponemal cytoplasmic filaments reveals novel bridging and anchoring components. *Mol. Microbiol.* **51,** 609–618.

Janosi, I. M., Chretien, D., and Flyvbjerg, H. (2002). Structural microtubule cap: Stability, catastrophe, rescue, and third state. *Biophys. J.* **83,** 1317–1330.

Jokelainen, P. T. (1967). The ultrastructure and spatial organization of the metaphase kinetochore in mitotic rat cells. *J. Ultrastruct. Res.* **19,** 19–44.

Kapoor, T. M., Lampson, M. A., Hergert, P., Cameron, L., Cimini, D., Salmon, E. D., McEwen, B. F., and Khodjakov, A. (2006). Chromosomes can congress to the metaphase plate before biorientation. *Science* **311,** 388–391.

Khodjakov, A., and Rieder, C. L. (2006). Imaging the division process in living tissue culture cells. *Methods* **38,** 2–16.

Kirschner, M., and Mitchison, T. (1986). Beyond self-assembly: From microtubules to morphogenesis. *Cell* **45,** 329–342.

Kline-Smith, S. L., and Walczak, C. E. (2004). Mitotic spindle assembly and chromosome segregation: Refocusing on microtubule dynamics. *Mol. Cell* **15,** 317–327.

Koster, A. J., Grimm, R., Typke, D., Hegerl, R., Stoschek, A., Walz, J., and Baumeister, W. (1997). Perspectives of molecular and cellular electron tomography. *J. Struct. Biol.* **120,** 276–308.

Kremer, J. R., Mastronarde, D. N., and McIntosh, J. R. (1996). Computer visualization of three-dimensional image data using IMOD. *J. Struct. Biol.* **116,** 71–76.

Li, Y., Bachant, J., Alcasabas, A. A., Wang, Y., Qin, J., and Elledge, S. J. (2002). The mitotic spindle is required for loading of the DASH complex onto the kinetochore. *Genes Dev.* **16,** 183–197.

Lucic, V., Foster, F., and Baumeister, W. (2005). Structural studies by electron tomography: From cells to molecules. *Annu. Rev. Biochem.* **74,** 833–865.

Maddox, P., Straight, A., Coughlin, P., Mitchison, T. J., and Salmon, E. D. (2003). Direct observation of microtubule dynamics at kinetochores in *Xenopus* extract spindles: Implications for spindle mechanics. *J. Cell Biol.* **162,** 377–382.

Maiato, H., Deluca, J., Salmon, E. D., and Earnshaw, W. C. (2004). The dynamic kinetochore-microtubule interface. *J. Cell Sci.* **117,** 5461–5477.

Maiato, H., Hergert, P., Pereira, S., Dong, Y., VandenBeldt, K. J., Rieder, C. L., and McEwen, B. F. (2006). The ultrastructure of the kinetochore and kinetochore fibre in *Drosophila* somatic cells. *Chromosoma* **115,** 469–480.

Mandelkow, E. M., Mandelkow, E., and Milligan, R. A. (1991). Microtubule dynamics and microtubule caps: A time-resolved cryo-electron microscopy study. *J. Cell Biol.* **114,** 977–991.

Mastronarde, D. N. (1997). Dual-axis tomography: An approach with alignment methods that preserve resolution. *J. Struct. Biol.* **120,** 343–352.

Mastronarde, D. N., McDonald, K. L., Ding, R., and McIntosh, J. R. (1993). Interpolar spindle microtubules in PTK cells. *J. Cell Biol.* **123,** 1475–1489.

McDonald, K. (1999). High-pressure freezing for preservation of high resolution fine structure and antigenicity for immunolabeling. *Methods Mol. Biol.* **117,** 77–97.

McDonald, K., Pickett-Heaps, J. D., McIntosh, J. R., and Tippit, D. H. (1977). On the mechanism of anaphase spindle elongation in *Diatoma vulgare*. *J. Cell Biol.* **74,** 377–388.

McDonald, K. L., O'Toole, E. T., Mastronarde, D. N., and McIntosh, J. R. (1992). Kinetochore microtubules in PTK cells. *J. Cell Biol.* **118,** 369–383.

McEwen, B. F., Heagle, A. B., Cassels, G. O., Buttle, K. F., and Rieder, C. L. (1997). Kinetochore fiber maturation in PtK1 cells and its implications for the mechanisms of chromosome congression and anaphase onset. *J. Cell Biol.* **137,** 1567–1580.

McEwen, B. F., Hsieh, C. E., Mattheyses, A. L., and Rieder, C. L. (1998). A new look at kinetochore structure in vertebrate somatic cells using high-pressure freezing and freeze substitution. *Chromosoma* **107,** 366–375.

McEwen, B. F., and Marko, M. (1999). Three-dimensional transmission electron microscopy and its application to mitosis research. *Methods Cell Biol.* **61,** 81–111.

McEwen, B. F., and Marko, M. (2001). The emergence of electron tomography as an important tool for investigating cellular ultrastructure. *J. Histochem. Cytochem.* **49,** 553–564.

McEwen, B. F., Marko, M., Hsieh, C. E., and Mannella, C. (2002). Use of frozen-hydrated axonemes to assess imaging parameters and resolution limits in cryoelectron tomography. *J. Struct. Biol.* **138,** 47–57.

McIntosh, J. R. (2005). Rings around kinetochore microtubules in yeast. *Nat. Struct. Mol. Biol.* **12,** 210–212.

McIntosh, J. R., and Euteneuer, U. (1984). Tubulin hooks as probes for microtubule polarity: An analysis of the method and an evaluation of data on microtubule polarity in the mitotic spindle. *J. Cell Biol.* **98,** 525–533.

McIntosh, J. R., and Landis, S. C. (1971). The distribution of spindle microtubules during mitosis in cultured human cells. *J. Cell Biol.* **49,** 468–497.

McIntosh, J. R., Nicastro, D., and Mastronarde, D. (2005). New views of cells in 3D: An introduction to electron tomography. *Trends Cell Biol.* **15,** 43–51.

McIntosh, J. R., Sisken, J. E., and Chu, L. K. (1979). Structural studies on mitotic spindles isolated from cultured human cells. *J. Ultrastruct. Res.* **66,** 40–52.

Meluh, P. B., and Koshland, D. (1995). Evidence that the MIF2 gene of Saccharomyces cerevisiae encodes a centromere protein with homology to the mammalian centromere protein CENP-C. *Mol. Biol. Cell.* **6,** 793–807.

Miranda, J. J., De Wulf, P., Sorger, P. K., and Harrison, S. C. (2005). The yeast DASH complex forms closed rings on microtubules. *Nat. Struct. Mol. Biol.* **12,** 138–143.

Mitchison, T., Evans, L., Schulze, E., and Kirschner, M. (1986). Sites of microtubule assembly and disassembly in the mitotic spindle. *Cell* **45,** 515–527.

Mitchison, T., and Kirschner, M. (1984). Dynamic instability of microtubule growth. *Nature* **312,** 237–242.

Mitchison, T. J., and Salmon, E. D. (1992). Poleward kinetochore fiber movement occurs during both metaphase and anaphase-A in newt lung cell mitosis. *J. Cell Biol.* **119,** 569–582.

Mitchison, T. J., and Salmon, E. D. (2001). Mitosis: A history of division. *Nat. Cell Biol.* **3,** E17–E21.

Niedermann, G., Grimm, R., Geier, E., Maurer, M., Realini, C., Gartmann, C., Soll, J., Omura, S., Rechsteiner, M. C., Baumeister, W., and Eichmann, K. (1997). Potential immunocompetence of proteolytic fragments produced by proteasomes before evolution of the vertebrate immune system. *J. Exp. Med.* **186,** 209–220.

O'Toole, E. T., McDonald, K. L., Mantler, J., McIntosh, J. R., Hyman, A. A., and Müller-Reichert, T. (2003). Morphologically distinct microtubule ends in the mitotic centrosome of *Caenorhabditis elegans*. *J. Cell Biol.* **163,** 451–456.

O'Toole, E. T., Winey, M., and McIntosh, J. R. (1999). High-voltage electron tomography of spindle pole bodies and early mitotic spindles in the yeast *Saccharomyces cerevisiae*. *Mol. Biol. Cell* **10,** 2017–2031.

Obuse, C., Iwasaki, O., Kiyomitsu, T., Goshima, G., Toyoda, Y., and Yanagida, M. (2004). A conserved Mis12 centromere complex is linked to heterochromatic HP1 and outer kinetochore protein Zwint-1. *Nat. Cell Biol.* **6,** 1135–1141.

Penczek, P., Marko, M., Buttle, K., and Frank, J. (1995). Double-tilt electron tomography. *Ultramicroscopy* **60,** 393–410.

Rath, B. K., Hegerl, R., Leith, A., Shaikh, T. R., Wagenknecht, T., and Frank, J. (2003). Fast 3D motif search of EM density maps using a locally normalized cross-correlation function. *J. Struct. Biol.* **144,** 95–103.

Rattner, J. B., Rao, A., Fritzler, M. J., Valencia, D. W., and Yen, T. J. (1993). CENP-F is a.ca 400 kDa kinetochore protein that exhibits a cell-cycle dependent localization. *Cell Motil. Cytoskeleton* **26,** 214–226.

Reipert, S., Fischer, I., and Wiche, G. (2004). High-pressure freezing of epithelial cells on sapphire coverslips. *J. Microsc.* **213,** 81–85.

Rieder, C. L. (1981). The structure of the cold-stable kinetochore fiber in metaphase PtK1 cells. *Chromosoma* **84,** 145–158.

Rieder, C. L., and Alexander, S. P. (1990). Kinetochores are transported poleward along a single astral microtubule during chromosome attachment to the spindle in newt lung cells. *J. Cell Biol.* **110,** 81–95.

Rieder, C. L., and Bowser, S. S. (1985). Correlative immunofluorescence and electron microscopy on the same section of epon-embedded material. *J. Histochem. Cytochem.* **33,** 165–171.

Rieder, C. L., Bowser, S. S., Cole, R., Rupp, G., Peterson, A., and Alexander, S. P. (1990). Diffuse kinetochores and holokinetic anaphase chromatin movement during mitosis in the hemipteran Agallia constricta (leafhopper) cell line AC-20. *Cell Motil. Cytoskeleton* **15,** 245–259.

Rieder, C. L., and Cassels, G. (1999). Correlative light and electron microscopy of mitotic cells in monolayer cultures. *Methods Cell Biol.* **61,** 297–315.

Rieder, C. L., and Salmon, E. D. (1994). Motile kinetochores and polar ejection forces dictate chromosome position on the vertebrate mitotic spindle. *J. Cell Biol.* **124,** 223–233.

Rieder, C. L., and Salmon, E. D. (1998). The vertebrate cell kinetochore and its roles during mitosis. *Trends Cell Biol.* **8,** 310–318.

Rieder, C. L., Schultz, A., Cole, R., and Sluder, G. (1994). Anaphase onset in vertebrate somatic cells is controlled by a checkpoint that monitors sister kinetochore attachment to the spindle. *J. Cell Biol.* **127,** 1301–1310.

Roos, U. P. (1973). Light and electron microscopy of rat kangaroo cells in mitosis. II. Kinetochore structure and function. *Chromosoma* **41,** 195–220.

Saitoh, H., Tomkiel, J., Cooke, C. A., Ratrie, H., 3rd, Maurer, M., Rothfield, N. F., and Earnshaw, W. C. (1992). CENP-C, an autoantigen in scleroderma, is a component of the human inner kinetochore plate. *Cell* **70,** 115–125.

Salmon, E. D. (1975). Pressure-induced depolymerization of spindle microtubules. I. Changes in birefringence and spindle length. *J. Cell Biol.* **65,** 603–614.

Sawin, K. E., and Mitchison, T. J. (1991). Poleward microtubule flux mitotic spindles assembled *in vitro*. *J. Cell Biol.* **112,** 941–954.

Shah, J. V., Botvinick, E., Bonday, Z., Furnari, F., Berns, M., and Cleveland, D. W. (2004). Dynamics of centromere and kinetochore proteins; implications for checkpoint signaling and silencing. *Curr. Biol.* **14,** 942–952.

Sillje, H. H., and Nigg, E. A. (2006). Purification of mitotic spindles from cultured human cells. *Methods* **38,** 25–28.

Skibbens, R. V., Skeen, V. P., and Salmon, E. D. (1993). Directional instability of kinetochore mobility during chromosome congression and segregation in mitotic newt lung cells: A push-pull mechanism. *J. Cell Biol.* **122,** 859–875.

Taylor, K. A., Schmitz, H., Reedy, M. C., Goldman, Y. E., Franzini-Armstrong, C., Sasaki, H., Tregear, R. T., Poole, K., Lucaveche, C., Edwards, R. J., Chen, L. F., Winkler, H., *et al.* (1999). Tomographic 3D reconstruction of quick-frozen, Ca^{2+}-activated contracting insect flight muscle. *Cell* **99,** 421–431.

Taylor, K. A., Tang, J., Cheng, Y., and Winkler, H. (1997). The use of electron tomography for structural analysis of disordered protein arrays. *J. Struct. Biol.* **120,** 372–386.

VandenBeldt, K. J., Barnard, R. M., Hergert, P. J., Meng, X., Maiato, H., and McEwen, B. F. (2006). Kinetochores use a novel mechanism for coordinating the dynamics of individual microtubules. *Curr. Biol.* **16,** 1217–1223.

Vandre, D. D., and Burry, R. W. (1992). Immunoelectron microscopic localization of phosphoproteins associated with the mitotic spindle. *J. Histochem. Cytochem.* **40,** 1837–1847.

Wadsworth, P., Shelden, E., Rupp, G., and Rieder, C. L. (1989). Biotin-tubulin incorporates into kinetochore fiber microtubules during early but not late anaphase. *J. Cell Biol.* **109,** 2257–2265.

Walczak, C. E., Gan, E. C., Desai, A., Mitchison, T. J., and Kline-Smith, S. L. (2002). The microtubule-destabilizing kinesin XKCM1 is required for chromosome positioning during spindle assembly. *Curr. Biol.* **12,** 1885–1889.

Walz, J., Tamura, T., Tamura, N., Grimm, R., Baumeister, W., and Koster, A. J. (1997). Tricorn protease exists as an icosahedral supermolecule *in vivo*. *Mol. Cell* **1,** 59–65.

Wang, H. W., and Nogales, E. (2005). Nucleotide-dependent bending flexibility of tubulin regulates microtubule assembly. *Nature* **435,** 911–915.

Warburton, P. E., Cooke, C. A., Bourassa, S., Vafa, O., Sullivan, B. A., Stetten, G., Gimelli, G., Warburton, D., Tyler-Smith, C., Sullivan, K. F., Poirier, G. G., and Earnshaw, W. C. (1997). Immunolocalization of CENP-A suggests a distinct nucleosome structure at the inner kinetochore plate of active centromeres. *Curr. Biol.* **7,** 901–904.

Wei, R. R., Sorger, P. K., and Harrison, S. C. (2005). Molecular organization of the Ndc80 complex, an essential kinetochore component. *Proc. Natl. Acad. Sci. USA* **102,** 5363–5367.

Westermann, S., Avila-Sakar, A., Wang, H. W., Niederstrasser, H., Wong, J., Drubin, D. G., Nogales, E., and Barnes, G. (2005). Formation of a dynamic kinetochore-microtubule interface through assembly of the Dam1 ring complex. *Mol. Cell* **17,** 277–290.

Westermann, S., Cheeseman, I. M., Anderson, S., Yates, J. R., III, Drubin, D. G., and Barnes, G. (2003). Architecture of the budding yeast kinetochore reveals a conserved molecular core. *J. Cell Biol.* **163,** 215–222.

Westermann, S., Wang, H. W., Avila-Sakar, A., Drubin, D. G., Nogales, E., and Barnes, G. (2006). The Dam1 kinetochore ring complex moves processively on depolymerizing microtubule ends. *Nature* **440,** 565–569.

Wigge, P. A., and Kilmartin, J. V. (2001). The Ndc80p complex from Saccharomyces cerevisiae contains conserved centromere components and has a function in chromosome segregation. *J. Cell Biol.* **152,** 349–360.

Wollman, R., Cytrynbaum, E. N., Jones, J. T., Meyer, T., Scholey, J. M., and Mogilner, A. (2005). Efficient chromosome capture requires a bias in the 'search-and-capture' process during mitotic-spindle assembly. *Curr. Biol.* **15,** 828–832.

Yao, X., Anderson, K. L., and Cleveland, D. W. (1997). The microtubule-dependent motor centromere-associated protein E (CENP-E) is an integral component of kinetochore corona fibers that link centromeres to spindle microtubules. *J. Cell Biol.* **139,** 435–447.

Zhai, Y., Kronebusch, P. J., and Borisy, G. G. (1995). Kinetochore microtubule dynamics and the metaphase-anaphase transition. *J. Cell Biol.* **131,** 721–734.

Ziese, U., Janssen, A. H., Murk, J. L., Geerts, W. J., Van der Krift, T., Verkleij, A. J., and Koster, A. J. (2002). Automated high-throughput electron tomography by pre-calibration of image shifts. *J. Microsc.* **205,** 187–200.

CHAPTER 12

Electron Microscopic Analysis of the Leading Edge in Migrating Cells

Tatyana Svitkina
Department of Biology
University of Pennsylvania, Philadelphia
Pennsylvania 19104

Actin polymerization is one of the major mechanisms that drives cellular motility. The most obvious application of this mechanism is the leading edge protrusion in motile cells. Knowledge of the structural organization of actin filaments in protrusive organelles, lamellipodia and filopodia, is instrumental for understanding the mechanisms of leading edge protrusion and of actin polymerization-based motility, in general. In this chapter, I review the contribution of electron microscopy (EM) to progress in this field and describe an approach to study the cytoskeleton at the leading edge of cultured cells by platinum replica EM after detergent extraction, triple fixation, critical point drying, and shadowing with platinum and carbon. This method provides a three-dimensional (3D) view of cytoskeletal architecture and is

0091-679X/07 $35.00
DOI: 10.1016/S0091-679X(06)79012-4

compatible with immunostaining and polarity determination of the actin filaments. Most importantly, the high quality and robustness of this EM technique make it easily compatible with light microscopy (LM) and allows one to correlate the organization of the cytoskeleton with both the behavior of a living cell and the localization of ectopically expressed proteins in the same cells.

I. Introduction

It is well established that cell locomotion plays vital roles in many aspects of the physiology of multicellular organisms such as development, morphogenesis, immunity, wound healing, and, adversely, cancer. Our understanding of the mechanisms of cell locomotion has progressed hand-in-hand with the technical advance of microscopy. Initially, the development of optical microscopes let us learn that cells actually can move, which happened quite early, almost simultaneously with the formulation of Schwann's cell theory in 1839. Electron microscopy (EM) subsequently provided information essential for understanding the mechanisms of motility and for developing a concept of the cytoskeleton, which we now know underlies all types of cell motility. However, the combination of light microscopy (LM) and EM remains the most powerful way to resolve many problems because the advantages and disadvantages of each microscopy technique are quite complementary. Thus, LM can be applied to live cells, but it is limited in resolution. Conversely, EM provides high resolution but is not applicable to living cells.

LM was obviously responsible for the first steps in studying cell motility. Initially, the most favorable objects were unicellular amoeboid organisms, which were large enough to be observed with the optics then available. Amoeboid motility of white blood cells from multicellular organism was discovered shortly thereafter, but these cells were too small for detailed studies by available technology. Felix Dujardin in 1838 described amoeboid motility as: "the extension of a pseudopod, its attachment at the tip, followed by its contraction to pull up the rest of the anima1,"[1] a concept that has survived until the present (Lauffenburger and Horwitz, 1996). Remarkably, Dujardin also gave a mechanistic explanation of amoeboid motility as being powered by the activity of a contractile and extensive substance (Abercrombie, 1977). It took over a century to understand the molecular nature of contractility and extensibility of this substance that was later termed protoplasm. The first step of the motility cycle—the extension of a pseudopod—and the role of EM in understanding the mechanisms of pseudopodia protrusion are the foci of this chapter.

[1] "RHIZOPODA." *LoveToKnow 1911 Online Encyclopedia.* © 2003, 2004 LoveToKnow. http://65.1911encyclopedia.org/R/RH/RHIZOPODA.htm

II. Understanding the Mechanisms of Leading Edge Protrusion: Complementary Roles of LM and EM

Intensive investigation of the cell locomotion in general and leading edge protrusion in particular was greatly stimulated by several significant technological accomplishments during first half of twentieth century. These included the invention of tissue culture by Ross Harrison (1907), which became a common technique by the late 1940s; development of time-lapse microscopy in late 1920s by Ronald Canti, which was later combined with phase-contrast microscopy by Arthur Hughes (Fell, 1958); and invention of an electron microscope by Ernst Ruska in 1933 followed by the development of commercial models by Siemens and RCA in 1939–1940. Further development and interaction of these key techniques brought us to the current state of understanding cell motility and the role of the cytoskeleton in this process.

If any individual should be credited with initiating cell motility research as a field, it would be Michael Abercrombie, who used time-lapse, phase-contrast microscopy to observe the motility of fibroblasts in cell culture and combined it with comprehensive, quantitative analysis to describe the behavior of individual cells and cell populations. His studies led to a principal conclusion that "a ruffled membrane tends to control the direction of whatever mechanism it is that moves the whole cell" (Abercrombie and Ambrose, 1958). Abercrombie later found that the term "ruffled membrane" does not adequately define the organelle directing the cell motility and introduced the term "lamellipodium," which combined flat anterior protrusions and ruffles, structures formed by the uplifting of flat protrusions or directly on the cell's dorsal surface (Abercrombie *et al.*, 1970). Another type of protrusive organelle, filopodia or microspikes, had been known since the mid-1800s, when the morphologies of pseudopodia characteristic of various Protozoa were classified.

Keith Porter was the first to bring EM to cell biology and to make the first informative electron micrograph of a cell (Porter *et al.*, 1945). This famous chick fibroblast was fixed with vapors of osmium tetroxide as a sole fixative, then dried and viewed as a whole cell. The use of whole mounts, however, was then put aside for a long time, as thin sectioning was a more universally applicable technique. Nonetheless, the osmium fixation in the early whole mounts clearly revealed diverse subcellular membranous organelles and allowed Porter *et al.* to image the nucleus, mitochondria, and endoplasmic reticulum, but the rest of the cell looked disappointedly empty until Karl Wohlfarth-Bottermann (1962) got an idea of applying additional heavy metals, such as phosphotungstic acid and uranium compounds, as contrasting and fixing agents. Using this approach, he discovered the first cytoskeletal structure, microfilaments bundles, and even showed their dynamic behavior by correlating the presence of fibrils with the contractile activity of an acellular slime mold *Physarum polycephalum* (Wohlfarth-Bottermann, 1964). The actin-containing nature of these filaments was immediately suspected and later proven using arrowhead decoration of filaments by heavy meromyosin (HMM)

(Ishikawa *et al.*, 1969). This technique, which was developed by Hugh Huxley (1963), allowed not only the recognition of actin filaments but also a determination of their polarity. The introduction of aldehyde fixatives, especially glutaraldehyde (Sabatini *et al.*, 1963), led to an explosion in publications describing fibrillar elements in the cytoplasm and permitted the discovery that microtubules were essentially ubiquitous in eukaryotic cells (Ledbetter and Porter, 1963). A few years later, intermediate filaments were recognized as a distinct type of filaments (Ishikawa *et al.*, 1968), thus completing the set of major cytoskeletal elements.

Cytoskeleton was a long-sought structural system for driving cell motility. The positioning of cytoskeletal fibers, especially actin, close to areas of cellular activity was consistent with this idea, but the actual mechanism of motility remained elusive for about another decade. During that time, the speculations circulated entirely around an idea of contractility because nobody could imagine that myosin might not be needed for actin to do its job. On this background, Norman Wessels and his group investigated thin sections of aldehyde/osmium-fixed cultured cells, where they recognized that protrusive organelles have an unusual cytoskeletal organization; they contain a nonoriented filament "lattice," rather than bundles, as in other regions of the cell (Wessels *et al.*, 1973; Yamada *et al.*, 1971). Since these data were essentially incompatible with the dominant belief that contractility drives actin-dependent motility, there was concern that the lattice might be a fixation artifact. Although osmium was later shown to disrupt actin filaments (Maupin-Szamier and Pollard, 1978; Small, 1981), and thin sectioning is definitely not extremely beneficial for obtaining a good view of the lamellipodial network, these findings pointed in a right direction that opened the recognition of another type of actin-based motility that might take place during protrusion.

A solution to these puzzles came shortly after, when Lewis Tilney studied the acrosome reaction in echinoderm sperm and discovered that actin polymerization per se can drive cell protrusion (Tilney *et al.*, 1973). This was a turning point in the field of cell motility and eventually made actin polymerization-driven motility a field on its own. During the acrosomal reaction, monomeric actin, which looked like amorphous material in EM sections, was reorganized within 30 sec into a 90-μm-long bundle of actin filaments. Remarkably, the same paper showed quite clear evidence about the polarity of the actin filaments in these bundles; their "barbed" or fast polymerizing ends were oriented toward the tip of the growing process. This finding definitely pointed toward the idea that actin might be inserted at the tip of the process during its polymerization, but the authors refrained from this far-reaching conclusion, probably, because the paper was already sufficiently heretical. Few years later, the point was nailed down and illustrated by EM images of incredible beauty, showing that actin bundles polymerize by adding subunits at the growing tip of the acrosomal process (Tilney and Kallenbach, 1979).

It remained an open question whether these findings could be applied to conventional tissue culture cells. Although filopodia resembled acrosomal processes in having an internal bundle of actin filaments, the structural organization of the lamellipodial cytoskeleton was largely irresolvable in thin-sectioned samples.

This prompted the field to return to whole-mount preparations, for which cells are grown on or transferred to EM grids and, after fixation and drying, viewed *in toto*. With this approach, thin regions of well-spread cells could be resolved, giving an idea of the cytoskeleton organization in its entirety (Buckley and Porter, 1967). The major problem for high-resolution studies by the whole-mount approach was the distortions caused by air-drying. For fragile cellular structures, surface tension is an enormous force that flattens them as the descending water/air interface goes through a cell during drying. The critical point drying procedure was introduced to help preserve three-dimensional (3D) cell structure (Anderson, 1951), and it allowed much better images of whole-mount cells (Buckley and Porter, 1975), but still the filament organization in lamellipodia was not resolved; it appeared as a feltlike meshwork.

A technique that allowed the clear visualization of individual actin filaments in lamellipodia and filopodia was developed by Vic Small; it involved detergent extraction of living cells followed by glutaraldehyde fixation, negative staining with uranyl acetate, and air-drying of the remaining cytoskeleton (Small *et al.*, 1978). Although it was a step backward, because of losing three-dimensionality, this approach revealed the diagonal network of actin filaments in lamellipodia and the parallel bundles in filopodia. These became especially clear when, in the later version of the technique (Hoglund *et al.*, 1980; Small, 1981), a mixture of Triton X-100 and glutaraldehyde was used instead of Triton X-100 alone as the first step of the procedure. Importantly, the polarity of actin filaments determined by myosin subfragment 1 (S1) decoration showed barbed ends facing forward in both lamellipodia and filopodia (Small *et al.*, 1978). These results strongly suggested that protrusion involves local polymerization of actin at the anterior leading edge. This idea was explicitly articulated by Uno Lindberg *et al.* (1981) and later proven by Yu-Li Wang, who observed actin dynamics using fluorescence microscopy to visualize labeled actin that had been microinjected into living cells and photobleached to mark already polymerized protein. He formulated the idea of actin filament treadmilling in lamellipodia (Wang, 1985), demonstrating *in vivo* the principle that had been so elegantly predicted by Albrecht Wegner (1977).

It still took quite a while to debate whether the force developed by actin polymerization could be sufficient to drive protrusion or activity of a motor is required for this process. A key to this puzzle was provided by an intracellular parasite, *Listeria monocytogenes*. Tilney and Portnoy (1989) found in 1989 that *Listeria* moves in a rocketlike fashion by polymerizing actin behind itself. This discovery eventually led to the reconstitution of *Listeria* motility in a mixture of pure proteins that lacked myosins, thus proving that actin polymerization per se is sufficient to drive motility (Loisel *et al.*, 1999). However, even now it is unclear whether any of numerous species of myosin contribute to protrusion in the context of a cell. Another puzzling aspect of the insertional actin polymerization, which delayed its acceptance, was to understand how actin can polymerize and push at the same time. This problem is not yet so much solved as simply familiar, as we have gotten accustomed to it. However, the Brownian ratchet model, proposed by

Moginer and Oster (1996), suggests that thermal fluctuations of the membrane and/or actin filaments can explain the phenomenon. These well-developed theories have made it much easier for us to think about the whole concept.

Initially, no distinction was made between the modes of protrusion in filopodia and lamellipodia. The actin filament arrays in both organelles appeared similar in a sense that they both contained long actin filaments, although they were situated at different orientation: parallel filaments in filopodia and a diagonal network in lamellipodia (Small, 1981; Small *et al.*, 1994). This view was strongly influenced by the negative-staining/air-drying EM technique, which specifically accentuated long, ordered actin filaments in the lamellipodial network, but because of sample flattening, it was less efficient in detecting short and randomly oriented actin filaments or filament–filament interactions. Alternative techniques preserving three-dimensionality, such as embedding, critical point drying, and freeze-drying, were better suited to the detection of short filaments and cross-links between filaments, but they suffered from other problems. In particular, the osmium fixation used for embedded samples led to actin filament degradation (Maupin-Szamier and Pollard, 1978; Small, 1981), while thin sectioning might deemphasize long filaments; dehydration in organic solvents, which was required for embedding and critical point drying, caused kinking of actin filaments (Small, 1981); critical point drying in conditions of incomplete water removal caused filaments to fuse (Ris, 1985); freeze-drying required a tricky balance between retaining too much or too little water to avoid the artifacts caused by ice crystals and air-drying, respectively (Small *et al.*, 1994). Because of these problems, any interactions between actin filaments in lamellipodia, even in very suggestive images (Hartwig and Shevlin, 1986; Heuser and Kirschner, 1980; Schliwa and van Blerkom, 1981), were considered artifacts and ignored. Consequently, the original model for protrusion, the filament treadmilling model, was based on the most visually appealing images of negatively stained cells and did not discriminate between filopodia and lamellipodia. This model proposed that all filaments elongated at the membrane, thus creating the pushing force, while the same filaments shortened at the rear, thus maintaining the steady state (Small, 1994).

A serious challenge to the filament treadmilling model came from LM. Theriot and Mitchison (1991) analyzed the dynamics of photoactivatable actin in lamellipodia of live cells. On one hand, they confirmed the filament treadmilling model by showing that the advance of the leading edge paralleled the assembly of actin at the cell front. On the other hand, they showed that the disassembly rate of F-actin in lamellipodia was too high to be consistent with long filaments. As an alternative, they proposed the nucleation-release model in which filaments are constantly nucleated at the leading edge and released after a short period of elongation, so the actin network in lamellipodia would consist of short cross-linked filaments; the numerous ends of such filaments would allow fast depolymerization. This proposal created a controversy between the structure observed by EM and the dynamics observed by LM of actin in lamellipodia. In contrast, the dynamics of actin in filopodia was quite slow (Mallavarapu and Mitchison, 1999) and totally consistent with the filament treadmilling model.

The development of an improved procedure for visualizing whole-mount cytoskeletal preparations (Svitkina and Borisy, 1998; Svitkina *et al.*, 1995) allowed us to resolve this controversy by showing that lamellipodia contain an extensively branched network of actin filaments (Svitkina and Borisy, 1999; Svitkina *et al.*, 1997). In this network, many filaments, especially those close to the leading edge, were rather short; they were bound to the sides of other filaments at their pointed ends and had their barbed ends facing the leading edge. An almost constant 70° angle of branching caused the majority of filaments to attain the diagonal orientation (Maly and Borisy, 2001), similar to that observed in negatively stained images (Small, 1981). Discovery that the Arp2/3 complex can nucleate actin filaments in branched configurations (Mullins *et al.*, 1998) and is present at the branch points in lamellipodia (Svitkina and Borisy, 1999) provided a molecular mechanism for actin filament branching. Identification of many other proteins involved in actin dynamics led eventually to the formulation of a dendritic nucleation model for lamellipodial protrusion (Pollard and Borisy, 2003). Some features of this model are unique, but others are a version of the nucleation-release hypothesis (Theriot and Mitchison, 1991).

III. Platinum Replica EM of the Leading Edge Cytoskeleton

In this section, I describe the platinum replica method for EM of critical point dried samples and present the application of this technique for studying the organization of the cytoskeleton in cultured cells. Much of the methodology results from the quite strict requirements regarding samples that can be introduced into an EM: they should be nonvolatile under the electron beam and must be sufficiently transparent to it, but they must have enough contrast to display features of interest as an image. An obviously important requirement is to maximally avoid artifacts. For analysis of the cytoskeleton in cultured cells, we employ a version of the whole-mount technique in which detergent extraction is used to decrease sample opacity; a three-step chemical fixation helps to preserve the cytoskeleton against distortions during subsequent procedures; dehydration by organic solvents removes cellular water, the major volatile component; critical point drying preserves the cell's 3D organization; and platinum coating creates contrast. None of these individual steps is novel; they have been developed by many great electron microscopists over the years. Our contribution has been to find a good combination of these steps (Svitkina *et al.*, 1984) and gradually to improve it by minimizing the common artifacts (Svitkina *et al.*, 1995) so that the resulting protocol is reliable, simple, and consistently produces excellent results with high yield (Svitkina and Borisy, 1998, 2006). Thus, the major advantage of this approach is its ability to facilitate large-scale EM analyses that are fast and relatively easy, resulting in reliable, high-quality images of the cytoskeleton. It should be noted that this method is not designed to

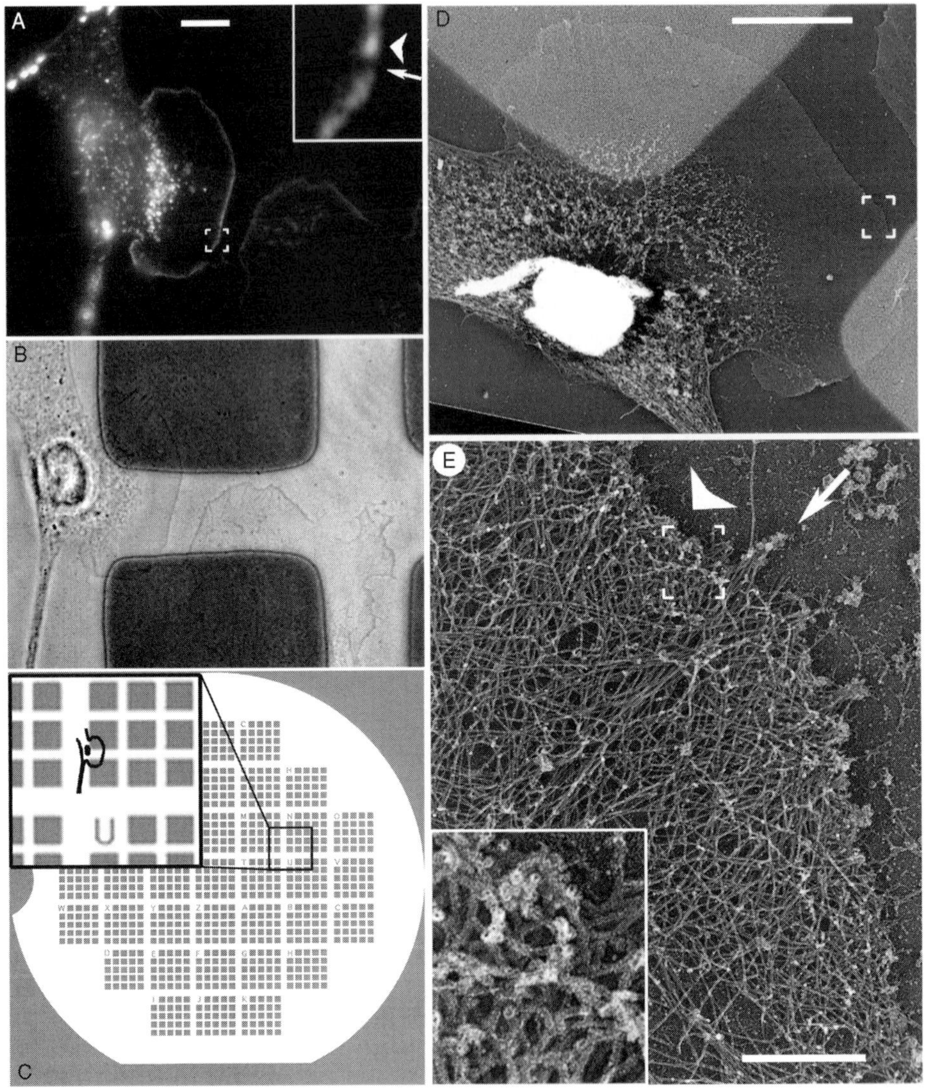

Fig. 1 Correlative fluorescence and EM of mouse melanoma B16F1 cell expressing EGFP-tagged β-subunit of heterodimeric capping protein, an important barbed end-capping protein in lamellipodia (Mejillano *et al.*, 2004). Transiently transfected cells were extracted with 1% Triton X-100 in PEM buffer supplemented by 2% PEG (MW = 35 kDa) and 2-μM phalloidin for 5 min at room temperature, rinsed with PEM buffer containing 1-μM phalloidin and, while in this buffer, fluorescence images of several cells were taken (~30 min total). Cells were subsequently fixed and processed for EM. (A) Fluorescence microscopy shows linear localization of EGFP-capping protein at the lamellipodial leading edge, as well as numerous puncta in the cell body. High magnification of the boxed region (inset) shows uneven distribution of capping protein along the leading edge with alternating bright (arrowhead) and dim (arrow) regions. (B) Phase-contrast image of the same cell growing on a labeled coverslip with dark regions of gold coating. (C) Diagram of the pattern generated by gold coating through a finder EM grid.

perform studies at molecular resolution; and it does not reveal, for example, individual subunits in cytoskeletal fibers or the native shape of individual proteins. Instead, the goal of this technique is to bridge the gap between low-resolution LM and high-resolution structural analysis of proteins or protein complexes. This technique is best suited to understanding the details of cytoskeleton architecture and their relation to live-cell activity.

The reliability of the protocol makes it easily compatible with light microscopic observations, so it becomes quite simple to correlate the living state directly with cytoskeletal architecture in the same cell (Figs. 1 and 2). Such correlative analysis is extremely important from at least two points of view. First, it provides an excellent way to control for some kinds of potential artifacts. For example, we have compared the distribution of fluorescently labeled cytoskeletal proteins in living cells and in the same cells in EM images and found that light and EM images are superimposable, suggesting that no significant distortions occurred during sample preparation for EM. This kind of analysis was performed for actin (Svitkina *et al.*, 2003; Verkhovsky *et al.*, 2003), myosin (Verkhovsky *et al.*, 1995), microtubules (Svitkina and Borisy, 1998), fascin (Svitkina *et al.*, 2003), and several other cytoskeletal proteins (unpublished data). Second, correlative microscopy provides an opportunity to establish functional connections between the cytoskeletal organization and the cell's motile behavior (Fig. 2) or the presence of certain cytoskeletal components (Fig. 4).

Step-by-step instructions on how to prepare cultured cells for platinum replica EM, and approaches for combining this EM protocol with LM, immunostaining, and actin filament polarity determination are described in following sections.

A. Cell Culture

The platinum replica technique requires cells growing on glass coverslips, which is an important advantage compared to other whole-mount techniques, both because glass is a much better support for cell spreading than the hydrophobic and pliable plastic films used in other techniques and because glass has the best optical characteristics, which is important for correlative LM and EM.

For routine EM, we grow cells on small ($\sim 7 \times 7\ mm^2$) trapezoidal glass coverslips cut from larger coverslips with a diamond pencil. This size is sufficient to fill

---

Inset shows position of the cell. (D) Low-magnification EM image of the cell (shown in inverse contrast). Boxed region is enlarged in E. (E) Actin organization at the leading edge of the cell. Lamellipodium with branched actin filament organization (arrowhead) corresponds to a bright region in the fluorescence image (arrowhead in A), whereas a filopodium (arrow) corresponds to a region with low fluorescence intensity (arrow in A) consistent with an idea that filopodial filaments elongate persistently without capping. Normally, filaments in filopodia are tightly bundled, but here they are separated, likely due to dissociation of a cross-linking protein, fascin, from the bundle during the imaging step (Svitkina *et al.*, 2003). Boxed region is enlarged in inset and shows branching of actin filaments. Scale bars = 10 μm (A, D) and 0.5 μm (E).

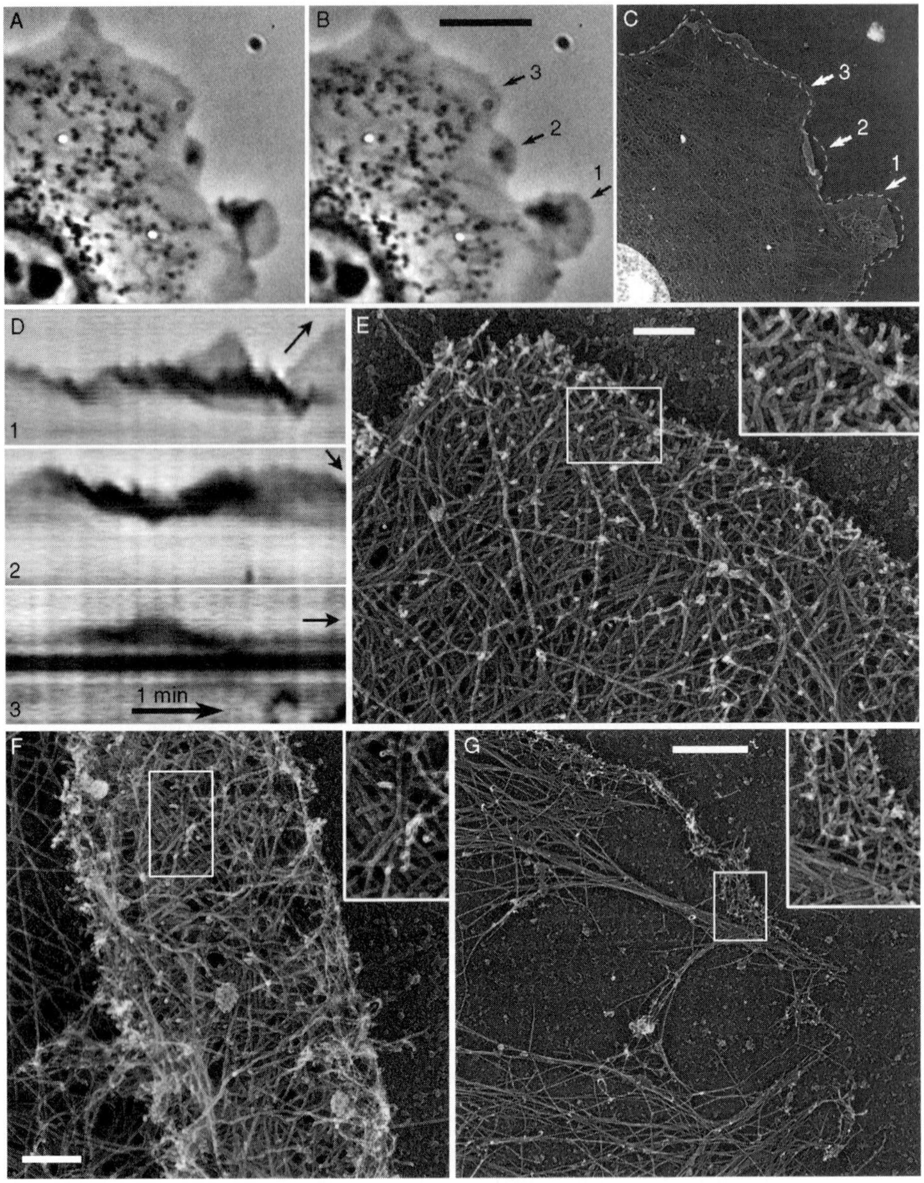

Fig. 2 Correlative microscopy of leading edge dynamics and cytoskeletal organization in Rat-2 rat embryo fibroblast. Cells were imaged by phase-contrast time-lapse microscopy with 4-sec time intervals. After the last frame, the medium was aspirated and replaced with 1% Triton X-100 in PEM buffer supplemented by 2-μM phalloidin; after 5-min extraction, cells were rinsed with PEM containing 1-μM phalloidin, fixed and processed for EM. (A, B) Frames from time-lapse sequence showing the last live-cell image (B) and an image 12 sec earlier (A). Numbered arrows point to protruded (1), retracted (2), and stationary (3) positions of the cell edge. (C) Low-magnification EM image of the same cell overlaid

at least six EM grids and even for a single cell population, it contains enough material for conclusive analysis. The trapezoidal shape of the coverslips (square with one oblique side) is helpful for easy determination of the cell-containing side because of the absence of mirror symmetry. Commercially available 12-mm round coverslips are also acceptable, but require extra time for dehydration, critical point drying, and the handling of replicas (see below).

For correlative microscopy, cells are cultured on $22 \times 22\ \text{mm}^2$ coverslips containing internal reference marks and mounted over a hole at the bottom of a plastic dish. Reference marks facilitate the relocalization of cells for subsequent EM. To produce marks, we evaporate gold onto a coverslip through an EM finder grid (Svitkina and Borisy, 2006), a variety of which is commercially available. The thickness of gold may be varied, but a thin layer, which gives a semitransparent, purple color, helps to give a clean working surface under the grid that is not contaminated with gold grains. Coated coverslips are baked at 160 °C overnight to stabilize the gold layer. Cells for correlative microscopy are selected within clear, uncoated glass areas, corresponding to the solid parts of the finder grid (Fig. 1). Etched coverslips, which are available commercially, are not suitable for this approach because the etched marks are lost when the replicas are released from the glass and placed on EM grids.

While various types of culture chambers can be used for live-cell observations, we use open dishes, which provide an opportunity for quick exchange of the culture medium to an extraction solution; this is important for correlative microscopy applications. However, this requires well-buffered media that prevent pH shift during observations. Perfusion chambers that allow for quick media exchange are also an option. We use marked coverslips mounted with silicon vacuum grease onto circular, 18-mm holes in the bottom of the 35-mm tissue culture dishes. Vacuum

---

with the cell outline, as in (B). Additional protrusion and retraction events occurred between acquisition of the last live image and extraction. Some shrinkage might also contribute to differences between phase and EM images. (D) Kymographs showing dynamics of cells' edges at numbered positions in (B) with distance and time plotted along *y*- and *x*-axis, respectively. At the time of extraction, the edge was protruding for ~42 sec (1), retracting for ~12 sec (2), or stationary for ~1.5 min (3), as evidenced by ascending, descending, or flat shape, respectively (arrows), of the rightmost parts of the kymographs. At position 1, protrusion was fast for ~40 sec, but decreased in the last 8 sec, suggesting that it might switch to retraction shortly; at position 2, the slight darkening at the end of a kymograph suggests beginning of ruffling; at position 3, no activity is seen on the kymograph, but comparison of the last live image and the EM image (C) suggests a slight advance of the edge at this position. (E) EM of the protruding edge (position 1) comprising a lamellipodium filled with dense branched network. A neighboring region (see C, next to arrow 1) shows slight retraction and, possibly, beginning of ruffle formation. (F) EM of the retracting edge (position 2) showing formation of a ruffle during the lapse between the last live image and extraction. Consistent with its formation from a lamellipodium, the ruffle is filled with the dendritic network. (G) EM of the stationary edge (position 3) containing very sparse cytoskeleton with stress fibers terminating at focal adhesions as dominant features. A narrow zone of branched network associated with the stress fiber (box) indicates beginning of a new protrusion. Boxed regions in (E–G) are enlarged in insets in corresponding panels. Scale bars = 5 μm (A–C), 0.2 μm (E, F), and 0.5 μm (G).

grease, in contrast to a permanent attachment, permits subsequent separation of coverslips to do EM processing. Avoid excessive amounts of grease, as it may complicate the subsequent excision of the central area that bears the selected cells.

B. Extraction

Platinum replica EM reveals only the surface of the sample. For application to cytoskeletal studies, the plasma membrane should therefore be removed to uncover the cell's internal organization. Detergent treatment is the most common way to remove the plasma membrane while retaining the cytoskeleton. Many versions of detergent extraction protocols have been developed (reviewed in Lindroth *et al.*, 1992), and the major incentive to vary the existing procedures was to find an extraction solution that would cause minimal distortion of the cytoskeleton while maximally removing the membrane and soluble proteins. The outcome of this search showed that nonionic detergents (most commonly, Triton X-100) in a slightly acidic buffer containing ethylene glycol bis(2-aminoethyl ether)-N,N,N′N′-tetraacetic acid (EGTA) and Mg^{2+} provide optimal conditions for preservation of the cytoskeleton. Other stabilizers are frequently added for better results. They include specific drugs, like phalloidin for actin or taxol for microtubules, or nonspecific chemicals, such as glycerol, polyethylene glycol (PEG), or various protein cross-linkers, for example, glutaraldehyde. While stabilizing supplements clearly help to preserve more material in the cytoskeleton, there is a concern whether they preserve the intrinsic cellular architecture or create something that did not exist in the living cell. These considerations are especially relevant regarding the use of phalloidin or taxol. This problem can be solved by comparing the organization of the cytoskeleton in the living state and after extraction in the same cells. This can be done using fluorescent cytoskeletal proteins and comparing the cytoskeleton organization *in vivo* and after the application of different extraction protocols. Using these approaches, we did not detect abnormal polymerization of actin or tubulin with our protocols when phalloidin or taxol was added to extraction solutions.

Depending on the particular application, we use one of three extraction protocols, all of which use Triton X-100 dissolved in PIPES-EGTA-magnesium (PEM) buffer, but they differ with respect to the stabilizers that are added (Table I and Fig. 3). Extraction at room temperature gives better results compared to 4 °C, when the plasma membrane is partially retained, or to 37 °C, when loss of material from the cytoskeleton is greater. Time of extraction is usually not very critical, and similar results are obtained over a 3- to 10-min range. Cytoskeletons should be rinsed after extraction two to three times with PEM buffer. Optionally, a quick rinse of cells with warm phosphate-buffered saline (PBS) can be done before extraction to remove serum proteins.

The best view of branched actin networks in lamellipodia can be achieved by extracting cells with 0.2–1% Triton X-100 in PEM buffer with 2-mM phalloidin as the only stabilizer (Fig. 3B). This procedure effectively removes all soluble and weakly associated proteins so that actin filaments look clean and are not obscured by associated proteins. However, this strong extraction may cause large-scale distortions

Table I
Protocol for Platinum Replica EM of Cytoskeletons

A. Extraction

1. Remove culture medium from a dish using a pipette or by pouring out
2. Optional: quickly rinse with prewarmed to 37°C PBS with Ca^{2+} and Mg^{2+}
3. Add one of the room temperature extraction solutions for 3–5 min, for example,
   (a) 1% Triton X-100, 4% PEG (MW 35 kDa), 2-μM phalloidin in buffer PEM (100-mM PIPES, pH 6.9; 1-mM EGTA; 1-mM $MgCl_2$); or
   (b) 1% Triton X-100, 2-μM phalloidin in buffer PEM; or
   (c) 0.5% Triton X-100, 0.25% glutaraldehyde in buffer PEM
4. Rinse cells with room temperature PEM buffer two or three times (for extraction "a," incubate 1 min in the last change)

B. Fixation

5. Add 2% glutaraldehyde in 0.1-M sodium cacodylate, pH 7.3. Fix at least 20 min at room temperature. If necessary, samples can be stored refrigerated at this stage
6. Remove glutaraldehyde and add 0.1% aqueous tannic acid (freshly prepared); incubate 20 min at room temperature. No washing is necessary before this step
7. Rinse specimens in five changes of distilled water and incubate 5 min in the third change
8. Add 0.2% aqueous uranyl acetate; incubate 20 min at room temperature
9. Replace uranyl acetate with distilled water

C. Dehydration

10. Place a specimen holder into a wide beaker with water
11. Load coverslips with cell side up into the holder alternating them with pieces of lens tissue fitting the size of the holder. Choose tissue with loose fibers, for example, from Kodak. Close a lid
12. Transfer holder through a series of graded ethanols:
    10%
    20%
    40%
    60%
    80%
    100% (twice)
    0.2% uranyl acetate in 100% ethanol
    100% (twice)
    100% ethanol dried over molecular sieve (twice). Incubate 5 min in each grade with stirring, except for uranyl acetate which incubate for 20 min without stirring.

of the weakly attached parts of the cell, especially if samples are not sufficiently gently handled, and it removes some intrinsic but dynamic cytoskeletal proteins. Another shortcoming of this procedure is that it frequently fails to dissolve scattered plasma membrane patches, even at 1% Triton X-100.

Better overall cytoskeleton preservation suitable for correlative microscopy is obtained if 1% Triton X-100 in PEM buffer is supplemented with high-MW PEG (20–40 kDa) at 1–4% concentration. This protocol produces minimal shrinkage, which is mostly restricted to the poorly attached cell body, and preserves many weakly associated cytoskeletal components. For example, the filopodia-specific, actin cross-linking protein fascin is rather weakly associated with the cytoskeleton; it completely dissociates from filopodia within 5 min of extraction in detergent-plus-buffer only,

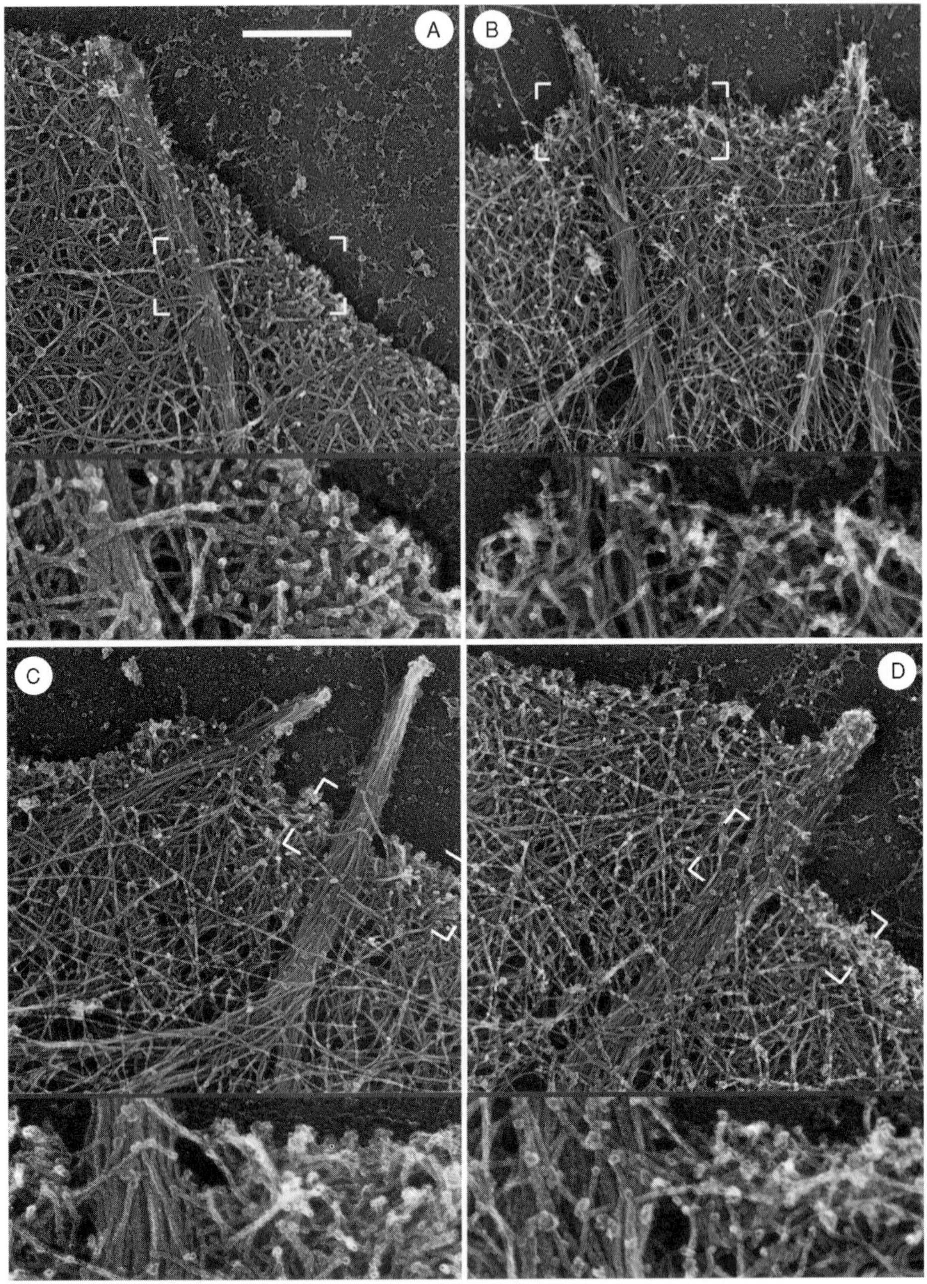

Fig. 3 Preservation of the leading edge cytoskeleton by different extraction protocols. (A) 1% Triton X-100, 4% PEG (MW 40 kDa), and 2-μM phalloidin in PEM buffer, 10 min at room temperature. Actin filaments in a filopodium form a tight bundle and in lamellipodia they form a dense dendritic network. (B) 1% Triton X-100 and 2-μM phalloidin in PEM buffer, 5 min at room temperature. Actin filaments

whereas in the presence of PEG, it is largely preserved for at least 30 min (Svitkina *et al.*, 2003). At the structural level, loss of fascin correlates with splaying of filopodial bundles (Figs. 1E and 3B), whereas conditions that preserve fascin produce tightly bundled filaments in filopodia (Fig. 3A, C, and D). By varying the concentration of PEG from 0% to 4% and/or its MW from 4 to 40 kDa, one can select a desirable degree of preservation, which increases with higher values of these parameters. With this procedure, the PEM wash after extraction should last ~1–3 min to surely remove the viscous PEG.

For very labile components of the cell protrusive machinery, especially those localized at the membrane/cytoskeleton interface, we apply a mixture of Triton X-100 and glutaraldehyde (Hoglund *et al.*, 1980; Small, 1981), optionally followed by 5-min extraction with 1% Triton X-100 in PEM. This approach helps, for example, to preserve a population of dynamic VASP molecules at the leading edge of lamellipodia, which is solubilized by the two other procedures (Fig. 4). This preservation correlates with the appearance of additional material at the leading edge as well as a cross-linking of the barbed ends of actin filaments at the sites where they apparently encountered the plasma membrane (Fig. 3C and D). It seems that proteins located close to the plasma membrane have better chances to be cross-linked and stabilized than proteins in the cell interior. The plasma membrane and soluble proteins are still readily dissolved by the combined glutaraldehyde/Triton X-100 treatment, so the view of the cytoskeleton is not blocked. Too high a ratio of glutaraldehyde to Triton X-100, however, leads to excessive protein cross-linking and the appearance of extra granular material, possibly captured soluble proteins (Fig. 3D). A combination of 0.25% glutaraldehyde and 0.5% Triton X-100 in this protocol gives a good balance between preserving the structural and molecular features of the cytoskeleton while not trapping too much irrelevant cytoplasmic material.

C. Fixation

Chemical fixation endows cell structures with a physical stability, and it minimizes the distortions caused by subsequent harsh procedures such as dehydration. Our fixation procedure includes consecutive treatments with 2% glutaraldehyde,

---

in the filopodial bundle are slightly separated, and in lamellipodia they also look less cross-linked and clearly display branches and filament ends. (C) 0.5% Triton X-100 and 0.25% glutaraldehyde in PEM buffer for 1 min followed by 1% Triton X-100 in PEM buffer for 5 min at room temperature. Filopodia are fairly tightly bundled, and lamellipodia display clear dendritic organization, but barbed ends of lamellipodial filaments at the very leading edge look cross-linked. (D) 0.25% Triton X-100 and 0.5% glutaraldehyde in PEM buffer for 1 min followed by 1% Triton X-100 in PEM buffer for 5 min at room temperature. Filopodial bundles and lamellipodial network are well preserved, but filament ends look highly cross-linked at this high glutaraldehyde/Triton X-100 ratio and filaments become associated with numerous particles, presumably, soluble or weakly associated proteins captured by glutaraldehyde. Boxed regions are enlarged at the bottom of each panel. Scale bar = 0.5 μm (for all panels).

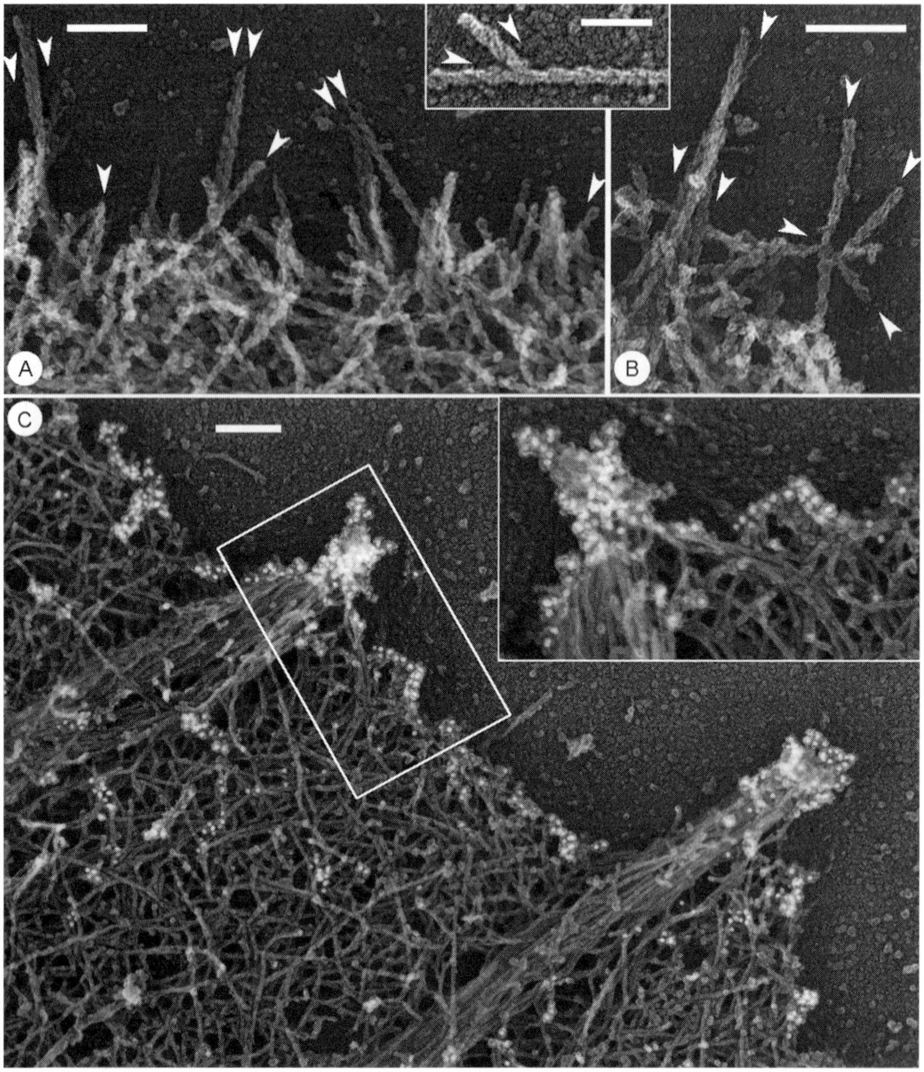

Fig. 4 Labeling of cytoskeletal components at the leading edge. (A, B) S1 decoration of actin filaments in lamellipodia of fish epidermal keratocytes (for methods see Svitkina *et al.*, 1997). Polarity of some filaments is indicated by arrowheads. Inset shows a branched filament at higher magnification. (C) Immunogold labeling of VASP at the leading edge of lamellipodia in B16F1 cell. VASP is a member of Ena/VASP family of actin-binding proteins, which preferentially associates with actin filament barbed ends and presumably protects them from capping by capping proteins (Bear *et al.*, 2002). Cells were extracted with 0.5% Triton X-100 and 0.1% glutaraldehyde in PEM buffer for 3 min followed by fixation with 2% glutaraldehyde for 20 min. After quenching with $NaBH_4$, cells were stained with rabbit anti-VASP antibody (courtesy Dr. F. Gertler) and secondary goat-anti-rabbit antibody conjugated with 18-nm colloidal gold (Jackson). Gold particles (white dots in the inversed contrast) are present at the known sites of VASP localization: they are highly concentrated at the filopodial tips; at lower level, they are present along the leading edge; and also are scattered in the lamellipodial interior. Boxed region is enlarged in inset. Scaled bars = 200 nm.

0.1% tannic acid, and 0.2% uranyl acetate. Since glutaraldehyde was introduced to the practice of EM preparation (Sabatini *et al.*, 1963), it has remained the most popular fixative for providing extensive cross-linking of proteins and excellent preservation of the structure in EM samples. Yet, glutaraldehyde alone is not sufficient, for example, to protect cells against some artifacts of dehydration (Small, 1981). Cofixation or postfixation with tannic acid has been shown to be beneficial in making samples more resistant to dehydration and other distortions, especially when combined with heavy metal treatment (Mizuhira and Futaesaku, 1972; Simionescu and Simionescu, 1976). Low-MW tannic acid (Cat. No. 1764) from Mallinckrodt, Inc. (Hazelwood, MO) gives the best results (Simionescu and Simionescu, 1976). This product is currently available from Fisher. Uranyl acetate works as a contrasting agent, but also as a fixative (Terzakis, 1968), apparently because it can form an extensive complex with tannic acid and thus protect structures from shrinkage. Therefore, during fixation, thorough washing is needed between tannic acid and uranyl acetate treatments to prevent formation of precipitates between the two compounds.

D. Dehydration and Critical Point Drying

The purpose of the critical point drying procedure is to transfer samples from a liquid to a gas without their having to cross the phase boundary with its intrinsic surface tension (Anderson, 1951). This is done by raising the temperature and the pressure above the critical point of the liquid, thus converting it homogeneously to a compressed gas. Subsequent release of the pressure leaves the sample dry and undamaged by surface tension. Water, however, is not a good liquid for critical point drying because its critical point has values of temperature and pressure that are too high for biological samples. A liquid with acceptable critical point parameters is CO_2. This material is not miscible with water and requires a transitional medium: ethanol or acetone, which can be mixed in any proportions with either water or CO_2. Thus, dehydration by one of these organic solvents is a necessary step preceding critical point drying. Ethanol is preferable because it is less harmful for both humans and equipment. However, dehydration removes not only free water but also water that is bound to proteins, thus significantly changing their conformation and causing shrinkage (Boyde and Boyde, 1980). The three-step fixation we use helps to minimize dehydration-related artifacts, especially for the peripheral, well-attached cell areas, as evidenced by good superimposition of light and EM images of the same cells. Some shrinkage still occurs in the weakly attached cell body.

Freeze-drying is an alternative to critical point drying for eliminating the liquid/gas boundary during drying. For this method, the cells are rapidly frozen, then cellular ice is sublimated under vacuum. Freeze-drying induces less shrinkage (Boyde and Boyde, 1980; Hartwig and Shevlin, 1986) and provides better preservation of fine structures, allowing one to see the subunit organization of cytoskeletal fibers (Heuser and Kirschner, 1980; Hirokawa *et al.*, 1988) and the shape of individual proteins (Heuser, 1989). However, freeze-drying is more challenging technically, and commonly produces quite a low rate of successful results, especially for imaging an entire cell. Thus, it is not practical for large-scale, routine analyses or for correlative microscopy. Conversely, critical point drying is a simple and reliable technique, which when applied with the necessary precautions produces images of excellent quality with high yield. The procedures most important for successful sample preparation are: (1) use of tannic acid-uranyl acetate fixation, which significantly reduces cell shrinkage during dehydration; (2) complete exchange of water to ethanol and then ethanol to liquid CO_2, as any remaining liquid other than CO_2 will eventually dry below its critical point and cause artifacts; (3) always keeping a layer of liquid above cells during sample handling, so the gas-liquid interface never passes through the specimen; and (4) protecting dried samples from absorbing moisture from air.

The process of dehydration through graded ethanols is the most sensitive procedure regarding the danger of air-drying because organic solvents are much more volatile than water. Therefore, after uranyl acetate fixation and before dehydration step, we place our specimens into a holder: a rectangular, wire-mesh basket that fits the chamber of a critical point drier (Svitkina and Borisy, 1998). Coverslips are loaded into a holder under water; they are kept separate with pieces of lens tissue. A wire-mesh lid is used to prevent the top lens tissue from flowing away. The lens tissue helps to retain a layer of liquid over the specimens. The coverslip/lens paper stack should be loose enough to permit easy liquid exchange. With this design, up to 12 coverslips with dimensions ~6–8 mm may be processed simultaneously without complications. Overloading the holder or increasing the coverslip size makes the exchange difficult and may result in poor quality of samples in the middle of the stack or at the center of a coverslip. Extended incubation times for dehydration and ethanol/CO_2 exchange are required to get successful result with larger coverslips.

The large, marked coverslips, which are used for correlative microscopy, should first be cleaned of immersion oil, using alcohol or ether, then detached from the bottom of the culture dish and quickly transferred into a wide, water-filled Petri dish. The remaining silicone grease helps to stabilize the coverslip in the dish, while the clean, central area with cells of interest (6–8 mm is the optimal size) is excised with a diamond pencil. It is helpful to use a razor blade as a guide while cutting off the greased edges of the coverslip. Use a sharp diamond pencil, and avoid glass crumbs around the cutting area to prevent coverslips from shattering.

Dehydration is done by transferring the holder with coverslips from beaker to beaker, going to increasing concentrations of ethanol and stirring for 5 min in each ethanol concentration. Double incubation is used for 100% ethanol. At this point, we use additional 20-min fixation with 1% uranyl acetate in 100% ethanol. This step apparently helps to stabilize structures in their new, dehydrated state; it improves the quality of the preparations. Two additional 5-min incubations in 100% ethanol with stirring are necessary to wash off the uranyl acetate. At the end of the dehydration procedure, it is better to use 100% ethanol dried over molecular sieves, which can be prepared as following. New molecular sieves (4 Å, 8–12 mesh) should be washed of dust with water, baked overnight at 160 °C, and after cooling, added to 100% ethanol.

After the last incubation in the dry ethanol, the holder with samples can be transferred to the chamber of the critical point drier filled with dried 100% ethanol. Critical point drying can be performed according to the recommendations of the manufacturer or using embedded automatic procedures, if available. Each critical point drying procedure first performs replacement of ethanol with CO_2, then a heating of the closed chamber to raise its pressure and temperature above the critical point for CO_2, and finally, a release of the pressure and cooling to room temperature. For the best quality of samples, the most critical point of this sequence is the first step, not the second, as one might think. Unfortunately, the manufacturer's protocols and automatic procedures do not always successfully remove ethanol from the samples. We use our own manual procedure to thoroughly remove all traces of ethanol from samples. This procedure includes 10 cycles of liquid exchange performed at 10–15 °C and consisting of 30 sec with CO_2 flushing through, followed by 5 min of incubation. During flushing, both IN and OUT valves are open and care should be taken to keep the level of CO_2 always above the samples. Stirring (if equipped) or manual shaking of the machine during flushing helps to further improve the ethanol removal. This exchange procedure can also be used with automatic driers after they are switched to a manual mode of operation.

Dried samples are extremely sensitive to humidity and should be kept in a desiccator until they have been coated. It is more difficult to protect samples from absorbing moisture while mounting them onto the stage of the vacuum evaporator, so this step should be done as fast as possible. If the relative humidity in the room is below 35%, the samples are quite safe during loading. Otherwise, running a powerful dehumidifier in the room solves a problem of moisture absorption.

E. Platinum/Carbon Coating

Metal coating provides contrast to biological specimens. In the replica technique, a thin layer of a metal is evaporated onto the sample from an oblique angle, then it is stabilized by carbon evaporation, released onto the surface of water, picked on a Formvar-coated EM grid, and viewed with transmission EM. The platinum replica technique was originally developed in conjunction with the freeze–fracture technique (Moor and Muhlethaler, 1963; Steere, 1957) and initially mostly used for analyses of internal membrane structure (reviewed in Fujimoto, 1997). John Heuser first used the platinum replica technique for cytoskeletons prepared by detergent extraction, quick-freezing, and freeze-drying (Heuser and Kirschner, 1980). His work demonstrated the excellent possibilities of the platinum replica approach for cytoskeletal studies. Compared to other whole mounts, replicas are more stable under the electron beam and provide an easily understandable 3D view of the sample. A disadvantage of replicas is that they reveal only the top layer of the sample, the one that is accessible to platinum shadowing. However, for thin objects, like lamellipodia of cultured cells, the method gives a sufficient depth of view. Platinum is a popular metal for coating because it represents a reasonable compromise between the melting temperature and the grain size, which are in inverse relationship with each other.

The quality of the image created by metal deposition depends on the angle and the thickness of shadowing. Rotary shadowing (Heuser and Kirschner, 1980) is superior to undirectional coating, as it does not create deep shadows, and it reveals more details. Higher angles of coating produce less contrast, but they better penetrate the depth of the sample. Increasing the thickness of metal layer increases contrast and improves the visualization of the cell interior, but it reduces resolution. We use platinum rotary shadowing at 45° angle with 2.5- to 2.8-nm thickness; this produces an adequate balance between contrast and depth of view.

Since the platinum layer is not cohesive, it is stabilized by subsequent evaporation of carbon, which can form a cohesive film and thus keep the platinum grains together. Carbon is practically transparent for electrons and therefore the thickness of the carbon layer is not very critical. Yet, too thick carbon becomes visible and this should be avoided. Insufficient carbon thickness may provide inferior stabilization and result in replica shattering. We use a 2- to 3-nm-thick carbon layer evaporated at 75°–90° angle.

F. Release and Mounting of Replicas

Platinum-carbon replicas must be released from the coverslip and mounted on EM grids for examination. Release of replicas from glass surfaces is achieved by placing a coated coverslip on the surface of ~10% hydrofluoric acid (HF) with the cell side up. In a few minutes or even seconds, the coverslip falls down, leaving the replica floating. It is helpful to cut the platinum/carbon layer while it is still on the coverslip; small pieces that will fit onto a grid are convenient. After separation from the coverslip by HF, the replica then falls apart along the introduced cuts, making the replica easier to handle. For correlative microscopy, the area of interest can be relocated under the dissection microscope before the replica is removed from the glass. After critical point drying and platinum coating, cells have good contrast and are clearly visible in a dissection microscope.

To wash the HF off the replicas, use a platinum loop to transfer replica pieces to the surface of water. However, the large difference in the surface tension between HF solution and pure water usually causes replicas to break. To prevent this, use highly diluted ($\sim10^{-6}$) detergent (Triton X-100 or household detergent are both fine) in water as a first wash, followed by a wash in pure water. An overdose of detergent may cause wrinkling or drowning of the replicas. It is recommended first to test a new detergent solution for wrinkling or breaking the replicas using a less valuable sample or its part before proceeding with the rest of the experiment. The washing detergent should be changed frequently because aged detergent solutions leave filmy contamination on cytoskeletons.

If cell areas of interest are thin and have low electron density, like lamellipodia and lamella in well-spread cultured cells, the above procedure is sufficient, so a replica can be picked up on Formvar-coated grids and investigated with the transmission EM. For thick and/or electron-dense samples, replicas can be treated with diluted household bleach to dissolve organic material. The quality of images, however, declines slightly after this treatment. Bleaching is also not compatible with immunogold labeling, since it removes gold particles from replicas.

IV. Identification of Cytoskeletal Components in Platinum Replica Preparations

An essential component of understanding the molecular machinery for motility is making connections between structures and molecules. Immunolabeling is the major approach to address such questions. However, it is not always possible to preserve the structural integrity and antigenicity of the sample at the same time (see also Chapter 19 by Morphew, this volume). An alternative approach is to perform correlative EM of cells ectopically expressing fluorescent fusion proteins (Fig. 1). For actin filaments, decoration with HMM or S1 myosin fragments has been instrumental for confirming their molecular identity (Huxley, 1963; Ishikawa *et al.*, 1969). At present, this technique remains an important tool for determining polarity of actin filaments. The use of these techniques in combination with the platinum replica EM of the cytoskeleton in cultured cells is presented in this section (Fig. 4).

A. Decoration of Actin Filaments with Myosin Subfragments

The polarity of actin filaments is a structural feature that is key to their function: it should be taken into account to understand the mechanisms of actin-dependent processes. In negatively stained (Huxley, 1963) or thin-sectioned (Ishikawa *et al.*, 1969) images, S1- or HMM-decorated actin filaments have a classic arrowhead pattern with barbed ends corresponding to fast growing ends (Bonder *et al.*, 1983). In platinum replicas, S1-decorated filaments have a helical, ropelike appearance, and polarity of actin filaments can be determined from the asymmetry of individual turns of the helix (Heuser and Cooke, 1983). With our approach, we are usually able to visualize the polarity of a substantial fraction of actin filaments even in areas of dense filament arrays, like lamellipodia or stress fibers (Fig. 4A and B; Svitkina *et al.*, 1997; Verkhovsky *et al.*, 1997).

Decoration of actin filaments in cytoskeletons is performed after detergent extraction and rinsing with the PEM buffer. We applied S1 to cytoskeletons at 0.25 mg/ml for 30 min at room temperature, rinsed quickly with PEM buffer, and fixed with glutaraldehyde. Methods for HMM (Margossian and Lowey, 1982) or S1 (Weeds and Taylor, 1975) preparation have been described. Alternatively, skeletal muscle HMM is available from Cytoskeleton, Inc. (Denver, CO).

B. Immunostaining

Colloidal gold-conjugated antibodies can be used together with platinum replica EM of the cytoskeleton, as gold has sufficient density to be visible on the platinum background (Fig. 4C). Particles as small as 5 nm may be difficult to discern, so a 10- to 20-nm size is preferable. The specific protocol for immuno-EM depends mostly on the antigen, the primary antibody, and their ability to react under particular conditions. For antigens that survive glutaraldehyde fixation, the primary antibody is applied after quenching the glutaraldehyde-fixed sample with $NaBH_4$ (2 mg/ml in PBS, two times 10 min each) and thorough washing with PBS. It helps sometimes to decrease the glutaraldehyde concentration to 0.2% at this stage, as 2% glutaraldehyde fixation will be applied anyway, later in the procedure. Formaldehyde or methanol fixation is deleterious to structure preservation and is not recommended for EM applications. Instead, staining of unfixed cytoskeletons with a primary antibody in PEM buffer is safe for many cytoskeleton structures and may be an option for antigens that are destroyed by glutaraldehyde fixation (Svitkina *et al.*, 1997; Verkhovsky *et al.*, 1995). Our buffer for diluting colloidal gold-conjugated antibodies contains 20-mM Tris–HCl (pH 8.0), 0.5-M NaCl, 0.05% Tween 20, and 1% BSA. The same buffer, but with 0.1% BSA, is used for washing off the antibody. In our experience, 1:5 to 1:10 dilutions are most useful for a number of commercially available gold-conjugated antibodies. After incubation with a secondary antibody, the samples should be fixed with 2% glutaraldehyde and then processed for EM according to the regular procedure.

Acknowledgments

I am grateful to Alexander Bershadsky and Vladimir Gelfand for suggesting the idea of platinum replica EM of the cytoskeleton very early on, Alexander Verkhovsky for his valuable help in improving the procedure along the way, and Gary Borisy for inspiring and supporting my work for many years, and acknowledge the current support from NIH grant GM 70898.

References

Abercrombie, M. (1977). Concepts in morphogenesis. *Proc. R. Soc. Lond. B Biol. Sci.* **199,** 337–344.

Abercrombie, M., and Ambrose, E. J. (1958). Interference microscope studies of cell contacts in tissue culture. *Exp. Cell Res.* **15,** 332–345.

Abercrombie, M., Heaysman, J. E., and Pegrum, S. M. (1970). The locomotion of fibroblasts in culture. II. "Ruffling". *Exp. Cell Res.* **60,** 437–444.

Anderson, T. F. (1951). Techniques for the preservation of three-dimensional structure in preparing specimens for the electron microscope. *Trans. NY Acad. Sci.* Series II **13,** 130.

Bear, J. E., Svitkina, T. M., Krause, M., Schafer, D. A., Loureiro, J. J., Strasser, G. A., Maly, I. V., Chaga, O. Y., Cooper, J. A., Borisy, G. G., and Gertler, F. B. (2002). Antagonism between Ena/VASP proteins and actin filament capping regulates fibroblast motility. *Cell* **109,** 509–521.

Bonder, E. M., Fishkind, D. J., and Mooseker, M. S. (1983). Direct measurement of critical concentrations and assembly rate constants at the two ends of an actin filament. *Cell* **34,** 491–501.

Boyde, A., and Boyde, S. (1980). Further studies of specimen volume changes during processing for SEM: Including some plant tissue. *Scan. Electron Microsc.* **132**(Pt. 2)**,** 117–124.

Buckley, I. K., and Porter, K. R. (1967). Cytoplasmic fibrils in living cultured cells. A light and electron microscope study. *Protoplasma* **64,** 349–380.

Buckley, I. K., and Porter, K. R. (1975). Electron microscopy of critical point dried whole cultured cells. *J. Microsc.* **104,** 107–120.

Fell, H. B. (1958). The cell in culture. *J. Clin. Pathol.* **11,** 489–494.

Fujimoto, K. (1997). SDS-digested freeze-fracture replica labeling electron microscopy to study the two-dimensional distribution of integral membrane proteins and phospholipids in biomembranes: Practical procedure, interpretation and application. *Histochem. Cell Biol.* **107,** 87–96.

Harrison, R. G. (1907). Observations on the living developing nerve fiber. *Anat. Rec.* **1,** 116–118.

Hartwig, J. H., and Shevlin, P. (1986). The architecture of actin filaments and the ultrastructural location of actin-binding protein in the periphery of lung macrophages. *J. Cell Biol.* **103,** 1007–1020.

Heuser, J. (1989). Protocol for 3-D visualization of molecules on mica via the quick-freeze, deep-etch technique. *J. Electron Microsc. Tech.* **13,** 244–263.

Heuser, J. E., and Cooke, R. (1983). Actin-myosin interactions visualized by the quick-freeze, deep-etch replica technique. *J. Mol. Biol.* **169,** 97–122.

Heuser, J. E., and Kirschner, M. W. (1980). Filament organization revealed in platinum replicas of freeze-dried cytoskeletons. *J. Cell Biol.* **86,** 212–234.

Hirokawa, N., Hisanaga, S., and Shiomura, Y. (1988). MAP2 is a component of crossbridges between microtubules and neurofilaments in the neuronal cytoskeleton: Quick-freeze, deep-etch immunoelectron microscopy and reconstitution studies. *J. Neurosci.* **8,** 2769–2779.

Hoglund, A. S., Karlsson, R., Arro, E., Fredriksson, B. A., and Lindberg, U. (1980). Visualization of the peripheral weave of microfilaments in glia cells. *J. Muscle Res. Cell. Motil.* **1,** 127–146.

Huxley, H. E. (1963). Electron microscope studies on the structure of natural and synthetic protein filaments from striated muscle. *J. Mol. Biol.* **77,** 281–308.

Ishikawa, H., Bischoff, R., and Holtzer, H. (1968). Mitosis and intermediate-sized filaments in developing skeletal muscle. *J. Cell Biol.* **38,** 538–555.

Ishikawa, H., Bischoff, R., and Holtzer, H. (1969). Formation of arrowhead complexes with heavy meromyosin in a variety of cell types. *J. Cell Biol.* **43,** 312–328.

Lauffenburger, D. A., and Horwitz, A. F. (1996). Cell migration: A physically integrated molecular process. *Cell* **84,** 359–369.

Ledbetter, M. C., and Porter, K. R. (1963). A "microtubule" in plant cell fine structure. *J. Cell Biol.* **19,** 239–250.

Lindberg, U., Hoglund, A. S., and Karlsson, R. (1981). On the ultrastructural organization of the microfilament system and the possible role of profilactin. *Biochimie* **63,** 307–323.

Lindroth, M., Bell, P. B., Jr., Fredriksson, B. A., and Liu, X. D. (1992). Preservation and visualization of molecular structure in detergent-extracted whole mounts of cultured cells. *Microsc. Res. Tech.* **22,** 130–150.

Loisel, T. P., Boujemaa, R., Pantaloni, D., and Carlier, M. F. (1999). Reconstitution of actin-based motility of Listeria and Shigella using pure proteins. *Nature* **401,** 613–616.

Mallavarapu, A., and Mitchison, T. (1999). Regulated actin cytoskeleton assembly at filopodium tips controls their extension and retraction. *J. Cell Biol.* **146,** 1097–1106.

Maly, I. V., and Borisy, G. G. (2001). Self-organization of a propulsive actin network as an evolutionary process. *Proc. Natl. Acad. Sci. USA* **98,** 11324–11329.

Margossian, S. S., and Lowey, S. (1982). Preparation of myosin and its subfragments from rabbit skeletal muscle. *Methods Enzymol.* **85**(Pt. B), 55–71.

Maupin-Szamier, P., and Pollard, T. D. (1978). Actin filament destruction by osmium tetroxide. *J. Cell Biol.* **77,** 837–852.

Mejillano, M. R., Kojima, S., Applewhite, D. A., Gertler, F. B., Svitkina, T. M., and Borisy, G. G. (2004). Lamellipodial versus filopodial mode of the actin nanomachinery; pivotal role of the filament barbed end. *Cell* **118,** 363–373.

Mizuhira, V., and Futaesaku, Y. (1972). New fixation method for biological membranes using tannic acids. *Acta Histochem. Cytochem.* **5,** 233–236.

Mogilner, A., and Oster, G. (1996). Cell motility driven by actin polymerization. *Biophys. J.* **71,** 3030–3045.

Moor, H., and Muhlethaler, K. (1963). Fine structure in frozen-etched yeast cells. *J. Cell Biol.* **17,** 609–628.

Mullins, R. D., Heuser, J. A., and Pollard, T. D. (1998). The interaction of Arp2/3 complex with actin: Nucleation, high affinity pointed end capping, and formation of branching networks of filaments. *Proc. Natl. Acad. Sci. USA* **95,** 6181–6186.

Pollard, T. D., and Borisy, G. G. (2003). Cellular motility driven by assembly and disassembly of actin filaments. *Cell* **113,** 549.

Porter, K. R., Claude, A., and Fullam, E. F. (1945). A study of tissue culture cells by electron microscopy. *J. Exp. Med.* **81,** 233–246.

Ris, H. (1985). The cytoplasmic filament system in critical point-dried whole mounts and plastic-embedded sections. *J. Cell Biol.* **100,** 1474–1487.

Sabatini, D. D., Bensch, K., and Barrnett, R. J. (1963). Cytochemistry and electron microscopy. The preservation of cellular ultrastructure and enzymatic activity by aldehyde fixation. *J. Cell Biol.* **17,** 19–58.

Schliwa, M., and van Blerkom, J. (1981). Structural interaction of cytoskeletal components. *J. Cell Biol.* **90,** 222–235.

Simionescu, N., and Simionescu, M. (1976). Galloylglucoses of low molecular weight as mordant in electron microscopy. I. Procedure, and evidence for mordanting effect. *J. Cell Biol.* **70,** 608–621.

Small, J. V. (1981). Organization of actin in the leading edge of cultured cells: Influence of osmium tetroxide and dehydration on the ultrastructure of actin meshworks. *J. Cell Biol.* **91,** 695–705.

Small, J. V. (1994). Lamellipodia architecture: Actin filament turnover and the lateral flow of actin filaments during motility. *Semin. Cell Biol.* **5,** 157–163.

Small, J. V., Herzog, M., Haner, M., and Abei, U. (1994). Visualization of actin filaments in keratocyte lamellipodia: Negative staining compared with freeze-drying. *J. Struct. Biol.* **113,** 135–141.

Small, J. V., Isenberg, G., and Celis, J. E. (1978). Polarity of actin at the leading edge of cultured cells. *Nature* **272,** 638–639.

Steere, R. L. (1957). Electron microscopy of structural detail in frozen biological specimens. *J. Cell Biol.* **3,** 45–60.

Svitkina, T. M., and Borisy, G. G. (1998). Correlative light and electron microscopy of the cytoskeleton of cultured cells. *Methods Enzymol.* **298,** 570–592.

Svitkina, T. M., and Borisy, G. G. (1999). Arp2/3 complex and actin depolymerizing factor/cofilin in dendritic organization and treadmilling of actin filament array in lamellipodia. *J. Cell Biol.* **145,** 1009–1026.

Svitkina, T. M., and Borisy, G. G. (2006). Correlative light and electron microscopy studies of cytoskeletal dynamics. *In* "Cell Biology: A Laboratory Handbook" 3rd ed. (J. Celis, ed.), Vol. 3, pp. 277–285. Elsevier.

Svitkina, T. M., Bulanova, E. A., Chaga, O. Y., Vignjevic, D. M., Kojima, S., Vasiliev, J. M., and Borisy, G. G. (2003). Mechanism of filopodia initiation by reorganization of a dendritic network. *J. Cell Biol.* **160,** 409–421.

Svitkina, T. M., Shevelev, A. A., Bershadsky, A. D., and Gelfand, V. I. (1984). Cytoskeleton of mouse embryo fibroblasts. Electron microscopy of platinum replicas. *Eur. J. Cell Biol.* **34,** 64–74.

Svitkina, T. M., Verkhovsky, A. B., and Borisy, G. G. (1995). Improved procedures for electron microscopic visualization of the cytoskeleton of cultured cells. *J. Struct. Biol.* **115,** 290–303.

Svitkina, T. M., Verkhovsky, A. B., McQuade, K. M., and Borisy, G. G. (1997). Analysis of the actin-myosin II system in fish epidermal keratocytes: Mechanism of cell body translocation. *J. Cell Biol.* **139,** 397–415.

Terzakis, J. A. (1968). Uranyl acetate, a stain and a fixative. *J. Ultrastruct. Res.* **22,** 168–184.

Theriot, J. A., and Mitchison, T. J. (1991). Actin microfilament dynamics in locomoting cells. *Nature* **352,** 126–131.

Tilney, L. G., Hatano, S., Ishikawa, H., and Mooseker, M. S. (1973). The polymerization of actin: Its role in the generation of the acrosomal process of certain echinoderm sperm. *J. Cell Biol.* **59,** 109–126.

Tilney, L. G., and Portnoy, D. A. (1989). Actin filaments and the growth, movement, and spread of the intracellular bacterial parasite, Listeria monocytogenes. *J. Cell Biol.* **109,** 1597–1608.

Tilney, L. G., and Kallenbach, N. (1979). Polymerization of actin. VI. The polarity of the actin filaments in the acrosomal process and how it might be determined. *J. Cell Biol.* **81,** 608–623.

Verkhovsky, A. B., Chaga, O. Y., Schaub, S., Svitkina, T. M., Meister, J. J., and Borisy, G. G. (2003). Orientational order of the lamellipodial actin network as demonstrated in living motile cells. *Mol. Biol. Cell* **14,** 4667–4675.

Verkhovsky, A. B., Svitkina, T. M., and Borisy, G. G. (1995). Myosin II filament assemblies in the active lamella of fibroblasts: Their morphogenesis and role in the formation of actin filament bundles. *J. Cell Biol.* **131,** 989–1002.

Verkhovsky, A. B., Svitkina, T. M., and Borisy, G. G. (1997). Polarity sorting of actin filaments in cytochalasin-treated fibroblasts. *J. Cell Sci.* **110**(Pt. 15), 1693–1704.

Wang, Y. L. (1985). Exchange of actin subunits at the leading edge of living fibroblasts: Possible role of treadmilling. *J. Cell Biol.* **101,** 597–602.

Weeds, A. G., and Taylor, R. S. (1975). Separation of subfragment-1 isoenzymes from rabbit skeletal muscle myosin. *Nature* **257,** 54–56.

Wegner, A. (1977). The mechanism of ATP hydrolysis by polymer actin. *Biophys. Chem.* **7,** 51–58.

Wessels, N. K., Spooner, B. S., and Luduena, M. A. (1973). Surface movements, microfilaments and cell locomotion. *In* "Locomotion of Tissue Cells," pp. 53–82. Associated Scientific Publishers, Amsterdam, New York.

Wohlfarth-Bottermann, K. E. (1962). Weitreichende. fibrillare Protoplasmadifferenzierungen und ihre Bedeutung fur die Protoplasmastromung. I. Elektronenmikroskopischer Nachweis und Feinstruktur. *Protoplasma* **54,** 514–539.

Wohlfarth-Bottermann, K. E. (1964). Differentiations of the ground cytoplasm and their significance for the generation of the motive force of ameboid movement. *In* "Primitive Motile Systems in Cell Biology" (R. D. Allen, and N. Kamiya, eds.), pp. 79–109. Academic Press, New York, London.

Yamada, K. M., Spooner, B. S., and Wessels, N. K. (1971). Ultrastructure and function of growth cones and axons of cultured nerve cells. *J. Cell Biol.* **49,** 614–635.

CHAPTER 13

Imaging Actomyosin *In Situ*

Kenneth A. Taylor,* Shenping Wu,* Mary C. Reedy,† and Michael K. Reedy†

*Institute of Molecular Biophysics
Florida State University
Tallahassee, Florida 32306

†Department of Cell Biology
Duke University Medical Center
Durham, North Carolina 27710

0091-679X/07 $35.00
DOI: 10.1016/S0091-679X(06)79013-6

Here we describe methods used in the analysis of electron tomograms of insect flight muscle from *Lethocerus* sp. Steps described include the preparation of the muscle fibers, emphasizing especially freezing with mechanical monitoring, fixation by freeze-substitution, tilt-series data collection, and merging of these data to produce a raw tomogram. Correspondence analysis of periodic subvolumes taken from the raw tomogram is particularly important for identifying similar forms from the very large distribution of individual structures found within each unit cell repeat. We describe schemes for aligning extracted repeat volumes, constructing classification masks, determining the number of factors to use for classification and the number of classes to compute. We also cover the fitting of atomic models of actin and myosin and the process of relating the high signal-to-noise ratio averages by "map back" to the original raw tomogram. Finally, we describe two applications where this methodology has been applied to nonmuscle specimens.

I. Introduction and Overview

A. Why Electron Tomography?

Muscle structure has been studied for many years using X-ray diffraction, electron microscopy (EM), and a variety of molecular probes. X-ray crystallography and NMR studies of purified proteins have provided atomic resolution structures of many components of the contractile assembly, that is, actin, myosin, and troponin. Cryo-EM has provided much of our knowledge of how these individual muscle proteins are assembled with one another (Woodhead *et al.*, 2005) and bound to filamentous actin (F-actin) (Holmes *et al.*, 2003; Jontes *et al.*, 1995; Milligan and Flicker, 1987; Narita *et al.*, 2001; Whittaker *et al.*, 1995). However, contractile proteins in muscle are assembled into a lattice of filaments, and during active contraction, they occupy different stages of the ATPase cycle. Under these circumstances, they are not amenable to crystallography, which requires individual proteins to be in a single state.

None of the above-mentioned techniques by itself is ideal for studying force production in muscle at the molecular level. *In vitro* mixtures of actin and myosin cannot show tension-bearing structures. Methods that measure fibers under tension yield averages of myosin crossbridges distributed over many different stages of the ATPase cycle; therefore, they average over many states. A different method is needed to provide three-dimensional (3D) images of individual myosin heads in a range of conformations from a muscle under tension.

Electron tomography (ET) can produce the requisite images, for example 3D images of single copies of a particular muscle structure. Thus, it offers a method by which to study biological structures that are not amenable to 3D imaging by any of the classical methods such as single-particle reconstruction (Frank, 2006), helical reconstruction (DeRosier and Moore, 1970; Egelman, 2000), or electron crystallography (Glaeser, 1999); all these methods rely on averaging large numbers of identical copies of a structure to obtain the 3D image. However, the capability of ET to produce a 3D image of a single copy of a muscle structure is offset by the low signal-to-noise ratio (SNR) of that image. Nevertheless, improvements in SNR in muscle tomograms can be obtained by subvolume averaging, provided that large numbers of identical structures can be found within the 3D image.

Striated muscle possesses many characteristics that make it difficult to image in 3D by the classical 3D EM methods mentioned above. It is a very well-ordered tissue, but from the standpoint of molecular imaging, it shows significant local disorder. Muscle is best characterized as a paracrystalline specimen (para—Greek prefix meaning faulty); in the present context, this means that some elements of the structure are highly ordered while other parts have intrinsic disorder or that the positioning of the highly ordered structural elements is imprecise. Striated muscle consists of hexagonal arrays of two types of well-ordered filaments: thick, myosin-containing filaments and thin, actin-containing filaments. In muscle these well-ordered filaments have a disordered arrangement within the lattice (Squire, 1981), and during muscle contraction, the interactions between filaments become highly variable.

The powerful methods of single-particle 3D reconstruction, which largely use single projections of individual molecules or assemblies, would be ineffective with muscle specimens. Single-particle methods are most effective when the molecules or "particles" are well separated and truly identical, even though they lie in random orientations. In the muscle lattice, the structures are not identical; they are closely apposed, and the filaments have a preferred orientation in thin, plastic sections and therefore require specimen tilting to obtain sufficient views to compute a 3D image. ET overcomes both preferred orientation and close apposition, and as will be shown below, it allows a single-particle approach to be applied to improve SNR. This method will be applied not to projections but to repeated volumes from within the reconstruction, each of which contains a single 3D motif.

One of the most important capabilities of modern 3D EM is its suitability for building pseudo-atomic models; atomic structures obtained by X-ray crystallography, NMR, or cryo-EM can be assembled into large macromolecular assemblies that have been reconstructed by single-particle cryo-EM or cryo-ET. Effective model building requires a 3D image with a high SNR and as much resolution as possible. Averaging to improve SNR is therefore an important part of ET if interpretation of heterogeneous molecular conformations is the ultimate goal. Averaging is most effective when the individual objects are as similar as possible and differ mainly in stochastic noise. Thus, the process of computing averaged images within the microheterogeneous muscle lattice is intimately based on identifying and grouping repeat-motif subvolumes into groups that have the same structure. The method of correspondence analysis (Frank and van Heel, 1982; Van Heel and Frank, 1981) has proven to be very effective at identifying structural motifs in the muscle lattice facilitating averaging of homogeneous repeat subvolumes. In addition, aligning and then averaging identical structures that have different orientations can help to correct for tilt angle limitations that exclude regions of data from the Fourier transform of the tomogram. This missing wedge, pyramid, or cone is discussed in detail in Chapter 29 by Förster and Hegerl, this volume.

One final concept is worth mentioning. Resolution in ET is largely limited by specimen thickness and the tilt angle increment between successive images. Regardless of the number of electrons used to record each image in the tilt series or the number of repeat subvolumes averaged, the reconstruction will only be valid to this limiting resolution even while SNR is improving. This is in contrast to single-particle cryo-EM methods, where resolution is limited largely by SNR; micrographs usually have information at higher resolution than the individual particle images contained within them. Resolution in single-particle reconstruction thus depends on how many particles or repeats that can be averaged and how well they can be aligned, all other parameters being equal.

B. Basic Features of Insect Flight Muscle

The asynchronous indirect flight muscles (IFMs) of the large water bug *Lethocerus* sp. are perhaps the best-ordered muscles in the animal kingdom; they have been a favorite of muscle structural biologists for decades (Reedy *et al.*, 1965). Their characteristic order makes them ideal for a study of muscle structure. IFM contains a hexagonal array of thick filaments, but in this case, the actin filaments interdigitate between thick filament pairs at pseudodyad positions in the unit cell. This filament arrangement, which differs from vertebrate striated muscle, where the actin filaments lie at trigonal positions within the unit cell, is significant because it facilitates the cutting of several types of thin section that have been extremely useful for imaging crossbridge structure in different muscle states (Reedy and Reedy, 1985). Two kinds of longitudinal sections, each 25-nm thick, can be cut parallel to the filament axis (Fig. 1A). "Myac" layers contain alternating *my*osin and *ac*tin filaments. Because the actin filament is placed between two myosin filaments, crossbridges approach the actin filament from each side and their full length is included in the section plane (Fig. 1A and B). Myac layers are ideal for studying crossbridge conformations in different muscle states because they contain all the features of the muscle, namely thick and thin filaments and the bridges that connect them, all arranged in a thin section that is easy to analyze. Another 25-nm-thick longitudinal section, the actin layer, contains only actin filaments alternating between two orientations that differ by an azimuthal rotation of 60° and an axial translation of 12.9 nm (Fig. 1A). The third section type is a 15-nm-thick section cut transverse to the filament lattice that includes only a single

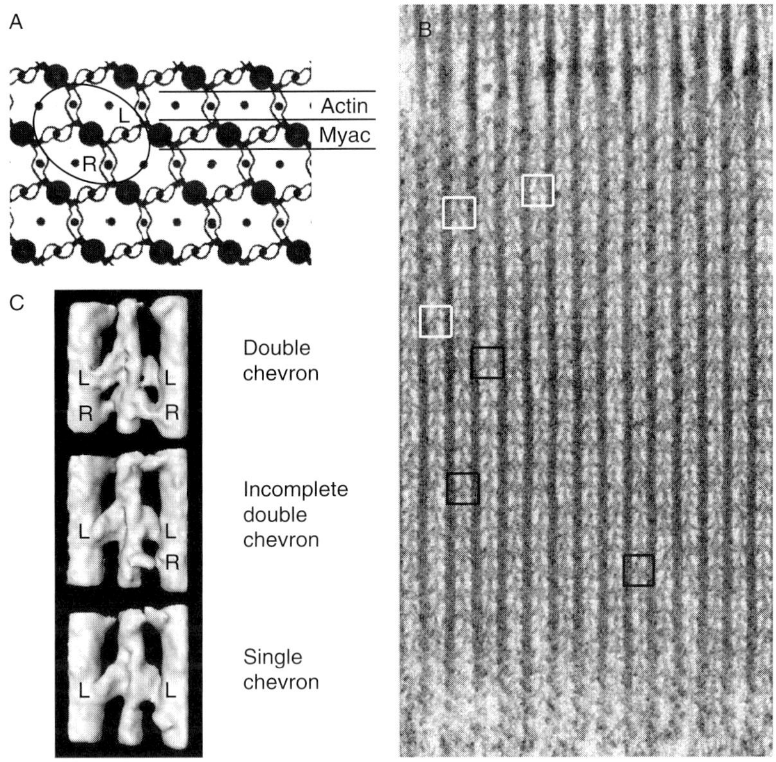

Fig. 1 Diagram of the muscle lattice and the myac layer. (A) Diagram showing the thick and thin filament lattice arrangement and the typical pattern of myosin crossbridges that occur in a 15-nm-thick transverse section. Two types of crossbridges, designated "lead" (L) and "rear" (R) are bound on the actin filaments. The "myac" layer is a 25-nm longitudinal section containing alternating thick and thin filaments. An "actin" layer is a 25-nm longitudinal section containing only actin filaments and their bound myosin crossbridges. A third motif, the "flared X" visible in 15-nm transverse sections is outlined by an ellipse. (B) Electron micrograph of a 25-nm longitudinal section of rigor insect flight muscle. Protein is black, embedding medium white. The rods of density running vertical are the thick filament backbones. Between thick filaments is an actin filament decorated with myosin heads that originate from the neighboring thick filaments. These are the crossbridges. (C) Examples of rigor repeats. At the top is a "double chevron," in the middle one of two versions of an incomplete double chevron and on the bottom a single chevron. The lattice of rigor crossbridges consists of a mixture of all these repeats types spread throughout the sarcomere. In addition, each repeat type can have a different structure for the two crossbridge types.

axial level of crossbridges. In rigor IFM, sections of this kind through the A-band display the "flared-X" motif (Fig. 1A).

Despite its regularity, IFM has several types of intrinsic disorder. IFM actin filaments have a highly regular helical structure of 28 actin subunits in 13 turns of the 5.9-nm genetic actin helix, but their orientation in the lattice can vary by a random 180° rotation about the filament axis (Holmes *et al.*, 1980). Alignment to correct this disorder is essential if actin monomers are to be resolved in the averages obtained from the tomogram. In the relaxed state, the thick filaments themselves are highly ordered and in helical register, but they can be offset axially by multiples of the 14.5-nm axial period (Schmitz *et al.*, 1994). In contracting muscle, neither axial nor helical thick filament registration is so regular (Taylor *et al.*, 1999).

Electron micrographs of rigor IFM myac layers (Fig. 1B) show a repeating pattern of crossbridges between the thick and thin filaments, known as the double chevron (Reedy *et al.*, 1965). This structure is produced by the binding of myosin heads to "target zones" on the thin filament that repeat every 38.7 nm (Reedy, 1967; Tregear *et al.*, 2004). The myosin and actin filaments of IFM have matched axial periodicities. IFM thick filaments display the 14.5-nm axial period typical of myosin filaments, but they also have a 38.7-nm helical period arising from the axial spacing of prominent helical tracks defining the surface array of myosin heads (AL-Khayat *et al.*, 2003; Morris *et al.*, 1991), seen as right-handed helices by direct imaging in thin sections (Reedy *et al.*, 1993). The actin filaments consist of a double helix of actin monomers with a pitch of 2×38.7 nm. Troponin, a key landmark on the thin filament, also has an axial repeat of 38.7 nm and thus repeats at the same azimuth along the entire length of the thin filament. The two filament systems in IFM have a common axial period of 232 nm (6×38.7 nm; 16×14.5 nm). The repeating motif of crossbridges and actin target zones every 38.7 nm provides large numbers of repeats that can be aligned, classified, and grouped according to subpopulation homogeneity and class-averaged for each subpopulation to yield high-quality images with improved SNR.

The term "repeat" in this discussion pertains to the 38.7-nm half-repeat of F-actin because this is a period that is invariant across many different catalytic and stable nucleotide-bound states. Thus defined, "repeat" can be equated with the term "single-particle" used in 3D reconstruction from projections. The main advantage of a paracrystalline specimen, such as muscle, is the high density of partially aligned repeats (particles) in the raw tomogram. However, these repeats are not all identical, even in an enzymatically homogeneous state such as rigor. Rigor IFM is dominated by four main classes of repeating motifs that can be defined based on the arrangement, rather than the structure, of individual crossbridges (Fig. 1C). The double chevron (Reedy and Reedy, 1985) contains a pair of opposed crossbridges, dubbed "lead" bridges (L), and an additional pair of opposed crossbridges, dubbed "rear" bridges (R). Variations on this theme include the single chevron, with no rear bridges and two types of incomplete double chevrons, with a single rear bridge that can be positioned on either side of the actin filament. The number of myosin heads (1 or 2) at both lead and rear bridges is variable. The myosin head in a single-headed crossbridge can be positioned on either of two actin monomers. Actin-bound heads may also differ in structure due to lattice constraints in the sarcomere; all myosin heads are linked to the thick filament shaft by the S2 domain. Contracting muscle has potentially many more possible combinations because the myosin heads are distributed over different steps in the catalytic cycle.

The goal of our ET of IFM has been to obtain molecular resolution of the crossbridges. Because of the noisy character of the raw tomograms, much effort has been expended to obtain averages to improve the SNR. Averaging also reduces the highly variable and very numerous structures of myosin crossbridges obtainable in a single tomogram to a more tractable number for pseudo-atomic model building. A single tomogram of a myac layer from a muscle half sarcomere usually contains ~500 repeats, far too many to analyze individually.

This chapter is ordered according to the separate steps that are used in electron tomographic analysis of IFM, which is the only muscle that has been imaged extensively in this way. Each heading will discuss one step, up to the last, which involves relating the averaged images obtained to the original specimen. Each heading will discuss the rationale for the approach, possible complications, and the particular steps used, many of which are generally applicable, but some of which may utilize unique features of IFM.

II. Preparing the Specimen

A. Materials

Muscle fibers were obtained from dorsal longitudinal flight muscles of *Lethocerus*. Live waterbugs were cold-stunned in ice–water slush for 10 min, then the muscles were exposed by dissection, and while still attached *in situ*, were detergent-permeabilized, taken through three 8- to 12-h cycles of glycerination/deglycerination (50% glycerol buffer; 5-mM Mg•ATP) on a mechanical rotator at 2 °C. After a further 8 h in 75% glycerol buffer (with Mg•ATP and DTT both 5 mM), whole or halved thoraces were cryo-stored at –80 to –100 °C in 75% glycerol for 2–24 months. These were briefly thawed to excise fibers, which were glued by their ends to stretching frames made from stainless steel wire, then oriented by an ~2% stretch and washed glycerol-free into suitable aqueous buffers, which were used as the vehicle for chemical fixation. Fixed fibers were Araldite-embedded for thin-section EM as described (Reedy and Reedy, 1985). Fibers for time-resolved studies were fast-frozen in various states and during various maneuvers, then freeze-substituted and embedded in Araldite (Liu *et al.*, 2004; Taylor *et al.*, 1999).

B. Mechanical Monitoring of Rapid Freezing

Earlier work with ET of IFM utilized rigor and stable, nucleotide-induced states that were chemically fixed (Chen *et al.*, 2001, 2002; Liu *et al.*, 2006b; Schmitz *et al.*, 1996, 1997), but recent efforts have worked to capture myosin head conformations in different stages of muscle contraction at known values of the fiber tension and stiffness (Liu *et al.*, 2004; Taylor *et al.*, 1999). It is largely through the efforts of Professors Yale Goldman and Clara Franzini-Armstrong at the University of

Pennsylvania that the device facilitating our millisecond-timed rapid freezing, length control and tension monitoring of single muscle fibers has reached its present advanced state (Fig. 2) (Hirose *et al.*, 1993, 1994; Lenart *et al.*, 1996; Liu *et al.*, 2004; Taylor *et al.*, 1999). This development started from a Heuser-type Med-Vac freeze-slammer, the freezing head of which was modified to integrate a miniaturized force transducer ($\pm 10\ \mu$N) and a piezo-bimorph length control lever ($\pm\Delta$length of 1–200 μm in 1–2000 msec) to control and monitor the length, tension, and stiffness of single fibers up to the moment of impact freezing. Two more changes important for our experiments were the gas bearings and the late-descent down-suction of ambient air into the freezing well just before impact freezing (Fig. 2). As detailed below, each run lasts for 0.5–3 h, permitting an output of 4–10 fibers/day.

The freezing head is detached and inverted to sit face up under a dissecting microscope. A standard Heuser-type planchette of thin, sheet-Al, its side-flaps bent to clasp the dovetail of the freezing anvil, carries a $2 \times 6 \times 0.5$ mm^3 slab of agar gel Superglued to the planchette, the free agar surface providing a moist, slick

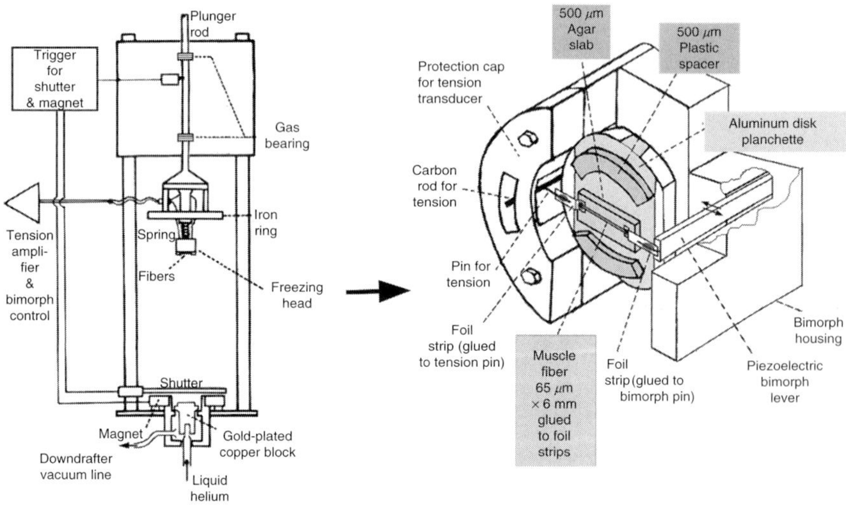

Fig. 2 Diagram of the Heuser Cryopress used for slam-freezing, as modified from Hirose *et al.* (1993). Modifications include: (1) support of glycerinated single fibers of IFM on a thin slab of agar and glued between force and length transducers; the Cu freezing anvil was machined to just clear both transducers as the fiber-loaded head hits the cold surface; (2) a piezoelectric bimorph lever to produce rapid fiber length changes; (3) replacement of plunge-rod pulleys with gas bearings to reduce noise in tension records to 10–20 μN and make 220-msec drop time reproducible to ± 1 msec; and (4) prevention of specimen precooling from updraft of cold helium by combination of a shutter lid, which opens 100 msec before impact, by a fan-driven duct that sucks ambient air horizontally across the top of the freezing chamber well and by a 100-msec down-suction pulse from within the freezing well initiated just 25 msec before impact to draw ambient air down just ahead of the specimen. Adapted from Liu *et al.* (2004), with permission.

support for the muscle fiber. Agar slabs are precision cast to 500-μm thickness, so that a typical 65- to 70-μm cylindrical fiber protrudes just 15–20 μm above the flanking nylon spacers of 550-μm thickness. Cylindrical fibers are compressed to an aspect ratio of 1:5 during impact freezing, a distortion we accept because the fibers remained straight without the frequent lateral shearing distortions that occur when supported on agar slabs 550- to 750-μm thick. We attribute the compression, which is larger than predicted by the difference between the agar + fiber slab thickness and the adjacent nylon spacer thickness, to freeze-trapping of a sudden elastic protrusion of agar + fiber during the abrupt deceleration of final impact. Although fibers are severely squashed during impact freezing, myofibrils on the surface remain cylindrical; the compression occurs in the poorly frozen regions deeper than 4–7 μm. However, a microwavy distortion was frequently visible even in the surface fibrils, evidenced by a 50–75% reduction in area of the normally extensive planar single filament layers typical in longitudinal sections obtained from fibers that were frozen more gently by liquid immersion. We hypothesize that this is likely due to shock waves generated on impact propagating through the specimen. Sound velocity, >1400 mm/msec in water, would exceed the estimated velocity of the freezing-front by 10× to 1000×.

A hand-sliced strip of Al foil, ~300-μm wide by ~15-mm long, is laid flat on the agar and its ends glued to the force and length pins. The 6-mm mid-segment from the foil strip is micro-scissored away to expose the underlying agar. A single insect fiber, still in 75% glycerol, is cut to 11-mm length and laid on the agar (moistened with 75% glycerol), and its ends are glued to the Al foil strips that remained as short extensions from force and length pins. The glue used for foil and fiber is cellulose nitrate in acetone. While the Cu freezing block is helium-cooled over 3 min to ~20°K, fiber and agar are irrigated drop wise with 10–50 drops each of the sequence of buffers needed to produce the desired final state, while being tested for tension and stiffness. (Optical effects show that 75% glycerol can be washed out of single permeabilized fibers in 10–60 sec by irrigation with 0% glycerol buffer, meanwhile timing the loss of visible refractile Schlieren-type striae at 10× to 60× under a dissecting microscope, or timing the change in phase-contrast appearance of intracellular structures at 100× to 400×.) At 5–15 sec before freezing, a video camcorder starts to record the action plus some commentary, while agar and fiber are microsuctioned free of surface liquid that might insulate the surface myofibrils from effective thermal contact with the Au-plated Cu freezing block. The freezing head is lifted, turned to face downward, and locked onto the plunger rod. When fiber, computer, and slammer are ready, a preprogrammed sequence is initiated that triggers start time for tension recording, start of the 220-msec drop-freeze event, and any length changes that are to occur during the drop.

After 15–30 sec of magnetically clamped contact against the Cu freezing block, the plunger rod is detached, the magnet is released, and the freezing head is manually lifted (in ~200 msec) into a cup of liquid nitrogen (LN_2). After transfer into a roomy Styrofoam box of LN_2, it is turned face up so the fiber can be cut free from force and length pins by scissoring the Al foil linkers. The force transducer is then pivoted clear to let the planchette slide off the freezing anvil into a storage vial.

C. Fixation by Freeze-Substitution

Slam-frozen samples are stored with the adhering agar block and planchette under LN_2 until freeze-substitution begins. Dried molecular sieve 3A is added to a freshly opened 500-ml bottle of absolute acetone and settled overnight at room temperature to give dried acetone, the solvent for the tannic acid (TA). TA was supplied by Electron Microscopy Sciences (Cat. No. 21720). A source of TA with consistently good quality is critical.

To begin freeze-substitution, planchettes are rapidly transferred from LN_2 into borosilicate glass scintillation vials at –80 to –93 °C that contain 10–15 ml of dried acetone and 0.2% TA dissolved $\leq$2 h before use. This vial is kept securely upright and loosely screw-capped in a Styrofoam box of dry ice (–80 °C) for $\geq$8 h. We get equally satisfactory results by freezing the TA–acetone mixture under LN_2, inserting planchette + specimen into the supernatant LN_2, placing the vial in dry ice, and leaving the planchette on the frozen surface to await immersion, once the LN_2 has evaporated and the TA–acetone has melted. Water ice dissolves in cryoacetone, and the freeze-trapped macromolecules are then chemically stabilized by TA. Unbound TA is then rinsed out. Next, muscle structure is further stabilized with uranyl acetate (UrAc) (0.5% in acetone, $\geq$2 h), here acting as a chemical fixative when used in combination with TA. As indicated by sample color change and final EM structure, this reaction appears to be complete within 10 min even if performed at –80 °C after rinsing out unbound TA. This encourages our belief that full chemical fixation can be effected by the TA and UrAc (TAURAC) sequence before warming above –80 °C. Nevertheless, we routinely follow the more convenient practice of warming the TA-acetone preparation to 0 °C after the initial 8+ h at –80 °C, then rinsing out TA (six changes dried acetone) and treating in UrAc-acetone at 0 °C for 2–12 h. Muscle fibers are now colored dark orange by the TAURAC sequence, so on warming to 23 °C, they are easily seen by stereomicroscopy. Each fiber + agar unit is now videotaped at 20× to 60× with commentary while under acetone in a deep glass dish (using an improvised glass-bottomed nosepiece for undersurface viewing) to record breaks, kinks, squishing, and so on, and to inspect the impact-frozen surface of the fiber, which always appears very finely pebbled by glancing reflection, rather than mirror-smooth as expected. The fiber + agar unit is then placed on a tilted rotor in a 60 °C oven for two 20-min sessions of infiltration with fresh changes of our epoxy resin mixture [Araldite 506 + DDSA + DER736 + DMP-30; 10 + 15 + 2 + 0.49 g (=1.8%) (Reedy and Reedy, 1985; Sosa *et al.*, 1994)], then sliced into four to eight segments of ~1-mm length, each separately laid fiber-down, agar-above on a smooth polypropylene surface where it is pre-cured 4–6 h at 60 °C in the thinnest possible (<50 μm) "puddle" of resin. This forces the semiflattened fiber to lie flat and parallel to the cured resin surface, at a depth of 5 μm or less. The sample is then covered with an inverted BEEM capsule filled with fresh resin, and cured a further 48 h at 80 °C. This orients the tangential cutting plane ideally for longitudinal thin sections that sample the shallow, vitrified zone. This zone is typically no more than 4–6 μm in thickness, a limitation that we think is worth accepting because we are convinced that whatever was freeze-vitrified without the use of cryoprotectant or high pressure to retard ice crystal nucleation must have undergone the fastest freezing that is physically attainable.

For longitudinal sections, blocks are trimmed to include only the ~1-mm fiber segment, oriented with its long axis parallel to the diamond knife-edge. Sections of 25 nm thickness are cut with a Diatome diamond knife and picked up on 200-mesh Cu grids coated with a thin carbon film. Grids are stained with 2% aqueous $KMnO_4$ for 15 min, rinsed with 10 drops of Pal's bleach (3 drops each of 1% Na sulfite and 1% oxalic acid freshly diluted in 10 ml of water), then with H_2O, followed by Sato lead stain for 1 min. This high-contrast $KMnO_4$/lead stain sequence can only be recommended for sections of our Araldite formulation; permanganate reacts with several other embedding resins, although Epon with DDSA only (no NMA) can be used with limited success. (The stain granularity is coarser than with Araldite.)

Sato Lead Stain (Sato, 1968). This method has been adapted according to Reedy. Chemicals consist of 1-g lead nitrate, 1-g lead citrate, 1-g lead acetate, and 2-g sodium citrate. Add all chemicals to 82 ml of distilled H_2O and shake well for one minute. This makes a milky solution. Add 18 ml of 4% NaOH freshly prepared from pellets. Mix and the solution will clear except for a few grains. Filter rapidly through Whatman No. 1 filter paper and store in a dark glass bottle with a minimum of air space above the solution. Solution will keep for a month or more at room temperature. When a white precipitate coats the bottom and sides, staining intensity decreases; discard stain when this occurs. Recent Sato lead stain modifications said to prolong shelf life have been described especially during 2006 by online contributors to MSA's Microscopy ListServer (http://www.microscopy.org/index.html) (search on "Sato" or "citrate").

D. Physical Limitations of Freezing

The conditions described above can induce and preserve a comparatively synchronous change in the tilt of the myosin crossbridges. Active IFM crossbridges subjected to a sudden stretch or release must be freeze-trapped within 2–6 msec, before detachment–reattachment begins to obscure the synchronized tilt change. Good freezing of rigor crossbridges requires less demanding protocols because the crossbridge attachments are longer lasting and sustain mechanical perturbations over seconds. Rapid stretch of rigor fibers (20 msec stretch, freeze at 21–23 msec) produces semiruptures at most Z-bands, while slower ramp stretches (1000 msec, freeze at 1200 msec) produced rare Z-band ruptures.

Slam-freezing against a liquid helium-cooled metal mirror is reported to produce a 5- to 10-μm vitrified surface zone (in our insect fiber work, usually 4–6 μm) frozen in 0.2–2 msec (Bald, 1986; Heuser *et al.*, 1979; Padrón *et al.*, 1988). Past estimates of the time required for freeze-vitrification were based on theory or on indirect measurements, but there are two more recent reports of freezing rates based on the time required to arrest progressive distortion of muscle dimensions. Baatsen (1993) reports 0.5 or 1.5 msec for the time needed to arrest sarcomere length (SL) elongation in a relaxed muscle fiber at depths of 0–10 μm, using a cryoblock moving transversely at 2.5 m/sec, thereby stretching the fiber at a known rate. His "absolute" value, based directly on frozen SL, is 0.5 msec; his larger 1.5 msec value incorporates a correction for "real" frozen SL by the 10–12% factor required to correct the 38.5-nm troponin periodicity for shrinkage during freeze-substitution.

More exciting progress, for those hoping for still shorter freezing times (0.02–0.06 msec; avg. 0.044 msec) comes from the myofilament lattice compression measured at a depth of 2.5 μm and extrapolated for a depth of 10 μm (0.7 msec) (Bennett, 1998). Using the LN_2-chilled Gatan CryoSnapper, Bennett found cryoprotected, 100-μm fiber bundles compressed to 20–40% of their original thickness in 0.2 msec. Compressive distortion of the near-surface filament lattice increased with depth. Since compression and freezing both commence at the "zero" instant of cryoblock contact with the muscle surface, the known rate of compression, based on a measured jaws-closing velocity of 400 μm/msec, provides a timescale. Each measured degree of freeze-trapped lattice compression can thus be read out as a measure of the time interval separating zero time and the instant of local freezing. Compressive distortion was undisturbed by formation and growth of ice crystals, thanks to Bennet's use of cryoprotectant. We also recognized progressively larger ice crystals and compression deformation deeper in our specimens, detectable by elliptical rather than circular profiles of the fibrils. However, compression distortion is minimal in the surface zone even in the absence of cryoprotectant.

We define vitrification as the absence of distortion by unstained ice microcrystal "ghosts" >2.5 nm that appear in the Z-band before showing in the A-band, and by the presence in cross sections of circular fibril profiles; the zone showing these features guaranteed the fastest possible freeze-trapping of native structure. The addition of cryoprotectants to this system can have two negative effects. First, it can alter actomyosin biochemistry and its coupling to mechanical events and states. Second, it is very likely to produce good freezing deeper in the specimen by retarding ice crystal growth at places where the freezing rate is slower. Thus, it removes the correspondence between good preservation and submillisecond speed of freezing.

One further question is whether temperature-sensitive structural states produced and sustained at 20–40 °C can be faithfully cryopreserved when freezing necessarily takes the specimen transiently through cold, prefreezing temperatures that would normally alter or terminate the warm-dependent state. This will clearly depend on the rates at which the warm-dependent processes collapse. Bennett *et al.* (2002) report success with a modified CryoSnapper in cryopreserving signature diffraction features of the "hot" relaxed thick filament state in mammalian skeletal muscle, which did not revert to the cold relaxed state of crossbridge helical disorder during the $\leq$1 msec freezing process.

E. Validation of Specimen Preservation

Both the fibers and fibrils of IFM are too thick for ET, even with the 300–400 keV microscopes available today. Sectioning of cryopreserved specimens is therefore a necessity to obtain molecular resolution 3D images. Cryosectioning of frozen-hydrated tissue offers considerable promise, but various artifacts, among these section compression and poor adhesion to the EM grid, make the application to muscle tissue problematic at the moment (see Chapter 1 by Dubochet on cryosectioning for more details). Thus, freeze-substitution and sectioning of plastic embedded specimens is currently the most useful approach. The question therefore arises as to the fidelity of preservation of molecular structure in freeze-substituted tissue.

Most of the literature documenting preservation comes from fixed and embedded protein crystals. However, surprisingly few papers make quantitative comparisons between data derived from sections and the native structure as observed by X-ray diffraction. Several examples report preservation of *order* in protein crystals, sometimes to beyond 1-nm resolution (Langer *et al.*, 1975). More common are examples reporting preservation to just beyond 2-nm resolution using TA as a fixative (Akey *et al.*, 1980; Winkelmann *et al.*, 1991). TA combined with UrAc provides the embedded protein crystal with negative contrasting and appears to make the protein crystals resistant to section compression , which is a major artifact in ultramicrotomy (Akey *et al.*, 1980). Use of TA in muscle specimens also gives enhanced structure preservation, as judged by the presence of the 5.9- and 5.1-nm layer lines from actin (Reedy and Reedy, 1985; Sosa *et al.*, 1994), which are notoriously difficult to retain in plastic-embedded specimens. In IFM, however, this treatment does not prevent section compression , which is usually ~10%. The reported order preserved in X-ray diffraction patterns from cryofixed, freeze-substituted rabbit muscle fibers includes meridional layer lines of "even higher resolution" than 5.9 nm (Sosa *et al.*, 1994). Axial-period layer lines seen by X-ray fiber diffraction extend to 1.3 nm in our TAURAC-fixed IFM (Sader *et al.*, 2007), but these same specimens present the curious anomaly that EM of sections cut from the same block show no order beyond 5 nm.

One quantitative comparison has been made between a 3D reconstruction by EM and X-ray diffraction patterns from native muscle (Winkler *et al.*, 1996). The largest observed changes were an enhancement in the intensity of the equatorial 1,0 reflection relative to the 2,0 intensity and a reduction in intensity of the meridional 14.5-nm reflection. The equatorial changes were attributed to a greater buildup of TA on the thick filament compared to the actin filament; the 14.5-nm reduction was attributed to image processing, which enhanced the actin order at the expense of the thick filament order. However, this comparison was done on chemically fixed specimens, not on freeze-substituted specimens.

III. Data Collection and Tomogram Calculation

A. Single- or Dual-Axis Tilt Series?

Selecting an approach for tilt-series data collection involves two issues: (1) single-axis or dual-axis tilting and (2) tilt angle increment. Limitations on the maximum angle to which a specimen grid can be tilted mean that some views of the specimen that are needed to give isotropic resolution cannot be obtained. The missing data are known as the "missing wedge." Single-axis tilt series usually have a missing wedge in the range of $\pm 20^\circ$ to $\pm 30^\circ$. This produces anisotropic resolution such that the resolution in the tomogram along the direction of the optical axis is less than the resolution in plane.

The missing wedge, which is most easily defined in the Fourier transform of the tomogram, can affect both resolution and the relative weighting of features in the image. Technically, all atoms in the structure contribute to all Fourier coefficients and because of this no feature should be entirely missing from a tomogram, even with a $\pm 30^\circ$ missing wedge. However, electron tomograms are typically low-resolution images where the contribution of some specimen features to the Fourier transform may be concentrated within a small region. The two best examples would be the density profile across a membrane and the mean radial density distribution of a filament.

The membrane profile is a projection of the density onto a line that runs perpendicular to the membrane surface. Its Fourier transform consists of a single line oriented in the same direction. When this line comes to lie within the missing wedge, the region of the membrane that contributed to it will disappear from the tomogram. Thus, tomograms of membrane-bound vesicles usually show the lipid bilayer along only their lateral edges and not on the top and bottom surfaces (Dierksen *et al.*, 1995). Proteins that extend out of the membrane, such as envelope protein spikes, remain visible, but without the adjoining lipid bilayer (Zhu *et al.*, 2006).

The mean radial density distribution of a filament is a density projection down its long axis. In Fourier space, this information is concentrated in a disk of intensity that is oriented perpendicular to the filament axis and is often referred to as the equator of the filament transform. When the filament axis is perpendicular to the tilt axis, and the filament lies roughly within the specimen plane, the equator of the filament transform will lie within the missing wedge. Filaments with this orientation will all but disappear in the tomogram (Mastronarde, 1997). These problems, which have a serious effect on tomograms of cells, have provided much of the incentive toward development of dual-axis (Mastronarde, 1997; Penczek *et al.*, 1995) and conical tilt (Lanzavecchia *et al.*, 2001) tomography. Dual-axis tomography measures more of the specimen transform; instead of a missing wedge, there is a missing pyramid. In the case of conical tilt geometry, a cone of data is missing.

The missing wedge effect can in some instances be completely eliminated, even from data collected in a single-axis tilt series. For example, when particles (molecules or macromolecular assemblies) are randomly oriented throughout the specimen, their images can be combined to produce an average (see Chapter 29 by Förster and Hegerl, this volume). Tomograms of enveloped viruses, which have glycoproteins extending out from the lipid bilayer, are one such specimen (Förster *et al.*, 2005; Zhu *et al.*, 2006). In an average of these particles, the missing wedge can be entirely eliminated because the data missing from the transform of one particle is provided by other particles in different orientations.

For tomograms of sections through muscle fibers, the mean radial density distribution of the filaments can be obtained largely through judicious choice of specimen orientation within the electron microscope. In our IFM work, it has been possible to collect multiple tomograms from sections with the tilt axis oriented approximately parallel to the filament axis, thereby collecting as much of the mean radial density distribution as possible in a single-axis tilt series. Were the filaments the only feature desired from the tomogram, this consideration would be sufficient to rule in favor of single-axis tilting. However, IFM provides a unique opportunity to visualize the interactions between myosin and actin and usually these interactions occur along an axis perpendicular to the filament axis. For example, the crossbridges are aligned in almost the worst possible orientation for obtaining the most accurate rendition of their structure. A second tilt series taken with the tilt axis perpendicular to the filament axis and combined with the first improves the definition of the crossbridges as well as the definition of the actin monomers within the thin filament.

Collecting dual-axis tilt series for plastic-embedded or negatively stained specimens is relatively simple because these specimens will usually tolerate the additional electron dose applied in the second tilt series [but see the work by Luther (2006) for a more detailed discussion]. In the absence of a rotating, high-tilt-angle holder, simply extricating the grid from the microscope, rotating it 90° in the holder, replacing it in the microscope, and then collecting a second tilt series is a simple solution. The separate tomograms can then be merged into a single tomogram (Mastronarde, 1997; Penczek *et al.*, 1995) or images from both tilt series can be merged simultaneously (Chen *et al.*, 2001). Most of the published work on ET of IFM has involved single-axis tilt series (Liu *et al.*, 2004; Schmitz *et al.*, 1996, 1997; Taylor *et al.*, 1999) but the early efforts on volume classification used a dual-axis tilt series (Chen *et al.*, 2001, 2002), and our current efforts are using exclusively dual-axis tilt series.

B. Tilt Angle Increment

The quality of a tomogram depends on collecting the right number of projections for the specimen thickness and resolution desired. In tomography, this translates into the tilt increment between successive members of the tilt series. The minimum number of images, N, needed to obtain a resolution, d, from an object of diameter, D, is given by the formula (Crowther *et al.*, 1970a):

$$N = \pi D/d \tag{1}$$

Thus, for an object 100-nm thick and a desired resolution of 2 nm, 157 images would be needed. Since these images are spread over 180°, the tilt increment would be 1.14°, but if tilting is restricted to ±60°, a total of 104 images would typically be needed.

Technically speaking, fixed increment tilt series are valid only for specimens that are spherical or cylindrical with the tilt axis oriented parallel to the cylinder axis. (In fact, the formula was derived as a rule of thumb for reconstructing spherical viruses from electron micrographs.) In this case, the specimen does not increase in thickness as the specimen is tilted. Most tomograms, however, are computed from "slabs" wherein the specimen is distributed over a lamina of some thickness. As these specimens are tilted, they become thicker with respect to the viewing direction, and thus the rate of change in the image increases as the tilt angle increases.

Most tilt series are collected using a fixed increment in the tilt angle between images in the tilt series. The advantage of using a fixed angular increment lies in the weighting of the back projection, which is simply multiplication by R, where R is the radial distance of points in the Fourier transform of the back projection. This weighting function is popularly known as an R-weighted back projection. However, generalized weighting functions have been developed for collections of images in any orientation, and these are widely used in "single-particle" 3D reconstructions (Harauz and van Heel, 1986; Radermacher, 2006). Thus, the relative advantages of R weighting are small.

A scheme for tilt angle increments that accounts for the increasing thickness of the specimen as a function of tilt angle is the so-called "cosine rule," also known as the "Saxton rule" (Saxton *et al.*, 1984). With cosine rule tilt increments, the tilt increment decreases as the tilt angle increases, exactly matching the rate that the specimen's apparent thickness increases. Cosine rule tilt increments also provide more data in the upper tilt angles, where it is most desired; images at high tilt provide the information on placement of features along the direction of the optical axis, that is, the z-axis. Use of cosine rule tilt increments also has important consequences for marker-free image alignment, as described below.

The best choice of a regimen for data acquisition, however, cannot be based solely on considerations of resolution. Radiation damage also affects the quality of the tomogram. With thin sections of plastic-embedded, fixed and stained tissue, however, this effect is commonly, and perhaps incorrectly, ignored. Nonetheless, one must ask whether cosine rule tilting requires more images. If the fixed tilt increment is determined according to the specimen thickness in the untilted specimen, then more images are required under the cosine rule. However, if the fixed increment is determined according to the projected specimen thickness in the highest tilt angle desired, the more correct value, then fewer images are needed under the cosine rule.

C. Tilt-Series Alignment

Another consideration is the method used to align the tilt series. The tilt-series images that constitute the raw data for ET must be aligned before a 3D map can be calculated. The vast majority of tilt series are aligned using added gold clusters to provide fiducial markers. IMOD (Kremer *et al.*, 1996) is a popular and powerful software tool for accomplishing this. An alternative method, commonly known as "marker-free" alignment, aligns the images based on features that are intrinsic to the specimen, not on extraneous features such as added gold markers. The latter are of proven utility for alignment, but they are irrelevant to the actual structure.

In our work with ET on IFM, the tilt series have been aligned *exclusively* using marker-free alignment. This is partly for historical reasons. The goal of our IFM imaging has always been to obtain molecular resolution images of the muscle lattice; thus our work is fundamentally a bottom-up approach. All our earlier IFM 3D reconstructions used spatial averaging (Taylor *et al.*, 1984, 1989a,b). Alignment for this kind of data analysis is usually done with a "phase origin search," which is a special kind of cross-correlation function, applied to just the phases of the Fourier coefficients obtained at reciprocal lattice positions. To obtain a 3D image without spatial averaging simply required aligning the images using all Fourier coefficients rather than the few data points on the reciprocal lattice. The methodology was thus heavily biased toward marker-free alignment from the very beginning. Our method of marker-free alignment has been described in detail in several publications (Taylor *et al.*, 1997; Winkler, 2006; Winkler and Taylor, 2006). Several other marker-free alignment methods have been described in the literature (Dengler, 1989; Guckenberger, 1982; Liu *et al.*, 1995; Owen and Landis, 1996), but there is not enough space in this chapter to describe or compare them. A more thorough discussion of various marker-free alignment methods can be found in the work by Brandt (2006).

Two images (projections) can be aligned using cross correlation only if they are relatively similar in appearance. In Fourier space, this constraint can be restated: they can be aligned only if their Fourier transforms have appreciable overlap. From the Fourier projection theorem, the Fourier transform of a projection of a volume is a single plane in Fourier space passing through the origin, the so-called "central section." Real objects have finite thickness so the central section has an "effective" thickness in Fourier space that is the inverse of the specimen thickness. This means that there is overlap beyond just the common line of intersection between central sections of different orientation. This fact forms the basis of the "cosine rule" tilt increment defined above, as well as its importance for cross-correlation-based alignment.

The software implementation for our method of marker-free tilt-series alignment (named PROTOMO) is downloadable as executables, shell scripts, and documentation from http://www.electrontomography.org/software/linux/. A flow chart for all the steps used in IFM image processing is illustrated in Fig. 3 and consists of the following steps:

1. Raw images are initially cleaned to remove density outliers.
2. Coarse alignment of the tilt series begins by aligning successive images with the 0° tilt and progressing successively to higher tilt using a reference that

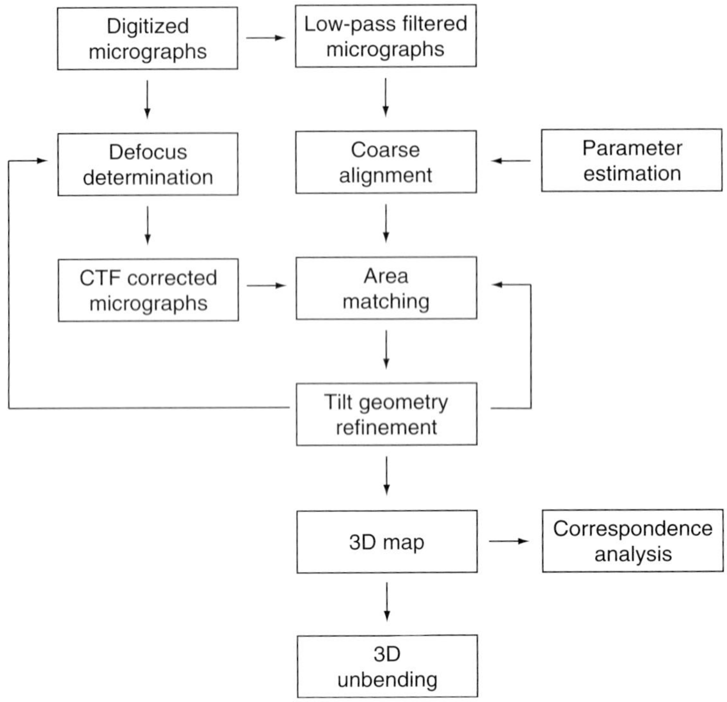

Fig. 3 Flow chart for marker-free alignment of tilt series. Although defocus determination and CTF gradient correction are possible (Winkler and Taylor, 2003), it provides minimal benefit for IFM sections when weighed against the computational expense. The steps of area matching and tilt geometry refinement are cycled until the data corrections are well below 1%. Adapted from Winkler and Taylor (2006), with permission.

represents the neighboring image at the adjacent lower tilt angle. This process proceeds until all the images in the tilt series are roughly aligned. Images are low-pass filtered to the first zero in the contrast transfer function (CTF). Parameter estimation in this step uses the tilt angles from the goniometer and the tilt azimuth obtained from previous tomograms collected at this magnification. Coarse alignment takes only minutes after the last image in the tilt series is collected.

3. From the first alignment of all images, a back-projection tomogram is calculated to produce a crude reconstruction. This reconstruction can be examined to determine whether the specimen has the features desired. If so, the tomogram is further refined.

4. An "area matching" refinement scheme is undertaken in which each member of the tilt series is aligned to a reference obtained by reprojecting the preliminary back projection. Area matching determines four parameters of a linear transformation (stretching factors) to match each image to the reference.

5. Geometry fitting takes the four parameters obtained from the area matching of each image and performs a least-squares fit to obtain the common tilt axis and the three Euler angles that define the specimen orientation within the coordinate frame of the tilt and optical axis. Although initial alignment of the tilt series is fast, the refinement of the tilt series is time consuming.

A key feature of this process is its use of a reprojection of a previously calculated back projection to obtain the reference image. The projection of the tomogram has the same SNR, based on electron counts, as the sum of all the images contributing to the tomogram (Hegerl and Hoppe, 1976). It is thus a much better reference, statistically speaking, than any single projection in the tilt series. Using a reprojection of the tomogram also prevents alignment errors from propagating through the alignment process. PROTOMO has the capability to calculate four different cross-correlation functions: (1) standard; (2) mutual (Van Heel *et al.*, 1992) (which is useful for searching for the correct peak when the cross-correlation function is generated from images of arrays, for example muscle); (3) phases only [in which the amplitudes in the transforms are set to 1.0 (Horner and Gianino, 1984); this produces especially sharp correlation peaks, even in spatially redundant specimens]; and (4) a phase-doubled correlation function (Saxton, 1994) (which increases the accuracy by doubling the magnitude of the shift).

In addition to determining stretching factors that match the individual images as accurately as possible, that is, area matching, PROTOMO refines four parameters that are constant across all images in the tilt series: the tilt azimuth of the specimen holder and three Euler angles that define the orientation of the specimen with respect to the specimen holder (Fig. 4). One of these amounts to the deviation of the nominal untilted image from 0°. The *increment* between successive tilt

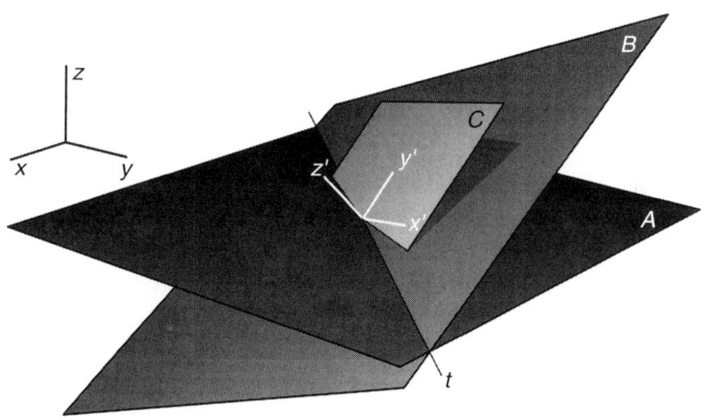

Fig. 4 Definition of orientation parameters determined during the area matching and geometry refinement procedure of marker-free alignment. Geometry refinement determines the tilt azimuth and three Euler angles that define the orientation of the specimen with respect to the tilt and optical axes. Adapted from Winkler and Taylor (2006), with permission.

angles is assumed to be accurate and is not adjusted because small changes in tilt angle give only small changes in individual images, whereas small changes in tilt azimuth and the specimen orientation produce measurable changes in all images. The quality of the alignment is assessed in two ways: (1) the shape and height of the cross-correlation peaks (Fig. 5A) and (2) by graphing the deviation

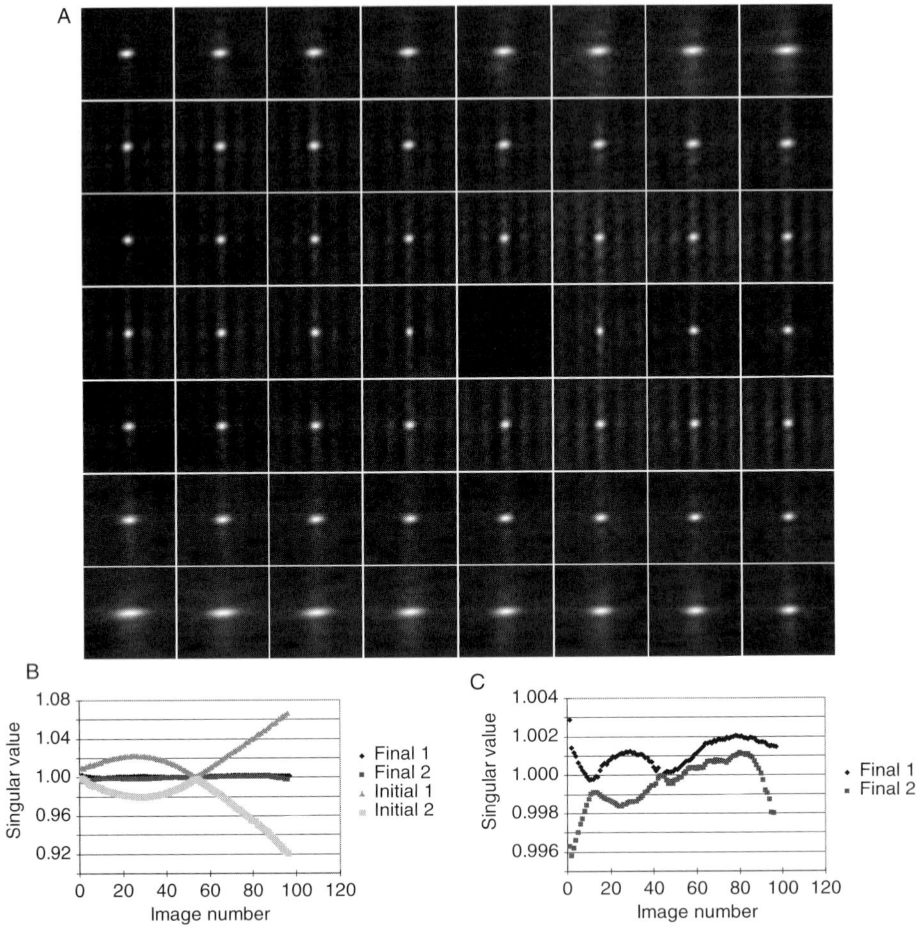

Fig. 5 Evaluation of tilt-series alignment. (A) Cross-correlation peaks. The tilt axis is oriented vertically which means that the stretching to resample each image occurs in the horizontal direction. Hence, the correlation peaks often appear stretched in this direction. The blank box corresponds to the nominal zero tilt, which is not modified in the area matching. (B) Plots of singular values which represent stretching factors from area matching prior to fitting the orientation parameters. Initially, the residuals can be quite large, but after area matching, are reduced to less than 1%. (C) Enlargement after final refinement. [Panel A courtesy of H. Winkler; panels B and C adapted from Winkler and Taylor (2006), with permission.]

obtained by area matching from the initial geometry (Fig. 5B). Cross-correlation peaks should be sharp and strong relative to the local background, but they usually broaden in a direction perpendicular to the tilt axis as tilt angle increases because of the foreshortening of the projection of the specimen (Fig. 5A). The deviations should be flat across the tilt series when alignment and area matching are completed. They are often, however, a bit worse at the high tilt angles (Fig. 5B).

Marker-free alignment works readily with thin specimens, like the sections of IFM that we have used for ET; they are 25- to 30-nm thick. Sections this thick give the best detail about the structures of the protein molecules in the specimen. Our method has also produced excellent alignment for different kinds of ice-embedded specimens up to ~100-nm thick (Liu *et al.*, 2006a; Zhu *et al.*, 2006). Thicker specimens seem to be more problematic, as are tilt series collected using fixed angle increments, unless the increments were defined according to the effective thickness at the highest tilt angle. The reason for this limitation is unclear, but one likely explanation lies in the very nature of the cross-correlation function. If two images are closely correlated in space, they will have high correlation coefficients; if not, the correlation will be poor. Relative to the area selected from the image at 0° tilt, as the specimen is tilted, portions of this area near the boundary are tilted outside the original boundary and neighboring regions outside the boundary are moved in. This results in steadily increasing differences between successive images with increasing tilt angle. This effect also increases with increasing specimen thickness. This change is smooth if the cosine rule is followed, but becomes more and more abrupt with fixed angle tilting. At the moment, there is no practical solution to this problem, which is not present when fiducials are used for alignment.

D. Computing and Displaying the Tomogram

The vast majority of tomograms are computed using weighted back projections. Alternative methods include Algebraic Reconstruction Techniques (Gordon *et al.*, 1970), Whittaker-Shannon interpolation in Fourier Space (Crowther *et al.*, 1970b), or reconstruction using orthogonal functions (Zeitler, 2006). In our work, we have used both weighted back projections and Whittaker-Shannon interpolation. Whittaker-Shannon interpolation is feasible only because when the images are aligned and area matched, accurate transformations are available to resample (i.e., stretch) each image of a tilt series onto the same grid. Then least-squares fitting to resample the transform can be done in one dimension (1D) along the equivalent of "lattice lines," similar to that used in two-dimensional (2D) crystal reconstruction. Resampling requires knowledge of the specimen thickness, but that is provided by the preliminary back projection computed during tilt-series alignment. There has yet been no detailed comparison between tomograms computed by the two methods. In our hands they are, for all practical purposes, indistinguishable. That said, in order to avoid artifacts in the image using Whittaker-Shannon interpolation, data must be collected according to the cosine rule, so the spacing between measurements near the resolution limit stays within the correct interval for the specimen thickness.

Interpreting the "raw" tomogram at the level of individual molecules is difficult, regardless of the specimen, because of problems with low SNR. Noise can be stochastic due to low electron counts, or in the case of fixed, embedded and stained sections, it can result from variations in the preservation and staining of the biological material. Regardless of its source, the display of the "raw" tomogram as a 3D volume is a frustrating exercise. A visualization tool such as IMOD that can pan through the actual density map is much more gratifying and has been used for analysis of IFM tomograms in one case (Tregear *et al.*, 2004). The noise in raw tomograms has provided much of the incentive for development of denoising algorithms (Frangakis and Hegerl, 2001; Jiang *et al.*, 2003; Stoschek and Hegerl, 1997). However, striated muscle offers an alternative method for improving the image SNR, the identification of multiple copies of equivalent structures, their alignment and averaging. This strategy has provided the possibility not only to improve the SNR in a tomogram, but also to perform operations such as quantifying the relative amounts of the different structures within the tomogram.

IV. Identifying Similar Structures in an Ensemble

Correspondence analysis is a powerful technique for analyzing patterns in data. It underlies all single-particle reconstruction methods. Correspondence analysis has been described in detail in an excellent monograph by Joachim Frank (Frank, 2006), so only the parts that are pertinent to the present discussion will be mentioned here. In EM, correspondence analysis has been applied largely to projection images. With tomograms, on the other hand, the process must be applied to volumes, so the missing wedge becomes an important issue (see Chapter 29 by Förster and Hegerl, this volume and Taylor *et al.*, 2006).

The four important aspects to correspondence analysis of volumes are: (1) selection of subvolumes that are appropriate for classification, (2) alignment of these subvolumes, (3) selection of the "classification mask," which defines the voxels that will be used in the classification, and (4) specifying the number of "classes" that will be computed.

A. Effect of the Missing Wedge

The complicating effect of the missing wedge on correspondence analysis of volumes is best described with reference to correspondence analysis of 2D projections in single-particle 3D reconstruction, for which this procedure is an integral component. The Fourier transform of a 2D projection from a 3D volume is a central section in the Fourier transform of that volume. Thus, the central section in Fourier space provided by the transform of a 2D projection can be thought of as the Fourier transform of the original 3D volume but with a ±90° missing wedge (as compared to the ±30° missing wedge of a typical tomographic volume). In single-particle reconstruction, correspondence analysis of a set of aligned projections groups them according to their orientation, and this is in fact the goal. In other words, images are grouped according to the orientation of their missing wedges. Once projections with the same orientation are determined and averaged, the 3D reconstruction will be obtained after determining the relative orientations between averaged projections.

In ET, each repeating volume is a 3D image by itself. Correspondence analysis can therefore be used to characterize structure variability from an aligned set of repeat volumes to improve SNR without sacrificing details that are heterogeneous. However, if correspondence analysis will classify a set of 2D projections according to the orientation of the "missing wedge" as described above, will it also classify a set of tomographic volumes according to the orientations of their missing wedges rather than according to the desired structure differences? At the moment, this is an open question, but with tomograms of IFM thin sections, the missing wedge, or pyramid if a dual-axis tomogram is used, is in virtually the same orientation for all repeat volumes and is therefore not expected to affect grouping according to structure.

There are a number of scenarios where the missing wedge might be expected to produce surprising and therefore unwanted effects on the classification of volume repeats. One such example is the classification of actomyosin interactions in a thick specimen. A longitudinal section of IFM, 50- to 60-nm thick, would contain all of the six thin filaments surrounding a thick filament. Myosin crossbridges would approach these actin filaments from six different directions with respect to the tilt axis, and therefore all six groups would show different effects of the missing wedge. Classification of the entire ensemble may cluster repeats according to missing wedge orientation rather than structure. However, this predicted effect has not yet been demonstrated, and once this kind of analysis is attempted, there may be no effect at all. A similar expectation would exist for classification of actomyosin repeat volumes obtained from tomograms of non-muscle cells. Experience so far with tomograms of 2D arrays on lipid monolayers is that classification is not strongly affected by missing wedge orientation (Liu *et al.*, 2006a). However, for these specimens, each repeat volume has in common the transform data from the in-plane projection, which may be dominating the grouping, although again this has not yet been shown. Finally, volume classification of proteins embedded in membrane vesicles produces reasonable results, so long as the proteins extracted from the top and bottom of the vesicle, where the membrane is invisible, are kept separate from those extracted from the vesicle sides, where the membrane is obvious (Zhu *et al.*, 2006).

In conclusion, although the missing wedge may have unwanted effects on repeat volume classification, this is not a reason to avoid the power of the method. It is only necessary to be aware that problems may occur and that initial results should be scrutinized carefully. Different kinds of specimens may show differing effects.

B. Selecting and Extracting Repeating Motifs

The first step in correspondence analysis is the extraction of individual repeats, that is the subvolumes, that will be classified. At the moment, there are few tools specifically designed to facilitate the manual picking of structures for 3D alignment and classification and fewer still for automated particle picking. In this regard, tomography lags behind the much more developed automated software for single-particle reconstruction from projections (Glaeser, 2004; Zhu *et al.*, 2004). Some consideration should be given to the most useful repeat segment. For example, the "target zone" on actin for actively contracting IFM is located midway between successive troponin complexes (Taylor *et al.*, 1999; Tregear *et al.*, 1998). In all IFM states except relaxed, troponin is easily identified in projections of longitudinal sections because of its size. Thus, we select repeats centered on the target zone of active contraction between these troponin densities.

For manual particle picking in tomograms, we have used several methods. Note that Unix shell scripts or executables used in the procedures are denoted by brackets [see tables in the work by Winkler (2007) for a more complete explanation].

1. EMAN (Ludtke *et al.*, 1999). This is an excellent particle-picking tool for projections and can be applied to either a projection of the tomogram or to an extracted *z*-section. Usually an extracted *z*-section has better contrast than the tomogram projection, but if the specimen is not flat, a projection may be better. Most of the repeats in a reconstruction of IFM are centered in the tomogram. Thus, the "*z*" coordinate can be set to 0.0 and the (x, y) center of each copy of a repeating structure identified by inspection and picked as if it were a single-particle and the list obtained from the EMAN output. With a list of centers for each repeat, a program [extract] is run to window the subvolumes.
2. We have also developed an "in-house" tool that facilitates manual particle picking within tomographic volumes (Winkler, 2007). This tool writes the x,y,z coordinate to a file that is used as input to the windowing script mentioned in step (1).
3. IMOD's image displaying and modeling program 3dmod can be used to pick up repeat coordinates (Kremer *et al.*, 1996). In the Human and Simian Immunodeficiency Virus (HIV/SIV) study (Zhu *et al.*, 2006), each envelope glycoprotein (Env) spike was treated as a separate object. After picking all points (the centers of the spikes), the model was saved as a.mod file. Then the model2point tool in IMOD was used to get the coordinates in ASCII format. Each line in the output consists of one (x,y,z) coordinate that is used as input into the windowing script in (1). Note that IMOD also has built-in volume extracting capability.

IFM is a paracrystalline object, so the repeats occur on a regular lattice. Thus, there are consistent features within each 38.7-nm repeat. A cross-correlation map is computed that determines the centers of each repeating unit according to a search template. A lattice can then be fit to either the 3D coordinates or the 2D coordinates obtained from the projected tomogram. Using these centers, the volumes of each repeat can be extracted. The axial length is usually longer than 38.7 nm to allow for axial shifts during alignment to correct for imprecise centers in the cross-correlation map.

The following procedure is used to compute the cross-correlation function, fit the centers of the peaks, and extract the individual repeats. It is modified from procedures used for the 3D unbending of tomograms obtained from paracrystalline specimens (Winkler and Taylor, 1996). Many of the steps can be accomplished with any of the lattice-fitting programs used to analyze 2D crystals. These typically fit a lattice to the "*x*, *y*" coordinates and write the results to a file that can be used as input to an extracting script (Henderson *et al.*, 1986; Schmid *et al.*, 1993).

1. Project the tomogram. This could also be an extracted central z-section.
2. Compute its Fourier transform.
3. Determine the reciprocal lattice vectors. We use a standard electron crystallography program, SPECTRA (Schmid *et al.*, 1993), to determine these vectors, from which the real space lattice vectors will be determined. SPECTRA also computes the cross-correlation map and fits a lattice to the centers of the correlation peaks if desired.
4. Average the map [tomoavg2.sh] using parameters obtained in the step (3).
5. Construct a reference from the averaged map [tomomot.sh]. The center, size, and mask of the reference can be modified by editing the script.
6. Compute a cross-correlation map [tomocorr.sh] using the reference obtained in step (5).
7. Peak search [tomosearch.sh]. Start position, searching vectors, index limits, and so on are modified in the script. Outputs include a peak position file. Depending on the quality of the tomogram, this step may be repeated using different start positions.
8. Verify peak positions by displaying the position file together with the cross-correlation map or the tomogram projection. It may be necessary to edit the position file manually.
9. Extract repeats using the position file [extract]. Size of repeats is defined by options in the command line. Outputs include the raw repeat stack and the corresponding position list.
10. The repeat volumes are contained within an "image stack," which is a four-dimensional (4D) array in which three dimensions define the individual volume size and the fourth specifies the number of volumes contained within the file.

With specimens that lack the preferred particle orientation present in muscle, an automated particle search is computationally demanding because it is necessary to rotate the reference over many orientations in fine enough increments that potential repeats are not missed because of a large orientation difference (see Chapter 29 by Förster and Hegerl, this volume). This process simultaneously defines the coarse particle orientation, which can later be refined. Several efforts at developing automated particle selection software for tomographic data have used the cross-correlation function as a basis for determining similarity (Bohm *et al.*, 2000; Frangakis *et al.*, 2002; Rath *et al.*, 2003). A potential pitfall of this approach is that the correlation peak height, which is the measure of similarity between the reference and the "raw" repeat, depends not only on the similarity in orientation and structure, but also on the local normalization. The missing wedge can affect the local normalization (Frangakis *et al.*, 2002), which is to say that high correlation due to similarities in particle structure and alignment could be offset if the local normalization is very different from the global normalization. The fast, local correlation function (Roseman, 2003) was developed to address this issue in automated particle selection for single-particle reconstruction, and it has been applied to 3D particle selection within tomograms of frozen-hydrated samples (Rath *et al.*, 2003).

C. Alignment of Repeating Units

Before applying correspondence analysis to IFM 3D repeats, each of the repeating elements must first be aligned to an appropriate reference. In projection alignment, only three parameters need be determined—x, y translation, and rotational around the z-axis. In volume alignment, on the other hand, there are six degrees of freedom: three translational and three rotational. This is computationally expensive. IFM repeats come from a lattice, so the alignment step is much simpler. However, there are still several types of disorder that must be detected and corrected if the actin monomers are to be resolved in the averages. With subvolumes defined by the 38.7-nm repeat of actin, two kinds of disorder should be examined. (1) The helical symmetry of the actin filament causes successive repeats along a single actin filament to be related by a 180° rotation about the filament axis. (2) Actin filaments within a half sarcomere are also rotated randomly by 180° about the filament axis.

To avoid reference bias in projection alignment, the initial classification is often done using derived functions that are invariant to translation and rotation. Such functions include the double autocorrelation function (Schatz and van Heel, 1990) and the double self-correlation function (Schatz and van Heel, 1992). For volume classification, generally applicable invariant functions have not been derived but some suitable for specific applications have been constructed.

A modified autocorrelation function for IFM classification (Winkler and Taylor, 1999) has been constructed, which was both invariant to translation and would distinguish the 180° ambiguity of the thin filament orientation (Holmes *et al.*, 1980). In its first application, a major limitation was the number of repeats; only about 400 were available with which to compute 16 classes. The SNR improvement was by only a factor of 4 on average in each class and resolution by the spectral SNR (Unser *et al.*, 1987) was only 7 nm.

D. Reference-Based Alignment Schemes

Addressing fundamental questions about muscle contraction requires a resolution of ~5 nm, so that the actin filament orientation can be defined independent of the myosin crossbridges. Invariant functions combined with multireference alignment of the actin filament and crossbridge, as described above, did not achieve this goal. Reference-based alignment schemes applied to projections have been very successful (Burgess *et al.*, 2004). The first application of reference-based alignment to volume data from IFM (Liu *et al.*, 2004) used the actin filament and the troponin complex as the reference. Reference-based alignment makes the implicit assumption that the structure of the reference, the actin filament, is invariant in each repeat. This assumption is reasonable at the resolution of electron tomograms of plastic-embedded specimens.

Our extraction, alignment, and classification procedures are performed using a data file, the "position file," which contains the voxel centers of each repeat referred to its original tomogram, plus the x,y,z shifts for translation alignment and the three Euler angles that define the optimal alignment (Winkler, 2007). This file is separate from the image stack defined above. Alignment of the repeats is usually done by searching over a small angular cone, whose angles are determined from a script. Transformations are applied to the original extracted repeats only once, thereby avoiding successive interpolations, which would lead to a loss of power at high resolution. The angular alignment is done over a grid that precesses around the axis determined by the previous best alignment. When completed, the best transformation is selected based on the highest correlation coefficient and the position file updated. Before classification, the best transformation is applied to each raw repeat to generate an aligned image stack.

The following procedure combines correspondence analysis with repeat alignment. At this stage the chief benefit of correspondence analysis is to detect errors in the alignment. Each cycle of correspondence analysis generates "class averages," which are the averages formed from the groups of individual repeats (classes). Although smaller in number, the repeats within the individual classes are typically more homogeneous than the full data set. For example, in the initial alignment, it is not uncommon to have two classes with the actin filament alignment differing by 180° rotation about the filament axis.

1. The raw repeats, already partially aligned thanks to the correlation map used to identify them, are aligned to a single actin filament selected from among the raw repeats to act as a reference. This alignment involves only a small angular search within a cone of ~4° width about the filament axis. This search removes any out-of-plane tilt of the filament segment and corrects for in-plane bends in the filament.

2. Several cycles of correspondence analysis followed by multireference alignment are carried out until the global average of all the actin filaments reveals actin subunits. Alignment and classification are done using a cylindrical mask that selects the actin filament and a small part of the crossbridges that cannot be windowed out efficiently. The mask also includes the troponin complex.

3. When classification reveals errors in the actin alignment, an improved reference is obtained by performing a cycle of alignment using the class averages themselves, followed by averaging. The changes in orientation that are determined from the alignment of the individual class averages to the reference can then be applied to the members of the class before performing the next alignment and classification cycle. Alternatively, all the volumes are realigned to an improved reference that is the average of the aligned class averages.

4. This process continues until one is satisfied with the alignment, which usually occurs when the shifts between successive cycles are small.

There is no guarantee that actin monomers will be resolved in the average of all repeats, the global average, since that is also a function of preservation and staining. Nonetheless, two features of the IFM muscle lattice provide cross-validation that the alignment is working (Fig. 6). One of these is provided by the presence of the large troponin complex of IFM (Bullard, 1984), which is enhanced in the averages when their alignment is good (Fig. 6A). The second is provided by the actin filaments themselves. In the actin target zone where "lead" crossbridges bind, the orientation of actin monomers is optimal for revealing the 2.75-nm offset between the two, long-pitch actin strands (Fig. 6B). The offset should become visible when the alignment is good enough and the SNR is high. It is not unusual, however, that the global average may fail to display any crossbridges if these structures are highly variable.

E. Mask Construction

The classification mask is a Boolean image (made up of 1's and 0's) that determines which voxels will be used in the correspondence analysis; thus, it has a strong effect on the result. The mask for correspondence analysis of volume data can be done in two ways with minimal subjectivity. The following procedure is adapted from mask construction in projection image processing (Frank, 2006):

1. Compute the global average of all aligned repeats.
2. Threshold the global average to produce a Boolean image.
3. Low pass filter the result. This will soften the edges.
4. Threshold again to produce the final mask.

Another method we have used is to form a Boolean image of the different class averages and combine them using a logical .or. operation. This gives a larger mask but one that is focused on the regions where the crossbridge features are occurring.

The alignment of repeats from IFM tomograms assumed that the actin filament was constant in structure. Thus, alignment on the actin filament should minimize its contribution to the interrepeat variance. The crossbridges, which are the desired part of the structure, would then be the source of highest interrepeat variance. Therefore, eliminating the actin filament from the mask focuses the correspondence

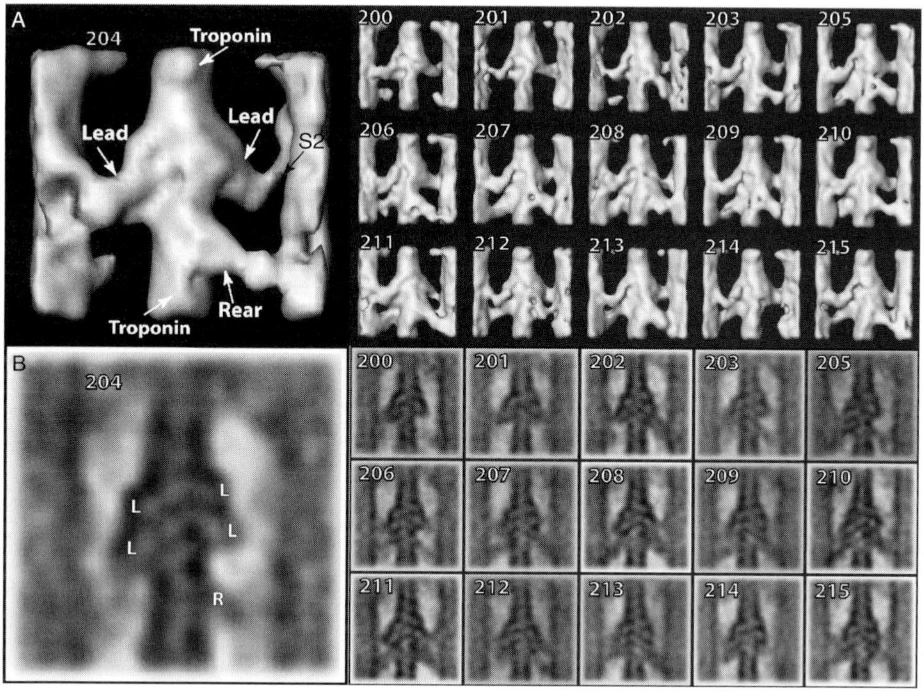

Fig. 6 Montage of 16 class averages calculated from 322 crossbridge repeats from a tomogram of rigor IFM fibers frozen after a ramp stretch. Thin filaments occupy the middle with thick filaments placed to either side. The orientation has Z-band at the bottom, M-line at the top. (A) Surface renditions. Lead bridges form the bridge pair in the middle; rear bridges form the pair toward the bottom. Numbers identify the class: single chevrons (classes 200 and 201) and incomplete double chevrons (classes 202, 204, and 214). Complete double chevrons compose the rest. In the left-hand panel are labeled the head region of troponin and a segment of S2 which has been pulled clear of the thick filament backbone. (B) Montage of z-sections through the center of the actin filament of each class averages shown in (A). The staggered arrangement of lighter density (actin monomers and crossbridges) indicates that alignment resolves actin monomers with a resolution of 5 nm. A distinct layer of stain surrounds the thin filament and attached myosin heads giving them a distinct "negatively stained" appearance and suggesting that the region corresponding to protein lies within the stain shell and also that the surface defined in (A) is actually within the stain shell. Only large structures appear negatively stained; smaller structures such as the crossbridge lever arm appear positively stained or dark. "L" and "R" denote motor domains of lead and rear crossbridges. Adapted from Liu *et al.* (2004), with permission.

analysis on the most variable parts of the structure. The actin filament can be eliminated from the mask by extracting the individual z-sections as tiff images, editing in Adobe Photoshop, and reassembling them into a volume. This procedure can be used to edit the mask in any way desired to emphasize the classification on any feature. This can be important if a volume representing the original tomogram is to be reassembled from high SNR averages of specific regions (see below).

Alternatively, the variance map of the global average can be substituted for the global average itself. A mask obtained from the variance map focuses the classification on those parts of the structure that exhibit the highest variability. The risk is that the variance map over a large population of highly heterogeneous repeats, such as those in IFM, may be featureless in the region of interest. This approach is more useful in those cases where the interrepeat variance is concentrated in a small number of features.

F. Correspondence Analysis of the Crossbridges

The previous section described an approach to optimize the alignment of the actin filaments. This is important because errors in alignment contribute to the structure variability in the class averages of the crossbridges. To obtain class averages with improved SNR the crossbridge patterns within the set of aligned repeats must be identified so that averaging is done only among "self-similar" repeats. The two classification schemes most widely used with 2D images are K-means and hierarchical ascendant classification (HAC), for example (Frank, 1990). The potential benefits of one of these over the other have not yet been explored for volume classification in IFM where we have used HAC exclusively. A neural net-based method, Self Organizing Maps (SOM) has also been shown to be effective in clustering crossbridge repeats (Pascual-Montano *et al.*, 2002).

G. Number of Factors to Use

Before grouping the collection of raw repeats, it is first necessary to determine the number of factors that represent the structure variance within the population of repeat volumes. This number will be only a small subset of the total number of factors determined; the rest will represent the noise component of the data. Without knowing what the factors represent in terms of structure, the number can nevertheless be determined in a straightforward fashion by a plot of the percentage of the variance contained in the factors as a function of factor number after ranking the factors in order of decreasing magnitude (Fig. 7). Factors that represent noise are readily identified because they usually decrease linearly, whereas the factors that represent actual structure in the data are usually the few high values at the beginning. By extending a line from the asymptote of the noise factors, those factors that contain signal are easily identified. In Fig. 7, no more than 13 factors would be needed to describe the structure variance in the data; the rest are all noise. However, a smaller number could easily be justified.

H. Number of Classes to Compute

Correspondence analysis of repeat volumes in IFM is used to bin what is literally a continuum of varying structures within the individual repeats. An important question, therefore, is the appropriate number of bins. The paired goals of SNR

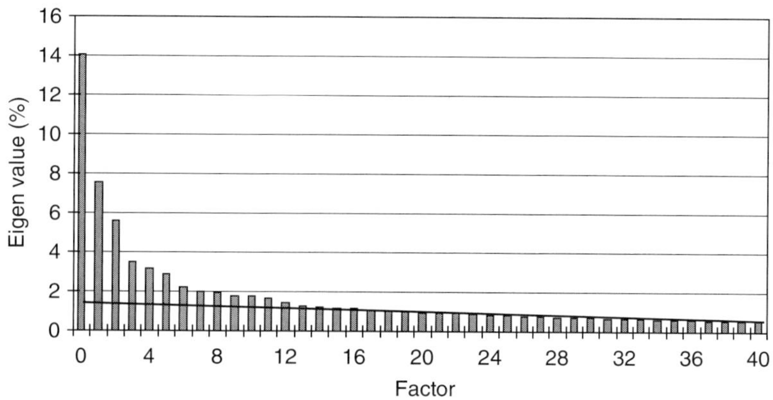

Fig. 7 Histogram of 40 factors obtained from the correspondence analysis of aligned swollen rigor repeats classified separately as left and right sides. Factors that represent the signal within the data are the strong ones near the origin. Factors that represent noise can be detected from the histogram because they decrease at a near-linear rate. The straight line is a manual fit to the "noise" factors. Factors that contain "signal" emerge as the ones that extend above the line. It is important to determine the number of factors that represent the "signal" portion of the data, as this defines the space in which clustering will be done.

improvement and the retention of structure variability are conflicting; SNR improvement is achieved by computing fewer classes, but retention of interimage variance is achieved by computing more classes. HAC permits different numbers of classes to be computed from a single classification but does not provide a ready estimate of the correct number of classes.

One possibility is an intelligent guess by considering the number of possible ways that crossbridges might plausibly attach to F-actin. In the first application of correspondence analysis to IFM, the number of possible *arrangements* of myosin heads in rigor repeats was estimated 9 and then approximately doubled to allow for some variation in *structure* among the different arrangements (Winkler and Taylor, 1999). Each of the 16 classes had between 9 and 29 repeats. Validation was provided by calculating the number of myosin heads represented by the classification. An average value of 2.7 myosin heads per repeat was obtained from the number of repeats per class and the number of myosin heads per repeat based on molecular modeling (Chen *et al.*, 2002). This result was in excellent agreement with experiment (Goody *et al.*, 1985; Lovell *et al.*, 1981; Thomas *et al.*, 1983).

A more systematic way is examination of the dendrogram. The dendrogram, which shows the relationship among classes, is a key component of HAC. By examining the dendrogram, one can identify classes that are closely related, thereby enabling the number of classes to be reduced. In Fig. 8, the top four classes are relatively similar double-chevron motifs, the next four classes are right-hand incomplete double chevrons, while the next three classes are single chevrons. The relationship between the bottom five classes is somewhat less distinctive.

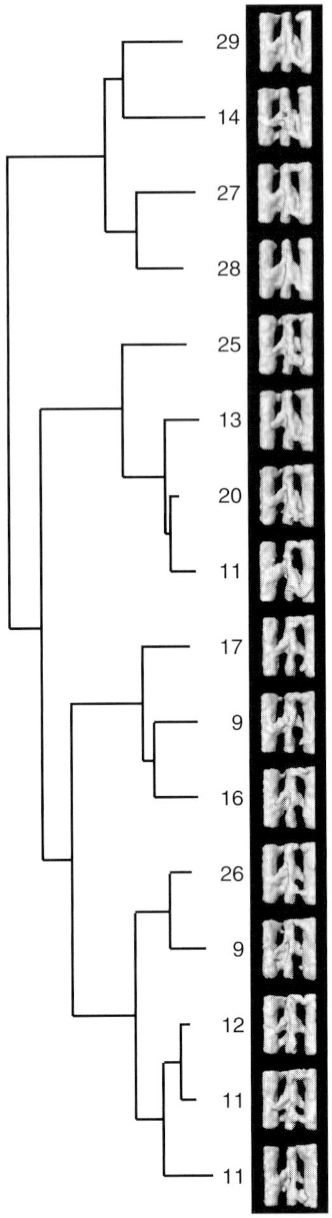

Fig. 8 Dendrogram from a classification of rigor IFM. The dendrogram contains information on the relationship between classes. Numbers adjacent to the pictures are the number of repeats in each class. Adapted from Winkler and Taylor (1999), with permission.

This process can be tedious if the suspected number of classes is large. It is not uncommon that 40–60 classes may be selected in classification of highly variable crossbridge repeats in IFM.

SOM (Pascual-Montano *et al.*, 2002) have the attractive feature of being fast and well adapted to the analysis of large, noisy data sets. They can also predict the number of particle classes. When applied to the same set of aligned repeats employed previously in an HAC classification (in fact the same data shown in Fig. 8), SOM predicted a smaller number of classes, but those predicted were nearly identical to ones obtained using HAC. SOM in fact verified what the dendrogram suggested, that is, fewer classes could be used to reveal the range of repeat variation.

Nevertheless, deciding *a priori* on the correct number of classes in highly heterogeneous data remains a difficult and important issue. One solution is separate classification for the different parts of the repeats. Until recently, classification of IFM repeats was done by including the crossbridges on both sides of the actin filament within the classification mask (myosin heads approach the actin filament from only two sides in IFM instead of from three sides in vertebrate striated muscle). We refer to this as two-sided classification because it assumes some correlation between structures on the left and right sides of the actin filament. However, several factors contribute to eliminating any correlation, if, indeed, it exists at all. First, the target zone in contracting muscle is three actin monomers long on one side of the actin filament and only two actins long on the other side (Tregear *et al.*, 2004). Second, aligning the repeats on the actin filament removes any registration along the thick filaments; successive repeats along an actin filament are rotated 180° so segments from all thick filaments are mixed on left and right sides of the aligned repeats. Third, in contracting muscle (Taylor *et al.*, 1999) and probably in rigor muscle, the axial registration of the thick filaments that is present in relaxed muscle (Schmitz *et al.*, 1994) is lost so that any correspondence between myosin head origins on the pair of thick filaments surrounding the actin filament does not exist in the aligned repeats. Thus, it is safe to assume that the myosin head distribution on the left side of the actin filament is independent of the head distribution on the right-hand side removing any justification for simultaneous classification of repeats using features on both sides; classification should be done on the left and right sides separately.

The following may illustrate how separate left and right side classification can significantly increase the amount of structure variance recoverable from image classification without sacrificing SNR improvement. Suppose there were nine different ways of arranging myosin heads on each side of the actin filaments within each 38.7-nm repeat. Simultaneous left–right classification would require 81 classes to recover the original arrangement. On the other hand, if left and right sides were classified separately, only nine classes would need to be computed on each side, resulting in a threefold increase in SNR for the same number of repeats. By recombining the separate left and right side class averages in all combinations, the original 81 classes can be obtained. Thus, a simple rule of thumb is to compute separate left and right side class averages corresponding to the square root of the total number of repeats being classified. Then each repeat can be reassembled using the left side and right side class averages. The benefits of this approach are easily apparent after describing the process of reassembling tomograms using the class averages (Section VII).

I. Alternative Averaging Strategies

For reasons outlined in Introduction, simple spatial averaging is likely to cause loss of more information than the improvement in SNR would justify, and this was the driving force for applying correspondence analysis to IFM tomograms. However, alternative spatial averaging methods exist that can in many instances give improved SNR without blurring out all the desired structures. Column averaging is a procedure that makes use of the highly regular structure of the individual filaments even in the presence of disordered placement within the filament lattice. Column averaging produces 1D averages along the filaments and was first used in IFM tomography on fast-frozen, isometrically contracting fibers (Taylor *et al.*, 1999).

Column averages are produced from a tomogram using the following protocol that is an extension of 3D unbending (Fig. 9) (Winkler and Taylor, 1996):

1. Compute a 3D cross-correlation map. It is often useful for this process to weight down the equatorial layer plane as it has two very strong spots that in effect broaden the cross-correlation peaks.
2. Fit a lattice to the cross-correlation map. This lattice will be used in step (3) to straighten the filaments and remove any wrinkles in the section.
3. Resample the tomogram so that the long direction of the filaments is aligned with one of the sampling axes and an integral number of voxels are included in each axial repeat. The integral sampling is important so that each order of diffraction from the filament structure is contained within a single layer plane in the 3D transform.
4. Compute the 3D transform of the resampled tomogram.
5. Apply a mask to the transform that allows only diffraction from the various orders of the filament periods to pass through the mask. The mask must allow passage of all points in the layer planes to avoid averaging over adjacent filaments.
6. Compute the inverse Fourier transform of the masked transform.

When done in this manner, averaging occurs only along the filament axis, with no averaging over adjacent filaments. If the actin and myosin filaments have different axial periods, this procedure can be applied twice using different masks and the filtered tomograms recombined with appropriate masking.

The amount of averaging, and thus SNR improvement, depends on the spacing of the layer planes. For example, if only layer planes corresponding to the 77.3-nm actin period are used, about 12-fold averaging is possible, and if the actin filament

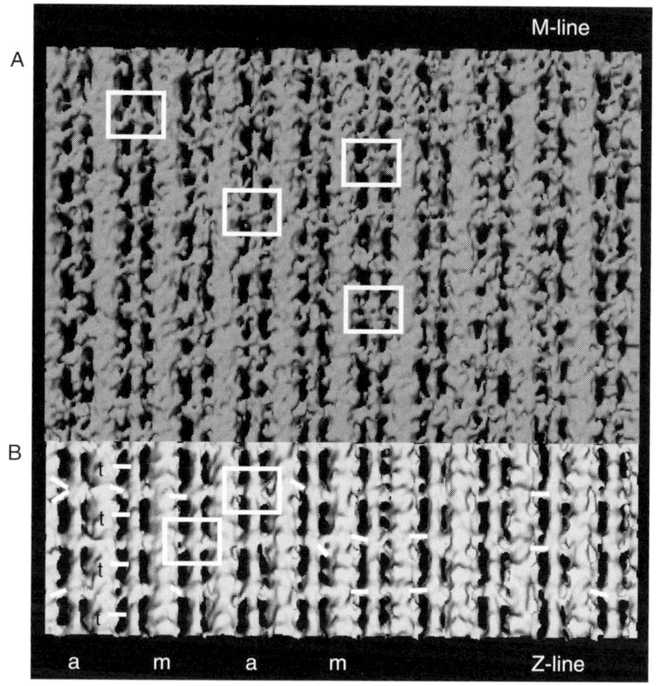

Fig. 9 Tomogram of IFM in the "high static tension" (HST) state, a condition that resembles an isometric contraction in vertebrate muscle. (A) Surface rendering of ~25% of one unaveraged tomogram. White boxes outline an important motif that has been dubbed the "mask motif." (B) Column average superimposed in phase with the unaveraged reconstruction. m = myosin filament, a = actin filament, t = troponin. White lines indicate crossbridge angle. Averaging has occurred axially over the 116-nm axial period that contains three 38.7-nm repeat periods so the short segment shown contains all the information of the column averaged tomogram. Adapted from Taylor *et al.* (1999), with permission.

structure is preserved, actin monomers may be resolved in the averages. If averaging over the 38.7-nm period, which can give up to 24-fold averaging, actin monomers will not be visible. The basic repeat of IFM is 232-nm, and if used, both actin and myosin filament periods can be recovered with one mask. However, there are less than four repeats of that period in a half sarcomere, so the more useful period is the 116-nm period (3 × 38.7 nm; 8 × 14.5 nm), of which there are about seven repeats but actin monomers will not be recovered in the average.

V. Building Atomic Models

Relating the tomogram to molecular function requires that molecular structures be fit to the molecular envelope of the class averages. Several methods have been used, but most involved manual building of the model from atomic structures using "O"—an x-ray crystallography model building program (Jones *et al.*, 1991). Usually model building begins in stages, starting with an atomic model of F-actin with an atomic model of the S1 fragment of myosin attached in the rigor configuration. The following procedure, referred to as "hybrid modeling" (Liu *et al.*, 2006b) has been used successfully to build atomic models into rigor IFM class averages and raw repeats:

1. Construct an actin filament model with the helical symmetry 28/13 and having 14 actin subunits. This would correspond in length to the 38.7-nm actin half period.

2. For active muscle modeling, decorate the central five actin monomers with the atomic model for S1 decorated actin (Holmes *et al.*, 2003; Rayment *et al.*, 1993). (These coordinates are available from the authors. The model from Holmes *et al.* can be downloaded from the *Nature* Web site. See also ftp://149.217.48.3/pub/holmes/.) For rigor muscle, decorate three additional actin monomers on the "barbed" end side of the target zone for active contraction.

3. Place the actoS1 atomic model in the global average of all repeats (Fig. 10A). This can be done manually or by generating a density map from the coordinates to fit to the global average. The transformation can then be used to transform the actoS1 model onto the global average. The pdb2mrc utility of EMAN can be used to generate this density map. By virtue of the alignment of all repeats onto the global average, this model can be placed without change into each class average or for that matter into each raw repeat.

4. Remove each S1 that does not lie within density. Usually at most six S1's will be needed for rigor IFM and no more than four for active contraction (Fig. 10B).

5. Remodel the S1's to fit the density (Fig. 10C). In "O" this is done using the "move_zone" utility. How the S1's are modified to fit the density is a matter of personal preference. For rigor muscle, we usually rebuild only the lever arm, which contains the light chains and the converter domain, leaving the motor domain docking on actin unchanged. Pivot points on the lever arm are usually, using the chicken S1 structure, heavy chain residues 710 (at the junction of motor domain and converter), 780 (at the beginning of the lever arm α-helix), and 806 (between the ELC and RLC). The heavy chain C-termini are angled slightly toward each other in two-headed crossbridges. For contracting muscle, in many instances, even the motor domain docking on actin has to be changed.

6. In some instances, classification fails to recover adequately some structures that are visible even in the raw repeats. The best example of this is swollen rigor, where even separate left–right classification could not produce accurately the position of the S2 domain which was pulled away from the thick filament surface (Liu *et al.*, 2006b). In this case, it was more useful to build the S2 domain into raw repeats (Fig. 10D).

There are a number of quantitative fitting programs that may successfully fit initial atomic models into tomogram volumes. These include SITUS (Wriggers *et al.*, 1999), COAN (Volkmann and Hanein, 1999), NMFF (Tama *et al.*, 2004),

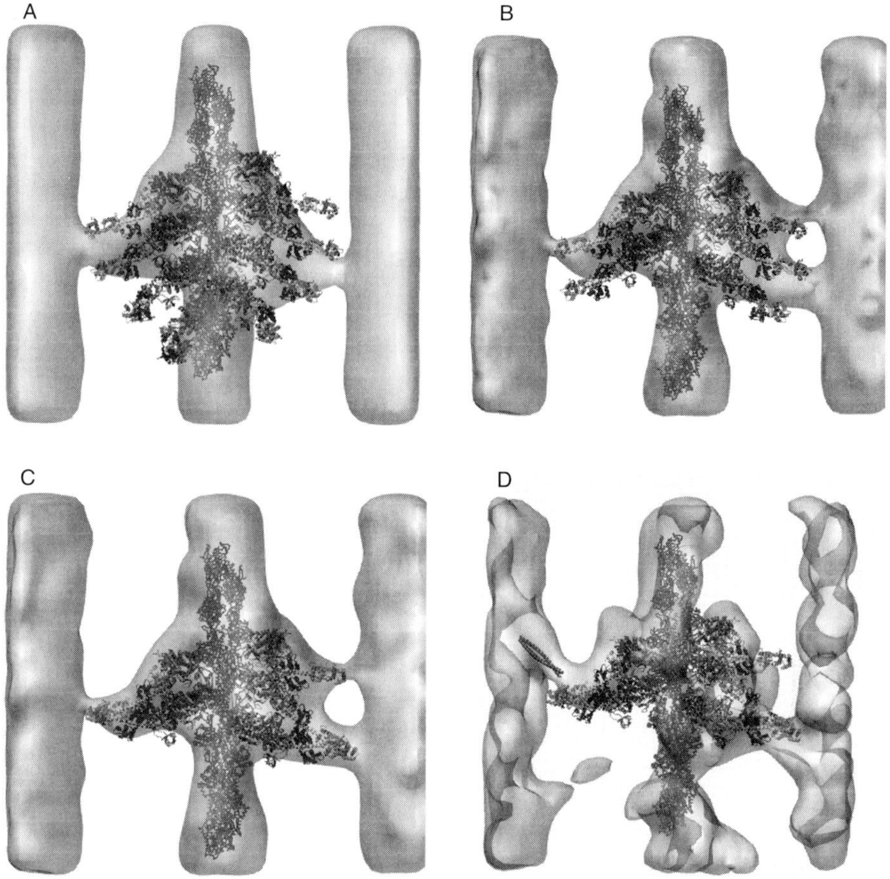

Fig. 10 Hybrid modeling of IFM class averages. (A) Global average in which a thin filament model containing 14 actin monomers decorated with 8 S1 structures in the rigor binding conformation has been placed. (B) The model in (A) placed into a class average obtained by reassembly of a left-hand class average and a right-hand class average. The bottom three S1's have been removed as they have no corresponding density. (C) Rebuilding of the model in (B) to bring the S1 lever arms into the envelope of the class average. (D) The model rebuilt for (C) placed unmodified into the "raw" repeat that corresponds to the class average. An S2 atomic model has been added to the crossbridge on the left. The S2 is more ambiguous on the right side. Note that S2 seems steeper in the "raw" repeat than in the class average, indicating that the correspondence was unable to efficiently classify feature that were close to the thick filament. (See Color Insert.)

and RSRef (Chapman, 1995). For IFM, we have used RSRef to refine a manually fit model to improve the geometry and to minimize poor contacts introduced by the manual fitting. We have not so far tried flexible fitting programs such as SITUS, COAN, and NMFF. Although both preferable and desirable, their use may be complicated by the negative stain effects of the TA fixative as described above.

VI. Relating Class Averages to the Specimen

Despite the order that is present in striated muscle, the specimen is heterogeneous from the standpoint of structural analysis, and it displays many kinds of disorder. The actin and myosin filaments are the major structures, and their interactions are a major source of heterogeneity. Disorder in filament positions is another. Obtaining high SNR requires that identical structures be located in the tomogram, then extracted, aligned, classified, and averaged. This process, however, removes the major structural features from their original context. This context is important for understanding function, but SNR improvement is necessary if molecular models are to be built. Can the original context be recovered after classification with improved SNR?

The alignment scheme used for the IFM repeats applies a single transformation for alignment of each raw repeat to the reference. The opposite process, aligning the class average to the raw repeats that are class members, requires only a single inverse transformation. Thus, replacement of the class members in the original tomogram by their class average can restore the context. This process, which we refer to as "mapping back," has been used to visualize the pattern of crossbridge repeats in tomograms of fast-frozen, freeze-substituted IFM rigor fibers that had been subjected to a ramp stretch (Fig. 11) (Liu *et al.*, 2004).

The process of mapping the class average back into the original repeat is done after all alignment, classification and averaging steps have been completed. The process can be done in following two ways:

1. A simple montage of the class averages placed in the proper sequence and with the correct separation as defined by the class members, that is, the raw repeats.
2. Replacement of the volume occupied by the raw repeat with the class average to which it is a member after application of the reverse transformation to the class average. A program [pasteimage] is used to perform this operation and runs from a Unix shell script.

Because the alignment is done on the actin filament, with no contribution by the myosin filaments themselves, this process sometimes produces discontinuities in the thick filament backbones. This effect can be reduced if only the actin filaments and crossbridges are pasted into the tomogram and the thick filaments left as they are.

It should be possible, using a judicious choice of alignment reference and classification mask, to separately average different structures in a multistructure sample such as muscle. Within the muscle lattice, actin filaments, myosin filaments, and crossbridges constitute three partially independent structures. A plausible scheme for reassembly of a tomogram using high SNR averages of the different components might be as follows:

1. Alignment on actin and classification using a mask that selects only the actin and troponin density. The alignment on actin minimizes one source of structure

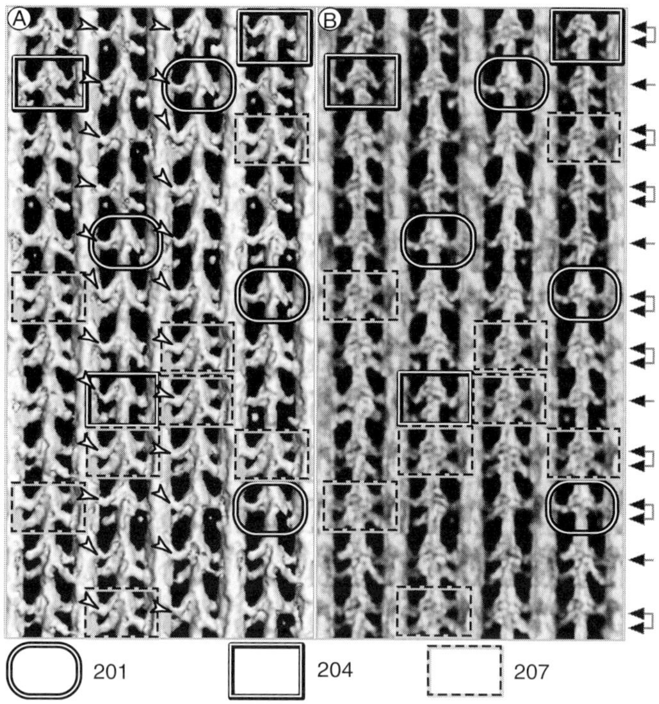

Fig. 11 A "map back" tomogram reassembled by replacing each "raw" repeat in the original tomogram with the class average to which it contributed. These classes produced by two-sided classification. In this figure "protein" is white. (A) Surface-rendered view of the "map back" tomogram obtained from rigor IFM fibers subjected to a ramp stretch and rapidly frozen and freeze-substituted. The "backbending" of many bridges in response to the stretch is clear as is the S2 segment of some bridges (arrows). Three class averages are outlined (201, 204, and 207). White arrowheads on two thick filaments mark the orientation of the S2 domain. Arrows on the right side of the figure show the levels in the 116-nm repeat that are predominantly double chevrons (paired arrows) or predominantly single chevrons (single arrows). (B) Volume rendering of the same "map back" area reveals the clear resolution of actin monomers in the thin filament and the individual motor domains attached to them. The S2 domain can be clearly seen in some cases. Adapted from Liu *et al.* (2004), with permission.

variability, and computation of a few classes would allow for some variation in actin features. This could be particularly important for the troponin density. Reassembly of the actin filaments would be done from these class averages.

2. Left and right side classification of the crossbridges using a mask that eliminates the actin filament and myosin filament backbone. This reduces the most variable part of the structure to a smaller number of class averages with high SNR.

3. Assemble an average for the raw actin repeat by combining the left and right side class averages for the crossbridges, together with the corresponding actin filament class average.

4. To obtain classes from the thick filament, repeat segments consisting of multiples of the 14.5-nm axial period would be extracted just as for the actin repeats and aligned to minimize one source of variance. The most obvious choice is alignment on the thick filament backbone and classification using a mask that includes only the myosin heads along the 14.5-nm axial period.

5. Reassembly of the "map back" tomogram would involve the following scheme, which has only been tested so far on the actin filament based repeats:

a. Replace the raw actin repeats, excluding structures associated only with the thick filaments, with the reassembled repeat from step (3).
b. Replace the 14.5-nm-thick filament repeats with the class average from step (4).
c. Low pass filter to the original resolution limit of the tomogram to reduce the appearance of any "seams" that develop at the boundaries where the individual components have been replaced.

Mapping back usually requires some decision on the disposition of repeats that are outliers. This can be done in three ways: (1) keep all raw repeats throughout the classification and alignment processes, discarding none regardless of how bad they may look; (2) replace outliers that were excluded from the class averages with the class average that they resemble best; and (3) simply use the raw repeat without replacement. Note that often in single-particle averaging, ~20% of the raw repeats are excluded from the averages for various reasons. Thus, the disposition of these outliers is not a trivial question for tomogram reassembly.

VII. Application to IFM

Initial applications of ET to IFM used film as the recording medium and with one exception was limited to single-axis tilt series (Chen *et al.*, 2001, 2002; Liu *et al.*, 2006b; Schmitz *et al.*, 1996, 1997; Taylor *et al.*, 1999). These tomograms were important for demonstrating the effect of adding nucleotide to rigor IFM fibers on the crossbridge structure (Schmitz *et al.*, 1996, 1997). A new crossbridge motif was identified for the first time, the "mask motif," in this work. Correspondence analysis and model building into class averages provided the basis for the first (and only) atomic models for two-headed rigor crossbridges (Chen *et al.*, 2001, 2002). Application of rapid freezing of fibers mounted on a tension transducer followed by freeze-substitution allowed for the first time crossbridges to be trapped and visualized during isometric active tension generation (Taylor *et al.*, 1999). The tomograms revealed crossbridges oriented over a wide range of axial and azimuthal angles and revealed "mask motifs" for the first time in an active muscle. However averaging at that time was limited to column averages, as we had not developed correspondence analysis to the stage that it would work on such a heterogeneous specimen so resolution even in the column averages was low.

The advent of CCD cameras, automated ET (Braunfeld *et al.*, 1994; Koster *et al.*, 1992) to reduce the electron dose accumulated over a tilt series, rapid-freezing/freeze-substitution methods for specimen preservation, and better tilt-series alignment methods (Winkler and Taylor, 2006) have improved the raw tomograms considerably. Improved tomograms combined with single-reference alignment and correspondence analysis of structural features have facilitated routine resolution of both actin monomers and myosin heads in rigor muscle (Fig. 6) (Liu *et al.*, 2004, 2006b). Class averages are routinely obtained with better than 5-nm resolution, which is good enough that atomic models can be built into the envelopes, using some structural constraints. From these atomic models, the effect of an imposed stretch on rigor IFM fibers could be determined at the level of the individual myosin heads (Liu *et al.*, 2004). The results showed that only a subset of crossbridges responded to the applied stress and did so by bending their lever arms toward the M-line. Recent work with swollen rigor fibers has provided an explanation of this effect in terms of the widely varying orientation of the S2 domain throughout the fibers (Liu *et al.*, 2006b). An imposed stretch will be sensed most efficiently by those crossbridges whose S2 domain is angled sharply toward the M-line.

In these new reconstructions, we clearly resolve a shell of stain that surrounds the actin filament, myosin motor domains, and the thick filament backbone (Fig. 6B) so these features appear negatively stained. On the other hand, additional features, such as the α-helical coiled-coil S2 domain, which links the myosin heads to the thick filament backbone, are now routinely seen in our raw tomograms and sometimes in class averages (Fig. 6A).

VIII. Application to Other Types of Specimen

The procedures described here have been adapted to the analysis of tomograms from other types of specimen with various types of intrinsic disorder and even to frozen-hydrated specimens. Correspondence analysis has been applied to studies of protein conformations in tomograms in only a few instances. The formation of protein arrays has two advantages for structure analysis. First, array formation reduces conformational variability, although this by itself does not automatically lead to arrays with high spatial order. Second, array formation places many molecules in similar, even if not identical environments so the methods of correspondence analysis described above can be applied to large numbers of molecular volumes, in fact, sometimes using numbers comparable to that used in single-particle analysis. A viable strategy for structure analysis would be 3D imaging by ET followed by correspondence analysis to determine the number of conformers that are present in the arrays.

A recent application of this approach involved determination of the structure of the inhibited state of myosin V from tomograms of ice-embedded, disordered 2D hexagonal arrays, formed on a lipid monolayer (Liu *et al.*, 2006a). A resolution of 2.4 nm was obtained from a class average with ~4000 repeats obtained from ~6000 that were extracted from the tomograms. This study was notable for its application of a focus gradient correction to the tilt-series images (Winkler and Taylor, 2003). With respect to the missing wedge of the repeats in these tomograms, the orientations varied only by rotation about the optical axis. Thus, the Fourier transforms of these repeats all had in common the in-plane projection information. Preliminary analysis suggests that classification is extracting actual structural variation, not missing wedge orientation. This suggests that under certain conditions, the missing wedge is not dominating and that correspondence analysis can be used to characterize structural variations.

A second application of these methods was recently reported (Zhu *et al.*, 2006) in which marker-free alignment, repeat volume alignment, correspondence analysis, and molecular model building were used to investigate the structure of Envs from HIV/SIV. Tilt series of these specimens are at the upper limit of thickness that we have attempted aligning with marker-free methods. Correspondence analysis was used in the repeat alignment to detect misalignment and was later used to investigate structural heterogeneity and symmetry in the Env particles. In this study, Env particles extracted from the top and bottom of virions, where the missing wedge causes the lipid bilayer to disappear were kept separate from those Env particles extracted from the sides where the lipid bilayer is visible. Both sets produced averages with complete coverage of Fourier space and except for the presence of the bilayer in one average, both groups of Env particles yielded the same structure.

IX. Summary

This chapter has presented the idea that ET of rapidly frozen, freeze-substitution fixed samples can be used to characterize structural variations in heterogeneous samples at the molecular level. We have also provided several suggestions for approaches that can work. Although there is reason for caution in characterizing structural differences when the missing wedge orientations are widely different, there is also reason to be encouraged that the powerful tool of correspondence analysis can be successfully applied to tomographic data under certain conditions. There is, therefore, promise that cryo-ET can be used to characterize large-amplitude motions in molecular motors in general as a series of snapshots.

Acknowledgments

The research on insect flight muscle and the software for marker-free alignment and volume classification was supported by NIH grants GM30598 (to K.A.T.), AR14317 (to M.K.R.) and by grant U54 GM64346 from NIGMS to the Cell Migration Consortium, which partially supported the marker-free alignment development. The rapid-freezing experiments would not have been possible without the help of the instrument maker, Joseph Pili, Tom Lenart, Keiko Hirose H. Sasaki, R. H. Edwards, and

R. T. Tregear. The skills of R. J. Edwards and H. Sasaki in inventing and mastering the Al foil strip adaptors that simplified fiber mounting and dismounting are worthy of note. Joe Pili's breakaway spring-loaded latch allowed us to slide the planchette off the freezing anvil dovetail and into a numbered vial for transfer to LN_2. We also thank Dr. Hanspeter Winkler for his comments on the chapter.

References

Akey, C. W., Moffat, K., Wharton, D. C., and Edelstein, S. J. (1980). Characterization of crystals of a cytochrome oxidase (nitrite reductase) from *Pseudomonas aeruginosa* by X-ray diffraction and electron microscopy. *J. Mol. Biol.* **136,** 19–43.

AL-Khayat, H. A., Hudson, L., Reedy, M. K., Irving, T. C., and Squire, J. M. (2003). Myosin head configuration in relaxed insect flight muscle: X-ray modelled resting crossbridges in a pre-powerstroke state are poised for actin binding. *Biophys. J.* **85,** 1063–1079.

Baatsen, P. H. (1993). Empirically determined freezing time for quick-freezing with a liquid-nitrogen-cooled copper block. *J. Microsc.* **172**(Pt. 1), 71–79.

Bald, W. B. (1986). On crystal size and cooling rate. *J. Microsc.* **143,** 89–102.

Bennett, P. M. (1998). Structural changes in samples cryofixed by contact with a cold metal block. *J. Microsc.* **192**(Pt. 3), 259–268.

Bennett, P. M., Tsaturyan, A., and Bershitsky, S. (2002). Rapid cryofixation of rabbit muscle fibres after a temperature jump. *J. Microsc.* **206,** 152–160.

Bohm, J., Frangakis, A. S., Hegerl, R., Nickell, S., Typke, D., and Baumeister, W. (2000). Toward detecting and identifying macromolecules in a cellular context: Template matching applied to electron tomograms. *Proc. Natl. Acad. Sci. USA* **97,** 14245–14250.

Brandt, S. S. (2006). Markerless alignment in electron tomography. *In* "Electron Tomography: Methods for Three-Dimensional Visualization of Structures in the Cell" (J. Frank, ed.), 2nd edn., pp. 187–216, Springer-Verlag.

Braunfeld, M. B., Koster, A. J., Sedat, J. W., and Agard, D. A. (1994). Cryo automated electron tomography: Towards high-resolution reconstructions of plastic-embedded structures. *J. Microsc.* **174,** 75–84.

Bullard, B. (1984). A large troponin in asynchronous insect flight muscle. *J. Muscle Res. Cell Motil.* **5,** 196.

Burgess, S. A., Walker, M. L., Thirumurugan, K., Trinick, J., and Knight, P. J. (2004). Use of negative stain and single-particle image processing to explore dynamic properties of flexible macromolecules. *J. Struct. Biol.* **147,** 247–258.

Chapman, M. S. (1995). Restrained real-space macromolecular atomic refinement using a new resolution-dependent electron-density function. *Acta Crystallogr. A* **51,** 69–80.

Chen, L. F., Blanc, E., Chapman, M. S., and Taylor, K. A. (2001). Real space refinement of acto-myosin structures from sectioned muscle. *J. Struct. Biol.* **133,** 221–232.

Chen, L. F., Winkler, H., Reedy, M. K., Reedy, M. C., and Taylor, K. A. (2002). Molecular modeling of averaged rigor crossbridges from tomograms of insect flight muscle. *J. Struct. Biol.* **138,** 92–104.

Crowther, R. A., Amos, L. A., Finch, J. T., De Rosier, D. J., and Klug, A. (1970a). Three dimensional reconstructions of spherical viruses by Fourier synthesis from electron micrographs. *Nature* **226,** 421–425.

Crowther, R. A., DeRosier, D. J., and Klug, A. (1970b). The reconstruction of a three-dimensional structure from projections and its application to electron microscopy. *Proc. R. Soc. Lond. A* **317,** 319–340.

Dengler, J. (1989). A multi-resolution approach to the 3d reconstruction from an electron microscope tilt series solving the alignment problem without gold particles. *Ultramicroscopy* **30,** 337–348.

DeRosier, D. J., and Moore, P. B. (1970). Reconstruction of three-dimensional images from electron micrographs of structures with helical symmetry. *J. Mol. Biol.* **52,** 355–369.

Dierksen, K., Typke, D., Hegerl, R., Walz, J., Sackmann, E., and Baumeister, W. (1995). Three-dimensional structure of lipid vesicles embedded in vitreous ice and investigated by automated electron tomography. *Biophys. J.* **68,** 1416–1422.

Egelman, E. H. (2000). A robust algorithm for the reconstruction of helical filaments using single-particle methods. *Ultramicroscopy* **85,** 225–234.

Förster, F., Medalia, O., Zauberman, N., Baumeister, W., and Fass, D. (2005). Retrovirus envelope protein complex structure in situ studied by cryo-electron tomography. *Proc. Natl. Acad. Sci. USA* **102,** 4729–4734.

Frangakis, A. S., and Hegerl, R. (2001). Noise reduction in electron tomographic reconstructions using nonlinear anisotropic diffusion. *J. Struct. Biol.* **135,** 239–250.

Frangakis, A. S., Bohm, J., Förster, F., Nickell, S., Nicastro, D., Typke, D., Hegerl, R., and Baumeister, W. (2002). Identification of macromolecular complexes in cryoelectron tomograms of phantom cells. *Proc. Natl. Acad. Sci. USA* **99,** 14153–14158.

Frank, J. (1990). Classification of macromolecular assemblies studied as 'single particles'. *Q. Rev. Biophys.* **23,** 281–329.

Frank, J. (2006). "Three-Dimensional Electron Microscopy of Macromolecular Assemblies—Visualization of Biological Molecules in Their Native State." Oxford University Press, New York.

Frank, J., and van Heel, M. (1982). Correspondence analysis of aligned images of biological particles. *J. Mol. Biol.* **161,** 134–137.

Glaeser, R. M. (1999). Review: Electron crystallography: Present excitement, a nod to the past, anticipating the future. *J. Struct. Biol.* **128,** 3–14.

Glaeser, R. M. (2004). Historical background: Why is it important to improve automated particle selection methods? *J. Struct. Biol.* **145,** 15–18.

Goody, R. S., Reedy, M. C., Hofmann, W., Holmes, K. C., and Reedy, M. K. (1985). Binding of myosin subfragment 1 to glycerinated insect flight muscle in the rigor state. *Biophys. J.* **47,** 151–169.

Gordon, R., Bender, R., and Herman, G. T. (1970). Algebraic reconstruction techniques. *J. Theor. Biol.* **29,** 471–482.

Guckenberger, R. (1982). Determination of a common origin in the micrographs of tilt series in three-dimensional electron microscopy. *Ultramicroscopy* **9,** 167–174.

Harauz, G., and van Heel, M. (1986). Exact filters for general geometry three dimensional reconstruction. *Optik* **73,** 146–156.

Hegerl, R., and Hoppe, W. (1976). Influence of electron noise on three-dimensional image reconstruction. *Z. Naturforsch* **31a,** 1717–1721.

Henderson, R., Baldwin, J. M., Downing, K. H., Lepault, J., and Zemlin, F. (1986). Structure of purple membrane from *Halobacterium halobium*: Recording, measurement and evaluation of electron micrographs at 3.5 Å resolution. *Ultramicroscopy* **19,** 147–178.

Heuser, J. E., Reese, T. S., Dennis, M. J., Jan, Y., Jan, L., and Evans, L. (1979). Synaptic vesicle exocytosis captured by quick freezing and correlated with quantal transmitter release. *J. Cell Biol.* **81,** 275–300.

Hirose, K., Franzini-Armstrong, C., Goldman, Y. E., and Murray, J. M. (1994). Structural changes in muscle crossbridges accompanying force generation. *J. Cell Biol.* **127,** 763–778.

Hirose, K., Lenart, T. D., Murray, J. M., Franzini-Armstrong, C., and Goldman, Y. E. (1993). Flash and smash: Rapid freezing of muscle fibers activated by photolysis of caged ATP. *Biophys. J.* **65,** 397–408.

Holmes, K. C., Angert, I., Kull, F. J., Jahn, W., and Schroder, R. R. (2003). Electron cryo-microscopy shows how strong binding of myosin to actin releases nucleotide. *Nature* **425,** 423–427.

Holmes, K. C., Tregear, R. T., and Barrington Leigh, J. (1980). Interpretation of the low angle X-ray diffraction from insect muscle in rigor. *Proc. R. Soc. Lond. B Biol. Sci.* **207,** 13–33.

Horner, J. L., and Gianino, P. D. (1984). Phase only matched filtering. *Appl. Optics.* **23,** 812–816.

Jiang, W., Baker, M. L., Wu, Q., Bajaj, C., and Chiu, W. (2003). Applications of a bilateral denoising filter in biological electron microscopy. *J. Struct. Biol.* **144,** 114–122.

Jones, T. A., Zou, J. Y., Cowan, S. W., and Kjeldgaard, M. (1991). Improved methods for building protein models in electron density maps and the location of errors in these models. *Acta Crystallogr. A* **47,** 110–119.

Jontes, J. D., Wilson-Kubalek, E. M., and Milligan, R. A. (1995). A 32 degree tail swing in brush border myosin I on ADP release. *Nature* **378,** 751–753.

Koster, A. J., Chen, H., Sedat, J. W., and Agard, D. A. (1992). Automated microscopy for electron tomography. *Ultramicroscopy* **46,** 207–227.

Kremer, J. R., Mastronarde, D. N., and McIntosh, J. R. (1996). Computer visualization of three-dimensional image data using IMOD. *J. Struct. Biol.* **116,** 71–76.

Langer, R., Poppe, C., Schramm, H. J., and Hoppe, W. (1975). Electron microscopy of thin protein crystal sections. *J. Mol. Biol.* **93,** 159–165.

Lanzavecchia, S., Cantele, F., and Bellon, P. L. (2001). Alignment of 3D structures of macromolecular assemblies. *Bioinformatics* **17,** 58–62.

Lenart, T. D., Murray, J. M., Franzini-Armstrong, C., and Goldman, Y. E. (1996). Structure and periodicities of cross-bridges in relaxation, in rigor, and during contractions initiated by photolysis of caged Ca^{2+}. *Biophys. J.* **71,** 2289–2306.

Liu, J., Reedy, M. C., Goldman, Y. E., Franzini-Armstrong, C., Sasaki, H., Tregear, R. T., Lucaveche, C., Winkler, H., Baumann, B. A. J., Squire, J. M., Irving, T. C., Reedy, M. K., *et al.* (2004). Electron tomography of fast frozen, stretched rigor fibers reveals elastic distortions in the myosin crossbridges. *J. Struct. Biol.* **147,** 268–282.

Liu, J., Taylor, D. W., Krementsova, E. B., Trybus, K. M., and Taylor, K. A. (2006a). 3-D structure of the myosin v inhibited state by cryoelectron tomography. *Nature* **442,** 208–211.

Liu, J., Wu, S., Reedy, M. C., Winkler, H., Lucaveche, C., Cheng, Y., Reedy, M. K., and Taylor, K. A. (2006b). Electron tomography of swollen rigor fibers of insect flight muscle reveals a short and variably angled s2 domain. *J. Mol. Biol.* **362,** 844–860.

Liu, Y., Penczek, P. A., McEwen, B. F., and Frank, J. (1995). A marker-free alignment method for electron tomography. *Ultramicroscopy* **58,** 393–402.

Lovell, S. J., Knight, P. J., and Harrington, W. F. (1981). Fraction of myosin heads bound to thin filaments in rigor fibrils from insect flight and vertebrate muscles. *Nature* **293,** 664–666.

Ludtke, S. J., Baldwin, P. R., and Chiu, W. (1999). EMAN: Semiautomated software for high-resolution single-particle reconstructions. *J. Struct. Biol.* **128,** 82–97.

Luther, P. K. (2006). Sample shrinkage and radiation damage of plastic sections. *In* "Electron Tomography: Methods for Three-Dimensional Visualization of Structures in the Cell" (J. Frank, ed.), 2nd edn., pp. 17–48, Springer-Verlag.

Mastronarde, D. N. (1997). Dual-axis tomography: An approach with alignment methods that preserve resolution. *J. Struct. Biol.* **120,** 343–352.

Milligan, R. A., and Flicker, P. F. (1987). Structural relationships of actin, myosin, and tropomyosin revealed by cryo-electron microscopy. *J. Cell Biol.* **105,** 29–39.

Morris, E. P., Squire, J. M., and Fuller, G. W. (1991). The 4-stranded helical arrangement of myosin heads on insect (*Lethocerus*) flight muscle thick filaments. *J. Struct. Biol.* **107,** 237–249.

Narita, A., Yasunaga, T., Ishikawa, T., Mayanagi, K., and Wakabayashi, T. (2001). Ca(2+)-induced switching of troponin and tropomyosin on actin filaments as revealed by electron cryo-microscopy. *J. Mol. Biol.* **308,** 241–261.

Owen, C. H., and Landis, W. J. (1996). Alignment of electron tomographic series by correlation without the use of gold particles. *Ultramicroscopy* **63,** 27–38.

Padrón, R., Alamo, L., Craig, R., and Caputo, C. (1988). A method for quick-freezing live muscles at known instants during contraction with simultaneous recording of mechanical tension. *J. Microsc.* **151,** 81–102.

Pascual-Montano, A., Taylor, K. A., Winkler, H., Pascual-Marqui, R. D., and Carazo, J. M. (2002). Quantitative self-organizing maps for clustering electron tomograms. *J. Struct. Biol.* **138,** 114–122.

Penczek, P., Marko, M., Buttle, K., and Frank, J. (1995). Double-tilt electron tomography. *Ultramicroscopy* **60,** 393–410.

Radermacher, M. (2006). Weighted back-projection methods. *In* "Electron Tomography: Three-Dimensional Imaging With the Transmission Microscope" (J. Frank, ed.), pp. 246–273. Elsevier, New York.

Rath, B. K., Hegerl, R., Leith, A., Shaikh, T. R., Wagenknecht, T., and Frank, J. (2003). Fast 3D motif search of EM density maps using a locally normalized cross-correlation function. *J. Struct. Biol.* **144,** 95–103.

Rayment, I., Holden, H. M., Whittaker, M., Yohn, C. B., Lorenz, M., Holmes, K. C., and Milligan, R. A. (1993). Structure of the actin-myosin complex and its implications for muscle contraction. *Science* **261,** 58–65.

Reedy, M. K. (1967). Cross-bridges and periods in insect flight muscle. *Am. Zool.* **7,** 465–481.

Reedy, M. K., and Reedy, M. C. (1985). Rigor crossbridge structure in tilted single filament layers and flared-x formations from insect flight muscle. *J. Mol. Biol.* **185,** 145–176.

Reedy, M. K., Holmes, K. C., and Tregear, R. T. (1965). Induced changes in orientation of the cross-bridges of glycerinated insect flight muscle. *Nature* **207,** 1276–1280.

Reedy, M. K., Lucaveche, C., Reedy, M. C., and Somasundaram, B. (1993). Experiments on cross-bridge action and filament sliding in insect flight muscle. *Adv. Exp. Med. Biol.* **332,** 33–44.

Roseman, A. M. (2003). Particle finding in electron micrographs using a fast local correlation algorithm. *Ultramicroscopy* **94,** 225–236.

Sader, K., Reedy, M. K., Studer, D., Popp, D., Lucaveche, C., and Trinick, J. (2007). Testing the structural preservation of sections cut with an oscillating ultramicrotome knife. *J. Struct. Biol.* (submitted for publication).

Sato, T. (1968). A modified method for lead staining of thin sections. *J. Electron Microsc. (Tokyo)* **17,** 158–159.

Saxton, W. O. (1994). Accurate alignment of sets of images. *J. Microsc.* **174,** 61–68.

Saxton, W. O., Baumeister, W., and Hahn, M. (1984). Three-dimensional reconstruction of imperfect two-dimensional crystals. *Ultramicroscopy* **13,** 57–70.

Schatz, M., and van Heel, M. (1990). Invariant classification of molecular views in electron micrographs. *Ultramicroscopy* **32,** 255–264.

Schatz, M., and van Heel, M. (1992). Invarient recognition of molecular projections in vitreous ice preparations. *Ultramicroscopy* **45,** 15–22.

Schmid, M., Dargahi, R., and Tam, M. (1993). Spectra: A system for processing electron images of crystals. *Ultramicroscopy* **48**(3), 251–264.

Schmitz, H., Lucaveche, C., Reedy, M. K., and Taylor, K. A. (1994). Oblique section 3-D reconstruction of relaxed insect flight muscle reveals the cross-bridge lattice in helical registration. *Biophys. J.* **67,** 1620–1633.

Schmitz, H., Reedy, M. C., Reedy, M. K., Tregear, R. T., and Taylor, K. A. (1997). Tomographic three-dimensional reconstruction of insect flight muscle partially relaxed by AMPPNP and ethylene glycol. *J. Cell Biol.* **139,** 695–707.

Schmitz, H., Reedy, M. C., Reedy, M. K., Tregear, R. T., Winkler, H., and Taylor, K. A. (1996). Electron tomography of insect flight muscle in rigor and AMPPNP at 23 °C. *J. Mol. Biol.* **264,** 279–301.

Sosa, H., Popp, D., Ouyang, G., and Huxley, H. E. (1994). Ultrastructure of skeletal muscle fibers studied by a plunge quick freezing method: Myofilament lengths. *Biophys. J.* **67,** 283–292.

Squire, J. M. (1981). "The Structural Basis of Muscle Contraction." Plenum Press, New York.

Stoschek, A., and Hegerl, R. (1997). Denoising of electron tomographic reconstructions using multi-scale transformations. *J. Struct. Biol.* **120,** 257–265.

Tama, F., Miyashita, O., and Brooks, C. L., III (2004). Normal mode based flexible fitting of high-resolution structure into low-resolution experimental data from cryo-EM. *J. Struct. Biol.* **147,** 315–326.

Taylor, K. A., Liu, J., and Winkler, H. (2006). Localization and classification of repetitive structures in electron tomograms of paracrystalline assemblies. *In* "Electron Tomography: Methods for Three-Dimensional Visualization of Structures in the Cell" (J. Frank, ed.), 2nd edn., pp. 417–440. Springer-Verlag.

Taylor, K. A., Reedy, M. C., Cordova, L., and Reedy, M. K. (1984). Three-dimensional reconstruction of rigor insect flight muscle from tilted thin sections. *Nature* **310,** 285–291.

Taylor, K. A., Reedy, M. C., Córdova, L., and Reedy, M. K. (1989a). Three-dimensional image reconstruction of insect flight muscle. I: The rigor myac layer. *J. Cell Biol.* **109,** 1085–1102.

Taylor, K. A., Reedy, M. C., Córdova, L., and Reedy, M. K. (1989b). Three-dimensional image reconstruction of insect flight muscle. II: The rigor actin layer. *J. Cell Biol.* **109,** 1103–1123.

Taylor, K. A., Schmitz, H., Reedy, M. C., Goldman, Y. E., Franzini-Armstrong, C., Sasaki, H., Tregear, R. T., Poole, K. J. V., Lucaveche, C., Edwards, R. J., Chen, L. F., Winkler, H., *et al.* (1999). Tomographic 3-D reconstruction of quick frozen, Ca^{2+}-activated contracting insect flight muscle. *Cell* **99,** 421–431.

Taylor, K. A., Tang, J., Cheng, Y., and Winkler, H. (1997). The use of electron tomography for structural analysis of disordered protein arrays. *J. Struct. Biol.* **120,** 372–386.

Thomas, D. D., Cooke, R., and Barnett, V. A. (1983). Orientation and rotational mobility of spin-labelled myosin heads in insect flight muscle in rigor. *J. Muscle Res. Cell. Motil.* **4,** 367–378.

Tregear, R. T., Edwards, R. J., Irving, T. C., Poole, K. J. V., Reedy, M. C., Schmitz, H., Towns-Andrews, E., and Reedy, M. K. (1998). X-ray diffraction indicates that active crossbridges bind to actin target zones in insect flight muscle. *Biophys. J.* **74,** 1439–1451.

Tregear, R. T., Reedy, M. C., Goldman, Y. E., Taylor, K. A., Winkler, H., Franzini-Armstrong, C., Sasaki, H., Lucaveche, C., and Reedy, M. K. (2004). Cross-bridge number, position, and angle in target zones of cryofixed isometrically active insect flight muscle. *Biophys. J.* **86,** 3009–3019.

Unser, M., Trus, B. L., and Steven, A. C. (1987). A new resolution criterion based on spectral signal-to-noise ratios. *Ultramicroscopy* **23,** 39–51.

Van Heel, M., and Frank, J. (1981). Use of multivariate statistics in analysing the images of biological macromolecules. *Ultramicroscopy* **6,** 187–194.

Van Heel, M., Schatz, M., and Orlova, E. (1992). Correlation functions revisited. *Ultramicroscopy* **46,** 307–316.

Volkmann, N., and Hanein, D. (1999). Quantitative fitting of atomic models into observed densities derived by electron microscopy. *J. Struct. Biol.* **125,** 176–184.

Whittaker, M., Wilson-Kubalek, E. M., Smith, J. E., Faust, L., Milligan, R. A., and Sweeney, H. L. (1995). A 35-Å movement of smooth muscle myosin on ADP release. *Nature* **378,** 748–751.

Winkelmann, D. A., Baker, T. S., and Rayment, I. (1991). Three-dimensional structure of myosin subfragment-1 from electron microscopy of sectioned crystals. *J. Cell Biol.* **114,** 701–713.

Winkler, H. (2007). 3D-reconstruction and processing of volumetric data in cryo-electron tomography. *J. Struct. Biol.* (in press).

Winkler, H. P., Reedy, M. C., Reedy, M. K., Tregear, R. T., and Taylor, K. A. (1996). Three-dimensional structure of nucleotide-bearing crossbridges *in situ.* Oblique section reconstruction of insect flight muscle in AMPPNP at 23 °C. *J. Mol. Biol.* **264,** 302–322.

Winkler, H., and Taylor, K. A. (1996). Three-dimensional distortion correction applied to tomographic reconstructions of sectioned crystals. *Ultramicroscopy* **63,** 125–132.

Winkler, H., and Taylor, K. A. (1999). Multivariate statistical analysis of three-dimensional cross-bridge motifs in insect flight muscle. *Ultramicroscopy* **77,** 141–152.

Winkler, H., and Taylor, K. A. (2003). Focus gradient correction applied to tilt series image data used in electron tomography. *J. Struct. Biol.* **143,** 24–32.

Winkler, H., and Taylor, K. A. (2006). Accurate marker-free alignment with simultaneous geometry determination and reconstruction of tilt series in electron tomography. *Ultramicroscopy* **106,** 240–254.

Woodhead, J. L., Zhao, F. Q., Craig, R., Egelman, E. H., Alamo, L., and Padron, R. (2005). Atomic model of a myosin filament in the relaxed state. *Nature* **436,** 1195–1199.

Wriggers, W., Milligan, R. A., and McCammon, J. A. (1999). SITUS: A package for docking crystal structures into low-resolution maps from electron microscopy. *J. Struct. Biol.* **125,** 185–195.

Zeitler, E. (2006). Reconstruction with orthogonal functions. *In* "Electron Tomography: Methods for Three-Dimensional Visualization of Structures in the Cell" (J. Frank, ed.), 2nd edn., pp. 275–306. Springer-Verlag.

Zhu, P., Liu, J., Bess, J. J., Chertova, E., Lifson, J. D., Grisé, H., Ofek, G., Taylor, K. A., and Roux, K. H. (2006). Distribution and three-dimensional structure of AIDS virus envelope spikes. *Nature* **441,** 847–852.

Zhu, Y., Carragher, B., Glaeser, R. M., Fellmann, D., Bajaj, C., Bern, M., Mouche, F., de Haas, F., Hall, R. J., and Kriegman, D. J. (2004). Automatic particle selection: Results of a comparative study. *J. Struct. Biol.* **145,** 3–14.

PART III

Imaging Frozen-Hydrated Cells and Cell Parts

If rapid freezing is the best way to preserve cell structure for study by electron microscopy (EM), then imaging these frozen cells without further treatment would appear to be the best method for subsequent structural study. The following chapters include an example of this approach for the study of a bacterial cell that can be imaged, at least at its periphery, without further treatment. Bigger cells require cryomicrotomy, and this method too is discussed. There are also chapters on the imaging of cell parts; these follow the principle of "divide and conquer" to get detailed and accurate information about subsystems from within a cell.

Each of these chapters demonstrates the power of cryo-EM for structural cell biology, but each also shows the difficulties that one encounters when following this apparently obvious route. Even when the problems of rapid freezing have been solved, and vitreous ice is assured, sample thickness is a critically important issue. It defines either a restrictive choice of sample, forcing the investigator to stick to those cells that are intrinsically thin, or the need for cryomicrotomy of vitrified samples, a task that is difficult in itself, but that also leads to its own set of problems in image acquisition and interpretation.

Once cryosamples of appropriate thickness are in hand, one must still confront two opposing problems. The contrast between ice and biological macromolecules is not great, so one must collect images that are based on enough electrons to give a statistically reliable distinction between the electron scattering by a given macromolecular complex and the solvent or cytoplasm that surrounds it. On the other hand, frozen-hydrated samples are sensitive to the electron beam, so dose must be limited to avoid destruction of the very objects one wishes to understand. This dichotomy is the principal problem for cryo-EM of biological materials; resolving it successfully requires good instrumentation, knowledge of the issues, technical skill, and experimental care.

The cells that are small enough to study directly by cryo-EM are few in number; even with something as small as a bacterium, the surface structures are far easier to image than the cell's interior. Thus, Subramaniam and colleagues (Chapter 14) have been able to learn important details about receptor molecules that inhabit the plasma membrane of a bacterium. Their work demonstrates both the importance of good images and the effectiveness of methods for averaging images from multiple copies of the same biological object; such averages provide the structural detail that illuminates the organization of component macromolecules.

Most cells require microtomy in preparation for cryo-EM; Chapter 15 by Dubochet *et al.* provides a guide to some of the issues that surround this difficult methodology. Other publications from this group describe methods for making cryosections, but this chapter instructs the reader on how to interpret the images that can be obtained from cryosections by cryo-EM. The good news is that image contrast is sufficient to display cellular structures with startling clarity and detail; even the protein subunits of single microtubules are obvious, and features of their shape that reveal their polar orientation can readily be seen. There is, however, a surprising range of subtle problems in "reading" cryo-electron micrographs, and this chapter will prove invaluable for people newly interested in the subject.

From there on, this section of the book deals with studies that have focused on parts of cells that are intrinsically small enough to be imaged by cryo-EM and/or cryo-electron tomography (cryo-ET). Serysheva *et al.* (Chapter 16) have used protein isolation and cryo-EM of single particles to build up an impressive view of the two kinds of calcium channels that are found in the pathway for excitation–contraction coupling in muscle cells.

The techniques for combining images from large numbers of single particles have been well developed for molecular microscopy, but they are treated in detail elsewhere (Frank, 2006). In essence, these methods use low doses of electrons to preserve the integrity of the sample, but the presence of many identical particles in isolation from the rest of the cell permits averaging of many structures to improve electron-counting statistics and thus the detail that can be seen. This method also uses a clever, model-based analysis of the images to estimate the orientation of each isolated molecule. Views from a given orientation can then be combined to improve image signal-to-noise for that particular orientation. Subsequently, the images from different views can be assembled to build up a three-dimensional (3D) view of the structure under study. Serysheva *et al.* have not only used these methods, they have also related their cryo-EM images to the atomic structures of protein pieces, generating a pseudo-atomic model of parts of their macromolecular complexes. Here, where our emphasis is on cells, they have also used classic images from cellular EM to place their high-resolution structures into a cellular context. The result shows how one can productively study biological material that has been isolated and frozen for cryo-EM, obtaining a significant improvement in resolution over conventional cellular methods, yet still placed the resulting objects in the context of their cellular function.

Hoenger and Nicastro (Chapter 17) have examined *in vitro*-assembled complexes of microtubules and their associated motor enzymes as well as a native, microtubule-based machine; they employ both cryo-EM and cryo-ET to determine mechanistically informative 3D structures. Their studies show how symmetry can be used to obtain averages of multiple structural units, increasing the image signal-to-noise ratio, and thus the detail that can be visualized in cryo-EM. For the samples assembled *in vitro* they can infer the biological relevance of each detail that their high-resolution imaging reveals. As an example of a native biological machine, they use sperm tails or flagella isolated from a biflagellate alga as samples where context is retained by the biological material itself. Here cryo-ET, together with image averaging, provides both the detail and the overview, so molecular structure is seen in its natural cellular context.

Both these studies show how one can combine images of a sample from multiple orientations of view and obtain 3D information from frozen-hydrated samples. Multiple copies of single particles, objects with helical symmetry, and asymmetric objects imaged with multiple specimen tilts can all be reconstructed in 3D. Each of these approaches has strengths and limitations, so the interested reader is well advised to understand them all. One strength in all these methods is that cryoimaging opens the possibility of using EM information to place or "dock" very high-resolution information from crystallography or NMR into the context provided by cryo-EM of a large macromolecular complex. These "pseudo-atomic" models provide valuable details about objects that are too big or ungainly to be crystallized. Such results often increase the meaning and value of the high-resolution structural work.

The final chapter of this part (Chapter 18) describes the use of atomic structures, single-particle methods, and cryo-ET to approach an atomic resolution understanding of an important cellular object: the clathrin-coated vesicle. This work by Cheng and Walz confronts the limitations of each method used by combining the best of each to help solve a cell biological problem. Atomic structures of parts of the clathrin molecule are docked into structural envelopes provided by a single-particle cryo-EM analysis of clathrin cages assembled *in vitro*. The next step on the road to a cellular structure is a population of coated vesicles isolated from cytoplasm However, the structural variation in these particles is great; single-particle methods will no longer allow the assembly of images from multiple objects because the object are just not enough alike. Now one must use tomography, the method that will build a 3D model from each object, making no assumptions about symmetry or similarity from one object to another. The resolution obtained with this method is less, though, because the allowable dose per image is so low. Nonetheless, a structural envelope obtained by tomography can be used to dock the higher resolution images from single-particle work, which in turn have served to dock the atomic models available from crystallography. This three-step process may turn out to be a model for how we can ultimately get pseudo-atomic models of many cellular structures.

References

Frank, J. (2006). "Three-Dimensional Electron Microscopy of Macromolecular Assemblies." Oxford University Press, Oxford.

CHAPTER 14

Electron Tomography of Bacterial Chemotaxis Receptor Assemblies

Peijun Zhang,* Robert M. Weis,† Peter J. Peters,‡ and Sriram Subramaniam[1],*

*Laboratory of Cell Biology, National Cancer Institute
NIH, Bethesda, Maryland 20892

†Department of Chemistry, University of Massachusetts
Amherst, Massachusetts 01003

‡Division of Tumor Biology, The Netherlands Cancer Institute
Amsterdam, The Netherlands

Bacteria sense many of the changes in their local chemical environment by the binding of ligands to a family of chemotaxis receptors, which in turn trigger the activation of a signaling pathway that ultimately regulates the rotation of the flagellar motor. We are using three-dimensional electron microscopy to describe the spatial architectures of chemotaxis receptor assemblies in bacterial cells. Here, we review our recent studies aimed at understanding the organization of the bacterial serine receptor Tsr in intact cells using electron tomography at room and cryogenic temperatures, coupled with imaging of thin vitreous sections obtained from rapidly frozen bacterial cells. We show that starting from tomograms of whole bacteria, the subvolumes corresponding to individual receptor units can be extracted and

[1] Corresponding author, email: ss1@nih.gov

0091-679X/07 $35.00
DOI: 10.1016/S0091-679X(06)79014-8

averaged to determine the three-dimensional structures of receptor assemblies in the physiologically relevant context of an intact cell.

I. Introduction

Emerging methods in three-dimensional (3D) biological electron microscopy offer the promise of bridging a critical gap in imaging in the biomedical size spectrum (Lucic *et al.*, 2005; McIntosh *et al.*, 2005; Subramaniam, 2005). This gap comprises a size range of considerable interest in biology and medicine that includes cellular protein machines, giant protein and nucleic acid assemblies, small subcellular organelles, and small bacteria. These objects are generally too large and/or too heterogeneous to be investigated by high-resolution x-ray and NMR methods; yet the level of detail afforded by conventional light and electron microscopy is often not adequate to describe their structures at resolutions high enough to be useful for understanding the chemical basis of biological function.

Electron tomography provides a powerful tool in the arsenal of cell biologists to investigate the 3D structures of large subcellular assemblies at resolutions one to two orders of magnitude higher than what is currently achieved using light microscopy. The fundamental principles underlying electron tomography and the strategy to use back-projection algorithms to get 3D information from a series of two-dimensional (2D) images recorded at different orientations were clearly articulated nearly four decades ago (DeRosier and Klug, 1968; Hart, 1968). However, this field has seen a significant burst of activity in the last 5 years with the availability of tools for automated data acquisition using modern computerized microscopes. Efforts at imaging complex assemblies at room temperature (using stained specimens) as well as cryogenic temperatures (using unstained specimens) have rapidly begun to provide many new insights into the 3D architecture of cells. In our laboratory, we are using 3D electron microscopy as a tool to discover and analyze the spatial architectures of receptors and membrane assemblies in viruses, cells, and tissues. The electron microscopic studies bridge the gap between cellular and molecular structure and provide a foundation to integrate the structural information with biochemical and genetic analyses.

In this review, we highlight our recent work on the architecture of receptor assemblies in intact *Escherichia coli* cells using complementary structural approaches. We first describe 2D electron microscopic studies of stained membrane extracts aimed at studying the morphologies of small, partially ordered receptor clusters in membranes isolated from cells overproducing the chemotaxis receptor Tsr. Next, we report the use of electron tomography at room temperature to describe the arrangement of these zipper-like assemblies in sections obtained from chemically fixed cells. This is followed by cryo-electron microscopic studies of vitreous sections prepared from cells frozen at high pressure which confirm and extend the findings from electron tomographic studies with stained sections at room temperature. Finally, we present findings from cryo-electron tomography of plunge frozen cells and show that receptor assemblies can be imaged in 3D in intact cells under low-dose conditions. Using averaging procedures similar to those used in electron crystallography, we demonstrate that *in situ* 3D structural information can be obtained by averaging as few as 300 individual Tsr receptor units.

II. Analysis of Tsr Assemblies in Membrane Extracts

The ability of bacterial cells to respond to very small changes of chemical concentration can be achieved by amplifying the signal of ligand binding through receptor clustering, involving extended receptor–receptor interactions in the membrane (Ames *et al.*, 2002; Parkinson *et al.*, 2005). However, direct structural evidence of these interactions is sparse. The serine receptor (Tsr) is one of the five methylatable chemotaxis proteins in *E. coli* that reside in the inner membrane. Receptors such as Tsr are part of a large class of transducers (Le Moual and Koshland, 1996) in the cell membrane that are sensitive to various environmental cues, and provide inputs to sensory pathways that bias cell movement toward favorable environments (Falke and Hazelbauer, 2001). Changes in the fraction of serine bound to Tsr initiate a cascade of biochemical events that ultimately regulate the rotation of the flagellar motor. The chemotaxis signaling pathway is one of many instances of the two-component superfamily of prokaryotic signal transduction pathways (Grebe and Stock, 1999; Stock and Da Re, 2000). Each pathway has a sensor that is usually an integral membrane protein, and one or more cytoplasmic phosphate-accepting response regulator proteins.

To elucidate the arrangement of Tsr in cell membranes, we began with structural studies of Tsr in membranes from cells overproducing Tsr. Membranes were isolated from these cells by lysis and subsequent fractionation, followed by treatment with a low concentration of Tween-80 detergent. The membrane preparations were stained with uranyl acetate and analyzed by transmission electron microscopy at room temperature (see Weis *et al.*, 2003 for a detailed description of these studies). As illustrated in Fig. 1A, overproduction of Tsr results in organization of the receptors into large assemblies within lipid bilayers. Three different morphologies of receptor assemblies can be identified, marked as m, z, and c in Fig. 1A for micellar, zipper-like, and crystalline assemblies, respectively. Front views of the zipper and micellar assemblies obtained in projection images are shown in more detail in Fig. 1B and C. Hybrid structures between zippers and micelles were also observed and all of the structural evidence indicates that these different morphologies simply represent different facets of receptor units that are packed with interacting cytoplasmic tails (Fig. 1D and E). The ordered crystalline patches represent top views of the zipper-like assemblies, and were analyzed using electron crystallographic methods to establish that they are organized in an approximately hexagonal lattice with unit cell dimensions of $\sim$75 Å, and unit area of $\sim$5000 Å^2. Each of the repeating units is large enough to accommodate six receptor molecules, consistent with models suggesting that receptors pack as trimers of dimer units. We note, however, that these images do not allow us to distinguish between trimer-of-dimer versus other packing arrangements.

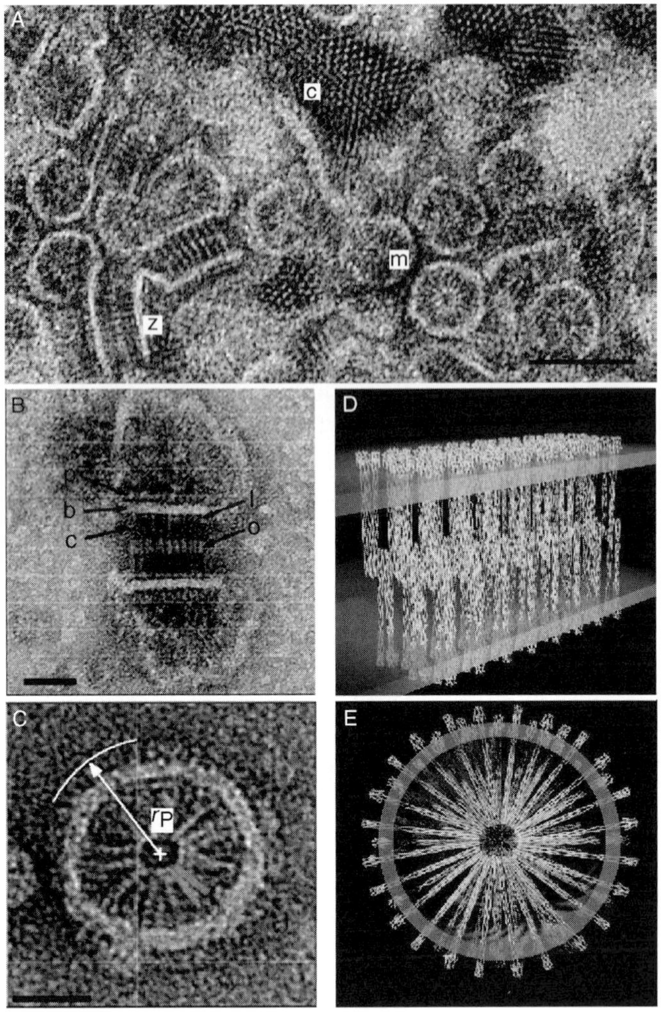

Fig. 1 Projection imaging of receptor assemblies. (A) Membranes were isolated from cells engineered to overproduce Tsr and stained with uranyl acetate. The different morphologies observed are indicated; receptors are organized either in zipper-like assemblies (labeled z) where the receptors are interdigitated at their cytoplasmic ends, and appear crystalline (labeled c) in the top view, or in rounded micellar assemblies (labeled m) where the cytoplasmic ends come together in a central region. Scale bar = 100 nm. (B) Expanded views of zipper-like assembly. One leaflet of the twin bilayer membrane is labeled to illustrate the topology of the receptor subunits: p, periplasmic ligand-binding domain; b, membrane bilayer; and c, cytoplasmic domain. Domain assignments in the other leaflet are apparent by symmetry. Two other features are labeled: a zone of interdigitation or overlap (o) and a feature that may represent the linker domain (l). Scale bar = 25 nm. (C) An enlarged micellar assembly of Tsr, illustrating how characteristic radial dimensions are determined (in this case r_P). Also, scaled silhouettes of two Tsr cytoplasmic domain dimers are superimposed on the assembly, patterned after the trimer-of-dimer structure (1*qu*7). The individual dimers are rendered as blue and red cigars in which the pair of dimers forms a "V"-shaped assembly. Scale bar = 25 nm. (D and E) Proposed molecular models for receptor arrangement in the zipper-like and micellar assemblies based on packing of an atomic structural model of Tsr. Figure adapted from Weis *et al.*, 2003.

III. Electron Tomography of Fixed, Cryosectioned *E. coli* Cells

To determine whether the receptor assemblies observed in the extracted membranes also exist in an intact cell, we obtained sections from frozen *E. coli* cells using the approach described first by Tokuyasu (1973). Specimens were prepared by sectioning at $-110\,^{\circ}$C, warmed to room temperature, treated with antibodies specific to Tsr, and embedded in methyl cellulose in the presence of uranyl acetate using established procedures (Peters and Hunziker, 2001). Electron tomographic studies of these sections were carried out at room temperature using a Tecnai F30 electron microscope operated at 300 kV (described in detail in Lefman *et al.*, 2004). Typically, a series of projection images at tilt range of $\pm 70^{\circ}$ was recorded on a Gatan 2K $\times$ 2K CCD camera mounted on a Tecnai F30 microscope, operating at 300 kV. The images in the tilt series were used to generate a 3D volume using weighted back-projection methods as implemented in the software package IMOD (Kremer *et al.*, 1996).

Tomograms constructed from 250-nm-thick sections confirm that the overproduction of Tsr leads to the formation of an extended internal membrane network composed of zipper-like and micelle-like structures (Fig. 2). These structures bear direct resemblance to the membrane assemblies observed in stained membrane extracts, but provide new insights into the organization of these assemblies in 3D. A tomographic slice (Fig. 2B) at a nominal thickness of 1 nm demonstrates the invagination of the membranes into the cytoplasm. The physical continuity of the invaginations with the inner membrane is especially clear in these images. To illustrate the overall topology of the invaginated membranes, a single slice from a small region of the tomogram is presented in Fig. 2C, accompanied by segmentation (Fig. 2D) to visualize the 3D arrangement of the membranes. The tomographic analysis demonstrates that the invaginations containing the zippered structures are not free-standing entities as viewed in the isolated membranes, but are continuous with the inner membrane. Tomograms obtained from either wild-type *E. coli* or control cells carrying deletions of Tsr do not display these internal membrane networks.

Using molecular models for these assemblies constructed from the isolated membrane preparations (Fig. 1D and E), we show an interpretation of the segmented structures in terms of the receptor-receptor interactions that result in the formation of the invaginated membranes. These observations are consistent with stabilization of the membrane network by interactions between the cytoplasmic ends of the receptors (wider spacing in Fig. 2D) as well as an interaction between the periplasmic ends of the receptors (narrow spacing in Fig. 2D). The presence of the periplasmic interaction that is presumably weaker than the interaction between interdigitated cytoplasmic ends of the receptors, could explain how a 3D network of membranes could be stabilized in the cell by combining the cytoplasmic and periplasmic pairs of interactions. We present a schematic representation of the formation and growth of the membrane invaginations in

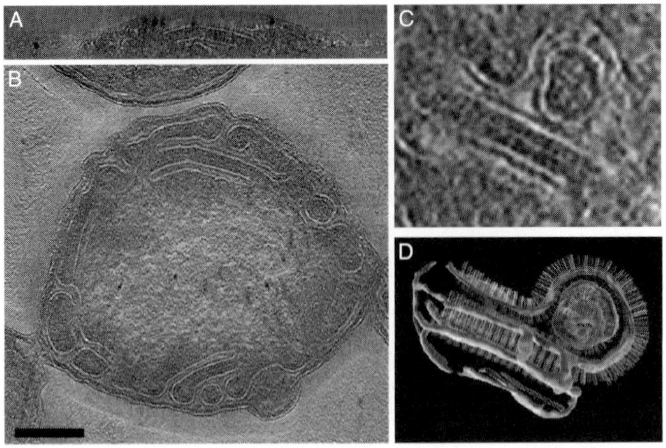

Fig. 2 3D imaging of 250-nm-thick cryosections at room temperature. Tomograms were obtained from immunolabeled sections of *E. coli* cells engineered to overproduce the chemotaxis receptor Tsr. (A and B) Vertical and horizontal slices (1-nm thick) from a tomogram illustrate the long membrane invaginations in these cells. The width of these invaginations corresponds to the length of the zipper formed by the end-to-end packing of receptors as shown in Fig. 1. Scale bar = 250 nm. (C and D) Enlarged view of a single tomographic slice and segmented representation of the 3D volume, respectively, showing the invaginated membrane architecture generated by the formation of zippered Tsr assemblies, shown along with a plausible molecular packing model for Tsr. Aside from the interaction between the cytoplasmic domains (yellow) in opposing bilayers, a complementary interaction between the periplasmic domains (blue) also appears to be likely. The uniformity in spacing between the rounded vesicular region and the invaginated membrane surrounded by it suggests that the entire membrane network is stabilized by interactions at both ends of the receptor. The segmented region is composed of two sets of stacked membranes. The lower membrane sheet is wrapped around a tubular invagination (see panel D in Fig. 3 for a schematic representation of the geometry). The segmentation was carried out to be faithful to the density in the tomogram and is therefore noisy, as seen, for example, in the patchy appearance of the rounded end of the tubular invagination. (See Color Insert.)

Fig. 3. The last panel of Fig. 3 is drawn to closely resemble specific structures actually observed in the tomogram and segmented in Fig. 2D.

IV. Cryo-Electron Microscopy of Frozen-Hydrated Sections of *E. coli* Cells

The combination of image information from negatively stained membrane extracts and tomograms from chemically fixed, cryosectioned cells has provided valuable structural information on the organization of Tsr assemblies in cells. However, it is also highly desirable to establish methods that allow imaging of assemblies of receptors and related molecules at high resolution, under conditions that are closer to the physiologically relevant state in living cells. One such approach is to carry out electron tomographic analyses of vitrified *E. coli* cells. However, the thickness

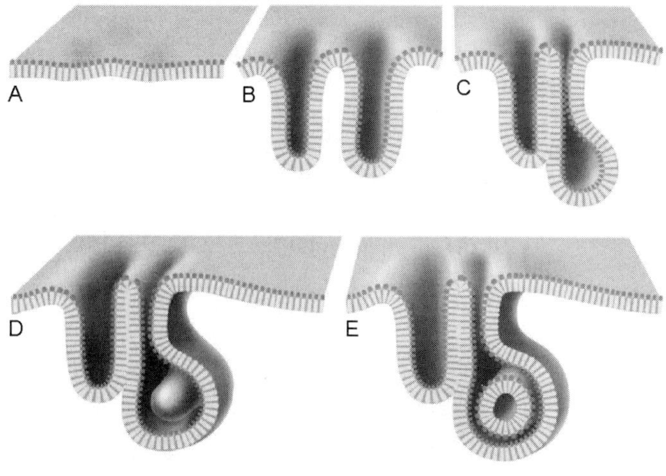

Fig. 3 A schematic illustration of the formation of invaginated membrane networks in cells overproducing Tsr. (A) Cytoplasmic membrane of wild-type cell or overproducing cell when the levels of Tsr expression are not very high. (B) Invaginations of the cytoplasmic membrane at different regions to accommodate the increased levels of Tsr by increasing the ratio of membrane surface to cytoplasmic volume. (C) Formation of two kinds of interactions between proximal invaginated membranes. The interactions between the cytoplasmic domains in adjacent membranes lead to a zipper-like structure with wider spacing (about 270 Å), while the interactions between the periplasmic domains lead to the zipper-like structure with narrower (about 150 Å) spacing. Note that the schematic is drawn to indicate that not all regions of the membrane are part of the zippered structures. (D) The rounded regions at the ends of the invaginations are occasionally large enough to accommodate an invagination from a different region of the membrane, as shown by the rounded tubule beginning to emerge from the backside. (E) The secondary invagination can grow to be so large that its growth is limited by the boundaries of other surrounding membranes. A cross-sectional view through such a region would have the appearance of a free-standing vesicular region. The cartoon in this panel has been drawn to correspond to the tomographic region shown in panel (C) of Fig. 2.

of *E. coli* cells (~1 μm) can limit the amount of useful information that can be recovered from whole-cell tomograms. To image the interior regions of cells, it is often necessary to generate sections of the cells at cryogenic temperatures and image them under low-dose conditions. While this can be technically challenging, it is nevertheless a powerful approach to validate and extend information that is obtained using 2D and 3D imaging carried out at room temperature and at higher doses.

We have used vitreous sectioning approaches to further confirm the formation of invaginated membranes in cells overproducing Tsr and to directly visualize the presence of Tsr arrays in the spacing between the membranes (described in more detail in Zhang *et al.*, 2004). For these experiments, wild-type *E. coli* or those overproducing Tsr were rapidly frozen under high pressure using a Leica EM PACT system. Thin frozen-hydrated samples (vitreous sections) were generated from frozen cells by sectioning at −150 °C using a diamond knife with a 25° knife angle, with nominal section thicknesses ranging from 20 to 60 nm (Zhang *et al.*, 2004). Figure 4A shows a projection image obtained from a 30-nm-thin section of a wild-type *E. coli* cell. In the sections generated from *E. coli* cells over producing Tsr receptor, linear receptor arrays and invaginated cell membranes are clearly visible (Fig. 4B and D). The presence of ordered receptor assemblies in cells overproducing Tsr provides a further test of the structural preservation in the frozen-hydrated sections. As indicated by the computed Fourier transform (Fig. 4B inset), the periodicity (~7.5 nm) is identical to that in membrane extracts

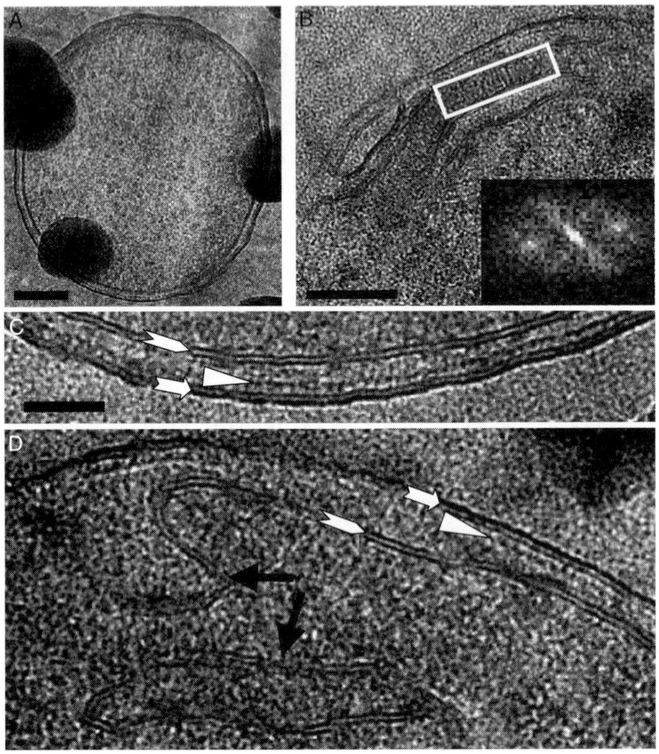

Fig. 4 Direct visualization of receptor arrays in frozen-hydrated sections of high-pressure frozen cells. (A) Low-dose electron microscopic image from a 30-nm-thick section of wild-type *E. coli* cell with identifiable densities corresponding to the outer membrane, the peptidoglycan layer, the inner membrane, and ribosome particles inside the cell. Scale bar = 0.2 μm. (B) Projection image of a frozen-hydrated section shows a zipper-like receptor linear array (boxed region) and its Fourier transform (inset). The diffraction spot that is visible corresponds to a frequency of ~1/7.5 nm^{-1}. Scale bar = 0.1 μm. (C) Both leaflets of inner and outer membranes (notched white arrows), as well as the peptidoglycan layer (white arrowhead) can be recognized clearly in images of frozen hydrated sections from wild-type *E. coli* cells. (D) The extensive invaginations observed in the tomograms of cells overproducing Tsr shown in Fig. 2 are also evident in low-dose images obtained from frozen-hydrated sections. Black arrows indicate the invaginated cytoplasmic membrane. Scale bars = 50 nm.

described earlier, indicating that there is excellent preservation of receptor packing in the sections. In both wild-type and Tsr-overproducing cell lines, the outer and inner membranes are clearly delineated, as are some regions of the intervening peptidoglycan layer. The densities corresponding to both leaflets of the inner and outer membranes are also well resolved in these images, which are recorded under low-dose conditions (Fig. 4C and D). While these sections are thin enough for visualizing structural detail, the use of thicker sections and imaging of whole cells can be expected to provide further insights into the 3D organization of the receptor assemblies in the cell membrane.

V. Structure of Chemotaxis Receptors Using Cryo-Electron Tomography

In an extension of the previous studies to study chemotaxis receptor arrangement in membrane extracts and cell sections, we have used electron tomography at cryogenic temperatures to obtain density maps of the receptor assemblies in plunge-frozen cells. One focus of these efforts has been to obtain density maps from plunge-frozen cells that have been engineered to overproduce Tsr, the serine chemotaxis receptor. Figure 5A shows an energy-filtered projection image from a portion of a single cell, recorded on a 2K $\times$ 2K CCD camera using a Tecnai Polara microscope operating at 300 kV, equipped with a Gatan 2000 imaging filter. As observed in the tomograms from fixed, cryosectioned cells and projection images from frozen-hydrated sections, the presence of high levels of Tsr leads to its accumulation in an extensively invaginated cytoplasmic membrane and to the formation of small ordered clusters in the plane of the membrane which can be readily seen in projection (Fig. 5A and B). These small patches of weakly ordered receptor clusters provide an interesting example of using a tomographic approach to determining 3D structure of the repeating units using methods pioneered by Ken Taylor and colleagues for myosin assemblies (Liu *et al.*, 2006). In order to obtain information on the 3D structure of the receptor, we recorded a series of images spanning a tilt range of $\pm 70°$, typically at composite doses in the range of 50–80 electrons per $Å^2$. The images in the tilt series were used to generate a 3D volume using weighted back-projection methods as implemented in the software package IMOD (Kremer *et al.*, 1996).

Figure 5C displays a 3-nm slice from a tomogram recorded from a cell such as the one shown in Fig. 5A. The partially ordered arrangement of receptors is readily visible. Each of the repeating regions of white density represents a slice through the central region of a partially ordered receptor cluster. Guided by determination of the local lattice vectors of the partially ordered patches, the volume corresponding to the small repeating units was boxed out and used to search the entire volume to generate a stack of subvolumes. These subvolumes were then subjected to several rounds of multireference alignment and finally averaged, first within a single cluster

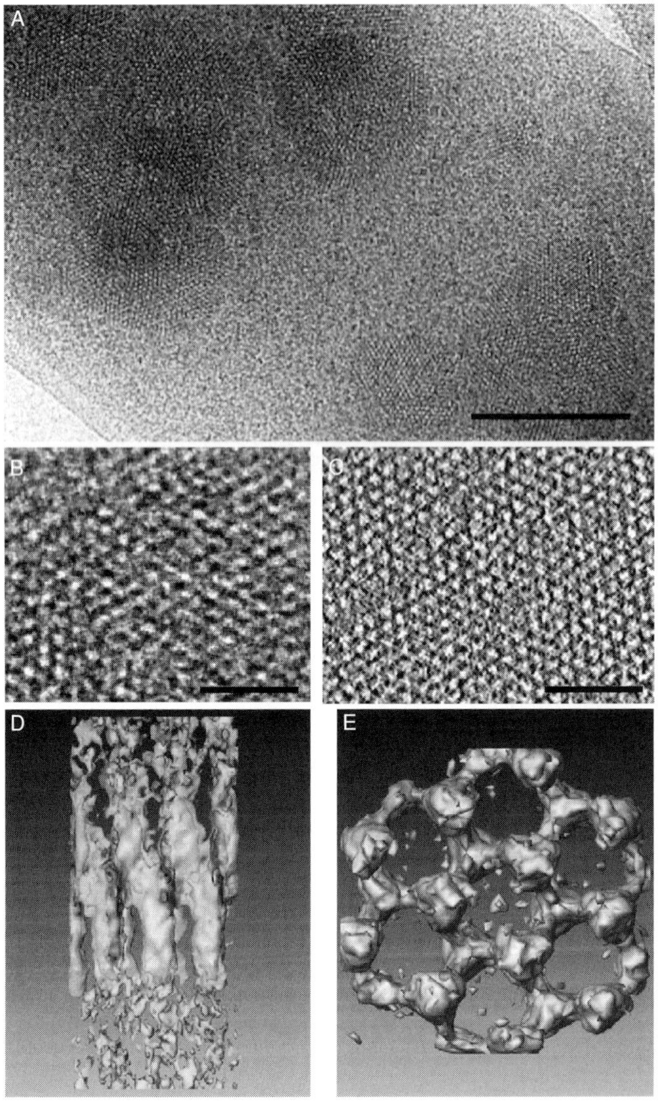

Fig. 5 Molecular architecture of Tsr receptor arrays. (A) A low-dose-projection image of a plunge-frozen cell shows domains of receptor arrays on the cell membrane. Scale bar = 200 nm. (B) An enlarged view of the receptor array as seen in a projection view down the long axis of the receptor. (C) A computational slice from a tomographic volume of the receptor array. Scale bars in panels (B) and (C) are 50 nm. (D) Surface representation (front view) of the 3D volume of a small cluster of receptors after 3D averaging of about 217 individual subvolumes. Each vertical rod of density has a size consistent with that expected from six receptor molecules. The orientation is the same as that in Fig. 1A, which shows a projection image of a zippered receptor assembly. (E) Top view of the assembly shown in panel (C) illustrating the hexagonal assembly formed by the receptor cluster. Each of the vertices in the hexagonally packed network corresponds to the top view from one repeating unit.

and subsequently over several clusters in multiple cells. A subvolume obtained from averaging about 217 of the small repeating units is presented in Fig. 5D and E, which shows front and top views of the structural motif. As discussed earlier (Lefman *et al.*, 2004; Weis *et al.*, 2003), the dimensions of this repeating unit are consistent with the size expected from a closely packed set of three receptor dimers, and provide an opportunity to generate structural models for receptor arrangement under physiological conditions. One such model that is plausible within the constraints determined from our analysis was already presented in Fig. 1C.

The density maps obtained by averaging information from such small numbers of molecules are still very noisy because of the low doses used and the small number of units averaged. Increasing the size of the data set can be expected to produce significant improvements. Comparison of the arrangement of receptors in these cells with cells expressing Tsr at wild-type levels in the presence of other signaling components will also be valuable, and such studies are underway in the laboratory. Despite the low signal-to-noise ratios in these images, the biologically interesting aspect of this study is the discovery that receptors are both laterally and axially clustered in these cells which are engineered to overproduce Tsr. Overall, these experiments demonstrate that useful structural information about receptor architecture can be obtained from cells in a near-native state, and are promising for *in vivo* structural analyses of other clustered membrane proteins in prokaryotic and eukaryotic cell membranes.

VI. Summary

Advances in instrumentation and image processing software are making electron tomography a routine tool that can be used to investigate the 3D architectures of viruses, cells, and tissues. Cryotomographic approaches have the potential to provide 3D information at resolutions that are good enough to directly recognize the shapes of large molecular assemblies, and combined with averaging, provide a new approach to *in situ* structure determination by averaging individual molecular tomograms. Tomography of stained specimens at room temperature and at high doses provides 3D structural information that reflects the distribution of stain in the specimen, and while the maps do not reflect the intrinsic distribution of proteins and nucleic acids, they provide extremely valuable information about cellular architecture. The use of dual-tilt strategies improves tomogram quality, at least in the case of imaging stained specimens at room temperature where beam damage is not a limiting factor. There is a great need for effective approaches to extract quantitative information from cellular tomograms, and this is an area where great advances may be expected in the coming years. We expect that the methods used here to reveal the organization of the bacterial membrane protein networks will be directly applicable to imaging similar structures in eukaryotic cells and that the strategy of combining tomographic, electron, and x-ray crystallographic approaches can be a tool that is generally applicable for constructing plausible 3D models for protein networks in cells at molecular resolution.

References

Ames, P., Studdert, C. A., Reiser, R. H., and Parkinson, J. S. (2002). Collaborative signaling by mixed chemoreceptor teams in *Escherichia coli*. *Proc. Natl. Acad. Sci. USA* **99,** 7060–7065.

DeRosier, D. J., and Klug, A. (1968). Reconstruction of three dimensional structures from electron micrographs. *Nature* **217,** 130–134.

Falke, J. J., and Hazelbauer, G. L. (2001). Transmembrane signaling in bacterial chemoreceptors. *Trends Biochem. Sci.* **26,** 257–265.

Grebe, T. W., and Stock, J. B. (1999). The histidine protein kinase superfamily. *Adv. Microb. Physiol.* **41,** 139–227.

Hart, R. G. (1968). Electron microscopy of unstained biological material: The polytropic montage. *Science* **159,** 1464–1467.

Kremer, J. R., Mastronarde, D. N., and McIntosh, J. R. (1996). Computer visualization of three-dimensional image data using IMOD. *J. Struct. Biol.* **116,** 71–76.

Le Moual, H., and Koshland, D. E., Jr. (1996). Molecular evolution of the C-terminal cytoplasmic domain of a superfamily of bacterial receptors involved in taxis. *J. Mol. Biol.* **261,** 568–585.

Lefman, J., Zhang, P., Hirai, T., Weis, R. M., Juliani, J., Bliss, D., Kessel, M., Bos, E., Peters, P. J., and Subramaniam, S. (2004). Three-dimensional electron microscopic imaging of membrane invaginations in *Escherichia coli* overproducing the chemotaxis receptor Tsr. *J. Bacteriol.* **186,** 5052–5061.

Liu, J., Taylor, D. W., Krementsova, E. B., Trybus, K. M., and Taylor, K. A. (2006). Three-dimensional structure of the myosin V inhibited state by cryoelectron tomography. *Nature* **442,** 208–211.

Lucic, V., Forster, F., and Baumeister, W. (2005). Structural studies by electron tomography: From cells to molecules. *Annu. Rev. Biochem.* **74,** 833–865.

McIntosh, R., Nicastro, D., and Mastronarde, D. (2005). New views of cells in 3D: An introduction to electron tomography. *Trends Cell Biol.* **15,** 43–51.

Parkinson, J. S., Ames, P., and Studdert, C. A. (2005). Collaborative signaling by bacterial chemoreceptors. *Curr. Opin. Microbiol.* **8,** 116–121.

Peters, P. J., and Hunziker, W. (2001). Subcellular localization of Rab17 by cryo-immunogold electron microscopy in epithelial cells grown on polycarbonate filters. *Methods Enzymol.* **329,** 210–225.

Stock, J., and Da Re, S. (2000). Signal transduction: Response regulators on and off. *Curr. Biol.* **10,** R420–R424.

Subramaniam, S. (2005). Bridging the imaging gap: Visualizing subcellular architecture with electron tomography. *Curr. Opin. Microbiol.* **8,** 316–322.

Tokuyasu, K. T. (1973). A technique for ultracryotomy of cell suspensions and tissues. *J. Cell Biol.* **57,** 551–565.

Weis, R. M., Hirai, T., Chalah, A., Kessel, M., Peters, P. J., and Subramaniam, S. (2003). Electron microscopic analysis of membrane assemblies formed by the bacterial chemotaxis receptor Tsr. *J. Bacteriol.* **185,** 3636–3643.

Zhang, P., Bos, E., Heymann, J., Gnaegi, H., Kessel, M., Peters, P. J., and Subramaniam, S. (2004). Direct visualization of receptor arrays in frozen-hydrated sections and plunge-frozen specimens of *E. coli* engineered to overproduce the chemotaxis receptor Tsr. *J. Microsc.* **216,** 76–83.

CHAPTER 15

How to "Read" a Vitreous Section

Jacques Dubochet,* Benoît Zuber,* Mikhail Eltsov,* Cédric Bouchet-Marquis,* Ashraf Al-Amoudi,[1],* and Françoise Livolant†

*Laboratory for Ultrastructural Analysis
Biophore, University of Lausanne
CH-1015 Lausanne, Switzerland

†Laboratoire de physique des solides
Université Paris sud
F-91 405 Orsay, Cedex, France

Cryo-electron microscopy of vitreous sections (CEMOVIS) makes it possible to observe biological specimen in a close-to-native state. There is no chemical fixation, no stain, and no aggregation of the biological material; all the water is present, immobilized in a vitreous state. An image obtained with CEMOVIS looks very different from a conventionally prepared dry section. The information it contains must be interpreted differently. First of all, the section looks smooth and homogeneous because, globally, the molecules are uniformly distributed in the cytosol.

[1] Present address: EMBL, Pf. 10.2209, D-69012 Heidelberg.

0091-679X/07 $35.00
DOI: 10.1016/S0091-679X(06)79015-X

Contrast appears at the level of macromolecules and their complexes, such as membranes, ribosomes and nucleosomes, fibers, and even single molecules, when they are large enough and appropriately arranged. The limitation on the amount of information that can be retrieved from an image is set by the plethora of overlapping structures in the two-dimensional image of the three-dimensional section. Computer electron tomography may overcome this limitation. It is, however, in the nature of vitreous sections to be deformed during the cutting process. This effect takes place without loss of information, as long as the deformation is homogeneous. Beam damage may be spectacular while one is imaging vitreous biological material, but it is probably less severe than in conventional preparations, as long as meaningful resolution is the limiting factor. A few observations, which require more elaborate interpretation, are presented.

I. Introduction

Electrons have a great advantage over photons; they interact strongly with matter. Even something as small as a single protein or a single DNA filament has a sufficient effect on the electron beam to produce an image from the interaction. Consequently, we electron microscopists can observe very thin specimens in exquisite detail. Unfortunately, we must observe them under vacuum and, consequently, our specimens must be prepared so they will not evaporate in the microscope. This simple fact is of fundamental importance for biologists studying material in which water is by far the most abundant constituent. Could it be that our biased view toward systematically dried specimens has impregnated and distorted our vision of cell ultrastructure? Could it be that a more realistic view, based on samples that retain a cell's normal hydration, would have a profound heuristic value and serve science better?

The age of dry electron microscopy is now over. Cryo-electron microscopy makes it possible to observe aqueous suspensions or fully hydrated specimens at a temperature at which water does not evaporate. The road to our current state was long; Fernandez-Moran (1960) initiated the approach and Taylor and Glaeser (1974, 1976) demonstrated that it works. Finding that vitrification of liquid water is possible and understanding its rationale was another important step (Dubochet and McDowall, 1981; Mayer and Brüggeller, 1980; and see also Chapter 1 by Dubochet, this volume). Instrument designers have fulfilled their essential role (Moor, 1987; Sitte, 1996). In the final analysis, though, it was only when all the small and big practicalities could be put together that cryo-electron microscopy became a feasible method. This took place in the 1980s for the observation of small biological particles suspended in a thin film of vitrified solution (Adrian *et al.*, 1984; Roos and Morgan, 1990). However, most cells and tissues are more difficult to deal with because a much larger volume must be vitrified and the latter must be cut into thin vitreous sections. It is only recently that cryo-electron microscopy of vitreous sections (CEMOVIS) was established as a practical method (Al-Amoudi *et al.*, 2004a,b, 2005a; Bouchet-Marquis *et al.*, 2006, 2007; Eltsov and Dubochet, 2005; Hsieh *et al.*, 2006; Leis *et al.*, 2005;

Matias and Beveridge, 2006; Nicastro *et al.*, 2005; Norlén and Al-Amoudi, 2004; Schwartz *et al.*, 2003; Zhang *et al.*, 2004; Zuber *et al.*, 2005). The method still requires some skill and dedication, so some time may pass before it becomes widely used. For the moment, it can be said that at least in favorable cases, CEMOVIS results in images that are much closer to the native state, and they provide higher resolution than those obtained from conventional sections. Contrary to what most people would guess, the major limitation of the new method does not arise from the lack of contrast of unstained structures suspended in water, but rather from the plethora of information overlapping in the thickness of the section. Computerized electron tomography (CET; see Chapter 25 by Best *et al.*, this volume) should overcome this problem.

In principle, CEMOVIS is as simple as possible. It consists in taking a small sample of biological material, vitrifying it, and cutting it into thin vitreous sections that can be observed directly in a cryo-electron microscope. In practice, there are a number of difficulties. Three of them deserve special attention. (1) Vitrification of pure water or of a dilute aqueous solution can only be achieved for objects whose size is measured in micrometers. By taking advantage of cryoprotectants and high-pressure freezing, it has been possible to extend this size by two orders of magnitude. Once vitrified, the sample must be kept at all times below the devitrification temperature of $-135\,^{\circ}\mathrm{C}$. (2) Cutting vitreous sections is a rather strange operation. It must be carried out in the dry state, since there is no known liquid on which sections will float at this temperature. Furthermore, a vitreous sample is, by definition, a liquid, even if its viscosity is very large. Consequently, deformation of the sample under the locally high pressure that is generated by the cutting process is inherent to vitreous sections. (3) Vitreous sections are sensitive to the electron beam. Here again, the liquid nature of vitreous sections manifests itself during irradiation by small deformations, which finally result in the complete destruction of the section.

The technique of CEMOVIS, namely how to vitrify, cut, and observe cryosections, is described in various articles (Al-Amoudi *et al.*, 2002, 2005b; Chang *et al.*, 1983; McDowall *et al.*, 1983; Richter, 1994a,b,c; Richter and Dubochet, 1989, 1993; Richter *et al.*, 1991; Sartori and Salamin Michel, 1994; Sartori Blanc *et al.*, 1998). Experimental protocols are also available (Dubochet *et al.*, 1988, 2007; Sartori Blanc and Salamin Michel, 1998). The present chapter is addressed to scientists confronted with CEMOVIS data. Our aim is to explain how to "read" micrographs, how to interpret them, and how to understand what they can and cannot provide.

II. Conventional Versus Cryo-Electron Microscopy: A Comparison

By conventional, we mean here every application of electron microscopy by which dry specimens are observed at room temperature, whereas cryo-electron microscopy concerns the observation of hydrated specimen at a temperature so low that water does not evaporate significantly in the microscope's vacuum.

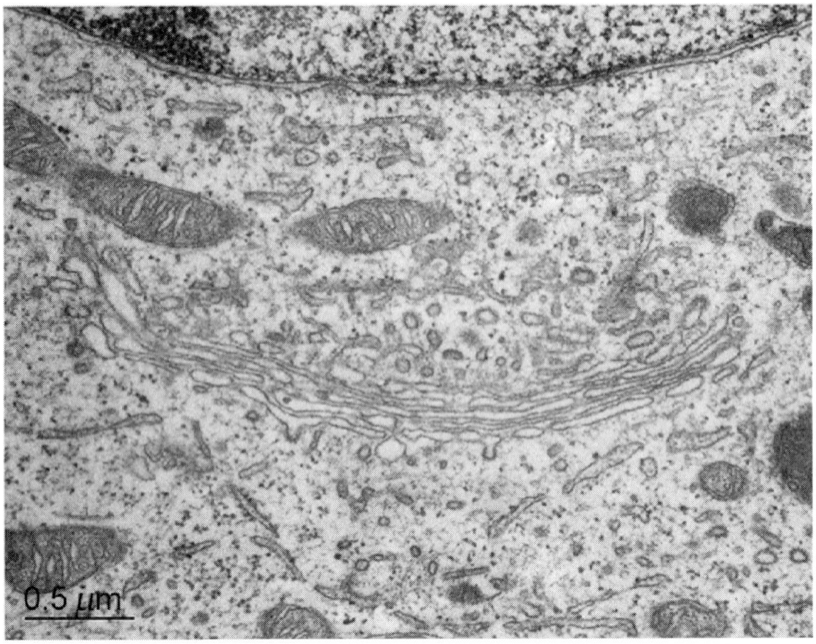

Fig. 1 Conventional, resin-embedded INS-1 cell (insulin secreting beta-cells). Cells in PBS were fixed in 2% glutaraldhyde for 30 min, washed, postfixed in 1% osmium reduced by 1.5% potassium ferrocyanide, washed, dehydrated in graded ethanol series (30%, 50%, 80%, 90%, 3 times 100%) and embedded in Epon 812, which was polymerized over night, cut at 50 nm and mounted on Formvar-coated grids. Contrast was enhanced by uranyl acetate and lead citrate. A large Golgi complex is visible in the center and a nucleus at the top. Courtesy M. Grabenbauer.

Figures 1–3 provide images of human culture cells with which to compare the results from these two methods. Figure 1 shows a conventional section of an Epon-embedded cell. It was chemically fixed and stained, water was replaced by a resin, and the section was stained once more after cutting. Figure 2 is a typical vitreous section of the kind that can now be routinely obtained from most cells and tissues. It is rich in valuable structural information, but cutting artifacts are also conspicuous. Figure 3 originates from regions of a mitotic cell, which is free of the organelles that modify the regularity of the cutting process (except at the bottom right). Other figures illustrate specific points discussed in the following sections.

A. Vitreous Sections Are Unstained and Nonaggregated

What is seen in conventional sections is not the native object but, primarily, the stain that is apposed to the biological structures. For example, a biological membrane frequently appears as a double layer because the stain is deposited on the

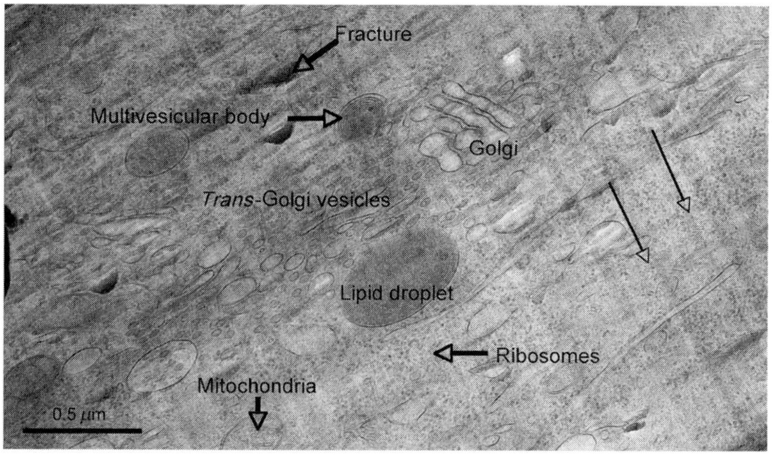

Fig. 2 Vitreous section of a HeLa cell. The cells where grown to confluence and mechanically harvested in PBS supplemented with 20% dextran 40 kDa as external cryoprotectant. The concentrated cell suspension was sucked into a 300-μm internal diameter sample tube for the EM PACT (Leica, Vienna, Austria) high-pressure freezer and immediately vitrified. The tube was trimmed and the specimen cut with a diamond knife (Diatome, Biel, Switzerland) in a Leica (Vienna, Austria) EM FC6 cryoultramicrotome and observed in an FEI (Eindhoven, The Netherlands) CM 100 electron microscope, equipped with a Gatan (Warrendale, PA) cryo-specimen holder. After vitrification, the temperature was always kept below −140 °C. Details of the preparation are given elsewhere (Bouchet-Marquis *et al.*, 2006). Various structures are marked. Note also the rich membranous system around the Golgi apparatus. The elliptical shape of all vesicles is due to cutting-induced deformation. Knife marks and the cutting direction are shown with long black arrows. There are some fractures due to irregular mechanical property of the medium. One of them is marked.

membrane's surfaces, while it scarcely penetrates to the interior. Similarly, we are accustomed to recognizing condensed chromatin because it generally appears dark (Fig. 1, upper left). This is not surprising, since the word "chromatin" means "stained substance," but it does not tell us whether the original material is particularly dense or not. Conventional preparations are generally stained twice. The whole block is stained first and the section is stained in the final stage of preparation. In most cases, section staining is even more abundant.

Aggregation is also in the "nature" of conventional preparations. Except for those that are fully hydrophobic, every biological structure or biomacromolecule interacts with water over part, or all, of its surface. These surfaces must rearrange on dehydration. This is most obviously the case for molecules floating in water. When water is removed, they have to bind to other molecules until, together, they form a new self-supporting aggregate (Kellenberger and Kistler, 1979). The same holds within any biological structure, whose regions, which were previously in contact with water, must find new contacts; typically they will do so by approaching other hydrophilic regions that were previously in water. In any case, the result is aggregation. It may correspond to large-scale rearrangement or only to local

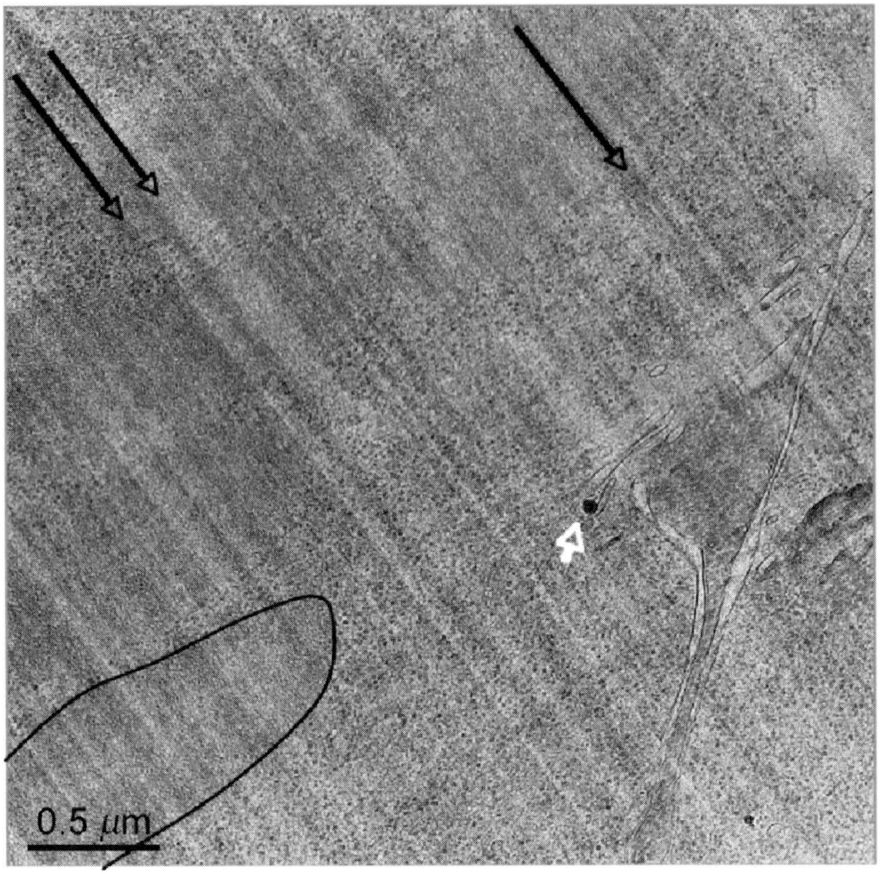

Fig. 3 Vitreous section of a mitotic HeLa cell prepared as for Fig. 2. Condensed chromatin corresponds to the smooth regions between the ribosome rich areas. One region of condensed chromatin is delineated (black line). Cutting direction and knife marks are marked by long black arrows. An ice crystal contaminating the surface of the section is marked with a white arrow.

compaction. The art of good specimen preparation does not consist in avoiding aggregation; it tries to minimize its effect. What do we know about the extent of rearrangements due to aggregation in conventional preparation? Not so much, but CEMOVIS will tell us more.

In the absence of any staining and with all the water present as immobilized liquid, CEMOVIS gives a faithful representation of the native state. What is seen on the image is the real distribution of the biological material within the thickness of the section. The global contrast between different regions of the micrograph is proportional to the density difference in the corresponding regions of the specimen.

Two remarks may be pertinent here. (1) Until now, CEMOVIS has been applied to unstained specimens because the inherent contrast of native material is sufficient for most observations, even at the molecular level. In the future, the use of labels for finely tuned specific contrast will certainly offer rich possibilities for characterizing and identifying structures and functions. (2) A word of caution about density and aggregation in the presence of vitreous water. Since the nature of vitreous water is not yet fully understood (Chapter 1 by Dubochet, this volume), changes in biological structures on vitrification cannot be excluded, although so far we know of no evidence in this direction. The density of vitreous water is thought to be 0.93 g/cm^3, which means that vitrification could correspond to a linear expansion of ca. 2%. Whether or not this could have a deleterious effect on biological structures is also unknown. We will see below that it probably produces some effects.

B. Vitreous Section Are Globally Uniform

From the above considerations we are not surprised that, when comparing conventional and vitreous sections (Figs. 1 with 2 and 3), the former appear to be formed from compact and well-separated features, whereas the latter are much more gray and uniform. The effect is more remarkable in Fig. 3, where mitotic chromosomes are dispersed in the ribosome-rich cytoplasm. One chromosomal region is marked at the bottom left. Others are recognized by the fact that they are free of ribosomes and by their fine, homogeneous, grainy texture, which is characteristic of condensed chromatin (Bouchet-Marquis *et al.*, 2006; McDowall *et al.*, 1986). If this figure is observed from a distance, it appears to be uniformly gray, except for the knife marks that are recognizable from their parallel orientation, and the region on the cell border at the bottom right. It may surprise many that condensed chromatin is not significantly denser than the rest of the cytoplasm. More precise measurements have shown that in mammalian cells, the density of condensed chromatin in the nucleus or in chromosomes does not exceed the surrounding nucleoplasm or cytoplasm by more than 3% (Sartori Blanc *et al.*, 1998). A similar observation concerns keratin filaments in the skin epithelium (Norlen *et al.*, 2003). Conventional electron microscopy shows intermediate filaments as compacted, dense bundles. In vitreous sections, they are no more compacted than the rest of the cytoplasm.

The global uniformity of the cell's interior is probably the simple consequence of thermodynamics; soluble material is uniformly distributed in a liquid. Of course, the rule does not apply when there are boundaries that define different compartments. In Fig. 2, there are several high-density multivesicular bodies and some stacks of the Golgi, the latter appearing with surprisingly low density. Mitochondria are generally among the densest compartments of cells.

C. Vitreous Sections Are Locally Rich in Fine Structures

The rule of global uniformity does not apply locally. Each individual structure, particle, or macromolecule has its own density and compaction. These finest details are present in a thin section. They are also present in the image, but not all of them can be recognized. What can be visualized and what cannot? The answer depends on several parameters. They include the conditions of image formation, the size and the nature of the particle, the surrounding structures, and the section thickness.

1. Phase Versus Amplitude Contrast

It is common knowledge that physicists invent strange theories. One of them is quantum mechanics, which claims that an event may be defined by how it is observed. Some electron microscopists do not realize that they are diligent users of this reality. An electron is either a particle or a wave, depending on how it is detected. A small objective aperture and sharp focus cause the intensity on the image to be reduced proportionally to the number of electrons scattered outside the objective aperture. In this situation, electrons are particles and the image is formed through amplitude contrast. The image is a fair representation of the *density* of the material in different regions of the specimen. In a section, this mode of image formation works well when there is a marked density difference between various regions. This is ideally the case in images of heavy metal accumulations, as in the case of stained sections. The structure must not be too small, though. If it is much smaller than the section thickness, the effect of its local high density is "diluted" over the whole section thickness and becomes negligible.

Amplitude contrast—where electrons behave as particles—is the ideal and dominant mode of image formation for conventional sections. Not so for vitreous sections! We have shown above that native biological material in the cell is surprisingly homogeneous. Amplitude contrast has little to depict under these conditions. For example, condensed chromatin can hardly be differentiated from its surrounding medium. At zero defocus, even a particle as big as a ribosome is practically invisible. Fortunately, an electron can also be a wave.

In order to observe the wave character of an electron, the objective aperture must be large and the defocus chosen appropriately. When this is the case, the contrast on the image, due to the small phase shift between the scattered and nonscattered electron waves, is proportional to the *variation of density* between adjacent regions. This effect is exquisitely sensitive and can reveal even the finest molecular details. A well-documented case is DNA (Dubochet *et al.*, 1994). A single molecule of DNA, suspended in a thin film of vitreous water, can be seen with good contrast and traced with precision because the density of DNA (ca. 1.5 g/cm^3) is much greater than that of the surrounding medium (0.93 g/cm^3 for vitreous water); this difference is visible even if it only extends over a 2-nm thickness. The presence or absence of DNA locally results in a very significant difference. It is sufficient to be detected by phase contrast, but certainly not by amplitude contrast. The observation of vitreous sections relies nearly exclusively on phase contrast.

The trouble with phase contrast is that it is selective. Depending on the degree of defocus, it favors the representation of some dimensions, whereas it blocks

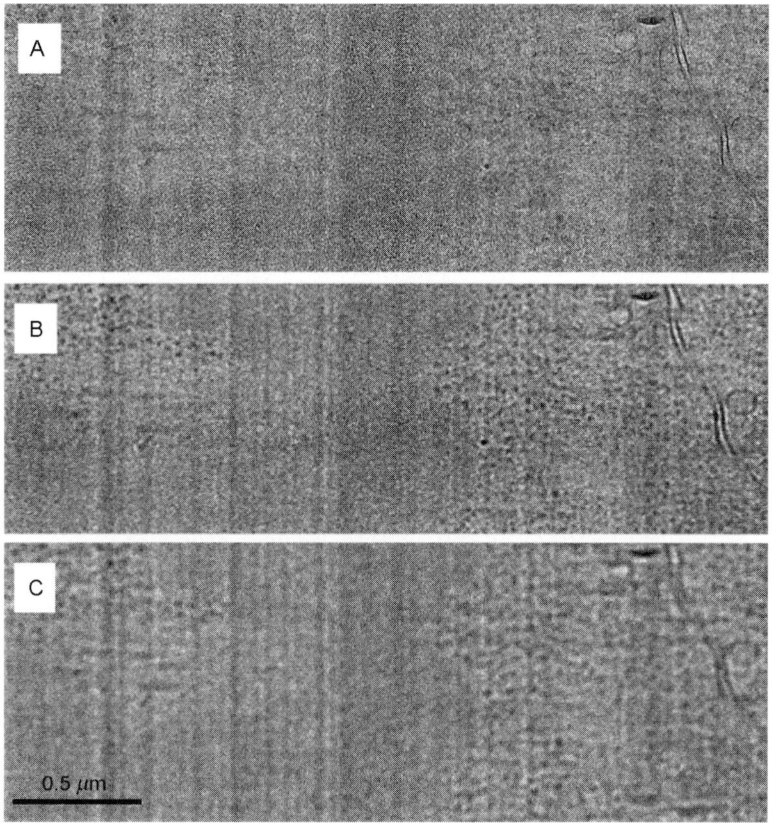

Fig. 4 Mitotic HeLa cell prepared as for Fig. 2. Focus series with images that are 6, 24, and 100 μm underfocus, shown in A, B, and C, respectively. These degrees of defocus corresponding, respectively, to optimal contrast transfer at 10, 20, and 40 nm.

others. For a given defocus, ΔZ, the dimension d that is best transferred to the image is given by the equation $d^2 = 2\lambda\Delta Z$ (where λ is the wavelength of the electron); the microscope blocks all information relating to the dimension $d^2 = \lambda\Delta Z$. Consequently, one photograph is not enough to give a full representation of the object. This effect must be dealt with very seriously when observations are aimed at very high resolution (nanometers) or at quantitative density determination. In practice and for the kind of observations that are at present achieved with CEMOVIS, as well as for those presented in this chapter, a fair representation of the specimen is obtained if the operator records different views over a broad range of defocus. The example presented in Fig. 4 shows conditions that favor the visualization of nucleosome-sized objects in (A); ribosomes are reinforced in (B), while 30- to 40-nm filaments, if present, would be best visualized in (C) [although none of these is visible, as is generally the case in vitrified images of condensed chromatin (Bouchet-Marquis *et al.*, 2006; McDowall *et al.*, 1986)].

2. Minimum Size of Visible Structure Is Affected by Section Thickness

What is the smallest structure that can be visualized in a vitreous section? It is now well established that phase contrast images of biological particles, vitrified in a thin layer of solution, can lead to atomic resolution under the most favorable conditions (Abbott, 2002). This can be the case when beam damage is overcome by summing information from many identical single particles or from a crystal (Baumeister and Steven, 2000). Our situation is less favorable; no two cells, two Golgi complexes, or two mitochondria are identical. Each biological structure must be visualized in its oneness. We discussed above the case of a single molecule of DNA floating in a thin film of solution, suggesting that reaching 2-nm resolution could be a realistic goal, but that situation is too easy for a fair comparison. Just as in light microscopy, one must distinguish between detection and resolution (viz., a single microtubule, which is easy to see as a fluorescent object but impossible to resolve from its neighbors when packed into clusters, as in the mitotic spindle). In any realistic cellular object, a macromolecule is not alone. Indeed, a cell is crowded (Goodsell, 1993). Many structures overlap over the thickness of a section. Here again, chromatin provides a useful illustration. We know that it is mostly formed of juxtaposed nucleosomes. Each nucleosome seen in isolation is clearly visible in a thin layer of vitrified buffer (Furrer *et al.*, 1995) or in a vitreous section (M.E., unpublished result). In the crowded state of condensed chromatin, however, there are perhaps 10 nucleosomes superimposed over the thickness of a vitreous section. As can be seen in Figs. 3 and 4, no specific structure emerges from this disordered "sea of nucleosomes," except the average close-packing distance of 10–12 nm, which is revealed in the grainy texture that characterizes condensed chromatin and chromosome structures.

Figure 5 shows a section through a dendrite in an organotypic slice of rat brain. Easily recognizable is a bundle of microtubules (μT) that lie approximately parallel to the viewing direction. Also readily seen, when they are parallel to the viewing direction, are the two membranes of a mitochondrion (Mit) and the cell membrane (Mb). These membranes are hardly distinguishable when they are seen obliquely (mb). A probable synapse (Sy) with synaptic vesicles (Sv) is marked on the right. Ribosomes (R) are also visible at bottom right. Many more filamentous and granular structures are visible all over the section, but they appear too faint to be interpreted. Some of the μT are exactly aligned with the viewing direction (μTa). They appear sharp and show high contrast. We note the chirality of the 13 protofilaments, reflected in the slew of the elliptical protofilament cross sections; these indicate that the μTs are seen from their minus toward their plus ends. Other MTs are slightly tilted (μTb) and less contrasted. We note also that several μTs contain a luminal density. A detailed study of these μT will be presented elsewhere (Bouchet-Marquis *et al.*, 2007; Garvalov *et al.*, 2006). It shows that the structure of μTs visible in vitreous sections coincides with the atomic model to better than 2 nm. Membranes are ca. 6-nm thick; some show a double-layered structure. Figure 6 illustrates how contrast depends on feature orientation.

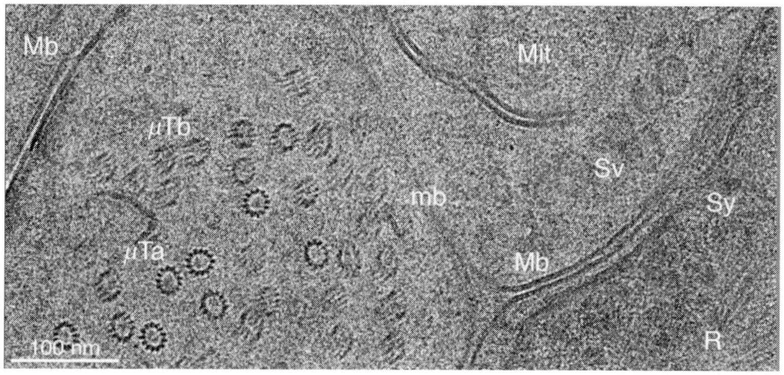

Fig. 5 Vitreous section in a dendrite from an organotypic slice of rat brain. See text for explanation. Details of the preparation are given in Zuber *et al.* (2005). μTa and μTb, microtubules seen along the axis and slightly tilted, respectively; Mit, mitochondrion; Mb, membrane appearing with high contrast when seen parallel to the viewing direction or with low contrast in other orientation (mb); Sy, probable synapse; Sv, synaptic vesicles; and R, ribosomes.

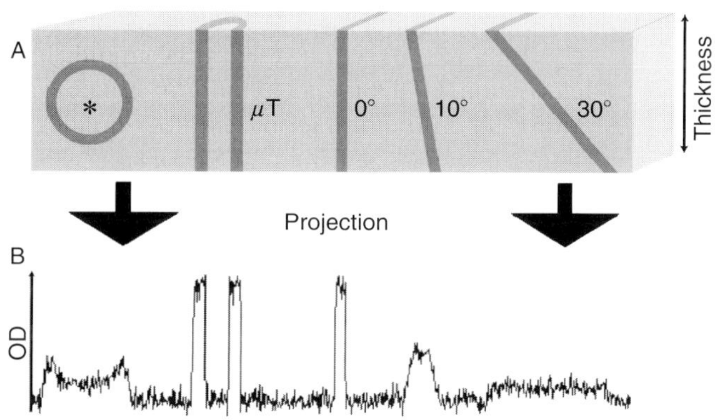

Fig. 6 Contrast in the image of a vitreous section. In this scheme, the 50-nm-thick section (A) contains a small vesicle (*), a microtubule (Mt) perpendicular to the section plan, and membranes perpendicular to the section or tilted by 10° or 30°. The density recorded on the image is shown in (B). It is apparent that straight or planar structures are seen with high contrast only when they are oriented along the viewing direction. Redrawn from Zuber *et al.* (2005), with permission.

It explains, for example, why a membrane visible at some place (Mb in Fig. 5) becomes invisible a bit further (mb).

From this and other images presented elsewhere, it is apparent that the image of a vitreous section contains a large amount of information. Those structures that are oriented along the viewing direction over the whole thickness of the section or a good part of it are seen with unprecedented detail. All other structures are less contrasted and are more difficult to interpret because they are superposed in the crowded environment of the cell interior. In other words, it is not the lack of visible structures or of contrast that is the limiting factor in CEMOVIS, but the plethora of information projected from the whole thickness of the section. CET will overcome this difficulty.

3. The Promise of CET

Tomography consists in reconstructing a 3D model of an object from a large number of 2D projections recorded from all possible directions. The beautiful theory of Hoppe (Hoppe *et al.*, 1974) further shows that recording this large number of projections does not mean that a very large cumulative electron dose is required for a tomographic reconstruction. It states that if one projection recorded with an electron dose D gives a 2-dimensions image at resolution δ, then a 3-dimensions map at the same resolution can be constructed from a large number of projections recorded with a total dose of $2D$ only. In principle, the reconstruction of a 3-dimensions model approaching the resolution of a 2-dimensions image is possible. In practice it faces impressive difficulties. The first one is how to record 100 or so tilted micrographs with a total dose that does not exceed the dose normally used for 2 micrographs, all perfectly focused and aligned. We note also that a 90° tilt will never be possible, but even approaching it presents a considerable technical challenge. The second difficulty is to assemble the information from all of these images when each of them is so underexposed that it is difficult to recognize anything on it. The third is to make sense of the 3-dimensions model, which contains so much information that the observer does not know how to look at it.

Nevertheless, CET works beautifully on thin vitrified layers containing squeezed cells or cell fragments (Medalia *et al.*, 2002; Plitzko *et al.*, 2002; see also the Chapter 3 by Hess, Chapter 16 by Sherysheva *et al.*, Chapter 17 by Hoenger and Nicastro, and Chapter 18 by Cheng and Walz). The combination of CEMOVIS with CET is just beginning (Bouchet-Marquis *et al.*, 2006; Hsieh *et al.*, 2002, 2006; Leis *et al.*, 2005; Nicastro *et al.*, 2005; Schwartz *et al.*, 2003; Zhang *et al.*, 2004), but considerable efforts are in progress to make these into efficient and widely used complementary techniques (3D EM, 2004).

D. Cutting-Induced Deformation Is Inherent to Vitreous Sections

During the cutting of thin sections, resin-embedded material is deformed, but it regains its original shape remarkably well after being floated on water and dried onto the support. It behaves like an elastic material. This does not hold for vitreous sections. By definition, vitreous material is a high-viscosity liquid, which means that it flows whenever a force is applied to it. It flows slowly when the force is small, but it flows rapidly when the force is great, as is the case during the sectioning process. What happens when a knife penetrates a block of vitreous material and how a vitreous section is formed has been studied in detail. A simple model describes the process (Al-Amoudi *et al.*, 2005b; Dubochet *et al.*, 2007) and accounts for all the major features of vitreous sections. The major aspects are the following:

Under optimal conditions, the knife progressing through the block with feed (f) compresses the material along the cutting direction so that dimensions are reduced by the compression factor (C), which can be anything between 15% to 60% or even more. This compression is compensated by a proportional increase of the section thickness (t), which thus becomes $t = f/(1 - \mathrm{C})$. To a first approximation, the compression factor is proportional to the sine of the cutting angle, but friction on the knife surface modulates this value; in the dimensions perpendicular to the cutting direction, sizes are conserved. Under these conditions the deformation is said to be homogeneous; dimensions are changed, but the transformation is the same at every point. A sphere in the block is deformed into an ellipsoid in the section but, very important, a plane is transformed into a plane and a straight line into a straight line. We insist here on the concept of homogeneous deformation because, as long as it holds, cutting-induced deformation takes place without loss of information. The original state can always be reconstructed from the measurable parameters of the deformation.

Some deformation is evident in every vitreous section. It results in circular structures appearing as ellipses with their small axes parallel to the cutting direction (Figs. 2 and 5). The remarkable homogeneity of the deformation can be tested in many cases. For example, the numerous membranes seen very sharply along the cutting direction indicate that their original planar structure remained a plane during cutting. In Fig. 5, the favorably oriented μT would not be sharp if the straight protofilaments had not stayed straight. The preservation of straight fragments of DNA in natural and model systems is the most stringent test of deformation homogeneity to date (Richter, 1994b; Sartori Blanc *et al.*, 2001).

In practice, the cryosectioning process does not always go so smoothly. Undesirable artifacts are frequent and severe. They are illustrated in Fig. 7. The most important are the following:

Crevasses: We have all been told that window glass is a vitrified liquid but, when a force is applied, it tends to break rather than flow. Sad! Similarly, when the force applied to a section by the knife is too great, it breaks where the shear stress is maximum, that is, at the surface of the section that is opposite to the knife-edge. It does not break only once; it forms a field of fractures that penetrate more or less deeply into the section. It happens also that a specific feature in the section may favor fracture formation. Those who are familiar with the remarkable preparation method of freeze-fracture will not be surprised that fractures in vitreous sections frequently take place through the middle of biological membranes.

Irregular compression: The rule of homogeneous compression holds good only as long as the material is itself homogeneous. Otherwise, the compression may vary from point to point in the section. Typically, the interior and exterior of a cell will frequently have different mechanical properties, and thus they deform differently. The effect can be quite complicated. For example, in Fig. 7, the deformation of the

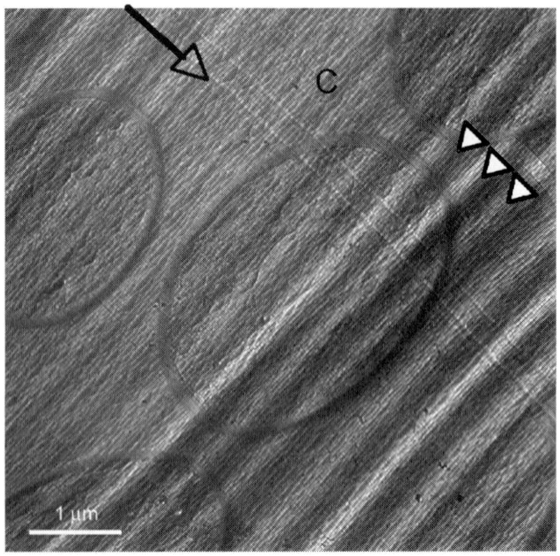

Fig. 7 Cutting artifacts. Yeast cells vitrified in 20% dextran. Cutting directions and knife marks: long black arrows; crevasses are all over, for example in region marked C. Chatter is the undulation marked by white arrow heads.

yeast envelope is not the same when the knife penetrates the cell as when it leaves it. A more surprising case is seen in Fig. 5. Many microtubules have an elliptical cross section with the small axis oriented along the cutting direction. But how is it possible that there is so much variation from one microtubule to the other? Why is the microtubule marked μTa perfectly circular? Is it because it was protected during cutting by the "shadow" of the mitochondrion? Could it be that microtubules are naturally noncircular or that they have irregular mechanical properties? Obviously, more must be known about the cutting process and about microtubules to answer these questions.

Chatter: One surprising form of nonhomogeneous compression is chatter. This oscillatory variation in section thickness along the cutting direction, well known from classical electron microscopy, is generally thought to be due to an oscillation of the block in relation to the knife-edge. In a vitreous section, the origin is generally different. It is due to the irregular friction of the forming section on the knife surface, which produces variable compression, causing section thickness to vary accordingly. This subtle effect points to the importance of the surface properties of the knife. The producer must know how to make it right.

Knife marks: Knife marks are partly due to damage at the cutting edge. When this is the case, a ridge on one surface of the section generally corresponds to a furrow on the other surface. More frequently, they are due to solid grains that contaminate the knife's surface. In this case they occur only on the bottom face of the section. Frequently, these grains consist of hard, ripped-off fragments of the sample (e.g., pieces of membrane).

It is the skill and the art of a good CEMOVISt to find the right cutting conditions in order to minimize deformation, to avoid crevasses, and to create conditions in which deformations are as homogeneous as possible. In order to fulfill these conditions, the operator can adjust the usual parameters of the microtome (speed, feed, knife angle, temperature, ion shower), but in addition he has much more to play with. The use of an oscillating knife has shown promise (Al-Amoudi *et al.*, 2003). The preparation of the section area by trimming and a good choice of section shape may play a major role in section quality. Of course, good vitrification is an essential prerequisite. Above all, the quality of the knife is vital. Unsurprisingly, the specimen itself influences cutting properties; for example, a nucleus is generally a good region to cut. Since some regions are better than others, it is a good idea to adjust the medium in which the specimen is bathing in order to attain optimal cutting properties. Experience has shown that for many cells and tissues, 20% 40-kDa dextran (a sugar polymer chosen for adequate viscosity and small osmotic effect) provides a medium in which the exterior and interior of the cell have equally favorable properties with respect to vitrification and cutting.

E. Vitreous Sections Are Beam Sensitive

A conventional section changes slightly at the beginning of the irradiation, then remains remarkably stable. There is not much difference between a minimally and a heavily irradiated specimen, but this difference denotes a proficient operator. For an irradiation dose that is typically used for a conventional record, a vitreous section flows gently at the beginning of the irradiation, then it starts to bubble, soon breaking away, leaving only a hole. A cryo-electron microscopist has no choice other than to keep the electron dose to a minimum.

1. Beam-Induced Drift

Charges are generated in the specimen during irradiation. They are rapidly eliminated if the specimen is a good electrical conductor—carbon film or metal coating helps—but they accumulate if the resistivity is high, as is the case in pure water or in biological vitreous specimens. The resulting forces, acting on the viscous material of the vitreous sample, induce drift according to the local charge distribution. In general, this drift can be limited so that it does not show up in a single micrograph. However, it becomes a serious challenge in CET where the information from many tilted views must be assembled. A more subtle effect is local. There are materials that specifically contract under the electron beam. Condensed chromatin is a characteristic example, but many nucleoprotein granules behave in the same way. As a consequence, the surface of condensed chromatin domains in a section shrink on irradiation, and their density increases accordingly (Sartori Blanc *et al.*, 1998). We know just about nothing of the underlying phenomena; probably they have to do with aggregation of entangled polymers on radiochemical cross-linking reactions.

2. Bubbling

Bubbling is a characteristic effect, which takes place in organic material embedded in vitreous water or hexagonal ice. As its name indicates, it describes the appearance of bubbles where the biological material is at its densest and when the electron dose is in the order of some ke/nm^2. The bubbles, which are very small at the beginning, grow and merge until big holes replace the irradiated region. Bubbles are indicative of what is going on in a biological specimen under the effect of electron irradiation. Of course, electron beam damage is not reserved exclusively for vitreous specimens and cryo-electron microscopy; in fact the basic phenomena are the same at room or at low temperatures. What makes a difference is what happens to the molecules formed during the irradiation process. Many of these molecules are of low molecular weight; they volatilize in a conventional specimen, thus resulting in mass loss. At low temperatures, they cannot escape because they are trapped in the vitreous water and cannot diffuse away. Bubbling starts when the concentration of gas exceeds what vitreous water can support. It is interesting to note that bubbling is subtly dependent on the nature of biological material. For example, the onset of bubbling is late in nucleic acids, probably because they are more radioresistant than proteins, but the contrary holds in glutaraldehyde-fixed specimens. Specialists in radiochemistry are not surprised at this.

III. Illustration: Two Pitfalls

CEMOVIS is arguably the most conservative way to visualize the native state of a biological specimen at high resolution. From the previous section, it is apparent however that an image of a vitreous section is only obtained at the end of a bumpy road. We have described the major steps, and this knowledge enables us to interpret rationally what we see on most images. There are exceptions, though, and sometimes features are observed which challenge our imagination. Are we witnessing strange biological structures or unexpected preparation artifacts? The zoo of unusual vitreous sections is rich. Finding out where the beasts come from is not always an easy task, but it contributes to the pleasure and intellectual agility of every CEMOVISt. We present here two unusual observations and their possible explanation.

A. Block Tectonics

Figure 8A is a strongly underfocused image of a yeast cell. As usual, the external medium has been complemented with 20% dextran in order to make vitrification possible and to improve the cutting properties. There are numerous crevasses in the external medium and some in the cell. The cutting direction (arrow) is parallel to the knife marks and perpendicular to the crevasses. The orientation of the arrow can be deduced from the banana-shape of the crevasses and from some asymmetry in

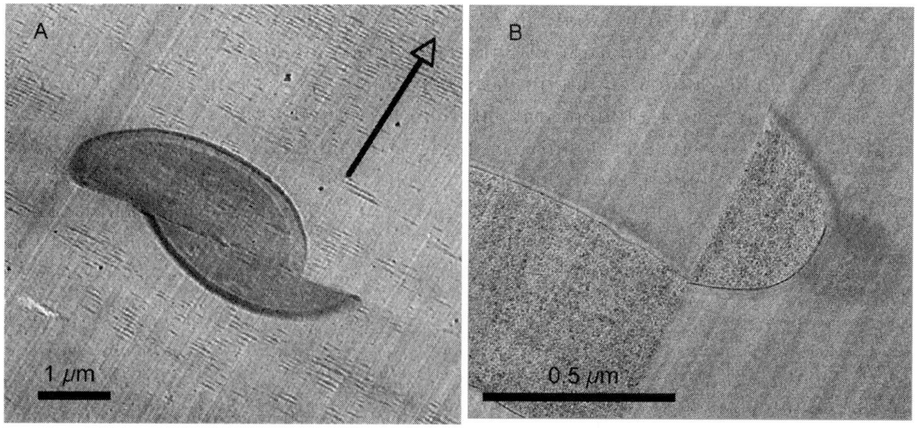

Fig. 8 Block tectonics. Vitreous sections of a yeast cell (A) and a bacterium (*E. gallinarum*; B) broken in a strange manner. The cutting direction is given by the black arrow.

the deformation of the cell envelope (Al-Amoudi *et al.*, 2005b). Obviously, there is a problem with this cell! Living yeast cells in suspension are not like this. At first, on the basis of this photograph, we thought that lateral stress during the cutting process might have induced a sudden slip of the knife with respect to the block along the knife-edge. Such an effect was considered in relation to an oscillating knife (Al-Amoudi *et al.*, 2003), but in fact this section was made with a static knife.

Other similar observations were made later. One of them is shown in Fig. 8B. Progress has been made in the mean time; this section is free of crevasses. The cutting direction is identified by the knife marks. The preparation is a suspension of the bacterium, *Enterococcus gallinarum*, again in 20% dextran. The fracture here is approximately oriented along the cutting direction. Altogether, a dozen similar images have been observed in our laboratory, and they show that the fracture can be in any direction, although with a preference around the cutting direction. How do they originate?

A hint is offered by the following observations. In the EM PACT high-pressure cooling apparatus (Leica, Vienna, Austria), the sample is vitrified in a 0.6-mm-diameter copper tube with 0.3-mm internal diameter. The first trimming of the specimen in the cryomicrotome results in a flat surface, formed by the section of the tube with the vitreous material in the middle. If the specimen is left in this situation/state, at $-140\,^{\circ}\mathrm{C}$ or below, it is then observed that the vitreous material is slowly expelled out of the tube, thus forming a bulge out of the copper surface. The effect is slow and is measured in micrometers, but it clearly indicates that tension is at work in the vitreous sample. Where does it originate?

Liquid water has a density of 1 g/cm^3. Ice has a density of 0.92. It is believed that vitreous water, obtained by rapid cooling at normal pressure, has nearly the same density as ice. However, direct measurements in the electron microscope are unreliable, since they are based on comparisons with reference material, such as polystyrene spheres, whose density variation with rapid cooling is uncertain. The situation is more complicated in our case since we cool at high pressure. High-density amorphous water (HDA) obtained by Mishima *et al.* (1984) at −194 °C and ca. 10,000 bar has a density of 1.27. This is certainly not what we obtain under our conditions of vitrification, but the actual density of vitreous material obtained by cooling under high pressure is likely to be higher than when vitrification takes place at low pressure.

What happens, then, when pressure is relaxed in the vitrified sample? The volume needs to expand, but the vitrified sample is hard. Pressure rises, the highly viscous vitreous material slowly deforms under stress . . . unless shear stress becomes too high and the sample breaks and subsides toward relaxation. This story is reminiscent of other events, alas too frequently reported in the news. The earth's mantle is also a fluid, moving under the action of forces generated by heat flow originating

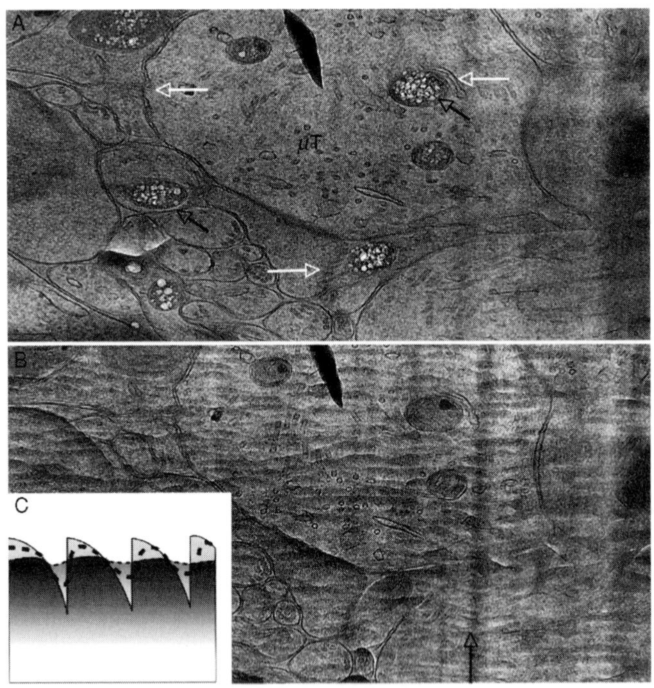

Fig. 9 Vitreous section of an organotypic slice of rat brain. See preparation details in Zuber *et al.* (2005). (A) The section appears smooth and without crevasses or knife mark. Membranes and microtubules in various orientations (μT) are clearly visible. However, bubbling is noticeable in dense regions (mitochondria; black arrow) and, at several places, membranes are discontinuous (white arrows). The total dose was 5 ke/nm^2. (B) Same section as in (A) recorded at low electron dose. There are numerous crevasses and an obvious knife mark (arrow). (C) Scheme of beam induced smoothing of a crevassed section. With electron irradiation, the sharp irregularities of the surface melt into a smooth surface (dotted line).

from the interior. Big tectonic plates are moving with respect to others, and the flow, sometimes exceeding the capabilities of viscous deformation, induces fractures and sudden displacement along fault lines. Apparently, some vitreous biological samples obtained by high-pressure cooling behave in just the same way.

B. Not Every Smooth Vitreous Section Is Admirable

Figure 9A shows a smooth vitreous section without crevasses and with very few shallow vertical knive marks. Should we admire it as an excellent section? We should not, because Fig. 9B shows the same section recorded at a lower electron dose before the 9A image was taken.

We propose the following explanation to account for what happens between Fig. 9B and A. It is a well-observed fact that the apparent viscosity of a vitreous sample decreases under the effect of an electron beam. We know that this effect is not primarily due to a general heating of the specimen (Dubochet *et al.*, 1988) but is probably related to local charge accumulations and chemical reactions involving radiochemical products. The consequence is that when force is applied, the specimen flows more than in the absence of irradiation. One force is surface tension. Figure 9C is a schematic representation of a cross section through a badly crevassed section. Crevasses are known to penetrate more or less deeply from the topside of a section, while the majority of knife marks are on the bottom side. Both are features that increase the surface area of a section. As is the case with any liquid, surface tension tends to minimize surface area, and consequently crevasses and knife marks are smoothed away on electron irradiation. Careful observation of Fig. 9A shows the real state of the section. We note in particular that (1) it is not a good sign when a specimen has only smooth knife marks, (2) biological structures fractured by crevasses are generally sealed with a visible seam, (3) the finest details of the biological structure are lost as a result of beam damage, and (4) bubbling in the densest regions is the certain sign of heavy irradiation.

Acknowledgments

We thank M. Grabenbauer (EMBL) for Fig. 1 and P. Buhayer for help with English. This work was supported by 3D-EM Network of Excellence within Research Framework Programme 6 of the European Commission. J.D. is the major author of this chapter. A.A.A., C.B.M., M.E., and B.Z. have provided most of the experimental data and have contributed to the redaction of this chapter; F.L. has contributed with the concept of block tectonics.

References

Abbott, A. (2002). The society of proteins. *Nature* **417,** 894–896.

Adrian, M., Dubochet, J., Lepault, J., and McDowall, A. W. (1984). Cryo-electron microscopy of viruses. *Nature* **308,** 32–36.

Al-Amoudi, A., Chang, J. J., Leforestier, A., McDowall, A., Salamin, L.-A., Norlén, L. P. O., Richter, K., Sartori Blanc, N., Studer, D., and Dubochet, J. (2004a). Cryo-electron microscopy of vitreous sections. *EMBO J.* **23,** 3583–3588.

Al-Amoudi, A., Dubochet, J., and Norlen, L. (2005a). Nanostructure of the epidermal extracellular space as observed by cryo-electron microscopy of vitreous sections of human skin. *J. Invest. Dermatol.* **124,** 764–777.

Al-Amoudi, A., Dubochet, J., and Studer, D. (2002). Amorphous solid water produced by cryosectioning of crystalline ice at 113 K. *J. Microsc.* **207,** 146–153.

Al-Amoudi, A., Dubochet, J., Gnaegi, H., and Lüthi, W. (2003). An oscillating cryo-knife reduces cutting-induced deformation of vitreous ultrathin sections. *J. Microsc.* **212,** 26–33.

Al-Amoudi, A., Norlén, L. P. O., and Dubochet, J. (2004b). Cryo-electron microscopy of vitreous sections of native biological cells and tissues. *J. Struct. Biol.* **148,** 131–135.

Al-Amoudi, A., Studer, D., and Dubochet, J. (2005b). Cutting artefacts and cutting process in vitreous sections for cryo-electron microscopy. *J. Struct. Biol.* **150,** 109–121.

Baumeister, W., and Steven, A. C. (2000). Macromolecular electron microscopy in the era of structural genomics. *Trends Biochem. Sci.* **25,** 624–631.

Bouchet-Marquis, C., Dubochet, J., and Fakan, S. (2006). Cryo-electron microscopy of vitrified sections: A new challenge for the analysis of functional nuclear architecture. *Histochem. Cell Biol.* **125,** 43–51.

Bouchet-Marquis, C., Zuber, B., Glynn, A. M., Eltsov, M., Grabenbauer, M., Goldie, K. N., Thomas, D., Frangakis, A. S., Dubochet, J., and Chrétien, D. (2007). Visualization of cell microtubules in their native state. *Biol. Cell* **99**.

Chang, J. J., McDowall, A. W., Lepault, J., Freeman, R., Walter, C. A., and Dubochet, J. (1983). Freezing, sectioning and observation artefacts of frozen hydrated sections for electron microscopy. *J. Microsc.* **132,** 109–123.

Dubochet, J., and McDowall, A. W. (1981). Vitrification of pure water for electron microscopy. *J. Microsc.* **124,** RP3–RP4.

Dubochet, J., Adrian, M., Chang, J.-J., Homo, J.-C., Lepault, J., McDowall, A. W., and Schultz, P. (1988). Cryo-electron microscopy of vitrified specimens. *Q. Rev. Biophys.* **21,** 129–228.

Dubochet, J., Al-Amoudi, A., Bouchet-Marquis, C., Eltsov, M., and Zuber, B. (2007). CEMOVIS. Cryo-electron microscopy of vitreous sections. *In* "Modern Cryo-Preparation Methods for Electron Microscopy" (A. Cavalier, B. M. Humbel, and D. Spehner, eds.). CRC Press, Boca Raton.

Dubochet, J., Bednar, J., Furrer, P., and Stasiak, A. (1994). Cryo-electron microscopy of DNA. *In* "Nucleic Acids and Molecular Biology" (F. EcksteinLilley, D. M. J., ed.), Vol. 8, pp. 41–55. Springer-Verlag, Berlin.

Eltsov, M., and Dubochet, J. (2005). Fine structure of the Deinococcus radiodurans nucleoid revealed by cryoelectron microscopy of vitreous sections. *J. Bacteriol.* **187,** 8047–8054.

Fernandez-Moran, H. (1960). Low-temperature preparation techniques for electron microscopy of biological specimens based on rapid freezing with liquid Helium II. *Ann. NY Acad. Sci.* **85,** 689–713.

Furrer, P., Bednar, J., Dubochet, J., Hamiche, A., and Prunell, A. (1995). DNA at the entry-excit of nucleosome observed by cryo-electron microscopy. *J. Struct. Biol.* **114,** 177–183.

Garvalov, B. K., Zuber, B., Bouchet-Marquis, C., Kudryashev, M., Gruska, M., Beck, M., Leis, A., Baumeister, W., Dubochet, J., Bradke, F., Frischknecht, F., and Cyrklaff, M. (2006). Luminal particles within native cellular microtubules. *J. Cell Biol.* **174,** 759–765.

Goodsell, D. S. (1993). "The Machinery of Life." Springer-Verlag, New York.

Hoppe, W., Gassmann, J., Hunsmann, N., Schramm, H. J., and Sturm, M. (1974). Three-dimensional reconstruction of individual negatively stained yeast fatty-acid synthetase molecules from tilt series in the electron microscope. *In* "Hoppe-Seyler's Zeitschrift fur physiologische Chemie," Vol. 355, pp. 1483–1487.

Hsieh, C. E., Leith, A., Mannella, C. A., Frank, J., and Marko, M. (2006). Towards high-resolution three-dimensional imaging of native mammalian tissue: Electron tomography of frozen-hydrated rat liver sections. *J. Struct. Biol.* **153,** 1–13.

Hsieh, C.-E., Marko, M., Frank, J., and Mannella, C. A. (2002). Electron tomographic analysis of frozen-hydrated tissue sections. *J. Struct. Biol.* **138,** 63–73.

Kellenberger, E., and Kistler, J. (1979). The physics of specimen preparation. *In* "Structure Research" (H. Mason, ed.), Vol. VII, pp. 1–14. Friedr. Vieweg & Son, Wiesbaden.

Leis, A., Andrees, L., Gruska, M., Al-Amoudi, A., Sartori, A., and Baumeister, W. (2005). Cryo-electron tomography and fluorescence microscopy of unicellular algae in vitreous cryosections. *Microsc. Microanal.* **11,** 330–331.

Matias, V. R. F., and Beveridge, T. J. (2006). Native cell wall organization shown by cryo-electron microscopy confirms the existence of a periplasmic space in staphylococcus aureus. *J. Bacteriol.* **188,** 1011–1021.

Mayer, E., and Brüggeller, P. (1980). Complete vitrification in pure liquid water and dilute aqueous solutions. *Nature* **288,** 569–571.

McDowall, A. W., Chang, J.-J., Freeman, R., Lepault, J., Walter, C. A., and Dubochet, J. (1983). Electron microscopy of frozen hydrated sections of vitreous ice and vitrified biological samples. *J. Microsc.* **131,** 1–9.

McDowall, A. W., Smith, J. M., and Dubochet, J. (1986). Cryo-electron microscopy of vitrified chromosomes *in situ*. *EMBO J.* **5,** 1395–1402.

Medalia, O., Weber, I., Frangakis, A. S., Nicastro, D., Gerisch, G., and Baumeister, W. (2002). Macromolecular architecture in eukaryotic cells visualized by cryoelectron tomography. *Science* **298,** 1209–1213.

Mishima, O., Calvert, L. D., and Whalley, E. (1984). "Melting ice" I at 77 K and 10 kbar: A new method of making amorphous solids. *Nature* **310,** 393–395.

Moor, H. (1987). Theory and practice of high pressure freezing. *In* "Cryotechniques in Biological Electron Microscopy" (R. A. Steinbrecht, and K. Zierold, eds.), pp. 175–191. Springer-Verlag, Heidelberg.

Nicastro, D., Austin, J., II, Pierson, J., Gaudette, R., Schwartz, C., Ladinsky, M., Staehelin, L. A., and McIntosh, J. R. (2005). Visualizing the macromolecular organization of chloroplast membranes using cryo-electron tomography. *Microsc. Microanal.* **11,** 150–151.

Norlén, L., and Al-Amoudi, A. (2004). Stratum corneum keratin structure, function and formation—the cubic rod-packing and membrane templating model. *J. Invest. Dermatol.* **123,** 715–732.

Norlen, L., Al-Amoudi, A., and Dubochet, J. (2003). A cryotransmission electron microscopy study of skin barrier formation. *J. Invest. Dermatol.* **120,** 555–560.

Plitzko, J. M., Frangakis, A. S., Foerster, F., Gross, A., and Baumeister, W. (2002). *In vivo veritas:* Electron cryotomography of intact cells with molecular resolution. *Trends Biotechnol.* **20**(8), 40–44.

Richter, K. (1994a). A cryoglue to mount vitreous biological specimens for cryoultramicrotomy at 110 K. *J. Microsc.* **173,** 143–147.

Richter, K. (1994b). Cutting artefacts on ultrathin cryosections of biological bulk specimens. *Micron* **25,** 297–308.

Richter, K. (1994c). High-density morphologies of ice in high-pressure frozen biological specimens. *Ultramicroscopy* **53,** 237–249.

Richter, K., and Dubochet, J. (1989). Electron microscopy of native specimen gluing of vitrified specimens for cryo-ultramicrotomy. *Experientia* **45,** A42.

Richter, K., and Dubochet, J. (1993). Ice II and ice III in partially vitrified high-pressure frozen biological samples (abstract). *Eur. J. Cell Biol.* **61,** 14.

Richter, K., Gnägi, H., and Dubochet, J. (1991). A model for cryosectioning based on the morphology of vitrified ultrathin sections. *J. Microsc.* **163,** 19–28.

Roos, N., and Morgan, A. J. (1990). "Cryopreservation of Thin Biological Specimens for Electron Microscopy: Methods and Applications." Oxford Science Publications, Oxford.

Sartori Blanc, N., and Salamin Michel, L. (1998). Cryotransmission electron microscopy of thin vitrified sections. *In* "A Laboratory Handbook" (E. Celis, ed.), pp. 323–332. Academic Press, Inc., London.

Sartori Blanc, N., Senn, A., Leforestier, A., Livolant, F., and Dubochet, J. (2001). DNA in human and stallion spermatozoa forms local hexagonal packing with twist and many defects. *J. Struct. Biol.* **134,** 76–81.

Sartori Blanc, N., Studer, D., Ruhl, K., and Dubochet, J. (1998). Electron beam-induced changes in vitreous sections of biological samples. *J. Microsc.* **192,** 194–201.

Sartori, N., and Salamin Michel, L. A. (1994). Cryo-transmission electron microscopy of thin vitrified sections. *In* "Cell Biology: A Laboratory Handbook" (E. Celis, ed.), pp. 177–185. Academic Press, Inc., London.

Schwartz, C., Nicastro, D., Ladinsky, M. S., Mastronarde, D., O'Toole, E., and McIntosh, J. R. (2003). Cryo-electron tomography of frozen-hydrated sections of eukaryotic cells. *Microsc. Microanal.* **9,** 1166–1167.

Sitte, H. (1996). Advanced instrumentation and methodology related to cryo-ultramicrotomy: A review. *Scanning Microsc. Suppl.* **10,** 387–466.

Taylor, K. A., and Glaeser, R. M. (1974). Electron diffraction of frozen, hydrated protein crystals. *Science* **186,** 1036–1037.

Taylor, K. A., and Glaeser, R. M. (1976). Electron microscopy of frozen hydrated biological specimens. *J. Ultrastruct. Res.* **55,** 448–456.

Zhang, P., Bos, E., Heymann, J., Gnaegi, H., Kessel, M., Peters, P. J., and Subramaniam, S. (2004). Direct visualization of receptor arrays in frozen-hydrated sections and plunge-frozen specimens of *E. coli* engineered to overproduce the chemotaxis receptor Tsr. *J. Microsc.* **216,** 76–83.

Zuber, B., Nikonenko, I., Klauser, P., Muller, D., and Dubochet, J. (2005). The mammalian central nervous synaptic cleft contains a high density of periodically organized complexes. *Proc. Natl. Acad. Sci. USA* **102,** 19192–19197.

CHAPTER 16

Single-Particle Electron Cryomicroscopy of the Ion Channels in the Excitation–Contraction Coupling Junction

Irina I. Serysheva, Wah Chiu, and Steven J. Ludtke

National Center for Macromolecular Imaging, Baylor College of Medicine
Houston, Texas 77030

Verna and Marrs McLean Department of Biochemistry and Molecular Biology
Baylor College of Medicine, Houston, Texas 77030

Single-particle electron cryomicroscopy (cryo-EM) has become the technique of choice for determining the structures of large molecular machines where crystallography is impractical or unsuccessful. Here, we review this methodology in its

0091-679X/07 $35.00
DOI: 10.1016/S0091-679X(06)79016-1

entirety, from specimen preparation to map interpretation, in the context of the membrane channel proteins essential to excitation–contraction (E–C) coupling in muscle cells. In the case of the Ca^{2+}-release channel in sarcoplasmic reticulum (SR) membranes, cryo-EM has been used to localize the binding sites of several effectors and to demonstrate conformational changes under different physiological conditions. The resolution of the most recent cryo-EM reconstruction of the Ca^{2+}-release channel in its closed conformation is sufficient to observe the α-helices forming the channel pore. For the L-type Ca^{2+} channel located in the membrane of striated muscle's T-tubule, two broadly similar, but not identical, low-resolution cryo-EM maps have been published. For both ion channels and the system as a whole, much work remains to be done. The resolution of both structures must be extended to 7–8 Å to better establish secondary structure throughout the entire structure. The rapid pace of improvement in the technology of single-particle reconstruction will soon permit improved resolution as well as the elucidation of more detailed, dynamic models of these important structures.

I. Introduction

A. Excitation–Contraction Coupling in Striated Muscle

E–C coupling consists of a series of physiological events that link the depolarization of striated muscle cell's plasma membrane to the release of Ca^{2+} from the sarcoplasmic reticulum (SR) into cytoplasm, resulting in muscle contraction. Muscle contraction is highly regulated by the intracellular free Ca^{2+} level through E–C coupling. The requisite set of events occur in specialized junctional regions between the invaginations of the plasma membrane, called transverse tubules (T-tubules), and the SR membrane (Fig. 1A). The major molecular components involved in E–C coupling are the voltage-gated "L-type" Ca^{2+} channel [aka, the dihydropyridine receptor (DHPR)], which is localized in the T-tubule membrane, and the Ca^{2+}-release channel [the ryanodine receptor (RyR)], located in the opposing SR junctional membrane. Defects in either of these two key proteins or modulation of their activity produce a number of human muscular genetic diseases (see reviews by Benkusky *et al.*, 2004; Brini, 2004; Hamilton, 2005; McCarthy *et al.*, 2000).

The groundwork for understanding the ultrastructural and spatial organization of both the DHPRs and RyRs was established by Franzini-Armstrong and coworkers using freeze-fracture methods (Block *et al.*, 1988; Ferguson *et al.*, 1984; Franzini-Armstrong and Kish, 1995; Protasi *et al.*, 1997). These elegant studies demonstrated that the physiological difference in the E–C coupling mechanism between skeletal and cardiac muscles may be attributed to the differences in the relative spatial arrangements of DHPR and RyR within the junctional regions. Electron microscopy of thin sections through the junctional regions revealed that DHPR organization is coordinated with RyR in both muscle types, though their overall patterns of distribution were substantially different. RyR appears as dense,

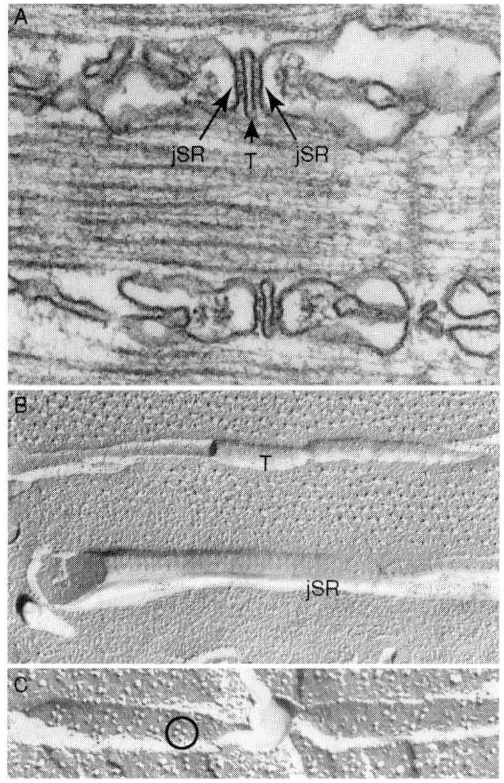

Fig. 1 (A) A cross section of two triads from fish muscle showing the T-tubule membrane flanked by two junctional SR (jSR) membranes. Feet structures (RyR1) are observed as bridging densities occupying the junctional gap between the opposed jSR and T-tubule membranes. Reproduced from the *J. Cell Biol.* (1970), **47,** 488–499, with permissions from the Rockefeller University Press and C. Franzini-Armstrong. (B) Freeze-fracture replica of a fracture face that grazes the junctional gap in fish muscle, showing a distinctive feature of the T-tubule membrane: the tall, intra-membranous particles that correspond to DHPRs. A double row of feet particles (RyR1) is seen on the surface of the jSR. (C) Freeze-fracture of a T-tubule from a fish showing evenly spaced tetrads of DHPRs. One tetrad is indicated with a circle. (B) and (C) are reproduced from the *J. Cell Biol.* (1988), **107,** 2587–2600, by copyright permission of the Rockefeller University Press and C. Franzini-Armstrong.

rectangular structures (junctional feet) in freeze-fracture replicas (Fig. 1B), forming two rows of regular arrays on the junctional SR membrane in both the skeletal and cardiac muscles. Freeze-fracture of T-tubules reveals dense and tall particles, which have been identified as DHPRs (Fig. 1B). In skeletal muscle, DHPRs are clustered into groups of four DHPRs, named "tetrads," which overlay every other RyR1 (Fig. 1C) (Block *et al.*, 1988). In cardiac muscle, DHPR tetrads are not observed. These observations provide the direct structural evidence suggesting that the Ca^{2+}-release triggering mechanism of skeletal muscle E–C coupling may involve direct physical contacts between the two channels. Understanding the structural basis for RyR–DHPR interactions remains a challenging and fascinating problem in the field of E–C coupling. Our current knowledge of the three-dimensional (3D) structures of DHPR and RyR is derived primarily from structural studies carried out using cryo-EM of isolated detergent-solubilized protein complexes (i.e., single particles). These studies have provided the best clues as to the structure–function relationships of these membrane proteins.

B. Cryo-EM of Single Particles

It is now well appreciated that most functions in the cell are carried out by protein complexes, often described as "molecular machines"; these generally comprise multiple subunits, each with a specific function. Ion channels are integral membrane proteins found in virtually all cellular membranes and represent one class of such molecular machines. *In vivo*, most ion channels function in close association with auxiliary proteins to endow each cellular compartment with specific biological functions.

The complex nature and large size of such molecular machines make the determination of their structure by standard techniques, like X-ray crystallography or NMR spectroscopy, very difficult. However, through use of cryo-EM, it is possible to study both large molecular machines and their conformational changes. The ice-embedding technique has been proven to preserve protein structure in its native, fully hydrated conformation (Chiu *et al.*, 1986; Dubochet *et al.*, 1988). Unlike electron or X-ray crystallography, cryo-EM single-particle reconstruction requires no crystals. Instead, images of individual molecular machines are computationally aligned to high precision, corrected for the effects of the microscope's contrast transfer function (CTF). Many images are then averaged and merged to produce a full, 3D reconstruction. Since this technique makes use of images of individual particles, it is also possible to apply statistical techniques and produce several 3D reconstructions from a single population of heterogeneous particles. This methodology requires images of only tens to hundreds of thousands of particles, so it requires substantially less material than crystallographic techniques, and impurities are more readily tolerated, though at a cost in structural resolution. Moreover, for some biological questions, such as determination of physiologically relevant conformations of molecules or study of interactions between molecules, packing forces present in a crystal may produce undesirable effects. The single-particle analysis avoids this problem, and in addition it provides the means to capture different conformational states of the molecular machines under solution conditions similar to those used for functional characterization (Chen *et al.*, 2006; Grob *et al.*, 2006; Orlova *et al.*, 1996; Penczek *et al.*, 2006; Tama *et al.*, 2003). Although most structure determinations of isolated molecular machines by cryo-EM are still far from atomic resolution, the computational and experimental technologies have gradually improved over the last decade, and subnanometer resolution is now frequently achieved (see reviews by Chiu, 2005; Henderson, 2004; Jiang and Ludtke, 2005; Orlova and Saibil, 2004). Studies from several groups have demonstrated that single-particle reconstructions from non-icosahedrally symmetric particles at 6- to 10-Å resolution are sufficient to accurately assign secondary structure (Fotin *et al.*, 2004; Gabashvili *et al.*, 2000; Ludtke *et al.*, 2004, 2005; Matadeen *et al.*, 1999).

Membrane protein assemblies remain the most challenging targets for structural analysis by single-particle cryo-EM for numerous reasons, including conformational heterogeneity of the sample due to different functional states, structural flexibility, detergent solubilization effects, and poor image contrast due to the presence of detergent. Nonetheless, RyR1 was among the first non-icosahedral structures to be solved by using single-particle cryo-EM (Radermacher *et al.*, 1994; Serysheva *et al.*, 1995). In this chapter, we describe structural studies of RyR1 and DHPR from rabbit skeletal muscle as well as the single-particle 3D reconstruction methodology used in the EMAN software suite (Ludtke *et al.*, 1999).

II. Single-Particle Cryo-EM Methodology

In single-particle reconstruction, a large set of images of individual molecular machines in random orientations is computationally combined to produce a complete, volumetric reconstruction of the machine. This technique is now capable of producing structures in the 5- to 10-Å resolution range. The underlying computational reconstruction techniques are still being actively developed, and detailed methodologies may vary significantly (Frank, 1996; Grigorieff, 1998; Ludtke *et al.*, 1999; Sorzano *et al.*, 2004; van Heel *et al.*, 1992). Nonetheless, the fundamental concepts remain the same, and within the limits of each, any of the techniques will produce equivalent reconstructions.

A. Specimen Preparation and Data Collection

The molecular machine to be studied must first be purified using specimen-specific techniques that preserve it in a chemically or functionally defined state. At the very least, prior to imaging, a functional assay on the purified protein should indicate a fully functional assembly, and purification assays should indicate very high purity, typically 99% or better. While purity requirements are quite stringent, very little specimen is required. To prepare a single electron microscopy grid, ~3 μl of solubilized specimen is required at concentrations ranging from ~0.1 to 1.0 mg/ml. That is, only micrograms of specimen are required for a complete reconstruction.

A small amount of purified specimen (3–5 μl) is deposited on a standard electron microscopy grid that has been coated with a holey carbon support film; then it is blotted with a piece of filter paper, leaving behind only a thin film of protein in aqueous solution that spans the holes in the carbon film (Fig. 2). The grid is then rapidly plunged into a bath of liquid ethane, freezing the water so rapidly that ice crystallization is impossible. The aqueous embedding medium is thus frozen in a glassy state known as vitreous ice (Dubochet *et al.*, 1988). Individual particles are thus preserved in a snapshot of their solution conformation

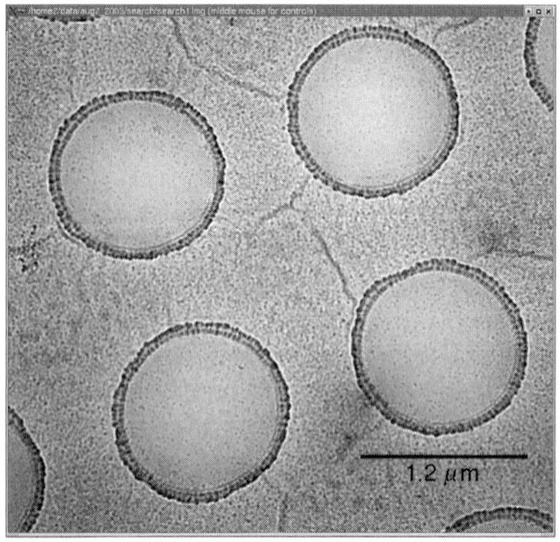

Fig. 2 Low-magnification TEM image of a holey carbon grid with ice-embedded RyR1 prepared by rapid plunge freezing. RyR1 can barely be observed as small dark specks within the holes and on the surface of the carbon between holes.

in random orientations within the ice (Fig. 3). It is worth noting that the correct choice of supporting film and the grid preparation procedure are critical for achieving proper vitrification of biological molecules.

The specimen is imaged on a transmission electron microscope using minimum dose conditions, designed to provide as much contrast as possible without damaging the specimen at the desired resolution (Unwin and Henderson, 1975). Dose levels typically used for imaging single particles are in the range of 5–25 electrons $(e^-)/Å^2$, although they vary with the voltage of the instrument being used and the expected resolution. Using low electron dose has the consequence that the image contrast is quite low, often with peak signal-to-noise ratios substantially less than one. In addition, single-particle cryo-EM is largely a phase contrast technique, meaning that overall contrast is enhanced by imaging far from focus, but with some cost in resolution. This type of imaging must deal with mutually exclusive goals: to go close-to-focus preserves information at higher resolution but provides little contrast; to go far-from-focus enhances the contrast sufficiently to facilitate the identification of particles and to accurately determine their orientation, but image resolution is worse. In addition, the CTF of the instrument is such that any single image is missing information at specific resolutions. To obtain complete information at all resolutions, it is necessary to collect data at a range of defocus values and properly combine the results using a CTF correction algorithm.

Images of ice-embedded specimen are recorded either on photographic film or on a charge-coupled device (CCD) detector. While CCD detectors produce accurate reconstructions (Booth *et al.*, 2004; Ludtke *et al.*, 2005; Mouche *et al.*, 2003) and

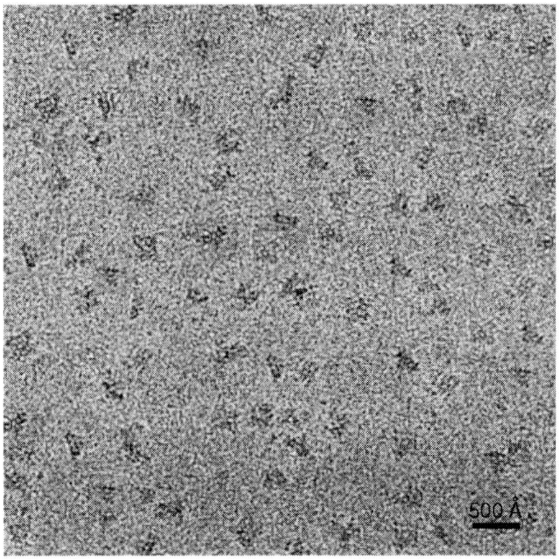

Fig. 3 A 200-kV electron image of ice-embedded RyR1, recorded under minimal-dose conditions (~15 e/Å^2) on JEM-2010F electron cryomicroscope with a Gatan 4K × 4K CCD camera at 40,000× magnification. Note the random orientations of channel particles in this specimen, which was prepared without a continuous carbon film.

require no developing or scanning, with the current generation of CCD detectors, photographic film still preserves higher resolution details than CCD frames at equivalent magnifications.

B. CTF Correction

Images of individual particles represent projections of the object being imaged, modulated by the microscope's CTF (Erickson and Klug, 1970). That is, the images represent sums of the density of the object being observed, integrated along lines parallel to the microscope's axis. They are not surface representations. Additionally, the complete imaging process produces several kinds of artifact, including a resolution-limiting envelope function, a high level of background noise, and the CTF of the instrument. The CTF artifacts in the image must be carefully corrected to avoid the presence of artifacts in the final reconstruction. This process involves characterization of each collected image, and a careful tracking of the parameters of each particle's image throughout the reconstruction procedure.

C. Image Data Quality Assessment

Only a fraction of the images collected will be usable in the final reconstruction. Prior to reconstruction, each image must be assessed for quality and suitability for

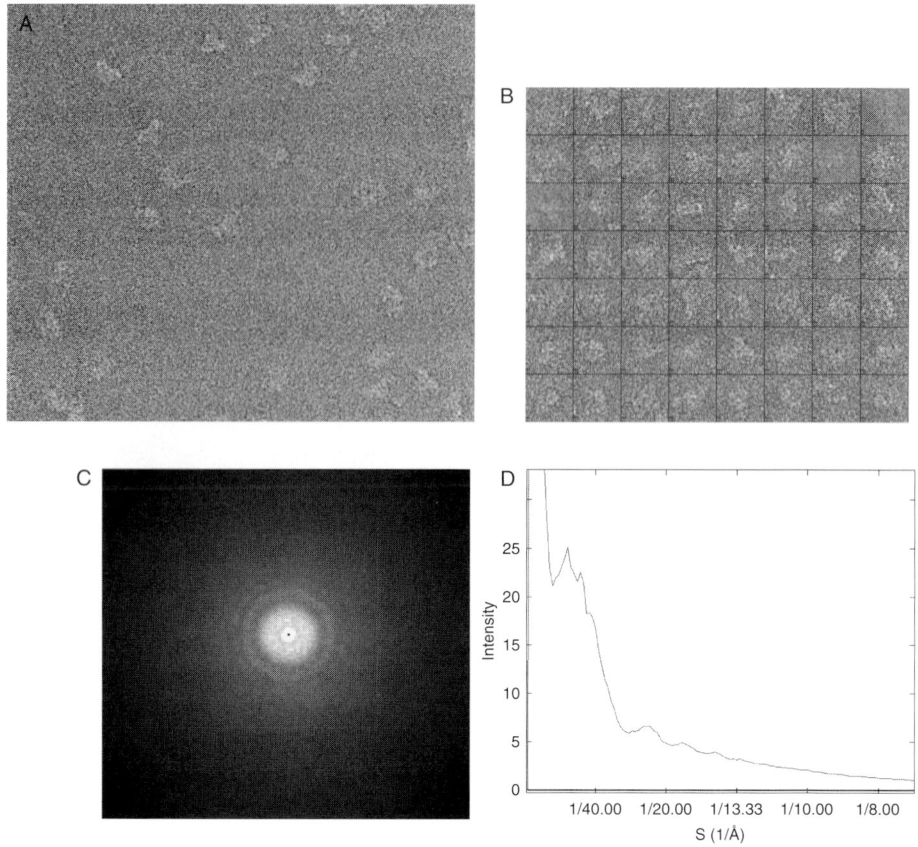

Fig. 4 Quality assessment of electron images. (A) Electron images of ice-embedded RyR; (B) images of individual RyR1 particles extracted from frames as in (A). (C) The incoherently averaged diffraction pattern computed for the particles extracted from one CCD frame. The Thon ring pattern is due to the CTF of the microscope. (D) The rotationally averaged power spectrum computed from (C). Structural information from the particles is embedded in the oscillatory part of the curve, which falls off at high resolution, due to the microscope's envelope function.

use in the final reconstruction (Fig. 4). This assessment process occurs in several stages. First, images are visually inspected. Images with obvious specimen drift, unusually low contrast, or severe contamination are discarded. Particles are then located in each of the remaining images (Fig. 4B). For each electron micrograph, the Fourier transforms of the individual particle images are calculated. These are used to form an incoherent average (Fig. 4C), which is then circularly averaged (Fig. 4D). The two- and one-dimensional (2D and 1D) power spectra (Fig. 4C and D) are then analyzed for drift, astigmatism, and resolution quality to decide if the image should be included in the reconstruction. The particles themselves, or a carbon support film if it is present, produce sufficient contrast to observe the Thon ring pattern, a consequence of the CTF and its associated envelope function. Finally, the CTF parameters of the final set of particles are determined on a per-micrograph basis and passed on with the particle images to the 3D reconstruction procedure.

D. Image Processing and 3D Reconstruction

As mentioned previously, several tools are available for performing single-particle reconstruction, including SPIDER (Frank *et al.*, 1996), IMAGIC (van Heel *et al.*, 1996), and FREALIGN (Grigorieff, 1998). We will discuss the methodology used to reconstruct RyR1 and DHPR as shown in Fig. 5, using the EMAN software suite (Ludtke *et al.*, 1999). In this methodology, a large number of particle images, ranging typically from a few thousand for a low-resolution reconstruction to tens of thousands for subnanometer resolution, are collected at a range of defocuses. In the first stage, particles from each micrograph or CCD frame are characterized, as discussed in Section II.C, to permit proper CTF correction to be

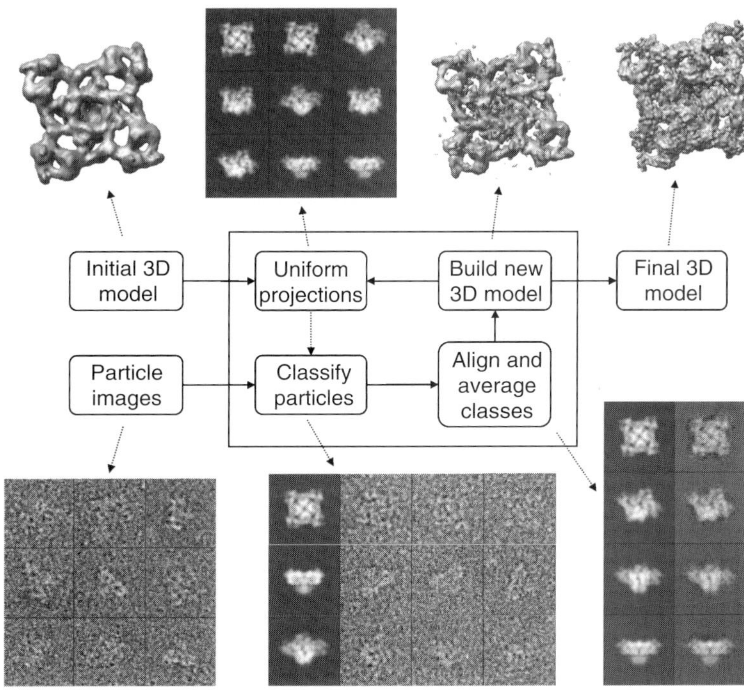

Fig. 5 Schematic diagram of the 3D reconstruction process in the EMAN software suite (Ludtke *et al.*, 1999), which was used to produce the subnanometer resolution reconstruction of RyR1 (Ludtke *et al.*, 2005).

carried out at a later stage of the reconstruction. Next, the particles are subjected to an iterative refinement loop, producing a 3D model that improves with each iteration. The refinement process continues until the model stabilizes. This iterative procedure is the key to obtaining an accurate model, even when there is little or no initial information about the structure.

The iterative refinement process begins with the complete set of characterized particle data and an initial 3D model. In current strategies, the initial 3D model is typically a featureless, asymmetric 3D Gaussian blob roughly the same size as the particles. Frequently, the entire refinement process is repeated with several different starting models to ensure that the final reconstruction is the same, irrespective of the initial model.

The first step of the iterative loop is the generation of a set of projections from the current 3D structure (initially, the starting model). These projections are generated at all possible orientations, constrained by any symmetry that is being imposed, and with some angular granularity. For a subnanometer resolution reconstruction of RyR1, for example, 2294 projections were generated with an approximate angular spacing of 1.4°. Each particle is then classified, based on the projection it is most similar to. That is, for each projection, we produce a list of particles that looked more similar to that specific projection than to any other projection. The particles within a single class, representing one specific view of the specimen, are then aligned with respect to each other in 2D. This alignment process is also iterative. Rather than simply aligning the particles to the corresponding projection, EMAN repeats the process several times. That is, particles are first aligned to a reference then averaged together. Particles are then realigned to the average and averaged together again, and so on. In each step of this iterative process, particles are compared to the previous average, and the least similar particles are excluded from the new average. This process produces a final class-average that represents the CTF-corrected consensus structure produced by the data, rather than the reference used to perform the classification. Once an average has been produced for each projection, the set of averages is used to construct a new 3D model and the overall process begins again. This iterative loop is repeated until the structure stabilizes, at which point the resolution is assessed by splitting the particle data into even- and odd-numbered halves and comparing the reconstructions produced independently from each half.

The reconstructions of each half of the data are compared by calculating their Fourier shell correlation (FSC) (van Heel, 1987) with a particular cutoff value used to define the resolution of the reconstruction (Fig. 7). Several different cutoff criteria have been proposed and are used, including 0.5 (Böttcher *et al.*, 1997), 0.33 (Penczek, 2002), 0.143 (Rosenthal and Henderson, 2003), and the so-called 3σ criterion (van Heel and Harauz, 1986). The smaller correlation thresholds produce resolution estimates that are higher (predict that the images include information about finer details) but are also more subject to artifacts in the FSC curves, due to incomplete independence of the structures or artifacts present in the raw data. Resolutions cited in this review are based on the 0.5 criterion.

E. Current Limitations

There are two primary limiting factors in single-particle reconstruction. The first is the high levels of noise present in the images. When the noise level is particularly high, model-bias in the reconstruction can become a problem (Stewart and Grigorieff, 2004). That is, it becomes possible for a feature in the initial model to dominate the alignment of some particles, thus preserving that feature in the final reconstruction, even though it is not real. There are a number of methods used to reduce this artifact, including starting with models that have no relationship to the actual structure and using the iterative class-averaging procedure described above.

The second limiting factor is compositional and/or conformational heterogeneity of the particle population itself. While there are several computational techniques to mitigate this problem, in general, any portion of the structure undergoing motion will be motion-averaged in the final structure, producing a blur as well as a local reduction in density. Due to the mechanism by which resolution is assessed in cryo-EM, these effects will have a minimal impact on assessed resolution, but they do produce local blurring in areas of the reconstruction.

F. Structural Feature Interpretation

Typically, cryo-EM reconstructions of macromolecular complexes are displayed using an isosurface representation of the underlying volumetric data. As the isosurface threshold is varied, the apparent connectivity of the map will also change. To properly interpret a map, it is necessary to consider the isosurface boundaries and the changes in connectivity over a range of isosurface thresholds. At low resolution, it is possible to estimate an appropriate isosurface threshold based on the known mass of the protein and the typical mean protein mass density of 1.3 g/cm^3. However, as resolution increases, this estimate no longer permits visualization of the structure's internal details. In addition, as mentioned above, any local motion or heterogeneity will produce both a local blurring and a local reduction in isosurface threshold. So, identifying all structural features in a map requires the realization that the appropriate threshold may be different in different parts of the structure; at higher resolutions, isosurface thresholds are selected based on how they represent specific structural details rather than based on a concept of molecular mass.

At resolutions worse than $\sim$10 Å, adjacent α-helices are not resolved but appear as a single larger structure. At 7- to 9-Å resolution, long helices can be resolved. In borderline situations such as the 9.6-Å resolution map of RyR1 (Fig. 7) (Ludtke *et al.*, 2005), care must be taken not to overinterpret the observed density. This map remains the highest resolution structure of a membrane protein yet resolved using single-particle analysis. This map was interpreted in two stages. First, the map was processed using the SSEhunter program (Baker *et al.*, 2006), which identifies secondary structure based on local density patterns and a set of scoring functions; this produces an estimate of the likelihood that a particular region contains α-helices or β-sheets. Second, the map is examined visually at varying isosurface thresholds, and helices are placed in the structure manually. The helices still present in the final interpretation are only those containing at least three turns that were identified as helices using *both* methods. Unambiguous identification of all secondary structure elements in a cryo-EM map, particularly of β-sheets, requires extending the resolution to 5–7 Å isotropically.

G. Interpretation of Cryo-EM Maps by Fitting X-ray Structures of Components or Domains

Intermediate-resolution (~10–20 Å) structures of macromolecular complexes often provide a 3D framework that is adequate for the assembly of crystal structures for individual protein subunits or domains when they are available. Another way to increase the information attainable from structural studies at this resolution range is to use complementary techniques such as homology modeling and bioinformatics (Baker *et al.*, 2002).

III. 3D Reconstruction of Proteins in Triad Junction

A. Ryanodine Receptor

RyRs mediate the gated release of Ca^{2+} from intracellular stores into the cytosol; they are the largest integral membrane channels known. In mammals, three different RyR isoforms have been identified and characterized, each encoded by a distinct gene located on a different chromosome. All three isoforms share ~70% sequence similarity. The RyR1 and RyR2 isoforms are the predominant Ca^{2+}-release channels in skeletal and cardiac muscles, respectively (Hakamata *et al.*, 1992; Takeshima *et al.*, 1989; Tunwell *et al.*, 1996; Zorzato *et al.*, 1990). RyR3 is primarily expressed in diaphragm, smooth muscle, and brain (Sorrentino and Reggiani, 1999; Sorrentino and Volpe, 1993; Sutko and Airey, 1996). The functional Ca^{2+}-release channel is a homotetramer composed of four RyR subunits, each of ~560 kDa; together, they constitute a single ion channel pore (Lai *et al.*, 1989). Hydropathy profiles of RyRs suggest that four-fifths of the protein from the N-terminus is exposed to cytoplasm and that transmembrane domains lie within the remaining C-terminal fifth of the protein (Takeshima *et al.*, 1989; Zorzato *et al.*, 1990). Thus, the RyR cytoplasmic region contains ~80% of the protein mass and accommodates multiple regulatory sites for a wide variety of endogenous molecules and pharmacological modifiers. The remaining RyR1 mass accounts for the membrane-spanning region, which comprises the Ca^{2+}-conduction pathway across the SR membrane.

1. Size and Shape of RyR

Cryo-EM produced the first clear 3D structures of RyR1 based on single-particle analysis (Radermacher *et al.*, 1994; Serysheva *et al.*, 1995; Wagenknecht and Radermacher, 1995). All of these structures exhibit largely the same characteristic features of RyR1. Additionally, 3D reconstructions of RyR1 compared to two other channel isoforms, RyR2 and RyR3, and demonstrate that the quaternary structure appears to be the same among isoforms at low-resolution level (Jeyakumar *et al.*, 1998; Sharma *et al.*, 1998).

Over the last decade, low-resolution structures of RyR have remained the primary basis for structure-function interpretations within the RyR, despite the fact that significant efforts have been made in several laboratories to improve the resolution beyond 20 Å. Substantial improvements in the resolution of the RyR1 reconstruction were achieved in two independent studies (Ludtke *et al.*, 2005; Samso *et al.*, 2005). Resolution improvements were made possible through advances in cryo-specimen preparation, which permitted exclusion of the carbon support film, and improvements to the image reconstruction techniques. The subnanometer resolution reconstruction of RyR1 in the closed conformation is shown in Fig. 6 (Ludtke *et al.*, 2005), with the corresponding resolution assessment in Fig. 7. This breakthrough structure offers more structural details than any previously published structure, and for the first time, secondary structure elements can be clearly distinguished (Figs. 10–11 and below).

The overall structure of RyR1 has a characteristic mushroom shape with fourfold symmetry. It comprises two square-shaped regions connected by four column-shaped densities. The large square-shaped region with overall dimensions of 280 × 280 × 120 Å represents the bulky, cytoplasmic region, which *in vivo* is exposed to the gap between the SR membrane and the T-tubule membrane. These regions are most likely involved in interdigitating with the neighboring molecules of RyR1, as seen *in situ* (Block *et al.*, 1988), or with modulatory auxiliary proteins, for example, calmodulin and FKBP12, a 12-kDa FK506-binding protein (Samso and Wagenknecht, 2002; Samso *et al.*, 2006; Sharma *et al.*, 2006;

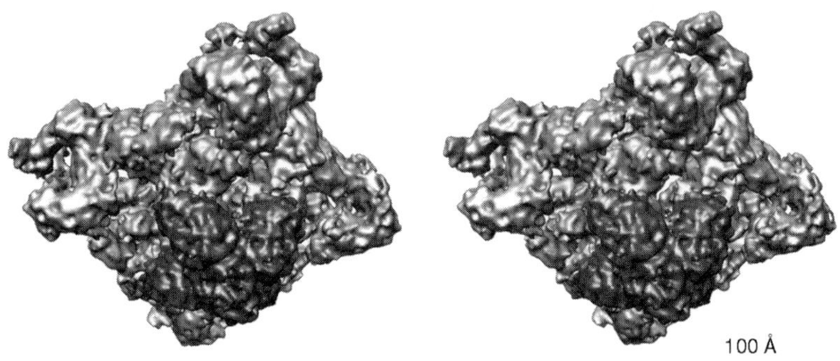

Fig. 6 The 9.6-Å resolution structure of RyR1 is shown as a stereo pair at an oblique angle. The structure is viewed from the SR lumen. The membrane-spanning region is shown in yellow. (See Color Insert.)

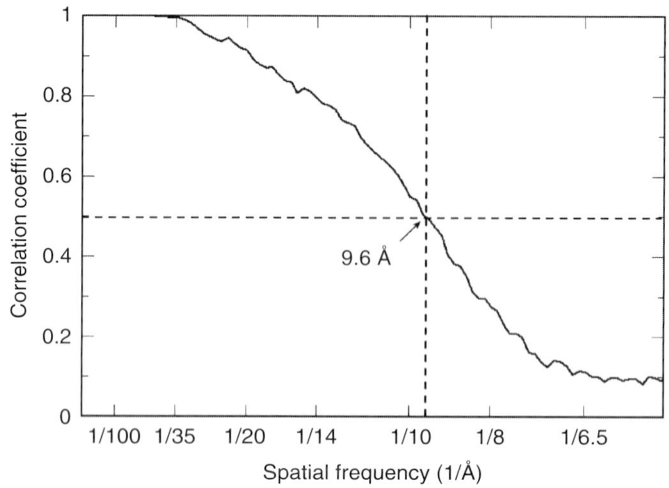

Fig. 7 Estimation of resolution of the 3D reconstruction using FSC between reconstructions from even and odd halves of the data set. The 0.5 threshold is used to assess the resolution in the final 3D reconstruction.

Wagenknecht *et al.*, 1997). The clamp-shaped regions are interconnected by "handle" domains (Fig. 8) and form a continuous network with the central rim and the column domains of the cytoplasmic region via several bridging densities. The small square-shaped structure with dimensions of 120 × 120 × 60 Å is rotated by ~40° with respect to the cytoplasmic assembly and represents the region that spans the SR membrane (Fig. 8). The proposed assignment of two major morphological regions within the 3D reconstruction of RyR is nicely consistent with the topological arrangement, which is predicted based on hydropathy analysis of the channel protein sequence (Takeshima *et al.*, 1989; Zorzato *et al.*, 1990).

2. Mapping Functional Domains in the Cryo-EM Map

Low resolution cryo-EM studies have also permitted localization of the binding sites for various macromolecular modulators of RyR1 such as calmodulin (Samso and Wagenknecht, 2002; Wagenknecht *et al.*, 1997) and FKBP12 (Samso *et al.*, 2006; Sharma *et al.*, 2006; Wagenknecht *et al.*, 1996, 1997). Mapping sequence onto 3D structure was done for some RyR regions by using sequence-specific antibodies (Baker *et al.*, 2002; Benacquista *et al.*, 2000). Another approach for assessing protein topology was based on structural analysis of an RyR-fusion protein that contained glutathione *S*-transferase (GST) fused to its N-terminus (GST-RyR3) (Liu *et al.*, 2001). Further studies by the same group employed the green fluorescent

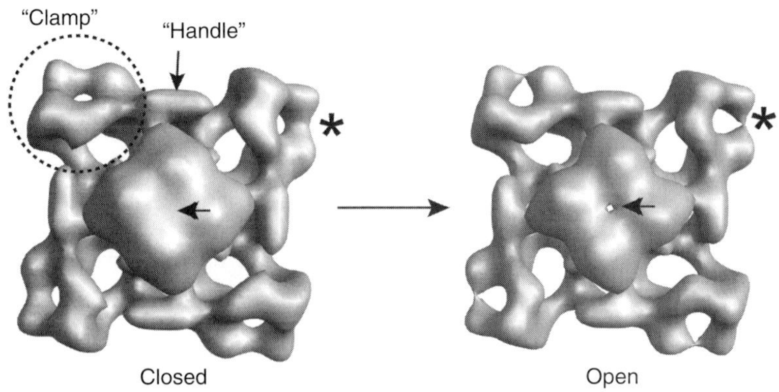

Fig. 8 The 30-Å resolution 3D maps of RyR1 in closed (~10-nM Ca^{2+}) and Ca^{2+}/ryanodine-open (~100-μM Ca^{2+}, 100-nM ryanodine) conformations (Orlova *et al.*, 1996; Serysheva *et al.*, 1999). Shown are views from the SR lumen, clamp-shaped and handle domains are indicated. Note differences in "clamp-shaped" domains (asterisks) and the presence or absence of the central opening in the TM domains (arrows).

protein (GFP) insertion for mapping sequence onto the 3D structure of RyR (Liu *et al.*, 2002, 2004, 2005; Zhang *et al.*, 2003).

The localizations of these domains are based on difference imaging between labeled and unlabeled RyR1. Their putative positions are summarized in Fig. 9. The accuracies of these localizations are often complicated by the size of the ligand (or the insert) and also the possible overall conformational differences between ligand-bound and ligand-free states.

3. Functional States of RyR1

During E–C coupling, the Ca^{2+}-release channel exists in several functional states, which can be characterized by specific channel-gating parameters. The functional channel transitions are regulated by a wide variety of endogenous molecules and pharmacological agents (Coronado *et al.*, 1994; Meissner, 1994; Ogawa, 1994). One of the greatest advantages of cryo-EM of single particles is its direct visualization of conformational changes, since the macromolecular complex can be trapped in solution under conditions similar to those used for electrophysiological or biochemical characterization. The structure of RyR has been analyzed under conditions that drive the channel population predominantly to its "open" (conducting) states (Orlova *et al.*, 1996; Serysheva *et al.*, 1999; Sharma *et al.*, 2000). Despite the low resolution of these reconstructions (~30 Å), global conformational changes were observed in both the cytoplasmic region and the putative membrane-spanning domain (Fig. 8). The clamp domains at the four corners of the cytoplasmic region appear in a more open conformation in the presence of Ca^{2+}/ryanodine or

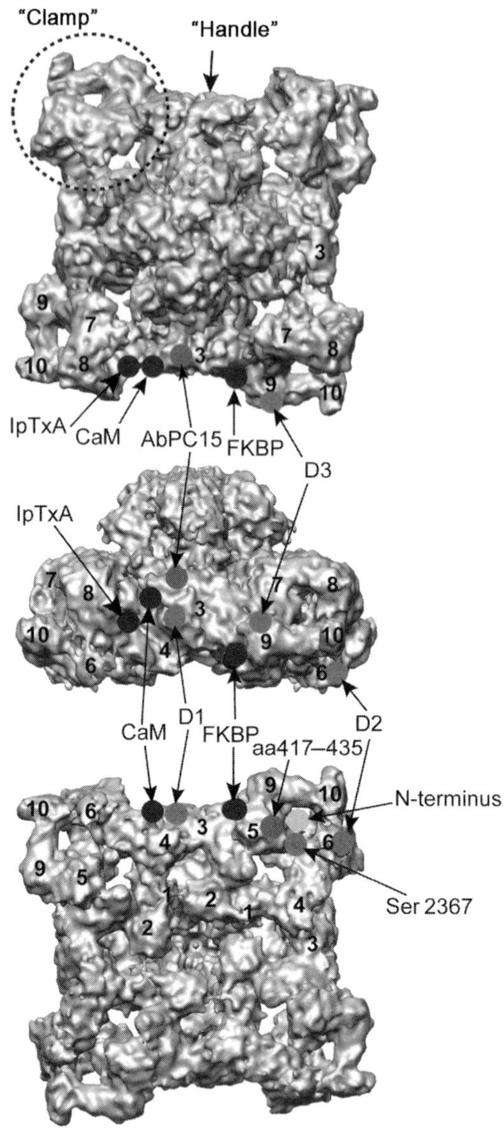

Fig. 9 The 9.6-Å resolution 3D map of the skeletal muscle Ca^{2+}-release channel in its closed conformation (Ludtke *et al.*, 2005) is shown in three different views: view from the SR lumen (top), side view (middle), and view from the cytoplasm (bottom). Numerals indicate putative structural domains as they were originally identified in low-resolution studies (Radermacher *et al.*, 1994). Red circles indicate locations of ligand-binding regions as determined by reconstructions of RyR-ligand complexes: calmodulin-binding region (CaM) (Samso and Wagenknecht, 2002; Wagenknecht *et al.*, 1997), binding region of the FK506-binding protein (FKBP12) (Samso *et al.*, 2006; Sharma *et al.*, 2006; Wagenknecht *et al.*, 1996, 1997), imperatoxin A (IpTxA)-binding region (Samso *et al.*, 1999). Green circles indicate locations of the RyR sequences as they were mapped by reconstructions of expressed

Ca^{2+}/ATP, which are reported to induce channel opening. In addition, central cavity and mass movements are detected in the transmembrane region. In the ryanodine-modified open channel, the transmembrane domain is twisted by $\sim 4°$ with respect to its position in the closed state, and a small hole appears in the center of the open state transmembrane domain. While the resolution in this structure is too low to characterize the hole, local thinning of the density at this point is apparent. On the basis of these observations, a mechanism for the channel opening similar to the opening-closing of the iris in a camera diaphragm was proposed (Serysheva *et al.*, 1999). Thus, low-resolution structural studies of RyR in different functional states suggest that channel activation is associated with significant mass rearrangements in the channel complex, implying highly allosteric regulation of channel gating. A number of other studies also support a model of long-range changes, rather than local conformational variation, in the quaternary architecture of the RyR1. These include fluorescence studies by Ikemoto *et al.* (1985). In addition, RyR1 mutations, associated with malignant hypothermia and central core disease, directly impact channel gating. However, the majority of these mutations have been mapped to two distinct regions (residues 35–614 and 1728–2504) within the N-terminal putative cytoplasmic portion of the channel protein (Benkusky *et al.*, 2004; Brini, 2004; Hamilton, 2005; McCarthy *et al.*, 2000), while the channel pore region is formed by the C-terminal portion of RyR. That is, the mutations affecting channel gating are not in close physical proximity to the channel.

4. Pore Structure

Currently, there are two central questions in the field of ion channel studies: how is the gating (opening–closing) process controlled by a wide variety of endogenous molecules and pharmacological modifiers, and how do channel pores discriminate between various (often very similar) ions? Knowledge of the detailed architecture of the ion channel pore is key to understanding channel function. Promising insights into the pore structure of the closed RyR1 channel have been gained through our cryo-EM studies at subnanometer resolution (Ludtke *et al.*, 2005). Using a combination of visualization and computational tools (Baker *et al.*, 2006), we have unambiguously identified five α-helices in the membrane-spanning region, including a bent inner helix and a short pore-associated helix that form the ion-conduction pathway across the membrane. There is also a helix parallel to the membrane-cytoplasmic

---

RyR2-GFP fusion proteins carrying the GFP inserted at different regions of RyR2 cDNA: divergent region 1 (D1) (Liu *et al.*, 2002), residues 4254–4631 (RyR1) and 4210–4562 (RyR2); D2 (Liu *et al.*, 2004), residues 1342–1403 (RyR1) and residues 4210–4562 (RyR2); D3 (Zhang *et al.*, 2003), residues 1872–1923 (RyR1) and residues 1852–1890 (RyR2); Ser 2365 (Liu *et al.*, 2005) is located within the central mutation region (residues 1728–2504 in RyR1 and residues 2246–2504 in RyR2). Binding sites for RyR1 sequence-specific antibodies are indicated with blue circles: antibody for residues 4425–4621 (AbPC15) (Benacquista *et al.*, 2000) and residues 416–434 (Baker *et al.*, 2002). The position of GST tag fused to the N-terminus of RyR3 (Liu *et al.*, 2001) is labeled with yellow circle. (See Color Insert.)

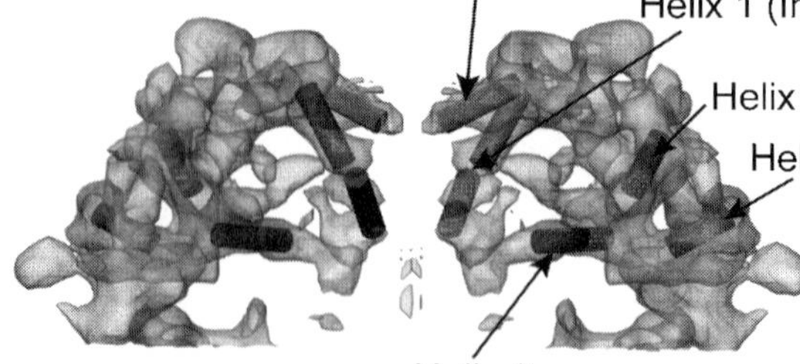

Fig. 10 Side view of the membrane-spanning region extracted from the 9.6-Å reconstruction of RyR1. Only two opposing subunits (of the four RyR1 subunits) are shown. A higher isosurface threshold was used to emphasize densities corresponding to five identified α-helices (annotated with cylinders).

interface (Fig. 10). It is worth emphasizing that these helices were independently identified using two different techniques: direct visual interpretation of the density and quantitative analysis using the *SSEhunter* program (an improved version of the *helixhunter* program; Jiang *et al.*, 2001). All proposed helices consist of continuous density, and, while difficult to represent in a static picture, these densities remain continuous over a wide range of thresholds. In fact, there are many continuous densities that we did not assign as helices, due to a low SSEhunter helical propensity score or unconvincing density patterns as the isosurface threshold was increased. Similar analysis has been used on other cryo-EM structures solved at similar resolutions and then later confirmed when high-resolution crystal structures of the protein or homologues became available (Chiu, 2005; Schmid *et al.*, 2004).

During the course of refining to increasingly high resolution, some features, such as the ring near the cytoplasmic side of the membrane-spanning region and "inner branches" connecting the ring densities with the column region as observed in Samso *et al.* (2005) appeared at 10- to 15-Å resolution. However, as the resolution of our reconstruction gradually improved, these ring-shaped densities resolved into the detailed pattern of rod-shaped densities shown in the final structure (Fig. 10). At lower isosurface thresholds, two ring-shaped densities can still be observed in the structure. It is tempting to speculate that the ring-shaped densities formed by inner helices correspond to the proposed "gating ring" near the cytoplasmic side of the membrane-spanning region in Samso *et al.* (2005).

5. Validation of Structural Features in the Pore Region

Comparison of the arrangement of the pore-forming helices in RyR1 with known channel structures reveals a striking similarity to the open MthK channel (Jiang *et al.*, 2002) (Fig. 11), although the RyR1 structure was determined under

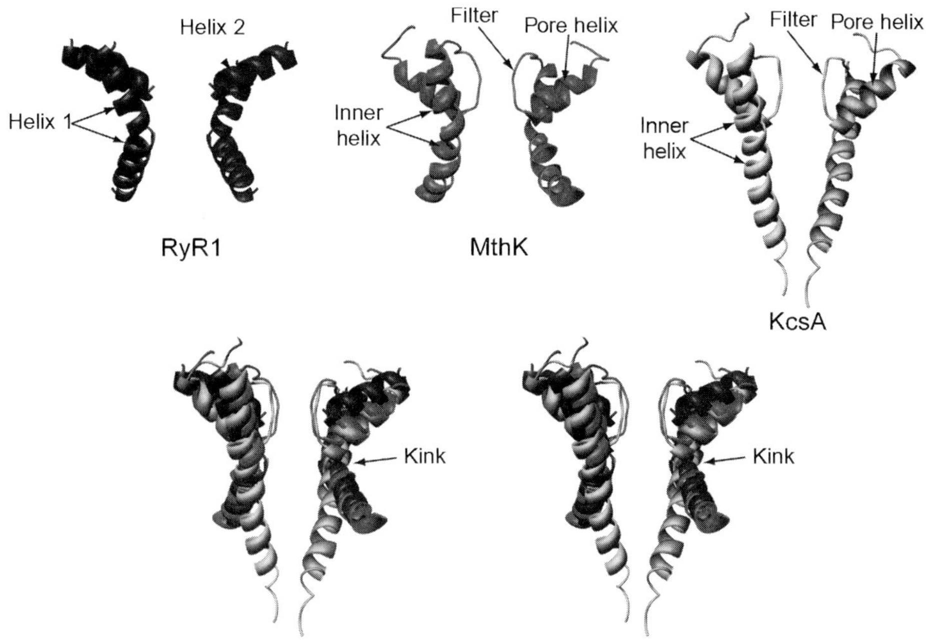

Fig. 11 Comparison of pore and inner helices from two opposing subunits of RyR1 and X-ray structures of the MthK channel, 1LNQ and of the KcsA channel, 1BL8. Stereo view of two subunits of RYR1, MthK, and KcsA pore-forming regions are superimposed and oriented with the cytoplasmic (out) side facing down.

conditions ($Ca^{2+} < 10$ nM) favoring the closed conformation (Serysheva *et al.*, 1995). This observation suggests that kinking of the inner helix is not essential for RyR1 opening. The membrane topology of RyR is not yet well defined. Predictions for the total number of transmembrane helices range from four (Takeshima *et al.*, 1989) to twelve (Zorzato *et al.*, 1990), with a model proposing eight (Du *et al.*, 2002). While some helices are resolved in the current structure, substantial mass in the transmembrane domain cannot yet be assigned unambiguously.

B. 3D Structure of DHPR

The DHPRs are called "L-type" or "slow," voltage-gated Ca^{2+} channels. They mediate Ca^{2+} influx across the muscle cell plasma membrane in response to its depolarization. The skeletal DHPRs are hetero-oligomeric membrane protein complexes with a total mass of ~430 kDa composed of five different subunits arranged in a 1:1:1:1:1 stoichiometry: the pore-forming α_1-subunit (190–212 kDa) is associated with auxiliary α_2-subunit (140 kDa), δ-subunit (25 kDa), β-subunit (53 kDa), and γ-subunit (33 kDa) (Catterall *et al.*, 1988; Ellis *et al.*, 1988; Glossmann and

Striessnig, 1988; Jay *et al.*, 1990). Each subunit is encoded by a separate gene, except the δ-subunit, which arises as a result of proteolysis from the C-terminus of the α_2/δ primary polypeptide. The α_2- and δ-subunits form a transmembrane, disulfide-linked glycoprotein dimer (Catterall, 1995). As predicted based on the primary sequence and the hydrophobicity profiles (Isom *et al.*, 1994), the α_1-subunit is organized into four transmembrane domain repeats, each of which contains six putative transmembrane segments (S1–S6). The S5 and S6 helices and the S5–S6 interlinking loop from each of repeat is believed to form the Ca^{2+}-channel pore. The S4 segment contains five to six positively charged amino acids, suggesting its essential role in the channel gating as the voltage sensor. Although the α_1-subunit has been shown to carry the characteristic pharmacological and functional properties of the Ca^{2+} channel for voltage sensing, ion permeability and drug binding, complete receptor function requires the presence of all subunits (Catterall, 2000; McDonald *et al.*, 1994; Melzer *et al.*, 1995; Tanabe *et al.*, 1988).

Cryo-EM has yielded two structures of the detergent-solubilized DHPR from skeletal muscle at ~30-Å resolution (Serysheva *et al.*, 2002; Wolf *et al.*, 2003). Although these reconstructions are broadly consistent in their shape, the detailed agreement between the structures is poor (Fig. 12). The overall size of the asymmetric channel structure determined by our group (Serysheva *et al.*, 2002) measures about $130 \times 115 \times 120$ Å^3 and consists of two major regions: a heart-shaped region connected at its widest end with a handle-shaped region (Fig. 12A). Wolf *et al.* (2003) described the structure as comprising two domains: the large globular domain exhibiting an ellipsoid shape, $165 \times 145 \times 80$ Å^3 in size, and a leg-shaped smaller domain attached off-center on one side of the globular domain (Fig. 12B). Although the transmembrane region of the α_1-subunit is predicted to be comprised of four homologous domains and, therefore, to be oriented in a pseudo-fourfold symmetric fashion in the membrane, no local symmetry was revealed in either of the reported 3D structures of DHPR. This is likely due to their low resolutions, combined with the presence of nonsymmetric auxiliary subunits within the channel complex.

Furthermore, these two studies disagree with respect to structural interpretations of densities within 3D maps of DHPR. Due to the low-resolution nature of both structures, the molecular boundaries of individual subunits could not be determined unambiguously, and mass densities were assigned primarily based on molecular mass and a proposed topological arrangement of the channel subunits in the T-tubule membrane. In our model, the heart-shaped region accounts for the main pore-forming α_1-subunit associated with the γ- and β-subunits, and the handle-shaped region comprises the α_2-δ complex. Therefore, the heart-shaped region spans the membrane, and its narrow part is exposed to the cytoplasm, while the major protein density comprising the handle-shaped region and the upper lobes of the heart-shaped structure is located on the extracellular side. This topology is consistent with the model proposed by Murata *et al.* (2001) based on antibody labeling.

In contrast, Grigorieff and coworkers (Wolf *et al.*, 2003) proposed that the major protein densities comprising the α_1-, γ-, δ-, and β-subunits are embedded within the membrane, placing the extracellular α_2-subunit within the smaller

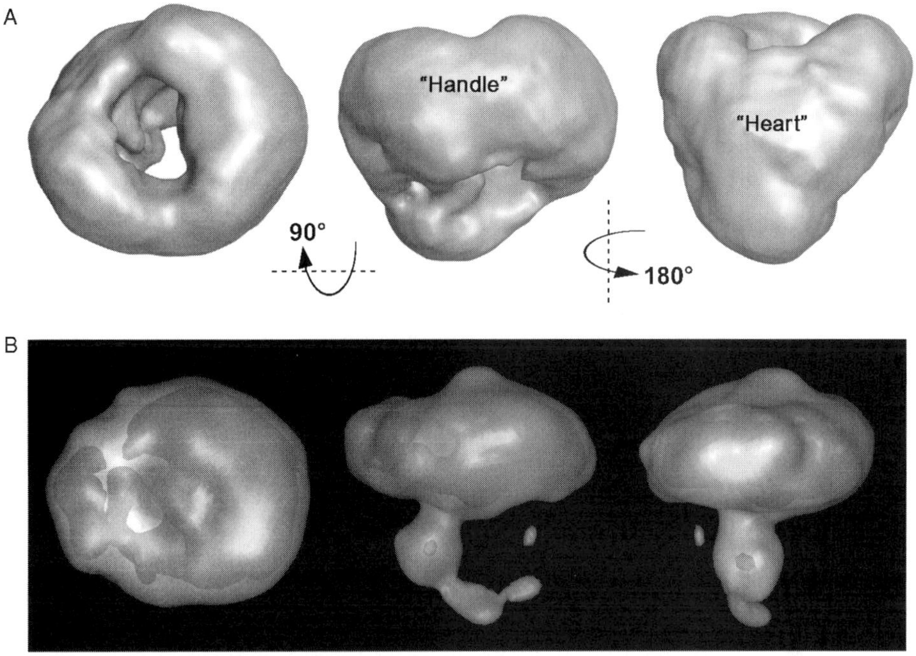

Fig. 12 The 30-Å resolution 3D reconstructions of DHPR obtained by single-particle cryo-EM. (A) 3D reconstruction of the DHPR determined by Serysheva *et al.* (2002) is shown in three different views. The reconstruction is displayed at the threshold level corresponding to a molecular mass of ~500 kDa. (B) Three orthogonal views, the 3D structure of the DHPR determined by Wolf *et al.* (2003). The map is displayed at two different density thresholds; the volume in gray encloses a molecular mass about 550 kDa. Reproduced with permission from *J. Mol. Biol.* (2003), **332,** 171–182. (See Color Insert.)

"leg"-shaped region. However, we argue that the β-subunit is a hydrophilic, peripheral protein located in the cytoplasm as was well established by many functional, biochemical, and structural studies (Opatowsky *et al.*, 2003; Pragnell *et al.*, 1991, 1994; Walker *et al.*, 1998). Moreover, recently reported crystal structures of the conserved core region of the β-subunit alone and in complex with the AID domain (α_1-interaction domain) (Chen *et al.*, 2004; Opatowsky *et al.*, 2004; Van Petegem *et al.*, 2004) provide insights into the molecular mechanism of the α_1-β interaction, confirming high-affinity binding between the intracellular domain of α_1-subunit and the β-functional core, which is most likely intracellular.

Another 27-Å resolution map of the DHPR was reported based on negative-stain electron microscopy and single-particle reconstruction (Wang *et al.*, 2002). This 3D reconstruction is strikingly different from either of the structures determined from frozen-hydrated channel protein (Serysheva *et al.*, 2002; Wolf *et al.*, 2003). It was proposed that negatively stained particles might correspond to a dimeric form of DHPR. The similar structure for the cardiac, voltage-gated, L-type Ca^{2+} channel in a dimeric form has been also reported by the same group (Wang *et al.*, 2004). It seems unlikely that differences between frozen-hydrated and negatively stained structures of DHPR would arise from image processing, since the reconstruction of the DHPR dimer was performed initially using the random conical tilt (Radermacher *et al.*, 1987) approach and was refined with the same EMAN software suite used for one of the vitrified structures (Ludtke *et al.*, 1999). The formation of channel dimers may be a consequence of different detergents used in these studies for solubilization of the channel complex, or it may be due to differences in specimen preparations for structural analysis (frozen-hydrated vs negatively stained). Clearly more experimentation is required to determine the causes of these discrepancies as well as to establish functional and physiological relevance of the DHPR dimer. Further structural analysis at higher resolution combined with specific antibody-labeling techniques should resolve the majority of these questions.

IV. Spatial Arrangement of DHPRs and RyRs in Triad Junctions

In both cardiac and skeletal muscles, DHPRs control activation of RyR1 by providing efficient coupling between depolarization of the plasma membrane and rapid Ca^{2+} release from the SR. DHPR is a voltage sensor, which undergoes conformational changes in response to depolarization of the plasmalemma (Rios and Brum, 1987). As one of the consequences of these changes, a slow L-type Ca^{2+} current is activated that in turn mediates entry of extracellular Ca^{2+} through the DHPR. In cardiac muscle, entry of Ca^{2+} through DHPRs activates RyR2s (cardiac isoform of RyR), causing them to release Ca^{2+} from the SR (the Ca^{2+}-induced Ca^{2+}-release mechanism) (Nabauer *et al.*, 1989). In skeletal muscle, however, entry of Ca^{2+} through the DHPR is not required for E–C coupling (Tanabe *et al.*, 1988). Instead, voltage-dependent conformational changes in DHPR produce a signal, which has been hypothesized to activate RyR1 allosterically. This signal is thought to be transmitted through a physical, possibly direct link between DHPR and RyR1 ("mechanical-link" mechanism) (Schneider and Chandler, 1973). Ca^{2+}-induced Ca^{2+} release was proposed to be supplementary to the direct molecular coupling in skeletal muscle (Pizarro *et al.*, 1991).

In addition, the interaction between the voltage-gated L-type Ca^{2+} channel and RyR1 is bidirectional: not only does L-type Ca^{2+} channel activate RyR1 ("orthograde" signaling from DHPR to RyR1), but RyR1 strongly effects L-type Ca^{2+} gating by transmitting a signal that enhances L-type Ca^{2+} current ("retrograde" signaling from RyR1 to DHPR) (Avila and Dirksen, 2000; Nakai *et al.*, 1996). The precise mechanism for the retrograde signaling in skeletal muscle is still not clear.

Thus, specific regions within both the DHPR and RyR proteins have been reported to mediate DHPR/RyR coupling and should be located in close opposition. Through a combination of observations from freeze-fracture studies and the molecular shapes of DHPR and Ca^{2+}-release channel determined by cryo-EM, a model for their interactions was proposed (Serysheva *et al.*, 2002). The handedness of the RyR array was inferred from a study by electron tomography of the frozen-hydrated entire triad junction isolated from skeletal muscle (Wagenknecht *et al.*, 2002) and from freeze-fracture studies (Paolini *et al.*, 2004). The site of inter-oligomeric contact within the RyR array corresponds to the clamp-shaped domain in single-particle 3D reconstruction of RyR (Fig. 8) (Serysheva *et al.*, 1999). These regions in the RyR structure undergo significant conformational changes on opening of the RyR1 (Orlova *et al.*, 1996; Serysheva *et al.*, 1999). Also, divergent regions 2 (residues 1342–1403 of RyR1) and 3 (residues 1872–1923 of RyR1), which were implicated to mediate interaction of RyR with DHPR, were localized to the clamp-shaped region (Liu *et al.*, 2004; Zhang *et al.*, 2003). Given the location of FKBP12 and calmodulin binding sites within the 3D structure of RyR (Samso *et al.*, 2006; Sharma *et al.*, 2006; Wagenknecht *et al.*, 1996, 1997), it should be pointed out that both binding sites are most likely not involved in the interaction between two adjacent RyRs.

Obviously in order to substantiate this model, it is necessary to perform further experiments to prove the sidedness of the observed structure of DHPR in the membrane and to correlate the linear sequences of RyR with the 3D structure.

V. Summary

In conclusion, while crystal structure determination of both Ca^{2+} channels has been in high demand for a along time, no high-resolution structures are yet available. To date, structures of DHPR and RyR based on single-particle cryo-EM have been providing the basis for interpreting structure–function relations for these proteins. Although our understanding of the molecular mechanism of E–C coupling is expanding rapidly, many essential questions on DHPR/RyR coupling cannot currently be answered, due to the absence of information about the arrangement of the primary sequences of DHPR and RyR in 3D. Continued progress in improving the resolution of both structures as well as elucidation of the structural changes associated with specific physiological states will gradually lead to a better understanding of this critical biological system.

The resolving power of single-particle cryo-EM has been gradually improving due to improvements in software, microscopes, and associated experimental equipment. This methodology has been demonstrated to be reliable and accurate for determining the structures of large molecular machines to 6-Å resolution without any use of information derived from X-ray crystallography, as demonstrated in the study of GroEL (Ludtke *et al.*, 2004). It is feasible to extend this approach toward 4-Å resolution, at which point it should be possible to trace the C_α protein backbone. As demonstrated in numerous physiological studies, RYR1 is a highly dynamic molecular machine. This heterogeneity is part of the normal functioning of the channel under physiological conditions. Therefore, this molecular machine represents an appropriate project to drive the development of novel image processing algorithms for determination of the structures of dynamic objects at high resolution.

Acknowledgments

We thank C. Franzini-Armstrong, T. Wagenknecht, and N. Grigorieff for their help in preparation of this chapter. We also acknowledge S. L. Hamilton for helpful discussions and support of studies presented in the chapter. Original research described in this chapter has been supported by NIH grants (P41RR02250, P01GM99116, R01GM072804, and R01AR44864) and a grant from the MDA.

References

Avila, G., and Dirksen, R. T. (2000). Functional impact of the ryanodine receptor on the skeletal muscle L-type calcium channel. *J. Gen. Physiol.* **115,** 467–480.

Baker, M. L., Ju, T., and Chiu, W. (2006). Identification of secondary structure elements in medium resolution density maps. *Structure* (in press).

Baker, M. L., Serysheva, I. I., Sencer, S., Wu, Y., Ludtke, S. J., Jiang, W., Hamilton, S. L., and Chiu, W. (2002). The skeletal muscle Ca^{2+} release channel has an oxidoreductase-like domain. *Proc. Natl. Acad. Sci. USA* **99,** 12155–12160.

Benacquista, B. L., Sharma, M. R., Samso, M., Zorzato, F., Treves, S., and Wagenknecht, T. (2000). Amino acid residues 4425–4621 localized on the three-dimensional structure of the skeletal muscle ryanodine receptor. *Biophys. J.* **78,** 1349–1358.

Benkusky, N. A., Farrell, E. F., and Valdivia, H. H. (2004). Ryanodine receptor channelopathies. *Biochem. Biophys. Res. Commun.* **322,** 1280–1285.

Block, B. A., Imagawa, T., Campbell, K. P., and Franzini-Armstrong, C. (1988). Structural evidence for direct interaction between the molecular components of the transverse tubule/sarcoplasmic reticulum junction in skeletal muscle. *J. Cell Biol.* **107,** 2587–2600.

Booth, C. R., Jiang, W., Baker, M. L., Hong Zhou, Z., Ludtke, S. J., and Chiu, W. (2004). A 9A single particle reconstruction from CCD captured images on a 200 kV electron cryomicroscope. *J. Struct. Biol.* **147,** 116–127.

Böttcher, B., Wynne, S. A., and Crowther, R. A. (1997). Determination of the fold of the core protein of hepatitis B virus by electron cryomicroscopy. *Nature* **386,** 88–91.

Brini, M. (2004). Ryanodine receptor defects in muscle genetic diseases. *Biochem. Biophys. Res. Commun.* **322,** 1245–1255.

Catterall, W. A. (1995). Structure and function of voltage-gated ion channels. *Annu. Rev. Biochem.* **64,** 493–531.

Catterall, W. A. (2000). Structure and regulation of voltage-gated Ca^{2+} channels. *Annu. Rev. Cell Dev. Biol.* **16,** 521–555.

Catterall, W. A., Seagar, M. J., and Takahashi, M. (1988). Molecular properties of dihydropyridine-sensitive calcium channels in skeletal muscle. *J. Biol. Chem.* **263,** 3535–3538.

Chen, D., Song, J. L., Chuang, D. T., Chiu, W., and Ludtke, S. J. (2006). A novel expanded conformation of single-ring GroEL-GroES complex encapsulates an 86-kDa substrate. *Structure* (in press).

Chen, Y. H., Li, M. H., Zhang, Y., He, L. L., Yamada, Y., Fitzmaurice, A., Shen, Y., Zhang, H., Tong, L., and Yang, J. (2004). Structural basis of the alpha1-beta subunit interaction of voltage-gated Ca^{2+} channels. *Nature* **429,** 675–680.

Chiu, W. (2005). Electron cryomicroscopy of cellular machines at subnanometer resolution. *Microsc. Microanal.* **11,** 1086–1087.

Chiu, W., Downing, K. H., Dubochet, J., Glaeser, R. M., Heide, H. G., Knapek, E., Kopf, D. A., Lamvik, M. K., Lepault, J., Robertson, J. D., Zeitler, E., and Zemlin, F. (1986). Cryoprotection in electron microscopy. *J. Microsc.* **141,** 385–391.

Coronado, R., Morrissette, J., Sukhareva, M., and Vaughan, D. M. (1994). Structure and function of ryanodine receptors. *Am. J. Physiol.* **266,** C1485–C1504.

Du, G. G., Sandhu, B., Khanna, V. K., Guo, X. H., and MacLennan, D. H. (2002). Topology of the Ca^{2+} release channel of skeletal muscle sarcoplasmic reticulum (RyR1). *Proc. Natl. Acad. Sci. USA* **99,** 16725–16730.

Dubochet, J., Adrian, M., Chang, J. J., Homo, J. C., Lepault, J., McDowall, A. W., and Schultz, P. (1988). Cryo-electron microscopy of vitrified specimens. *Q. Rev. Biophys.* **21,** 129–228.

Ellis, S. B., Williams, M. E., Ways, N. R., Brenner, R., Sharp, A. H., Leung, A. T., Campbell, K. P., McKenna, E., Koch, W. J., Hui, A., Scwartz, A., and Harpold, M. M. (1988). Sequence and expression of mRNAs encoding the alpha 1 and alpha 2 subunits of a DHP-sensitive calcium channel. *Science* **241,** 1661–1664.

Erickson, H. P., and Klug, A. (1970). The Fourier transform of an electron micrograph: Effects of defocussing and aberrations, and implications for the use of underfocus contrast enhancement. *Philos. Trans. Roy. Soc. Lond. B* **261,** 105–118.

Ferguson, D. G., Schwartz, H. W., and Franzini-Armstrong, C. (1984). Subunit structure of junctional feet in triads of skeletal muscle: A freeze-drying, rotary-shadowing study. *J. Cell Biol.* **99,** 1735–1742.

Fotin, A., Cheng, Y., Sliz, P., Grigorieff, N., Harrison, S. C., Kirchhausen, T., and Walz, T. (2004). Molecular model for a complete clathrin lattice from electron cryomicroscopy. *Nature* **432,** 573–579.

Frank, J. (1996). "Three-Dimensional Electron Microscopy of Macromolecular Asemblies." Academic Press, Inc., San Diego.

Frank, J., Radermacher, M., Penczek, P., Zhu, J., Li, Y., Ladjadj, M., and Leith, A. (1996). SPIDER and WEB: Processing and visualization of images in 3D electron microscopy and related fields. *J. Struct. Biol.* **116,** 190–199.

Franzini-Armstrong, C., and Kish, J. W. (1995). Alternate disposition of tetrads in peripheral couplings of skeletal muscle. *J. Muscle Res. Cell Motil.* **16,** 319–324.

Gabashvili, I. S., Agrawal, R. K., Sahn, C. M., Grassucci, R. A., Svergun, D. I., Frank, J., and Penczek, P. (2000). Solution structure of the *E. Coli* 70S ribosome at 11.5 A resolution. *Cell* **100,** 537–549.

Glossmann, H., and Striessnig, J. (1988). Calcium channels. *Vitam. Horm.* **44,** 155–328.

Grigorieff, N. (1998). Three-dimensional structure of bovine NADH: Ubiquinone oxidoreductase (complex I) at 22 A in ice. *J. Mol. Biol.* **277,** 1033–1046.

Grob, P., Cruse, M. J., Inouye, Peris, M., Penczek, P. A., Tjian, R., and Nogales, E. (2006). Cryo-electron microscopy studies of human TFIID: Conformational breathing in the integration of gene regulatory cues. *Structure* **14,** 511–520.

Hakamata, Y., Nakai, J., Takeshima, H., and Imoto, K. (1992). Primary structure and distribution of a novel ryanodine receptor/calcium release channel from rabbit brain. *FEBS Lett.* **312,** 229–235.

Hamilton, S. L. (2005). Ryanodine receptors. *Cell Calcium* **38,** 253–260.

Henderson, R. (2004). Realizing the potential of electron cryo-microscopy. *Q. Rev. Biophys.* **37,** 3–13.

Ikemoto, N., Antoniu, B., and Meszaros, L. G. (1985). Rapid flow chemical quench studies of calcium release from isolated sarcoplasmic reticulum. *J. Biol. Chem.* **260,** 14096–14100.

Isom, L. L., De Jongh, K. S., and Catterall, W. A. (1994). Auxiliary subunits of voltage-gated ion channels. *Neuron* **12,** 1183–1194.

Jay, S. D., Ellis, S. B., McCue, A. F., Williams, M. E., Vedvick, T. S., Harpold, M. M., and Campbell, K. P. (1990). Primary structure of the gamma subunit of the DHP-sensitive calcium channel from skeletal muscle. *Science* **248,** 490–492.

Jeyakumar, L. H., Copello, J. A., O'Malley, A. M., Wu, G. M., Grassucci, R., Wagenknecht, T., and Fleischer, S. (1998). Purification and characterization of ryanodine receptor 3 from mammalian tissue. *J. Biol. Chem.* **273,** 16011–16020.

Jiang, W., and Ludtke, S. J. (2005). Electron cryomicroscopy of single particles at subnanometer resolution. *Curr. Opin. Struct. Biol.* **15,** 571–577.

Jiang, W., Baker, M. L., Ludtke, S. J., and Chiu, W. (2001). Bridging the information gap: Computational tools for intermediate resolution structure interpretation. *J. Mol. Biol.* **308,** 1033–1044.

Jiang, Y., Lee, A., Chen, J., Cadene, M., Chait, B. T., and MacKinnon, R. (2002). The open pore conformation of potassium channels. *Nature* **417,** 523–526.

Lai, F. A., Misra, M., Xu, L., Smith, H. A., and Meissner, G. (1989). The ryanodine receptor-Ca^{2+} release channel complex of skeletal muscle sarcoplasmic reticulum. Evidence for a cooperatively coupled, negatively charged homotetramer. *J. Biol. Chem.* **264,** 16776–16785.

Liu, Z., Wang, R., Zhang, J., Chen, S. R., and Wagenknecht, T. (2005). Localization of a disease-associated mutation site in the three-dimensional structure of the cardiac muscle ryanodine receptor. *J. Biol. Chem.* **280,** 37941–37947.

Liu, Z., Zhang, J., Li, P., Chen, S. R., and Wagenknecht, T. (2002). Three-dimensional reconstruction of the recombinant type 2 ryanodine receptor and localization of its divergent region 1. *J. Biol. Chem.* **277,** 46712–46719.

Liu, Z., Zhang, J., Sharma, M. R., Li, P., Chen, S. R., and Wagenknecht, T. (2001). Three-dimensional reconstruction of the recombinant type 3 ryanodine receptor and localization of its amino terminus. *Proc. Natl. Acad. Sci. USA* **98,** 6104–6109.

Liu, Z., Zhang, J., Wang, R., Wayne Chen, S. R., and Wagenknecht, T. (2004). Location of divergent region 2 on the three-dimensional structure of cardiac muscle ryanodine receptor/calcium release channel. *J. Mol. Biol.* **338,** 533–545.

Ludtke, S. J., Baldwin, P. R., and Chiu, W. (1999). EMAN: Semiautomated software for high-resolution single-particle reconstructions. *J. Struct. Biol.* **128,** 82–97.

Ludtke, S. J., Chen, D. H., Song, J. L., Chuang, D. T., and Chiu, W. (2004). Seeing GroEL at 6 A resolution by single particle electron cryomicroscopy. *Structure (Camb.)* **12,** 1129–1136.

Ludtke, S. J., Serysheva, I. I., Hamilton, S. L., and Chiu, W. (2005). The pore structure of the closed RyR1 channel. *Structure (Camb.)* **13,** 1203–1211.

Matadeen, R., Patwardhan, A., Gowen, B., Orlova, E. V., Pape, T., Cuff, M., Mueller, F., Brimacombe, R., and van Heel, M. (1999). The *Escherichia coli* large ribosomal subunit at 7.5 A resolution. *Struct. Fold. Des.* **7,** 1575–1583.

McCarthy, T. V., Quane, K. A., and Lynch, P. J. (2000). Ryanodine receptor mutations in malignant hyperthermia and central core disease. *Hum. Mutat.* **15,** 410–417.

McDonald, T. F., Pelzer, S., Trautwein, W., and Pelzer, D. J. (1994). Regulation and modulation of calcium channels in cardiac, skeletal, and smooth muscle cells. *Physiol. Rev.* **74,** 365–507.

Meissner, G. (1994). Ryanodine receptor/Ca^{2+} release channels and their regulation by endogenous effectors. *Annu. Rev. Physiol.* **56,** 485–508.

Melzer, W., Herrmann-Frank, A., and Luttgau, H. C. (1995). The role of Ca^{2+} ions in excitation-contraction coupling of skeletal muscle fibres. *Biochim. Biophys. Acta* **1241,** 59–116.

Mouche, F., Zhu, Y., Pulokas, J., Potter, C. S., and Carragher, B. (2003). Automated three-dimensional reconstruction of keyhole limpet hemocyanin type. *J. Struct. Biol.* **144,** 301–312.

Murata, K., Odahara, N., Kuniyasu, A., Sato, Y., Nakayama, H., and Nagayama, K. (2001). Asymmetric arrangement of auxiliary subunits of skeletal muscle voltage-gated l-type Ca(2+) channel. *Biochem. Biophys. Res. Commun.* **282,** 284–291.

Nabauer, M., Callewaert, G., Cleemann, L., and Morad, M. (1989). Regulation of calcium release is gated by calcium current, not gating charge, in cardiac myocytes. *Science* **244,** 800–803.

Nakai, J., Dirksen, R. T., Nguyen, H. T., Pessah, I. N., Beam, K. G., and Allen, P. D. (1996). Enhanced dihydropyridine receptor channel activity in the presence of ryanodine receptor. *Nature* **380,** 72–75.

Ogawa, Y. (1994). Role of ryanodine receptors. *Crit. Rev. Biochem. Mol. Biol.* **29,** 229–274.

Opatowsky, Y., Chen, C. C., Campbell, K. P., and Hirsch, J. A. (2004). Structural analysis of the voltage-dependent calcium channel beta subunit functional core and its complex with the alpha 1 interaction domain. *Neuron* **42,** 387–399.

Opatowsky, Y., Chomsky-Hecht, O., Kang, M. G., Campbell, K. P., and Hirsch, J. A. (2003). The voltage-dependent calcium channel beta subunit contains two stable interacting domains. *J. Biol. Chem.* **278,** 52323–52332.

Orlova, E. V., and Saibil, H. R. (2004). Structure determination of macromolecular assemblies by single-particle analysis of cryo-electron micrographs. *Curr. Opin. Struct. Biol.* **14,** 584–590.

Orlova, E. V., Serysheva, I. I., van Heel, M., Hamilton, S. L., and Chiu, W. (1996). Two structural configurations of the skeletal muscle calcium release channel. *Nat. Struct. Biol.* **3,** 547–552.

Paolini, C., Protasi, F., and Franzini-Armstrong, C. (2004). The relative position of RyR feet and DHPR tetrads in skeletal muscle. *J. Mol. Biol.* **342,** 145–153.

Penczek, P. A. (2002). Three-dimensional spectral signal-to-noise ratio for a class of reconstruction algorithms. *J. Struct. Biol.* **138,** 34–46.

Penczek, P. A., Yang, C., Frank, J., and Spahn, C. M. (2006). Estimation of variance in single-particle reconstruction using the bootstrap technique. *J. Struct. Biol.* **154,** 168–183.

Pizarro, G., Csernoch, L., Uribe, I., Rodriguez, M., and Rios, E. (1991). The relationship between Q gamma and Ca release from the sarcoplasmic reticulum in skeletal muscle. *J. Gen. Physiol.* **97,** 913–947.

Pragnell, M., De Waard, M., Mori, Y., Tanabe, T., Snutch, T. P., and Campbell, K. P. (1994). Calcium channel beta-subunit binds to a conserved motif in the I-II cytoplasmic linker of the alpha 1-subunit. *Nature* **368,** 67–70.

Pragnell, M., Sakamoto, J., Jay, S. D., and Campbell, K. P. (1991). Cloning and tissue-specific expression of the brain calcium channel beta-subunit. *FEBS Lett.* **291,** 253–258.

Protasi, F., Franzini-Armstrong, C., and Flucher, B. E. (1997). Coordinated incorporation of skeletal muscle dihydropyridine receptors and ryanodine receptors in peripheral couplings of BC3H1 cells. *J. Cell Biol.* **137,** 859–870.

Radermacher, M., Rao, V., Grassucci, R., Frank, J., Timerman, A. P., Fleischer, S., and Wagenknecht, T. (1994). Cryo-electron microscopy and three-dimensional reconstruction of the calcium release channel/ryanodine receptor from skeletal muscle. *J. Cell Biol.* **127,** 411–423.

Radermacher, M., Wagenknecht, T., Verschoor, A., and Frank, J. (1987). Three-dimensional structure of the large ribosomal subunit from *Escherichia coli*. *EMBO J.* **6,** 1107–1114.

Rios, E., and Brum, G. (1987). Involvement of dihydropyridine receptors in excitation-contraction coupling in skeletal muscle. *Nature* **325,** 717–720.

Rosenthal, P. B., and Henderson, R. (2003). Optimal determination of particle orientation, absolute hand, and contrast loss in single-particle electron cryomicroscopy. *J. Mol. Biol.* **333,** 721–745.

Samso, M., and Wagenknecht, T. (2002). Apocalmodulin and Ca^{2+}-calmodulin bind to neighboring locations on the ryanodine receptor. *J. Biol. Chem.* **277,** 1349–1353.

Samso, M., Shen, X., and Allen, P. D. (2006). Structural characterization of the RyR1-FKBP12 interaction. *J. Mol. Biol.* **356,** 917–927.

Samso, M., Trujilo, R., Gurrola, G. B., Valdivia, H. H., and Wagenknecht, T. (1999). Three-dimensional location of the imperatoxin a binding site on the ryanodine receptor. *J. Cell Biol.* **146,** 493–499.

Samso, M., Wagenknecht, T., and Allen, P. D. (2005). Internal structure and visualization of transmembrane domains of the RyR1 calcium release channel by cryo-EM. *Nat. Struct. Mol. Biol.* **12,** 539–544.

Schmid, M. F., Sherman, M. B., Matsudaira, P., and Chiu, W. (2004). Structure of the acrosomal bundle. *Nature* **431,** 104–107.

Schneider, M. F., and Chandler, W. K. (1973). Voltage dependent charge movement of skeletal muscle: A possible step in excitation-contraction coupling. *Nature* **242,** 244–246.

Serysheva, I. I., Ludtke, S. J., Baker, M. R., Chiu, W., and Hamilton, S. L. (2002). Structure of the voltage-gated L-type Ca^{2+} channel by electron cryomicroscopy. *Proc. Natl. Acad. Sci. USA* **99,** 10370–10375.

Serysheva, I. I., Orlova, E. V., Chiu, W., Sherman, M. B., Hamilton, S. L., and van Heel, M. (1995). Electron cryomicroscopy and angular reconstitution used to visualize the skeletal muscle calcium release channel. *Nat. Struct. Biol.* **2,** 18–24.

Serysheva, I. I., Schatz, M., van Heel, M., Chiu, W., and Hamilton, S. L. (1999). Structure of the skeletal muscle calcium release channel activated with Ca^{2+} and AMP-PCP. *Biophys. J.* **77,** 1936–1944.

Sharma, M. R., Jeyakumar, L. H., Fleischer, S., and Wagenknecht, T. (2000). Three-dimensional structure of ryanodine receptor isoform three in two conformational states as visualized by cryo-electron microscopy. *J. Biol. Chem.* **275,** 9485–9491.

Sharma, M. R., Jeyakumar, L. H., Fleischer, S., and Wagenknecht, T. (2006). Three-dimensional visualization of FKBP12.6 binding to an open conformation of cardiac ryanodine receptor. *Biophys. J.* **90,** 164–172.

Sharma, M. R., Penczek, P., Grassucci, R., Xin, H. B., Fleischer, S., and Wagenknecht, T. (1998). Cryoelectron microscopy and image analysis of the cardiac ryanodine receptor. *J. Biol. Chem.* **273,** 18429–18434.

Sorrentino, V., and Reggiani, C. (1999). Expression of the ryanodine receptor type 3 in skeletal muscle. A new partner in excitation-contraction coupling? *Trends Cardiovasc. Med.* **9,** 54–61.

Sorrentino, V., and Volpe, P. (1993). Ryanodine receptors: How many, where and why? *Trends Pharmacol. Sci.* **14,** 98–103.

Sorzano, C. O., Marabini, R., Velazquez-Muriel, J., Bilbao-Castro, J. R., Scheres, S. H., Carazo, J. M., and Pascual-Montano, A. (2004). XMIPP: A new generation of an open-source image processing package for electron microscopy. *J. Struct. Biol.* **148,** 194–204.

Stewart, A., and Grigorieff, N. (2004). Noise bias in the refinement of structures derived from single particles. *Ultramicroscopy* **102,** 67–84.

Sutko, J. L., and Airey, J. A. (1996). Ryanodine receptor Ca^{2+} release channels: Does diversity in form equal diversity in function? *Physiol. Rev.* **76,** 1027–1071.

Takeshima, H., Nishimura, S., Matsumoto, T., Ishida, H., Kangawa, K., Minamino, N., Matsuo, H., Ueda, M., Hanaoka, M., Hirose, T. and Numa, S. (1989). Primary structure and expression from complementary DNA of skeletal muscle ryanodine receptor. *Nature* **339,** 439–445.

Tama, F., Valle, M., Frank, J., and Brooks, C. L., III. (2003). Dynamic reorganization of the functionally active ribosome explored by normal mode analysis and cryo-electron microscopy. *Proc. Natl. Acad. Sci. USA* **100,** 9319–9323.

Tanabe, T., Beam, K. G., Powell, J. A., and Numa, S. (1988). Restoration of excitation-contraction coupling and slow calcium current in dysgenic muscle by dihydropyridine receptor complementary DNA. *Nature* **336,** 134–139.

Tunwell, R. E., Wickenden, C., Bertrand, B. M., Shevchenko, V. I., Walsh, M. B., Allen, P. D., and Lai, F. A. (1996). The human cardiac muscle ryanodine receptor-calcium release channel: Identification, primary structure and topological analysis. *Biochem. J.* **318,** 477–487.

Unwin, N., and Henderson, R. (1975). Molecular structure determination by electron microscopy of unstained crystalline specimens. *J. Mol. Biol.* **94,** 425–440.

van Heel, M. (1987). Similarity measures between images. *Ultramicroscopy* **21,** 95–100.

van Heel, M., and Harauz, G. (1986). Resolution criteria for three dimensional reconstructions. *Optik* **73,** 119–122.

van Heel, M., Harauz, G., and Orlova, E. V. (1996). A new generation of the IMAGIC image processing system. *J. Struct. Biol.* **116,** 17–24.

van Heel, M., Winkler, H., Orlova, E. V., and Schatz, M. (1992). Structure analysis of ice-embedded single particles. *Scanning Microsc. Suppl.* **6,** 23–42.

Van Petegem, F., Clark, K. A., Chatelain, F. C., and Minor, D. L., Jr. (2004). Structure of a complex between a voltage-gated calcium channel beta-subunit and an alpha-subunit domain. *Nature* **429,** 671–675.

Wagenknecht, T., and Radermacher, M. (1995). Three-dimensional architecture of the skeletal muscle ryanodine receptor. *FEBS Lett.* **369,** 43–46.

Wagenknecht, T., Grassucci, R., Berkowitz, J., Wiederrecht, G. J., Xin, H. B., and Fleischer, S. (1996). Cryoelectron microscopy resolves FK506-binding protein sites on the skeletal muscle ryanodine receptor. *Biophys. J.* **70,** 1709–1715.

Wagenknecht, T., Hsieh, C. E., Rath, B. K., Fleischer, S., and Marko, M. (2002). Electron tomography of frozen-hydrated isolated triad junctions. *Biophys. J.* **83,** 2491–2501.

Wagenknecht, T., Radermacher, M., Grassucci, R., Berkowitz, J., Xin, H. B., and Fleischer, S. (1997). Locations of calmodulin and FK506-binding protein on the three-dimensional architecture of the skeletal muscle ryanodine receptor. *J. Biol. Chem.* **272,** 32463–32471.

Walker, D., Bichet, D., Campbell, K. P., and De Waard, M. (1998). A beta 4 isoform-specific interaction site in the carboxyl-terminal region of the voltage-dependent Ca^{2+} channel alpha 1A subunit. *J. Biol. Chem.* **273,** 2361–2367.

Wang, M. C., Collins, R. F., Ford, R. C., Berrow, N. S., Dolphin, A. C., and Kitmitto, A. (2004). The three-dimensional structure of the cardiac L-type voltage-gated calcium channel: Comparison with the skeletal muscle form reveals a common architectural motif. *J. Biol. Chem.* **279,** 7159–7168.

Wang, M. C., Velarde, G., Ford, R. C., Berrow, N. S., Dolphin, A. C., and Kitmitto, A. (2002). 3D structure of the skeletal muscle dihydropyridine receptor. *J. Mol. Biol.* **323,** 85–98.

Wolf, M., Eberhart, A., Glossmann, H., Striessnig, J., and Grigorieff, N. (2003). Visualization of the domain structure of an L-type Ca^{2+} channel using electron cryo-microscopy. *J. Mol. Biol.* **332,** 171–182.

Zhang, J., Liu, Z., Masumiya, H., Wang, R., Jiang, D., Li, F., Wagenknecht, T., and Chen, S. R. (2003). Three-dimensional localization of divergent region 3 of the ryanodine receptor to the clamp-shaped structures adjacent to the FKBP binding sites. *J. Biol. Chem.* **278,** 14211–14218.

Zorzato, F., Fujii, J., Otsu, K., Phillips, M., Green, N. M., Lai, F. A., Meissner, G., and MacLennan, D. H. (1990). Molecular cloning of cDNA encoding human and rabbit forms of the Ca^{2+} release channel (ryanodine receptor) of skeletal muscle sarcoplasmic reticulum. *J. Biol. Chem.* **265,** 2244–2256.

CHAPTER 17

Electron Microscopy of Microtubule-Based Cytoskeletal Machinery

Andreas Hoenger and Daniela Nicastro

Department of Molecular, Cellular, and Developmental Biology
University of Colorado, Boulder, Colorado 80309

Most intracellular motility and transport processes are based either on myosin motors interacting with the actin network, or on dyneins and kinesins operating on microtubules. This chapter focuses on electron microscopy-based methods to assess structural and functional aspects of kinesins and dyneins in complex with microtubules. To this end, we will describe two fundamentally different approaches to the study of motor–microtubule complex structure in three dimensions (3D). Both methods have been applied very successfully in the past, but they will require further refinement to generate a comprehensive model of the intracellular transport machinery at near-atomic resolution. The more established and older 3D reconstruction methods apply averaging of image elements to regenerate a 3D data set from many different projections of apparently identical particles. More recently, advances in image recording technology and the availability of dramatically increased computing power has initiated new approaches that now apply tomographic 3D reconstruction

0091-679X/07 $35.00
DOI: 10.1016/S0091-679X(06)79017-3

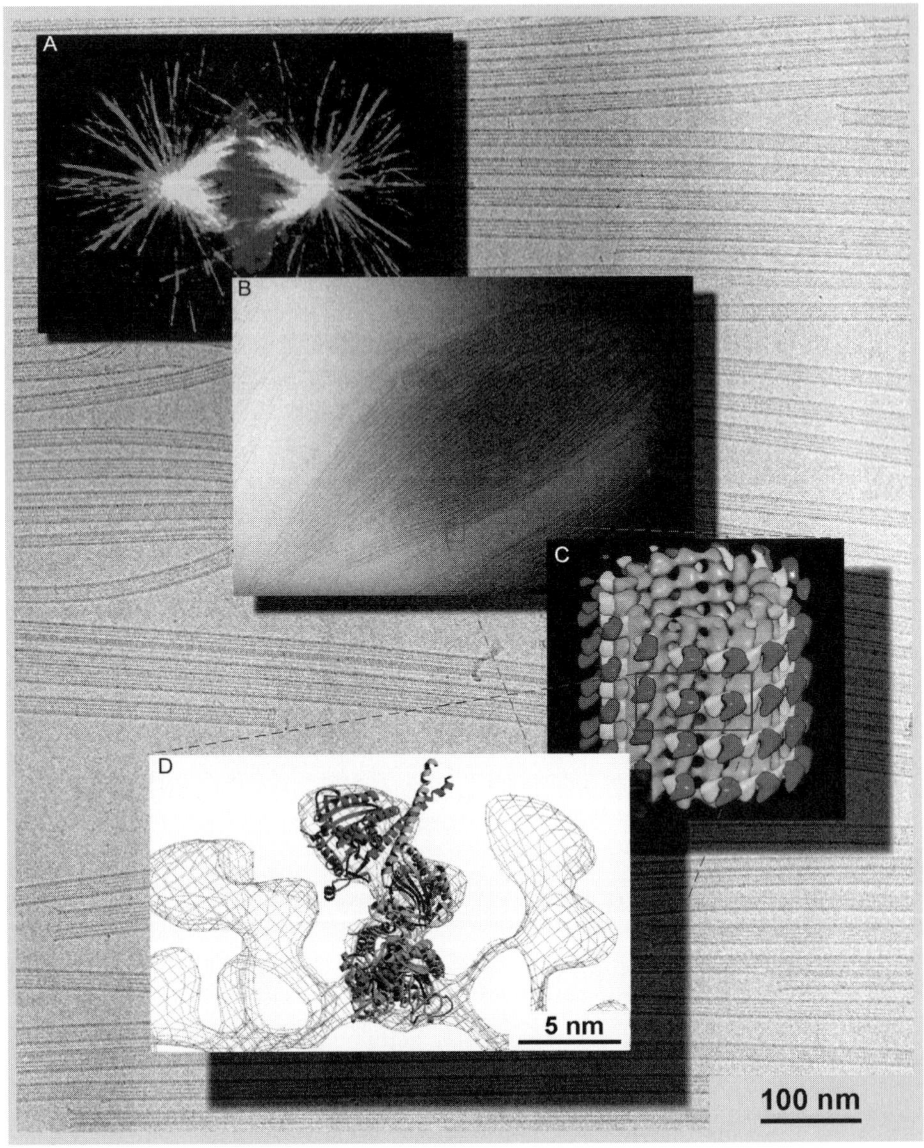

Fig. 1 Correlative aspects of structural techniques that are used to study microtubule-based motility, ranging from large complex assemblies to atomic detail. Rapid freezing preserves the details of biological structures. Cryo-EM can image them, and averaged images from cryo-micrographs enabled the alignment of atomic resolution data with structures that cannot be crystallized. The Background: cryo-EM image of microtubules. Insets from top to bottom: (A) Fluorescence image of a mitotic spindle obtained by light microscopy: Andrei Popov, EMBL-Heidelberg. Microtubules are detectable as green bars, but they are not resolved from one another in such a dense, fibrous array. (B) A minimal spindle arrangement of microtubules artificially induced by the spindle protein, TPX2 (Schatz *et al.*, 2003).

to essentially any type of molecular and cellular structure without the need for averaging. In both cases, 3D scaffolds are generated at resolutions between 1–3 nm on thin objects if averaging is applied, and 3–4 nm for thick-section tomography. These scaffolds allow the interpretation of molecular details by supplementing them with near-atomic structural data obtained by x-ray and electron crystallography or NMR spectroscopy.

I. Introduction

The majority of eukaryotic cells contain three filamentous cytoskeletal systems, namely actin filaments, microtubules, and intermediate filaments. Actin filaments and microtubules are polar structures, while intermediate filaments are apolar as a result of head-to-head assembly of their structural subunits. This lack of polarity is the reason why molecular motors travel exclusively along actin filaments and microtubules, not on the intermediate filaments. In this chapter, we will focus on structural and functional properties of the interactions between dynein and kinesin motor molecules and microtubules. In particular, this chapter describes three-dimensional (3D) cryo-electron microscopy (cryo-EM) methods that create structural data in a resolution range from 1 to 4 nm, which is suitable for further detailed interpretation in conjunction with x-ray and electron crystallography or NMR spectroscopy (Fig. 1).

Microtubules are highly dynamic tubular structures composed of $\alpha\beta$-tubulin dimers that polymerize head-to-tail into so-called protofilaments under the control of GTP. The $\alpha\beta$-tubulin structure has been solved to near-atomic resolution by electron crystallography (Löwe *et al.*, 2001; Nogales *et al.*, 1998). So far, the best resolved structure of an intact microtubule has been reported by Li *et al.* (2002; see also Krebs *et al.*, 2004). *In vivo*, protofilaments associate laterally and form a hollow tube. Most commonly, microtubules found *in vivo* are composed of 13 protofilaments (Fig. 2). These microtubules do not exhibit helical symmetry because they contain so-called seams that interrupt the lattice at one or more lateral contacts between protofilaments [visualized by Kikkawa *et al.*, 1994 (freeze-fracturing); Sosa and Milligan, 1996 (back-projection reconstruction)]. At the seam, protofilaments interact laterally in a so-called A-lattice conformation, while the rest of the tube forms a B-lattice (Amos and Klug, 1974; Song and Mandelkow, 1993, 1995). *In vitro* it is possible to form tubulin polymers that

(C) Helical 3D reconstruction of a motor–microtubule complex. (D) Molecular docking of x-ray data from Nogales *et al.* (1998) for tubulin and the x-ray data from the dimeric kinesin, ncd (Sablin *et al.*, 1998; Wendt *et al.*, 2002). Helical averaging is most successful on artificially reconstituted microtubule-motor complexes. Cryo-ET now creates new possibilities for the study of large, irregular assemblies that represent *in vivo* conditions more accurately (see Fig. 5). (See Color Insert.)

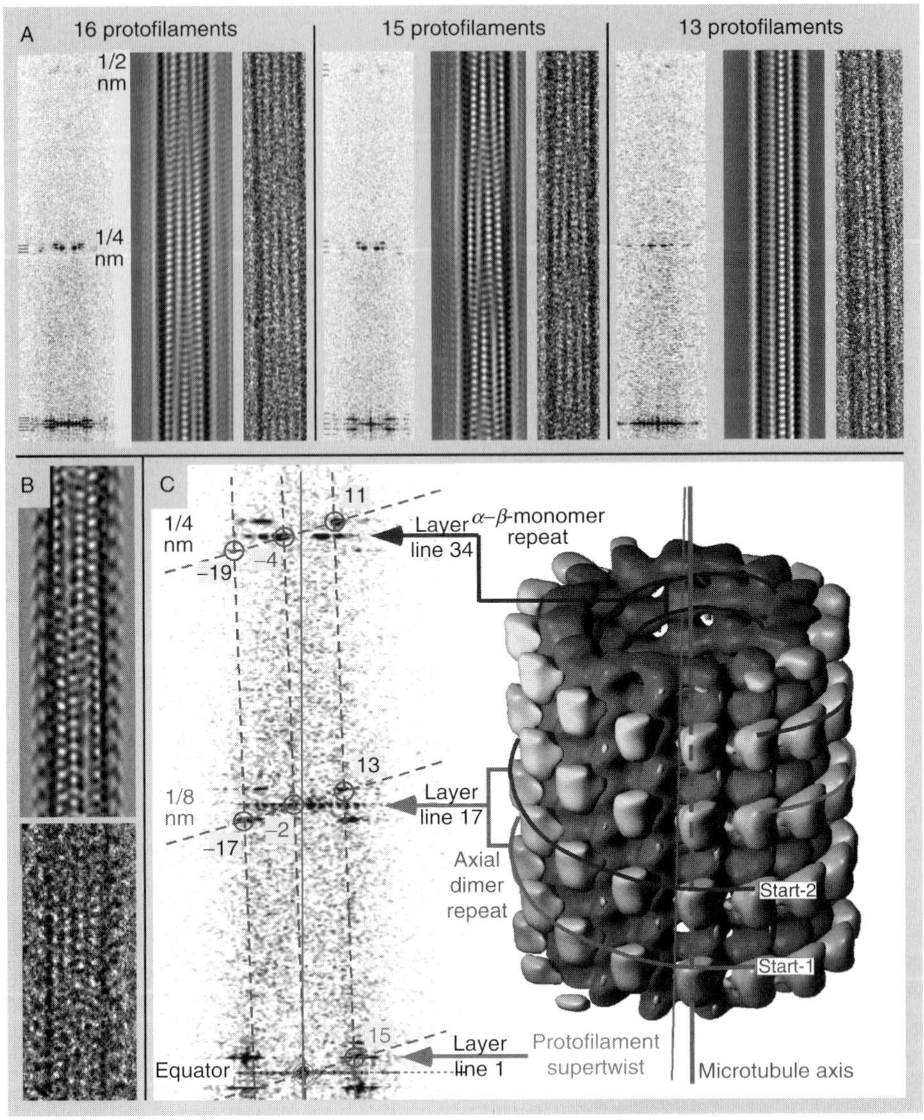

Fig. 2 Structural variations in microtubules and 3D reconstruction of microtubule–motor complexes by helical averaging. (A) Microtubules may be composed of different numbers of protofilaments. Examples shown here are composed of 16, 15, and 13 protofilaments (panels from the left). Each panel shows raw data image (right), a Fourier-filtered image (center), and the corresponding diffraction pattern (left). Only microtubules with 15 or 16 protofilaments are truly helical assemblies. Microtubules without any associated proteins show a distinct layer-line cluster at 1/4 nm that corresponds to the axial $\alpha-\beta-\alpha-\beta$ tubulin monomer repeat; weaker clusters of diffracted intensity that lie further out (i.e., 1/2 nm) are often directly visible. The increments between layer lines are given by the length of the protofilament supertwist illustrated in (C). The longer the supertwist the narrower the layer-line pattern (red lines at each cluster of the diffraction pattern: compare 16- with 15-protofilament microtubules). 13-protofilament

contain between 9 and ~16 protofilaments (Chrétien *et al.*, 1996). This variability in protofilament number can be controlled to some extent by modifying the tubulin polymerization buffers. Arnal *et al.* (1996) were the first to generate 15-protofilament microtubules with a right-handed protofilament supertwist (Fig. 2). These microtubules are truly helical structures and therefore suitable for 3D reconstruction based on helical averaging (Fig. 2; DeRosier and Moore, 1970). Since their introduction, they have been widely used for many structural studies of motors and microtubule-associated proteins (MAPs) as they interact with microtubules (Kinesins: Sosa *et al.*, 1997; Hoenger *et al.*, 1998, 2000a; Hirose *et al.*, 1998, 2000; Kikkawa *et al.*, 2001; Tau: Al-Bassam *et al.*, 2002; Kar *et al.*, 2003; Santarella *et al.*, 2004; Ncd-tail: Wendt *et al.*, 2003; Dynein microtubule-binding domain: Mizuno *et al.*, 2004).

Microtubule-based motility is driven by motor proteins of either the dynein family (Gibbons and Gibbons, 1972, 1973; Koonce *et al.*, 1992; reviewed in Vallee *et al.*, 2004; for the newest nomenclature on cytoplasmic dynein see Pfister *et al.*, 2005) or the kinesin superfamily (Hirokawa, 1998; for the newest nomenclature see Lawrence *et al.*, 2004). Both motor families interact with microtubules in an ATP-driven process, but the mechanisms of force transduction are quite different. Dyneins occur either in the cytoplasm, where they mediate retrograde microtubule transport, or as an integral structure of axonemes, where they generate the forces that drive flagellar motion. Dyneins are composed of several different polypeptides that form large complexes with one, two, or three motion-causing heads (reviewed in Vallee *et al.*, 2004 and Pfister *et al.*, 2006). The core element of all dyneins is an ~530-kDa heavy chain that includes 6 AAA-ATPase domains and a stalk domain that mediates the actual mechanochemical connection to the microtubule surface at roughly the same location as kinesin motor domains (Mizuno *et al.*, 2004). Dynein's tail domain interacts with cargo, which may be a vesicle, some other cytoplasmic cargo, or another microtubule. Dyneins range in molecular mass from ~700 kDa to well over 1 MDa, depending mostly on the number of heavy chains that they include but also on their content of both light (12–18 kDa) and intermediate (~40–150 kDa) chains.

---

microtubules lack a supertwist, so the layer-line clusters at the equator and at 1/4-nm overlay into a single layer line. (B) Raw (bottom) and filtered image (top) of a 15-protofilament microtubule decorated with kinesin motor domains at a 1:1 ratio between motors and tubulin dimers. The motor densities are visible at the outer microtubule surface as little blobs that protrude outward from the microtubule surface. (C) Layer-line indexing on a diffraction pattern from a motor-decorated 15-protofilament microtubule. Due to the added motor density at each tubulin dimer, a strong layer-line cluster shows up at 1/8 nm, enhancing the structural features of the $\alpha\beta$-tubulin dimer along the protofilament axis (red arrow in 3D model). This helical path actually forms a two-start helix with two separate, left-handed helical paths (red and burgundy lines drawn on the model of the microtubule); therefore the corresponding Bessel order is −2. The protofilament supertwist creates layer line 1 in close proximity to the equator due to the reciprocal principles of Fourier space (purple line = Bessel order + 15). The microtubule axis (green line in model) is parallel to the meridian (green line in the diffraction image). Blue dashed lines mark one of the two lattices that correspond to the so-called near and far side of the microtubule surface.

Kinesins are much smaller than dyneins, but they are structurally and functionally more diverse. Depending on their location and composition, kinesin family members interact with a myriad of different cargo structures, which range from little vesicles to microtubules and other cellular components to entire mitochondria. The majority of kinesins are microtubule plus-end-directed dimers that form complexes between so-called "heavy chains," and some of them interact with "light chains" that mediate cargo binding. The heavy chains include a conserved motor domain (~350 residues), which contains the ATPase pocket and microtubule-binding site; this is commonly located at the N-terminal end of the polypeptide chain. Near to the motor domain, the primary structure of most kinesins includes a segment of heptad repeats that predicts the presence of a coiled coil. This leads many kinesin heavy chains to dimerize, a feature that is important for their mechanism of motility (see below). These dimers either interact directly with cargo or they bind additional polypeptides, the light chains that can contribute either to cargo binding or to regulation of the kinesin motor domain. Commonly, these functions are mediated by the C-terminal part of the complex. There are, however, many exceptions, such as kinesin-14 members, where the head domain locates to the C-terminal end of the dimeric complex (e.g., Ncd: Chandra *et al.*, 1993; McDonald *et al.*, 1990; Walker *et al.*, 1990), kinesin-2 members that form heterodimeric complexes (Scholey, 1996), and kinesin-3 members that operate as monomers (e.g., KIF1A: Kikkawa *et al.*, 2001), or in transient states between monomeric and dimeric forms (e.g., Unc104: Al-Bassam *et al.*, 2003).

II. Rationale

This chapter focuses on structural examinations of kinesins and dyneins performed by cryo-EM and 3D image analysis. Typically, these 3D EM data have been interpreted using near-atomic resolution x-ray structures (Fig. 1), as well as biochemical and biophysical measurements of various kinds. Generally, EM is an ideal tool with which to integrate structural data into cellular and physiological studies, as it bridges nicely the resolution gap between near-atomic x-ray crystallography or NMR spectroscopy data and light microscopy. 3D data from cryo-EM at sufficient resolution can provide molecular detail about a complex tertiary arrangement that is not suited for 3D crystallization (e.g., entire microtubules) or too large for NMR spectroscopy. The interpretation of cryo-EM 3D data by docking near-atomic resolution components into an EM-derived scaffold has evolved over the past decade or so from an often quite ambiguous, hand-guided process to a more quantitative discipline supported by several refined computational approaches (Rossmann *et al.*, 1994; Volkmann and Hanein, 1999; Wriggers *et al.*, 1999). In the kinesin field, the resulting images have provided important insights into the behavior of kinesin motor domains under different nucleotide conditions. Molecular docking into EM-scaffolds has allowed the reconstruction of both processive and nonprocessive walking processes by monomeric (Kikkawa *et al.*, 2001) and dimeric kinesins (Hoenger *et al.*, 2000a;

Rice *et al.*, 1999; Skiniotis *et al.*, 2003; Wendt *et al.*, 2002), as well as the determination of the oligomerization properties of transiently monomeric motors, such as Unc104 (Al-Bassam *et al.*, 2003), or the microtubule-depolymerization activities of KinI (Moores *et al.*, 2002).

So far, most cryo-EM 3D data about microtubules, kinesins, and dyneins have been obtained from structures that either exhibit helical symmetry or form complexes that are large but structurally integer, and are therefore suitable for single-particle, 3D analysis. While kinesin head domains have been approached mostly using helical methods to study the motor while it is complexed to helical microtubules (Fig. 2), the bulk structure of dynein has been investigated by single-particle reconstruction (Burgess *et al.*, 2003; Samso and Koonce, 2004). Now, however, we have a third technique available, that of cryo-electron tomography (cryo-ET), which allows the reconstruction of polymorphic and large complexes such as the dyneins found in axonemes (Fig. 5). Cryo-ET circumvents averaging procedures by collecting tilt series from individual structures, so these structures do not need to be reproducible in size and shape. Hence, cryo-ET opens an avenue toward 3D reconstruction of essentially any cellular component, providing it complies with the requirements of image acquisition in an electron microscope. In addition, when repetitive particles are present, cryo-ET can be combined with 3D volume averaging to increase the resolving power of this imaging technique (Fig. 5).

By now there is a large amount of cryo-EM 3D data available that concentrates on kinesin motor domains complexed with microtubules (initiated by Kikkawa *et al.*, 1995) or tubulin sheets (Hoenger and Milligan, 1997; Hoenger *et al.*, 1995). The small size of even full-sized kinesins essentially precludes cryo-EM studies on frozen-hydrated particles in isolation. Therefore, structural data on full-sized kinesin complexes have mostly been obtained by low-angle rotary shadowing that enhances the low-resolution features of such a molecule by depositing large metal clusters along the outer molecule boundary. Such work does not, however, reveal any high-resolution, 3D structural details. Due to their structural variability and conformational flexibility, few structural data are available to date on kinesin tail domains. So far, single events of kinesin–tubulin interactions (i.e., elucidating conformational details independent from averaging procedures) have been obtained most reliably with high-resolution, unidirectional shadowing (Fig. 3; Krzysiak *et al.*, 2006). However, recent progress in cryo-ET is promising and should provide further insight into the structural properties of entire kinesin molecules from work under (or close to) native conditions (Fig. 4).

III. Methods

The first 3D reconstructions of microtubule protofilaments were achieved by Amos and Baker (1979) and Tamm *et al.* (1979). Amos and Baker (1979) proposed a 3D microtubule structure that compares fairly well with more recent helical reconstructions obtained from cryo-EM data, although these early reconstructions were

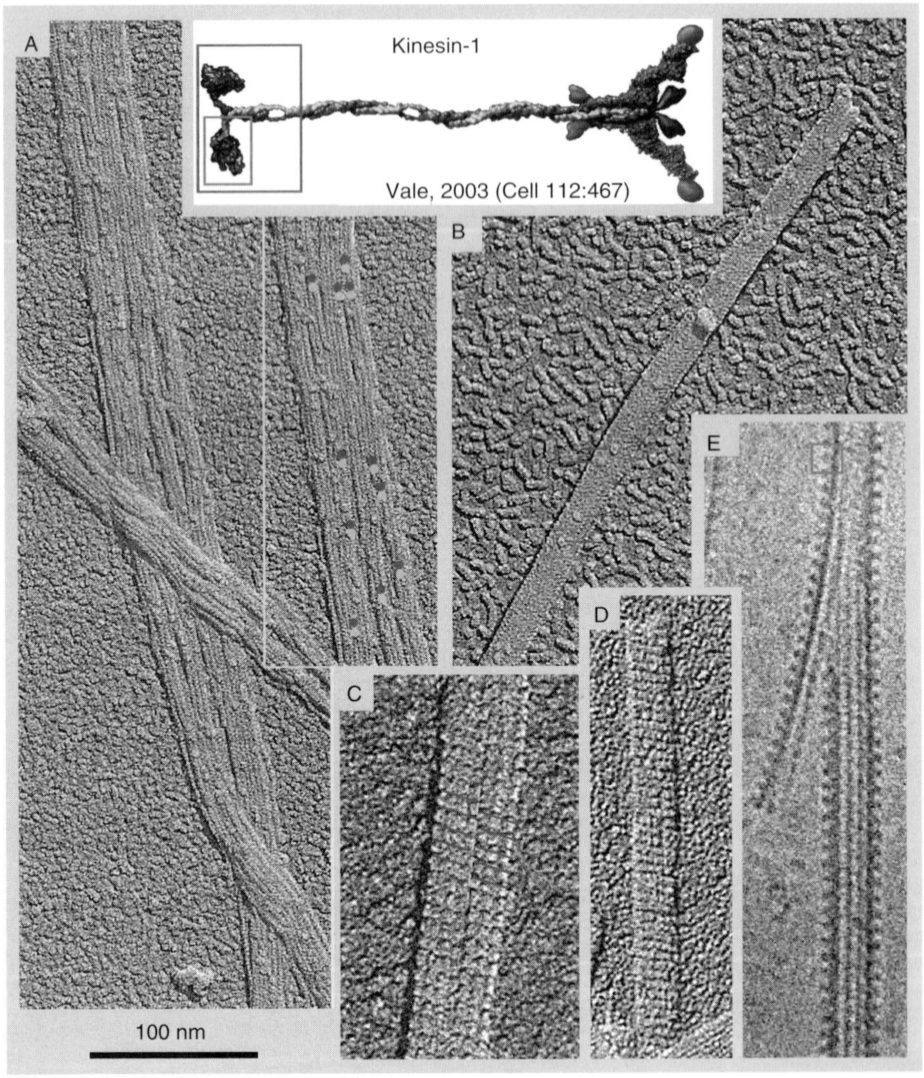

Fig. 3 High-resolution surface shadowing of microtubule–kinesin complexes. High-resolution, unidirectional metal shadowing is a powerful structural method that is complementary to cryo-EM and cryo-ET. Depending on the length of the genetic constructs used to express pieces of the kinesin-1 motor domain, the resulting polypeptides either remain monomeric (blue box on the model of kinesin's structure) or, if sufficient neck-region is left, dimerize via a coiled-coil interaction (red box). (A) Surface shadowing of microtubules complexed with substoichiometric amounts of dimeric Eg5 (Krzysiak *et al.*, 2006). The dimers are readily visible on the tubulin polymer's surface as little double dots. The area in the red-dotted frame in Fig. 3A is repeated as an inset with the dimers highlighted with red and yellow markers. (B) Microtubules complexed with substoichiometric amounts of a shorter Eg5 construct, illustrating the monomeric nature of this construct. (C) Codecoration of the *Schizosaccharomyces*

performed on negatively stained, zinc-induced two-dimensional (2D) layers of tubulin. Here we will focus on methods suitable for cryo-EM data obtained with intact microtubules. Hence, this chapter will describe the most popular 3D reconstruction methods that are currently used for the analysis of cryo-EM data on microtubules and their associated proteins with a particular emphasis on helical approaches. Typically, microtubule–kinesin motor domain complexes are investigated by helical 3D reconstruction and image averaging (Fig. 2). This approach

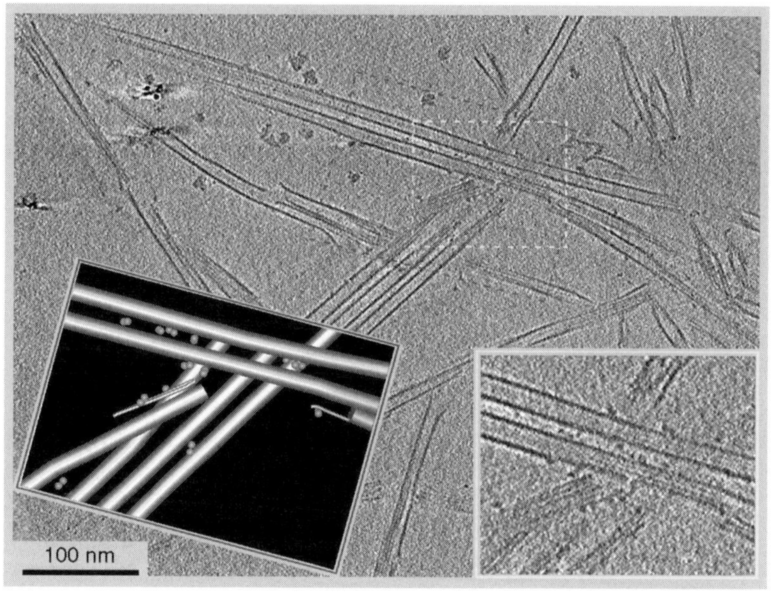

Fig. 4 Microtubule bundling by the fission yeast kinesin-like protein, Klp6p: motor-domain plus-neck (Klp6MDN) visualized by cryo-ET. 3D reconstructions of rapidly frozen specimens show that Klp6MDN forms dimers and can link taxol-stabilized microtubules into bundles *in vitro*. In the depicted tomographic slice bundles of two (running almost horizontally) and three microtubules, respectively, are visible. The distance between linked microtubules ranges from 10 to 13 nm and diagonal linkers that bridge between neighboring microtubules are visible. Right inset: Tomographic slice showing a higher magnification of several microtubules and the Klp6MDN particles, visible as small, dark disks (~4 nm in diameter). Left inset: 3D graphic model of a tomographic subvolume showing microtubules and a few single protofilaments and Klp6MDN molecules. Note that most of the motor proteins are arranged in pairs, indicating that they form dimers under these conditions. Modified after Fiedler *et al.*, in preparation.

---

pombe EB1-homologue Mal3p with kinesin motor domains makes it possible to visualize microtubule lattice seams directly (Sandblad *et al.*, 2006). (D) In the absence of Mal3p, the lattice seams are the least stable lateral interaction in a microtubule; the polymers typically collapse along these seams. (E) Same specimen as in (D), but imaged by cryo-EM; note that the previously described features remain invisible in cryo-EM images. (From Vale, 2003 with permission from Elsevier.) (See Color Insert.)

is based on a helical microtubule template that is only possible with some specific microtubule configurations (see above). However, averaging methods are now complemented with cryo-ET, which is becoming increasingly powerful (Fig. 4). Microtubules complexed with nonmotor MAPs have been reconstructed with helical methods, but often with somewhat limited success as these molecules exhibit a high degree of structural flexibility. In our laboratory, we often complement these data with surface structural data that are generated by high-resolution metal shadowing (HRMS) (Fig. 3; Santarella *et al.*, 2004). Sosa and Milligan (1996) developed a 3D reconstruction method for microtubule–kinesin complexes that is based on weighted back projection. This method was developed as a solution for work with nonhelical microtubules. Larger complexes, such as dynein motors, are not suitable for helical methods in spite of their binding affinity for microtubules (except for their microtubule-binding domain; Mizuno *et al.*, 2004); no one has yet achieved stoichiometric decoration with these complexes, an arrangement that would follow the helical template. However, dynein is large enough for single-particle averaging approaches (Burgess *et al.*, 2003; Samso and Koonce, 2004). Furthermore, dynein complexes and even larger structures, for example entire axonemes, have now been successfully reconstructed by cryo-ET to about 4-nm resolution (Fig. 5; Nicastro *et al.*, 2005, 2006).

A. Helical 3D Reconstruction

Helical arrays are either naturally occurring or sometimes artificially produced repetitive arrangements of macromolecules or complexes of macromolecules; they are often of great use to structural biologists. Helical reconstruction approaches have yielded near-atomic resolution on either naturally occurring helices, such as bacterial flagella (Yonekura *et al.*, 2003), or artificially reconstituted ones such as tubes formed by the acetylcholine receptor (Miyazawa *et al.*, 2003). Under certain conditions, as outlined in the following sections, microtubules may also form true helices, but often their lattices show one or more discontinuities by so-called seams (Fig. 3).

1. Special Properties of Helical Reconstruction

Helical reconstruction methods combine features of tilt-series reconstruction from 2D crystalline arrays with methods used for icosahedral and single-particle reconstruction (DeRosier and Moore, 1970). The regularity of the helical packing generates distinct layer-line patterns in Fourier space that can be analyzed by methods resembling those used for crystallography. These methods take advantage of the discrete layer-line spacing that reflect the axial repeats of helical turns, analogous to the patterns of distinct spots generated by 2D crystals. The rotation along a helical path projects the repeating building blocks at all the angles that are required for an isotropic 3D reconstruction. In the real part of Fourier space (the visible diffraction pattern: Fig. 2A left panels and C) this information is reflected in the appearance of continuous layer lines, running perpendicular to the helix

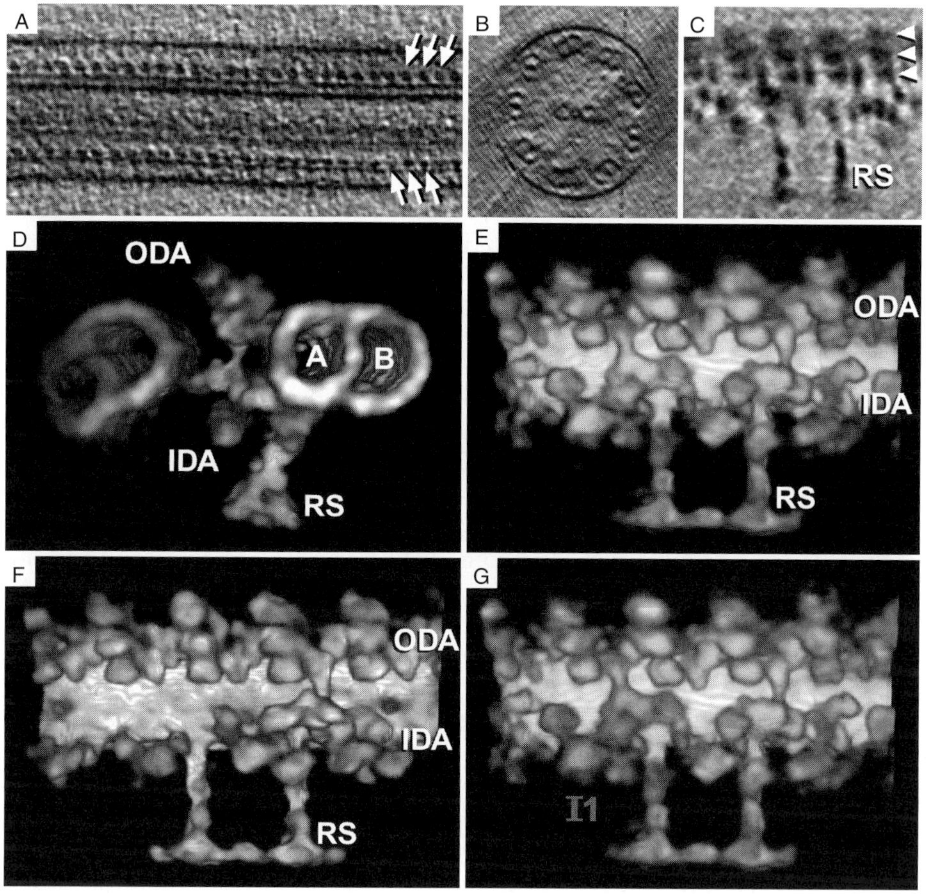

Fig. 5 Cryo-ET and volume averaging of eukaryotic flagella. (A and B) Longitudinal (A) and cross-sectional (B) tomographic slices of a sea urchin sperm flagellum; the dotted line in (B) indicates the location of the slice through the tomogram shown in (A); in (A) note the motor domains in the outer dynein arm rows on both sides of the axoneme (white arrows). (C) Tomographic slice showing the proteins associated with one microtubule doublet from a *Chlamydomonas* axoneme in longitudinal orientation viewed from the B-tubule of the adjacent doublet; average of 150-axonemal repeats. Note the three motor domains of the outer dynein arms in this species (white arrows) and the radial spokes (RS). (D and E) Volume-rendered representations that show an average of a *Chlamydomonas* doublet microtubule, which is comprised of an A-tubule and a B-tubule, seen in cross-sectional (D) and longitudinal (E) orientation; note the many microtubule-associated structures, such as inner and outer dynein arms (IDA, ODA), and the RS. (F) As in (E), but the image shows an averaged microtubule doublet from a strain that carries a loss of function allele of the pf9 gene, which encodes a heavy chain for an inner dynein arm Comparisons between (E) and (F) show clear differences: the inner dynein arm complex I1 (G) is missing in the mutant (F). Modified after Nicastro *et al.* (2006).

axis, rather than rows of distinct spots, as in 2D crystal diffraction patterns. A 2D projection of a helix is a one-dimensional (1D) crystal that features repeating elements along the helical axis, but perpendicular to the axis the image features are continuous, so in Fourier space the information that corresponds to these features is not sampled in discrete spots. Reaching isotropic resolution requires projecting an image element at small intervals between 0° and 180°. Hence, the projected image of a helical array that includes at least one-half of a helical turn is theoretically sufficient for two independent isotropic helical 3D reconstructions. In practice, however, the low contrast in cryo-EM micrographs requires that several helical repeats must be averaged together to increase the signal-to-noise ratio to a useful level.

2. Specific Difficulties Associated with Helical Reconstruction Methods

a. Inherent to all procedures that rely on improving the signal-to-noise ratio by averaging over repetitive image elements and imposing an assumed symmetry on a structure, helical reconstruction depends on the structural integrity and precise registration of such image elements. In the case of a microtubule decorated with monomeric motor head domains, the repetitive element is one motor domain complexed to one tubulin dimer. When dimeric motor constructs are used, the repeating element may be two motor heads complexed to two tubulin dimers. This is often problematic because the tubulin dimers do not bind to each other in perfect register. In cases where decoration of the helical template is incomplete, the results are often barely interpretable.

b. Solely based on cryo-EM projection data, it is often difficult to resolve the helical parameters of an unknown structure. For example, helices may be left- or right-handed, and cryo-EM projections do not reveal that handedness unless the helix is tilted perpendicular to its axis. Therefore, in our laboratory we collect additional data by complementing techniques to determine crucial parameters such as helical handedness and the number of starts in a helix. One of the safest approaches to determine handedness is surface metal shadowing. A helical assembly may combine several identical helices intertwined with each other: for example, actin filaments can be described as two-start helices with two identical strands of actin molecules intertwined. Each perpendicular plane through the helix produces a density distribution according to a point-symmetry through the helix axis. Only then are the symmetry requirements of a two-start helix met. Accordingly multiple-start helices of any number must comply with this requirement, according to the number of starts. In contrast, the symmetry of double-stranded DNA, despite its name, creates a one-start helix, forming major and minor grooves. Only if these grooves were identical one could speak of a two-stranded helix. DNA differs from a true two-start helix because each of its strands is polar and they are arranged antiparallel to one another.

c. Helical arrays are often quite flexible. Even tubes with a large diameter tend to bend and the wider a tube the more severe are the flattening effects. While an actin filament behaves almost like a string of pearls, tubular assemblies like microtubules may be distorted in ways that are often difficult to handle. For example, a straight tube shows a perfectly circular cross section, while a bent tube shows an elliptic cross section; in the latter case, the layer lines smear out at higher spatial frequencies. In addition, tubes may tilt out of the plane perpendicular to the beam axis, and this distorts the sinus-shaped wave character of a projected helix.

3. Geometric Properties of 15-Protofilament Microtubules

As mentioned above, microtubules are not necessarily helical. Chrétien and Wade (1991) investigated the microtubule lattice with respect to protofilament number and their probability of forming seams. They found that either microtubules with 9–11 or >15 protofilaments are the most likely ones to exhibit a truly helical lattice. Kikkawa *et al.* (1995) were the first to reconstruct a kinesin-decorated 10-protofilament microtubule. However, for all practical purposes, 10-protofilament microtubules proved unreliable, occasionally exhibiting multiple seams (Dias and Milligan, 1999). Hence, the community of scientists using microtubules as helical supports for the study of associating proteins has switched to 15-protofilament microtubules (Arnal *et al.*, 1996).

Detailed protocols describing the preparation of 15-protofilament microtubules, quick-freezing, and recording in an electron microscope have been published previously in Beuron and Hoenger (2001). Here is a brief summary: The fraction of microtubules that contain 15 protofilament after polymerization *in vitro* is enhanced by adding ~5% dimethyl sulfoxide (DMSO) to the polymerization buffer (BRB80) that typically consists of 80-mM PIPES pH 6.8, 2-mM GTP, and 100-mM NaCl. Under these conditions, we consistently obtain between 5% and 10% 15-protofilament microtubules. Figure 2A compares projection images of 13-, 15-, and 16-protofilament microtubules. The summing of densities from the top and bottom half of the tube in 2D projections is responsible for the Moiré patterns observed (Fig. 2A and B; Chrétien *et al.*, 1992; Mandelkow and Mandelkow, 1985; Mandelkow *et al.*, 1986; Sosa and Chrétien, 1998). 13-protofilament microtubules are recognizable by the straight axial alignment of their protofilaments. Any other number of protofilaments results in an axial protofilament supertwist. Generally, projections of microtubules with odd numbers of protofilament show a left–right asymmetry about their axis, while even numbers of protofilaments create a symmetrical left–right pattern (see Fig. 2A). In all cases, the Moiré pattern is specific for the protofilament number and handedness of the microtubule; as such it provides a very practical tool for the identification of different microtubules. It may even be used for polarity determinations (Chrétien *et al.*, 1992; Sosa and Chrétien, 1998). In 15-protofilament microtubules this supertwist is normally right-handed.

a. Indexing Helical Diffraction Patterns: Axial Supertwists of Protofilaments

In supertwisted microtubules, the protofilament supertwist is the longest-pitched helix and is therefore responsible for the layer-line increment in diffraction patterns (Fig. 2B). For those unfamiliar with diffraction and the Fourier Transform, which is the mathematics most useful for understanding it, the following brief descriptions may be useful. The object under study, for example a motor domain bound to a tubulin dimer, is an asymmetric structure that can be described by a continuous function in 3D space that defines where all its component structures lie. The Fourier transform of such continuous information is also continuous, although its dimensions are "reciprocal" in the sense that big features in real space correspond to details in "Fourier space" and vice versa. Regularly repeated structures, however, generate a Fourier transform that is sampled only discontinuously because destructive and constructive interferences of phases that occur repetitively in the regular array in real space. Helices repeat themselves along their axis (series of identical helical pitches), but helices show continuous image features perpendicular to their axis. These geometrical facts lead to the formation of sampled lines in the diffraction pattern (called layer lines). These are periodically repeated along the axis in Fourier space due to the repetitive nature of helical turns, but the lines themselves are continuous, as there is no truly repetitive structure perpendicular to the helical axis. The layer line formed by the supertwist represents layer line 1 (right above the equator), marked with the number 15 in Fig. 2C. This number, and others that appear below, derive from the different orders of "Bessel functions," which are quasiperiodic functions that behave a bit like sine and cosine functions wrapped around a cylinder. These enter the picture because they best describe the Fourier transform of cylindrical objects, including helices. The "Bessel order" indicates the numbers of parallel starts of that particular helix, and since there are 15 protofilaments, the corresponding Bessel order is 15, as each individual helix follows one protofilament in this case. The Bessel order is (+) for right-handed and (−) for left-handed helical paths. In helical 15-protofilament microtubules (Fig. 2A center, B), the supertwist has to be right-handed (Fig. 2C). The supertwist of 15-protofilament microtubules can also be left-handed, but these microtubules are not helical (see Sosa *et al.*, 1997). The longer the supertwist, the closer its corresponding layer line is to the equator (compare 15- and 16-protofilament microtubules in Fig. 2A), and the finer the axial layer-line sampling will be. If the supertwist is indefinitely long, that is nonexistent as in 13-protofilament microtubules, that layer line coincides with the equator (Fig. 2A left panel), and the layer-line spacing in the diffraction pattern is determined by the next shortest helix, as described in the following section.

b. Indexing: Short-Pitched Helices

Correctly indexing a helical diffraction pattern is crucial for a successful 3D reconstruction by helical methods. If the helical parameters such as handedness and Bessel order are unknown, extracting the 3D information from a 2D projection according to the object's helical symmetry is impossible. However, even if these parameters are not retrievable for some reason, 3D information may still be extracted from helices by a process called "iterative helical real space reconstruction" (IHRSR: Egelman, 2000; see also Chapter 23 by Sosinsky *et al.*, this volume).

Motor-decorated, 15-protofilament right-handed microtubules show another strong helical feature, and that is the helical path defined by the B-lattice arrangement of the tubulin dimers (Amos and Klug, 1974; Song and Mandelkow, 1993). The B-lattice helix is a left handed, two-start helix that is formed by a lateral staggering of about 0.9 nm between each dimer and the one to its left (looking in toward the helix axis). The elements of this two-start helix are marked with red and burgundy lines in Fig. 2C. Diffraction from this helix gives rise to a cluster of layer lines (Fig. 2C) at 1/8 nm from the origin in Fourier space (corresponding to the 8-nm axial repeat of motor–tubulin complexes along the microtubule axis). On cryo-EM projection images, as shown in Fig. 2B, the left-handedness of the helix is not directly visible, and neither is the protofilament supertwist. However, high-resolution surface shadowing reveals the short-pitched handedness in a very straightforward way (striations that are almost perpendicular to the polymer axis in Fig. 3C and D). The hand of the supertwist in a given frozen-hydrated microtubule can, however, be determined from the diffraction pattern of a cryo-EM image, according to the geometrical relationships of the layer lines to each other. A helical diffraction pattern always shows two overlaid lattices, one from the so-called near side (the upper half of the helical tube) and one from the far side (the lower half). Both constitute two individual and complete 3D data sets. One of these lattices is marked with blue dashed lines on the diffraction pattern in Fig. 2C. For each individual layer line, reflections from the near and far sides of the microtubule correspond to diffracted intensities lying on either the left or the right side of the meridian. If the left-handed helical paths are on the left side (i.e., −2, −4 in Fig. 2C), the corresponding right-handed paths will be on the right (+15 in Fig. 2C) and vice versa. Layer lines are mirror-symmetric to the meridian with regards to intensity of diffraction (=[Fourier amplitudes]^2), but phases are only identical on even Bessel orders. They differ by 180° for odd Bessel orders.

The procedure for indexing a helical lattice now works as follows: The higher a Bessel order, the further its peaks lie from the meridian. From the number of protofilaments (15) we know that the Bessel order of the supertwist peak has to be +15 or −15. By metal-shadowing experiments (Fig. 3D) we have previously determined the Bessel order of the short-pitched helix in the 1/8-nm cluster as −2 (left-handed two-start helix, see Fig. 2C). Unlike the supertwist, the short-pitched helices do not change hand according to protofilament number. Hence their handedness (but not necessarily their Bessel order) applies for every microtubule. The lattice tells us that the corresponding supertwist peak should locate to the other side of the meridian. That makes the handedness sign to be opposite of that for the −2 helix. Hence, the Bessel order of the supertwist has to be +15. With this knowledge we can now calculate the Bessel orders of the remaining layer lines to complete the indexing (start counting along a lattice line from 0 on the meridian outward: if minus is left, plus must be right and vice versa). Accordingly, the layer-line cluster at 1/4 nm corresponds to individual α- and β-tubulin monomers, as if they were structurally identical. In fact, at this resolution the structural features of α- and β-tubulin are not identical, but they are indeed very similar (Krebs *et al.*, 2004); this is why the 4-nm repeat shows up as such a strong feature of the diffraction pattern. Hence, if the dimer forms a two-start helix, the monomers (which are nearly identical) form a four-start helix (as there are two for each dimer); therefore, the Bessel order that is twice as far out in reciprocal space has to be -4. The other layer lines in the two clusters are convolutions of the short-pitched helix with the supertwist, and the indexing works accordingly (i.e., $-2 \pm 15 = -17$ and 13).

4. Software for Helical 3D Reconstruction

To date, there are several helical-processing packages available as public domain software to academic institutions. The package we have routinely used is called PHOELIX (Carragher *et al.*, 1996; Whittaker *et al.*, 1995). This package combines routines from SUPRIM (Schroeter and Bretaudiere, 1996) and MRC (DeRosier and Moore, 1970; Moore *et al.*, 1970). Probably the most versatile package for helical reconstruction is the so-called Brandeis package (Morgan and DeRosier, 1992) that has recently been revised and made available for several computer platforms.

B. Back Projection

Back projection is a generally applicable approach for restoring the 3D information of an object from a set of 2D projections, provided the angular relations between the projections are known (Rademacher, 1992). This is the basic principle of single-particle 3D reconstruction and of tomography. The back-projection approach described here has been designed specifically for nonhelical microtubules complexed with kinesin motor domains. Motivated by the apparent symmetry problems that result from microtubule lattice seams, Sosa and Milligan (1996) and Metoz *et al.* (1997) developed a 3D reconstruction procedure for microtubule–motor complexes, which was also named "tomography without tilt" (Metoz *et al.*, 1997; see also Wade *et al.*, 1995). The method has some limitations, and the discovery of 15-protofilament microtubules diminished its importance for the field. However, back projection has a few particular features that are nevertheless very attractive for structural investigations into microtubule–motor complexes. This is the reason why we believe it is important to mention this technique here.

For specimens that have highly variable helical parameters, the classical Fourier–Bessel algorithm is either difficult or impossible to implement. In these cases, several approaches based in real space reconstruction or back projection have been implemented (Bluemke *et al.*, 1988; Egelman, 2000; Li *et al.*, 2002; Metoz *et al.*, 1997; Sosa and Milligan, 1996). If the helical parameters of a specimen are known (radius, axial rise, and rotation per subunit), then it is possible to obtain a 3D reconstruction by back-projecting slices taken at each subunit, given its axial position and angle, according to helical symmetry. In principle, a 3D reconstruction made by using either the back projection or the Fourier–Bessel algorithm should give equivalent results. One advantage of the back-projection approach is that the algorithm makes it relatively easy to account for changes in helical parameters along the filament. Another advantage is that it is also relatively easy to redefine the asymmetric unit to avoid averaging subunits that may not be equivalent.

Back projection as used in the context of microtubules and associated proteins is the only averaging-based 3D reconstruction method that allows for a 3D analysis of microtubules that exhibit one or more lattice seams. However, it only works on microtubules with supertwisted protofilaments and cannot be applied to the most commonly occurring 13-protofilament microtubules. A complete reconstruction with isotropic resolution requires a microtubule whose image includes sufficient length to provide a 180° rotation of its supertwist. On a right-handed 15-protofilament microtubule the required minimum length is about 800 nm. On 14-protofilament microtubules a 180° turn of the supertwist is almost 2-μm long, which is the major drawback of this approach for practical applications. At ~30,000× magnification, this length requires the width of an entire photographic plate. On top of that, microtubules polymerized *in vitro* are often not structurally well preserved over such a long distance. Alternatively, they often exhibit one or more internal distortions; these can easily be eliminated in helical reconstructions, but they disrupt the back-projection approach. The repetitive structural feature used for back projection is not the $\alpha\beta$-tubulin–motor complex as for helical analysis, but an entire 360° disk or slice of a microtubule with a minimum length of one $\alpha\beta$-tubulin dimer, or any multiple of this. The shorter the disk, the more effective the averaging will be, and the finer the rotation increments for the back projection will be. As an example: a 180° right-handed supertwist in a 15-protofilament is about 1200 nm. With 8-nm intervals (the length of a tubulin dimer) that amounts to ~150 disks and each disk rotates about 1.2° relative to the previous one. With a disk the size of a four-dimer interval, each rotation would be four times that (=4.8°) and there would be four times fewer units to be averaged. Helical and back-projection reconstructions have been directly compared in Sosa *et al.* (1997).

C. High-Resolution Metal Shadowing (HRMS)

Metal shadowing has long been used to see the overall shape of macromolecules. It has been used to study the tertiary structure of kinesin motors (Hirokawa, 1982). The surface of microtubules has been investigated by freeze-fracturing methods, showing for the first time the characteristics of both inner and outer microtubule surfaces without the need for averaging (Heuser, 1986). Similar methods have been used to look at microtubules decorated with kinesin motor domains stoichiometrically bound to the tubulin dimers (Kikkawa *et al.*, 1994). These beautiful images revealed the lattice seams of normal microtubules.

Our laboratory has used a different high-resolution shadowing approach on freeze-dried specimens. HRMS is a preparative method that falls into the general category of cryomethods for 3D reconstruction, and it delivers structural data that are complementary to 3D cryo-EM and cryo-ET. My laboratory has a long-standing collaboration with Dr. H. Gross at the ETH of Zuerich, and this collaboration has produced a series of publications that illustrate the power of HRMS for microtubules and microtubules complexed with kinesin motor domains or nonmotor MAPs (Hoenger *et al.*, 2000b; Krzysiak *et al.*, 2006; Santarella *et al.*, 2004; Fig. 3). HRMS visualizes exclusively the surface features of large macromolecular assemblies or cellular organelles. It delivers a very high signal-to-noise ratio that allows direct interpretations of 2- to 3-nm details without further image enhancements or averaging. The restriction to surface feature image information enhances structures that are otherwise difficult to detect with cryo-EM, such as individual, stochastically dispersed monomeric and dimeric motor constructs on a microtubule surface (Fig. 3A dimeric Eg5; Fig. 3B monomeric Eg5), or microtubule lattice seams on motor saturated tubes (Fig. 3C). In the form described here, HRMS is not common in the EM community, and its potential is generally underestimated and not fully recognized. We regularly use surface shadowing as a method that is complementary to 3D reconstruction. Shadowing allows direct visualization of helical lattice parameters such as handedness and sometimes Bessel orders. These parameters would be much harder to obtain from cryo-EM micrographs (compare Fig. 3D with E).

The initial step of specimen preparation for shadowing is comparable to that of cryo-EM. Specimens are quick-frozen in liquid ethane or nitrogen. However, this sample is then allowed to freeze-dry at $-80\,^{\circ}$C and an ambient pressure of ca. $\leq 10^{-5}$ Pa for $\sim$90 min. This sublimates the ice and exposes the outer surface of the specimen's component macromolecules. Freeze-drying does not flatten internal protein structures, but, as shown in Fig. 3, weak cylindrical structures like microtubules flatten during this process. HRMS then applies a fine layer of tantalum/tungsten (Ta/W). This alloy reproduces the surface much better than platinum/carbon (Pt/C), which is routinely used for freeze-fractured specimens. However, Ta/W oxidizes quickly under atmospheric conditions; with this method it is therefore necessary to use a vacuum/cryo-transfer system. Unlike cryo-EM that generates a "see through" 2D projection of an object, shadowing exclusively reproduces the outer surface to a resolution of $\sim$2 nm (and better if surface reconstruction methods are applied; Walz *et al.*, 1996, 1.06 nm resolution). These details are shown at a very high signal-to-noise ratio, while internal densities reproduce at a much lower contrast and are therefore not relevant for image formation.

D. 3D Reconstructions from Single Particles

3D analysis of macromolecular complexes by single-particle averaging is the most generally applicable averaging-based 3D reconstruction method for EM data (Frank, 2002; Frank and van Heel, 1982; Van Heel, 1984; Van Heel and Frank, 1981). As expressed in its name, single-particle averaging does not rely on regular arrangements such as helices or 2D crystalline arrays. However, due to the low contrast in cryo-EM it is limited to relatively large particles, typically with a molecular weight >300 kDa; therefore, single-particle approaches are not applicable to structural investigations of kinesin motors. Single-particle reconstruction has, however, been used in two cases for structural studies on microtubules and molecular motors: (1) for a high-resolution map of microtubules (Li *et al.*, 2002) and (2) for the structure of dynein motor domains (most recent works: Burgess *et al.*, 2003; Samso and Koonce, 2004). Li *et al.* (2002) reconstructed 13-protofilament microtubules from small tubule fragments that were treated as single particles; they obtained a resolution of 0.8 nm. Unfortunately, the method was not sensitive enough to detect the microtubule lattice seam and therefore α- and β-tubulin monomers were averaged. As a result, the structure resolved the common parts in both subunits to the highest resolution ever achieved on an intact microtubule, but the process lost the specific structural elements that are different between the two monomers. Both recent publications on dynein structure obtained by single-particle averaging were performed in conjunction with negative staining; they achieved a resolution of ~2.5 nm (Burgess *et al.*, 2003; Samso and Koonce, 2004).

E. Cryo-ET of Microtubules Complexed with Kinesin or Axonemal Dynein

Tomography is the most versatile 3D reconstruction approach for specimens viewed by EM because, unlike averaging methods, it is not limited to molecules that are regularly arranged or isolated. Thus, ET allows the reconstruction of large and polymorphic structures like entire organelles and cells. To this end tomography does not combine different angular views from many identical particles, it combines a set of 2D projections recorded from different viewing angles of a single 3D object. Thus, ET is uniquely suited to study macromolecular complexes in their native environment. The principles and technical details of tomography are explained in more detail in several chapters in this book (see also reviews by Baumeister *et al.*, 1999; McIntosh *et al.*, 2005). Here we will focus on the particular benefits of cryo-ET for structural investigations of microtubules and their associated proteins.

Currently the resolution limit of 3D reconstructions obtained by cryo-ET is on the order of 6–8 nm for raw cryotomograms and 3–4 nm after volume-averaging of repetitive particles extracted from these tomograms. One limitation of 3D reconstructions from single-axis tilt series is the so-called "missing wedge" of data in Fourier space (for a complete discussion, see Chapter 29 by Förster and Hegerl, this volume). In 3D Fourier space, isotropic resolution at a certain value describes a sphere. Isotropic resolution in tomograms would require the acquisition of continuous projections from $-90°$ to $+90°$; however, due to geometric constraints in the support grids a tilt range up to about $\pm 70°$ is realistic; beyond that things get problematic. The resulting lack of high-tilt data translates into a wedge from this sphere of data in the Fourier space that is not sampled; therefore, the resolution in the reconstruction is anisotropic; it is highest in the x–y plane of a tomogram, but it is lower in z-direction. Accordingly, on a spherical object with a thin wall this wall may be very clear within the x–y plane of the tomogram, but smeared out on the top and bottom of the sphere. Substantial future developments on various frontlines will be required to push the resolution of ET toward 2 nm, which would then provide enough detail to assign geometric parameters to small molecules such as kinesin motor domains. However, even at 3- to 4-nm resolution kinesin motors can be imaged and identified in complex with microtubules, allowing a study of their microtubule and cargo interaction properties under close to *in vivo* conditions. Figure 4 shows vitrified microtubules that are cross-linked by a portion of the kinesin-like protein from fission yeast, Klp6p (expressed in bacteria), which dimerizes under the chosen conditions (Fiedler *et al.*, in preparation).

For structural studies at higher resolution, the kind of data shown in Figs. 4 and 5A and B constitutes the basis for 3D averaging procedures. This approach does not begin at the level of 2D projections, as is the case in the previously described methods; it goes a step further by extracting repetitive structures in 3D directly from a tomogram. These structures are then averaged to enhance the signal-to-noise ratio. In addition, the problem of anisotropic resolution, which is due to the missing wedge, can be overcome if the individual particles included in the average have different orientations within the tomogram and thus allow filling in the missing spatial frequencies (see Chapter 29 by Förster and Hegerl). The procedure we have applied in the following example (see below and Fig. 5) is essentially identical to that protocol. It demonstrates the potential of posttomographic averaging using small 3D volumes that were extracted from tomograms.

Eukaryotic flagella and cilia are important organelles for cell motility. They are composed of an ordered array of microtubules and associated proteins called the axoneme. Typical axonemes contain two singlet microtubules that are surrounded by nine microtubule doublets (the familiar 9 + 2). Each doublet is connected to its neighbors by an inner and an outer row of dynein arms, which are the force-producing motors responsible for flagellar motion (reviewed in Gibbons, 1981; for tomographic reconstruction: Nicastro *et al.*, 2005). Several characteristics make frozen-hydrated axonemes an ideal testing ground for the combination of cryo-ET and volume averaging: the size and complexity of the structure both allows and warrants the use of cryo-ET. Meanwhile, the highly repetitive arrangement of 96-nm axial repeats allows for averaging. By making use of all nine microtubule doublets one can achieve isotropic resolution, since they all project onto the image plan at different angles. This kind of study has now provided important insights into the structural and functional organization of axonemes. It has revealed several novel structures such as proteins binding to the inner side of the doublet microtubule walls (microtubule inner proteins, MIPs) and links between dynein complexes. These findings have important implications for microtubule stability and the regulation of dynein activity (Fig. 5C–E; Nicastro *et al.*, 2006). The high reproducibility and sensitivity of this technique, currently producing ~4-nm resolution, holds promise for future studies of macromolecular complexes *in situ*, as well as comparisons between wild-type and mutant axonemes for the localization of protein components (compare Fig. 5F and G; Nicastro *et al.*, 2005).

IV. Discussion

This chapter has described the application of cryo-EM and 3D image reconstruction methods, including cryo-ET, for studying microtubules complexed with kinesin and dynein motor proteins and nonmotor MAPs. With cryo-ET becoming an increasingly powerful tool, the method should soon reach a resolution level on the order of 2 nm, which is comparable to routine helical or single-particle averaging. The scientific driving force behind cryo-ET is the possibility of generating 3D images from large, complex, and irregular structures that are beyond the capabilities of averaging procedures. Cryopreparation methods, such as vitrification and sectioning of vitrified cells and tissue, may render cryo-ET a prime tool for a unifying approach to structural and cellular biology, connecting atomic structures determined by crystallography directly to the high-resolution end of light microscopy on living cells and tissue. This development will be driven by several technical advances such as (1) improved imaging with electron-energy filters and refined high-contrast lens technology, (2) data recording with high-sensitivity and noise-reduced CCD cameras, and (3) data handling, such as alignment of individual projections and 3D analysis of huge volumes, as a result of dramatically enhanced computational speed and economical mass data storage capabilities (tomographic data sets easily reach several gigabytes).

With these exciting prospects for cryo-ET in cell biology, one can ask what remains as the role of averaging procedures such as helical and single-particle approaches? Averaging is only applicable to image data derived from particles that are structurally reproducible and controllable with regards to environmentally induced conformational states. Hence, while essential for high-resolution studies at molecular and near-atomic detail, the scope for averaging-based 3D reconstruction techniques in cell biology seems limited. Nevertheless, averaging methods will not be abandoned that quickly, even for cell biology applications, and there are several reasons for keeping them around. One is that cryo-ET will most likely not attain or surpass 2-nm resolution in the immediate future. Therefore, averaging methods will continue to be valuable for generating structural and functional data on a detailed molecular level. Second, yet another very important role for averaging methods is creating detailed structural models of biological complexes at resolutions better than 2 nm. These will act as templates and help the process of localizing structures of interest in the crowded environment displayed by tomograms of intact cells or other large structures. Finally, the goal of reaching 2-nm resolution in cryo-ET aims ultimately at the docking of x-ray determined structures directly into tomographic data. However, cryo-ET often produces continuous maps of electron densities that smear over the space between large protein assemblies in a way that may diffuse their outer boundaries (e.g., cytoskeletal structures complexed with various proteins). Therefore, preliminary docking of atomic structures onto isolated models will certainly facilitate the process and enhance the chance for successful molecular and structural interpretations of tomographic data.

In the past structural and functional studies by cryo-EM of microtubules and associated proteins have been dominated by averaging procedures, which have all the strengths and limitations described here. The new possibilities, based on cryo-ET, will now enable structural and functional studies directly in a cellular environment and, thus, will direct the study of these structures to conditions that are much closer to *in vivo* situations. Meanwhile, the number of high-resolution structures of microtubule-interacting proteins is constantly growing, and the combination of technologies available today should soon produce a comprehensive picture of the microtubule cytoskeleton and its interacting components in various cells and tissues.

Acknowledgments

We would like to thank Dr. Hernando Sosa (AECOM, New York) for his comments on the back-projection paragraph. We are grateful to Linda Batlin and Regula Mueller Hoenger (both CU at Boulder, CO) for their help with the chapter.

References

Al-Bassam, J., Cui, Y., Klopfenstein, D., Carragher, B. O., Vale, R. D., and Milligan, R. A. (2003). Distinct conformations of the kinesin Unc104 neck regulate a monomer to dimer motor transition. *J. Cell Biol.* **163,** 743–753.

Al-Bassam, J., Ozer, R. S., Safer, D., Halpain, S., and Milligan, R. A. (2002). MAP2 and tau bind longitudinally along the outer ridges of microtubule protofilaments. *J. Cell Biol.* **157,** 1187–1196.

Amos, L. A., and Baker, T. S. (1979). The three-dimensional structure of tubulin protofilaments. *Nature* **279,** 607–612.

Amos, L. A., and Klug, A. (1974). Arrangement of subunits in flagellar microtubules. *J. Cell Sci.* **14,** 523–549.

Arnal, I., Metoz, F., DeBonis, S., and Wade, R. H. (1996). Three-dimensional structure of functional motor proteins on microtubules. *Curr. Biol.* **6,** 1265–1270.

Baumeister, W., Grimm, R., and Walz, J. (1999). Electron tomography of molecules and cells. *Trends Cell Biol.* **9,** 81–85.

Beuron, F., and Hoenger, A. (2001). "Structural Analysis of the Microtubule-Kinesin Complex by Cryo-Electron Microscopy," Vol. 164. Humana Press, Totowa, NJ.

Bluemke, D. A., Carragher, B., and Josephs, R. (1988). The reconstruction of helical particles with variable pitch. *Ultramicroscopy* **26,** 255–270.

Burgess, S. A., Walker, M. L., Sakakibara, H., Knight, P. J., and Oiwa, K. (2003). Dynein structure and power stroke. *Nature* **421,** 715–718.

Carragher, B. O., Whittaker, M., and Milligan, R. A. (1996). Helical processing using PHOELIX. *J. Struct. Biol.* **116,** 107–112.

Chandra, R., Salmon, E. D., Erickson, H. P., Lockhart, A., and Endow, S. A. (1993). Structural and functional domains of the *Drosophila* ncd microtubule motor protein. *J. Biol. Chem.* **268,** 9005–9013.

Chrétien, D., and Wade, R. H. (1991). New data on the microtubule surface lattice. *Biol. Cell* **71,** 161–174.

Chrétien, D., Kenney, J. M., Fuller, S. D., and Wade, R. H. (1996). Determination of microtubule polarity by cryo-electron microscopy. *Structure* **4,** 1031–1040.

Chrétien, D., Metoz, F., Verde, F., Karsenti, E., and Wade, R. H. (1992). Lattice defects in microtubules: Protofilament numbers vary within individual microtubules. *J. Cell Biol.* **117,** 1031–1040.

DeRosier, D., and Moore, P. B. (1970). Reconstruction of three-dimensional images from electron micrographs of structures with helical symmetry. *J. Mol. Biol.* **52,** 355–369.

Dias, D. P., and Milligan, R. A. (1999). Motor protein decoration of microtubules grown in high salt conditions reveals the presence of mixed lattices. *J. Mol. Biol.* **287,** 287–292.

Egelman, E. H. (2000). A robust algorithm for the reconstruction of helical filaments using single-particle methods. *Ultramicroscopy* **85,** 225–234.

Frank, J. (2002). Single-particle imaging of macromolecules by cryo-electron microscopy. *Annu. Rev. Biophys. Biomol. Struct.* **31,** 303–319.

Frank, J., and van Heel, M. (1982). Correspondence analysis of aligned images of biological particles. *J. Mol. Biol.* **161,** 134–137.

Gibbons, B. H., and Gibbons, I. R. (1972). Flagellar movement and adenosine triphosphatase activity in sea urchin sperm extracted with triton X-100. *J. Cell Biol.* **54,** 75–97.

Gibbons, B. H., and Gibbons, I. R. (1973). The effect of partial extraction of dynein arms on the movement of reactivated sea-urchin sperm. *J. Cell Sci.* **13,** 337–357.

Gibbons, I. R. (1981). Cilia and flagella of eukaryotes. *J. Cell Biol.* **91,** 107s–124s.

Heuser, J. E. (1986). Different structural states of a microtubule cross-linking molecule, captured by quick-freezing motile axostyles in protozoa. *J. Cell Biol.* **103,** 2209–2227.

Hirokawa, N. (1982). Cross-linker system between neurofilaments, microtubules, and membranous organelles in frog axons revealed by the quick-freeze, deep-etching method. *J. Cell Biol.* **94,** 129–142.

Hirokawa, N. (1998). Kinesin and dynein superfamily proteins and the mechanism of organelle transport. *Science* **279,** 519–526.

Hirose, K., Cross, R. A., and Amos, L. A. (1998). Nucleotide-dependent structural changes in dimeric NCD molecules complexed to microtubules. *J. Mol. Biol.* **278,** 389–400.

Hirose, K., Henningsen, U., Schliwa, M., Toyoshima, C., Shimizu, T., Alonso, M., Cross, R. A., and Amos, L. A. (2000). Structural comparison of dimeric Eg5, Neurospora kinesin (Nkin) and Ncd head-Nkin neck chimera with conventional kinesin. *EMBO J.* **10,** 5308–5314.

Hoenger, A., Doerhoefer, M., Woehlke, G., Tittmann, P., Gross, H., Song, Y.-H., and Mandelkow, E. (2000b). Surface topography of microtubule walls decorated with monomeric and dimeric kinesin constructs. *Biol. Chem.* **381,** 1001–1011.

Hoenger, A., and Milligan, R. A. (1997). Motor domains of kinesin and ncd interact with microtubule protofilaments with the same binding geometry. *J. Mol. Biol.* **265,** 553–564.

Hoenger, A., Sablin, E. P., Vale, R. D., Fletterick, R. J., and Milligan, R. A. (1995). Three-dimensional structure of a tubulin-motor-protein complex. *Nature* **376,** 271–274.

Hoenger, A., Sack, S., Thormählen, M., Marx, A., Muller, J., Gross, H., and Mandelkow, E. (1998). Image reconstructions of microtubules decorated with monomeric and dimeric kinesins: Comparison with x-ray structure and implications for motility. *J. Cell Biol.* **141,** 419–430.

Hoenger, A., Thormählen, M., Diaz-Avalos, R., Doerhoefer, M., Goldie, K. N., Muller, J., and Mandelkow, E. (2000a). A new look at the microtubule binding patterns of dimeric kinesins. *J. Mol. Biol.* **297,** 1087–1103.

Kar, S., Fan, J., Smith, M. J., Goedert, M., and Amos, L. A. (2003). Repeat motifs of tau bind to the insides of microtubules in the absence of taxol. *EMBO J.* **22,** 70–77.

Kikkawa, M., Ishikawa, T., Nakata, T., Wakabayashi, T., and Hirokawa, N. (1994). Direct visualization of the microtubule lattice seam both *in vitro* and *in vivo*. *J. Cell Biol.* **127,** 1965–1971.

Kikkawa, M., Ishikawa, T., Wakabayashi, T., and Hirokawa, N. (1995). Three-dimensional structure of the kinesin head-microtubule complex. *Nature* **376,** 274–277.

Kikkawa, M., Sablin, E. P., Okada, Y., Yajima, H., Fletterick, R. J., and Hirokawa, N. (2001). Switch-based mechanism of kinesin motors. *Nature* **411,** 439–445.

Koonce, M. P., Grissom, P. M., and McIntosh, J. R. (1992). Dynein from Dictyostelium: Primary structure comparisons between a cytoplasmic motor enzyme and flagellar dynein. *J. Cell Biol.* **119,** 1597–1604.

Krebs, A., Goldie, K. N., and Hoenger, A. (2004). Complex formation with kinesin motor domains affects the structure of microtubules. *J. Mol. Biol.* **335,** 139–153.

Krzysiak, T. C., Wendt, T., Sproul, L. R., Tittmann, P., Gross, H., Gilbert, S. P., and Hoenger, A. (2006). A structural model for monastrol inhibition of dimeric kinesin Eg5. *EMBO J.* **25,** 2263–2273.

Lawrence, C. J., Dawe, R. K., Christie, K. R., Cleveland, D. W., Dawson, S. C., Endow, S. A., Goldstein, L. S., Goodson, H. V., Hirokawa, N., Howard, J., Malmberg, R. L., McIntosh, J. R., *et al.* (2004). A standardized kinesin nomenclature. *J. Cell Biol.* **167,** 19–22.

Li, H., DeRosier, J., Nicholson, W. V., Nogales, E., and Downing, K. H. (2002). Microtubule structure at 8 Å resolution. *Structure* **10,** 1317–1328.

Löwe, J., Li, H., Downing, K. H., and Nogales, E. (2001). Refined structure of alpha beta-tubulin at 3.5 A resolution. *J. Mol. Biol.* **313,** 1045–1057.

Mandelkow, E. M., and Mandelkow, E. (1985). Unstained microtubules studied by cryo-electron microscopy. Substructure, supertwist and disassembly. *J. Mol. Biol.* **181,** 123–135.

Mandelkow, E. M., Schultheiss, R., Rapp, R., Muller, M., and Mandelkow, E. (1986). On the surface lattice of microtubules: Helix starts, protofilament number, seam and handedness. *J. Cell Biol.* **102,** 1067–1073.

McDonald, H. B., Stewart, R. J., and Goldstein, L. S. (1990). The kinesin-like ncd protein of *Drosophila* is a minus end-directed microtubule motor. *Cell* **63,** 1159–1165.

McIntosh, R., Nicastro, D., and Mastronarde, D. (2005). New views of cells in 3D: An introduction to electron tomography. *Trends Cell Biol.* **15,** 43–51.

Metoz, F., Arnal, I., and Wade, R. H. (1997). Tomography without tilt: Three-dimensional imaging of microtubule/motor complexes. *J. Struct. Biol.* **118,** 159–168.

Miyazawa, A., Fujiyoshi, Y., and Unwin, N. (2003). Structure and gating mechanism of the acetylcholine receptor pore. *Nature* **423,** 949–955.

Mizuno, N., Toba, S., Edamatsu, M., Watai-Nishii, J., Hirokawa, N., Toyoshima, Y. Y., and Kikkawa, M. (2004). Dynein and kinesin share an overlapping microtubule-binding site. *EMBO J.* **23,** 2459–2467.

Moore, P. B., Huxley, H. E., and DeRosier, D. J. (1970). Three-dimensional reconstruction of F-actin, thin filaments and decorated thin filaments. *J. Mol. Biol.* **50,** 279–295.

Moores, C. A., Yu, M., Guo, J., Beraud, C., Sakowicz, R., and Milligan, R. A. (2002). A mechanism for microtubule depolymerization by KinI kinesins. *Mol. Cell* **9,** 903–909.

Morgan, D. G., and DeRosier, D. (1992). Processing images of helical structures: A new twist. *Ultramicroscopy* **46,** 263–285.

Nicastro, D., McIntosh, J. R., and Baumeister, W. (2005). 3D structure of eukaryotic flagella in a quiescent state revealed by cryo-electron tomography. *Proc. Natl. Acad. Sci. USA* **102,** 15889–15894.

Nicastro, D., Schwartz, C., Pierson, J., Gaudette, R., Porter, M. E., and McIntosh, J. R. (2006). Cryotomography of axonemes: Insights into the regulation of dynein activity and microtubule stability. *Science* **313,** 944–948.

Nogales, E., Wolf, S. G., and Downing, K. H. (1998). Structure of the alpha beta tubulin dimer by electron crystallography. *Nature* **391,** 199–203.

Pfister, K. K., Fisher, E. M., Gibbons, I. R., Hays, T. S., Holzbaur, E. L., McIntosh, J. R., Porter, M. E., Schroer, T. A., Vaughan, K. T., Witman, G. B., King, S. M., and Vallee, R. B. (2005). Cytoplasmic dynein nomenclature. *J. Cell Biol.* **171** 411–413.

Pfister, K. K., Shah, P. R., Hummerich, H., Russ, A., Cotton, J., Annuar, A. A., King, S. M., and Fisher, E. M. (2006). Genetic analysis of the cytoplasmic dynein subunit families. *PLoS Genetics* **2,** e1.

Rademacher, M. (1992). "Weighted Back-Projection Methods." Plenum Press, New York.

Rice, S., Lin, A. W., Safer, D., Hart, C. L., Naber, N., Carragher, B. O., Cain, S. M., Pechatnikova, E., Wilson-Kubalek, E. M., Whittaker, M., Pate, E., Cooke, R., *et al.* (1999). A structural change in the kinesin motor protein that drives motility. *Nature* **402,** 778–784.

Rossmann, M. G., Olson, N. H., Kolatkar, P. R., Oliveira, M. A., Cheng, R. H., Greve, J. M., McClelland, A., and Baker, T. S. (1994). Crystallographic and cryo EM analysis of virion-receptor interactions. *Arch. Virol. Suppl.* **9,** 531–541.

Sablin, E. P., Case, R. B., Dai, S. C., Hart, C. L., Ruby, A., Vale, R. D., and Fletterick, R. J. (1998). Direction determination in the minus-end-directed kinesin motor Ncd. *Nature* **395,** 813–816.

Samso, M., and Koonce, M. P. (2004). 25 Å resolution structure of a cytoplasmic dynein motor reveals a seven-member planar ring. *J. Mol. Biol.* **340,** 1059–1072.

Sandblad, L., Busch, K. E., Tittmann, P., Gross, H., Brunner, D., and Hoenger, A. (2006). The *Schizosaccharomyces pombe* EB1 homolog Mal3p localizes preferentially to the Microtubule Lattice Seam. *Cell,* in press.

Santarella, R. A., Skiniotis, G., Goldie, K. N., Tittmann, P., Gross, H., Mandelkow, E. M., Mandelkow, E., and Hoenger, A. (2004). Surface-decoration of microtubules by human tau. *J. Mol. Biol.* **339,** 539–553.

Schatz, C. A., Santarella, R. A., Hoenger, A., Karsenti, E., Mattaj, I. W., Gruss, O. J., and Carazo-Salas, R. E. (2003). Importin alpha-regulated nucleation of microtubules by TPX2. *EMBO J.* **22,** 2060–2070.

Scholey, J. M. (1996). Kinesin-II, a membrane traffic motor in axons, axonemes, and spindles. *J. Cell Biol.* **133,** 1–4.

Schroeter, J. P., and Bretaudiere, J. P. (1996). SUPRIM: Easily modified image-processing software. *J. Struct. Biol.* **116,** 131–137.

Skiniotis, G., Surrey, T., Altmann, S., Gross, H., Song, Y.-H., Mandelkow, E., and Hoenger, A. (2003). Nucleotide-induced conformations in the neck region of dimeric kinesin. *EMBO J.* **22,** 1518–1528.

Song, Y.-H., and Mandelkow, E. (1993). Recombinant kinesin motor domain binds to $\alpha\beta$-tubulin and decorates microtubules with a B surface lattice. *Proc. Natl. Acad. Sci. USA* **90,** 1671–1675.

Song, Y.-H., and Mandelkow, E. (1995). The anatomy of flagellar microtubules: Polarity, seam, junctions and lattice. *J. Cell Biol.* **128,** 81–94.

Sosa, H., and Chrétien, D. (1998). Relationship between moiré patterns, tubulin shape, and microtubule polarity. *Cell Motil. Cytoskeleton* **40,** 38–43.

Sosa, H., Dias, D. P., Hoenger, A., Whittaker, M., Wilson-Kubalek, E., Sablin, E., Fletterick, R. J., Vale, R. D., and Milligan, R. A. (1997). A model for the microtubule-Ncd motor protein complex obtained by cryo-electron microscopy and image analysis. *Cell* **90,** 217–224.

Sosa, H., and Milligan, R. A. (1996). Three-dimensional structure of ncd-decorated microtubules obtained by a back-projection method. *J. Mol. Biol.* **260,** 743–755.

Tamm, L. K., Crepeau, R. H., and Edelstein, S. T. (1979). Three-dimensional reconstruction of tubulin in zinc-induced sheets. *J. Mol. Biol.* **130,** 473–492.

Vale, R. D. (2003). The molecular motor toolbox for intracellular transport. *Cell* **112,** 467–480.

Vallee, R. B., Williams, J. C., Varma, D., and Barnhart, L. E. (2004). Dynein: An ancient motor protein involved in multiple modes of transport. *J. Neurobiol.* **58,** 189–200.

Van Heel, M. (1984). Multivariate statistical classification of noisy images (randomly oriented biological macromolecules). *Ultramicroscopy* **13,** 165–183.

Van Heel, M., and Frank, J. (1981). Use of multivariate statistics in analyzing the images of biological macromolecules. *Ultramicroscopy* **6,** 187–194.

Volkmann, N., and Hanein, D. (1999). Quantitative fitting of atomic models into observed densities derived by electron microscopy. *J. Struct. Biol.* **125,** 176–184.

Wade, R. H., Horowitz, R., and Milligan, R. A. (1995). Toward understanding the structure and interactions of microtubules and motor proteins. *Proteins Struct. Funct. Genet.* **23,** 502–509.

Walker, R. A., Salmon, E. D., and Endow, S. A. (1990). The *Drosophila* claret segregation protein is a minus-end directed motor molecule. *Nature* **347,** 780–782.

Walz, T., Tittmann, P., Fuchs, K. H., Muller, D. J., Smith, B. L., Agre, P., Gross, H., and Engel, A. (1996). Surface topographies at subnanometer-resolution reveal asymmetry and sidedness of Aquaporin-1. *J. Mol. Biol.* **264,** 907–918.

Wendt, T., Karabay, A., Krebs, A., Gross, H., Walker, R. A., and Hoenger, A. (2003). A structural analysis of the interaction between ncd tail and tubulin protofilaments. *J. Mol. Biol.* **333,** 541–552.

Wendt, T. G., Volkmann, N., Skiniotis, G., Goldie, K. N., Muller, J., Mandelkow, E., and Hoenger, A. (2002). Microscopic evidence for a minus-end directed power stroke in the kinesin motor ncd. *EMBO J.* **21,** 5969–5978.

Whittaker, M., Carragher, B. O., and Milligan, R. A. (1995). PHOELIX: A package for semi-automated helical reconstruction. *Ultramicroscopy* **58,** 245–259.

Wriggers, W., Milligan, R. A., and McCammon, J. A. (1999). Situs: A package for docking crystal structures into low-resolution maps from electron microscopy. *J. Struct. Biol.* **125,** 185–195.

Yonekura, K., Maki-Yonekura, S., and Namba, K. (2003). Complete atomic model of the bacterial flagellar filament by electron cryomicroscopy. *Nature* **424,** 643–650.

CHAPTER 18

Reconstructing the Endocytotic Machinery

Yifan Cheng* and Thomas Walz†

*Department of Biochemistry and Biophysics
University of California-San Francisco
San Francisco, California 94143

†Department of Cell Biology, Harvard Medical School
Boston, Massachusetts 02115

Most biological processes are too complex to be studied by a single structural technique because they are carried out by dynamic protein assemblies that function as macromolecular machines. Cellular iron uptake is a typical example for such a complex biological process. In this chapter, we describe the "divide and conquer" approach that we used to study clathrin-mediated endocytosis of the transferrin–transferrin receptor complex, the machinery used by cells to acquire iron. We first used single-particle electron microscopy of vitrified samples to produce density maps at subnanometer resolution for the transferrin–transferrin receptor complex and for *in vitro*-assembled clathrin cages. We could interpret these density maps by docking

0091-679X/07 $35.00
DOI: 10.1016/S0091-679X(06)79018-5

crystal structures of the subunits into the density maps, thus producing pseudo-atomic models for both the transferrin–transferrin receptor complex and the D6 barrel clathrin lattice. In a next step, we used cryo-electron tomography of vitrified samples to produce density maps for *in vivo*-assembled clathrin-coated vesicles, showing that the clathrin lattice, the membrane vesicle, and the connections between them can be studied in individual particles.

I. Introduction

Almost all biological processes are carried out by complex and dynamic assemblies of biological macromolecules. The functions of such macromolecular machines are based on the regulated assembly and disassembly of their protein and nucleic acid components during different stages of the process. Another important mechanism that contributes to the functions of protein complexes are energy-dependent conformational changes powered by the conversion of chemical energy into mechanical work. A full understanding of biological processes thus requires structural information at a wide range of resolutions, ranging from the cellular context all the way to atomic details of the involved macromolecules. Light microscopy, especially live-cell imaging, is a powerful tool for characterizing a process on the cellular level and to provide information about its dynamics. X-ray crystallography and NMR spectroscopy are the traditional approaches used in structural biology and can be used to determine the structure of macromolecules at atomic resolution. Electron microscopy (EM) can, however, be of great value in developing our understanding of biological structure because it can be used to study processes at a wide range of resolutions. EM can thus integrate the information obtained by light microscopy, x-ray crystallography and NMR spectroscopy, in effect merging the disciplines of cell biology and structural biology.

Conventional EM on thin sections has been used for many decades to visualize the ultrastructure of cells and tissues. Especially the combination of EM with immunocytochemistry has proven extremely powerful in the elucidation of cellular processes, and it became one of the techniques that ushered in modern cell biology. More recently, with the development of sample vitrification, EM has also established a firm role for itself in the structure determination of proteins and macromolecular complexes. Molecular EM, so called because it is concerned with the structure of molecules rather than the ultrastructure of cells, now complements structure determination by x-ray crystallography and NMR spectroscopy. Electron crystallography can be used to determine the structure of proteins, in particular membrane proteins, to near-atomic resolution. This approach requires, however, that the proteins can be assembled into two-dimensional (2D) crystals. EM of single particles has been developed to cope with macromolecules that are hard or impossible to crystallize. In this method, images of thousands of individual molecules are combined to obtain, in favorable cases, a density map at subnanometer resolution. While single-particle EM can be applied to heterogeneous molecule populations, the technique requires that a large number of molecules have the same structure so that their images can be averaged. Finally, electron tomography, which is most often used to obtain three-dimensional (3D) maps of sections through tissues or even entire cells, can also be used to visualize the structure of large macromolecular complexes or isolated organelles. The resolution obtained with electron tomography is currently still limited to a few nanometers, but since it does not depend on image averaging, this approach makes it possible to determine the structure of objects with unique structures.

We are combining information from several structural methods to study the molecular machinery for receptor-mediated endocytosis. Using transferrin-dependent iron uptake by clathrin-mediated endocytosis as an example, we illustrate in this chapter how the fitting of crystal structures into molecular envelopes of protein complexes obtained by EM can be used to study the structural basis of a cell biological process. While other chapters in this book are concerned with the combination of electron tomographic reconstructions with light microscopy data, this chapter focuses on how molecular EM can bring high-resolution structures obtained with x-ray crystallography or NMR spectroscopy into the context of protein complexes and small organelles.

II. Rationale

From archaea to human, almost every living organism depends on iron. Because of its easy conversion between the ferrous (Fe^{2+}) and ferric (Fe^{3+}) states, many organisms utilize iron as a cofactor for the catalysis of redox reactions. However, under physiological conditions ferric iron forms highly insoluble hydroxide salts, and in the presence of oxygen ferrous iron catalyzes the generation of hydroxyl radicals, a major source for oxidative damage in cells. Because of its insolubility and toxicity, iron found in vertebrates is transported through the circulation bound to a carrier protein, plasma transferrin (Tf); this complex is taken up by cells through a clathrin-mediated endocytotic pathway involving the Tf receptor (TfR) (Enns *et al.*, 1996; reviewed in Aisen *et al.*, 2001). Figure 1A illustrates the different steps in the Tf cycle. At the extracellular pH of 7.4, Tf can bind two ferric ions. Two diferric Tf (dTf) molecules can then bind to the dimeric TfR, which is localized on the surface of iron-dependent cells; this forms the dTf–TfR complex. dTf–TfR complexes are recruited to clathrin-coated pits, which invaginate and bud off from the plasma membrane to form clathrin-coated vesicles. After losing their coat, the vesicles fuse into endosomes, where the acidic pH of 5.6 in the lumen causes the Tf molecules to change their conformation, resulting in the release of the bound iron. The divalent metal ion transporter 1 (DMT1) transports the released iron across the endosomal membrane into the cytoplasm. Iron-free Tf (apoTf) remains tightly bound to the receptor at the endosomal pH, and the complex cycles back to the cell surface, where the extracellular pH causes apoTf to dissociate from TfR.

Tf–TfR complex-dependent iron uptake by clathrin-mediated endocytosis is a typical example of a cellular process, which often involves multiple steps in which protein assemblies, such as Tf–TfR complexes and clathrin coats, form and disassemble in a dynamic fashion. Conventional EM played a central role in elucidating

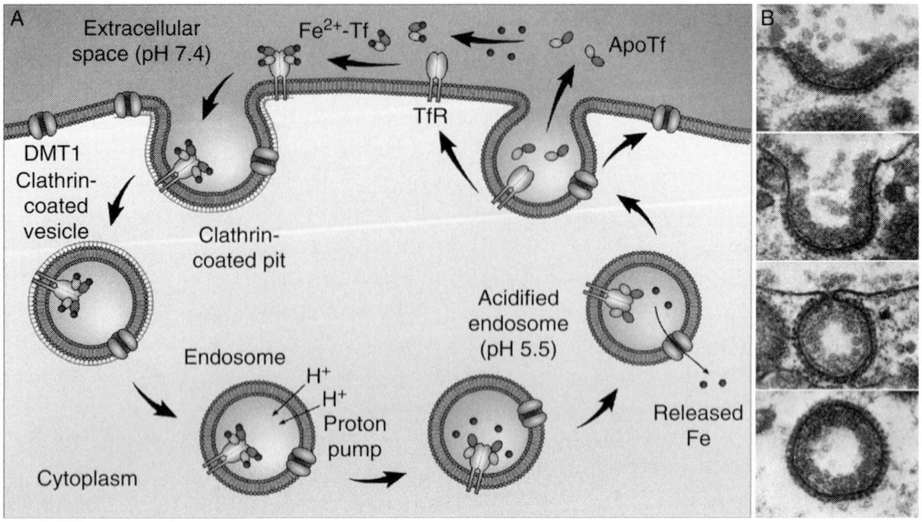

Fig. 1 Tf-dependent iron uptake by clathrin-mediated endocytosis. (A) Schematic diagram of the Tf cycle. (B) EM images of thin-sectioned oocyte membranes showing different stages in the formation of clathrin-coated vesicles from the coated pit to the fully formed coated vesicle. Reprinted from Perry and Gilbert (1979), with permission from Company of Biologists.

the Tf cycle (Willingham *et al.*, 1984), and it provided many structural insights into this process on the level of cellular ultrastructure such as the discovery of coated vesicles (Roth and Porter, 1964), the assembly of clathrin coats on the inside of the plasma membrane (Heuser, 1980), and the localization of TfRs to clathrin-coated vesicles (van Dam and Stoorvogel, 2002). Figure 1B shows typical images of the endocytotic process captured by conventional EM of a tissue section (Perry and Gilbert, 1979). More recently, x-ray crystallography provided atomic structures for many of the proteins and protein domains involved in iron uptake, including crystal structures of various members of the Tf family (reviewed in Baker *et al.*, 2003), the ectodomain of TfR (Lawrence *et al.*, 1999), the core of the adaptor protein complex 2 (AP-2) (Collins *et al.*, 2002), as well as the proximal region (Ybe *et al.*, 1999) and the N-terminal domain of the clathrin heavy chain (ter Haar *et al.*, 1998). Molecular EM was then the technique of choice to understand how these proteins interact with each other to form the assemblies that carry out the cell biological processes previously visualized by conventional EM.

In this chapter, we discuss how single-particle EM has been used to determine molecular envelopes of the dTf–TfR complex (Cheng *et al.*, 2004) and *in vitro*-assembled clathrin coats (Fotin *et al.*, 2004b) at subnanometer resolution. We then describe how electron tomography has been used to obtain low-resolution structures of individual clathrin-coated vesicles isolated from calf brains (Cheng *et al.*, 2006). We also describe how crystal structures were fit into the EM density maps to generate pseudo-atomic models for these protein assemblies, thus helping to understand how the component proteins work in concert to fulfill their biological function.

III. Methods

A. Vitrification of Protein Assemblies

In conventional EM of cells, the most commonly used specimen preparation technique involves the fixation, dehydration, embedding, thin sectioning, and staining of the tissue or cell sample (reviewed in Wright, 2000). These fixation/staining techniques do not preserve high-resolution structural detail in the samples. More than 30 years ago, Taylor and Glaeser (1974) showed that a sample prepared in the frozen-hydrated state can preserve the native structure of biological specimens to the level of atomic resolution. Only recently, however, have sample vitrification and cryosectioning developed into techniques that can routinely be used in conventional EM (see Chapter 1 by Dubochet, this volume).

Although the techniques and instruments used for preparing frozen-hydrated specimens are somewhat different for conventional and molecular EM, the principle underlying specimen vitrification is the same. Vitrification of proteins or macromolecular complexes in buffer solution has become a routine specimen preparation technique in molecular EM. The quick-freezing technique in common use is still essentially the same as that initially developed by Dubochet and colleagues for the vitrification of viruses (Adrian *et al.*, 1984). Briefly, a drop of sample solution is applied to a holey carbon grid. The grid is blotted with a filter paper to remove most of the sample solution, while the remaining solution forms a thin layer of liquid across the holes in the carbon film. The grid is then quick-frozen by plunging it rapidly into liquid ethane cooled by liquid nitrogen to a temperature of about $-180\,^{\circ}\mathrm{C}$. The cooling rate in this process is so high that the water transitions from its fluid to a solid state without forming ice crystals, which would damage the specimen. The resulting layer of vitreous (or amorphous) ice embeds the proteins and preserves the specimen virtually artifact-free in a near-native environment.

The thickness of the vitreous ice layer is the most critical factor in the preparation of frozen-hydrated specimens. The thickness of this layer must fit the size of the specimen and is ideally only a little bit thicker than the specimen itself. If the ice layer is much thicker than the size of the specimen, the ice will unnecessarily reduce the contrast of the protein complexes in the images. This is problematic because no artificial contrasting agents are used in the preparation of frozen-hydrated samples, so the contrast of the sample is very low to begin with. In the case of small protein complexes, such as the 300-kDa Tf–TfR complex, the ice layer has to be very thin in order for the small particles to still generate sufficient contrast to be visible at all (Cheng *et al.*, 2004) (Fig. 3). On the other hand, if the ice layer is too thin for a particular specimen, the surface tension of the liquid layer, which thins during the blotting of the grid, will crush the protein assemblies and distort their structure. For example, for the large, *in vitro*-assembled clathrin coats, the ice layer had to be more than 100-nm thick to preserve the symmetries of the clathrin lattices (Fotin *et al.*, 2004b) (Fig. 7A). An even thicker ice layer was necessary for clathrin-coated vesicles purified from calf brain, which range in diameter from 50 to over 200 nm (Fig. 9).

The thickness of the ice layer can be adjusted in two ways: by varying the times for blotting or for drying between the blotting and freezing of the grid. The longer these times, the thinner the resulting ice layer. While prolonged blotting of the grid removes sample with the solution, prolonged drying removes only water, thus concentrating the sample, including buffer components. Once the best blotting and drying times have been found, grids can be prepared reproducibly using commercially available, semiautomated vitrification machines such as the Vitrobot (FEI Company, Hillsboro, OR).

B. Single-Particle EM

Single-particle EM, a technique pioneered by Joachim Frank and well explained in his revised book (Frank, 2006), is probably the most versatile technique currently available to study the structure of large, dynamic protein assemblies. The technique has made dramatic advances over the last decade, mostly due to the development of (1) electron microscopes equipped with field emission electron sources that produce more coherent electron beams, (2) specimen vitrification that preserves the atomic detail of the samples, and (3) innovative software packages that are used to process and average image data, such as SPIDER (Frank *et al.*, 1996), IMAGIC (van Heel *et al.*, 1996), EMAN (Ludtke *et al.*, 1999), and FREALIGN (Stewart and Grigorieff, 2004). Today, single-particle cryo-EM can produce 3D density maps at subnanometer resolution not only for large and highly symmetric specimens, such as viruses (Bottcher *et al.*, 1997; Zhou *et al.*, 2000), but also for smaller protein complexes with low or no symmetry (Cheng *et al.*, 2004; Jiang *et al.*, 2006; Ludtke *et al.*, 2004).

To prevent excessive radiation damage by the high-energy electron beam, images of frozen-hydrated specimens are collected following strict low-dose procedures. As a result of the limited electron dose used to image the sample, however, the raw images are based on too few electrons to provide a good representation of the electron scattering properties of each region in the specimen. The resulting images thus have a very low signal-to-noise ratio. Since EM images are projections of the imaged molecule and since these projection images have very low contrast, it is impossible to understand the 3D structure of the specimen directly from the raw images. Molecular EM thus relies on extensive image processing and image averaging to extract the significant information from the images and to reconstruct 3D density maps of the imaged molecules. The result of this image processing depends strongly on the specific procedures that were used and on whether the algorithms worked properly to find the correct solutions. 3D reconstructions obtained through extensive image processing should therefore always be regarded with caution. Whenever possible, interpretations of EM density maps should be verified by independent means such as the docking of known protein structures into the EM map, mutagenesis, and gold or antibody labeling. It has proved similarly difficult to assess the exact resolution of a density map obtained by single-particle EM. In single-particle cryo-EM, the resolution of a 3D reconstruction is often measured by Fourier shell correlation (FSC) curves (Harauz and van Heel, 1986), which in the case of large data sets has a tendency to overestimate the resolution of 3D maps (Stewart and Grigorieff, 2004). It is therefore important to verify that the resolution that was determined from an FSC curve is reflected in the features seen in the density maps. For example, a density map at subnanometer resolution should resolve individual α-helices (Cheng *et al.*, 2004; Fotin *et al.*, 2004b; Ludtke *et al.*, 2004).

C. Cryo-Electron Tomography

The same electron tomographic technique described in other chapters of this book for the 3D reconstruction of thin sections can also be applied to isolated protein assemblies. In contrast to the single-particle approach just described, electron tomography does not rely on image averaging, so 3D reconstructions can be obtained for objects with unique structures. The disadvantage of electron tomography is that the same sample has to be imaged many times, resulting in a high accumulated electron dose and concomitant beam damage to the sample. In addition, since images are not being averaged, electron tomographic reconstructions suffer from extremely low signal-to-noise ratios, especially when vitrified samples are being imaged. As a result, the resolution of electron tomograms is usually limited to a few nanometers.

In addition to the accumulated dose problem, cryo-electron tomography battles additional technical challenges. The same specimen area of the frozen-hydrated specimen has to be imaged through a large tilt range, but it is difficult to image highly tilted specimens because of the lower stability of side-entry specimen stages under tilted conditions and because of the effects of specimen charging. Data collection by manual tilting of the specimen is a very tedious and time-consuming process, and it is difficult to control the total electron dose in this way. Alternatively, a number of software packages can be used to automatically collect tomographic tilt series such as UCSF Tomography (Zheng *et al.*, 2004), Xplore3D of FEI software, and SerialEM (Mastronarde, 2005). All cryo-electron tomographic data presented in this chapter were collected by manual data collection.

Interpretation of electron tomograms is another challenge in cryo-electron tomography because the density maps are usually very noisy. In addition, the tomograms are distorted due to the inevitable missing wedge effect caused by the physical limitation that samples usually cannot be tilted to 90° in the electron microscope, typically limiting data collection to a maximum angle of 60° or 70°. Computational de-noising procedures, often involving nonlinear algorithms such as nonlinear anisotropic diffusion (Frangakis and Hegerl, 2001; Frangakis *et al.*, 2001), have been introduced to reduce the noise level in tomograms. In addition, segmentation procedures, either interactive or automated, are often used to generate surface representations of objects in the density maps. While these procedures make it easier to distinguish large structural features, they also carry the risk of altering more subtle features in the raw 3D density maps. Extreme caution should therefore be taken when interpreting small structural details in electron tomograms that were subjected to nonlinear de-noising and segmentation. Again, verification by independent means would augment confidence in such interpretations. The surface representations of the clathrin-coated vesicles presented in this chapter were produced by isosurface contouring of electron tomograms that were de-noised using a median filter (Winkler and Taylor, 2006).

D. Docking Crystal Structures into EM Density Maps

A commonly used way to interpret EM density maps of macromolecular complexes is the docking of known atomic structures of protein components or their domains into the EM density map. The fit of known atomic structures makes it often possible to localize specific subunits within the complex and, if the structures of all the components are known, to build a pseudo-atomic model of the assembly. The docking procedure usually starts with the interactive placement of the atomic structures into the reconstructed EM density map to obtain visually a good fit. While it is sometimes not possible to do more, visual docking is often followed by computational refinement of the fit. A number of programs and methods have been developed for the refinement of docking, including foldhunter (Jiang *et al.*, 2001), Situs (Chacon and Wriggers, 2002), real space refinement (RSRef; Korostelev *et al.*, 2002), and normal mode-based flexible fitting (Tama *et al.*, 2004). Docking results become more reliable with increasing resolution of the density map, and at subnanometer resolution, where α-helices become resolved in density maps, the fit of the atomic structures is usually very accurate. For the examples presented here, where crystal structures were docked into single-particle EM reconstructions, the dockings were all done by visual placement and subsequent computational refinement of the fit, treating the crystal structures as rigid bodies. In the case of density maps obtained by cryo-electron tomography, where the resolution is comparably low, atomic structures were only visually placed into the density maps without further computational refinement.

IV. Results

A. The Human Tf–TfR Complex

Members of the Tf family are highly conserved, both in primary sequence and protein fold (Fig. 2A). The 3D structure of Tf proteins has been extensively studied by x-ray crystallography (reviewed in Baker *et al.*, 2003), and crystal structures for a variety of Tf molecules from different species and harboring different mutations

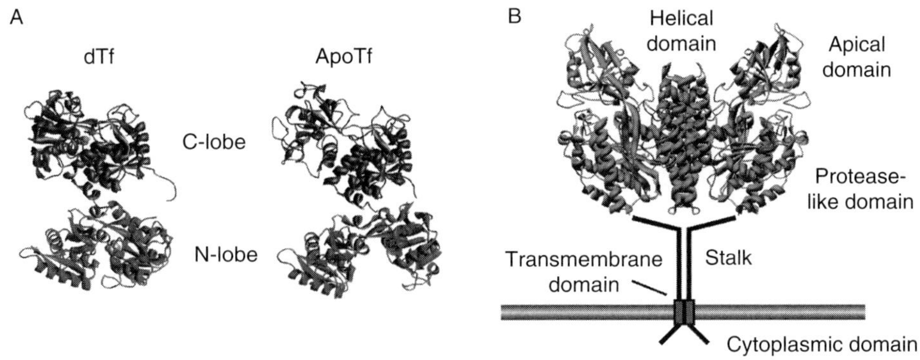

Fig. 2 Crystal structures of Tf and the TfR ectodomain. Reprinted from Supplementary Figure 1 of Cheng *et al.* (2005), with permission from Elsevier. (A) Comparison of iron-bearing transferrin (dTf) with iron-free transferrin (apoTf) shows the jawlike opening of the two domains in each of the two lobes. (B) The crystal structure of the TfR contains only the ectodomain consisting of the protease-like, apical and helical domains. The positions of the stalk, transmembrane domain and cytoplasmic domain as well as the lipid bilayer are indicated.

have been deposited in the protein data bank (www.pdb.org). Tf molecules consist of two homologous lobes termed the N- and C-lobes, for N-terminal and C-terminal lobes. The two lobes are connected by a single loop. Each lobe is composed of two domains (N1, N2 and C1, C2 domains), which form at their interfaces a high-affinity Fe^{3+}-binding site. At the extracellular pH of 7.4, each lobe can bind a Fe^{3+} ion. Coordination of Fe^{3+} requires amino acid ligands from both domains, locking Fe^{3+}-bearing lobes in a closed conformation (Fig. 2A, left side). Release of the Fe^{3+} under slightly acidic conditions, such as in the endosomal lumen, releases the domains and the Tf lobes appear to be in a dynamic conformational equilibrium between their closed and open states (reviewed in Baker *et al.*, 2003) (Fig. 2A, right side).

The structure of the ectodomain of human TfR has also been determined by x-ray crystallography (Lawrence *et al.*, 1999). This work revealed that each subunit of the dimeric receptor consists of an apical domain, a protease-like domain, and a helical domain responsible for dimer interactions (Fig. 2B). Not included in this atomic structure were the stalk region, which connects the TfR ectodomain to the membrane, the single-span transmembrane domain, and the short cytoplasmic N-terminal domain that contains the internalization sequence. On the basis of the crystal structures of Tf and the TfR ectodomain, as well as on available biochemical data, a model was generated for the dTf–TfR complex in which the Tf C-lobe was proposed to bind to the protease-like domain and the Tf N-lobe to the apical domain of the receptor (Lawrence *et al.*, 1999). Attempts to confirm this model by x-ray studies were thwarted by the inability of the dTf–TfR complex so far to form well-ordered crystals.

Single-particle cryo-EM has provided an alternative way to obtain direct structural information for the dTf–TfR complex; this method, of course, does not depend on the dTf–TfR complex forming 3D crystals. Because of its small size (300 kDa) and low symmetry (twofold), the dTf–TfR complex is a challenging object for structure analysis by single-particle cryo-EM. Great care was therefore taken to prepare frozen-hydrated specimens in a very thin ice layer to obtain the best possible image contrast. Furthermore, the work has taken advantage of the high coherence of the electron beam produced by a field emission electron source and of recently developed software to correct for the contrast transfer function (CTF) of the electron microscope in the processing of single-particle images (Grigorieff, 2006). Finally, the images of vitrified dTf–TfR complexes were taken using large defocus values to increase image contrast. Figure 3A shows a typical spread of dTf–TfR complexes embedded in a thin layer of vitreous ice and imaged at -5-μm defocus. The enlarged views shown in Fig. 3B and C reveal that the images contain sufficient structural information that the molecular organization of the dTf–TfR complex can already be seen in the raw images. About 36,000 such particles were manually picked from images of frozen-hydrated dTf–TfR complexes and used to calculate an initial model of the complex, following standard image-processing procedures (described in detail in van Heel *et al.*, 2000). The initial density map was further refined with the program FREALIGN (Grigorieff, 2006), which was also used to correct the CTF of the particle images. The final 3D reconstruction of the dTf–TfR complex was calculated to a nominal resolution of 7.5 Å (Cheng *et al.*, 2004).

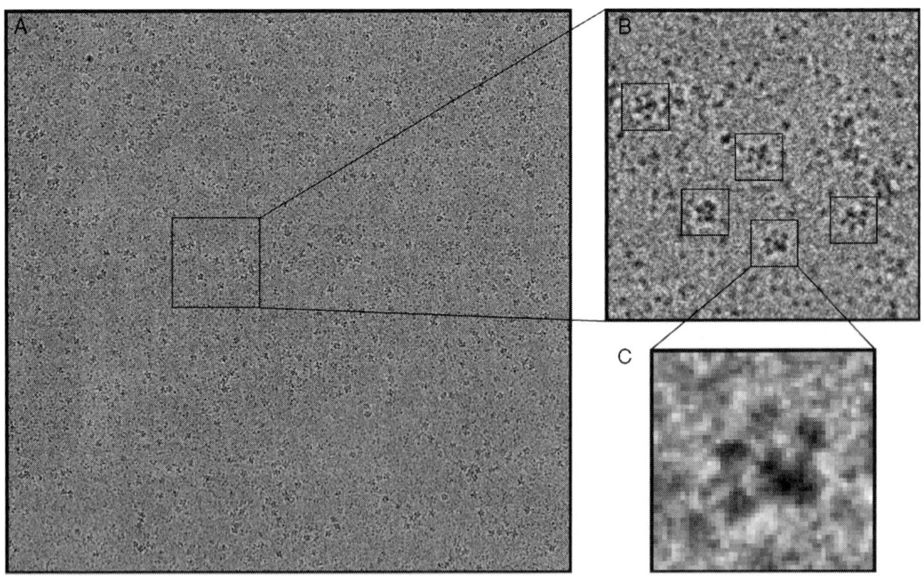

Fig. 3 Cryo-EM of frozen-hydrated Tf–TfR complex. (A) Overview image in which individual complexes are visible as dark speckles. (B) Enlarged image of the area shown in A with individual complexes marked by square frames. (C) Enlarged image of the Tf–TfR complex marked in B, which reveals the individual domains of the complex.

Since crystal structures were available for both dTf and the TfR ectodomain, these structures were docked into the EM density map (Fig. 4A). At the subnanometer resolution of the density map, individual α-helices were resolved, which made it possible to dock the atomic structures very accurately into the EM map. Shown in Fig. 4B is a section through the density map (green) of the helical and protease-like domains of the TfR ectodomain with the docked crystal structure (Cα trace in yellow). Docking the crystal structures of both Tf and the TfR ectodomain into the

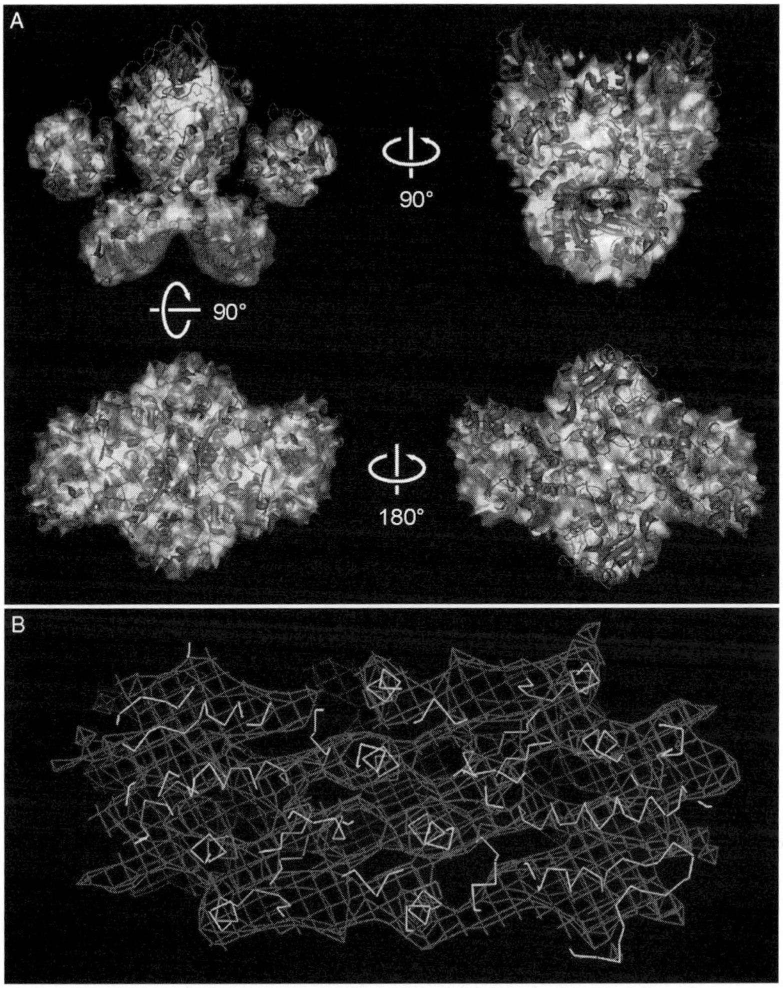

Fig. 4 Docking of the Tf and TfR crystal structures into the EM density map. (A) The Tf and TfR crystal structures (red) were fit into the subnanometer resolution 3D reconstruction of the complex (gold). (B) Section through the EM density map (green) of the helical domain of the TfR. The fit of the Cα trace of the TfR crystal structure (yellow) shows that the density map resolves individual helices. (See Color Insert.)

EM density map produced a pseudo-atomic model for the human dTf–TfR complex (Fig. 5A). The model revealed that the Tf molecules do not bind to the most exposed, membrane-distal domains of the receptor, but instead bind to TfR from the side and bottom. The Tf C-lobe binds to the side of the TfR ectodomain through a single, large interaction surface formed between the Tf C1 domain and the TfR helical domain. The N-lobe inserts itself into a gap between the TfR ectodomain and the membrane surface, and the N1 as well as the N2 domain interact with TfR involving the protease-like and helical domains of the receptor.

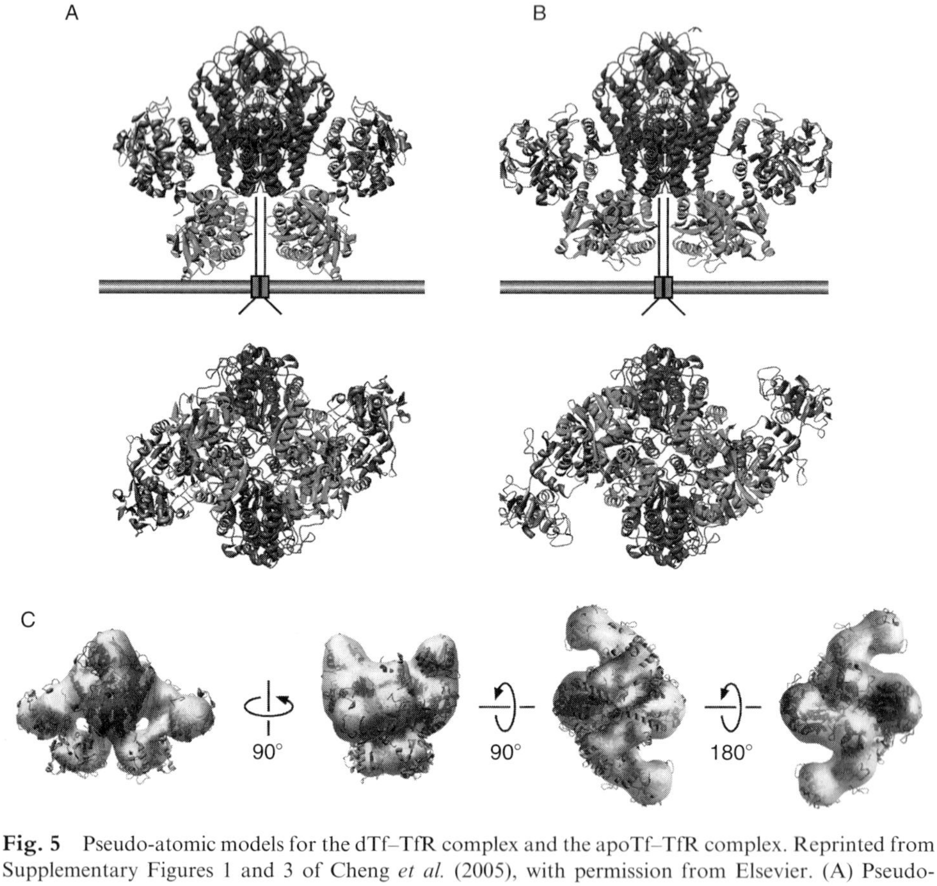

Fig. 5 Pseudo-atomic models for the dTf–TfR complex and the apoTf–TfR complex. Reprinted from Supplementary Figures 1 and 3 of Cheng *et al.* (2005), with permission from Elsevier. (A) Pseudo-atomic model for the complex seen from the side (upper panel) and the bottom (lower panel) obtained by docking the dTf and TfR crystal structures into the EM density map of the complex. (B) Proposed atomic model for the apoTf–TfR complex seen from the side (upper panel) and the bottom (lower panel) obtained by substituting the dTf for the apoTf crystal structure. (C) The good fit of the proposed model for the apoTf–TfR complex into the experimental single-particle EM density map (gold) confirmed the predicted structure of the iron-free complex. (See Color Insert.)

The resolution of the single-particle reconstruction of the human dTf–TfR complex was sufficiently high to resolve some of the α-helices in dTf and TfR, so that docking of the atomic structures into the EM density map could be done with a high level of accuracy and confidence. Even so, because of the very high sequence homology and structural similarity between the Tf N- and C-lobes and because of the unexpected binding mode between Tf and its receptor, it was important to confirm the correctness of the pseudo-atomic model. This was accomplished by studies of negatively stained complexes formed between the TfR ectodomain and isolated Tf N- and C-lobes, which confirmed the relative location of the two Tf lobes with respect to the TfR (Cheng *et al.*, 2004). Furthermore, substitutions of Tf residues identified by the pseudo-atomic model to interact with the TfR ectodomain indeed reduced the binding affinity of mutant Tf to TfR (Cheng *et al.*, 2004). Conversely, many of the TfR residues identified by the pseudo-atomic model to interact with Tf were also identified as important for Tf binding in an independent mutagenesis analysis of TfR (Giannetti *et al.*, 2003).

On the basis of the assumption that the interaction between Tf and TfR remains the same after iron release from the Tf lobes, a model was built for the apoTf–TfR complex simply by substituting the dTf crystal structure by that of apoTf (Fig. 5B) (Cheng *et al.*, 2004). To confirm the resulting model for the apoTf–TfR structure, single-particle cryo-EM was used to produce an experimental density map for the complex. The resolution of the 3D reconstruction of the apoTf–TfR complex could not be improved beyond 17 Å, but the good fit of the pseudo-atomic model of the apoTf–TfR complex into the EM density map (Fig. 5C) confirmed the proposed structural change in the Tf–TfR complex on iron release (Cheng *et al.*, 2005). The reason for the substantially lower resolution of the apoTf–TfR reconstruction as compared to the dTf–TfR reconstruction is most likely due to the structural heterogeneity in the iron-free Tf lobes, which are thought to exist in a dynamic equilibrium between open and closed conformations (Baker *et al.*, 2003).

B. Clathrin Coats

In vertebrates the Tf–TfR complex is taken up by clathrin-mediated endocytosis. Clathrin-coated vesicles are found in all eukaryotic cells in which they transport lipids and cargo proteins from the plasma membrane and the *trans*-Golgi network to early endosomes (Kirchhausen, 2000). The building block of clathrin coats is a trimer of clathrin molecules referred to as triskelion (Kirchhausen and Harrison, 1981; Ungewickell and Branton, 1981). The clathrin triskelion consists of three elongated legs formed by clathrin heavy chains that radiate from a central hub. Each leg of the triskelion can be divided into domains, namely the proximal and distal leg regions, the ankle, the linker, and the N-terminal domain (Fig. 6, top panel) (Fotin *et al.*, 2004b). In addition, each of the three clathrin heavy chains has a light chain bound. In the presence of organelle-specific APs, the clathrin legs intertwine and the triskelions assemble into lattices with mostly pentagonal and hexagonal facets. Clathrin coats without membrane vesicles can be assembled

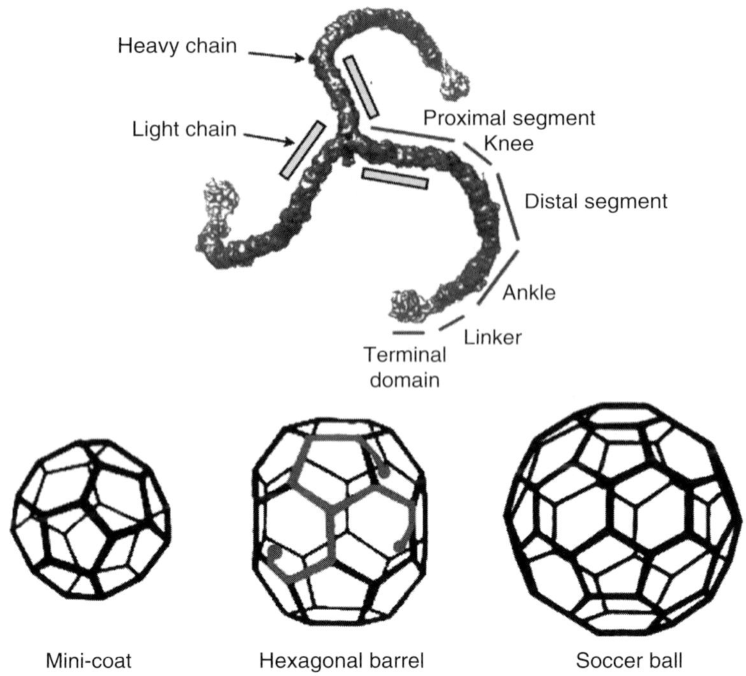

Fig. 6 Clathrin structure. Reprinted from Fotin *et al.* (2004b). The top panel shows a clathrin triskelion with the different domains of the heavy chain marked and the light chain indicated in yellow. The bottom panels show schematic representations of three types of *in vitro*-assembled clathrin coats; the mini-coat with tetrahedral symmetry, the hexagonal barrel with D6 symmetry, and the soccer ball with icosahedral symmetry. (See Color Insert.)

in vitro using clathrin and AP-2 purified from calf brain. *In vitro* assembly conditions can be chosen such that mostly three types of clathrin coats with symmetric lattices form, namely icosahedral coats, D6 barrels, and tetrahedral coats (Fig. 6, bottom panels). Because of its high symmetry and stable structure and because under favorable *in vitro* assembly conditions it is the predominant clathrin coat (Fotin *et al.*, 2004b), to date almost all single-particle cryo-EM studies focused on the structure of the D6 barrel (Fotin *et al.*, 2004b; Heymann *et al.*, 2005; Smith *et al.*, 1998, 2004; Vigers *et al.*, 1986).

In contrast to the Tf–TfR complex, which required the vitreous ice layer to be very thin, clathrin coats are large. A thick ice layer was therefore necessary to prevent deformations of the clathrin lattices (Fig. 7A). Using the single-particle cryo-EM approach with some additional image-processing steps (Fotin *et al.*, 2006), a 3D reconstruction of the D6 barrel coat could be calculated at a resolution of 7.9 Å (Fotin *et al.*, 2004b) (Fig. 7B). Unlike the situation with the Tf–TfR complex, where the atomic structures of the two components of the complex had been solved by X-ray

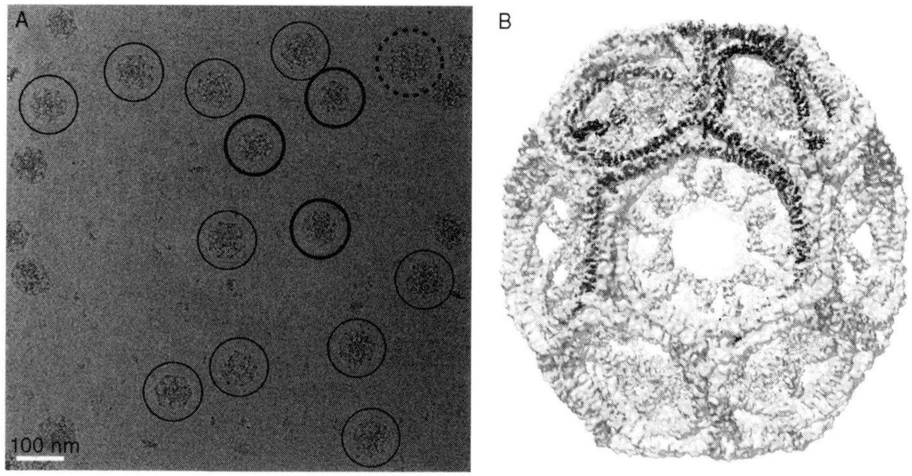

Fig. 7 Cryo-EM of *in vitro*-assembled clathrin coats. (A) Raw image of clathrin coats in vitreous ice. Reprinted from Fotin *et al.* (2004b). The thin circles mark D6 barrels, the thick circles mini-coats, and the dashed circle indicates a soccer ball. (B) 3D reconstruction of a D6 barrel clathrin coat (gold) with docked pseudo-atomic models of three clathrin triskelions forming an asymmetric unit (red, green, and blue). (See Color Insert.)

crystallography, the atomic structure of the clathrin triskelion is not yet known. Instead the structures of two clathrin fragments were available, namely the structures of the proximal leg region and the N-terminal domain of the clathrin heavy chain (ter Haar *et al.*, 1998; Ybe *et al.*, 1999), which could both be docked into the EM density map of the D6 barrel. Comparison of the density map calculated from the crystal structure of the proximal leg region limited to a resolution of 7.9 Å with the corresponding density in the EM map confirmed that the 3D reconstruction had subnanometer resolution (Fotin *et al.*, 2004b). Most of the clathrin heavy chain consists of tandem repeats, known as clathrin heavy chain repeats (CHCRs) (Ybe *et al.*, 1999). The crystal structure of the proximal region contained the entire CHCR6 and revealed that the repeats consist of an α-helical zigzag. On the basis of the crystal structure of CHCR6 and the amino acid sequences of the other repeats, homology models were generated for the remaining CHCRs and docked into the density map. Some density at the hub of the triskelion remained unaccounted for, but could be modeled as an extended α-helix, in accordance with the high probability of the C-terminal region of the clathrin heavy chain to adopt α-helical secondary structure (Fotin *et al.*, 2004b). In this way a pseudo-atomic model was generated for an entire clathrin heavy chain, based on which the structures of a clathrin triskelion and finally of an entire D6 barrel could be built (Fig. 12C).

In a next step, 3D reconstructions were calculated for D6 barrel coats assembled from clathrin triskelions containing or lacking light chains. The difference map (Fig. 8A) visualized the light chain as a rod-shaped density, about 100-Å long,

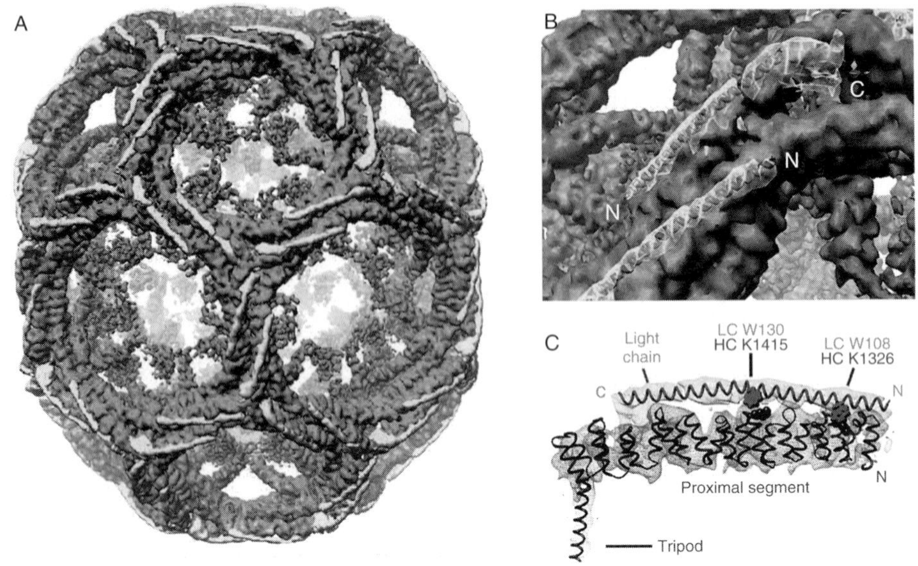

Fig. 8 Interaction between the clathrin light and heavy chains. Reprinted from Fotin *et al.* (2004b). (A) 3D reconstruction of a D6 barrel *in vitro*-assembled from clathrin triskelions lacking light chains (blue). The yellow densities are the difference peaks between density maps of D6 barrels with and without light chains. (B) A close-up look on the clathrin light chain (yellow) with a model for the helical region of the light chain (red) docked into it. (C) Side chains in the clathrin heavy and light chains known from mutagenesis studies to interact with each other, thus defining the orientation of the light chain relative to the heavy chain. (See Color Insert.)

attached to the outer surface of the clathrin heavy chain with a more globular density near the vertices of the clathrin lattice (Fig. 8B). On the basis of its amino acid sequence, the central region of the light chain was predicted to adopt an α-helical secondary structure, and therefore a model for the α-helical central region of the light chain was docked into the difference density (Fig. 8B). The polarity of the light chain was already known from antibody labeling experiments that localized the C-terminus of the light chain near the vertices of the clathrin lattice (Kirchhausen and Toyoda, 1993). The docking was further guided by mutagenesis studies that showed that two lysine residues in the heavy chain, Lys 1326 and Lys 1415, interacted with two tryptophan residues in the light chain, Trp 108 and Trp 130 (Chen *et al.*, 2002). The combination of all this information led to a uniquely defined model for the binding of the clathrin light chain to the heavy chain (Fig. 8C) (Fotin *et al.*, 2004b). Finally, D6 barrel coats were decorated with a fragment of auxilin, a protein involved in the disassembly of clathrin coats. A difference map between the 3D reconstructions with and without bound auxilin localized the clathrin-binding auxilin fragment to the vertices of the clathrin lattice (Fotin *et al.*, 2004a) (Fig. 12C).

C. Clathrin-Coated Vesicles

With the pseudo-atomic model of a complete D6 barrel clathrin coat in hand, the next question concerned the interactions between clathrin coats and the encapsulated lipid vesicles. In particular, it was of interest to understand how clathrin lattices assemble to accommodate vesicles of different sizes and how the APs anchor the vesicle to the clathrin lattice. To address these questions, the structure of entire clathrin-coated vesicles had to be visualized.

Clathrin-coated vesicles were isolated from calf brain. Most of the brain-derived vesicles are precursors of synaptic vesicles, and the vesicles are therefore small and quite homogeneous in size compared to clathrin-coated vesicles isolated from other tissues (Maycox *et al.*, 1992). Figure 9A shows an image of clathrin-coated vesicles in a layer of vitreous ice. In spite of their apparent similarity, close examination shows that all the vesicles in this image differ in size, shape, and protein composition. Moreover, some of the vesicles have already lost their clathrin coats, whereas some clathrin coats do not contain a vesicle and may in fact only have assembled during the purification procedure. The obvious structural heterogeneity made it impossible to use single-particle EM because this approach depends on the averaging of images of structurally identical complexes. Electron tomography was thus the only approach that could be used to study the structure of *in vivo*-assembled clathrin-coated vesicles.

Figure 9B–D show three raw images from a tilt series (–62.6° to +51.0° recorded with 2° increments) of clathrin-coated vesicles embedded in vitreous ice. Figure 10A shows a central section through the tomogram of the vitreous ice slab reconstructed from this tilt series. The section clearly reveals membrane vesicles in many of the imaged clathrin coats. Shown in Fig. 10B is the 3D density map of one of the coated vesicles boxed out from the entire tomogram after the noise was reduced by applying a 3×3 pixel median filter (Winkler and Taylor, 2006). The pentagons and hexagons of the clathrin lattice are clearly resolved in this density map, as is the membrane of the encapsulated vesicle. Because of the missing wedge effect (Baumeister, 2005), the density of the vesicle membrane is only continuous around the equatorial area; it is broken at the top and bottom of the tomogram. A central section through the density map in the equatorial area (Fig. 10C) shows that the lipid vesicle is surrounded by two layers of density. The pseudo-atomic model obtained from the single-particle reconstruction of *in vitro*-assembled clathrin coats was used to model the clathrin lattice of the coated vesicles (red in Fig. 10C). The pentagons and hexagons formed by the intertwining legs of the clathrin triskelions account well for the outermost layer of density. The weaker and less well-defined layer of densities in between the outer density of the clathrin legs and the inner density of the vesicle membrane represents the N-terminal domains of the clathrin heavy chain. There are also some additional densities that connect the N-terminal domains of the clathrin triskelions to the vesicle membrane; these might represent adaptor complexes. Finally, densities inside the vesicle may be due to the presence of cargo molecules trapped inside the vesicle.

All coated vesicles in the tomogram have different sizes and shapes. Figure 11A shows central sections through eight clathrin-coated vesicles that were boxed out

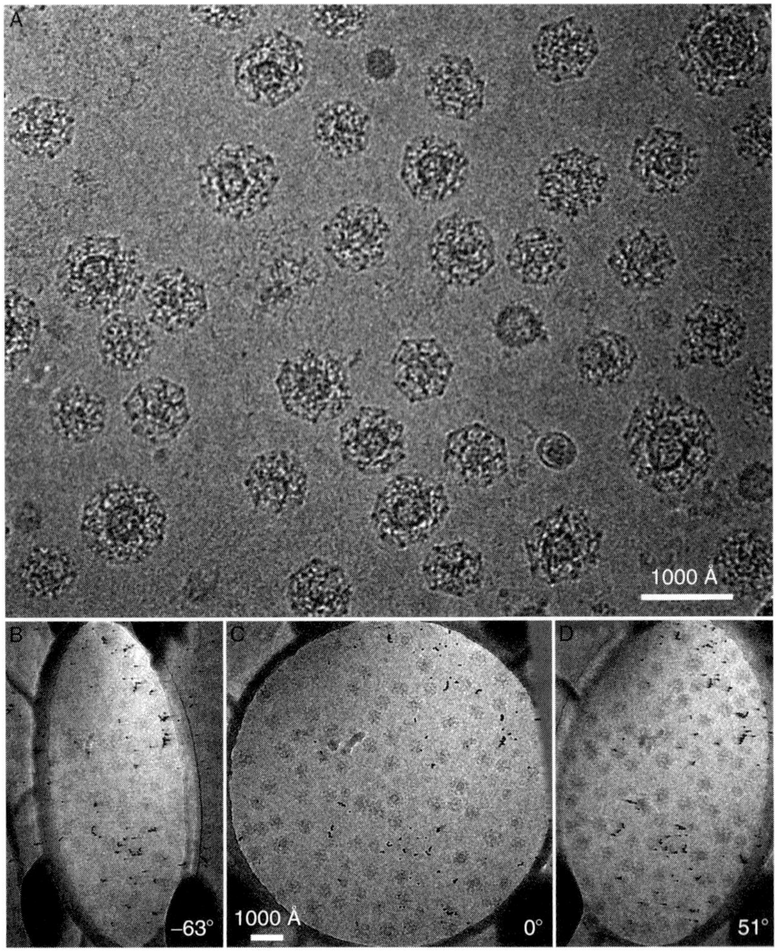

Fig. 9 Electron tomography of clathrin-coated vesicles in vitreous ice. Reprinted from Figure 1 of Cheng *et al.* (2006), with permission from Elsevier. (A) High electron dose cryo-EM image of clathrin-coated vesicles embedded in vitreous ice. (B–D) Three low electron dose cryo-EM images of a tomographic tilt series. The black dots are colloidal gold particles used for aligning the images of the tilt series to each other.

from the tomogram, arranged in size from the largest (1) to the smallest coated vesicle (8). The vesicles inside the clathrin lattice have a spherical shape, and their sizes change with those of the coats. The clathrin lattices of almost all the vesicles show no obvious symmetry. Nonetheless, the organization of the individual clathrin coats could unambiguously be determined by docking straight lines (polyalanine α-helices) into the densities forming the edges of the clathrin lattices. The resulting lattice designs were all different (Fig. 11B). All the clathrin lattices contained predominantly pentagons and hexagons, but occasionally heptagons were also present (vesicles 1 and 4). While different from each other, all these clathrin lattices

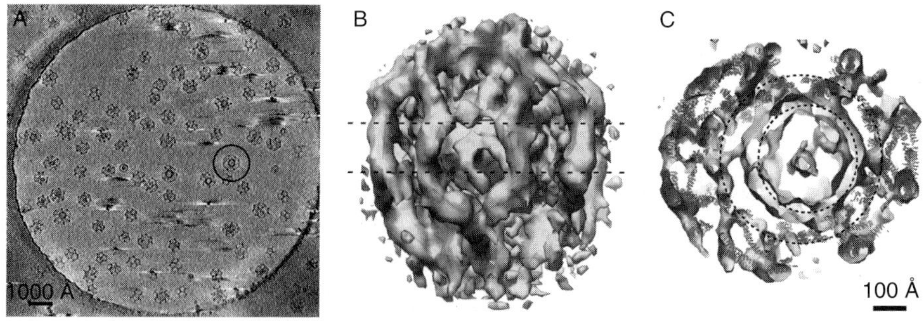

Fig. 10 Electron tomographic reconstruction of a clathrin-coated vesicle. Reprinted from Figures 2 and 5 of Cheng *et al.* (2006), with permission from Elsevier. (A) Central section (15.5-nm thick) through a cryo-electron tomogram of a hole in a carbon film filled with vitrified ice containing coated vesicles. (B) Surface-rendered representation of the clathrin-coated vesicle marked in panel A. (C) Slice through the clathrin-coated vesicle at the position indicated in B. The density map was interpreted using the pseudo-atomic model of the clathrin heavy chain (red). The dotted lines represent the positions of the lipid bilayer (inner circle) and the adaptor complexes (outer circle). (See Color Insert.)

followed Euler's law governing the design of regular polyhedrons: $E = F + V - 2$, where E is the number of edges, V is the number of vertices, and F is the number of faces of the polyhedron. Euler's law demands that all closed lattices consisting exclusively of hexagons and pentagons will contain 12 pentagons, although they can have different numbers of hexagons. All lattices containing a heptagon can have varying numbers of hexagons but must have 13 pentagons. Only the lattice of the smallest coated vesicle observed (vesicle 8 in Fig. 11B) was symmetric. While its size was that of a D6 barrel, the lattice had D2 symmetry, constituting a so-called tennis ball design first described by Crowther (Crowther *et al.*, 1976). The lattices of all other observed coated vesicles did not display symmetry. The coated vesicle with the tennis ball design contained a vesicle with a mean diameter of about 27 nm, while the mean diameter of the surrounding clathrin coat was about 70 nm. This is the smallest, vesicle-containing clathrin lattice observed so far.

The vesicles in the coated vesicles were all spherical in shape, but they were not located exactly in the center of the clathrin lattices. This is indicated by the two dashed circles in Fig. 10C, where the larger circle marks the location of the N-terminal domains of the clathrin triskelions and the smaller circle delineates the membrane of the vesicle. The N-terminal domains on one side of the coat are in close contact with the vesicle, while there is a 50- to 100-Å-wide gap on the opposite side. This asymmetry was seen in all analyzed vesicles (Fig. 11). The observed eccentricity of the vesicles within the clathrin lattices may be explained by differences between the initiation and finishing points in the formation of clathrin coats. While there have to be interactions between clathrin triskelions and the plasma membrane at the initiation site of clathrin-coated pits, such interactions may not be required once the vesicles are about to be pinched off from the plasma membrane and the clathrin coats close up. This interpretation is in agreement with the current picture of initiation and growth of clathrin-coated vesicles, based on many other studies (Ehrlich *et al.*, 2004).

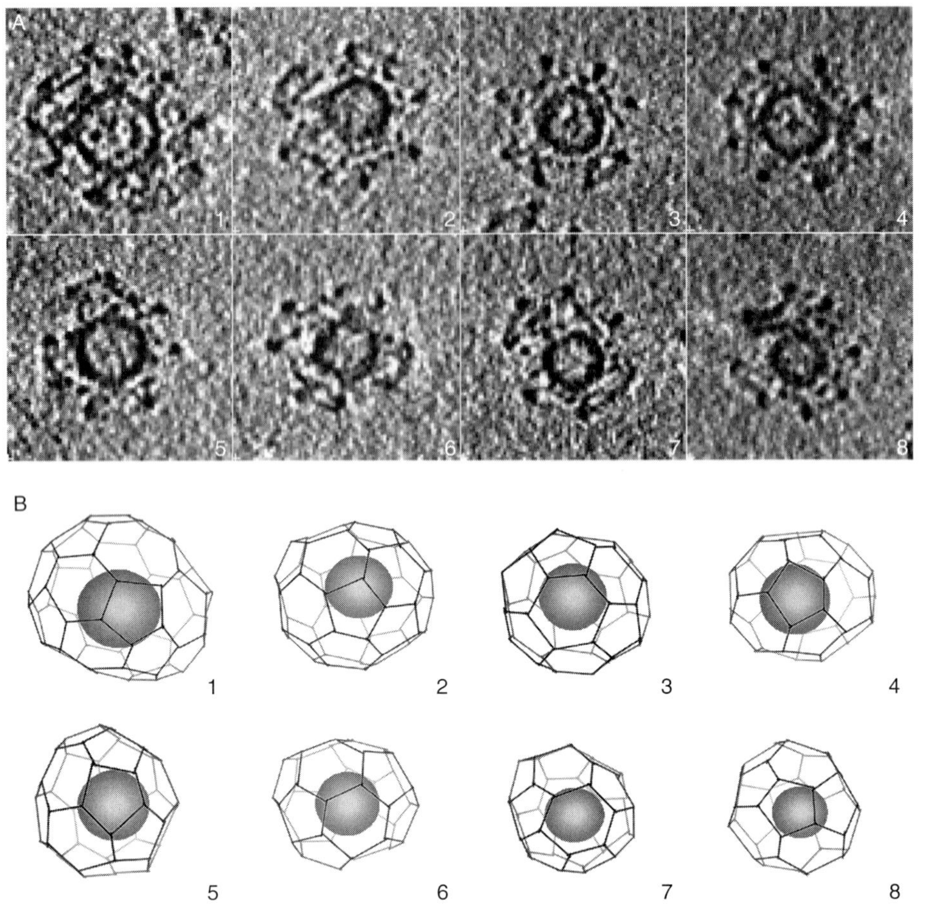

Fig. 11 Design of clathrin-coated vesicles. Reprinted from Figure 4 of Cheng *et al.* (2006), with permission from Elsevier. (A) Central sections of eight coated vesicles boxed out from a cryo-electron tomogram. (B) Schematic representations of the design of the clathrin lattices and the positions of the vesicles relative to the lattices.

V. Discussion

In the example we present here, two different molecular EM techniques, single-particle EM and electron tomography, were used to study the structure of protein assemblies involved in the process of iron uptake, which is based on the clathrin-mediated endocytotic cycle of the Tf–TfR complex. By applying single-particle EM to vitrified samples, density maps at subnanometer resolution could be produced that resolved α-helices. The technique relies, however, on the averaging of the images of a large number of structurally identical proteins or protein complexes. Because of the requirement to average images, the structural homogeneity of the sample is a critical factor for the success of this technique. By contrast, electron tomography does not rely on image averaging, although in some instances it is possible to average objects that have been visualized in a tomogram. Electron tomography can thus be used to study the structure of large protein assemblies with structural heterogeneities. The resolution of electron tomograms is, however, considerably lower than the resolution that can be obtained with single-particle cryo-EM, and the interpretation of electron tomographic density maps can therefore be quite challenging.

As shown in Fig. 12A, cellular iron uptake is a complex process that involves many steps such as initiation of clathrin coat formation, cargo selection (in this case the Tf–TfR complex), growth of the clathrin coat, pinching off of the coated vesicle, disassembly of the clathrin coat, fusion of the transport vesicle with the endosome, iron release by the Tf–TfR complex, and many others. All these steps are carried

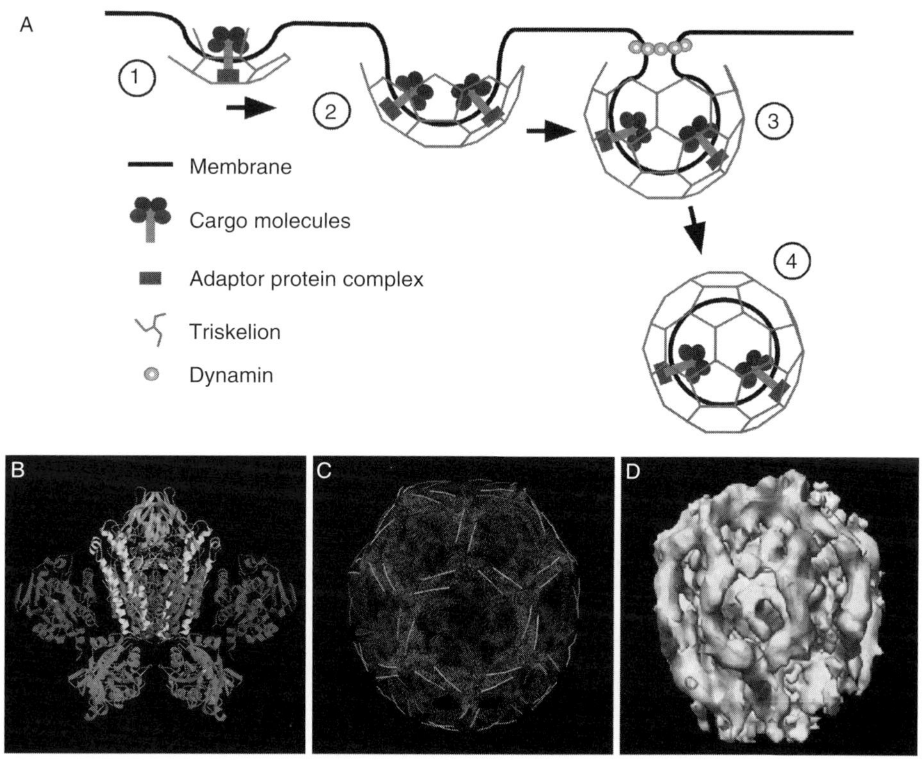

Fig. 12 Structures of some of the protein assemblies involved in iron uptake by clathrin-mediated endocytosis. (A) Schematic drawing showing the formation of clathrin-coated vesicle containing Tf–TfR complexes. (B) Pseudo-atomic model of human dTf–TfR complex. (C) Pseudo-atomic model of a complete D6 barrel clathrin coat (blue) with bound light chain (yellow) and density representing the clathrin-binding fragment of auxilin (red density). (D) 3D reconstruction of a clathrin-coated vesicle obtained with cryo-electron tomography. (See Color Insert.)

out by protein complexes that assemble and disassemble at different stages of the process.

As a whole, the process is much too complicated to be studied by any single approach currently used in structural biology, and it is thus inevitable to employ a "divide and conquer" approach. Typically, the first step in the structural analysis of a cellular process is to isolate individual proteins or protein domains and to determine their structure by x-ray crystallography or NMR spectroscopy. In a next step, *in vivo*-assembled complexes are isolated or isolated proteins are assembled *in vitro* to form a complex. In favorable cases the structures of protein complexes can also be determined by x-ray crystallography, but in some cases it proved difficult to grow crystals of complexes as in the case of the Tf–TfR complex. Obtaining 3D crystals becomes more difficult with less stable and/or more complex protein assemblies such as clathrin coats, and is impossible for nonstoichiometric assemblies such as clathrin-coated vesicles. Molecular EM is then the approach with which to visualize the structure of complexes: single-particle EM when the particle images can be averaged and electron tomography for objects with unique structures. Finally, docking of crystal structures into the resulting EM density maps generates pseudo-atomic models for macromolecular assemblies as demonstrated for the Tf–TfR complex (Fig. 12B), the D6 barrel clathrin coat (Fig. 12C), and to some degree even for clathrin-coated vesicles (Fig. 12D).

As mentioned before, EM density maps, obtained either by single-particle EM or by electron tomography, and pseudo-atomic models based on these EM maps have to be interpreted with caution. As illustrated with the presented examples, molecular interpretations should whenever possible be verified by independent means. Should this not be possible, it is important to draw attention to the speculative nature of unconfirmed models, since atomic models, irrespective of how they were obtained, are often taken for the ultimate truth and can thus be misleading. Therefore, researchers outside the molecular EM field have to be made aware of the limitations of the techniques used to create pseudo-atomic models.

There are still many components of the clathrin-mediated pathway for endocytosis of the Tf–TfR complex that have not yet been integrated into the big "structural" picture. However, similar studies that follow the divide and conquer approach presented here should be able to produce mechanistic insights into how Hsc70 and auxilin initiate the disassembly of clathrin coats, how dynamin pinches off membrane vesicles, and the many other steps of the Tf cycle. The ultimate goal would be to integrate this molecular information with the dynamic information provided by live-cell light imaging to produce molecular movies of cellular processes.

VI. Summary

Using clathrin-mediated endocytosis of the Tf–TfR complex as an example, we described how multistep biological processes performed by dynamic protein assemblies can be studied by a "divide and conquer" approach that combines different structural techniques. By fitting crystal structures of the subunits into density maps obtained by single-particle EM for objects with identical structures or by electron tomography for objects with unique structures, it is possible to produce pseudo-atomic models for the macromolecular machineries that carry out the various processes that occur in living cells.

References

Adrian, M., Dubochet, J., Lepault, J., and McDowall, A. W. (1984). Cryo-electron microscopy of viruses. *Nature* **308,** 32–36.

Aisen, P., Enns, C., and Wessling-Resnick, M. (2001). Chemistry and biology of eukaryotic iron metabolism. *Int. J. Biochem. Cell Biol.* **33,** 940–959.

Baker, H. M., Anderson, B. F., and Baker, E. N. (2003). Dealing with iron: Common structural principles in proteins that transport iron and heme. *Proc. Natl. Acad. Sci. USA* **100,** 3579–3583.

Baumeister, W. (2005). From proteomic inventory to architecture. *FEBS Lett.* **579,** 933–937.

Bottcher, B., Wynne, S. A., and Crowther, R. A. (1997). Determination of the fold of the core protein of hepatitis B virus by electron cryomicroscopy. *Nature* **386,** 88–91.

Chacon, P., and Wriggers, W. (2002). Multi-resolution contour-based fitting of macromolecular structures. *J. Mol. Biol.* **317,** 375–384.

Chen, C. Y., Reese, M. L., Hwang, P. K., Ota, N., Agard, D., and Brodsky, F. M. (2002). Clathrin light and heavy chain interface: Alpha-helix binding superhelix loops via critical tryptophans. *EMBO J.* **21,** 6072–6082.

Cheng, Y., Boll, W., Kirchhausen, T., Harrison, S. C., and Walz, T. (2006). Cryo-electron tomography of clathrin coated vesicles—structural implications for coat assembly. *J. Mol. Biol.* (in press).

Cheng, Y., Zak, O., Aisen, P., Harrison, S. C., and Walz, T. (2004). Structure of the human transferrin receptor-transferrin complex. *Cell* **116,** 565–576.

Cheng, Y., Zak, O., Aisen, P., Harrison, S. C., and Walz, T. (2005). Single particle reconstruction of the human apo-transferrin-transferrin receptor complex. *J. Struct. Biol.* **152,** 204–210.

Collins, B. M., McCoy, A. J., Kent, H. M., Evans, P. R., and Owen, D. J. (2002). Molecular architecture and functional model of the endocytic AP2 complex. *Cell* **109,** 523–535.

Crowther, R. A., Finch, J. T., and Pearse, B. M. (1976). On the structure of coated vesicles. *J. Mol. Biol.* **103,** 785–798.

Ehrlich, M., Boll, W., Van Oijen, A., Hariharan, R., Chandran, K., Nibert, M. L., and Kirchhausen, T. (2004). Endocytosis by random initiation and stabilization of clathrin-coated pits. *Cell* **118,** 591–605.

Enns, C. A., Rutledge, E. A., and Williams, A. M. (1996). The transferrin receptor. *Biomembranes* **4,** 255–287.

Fotin, A., Cheng, Y., Grigorieff, N., Walz, T., Harrison, S. C., and Kirchhausen, T. (2004a). Structure of an auxilin-bound clathrin coat and its implications for the mechanism of uncoating. *Nature* **432,** 649–653.

Fotin, A., Cheng, Y., Sliz, P., Grigorieff, N., Harrison, S. C., Kirchhausen, T., and Walz, T. (2004b). Molecular model for a complete clathrin lattice from electron cryomicroscopy. *Nature* **432,** 573–579.

Fotin, A., Kirchhausen, T., Grigorieff, N., Harrison, S. C., Walz, T., and Cheng, Y. (2006). Structure determination of clathrin coats to subnanometer resolution by single particle cryo-electron microscopy. *J. Struct. Biol.* (in press).

Frangakis, A. S., and Hegerl, R. (2001). Noise reduction in electron tomographic reconstructions using nonlinear anisotropic diffusion. *J. Struct. Biol.* **135,** 239–250.

Frangakis, A. S., Stoschek, A., and Hegerl, R. (2001). Wavelet transform filtering and nonlinear anisotropic diffusion assessed for signal reconstruction performance on multidimensional biomedical data. *IEEE Trans. Biomed. Eng.* **48,** 213–222.

Frank, J. (2006). "Three-dimensional Electron Microscopy of Macromolecular Assemblies." Oxford University Press, Cary.

Frank, J., Rademacher, M., Penczek, P., Zhu, J., Ladjadj, M., and Leith, A. (1996). SPIDER and WEB: Processing and visulization of images in 3D electron microscopy and related fields. *J. Struct. Biol.* **116,** 190–199.

Giannetti, A. M., Snow, P. M., Zak, O., and Bjorkman, P. J. (2003). Mechanism for multiple ligand recognition by the human transferrin receptor. *PLoS Biol.* **1,** 341–350.

Grigorieff, N. (2006). FREALIGN: High-resolution refinement of single particle structures. *J. Struct. Biol.* (in press).

Harauz, G., and van Heel, M. (1986). Exact filters for general geometry 3-dimensional reconstruction. *Optik* **73,** 146–156.

Heuser, J. (1980). Three-dimensional visualization of coated vesicle formation in fibroblasts. *J. Cell Biol.* **84,** 560–583.

Heymann, J. B., Iwasaki, K., Yim, Y. I., Cheng, N., Belnap, D. M., Greene, L. E., Eisenberg, E., and Steven, A. C. (2005). Visualization of the binding of Hsc70 ATPase to clathrin baskets: Implications for an uncoating mechanism. *J. Biol. Chem.* **280,** 7156–7161.

Jiang, W., Baker, M. L., Ludtke, S. J., and Chiu, W. (2001). Bridging the information gap: Computational tools for intermediate resolution structure interpretation. *J. Mol. Biol.* **308,** 1033–1044.

Jiang, W., Chang, J., Jakana, J., Weigele, P., King, J., and Chiu, W. (2006). Structure of epsilon15 bacteriophage reveals genome organization and DNA packaging/injection apparatus. *Nature* **439,** 612–616.

Kirchhausen, T. (2000). Clathrin. *Annu. Rev. Biochem.* **69,** 699–727.

Kirchhausen, T., and Harrison, S. C. (1981). Protein organization in clathrin trimers. *Cell* **23,** 755–761.

Kirchhausen, T., and Toyoda, T. (1993). Immunoelectron microscopic evidence for the extended conformation of light chains in clathrin trimers. *J. Biol. Chem.* **268,** 10268–10273.

Korostelev, A., Bertram, R., and Chapman, M. S. (2002). Simulated-annealing real-space refinement as a tool in model building. *Acta Crystallogr. D Biol. Crystallogr.* **58,** 761–767.

Lawrence, C. M., Ray, S., Babyonyshev, M., Galluser, R., Borhani, D. W., and Harrison, S. C. (1999). Crystal structure of the ectodomain of human transferrin receptor. *Science* **286,** 779–782.

Ludtke, S. J., Baldwin, P. R., and Chiu, W. (1999). EMAN: Semiautomated software for high-resolution single-particle reconstructions. *J. Struct. Biol.* **128,** 82–97.

Ludtke, S. J., Chen, D. H., Song, J. L., Chuang, D. T., and Chiu, W. (2004). Seeing GroEL at 6 Å resolution by single particle electron cryomicroscopy. *Structure* **12,** 1129–1136.

Mastronarde, D. N. (2005). Automated electron microscope tomography using robust prediction of specimen movements. *J. Struct. Biol.* **152,** 36–51.

Maycox, P. R., Link, E., Reetz, A., Morris, S. A., and Jahn, R. (1992). Clathrin-coated vesicles in nervous tissue are involved primarily in synaptic vesicle recycling. *J. Cell Biol.* **118,** 1379–1388.

Perry, M. M., and Gilbert, A. B. (1979). Yolk transport in the ovarian follicle of the hen (*Gallus domesticus*): Lipoprotein-like particles at the periphery of the oocyte in the rapid growth phase. *J. Cell Sci.* **39,** 257–272.

Roth, T. F., and Porter, K. R. (1964). Yolk protein uptake in the oocyte of the mosquito *Aedes Aegypti. L. J. Cell Biol.* **20,** 313–332.

Smith, C. J., Dafforn, T. R., Kent, H., Sims, C. A., Khubchandani-Aswani, K., Zhang, L., Saibil, H. R., and Pearse, B. M. (2004). Location of auxilin within a clathrin cage. *J. Mol. Biol.* **336,** 461–471.

Smith, C. J., Grigorieff, N., and Pearse, B. M. (1998). Clathrin coats at 21 Å resolution: A cellular assembly designed to recycle multiple membrane receptors. *EMBO J.* **17,** 4943–4953.

Stewart, A., and Grigorieff, N. (2004). Noise bias in the refinement of structures derived from single particles. *Ultramicroscopy* **102,** 67–84.

Tama, F., Miyashita, O., and Brooks, C. L., III (2004). Flexible multi-scale fitting of atomic structures into low-resolution electron density maps with elastic network normal mode analysis. *J. Mol. Biol.* **337,** 985–999.

Taylor, K. A., and Glaeser, R. M. (1974). Electron diffraction of frozen, hydrated protein crystals. *Science* **186,** 1036–1037.

ter Haar, E., Musacchio, A., Harrison, S. C., and Kirchhausen, T. (1998). Atomic structure of clathrin: A beta propeller terminal domain joins an alpha zigzag linker. *Cell* **95,** 563–573.

Ungewickell, E., and Branton, D. (1981). Assembly units of clathrin coats. *Nature* **289,** 420–422.

van Dam, E. M., and Stoorvogel, W. (2002). Dynamin-dependent transferrin receptor recycling by endosome-derived clathrin-coated vesicles. *Mol. Biol. Cell* **13,** 169–182.

van Heel, M., Gowen, B., Matadeen, R., Orlova, E. V., Finn, R., Pape, T., Cohen, D., Stark, H., Schmidt, R., Schatz, M., and Patwardhan, A. (2000). Single-particle electron cryo-microscopy: Towards atomic resolution. *Q. Rev. Biophys.* **33,** 307–369.

van Heel, M., Harauz, G., Orlova, E. V., Schmidt, R., and Schatz, M. (1996). A new generation of the IMAGIC image processing system. *J. Struct. Biol.* **116,** 17–24.

Vigers, G. P., Crowther, R. A., and Pearse, B. M. (1986). Three-dimensional structure of clathrin cages in ice. *EMBO J.* **5,** 529–534.

Willingham, M. C., Hanover, J. A., Dickson, R. B., and Pastan, I. (1984). Morphologic characterization of the pathway of transferrin endocytosis and recycling in human KB cells. *Proc. Natl. Acad. Sci. USA* **81,** 175–179.

Winkler, H., and Taylor, K. A. (2006). Accurate marker-free alignment with simultaneous geometry determination and reconstruction of tilt series in electron tomography. *Ultramicroscopy* **106,** 240–254.

Wright, R. (2000). Transmission electron microscopy of yeast. *Microsc. Res. Tech.* **51,** 496–510.

Ybe, J. A., Brodsky, F. M., Hofmann, K., Lin, K., Liu, S. H., Chen, L., Earnest, T. N., Fletterick, R. J., and Hwang, P. K. (1999). Clathrin self-assembly is mediated by a tandemly repeated superhelix. *Nature* **399,** 371–375.

Zheng, Q. S., Braunfeld, M. B., Sedat, J. W., and Agard, D. A. (2004). An improved strategy for automated electron microscopic tomography. *J. Struct. Biol.* **147,** 91–101.

Zhou, Z. H., Dougherty, M., Jakana, J., He, J., Rixon, F. J., and Chiu, W. (2000). Seeing the herpesvirus capsid at 8.5 Å. *Science* **288,** 877–880.

PART IV

Localizing Macromolecules in Cells

Learning the distribution of a cell's structural components, like its membranes and chromatin, is informative; the results describe aspects of cell organization, which in turn can be used to exclude specific models and clarify cellular mechanisms. Such knowledge is, however, much less information than we would like to obtain, because it usually tells us nothing about the specific macromolecules that are components of the structures we can see. Particularly in the context of genomic and proteomic information, it is desirable to know just which molecular players are contributing to each of the objects visualized. From another point of view, an effort to understand the function of a macromolecule is often abetted by knowledge about that molecule's cellular localization as a function of time and experimental condition. Thus, it is important to develop techniques that can place macromolecules into maps of cell structure.

The most common method for this task uses antibodies to affix a visible tag to a given macromolecule, displaying its localization with whatever optics are used. Immunofluorescence and immuno-electro microscopy (immuno-EM) are familiar techniques, and several chapters in this section describe some of the best methods currently available for traveling this useful road. Morphew (Chapter 19) has reviewed methods for labeling structures of interest with small particles of colloidal gold, applying this approach in combination with epitope tagging or with chimeras that include a protein of interest and the green fluorescent protein. She also shows how serial section EM and electron tomography (ET) can be used to get some level of immunolocalization in three dimensions (3D). Pelchen-Mathews and Marsh (Chapter 20) show how immuno-EM can be invaluable for the study of virus development because antigenicity can distinguish between intermediates in virus formation and other cellular particles. Ladinsky and Howell (Chapter 21) describe ways in which antibodies can be usefully applied to frozen sections cut from aldehyde-fixed samples that were infiltrated with sucrose, then frozen to permit their sectioning without dehydration or embedding (Tokuyasu, 1980). Here again, some 3D localization of label is possible, thanks to tomographic reconstructions of the labeled sections.

Cheutin *et al.* (Chapter 22) show how one can combine chemical fixation with gentle cellular lysis to make preparations that retain many aspects of cell structure but allow the application of antibodies labeled with very small tags before the cell is further processed for EM. The two great advantages of this approach are that the antibodies have wide-ranging access to their antigens within the lysed cell, and the resulting distributions of label are readily seen in 3D, either by ET or serial sections. The concern with this approach is that a light fixation, combined with a lysis that is sufficient to allow labeled antibody entry, may not retain the antigen's position from the living cell. As these authors discuss, the suitability of the method depends on the structure that is to be studied.

Sosinsky *et al.* (Chapter 23) have developed a method that again relies on aldehyde-fixation but employs probes that are so small that no special permeablization is required. A protein motif that can be engineered into many if not all proteins is used to bind a biarsenical group developed in the laboratory of Roger Tsien (Gaietta *et al.*, 2002). This combining group can be attached to various labels: either a small fluorescent dye or a fluorophore that is readily bleached, generating free radicals that can go on to oxidize a reporter like diaminobenzidine (DAB). The resulting precipitate binds enough OsO_4 to make a readily visible EM stain. This promising method is compared with several others and provides an important new way to think about placing labels on proteins in cells.

Leapman and Aronova (Chapter 24) show the value of an electron energy filter for detecting certain kinds of labels. During the imaging process, some beam electrons are scattered "inelastically," meaning that they lose a small fraction of their energy to the atoms in the sample. This energy loss is often associated with an electronic transition in the atom that did the scattering, so energy loss is quantized. An electron energy loss spectrometer can measure these losses and provide evidence about the atom that did the scattering. An imaging, energy loss spectrometer can place each inelastic event within an image of the specimen, revealing the cellular location of the scattering atom. This approach can be used to determine the distribution of some naturally occurring atoms, like the phosphorous in nucleic acids, and a macromolecule, like ferritin, which contains iron. The method may also work for other atomic labels that give characteristic electron energy loss spectra.

Finally, Best *et al.* (Chapter 25) describe methods that are under development in Wolfgang Baumeister's laboratory to use correlations between 3D images of frozen-hydrated cells and databases of 3D protein structures to place specific macromolecular complexes in cells on the basis solely of their appearance; this method circumvents the need for added labels of any kind (Bohm *et al.*, 2000; Medalia *et al.*, 2002). In principle, this method should beat all others, since genes and gene products could be localized without tagging. In practice, there are many problems still to be solved before this approach can be widely applied to a protein of choice. The evidence presented here and in the literature shows that large protein complexes can be identified by shape alone, both in simple mixtures with a few other complexes and for some cases *in vivo*. The method is computationally intensive, yet it may ultimately be applicable to a wide variety of structures. It is exciting to think about the possibility of placing all proteins whose atomic structures have been determined within a 3D map of a frozen cell! Readers can assess for themselves how progress, both past and future, will bring us toward this goal.

References

Bohm, J., Frangakis, A. S., Hegerl, R., Nickell, S., Typke, D., and Baumeister, W. (2000). Toward detecting and identifying macromolecules in a cellular context: Template matching applied to electron tomograms. *Proc. Natl. Acad. Sci. USA* **97,** 14245–14250.

Gaietta, G., Deerinck, T. J., Adams, S. R., Bouwer, J., Tour, O., Laird, D. W., Sosinsky, G. E., Tsien, R. Y., and Ellisman, M. H. (2002). Multicolor and electron microscopic imaging of connexin trafficking. *Science* **296,** 503–507.

Medalia, O., Weber, I., Frangakis, A. S., Nicastro, D., Gerisch, G., and Baumeister, W. (2002). Macromolecular architecture in eukaryotic cells visualized by cryoelectron tomography. *Science* **298,** 1209–1213.

Tokuyasu, K. T. (1980). Immunochemistry on ultrathin frozen sections. *Histochem. J.* **12,** 381–403.

CHAPTER 19

3D Immunolocalization with Plastic Sections

Mary K. Morphew

Boulder Laboratory for 3D Electron Microscopy of Cells
Department of Molecular, Cellular and Developmental Biology
University of Colorado, Boulder, Colorado 80309

Immunolocalization of antigens on serial plastic sections cut for electron microscopy (EM) has the potential to map the 3D distribution of macromolecules in the context of relatively large volumes of biological material. The challenges of this technique involve the necessary compromise between the preservation of cellular ultrastructure and the preservation of antigen reactivity. The best approach to these challenges will depend on the scientific questions under investigation. Many antibody localization studies on insoluble and reactive proteins are suited to methods that use permeabilization and extraction to allow penetration of EM-traceable markers into the cell (Chapter 22 by Cheutin *et al.*, this volume), while others require the retention and preservation of the entire cell matrix. Once a suitable compromise is achieved, the reconstructing and modeling of labeled samples can be performed in ways that address specific questions of the investigation. Here, we discuss several options for labeling techniques, protocols for fixation and embedding, antibody options and types of 3D reconstructions for EM immunocytochemistry. We also provide several

0091-679X/07 $35.00
DOI: 10.1016/S0091-679X(06)79019-7

examples of successful applications of these methods. Finally, we will suggest new methods that may bridge the gap between the compromise of good structure and antigen reactivity through the use of low-temperature methods and other forms of labeling.

I. Introduction

One important goal of cellular electron microscopy (EM) is to correlate a biological structure visualized in the EM with a particular function of this cell or tissue. Structural cell biology has seen major progress in the past 20 years, thanks to the development of new technologies, which have improved both the preservation of specimens and the capabilities of the EM. The use of rapid freezing to immobilize cells has become routine in many laboratories, though improvements in the relevant methods are still being made. Samples whose primary fixation is by cryoimmobilization can be imaged directly in ice following either thin-film plunge freezing or cryomicrotomy of frozen-hydrated samples. Alternatively, frozen samples may be freeze-substituted into organic solvents containing either aggressive fixatives [like osmium tetroxide (OsO_4)] followed by embedding in epoxy resins for elucidating pure structure in the EM or light fixatives followed by embedding in acrylic or methacrylate resins for antigen localization. The recent popularity of live confocal imaging of cells transfected with GFP-tagged proteins has given new importance to immuno-EM, with the hope of improving the resolution at which these GFP chimeras are localized.

Immuno-EM for localizing macromolecules *in situ* involves two basic steps. First, the cell or tissue must be stabilized in such a way as to preserve ultrastructure in as close to the native state as possible, while still retaining necessary antigenicity. This step is demanding because well-preserved ultrastructure and immunoreactivity are often mutually exclusive. The preparation of tissues for microscopic evaluation requires procedures that alter the chemical properties of most components. On the other hand, techniques that allow for the retention of biological activity usually result in poor structural preservation (Bendayan, 1987). Many EM localization studies begin only after preliminary work in the light microscope (LM). But since the resolution of the EM is more than an order of magnitude higher than that of the LM, there are different criteria for morphological preservation in these two imaging modalities. Sample preparations for LM are usually unacceptable in the EM because of the poor preservation of the structural detail that is visible with the latter kind of microscope; they generally rely on a relatively high degree of permeabilization and only a "light" fixation. The cell's antigens remain, however, more immunoreactive than those in samples prepared by the more rigorous methods required for EM. Usually, a variety of fixation protocols must be combined with an equal variety of embedding regimens to obtain an acceptable sample for EM immunolocalization.

The next basic step for EM immunolabeling is the application of the antibody and the visualization of the antibody–antigen complex *in situ*, either as a two-dimensional (2D) projection or as a three-dimensional (3D) localization through a volume of biological material. Again, the success of these steps requires the testing of different approaches, as some antibodies work well for one method but not as well (or not at all) for others. For instance, some antibodies may only recognize the antigen in its native conformation. Strong fixation can either destroy this structure or cluster other proteins around the antigen of interest, making it inaccessible to the antibody. In addition, the antibody may recognize one or a few sequences on the target molecule, so antibody binding will depend on the position and orientation of these sequences in the protein structure (Webster, 1999). The ability to localize macromolecules in 3D also depends on the accessibility and orientation of the epitope and the ease with which the sample can be used for 3D reconstructions. This chapter will address a number of factors that contribute to successful EM immunolocalization, specifically fixation, embedding, antibody selection, and 3D reconstruction of immunolabeled cells for localization of specific molecules. Also, a variety of protocols will be discussed that have proven useful for a wide range of antibody–antigen complexes in samples used to date.

II. Rationale for the Various Approaches to Immuno-EM

3D EM immunolocalization involves labeling a particular macromolecule within a relatively large volume of biological material. Such labeling may be of three distinct types: negative-stain labeling, preembed labeling, and postembed labeling. Negative-stain labeling is the combination of immunological labeling of specific antigens in various preparations (e.g., isolated protein molecules, viruses, cellular membrane fractions, intact cytoskeleton, and cytoskeletal proteins) with the use of negative stain. The method offers considerable possibilities for antigen/epitope localization at comparatively high resolution (Harris, 1999), but much of the cellular context is lost by the preparative procedure. Therefore, this method will not be reviewed here. In preembed labeling of cells, immunoreagents are allowed to penetrate into the specimen prior to embedding (either before or after fixation), producing a 3D distribution of probe (Stirling, 1990; Chapter 22 by Cheutin *et al.*, this volume). For preembed methods, the cells must be permeabilized either mechanically by cutting frozen sections or freezing/thawing, or chemically by pore-forming proteins, detergents, or organic solvents; only then do the markers have access to cytoplasmic or nuclear antigens (Humbel *et al.*, 1998). During permeabilization, the cellular membranes are perforated or completely removed, which means that not only can the antibody enter, soluble cellular proteins may be redistributed or washed out of the cell (Melan and Sluder, 1992). This method is useful for the cases in which the proteins of interest remain in their proper places. Not only will label be distributed in 3D for subsequent study, epitopes may be unmasked by the permeabilization solutions, and sensitive antigens that might have been affected by embedding in plastic can still be studied. The difficulty with this method resides, of course, in knowing whether the antigens that are localized are indeed in their physiological position after all the treatments they have received. We will describe a few of many protocols here, but more thorough review of this technique can be found (Humbel *et al.*, 1998; a specific example is seen in Chapter 22 by Cheutin *et al.*, this volume).

Postembed labeling refers to techniques where antibodies are applied to samples after processing, embedding, and sectioning. This is usually the preferred technique for immunolocalization on plastic sections because the biological material can be prepared with special care to preserve good ultrastructure (see Chapter 3 by Hess and Chapter 20 by Pelchen-Matthews and Marsh, this volume). Postembed immunolabeling can be performed on tissues that have been immobilized by conventional chemical fixation (i.e., paraformaldehyde, glutaraldehyde) or preferably by rapid freezing and freeze-substitution fixation (RF-FSF) in organic solvents, often containing a secondary fixative. The resulting sample can be embedded in a number of different immunocompatible resins such as LR White, LR Gold, or the Lowicryls. Occasionally, if the epitope is particularly robust, immunoreactivity is retained in epoxy resin, but usually etching and antigen unmasking are required with this embedment.

III. Methods for Immuno-EM

A. Preembedding Methods

Methods for preembed immunolabeling require that the cells be permeabilized prior to immunolabeling, either before or after fixation. Common reagents for permeabilization include Triton X-100 (de Graaf *et al.*, 1992), Saponin (Macville *et al.*, 1995), sodium borohydride (Van Lookeren Campagne *et al.*, 1992), streptolysin-O (Dundr and Raska, 1993), and Brij-opening (Schliwa *et al.*, 1987). For a thorough comparison of these methods, see Humbel *et al.* (1998). Some studies have shown that the method of cell preparation can be sufficient to permeabilize cells. Adams and Kilmartin (1999) found that the yeast *Saccharomyces cerevisiae* fractured during their processing procedure for EM. This was advantageous in avoiding the need detergent permeabilization; their approach preserved the nuclear membrane, allowing for the proper identification of spindle pole body components.

Following permeabilization, samples are fixed with formaldehyde, often containing low amounts of glutaraldehyde (0.1–0.5%). Again, the choice of fixatives and their strength depends on the epitope being labeled. One advantage of the preembed label technique is that fixation and immunostaining can often be evaluated in the LM, using fluorescent probes to confirm antigen survival prior to EM. Fixed samples are then incubated in primary antibody, rinsed, and finally incubated in secondary antibodies conjugated to gold particles. Small gold clusters are recommended for this task because of their higher rate of diffusion (see section on gold probes). If nanogold particles are used, silver enhancement is required for visualization in the EM. Colloidal gold particles measuring 5 nm or greater require no enhancement, but their greater size can inhibit their free permeation of reasonably well-preserved cytoplasm. Labeled samples may then be dehydrated and embedded in epoxy resin for microtomy and microscopy.

B. Postembedding Methods

Postembed labeling requires sample preparation and sectioning prior to the application of antibodies. As with the preembed method, the fixation should be chosen to be as rigorous as possible for preserving ultrastructure, but it must still provide sufficient retention of antigen activity and availability to yield a useable labeling signal. Samples that are chemically fixed (usually paraformaldehyde with or without glutaraldehyde) may be dehydrated and embedded at room temperature or they may be dehydrated using either a progressive lowering of temperature (PLT) protocol (Kellenberger *et al.*, 1980) or cryosubstitution dehydration (Oprins *et al.*, 1994). They are then embedded at low temperature in acrylic or methacrylate resin. Samples immobilized by rapid freezing are freeze-substituted into organic solvent (with or without the addition of glutaraldehyde or other fixatives) (see Part I, this volume), then embedded at low temperature (–20 to $-50\,^{\circ}$C) in acrylic or methacrylate resin. Alternatively, samples may be fixed with aldehydes, followed by infusion with sucrose (without the need for dehydration) and then freezing in liquid nitrogen for cryosectioning (Tokuyasu, 1980; Chapter 21 by Ladinsky and Howell, this volume). Tissue sections resulting from this method are often very immunoreactive and membranes are well contrasted. The details of fixation, dehydration, and embedding options for plastic section immunolocalization are discussed below.

C. Sample Preparation

1. Fixation

The methods for fixation of cells and tissues for immuno-EM vary, depending on the structure to be examined, the type of embedding medium to be used, and the robustness of the epitope. In general, it is best to use the most rigorous fixation available (i.e., that which best preserves the structure of interest) while still retaining the antigenicity of the epitope. One can usually begin with a cocktail of formaldehyde (1–4%) and glutaraldehyde (0.1–0.5%) in either phosphate buffer or a cytoskeleton-stabilizing buffer, for example, one that contains PIPES, HEPES, EGTA, and magnesium. If one is fixing by freeze-substitution, uranyl acetate may be added to the aldehyde mixture (0.01–0.5%) or may be used alone as a primary fixative. The addition of uranyl acetate improves membrane contrast, usually without loss of antigenicity. Secondary fixation with OsO_4 is often not possible because it destroys or masks most antigens and also because the tissue darkening inherent to fixation in OsO_4 may prevent complete UV-induced

Table I
Sample Protocols for Fixation, Embedding, and Immunolabeling

A. Preembed labeling	
1. Buffer rinse	
2. Fix 2% paraformaldehyde, 0.1% glutaraldehyde in buffer	15 min
3. Buffer rinse	3 × 5 min
4. Permeabilize 0.5% Triton X-100	5 min
5. Buffer rinse	3 × 5 min
6. Treat sodium borohydrate (1 mg/ml in 1:1 PBS:ETOH)	5 min
7. Buffer rinse	3 × 5 min
8. Block buffer	30 min
9. Primary antibody in block buffer	2 h at RT
10. Rinse PBS-T	3 × 10 min
11. Secondary gold conjugate in block buffer	2 h at RT
12. Rinse PBS-T	3 ×10 min
13. Rinse PBS	2 × 5 min
14. Silver enhance (if nanogold is used)	
15. Dehydrate acetone	
16. Embed EPOM/Araldite	
B. Postembed fixation for chemical/PLT	
1. Buffer rinse	
2. Fix 2% paraformaldehyde, 0.1% glutaraldehyde	15 min at RT
3. Buffer rinse	3 × 10 min
4. 30% ETOH at 0°C	30 min
5. 50% ETOH at −20°C	60 min
6. 70% ETOH at −35°C	60 min
7. 95% ETOH at −35°C	60 min
8. 100% ETOH at −35°C	60 min
9. 100% ETOH at −35°C	60 min
10. ETOH:resin mix (1:1) at −35°C	60 min
11. ETOH: resin mix (1:2) at −35°C	60 min
12. Pure resin at −35°C	60 min
13. Pure resin at −35°C	overnight
14. Polymerize at −36°C under UV light	2 days
15. Polymerize while warming to RT	1 day
C. Rapid freeze/freeze-substitution embedding	
1. Rapid freeze by plunge or high-pressure methods	
2. Substitute into acetone ±0.1% glutaraldehyde, 0.1% uranyl acetate, (0.01% OsO_4) at −90°C	2–4 days
3. Warm to −50°C	12–24 h
4. Rinse in acetone at −50°C	3 × 20 min
5. Acetone:HM20 mix (1:1)	1 day
6. Acetone:HM20 mix (1:2) at −50°C	1 day
7. Pure HM20	3 × 2 days
8. Polymerize UV light at −50°C	2 days
9. Polymerize while warming to 0°C	1 day

RT, room temperature.

polymerization of resin. However, in some cases with highly robust epitopes, low percentages of OsO_4 (0.01%) can be used to produce superior preservation with antigen retention (McDonald and Morphew, 1993; Müller-Reichert, 2003). For sample protocols, including buffers and fixation recipes, see Tables I and II.

2. Dehydration

The preferred method of sample preparation for postembed labeling is by rapid freezing followed by freeze-substitution dehydration, but this is often not possible, particularly if the necessary equipment is unavailable. In cases where chemical fixation is used, one can still take advantage of the benefits of low-temperature dehydration, which reduces the extraction of soluble tissue components and preserves both ultrastructure and the reactivity of antigen sites, by using one of two powerful methods. (1) PLT employs the stepwise decrease of specimen temperature and a simultaneous increase in the concentration of the dehydrating agent, taking care to prevent the sample from freezing. (2) Fixed tissue can also be dehydrated under conditions of cryosubstitution (CS) (Oprins *et al.*, 1994) by which fixed samples are frozen in liquid nitrogen and then freeze-substituted into solvents at low temperature.

Ethanol, methanol, or acetone can all be used as dehydrating agents for immunocompatible resins, either during PLT or freeze-substitution. We have found that acetone is the most useful dehydrating solvent for fine structure studies. This was determined by dehydrating samples in a number of different solvents and comparing them by the criteria of their having smooth, continuous contours of membranes, straight or smooth cytoskeletal elements, and a ground cytoplasm with little or no

Table II
Suggested Recipes for Immunocytochemistry

A. Buffer rinse for preembed ICC
60-mM PIPES
25-mM HEPES
10-mM EGTA
2-mM $MgSO_4$
B. Buffer rinse for on-section labeling
Phosphate buffered saline
0.01% Tween-20[a]
C. Blocking buffer
0.1% Coldwater fish gelatin
0.8% Bovine serum albumin
0.01% Tween-20

[a] Tween-20 is used to help rinse away unbound antibody. Excessive exposure to Tween may cause loss of membrane structure. This can be prevented by postfixation of sections with 0.5% formaldehyde.

extraction (McDonald and Morphew, 1993). Following dehydration in acetone, care must be taken to remove as much of the solvent as possible, as any residual solvent can prevent the polymerization of acrylic resins. Excess solvent is eliminated by a number (at least 3) of changes of fresh resin prior to polymerization.

3. Embedding Resins

3D localization of antigens may be obtained from tomographic reconstructions of thick sections (~250 nm) of preembed labeled material, by serial thin sections (20–70 nm) that have been postembed labeled with antibodies, or by tomographic reconstruction of serial thick sections labeled with antibodies. When choosing an embedding resin for these applications, two factors should be considered: the sectioning characteristics of the resin block and the stability of the resin section in the electron beam. For an evaluation of resin properties, see Table III.

a. Epoxy Resin

Epoxy resins are the easiest to section, especially for serial applications. They are also the most stable in the electron beam, and they usually reveal well-preserved ultrastructure. However, because of their highly cross-linked, hydrophobic character and their ability to copolymerize with biological material, they are usually discounted for postembedding studies (Stirling, 1990). Some very hardy antigens may, however, be labeled in epoxy resins, and the likelihood of such labeling can be increased by using strategies that revive antigenic activity. "Deosmification" of epoxy-embedded material is carried out by incubating the section on a solution saturated with sodium metaperiodate (Bendayan and Zollinger, 1983) followed by a treatment to etch and remove resin components with oxidizing agents such as hydrogen peroxide, potassium permanganate, and periodic acid (Causton, 1984).

b. Immunocompatible Resins

Resins formulated for immunocytochemistry are unique in their properties of polarity, hydrophilicity, and degrees of cross-linking. Polar resins mimic the

Table III
Comparison of Different Embedding Resins

	Ease of use	Temperature of polymerization	Immunoreactivity	Stability in EM beam	Ultrastructure preservation	Ease of sectioning
Epoxy	+++	60°C w/accelerator	possible w/etching	+++	+++	+++
LR White	++	>−20°C w/UV	+	+	+	+
LR Gold	++	>−20°C w/UV	+	+	+	+
Lowicryl K4M	+	>−35°C w/UV	+	+	+	−
Lowicryl HM20	+	>−50°C w/UV	+	++	++	++

environment that proteins or other biological structures had *in vivo*. These resins are not as heavily cross-linked as the epoxy resins and absorb water readily, which is said to decrease the attraction of the plastic for immunoreagents. This reduces background and can enhance sensitivity (Newman and Hobot, 1987; Stirling, 1990). A further factor affecting labeling efficiency is the specimen-related surface relief of the section. This appears to depend on retention of hydration shells of the constituent tissue macromolecules and is best achieved by low-temperature embedding. (Acetarin *et al.*, 1986).

LR White and LR Gold. LR White (London Resin Company, Reading, Berkshire, England) is a hydrophilic resin which tolerates ~12% water (Newman and Hobot, 1987). It can be polymerized by heat at room temperature, or at −20 °C using UV light. LR White is easy to use and is a good choice if samples are fixed and dehydrated at room temperature by standard protocols. The sectioning properties of LR White are fair, however, and cells embedded in this resin often exhibit some cytoplasmic extraction. LR Gold is a polar acrylic plastic which is suitable for low-temperature embedding of fixed or unfixed tissues (Thorpe, 1999). It is easy to use and can be polymerized with UV light at −20 °C or at room temperature.

Lowicryls. Lowicryl resins were introduced in 1980 (Carlemalm *et al.*, 1982; Kellenberger *et al.*, 1980) and were developed to improve the preservation of cellular ultrastructure. They are methacrylate resins whose properties are particularly well suited for use at low temperature: miscibility with various organic solvents, low viscosity, and the ability to be polymerized by UV light. Lowicryl K4M is a polar medium and is suitable for infiltration and polymerization at temperatures down to −35 °C. K4M gives good results for structural preservation and immunostaining, but it is challenging to section, especially when one wants ribbons of serial sections. Lowicryl HM20 is a nonpolar and comparatively hydrophobic resin that can be used down to −50 °C. It produces blocks with excellent sectioning properties and good ultrastructure. Surprisingly, immunoreactivity is quite well preserved, in spite of the more hydrophobic properties of this medium. Because this resin is useful at lower temperatures, has superior sectioning qualities, and is stable in the electron beam; it is often the best choice for 3D immuno-EM. It is important to note, however, that these methacrylate resins are volatile and toxic. Fume hoods and ventilated apparatus are necessary, and gloves must always be worn.

4. Microtomy

3D immunocytochemistry generally requires that the embedding resin can be serially sectioned. For convenient viewing of these serial sections, it is advisable to pick the sections up on Formvar-coated slot grids, so grid bars do not obscure the structures that are being serially imaged. Nickel grids are often used for EM cytochemistry, because they do not oxidize while they are exposed to buffers during the labeling procedure. However, in some microscopes the magnetic properties of nickel makes them impossible to image well. We have found copper/rhodium grids suitable for immunocytochemistry provided the copper side of the grid is coated with a Formvar film to prevent oxidizing copper from contamination of the antibody solutions. It is often useful to trim block faces so their height is no more than half their length to increase the number of sections on a grid and to ease the tracking of cells through a series.

5. Block Buffers and Primary Antibodies

Prior to antibody labeling, sections should be treated with reagents that reduce nonspecific antibody binding. A number of different mixtures may be used for this purpose, depending on the possible sources of background. To prevent the binding of antibodies to cellular components through unreacted fixatives and tissue aldehydes, the surface of the section may be treated with 50-mM glycine or bovine serum albumin (BSA). When more specific receptors in tissue bind secondary antibody nonspecifically, the section's surface can be blocked prior to incubation with specific antibodies by using a normal, unlabeled serum from the same animal source that will be used for the secondary antibody. Cold-water fish gelatin and/or nonfat milk may also be used to block nonspecific binding. It is also useful to add a surfactant, such as Tween-20, to the blocking buffer and to subsequent washes; this will aid in rinsing antibody from nonspecific, hydrophobic bonds. We have found that the best universal blocking buffer contains BSA, fish gelatin, and Tween-20 (Table II). However, prolonged incubation in Tween-containing buffers can degrade the quality of membrane preservation; a 10-min postfixation of sections in 0.5% formaldehyde prior to blocking and antibody incubation will reduce this degradation of structural preservation.

Primary antibodies and their reactivity to epitopes are likely the most idiosyncratic yet important consideration in immunolabeling. It is, however, one that must be evaluated empirically. In general, monoclonal antibodies or affinity-purified, polyclonal antibodies are most effective, producing good localization with little background labeling. Dilutions of antibodies must be tested through a series of protein concentrations, and antibodies, both primary and secondary, should be diluted in blocking buffer.

6. Gold Probes

Faulk and Taylor first introduced colloidal gold as a label for EM in 1971. There are two major considerations when choosing a gold probe: type and particle size. Gold probes can be made in-house (Slot and Geuze, 1989) or purchased preconjugated to either protein A (pA), protein G (pG), or IgG. Staphylococcal pA binds the Fc portion of IgG molecules from many animal species, with highest affinity for human, rabbit, guinea pig, swine, monkey, and dog (Biberfeld *et al.*, 1975). Streptococcal pG binds strongly to all human IgG subclasses and rabbit, cow, horse, goat, sheep, and swine IgG. IgG-bound gold probes, which are species specific, have a higher sensitivity for detecting low levels of antigens, but they may show higher nonspecific binding than pA (Slot *et al.*, 1989).

The usual size range for standard colloidal gold probes, which can be visualized in the EM without enhancement, is 5–20 nm. Smaller particles (5 nm) have the advantage of improved spatial resolution of the labeling. Moreover, with larger gold probes steric hindrance may decrease the observed immunolabel density. Generally, the smaller the gold particle, the more sites are labeled. Smaller particles also do not obscure adjacent antigens, and they diffuse better to their sites of reaction (Takizawa and Robinson, 1994; Thorpe, 1999). The application of nanogold particles has become useful, especially for 3D preembed techniques (see Chapter 22 by Cheutin *et al.*, this volume). Nanogold probes are 1.4-nm gold clusters (~67 atoms) that can be covalently linked to Fab antibody fragments, making for small and stable probes. The Fab fragment is about one-third the size of a full IgG molecule. Gold clusters can be site-specifically attached to the hinge region cysteine, which is the end opposite the antigen-binding region; this helps to preserve immunoreactivity (Hainfeld and Powell, 2000). Since nanogold probes are at or beneath the resolution of most EM of cells, silver enhancement is necessary to visualize them, unless HADEF-STEM is used to image small gold. Silver enhancement kits are available from commercial vendors (Nanoprobes, Inc., Yaphank, NY) or solutions can be made in-house (Danscher *et al.*, 1995; Lah *et al.*, 1990). For an evaluation of particle size, see Table IV.

Multiple labeling on plastic sections, using two different sizes of gold probes, is often extremely useful to colocalize components thought to belong to a particular structure or to confirm labeling of one protein (i.e., one antibody to a protein and another to an epitope tag, such as myc, HA, or Pk, which has been engineered into another allele of that same protein). It is, however, important to insure that the two antibodies and the secondary antibodies being used do not cross-react with each other during the labeling procedure. The use of primary antibodies from different species (rabbit and mouse) usually solves this problem, so long as appropriate, non-cross-reacting secondary antibodies conjugated to an EM visible tag are available. If the labeling of the second antigen greatly overlaps that of the first, it is likely that the labeling density will be lower than that observed in a single labeling, due to steric hindrance of the first antibody–gold conjugate (Rabouille, 1999). Some useful Web sites for resin embedding and immunocytochemistry may be found in Table V.

Table IV
Gold Probes (Commercially Available) Useful for Immunocytochemistry

Gold probe	Silver enhance	Spatial resolution
0.8 nm	+	+++
1.4 nm	+	++
5 nm	–	+
10 nm	–	–
15 nm	–	– – –

Table V
Useful Web Sites for Resin Embedding and Immunocytochemistry

http://www.emsdiasum.com/microscopy/technical/datasheet/14330c.aspx
http://www.2spi.com/catalog/chem/lr-elec_micro.htm
http://www.nanoprobes.com/LabRgts.html
http://www.nanoprobes.com/TechSE.html
http://bio3d.colorado.edu/imod/

7. 3D Reconstruction of Cells with Immunolabels

If 3D data is desired, one may collect images of serial thin sections to which label has been applied and map the 3D distribution of antibody by aligning those sections to create a representative volume. Serial section reconstruction for on-section labeling has two distinct advantages over a simple 2D projections of individual sections. First, one determines the distribution of antigen through a larger volume or structure. Second, this method allows one to evaluate the specificity of antibody binding and to assess background labeling through a larger volume of tissue (Müller-Reichert *et al.*, 2003) using quantitative analysis. Since label is only bound to the section surface where antigen is presented, it is advisable to cut the thinnest sections possible that will still provide adequate contrast (usually 40–50 nm); this maximizes the surface-to-volume ratio of the sample and increases the amount of labeling. Alternatively, one may use EM tomography to determine the 3D structure of stained material within the section and compare this with the antibody distribution on the section's surface. In addition, these two approaches may be combined to obtain an even larger volume through serial tomographic reconstructions of thick sections, which have on-section labeling (Donohoe *et al.*, 2006). Labeled serial thick sections do not give the same advantage of a high surface-to-volume labeling ratio, as is available from thin section reconstructions, but it does benefit from the ability to correlate high-resolution structure obtained from the tomogram with antigen detected at the surface of the rendered volume. In either case, serial images are accurately aligned and 3D models constructed using computer modeling tools available (McDonald and Morphew, 1993; Winey *et al.*, 1995). For additional comments on both 3D modeling and tomographic reconstruction, see Parts II and V, this volume.

IV. Comparison of the Methods

The goal of 3D immunocytochemistry is to determine the position of specific macromolecules within the volume of a cell or organelle relative to other molecules or markers using material that is an accurate representation of the living cell. The problems in achieving this goal derive from the fact that one cannot do EM on live material, so samples must be processed in such a way that they may be sectioned in a microtome and are suitable for imaging under vacuum conditions. If samples are extremely well preserved, they are often difficult to immunolabel, so a successful method requires compromise.

Two classes of compromise-based methods are preembed labeling and postembed labeling. Preembed is a useful technique for studying monolayers of cultured cells or isolated cells in suspension, particularly where the specific macromolecules to be labeled are not solubilized or displaced by the treatments necessary to permeabilize the cells and allow the entry of antibody probes. Many components of the cytoskeleton and components of the kinetochore plate of cultured mammalian cells retain their structure and are reactive to antibodies after treatment with 0.2% Triton X-100 prior to fixation and antibody labeling (Fig. 1). While the treatment with detergents is relatively harsh and removes a substantial pool of soluble proteins and membranous structures, one can obtain a strong signal with good spatial resolution of some insoluble cellular components. An important concern about such methods, though, is that one has no control for the possibility that the conditions that extracted the cell's soluble components may also have rearranged some of the components that are left. Thus, there is a fundamental uncertainty in the reliability of this approach to component localization. Preembed labeling of the yeast *Saccharomyces cerevisiae* can be achieved by treating the cells with enzymes (glusulase and zymolyase) following formaldehyde fixation, thus avoiding detergent permeabilization and allowing for the preservation of the nuclear membrane (Adams and Kilmartin, 1999). In either case, small gold probes (0.8–1.4 nm) are used, both to increase the efficiency of gold entry into the cell and to decrease steric hindrance of labeling.

Postembed methods are focused on maintaining higher levels of structural preservation than preembed methods; they take advantage of methods for rapid freezing and low-temperature embedding, then use the subsequent processes of cutting sections to solve the problem of probe entry into cells. Low-temperature methods may also help to retain antigen activity and availability, but there is the concern that resin embedding will reduce both the access and the strength of interaction for antibodies and their respective antigens. Special resins have been developed to help minimize this problem, so even if low-temperature methods are not used, the resins available for these protocols can provide for superior antibody labeling because their hydrophilicity, their degree of cross-linking, and their specimen-related surface relief that develops during sectioning. LR White resin has proven successful when low-temperature methods are not available as it is of low viscosity, relatively low toxicity, easy to section, and sensitive to label (Fig. 2A). The disadvantages of using LR White are that membranes are poorly preserved and sometimes appear swollen or extracted. In addition, it is difficult to use LR White in flat-embedding applications, such as for cell monolayers, because its cross-linking ability is highly sensitive to oxygen.

The Lowicryl family of resins are the most widely used for postembed immunolabeling because they are compatible with low-temperature methods and reveal superior ultrastructure after sectioning. Lowicryl K4M is often a good choice

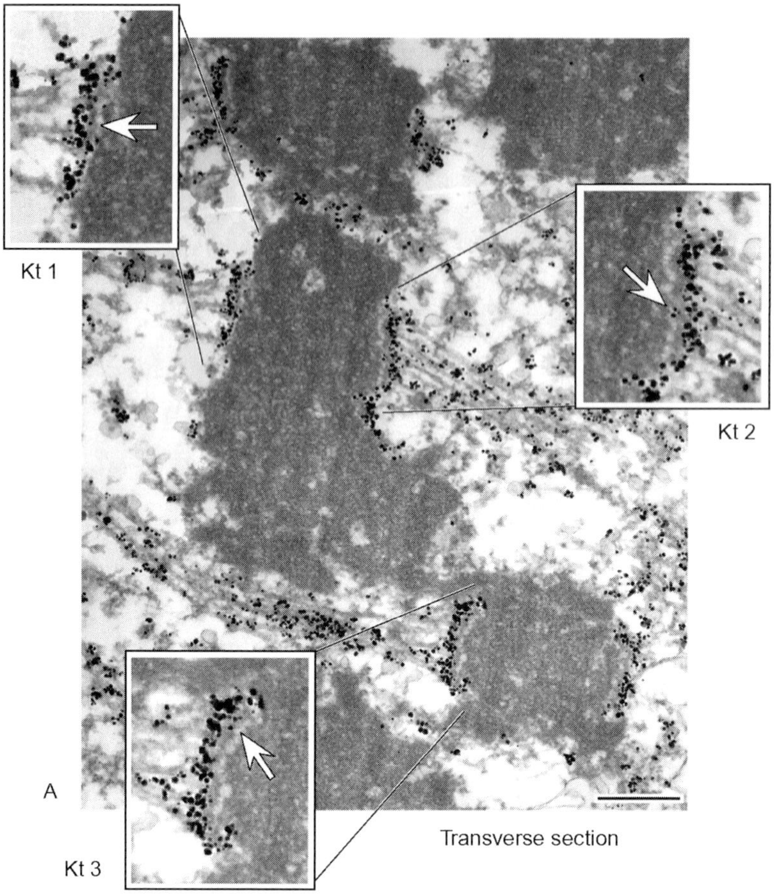

Fig. 1 Preembed labeling of Indian Muntjac fibroblasts permeabilized with 0.2% Triton X-100 in PHEM buffer and fixed with 0.5% glutaraldehyde. Permeabilized cells were labeled with affinity-purified rabbit anti-CENP-E and incubated with 0.8-nm goat anti-rabbit gold followed by silver enhancement. CENP-E is seen located in the fibrous corona and outer plate of kinetochores at metaphase. Insets show higher magnification views of three of the kinetochores. Examples of enhanced gold particles clearly labeling the outer plate are indicated by white arrows. Reprinted with kind permission from Springer Science and Business Media, Cooke *et al.* (1997), Fig. 3A.

when room temperature fixation is followed by PLT during dehydration (Fig. 2B). It is especially useful for flat-embedding cultured monolayers because the resin retains viscosity for this methodology at −35 °C. The disadvantage of K4M is its sectioning characteristics. The strongly hydrophobic nature of K4M makes collecting ribbons of serial sections difficult and frustrating. It is advisable to pull water in the knife boat as low as possible and to avoid wetting of the block face as it passes by the knife-edge.

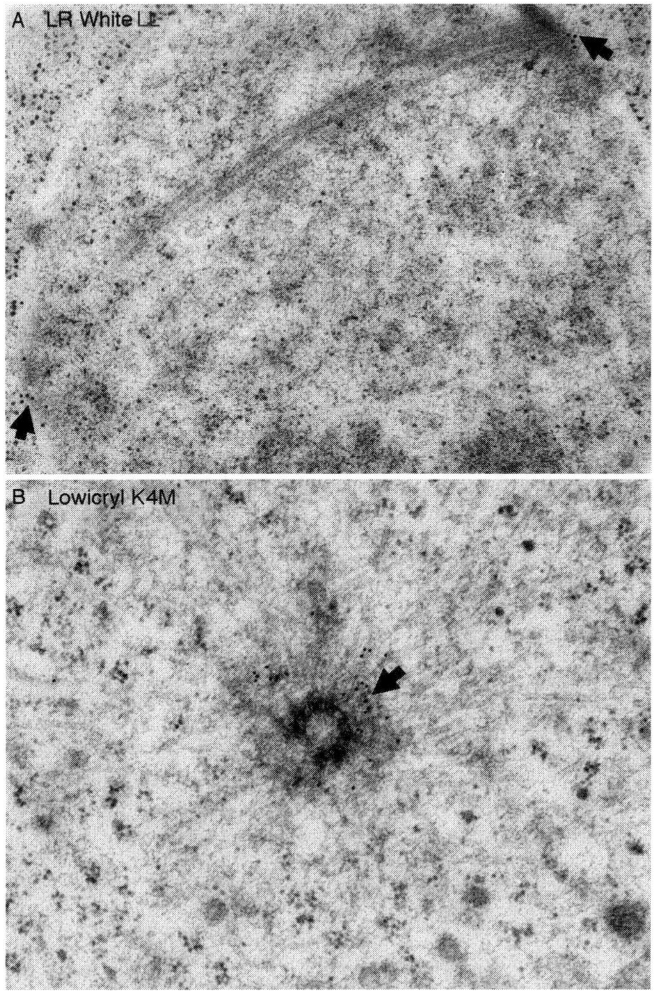

Fig. 2 (A) Section through the mitotic spindle of yeast *Schizosaccharomyces pombe* tagged with Cut11:GFP (West *et al.*, 1998). Cells were embedded in LR White and polymerized at 45 °C. Sections were stained with antibodies to GFP and secondary antibodies conjugated to 10-nm gold. Gold label is located at the spindle pole bodies (black arrows). Note the swollen appearance around the nucleus and the lack of contrast in the nuclear envelope. (B) A section through the centriole of a HeLa cell prepared by PLT dehydration followed by embedding and UV polymerization in Lowicryl K4M at −35 °C. Sections were labeled with antibodies to δ-tubulin and secondary antibodies conjugated to 10-nm gold. Black arrow indicates gold labeling at gold label at the dense area surrounding the centriole.

RF-FSF and low-temperature embedding with UV polymerization in Lowicryl HM20 has become the most popular and successful method for processing samples for 3D immuno-EM. When cells and tissues are well frozen (without ice crystal damage) and when freeze-substitution and embedding are successful at preserving both cell structure and antigen activity, these samples are good candidates for serial sectioning and imaging, even during the large beam dose necessary for tomographic reconstruction data collection.

3D reconstruction of immunolabel on the surface of serial sections is a useful method to determine the position of markers in cells or tissues relative to other macromolecules or structures throughout a large volume. Thin sections of RS-FSF material in HM20 reveal clearly contrasted microtubules and distribution of gold label on kinetochore protein Ndc10p in yeast *Saccharomyces cerevisiae* (Fig. 3). Serial sections measuring 40–50 nm are recommended for reconstructions of mitotic spindles, as sections of this thickness are easy to obtain, contrast in the EM is reasonable and surface to volume ratios of immunolabel allows one to build some statistical analysis of label versus structure (Müller-Reichert *et al.*, 2003). When the desired localization is on a component of a relatively small structure, such as the spindle pole body of the yeast *Schizosaccharomyces pombe*, thinner sections are necessary (20–30 nm) to obtain the required spatial resolution (Fig. 4).

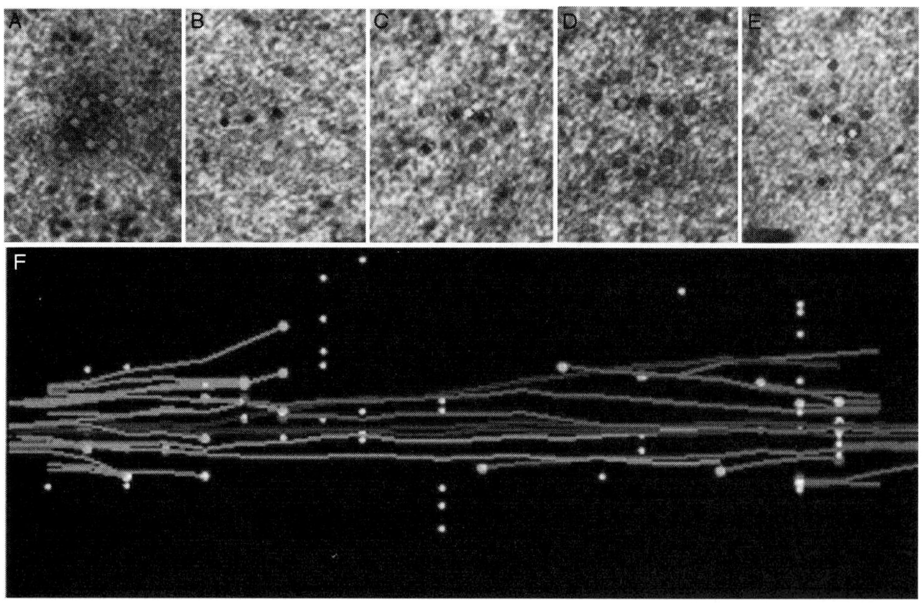

Fig. 3 3D immunolocalization of the Ndc10 antigen in *Saccharomyces cerevisiae*. Cells prepared by high-pressure freezing, followed by freeze-substitution and Lowicryl-HM20 embedding show excellent preservation of fine structure. Serial 50-nm-thick sections are immunolabeled with anti-Ndc10 antibody and secondary antibody conjugated with colloidal gold. Complete sections through a mitotic spindle were imaged and spindle microtubule cross sections tracked. Sample cross sections are shown (A–E) and the approximate position of each cross section lies above the 3D model (F). The positions of the gold particles are marked with a yellow dot and the 3D distribution of the antigen is shown (F). Courtesy Dr. Thomas Müller-Reichert. Max Plank Institute for Molecular Cell Biology and Genetics, Dresden, Germany. (See Color Insert.)

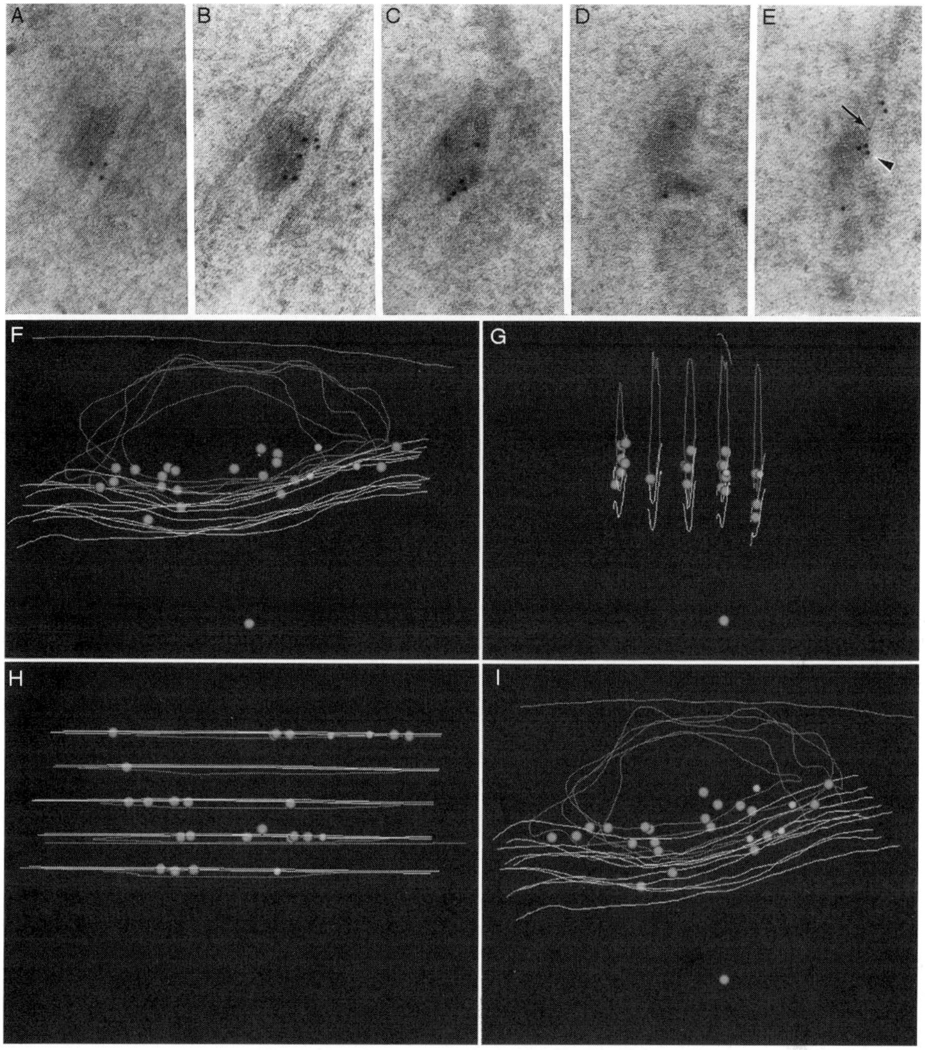

Fig. 4 Five-serial sections through the spindle pole body (SPB) of the yeast *Schizosaccharomyces pombe* strain which carries the Pk epitope tag on Cut12. Cells have been high-pressure frozen, embedded, and UV polymerized in Lowicryl HM20, and serial sections (30 nm) were collected. Sections were double-labeled with monoclonal antibodies to the Pk epitope tag and polyclonal anti-Cut12 antibody. Micrographs of serial sections (A–E) were aligned and components of the SPB were modeled. The 10-nm gold particles (anti-Cut12, green dots), 5-nm gold particles (anti-Pk, blue dots) nuclear envelope (yellow contours), SPB (red contours), and cytoplasmic microtubule (blue line) are seen in four views of the resulting model (F–I). Reprinted with permission from Genes and Development, Bridge *et al.* (1998). (See Color Insert.)

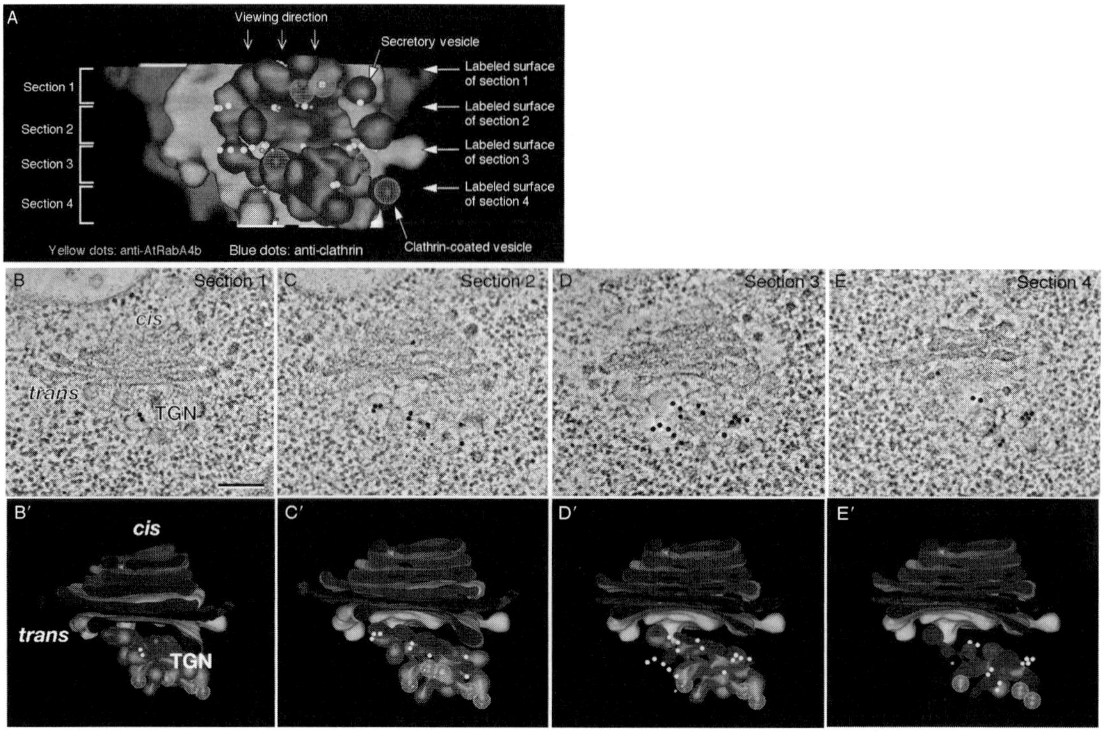

Fig. 5 3D distribution of anti-AtRabA4b and anti-clathrin gold particles associated with a *trans*-Golgi network in an Arabidopsis root meristem cell (A). Face-on view of a 3D model of the TGN (red) and *trans*-Golgi cisternae (blue, purple) generated from four consecutive sections (1–4 from top to bottom) that were immunolabeled with AtRabA4b and clathrin heavy-chain antibody. Four layers of immunogold particles on the section surfaces are seen. Side views (from the viewing direction indicated) of the four layers of the gold particles in the Golgi model are shown below. Tomographic slices of the surfaces of the four sections (B–E) and corresponding Golgi models (B′–E′) viewed from the direction indicated by three arrows in the top panel. Immunogold particles (yellow and blue dots) in the models can be readily matched with the gold particles in the upper micrographs. AtRabA4b is found on the cytoplasmic side of the TGN membrane in the regions producing secretory vesicles. Scale bar = 200 nm. Courtesy Byung-Ho Kang and L. Andrew Staehelin, Department MDCB, University of Colorado, Boulder, CO. (See Color Insert.)

Tomographic reconstructions of relatively thick (250 nm) sections imaged through serial tilts allow one to obtain high-resolution information (~7 nm) through the volume of the section (see Chapter 6 by Höög and Antony, Chapter 8 by Marsh *et al.*, and Chapter 7 by Robertson, this volume). While this technology can be applied to preembed labeled samples, their extracted nature does not lend itself nicely to study with high-resolution detail. The application of antibody–gold conjugates to the surface of RF-FSF, HM20 sections followed by serial tilt imaging and tomographic reconstruction is recently becoming a strong technique for identifying protein distribution corresponding to a well-resolved structure under the surface of label. As with serial thin section reconstruction, these thick tomographic slices can be serially reconstructed combined with antigen labels at their surface (Fig. 5). This technique is particularly useful for antigen localization through large volumes of membrane compartments.

V. Summary

EM tomography combined with immunolabeling is an important first step toward localizing proteins in 3D volumes of well-fixed material at high resolution, but the method has important limitations. With the well-fixed tissue one wants for high-resolution reconstruction, antigens are only labeled at the surface of the serial sections, though probe permeation can occur in some special cases (see Chapter 21 by Ladinsky and Howell, this volume). The ideal situation would arise if macromolecules could be localized within a cell or tissue without the need for permeabilization to accomplish antibody access. One possible approximation to this idea includes the photooxidation of fluorescent markers such as GFP. While this idea has been accomplished with fluorophores in aqueous solutions at ambient temperatures (see Chapter 23 by Sosinsky *et al.*, this volume), these techniques have not yet been combined with protocols for RF-FSF, so there are concerns about the issues that are always associated with aldehyde primary fixation. Other possible improvements in the ability to localize macromolecules in 3D may be in the use of clonable tags. Nanogold-tagged aptamers that can bind to proteins bearing clonable tags are worth exploration, especially if they can be used at low temperature in organic solvents, as during freeze-substitution. Further, the ability to silver enhance nanogold during freeze-substitution protocols would be a critical step in visualizing small gold tags RF-FSF preparations.

Acknowledgments

The author would like to thank all those who contributed material used in this chapter. Their names and affiliations are listed with the appropriate figures. Much of this work was supported by grant RR00592 to J. R. McIntosh from the National Center for Research Resources of the National Institutes of Health.

References

Acetarin, J. D., Carlemalm, E., and Villiger, W. (1986). Development of new Lowicryl resins for embedding biological specimens at even lower temperatures. *J. Microsc.* **143,** 81.

Adams, I. R., and Kilmartin, J. V. (1999). Localization of core spindle pole body (SPB) components during SPB duplication in *Saccharomyces cerevisiae*. *J. Cell Biol.* **145,** 809–823.

Bendayan, M. (1987). Effect of tissue processing on colloidal gold cytochemistry. *J. Histochem. Cytochem.* **35,** 983.

Bendayan, M., and Zollinger, M. (1983). Ultrastructural localization of antigenic sites on osmium-fixed tissues applying the protein-A gold approach. *J. Histochem. Cytochem.* **31,** 101.

Biberfeld, B., Ghetie, V., and Sjoquist, J. (1975). Demonstration and assaying of IgG antibodies in tissues and on cells by labeled Staphyloccal protein A. *J. Immunol. Methods* **6,** 249–259.

Bridge, A., Morphew, M., Bartlett, R., and Hagan, I. (1998). The fission yeast SPB component Cut12 links bipolar spindle formation to mitotic control. *Genes Dev.* **12,** 927–942.

Carlemalm, E., Garavito, R. M., and Villiger, W. (1982). Resin development for electron microscopy and an analysis of embedding at low temperature. *J. Microsc.* **126,** 123–143.

Causton, B. E. (1984). The choice of resins for electron immunochemistry. *In* "Immunolabeling for Electron Microscopy" (J. M. Polak and I. M. Varnell, eds.), p. 29. Elsevier, Amsterdam.

Cooke, C. A., Schaar, B., Yen, T. J., and Earnshaw, W. C. (1997). Localization of CENP-E in the fibrous corona and outer plate of mammalian kinetochores from prometaphase through anaphase. *Chromosoma* **106,** 446–455.

Danscher, G., Hacker, G. W., Hauser-kronberger, C., and Grimelius, L. (1995). Trends in autometallographic silver amplification of colloidal gold particles. *In* "Immuno-Gold Staining: Principles, Methods, and Applications" (M. A. Hyatt, ed.), pp. 11–18. CCR Press, Boca Ratom, FL.

de Graaf, A., Humbel, B., Stuurman, N., van Bergen en Henegouwen, P., and Verkleij, A. (1992). Three-dimensional immunogold labeling of nuclear matrix proteins in permeabilized cells. *Cell Biol. Int. Rep.* **16,** 827–836.

Donohoe, B. S., Kang, B.-H., Nebenführ, A., and Staehelin, L. A. (2006). Identification and characterization of COPIa- and COPIb-type vesicle classes associated with plant and algal Golgi. PNAS, in press.

Dundr, M., and Raska, I. (1993). Nonisotopic ultrastructural mapping of transcription sites within the nucleolus. *Nature* **208,** 275–281.

Faulk, W. P., and Taylor, G. M. (1971). An immunocolloid method for the electron microscope. *Immunochemistry* **8,** 1081–1083.

Hainfeld, J., and Powell, R. (2000). New frontiers in gold labeling. *J. Histochem. Cytochem.* **48,** 471–480.

Harris, J. R. (1999). Negative staining of thinly spread biological particulates. *In* "Electron Microscopy Methods and Protocols" (N. Hajibagheri, ed.), pp. 13–30. Humana Press, NJ.

Humbel, B., De Jong, M., Muller, W., and Verkleij, A. (1998). Pre-Embedding immunolabeling for electron microscopy: An evaluation of permeabilization methods and markers. *Microsc. Res. Tech.* **42,** 43–58.

Kellenberger, E., Carlamalm, E., villinger, W., Roth, J., and Garavito, R. M. (1980). Low denaturation embedding for electron microscopy of thin sections. Chemische Werke Lowi GmbH, Postfach, D-8264 Waldkraiburg Federal Republic of Germany.

Lah, J. J., Hayes, D. M., and Burry, R. W. (1990). A neutral pH silver development method for the visualization of 1-nanometer gold particles in preembedding electron microscopic immunocytochemistry. *J. Histochem. Cytochem.* **38,** 503–508.

Macville, M., Wiesmeijer, K., Dirks, R., Fransen, J., and Raap, A. (1995). Saponin pre-treatment in pre-embedding electron microscopic *in situ* hybridization for the detection of specific RNA sequences in cultured cells: A methodological study. *J. Histochem. Cytochem.* **43,** 1005–1018.

McDonald, K., and Morphew, M. (1993). Improved preservation of ultrastructure in difficult-to-fix organisms by high pressure freezing and freeze substitution: I. *Drosophila melanogaster* and *Strongylocentrotus purpuratus* embryos. *Microsc. Res. Tech.* **24,** 465–473.

Melan, M., and Sluder, G. (1992). Redistribution and differential extraction of soluble proteins in permeabilized cultured cells: Implications for immunofluorescence microscopy. *J. Cell Sci.* **101,** 731–743.

Müller-Reichert, T., Sassoon, I., O'Toole, E., Romao, M., Ashford, A. J., Hyman, A. A., and Antony, C. (2003). Analysis of the distribution of the kinetochore protein Ndc10p in *Saccharomyces cerevisiae* using 3-D modeling of mitotic spindles. *Chromosoma* **111,** 417–428.

Newman, G. R., and Hobot, J. A. (1987). Modern acrylics for post-embedding immunostaining techniques. *J. Histochem. Cytochem.* **35,** 971.

Oprins, A., Geuze, H. J., and Slot, J. W. (1994). Cryosubstitution dehydration of aldehyde-fixed tissue: A favorable approach to quantitative immunocytochemistry. *J. Histochem. Cytochem.* **42,** 497–503.

Rabouille, C. (1999). Quantitative aspects of immunogold labeling in embedded and nonembedded sections (1999). *In* "Electron Microscopy Methods and Protocols" (N. Hajibagheri, ed.), pp. 125–144. Humana Press, NJ.

Slot, J. W., and Geuze, H. J. (1989). A new method for preparing gold probes for multi-labeling cytochemistry. *Eur. J. Cell Biol.* **38,** 87–93.

Slot, J. W., Posthuma, G., Crapo, J. D., and Geuze, H. J. (1989). Quantitative aspects of immunogold labeling in embedded and in nonembedded sections. *Am. J. Anat.* **185,** 271–281.

Takizawa, T., and Robinson, J. M. (1994). Use of 1.4 nm immunogold particles for immunocytochemistry on ultrathin cryosections. *J. Histochem. Cytochem.* **42,** 1616–1623.

Thorpe, J. R. (1999). The application of LR Gold resin for immunogold labeling. *In* "Electron Microscopy Methods and Protocols" (N. Hajibagheri, ed.), pp. 99–110. Humana Press, NJ.

Tokuyasu, K. T. (1980). Immunochemistry on ultrathin frozen sections. *Histochem. J.* **12,** 381–403.

Schliwa, M., Euteneuer, U., and Porter, K. (1987). Release of enzymes of intermediary metabolism from permeabilized cells: Further evidence in support of a structural organization of the cytoplasmic matrix. *Eur. J. Cell Biol.* **44,** 214–218.

Stirling, J. (1990). Immuno- and affinity probes for electron microscopy: A review of labeling and preparation techniques. *J. Histochem. Cytochem.* **38,** 145–157.

Van Lookeren Campagne, M., Oestreicher, A. B., Van der Krift, T. P., Gispen, W. H., and Verkleij, A. J. (1992). Freeze-substitution and Lowicryl HM20 embedding of fixed rat brain: Suitability for immunogold ultrastructural localization of neural antigens. *J. Histochem. Cytochem.* **39,** 1267.

Webster, P. (1999). The production of cryosections through fixed and cryoprotected biological material and their use in immunocytochemistry. *In* "Electron Microscopy Methods and Protocols" (N. Hajibagheri, ed.), pp. 49–76. Humana Press, NJ.

West, R. R., Viasberg, E. V., Ding, R., Nurse, P., and McIntosh, J. R. (1998). *cut11*: A gene required for cell-dependent spindle pole body anchoring in the nuclear envelope and bipolar spindle formation in *Schizosaccharomyces pombe*. *Mol. Biol. Cell* **9,** 2839–2855.

Winey, M., Yarar, D., Giddings, T., and Mastronarde, D. (1995). Nuclear pore complex number and distribution throughout the *Saccharomyces cerevisiae* cell cycle by three-dimensional reconstruction from electron micrographs of nuclear envelopes. *Mol. Biol. Cell* **8,** 2119–2132.

CHAPTER 20

Electron Microscopy Analysis of Viral Morphogenesis

Annegret Pelchen-Matthews and Mark Marsh
Cell Biology Unit, MRC Laboratory for Molecular Cell Biology and
Department of Biochemistry and Molecular Biology
University College London, London WC1E 6BT, United Kingdom

Due to their small size, viruses can only be clearly visualized by electron microscopy (EM). Consequently, our understanding of the replication of many viruses has been greatly enhanced by high-resolution EM studies. Here we describe how transmission EM of plastic-embedded material and immunolabeling studies can be used to analyze the interactions of viruses with their host cells. We will

0091-679X/07 $35.00
DOI: 10.1016/S0091-679X(06)79020-3

focus particularly on the assembly of two types of enveloped viruses: the beta-herpesvirus, human cytomegalovirus (HCMV), and the primate lentiviruses, the simian and human immunodeficiency viruses (SIV and HIV, respectively).

I. Introduction

Viruses are responsible for numerous diseases ranging from the common cold to smallpox, from childhood infections such as measles and chicken pox to major epidemics like AIDS, which now affects more than 40 million people worldwide and kills more than 3 million a year.[1] Outbreaks of highly pathogenic viruses such as Ebola, which causes a lethal hemorrhagic fever, the SARS coronavirus, or, most recently, avian influenza have raised widespread public concern. Other viruses are implicated as causative agents for some cancers.

The virus particles themselves are tiny, with the sizes of most virions at the limit of the resolution of even the best light microscopes. Most viruses are on the order of 50–200 nm in diameter, though the smallest parvoviruses can measure less than 30 nm, while poxviruses or rhabdovirus particles like the vesicular stomatitis virus can be 0.3- to 0.4-μm long. Filoviruses such as Ebola or Marburg virus or filamentous forms of the influenza virus have diameters of about 80–100 nm, but can reach lengths of several μm. Viruses can be localized by fluorescence methods using specific antibodies directed against viral proteins, or viral components coupled to the green fluorescent protein (GFP) or its derivatives. This allows virus-infected cells to be analyzed by FACS methods or by immunofluorescence staining to reveal the distribution of major viral components, but these techniques cannot reveal much about structure. Viruses can be visualized by fluorescence microscopy as small spots, though the resolution is rarely sufficient to determine whether fluorescent spots represent virions, aggregates of viral proteins, or subviral particles. When carefully controlled, fluorescence methods can sometimes allow virus particles to be counted (Pizzato *et al.*, 1999), while more recent multicolor fluorescence coincidence detection methods allow quantitative measurements of the antigenic composition of individual virions (Li *et al.*, 2004). However, a detailed analysis virus structures requires higher resolution methods such as electron microscopy (EM). At the simplest level, negative staining of virus preparations can reveal some features of virus structure, size, and heterogeneity, and this can be of diagnostic value (Curry *et al.*, 2006). More detail is revealed when virus preparations are rapidly frozen and the vitrified specimens examined by cryo-EM. When combined with data from X-ray diffraction studies, or with electron tomography or single-particle analysis of isolated virions, highly detailed virus structures can be derived at near atomic resolution (Briggs *et al.*, 2006; Grunewald *et al.*, 2003).

[1] UNAIDS Epidemic Update, November 2005.

As obligate intracellular parasites, viruses are dependent on living host cells for their replication. The viruses must first identify and adhere to a suitable host cell, which occurs through interactions of viral surface components or envelope glycoproteins with specific cellular receptors. The viral DNA or RNA genome must then access the relevant replication machineries of the host cells (Marsh and Helenius, 2006). Enveloped viruses have to fuse with a cellular membrane, allowing a naked virus core to be released into the cytoplasm. Further uncoating reactions and/or transport of the core to the cell's nucleus may also be required. Even the complex poxviruses, which carry their own machineries enabling them to carry out transcription and replication in the cytoplasm, need to access the appropriate location inside the cell, for example, in the vicinity of the endoplasmic reticulum (Schramm and Krijnse Locker, 2005). Once access to the replicative machineries has been established, the viral genetic material induces the host cell to synthesize the components required for the production of new virus particles, as well as virus-specific enzymes or regulatory proteins that may reprogram the cells' activities. New virus particles are assembled, and, for enveloped viruses, acquire a lipid membrane by budding through a cellular membrane. Many cellular organelles have been shown to be involved in viral replication and assembly, including the plasma membrane, components of the biosynthetic pathway [endoplasmic reticulum, the Golgi apparatus or the *trans*-Golgi network (TGN)], and endosomes. These compartments are exploited in different ways and to varying extents by different viruses. Some viruses subvert the host cells' normal functions, perhaps generating new virus-induced structures or modifying normal cellular organelles to produce extensive "virus factories." These processes are most effectively studied with the electron microscope. The characteristic morphologies allow the easy identification of mature and assembling virus particles in cells, and the effects on the cell's own structure can be detected alongside. The EM techniques are even more powerful when combined with immunolabeling protocols, in particular immunolabeling of cryosections.

Here we review the use of standard transmission electron microscopy (TEM) of plastic-embedded material, as well as protocols for immunolabeling of cryosections, in the analysis of viral interactions with cells. As examples, we will focus particularly on the assembly of the human and simian immunodeficiency viruses (HIV and SIV, respectively) and on the more complex herpesvirus, the human cytomegalovirus (HCMV).

II. Analysis of Virus Assembly on Plastic Sections

Since many viruses have characteristic structures that are easily identified in the electron microscope, studies of ultrathin sections of plastic-embedded preparations of virus infected cells by TEM can be remarkably informative. Indeed, most of the early investigations into viral life cycles used plastic-embedded specimens (Dales, 1973). Various stages of the viral life cycles have been documented by EM on plastic sections, including virus entry pathways using clathrin-coated pits (Helenius *et al.*, 1980) or caveolae (Kartenbeck *et al.*, 1989; Stang *et al.*, 1997) and transport of viruses to the nucleus (Sodeik *et al.*, 1997). Such studies require very high input numbers of virus particles, and since often only a fraction of the visible virus particles are actually infectious, nonproductive pathways might also be visualized. The EM data therefore have to be interpreted alongside biochemical and pharmacological investigations.

TEM has also been used to study the generation of new virions, and the most striking images are obtained during analysis of virus assembly in infected cells. Many viruses induce spectacular changes in the host cell's morphology, for example modifying cellular structures to establish specialized replication structures or "viral factories" (Novoa *et al.*, 2005). EM can be used to identify morphological features of the various stages in the assembly of virus particles, distinguish immature and mature particles, or analyze steps in the acquisition of lipid membranes by enveloped viruses. For example, a range of EM techniques, including analysis of plastic-embedded material, are being used to determine the origin of the membranes wrapping the different intracellular forms of vaccinia virus (Heuser, 2005; Schramm and Krijnse Locker, 2005; Sodeik and Krijnse-Locker, 2002). In addition, EM can be used to study the organelles involved in virus production and to determine whether and how viral infection modifies cellular compartments or generates new structures.

A. Epon Embedding of Infected Cell Preparations: Technical Considerations

As for all EM studies, cell preparation techniques must aim to preserve the structure of the virus-infected cells as close to the living state as possible. Standard fixation protocols employ aldehyde fixatives to immobilize proteins and osmication to stabilize membrane structures. We now use a mixture of 2% formaldehyde (from paraformaldehyde), which is small and penetrates easily into cells, and 1.5% glutaraldehyde, a more potent fixative that cross-links cell components, in 0.1-M phosphate buffer, pH 7.4. To minimize disturbance to the cells, it is best to prepare a double-strength fixative (i.e., 4% formaldehyde/3% glutaraldehyde) at the temperature of the cells and add this directly into the culture medium. High-pressure freezing, followed by freeze-substitution with aldehyde fixatives may give even better preservation (Murk *et al.*, 2003; Studer *et al.*, 2001). However, many viruses are pathogenic and must be inactivated by fixation before the infected cells can be handled safely outside safety hoods or containment laboratories.

For enveloped viruses, the preservation of membrane structures in the preparations is also important. We obtain good results using a post-fixation step in reduced osmium (1% osmium tetroxide, OsO_4, in 1.5% potassium ferricyanide, $K_3Fe(CN)_6$, for 1 h on ice). Membrane structures are especially well preserved if the cells are also treated with tannic acid, which acts as a mordant between osmium-treated structures and lead stains, improving the contrast and fine delineation of various cellular structures, but especially membranes (Simionescu and Simionescu, 1976). Cells are washed and incubated in a low molecular weight tannic acid (gallotannin, 1%)

in 0.05-M sodium cacodylate buffer for 45 min at room temperature, followed by a brief treatment (5–10 min) with 1% sodium sulfate, dehydrated and embedded in an epoxy resin such as Epon 812. Ultrathin sections are contrasted with a lead citrate stain before viewing in the EM.

B. Assembly of the Human Cytomegalovirus

The beta-herpesvirus HCMV is a complex double-stranded DNA virus whose 230 kilo base-pair (kB) genome encodes around 200–250 open reading frames, depending on the virus strain (Britt and Boppana, 2004). Virus particles consist of an inner icosahedral capsid enclosing the DNA genome. This is coated by a layer of at least 15 different tegument proteins and wrapped in an outer lipid envelope containing the viral membrane proteins. The HCMV genome encodes around 40 proteins with features of transmembrane glycoproteins, ranging from the abundantly expressed glycoprotein complexes (gM/gN, gB, and gH/gO/gL) to regulatory proteins, viral Fc receptors, and chemokine receptor-like proteins (Britt and Boppana, 2004). The gB and gH/gO/gL complexes are essential for viral infectivity, functioning in virus attachment to the host cell and membrane fusion (Britt and Mach, 1996).

The assembly of HCMV takes place at specific replication sites both in the nucleus and in the cytoplasm of infected cells, often leading to large increases in cell size (cytomegaly). Distinct stages in the replication of HCMV have been documented by EM, as illustrated in Fig. 1. The viral capsid proteins move into the nucleus, where they assemble into 100-nm-diameter icosahedral cores that are then filled with the viral DNA to produce nucleocapsids (Fig. 1A). The nucleocapsids are thought to leave the nucleus by budding through the inner membrane of the nuclear envelope, before fusing and de-enveloping at the outer nuclear membrane, so that naked capsids are released into the cytoplasm. This idea is supported by EM analysis, where enveloped capsids can sometimes be observed in dilations of the nuclear envelope (Fig. 1B). In the cytoplasm, the particles acquire a layer of tegument proteins, a process that often occurs at characteristic cytoplasmic virus factories (Fig. 1C), identified as accumulations of electron-dense tegument proteins in the vicinity of the centrioles/microtubule organizing center. The virus factories also tend to contain various membrane vesicles and are often surrounded by enlarged mitochondria. The virus particles acquire their final envelope by budding into characteristic crescent-shaped membrane vesicles (Fig. 1D), and simultaneously become enclosed in small membrane-limited vacuoles. Fusion of these vacuoles with the plasma membrane leads to the release of HCMV virions into the extracellular space. The morphology of the fully assembled extracellular virus particle (150- to 200-nm diameter) shows the viral envelope, coated with a layer of the viral glycoprotein complexes (Fig. 1E). Three types of particles can be distinguished on HCMV-infected cells, including enveloped nucleocapsids, empty capsid particles that lack the DNA genome [giving rise to noninfectious enveloped particles (NIEPs)], and cytoplasmic aggregates of tegument proteins, which produce enveloped "dense bodies" (see Fig. 1F). On infected human fibroblasts, we have also occasionally observed HCMV particles within or

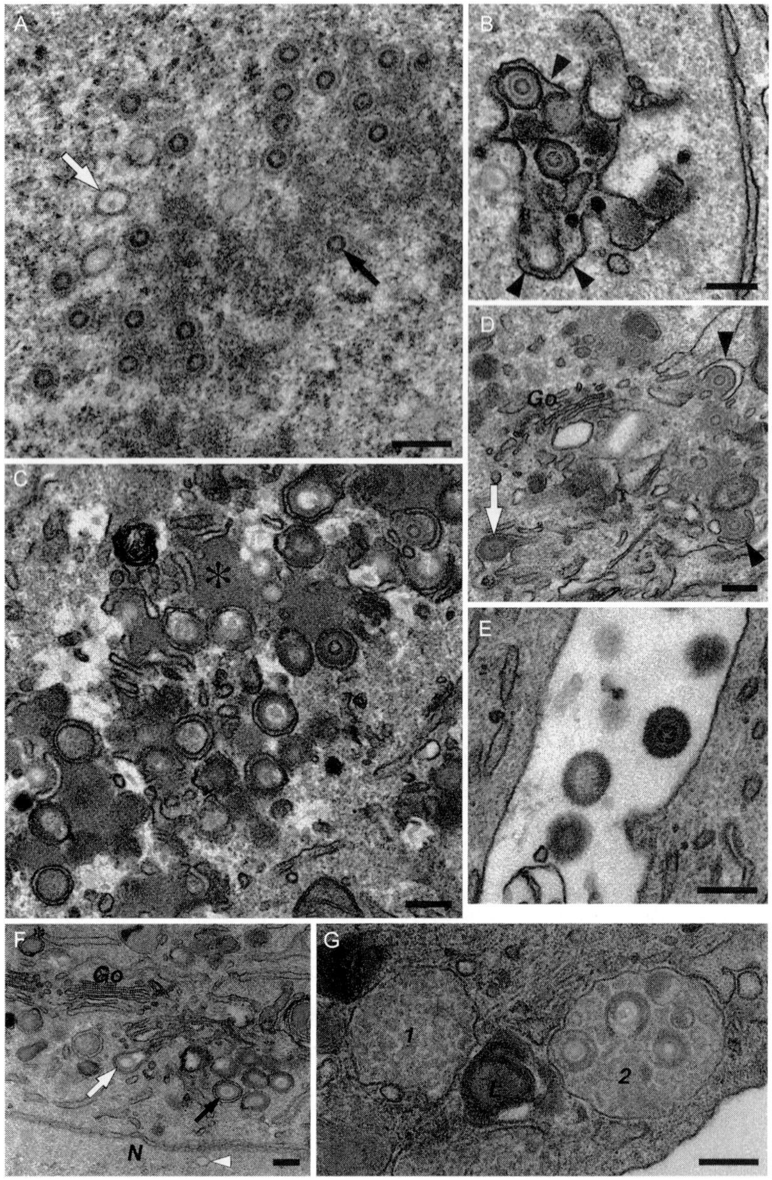

Fig. 1 Stages of HCMV assembly. HCMV assembly was documented in human embryonic lung or foreskin fibroblast cells fixed in 2% formaldehyde/1.5% glutaraldehyde and postfixed with reduced OsO_4 and tannic acid. (A–E) Human embryonic lung cells. (A) The nucleus contains empty icosahedral capsids (e.g., white arrow) or nucleocapsids enclosing a darker ring composed of the viral DNA (black arrow). (B) The capsids bud into the lumen of the nuclear envelope, acquiring a first membrane from the inner nuclear envelope (black arrowheads). This membrane is lost during fusion with the outer membrane of the nuclear envelope, so that naked capsids are released into the cytoplasm. (C) A cytoplasmic "virus

budding into multivesicular late endosomes (Fig. 1G). Various HCMV glycoproteins, including several viral chemokine receptor-like proteins are also located on endosomes and multivesicular bodies and may be incorporated into the viral membrane during the budding step (Fraile-Ramos *et al.*, 2001, 2002).

C. Assembly of SIV Virions and Virus-like Particles

Retroviruses, including the primate lentiviruses HIV and SIV, are single-stranded RNA viruses with a much smaller genome and simpler structure than the herpesviruses. The two copies of the unspliced 9.8-kB RNA genome contain three major genes, *gag*, *pol*, and *env*; these code for the structural proteins, viral enzymes (reverse transcriptase, integrase, and protease), and the envelope glycoprotein (Env), respectively. There are also sequences for a small number of accessory proteins (Freed, 2004). Like other enveloped viruses, HIV and SIV acquire their membrane by budding through a cellular membrane (Fig. 2A). Hence all of the components that make up an infectious virus particle need to be brought together at this membrane budding site in a coordinated manner (Pelchen-Matthews *et al.*, 2004). Virus assembly is driven primarily by the Gag polyprotein that contains all of the structural proteins of the virus, namely the matrix protein (MA) that forms a protein shell immediately beneath the viral membrane, the capsid protein (CA) that condenses into the truncated cone-shaped core that is characteristic of HIV and SIV particles (see Fig. 2A), and the nucleocapsid (NC) that stabilizes the viral RNA genome. Gag alone is able to bind membranes, to deform them into a bud and later a stalked "lollipop"-type structure, and the presence of Gag can be visualized as a thick, membrane-associated, electron-dense layer at the budding site (Fig. 2A). Eventually, a scission reaction releases the budding virion from the host cell, and, concomitant with or shortly after release, the viral protease cleaves the Gag polyprotein, allowing CA and NC to condense into the cone-shaped viral core of mature virions. Indeed, cells transfected with the *gag* gene alone will produce virus-like particles (VLPs) with morphological features identical to immature SIV buds (Fig. 2B), but in the absence of the protease these particles cannot mature.

---

factory" composed of electron-dense accumulations of viral tegument proteins (asterisk) and various membrane vesicles. HCMV capsid particles acquire a layer of tegument and then proceed to wrap into the membrane vesicles. (D) Detail of the envelopment of tegumented capsids at cytoplasmic membranes near the Golgi complex (*Go*). The membrane vesicles are deformed into characteristic crescents (black arrowheads). This leads to the production of mature enveloped HCMV particles within cytoplasmic vacuoles (white arrow). (E) Fusion of these vesicles with the plasma membrane releases HCMV particles into the extracellular space. (F and G) Human foreskin fibroblasts. (F) Three types of enveloping particles in the cytoplasm. The white and black arrows indicate envelopment of an empty capsid and a DNA-filled nucleocapsid, respectively, while the asterisk marks an enveloped tegument aggregate or "dense body." Note also the empty capsid (white arrowhead) in the nucleus (*N*), which lacks the tegument layer. (G) Two multivesicular late endosomes (*1, 2*) and a lysosome containing multilamellar membranes (*L*). Endosome 2 contains several assembled HCMV virions. Images were taken from cells infected for 72 h (A and B), 96 h (E, F, and G), or 120 h (C and D). Scale bars = 200 nm.

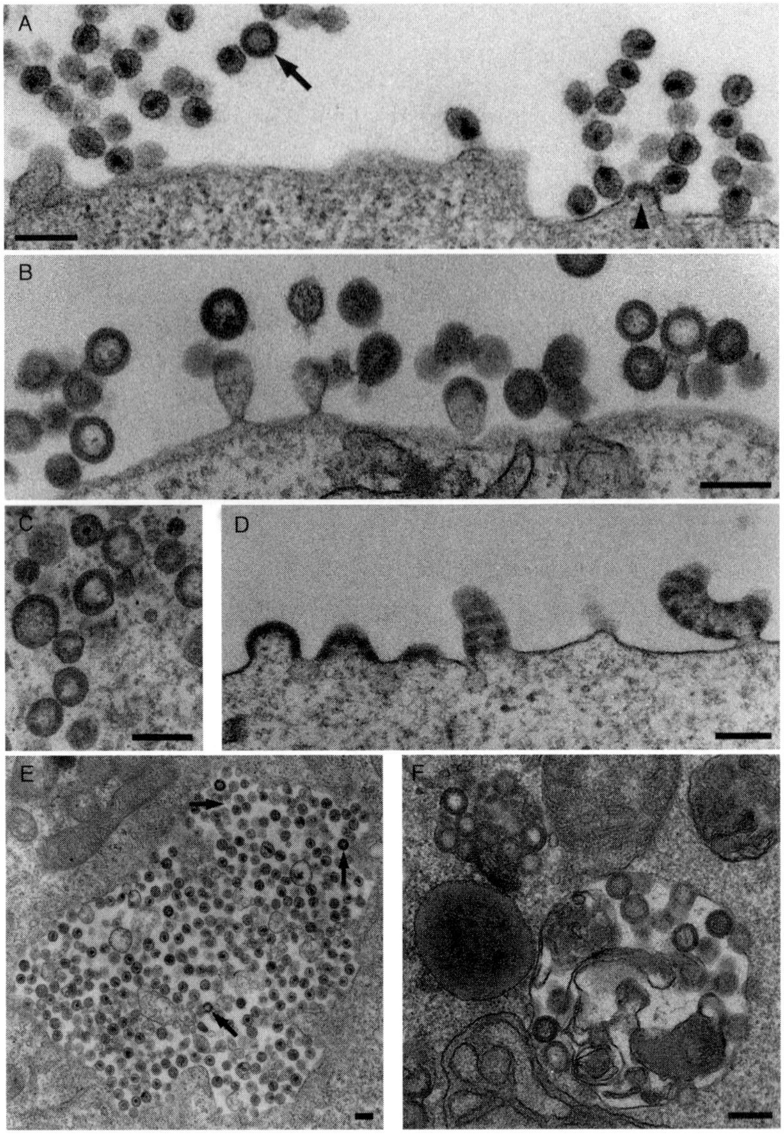

Fig. 2 Assembly of SIV viruses and VLPs. (A) CEMx174 T2 cells acutely infected with a SIVmac251-derived virus produce many virions at the cell surface. A budding virus can be recognized by the electron-dense Gag layer accumulating under the membrane (arrowhead). The Gag layer is also visible in the immature virus particle (arrow). (B) VLPs with the morphology of immature viruses and budding figures accumulate at the surface of COS cells that have been transfected with the SIV Gag protein. (C and D) COS cells expressing a chimeric SIV Gag-GFP show more irregular VLPs (C) or cell surface-budding figures (D). The particles lack the thin electron-dense line next to the VLP lumen, and the electron-dense Gag protein layer is frequently interrupted due to steric interference by the GFP moiety. (E and F) Intracellular vacuoles containing viruses or VLPs can occasionally be found in CEMx174 cells infected with SIV NC-MAC (LaBranche *et al.*, 1995) (E) or COS cells expressing SIV Gag (F), respectively. Arrows in (E) indicate virus particles with immature morphology. Scale bars = 200 nm.

We have analyzed VLP assembly driven by the Gag protein in COS cells transfected with the SIVmac239 *gag* gene, or with a chimeric Gag-GFP construct. In both cases, the cells released VLPs, which could be isolated and purified on sucrose gradients and which contained intact membrane envelopes as judged by their resistance to digestion with proteinase K. The Gag-GFP particles could be visualized by fluorescence microscopy as brightly fluorescent spots at the cell surface and as larger aggregates at intracellular sites.

On Gag-transfected cells, VLPs were easily detected by EM, both budding from the plasma membrane and free in the medium surrounding the cells. Detailed analysis revealed a number of differences between particles budding from Gag or Gag-GFP-transfected cells. VLPs produced by Gag-transfected cells had a fairly homogeneous size distribution of 110–125 nm and consisted of a membrane envelope with an internal electron-dense protein layer and an electron-lucent center (Fig. 2B). Characteristically, the protein layer ended with a sharp line of electron density next to the lumen, similar to the appearance of immature virions budding from SIV-infected cells (cf. Fig. 2A, arrow). Particles from Gag-GFP-transfected cells showed a wider size distribution (110–160 nm, with occasional even larger particles) and significant shape heterogeneity, including oval or elongated particles (Fig. 2C). The electron-dense layer at the budding sites appeared thicker (31 ± 7 nm as compared to 23 ± 3 nm for Gag buds) and lacked the electron-dense line next to the particle lumen. In addition, the protein coat often showed a banded pattern with a regular periodicity of about 50–60 nm, both under the plasma membrane and in the budding or budded particles (Fig. 2C and D). Such interruptions in the protein layer were never seen for Gag VLPs and are likely to result from steric interference to Gag packing by the GFP moiety, which represents approximately one-third of the chimeric Gag-GFP protein. Aberrant Gag-GFP-budding structures have also been observed for HIV (Larson *et al.*, 2005; Muller *et al.*, 2004), though budding particles of normal appearance were seen in cells coexpressing Gag-GFP together with unmodified Gag. Such cells still release green fluorescent VLPs, allowing correlative light microscopy/EM analysis (Larson *et al.*, 2005).

As for other retroviruses, the budding of SIV and HIV Gag is dependent on an intact "late" or L-domain (Freed, 2002) that recruits cellular factors required for the scission and release of the budding particles (Morita and Sundquist, 2004). Again EM analysis has been useful in defining the stages of budding arrest, and the effects of mutations in the viral late domains or of dominant negative versions of the cellular ESCRT (endosomal sorting complexes required for transport) proteins (Pornillos *et al.*, 2003; von Schwedler *et al.*, 2003).

The distribution of virus buds can also be useful in defining the sites of virus assembly. For example, SIV and HIV assemble primarily at the plasma membrane of T cells, T-cell lines, or a variety of model tissue culture cell lines. However, intracellular vacuoles filled with virus particles can also occasionally be observed, both on SIV-infected T cells (Fig. 2E) and on COS cells expressing high levels of the SIV Gag protein (Fig. 2F). This is similar to the assembly of HIV in primary human monocyte-derived macrophages, which accumulate large vacuoles packed with virus particles (Gendelman *et al.*, 1988). But on these static EM images, it is unclear where these intracellular viruses have assembled: virions near the cell surface might have been formed at intracellular locations and secreted, while intracellular virions could have been taken up by the cell, for example, by endocytosis. The presence of budding profiles or immature virions (Fig. 2E) can give an indication that viruses are actually assembling at a particular site, but such images are comparatively rare. Although the oligomerizing Gag protein can easily be identified on plastic sections, the presence of other viral proteins, such as Env, in the viruses is more difficult to establish. A further limitation of TEM of Epon sections is that it can be hard to identify the nature of the intracellular virus-containing structures, for example, whether they are part of the endosome/lysosome system, or whether they represent structures modified by the virus, such as swollen Golgi cisternae or enlarged TGN vesicles. These questions can in part be addressed by immunolabeling of infected cells with antibodies against various intracellular markers. At least in the case of HIV assembly in macrophages, immunolabeling methods have demonstrated that the intracellular virus-containing vacuoles contain some markers for late endosomes and lysosomes (Pelchen-Matthews *et al.*, 2003; Raposo *et al.*, 2002).

III. Immunolabeling of Ultrathin Cryosections: Applications of the Tokuyasu Technique to Study Virus Assembly

Many viral structural proteins are expressed at high levels and tend to be incorporated into virus particles at high copy number. For example, HIV may contain up to 5000 Gag molecules per particle (Briggs *et al.*, 2004). In addition, as these proteins tend to be highly immunogenic, a range of antibodies or sera against viral components are available commercially or through research facilities such as the NIH AIDS Research and Reference Reagent Program. Many antibodies recognize the viral proteins in their native state and give strong signals in immunolabeling reactions. To identify cellular compartments, it is necessary to perform double labeling experiments with antibodies directed against specific marker proteins, and this requires postembedding immunolabeling protocols.

Some highly expressed viral proteins have been immunolocalized on Araldite-embedded sections (Landini *et al.*, 1987, 1991) or on Epon etched by treatment with, for example, hydrogen peroxide (Hensel *et al.*, 1995), but the harsh conditions required during cell preparation (fixation in high concentrations of aldehyde fixatives, osmication, dehydration, and embedding in plastic) and etching destroy many of the antibody-binding sites. Hence this method is not generally recommended. Embedding in more hydrophilic resins such as Lowicryl or LR White retains more antigenic epitopes, but the preservation of cellular membranes in these resins is poor. This is a particular disadvantage if enveloped viruses are studied, as the interaction with the cell's membrane systems is critical to virus production. By far the best option for antigen preservation is to maintain the biological material in its hydrated state, which is achieved by using vitreous ice as the embedding medium for cryosectioning. Cellular membranes are well preserved in this technique—they tend to be negatively stained and appear as white lines.

A. Immunolabeling of Cryosections

The use of cryosections for immunolocalization studies was originally proposed by Tokuyasu (Tokuyasu and Singer, 1976) and is described in more detail in Tokuyasu (1997). Subsequently, the method has been further improved and modified and is now routinely used in many laboratories. Good descriptions of the technique can be found in Liou *et al.* (1996) and Raposo *et al.* (1997). The following section focuses and expands on some of the steps involved.

1. Preparation of Cryosections

For the standard immunolabeling of cryosections, it is still necessary to fix the infected cells and tissues with aldehyde fixatives to immobilize proteins, though milder fixations largely avoiding glutaraldehyde can also give good results. Methods for high-pressure freezing followed by freeze-substitution with aldehyde fixatives and rehydration before processing for cryosectioning are also being developed and may give still better cell preservation (Koster and Klumperman, 2003). However, when working with infected cell preparations, the fixation step also provides a useful way to inactivate the viruses. We routinely use a fixative containing 4% formaldehyde (prepared from paraformaldehyde and added to the cells as a double strength, i.e., 8%, fixative as detailed above). As glutaraldehyde is a cross-linking fixative, it can improve cell preservation and the sectioning properties of the embedded, frozen cells, even at low concentrations ($\leq$0.2%), but even these low amounts can destroy some antigenic epitopes. Fixed cells are rinsed in a glycine-containing buffer before embedding in gelatin, and embedded cell pellets are dissected into blocks suitable for sectioning, infiltrated with 2.3-M sucrose as a cryoprotectant and frozen onto specimen pins by plunging into liquid N_2. Specimen blocks can be stored in a nitrogen bank for extended times, allowing further analysis at later dates or when new antibodies become available.

With the availability of stable nitrogen-cooled chambers for the new generation of ultramicrotomes (Leica Microsystems, Vienna, Austria) and the introduction of diamond knives specifically for cryosectioning (e.g., the Diatome cryoimmuno 35° knife with its large diamond pick-up platform), ultrathin sectioning for EM has become feasible in any EM laboratory. Sections with a nominal thickness of 50–70 nm with silver to pale gold interference colors tend to be the most useful. Further improvements have been introduced in the methods for section retrieval with mixtures of sucrose and methylcellulose (Liou *et al.*, 1996), while the ability to store sections for up to a year on Formvar and carbon-coated grids (Griffith and Posthuma, 2002) makes immunolabeling experiments more productive.

2. Immunolabeling

For immunolabeling, grids are removed from the slides on which they were stored and placed—sections down—on buffered 2% gelatin, which is then melted. From the gelatin, grids are successively floated on droplets of the various incubation solutions in phosphate-buffered saline (PBS), comprising:

- Quench buffer (50-mM glycine buffer, perhaps containing 50-nM NH_4Cl) to inactivate any residual fixative. Rinse in PBS.
- A protein solution to block nonspecific sites. We normally use 1–2% bovine serum albumin (BSA), but fish-skin gelatin or fetal calf serum have also been used.
- *The primary antibody*: Droplets of 5–25 μl of antibody in a 1–2% BSA-containing carrier buffer. For some antibodies, nonspecific binding can be reduced by the inclusion of acetylated BSA (Aurion, Wageningen, The Netherlands) in this step only.
- Extensive washing in buffers containing BSA.
- The colloidal *gold probe* to react with the primary antibody, followed by more washes.
- A brief fixation (5–10 min) with 1–2% glutaraldehyde to stabilize the immune complexes.
- Finally, samples are washed extensively with distilled water, stained briefly in neutral uranyl acetate (Tokuyasu, 1978), rinsed in water, and embedded in uranyl acetate–methylcellulose (Raposo *et al.*, 1997).

3. Gold Probes for Immunolabeling

While it is possible to stabilize colloidal gold directly with the desired primary antibody, gold-labeled second antibody reagents are more convenient to use, and many are now commercially available. Gold probes conjugated with second antibodies against a variety of species are available, as are colloidal gold particles stabilized with protein A, protein G, or streptavidin (to detect biotinylated compounds) or with lectins such as concanavalin A or wheat germ agglutinin.

We prefer to use protein A-gold (PAG) for localization studies as it is very versatile and allows double-staining immunolabeling even with primary antibodies from the same species (see below). Although protein A has up to four immunoglobulin binding sites, in practice, only one site tends to be available in a suitable orientation to bind the IgG Fc region. Protein A reacts strongly with antibodies from a number of species, including rabbit, human (but not human IgG3), and some mouse IgGs. However, for rat or goat primary antibodies, and for most mouse antibodies, a bridging antibody is required. This can give the appearance of clustered or clumpy labeling. Protein A conjugates can also be used to stain cells expressing Fc receptors: as the protein A-binding site on antibodies is located in the Fc region, PAG will only bind to those antibody molecules bound to their specific antigens (with their protein A-binding site exposed) and not those that have been captured by Fc receptors, where this site is hidden.

Gold particles suitable for EM are in the range of 5–20 nm, while ultrasmall particles can be visualized after silver enhancement (see Chapter 22 by Cheutin *et al.*, this volume). Labeling efficiencies tend to be higher for the smaller gold particles (Robinson *et al.*, 2000), while large probes (15–20 nm) should only be used with high-affinity antibodies directed against abundant epitopes. On the other hand, the larger 15-nm-gold particles are easier to see in the EM and this can be used to focus the investigator's attention to features of interest such as virus-infected cells or virus assembly compartments within these cells; modern digital cameras can be calibrated to enhance the contrast so that accumulations of gold particles can be detected even at low magnification and used to "home in" on the labeled structures. By contrast, 5-nm-gold particles can only be seen clearly at primary magnifications of >25,000×, they are easier to see on weakly contrasted sections, and can be missed if the contrast is high or if they are located near more electron-dense structures.

B. An Example of Immunolabeling for EM: Assembly of SIV VLPs and Virions

Some of our own studies on the assembly of VLPs from the SIV Gag protein illustrate applications of immunolabeling of cryosections (Fig. 3). The VLP-budding profiles in COS cells transfected with the Gag-GFP chimera were labeled with antibodies directed against the SIV MA or CA proteins (Fig. 3A and B, respectively), or with an antiserum against GFP (Fig. 3C). These antibodies show finely distinct labeling patterns, demonstrating the resolution of the EM-immunolabeling method: antibodies against MA label the buds immediately beneath the membrane, consistent with the location of MA at the periphery of the VLPs, while anti-CA antibodies stain more centrally over the electron-dense Gag layer that is polymerizing under the membrane. Both the anti-MA and anti-CA antibodies are mouse IgG1s requiring a rabbit anti-mouse IgG-bridging antibody (an extra incubation step between the primary antibody and PAG in the protocol above). This gives the possibility to amplify the signal, as several bridging antibody molecules may bind to the primary antibody, and results in a more clustered distribution of the gold particles. The rabbit anti-GFP antibody interacts directly with PAG, and no bridging antibody was used (Fig. 3C). Labeling here is by single gold particles. This is particularly striking over an area of flat polymerizing Gag-GFP, where the striations in the Gag-GFP layer are decorated with evenly spaced gold particles, located just below the electron-dense Gag layer. Since the length of an antibody molecule is ~9 nm, gold particles would be expected to be located within 10 nm of the antigen in labeling reactions with a single primary antibody. If a bridging antibody is used, this will reduce the space resolution. While the orientation of the primary and bridging antibodies relative to the section may vary, for quantitative analysis it is generally assumed that the antigen of interest is located within ~20 nm of the surface of the gold particle.

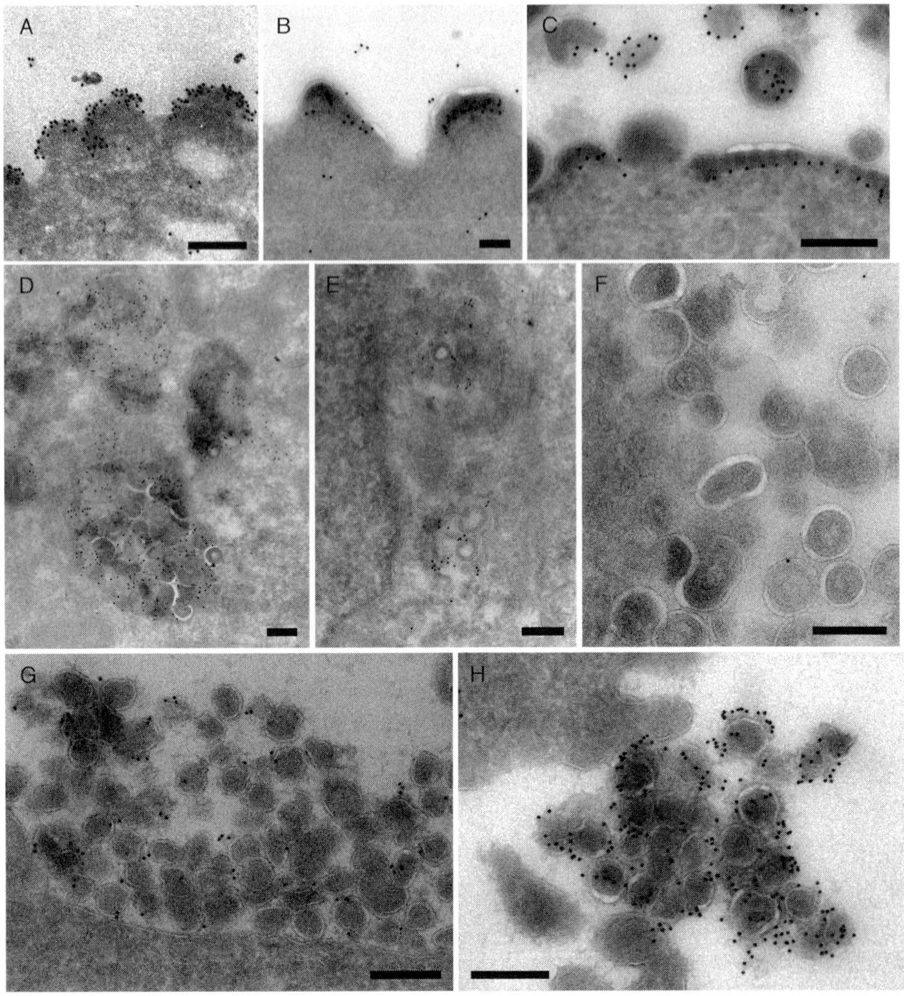

Fig. 3 Analysis of SIV assembly by immunolabeling of cryosections. (A, B, and C) SIV Gag-GFP VLPs budding from the surface of transfected COS cells. The buds were immunolabeled with mouse monoclonal antibodies against SIV MA (mAb KK59 in A) or CA (mAb KK64 in B, antibodies provided by Dr. K. Kent through the NIBSC Centralised Facility for AIDS Reagents, Potters Bar, UK) and a rabbit anti-mouse IgG-bridging antibody (Dako UK Limited, Ely, UK), or with a rabbit polyclonal antiserum recognizing GFP (C; Living Colors™ peptide antibody, Clontech Laboratories, Inc., Mountain View, CA, USA). All antibodies were detected with 10-nm PAG. (D, E, and F) COS cells expressing SIV Gag and assembling VLPs in intracellular vacuoles resembling multivesicular bodies were immunolabeled with mouse monoclonal antibodies against CD63 (Fraile-Ramos *et al.*, 2001) or against LBPA (Kobayashi *et al.*, 1998, provided by Dr. J. Gruenberg, University of Geneva, Switzerland) and 10-nm PAG. The LBPA antibody labels intracellular (E), but not cell surface (F) VLPs. (G and H) Virus particles at the surface of CEMx174 cells chronically infected with SIVmac239 expressing an Env glycoprotein with a long cytoplasmic domain (G) or SIVmac239/251T with a cytoplasmically truncated Env protein (H) were stained with an antibody against SIV Env SU (Edinger *et al.*, 2000, provided by Dr. R. W. Doms, University of Pennsylvania, Philadelphia) and 10-nm PAG. The immunolabeling suggests that these two virus strains incorporate different amounts of the Env protein. PAG reagents were purchased from The Cell Microscopy Center, University Medical Center, Utrecht, The Netherlands. Scale bars = 200 nm.

Because of their characteristic structures, VLPs can be recognized in transfected cells even without immunolabeling, especially for the Gag VLPs, where more regular particles are formed. Since VLPs and budding figures are observed both at the plasma membrane and in some intracellular vacuoles, immunolabeling with antibodies against cellular markers can be used to identify these intracellular structures. Here, the VLP-containing vacuoles were labeled by antibodies against the late endosomal protein CD63 (Fig. 3D) or the lipid lysobisphosphatidic acid (LBPA) (Kobayashi *et al.*, 1998), which accumulates on the internal membranes of late endosomes and lysosomes (Fig. 3E). The presence of these markers identifies the intracellular vacuoles as late endosomes. Some VLPs appear to be directly labeled with gold particles, suggesting that CD63 or LBPA can themselves be incorporated into the budding virion membrane. By contrast, VLPs budded at the cell surface are not stained by the anti-LBPA antibody (Fig. 3F), indicating that there may be biochemical differences between VLPs budded at the cell surface or at the membrane of endosomes.

We have also examined the assembly of various SIV strains that differ only in their cell surface Env. As for other retroviruses, the SIV Env glycoprotein is composed of two subunits, the surface glycoprotein (SU, gp120), which is noncovalently attached to the transmembrane subunit (TM, gp41). TM contains a long cytoplasmic domain of around 150–180 amino acids, depending on the virus strain. This long cytoplasmic domain is required for virus transmission *in vivo*, but when some SIV isolates, such as SIVmac239, are passaged in human tissue culture cell lines, they spontaneously truncate the cytoplasmic domain and express an Env protein with a shorter cytoplasmic tail of only around 20 amino acids. Truncation of the Env TM cytoplasmic domain can increase Env incorporation into virions (Johnston *et al.*, 1993; Zingler and Littman, 1993). We have examined CEMx174 cells chronically infected with SIVmac239 or with a variant expressing an Env protein with a truncated TM protein (SIVmac239/251Tail). Ultrathin cryosections were stained with an antibody against the V3 loop on the SIV SU protein (Edinger *et al.*, 2000). As shown in Fig. 3G and H, there was only sparse gold labeling of the budded SIVmac239 particles (Fig. 3G), while virions with the truncated Env TM protein were much more strongly labeled (Fig. 3H). Consistent with the exposure of Env at the surface of the particles, the virions were decorated with PAG on the outside of the virions. Indeed, the abundant Env glycoproteins on these particles was visualized as a fuzzy electron-dense layer (Fig. 3H). Since gold particles appeared as discrete spots on the images, they could be counted to give an estimate of the Env densities on virions. This suggested that, while SIVmac239 viruses were labeled with only 1.2 gold particles per virion (Fig. 3G), the short tail variant (SIVmac239/251Tail) contained 8.7 gold particles per virion (Fig. 3H), suggesting that this SIV variant incorporates more than seven times as much Env as the wild-type SIVmac239, in agreement with the analysis of purified SIV particles by HPLC or electron tomography studies, which reveal ~8- to 10-fold higher incorporation of truncated Env proteins into SIV (Chertova *et al.*, 2002; Zhu *et al.*, 2003). However, immunolabeling of cryosections offers the possibility of analyzing the composition of individual virus particles, rather than bulk populations, and may be used to compare, for example, viruses budded at intracellular sites or at the cell surface.

The gold particle counts also give a rough indication of the efficiency of antigen detection on immunolabeled cryosections: if SIVmac239 virions with a truncated cytoplasmic domain, such as the SIVmac239/251Tail, contain about 79 Env trimers per virion (Zhu *et al.*, 2003) and virus particles are, approximately, cut in half on 50- to 70-nm cryosections, then the gold particle counts above represent a labeling efficiency of around 7%. This is in broad agreement with the values previously determined for the labeling of the Semliki Forest virus spike protein complex in infected baby hamster kidney cells (Griffiths and Hoppeler, 1986), or for labeling of 5′-nucleotidase at the cell surface (Howell *et al.*, 1987). The high labeling of virions here probably results from the use of a rabbit anti-mouse bridging antibody in the staining reaction, which gives some amplification, and from the fact that the stained virus particles are located in the extracellular space rather than embedded in the cytoplasmic matrix. The penetration of gold probes into cryosections depends on the density of the organelles, allowing reagents to penetrate well into some structures, while access to dense structures such as secretory granules may be severely restricted (Slot *et al.*, 1989; Chapter 21 by Ladinsky and Howell, this volume). Labeling efficiencies vary for different antibody/antigen combinations, and depend on the size of the gold probes (higher labeling efficiencies are obtained with smaller gold probes), while steric effects may reduce the access of antibodies or gold probes at sites of high antigen density. While labeling efficiencies of around 10% or less can be acceptable for the study of viruses or virus-infected cells, where viral proteins are expressed in abundance, inefficient detection can be a problem in the study of cellular antigens that may be present at much lower copy number.

C. Protocols for Double-Staining Immunolabeling

More information can be obtained in double-labeling studies using different sized gold particles to detect several distinct antigens. In studies of virus assembly, it can be useful to localize a viral structural protein and a cellular marker to define the virus assembly compartments or to detect multiple viral antigens. For example, in HIV the levels of CA and Env can be compared and the composition of individual virus particles at different locations in the cell analyzed. To ensure specificity of the labeling, these studies have to be more carefully designed and controlled than single antibody immunolabeling reactions.

1. Immunofluorescence Staining of Semithin Cryosections

It is often useful to screen the samples by immunofluorescence staining before embarking on more complex investigations. This is conveniently done with semithin (0.5 μm) cryosections that are cut with glass knives at −80 °C, picked up on a droplet of 2.3-M sucrose, and placed on glass slides. The slides should be marked by scoring small (5-mm diameter) circles to indicate the area where the sections will be placed. A number of sections from different specimens can be placed side-by-side and screened together with the same antibody. Immunolabeling is performed on the slides by placing drops of solutions over the sections. Fixative is quenched by incubation in a solution of glycine and NH_4Cl as above. For sections fixed in glutaraldehyde fixatives, a brief incubation in a freshly prepared solution of sodium borohydride is also required to reduce the autofluorescence of glutaraldehyde-fixed tissue. Normally, there is no need to permeabilize the cells, as many organelles are cut open during the sectioning; however, if desired, buffers containing detergents such as Triton X-100 or Saponin may be used. When these samples are stained, good images without "out-of-focus" fluorescence from adjacent layers in the cell can be obtained, even with a standard fluorescence microscope. Immunofluorescence thus allows a rapid assessment of the specimens: it can give an indication of the proportion of virus-infected cells and, in situations where virus assembly occurs in a restricted location, of the number of cell profiles sectioned through the area of interest. The immunofluorescence patterns will give a first indication of the location of structures of interest, antibody reactivities can be conveniently assessed and antibodies titered, while double-staining studies can give a first indication of the compartments involved.

An example of screening by immunofluorescence is shown in Fig. 4. Here, sections of human primary monocyte-derived macrophages infected for 13 days with HIV-1_{BaL} were stained with a rabbit serum to the HIV MA protein and a fluorescent anti-rabbit antibody. The staining reveals that some of the cells contain

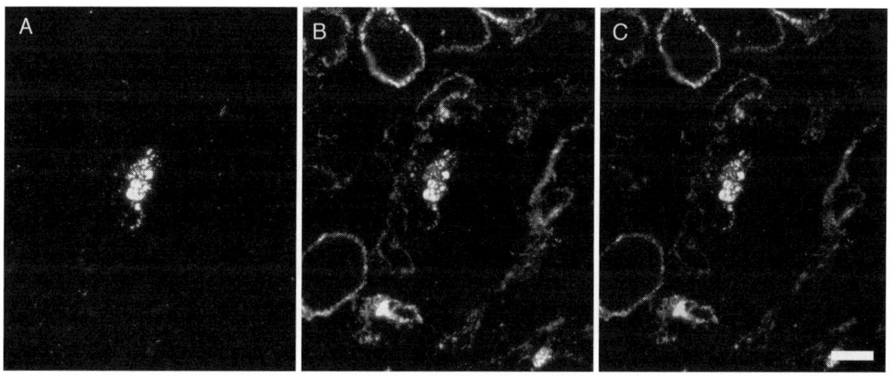

Fig. 4 Screening semithin cryosections by immunofluorescence. Cryosections (0.5-μm thick) from macrophages infected for 13 days with HIV-1_{BaL} were immunolabeled with a rabbit antiserum against HIV-1 MA (UP595, provided by Dr. M. Malim, King's College London) and a monoclonal antibody against CD9 (mouse monoclonal anti-CD9, Serotec, Kidlington, Oxford, UK), and the bound antibodies were detected with goat anti-rabbit Alexa Fluor-594 and goat anti-mouse Alexa Fluor-488 secondary antibodies (Invitrogen Molecular Probes, Paisley, UK). (A) Red fluorescence. MA staining reveals an accumulation of HIV viruses in one of the cells. (B) Green fluorescence. CD9 staining. (C) Merged image shows colocalization in yellow. Scale bar = 10 μm. (See Color Insert.)

intracellular accumulations of the virus (Fig. 4A). In addition, the sections were incubated with a monoclonal antibody recognizing the tetraspanin CD9 and anti-mouse Alexa Fluor-488 (Fig. 4B). CD9 is found primarily at the plasma membrane and is here seen to outline a number of cells. In addition, some cells show granular deposits of intracellular CD9 staining. In the HIV-infected cell, the intracellular virus accumulation (MA-stained) colocalizes with the CD9 marker.

2. Double-Staining Protocols with PAG Reagents

The simplest double immunolabeling protocols are achieved when staining with two primary antibodies from different species is followed by the relevant gold-labeled secondary antibodies; this can sometimes be performed in just two incubations with mixed primary and secondary reagents, respectively. Double staining with PAG reagents requires sequential incubations, but it does sometimes permit double staining even with primary antibodies from the same species. These double-staining protocols rely on the fact that the protein A-binding site on antibodies is easily inactivated by a brief incubation in glutaraldehyde (5–10 min in 1–2% glutaraldehyde; we have found that a similar incubation in formaldehyde fixatives is not sufficient to block later cross-reactions). With this in mind, a number of double- or even triple-staining protocols can be devised (Table I). Antibodies that

Table I
Double-Staining Protocols with PAG

Four-step staining
(two primary antibodies, both able to bind PAG directly[a])
First primary antibody/PAG-1 -GA-[b] Second primary antibody/PAG-2
Five-step staining
(one primary antibody requiring a bridging antibody, the second, from a different species able to bind PAG directly)
First primary antibody/Bridging antibody/PAG-1 -GA- Second primary antibody/PAG-2
OR First primary antibody/PAG-1 -GA- Second primary antibody/Bridging antibody/PAG-2
Six-step staining
(two primary antibodies from different species, both requiring a bridging antibody)
First primary antibody/Bridging antibody/PAG-1 -GA- Second primary antibody/Bridging antibody/PAG-2
(two mouse primary antibodies, the first requiring a rabbit anti-mouse IgG-bridging antibody)
First primary antibody/Bridging antibody/PAG-1 -GA- Isotype-matched blocking antibody[c] -GA- Second primary antibody/PAG-2

[a]Strong protein A-binding antibodies include (among others) rabbit antibodies, human IgG1 and 2 (not IgG3) and some mouse IgG 2a, 2b. Antibodies from rat, goat, or sheep do not generally bind to protein A.

[b]GA denotes the fixation to inactivate protein A-binding sites. This consists of 5–10 min in 1–2% glutaraldehyde, followed by washing in PBS, quenching in 50-mM glycine buffer, as well as a preincubation in BSA-containing blocking buffer.

[c]Same isotype as the first primary antibody.

are able to bind protein A directly allow the simplest staining sequences, but protocols for labeling two mouse antibodies, even if one requires a bridging antibody, can also be developed. For example, labeling with a mouse IgG1, a rabbit anti-mouse bridging antibody and a small PAG particle can be followed by glutaraldehyde fixation, quenching and then a blocking incubation with an isotype-matched (again IgG1) irrelevant antibody to saturate any remaining sites on the bridging antibody that have affinity for mouse antibodies. Following another fixation step to inactivate PAG-binding sites on the blocking antibody, the sections can be labeled with a mouse IgG2a or IgG2b followed by direct staining with a larger PAG particle. It is still not possible to double stain with two antibodies from the same species if they both require a bridging antibody (e.g., two weakly binding mouse IgG1s).

The choice of the size of gold particles used is central to the success of a double-staining immunolabeling experiment. Since all colloidal gold preparations contain some spread of particle sizes, combinations of 5- and 15-nm gold particles give the most unambiguous distinction between the labels. Commercial preparations of 10- and 15-nm gold both contain a small proportion of 12-nm particles, so it may be difficult to identify the antibody associated with some individual particles. Given the ease with which large gold particles can be located at the EM, the biggest gold particles should be used to label structures of particular interest, for example, virus particles or a marker for a rare compartment. Due to steric interference and masking of sites, the efficiency of immunolabeling is reduced for the second or later antibody incubations, and hence the most critical staining reactions should be done first. For steric reasons, there is also less cross-reaction if the smallest gold particles are used before larger conjugates in the staining sequence. Further restrictions arise if one antibody is sensitive to the glutaraldehyde fixation step: such an antibody then has to be used as the first primary antibody. To allow interpretations of the labeling patterns, antigen distributions should be carefully documented in single-staining experiments, and all staining experiments need to be carefully controlled, for example, by using the "wrong" first antibodies or omitting reagents. The most informative controls are provided by cell lines lacking an antigen of interest, for example, after depletion by small interfering RNAs (siRNAs) or, when studying virus assembly, by uninfected cells, by pre-absorbing antibodies with their antigen, and by omitting the second primary antibody.

D. Double- and Triple-Labeling Studies of HIV Assembly in Macrophages

As already mentioned, HIV assembly in primary human monocyte-derived macrophages occurs in large, apparently intracellular, vacuoles with various complex morphologies. Some vacuoles are filled with viruses, while others contain scattered virus particles in irregular or interconnected structures. These structures can be stained with antibodies against markers for late endosomes (Pelchen-Matthews *et al.*, 2003) or for the major histocompatibility class II complex (Raposo *et al.*, 2002), suggesting that they may be related to late endosomes or equivalent to the

MIIC, the compartment where major histocompatibility class II molecules are loaded with peptides. Often only a fraction of the macrophages in a culture are infected with virus, which may be found in a restricted location in the cell, so screening samples by immunofluorescence (see Fig. 4 and above) is essential. Here, the preliminary immunofluorescence studies suggested that the virus vacuoles contain the tetraspanin CD9. This was further investigated in EM double-labeling experiments. CD9 is located primarily at the plasma membrane, with rare staining also of intracellular vacuoles (cf. Fig. 4B), hence we wanted to mark the location of CD9 with large gold particles (PAG 15 nm). However, the CD9 antibody was extremely sensitive to glutaraldehyde fixation, with even a 5-min incubation in 1% glutaraldehyde eliminating most of the signal. Thus, we used a protocol where staining with anti-CD9/rabbit anti-mouse bridging antibody/15-nm PAG was followed by a rabbit anti-HIV MA antiserum and 5-nm PAG. Virus-containing vacuoles were strongly labeled for CD9, which was frequently located over the virus particles in the lumen of the vacuole or over small vesicles resembling the intralumenal vesicles of multivesicular bodies or "exosomes" (Fig. 5A). The identity of the virus particles was confirmed by their staining with 5-nm PAG.

In this double-staining protocol, we observed some cross-reaction of the second gold reagent (5-nm PAG) with the first antibody (anti-CD9). This was particularly noticeable when we examined CD9 labeling at the cell surface of uninfected macrophages (which lacked the MA antigen). A number of 5-nm gold particles were still observed, and most of these were located close to (within 20 nm of) a 15-nm PAG particle (Fig. 5A, arrows). Such cross-reaction would not occur if the smaller gold were used first, as a larger gold identifying the second antigen would be sterically blocked from accessing any surviving PAG-binding sites in the first antibody layer. Given the distinct pattern with which the anti-MA antibody labels virus particles, the cross-reacting gold particles could easily be identified in this study. However, this example serves to illustrate the importance of careful controls in the immunolabeling reactions.

Figure 5C shows an example of a triple labeling with three sizes of gold particles. Here sections from HIV-1_{BaL}-infected macrophages were stained sequentially with three antibodies that are all able to bind PAG directly. Virus particles are identified with a mouse monoclonal antibody against HIV-1 MA and 5-nm PAG, the 10-nm PAG marks the late endosome protein CD63, while a rabbit antiserum and 15-nm PAG were used to detect the lysosomal marker, LAMP-1. Virus particles are clearly decorated with the 5-nm gold particles, and often also labeled with the 10-nm PAG identifying CD63. Indeed, biochemical studies have shown that CD63 is incorporated into the envelope of infectious HIV virions produced by macrophages (Nguyen *et al.*, 2003; Pelchen-Matthews *et al.*, 2003). This is important because for many viruses, including HIV, only a fraction of the visible particles are infectious, and EM techniques alone cannot distinguish infectious from morphologically normal noninfectious virus particles. Cryosection immunolabeling allows the analysis of virus components, for example, the content of Envs that are essential for infectivity. Nonetheless, it is important to study statistically relevant numbers of

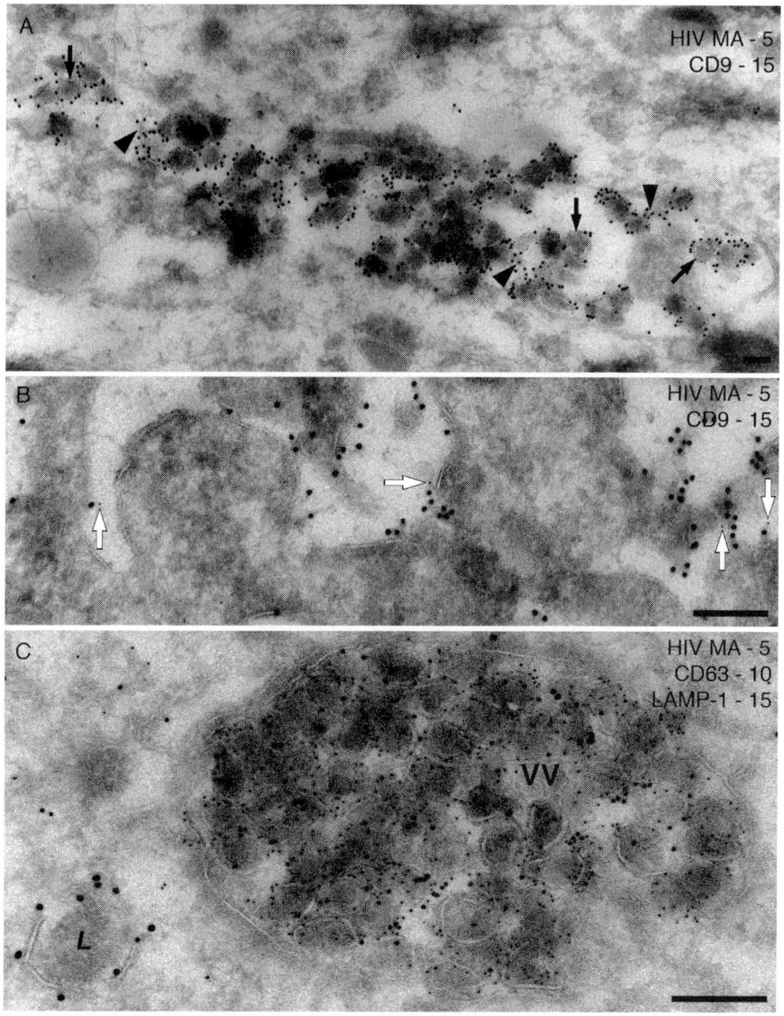

Fig. 5 Double- and triple-staining immunolabeling with PAG. (A and B) Ultrathin cryosections of 7-day HIV-1_{BaL} infected (A) or uninfected macrophages (B) were stained for CD9 (mouse monoclonal anti-CD9, rabbit anti-mouse bridging antibody and 15-nm PAG) followed by the rabbit antiserum against HIV-1 MA and 5-nm PAG. (A) An intracellular vacuole containing many virus particles is stained strongly for CD9, both on the virus particles (black arrows) and on smaller intralumenal vesicles (black arrowheads). (B) CD9 labeling at the cell surface of uninfected macrophages. A small number of 5-nm PAG particles are seen close to 15-nm PAG (white arrows), indicating a minor amount of cross-reaction in this double-labeling protocol. (C) Triple staining with a mouse monoclonal anti-HIV MA antibody (4C9, Ferns *et al.*, 1987, provided by Drs. R. B. Ferns and R. S. Tedder through the NIBSC Centralised Facility for AIDS Reagents) and 5-nm PAG, the mouse monoclonal anti-CD63 and 10-nm PAG and a rabbit anti-LAMP-1 (provided by Dr. S. Carlsson, Umea University, Sweden) and 15-nm PAG. The large vacuole is packed with virus particles that are also labeled for CD63, but there is little labeling for LAMP-1. By contrast, a lysosome (*L*) nearby is strongly labeled for LAMP-1 on its limiting membrane. Scale bars = 200 nm.

virus particles, and the data are best analyzed in conjunction with complementary virological and biochemical studies. Finally in Fig. 5C, the 15-nm gold particles identifying LAMP-1 are only rarely seen in the virus vacuole, but nearby an electron-dense vacuole, probably a lysosome, is clearly labeled on its limiting membrane. These studies support the idea that the compartment where HIV assembles in macrophages is related to late endosomes but is distinct from lysosomes (Pelchen-Matthews *et al.*, 2003; Raposo *et al.*, 2002).

E. Localization of Other Cellular and Viral Components

The immuno-EM method described so far focused mainly on detecting protein antigens with suitable antibodies. Some antibodies are also available against lipid antigens, as shown above for LBPA (Kobayashi *et al.*, 1998) (see Fig. 3). The standard aldehyde fixatives described adequately immobilize proteins, but some lipids may relocate during the sectioning and labeling procedures. Hence lipid localization studies can benefit from high-pressure freezing to prepare samples (Mobius *et al.*, 2002). Cryosections are cut from the frozen specimens and collected with pick-up solutions supplemented with aldehyde fixatives and uranyl acetate to stabilize the membranes. For lipid localizations, the cryosections should be labeled immediately, and it is also advisable to perform the staining incubations on ice to prevent or at least slow the relocation of the lipid components (Downes *et al.*, 2005). In addition to antibodies, various lipid-binding probes have been described that are useful for EM localization studies, including toxins such as perfringolysin O, which binds cholesterol (Mobius *et al.*, 2002). Other probes have been developed using protein domains that bind specific phosphoinositides (Downes *et al.*, 2005).

In addition, *in situ* hybridization techniques are being adapted to detect viral DNA or RNA genomes at the EM level (Risco and Carrascosa, 1999). Although EM *in situ* hybridization reactions on ultrathin frozen sections appear to be the most sensitive (Le Guellec, 1998), the ultrastructural preservation can be poor, and many investigators favor postembedding methods with Lowicryl or LR White resin-embedded material. *In situ* hybridization studies localizing adenovirus or Herpes Simplex virus DNA have been described (Goto *et al.*, 2001; Puvion-Dutilleul and Puvion, 1989; Puvion-Dutilleul *et al.*, 1998). Similarly, Lowicryl-embedded preparations of infected cells were used to detect HIV or SIV genomes in virus particles or budding figures and at cytoplasmic sites near the plasma membrane of infected cells (Canto-Nogues *et al.*, 2001). Viral RNA was also detected in the nucleus of peripheral blood mononuclear cells even before they exhibited morphological evidence of infection.

F. Quantification of Gold Particle Distributions

Immunolabeling is extremely useful as an aid to the identification of structures and to establish patterns of colocalization with marker proteins. In addition, gold particles appear as discrete spots on electron micrographs and so can be counted and their distributions analyzed for a more quantitative description of the experimental data.

This is particularly useful for poorly expressed antigens or in situations where the gold labeling is sparse and dispersed. When gold particle distributions on a set of photographs are compared with the distribution of random points (e.g., intersections on an overlaid lattice), relative labeling indexes can be calculated and used to identify significantly labeled compartments (Mayhew *et al.*, 2002, 2004), and different experimental samples can be compared under various conditions.

Many of the new generation of CCD cameras that are being developed for EM are equipped with particle identification programs that can facilitate the recognition and counting of gold particles of various sizes and the analysis of particle distributions and densities. This can again be useful to compare labeling over different compartments or on specimens prepared under different conditions. More complex methods of analyzing particle distributions using, for example, nearest neighbor algorithms and clustering analysis are also being developed (Hancock and Prior, 2005).

Given the ease with which gold particles can be detected and counted, it is important to be aware of the pitfalls associated with gold particle analyses. Nonspecific labeling needs to be clearly defined, for example on preparations lacking the relevant antigen, and quantified. Variations in the labeling efficiencies for different antibody antigen combinations or gold particle sizes must be accounted for. In particular, variations in the ability of different sized gold particles to penetrate cryosections (Slot *et al.*, 1989), which may depend on the density of the structure to be analyzed, can give rise to different labeling efficiencies over different structures. For example, an antigen at the cell surface may be more accessible, and hence more strongly labeled, than the same molecule within an intracellular organelle, leading to large variations in labeling efficiencies (Griffiths and Hoppeler, 1986); hence, in some situations, labeling densities can only be used to compare variations in antigen concentration in similar organelles, and EM studies should be supported by, for example, quantitative biochemical analysis from cell fractionation.

IV. Future Developments

As illustrated above, the traditional methods of analyzing plastic-embedded preparations of virus-infected cells, together with the more powerful application of immunolabeling of ultrathin cryosections can provide detailed information on the sites and mechanisms of assembly of viruses. Often, limitations arise due to small numbers of infected cells, which make finding areas of interest laborious and time consuming. This task may be facilitated when a CCD camera is available to allow systematic scanning across the specimens.

The development of stable specimen holders and tilt stages and the associated computer software is making EM tomography more widely available (see Part V, this volume), and these methods should also be applied to the analysis of virus-infected cells. Tomography can reveal membrane connections and might be particularly revealing in studies of surface connections of budding viruses or of budding mutants. Tomography methods have also been used on thick cryosections (Koster and

Klumperman, 2003; Chapter 21 by Ladinsky and Howell, this volume), revealing more structural details. However, colloidal gold probes do not penetrate into thick cryosections. Ultrasmall gold reagents are available, which can penetrate much further into sections and can be visualized by silver enhancement. A range of ultrasmall gold reagents against various antibodies is already available. Some of these are linked to Fab fragments, while reagents based on even smaller antibody fragments such as camelid heavy-chain variable domains (Spinelli *et al.*, 2000) are being developed. A related technology is the development of gold nanoparticles with defined chemical structures and therefore more uniformity than gold colloids. These particles can be covalently linked to various binding reagents, including antibodies, Fab fragments, ligands, or haptens, and show good stability and a long shelf life. Like ultrasmall gold reagents, gold nanoparticles can penetrate more deeply into sections, and can be visualized by silver enhancement or using gold-based autometallography reactions that may be compatible with biological materials.

A particularly exciting development is correlative light microscopy and EM. Already applications have been described for GFP-labeled HIV (Larson *et al.*, 2005) or HCMV (Sampaio *et al.*, 2005). Cells growing on gridded coverslips can be transfected, infected, or microinjected with GFP chimeras, or engineered virus strains. After imaging in a fluorescent microscope, possibly with live-cell recording of the fluorescent constructs, cells are fixed and embedded, and the same cells can be relocated in the EM and may be imaged, even by tomography. Similar techniques are being developed for preparing cryosections from monolayers of cells (Oorschot *et al.*, 2002), and for relocating the fluorescent cells on immunolabeled cryosections (Koster and Klumperman, 2003). Correlative methods may also make use of reagents such as FluoroNanogold® (nanogold particles as above, that have been modified with a fluorescent probe, either fluorescein or one of the Alexa Fluor® dyes). Such reagents can be used to label cells, or even cryosections, for analysis both at the fluorescent light microscope level and subsequently, after silver or gold enhancement, at the EM level. Indeed, ultrathin cryosections may be stained with FluoroNanogold® and then silver-enhanced and viewed at the EM, with significantly higher labeling efficiencies than can be obtained with colloidal gold probes (Takizawa and Robinson, 2003). Fluorescent quantum dots are also reported to be easily visible in the EM and may have similar applications (Giepmans *et al.*, 2005).

V. Summary and Conclusions

Here we have demonstrated how EM, and particularly immunolabeling of cryosections, constitute powerful methods to analyze virus assembly. Particles with characteristic structures are easily identified, while immunolabeling allows quantification of viral components, even in individual virus particles, and comparisons between particles at different locations in the cell or at different stages in viral assembly. For many viruses, a range of specific antibodies are available, while reagents against cellular markers are widely commercially available. Together with the newly developed methods for electron tomography and correlative immunofluorescence studies and EM, huge potential exists to unravel more details of virus assembly in the near future.

Acknowledgments

We thank our collaborators Dr. Thomas Kledal, Dr. Jim Hoxie, and Dr. Wolfgang Ballensiefen for providing infected cell preparations, Dr. Lindsay Hewlett for help with the HCMV EM experiments, and Dr. Lucy Collinson for critical comments on the manuscript. We are grateful to the NIBSC Centralised Facility for AIDS Reagents, Potters Bar, UK and to Drs. Jean Gruenberg, Bob Doms, Michael Malim, and Sven Carlsson for providing antibody reagents. This work was supported by the UK Medical Research Council.

References

Briggs, J. A., Grunewald, K., Glass, B., Forster, F., Krausslich, H. G., and Fuller, S. D. (2006). The mechanism of HIV-1 core assembly: Insights from three-dimensional reconstructions of authentic virions. *Structure* **14,** 15–20.

Briggs, J. A., Simon, M. N., Gross, I., Krausslich, H. G., Fuller, S. D., Vogt, V. M., and Johnson, M. C. (2004). The stoichiometry of Gag protein in HIV-1. *Nat. Struct. Mol. Biol.* **11,** 672–675.

Britt, W. J., and Boppana, S. (2004). Human cytomegalovirus virion proteins. *Hum. Immunol.* **65,** 395–402.

Britt, W. J., and Mach, M. (1996). Human cytomegalovirus glycoproteins. *Intervirology* **39,** 401–412.

Canto-Nogues, C., Hockley, D., Grief, C., Ranjbar, S., Bootman, J., Almond, N., and Herrera, I. (2001). Ultrastructural localization of the RNA of immunodeficiency viruses using electron microscopy *in situ* hybridization and *in vitro* infected lymphocytes. *Micron* **32,** 579–589.

Chertova, E., Bess, J. W., Jr, Crise, B. J., Sowder, I. R., Schaden, T. M., Hilburn, J. M., Hoxie, J. A., Benveniste, R. E., Lifson, J. D., Henderson, L. E., and Arthur, L. O. (2002). Envelope glycoprotein incorporation, not shedding of surface envelope glycoprotein (gp120/SU), is the primary determinant of SU content of purified human immunodeficiency virus type 1 and simian immunodeficiency virus. *J. Virol.* **76,** 5315–5325.

Curry, A., Appleton, H., and Dowsett, B. (2006). Application of transmission electron microscopy to the clinical study of viral and bacterial infections: Present and future. *Micron* **37,** 91–106.

Dales, S. (1973). Early events in cell-animal virus interactions. *Bacteriol Rev.* **37,** 103–135.

Downes, C. P., Gray, A., and Lucocq, J. M. (2005). Probing phosphoinositide functions in signaling and membrane trafficking. *Trends Cell Biol.* **15,** 259–268.

Edinger, A. L., Ahuja, M., Sung, T., Baxter, K. C., Haggarty, B., Doms, R. W., and Hoxie, J. A. (2000). Characterization and epitope mapping of neutralizing monoclonal antibodies produced by immunization with oligomeric simian immunodeficiency virus envelope protein. *J. Virol.* **74,** 7922–7935.

Ferns, R. B., Tedder, R. S., and Weiss, R. A. (1987). Characterization of monoclonal antibodies against the human immunodeficiency virus (HIV) gag products and their use in monitoring HIV isolate variation. *J. Gen. Virol.* **68,** 1543–1551.

Fraile-Ramos, A., Kledal, T. N., Pelchen-Matthews, A., Bowers, K., Schwartz, T. W., and Marsh, M. (2001). The human cytomegalovirus US28 protein is located in endocytic vesicles and undergoes constitutive endocytosis and recycling. *Mol. Biol. Cell* **12,** 1737–1749.

Fraile-Ramos, A., Pelchen-Matthews, A., Kledal, T. N., Browne, H., Schwartz, T. W., and Marsh, M. (2002). Localization of HCMV UL33 and US27 in endocytic compartments and viral membranes. *Traffic* **3,** 218–232.

Freed, E. O. (2002). Viral late domains. *J. Virol.* **76,** 4679–4687.

Freed, E. O. (2004). HIV-1 and the host cell: An intimate association. *Trends Microbiol.* **12,** 170–177.
Gendelman, H. E., Orenstein, J. M., Martin, M. A., Ferrua, C., Mitra, R., Phipps, T., Wahl, L. A., Lane, H. C., Fauci, A. S., Burke, D. S., Skillman, D., and Meltzer, M. S. (1988). Efficient isolation and propagation of human immunodeficiency virus on recombinant colony-stimulating factor 1-treated monocytes. *J. Exp. Med.* **167,** 1428–1441.
Giepmans, B. N., Deerinck, T. J., Smarr, B. L., Jones, Y. Z., and Ellisman, M. H. (2005). Correlated light and electron microscopic imaging of multiple endogenous proteins using quantum dots. *Nat. Methods* **2,** 743–749.
Goto, T., Kohno, T., Nakano, T., Fujioka, Y., Morita, C., and Sano, K. (2001). An improved procedure of electron microscopic *in situ* hybridization for detecting adenovirus DNA. *J. Electron Microsc. (Tokyo)* **50,** 73–76.
Griffiths, G., and Hoppeler, H. (1986). Quantitation in immunocytochemistry: Correlation of immunogold labeling to absolute number of membrane antigens. *J. Histochem. Cytochem.* **34,** 1389–1398.
Griffith, J. M., and Posthuma, G. (2002). A reliable and convenient method to store ultrathin thawed cryosections prior to immunolabeling. *J. Histochem. Cytochem.* **50,** 57–62.
Grunewald, K., Desai, P., Winkler, D. C., Heymann, J. B., Belnap, D. M., Baumeister, W., and Steven, A. C. (2003). Three-dimensional structure of herpes simplex virus from cryo-electron tomography. *Science* **302,** 1396–1398.
Hancock, J. F., and Prior, I. A. (2005). Electron microscopic imaging of Ras signaling domains. *Methods* **37,** 165–172.
Helenius, A., Kartenbeck, J., Simons, K., and Fries, E. (1980). On the entry of Semliki forest virus into BHK-21 cells. *J. Cell Biol.* **84,** 404–420.
Hensel, G., Meyer, H., Gartner, S., Brand, G., and Kern, H. F. (1995). Nuclear localization of the human cytomegalovirus tegument protein pp150 (ppUL32). *J. Gen. Virol.* **76,** 1591–1601.
Heuser, J. (2005). Deep-etch EM reveals that the early poxvirus envelope is a single membrane bilayer stabilized by a geodetic "honeycomb" surface coat. *J. Cell Biol.* **169,** 269–283.
Howell, K. E., Reuter-Carlson, U., Devaney, E., Luzio, J. P., and Fuller, S. D. (1987). One antigen, one gold? A quantitative analysis of immunogold labeling of plasma membrane 5′-nucleotidase in frozen thin sections. *Eur. J. Cell Biol.* **44,** 318–327.
Johnston, P. B., Dubay, J. W., and Hunter, E. (1993). Truncations of the simian immunodeficiency virus transmembrane protein confer expanded virus host range by removing a block to virus entry into cells. *J. Virol.* **67,** 3077–3086.
Kartenbeck, J., Stukenbrok, H., and Helenius, A. (1989). Endocytosis of simian virus 40 into the endoplasmic reticulum. *J. Cell Biol.* **109,** 2721–2729.
Kobayashi, T., Stang, E., Fang, K. S., de Moerloose, P., Parton, R. G., and Gruenberg, J. (1998). A lipid associated with the antiphospholipid syndrome regulates endosome structure and function. *Nature* **392,** 193–197.
Koster, A. J., and Klumperman, J. (2003). Electron microscopy in cell biology: Integrating structure and function. *Nat. Rev. Mol. Cell Biol.* Suppl. SS6–S10.
LaBranche, C. C., Sauter, M. M., Haggarty, B. S., Vance, P. J., Romano, J., Hart, T. K., Bugelski, P. J., Marsh, M., and Hoxie, J. A. (1995). A single amino acid change in the cytoplasmic domain of the simian immunodeficiency virus transmembrane molecule increases envelope glycoprotein expression on infected cells. *J. Virol.* **69,** 5217–5227.
Landini, M. P., Severi, B., Cenacchi, G., Lazzarotto, T., Lindenmeier, W., and Necker, A. (1991). Human cytomegalovirus structural components: Intracellular and intraviral localization of p38. *Virus Res.* **19,** 189–198.
Landini, M. P., Severi, B., Furlini, G., and Badiali De Giorgi, L. (1987). Human cytomegalovirus structural components: Intracellular and intraviral localization of p28 and p65–69 by immunoelectron microscopy. *Virus Res.* **8,** 15–23.
Larson, D. R., Johnson, M. C., Webb, W. W., and Vogt, V. M. (2005). Visualization of retrovirus budding with correlated light and electron microscopy. *Proc. Natl. Acad. Sci. USA* **102,** 15453–15458.

Le Guellec, D. (1998). Ultrastructural *in situ* hybridization: A review of technical aspects. *Biol. Cell* **90,** 297–306.

Li, H., Zhou, D., Browne, H., Balasubramanian, S., and Klenerman, D. (2004). Molecule by molecule direct and quantitative counting of antibody-protein complexes in solution. *Anal. Chem.* **76,** 4446–4451.

Liou, W., Geuze, H. J., and Slot, J. W. (1996). Improving structural integrity of cryosections for immunogold labeling. *Histochem. Cell Biol.* **106,** 41–58.

Marsh, M., and Helenius, A. (2006). Virus entry: Open Sesame. *Cell* **124,** 729–740.

Mayhew, T. M., Griffiths, G., and Lucocq, J. M. (2004). Applications of an efficient method for comparing immunogold labelling patterns in the same sets of compartments in different groups of cells. *Histochem. Cell Biol.* **122,** 171–177.

Mayhew, T. M., Lucocq, J. M., and Griffiths, G. (2002). Relative labelling index: A novel stereological approach to test for non-random immunogold labelling of organelles and membranes on transmission electron microscopy thin sections. *J. Microsc.* **205,** 153–164.

Mobius, W., Ohno-Iwashita, Y., van Donselaar, E. G., Oorschot, V. M., Shimada, Y., Fujimoto, T., Heijnen, H. F., Geuze, H. J., and Slot, J. W. (2002). Immunoelectron microscopic localization of cholesterol using biotinylated and non-cytolytic perfringolysin O. *J. Histochem. Cytochem.* **50,** 43–55.

Morita, E., and Sundquist, W. I. (2004). Retrovirus budding. *Annu. Rev. Cell Dev. Biol.* **20,** 395–425.

Muller, B., Daecke, J., Fackler, O. T., Dittmar, M. T., Zentgraf, H., and Krausslich, H. G. (2004). Construction and characterization of a fluorescently labeled infectious human immunodeficiency virus type 1 derivative. *J. Virol.* **78,** 10803–10813.

Murk, J. L., Posthuma, G., Koster, A. J., Geuze, H. J., Verkleij, A. J., Kleijmeer, M. J., and Humbel, B. M. (2003). Influence of aldehyde fixation on the morphology of endosomes and lysosomes: Quantitative analysis and electron tomography. *J. Microsc.* **212,** 81–90.

Nguyen, D. G., Booth, A., Gould, S. J., and Hildreth, J. E. (2003). Evidence that HIV budding in primary macrophages occurs through the exosome release pathway. *J. Biol. Chem.* **278,** 52347–52354.

Novoa, R. R., Calderita, G., Arranz, R., Fontana, J., Granzow, H., and Risco, C. (2005). Virus factories: Associations of cell organelles for viral replication and morphogenesis. *Biol. Cell* **97,** 147–172.

Oorschot, V., de Wit, H., Annaert, W. G., and Klumperman, J. (2002). A novel flat-embedding method to prepare ultrathin cryosections from cultured cells in their *in situ* orientation. *J. Histochem. Cytochem.* **50,** 1067–1080.

Pelchen-Matthews, A., Kramer, B., and Marsh, M. (2003). Infectious HIV-1 assembles in late endosomes in primary macrophages. *J. Cell Biol.* **162,** 443–455.

Pelchen-Matthews, A., Raposo, G., and Marsh, M. (2004). Endosomes, exosomes and Trojan viruses. *Trends Microbiol.* **12,** 310–316.

Pizzato, M., Marlow, S. A., Blair, E. D., and Takeuchi, Y. (1999). Initial binding of murine leukemia virus particles to cells does not require specific Env-receptor interaction. *J. Virol.* **73,** 8599–8611.

Pornillos, O., Higginson, D. S., Stray, K. M., Fisher, R. D., Garrus, J. E., Payne, M., He, G. P., Wang, H. E., Morham, S. G., and Sundquist, W. I. (2003). HIV Gag mimics the Tsg101-recruiting activity of the human Hrs protein. *J. Cell Biol.* **162,** 425–434.

Puvion-Dutilleul, F., Besse, S., Pichard, E., and Cajean-Feroldi, C. (1998). Release of viruses and viral DNA from nucleus to cytoplasm of HeLa cells at late stages of productive adenovirus infection as revealed by electron microscope *in situ* hybridization. *Biol. Cell* **90,** 5–38.

Puvion-Dutilleul, F., and Puvion, E. (1989). Ultrastructural localization of viral DNA in thin sections of herpes simplex virus type 1 infected cells by *in situ* hybridization. *Eur. J. Cell Biol.* **49,** 99–109.

Raposo, G., Kleijmeer, M. J., Posthuma, G., Slot, J. W., and Geuze, H. J. (1997). Immunogold labeling of ultrathin cryosections: Application in immunology. *In* "Handbook of Experimental Immunology" (L. A. Herzenberg, D. M. Weir, L. A. Herzenberg, and C. Blackwell, eds.), pp. 1–11. Blackwell Science, Inc., Oxford, UK.

Raposo, G., Moore, M., Innes, D., Leijendekker, R., Leigh-Brown, A., Benaroch, P., and Geuze, H. (2002). Human macrophages accumulate HIV-1 particles in MHC II compartments. *Traffic* **3,** 718–729.

Risco, C., and Carrascosa, J. L. (1999). Visualization of viral assembly in the infected cell. *Histol. Histopathol.* **14,** 905–926.

Robinson, J. M., Takizawa, T., and Vandre, D. D. (2000). Enhanced labeling efficiency using ultrasmall immunogold probes: Immunocytochemistry. *J. Histochem. Cytochem.* **48,** 487–492.

Sampaio, K. L., Cavignac, Y., Stierhof, Y. D., and Sinzger, C. (2005). Human cytomegalovirus labeled with green fluorescent protein for live analysis of intracellular particle movements. *J. Virol.* **79,** 2754–2767.

Schramm, B., and Krijnse Locker, J. (2005). Cytoplasmic organization of POXvirus DNA replication. *Traffic* **6,** 839–846.

Simionescu, N., and Simionescu, M. (1976). Galloylglucoses of low molecular weight as mordant in electron microscopy. I. Procedure, and evidence for mordanting effect. *J. Cell Biol.* **70,** 608–621.

Slot, J. W., Posthuma, G., Chang, L. Y., Crapo, J. D., and Geuze, H. J. (1989). Quantitative aspects of immunogold labeling in embedded and in nonembedded sections. *Am. J. Anat.* **185,** 271–281.

Sodeik, B., Ebersold, M. W., and Helenius, A. (1997). Microtubule-mediated transport of incoming herpes simplex virus 1 capsids to the nucleus. *J. Cell Biol.* **136,** 1007–1021.

Sodeik, B., and Krijnse-Locker, J. (2002). Assembly of vaccinia virus revisited: *De novo* membrane synthesis or acquisition from the host? *Trends Microbiol.* **10,** 15–24.

Spinelli, S., Frenken, L. G., Hermans, P., Verrips, T., Brown, K., Tegoni, M., and Cambillau, C. (2000). Camelid heavy-chain variable domains provide efficient combining sites to haptens. *Biochemistry* **39,** 1217–1222.

Stang, E., Kartenbeck, J., and Parton, R. G. (1997). Major histocompatibility complex class I molecules mediate association of SV40 with caveolae. *Mol. Biol. Cell* **8,** 47–57.

Studer, D., Graber, W., Al-Amoudi, A., and Eggli, P. (2001). A new approach for cryofixation by high-pressure freezing. *J. Microsc.* **203,** 285–294.

Takizawa, T., and Robinson, J. M. (2003). Ultrathin cryosections: An important tool for immunofluorescence and correlative microscopy. *J. Histochem. Cytochem.* **51,** 707–714.

Tokuyasu, K. T. (1978). A study of positive staining of ultrathin frozen sections. *J. Ultrastruct. Res.* **63,** 287–307.

Tokuyasu, K. T. (1997). Immunocytochemistry on ultrathin cryosections. *In* "Cells, a Laboratory Manual: Volume 3: Subcellular Localization of Genes and Their Products" (D. L. Spector, R. D. Goodman, and L. A. Leinwand, eds.), pp. 131.131–131.127. Cold Spring Harbor Laboratory Press, Woodbury, New York.

Tokuyasu, K. T., and Singer, S. J. (1976). Improved procedures for immunoferritin labeling of ultrathin frozen sections. *J. Cell Biol.* **71,** 894–906.

von Schwedler, U. K., Stuchell, M., Muller, B., Ward, D. M., Chung, H. Y., Morita, E., Wang, H. E., Davis, T., He, G. P., Cimbora, D. M., Scott, A., Krausslich, H. G., *et al.* (2003). The protein network of HIV budding. *Cell* **114,** 701–713.

Zhu, P., Chertova, E., Bess, J., Jr., Lifson, J. D., Arthur, L. O., Liu, J., Taylor, K. A., and Roux, K. H. (2003). Electron tomography analysis of envelope glycoprotein trimers on HIV and simian immunodeficiency virus virions. *Proc. Natl. Acad. Sci. USA* **100,** 15812–15817.

Zingler, K., and Littman, D. R. (1993). Truncation of the cytoplasmic domain of the simian immunodeficiency virus envelope glycoprotein increases env incorporation into particles and fusogenicity and infectivity. *J. Virol.* **67,** 2824–2831.

CHAPTER 21

Electron Tomography of Immunolabeled Cryosections

Mark S. Ladinsky* and Kathryn E. Howell†

*Boulder Laboratory for 3D Electron Microscopy of Cells
Department of MCD Biology, University of Colorado
Boulder, Colorado 80309

†Department of Cellular and Structural Biology
University of Colorado Medical School
Aurora, Colorado 80045

Electron tomography (ET) and immunolabeling of cells and tissues are both established techniques that give the cell biologist important insights into structure–function relationships. While immunolabeling provides localization of specific macromolecules to a structure of interest, ET provides a high-resolution, three-dimensional (3D) perspective that allows one to characterize fine-structural detail and connectivity between compartments that often cannot be seen by other approaches. Combining the two techniques by means of preparing tomographic reconstructions of immunolabeled samples results in a powerful method for localizing macromolecules within the 3D context of a cell. Several different immunolabeling techniques can provide samples that are suitable for tomography. Most of these techniques favor certain structures or classes of antibodies. Here we describe a

0091-679X/07 $35.00
DOI: 10.1016/S0091-679X(06)79021-5

method by which the Golgi complex and other organelles in mammalian cells are immunolabeled with the technique pioneered by Tokuyasu, and subsequently analyzed by ET. The data generated by this method not only gives a 3D perspective on the labeled structures, they can also be compared directly with similar samples, prepared by rapid-freezing and freeze-substitution, to give a broader, high-resolution view of the cell.

I. Introduction

Electron tomography (ET) provides an informative view of cellular fine structure in three dimensions (3D). Our own studies with ET have focused mostly on the Golgi complex in cells or tissues that were immobilized by rapid freezing, then fixed by freeze-substitution and subsequently embedded in plastic (Ladinsky *et al.*, 1999, 2002; Marsh *et al.*, 2001). This methodology has helped in deciphering the complex structure of the Golgi with a high degree of confidence. However, ET on its own provides only part of an important story, as it generally does not permit the localization of specific macromolecules, a process that is crucial for addressing structure–function relationships in all organelles, including the Golgi complex. Equally important, the absence of a 3D perspective from studies that use traditional EM immunolabeling limits what can be learned about complex and dynamic organelles. In principle, however, one can couple tomography with techniques for immunolabeling, generating a powerful approach for localizing specific molecules within the context of 3D cell structure at a resolution that surpasses anything available from conventional light microscopy. Nonetheless, the method as currently practiced has limitations that help to define ways in which new techniques might advance this aspect of cell biology.

In light microscopy, localization of specific molecules is commonly achieved by indirect immunofluorescence, which employs a primary antibody that recognizes a specific macromolecule and a fluorescently conjugated secondary antibody against a host-specific domain of the primary IgG. This approach can also be applied in electron microscopy, except that electron-dense probes, such as colloidal gold, are used in place of fluorescent probes (Goodman *et al.*, 1979).

A histochemical method, based on horseradish peroxidase conjugated either to an antibody or directly to the protein of interest, can also be used to develop a dense reaction product. This method has been used for cytochemistry and immunocytochemistry since 1951 (Novikoff, 1951; Singer, 1959), and it has also been adapted for electron microscope studies of Golgi structure and function (Brown and Farquhar, 1989; Ladinsky *et al.*, 1994; Thorne-Tjomsland *et al.*, 1991). More recently this technique has been redesigned to employ photooxidation of a fluorescent-tagged protein (Pagano *et al.*, 1989) into an electron-dense product, which can subsequently be analyzed by ET (Grabenbauer *et al.*, 2005; Ladinsky *et al.*, 1994). The technique is limited, however, because the precipitated reaction product diffuses and tends to fill entire compartments. As such it is inaccurate when compared with colloidal gold probes.

One can imagine a powerful, modern approach to protein localization by ET that would be analogous to the green fluorescent protein (GFP) for light microscopy. In this model, a small piece of DNA could be cloned along with the sequence encoding the protein of interest, allowing the protein product to be visible in the EM but without altering its behavior *in vivo*. At present, there is no such tag, though several laboratories are working to generate the best approximation possible (Gaietta *et al.*, 2002). Instead, a method that is in common use relies on fluorescently tagged molecules that can be expressed *in vivo* and then localized with gold-conjugated antibodies that are specific for the fluorescent tag itself (see Chapter 19 by Morphew, this volume). In spite of recent progress with clonable groups that can bind labels (see Chapter 23 by Sosinsky *et al.*, this volume) and other novel methods that are under development, tagging structures of interest with antibodies labeled with heavy metals remains one of the most widely used methods for localizing macromolecules by EM.

There are several methods for immunolabeling samples for EM studies, some of which are discussed in other chapters of this book. Each method has different virtues with respect to the key issues of sample preservation, contrast, effectiveness of antibody labeling (the signal-to-noise ratio), and suitability for the cell type or structure being studied. We have found that a modified version of the method developed by Tokuyasu for immunolabeling cryosections (Tokuyasu, 1971, 1986) is valuable for studies of the Golgi complex and associated vesicles, tubules, and organelles. Instead of plastic embedding, Tokuyasu's method employs a light fixation followed by sucrose infiltration to facilitate subsequent freezing without incurring ice crystal damage. This approach preserves antigen structure and promotes its accessibility for antibody binding, thereby increasing the likelihood that a wide array of antibodies will give accurate and repeatable results. Negative staining of the labeled sections is sufficient for contrasting the membrane structures of interest with an acceptable signal-to-noise ratio. Tomography of the resulting sections gives 3D data on the localization of labeled proteins that can be compared with similar samples prepared for optimal preservation of structure to more completely characterize the structure of interest, and its function, within the cell.

II. Methods

The methods described below are routinely used on mammalian tissue (rat liver and mammary gland) and cultured cell lines [normal rat kidney (NRK) and HeLa]. Both cellular systems are effectively labeled and analyzed by these methods. There are many published variations of this technique. The most complete listing of them can be found in *Fine Structure Immunocytochemistry* (Griffiths, 1993).

A. Specimen Fixation and Freezing

Cultured cells are grown to confluence on 100-mm Petri dishes in a medium appropriate to the cell type (e.g., Dulbecco's modified Eagle's medium supplemented with 10% calf serum for NRK cells), using a 37 °C incubator with 5% CO_2.

Typically two dishes are prepared; these yield enough cells for at least five cryosectioning samples, which is sufficient for a large number of experiments. The monolayer of cells is removed from the dish with trypsin-EDTA (Sigma-Aldrich, Inc., St. Louis, MO) and transferred to a 50-ml centrifuge tube. The cells are then pelleted and the trypsin-containing supernatant is discarded and replaced with fresh medium. The resuspended cells are then returned to the incubator for 1 h to recover from the trypsin treatment. In our experience this approach is less damaging to the cells than fixing them while they are still attached to the Petri dish and subsequently removing them by scraping with a rubber policeman. However, for specific situations, for example, when cell adherence or junction formation are being studied, scraping of the cells is required.

Following removal from the dish, the cells are gently pelleted, the medium is removed, and the pellet is resuspended in 1 ml of fixative solution. Cells are fixed with 4% EM-grade paraformaldehyde in phosphate-buffered saline (PBS) containing 5% sucrose and 100-mM HEPES, pH 7.2. A common problem with chemically fixed EM samples is a noticeable dilation of one or more Golgi cisternae, usually on the *trans*-sides of the Golgi stack. This dilation is usually absent in samples that have been prepared by rapid-freezing and freeze-substitution fixation (RF-FSF), so this structural change can confidently be attributed to experimental conditions (Ladinsky *et al.*, 2002). The addition of sucrose and HEPES to the fixation solution decreases artifacts in these highly labile membrane systems that can result from rapid changes in the osmotic environment and pH. Once resuspended in fixative, the cells are transferred to a 1.5-ml Eppendorf centrifuge tube and promptly pelleted (3000 rpm for 2 min) again. This step allows for even exposure of the cells to the fixative and simplifies manipulation of the sample in subsequent steps, eliminating the need for multiple rounds of centrifugation. The supernatant is replaced with fresh fixative solution and the cells are fixed overnight at 4°C.

Fresh tissue is prepared similarly to cultured cells except that the tissue, once removed from the sacrificed animal, is minced into blocks no larger than 2 mm^3 and placed into a tube of fixative solution without centrifugation. Tissue from the same preparation is almost always rapidly frozen in a high-pressure freezer, freeze-substituted, and plastic embedded in parallel with immunolabeling procedure, facilitating accurate comparison of material that was prepared for EM by both approaches.

Following fixation, the samples are warmed to room temperature. Supernatant fixative is removed without disturbing the cell pellet or tissue blocks and is replaced with PBS containing 2.1-M sucrose. The sugar infiltrates the cells, making them easier to freeze without the formation of damaging ice crystals. The sucrose also provides an ice–solute mixture that facilitates sectioning at low temperature. Infiltration proceeds over 8–10 h, during which the sucrose solution is changed every hour without disturbing the sample. This infiltration regimen, albeit slow, results in far less sample loss than other approaches that call for physically transferring the sample into pools of increasing concentrations of sucrose.

Once the cell pellets are infiltrated with sucrose, the supernatant is mostly removed, leaving ~25 μl of solution above the pellet. Under a dissecting microscope, 5–10 μl of the sample is drawn into a glass capillary tube or pulled Pasteur pipette and transferred to the tip of an aluminum cryoultramicrotomy peg (available from Ted Pella Inc., Redding, CA or Electron Microscopy Sciences, Hatfield, PA). After the peg has been loaded, it is quickly submerged in liquid nitrogen to freeze the sample onto the peg. This approach results in a rounded sample block that is strongly affixed to the peg and is large enough for repeated rounds of trimming and sectioning. For tissue, a single 1–2 mm^3 tissue block is placed onto the peg, and excess fluid is carefully removed with a strip of Whatman No. 1 filter paper prior to freezing. Pegs with frozen samples attached are stored in cryotubes (Nalge Nunc International, Rochester, NY) under liquid nitrogen.

B. Cryoultramicrotomy

Cryosections are cut with a cryomicrotome like the UltraCut UCT equipped with a FCS cryostage (Leica Microsystems, Wetzlar, Germany). Typically sections are cut at $-110°C$. The block is trimmed with a diamond cryotrimming tool (Diatome US, Hatfield, PA) to a rectangle with dimensions of $\sim 100 \times 150\ \mu m^2$. This is a relatively large blockface, but it is necessary to provide a large sample size in a given section or grid. Large areas are often important for obtaining quantifiable results or for efficiently selecting good regions for tilt-series imaging. Sections are cut with a Cryo-P diamond knife that has a 35° included angle (Diatome US, Hatfield, PA). Usually the microtome is set for 90-nm nominal advance and a cutting speed of 1 mm/sec, but these values are almost always adjusted to optimize the quality of the sections from a specific block. On particularly dry days, when static is a problem, a "Static Line" active ionizer (Diatome) is placed in the cutting chamber to control static-induced movement or compression of the sections. Ribbons of sections are collected from the knife-edge with a small wire loop containing a drop of 2.3-M sucrose in PBS. The drop is allowed to partially freeze prior to contact with the sections. The frozen sections are then quickly transferred to the surface of a Formvar-carbon-coated 100-mesh copper/rhodium grid, which is subsequently floated on the surface of a pool of PBS in a Petri dish. The transfer of sections to the grid surface is accomplished as quickly as possible to minimize melting of the drop or condensation of water on the sections prior to their attachment to the grid. The sections now melt, but since the tissue was fixed before sucrose impregnation, they are sufficiently robust that cell and tissue structure remain well preserved.

C. Immunolabeling of Cryosections

Prior to labeling the sections with antibodies, nonspecific binding sites are blocked by floating the grids on a drop of 10% calf serum in PBS for 30 min within a humidified chamber. The grid is then carefully drained of excess fluid by touching its edge with Whatman No. 1 filter paper and floated on a 10-μl drop of primary antibody diluted in PBS containing 5% calf serum. The appropriate dilution, defined as a concentration at which specific labeling is strong and nonspecific labeling is weak, must be determined for each antibody through a series of dilution experiments. With a good antibody, there should be at least a 100-fold difference between the concentration that gives adequate staining with low background and the concentrations that either give no staining at all or extensive nonspecific staining. Once the right antibody concentration is found, the samples are incubated on this antibody preparation for 2 h at room temperature. The grid is then rinsed by incubating it on four successive drops of PBS for 10 min each. The labeling process is repeated with secondary antibodies specific for the IgG of the primary antibody's host animal, conjugated to colloidal gold (5, 10, or 15 nm) (Ted Pella Inc., Redding, CA). After the final PBS rinse, the grids are washed twice in dH_2O then negatively stained with 1% uranyl acetate in dH_2O and 1% methylcellulose on ice for 30 min. The grid is picked up in a wire loop, the diameter of which is slightly larger than that of the grid, and excess stain is removed with Whatman No. 1 filter paper. Proper draining of the grid is important for minimizing the thickness of the methylcellulose embedment and for insuring even drying. Inconsistent drying often results in distortion or tearing of the sample. Holding the loop at a slight angle while passing it across the sheet of filter paper at a slow and consistent speed will help provide a methylcellulose film that is of even thickness across the grid. The grids are allowed to air-dry overnight within the wire loops. A pair of sharp forceps is used to remove the finished grid from the loop for storage in a standard grid box. Best results are obtained when the grids are examined within a week of preparation. For long-term storage of methylcellulose-embedded grids, especially in areas prone to high humidity, it is recommended that they should be stored in a desiccator.

D. Immuno-Electron Microscopy and Tomography

With few exceptions (one of which is discussed later in this chapter), colloidal gold-conjugated antibodies label antigens on only the surfaces of sucrose-impregnated cryosections. However, 3D analysis allows for the structures that are labeled at the surfaces to be followed through the volume of the section. As such, tomography of immunolabeled sections is a powerful tool for studying structural–functional relationships between adjacent membranes.

We utilize the same procedure for tomographic imaging of labeled sections as we do for plastic-embedded, osmicated samples. The grid is placed in a Model 925 (Gatan, Inc., Pleasanton, CA) tilt-rotate holder and imaged in a Tecnai TF-20 electron microscope (FEI company, Hillsboro, OR) operating at 200 KeV. When a suitable region is selected, the grid is tilted through a range of $\pm 60°$ with images recorded digitally at 1° intervals. The tilting and imaging of the sample is controlled automatically with the SerialEM software package (Mastronarde, 2005). Once recorded, the images are aligned, back-projected to make a tomogram, analyzed, and modeled using the IMOD software package (Kremer *et al.*, 1996) operating on a Macintosh G4 or G5 computer. Gold particles denoting labeled antigens can often be used as fiducial markers for image alignment.

III. Results and Discussion

We favor a procedure that examines a given structure of interest using tomographic analyses on two kinds of samples: one that optimizes cellular preservation and another that optimizes immunolocalization of specific molecules. This is a powerful way to study complex organelles, like the Golgi. The Tokuyasu method for immunolabeling of cryosections gives a reliable and repeatable localization for many molecules of interest to the membrane traffic community, and it has contributed many insights into structural–functional relationships in the Golgi (Griffiths, 1993; Griffiths *et al.*, 1983, 1984; Ladinsky and Howell, 1992; Ladinsky *et al.*, 2002; Snyder *et al.*, 2006; Wu *et al.*, 2000). However, this technique, which features chemical fixation and sucrose infiltration, does not adequately preserve the cell for high-resolution structural studies. Rapid freezing followed by freeze-substitution and embedding in Epon resin gives excellent preservation of a cell's fine structure but is rarely ever compatible with successful immunolabeling.

Figure 1 shows similar regions of Golgi complexes in lactating rat mammary epithelial cells, prepared either by high-pressure freezing and freeze-substitution (A) or by the Tokuyasu method (B). In the rapidly frozen sample, the Golgi cisternae and associated tubules and vesicles are well preserved and contrasted. Tomographic reconstruction of this region allows for unambiguous tracking and identification of many individual structures through the volume of the section. For example, endoplasmic reticulum (ER)-derived, pleomorphic compartments that concentrate on the *cis*-side of the Golgi stack (*), known as vesicular-tubular clusters (VTCs) (Bannykh *et al.*, 1996; Saraste and Svensson, 1991) can be followed in 3D and shown to be fusing with one another. These VTC "complexes" subsequently fuse with existing *cis*-Golgi cisternae or with each other to form a new cisterna (Bonfanti *et al.*, 1998). Cryosections of similar lactating mammary epithelial cells prepared by the Tokuyasu method, immunolabeled with antibodies against the *cis*-Golgi matrix protein p115, and subsequently imaged by 3D tomography, show localization of p115 to *cis*-Golgi cisternae and to structures on the *cis*-side of the stack that resemble those identified as VTCs in rapidly frozen samples (* in B). Comparison of the labeling pattern in (B) with the fine-structural detail in (A) strengthens the theory that ER-derived VTCs form into new *cis*-Golgi cisternae.

Although tomography of immunolabeled Tokuyasu cryosections cannot yield the finely detailed and reliable structural insights afforded by rapidly frozen/freeze-substituted specimens, it can be used to gain a clearer picture of the specimen than is afforded by single, two dimensional (2D) projections and to distinguish and follow immunologically distinct structures through a volume. Since usually only antigens that are exposed on the surfaces of cryosections can be labeled with antibodies, and since the compartments of the Golgi complex are highly convoluted in 3D space, it is difficult to distinguish different organelles (i.e., Golgi cisternae and ER) when they are in close proximity, and even when they are specifically labeled.

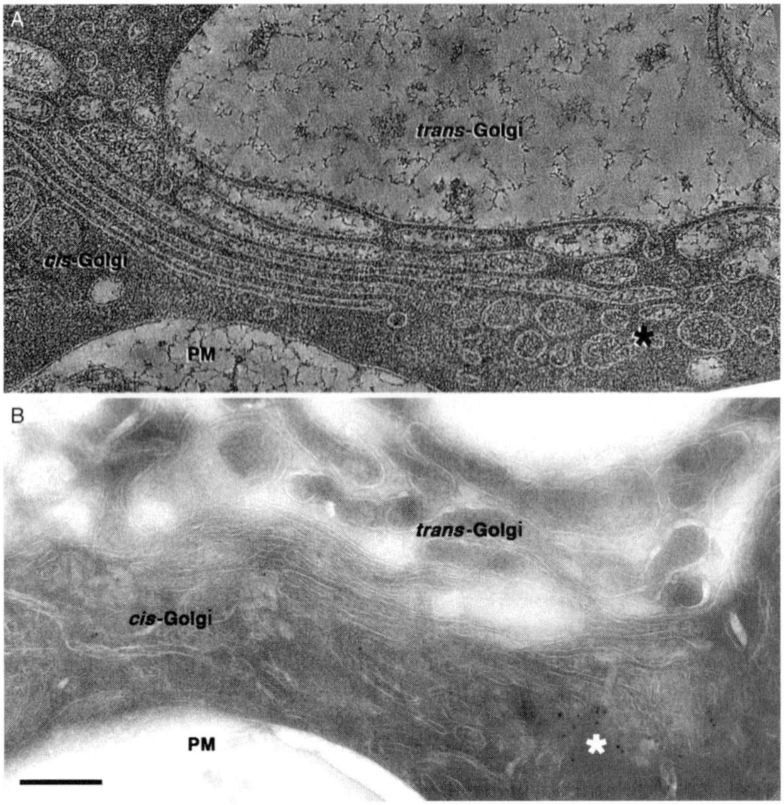

Fig. 1 Data from immunolabeled tomograms can often be correlated with similar tomographic data from rapidly frozen, freeze-substitution fixed material. The comparison helps to localize molecules on or within specific structures. (A) A tomographic slice (0.9-nm thick) showing part of the Golgi in a plastic-embedded lactating mammary epithelial cell. The *cis-* and *trans*-sides of the Golgi stack are easily discernable. The asterisk (*) indicates a region of vesicular-tubular clusters (VTCs) near the *cis*-face of the Golgi stack. (B) Shows a similar Golgi region in a thin (90 nm) cryosection from a mammary cell that was prepared for Tokuyasu immunolabeling and stained with antibodies against the *cis*-Golgi matrix protein p115. The asterisk (*) indicates the localization of p115. Comparison of the two images suggests that p115 localizes to the VTCs. Correlation of samples optimally prepared for immunocytochemistry with equivalent samples optimally prepared for preservation of fine structure allows one to confidently localize specific molecules within the complex 3D structure of a cell. Scale bar = 0.2 μm.

Figure 2 demonstrates how tomographic analysis of a Tokuyasu cryosection aids in solving this problem. Panel A shows a traditional 2D projection of a ~100-nm cryosection from a HeLa cell, transfected to express a GFP-tagged variant of the medial-Golgi matrix protein GRASP55. The section is double-labeled with antibodies against GFP (10-nm gold particles) and the ER marker, protein disulfide isomerase (PDI; 15-nm gold). The single image shows a Golgi stack (G) that is extensively labeled with GRASP55–GFP. A pronounced ER cisterna, labeled with

PDI (ER) sits to the right of the Golgi stack. There is little detail visible within individual Golgi cisternae or their associated tubules and vesicles. Also, large gold particles indicating PDI labeling are present within the Golgi stack, but it is difficult to distinguish the underlying compartment from adjacent Golgi membranes. Panel B shows a single tomographic slice from a reconstruction of the same area shown in A. This slice is taken from the middle of the section's volume. The resolution and detail are now sufficient to follow individual membrane-bounded structures, revealing their conformations and interactions with their neighbors. Since only antigens exposed at the surfaces of a section are labeled by this technique, these tomograms are analyzed by noting the positions of gold particles in tomographic slices near the top of (C), and bottom (D) of the cryosection, and then following the structures labeled by those particles through the volume of the sample. By this method, compartments that appear very close to one another and cannot be distinguished by 2D EM can be identified both by their specific markers and by their structure. An example of this is shown in Fig. 2E and F.

Specialized ER that closely associates with *trans*-Golgi cisternae (Ladinsky *et al.*, 1999) is often difficult to distinguish in traditional EM views, but it can be easily detected by the specific PDI labeling and subsequent tracking through the tomographic volume. Panel E shows a detail from the reconstruction, indicating a piece of *trans*-ER within the Golgi stack, identified by PDI label (arrow). The boundaries of the ER membrane can then be tracked through the volume of the section and modeled using the IMOD software (F). This shows that tomography is a powerful enhancement to a technique that often suffers from its inherently poor structural preservation and display.

It is important to note that the labeling pattern and strength of a particular antibody may vary considerably among different cell types, even among different cells from the same host organism. For example, the distribution of a Golgi antigen can differ depending on whether the cell under study is extensively secreting Golgi-processed molecules. Figure 3 shows the labeling pattern of the *trans*-Golgi, *trans-membrane* protein, TGN38 in two different cells from the rat: (A) shows labeling in a nonsecretory NRK cell, and labeling in a lactating mammary epithelial cell is shown in (B). Since *trans*-Golgi cisternae in lactating mammary cells contain large amounts of casein and other milk proteins, they are grossly expanded due to the osmolarity of the secreted fluid. This expansion results in greater accessibility to the antigen, and thus a comparatively higher density of gold particles.

By imaging immunolabeled sections with 3D tomography, we have found that certain antibodies applied to certain specimens can label antigens within the section, not just on its surface. This kind of labeling is common in mammary epithelial cells and other tissues that contain large, open structures such as lipid droplets. At least some of the internal labeling can be attributed to incomplete fixation of the lipid-containing bodies, which results in loss of material from those regions during sectioning or in subsequent processing steps. Mature lipid droplets are often larger in diameter than the thickness of the section, so antigens on the periphery of the open spaces become exposed to the antibodies. In other cases, however, tomography

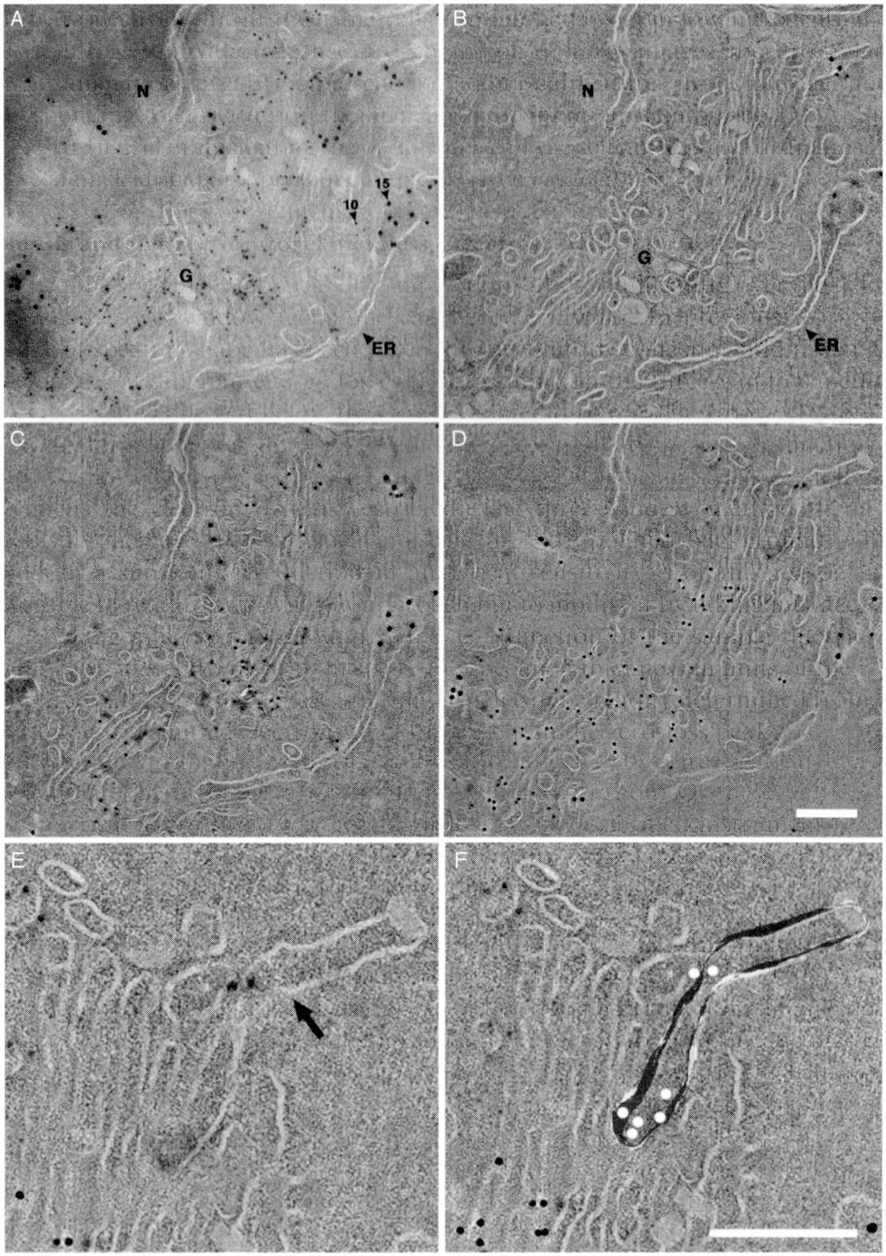

Fig. 2 Tomographic reconstruction and 3D analysis of an immunolabeled cryosection. Cultured HeLa cells were transfected to overexpress GFP-tagged GRASP55, a medial-Golgi matrix protein. The cells were fixed, cryosectioned, and double-immunolabeled with antibodies against GFP (10-nm gold secondary) and a marker for the endoplasmic reticulum (ER), protein disulfide isomerase (PDI;

reveals penetration of antibodies into regions of the cell that appear well preserved. An example of this staining is shown in Fig. 4. HEK-293 cells, expressing a chimera of the lipid droplet protein, adipophilin (ADPH) with GFP, were prepared by this technique, and sections labeled with antibodies against GFP, followed by gold-conjugated secondary antibodies. A single, 2D projection image of a lipid droplet (A) shows considerable ADPH labeling at, and extending away from, the droplet's edge. A tomographic reconstruction of the same region reveals that some of the labeling is within the section. Panels B through E are individual tomographic slices (0.9-nm thick) taken at 9-nm intervals through the volume of the section. Arrowheads point to a region that is devoid of gold particles at one surface (B), but that displays a number of particles at points deeper in the section (C and D). Still other particles are visible only on the far surface (E). A tomographic view of a slice taken perpendicular to the plane of the section (F) clearly shows gold particles within the section (arrows).

It is important to note that there are numerous other methods for immunolabeling cells for electron microscopy, some of which are discussed in other chapters of this book. These different EM immunolabeling techniques have both advantages and disadvantages and are selected with specific samples or labeling goals in mind (compare Chapter 3 by Hess, Chapter 22 by Cheutin *et al.*, Chapter 20 by Pelchen-Matthews and Marsh, Chapter 19 by Morphew, and Chapter 23 by Sosinsky *et al.*, this volume).

A common method is embedding samples in a methacrylate resin such as Lowicryl HM-20 or K4M. These resins are miscible with solvents like acetone or methanol and remain liquid at low temperature, meaning they can be used with rapidly frozen samples and applied during freeze-substitution. This procedure results in preservation that is nearly as good as that obtained by rapidly frozen, osmicated samples that have been embedded in Epon. Also, once Lowicryl is polymerized it may be sectioned

---

15-nm gold secondary). (A) A traditional 2D view of a labeled Golgi region from the ~100-nm thick cryosection, showing extensive GRASP55 labeling within the Golgi stack (G). ER is present to the right of the Golgi, distinguished by specific PDI labeling. The large gold particles (15) representing PDI can easily be distinguished from the smaller particles (10) that represent GFP. (B) A single slice (0.9-nm thick) from a tomographic reconstruction of the same region is shown in (A). This slice is taken from the middle of the section's volume. The resolution is now sufficient to follow individual structures. Since only antigens exposed at the surfaces of a section are typically labeled by this technique, these tomograms are analyzed by noting the positions of gold particles in tomographic slices near the top (C) or bottom (D) of the cryosection, and then following the structures labeled by those particles through the volume of the sample. By this method, organelles that appear to be very close to one another can be identified both by their specific markers and by their structure. For example, specialized ER that closely associates with *trans*-Golgi cisternae is often difficult to distinguish in traditional EM views but can easily be detected by the specific PDI label and subsequent tracking through the tomographic volume. (E) Detail from the reconstruction, indicating a piece of *trans*-ER within the Golgi stack, identified by PDI label (arrow). The boundaries of the ER membrane can then be tracked through the volume and modeled using the IMOD software (F). Scale bars in D and F = 0.2 μm.

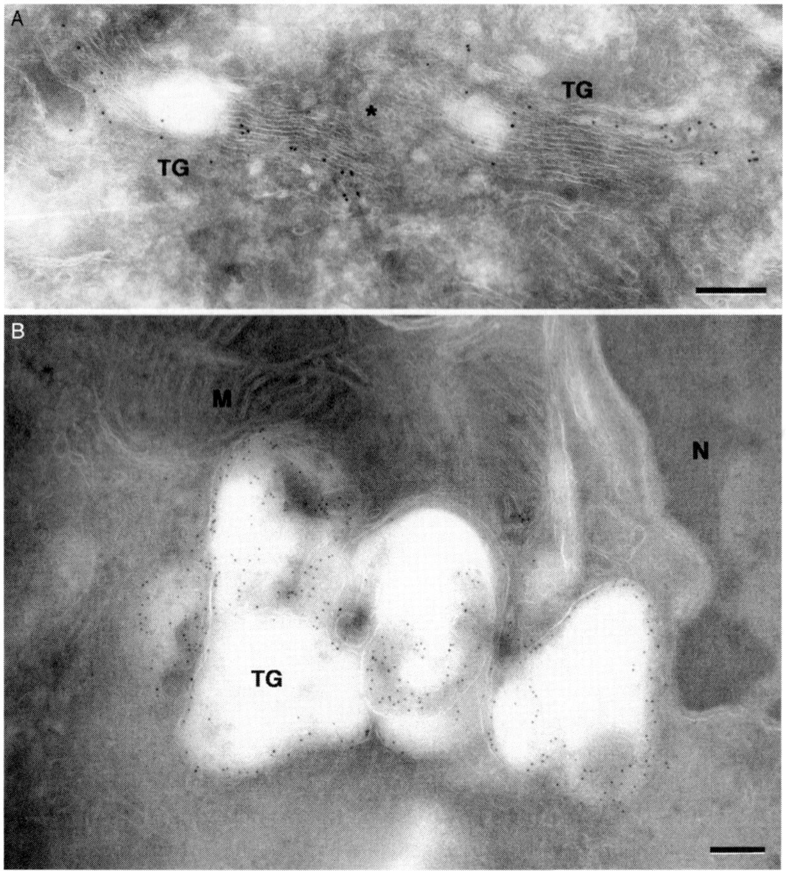

Fig. 3 Labeling strength of a particular antibody may vary considerably in different cell types, even those from the same host organism. Shown are Golgi regions from (A) a normal rat kidney cell and (B) a rat lactating mammary epithelial cell, labeled with antibodies against the transmembrane *trans*-Golgi network protein, TGN38. The samples were fixed by the same method, and the concentrations of the primary and secondary antibodies were the same for both preparations. TGN38 is localized to multiple transcisternae (TG) in each cell. The Golgi ribbon in (A) is rotated on either side of a noncompact region (*), so the *trans*-side of each stack faces in the opposite directions. *Trans*-Golgi cisternae in lactating mammary cells contain large amounts of casein and other milk proteins and are grossly expanded due to the osmolarity of the secreted fluid. This expansion results in greater accessibility to the antigen, which explains the comparatively higher density of gold particles. Scale bars = 0.2 μm.

in a standard microtome at room temperature, which is more straightforward than cryomicrotomy (see Chapter 19 by Morphew, this volume).

Although the overall preservation of Lowicryl-embedded samples is very good, there is often variation in preservation and contrasting among different components of the cell, due at least in part to the extractive nature of the resin. Microtubules are particularly well preserved with Lowicryl embedding, which makes this approach

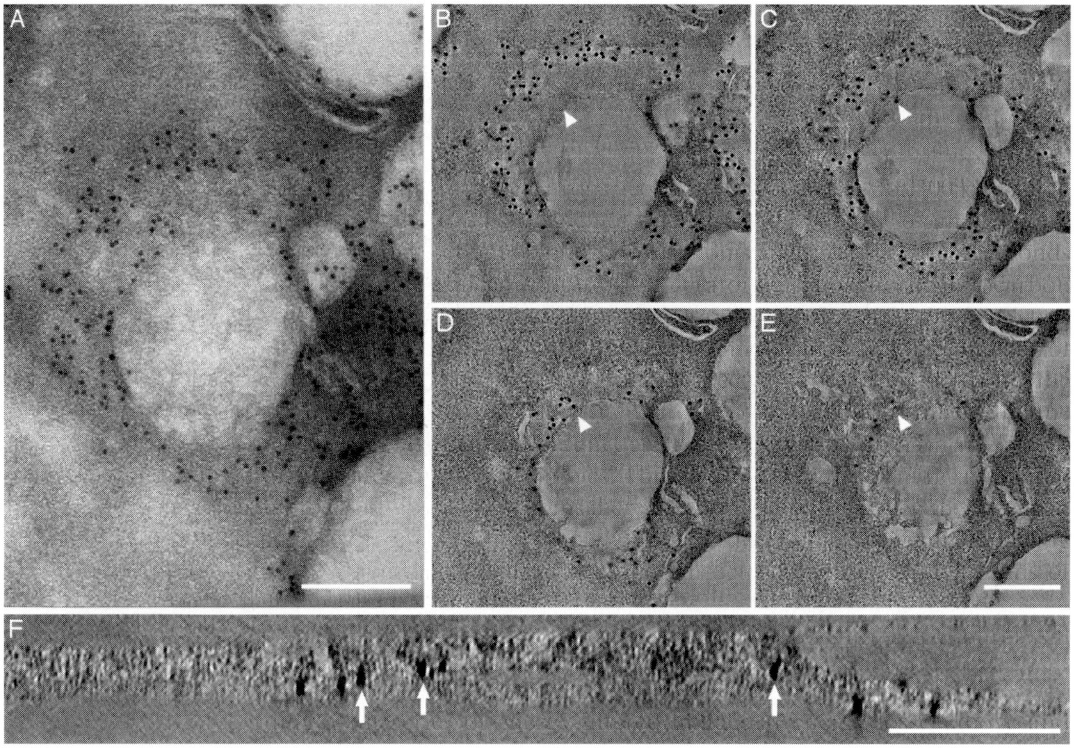

Fig. 4 Tomography reveals that in certain samples, antibodies can penetrate and label antigens within the section. Shown is a detail from a HEK-293 cell expressing a chimera of the lipid droplet protein, ADPH, and GFP. The cell was prepared as described in the text and labeled with antibodies against GFP. (A) A 2D projection image showing extensive ADPH labeling at and near the periphery of a lipid droplet. It is impossible, however, to ascertain the 3D position of the gold particles in the section from this single view. Tomographic reconstruction of the same area clearly shows that the gold particles are situated on both surfaces of the section as well as within it. Panels (B–E) are individual tomographic slices (0.9-nm thick) taken parallel to the section plane and separated by ~10 nm. Arrowheads indicate an area where gold particles are absent at one level of the section (B), but appear in slices from within the section's volume (C and D), and are again mostly absent in a slice taken near the edge (E). (F) A tomographic slice cut perpendicular to the section plane shows gold particles embedded within the volume (arrows). The entry of labeled antibodies into this section seems to be associated with places where there are rather large, empty places, like vesicles or lipid droplets, from which the biological material has been extracted during preparation. Scale bars = 0.2 μm.

ideal for immunolocalization of molecules associated with the cytoskeleton. However, membranous organelles, such as ER or the Golgi, are less well preserved, and antibodies against molecules in those structures are often less effective, or do not work at all. Figure 5 shows Golgi regions in cells that have been prepared in Lowicryl HM-20 (A) or by the Tokuyasu method (B). The general cytoplasm is dense and well contrasted in the Lowicryl sample, but the Golgi stack (GS) itself is defined less-well than other organelles, such as ribosomes and particularly

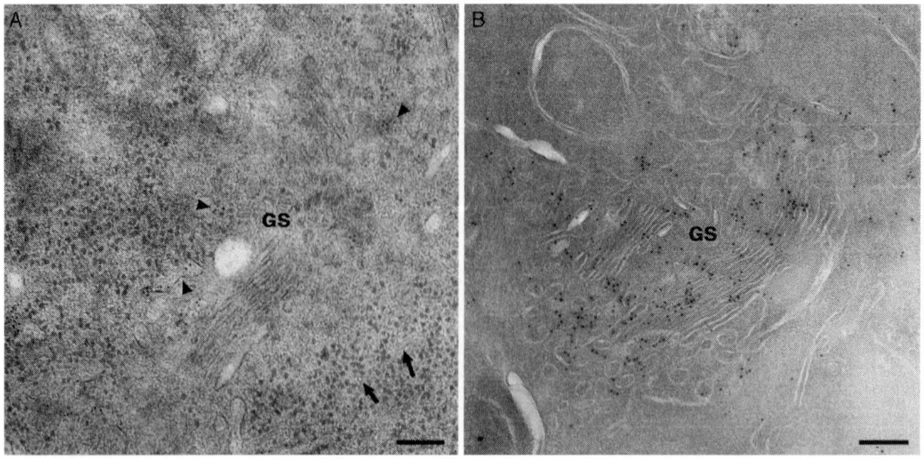

Fig. 5 Different EM immunolabeling techniques are often selected with specific samples or labeling goals in mind. (A) A Golgi region from a cultured cell, rapidly frozen, and embedded in Lowicryl HM-20 resin compared to a similar region from a Tokuyasu prepared sample (B). Note that the region shown in (B) is complex, because two Golgi stacks are present. Embedding in Lowicryl allows for the sample to be rapidly frozen and both fixed and embedded by freeze-substitution at low temperatures. Sectioning, on the other hand, can be done at room temperature. Lowicryl generally yields very good structural preservation, and it maintains the cytoskeleton particularly well. It is also an improvement for immunolabeling over more hydrophobic resins, like Epon-Aralite. As such, Lowicryl-embedded samples are ideal for labeling microtubules (arrows in A) and their associated proteins. It is less well suited for study of organelles such as the Golgi or ER, as membranes are often extracted or not as well contrasted, compared to equivalent samples prepared by the Tokuyasu method. Furthermore, antibodies against membrane proteins such as PDI (arrowheads in A) do not label as strongly or consistently as they do with the Tokuyasu methods, and often they do not work at all. Scale bars = 0.2 μm.

microtubules (arrows). The definition and contrast of individual Golgi cisternae and associated membranes is much higher in the Tokuyasu image, but ribosomes, microtubules, and other cytoplasmic elements are virtually undetectable. Also, antibodies against a common ER marker, PDI (arrowheads in A), label with far lower efficiency on Lowicryl HM-20 than on a Tokuyasu sample (compare with Fig. 2A). Antibodies against cytoskeletal proteins, on the other hand, often work with efficiency and repeatability in these preparations (see Chapter 19 by Morphew, this volume). While not all Golgi antibodies work for Tokuyasu sections, those that do generally label the surface of the section efficiently. As with all other EM immunolabeling techniques that apply the antibody to a slice of the sample, there is little if any penetration of the probe beyond the surface of the section. This point is demonstrated in Fig. 2, where a tomographic slice from the middle of the section (B) contains no gold particles, while slices from either surface of the section (C and D) show extensive gold labeling. It may be helpful to use multiple approaches to balance the advantages and disadvantages of each technique.

A final point to emphasize is that there are many factors that can affect the efficiency with which an antibody recognizes its antigen, regardless of the EM immunolabeling method that is used. For example, the amount and type of chemical fixative used can render an epitope unrecognizable by the antibody; stains or embedments may cover the antibody binding sites (Slot *et al.*, 1989), or they may be "hidden" because the protein of interest is posttranslationally modified or is complexed with other molecules. These factors, combined with the fact that usually only antigens at the surfaces of sections can be labeled, suggest that the ratio of antigens that get labeled, compared to the actual number of antigens in a sample, is quite low. One estimate for labeling efficiency, based on labeling of a nonabundant plasma membrane protein, is 15% (Howell *et al.*, 1987). One must therefore keep in mind that while immunolabeling at the EM level is an important tool for reporting the relative location of molecules, it cannot report the actual quantity of a particular molecule within a given sample.

We have described one of several methods for immunolabeling at the EM level and how its usefulness is enhanced by coupling it with electron microscope tomography. Although the Tokuyasu method is somewhat limited by its use of chemical fixation, it has proven most efficient for localization of molecules in membranous compartments, such as the Golgi complex and ER. Many of the problems associated with this approach can be reduced or eliminated by tomographic reconstruction of immunolabeled cryosections and by comparing those data to similar samples, prepared by state-of-the-art rapid freezing/freeze-substitution techniques.

References

Bannykh, S. I., Rowe, T., and Balch, W. E. (1996). The organization of endoplasmic reticulum export complexes. *J. Cell Biol.* **135,** 19–35.

Bonfanti, L., Mironov, A. A., Jr., Martinez-Menarguez, J. A., Martella, O., Fusella, A., Badassarre, M., Buccione, R., Heuze, H. J., Mironov, A. A., and Luini, A. (1998). Procollagen traverses the Golgi stack without leaving the lumen of cisternae: Evidence for cisternal maturation. *Cell* **95,** 993–1003.

Brown, W. J., and Farquhar, M. G. (1989). Immunoperoxidase methods for the localization of antigens in cultured cells and tissue sections by electron microscopy. *Methods Cell. Biol.* **31,** 553–569.

Gaietta, G., Deerinck, T. J., Adams, S. R., Bouwer, J., Tour, O., Laird, D. W., Sosinsky, G. E., Tsien, R. Y., and Ellisman, M. H. (2002). Multicolor and electron microscopic imaging of connexin trafficking. *Science* **296,** 503–507.

Goodman, S. L., Hodges, G. M., Trejdosiewicz, L. K., and Livingston, D. C. (1979). Colloidal gold probes—a further evaluation. *Scan Electron Microsc.* **3,** 619–628.

Grabenbauer, M., Geerts, W. J., Fernandez-Rodriguez, J., Hoenger, A., Koster, A. J., and Nilsson, T. (2005). Correlative microscopy and electron tomography of GFP through photooxidation. *Nat. Methods* **2,** 857–862.

Griffiths, G. (1993). "Fine Structure Immunocytochemistry." Springer Verlag, Heidelberg and Berlin.

Griffiths, G., McDowall, A., Back, R., and Dubochet, J. (1984). On the preparation of cryosections for immunocytochemistry. *J. Ultrastruct. Res.* **89,** 65–78.

Griffiths, G., Simons, K., Warren, G., and Tokuyasu, K. T. (1983). Immunoelectron microscopy using thin, frozen sections: Application to studies of intracellular transport of Semliki Forest virus spike glycoproteins. *Methods Enzymol.* **96,** 466–485.

Howell, K. E., Reuter-Carlson, U., Devaney, E., Luzio, J. P., and Fuller, S. D. (1987). One antigen, one gold? A quantitative analysis of immuno-gold labeling of plasma membrane 5′ nucleotidase in frozen-thin sections. *Eur. J. Cell Biol.* **44,** 318–327.

Kremer, J. R., Mastronarde, D. N., and McIntosh, J. R. (1996). Computer visualization of three-dimensional image data using IMOD. *J. Struct. Biol.* **116,** 71–76.

Ladinsky, M. S., and Howell, K. E. (1992). The trans-Golgi network can be dissected structurally and functionally from the cisternae of the Golgi complex by brefeldin-A. *Eur. J. Cell Biol.* **59,** 92–105.

Ladinsky, M. S., Kremer, J. R., Furcinitti, P. S., McIntosh, J. R., and Howell, K. E. (1994). HVEM tomography of the *trans*-Golgi network: Structural insights and identification of a lace-like vesicle coat. *J. Cell Biol.* **127,** 29–38.

Ladinsky, M. S., Mastronarde, D. N., McIntosh, J. R., Howell, K. E., and Staehelin, L. A. (1999). Golgi structure in 3 dimensions: Functional insights from the NRK cell. *J. Cell Biol.* **144,** 1135–1149.

Ladinsky, M. S., Wu, C. C., McIntosh, S., McIntosh, J. R., and Howell, K. E. (2002). Structure of the Golgi and distribution of reporter molecules at 20°C reveals the complexity of the exit compartments. *Mol. Biol. Cell* **13,** 2810–2825.

Marsh, B. J., Mastronarde, D. N., Buttle, K. F., Howell, K. E., and McIntosh, J. R. (2001). Organellar relationships in the Golgi region of the pancreatic beta cell, line, HIT-T15, visualized by high-resolution electron tomography. *Proc. Natl. Acad. Sci. USA* **98,** 2399–2406.

Mastronarde, D. N. (2005). Automated electron microscope tomography using robust prediction of specimen movements. *J. Struct. Biol.* **152,** 36–51.

Novikoff, A. B. (1951). The validity of histochemical phosphatase methods on the intracellular level. *Science* **113,** 320–325.

Pagano, R. E., Sepanski, M. A., and Martin, O. C. (1989). Molecular trapping of a fluorescent ceramide analogue at the Golgi apparatus of fixed cells: Interaction with endogenous lipids provides a *trans*-Golgi marker for both light and electron microscopy. *J. Cell Biol.* **113,** 1267–1279.

Saraste, J., and Svensson, K. (1991). Distribution of the intermediate elements operating in ER to Golgi transport. *J. Cell Sci.* **100,** 415–430.

Singer, S. J. (1959). Preparation of an electron-dense antibody conjugate. *Nature* **183,** 1523–1524.

Slot, J. W., Posthuma, G., Chang, L. Y., Crapo, J. D., and Geuze, H. J. (1989). Quantitative aspects of immunogold labeling in embedded and in nonembedded sections. *Am. J. Anat.* **185,** 271–281.

Snyder, C. M., Mardones, G., Ladinsky, M. S., and Howell, K. E. (2006). GMx33 associates with the *trans*-Golgi matrix in a dynamic manner and sorts within tubules exiting the Golgi. *Mol. Biol. Cell* **17,** 511–524.

Thorne-Tjomsland, G., Clermont, Y., and Tang, X. (1991). Glucose-6-phosphate activity of endoplasmic reticulum and Golgi apparatus in spermatocytes and spermatids of the rat: An electron microscopic cytochemical study. *Biol. Cell* **71,** 33–41.

Tokuyasu, K. T. (1971). A technique for ultramicrotomy of cell suspensions and tissues. *J. Cell Biol.* **57,** 551–565.

Tokuyasu, K. T. (1986). Application of cryoultramicrotomy to immunocytochemistry. *J. Microsc.* **143,** 139–149.

Wu, C. C., Taylor, R. S., Lane, D. R., Ladinsky, M. S., Weisz, J. A., and Howell, K. E. (2000). GMx33: A novel family of trans-Golgi proteins identified by proteomics. *Traffic* **1,** 963–975.

CHAPTER 22

Visualizing Macromolecules with Fluoronanogold: From Photon Microscopy to Electron Tomography

T. Cheutin,* C. Sauvage,† P. Tchélidzé,† M. F. O'Donohue,† H. Kaplan,‡ A. Beorchia,§ and D. Ploton†

*Institut de Génétique Humaine, CNRS UPR 1142
34396 Montpellier Cédex 5, France

†Unité MéDIAN, CNRS UMR 6142, UFR de Pharmacie
51096 Reims Cédex, France

‡IFR53, 51096 Reims Cédex, France

§DTI, UMR 6107, UFR de Sciences, BP 1039
51687 Reims Cédex, France

Fluoronanogold (FNG) contains a fluorochrome molecule (FITC, Alexa, and so on) and a 1.4-nm cluster of gold atoms; these permit visualization both by photon microscopy (direct visualization) and by electron microscopy (EM) (after silver or

0091-679X/07 $35.00
DOI: 10.1016/S0091-679X(06)79022-7

gold enhancement). Classical applications of FNG concern immunolabeling of any antigen and its visualization with cellular imaging in two dimensions (2D), that is, using wide-field optical microscopy and EM (ultrathin sections). In the present chapter, we show that FNG can be localized in three dimensions (3D), both by confocal light microscopy and electron tomography. These approaches are useful for investigating the volumetric organization of proteins within well-preserved organelles at different levels of resolution (200 nm in confocal microscopy and 10 nm in electron tomography), as shown with two different nucleolar proteins.

I. Introduction

The nucleus is a highly organized cell compartment where numerous functions take place, such as modulation of chromatin compaction, synthesis of DNA, transcription, and maturation of RNA (reviewed in Cremer and Cremer, 2001; O'Brien *et al.*, 2003; Thiry and Goessens, 1996). Very recent work has demonstrated that nuclear proteins diffuse rapidly throughout the volume of the nucleus, and that their local concentration is very high within the sub-volumes where they are active (Bubulya and Spector, 2004; Gorski and Misteli, 2005; Misteli, 2005). To study the highly dynamic organization of the nucleus, new approaches are required in order to investigate how the proteins of various macromolecular machines are spatially arranged, both within living and fixed cells. In the latter case, it is now necessary:

1. to use sensitive probes with a high penetration power
2. to perform electron tomography, which is a powerful way to study the spatial supramolecular organization of molecular machines (Baumeister, 2005; Lucic *et al.*, 2005; McEwen and Marko, 2001; McIntosh *et al.*, 2004)
3. to investigate the same immunolabeled proteins at different levels of resolution by using both photon and electron microscopy (EM).

Here, we show that fluoronanogold (FNG) allows the localization of any given antigen at different levels of imaging by using multimodal microscopy (i.e., photon and EM). The results presented here deal with two nucleolar proteins, pKi-67 (Scholzen and Gerdes, 2000) and RNA polymerase I (RNAP I) (Gilbert *et al.*, 1995; Scheer and Rose, 1984), both of which have been localized by using FNG in the same cell line.

II. Experimental Approach for Immunostaining of pKi-67 and RNAP I

A. Cells

Human A549 cells were grown in Ham's F12 supplemented with 10% fetal calf serum, 100-U/ml penicillin, 100-μg/ml streptomycin and 2-mM l-glutamine (Sigma,

L'Isle d'Abeau, France) at 37 °C in an atmosphere enriched with 5% CO_2. They were seeded at 5×10^4 cells/cm^2 and grown for 48–72 h on glass coverslips.

B. Fixation

The two following protocols gave similar results: (1) cells were fixed for 10 min in 3% paraformaldehyde diluted in phophate-buffered saline (PBS), then permeabilized with 0.3% Triton X-100 in PBS (140-mM NaCl, 6-mM Na_2HPO_4, 4-mM KH_2PO_4, pH 7.2) (Sigma) for 2 min; or (2) cells were simultaneously fixed and permeabilized for 4 min in 3% paraformaldehyde and 1% Triton X-100 diluted in PBS. After fixation, cells were incubated for 30 min in PBS containing 3% bovine serum albumin (BSA), 1-mM $CaCl_2$, and 0.5-mM $MgCl_2$ to block subsequent nonspecific binding of immunoglobulins.

C. Immunostaining

Cells were incubated for 30 min in MM1 antibodies against human pKi-67 (Novocastra, Newcastle upon Tyne, UK), diluted 1:50 in PBS containing 1% BSA, 1-mM $CaCl_2$, and 0.5-mM $MgCl_2$, and rinsed three times for 5 min in PBS. Alternatively, they were incubated for 4 h with a rabbit polyclonal antibody directed against the large subunit of RNAP I (courtesy Dr. K. Rose, Houston, TX) diluted 1:400. A biotinylated goat anti-mouse antibody (Jackson Immuno Research, West Grove, PA) (or anti-rabbit, Sigma) was applied for 30 min and then stained for 15 min with a streptavidin-fluoronanogold, diluted 1:20 (Nanoprobes, Inc., New York). This reagent consisted of streptavidin conjugated both to a fluorescein (FITC) molecule and a 1.4-nm Nanogold® particle (Nanoprobes, Inc., New York) (Cheutin *et al.*, 2002, 2003).

D. Confocal Microscopy

Confocal microscopy was performed with an MRC 1024ES System (Bio-Rad, Hercules, CA), mounted on an IX70 optical microscope (Olympus, Rungis, France), equipped with a plan apochromat 63× oil immersion objective, numerical aperture = 1.4. The FITC contained within FNG was excited with the 488-nm line of an argon ion laser. Depending on the cell's thickness, 20–30 optical sections per cell were recorded with a 0.2-μm z step. Processing of the z-series was performed with Amira software (Mercury®, Merignac, France) in order to obtain surface and volumetric visualization. The inner parts of the volumes were investigated by presenting sections computed to display any chosen plane.

E. Electron Microscopy

Cells immunolabeled as described above with FNG were postfixed for 12 min with 1.6% glutaraldehyde in PBS. After rinsing in deionized water, HQ silver enhancement (Nanoprobes, Inc., New York) was performed for 8–10 min at room temperature. This developed the nanogold into particles ~10 nm in diameter. Cells were then harvested by scraping, dehydrated in graded alcohols, and embedded in Epikotte 812.

1. Classical Electron Microscopy

Ultrathin sections (80 nm) mounted on copper grids were counterstained with uranyl acetate and lead citrate before examination using a 200 CX microscope at 80 kV (Jeol, Croissy-sur-Seine, France).

2. Electron Tomography

Sections, 500–2000 nm in thickness, were observed in an intermediate-voltage CM30 electron microscope operating at 250 kV in the STEM mode with a spot size of 8.4 nm and a scanning time of 0.5 msec per pixel (Philips, Eindhoven, The Netherlands) (Beorchia *et al.*, 1993). To see the immunolabels with high contrast, these sections were not counterstained. Before initiating a tilt series, the section was stabilized under the electron beam at a dose of 100 e/(Å^2 sec) for 10 min to limit anisotropic thinning of the specimen during data collection. In these conditions, total dose received per pixel during 10 min is 1270 electrons. Each section was tilted every 2° from −50° to +50°. The 51 corresponding images (512 × 512 pixels) were recorded directly on a disk-type scintillator-photomultiplier detector system and digitized online using Orion hardware (ELI sprl, Charleroi, Belgium). Images were then aligned individually by using a sinogram technique (Tchelidze *et al.*, 2006). Tomographic reconstructions were performed with an extended field-additive algorithm-reconstruction technique on aligned images, as previously described (Cheutin *et al.*, 2002, 2003). Extensive investigations on the reconstructed volumes were performed by using the Amira software (Mercury®). Surface or volumetric visualizations with a high depth of field were calculated to obtain relevant images at specific angles of view. Finally, the inner parts of the volumes were investigated by performing sections in any chosen plane.

III. Applications

The distribution of pKi-67 or RNAP I was revealed with FNG and investigated by using confocal microscopy. Another part of the same cell culture was kept in order to amplify the FNG with silver. The ultrastructural localization of the labeling was then investigated within ultrathin sections (classical EM) and thick sections (medium voltage EM) to perform electron tomography.

A. pKi-67

1. Confocal Microscopy

Optical sections (z-series) were acquired by confocal microscopy within the whole nuclear volume by exciting FITC molecules of FNG at 488 nm. The labeling appeared as several clusters of bright spots for optical sections collected at the bottom (Fig. 1A) or at the top of the volume (Fig. 1E). In between the labeling formed a complex surface, ~300-nm thick, that looks like many connected strands surrounding the nucleolus (Fig. 1B–D). The projection of a z series (Fig. 1F) revealed that pKi-67 was mainly distributed on the periphery of the nucleolus, but that discrete spots were also located in the nucleoplasm and along the nuclear envelope. Orthogonal sections were cut to investigate the distribution of pKi-67 within the nuclear thickness (Fig. 1G). These confirmed that the labeling was distributed among several cords (~300 nm in thickness), surrounding unlabeled regions. A few protrusions, emanating either from the top (arrows) or the bottom (arrowheads) of the cords, were clearly seen (Fig. 1G). These corresponded to the bright spots observed in the distal sections (Fig. 1A and E). The protrusions appear to make direct contact with the nuclear envelope (arrows and arrowheads, Fig. 1G). Orthogonal sections, performed in the (x, z) plane (Fig. 1H–J),

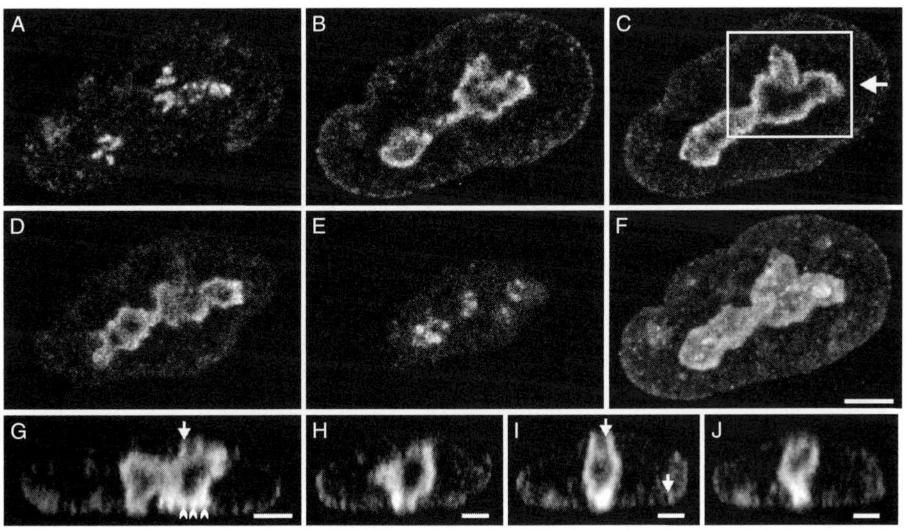

Fig. 1 Immunolabeling of pKi-67 antigen detected with a specific antibody revealed by FluoroNanogold. The fluorescent dye of the probe (FITC) was visualized by using confocal microscopy. (A–E) Gallery of five optical sections taken from a z series. The labeling is distributed either in strands, ~300 nm in width, or in small perinucleolar dots. (F) Projection of all the sections contained in the z series. (G) Orthogonal section of the reconstructed volume, performed at the level of the arrow in Fig. 1C. (H–J) Transversal sections performed at the level of the three arrowheads in Fig. 1G. Scale bars = 2 μm. With permission from Cheutin *et al.* (2003).

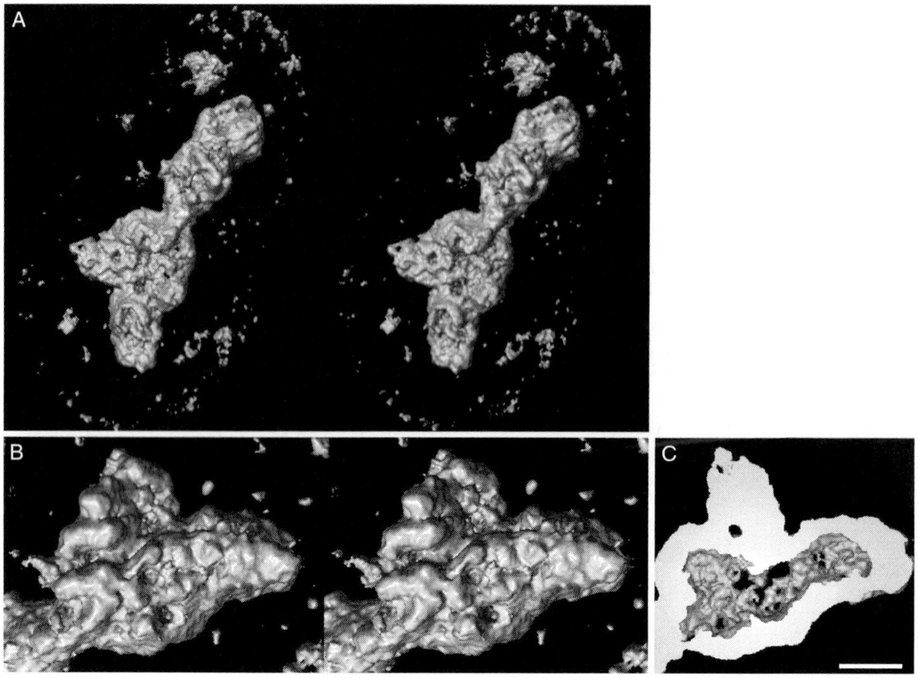

Fig. 2 Surface visualization of pKi-67. A surface-rendering method was applied to the data presented in Fig. 1. (A) Stereo pair of a back view of the volume presented in Fig. 1F. (B) Stereo pair of a front view, corresponding to the area framed in Fig. 1C. (C) Split view of the stereo pair shown in (B). Scale bar = 1 μm. With permission from Cheutin *et al.* (2003).

confirmed that pKi-67 forms a set of almost continuous cords from which some protrusions are directed toward the nuclear envelope (arrows, Fig. 1I).

A surface visualization of the whole volume, displayed as a stereo pair, revealed the external organization of pKi-67 (Fig. 2A). A focus on the area framed in Fig. 1C showed a complex and irregular surface, composed of many intertwined fibers and protrusions (Fig. 2B). Cutting this volume in half revealed that the labeling was organized as fibers surrounding a large cavity (Fig. 2C). Overall, these tomographic results show that pKi-67 was arranged in one or more contorted strands that form a shell around the nucleoli.

2. Electron Microscopy

Immunolocalization of pKi-67 revealed with FNG was then performed at the ultrastructural level on the remaining part of the cell culture. The gold clusters of FNG were amplified with silver in order to obtain 10-nm silver–gold particles visible in EM.

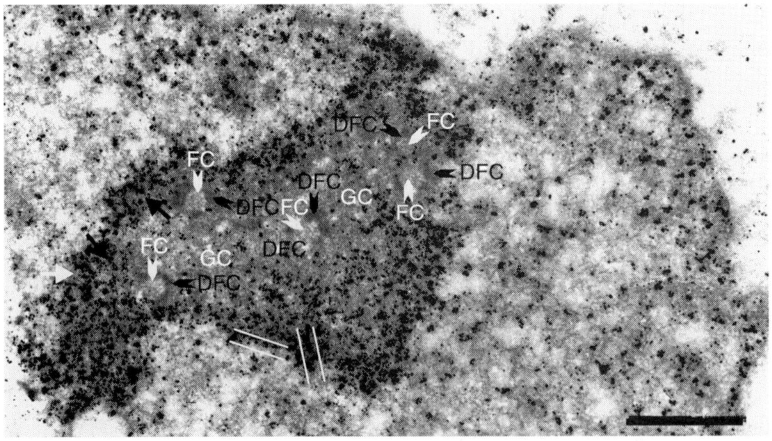

Fig. 3 Ultrastructural localization of pKi-67 in ultrathin sections. After silver amplification of the gold clusters contained in FNG, cells were embedded in Epikotte 812. Ultrathin sections (80 nm) were counterstained and observed at 80 kV. The nucleolar components are clearly identified, demonstrating the excellent preservation of the structures: the fibrillar centers (FCs, white arrowheads), surrounded with the dense fibrillar component (DFC, black arrowheads) and the granular component (GC). The labeling appears as numerous dots, ~10–15 nm in diameter, mainly localized in the cortical part of the nucleolus, in the vicinity of perinucleolar chromatin. In some areas, the labeling is organized as thinner strands, 50 nm in diameter (arrows and parallel lines). Labeling localized in the nucleoplasm and clumps of chromatin is not interpreted as unspecific background. Scale bar = 1 μm. With permission from Cheutin *et al.* (2003).

a. Identification of the Nuclear Components Containing pKi-67

Ultrathin sections (80 nm) were cut and counterstained with uranyl acetate and lead citrate. This allowed a clear identification of the three typical nucleolar components: the fibrillar centers (FCs), surrounded by the dense fibrillar component (DFC) and the granular component (GC) (see arrows and arrowheads on Fig. 3). These findings attested to the overall preservation of the cells after labeling and amplification with silver.

The number of black dots that represent labeling was low in the cytoplasm and high within the nucleoplasm, where most of it was found in close association with condensed chromatin. Within the nucleolus, the FCs, DFC, and GC were almost totally devoid of pKi-67. Most labeling was found at the periphery of the nucleoli, where the perinucleolar-condensed chromatin was located. Noticeably, the particles were grouped in small clusters, which in place seemed to form short threads, ~50 nm in diameter (arrows and parallel lines) (Fig. 3).

b. Electron Tomography of pKi-67 in 2-μm-Thick Sections

In order to investigate the three-dimensional (3D) organization of pKi-67 within the periphery of the nucleolus, electron tomography of the labeling was performed by using a STEM working at 250 kV (Beorchia *et al.*, 1992). Sections

(2 μm in thickness) were cut from the same blocks as those used for ultrathin sectioning, tilted, and then imaged at different angles of view. Figure 4 (A–E) shows a nucleolus imaged with inverted contrast at different angles of view as the goniometer stage was tilted from −40° to +40°. The labeling, which was more

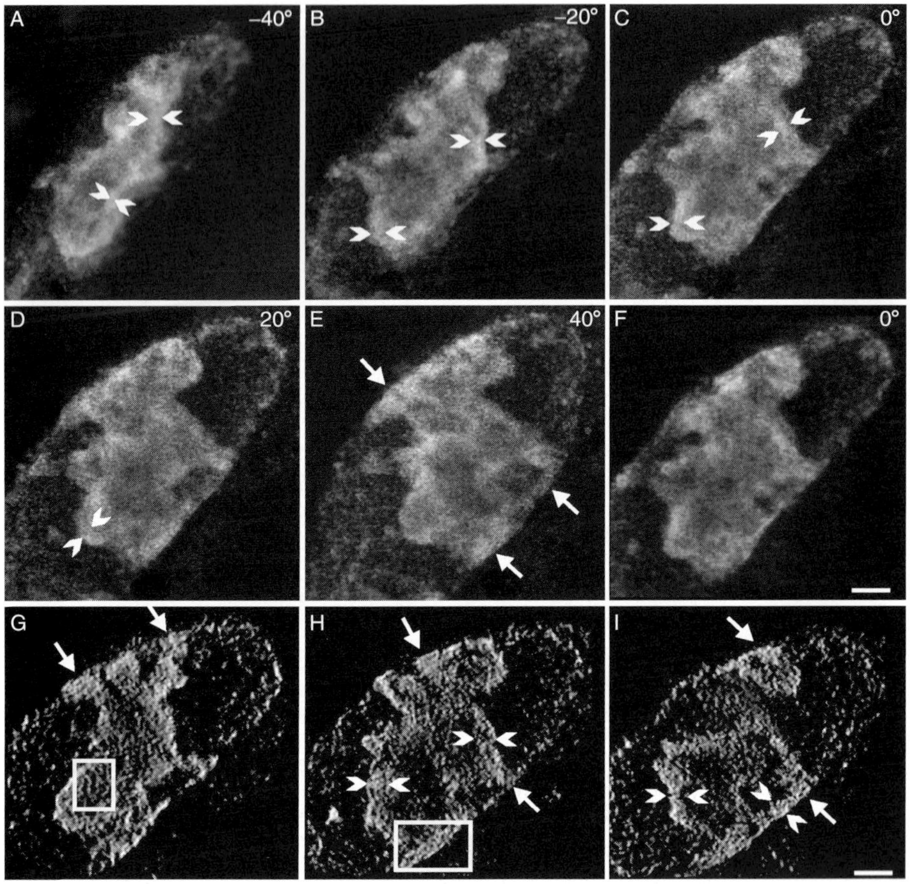

Fig. 4 Electron tomography analysis of pKi-67 in 2-μm-thick sections. After silver amplification of the gold clusters contained in FNG, cells were embedded in Epikotte 812. Thick sections (2 μm in thickness) were imaged with a STEM working at 250 kV. Image contrast was reversed for convenience. (A–E) Gallery of five projections taken from the same tilt series (−40° to +40°). The labeling appears as a strand, 250–300 nm in thickness (bracketed by arrowheads). Contacts with the nuclear envelope are evidenced (arrows). (F) Projections of an entire tilt series were used to compute a tomogram. The 0° projection shows a similar organization to that observed on the initial image (compare C and F), indicating that the reconstruction process introduces no distortion. (G and H) Three 50-nm-thick digital sections were taken from the top (G), the middle (H), and the bottom (I) of the tomogram. The labeling is distributed in strands (bracketed by arrowheads), 250–300 nm in diameter, disposed around a central, unstained area. The strands consist of thinner-ones, 40–50 nm in diameter (frames in G and H). Arrows point to contacts with the nuclear envelope. Scale bar = 0.5 μm. With permission from Cheutin *et al.* (2003).

intense near the nucleolar periphery, was clearly organized as one or more strands, 250–300 nm in diameter (bracketed by arrowheads, Fig. 4A–D). Some labeled areas were in close proximity to the nuclear envelope (arrows, Fig. 4E) and a weak perinuclear labeling, corresponding to small nucleoplasmic clusters, was also observed. All the projections of the tilt series were used to compute a tomogram. The computed projection at 0° (Fig. 4F) is similar to the initial projection at 0° (Fig. 4C), which demonstrates the validity of the reconstruction. Digital sections were then cut from the tomogram to study the internal distribution of pKi-67 (Fig. 4G–I). The labeled proteins were organized as strands (bracketed with arrowheads, Fig. 4H and I) surrounding an unlabeled, central area, as already observed in optical sections (Fig. 1). The thickness of the strands was relatively constant (250–300 nm) and many protrusions emanating from them were in direct contact with the nuclear envelope (arrows). A closer examination revealed that the strands and the protrusions consisted of fibers 30–50 nm in thickness (framed in Fig. 4G and H). These fibers were straight (500 nm in length) (frame, Fig. 4G) or appeared as twisted segments (frame, Fig. 4H).

c. Electron Tomography of pKi-67 in 0.5-μm-Thick Sections

Due to the complexity of the labeling evidenced within 2-μm-thick section, the 3D organization of pKi-67 fibers was reinvestigated within thinner sections (~0.5 μm). The resulting improvement in resolution revealed a fibrillar organization within the perinucleolar strands (frame, Fig. 5C), which was further investigated at higher magnification by computing new tomograms. Observation of this region at different angles of view showed that particles were grouped to constitute thin strands (Fig. 6A–E; semicircles, Fig. 6A), whose thickness (~30–50 nm) appeared more clearly on transversal sections performed at different levels (circles, Fig. 6F–H) and on stereo pairs viewed

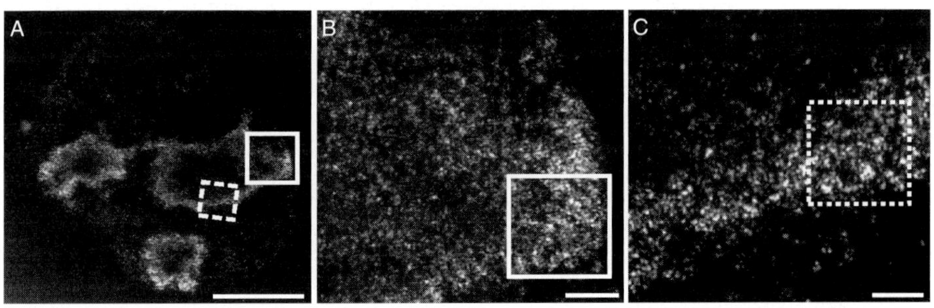

Fig. 5 Electron tomography analysis of pKi-67 in 0.5-μm-thick sections (A–C). The two framed areas in (A) were observed at a higher magnification. (B) A detailed view of a protrusion in close contact with the nuclear envelope. (C) A view of the perinucleolar strand. In both cases, the labeling appears as a complex network of intertwined thinner strands, whose compaction is higher in the protrusion (B) than in the perinucleolar cord (C). Scale bars = 2 μm (A) and 200 nm (B and C). With permission from Cheutin *et al.* (2003).

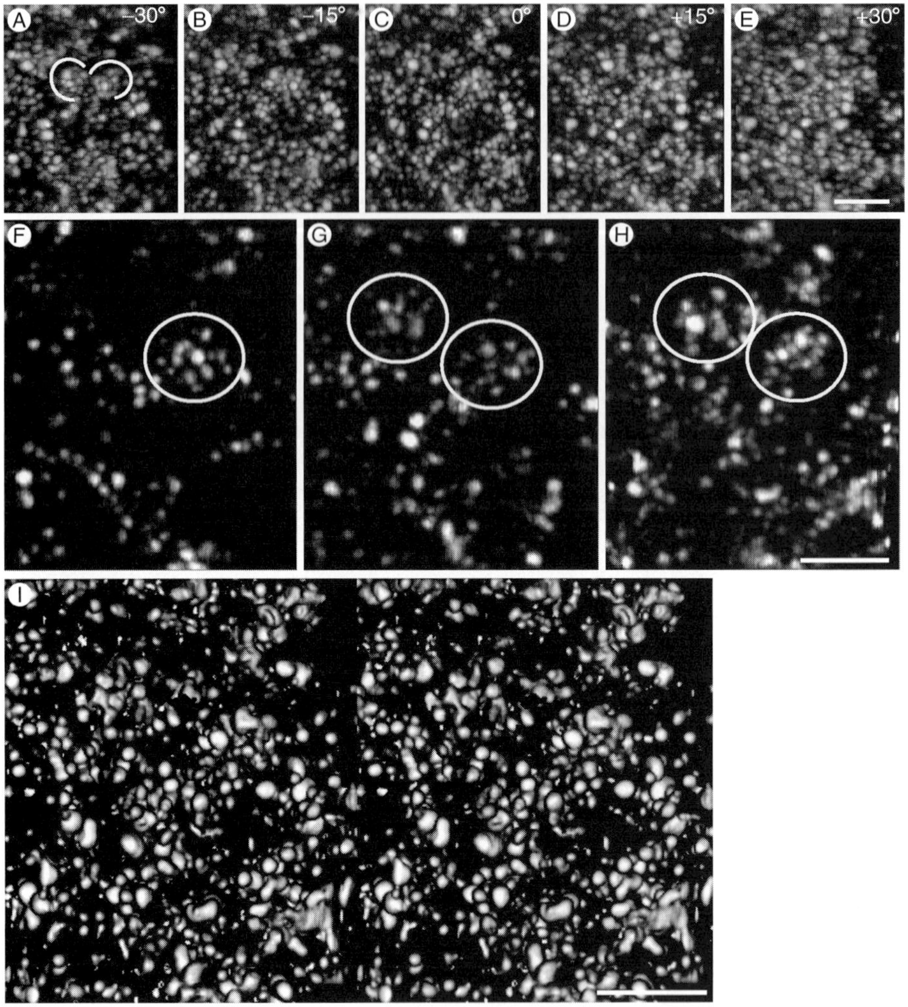

Fig. 6 Internal organization of the perinucleolar area analyzed by electron tomography. (A–E) Projections of the tomogram reconstructed from the area framed in Fig. 5C were computed at five different angles. At −30°, tangential sections of the 50-nm strands clearly appear (semicircles). (F–H) Three 50-nm-thick digital sections were obtained from the tomogram shown in (C). Circles indicate tangential sections of two different strands, 50 nm in diameter. (I) A stereo pair, presented with a surface visualization mode, shows the constant diameter of the strands and their regular spacing. Scale bars = 100 nm. With permission from Cheutin *et al.* (2003).

with a surface visualization (Fig. 6I). Finally, a systematic study of these regions by using electron tomography showed that the fibers were homogeneously positioned throughout the whole thickness of the labeling and that they were regularly spaced (∼100-nm distance between fibers).

B. RNAP I

1. Confocal Microscopy

Projection of a z-series, acquired at 488 nm in the whole nuclear volume by confocal microscopy, revealed that the labeling appeared as numerous dots, ~0.5–1 μm in diameter (Fig. 7A). Frequently, these dots were organized as contorted necklaces (Fig. 7B), as previously revealed by using classical FITC-labeled secondary antibodies (Gilbert *et al.*, 1995; Thiry *et al.*, 2000).

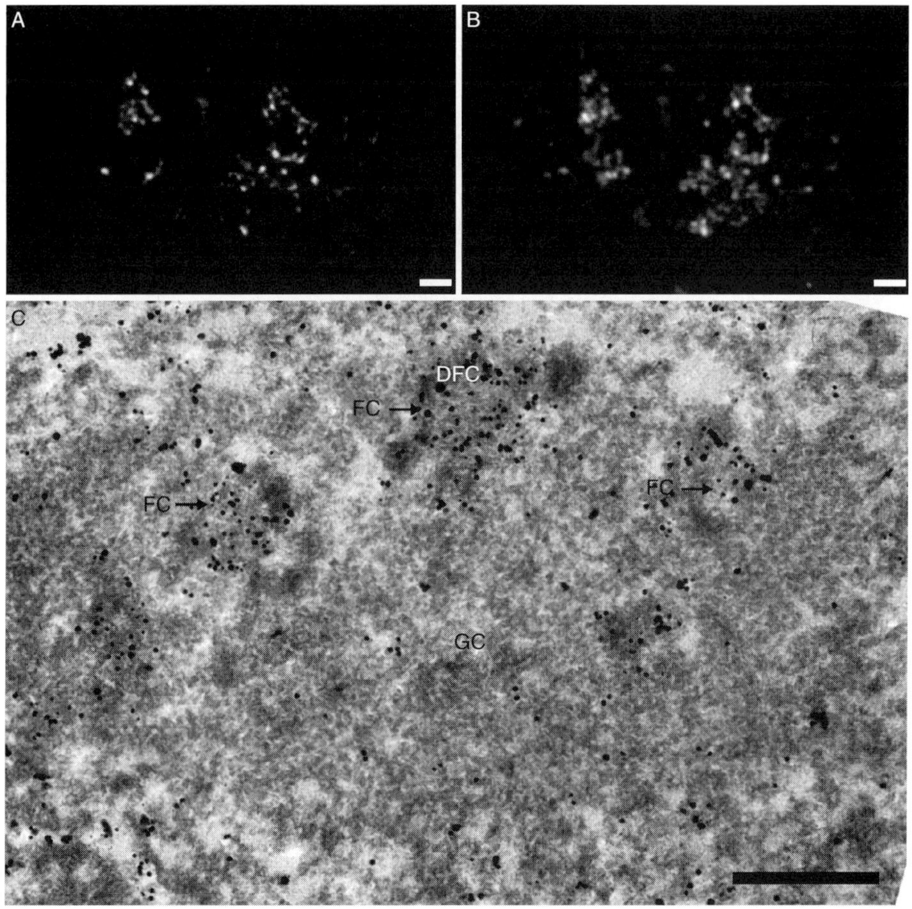

Fig. 7 Localization of RNAP I within A549 cells. (A) An optical section obtained by using confocal microscopy. (B) All the optical sections of a z-series were used to compute a projection showing that the labeling is organized as numerous spots disposed as several necklaces. Scale bars = 1 μm. (C) After silver enhancement and Epikotte embedding, ultrathin sections (80 nm) were cut, counterstained and observed with an electron microscope at 100 kV. The main nucleolar components are identified (FC, DFC, and GC). A high density of particles is observed within the FCs. Scale bar = 0.5 μm. With permission from Cheutin *et al.* (2002).

2. Electron Microscopy

Localization of RNAP I at the ultrastructural level was performed as described for pKi-67.

a. Identification of Nuclear Components Containing RNAP I

Ultrathin sections (80 nm) were cut and counterstained with uranyl acetate and lead citrate. Typical nucleolar components such as the FCs (~270 nm in diameter),

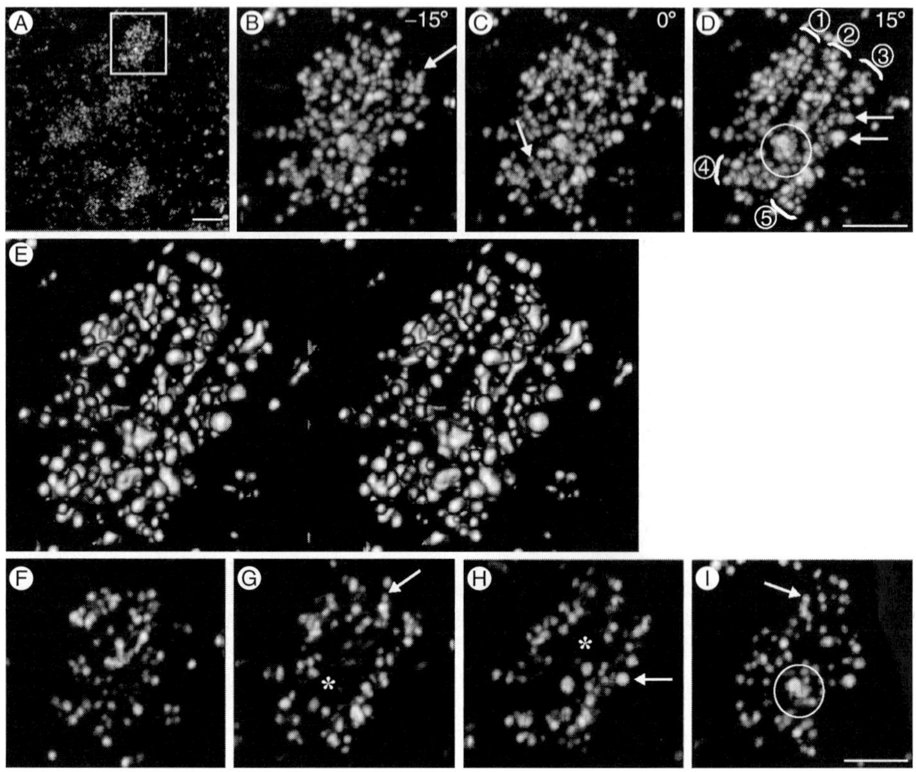

Fig. 8 Electron tomography of A549 cells, immunolabeled with an anti-RNAP I antibody, revealed with FNG and amplified with silver. Contrast was inverted so that silver particles appear white. A 500-nm-thick section observed using a STEM working at 250 kV is displayed: several independent clusters, 270 nm in diameter, are seen (A). (B–D) Different projections at −15°, 0°, and +15° of the tomogram were calculated after tomographic reconstruction of the cluster framed in (A). At +15° (D), five 60-nm coils are evidenced as indicated by brackets (#1–5). The large circle shows the area where the coils are fused together. The arrows point to twines, 20 nm in thickness. (E) A stereo pair of the tomogram presented in the same orientation as in (D) was calculated thanks to a surface-rendering mode. (F–I) Four successive 30-nm-thick sections were achieved within the tomogram presented in (D). Asterisks (G and H) indicate the internal part of the cluster, devoid of labeling, and arrows (G–I) refer to twines. Scale bar = 200 nm (A) or 100 nm (B–I). With permission from Cheutin *et al.* (2002).

surrounded by the DFC and the GC, were clearly identified (Fig. 7C). This showed the excellent ultrastructural preservation of the cells, as also observed for pKi-67 (Fig. 3). RNAP I molecules, identified by 10-nm silver–gold particles, were mainly localized within the FCs and were rare within the DFC and the GC.

b. Electron Tomography of RPAP I in 500-nm-Thick Sections

In order to investigate the 3D organization of RNAP I molecules within the FCs, electron tomography was performed by using a STEM working at 250 kV. The labeling (Fig. 8A) appeared as discrete clusters of particles in which the density of the labeling was 10 times higher than that of the nucleoplasm (data not shown). The clusters were organized as spheroids, ~270 nm in diameter, in which the particles presented an organized structure, as revealed by a rotation (framed in Fig. 8A) from −15° to +15° (Fig. 8B–D). This cluster was composed of several coils, 60 nm in diameter, as is also seen on a stereo pair (Fig. 8E). Rotations were performed to determine the number of coils, to measure their length, and to investigate their spatial organization (see Cheutin *et al.*, 2002 for details). Three curved coils (150–200 nm in length, #1–3) and two shorter ones (30–50 nm in length, #4 and 5) were observed (Fig. 8D). These coils shared a common origin (Fig. 8D and I, circles), but displayed separate extremities (brackets).

The internal arrangement of the labeling within the FC was investigated by computing 30-nm-thick sections within the tomogram displayed in Fig. 8D (Fig. 8F–I). This analysis demonstrates that the central region of the FC is devoid of particles (asterisks). In the coils, the particles are frequently aligned to form 20-nm-thick twines with a free end (Fig. 8G–I, arrows). To investigate their orientation, the volume was rotated to place the coils in a vertical orientation, and a sectioning plane parallel to the axis of coil #2 was cut (Fig. 9A and B). Two successive 1.5-nm-thick sections revealed seven twines, which were orthogonally disposed relative to the axis of the coil and were parallel to each other (Fig. 9A and B, arrows). When 1.5-nm-thick sections were cut perpendicular to the axis of the coil, it appeared that the twines formed full or partially open circles, 60 nm in diameter (Fig. 9C and D, white circles). These features were also evidenced on a stereo pair, imaging part of the volume (Fig. 9E).

IV. Discussion

Since FNG is a probe containing two different markers, it opens the possibility of imaging the same sample at both the optical and the EM level (after silver amplification). Thus, it is easy to perform a multimodal investigation, either on different cells or on the same cell visualized by correlative microscopy (Robinson and Vandré, 1997; Robinson *et al.*, 2000). The very small size of this gold atom cluster allows one to perform an efficient immunolabeling within fixed cells prior to embedding. This approach leads to a much greater sensitivity comparatively to immunolabeling based on 5- to 10-nm gold particles (Hainfeld and Powell, 2000;

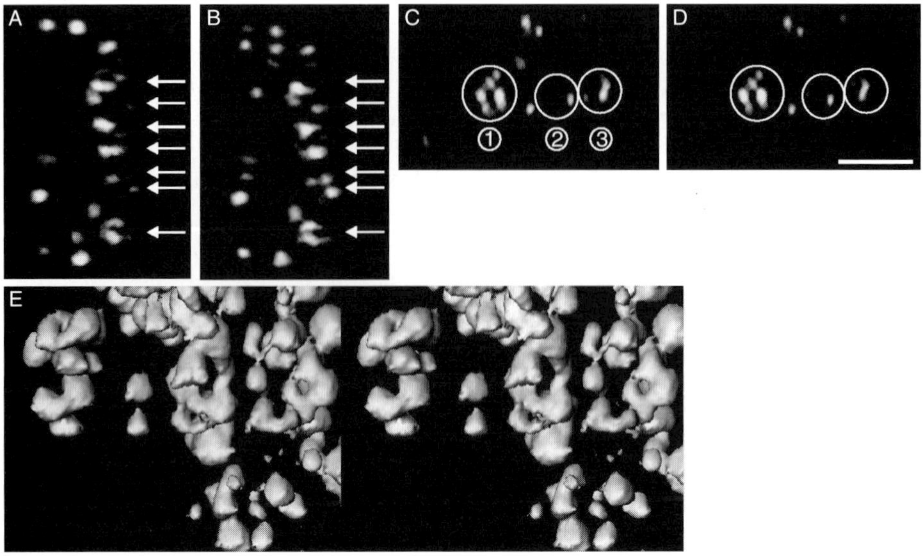

Fig. 9 Organization of the coils. (A and B) Two successive 1.5-nm-thick sections were computed parallel to the long axis of coil #2. The positions of seven twines, orthogonal to the long axis of coil #2, are indicated by arrows. (C and D) Two successive 1.5-nm-thick sections computed perpendicularly to the long axis of coils #1–3. Twines (large circles) are bent and constitute closed or partially open circles. Scale bar = 100 nm. (E): Stereo pair showing the twines observed in (A–D). Several open and closed circles corresponding to bent twines are evidenced. With permission from Cheutin *et al.* (2002).

Robinson *et al.*, 2000; Takizawa and Robinson, 2000). This increased efficiency is a prerequisite both for a precise localization of any antigen within ultrathin sections and for the investigation of its 3D organization by electron tomography.

In the present chapter, we stress on the fact that the multimodal investigation offered by FNG can be widened to 3D visualization by confocal microscopy and electron tomography (Cheutin *et al.*, 2002, 2003). Consequently, the spatial organization of a protein can be investigated through a single probe by performing a virtual zoom, with a resolution increasing from 200–500 nm (3D confocal microscopy) to ~10 nm (electron tomography). This approach can be extremely useful for the molecular imaging of the cell. It has been applied to analyze the 3D organization of complex molecular machines such as the site for ribosomal synthesis and processing localized in the cell nucleoli (Thiry and Goessens, 1996; Thiry *et al.*, 2000).

For the two nucleolar proteins presented here, this approach brings a new vision of their localization at the ultrastructural level and allows us to visualize their 3D organization at the level of the nucleolus. As compared to other pre- and post-embedding protocols, a much stronger labeling is obtained with FNG. This is due the very high sensitivity of the method. Thus, by using this approach, it is possible to evidence different levels of nucleolar proteins concentration, that is, low concentration of RNAP I within the nucleoplasm and high concentration of this protein within the FCs in which rRNA genes and their associated transcriptional machinery are localized. Thus, the "dynamic range" of the labeling with FNG is much larger than that of classical postembedding methods, which only reveal proteins within the nucleolus due to their higher concentration in this subdomain. Our results are in good agreement with the study of GFP-tagged nucleolar proteins as seen in FRAP experiments. Indeed, this technique demonstrates the presence of rapidly diffusing, but inactive nucleolar proteins within the nucleoplasm before (or after) they reach the nucleolar sites (Misteli, 2005). On the contrary, these proteins are highly concentrated at the active nucleolar sites, but their mobility is then significantly reduced.

Finally, due to its high sensitivity and excellent penetration within cells, FNG in particular and nanoprobes in general are excellent tools with which to investigate the 3D organization of complex machines by electron tomography. With this approach, we have visualized a nonrandom disposition of silver/gold particles (e.g., aligned as fibers or loops) which trace given proteins and reveal their 3D disposition within part of an organelle. Such applications on the localization of pKi-67 (Cheutin *et al.*, 2003) and RNAP I have been very fruitful (Cheutin *et al.*, 2002), allowing us to propose the first model of the 3D organization of rDNA genes within the cortex of the FCs (Cheutin *et al.*, 2002; Ploton *et al.*, 2004). This model is in good agreement with results obtained after specific labeling of rDNA (Derenzini *et al.*, 2006), of elongating rRNA (Cheutin *et al.*, 2002), and of immunolocalized GFP-tagged proteins (Tchelidze *et al.*, 2006).

Acknowledgments

This work was supported by grants from ARC (contract No. 4497) and Ligue contre le Cancer (Départements de l'Aube, de la Marne et des Ardennes).

References

Baumeister, W. (2005). A voyage to the inner space of cells. *Protein Sci.* **14,** 257–269.

Beorchia, A., Héliot, L., Ménager, M., Kaplan, H., and Ploton, D. (1993). Applications of medium-voltage STEM for the 3-D study of organelles within thick sections. *J. Microsc.* **170,** 247–258.

Bubulya, P. A., and Spector, D. L. (2004). On the movements of nuclear components in living cells. *Exp. Cell Res.* **296,** 4–11.

Cheutin, T., O'Donohue, M. F., Beorchia, A., Klein, C., Kaplan, H., and Ploton, D. (2003). Three-dimensional organization of pKi-67: A comparative fluorescence and electron tomography study using fluoronanogold. *J. Histochem. Cytochem.* **51,** 1411–1423.

Cheutin, T., O'Donohue, M. F., Beorchia, A., Vandelaer, M., Kaplan, H., Deféver, B., Ploton, D., and Thiry, M. (2002). Three-dimensional organization of active rRNA genes within the nucleolus. *J. Cell Sci.* **115,** 3297–3307.

Cremer, T., and Cremer, C. (2001). Chromosome territories, nuclear architecture and gene regulation in mammalian cells. *Nat. Rev. Genet.* **2,** 292–301.

Derenzini, M., Pasquinelli, G., O'Donohue, M. F., Ploton, D., and Thiry, M. (2006). Structural and functional organization of ribosomal genes within the mammalian cell nucleolus. *J. Histochem. Cytochem.* **54,** 131–145.

Gilbert, N., Lucas, L., Klein, C., Ménager, M., Bonnet, N., and Ploton, D. (1995). Three-dimensional co-location of RNA polymerase I and DNA during interphase and mitosis by confocal microscopy. *J. Cell Sci.* **108,** 115–125.

Gorski, S., and Misteli, T. (2005). Systems biology in the cell nucleus. *J. Cell Sci.* **118,** 4083–4092.

Hainfeld, J. F., and Powell, R. D. (2000). New frontiers in gold labeling. *J. Histochem. Cytochem.* **48,** 471–480.

Lucic, V., Förster, F., and Baumeister, W. (2005). Structural studies by electron tomography: From cells to molecules. *Annu. Rev. Biochem.* **74,** 833–865.

McEwen, B. F., and Marko, M. (2001). The emergence of electron tomography as an important tool for investigating cellular ultrastructure. *J. Histochem. Cytochem.* **49,** 553–563.

McIntosh, R., Nicastro, D., and Mastronarde, D. (2004). New views of cells in 3D: An introduction to electron tomography. *Trends Cell Biol.* **15,** 43–51.

Misteli, T. (2005). Concepts in nuclear architecture. *BioEssays* **27,** 477–487.

O'Brien, T. P., Bult, C. J., Cremer, C., Grunze, M., Knowles, B. B., Langowski, J., McNally, J., Pederson, T., Politz, J. C., Pombo, A., Schmahl, G., Spatz, J. P., *et al.* (2003). Genome function and nuclear architecture: From gene expression to nanoscience. *Genome Res.* **13,** 1029–1041.

Ploton, D., O'Donohue, M. F., Cheutin, T., and Thiry, M. (2004). Three-dimensional organization of rDNA and transcription. *In* "The Nucleolus" (Mark Olson, ed.), pp. 154–169. Landes Biosciences, Georgetown, Texas.

Robinson, J. M., Takizawa, T., and Vandré, D. D. (2000). Applications of gold cluster compounds in immunocytochemistry and correlative microscopy: Comparison with colloidal gold. *J. Microsc.* **199,** 163–179.

Robinson, J. M., and Vandré, D. D. (1997). Efficient immunocytochemical labeling of leukocyte microtubules with FluoroNanogold: An important tool for correlative microscopy. *J. Histochem. Cytochem.* **45,** 631–642.

Scheer, U., and Rose, K. M. (1984). Localization of RNA polymerase I in interphase cells and mitotic chromosomes by light and electron microscopic immunocytochemistry. *Proc. Natl. Acad. Sci. USA* **81,** 1431–1435.

Scholzen, T., and Gerdes, J. (2000). The Ki-67 protein: From the known and the unknown. *J. Cell Physiol.* **182,** 311–322.

Takizawa, T., and Robinson, J. M. (2000). FluoroNanogold is a bifunctional immunoprobe for correlative fluorescence and electron microscopy. *J. Histochem. Cytochem.* **48,** 481–486.

Tchelidze, P., Sauvage, C., Bonnet, N., Kilian, L., Beorchia, A., O'Donohue, M. F., Ploton, D., and Kaplan, H. (2006). Electron tomography of amplified nanogold immunolabelling: Improvement of quality based on alignment of projections with sinograms and use of post-reconstruction deconvolution. *J. Struct. Biol.* In press.

Thiry, M., Cheutin, T., O'Donohue, M. F., Kaplan, H., and Ploton, D. (2000). Dynamics and three-dimensional localization of ribosomal RNA within the nucleolus. *RNA* **6,** 1750–1761.

Thiry, M., and Goessens, G. (1996). The nucleolus during the cell cycle. *In* "Molecular Biology Intelligence Unit" (R. G. Landes, ed.), pp. 1–144. Springer-Verlag, Heidelberg.

CHAPTER 23

Markers for Correlated Light and Electron Microscopy

Gina E. Sosinsky, Ben N. G. Giepmans, Thomas J. Deerinck, Guido M. Gaietta, and Mark H. Ellisman

National Center for Microscopy and Imaging Research and Center for Research in Biological Systems
University of California, San Diego, La Jolla, California 92093

The dynamic behavior of cells is a consequence of the coordinated and elaborate interactions between complexes of macromolecules that constitute their formed structures or organelles. Live-cell imaging and high-resolution determination of the location of nucleic acids, proteins, metabolites, and ions can now be made within subcellular domains down to the molecular level, revealing the important information required to understand complex cellular functioning. Light and electron microscopic methods have proved extremely powerful in providing information about cells, and techniques such as immunolabeling have been applied successfully to identify molecules *in situ*. Recently, developments in recombinant methods of modern molecular biology and synthetic chemistry have lead to new probes and targeting techniques for these

0091-679X/07 $35.00
DOI: 10.1016/S0091-679X(06)79023-9

correlated modalities. Here we review whole cell, organelle, and protein labeling methods for correlated observations in the spatial domains of both light and electron microscopy.

I. Introduction

A continuing challenge to structural biologists is the elucidation of large macromolecular complexes and organelles, including multicomponent combinations of proteins, membranes, and/or nucleic acids. We have coined and have popularized the term "mesoscale" (from the Greek "mesos," middle) as the imaging range we mine to obtain new knowledge about the structure and dynamics of the macromolecular aggregates that compose the functional complexes within cells, tissues, and organs. These mesoscale structures are imaged with light microscopy (LM) and with more detail with electron microscopy (EM) at moderate resolutions, achievable with numerous preparatory methods, as opposed to high resolution, atomic-level, 3D structural methods, such as nuclear magnetic resonance (NMR), x-ray crystallography, and molecular microscopy (e.g., electron cryomicroscopy). In addition, along with the task of creating systems biology computational approaches to protein–protein interactions (the "interactomes" Rual *et al.*, 2005), one would ideally like to place macromolecular complexes within their cellular environments in order to fully understand their functional interactions ("visual proteomics," Nickell *et al.*, 2006). The focus of this chapter is the visualization of cells and organelles, and the identification of molecules for correlative LM and EM.

II. How Do LM and EM Complement Each Other?

One way of highlighting proteins of interest is to attach probes to them. For LM, this most often involves the tagging of molecules with fluorophores (Giepmans *et al.*, 2006) or nonfluorescent stains, such as peroxidase-mediated precipitates. EM markers must be highly electron dense in order to be distinguished from background constituents. Such markers can be visualized by EM tomography to provide information about the 3D ultrastructure of tissues, cells, and macromolecular complexes (as reviewed in Baumeister *et al.*, 1999; Frey *et al.*, 2006; McEwen and Marko, 2001; Sosinsky and Martone, 2003; Subramaniam, 2005). The combination of EM tomographic reconstructions with labeling of specific components provides a powerful method for localizing proteins within their cellular components.

The application of correlated LM and EM is leading the way in revolutionizing our understanding of the structural basis of cell biology. Better fluorescent probes have been and are being developed that highlight a particular protein of interest, and better LM instrumentation allows us to monitor these probes in three dimensions (3D) and real time (4D-imaging). Developments in fluorescence LM provide great opportunities to determine protein colocalization, turnover, and function (reviewed in

Giepmans *et al.*, 2006). However, imaging at the resolution of the LM provides only a general cellular location of the probe, a size and shape, and a sense of how the probe moves during sequential data acquisition. The relatively low resolution it provides might lead to a conclusion of colocation of two probes or the localization of a marker to a specific organelle that does not reflect the actual *in situ* situation. For example, a molecule or complex that sits just under the plasma membrane could be misinterpreted in LM imaging as being localized in the plasma membrane. While technological advances in LM imaging are pushing the resolution beyond the classical limit of 0.1–0.2 μm (Garini *et al.*, 2005), there are great additional benefits from probes that can also be imaged at the EM level. The shorter wavelength of the electron particle beam offers a substantial increase in resolution, and it also provides valuable information about the location of probes within the full cellular context; in addition to the probe, one sees many subcellular structures (e.g., organelles), membrane systems (e.g., plasma membrane), and macromolecular complexes (e.g., ribosomes), all at high resolution. Theoretically, this resolution increase when based only on instrumentation parameters is $>$1000-fold, but practically speaking, for biological specimens this ratio is $\sim$40- to 100-fold improvement. EM samples are inanimate, static structures. A well-known neuroscientist, Floyd Bloom, once went so far as to say "the gain in the brain lies mainly in the stain" (personal communication), but it is worth noting that this concept of improved protein detection is applicable to many fields of biological EM. Therefore, new probes to localize molecules at both the LM and EM levels have great potential to combine the benefits of the two techniques.

In this chapter, we divide the correlative LM and EM probes aimed at obtaining mesoscale structure and dynamics into three categories: (1) fluorescence photoconversion, (2) enzymatic-based detection, and (3) particle-based methods for protein localization. Each of these techniques has its own advantages and drawbacks; in reality, an investigator needs to decide ahead of the experiment as to the kind of information that s/he wishes to obtain. A flowchart of the experimental approaches employed in determining the localization of proteins in different samples and conditions is shown in Fig. 1.

III. Fluorescence Photooxidation

A. Principles

Photooxidation is a powerful and versatile technique for LM and EM whereby the oxygen radicals generated by a fluorescent compound under illumination are used to drive the oxidation of diaminobenzidine (DAB; *see schematic in* Fig. 2). Oxidized DAB can be made electron dense through osmification. Table I summarizes the photooxidation experiments with small fluorophores that are described in the following sections. Photooxidation of DAB was first shown with Lucifer Yellow (Maranto, 1982) in whole cells. Photooxidation of highly permeant

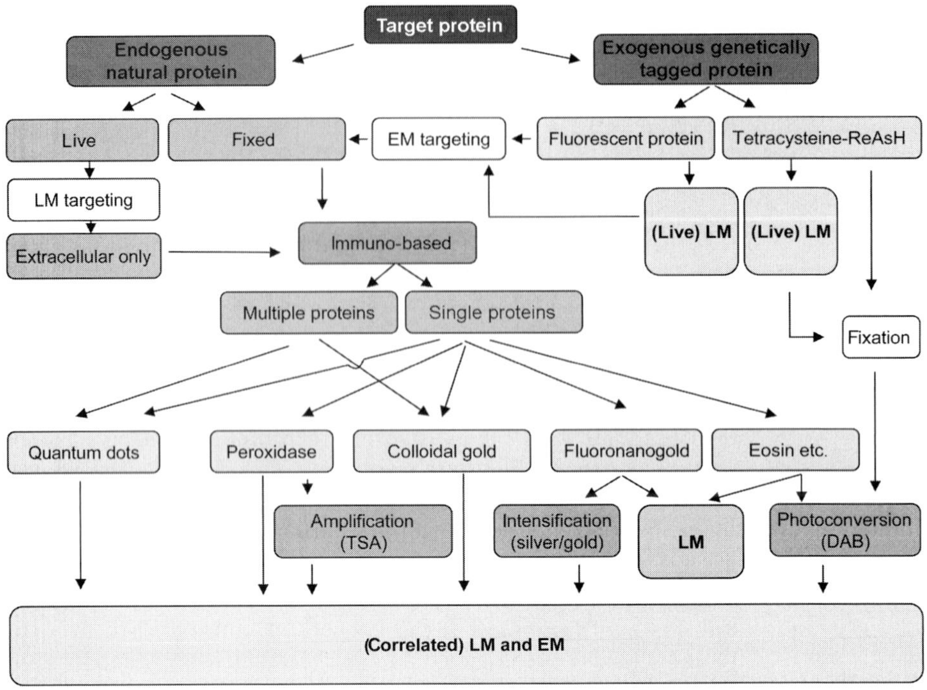

Fig. 1 Flowchart of the methods for correlative LM and EM described in this chapter. These include photooxidation, enzymatic-based detection, and particle-based methods. See text for details. (See Color Insert.)

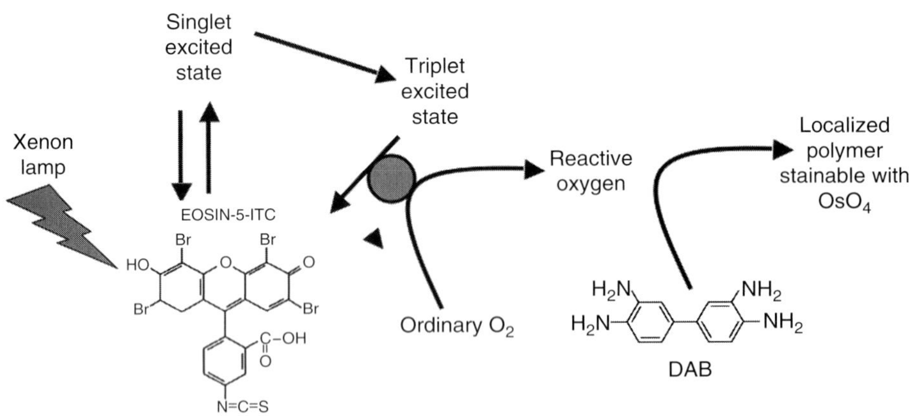

Fig. 2 Schematic of the photooxidation process. The process reads left to right, first with illumination by a xenon source that excites the photooxidizer to create reactive oxygen species that reacts with DAB to form an osmiophilic polymer.

Table I
Major Classes of Small Fluorophores for Photoconversion and LM and EM Analysis

Dye	Target	Reference(s)
Lucifer Yellow	Entire cell (microinjection)	Maranto, 1982
		Bushong *et al.*, 2002
Fluoro-Ruby	Anterograde axon transport	Schmued and Snavely, 1993
DiI	Membranes	von Bartheld *et al.*, 1990
PI	Nucleic acids	Balercia *et al.*, 1992
BODIPY	Conjugates (like BODIPY ceramide: Golgi)	Pagano *et al.*, 1989; Takizawa *et al.*, 1993
FM1-43	Plasma membrane and transport vesicles	Brumback *et al.*, 2004
		Henkel *et al.*, 1996
		Harata *et al.*, 2006
		Richards *et al.*, 2000
		Schikorski and Stevens, 2001
		Fomina *et al.*, 2003
Eosin	Immuno-conjugates to identify proteins	Deerinck *et al.*, 1994; Hand *et al.*, 2001
	Oligonucleotides conjugates to identify nucleic acids	Huang *et al.*, 1994
	Ligand conjugates:	
	Phalloidin-conjugate to identify F-actin	Capani *et al.*, 2001
	α-Bungarotoxin to nictonic acetylcholine receptor	Martone *et al.*, 1999
ReAsH	Genetically targeted to identify proteins	Gaietta *et al.*, 2002
		Sosinsky *et al.*, 2005
		Martin *et al.*, 2005
		Beahm *et al.*, 2006

Typical applications of the different dyes are summarized. See text for further details.

fluorophores (Lucifer Yellow, BODIPY ceramide) or reagents (FlAsH/ReAsH), in combination with DAB and reactive oxygen, possesses several advantages over other methods for high-resolution protein localization studies: (1) due to the small size of the reagents, the penetration of labels into the tissue is substantial, even without the use of harsh pretreatments such as detergents; (2) it can be used for correlated LM and EM; (3) due to the limited spread of reaction product, the spatial resolution under the EM is vastly superior to enzymatic methods; and (4) whereas photoconversion specifically stains the target protein, particle methods [gold or quantum dots (QDs)] decorate it. Moreover, in case of tetracysteine (Cys4) labeling the photooxidation product is directly on the protein. For gold labeling or eosin-mediated photoconversion, the "tree" of antibodies or other targeting reagents is a complex typically >10 nm in diameter.

Following treatment with osmium, the photooxidized DAB reaction product exhibits uniform staining with little granularity. The positively stained, photoconverted material often has a "negative-stain"-like appearance such that macromolecules appear light, but are surrounded by electron density. This technique provides specifically labeled, thick (>0.1 μm) specimens for high-resolution, 3D correlated confocal and high voltage microscopies (e.g., EM tomography). Standard EM techniques such as colloidal gold (CG) immunolabeling or immunoperoxidase are usually suboptimal for thick sections, because the label cannot penetrate thick sections unless the cell ultrastructure is highly compromised.

B. Small, Organic Fluorophores

Small organic dyes and compounds that either diffuse easily, or naturally intercalate into membranes or organelles, have been useful for generating the reactive oxygen species critical to good photooxidation (Table I). Maranto (1982) demonstrated the usefulness of photooxidation with neurons injected with the fluorescent tracer dye, Lucifer Yellow. In that study, horseradish peroxidase (HRP) was used to label synapses in combination with Lucifer Yellow. Lucifer Yellow (molecular weight 457 D) diffuses easily through entire cells, and thus delineates the cell's expanse. This technique was recently used to show that astrocyte processes establish exclusive territories and are much more expansive than had been shown with standard immunolabeling of an astrocyte marker (Bushong *et al.*, 2002). Similarly, Fluoro-Ruby, a fluorescent tetramethylrhodamine-dextran amine, used to demonstrate anterograde axon transport, has been successfully photoconverted and examined by EM (Schmued and Snavely, 1993). The membrane stain, 1,1′-dioctadecyl-3,3,3′,3′-tetramethylindocarbocyanine perchlorate (DiI), is a fluorescent dye that diffuses within cell membranes and can be photoconverted (von Bartheld *et al.*, 1990). Propridium iodide (PI), a fluorescent compound that intercalates into nucleic acid structures stains the nucleus and can be photoconverted (Balercia *et al.*, 1992). Fluorescent conjugates of ceramide, such as BODIPY ceramide, accumulate at the Golgi apparatus of fixed cells and can be photoconverted to selectively stain this organelle (Pagano *et al.*, 1989; Takizawa *et al.*, 1993).

Some classes of fluorescent lipid dyes intercalate into smaller vesicles and can then be photoconverted for EM analysis. One such dye is FM1-43, a dye that becomes highly fluorescent when partitioning into a hydrophobic environment, such as membranes; it has therefore been instrumental in monitoring exocytosis, endocytosis, and endosomal traffic (Brumback *et al.*, 2004; Henkel *et al.*, 1996). The charged FM1-43 is applied to the outside of the cell; it partitions into the outer leaflet of the plasma membrane, but it cannot flip-flop or diffuse across the membrane. Because surface-exposed FM1-43 can be washed out, the endosome-trapped dye can be used to specifically stain endosomes at the LM level and, following photoconversion, at the EM level. This technique has proven useful for identifying intermediates of the synaptic vesicle cycle and in the recycling of these membrane components (Harata *et al.*, 2006; Henkel *et al.*, 1996; Richards *et al.*, 2000; Schikorski and Stevens, 2001). FM1-43 has also been used to investigate the kinetics of constitutive endocytosis and to visualize the fate of endocytic organelles in resting and activated human T lymphocytes. Internalized cargo was carried to lysosome-like compartments in resting T cells and to multivesicular bodies in activated T cells (Fomina *et al.*, 2003).

C. Eosin-Based Fluorescent Probes

Eosin, a tetrabrominated derivative of fluorescein, is a potent generator of singlet oxygen while still possessing reasonable fluorescence. Eosin has a quantum yield of 0.57 for singlet oxygen generation, however the fluorescence quantum yield of eosin is typically only about 10–20% that of fluorescein (Deerinck *et al.*, 1994). Eosin can also be conjugated to various molecules, such as antibodies, phalloidin, and α-bungarotoxin, to achieve high-resolution protein localization at the fluorescence LM and EM levels (Martone *et al.*, 1999). Capani *et al.* (2001) used phalloidin conjugated to eosin and photooxidation to investigate the actin cytoskeleton in dendritic spines. Phalloidin selectively distinguished among different populations of dendritic spines, both within and between brain regions (Capani *et al.*, 2001). In addition, streptavidin–eosin conjugates enabled the photoconversion of RNA probes (Huang *et al.*, 1994) for EM analysis of the distribution of poly(A)-RNA in mammalian cells.

Despite its advantages, photoconversion suffers from a lack of sensitivity compared to enzyme-based methods—which "amplify" the signal by churning out an often diffusible reaction product. In fact, for most immunolabeling procedures, the standard fluorescent dyes fluorescein isothiocyanate (FITC) and rhodamine cannot be used because they are poor generators of singlet oxygen and thus are too inefficient at driving the polymerization of a reactive compound like DAB (Deerinck *et al.*, 1994). As outlined above, eosin offers a reasonable compromise between fluorescence yield and oxidation of DAB. However, even with this more potent oxidizer, the fluorescent signal (local concentration of target molecules) may not be strong enough to drive adequate photooxidation when stronger aldehyde fixation is used for maintaining ultrastructure for EM imaging (Hand *et al.*, 2001). We found that the immunofluorescence of many commercial connexin43 antibodies decreased dramatically even with small amounts of glutaraldehyde. Using tyramide signal amplification methods can modulate some of the antibody fixation sensitivity. These tyramide amplification techniques have also been successfully implemented for enhanced postembedding immuno-based CG labeling of proteins in pancreatic and liver tissues (Mayer and Bendayan, 1999). While there is a small decrease in the precise localization of labeling, due to the enzymatic nature of the amplification and the larger size of the immunolabeling reagents, we demonstrated that the labeling for connexin43 gap junctions was specific and provided specimens suitable for 3D localization by EM tomography (Hand *et al.*, 2001).

D. Genetic-Based Tags

A focus of further development for the fluorescence photooxidation technology is the application and invention of fluorescent genetic probes with photooxidation capabilities that are appended to or inserted into proteins of interest (see also Giepmans *et al.*, 2006). Fluorescent proteins (FPs) have provided a wealth of LM data, but genetically encoded tags for EM are just being developed.

Green fluorescent protein (GFP) is the most widely used fluorescent genetic tag, but GFP has been used for photoconversion of DAB in only a couple of situations (Grabenbauer *et al.*, 2005; Monosov *et al.*, 1996) because the reactive oxygen generation is much lower than that of ReAsH and may only work for situations where it is expressed at extremely high concentrations and by forced packing, changing its own properties, such that singlet oxygen can be generated. In addition, FPs (~26 kDa) are often larger than the protein of interest. While not necessarily a significant limitation in cell biology, the large size of FPs can interfere with the distribution, function, and fate of some recombinant proteins (reviewed in Giepmans *et al.*, 2006).

Recombinant proteins in intact cells can also be labeled by genetically appending or inserting a small motif (6–20 residues) that encodes a Cys4 sequence -Cys-Cys-Pro-Gly-Cys-Cys-, then exposing the cells to a membrane-permeant, nonfluorescent, biarsenical derivative of fluorescein, FlAsH. Roger Y. Tsien and his colleagues (Adams *et al.*, 2002; Griffin *et al.*, 1998) developed this system originally for live-cell imaging. FlAsH binds with high affinity and specificity to the Cys4 motif and thereby becomes strongly fluorescent, emitting green light when excited with blue light. These fluorescent complexes can survive for days in the absence of excessive (mM) concentrations of competing 1,2-ethanedithiol (EDT) (Martin *et al.*, 2005).

A biarsenical resorufin derivative, ReAsH, has been shown to photoconvert DAB (Gaietta *et al.*, 2002). Because the labeling is carried out in live cells, as described above for FlAsH, preparation for EM allows stringent fixation and does not require permeabilization. This preserves the ultrastructural aspects of the specimen. In principle, application of cryofixation methods to these labeled specimens would be highly advantageous. In practice, however, cells or tissues being photoconverted cannot be fast frozen and then photooxidized. Nonetheless, we have demonstrated that aldehyde fixation used prior to rapid freezing provides very good ultrastructural preservation (Sosinsky *et al.*, 2005). The combination of aldehyde fixation prior to high pressure freezing is one aspect of future development of photoconversion methods.

The Cys4/biarsenical ligand technique allows pulse-chase labeling at the fluorescence LM and EM level (Gaietta *et al.*, 2002) to address questions about how proteins move through the various cell compartments during their life cycle. Examples of optical pulse-chase applications include the trafficking of the gap junction protein, connexin43, (Gaietta *et al.*, 2002) or the localization of protein synthesis for the α-amino-5-hydroxy-3-methyl-4-isoxazole propionic acid (AMPA) receptor, glutamate receptor 1, GluR1 (Ju *et al.*, 2004). AMPA receptors are named for their synthetic agonist; they include glutamate receptors (GluR). This class of receptors is the most commonly found receptor in the nervous system, and therefore, proper trafficking of their subunits (e.g., GluR1) is an important consideration in synaptic remodeling. The proper trafficking of gap junction proteins has great relevance to the various connexin diseases, because many of these mutations cause mistrafficking of the protein.

A peptide with higher affinity for ReAsH (Martin *et al.*, 2005) and GFP combined with ReAsH have both been used to increase the specificity of ReAsH labeling to Cys4 and to increase specific photooxidation at the ReAsH-labeled chimeric protein by excitation of ReAsH through FRET from the donor GFP (Beahm *et al.*, 2006; Martin *et al.*, 2005). These samples are amenable to EM tomographic imaging (Sosinsky *et al.*, 2003) and can be used to follow Golgi components through mitosis (Gaietta *et al.*, 2006). Whichever method is used for photoconversion, DAB precipitation is most easily detectable on structural compartments or proteins that are concentrated in the cell. While there are no published quantifications of the concentration of label and size, we have some estimates based on the Cys4-tagged specimens we have developed. For example, 180 copies of a viral protein tagged with a Cys4 domain and labeled with ReAsH are enough for good photoconversion of a ~30-nm diameter virus (Lanman *et al.*, personal communication). In addition, individual actin filaments made from native and Cys4-tagged actin monomers are also easily photoconverted (Giepmans *et al.*, personal communication). Since this technique does "paint" individual labeled macromolecules like hemichannels with as few as six Cys4–ReAsH molecules, making them directly visible in the TEM—there is hope for further improvement in sensitivity, if the signal-to-noise ratio (SNR) can be improved. FRET-based photoconversion of ReAsH has promise here (Martin *et al.*, 2005). For more diffuse proteins, enzymatic or particle-based markers might be preferable.

IV. Enzymatic-Based Methods

A more traditional technique for visualizing proteins within cells involves the use of enzyme-based systems, typically in combination with immunolabeling (Etemadi, 1980; Farr and Nakane, 1981; van Leeuwen, 1981). Enzyme-based methods were first introduced in the mid-1960s and continue to be used today. The benefits of these methods are that they are straightforward and cheap, they penetrate well into the sample, and they have higher sensitivity than eosin-based methods. Commercial kits are available for ease of use (e.g., from Pierce, Clinisciences, or Molecular Probes, Inc.). These methods are still a valuable complementary approach to fluorescence photooxidation.

The two enzymes most commonly used for this work are HRP and alkaline phosphatase. These enzymes can be targeted to specific proteins by antibodies or by ligands conjugated to an enzyme. HRP catalyzes a reaction that uses DAB and H_2O_2 as substrates; as with fluorescence photooxidation, the DAB forms an amorphous polymer that is readily stained with OsO_4 (see above). With this method, one can observe the same preparation with nonfluorescent LM and EM. However, the diffusion of the reaction product away from the reaction site and onto membranous structures is a significant drawback, decreasing the resolution of the labeling (Deerinck *et al.*, 1994). The enhanced resolution of the photoconversion reaction product versus that generated by peroxidase is probably the result of two factors.

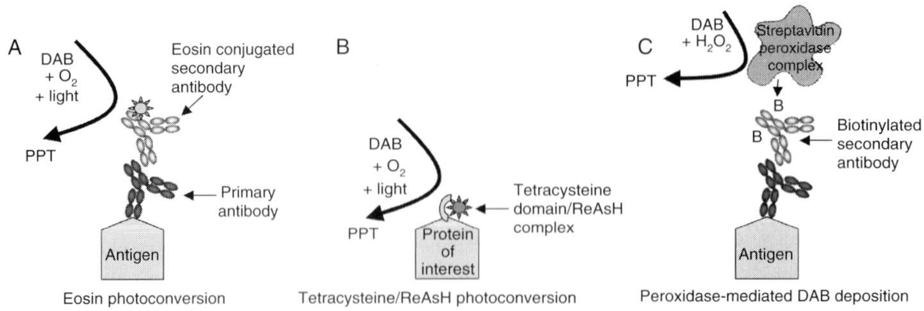

Fig. 3 Comparison of relative labeling sizes for three methods discussed in this chapter that use DAB polymerization as a labeling strategy. In each of these labeling procedures, DAB is used in combination with reactive oxygen species to form an osmiophilic precipitate that in combination with osmium tetroxide forms an electron dense substance. (A) Eosin photoconversion involves the binding of an eosin-conjugated ligand. In this case, eosin is bound to a primary antibody, but this method can also be used with eosin-conjugated toxins or ligands. (B) A Cys4 domain is genetically appended to a protein of interest and the biarsenical ligand, ReAsH, is bound to the Cys4 domain. The ReAsH serves as the singlet oxygen generator. (C) Schematic of peroxidase-mediated DAB polymerization (ABC methods) involving the use of biotin, streptavidin, and HRP. While this method may provide more sensitivity than the photooxidation methods (A or B), the resolution (size of the precipitate) is not nearly as precise.

First, the photoconversion reaction is carried out after postfixation with high concentrations of glutaraldehyde. This fixation increases the relative cross-linking of the cytoplasm and reduces the ability of the reaction product to diffuse. In addition, when singlet oxygen is generated by a point source (the fluorophore), the radius of influence is limited in our estimations to the single nanometer range, and the DAB appears to form a polymer in that limited domain. This was shown with eosin-based photoconversion of antibody labeled microtubules in 1994 (Deerinck *et al.*, 1994). Finally, photoconversion is done on ice. This also reduces the likelihood of diffusion of the reaction product. These interpretations are supported by the images shown in Fig. 4C and H. Figure 3 contains a schematic comparing the relative sizes and EM staining processes for (A) eosin-based photooxidation (B) Cys4/ReAsH-based photooxidation and (C) peroxidase (enzymatic)-based reactions.

V. Particle-Based Methods for Protein Localization

A. Gold Particles

Colloidal Gold (CG) bound to antibodies, soluble proteins, or ligands remains the standard method for particle-based methods for protein localization at present. As such, there are many good references for the use of CG as an EM marker (see Burry, 2000; Hayat, 1990; Roth, 1996). Here, we will discuss only recent technological improvements in CG labeling for correlative fluorescent LM and EM. Depending on the protein to be imaged and detectability required for EM, the size of CG typically used is 5–20 nm. While such gold particles are instantly recognizable in the EM, they are difficult to introduce into cells and tissues unless cells are permeabilized with detergents, whereupon the cell's ultrastructure is compromised. Exceptions to this are cell surface receptors, where permeabilization is not required for ligand binding. An example of this approach was an electron tomographic reconstruction of the ciliary ganglion in chick lens after marking of the cell surface for specific acetylcholine receptors (Shoop *et al.*, 1999). Because of these problems, most published studies that use CG for intracellular labeling apply the tagged probe after embedding rather than using a preembedding protocol; they are stand-alone EM experiments. A major benefit of this approach is the ability to use different sizes of gold simultaneously, allowing for localization of multiple distinct targets with clearly defined labels in the same specimen.

Progress in gold probes for correlative microscopy involves small gold labels (~1–3 nm) that can be increased in size by gold "toning" or silver "enhancing" procedures on thin sections for better detection (Baschong and Stierhof, 1998; Hainfeld and Furuya, 1995; Robinson *et al.*, 1998). These labels have also been made fluorescent by covalently binding FITC or Cy3 to the Fab. Probes like this, referred to as "Fluoronanogold," use an ~2-nm gold particle (Nanogold) (Powell and Hainfeld, 2002; Takizawa and Robinson, 2000a,b; Takizawa *et al.*, 1998). Nanogold is a commercial product of Nanoprobes, Inc. (Yapank, NY, http:www.nanoprobes.com; Hainfeld and Powell, 2000; Powell *et al.*, 1997, 1998). Applications of this technology for correlated LM and EM or EM tomography include that labeling of microtubules in leukocytes (Robinson and Vandre, 1997), localization of pKi-67, a putative chromatin-binding protein of unknown function (Cheutin *et al.*, 2003), release of interleukin-4 from granule distinct vesicular compartments (Melo *et al.*, 2005), and localization of the voltage-gated K^+ Channel, Kv4.2, in T-tubules of rat myocytes (Takeuchi *et al.*, 2000). However, nanogold is not nearly as visible in the EM as larger sized gold particles; it typically requires an amplification step, such as silver or gold intensification, that no longer allows for precise discriminative labeling of multiple distinct of the protein targets. QDs show great promise in their preembedding multiprotein labeling abilities (see below).

B. Quantum Dots

QDs are inorganic nanocrystals that fluoresce at distinct wavelengths depending on their size (~4–10 nm) and shape. The core, typically a CdSe or CdTe crystal, is electron dense and enables discrimination of the distinct QDs at the EM level (Giepmans *et al.*, 2005). A water-soluble coating allows for conjugation to targeting molecules, such as antibodies. While QDs are excellent fluorophores for LM, they are also a powerful marker for protein detection at the EM level (reviewed in Giepmans *et al.*, 2006). The drawback of antibody-mediated targeting of QDs to cytoplasmic epitopes, as with all preembedding immuno-based techniques, is that

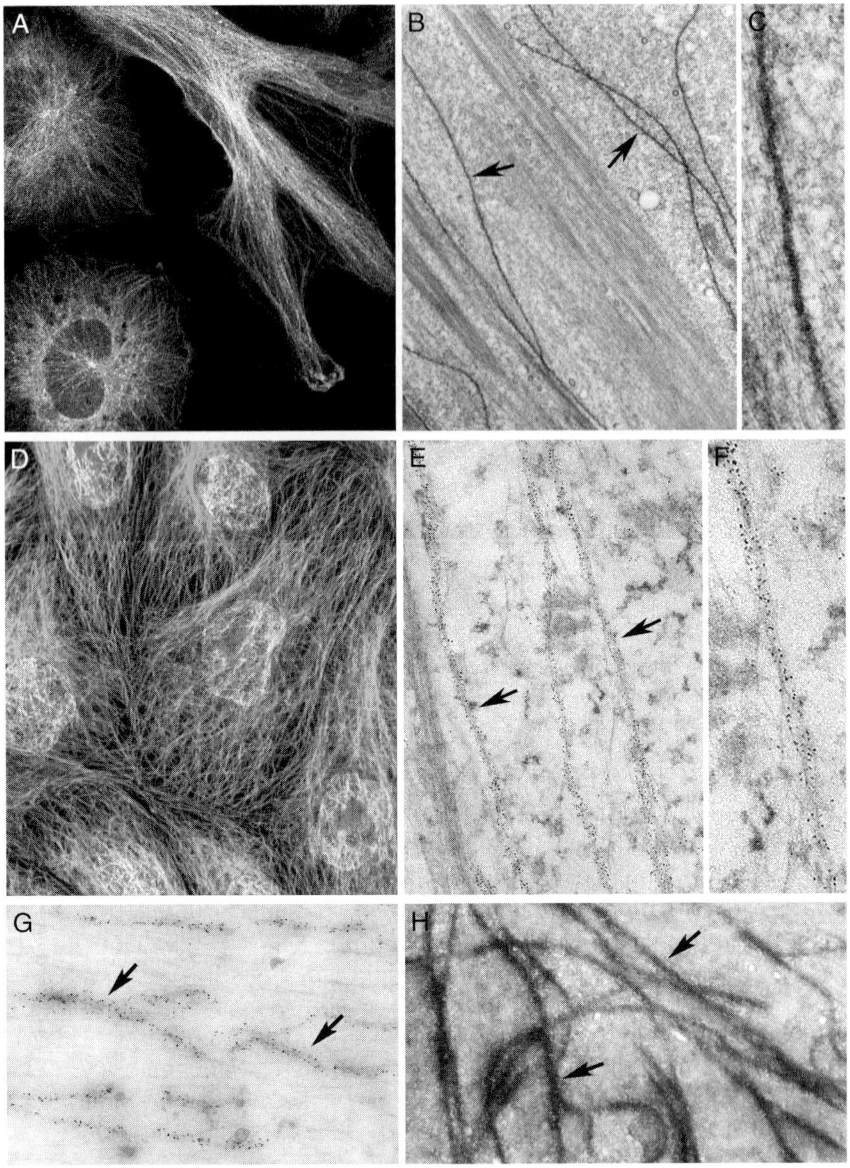

Fig. 4 Comparison of immunolabeled microtubules stained by photooxidation, particle-based methods, and enzymatic methods. Monolayers of cultured cells were immunolabeled for the microtubule protein tubulin followed by eosin-conjugated secondary antibodies (A–C) or QD-conjugated secondary antibodies (D–F). Fluorescence LM imaging (A, D) reveals many microtubules (A: green, D: red; nuclei are counterstained in blue). (B) Eosin-mediated photooxidation of DAB leads to a fine, osmiophilic precipitate that evenly increases the density of the microtubules in EM images. A higher magnification view is shown in (C). (E) QDs are large and electron dense enough to image at the EM at near-molecular

permeabilization is required, and these treatments have a negative impact on the cellular ultrastructure. Compared to CG, QDs have a relative low electron density, which can be increased by silver enhancement, as has been successfully applied to detect glycine receptors in synapses (Dahan *et al.*, 2003), but now multilabeling as outlined earlier is no longer possible. Postsectioning labeling of a nuclear protein has also been shown with QD and correlated microscopies (Nisman *et al.*, 2004). We recently reported a generic application to simultaneously label multiple endogenous proteins for correlated LM–EM in cells and tissues using preembedding immunolabeling (Giepmans *et al.*, 2005). Proteins in cells and tissues were immuno-double or immuno-triple labeled using QDs and visualized in the LM and EM. Preembedding labeling using QDs penetrated tissues more efficiently than conventional 5-nm CG, but not nearly as well as small dyes such as FITC that are suitable solely for LM. QDs allow for labeling with several micrometer penetration, as opposed to immunogold (Giepmans *et al.*, 2005, 2006). Therefore, QDs can be successfully used to label proteins in thick EM-sections for immuno-tomography (Giepmans *et al.*, our unpublished observations). Since the different fluorescent emissions of QDs can be correlated with different sizes or shapes of QDs, which are discernable in the EM, double or triple staining of proteins followed by correlated microscopy is relatively straightforward. Moreover, labeling efficiency can be assessed by fluorescence at the LM level before continuing with more laborious EM. For high-throughput EM, large sections of labeled tissues can be screened at LM level, followed by a subsequent EM analysis of regions of interest. As an alternative to immuno-based techniques, QDs can be targeted to molecules of interest using other conjugates such as streptavidin (reviewed in Giepmans *et al.*, 2006).

VI. Concluding Remarks

In summary, new tools have been developed for correlated LM and EM (e.g., photooxidation of fluorescent probes and immuno-QD labeling); these complement more established techniques, such as peroxidase or CG labeling. A comparison of these methods has been achieved by using a standard structure, like microtubules, that were immunolabeled for tubulin and then detected by photooxidation (Fig. 4A–C), QDs (Fig. 4D–F), CG (Fig. 4G), and enzymatic labeling (Fig. 4H). Such images show the different appearances of these labeling strategies. In our experiences, the SNR for enzymatic methods is better than for photooxidation, and QDs are better than standard CG, because of better penetration of the particles. However, to compare the SNR of QDs with Cys4 photooxidation is not a valid comparison, because while QDs give far better fluorescence than eosin or ReAsH, but

resolution. The image in (F) is a higher magnification view. Older methods: a 5-nm immunogold microtubule labeling (G) and peroxidase-mediated enzymatic labeling (H) are shown for comparison. Black arrows in this figure point to individual microtubules. For scale, the diameter of a native microtubule is ~25 nm. (See Color Insert.)

they do not label the tagged protein complex with reaction product. The applicability and implementation of these techniques to any particular biological problem depends on the exact settings and goals, like the viewing of a single versus multiple distinct targets, the desired resolution, the ease of use, desired integrity of cell architecture, and the detectability of features. While each of these techniques has advantages and disadvantages relative to the others, each method can provide important and significant information about the dynamics, localizations, and arrangements of proteins within their cellular context.

Acknowledgments

We are grateful to Drs. Maryann E. Martone and Roger Y. Tsien for invaluable discussions. Some of the work described in this review was supported by grants from the National Institutes of Health grant NIH-RR004050, NS 14718, RR019701, DK54441 (M.H.E.), GM072033 (Roger Y. Tsien), GM65937 (G.E.S) GM72881 (G.E.S.), and NSF grant MCB-0543934 (G.E.S.).

References

Adams, S. R., Campbell, R. E., Gross, L. A., Martin, B. R., Walkup, G. K., Yao, Y., Llopis, J., and Tsien, R. Y. (2002). New biarsenical ligands and tetracysteine motifs for protein labeling *in vitro* and *in vivo*: Synthesis and biological applications. *J. Am. Chem. Soc.* **124,** 6063–6076.

Balercia, G., Chen, S., and Bentivoglio, M. (1992). Electron microscopic analysis of fluorescent neuronal labeling after photoconversion. *J. Neurosci. Methods* **45,** 87–98.

Baschong, W., and Stierhof, Y. D. (1998). Preparation, use, and enlargement of ultrasmall gold particles in immunoelectron microscopy. *Microsc. Res. Tech.* **42,** 66–79.

Baumeister, W., Grimm, R., and Walz, J. (1999). Electron tomography of molecules and cells. *Trends Cell Biol.* **9,** 81–85.

Beahm, D. L., Oshima, A., Gaietta, G. M., Hand, G. M., Smock, A. E., Zucker, S. N., Toloue, M. M., Chandrasekhar, A., Nicholson, B. J., and Sosinsky, G. E. (2006). Mutation of a conserved threonine in the third transmembrane helix of alpha- and beta-connexins creates a dominant-negative closed gap junction channel. *J. Biol. Chem.* **281,** 7994–8009.

Brumback, A. C., Lieber, J. L., Angleson, J. K., and Betz, W. J. (2004). Using FM1-43 to study neuropeptide granule dynamics and exocytosis. *Methods* **33,** 287–294.

Burry, R. W. (2000). Specificity controls for immunocytochemical methods. *J. Histochem. Cytochem.* **48,** 163–166.

Bushong, E. A., Martone, M. E., Jones, Y. Z., and Ellisman, M. H. (2002). Protoplasmic astrocytes in CA1 stratum radiatum occupy separate anatomical domains. *J. Neurosci.* **22,** 183–192.

Capani, F., Ellisman, M. H., and Martone, M. E. (2001). Filamentous actin is concentrated in specific subpopulations of neuronal and glial structures in rat central nervous system. *Brain Res.* **923,** 1–11.

Cheutin, T., O'Donohue, M. F., Beorchia, A., Klein, C., Kaplan, H., and Ploton, D. (2003). Three-dimensional organization of pKi-67: A comparative fluorescence and electron tomography study using FluoroNanogold. *J. Histochem. Cytochem.* **51,** 1411–1423.

Dahan, M., Levi, S., Luccardini, C., Rostaing, P., Riveau, B., and Triller, A. (2003). Diffusion dynamics of glycine receptors revealed by single-quantum dot tracking. *Science* **302,** 442–445.

Deerinck, T. J., Martone, M. E., Lev-Ram, V., Green, D. P. L., Tsien, R. Y., Spector, D. L., Huang, S., and Ellisman, M. H. (1994). Fluorescence photooxidation with eosin: A method for high resolution immunolocalization and *in situ* hybidization detect for light and electron microscopy. *J. Cell Biol.* **126,** 901–910.

Etemadi, A. H. (1980). Membrane asymmetry. A survey and critical appraisal of the methodology. I. Methods for assessing the asymmetric orientation and distribution of proteins. *Biochim. Biophys. Acta* **604,** 347–422.

Farr, A. G., and Nakane, P. K. (1981). Immunohistochemistry with enzyme labeled antibodies: A brief review. *J. Immunol. Methods* **47,** 129–144.

Fomina, A. F., Deerinck, T. J., Ellisman, M. H., and Cahalan, M. D. (2003). Regulation of membrane trafficking and subcellular organization of endocytic compartments revealed with FM1-43 in resting and activated human T cells. *Exp. Cell Res.* **291,** 150–166.

Frey, T. G., Perkins, G. A., and Ellisman, M. H. (2006). Electron tomography of membrane-bound cellular organelles. *Annu. Rev. Biophys. Biomol. Struct.* **35,** 199–224.

Gaietta, G., Deerinck, T. J., Adams, S. R., Bouwer, J., Tour, O., Laird, D. W., Sosinsky, G. E., Tsien, R. Y., and Ellisman, M. H. (2002). Multicolor and electron microscopic imaging of connexin trafficking. *Science* **296,** 503–507.

Gaietta, G. M., Giepmans, B. N. G., Deerinck, T. J., Smith, W. B., Ngan, L., Llopis, J., Adams, S. R., Tsien, R. Y., and Ellisman, M. H. (2006). Golgi twins in late mitosis revealed by genetically encoded tags for live cell imaging and correlated electron microscopy. *Proc. Natl. Acad. Sci. USA.* In press.

Garini, Y., Vermolen, B. J., and Young, I. T. (2005). From micro to nano: Recent advances in high-resolution microscopy. *Curr. Opin. Biotechnol.* **16,** 3–12.

Giepmans, B. N., Adams, S. R., Ellisman, M. H., and Tsien, R. Y. (2006). The fluorescent toolbox for assessing protein location and function. *Science* **312,** 217–224.

Giepmans, B. N., Deerinck, T. J., Smarr, B. L., Jones, Y. Z., and Ellisman, M. H. (2005). Correlated light and electron microscopic imaging of multiple endogenous proteins using Quantum dots. *Nat. Methods* **2,** 743–749.

Grabenbauer, M., Geerts, W. J., Fernadez-Rodriguez, J., Hoenger, A., Koster, A. J., and Nilsson, T. (2005). Correlative microscopy and electron tomography of GFP through photooxidation. *Nat. Methods* **2,** 857–862.

Griffin, B. A., Adams, S. R., and Tsien, R. Y. (1998). Specific covalent labeling of recombinant protein molecules inside live cells. *Science* **281,** 269–272.

Hainfeld, J. F., and Furuya, F. R. (1995). Silver enchancement of Nanogold and undecagold. *In* "Immunogold-Silver Staining" (M. A. Hayat, ed.), pp. 71–96. CRC Press, Boca Raton, FL.

Hainfeld, J. F., and Powell, R. D. (2000). New frontiers in gold labeling. *J. Histochem. Cytochem.* **48,** 471–480.

Hand, G. M., Martone, M. E., Stelljes, A., Ellisman, M. H., and Sosinsky, G. E. (2001). Specific labeling of connexin43 in NRK cells using tyramide-based signal amplification and fluorescence photooxidation. *Microsc. Res. Tech.* **52,** 331–343.

Harata, N. C., Choi, S., Pyle, J. L., Aravanis, A. M., and Tsien, R. W. (2006). Frequency-dependent kinetics and prevalence of kiss-and-run and reuse at hippocampal synapses studied with novel quenching methods. *Neuron* **49,** 243–256.

Hayat, M. A. (1990). "Colloidal Gold: Principles, Methods, and Applications." Academic Press, San Diego, CA.

Henkel, A. W., Lubke, J., and Betz, W. J. (1996). FM1-43 dye ultrastructural localization in and release from frog motor nerve terminals. *Proc. Natl. Acad. Sci. USA* **93,** 1918–1923.

Huang, S., Derrinck, T. J., Ellisman, M. H., and Spector, D. L. (1994). *In vivo* analysis of the stability and transport of nuclear poly(A)$^+$RNA. *J. Cell Biol.* **126,** 877–899.

Ju, W., Morishita, W., Tsui, J., Gaietta, G., Deerinck, T. J., Adams, S. R., Garner, C. C., Tsien, R. Y., Ellisman, M. H., and Malenka, R. C. (2004). Activity-dependent regulation of dendritic synthesis and trafficking of AMPA receptors. *Nat. Neurosci.* **7,** 244–253.

Maranto, A. (1982). Neuronal mapping: A photooxidation reaction makes Lucifer Yellow useful for electron microscopy. *Science* **217,** 953–955.

Martin, B. R., Giepmans, B. N., Adams, S. R., and Tsien, R. Y. (2005). Mammalian cell-based optimization of the biarsenical-binding tetracysteine motif for improved fluorescence and affinity. *Nat. Biotechnol.* **23,** 1308–1314.

Martone, M. E., Deerinck, T. J., Young, S. J., and Ellisman, M. E. (1999). Three-dimensional protein localization using high voltage electron microscopy. *Acta Histochem. Cytochem.* **32,** 35–43.

Mayer, G., and Bendayan, M. (1999). Immunogold signal amplification: Application of the CARD approach to electron microscopy. *J. Histochem. Cytochem.* **47,** 421–429.

McEwen, B. F., and Marko, M. (2001). The emergence of electron tomography as an important tool for investigating cellular ultrastructure. *J. Histochem. Cytochem.* **49,** 553–564.

Melo, R. C., Spencer, L. A., Perez, S. A., Ghiran, I., Dvorak, A. M., and Weller, P. F. (2005). Human eosinophils secrete preformed, granule-stored interleukin-4 through distinct vesicular compartments. *Traffic* **6,** 1047–1057.

Monosov, E. Z., Wenzel, T. J., Luër, G. H., Heyman, J. A., and Subramani, S. (1996). Labeling of peroxisomes with green fluorescent protein in living *P. pastoris* cells. *J. Histochem. Cytochem.* **44,** 581–589.

Nickell, S., Kofler, C., Leis, A. P., and Baumeister, W. (2006). A visual approach to proteomics. *Nat. Rev. Mol. Cell Biol.* **7,** 225–230.

Nisman, R., Dellaire, G., Ren, Y., Li, R., and Bazett-Jones, D. P. (2004). Application of quantum dots as probes for correlative fluorescence, conventional, and energy-filtered transmission electron microscopy. *J. Histochem. Cytochem.* **52,** 13–18.

Pagano, R. E., Sepanski, M. A., and Martin, O. C. (1989). Molecular trapping of a fluorescent ceramide analogue at the Golgi apparatus of fixed cells: Interaction with endogenous lipids provides a trans-Golgi marker for both light and electron microscopy. *J. Cell Biol.* **109,** 2067–2079.

Powell, R. D., and Hainfeld, J. F. (2002). Combined fluorescent and gold probes for microscopic and morphological investigations. *In* "Gold and Silver Staining: Techniques in Molecular Morphology" (G. W. Hacker, and J. Gu, eds.), pp. 107–118. CRC Press, Boca Raton, FL.

Powell, R. D., Halsey, C. M., and Hainfeld, J. F. (1998). Combined fluorescent and gold immunoprobes: Reagents and methods for correlative light and electron microscopy. *Microsc. Res. Tech.* **42,** 2–12.

Powell, R. D., Halsey, C. M., Spector, D. L., Kaurin, S. L., McCann, J., and Hainfeld, J. F. (1997). A covalent fluorescent-gold immunoprobe: Simultaneous detection of a pre-mRNA splicing factor by light and electron microscopy. *J. Histochem. Cytochem.* **45,** 947–956.

Richards, D. A., Guatimosim, C., and Betz, W. J. (2000). Two endocytic recycling routes selectively fill two vesicle pools in frog motor nerve terminals. *Neuron* **27,** 551–559.

Robinson, J. M., Takizawa, T., Vandre, D. D., and Burry, R. W. (1998). Ultrasmall immunogold particles: Important probes for immunocytochemistry. *Microsc. Res. Tech.* **42,** 13–23.

Robinson, J. M., and Vandre, D. D. (1997). Efficient immunocytochemical labeling of leukocyte microtubules with FluoroNanogold: An important tool for correlative microscopy. *J. Histochem. Cytochem.* **45,** 631–642.

Roth, J. (1996). The silver anniversary of gold: 25 years of the colloidal gold marker system for immunocytochemistry and histochemistry. *Histochem. Cell Biol.* **106,** 1–8.

Rual, J. F., Venkatesan, K., Hao, T., Hirozane-Kishikawa, T., Dricot, A., Li, N., Berriz, G. F., Gibbons, F. D., Dreze, M., Ayivi-Guedehoussou, N., Klitgord, N., Simon, C., *et al.* (2005). Towards a proteome-scale map of the human protein-protein interaction network. *Nature* **437,** 1173–1178.

Schikorski, T., and Stevens, C. F. (2001). Morphological correlates of functionally defined synaptic vesicle populations. *Nat. Neurosci.* **4,** 391–395.

Schmued, L. C., and Snavely, L. F. (1993). Photoconversion and electron microscopic localization of the fluorescent axon tracer fluoro-ruby (rhodamine-dextran-amine). *J. Histochem. Cytochem.* **41,** 777–782.

Shoop, R. D., Martone, M. E., Yamada, N., Ellisman, M. H., and Berg, D. K. (1999). Neuronal acetylcholine receptors with alpha7 subunits are concentrated on somatic spines for synaptic signaling in embryonic chick ciliary ganglia. *J. Neurosci.* **19,** 692–704.

Sosinsky, G. E., Deerinck, T. J., Greco, R., Buitenhuys, C. H., Bartol, T. M., and Ellisman, M. H. (2005). Development of a model for microphysiological simulations: Small nodes of ranvier from peripheral nerves of mice reconstructed by electron tomography. *Neuroinformatics* **3,** 133–162.

Sosinsky, G. E., Gaietta, G. M., Hand, G., Deerinck, T. J., Han, A., Mackey, M., Adams, S. R., Bouwer, J., Tsien, R. Y., and Ellisman, M. H. (2003). Tetracysteine genetic tags complexed with biarsenical ligands as a tool for investigating gap junction structure and dynamics. *Cell Commun. Adhes.* **10,** 181–186.

Sosinsky, G., and Martone, M. (2003). The nuts and bolts of electron tomography: Imaging of big and messy biological structures. *In* "Microscopy Today," Vol. 11, pp. 7–11.

Subramaniam, S. (2005). Bridging the imaging gap: Visualizing subcellular architecture with electron tomography. *Curr. Opin. Microbiol.* **8,** 316–322.

Takeuchi, S., Takagishi, Y., Yasui, K., Murata, Y., Toyama, J., and Kodama, I. (2000). Voltage-gated K(+)Channel, Kv4.2, localizes predominantly to the transverse-axial tubular system of the rat myocyte. *J. Mol. Cell Cardiol.* **32,** 1361–1369.

Takizawa, P. A., Yucel, J. K., Veit, B., Faulkner, D. J., Deerinck, T., Soto, G., Ellisman, M., and Malhotra, V. (1993). Complete vesiculation of Golgi membranes and inhibition of protein transport by a novel sea sponge metabolite, ilimaquinone. *Cell* **73,** 1079–1090.

Takizawa, T., and Robinson, J. M. (2000a). Analysis of antiphotobleaching reagents for use with FluoroNanogold in correlative microscopy. *J. Histochem. Cytochem.* **48,** 433–436.

Takizawa, T., and Robinson, J. M. (2000b). FluoroNanogold is a bifunctional immunoprobe for correlative fluorescence and electron microscopy. *J. Histochem. Cytochem.* **48,** 481–486.

Takizawa, T., Suzuki, K., and Robinson, J. M. (1998). Correlative microscopy using FluoroNanogold on ultrathin cryosections. Proof of principle. *J. Histochem. Cytochem.* **46,** 1097–1102.

van Leeuwen, F. W. (1981). An introduction to the immunocytochemical localization of neuropeptides and neurotransmitters. *Acta Histochem. Suppl.* **24,** 49–77.

von Bartheld, C. S., Cunningham, D. E., and Rubel, E. W. (1990). Neuronal tracing with DiI: Decalcification, cryosectioning, and photoconversion for light and electron microscopic analysis. *J. Histochem. Cytochem.* **38,** 725–733.

CHAPTER 24

Localizing Specific Elements Bound to Macromolecules by EFTEM

Richard D. Leapman and Maria A. Aronova

National Institute of Biomedical Imaging and Bioengineering
National Institutes of Health
Bethesda, Maryland 20892

This chapter contains a description of the basic principles of energy filters and their application for mapping elements in sectioned cells. Quantitative elemental distributions are obtained by subtracting a scaled image acquired at an energy loss just below a core-edge from an image acquired just above the edge. The net signal is then normalized by the zero-loss intensity and divided by the ionization cross section to give the number of atoms per unit area. An expression is also derived for the statistical errors in the measured number of atoms for a particular set of imaging conditions, accelerating voltage, and probe current. Energy filtered imaging of phosphorus, which serves as an intrinsic label for nucleic acid, reveals the distribution of DNA in the cell nucleus. However, it is shown that determining the phosphorus distributions in specimens of variable thickness requires a correction

0091-679X/07 $35.00
DOI: 10.1016/S0091-679X(06)79024-0

for plural inelastic scattering. An important application of electron spectroscopic imaging is its ability to image specific metal atoms bound to macromolecules. As an example, iron maps are acquired from sections containing the iron storage protein, ferritin. Finally, by acquiring electron spectroscopic images in tilt series and employing established methods of electron tomography, it is possible to reconstruct the distributions of specific elements in three dimensions.

I. Introduction

Over the past decade, there have been significant advances in energy filtered transmission electron microscopy (EFTEM), which can now provide two-dimensional (2D) distributions of specific chemical elements with a higher sensitivity and higher spatial resolution than was achievable before (Egerton, 2003; Hofer and Warbichler, 2004; Krivanek *et al.*, 1995; Reimer, 1995). Progress has involved improvements in the electron optical design of imaging filters, development of efficient array detectors with large numbers of pixels, and sophisticated digital control of image acquisition. Furthermore, by using tomographic methods, it has recently become possible to extend elemental mapping into 3D, although this technique is quite new and is still under development (Leapman *et al.*, 2004; Midgley and Weyland, 2003).

In the EFTEM electrons that have been transmitted through a thin specimen enter the field of magnetic prism spectrometer, which disperses the electrons according to their kinetic energy or velocity. A slit placed at the dispersion plane of the magnetic prism selects electrons of a given energy loss, which then travel through a series of lenses to form a high resolution image of the specimen on a 2D array detector (Reimer, 1995). As already discussed in an earlier Chapter, imaging filters can be used to eliminate blurring of the image in thick frozen-hydrated specimens due to chromatic aberration of the objective lens, which fails to focus properly those electrons that have lost energy by inelastic scattering (Beck *et al.*, 2004; Hsieh *et al.*, 2006; Koster *et al.*, 1997; Nicastro *et al.*, 2005; Steven and Aebi, 2003; Zhang *et al.*, 2004). Improvement in contrast is achieved by setting the energy selecting slit to pass only elastically scattered as well as unscattered electrons, which have not suffered any energy loss. In such applications, the energy-loss electrons are considered to be a nuisance and are simply excluded altogether from the image. However, the inelastic scattering also carries valuable information about the composition of the specimen and, in particular, those electrons that have excited atomic inner shell electrons provide information about the distributions of specific chemical elements (Bazett-Jones *et al.*, 1994, 1996; Ottensmeyer, 1984). Inner-shell or core-level excitations that are relevant to biological systems typically occur with energy losses in the range of 100–1000 eV and are very weak compared to the total elastic or inelastic signals. Nevertheless, the signal is still strong enough to map small numbers of atoms provided that a sensitive detector is used to record the energy-selected images and that a sufficiently high electron dose can be applied to the sample (Leapman, 2003; Leapman and Rizzo, 1999).

For most applications, it is necessary to use freeze-substituted plastic-embedded sections that are unstained (Müller-Reichert *et al.*, 2003) or freeze-dried cryosections (Leapman and Ornberg, 1988; Somlyo *et al.*, 1979); frozen-hydrated specimens are generally unable to withstand the required electron dose unless only very low spatial resolution information is sought. By using EFTEM, it is possible to detect physiologically relevant levels of phosphorus, sulfur, and certain other elements in cellular regions that are a few nanometers in size (Bazett-Jones *et al.*, 1994; Goping *et al.*, 2003; Leapman *et al.*, 1997). For example, by imaging phosphorus it is feasible to map quantitatively the distribution of nucleic acid with a detection limit (standard deviation) of ~20 base pairs. Bound metal atoms, such as iron, can be imaged too with potential applications to the biology of metalloproteins or to the identification of specific metal labels attached to proteins. In this chapter, we describe the basic principles of elemental mapping in 2D as well as the much more recent extension to 3D and demonstrate how these techniques can reveal the distributions of metals and other bound elements in cells on a macromolecular scale.

II. General Principles

A. Electron Energy Loss Spectroscopy and Imaging

Imaging filters can be incorporated into the column of a conventional transmission electron microscope before the projector lenses (in-column filters) or they can be mounted at the end of the column (post-column filters) (Krivanek *et al.*, 1991, 1995; Lanio, 1986; Reimer, 1995). Although in-column and post-column filters look quite different (Fig. 1), their principles of operation are basically the same. Each comprises four main parts: (1) an energy-dispersive component, which separates electrons that have lost different amounts of energy by inelastic scattering in the specimen; (2) an energy-selecting slit that defines a band of energies to be transmitted to the image; (3) a series of lenses that focuses the transmitted electrons into an image that is conjugate to the object plane of the filter, which is situated at the final image plane of the microscope for post-column filters or after the intermediate lens for in-column filters; and (4) a 2D array detector, usually a cooled charge couple device (CCD) camera with near single electron sensitivity. The in-column filter consists of four magnetic prisms in the shape of a Greek omega and is often called an Ω-filter. This type of filter uses the projector lenses of the microscope to form a final image on a CCD camera mounted below the camera chamber (Reimer, 1995). The post-column filter has a single magnetic prism, which bends the electrons through an angle of 90°. A series of magnetic quadrupole and sextupole lenses enables an energy-filtered image of high quality

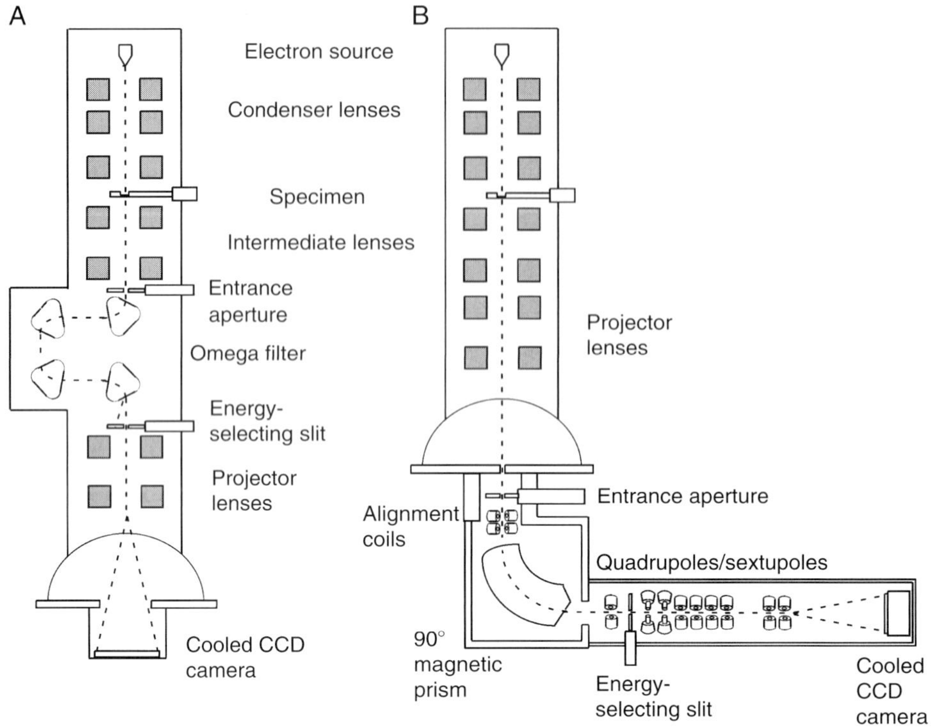

Fig. 1 Simplified diagrams of two different types of imaging filters used with transmission electron microscopes: (A) in-column Ω-filter consisting of four 90° magnetic sectors; (B) post-column filter, consisting of one 90° magnetic sector mounted as an attachment below the microscope camera chamber.

to be formed without introducing additional image rotations that occur with round magnetic lenses (Krivanek *et al.*, 1995).

To understand how to measure the distribution of elements in a biological specimen, it is helpful to examine briefly the form of the electron energy loss spectrum (EELS) (Egerton, 1996; Leapman, 2004), as shown in Fig. 2 for a sample of pure DNA supported on a carbon film. The most intense feature in the spectrum is the zero-loss peak corresponding to incident electrons that are either unscattered and pass undeviated through the specimen, or are elastically scattered by atomic nuclei with a change in direction but with negligible energy loss. The most probable inelastic scattering events correspond to excitations of atomic outer shell electrons and typically involve energy losses of 10–50 eV. This low-loss spectrum contains a maximum at around 20 eV attributable to a plasmon or collective excitation of the valence electrons; it also includes a tail of decreasing intensity that extends to higher energy losses, due to ejection of valence electrons with increasing amounts of kinetic energy. Superposed on this tail due to valence excitations are weak signals or "edges" originating from excitation of inner-shell or core levels of specific atoms

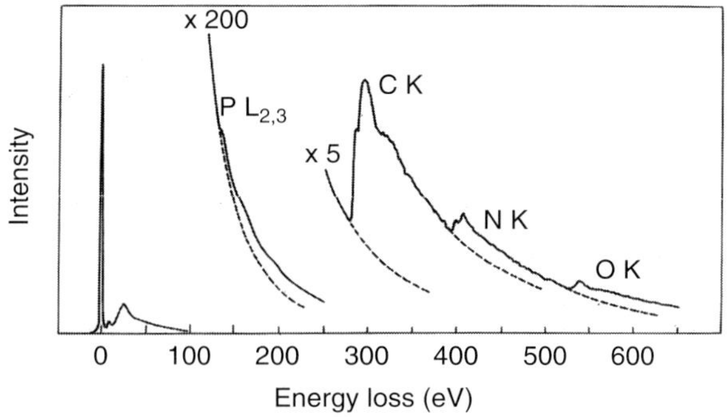

Fig. 2 Electron energy-loss spectrum from a sample of DNA deposited on a carbon film, showing the zero-loss peak, excitation of the valence electrons (peak at around 20 eV), and core-edges at 132 (phosphorus), 285 (carbon), 402 (nitrogen), and 531 eV (oxygen). The intensity scale at the P $L_{2,3}$ edge is increased by a factor of 200, and by a further factor of 5 at the carbon K edge as shown. The background intensity is shown beneath each edge, extrapolated according to an inverse power law.

in the specimen. Excitations of 1s electrons give rise to potassium (K) edges, whereas excitation of 2p electrons give rise to $L_{2,3}$ edges. Core-level excitations are typically a factor of 10^3 or 10^4 weaker than the total inelastic or elastic signals. Therefore to map the distribution of an element successfully usually requires $>10^6$ beam electrons to pass through each area of the specimen that corresponds to a single pixel in the image. For a nanometer-sized pixel, the corresponding dose would be at least 10^6 electrons/nm^2.

To obtain element-specific maps from sectioned cells, it is typically necessary to record two images, one at an energy loss just above the atomic core level excitation threshold (also called a "core-edge") and one at an energy loss just below the threshold (Leapman, 1986; Ottensmeyer, 1984). By carefully subtracting the pre-core-edge background intensity from the post-core-edge intensity, it is possible not only to determine the net core-edge signal intensity but also to quantify the number of atoms of a particular element that is present in an image feature by using the theoretical scattering cross section for the core level excitation (Egerton, 1996; Isaacson and Johnson, 1975).

The number of atoms n_x of an element x in a pixel of dimension d can be determined from the net signal at energies above the core-edge $S_x(\Delta, \beta)$ measured in a band of width Δ defined by the energy-selecting slit and for a collection semiangle β, which is defined by the objective aperture.

$$n_x = \frac{S_x(\Delta, \beta) d^2}{I_0(\Delta, \beta) \cdot \sigma_x(\Delta, \beta)} \tag{1}$$

where $I_0(\Delta, \beta)$ is the measured intensity in the transmitted electron beam, including the zero-loss peak for the same energy band and objective aperture, and $\sigma_x(\Delta, \beta)$ is the inelastic scattering cross section for the core level excitation for the energy band Δ and collection semiangle β (Egerton, 1979).

Usually the number of atoms of element x to be analyzed is much smaller than the total number of atoms in the volume of the specimen corresponding to a single pixel in the image, so there is a strong background intensity underlying a relatively weak signal. Therefore, it becomes essential to subtract the background very carefully from the post-edge energy-selected image. Several methods have been employed to take account of the background intensity (Hofer *et al.*, 1997; Ottensmeyer, 1984; Thomas and Midgley, 2001). The simplest approach is ratio mapping, where the post-edge image intensity I_{post} is divided by the pre-edge intensity I_{pre}. If the shape of the background remains constant across the entire image, the ratio value k_x gives simple representation of the distribution of element x.

$$k_x = \frac{I_{\mathrm{post}}}{I_{\mathrm{pre}}} \tag{2}$$

It is useful, however, to obtain a modified ratio, k_x', by subtracting k_0, which corresponds to the measured ratio value in a region of the specimen where the element x is known to be absent.

$$k_x' = \frac{I_{\mathrm{post}}}{I_{\mathrm{pre}}} - k_0 \tag{3}$$

Multiplication of this equation by the pre-edge intensity gives the so-called "two-window" image intensity, which is an accurate measure of the core-edge signal, provided that the shape of the background remains constant over the field of interest.

$$S_x = I_{\mathrm{post}} - k_0 I_{\mathrm{pre}} \tag{4}$$

If the specimen thickness is uniform over the field of interest, the shape of the spectral background often remains sufficiently constant for two-window mapping to be valid. Two-window imaging is illustrated in Fig. 3, where phosphorus and nitrogen maps are acquired at low magnification from a thymocyte in an unstained, thin plastic section. These images were recorded using an FEI CM120 TEM equipped with a Gatan GIF100 post-column imaging filter. The schematic spectrum in Fig. 3A indicates the pre-edge and post-edge energy windows at 120 ± 10 and 152 ± 10 eV, respectively, which were selected to map phosphorus using the $L_{2,3}$ edge signal. The phosphorus $L_{2,3}$ core loss spectrum contains a weak peak at the edge threshold and a broad maximum at 152 eV. Approximately 20% of the core-edge intensity is included by selecting a post-edge window of 152 ± 10 eV, which avoids the overlapping sulfur $L_{2,3}$ edge at 165 eV. The pre-edge window is selected at 120 ± 10 eV, which is as close as possible to the phosphorus (P) $L_{2,3}$ edge. The pre-edge, post-edge, and net phosphorus images in Fig. 3B–D provide a good approximation of the distribution of DNA in the dense chromatin of the nucleus. While there are other phosphorylated macromolecules present, such as histones and other chromatin-associated proteins, their contributions to net phosphorous per unit volume is negligible, in comparison with that of DNA. By increasing the energy-loss offset, images were recorded below and above the nitrogen (N) K edge, at 380 ± 10 and 420 ± 10 eV (Fig. 3E

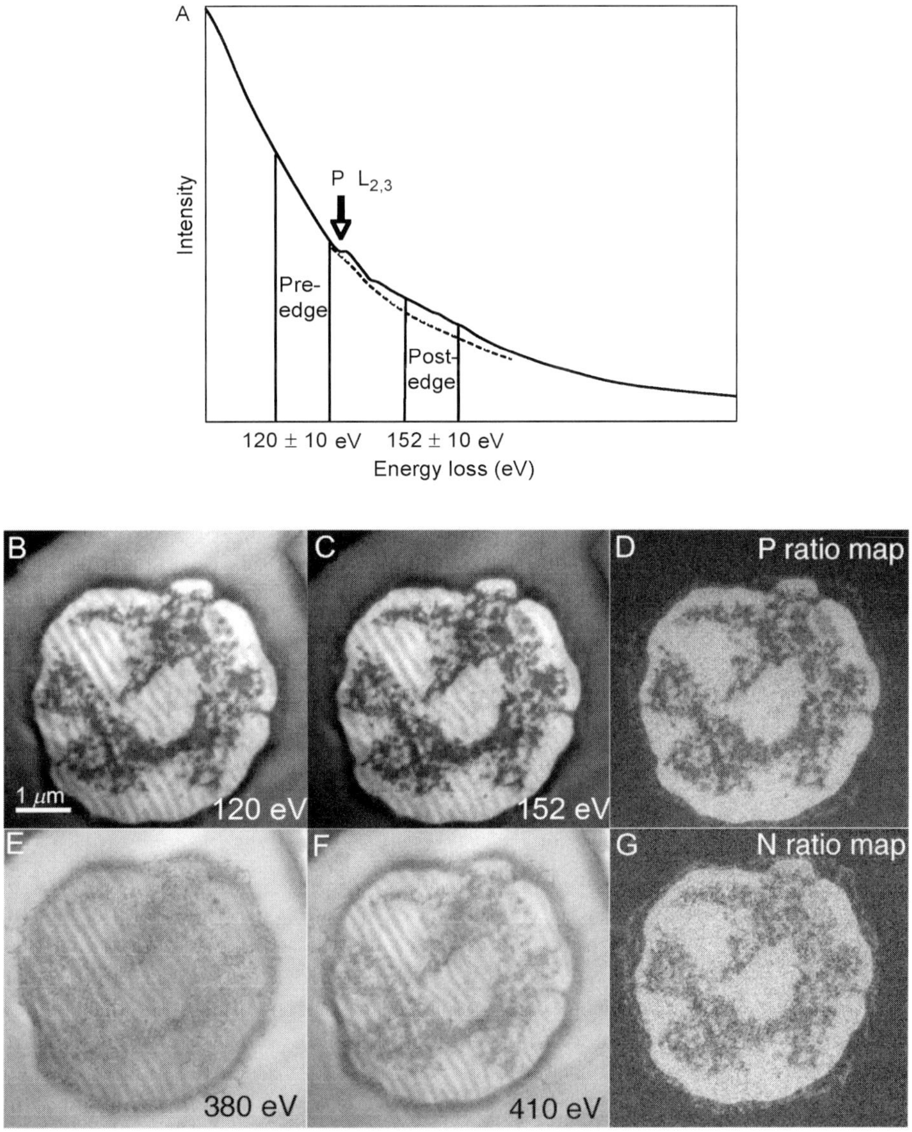

Fig. 3 Example of elemental mapping of an unstained plastic-embedded thin section of a thymocyte: (A) schematic diagram of EELS in the vicinity of the P $L_{2,3}$ edge indicating the pre-edge and post-edge energy windows; (B) pre-P $L_{2,3}$ edge image; (C) post-P $L_{2,3}$ edge image; (D) net phosphorus ratio map; (E) pre-N K edge image; (F) post-N K edge image; (G) net nitrogen ratio map. The strong signal from phosphorus and nitrogen reflects the dense nuclear chromatin, and subtle differences in contrast between the two images are attributed to variation in the amounts of proteins and nucleic acids.

and F, respectively) from the same cell; and the net two-window nitrogen map is shown in Fig. 3G. Although the nitrogen and phosphorus images appear similar, a close examination reveals the presence of additional features in the nitrogen map, corresponding to structures in the nucleus that contain protein but no DNA. Such images are useful in elucidating the organization of the cell nucleus and distinguishing between nucleic acid and protein complexes (Ren *et al.*, 2003).

B. Quantitative Analysis and Correction for Thickness Effects

When the specimen thickness or composition varies significantly over the field of interest, the two-window technique can become invalid. In previous work, this change of spectral shape has been taken into account by collecting additional pre-edge images at different energy losses. For example, in the three-window method, two images are collected below the core-edge in addition to one above (Leapman, 1986). By fitting the pre-edge background intensity as an inverse power function, its shape is allowed to vary across the image. However, for beam-sensitive biological specimens the electron dose is restricted. The three-window method therefore drastically increases the statistical noise in the extrapolated background, which results in a substantial reduction in analytical sensitivity. For this reason, it is preferable to retain a two-window approach but to modify it by taking into account subtle changes in the background shape as a function of the sample thickness t. In this case, it is convenient to measure the specimen thickness in units of the total mean free path λ_i for inelastic scattering, which can be readily determined from the fraction of transmitted electrons that are present in the zero-loss peak:

$$\frac{t}{\lambda_i} = \ln\left(\frac{I_{\text{tot}}}{I_0}\right) \tag{5}$$

where I_{tot} is the total unfiltered image intensity and I_0 is the zero-loss intensity (Egerton, 1996). The inelastic mean free path for plastic embedding material is typically 150–200 nm for a microscope accelerating voltage of 120 kV and 300–400 nm for a microscope operating at 300 kV. During irradiation, the embedding material tends to collapse, which can alter the effective inelastic mean free path. Electron doses for elemental mapping are typically $>10^6$–10^7 e/nm^2, which are much greater than the dose for mass loss due to radiolytic damage. This means that terminal mass loss has already occurred soon after acquisition is started. However, at even greater doses ($>10^8$ e/nm^2) and at higher accelerating voltages (e.g., 300 kV), displacement or "knock-on" damage becomes a relevant damage mechanism, which is caused by direct ejection of atoms from the specimen by rare high angle elastic scattering events (Isaacson, 1977). Under those extreme conditions, there might be some continuous mass loss during acquisition of elemental maps.

The variation in shape of the spectral background is determined by modeling the value of the ratio k_0 as a function of relative specimen thickness (Aronova *et al.*, 2006). First, the single scattering intensity is determined for pure embedding medium, for example, epon, by recording the energy-loss spectrum from a uniformly thin section of the material. The single scattering intensity $Q(E)$ is then obtained by using the logarithmic Fourier deconvolution method (Egerton, 1996; Leapman, 2004):

$$Q(E) = \left[\frac{t}{\lambda_i}\right]^{-1} FT^{-1}\{\ln[\tilde{I}(\tilde{E})] - \ln I_0\} \tag{6}$$

where $I(E)$ is the intensity of the energy loss spectrum except with the zero loss peak $I_0(E)$ replaced by a δ-function of magnitude equal to its integrated area I_0, and $\tilde{I}(\tilde{E})$ is the Fourier transform of the energy loss spectrum. The spectrum, with the addition of plural scattering, can then be modeled by the equation for $I(E)$ in terms of $Q(E)$ and t/λ_i:

$$I(E) = I_0 FT^{-1}\left\{\exp\left[\left(\frac{t}{\lambda_i}\right)\tilde{Q}(\tilde{E})\right]\right\} \tag{7}$$

where $\tilde{Q}(\tilde{E})$ is the Fourier transform of the single scattering distribution $Q(E)$. By computing the spectral intensity as a function of t/λ_i, it is then possible to calculate a k-curve (i.e., ratio of post-edge to pre-edge intensities, see Eq. 2), which can be used to determine a more accurate two-window map. Figure 4 shows the calculated k-curve for mapping phosphorus ($L_{2,3}$ edge at 132 eV) in a matrix of epon embedding resin. Post-edge and pre-edge windows were selected at 152 and 120 eV, respectively, each with a width of 20 eV. A fifth order polynomial function of t/λ_i was sufficient to fit through the k-curve to provide a set of parameters that can be used to adjust the k-value at each image pixel. The k-curve displays a minimum at a t/λ_i value of about 0.7, with k varying by around 10% for a typical specimen.

From this curve it is evident that artifacts can occur in two-window phosphorus maps calculated using a constant k_0-value. In such a specimen, it is possible to correct for thickness effects by recording unfiltered and zero-loss images, from which a t/λ_i map can be calculated using Eq. (5). It is then straightforward to compute a thickness-dependent k_0-map of the specimen from the parameterized k-curve together with the t/λ_i map, and the k_0-map can be used to generate a thickness-corrected two-window phosphorus map as given by:

$$S_x = I_{\text{post}} - k_0\left(\frac{t}{\lambda_i}\right)I_{\text{pre}} \tag{8}$$

Then combining Eq. (1) with Eq. (8), we can write the following expression for the number of atoms in a pixel of dimension d.

$$n_x = \left[\frac{I_{\text{post}}(\Delta,\beta) - k_0(t/\lambda_i)I_{\text{pre}}(\Delta,\beta)}{I_0(\Delta,\beta)\cdot\sigma_x(\Delta,\beta)}\right]d^2 \tag{9}$$

This ability to correct for thickness effects in elemental maps by using a t/λ_i dependent k-ratio enables two-window elemental mapping to be performed in specimens that are as thick as one inelastic mean free path, without artifacts due to plural inelastic scattering.

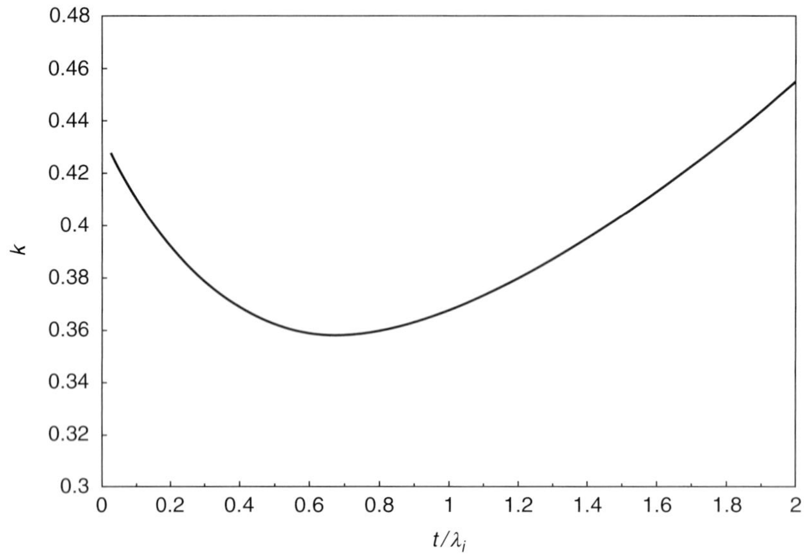

Fig. 4 Post-edge to pre-edge intensity ratio *k* for epon at the phosphorus edge as a function of relative specimen thickness, t/λ_i. Energies of pre-edge and post-edge windows are 120 and 152 eV, respectively, and the width of the energy-selecting slit is 20 eV. Variation in *k* is due to plural inelastic scattering, which can result in systematic errors in mapping phosphorus using the two-window method.

C. Detection Limits

To estimate the minimum detectable number of atoms in a given pixel of an elemental map, it is necessary to determine the variance of the signal due to statistical errors in the background. The estimated variance of the signal, var(S_x), is given by the sum of the variances of the post- and pre-edge intensities, var(I_{post}) and var(I_{pre}), respectively, scaled by a factor of k.

$$\begin{aligned} \mathrm{var}(S_x) &= \mathrm{var}(I_{post}) + \mathrm{var}(kI_{pre}) = \mathrm{var}(I_{post}) + k^2\ \mathrm{var}(I_{pre}) \\ \mathrm{var}(S_x) &\approx I_{post} + k^2 \frac{I_{post}}{k} \approx (1+k)I_{post} \end{aligned} \tag{10}$$

The statistical noise as represented by the standard deviation in the signal is then approximated by:

$$\mathrm{st\ dev}(S_x) \approx \sqrt{(1+k)I_{post}} \tag{11}$$

The noise is therefore determined largely by the intensity (electron counts) in the post-edge window, and from Eq. (1) the estimated standard deviation in the number of atoms within a given pixel of dimension d, that is, the detection limit, is given by:

$$\text{st dev}(n_x) \approx \frac{d^2\sqrt{(1+k)I_{\text{post}}}}{I_0(\Delta,\beta)\sigma_x(\Delta,\beta)} \tag{12}$$

For example, with a flux of 10^7 electrons accelerated to an energy of 300 keV into a nanometer-sized pixel and for a specimen of thickness $t/\lambda_i = 0.5$, the estimated noise in one pixel corresponds to about 60 phosphorus atoms. This noise level corresponds to a minimum detectable concentration of 0.5% (atomic) phosphorus in a carbon specimen of 100-nm thickness, which approaches a molar concentration value. It is possible to improve the sensitivity in terms of the minimum detectable number of phosphorus atoms per unit area by increasing the electron dose or by increasing the pixel size. Because the background intensity increases rapidly with thickness and the zero-loss intensity decreases, the minimum detectable number of atoms in an elemental map is much greater for thicker samples. Thus, Eq. (12) predicts that the minimum detectable number of phosphorus atoms increases by a factor of 6 when the specimen thickness increases from $t/\lambda_i = 0.5$ to $t/\lambda_i = 1.0$. The same spectrum modeling approach can be applied to predict the minimum detectable numbers of atoms of other elements in EFTEM mapping.

D. Mapping Elemental Distributions in 3D

By recording a tilt series of energy-filtered images at energy losses above and below the core-edge of an element and by using established methods for tomographic reconstruction (Frank, 1992), it is possible to compute the 3D distribution of that element in the specimen (Leapman *et al.*, 2004; Midgley and Weyland, 2003). Automated elemental tomography is more complicated to perform than standard tomography and some special procedures must be added to collect the data (Harmon *et al.*, 2005). It is often impractical to correct for spatial drift and focus simply by using the pre-edge and post-edge images because the gold fiducial particles deposited on the specimen surfaces for alignment might not be visible or the particles might undergo contrast reversal over the tilt series. To overcome this difficulty, a zero-loss image is recorded for each tilt angle. As indicated earlier, the zero-loss image is needed for several other purposes: to calculate the relative specimen thickness for plural scattering correction, to enable a quantitative determination of the number of atoms, and to enable the final tilt-series alignment in the tomographic reconstruction. In practice, a high-speed shutter must be employed to acquire the zero-loss image without saturating the CCD detector while maintaining the same illumination on the specimen that is used to collect the weak signal in the core-edge maps.

To perform elemental tomography, it is necessary for the acquisition computer to control the spectrum energy offset (via the microscope accelerating voltage) and the width of the energy-selecting slit. In addition, the acquisition system must control the tilt of the specimen goniometer and the excitation of the image

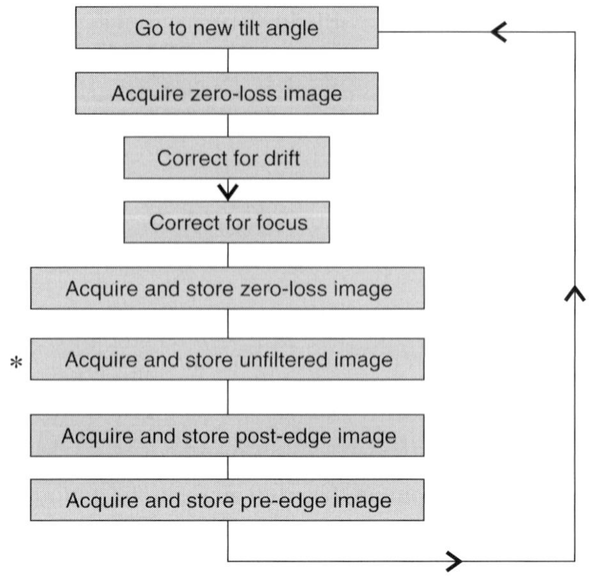

Fig. 5 Sequence for acquiring 3D elemental maps by EFTEM tomography using the two-window mapping method; drift and focus correction is performed using the zero-loss image. Acquisition of the unfiltered image (*) is only necessary when a variable *k*-factor is applied for thickness correction.

deflection coils and objective lens, just as in standard electron tomography. The acquisition sequence for automated elemental tomography in Fig. 5 illustrates the collection of zero-loss, post-edge, and pre-edge images from an unstained sectioned cell on which gold nanoparticles have been deposited for alignment during acquisition and subsequent 3D reconstruction. Since the pre-edge and post-edge images are collected immediately after acquisition of the corresponding zero-loss image (and after acquisition of the unfiltered image, if the thickness correction is needed), it can be assumed that each group of images is perfectly aligned. This type of acquisition sequence has been implemented in the energy-filtered tomography software designed by Gatan Inc. (Harmon *et al.*, 2005); examples of its application are illustrated below.

III. Applications

A. Visualizing Metals

There is considerable interest in being able to map metal clusters within the 3D volume of a cell. Such a capability provides information about the distributions of metalloproteins as well as the presence of nanoparticles that might be used to label specific biomolecules. A particularly important metalloprotein is ferritin, an iron storage protein that contains a dense 8-nm diameter core with up to 4000 Fe atoms surrounded by a thin protein shell called apoferritin, whose outer diameter is about 12 nm.

The capability of mapping the 3D distribution of iron in ferritin cores, is illustrated in Fig. 6, which shows electron tomographic data recorded from a plastic section with ferritin molecules deposited on both its top and bottom surfaces, along with 10-nm gold particles that serve as alignment markers. The data were acquired in an FEI Tecnai TF30 electron microscope operating at an accelerating voltage of 300 kV and equipped with a Gatan Tridiem imaging filter. Figure 6A and B display energy-selected images recorded in an energy window of 670 ± 7.5 eV below the iron (Fe) $L_{2,3}$ edge and in an energy window of 715 ± 7.5 eV above the Fe $L_{2,3}$ edge at 707 eV for a 0° specimen tilt. Although the total integration time was 16 sec per image, the signal was still relatively weak. This is evident from the calculated two-window map in Fig. 6C, which also shows that the bright features corresponding to the gold particles are removed by the background subtraction. The images were subjected to 4×4 binning to give an image size of 512×512 pixels with a pixel size of 2.7 nm. After reconstruction of 62 projections over an angular range of ±60° with 4° increments and two orthogonal tilt axes, the resulting iron distribution was projected back onto 0° to give the image in Fig. 6D. The significantly lower noise level illustrates the principle of dose fractionation as applied to elemental tomography. Dose fractionation is based on the theoretical prediction that the total electron dose required to visualize structural details close to the resolution limit in a 3D image is about the same as that for a 2D image, except that in 3D it is possible to separate the overlapping features. Therefore, it is possible to fractionate the dose between different projections of a 3D elemental distribution, yet build up a strong signal-to-noise ratio in the reconstructed volume (McEwen *et al.*, 1995). Figure 6E–G show three ortho-slices of the reconstructed volume of the specimen, through a layer of ferritin molecules situated on the top surface (Fig. 6E), through a plane in the middle of the plastic section (Fig. 6F), and through a layer of ferritin molecules situated on the bottom surface (Fig. 6G). The ferritin cores are visible on the surface layers, whereas they are absent in the slice through the center of the section. Intensity profiles across three ferritin cores are displayed as insets in Fig. 6E and G, which show that the signal is easily detectable above the noise level. The 3D volume was reconstructed using the IMOD software package from the University of Colorado (Kremer *et al.*, 1996; Marsh *et al.*, 2001; Mastronarde, 1997, 2005). The 3D reconstructions were visualized by volume rendering with software from Amira (Mercury Computer Systems, San Diego, CA). The 3D visualization in Fig. 6H again shows the layer of ferritin molecules on the specimen surfaces and reveals that the specimen has collapsed down to a thickness of about 70 nm under electron dose required to record the tilt series. In this specimen there was no heavy metal staining, so strong scattering from the dense cores would have made it possible to image the ferritin molecules directly by standard bright field tomography. However, in stained specimens, elemental tomography allows discrimination between iron in the ferritin core and other electron-dense regions of stain, which could contain uranium,

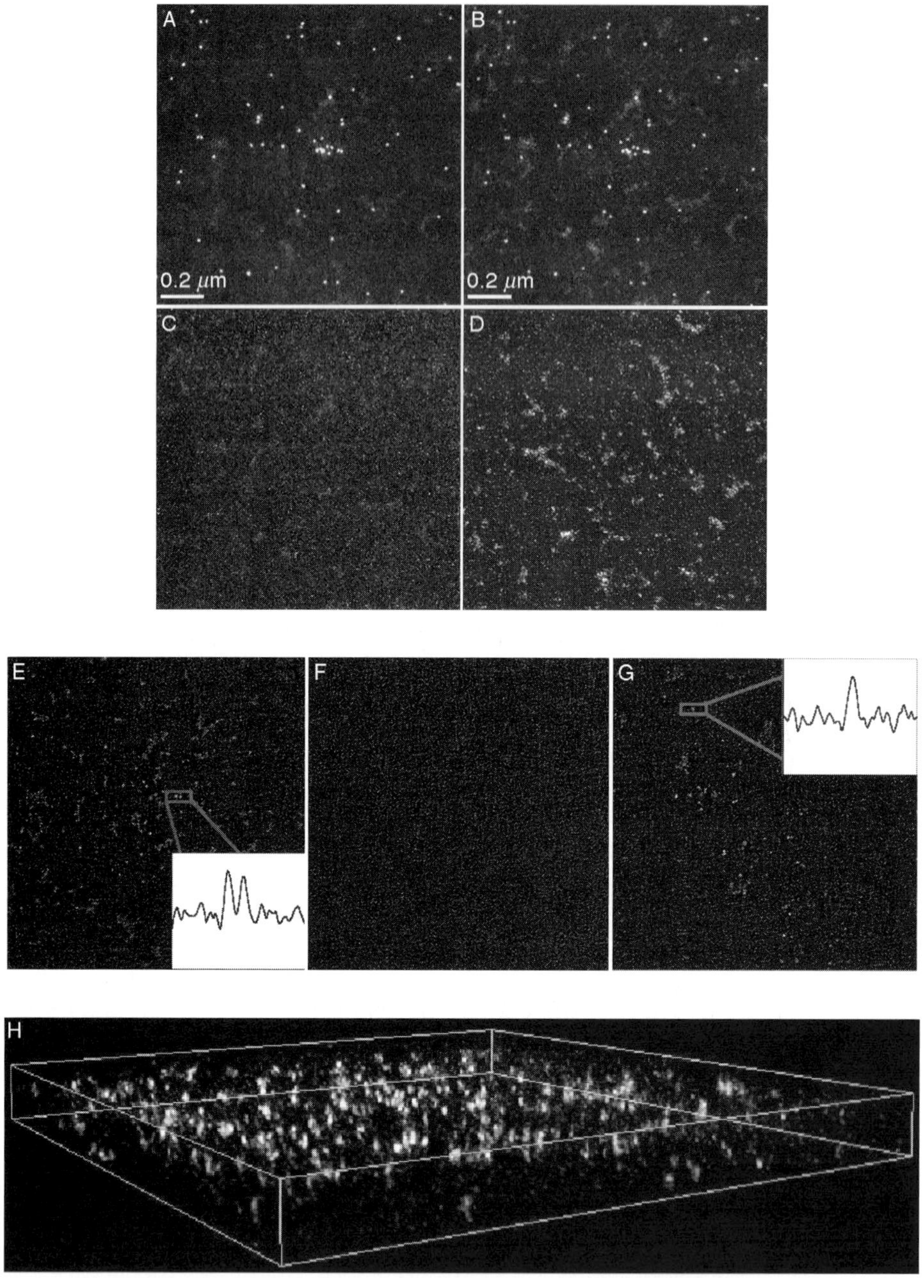

Fig. 6 EFTEM tomography of a 70-nm-thick plastic section on which bovine ferritin molecules have been deposited along with 10-nm diameter gold fiducial nanoparticles: (A) pre-Fe $L_{2,3}$ edge image at 0° tilt; (B) post-Fe $L_{2,3}$ edge image at 0° tilt; (C) net iron image at 0° tilt; (D) projection of the reconstructed tomogram onto the 0° tilt plane. It is evident that the signal-to-noise ratio is much higher

lead or other elements. Quantitative EFTEM mapping also has the advantage of providing an estimate for the number of iron atoms contained in each ferritin core.

A number of studies have described methods for determining the 2D distributions of ferritin molecules in various cell types. One application where information about the 3D distribution of ferritin has proved to be important is in the study of iron regulation in brain, where iron is needed for many important biochemical processes, but where the element needs to be tightly regulated to avoid neurotoxicity (Zhang *et al.*, 2005). The regulation of iron by ferritin can be affected in knockout mice when the expression of iron regulatory proteins (IRPs) is absent or reduced. To map the distribution of ferritin in neuronal tissue, it has proved useful to combine techniques: EELS analysis to identify iron in the ferritin cores, and conventional bright-field electron tomography to determine their 3D distribution. In fact, it is not necessary to measure the elemental signal in 3D, provided that the electron-dense nanoparticles can be identified in a 2D projection by their bright-field contrast produced by elastic scattering. Figure 7A shows a region from a plastic section of glutaraldehyde-fixed, osmicated, and lightly stained IRP-knockout mouse cerebellar tissue that contains myelinated axons and surrounding oligodendrocytes. The micrograph was recorded using a VG Microscopes HB501 scanning transmission electron microscope (STEM) equipped with a high-brightness field-emission source and operating at an accelerating voltage of 100 kV (Leapman, 2003). The image was obtained using a STEM annular dark-field detector, and the contrast was inverted to resemble a bright-field TEM image. Subsequently, iron maps were recorded with the STEM-EELS spectrum-imaging technique, which is complementary to EFTEM spectroscopic imaging (Hunt *et al.*, 1999; Jeanguillaume and Colliex, 1989). In STEM spectrum-imaging, a finely focused electron probe is digitally scanned across a specimen, and spectra are recorded at each pixel, which allows a more accurate quantitative analysis, but the number of pixels from which such data can practically be obtained is limited. In this case, elemental maps were obtained by using an electron spectrometer (Gatan Enfina EELS) interfaced to the STEM, and spectrum-images were acquired from small regions of cerebellum containing myelinated axons and surrounding oligodendrocytes. The spectrum in Fig. 7B was extracted from a single electron dense particle in a nonaxonal region of the tissue; the presence of a strong Fe $L_{2,3}$ signal identifies the particle as a ferritin molecule. Pre-edge, post-edge, and net iron maps extracted from the spectrum-image (Fig. 7C–E, respectively) show a cluster of ferritin molecules. Such images

in (D) than in (C), illustrating the principle of dose fractionation; slices through the 3D reconstruction parallel to the plane of the plastic section and situated (E) at the top surface layer, (F) at the center of specimen, and (G) at the bottom surface layer. Ferritin cores are visible only at the two surfaces; line density profiles across ferritin molecules in (E) and (F) show that the iron signal in the core is well above the noise level; (H) volume-rendered reconstruction of the specimen showing surface layers of ferritin molecules.

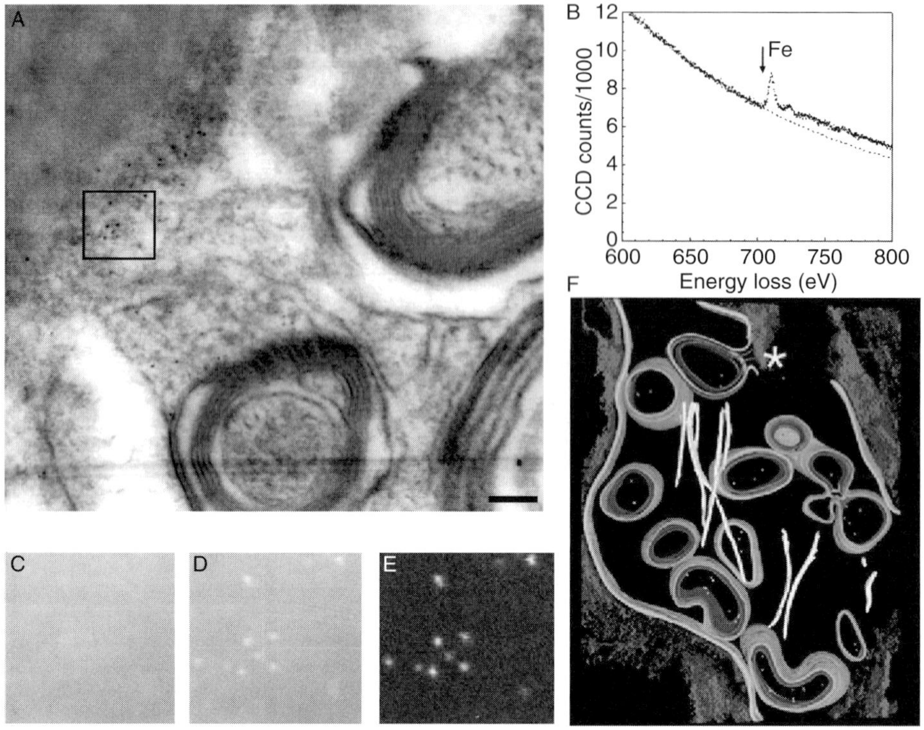

Fig. 7 Mapping the 3D distribution of ferritin in brain using a combination of 2D EELS imaging and 3D bright-field electron tomography. (A) Contrast-reversed dark-field STEM image of thin plastic section of fixed and stained cerebellum from an IRP knockout mouse; scale bar = 100 nm. (B) Spectrum extracted from STEM-EELS image in a 6-nm diameter region containing a single ferritin molecule. (C) Fe $L_{2,3}$ pre-edge image and (D) Fe $L_{2,3}$ post-edge image from boxed region in Fig. 7A. (E) Net Fe $L_{2,3}$ edge image after subtracting background by fitting to inverse power law at each pixel. (F) Reconstructed tomogram recorded at an accelerating voltage of 120 kV in an FEI CM120 TEM, showing volume-rendered visualization of another area of the section: axonal membrane is displayed in orange, oligodendrocyte membrane in blue, and invaginated double-walled vesicles containing oligodendrocytic cytoplasm and ferritin molecules in white; asterisk indicates double-walled vesicle emerging from junction of two membranes; full width of image is 1.5 μm. Reprinted from P. Zhang *et al.*, Electron tomography of degenerating neurons in mice with abnormal regulation of iron metabolism. *J. Struct. Biol.*, **150,** 144–153, Copyright 2005; with permission from Elsevier. (See Color Insert.)

reveal that more ferritin is present in oligodendrocytes than in the axons of the knockout mice. Bright-field electron tomograms from the same specimen indicate that myelinated axons can invaginate processes of oligodendrocytes containing ferritin, as suggested by the reconstruction in Fig. 7F. In 2D projections, it appears that ferritin is situated in the axons, whereas a 3D analysis of many such tomograms reveals that the ferritin-containing compartment is topologically separate from the axons.

B. Visualizing Nucleic Acid

Although structural biology of cellular proteins has progressed rapidly in recent years, the problem of determining the organization of chromatin in the cell nucleus has proved more difficult to address. Several models for the arrangement of DNA in chromatin have been proposed, including a hierarchical assembly of 10- and 30-nm fibrils, containing DNA wound around histones. It has been difficult to apply conventional methods of specimen preparation involving osmium tetroxide fixation and heavy metal staining, because the contrast is hard to interpret. For example, it may not be possible to distinguish between negative staining and positive staining of chromatin structures. Although cryo-electron microscopy does provide native high-resolution structure of cells at the macromolecular scale, it is difficult to apply this technique to determine chromatin structure in intact cells, in part due to the technical problem of cryosectioning eukaryotic cells. Since DNA contains one phosphate for every base in the structure, EFTEM imaging of phosphorus provides an intrinsic molecular label for nucleic acid, and by combining phosphorus imaging with nitrogen imaging, the technique also provides a means of distinguishing between DNA and protein complexes. It is not feasible to map the distribution of phosphorus in frozen cells, except at low spatial resolution, because of the high sensitivity of frozen-hydrated specimens to damage in the electron microscope. However, the EFTEM can be used to image phosphorus in cells or tissues that have been prepared by high-pressure freezing, freeze-substitution, embedding, and sectioning. In such preparations, it is important to avoid heavy metal fixatives and stains that could interfere with the P $L_{2,3}$ edge in the energy loss spectrum.

Until recently, EFTEM phosphorus mapping of cells in unstained plastic sections has involved the collection of 2D maps. By applying energy-filtered tomography, it is now possible to obtain 3D phosphorus distributions of cellular components such as ribosomes (Leapman *et al.*, 2004). The complicated overlapping structure of chromatin in 2D projections of a cell also makes EFTEM an ideal approach for studying the organization of the nucleus. The feasibility of this approach is demonstrated by mapping the 3D distribution of DNA in a sectioned nucleus of a *Drosophila* larva prepared by freezing in a Baltec HPM10 high-pressure freezing machine. The frozen larvae were freeze-substituted in 0.2% glutaraldehyde in acetone at a temperature of $-90\,^{\circ}$C for 3 days; they were then warmed to $20\,^{\circ}$C, rinsed several times in acetone, and infiltrated with Epon-Araldite and polymerized for 2 days at $60\,^{\circ}$C. Sections were cut to a thickness of about 100 nm and were picked up on 400 mesh copper grids. EFTEM tilt series were acquired at energy losses of 0, 120, and 152 eV for a tilt angular range of $\pm 66^{\circ}$ with increments of 3°. The specimen grid was rotated by 90° and a second orthogonal tilt series was recorded with the same parameters as for the first series. The total electron dose was $\sim 10^7$ e/nm^2 for the dual-axis series. Images were binned by a factor of 4 to give 512×512 pixels with an acquisition time of 0.1 sec for the zero-loss image and 4 sec for the pre-edge and post-edge images, and with four readouts for each acquisition. Each voxel in the reconstruction was 2.7 nm on a side.

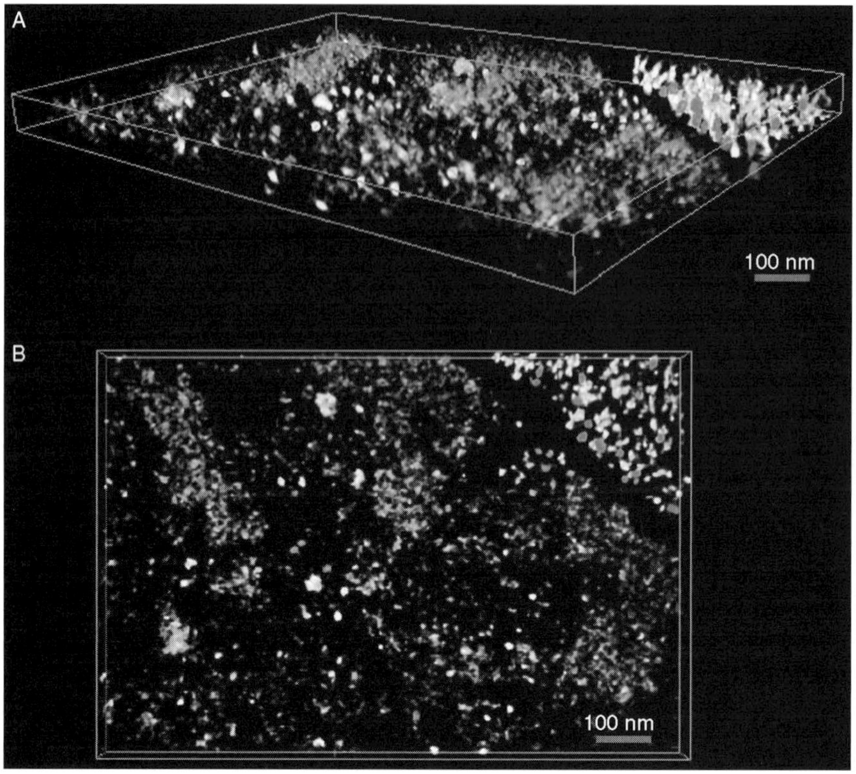

Fig. 8 3D phosphorus distribution in unstained plastic section of high-pressure frozen and freeze-substituted *Drosophila* larva, showing part of cell nucleus and a region of cytoplasm containing ribosomes (top right); the phosphorus distribution reflects the distribution of nucleic acid; (A) side view and (B) top view. Ribosomes colored green were used to quantify the phosphorus concentration in the nucleus, since ribosomes are known to contain about 7000 P atoms in their RNA. Another series phosphorus-containing dense particles of unknown origin are present in the nucleus. Short segments of fibrils occur throughout the cytoplasm are attributed to DNA. (See Color Insert.)

The tilt series were aligned using the IMOD program by tracking the fiducial gold particles in the zero-loss image and the 3D volume was reconstructed. Figure 8A and B are two views of the volume-rendered phosphorus distribution of a nucleus in a sectioned cell. The chromatin appears as a dense array of fiber segments, which reflects the distribution of DNA. Relatively clear spaces are also visible, which could correspond to a sparser density of DNA in regions of euchromatin or to the location of large protein complexes that do not contain high concentrations of phosphorus. Dense particles containing several thousand phosphorus atoms are also visible, and these are colored magenta in the top view in Fig. 8B. A region of cytoplasm is evident at the top right corners of Fig. 8A and B, where a clear space separates the boundary of the nucleus from a cluster of dense phosphorus containing particles. These particles, some of which are colored green in Fig. 8A and B, can be readily identified as ribosomes, each of which contains about 7000 phosphorus atoms in its RNA.

Phosphorus distributions such as the one in Fig. 8, in combination with structural data obtained using other methods, might eventually be able to provide information about the organization of DNA in the cell nucleus. Although spatial resolution will be limited by sample preparation and beam damage, the EFTEM approach is quantitative and could enable measurement of phosphorus content per unit volume, and hence the variation in packing density of DNA within the nucleus.

IV. Summary and Future Directions

Quantitative 3D imaging of elemental distributions in the EFTEM provides a new approach for localizing specific components in sectioned freeze-substituted cells. It is anticipated that EFTEM tomography will provide information about a number of important elements that are intrinsic to cells. 3D phosphorus distributions indicate the organization of nucleic acids and other phosphate-containing molecules; sulfur reveals the distribution of proteins that have high concentrations of cysteine or methionine residues; nitrogen reflects the distribution of proteins as well as nucleic acids; and metals such as iron indicate the location of metalloproteins. For example, it has been demonstrated that single molecules of ferritin containing between 2000 and 4000 Fe atoms can be visualized in 3D in specimens that are around 100-nm thick. As EFTEM techniques and reconstruction methods improve, it is likely that the minimum detectable number of atoms will approach theoretical limits (Egerton, 1996; Isaacson and Johnson, 1975; Leapman, 2004). EFTEM can not only detect intrinsic concentrations of biological elements, it is also a useful technique for imaging molecular labels, such as nanoparticles containing synthetic metal clusters that bind to specific cellular proteins. In particular, EFTEM imaging can be used to image the distribution of novel labels synthesized for correlative optical and electron microscopy. For example, quantum dots that provide a strong fluorescent signal in the optical microscope contain metals like cadmium that can be imaged by EFTEM (Dellaire *et al.*, 2004; Nisman *et al.*, 2004). In the future, it would be useful to explore other types of nanoparticle labels containing transition elements or rare earths that provide signals of suitable strength for detection in the EFTEM.

Acknowledgments

The authors thank Dr. Robin Harmon, Dr. Steffen Mayer, and Dr. John Hunt of Gatan Inc. for providing the energy filtered tomography program in the Gatan Microscopy Suite 1.5 software. They thank Lori Dorward for help with the Amira software, Dr. Guofeng Zhang for providing the specimen of *Drosophila* larvae, and Dr. Michael Kruhlak for providing the thyomocyte sample. This research was supported by the Intramural Research Program of the NIH.

References

Aronova, M. A., Kim, Y. C., Zhang, G., and Leapman, R. D. (2006). Quantification and thickness correction of EFTEM phosphorus maps. *Ultramicroscopy* (in press).

Bazett-Jones, D. P., Leblanc, B., Herfort, M., and Moss, T. (1994). Short-range DNA looping by the *Xenopus* HMG-box transcription factor, xUBF. *Science* **264,** 1134–1137.

Bazett-Jones, D. P., Mendez, E., Czarnota, G. J., Ottensmeyer, F. P., and Allfrey, V. G. (1996). Visualization and analysis of unfolded nucleosomes associated with transcribing chromatin. *Nucleic Acids Res.* **24,** 321–329.

Beck, M., Förster, F., Ecke, M., Plitzko, J. M., Melchior, F., Gerisch, G., Baumeister, W., and Medalia, O. (2004). Nuclear pore complex structure and dynamics revealed by cryoelectron tomography. *Science* **306,** 1387–1390.

Dellaire, G., Nisman, R., and Bazett-Jones, D. P. (2004). Correlative light and electron spectroscopic imaging of chromatin *in situ*. *Methods Enzymol.* **375,** 456–478.

Egerton, R. F. (1979). K-shell ionization cross-sections for use in microanalysis. *Ultramicroscopy* **4,** 169–179.

Egerton, R. F. (1996). "Electron Energy Loss Spectroscopy," 2nd Edn. Plenum, New York.

Egerton, R. F. (2003). New techniques in electron energy-loss spectroscopy and energy-filtered imaging. *Micron* **34,** 127–139.

Frank, J. (1992). "Electron Tomography: Three-Dimensional Imaging with the Transmission Electron Microscope." Plenum, New York.

Goping, G., Pollard, H. B., Srivastava, M., and Leapman, R. (2003). Mapping protein expression in mouse pancreatic islets by immunolabeling and electron energy loss spectrum-imaging. *Microsc. Res. Tech.* **61,** 448–456.

Harmon, R. T., Meyer, S., and Hunt, J. A. (2005). "Software for EFTEM Tomography Incorporated into Digital Micrograph/GMS." Gatan, Inc., Pleasanton, CA.

Hofer, F., Grogger, W., Kothleitner, G., and Warbichler, P. (1997). Quantitative analysis of EFTEM elemental distribution images. *Ultramicroscopy* **67,** 83–103.

Hofer, F., and Warbichler, P. (2004). Elemental mapping using energy filtered imaging. *In* "Transmission Electron Energy Loss Spectrometry in Materials Science and the EELS Atlas" (C. Ahn, ed.), 2nd Edn., Chapter 6, pp. 159–222. Wiley-VCH, Berlin.

Hsieh, C., Leith, A., Mannella, C. A., Frank, J., and Marko, M. (2006). Towards high-resolution three-dimensional imaging of native mammalian tissue: Electron tomography of frozen-hydrated rat liver sections. *J. Struct. Biol.* **153,** 1–13.

Hunt, J. A., Kothleitner, G., and Harmon, R. (1999). Comparison of STEM EELS spectrum imaging vs. EFTEM spectrum imaging. *Microsc. Microanal.* **5**(Suppl. 2), 616–617.

Isaacson, M., and Johnson, D. (1975). Low Z elemental analysis using energy loss electrons. *Ultramicroscopy* **1,** 33–52.

Isaacson, M. S. (1977). Specimen damage in the electron microscope. *In* "Principles and Techniques of Electron Microscopy VII" (M. A. Hayat, ed.), pp. 1–78. Van Nostrand Reinhold, New York.

Jeanguillaume, C., and Colliex, C. (1989). Spectrum-imaging: The next step in EELS digital acquisition and processing. *Ultramicroscopy* **28,** 252–257.

Koster, A. J., Grimm, R., Typke, D., Hegerl, R., Stoschek, A., Walz, J., and Baumeister, W. (1997). Perspectives of molecular and cellular electron tomography. *J. Struct. Biol.* **120,** 276–308.

Kremer, J. R., Mastronarde, D. N., and McIntosh, J. R. (1996). Computer visualization of three-dimensional image data using IMOD. *J. Struct. Biol.* **116,** 71–76.

Krivanek, O. L., Gubbens, A. J., Dellby, N., and Meyer, C. E. (1991). Design and first applications of a post-column imaging filter. *Microsc. Microanal. Microstr.* **3,** 187–199.

Krivanek, O., Friedman, S., Gubbens, A., and Kraus, B. (1995). An imaging filter for biological applications. *Ultramicroscopy* **59,** 267–282.

Lanio, S. (1986). High-resolution imaging magnetic energy filters with simple structure. *Optik* **73,** 56–68.

Leapman, R. D. (1986). STEM elemental mapping by electron energy loss spectroscopy. *Ann. N. Y. Acad. Sci.* **483,** 326–338.

Leapman, R. D. (2003). Detecting single atoms of calcium and iron in biological structures by electron energy-loss spectrum-imaging. *J. Microsc.* **210,** 5–15.

Leapman, R. D. (2004). EELS quantitative analysis. *In* "Transmission Electron Energy Loss Spectrometry in Materials Science and the EELS Atlas" (C. Ahn, ed.), 2nd Edn., Chapter 3, pp. 49–94. Wiley-VCH, Berlin.

Leapman, R. D., Jarnik, M., and Steven, A. C. (1997). Spatial distributions of sulfur-rich proteins in cornifying epithelia. *J. Struct. Biol.* **120,** 168–179.

Leapman, R. D., Kocsis, E., Zhang, G., Talbot, T. L., and Laquerriere, P. (2004). Three-dimensional distributions of elements in biological samples by energy-filtered electron tomography. *Ultramicroscopy* **100,** 115–125.

Leapman, R. D., and Ornberg, R. L. (1988). Quantitative electron energy loss spectroscopy in biology. *Ultramicroscopy* **24,** 251–268.

Leapman, R. D., and Rizzo, N. W. (1999). Towards single atom analysis of biological structures. *Ultramicroscopy* **78,** 251–268.

McEwen, B. F., Downing, K. H., and Glaeser, R. M. (1995). The relevance of dose fractionation in tomography of radiation-sensitive specimens. *Ultramicroscopy* **60,** 357–373.

Marsh, B. J., Mastronarde, D. N., Buttle, K. F., Howell, K. E., and McIntosh, J. R. (2001). Organellar relationships in the Golgi region of the pancreatic beta cell line, HIT-T15, visualized by high resolution electron tomography. *Proc. Natl. Acad. Sci. USA* **98,** 2399–2406.

Mastronarde, D. N. (1997). Dual axis tomography: An approach with alignment methods that preserve resolution. *J. Struct. Biol.* **120,** 343–352.

Mastronarde, D. N. (2005). Automated electron microscope tomography using robust prediction of specimen movements. *J. Struct. Biol.* **152,** 36–51.

Midgley, P. A., and Weyland, M. (2003). 3D electron microscopy in the physical sciences: The development of Z-contrast and EFTEM tomography. *Ultramicroscopy* **96,** 413–431.

Müller-Reichert, T., Hohenberg, H., O'Toole, E. T., and McDonald, K. (2003). Cryoimmobilization and three-dimensional visualization of *C. elegans* ultrastructure. *J. Microsc.* **212,** 71–80.

Nicastro, D., McIntosh, J. R., and Baumeister, W. (2005). 3D structure of eukaryotic flagella in a quiescent state revealed by cryo-electron tomography. *Proc. Natl. Acad. Sci. USA* **102,** 15889–15894.

Nisman, R., Dellaire, G., Ren, Y., Li, R., and Bazett-Jones, D. P. (2004). Application of quantum dots as probes for correlative fluorescence, conventional, and energy-filtered transmission electron microscopy. *J. Histochem. Cytochem.* **52,** 13–18.

Ottensmeyer, F. P. (1984). Electron spectroscopic imaging: Parallel energy filtering and microanalysis in the fixed-beam electron microscope. *J. Ultrastruct. Res.* **88,** 121–134.

Reimer, L. (1995). "Energy-Filtering Transmission Electron Microscopy." Springer, Berlin.

Ren, Y., Kruhlak, M. J., and Bazett-Jones, D. P. (2003). Same serial section correlative light and energy-filtered transmission electron microscopy. *J. Histochem. Cytochem.* **51,** 605–612.

Somlyo, A. P., Somlyo, A. V., and Shuman, H. (1979). Electron probe analysis of vascular smooth muscle. *J. Cell Biol.* **81,** 316–335.

Steven, A. C., and Aebi, U. (2003). The next ice age: Cryo-electron tomography of intact cells. *Trends Cell Biol.* **13,** 107–110.

Thomas, P. J., and Midgley, P. A. (2001). Image-spectroscopy–I. The advantages of increased spectral information for compositional EFTEM analysis; and Image-spectroscopy–II. The removal of plural scattering from extended energy-filtered series by Fourier deconvolution. *Ultramicroscopy* **88,** 179–194.

Zhang, P., Bos, E., Heymann, J., Gnaegi, H., Kessel, M., Peters, P. J., and Subramaniam, S. (2004). Direct visualization of receptor arrays in frozen-hydrated sections and plunge-frozen specimens of E. coli engineered to overproduce the chemotaxis receptor Tsr. *J. Microsc.* **216,** 76–83.

Zhang, P., Land, W., Lee, S., Juliani, J., Lefman, J., Smith, S. R., Germain, D., Kessel, M., Leapman, R., Rouault, T. A., and Subramaniam, S. (2005). Electron tomography of degenerating neurons in mice with abnormal regulation of iron metabolism. *J. Struct. Biol.* **150,** 144–153.

CHAPTER 25

Localization of Protein Complexes by Pattern Recognition

Christoph Best, Stephan Nickell, and Wolfgang Baumeister

Max Planck Institute of Biochemistry
Am Klopferspitz 18, D-82152 Martinsried, Germany

As cryo-electron tomography of whole cells or cell sections approaches molecular resolution, it becomes feasible to locate protein complexes in cells by their distinctive three-dimensional (3D) structures alone, without the need of markers. This opens the way to creating a molecular map of the cellular proteome and its interactions under near-life conditions. However, the process requires sophisticated computational methods to create large libraries of molecular templates and search for them in cellular tomograms, taking into account the low signal-to-noise ratio and limited resolution of cryo-electron tomography (cryo-ET), as well as the problem of the missing wedge.

0091-679X/07 $35.00
DOI: 10.1016/S0091-679X(06)79025-2

I. Introduction

Cryo-electron tomography (cryo-ET) is unique among biological imaging methods in being able to produce three-dimensional (3D) representations of organelles and cells in near-life conditions with a resolution high enough to identify and locate individual macromolecules. This opens the door to the creation of molecular maps of cellular proteomes (Baumeister, 2004; Nickell *et al.*, 2006). It is now widely recognized that cells are not just membrane-bound reaction compartments in which macromolecules diffuse freely and interact randomly. Although there is a strong element of stochasticity, macromolecules must form functional modules, at least transiently, to execute cellular processes. Thus, a molecular map of a cell would provide significant insights into the functional associations between macromolecular complexes that make up the operating modules of the cellular "factory."

Drawing such a molecular map must rely on the ability to localize protein complexes systematically and objectively. One could imagine introducing labels that are specific for each macromolecule of interest, but it is a significant challenge to get markers that are visible by electron microscopy (EM) into cells without perturbing them. We are therefore pursuing an alternative approach. Cryo-ET of cells combines an optimal preservation of the sample (see Chapter 1 by Dubochet, this volume) with 3D imaging at molecular resolution (<5 nm), and it provides faithful representations of the density landscapes inside cells. However, interpretation of the tomograms is hampered by the low signal-to-noise ratio, which necessitates the use of powerful statistical pattern recognition techniques to map objects of interest in the images. Template matching, in which the content of the sample volume is compared to templates representing the molecules under consideration, has shown great promise to achieve this goal.

II. Template Matching

A. Template Libraries for Visual Proteomics

The database of high-resolution protein structures has already grown to an impressive size in recent years, and structural genomics initiatives have accelerated the pace with which new structures are added to it. We may anticipate that within the next decade this database will become eventually complete, that is, it will contain structures representing most of the proteomic inventory of cells. On the other hand, our knowledge of how, when, and where these proteins interact to perform their functions is limited (for review see Aloy and Russell, 2006 and Sali *et al.*, 2003). Genome-scale interaction networks that have been more or less successfully generated by techniques such as the yeast-two-hybrid system tell mainly about possible physical interactions between two proteins; less is known about whether these interactions actually take place and have a functional role in living cells. Characterization of such interactions is a challenge for classical biochemical techniques, which rely on isolating molecules from within the cell and characterizing them in isolation. EM, on the other hand, has the potential to provide a whole-cell picture: a high-resolution cryo-electron tomogram is essentially a picture of the cell's entire proteome, and it should, in principle, allow us to map macromolecular interaction patterns in a fairly comprehensive manner ("visual proteomics," Nickell *et al.*, 2006; see Lucic *et al.*, 2005, for review).

The biggest challenge in achieving this goal, besides the sheer size of the data sets and the high level of noise within them, is that EM cannot take advantage of the labeling techniques that allow easy discrimination between different macromolecules. In light microscopy, fluorescence labeling has provided a powerful route to molecular identification and localization, thanks in particular to the construction and expression of chimeras with fluorescent proteins. EM currently lacks effective clonable labels, though efforts have been undertaken to label specific proteins with gold nanoparticles that are clearly visible in the images. Unfortunately, these techniques are limited, both in their effectiveness for cellular work and in their usefulness for proteomic applications. Not only is it not simple to insert such labels into a molecularly crowded cell without running the risk of introducing artifacts, nanoparticle labeling usually is restricted to one or maybe a few molecular species at a time.

Computational identification of macromolecules in cellular tomograms based on template matching is an alternative approach that aims at identifying macromolecules solely by their inherent structural features as seen in their 3D density distribution.

Central to this approach is the availability of 3D structures to be used as templates. Ideally, a template library would contain structures from the entire proteomic inventory of the cellular system under scrutiny. If high-resolution structures from x-ray crystallography or NMR spectroscopy are not available, as is often the case for membrane proteins or large and labile complexes, medium-resolution (1–2 nm) structures can also be used as templates. EM *single-particle analysis* (Frank, 1996) and hybrid methods that combine high-resolution structures of components (subunits or domains) with the lower-resolution structures of large complexes (Baumeister and Steven, 2000) will undoubtedly contribute to the template libraries in a major way. Currently a slow and specialized craft, EM single-particle analysis is expected to become a high-throughput method in the near future, thanks to automated data acquisition and analysis.

Once the challenge of creating a comprehensive template library is met, it can be used to create a proteomic map of a cell through the comparison of the densities observed in a tomogram to the templates in the library. In practice, several technical problems must be overcome for such an approach to be successful. First, the technique is based on the proposition that different macromolecules appear sufficiently different in electron tomograms that they can be reliably distinguished from each other. It also requires tomograms of cells that have a sufficiently high resolution and acceptable signal-to-noise ratios to allow meaningful comparisons with the template library to be made. Since there are limits as to how far these issues can be pushed with current and near-future electron microscopes, improvements in the computational processing are needed that allow one to identify template patterns even in the presence of relatively high noise and at borderline resolutions. Luckily, similar problems of pattern recognition in the presence of high noise have been encountered in a variety of fields from computer vision through remote sensing to bioinformatics. Progress in these fields has led to the development of advanced pattern recognition algorithms that are based on statistical learning techniques.

Another problem is the limitation in specimen thickness, which precludes tomographic imaging of eukaryotic cells in their entirety. Here cryosectioning techniques that physically cut a frozen-hydrated (vitrified) cell into slices of about 400-nm thickness show considerable promise (see Chapter 15 by Dubochet *et al.*, this volume). Tomograms of these slices could then in principle be computationally reassembled into a complete image of the original cell.

Finally, imaging and analyzing single cells is not enough to make general statements about the proteome. Sufficient statistics about the molecular distribution in single cells will have to be achieved. However, advances in automated EM will facilitate the acquisition of large numbers of images and tomograms in the near future.

Taken together, these techniques have the potential to make cryo-EM into a valuable proteomics tool with the unique capability of providing insight into the actual molecular interaction network in living cells.

B. Principles of Machine Learning and Pattern Recognition

In the theory of machine learning (Duda *et al.*, 2001; Hastie *et al.*, 2003), the term "pattern recognition" loosely refers to techniques for identifying repetitive features in a set of objects and for designing machines (i.e., computer programs) that are capable of recognizing these features. One distinguishes here between the problem of *unsupervised learning*, in which the feature is not known a priori and has to be identified by the program, and *supervised learning*, in which the feature is known and can be given to the program as an example. This general theory is agnostic about what constitutes a feature, and it can be applied equally well to computer vision as well as to mining credit card data.

When trying to identify macromolecules in ET, we can generally assume that we know the shape of the molecule and can therefore produce a "template." The problem of pattern recognition then reduces to finding occurrences of this template in the sample volume, while accounting for noise and imaging artifacts. Machine learning theory offers two alternative approaches to this problem: in the probabilistic approach, physical knowledge of the imaging process is used to produce a noise model of the image, which can then be used to calculate the probability that a subvolume arises from the template under the influence of the noise. While this is the classical approach used in statistics, it depends on a good characterization of the noise and it is often only feasible for specialized noise distributions such as the Gaussian or the Poisson distribution.

A more pragmatic alternative is the pure machine learning approach in which one seeks a mathematical function, the *discriminant*, that allows empirical discrimination between the occurrence and nonoccurrence of the template. This approach makes use of two training sets of images, one representing the object that is sought and another representing observations of noise or other objects. A function is then devised pragmatically that has a large value on the positive set and a small value on the negative set; thus it serves to discriminate between the two sets. This approach has been shown to be very effective for many pattern recognition problems.

An alternative to template matching is often used in segmentation problems. Here the focus is not on the detection of the overall shape of a known object, but on the recognition of typical features associated with macromolecules like their surface curvatures and sizes. Methods such as anisotropic diffusion, mean-curvature motion (MCM; Böhm *et al.*, 2000), and scaling indices (Ortiz *et al.*, 2006), which are ordinarily used for segmentation, can be tuned to detect subvolumes that contain macromolecular complexes. This approach is often useful for the initial identification of target volumes. It can in principle be combined with unsupervised learning methods (clustering) to identify macromolecular complexes without having prior knowledge of their structure.

C. Templates

A *template* is an idealized image of the macromolecular complex under consideration, scaled to the resolution of the target volume and appropriately filtered to account for the physical imaging process. It is usually taken from an atomic-scale model of the complex, derived either by x-ray or electron crystallography, by single-particle EM, or from any combination of these methods. To construct the idealized image of the complex, as it would appear in an electron microscope, the electrostatic potential is calculated approximately from an atomic map by summing the total atomic numbers Z in each volume element. At the resolution afforded by current EM, this approximation is reasonable. The resulting density is then convoluted with the appropriate contrast transfer function of the microscope and low-pass filtered.

The resulting template is a 3D volume V_{templ}, typically with a linear extension with size on the order of 10^2 volume elements (*voxels*). To identify occurrences of the template in a larger sample volume V, with linear extensions around 10^3 voxels, the template is compared to all possible subvolumes having the size of the template. Since the rotational orientation of the particles is also unknown, all possible orientations of the template have to be considered. The comparison itself has to take into account that the absolute magnitude (the *scale*) and the zero offset (the *bias*) of the voxel values are often unknown, as they depend on the details of the image acquisition process and the postprocessing. Note that when we refer to "scale" here, it is in reference to the magnitude of the individual voxel *values*, not the 3D size of the template.

Mathematically, this comparison can be expressed by the approximate equality:

$$f(x) \sim \alpha(T_y R_{\phi,\theta,\psi} c)(x) + b \qquad \text{for } x \in V_y \tag{1}$$

Here $c(x)$ is the template function, $R_{\phi,\theta,\psi}$ is the rotation operator with Euler angles ϕ, θ, and ψ, T_y indicates a translation of the template to the position y, and α and b are the scale and bias of the voxel values as mentioned above. Equation (1) expresses that a translated and rotated copy of the template, after approximate rescaling of the voxel values, should be (approximately) identical to the subvolume of the sample volume under consideration (Fig. 1). This is the basic operation in template matching. In the following we will, in the guise of different correlation functions, lay out the mathematical procedure to efficiently test for this approximate equality in large sample volumes.

The comparison is performed at all elements of the subvolume V_y, which has the size and shape of the template and is located at the reference point y. The parameters $y, \phi, \theta, \psi, \alpha$, and b are initially unknown. The correct values for some of them, in particular the scale α and bias b, can be intelligently estimated from the characteristics of the sample volume or calculated efficiently by special transformations, such as the offset y. However, at least the rotation parameters ϕ, θ, and ψ are usually determined by complete enumeration, that is, by considering all independent combinations of the Euler angles, typically in increments of 5° or 10°, looking for the orientation that provides the best match. This is often followed by a refinement step in which only angles immediately around the coarsely determined maximum are scanned. Even so, on the order of 10^4 different rotations of the template must be compared to each target volume.

The idea of template matching using the correlation function (Frank, 1972) occurred initially in EM in relation to the problem of picking particles (Frank and Wagenknecht, 1984; Huang and Penczek, 2004) for single-particle analysis (see Frank, 2002 for a review on single-particle methods, and Zhu *et al.*, 2004 for a review of current particle picking methods). Its application to 3D objects was introduced by Walz *et al.* (1997), as a tool to align manually selected particles for averaging, and by Volkmann and Hanein (1999), and Roseman (2000), for the problem of docking high-resolution crystallographic structures into low-resolution EM density maps. Böhm *et al.* (2000) first made use of a combination of segmentation and cross-correlation techniques to localize macromolecules in tomograms.

D. The Cross-Correlation Function

The most popular discriminant function for identifying the occurrence of templates in subvolumes is the cross-correlation function. It is defined as the voxel-wise sum of the products of the translated template and the subvolume:

$$\mathrm{CCF}[f, c](y) = \sum_{x \in V_y} f(x) c(x - y) \tag{2}$$

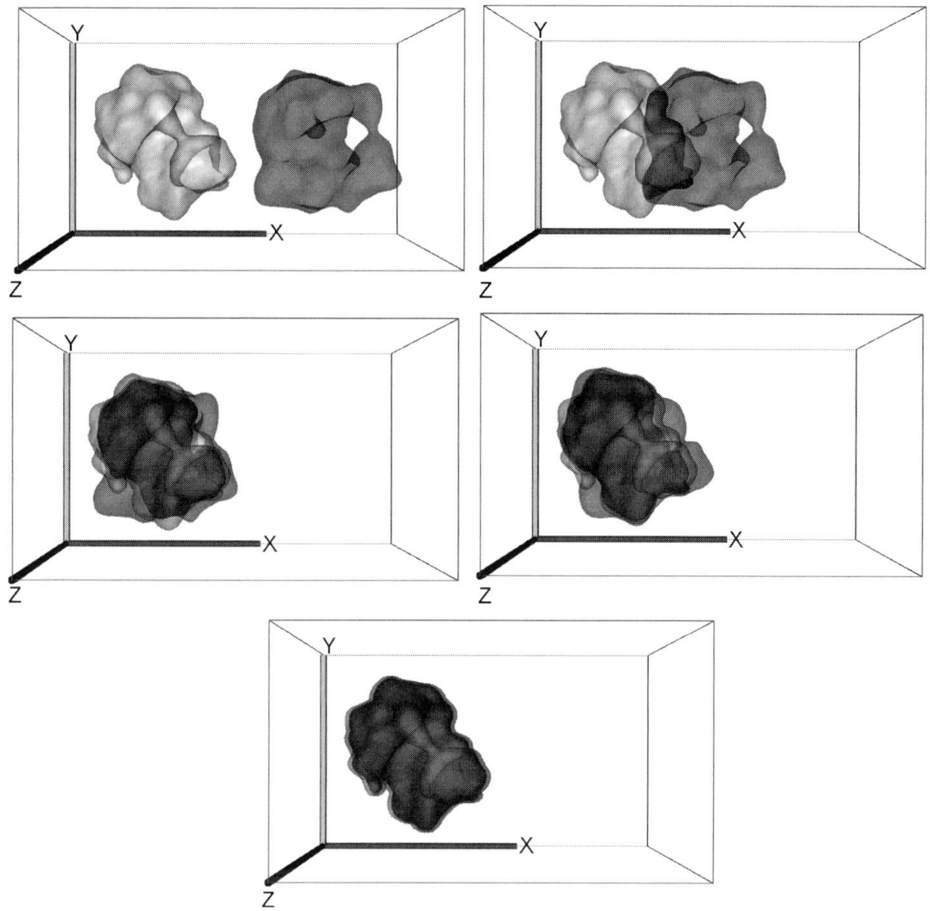

Fig. 1 Three-dimensional (3D) template matching between a sample volume (gray) and a template (green). The correlation function is determined by the amount of overlap between both (red). In the template matching process, the template is translated and rotated until maximum overlap is achieved. (See Color Insert.)

It is a function of the offset y of the template in the volume. Its significance is most easily explained in terms of the concept of mean-square deviation, which is central to the theory of normal distributions. The deviation between sample volume f and template c can be expressed as the difference image:

$$f - T_y c \tag{3}$$

where the operator T_y signifies that the template c has been translated to the point y, mathematically expressed as:

$$(T_y c)(x) = c(x - y) \tag{4}$$

The closer this difference image is to zero, the less f and c differ from each other. A natural measure of this deviation is the square norm of the difference image, that is, the sum over its squared voxel values:

$$|f - T_y c|^2 = \sum_{x \in V_y} [f(x) - c(x-y)]^2 \tag{5}$$

This sum is taken over the subvolume V_y, which has the size and shape of the template volume and is translated to the reference point y. In signal processing, the sum over squared values is often referred to as total power in an image. Since the difference image may have both positive and negative voxel values, it is important that the voxel values are squared, as this makes all contributions to the sum nonnegative. The norm of the difference image is therefore zero if and only if both images agree completely. It is thus a good measure of the total deviation between the two images. In Section II.H, we will discuss in more details the character of this measure.

Expanding the square on the right-hand side leads to the expression:

$$|f - T_y c|^2 = \sum_{x \in V_y} f(x)^2 + \sum_{x \in V_{\text{templ}}} c(x)^2 - 2 \sum_{x \in V_y} f(x)c(x-y) \tag{6}$$

In the second term, we made use of the identity:

$$\sum_{x \in V_y} c(x-y)^2 = \sum_{x \in V_{\text{templ}}} c(x)^2 \tag{7}$$

as V_y is just the template volume V_{templ} translated to the point y. The last term in Eq. (6), the cross-term in the sense that it contains both the template and the subvolume, is the cross-correlation function. The two other terms only contain either the subvolume or the template. In signal-processing terminology, they give the total power of the signal in the subvolume V_y and in the template and thus set the overall scale of the comparison. As the cross-correlation function enters Eq. (6) with a negative term, the higher the value of the cross-correlation function, the lower the mean-square deviation.

In general, we evaluate Eq. (6) at all (or a large number of) points y. If the volume V_{templ} (and correspondingly V_y) is fairly large, the first two terms change only slowly. Since they are sums of squares, and thus each voxel contributes positively, shifting the volume V_y by a small amount only changes a small number of contributing voxels in the sum (namely the edges of the volume), and most of the time this will only result in a small change in the result. This approach rests, of course, on certain assumptions about the statistical nature of the voxel value distributions, but these assumptions hold in most cases.

The cross-term, that is, the cross-correlation function, however, receives both positive and negative contributions, which may cancel or reinforce each other. Small changes in the contributing voxels can then result in large changes in the cross-correlation function. Therefore, most of the variation of Eq. (6) comes from the cross-term, so it is generally justified to consider exclusively the cross-correlation function as an indicator for template similarity. This holds especially when only local maxima of the correlation function are sought, and its absolute value is disregarded.

From a pure machine learning point-of-view, the cross-correlation function is one of the simplest possible discriminant functions, namely a linear combination of the voxel values $f(x)$ with weights given by the template function. In signal processing, such a function is called a *finite-response filter* operating on the voxel data and producing a detection signal, and techniques from digital filter design can be applied to improve the filter beyond the cross-correlation function.

E. Normalization of the Correlation Function

The mean-square deviation, Eq. (5), does not account for a possible scale and bias difference between the voxel values in the template and the sample volume. One way to deal with such a difference is to normalize both functions to the same scale before calculating the cross-correlation function. We first discuss this concept in the general setting of comparing two functions $f(x)$ and $g(x)$ and will show its application to template matching in the form of the local correlation function in the next section.

In a first step, a bias, that is, an additive constant or a constant background, can be removed from a function $f(x)$ by subtracting the respective average from both functions, with the average defined in general by:

$$\bar{f} = \frac{1}{N}\sum_x f(x) \tag{8}$$

(N gives the number of voxels over which the function f is defined). Then an overall scale, that is, a multiplicative constant, can be removed by dividing both functions by their square norm, defined as:

$$|f| = \sqrt{\frac{1}{N}\sum_x \left[f(x) - \bar{f}\right]^2} \tag{9}$$

Note that this quantity, up to the factor of $1/N$, is just the squared term found in Eq. (6). The normalized function $\hat{f}$ corresponding to f is then given by:

$$\hat{f} = \frac{f - \bar{f}}{|f|} \tag{10}$$

Going back to Eq. (6) for the mean-square deviation of two functions, we then find that the square deviation of two functions $\hat{f}$ and $\hat{g}$ normalized in this way is simply:

$$|\hat{f} - \hat{g}|^2 = 2 - 2\sum_x \hat{f}(x)\hat{g}(x) = 2 - 2r \tag{11}$$

with the *Pearson correlation coefficient* r defined as:

$$r = \frac{\sum_x [f(x) - \bar{f}][g(x) - \bar{g}]}{\sqrt{\sum_x [f(x) - \bar{f}]^2}\sqrt{\sum_x [g(x) - \bar{g}]^2}} \tag{12}$$

While this expression may look fairly complicated, it can be calculated in two simple steps by first applying the normalization transformation, Eq. (10), and then calculating the cross-correlation function:

$$r = \sum_x \hat{f}(x) \cdot \hat{g}(x) \tag{13}$$

This expression measures the similarity or dissimilarity of the functions $f(x)$ and $g(x)$ if bias and scale of the voxel values are ignored. The normalization factors in the denominator make sure that the correlation coefficient is confined to values between $+1$ and -1, where $+1$ indicates complete correlation and -1 complete anticorrelation (i.e., the two subvolumes are proportional to each other with a negative proportionality constant). At a value $r = 0$, the vectors are said to be *orthogonal to each other*, and thus completely uncorrelated.

The correlation coefficient has a natural interpretation in the setting of high-dimensional vector spaces. In such a vector space, both the template and the subvolume are represented as vectors. Just as in 3D Euclidean geometry, a vector is a composite quantity whose components are (usually) real numbers. The number of components is called the dimension of the vector space. To represent subvolumes as vectors, we choose the dimension to be the number of voxels in each subvolumes, and assign vectors to volumes such that each component of the vector represents one voxel in the volume. It must be emphasized that the spatial organization of the voxels is of no import here; they are simply lined up in the vector.

The reason for thinking of the cross-correlation function in terms of vectors lies in the peculiar interpretation of Eq. (12) in geometry. The sum of the componentwise products of two vectors f and g is called the *scalar product* and frequently written as $\langle f, g \rangle$. In complete analogy to ordinary geometry, it can be written using the length $|f|$ and $|g|$ of the vectors, as given by their square norm, Eq. (9), and the angle ϕ between them:

$$\langle f, g \rangle = |f||g| \cos \phi \tag{14}$$

In elementary geometry, the angle ϕ is a quantity independent of the lengths of the vectors. Rewriting Eq. (14) as:

$$\cos \phi = \frac{\langle f, g \rangle}{|f||g|} \tag{15}$$

we see that we can identify cos ϕ with the correlation coefficient, Eq. (12), for the case $\bar{f} = \bar{g} = 0$. In a vector space, the correlation coefficient characterizes the angle between the vectors f and g, with $+1$ corresponding to $\phi = 0$ (two parallel

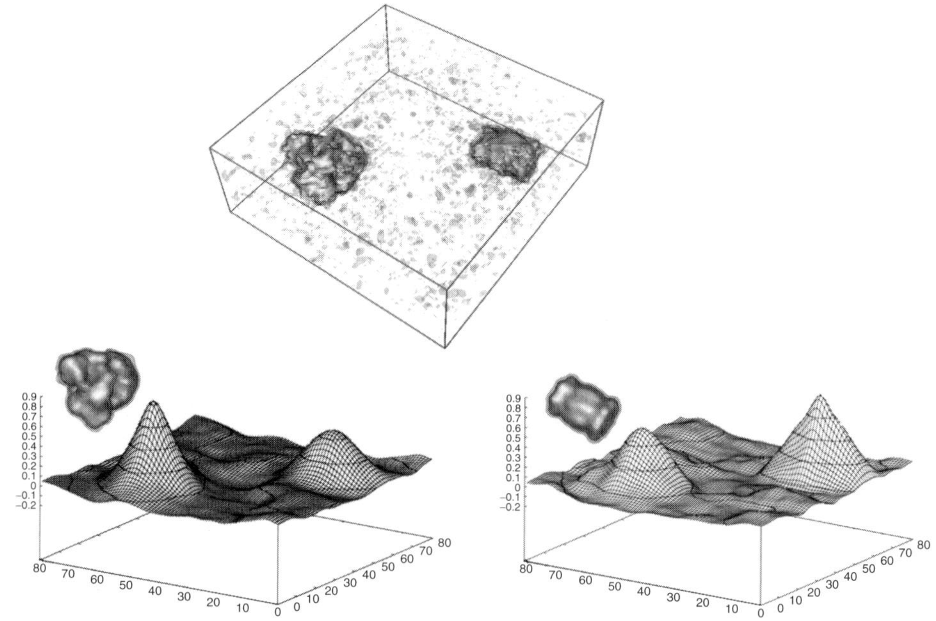

Fig. 2 Localization and discrimination of objects using the cross-correlation coefficient. The top image shows the sample volume containing two different objects in a noisy environment. Templates matching either object were used to search for the objects in the volume. The result is shown, for two templates matching either object, in the surface plots at the bottom. Peaks signify the locations of the objects in the sample volume. While a signal can be observed for both objects with either template, the signal is strongest when the correct template is used.

vectors), -1 to $\phi = 180°$ (antiparallel vectors), and 0 to $\phi = 90°$ (orthogonal vectors).

F. Local Correlation Functions

In the case of template matching, the correlation coefficient is computed between a subvolume of the sample volume $f(x)$ and a translated copy of the template, given by $g(x) = c(x - y)$ (Fig. 2). The two functions are compared in a small volume V_y, corresponding to the template volume translated to the reference point y. The resulting normalized cross-correlation function then reads:

$$r(y) = \frac{\sum_{x \in V_y} \left[f(x) - \bar{f}\right] c(x - y)}{\sqrt{\sum_{x \in V_y} \left[f(x) - \bar{f}\right]^2} \sqrt{\sum_{x \in V_{\text{templ}}} c(x)^2}} \tag{16}$$

We assume that $c(x)$ is already properly normalized, that is $\bar{c} = 0$, which is possible without restriction. One problem arising here, however, is that the normalization of the sample $f(x)$ now depends on the reference point y, that is, $\bar{f}$ and $|f|$ are derived from the sample in the subvolume V_y, not in the complete volume V. This is called the local correlation function. Unfortunately, this implicit dependence makes it impossible to calculate $r(y)$ by the simple two-step procedure, Eq. (13).

Alternatively, one can introduce a global correlation function in which the normalization of f is performed over the whole volume V, that is, using the average $\bar{f}$ and square norm $|f|$ calculated over all voxels of the sample volume. Now the normalization can be performed by preprocessing the volume and the template using Eq. (10), and during the matching procedure, only the unnormalized cross-correlation function, that is, the numerator, has to be computed. While this is an approximation, it has a certain justification, as discussed above, in the small variation of the squared terms as compared to the cross-term.

To illustrate the difference between the local and the global correlation function, let us consider the following situation: instead of considering a small template volume V_{templ} and a large sample volume V, we extend the template to the sample volume by padding it with zeros, creating a function $c(x)$ that is now defined everywhere, but zero outside the original template volume. We further stipulate that, if the value $x - y$ falls outside the volume, it "wraps around," that is, it is periodically continued. This periodical continuation is a technical detail mandated by the use of the Fourier transform (see next section). The global cross-correlation coefficient of this extended template, shifted to position y, then reads:

$$r_{\mathrm{glob}}(y) = \frac{\sum_{x \in V} \left[f(x) - \bar{f} \right] c(x - y)}{\sqrt{\sum_{x \in V} \left[f(x) - \bar{f} \right]^2 \cdot \sum_{x \in V} c(x)^2}} \tag{17}$$

Since multiplying any number with zero gives zero, the numerator here, that is, the cross-correlation function, Eq. (2), is unchanged as compared to the local function (16). In the denominator, the second term again remains unchanged whether we sum over V_{templ} or V, but in the first term, the summation now includes the full volume V rather than the shifted template volume V_y: this slight change vastly simplifies the calculation, as the numerator is now completely independent of the template position y and needs only to be calculated a single time, if at all, as the relative magnitude of the correlation coefficient at different y is completely determined by the correlation function in the numerator.

The difference between the global and the local cross-correlation function can be seen by considering the different interpretations: the global cross-correlation coefficient measures the probability that the full volume is identical, *modulo* noise, to a volume that contains a single copy of the template at position y, and zeros otherwise. The second part of this statement is, of course, wrong in general, as there will be other objects and noise in the volume. The local cross-correlation function only considers information in a volume of the size of the template volume around the position y. By choosing the correct, localized normalization in the denominator of the correlation coefficient, we ensure that the comparison between template and volume is only performed over the appropriate subvolume.

Note that again an equivalent way of writing Eq. (16) is by applying a mask to the sample volume, zeroing out all components outside a template volume around y:

$$r(y) = \frac{\sum_{x \in V} [f(x) - \bar{f}] c(x - y)}{\sqrt{\sum_{x \in V} [f(x) - \bar{f}]^2 w_y(x)} \sqrt{\sum_{x \in V} c(x)^2}} \tag{18}$$

Here $w_y(x)$ is a function that is unity only if $x \in V_y$, and zero otherwise. After applying the mask, and subtracting the local average $\bar{f}$, global and local correlation functions coincide. However, the position of the mask depends on y, so it must be applied for every y individually. This general principle that a locally significant comparison of two functions can be achieved by applying the same mask to both, and comparing only the information remaining after masking, will be applied in the section on dealing with the missing wedge below.

G. Fourier-Space Representation

An attractive feature of the cross-correlation function is that it can be efficiently computed using the Fourier transform. From Eq. (2), the calculation of the cross-correlation function for a spindle point y requires $N_0 = |V_y| = |V_{\text{templ}}|$ operations, that is, as many as there are voxels in the template volume. To calculate it for all possible y thus requires $N \cdot N_0$ operations, where $N = |V|$ is the number of voxels in the sample volume being searched. With volumes of the order of 10^9 voxels and templates of size 10^6, the resulting 10^{15} operations would be prohibitively large even with today's computer speeds that approach 10^{10} operations per second and per processor.

Luckily, the convolution theorem of harmonic analysis allows us to turn the calculation of the correlation function into a simpler operation on the Fourier transforms of the volume and the template. It states that the Fourier transform of the correlation function:

$$C_{f,g}(y) = \sum_x f(x) g(x - y) \tag{19}$$

can be computed as the direct product of the Fourier transforms of f and the complex conjugate of g:

$$\mathcal{F}C_{f,g}(k) = \mathcal{F}f(k) \cdot \mathcal{F}g(k)^* \tag{20}$$

where k runs over all Fourier frequencies. For the Fourier transform to be applicable, the functions f and g must be defined in the same volume, and x runs over all voxels in the volume. Since $y - x$ may then fall onto a point outside the volume, the volume must be extended periodically, that is, the coordinates "wrap over" to the other side at the edge of the volume. Usually, this is a technical detail that can be satisfied by sufficiently padding and extending the sample and template volumes, and/or applying appropriate filters at the edges.

The calculation thus requires two Fourier transforms, one multiplication and one inverse Fourier transform. Since the complexity of a fast Fourier transform (FFT, a specially optimized algorithm to calculate the Fourier transform) is proportional to $N \log N$, and the multiplication just to N, the total number of operations does not grow faster than $N \log N$, which is significantly better than the $N{\cdot}N_0$ number of operations in the original problem, at least if N and N_0 are large. This is, of course, an estimate, and the total efficiency of the algorithm depends on the size of the volume and the template. In most cases, however, the calculation in Fourier space is vastly more efficient.

The calculation of the locally normalized cross-correlation function can also be sped up using the Fourier transform (Roseman, 2003). If we define a function $w_y(x)$ that is unity if x is inside the volume V_y, and zero otherwise, we can write the local power of $f(x)$ as:

$$\sum_{x\in V_y} f(x)^2 = \sum_x f(x)^2 w_y(x) = \sum_x f(x)^2 w_0(x-y) \tag{21}$$

This again has the form of a correlation function and can be calculated by multiplying the Fourier transform of $f(x)^2$ and of the function $w_0(x)$:

$$\sum_{x\in V_y} f(x)^2 = \mathcal{F}^{-1}\left[\mathcal{F}f^2 \cdot \mathcal{F}w_0^*\right] \tag{22}$$

H. Probabilistic Interpretation

Up to this point, we have just posited that Eqs. (5) or (17) are good measures of the similarity between a template and a subvolume. The theoretical underpinning of this assumption lies in the theory of normally distributed errors and the stochastic model of the imaging process. In this theory, it is assumed that the observed image f can be mathematically expressed as a combination of an ideal value f_0 and some unknown noise ε:

$$f(x) = f_0(x) + \varepsilon(x) \tag{23}$$

The actual value of the noise $\varepsilon(x)$ is of course unknown, but we can assume that its probability distribution is known. In some cases, it may be reasonable that the noise is distributed according to a normal (or Gaussian) distribution with certain characteristics. Knowing this probability distribution allows us to ask how probable it is that a certain observed image $f(x)$ has arisen from an idealized image $f_0(x)$ by calculating the difference image and evaluating it according to the probability distribution of the noise. In the theory of normal distributions, for example, a central role is given to the mean-square deviation, Eq. (5), generally called χ^2.

It is, in this theory, a direct measure of the probability that the difference between two functions can be attributed to chance.

In general, the idealized image $f_0(x)$ is not exactly known. In template matching, it depends on the displacement y of the template, the Euler rotation angles, and also an overall proportionality factor and shift in the voxel values. Probabilistic modeling offers one technique to systematically deal with this ignorance by means of the *maximum likelihood principle*.

The maximum likelihood principle states that the unknown transformation parameters of the image can be estimated by choosing them in such a way that the image they produce is the most probable match for the actual image. It is a very general principle and underlies many parameter estimation problems in statistics, such as linear fits. In principle, it requires us to systematically consider all values of the unknown parameters, apply them to the template and compare the resulting template image to the actual image, and then choose the best correspondence. Luckily, in some cases (such as linear fits) this problem can be solved analytically without actually enumerating all values, but in other cases (such as the rotation and translation parameters), this is not possible, and enumeration is the only viable way to proceed.

While the maximum likelihood method provides us with the most likely fit of our unknown parameters (position and orientation of the template) to the observed image, it does not make any statement whether this fit is the only likely one or one of many similar ones. It is possible to calculate the probability that the observed image arose from the template, but it is much more difficult to make any statement about the probability of this fit among all possible fits in the maximum-likelihood framework. Such questions are especially important when different templates have to be considered under noisy conditions. Such issues are beyond the scope of this chapter, and they often require more advanced statistical techniques such as Bayesian estimation and Monte Carlo analysis.

I. Practical Implementation

In the preceding sections, we have introduced the correlation function, its correct normalization, and its probabilistic interpretation. Putting the pieces together, template matching in this approach requires the calculation of the locally normalized cross-correlation function, Eq. (16), for all (or a sufficiently large number) of Euler rotations (ϕ, θ, ψ) of the template. The maxima of the resulting cross-correlation coefficient:

$$r_{(\phi,\theta,\psi)}(y) = \frac{\sum_{x \in V_y} \left[f(x) - \bar{f}\right] R_{(\phi,\theta,\psi)} c(x-y)}{\sqrt{\sum_{x \in V_y} \left[f(x) - \bar{f}\right]^2 \cdot \sum_{x \in V_{\text{templ}}} R_{(\phi,\theta,\psi)} c(x)^2}} \tag{24}$$

give the most probable locations of the template in the volume V. For any given set of Euler angles, both the numerator and the denominator can be efficiently calculated in Fourier space, resulting in one 3D function per Euler angle

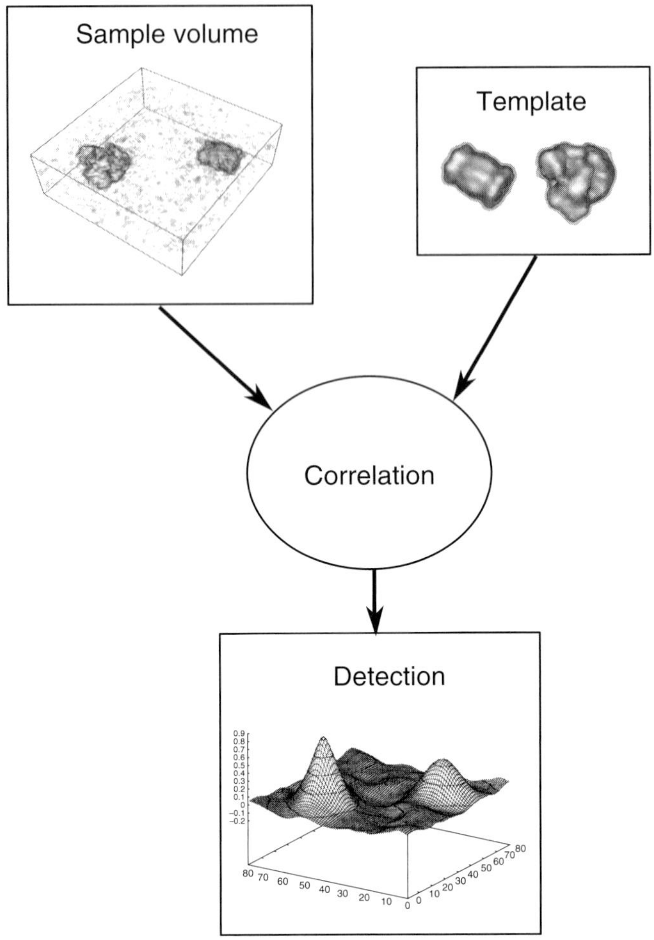

Fig. 3 The template matching process for the detection of macromolecules. Possible locations of molecules can be identified by looking for peaks in the cross-correlation functions between a set of template volumes and the sample volume.

combination. The local peaks of this function indicate likely locations of the template in the orientation given (Fig. 3).

If the template volume V_{templ} is rotationally symmetric, the local normalization factor in the denominator does not depend on the Euler angles, and one calculation of the localized normalization factors is sufficient for all Euler angles, significantly speeding up the calculation.

Typically, the space of Euler angles is sampled at 10° or 5° increments, resulting in 23,328 and 186,624 different combinations, respectively, and making it infeasible to store the complete function $r_{(\phi,\theta,\psi)}(y)$ in core memory. Rather, one computes it consecutively for all Euler angles, keeping only the maximum values at each location and the corresponding values of the Euler angles in memory. Local peaks in the maximum values for each location then identify possible locations of the template in the sample volume in any orientation.

As usual in image processing, suitable filtering is applied to the image before the correlation coefficients are calculated. This filtering removes Fourier components that carry little or no information of relevance. Usually a band-pass filter is applied that removes both high-frequency information that are often noise, and low-frequency areas that correspond to features significantly larger than the objects under consideration.

III. The Missing-Wedge Problem

A. Tomographic Imaging

In tomographic imaging, a series of projection images is acquired by mechanically tilting the specimen around a fixed axis (see Leis *et al.*, 2006, for a recent review of the image-processing problems associated with this technique; Chapter 29 by Förster and Hegerl, this volume). The images acquired in this process show the projections of the electrostatic density in the object, that is, each pixel represents the total density along a straight line, or *ray*, perpendicular to the imaging plane. Features of the object therefore appear superimposed on the individual images, but a mathematical procedure based on the Radon transform allows one to disentangle the individual features in the object and recover the 3D locations by combining the information from different tilt angles.

The biggest artifact in tomographic reconstruction comes from the limited tilt range that is accessible experimentally. This limit is caused simply by the thickness of the vitreous ice layer and the holder in which the specimen is suspended and which, at some inclination, becomes impenetrable to the electrons. Tilt angles beyond 70° are rarely accessible, and the information corresponding to these directions is not included in the reconstruction. In Fourier space, this corresponds to a cone- or wedge-like area around the positive and negative z-axis where no information is available, the so-called *missing wedge*. In ordinary space, this is manifest as an elongation or smearing of the object along the z-axis, caused by the fact that these Fourier components are missing and features in this direction lack definition.

This missing wedge causes a serious problem for template matching, as it distorts the shape of the observed macromolecules. To understand the effect of the missing wedge on the correlation function, one should remember that it does not matter whether the objects are compared in real space or in Fourier space. The missing wedge is a very pronounced feature in Fourier space, and a pattern-matching algorithm, such as the correlation function, will have a tendency to fit this feature to some similar feature of the template. This process is in competition to the alignment with the actual distinctive features of the molecule, and if it wins, the molecule will be misaligned and/or misidentified.

B. Restricted Cross-Correlation Coefficients

The missing wedge can be dealt with in a similar way as with the local normalization discussed in Section II.F (Förster, 2005). There, we dealt with the problem that the comparison between the template and the volume should only be performed inside a localized subvolume in which the template is defined. This was achieved by choosing as the correct normalization of the correlation coefficient the total power contained in the subvolume under consideration. With the missing wedge, we have the similar situation that the comparison should be limited to the information outside the missing wedge.

In the definition of the correlation coefficient, Eq. (16), correct normalization was achieved by summing over only these elements x for which information is available, that is, the shifted subvolume V_y. Another way of including this weight function in the calculation is to set all elements of the sample volume and the template, for which no information is available, to zero and applying this transformation before the numerator and denominator of Eq. (16) are calculated.

In the case of the missing wedge, the unavailable information is a volume in Fourier space. Applying the previous prescription, this can be taken into account by setting the Fourier coefficients of both the template and the sample volume inside the missing wedge to zero. Denoting the Fourier transform of the template by $\mathcal{F}c(k)$, the template is replaced by a smeared template:

$$c'(x) = \mathcal{F}^{-1}[w(k)\mathcal{F}c(k)] \tag{25}$$

where $w(k)$ is the wedge function, which has the value one inside the experimentally accessible cone and zero outside it. This operation is a multiplication in Fourier space that can be translated to a convolution in position space:

$$c'(x) = \sum_y \mathrm{PSF}(x - y)\dot{c}(y) = (\mathrm{PSF} * c)(x) \tag{26}$$

The *point-spread function* $\mathrm{PSF}(x - y)$ is the Fourier transform of the wedge function; it describes how a single bright point would appear under the influence of the missing wedge.

Applying the previous recipe for accounting for unknown information, it is sufficient to replace the template $c(x)$ by the smeared-out template $c'(x)$, both in the numerator and the denominator of the locally normalized cross-correlation function, Eq. (24). The smearing operation must be performed after the template has been rotated:

$$r_{(\phi,\theta,\psi)}(y) = \frac{\sum_{x\in V_y}\left[f(x)\mathrm{PSF} * R_{(\phi,\theta,\psi)}c\right](x - y)}{\sqrt{\sum_{x\in V_y} f(x)^2 \cdot \sum_{x\in V_{\mathrm{templ}}}\left[\mathrm{PSF} * R_{(\phi,\theta,\psi)}c\right](x)^2}} \tag{27}$$

This means that now, even if the template volume is spherical, the second term in the denominator will depend on the rotation angles, reflecting the fact that due to the missing wedge, different amounts of information are available in different orientations of view.

IV. Applications

The usefulness of template matching in 3D to the identification of macromolecules in ET was first explored by Böhm *et al.* (2000). This study investigated the power of the cross-correlation coefficient to discriminate between different macromolecules, namely the 20S proteasome, the GroEL chaperonin, and the thermosome using both synthetic data and real electron-tomographic images of isolated macromolecules. Templates for the three molecules were generated either from atomic models available in the Protein Data Base (for the 20S proteasome and GroEL) or from pseudo-atomic models derived from EM (for the thermosome).

Tomographic sample volumes were generated both synthetically, by filtering and adding colored noise, and from tomographic reconstructions of actual tilt series obtained from purified samples embedded in vitreous ice. The different templates were then compared to the sample volumes using the cross-correlation coefficient. Histograms of the coefficient show that it generally takes on a larger value whenever the correct template is used. The quantitative accuracy of the particle identification depends on the combination of template and sample molecule in use, as well as on the resolution of the images. However, it was found that, in this setting, the cross-correlation function was highly significant, at several standard deviations, in discriminating between the different objects. Performance was especially good for the 20S proteasomes, which differ markedly in size from other two species, while the discrimination between GroEL and the thermosome required a resolution better than 4 nm.

This chapter also considered the use of an alternative approach to particle identification based on segmentation of the volume using an anisotropic diffusion technique called MCM. In this approach, particles are not identified according to a template, but subvolumes exhibiting surfaces of low curvature such as would by typical for macromolecules or cellular compartments. This is used to find interesting regions of the tomogram that contain macromolecules before the correlation function is calculated, thus saving computing time.

To demonstrate the performance of the method in a cellular context, Frangakis *et al.* (2002) have applied template matching to the detection of macromolecules in "phantom cells" constructed from liposome vesicles that mimic real prokaryotic cells in size and shape, but whose content is known (Fig. 4). Again, the thermosome and the 20S proteasome were used as sample macromolecules. A solution of purified proteins was encapsulated in lipid vesicles with an average diameter of ≈400 nm

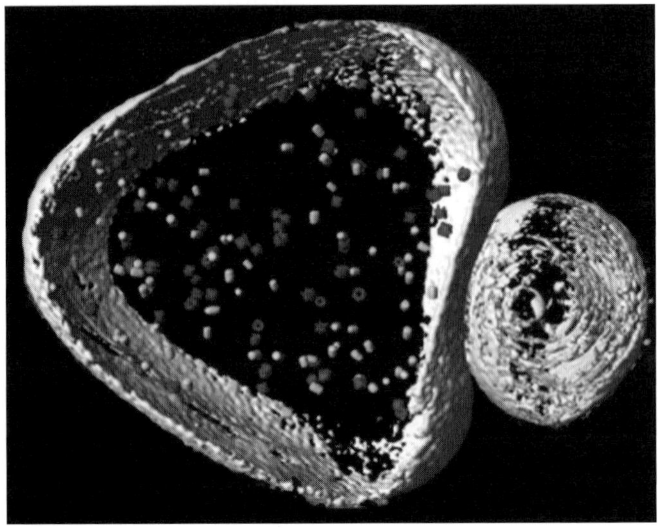

Fig. 4 Detection of macromolecules in a phantom cell. The image shows a liposome vesicle (white) in which the locations of macromolecules have been identified by cross-correlation with templates. The molecule locations and orientations are marked by color-coded templates pasted into the volume. Adapted from Frangakis *et al.* (2002). (See Color Insert.)

and embedded in vitreous ice. The resulting tomograms were investigated using the local correlation coefficient. The template was constructed from atomic-scale models of the macromolecules. To optimize the template, different band-bass filters were applied and the cross-correlation function inspected manually. Locally normalized cross-correlation coefficients were calculated with both the correct and the incorrect template for the tomogram under consideration. The authors found that there was a clear separation in the values of the correlation coefficients between both templates, leading to identification accuracies of 96% for the 20S proteasomes and 87% for the thermosomes, at a resolution of about 4 nm. Applying the algorithm to phantom cells that contain a mixture of both species, the molar ratio was retrieved with a deviation of less than 10%. This led to the conclusion that macromolecular complexes in the size range of 0.5–1 MDa can be identified with satisfactory fidelity. There remains, however, the question of how this identification algorithm is affected by molecular crowding and the presence of other cellular structures in the tomogram.

Recent applications of the template matching technique have focused on particle averaging, which is described elsewhere in this book (see Chapter 29 by Förster and Hegerl, this volume). Here the template matching is used to locate and align subvolumes containing the macromolecule under consideration which are then averaged to produce a density map with improved resolution. This is similar to the single-particle method without the problem of assigning projection angles to the images.

Förster *et al.* (2005) studied the structure of envelope glycoprotein (Env) complexes on intact Moloney murine leukemia retrovirus particles using the restricted cross-correlation coefficient. These trimeric transmembrane complexes stud the surface of the retrovirus. They were located manually on the surface of the virions, and the restricted cross-correlation function was used to align the particles to a common template. The authors could take advantage of the fact that the complexes are oriented radially on the spherical surface of the virus, so that only one rotational degree of freedom had to be considered initially. Averaging of several thousand particles resulted in a reconstruction with about 2.5-nm resolution.

Beck *et al.* (2004) used a similar approach to the nuclear pore complex in *Dictyostelium discoideum*. Again, the restricted cross-correlation function was used to align manually identified copies of the complex, and an improved density map was found by averaging.

Rath *et al.* (2003) explored the application of the locally normalized cross-correlation function to identify specific ribosomal proteins in a tomogram of the *Escherichia coli* ribosome. They also looked at ryanodine receptors in a suspension of vesicles derived from skeletal muscle sarcoplasmic reticulum and transverse tubules. In the first case, the results could be compared against the results of real-space refinement in which the known structures of the ribosomal proteins had been fitted in the cryo-EM density. Here, the cross-correlation function was able to identify the location of the proteins to within two pixels, and the angular orientation was specified to 8°. In the second case, a template volume was picked manually from the tomogram and used to identify other occurrences of this motif. Most particles were found in plausible locations on the vesicle membranes.

Ortiz *et al.* (2006) applied the restricted cross-correlation function to identify 70S ribosomes in *Spiroplasma melliferum* (Fig. 5). The authors were able to identify several hundred of these macromolecular complexes in a single tomogram and average the aligned subtomograms to create a density map with a resolution of 4.7 nm that matched the structure of an *E. coli* 70S ribosome. Similar to Böhm *et al.* (2000), an alternative method of particle identification based on the scaling index segmentation method was considered. This method identifies possible particles based on fractal dimensional properties of the subvolumes; since it does not make use of a template, the method is not able to discriminate between different particles. Still an overlap of 70% was observed between the two methods.

V. Conclusions

Improved methods for recognizing patterns in highly noisy 3D tomographic reconstructions are helping to develop cryo-ET into a useful tool for mapping the macromolecular content and interactions of cells in near-life conditions. At the

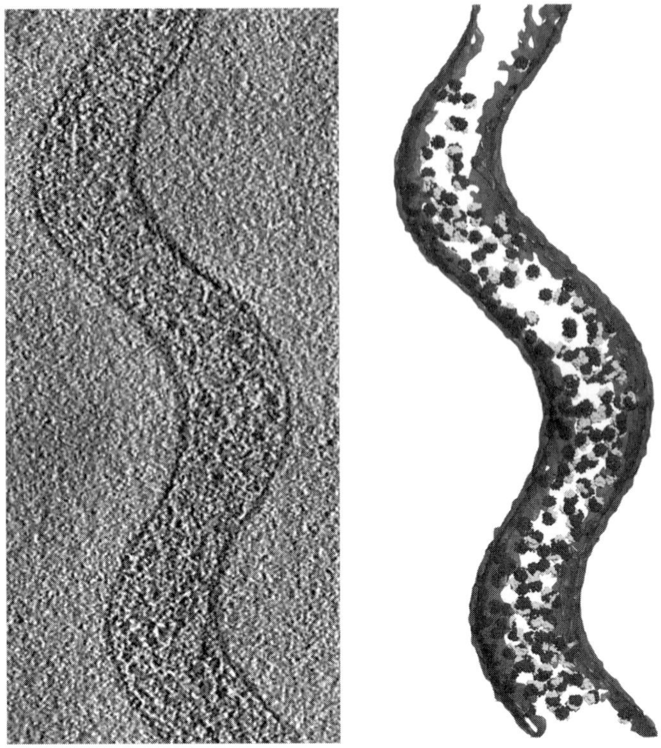

Fig. 5 Identification of ribosomes in *Spiroplasma melliferum*. The left-hand side shows a reconstructed tomographic slice with macromolecular complexes visible as dark regions. In the 3D view on the right-hand side, locations and orientations of 70S ribosomes identified by template matching are indicated by copies of the templates (in blue and yellow) (image by J. Ortiz, see Ortiz *et al.*, 2006). (See Color Insert.)

resolutions attainable today, large macromolecular complexes can already be reliably identified and discriminated from each other both in isolation and in phantom cells. Recent progress indicates that, at least in relatively simple cells, satisfactory results can be obtained with intact cells as well.

Further improvements are under way in the cryo-EM pipeline. These range from sample preparation and sectioning to the automation of the image acquisition and postprocessing process to intelligent pattern recognition algorithms. Meanwhile, large libraries of templates are becoming available from, among others, the various structural genomics efforts. Given the progress we may reasonably expect from all this work, the method described here holds the promise of becoming a viable tool for the study of biologically significant protein interactions that underly the processes of life.

References

Aloy, P., and Russell, R. B. (2006). Structural systems biology: Modelling protein interactions. *Nat. Rev. Mol. Cell. Biol.* **7,** 188.

Baumeister, W. (2004). Mapping molecular landscapes inside cells. *Biol. Chem.* **385**(10), 865–872.

Baumeister, W., and Steven, A. C. (2000). Macromolecular electron microscopy in the era of structural genomics. *Trends Biochem. Sci.* **25**(12), 624–631.

Beck, M., Förster, F., Ecke, M., Plitzko, J. M., Melchior, F., Gerisch, G., Baumeister, W., and Medalia, O. (2004). Nuclear pore complex structure and dynamics revealed by cryoelectron tomography. *Science* **306**(5700), 1387–1390.

Böhm, J., Frangakis, A. S., Hegerl, R., Nickell, S., Typke, D., and Baumeister, W. (2000). Toward detecting and identifying macromolecules in a cellular context: Template matching applied to electron tomograms. *Proc. Natl. Acad. Sci. USA* **97**(26), 14245–14250.

Duda, R. O., Hart, P. E., and Stork, D. G. (2001). "Pattern Classification." John Wiley & Sons, New York.

Förster, F. (2005). Quantitative Analyse von Makromolekülen in Kryoelektronentomogrammen mittels Korrelationsmethoden. Ph.D. Thesis, Technische Universität München.

Förster, F., Medalia, O., Zauberman, N., Baumeister, W., and Fass, D. (2005). Retrovirus envelope protein complex structure *in situ* studied by cryo-electron tomography. *Proc. Natl. Acad. Sci. USA* **102**(13), 4729–4734.

Frangakis, A. S., Böhm, J., Förster, F., Nickell, S., Nicastro, D., Typke, D., Hegerl, R., and Baumeister, W. (2002). Identification of macromolecular complexes in cryoelectron tomograms of phantom cells. *Proc. Natl. Acad. Sci. USA* **99**(22), 14153–14158.

Frank, J. (1972). Two-dimensional correlation functions in electron microscope image analysis. *In* "Proceedings of the Fifth European Congress on Electron microscopy," pp. 622–623. Institute of Physics, London, UK.

Frank, J. (1996). "Three-Dimensional Electron Microscopy of Macromolecular Assemblies." Academic Press, San Diego, CA.

Frank, J. (2002). Single-particle imaging of macromolecules by cryo-electron microscopy. *Annu. Rev. Biophys. Biomol. Struct.* **31,** 303–319.

Frank, J., and Wagenknecht, T. (1984). Automatic selection of molecular images from electron micrographs. *Ultramicroscopy* **12,** 169–176.

Hastie, T., Tibshirani, R., and Friedman, J. H. (2003). "The Elements of Statistical Learning." Springer-Verlag, New York, Heidelberg, Berlin.

Huang, Z., and Penczek, P. A. (2004). Application of template matching technique to particle detection in electron micrographs. *J. Struct. Biol.* **145**(1–2), 29–40.

Leis, A. P., Beck, M., Gruska, M., Best, C., Hegerl, R., Baumeister, W., and Leis, J. W. (2006). Cryo-electron tomography of biological specimens. *IEEE Signal Process. Mag.* **23**(3), 95.

Lucic, V., Förster, F., and Baumeister, W. (2005). Structural studies by electron tomography: From cells to molecules. *Ann. Rev. Biochem.* **74,** 833–865.

Nickell, S., Kofler, C., Leis, A. P., and Baumeister, W. (2006). A visual approach to proteomics. *Nat. Rev. Mol. Cell. Biol.* **7**(3), 225–230.

Ortiz, J., Förster, F., Kurner, J., Linaroudis, A., and Baumeister, W. (2006). Mapping 70S ribosomes in intact cells by cryo-electron tomography and pattern recognition. (In press).

Rath, B. K., Hegerl, R., Leith, A., Shaikh, T. R., Wagenknecht, T., and Frank, J. (2003). Fast 3D motif search of EM density maps using a locally normalized cross-correlation function. *J. Struct. Biol.* **144**(1–2), 95–103.

Roseman, A. M. (2000). Docking structures of domains into maps from cryo-electron microscopy using local correlation. *Acta Crystallogr. D Biol. Crystallogr.* **56,** 1332–1340.

Roseman, A. M. (2003). Particle finding in electron micrographs using a fast local correlation algorithm. *Ultramicroscopy* **94**(3–4), 225–236.

Sali, A., Glaeser, R., Earnest, T., and Baumeister, W. (2003). From words to literature in structural proteomics. *Nature* **422,** 216–225.

Volkmann, N., and Hanein, D. (1999). Quantitative fitting of atomic models into observed densities derived by electron microscopye. *J. Struct. Biol.* **125,** 176–184.

Walz, J., Typke, D., Nitsch, M., Koster, A. J., Hegerl, R., and Baumeister, W. (1997). Electron tomography of single ice-embedded macromolecules: Three-dimensional alignment and classification. *J. Struct. Biol.* **120**(3), 387–395.

Zhu, Y., Carragher, B., Glaeser, R. M., Fellmann, D., Bajaj, C., Bern, M., Mouche, F., de Haas, F., Hall, R. J., Kriegman, D. J., Ludtke, S. J., Mallick, S. P., *et al.* (2004). Automatic particle selection: Results of a comparative study. *J. Struct. Biol.* **145,** 3–14.

PART V

Aspects of Data Collection and Analysis

Chapters in the previous parts of this volume have shown how current methodologies empower cell biologists to use electron microscopy (EM) in many informative ways. This part looks more toward the future during which existing methodologies will evolve to give us even more powerful ways to study cell structure. Two areas that seem ripe for improvement are the technologies for image acquisition and the ways in which acquired images can be analyzed for the fruitful extraction of information.

Instrumentation for image acquisition is important because currently available electron lenses are not well corrected for chromatic aberration, that is, variation in focal length as a function of electron energy. Each beam electron that interacts with the sample, and therefore can convey information about sample structure, has a chance of having experienced an "inelastic scattering event," which means that its energy has changed relative to that of others, so it will not be properly handled by the imaging lenses. The result is that it crosses the image plane at a point that is not conjugate to the specimen point from which it was scattered. Elastic scattering is essential for getting information about the sample, but inelastic scattering produces beam electrons that blur the image rather than contributing to it.

Ellisman's laboratory has been experimenting with an energy filter to remove inelastically scattered electrons from the beam before the image is formed. This approach opens the possibility of increasing image clarity, particularly for images of very thick samples. Energy filters have been used for several years to image frozen-hydrated samples (Grimm *et al.*, 1998), where electron phase contrast is important in the imaging process. In this case, there is clear evidence for improvement in image quality from energy filtering. Chapter 26 by Bouwer *et al.* shows how a similar strategy can also improve images of very thick samples of fixed, stained, and embedded samples, particularly if they are selectively stained. This technology may open up electron tomography (ET) for cellular samples that are literally micrometers thick, though the applicability of the method to cells that are simply well fixed and not selectively stained remains to be demonstrated.

Mooney (Chapter 27) reviews the electronic cameras that are currently available and shows how their development has enhanced progress in EM and ET. He emphasizes the importance of signal-to-noise ratio (SNR) in evaluating imaging devices, drawing particular attention to methods that consider SNR as a function of the spatial frequency, or spectral signal to noise. He also discusses currently active routes toward the improvement of cameras that are based largely on the detection of light that is produced from beam electrons by a scintillator and imaged on a charge-coupled device (CCD). He shows how attention to all the details of electron detection could lead to significantly improved images, even with current technology. The group led by Xuong (Chapter 28) has also addressed the difficulties imposed by current CCD technology. Their chapter reviews the various approaches that have been following in San Diego and goes on to describe their recent work on alternative camera technologies. A silicon-based detector can be used directly in the electron beam to produce images that report on specimen structure with even fewer beam electrons than are required by a sensitive CCD camera.

Taken together, these chapters alert us to the importance of supporting research on image generation and acquisition. They also encourage us to keep abreast of developments in this field. Such advances may well provide us with even better ways of imaging cells in the near future.

Methods for the analysis of EM data are also important for improving the quality and quantity of information we can obtain from images of biological material. Given the tremendous strides that have been made in the speed and quality of computing machines, this is again a fruitful area for current and future work on improving our understanding of cell structure. This part begins with Chapter 29 by Föster and Hegerl, which addresses the key problem of SNR in EM images of frozen-hydrated samples. As mentioned many times in earlier parts of this book, the sensitivity of such samples to damage by the electron beam requires that electron doses be kept low. When they are low, however, there are not enough electrons being scattered by each region of the sample to give a good measure of the local scattering power (local electron density). An image that is underexposed in this way may contain information at very high resolution, but the image is painfully uninformative because the counting statistics for each region of the image are so poor. Better statistics would appear to require more dose, which means more damage and thus lower sample resolution, defeating the purpose of the whole exercise. These authors show how images of identical objects in a cryotomogram can be identified, cut out, aligned, and averaged with one another so as to increase image SNR without using more dose per unit volume. When multiple copies of a structure can be found at diverse orientations relative to the axis of tilt, the averaging can also be used to eliminate the problem of the missing wedge, which commonly gives ET an anisotropic resolution. This chapter points the reader, then, toward ways of thinking about image processing that may provide greatly improved information from a given electron tomogram.

Once a good-quality image is in hand, from either EM or ET, one is then faced with visualizing, extracting, and studying the features of interest. Earlier chapters in the volume have shown several examples of scientists using their eyes to identify features of interest (membranes or cytoskeletal components) and a drawing tool to generate graphic representations of those features for subsequent study. This method works well, but it is time consuming and labor intensive. Scientists using cryo-EM of frozen-hydrated samples imaged in isolation often identify the boundaries of their structures automatically, simply by setting a threshold in density or contrast, so everything above that value is included and everything else is excluded. In a cell, however, this method is problematic because so many cellular structures are packed densely with one another; finding boundaries through simple thresholding is generally impossible. Sandberg (Chapter 30) describes several computational methods that have been used to solve this problem of "segmenting" features of interest away from everything else in a complex structure like the cell. He also introduces a new algorithm of his own that identifies cellular membranes quite well.

Once images are in hand and the data are ready for analysis, there are still computational problems that confront us. Electron images of cells are complex, and their accurate representation requires huge amounts of data; tomograms are even worse, often comprising gigabytes of information. How should all this information be organized and stored, so it will be available for detailed study? Martone *et al.* (Chapter 31) describe important efforts to build a database for structural information about cells. This is a daunting task, given both the size of the data sets in use and the complexity of the information they contain. While one can represent human hemoglobin with three-dimensional models of only a few objects, representing a human liver cell mitochondrion is much more difficult; these structures and organelles like them are pleomorphic. Representing a given cell type is even harder. Thus, this chapter addresses a key problem that will be with us for a long time. The task of setting up useful databases that are constructively linked with gene and protein databases and can serve the cell or developmental biologist is an important effort whose time is ripe.

References

Grimm, R., Singh, H., Rachel, R., Typke, D., Zillig, W., and Baumeister, W. (1998). Electron tomography of ice-embedded prokaryotic cells. *Biophys. J.* **74,** 1031–1042.

CHAPTER 26

The Application of Energy-Filtered Electron Microscopy to Tomography of Thick, Selectively Stained Biological Samples

James C. Bouwer, Mason R. Mackey, Albert Lawrence, Tom J. Deerinck, Ying Z. Jones, Masako Terada, Maryann E. Martone, Steven T. Peltier, and Mark H. Ellisman

University of California, San Diego
La Jolla, California 92093

We describe the use of an electron energy filter to enhance contrast and reduce chromatic aberration in the imaging of thick, selectively stained specimens. These efforts have been driven by the need for scientists investigating biological structure at scales just below the 0.3-μm resolution achievable with a light microscope, which we term the "mesoscale," to produce data sets that represent complete models of large structures at the resolution of the electron microscope. For thick specimens, however, the resolution of electron microscopy is severely limited by chromatic aberration that results from the inability of current electron lenses to deal uniformly with beam electrons that have a distribution of energies as a result of inelastic scattering in the

0091-679X/07 $35.00
DOI: 10.1016/S0091-679X(06)79026-4

samples. The technique we describe is that of automated, most-probable loss (MPL) tomography. We show that for thick, selectively stained biological specimens, this method produces a dramatic increase in the resolution of the projected images and the resulting computed volumes versus unfiltered transmission electron microscopy (TEM) methods on the JEOL JEM-3200 omega-filtered microscope. These improvements are particularly evident at the large tilt angles required to improve tomographic resolution in the z-direction. MPL tomography effectively increases the usable thickness of selectively stained samples that can be imaged at a given accelerating voltage by improving resolution relative to unfiltered TEM; it also increases the signal-to-noise in comparison to zero-loss imaging of thick sections. The method therefore expands the utility of the intermediate voltage electron microscope (IVEM) for delivering information from specimens up to 3-μm thick.

I. Introduction

Energy filtering has been used for a wide variety of applications in biology. Some of these include: improved contrast through zero-loss imaging of samples embedded in vitreous ice (Grimm *et al.*, 1996, 1997), imaging of thick specimens (Kortje *et al.*, 1996; Reimer *et al.*, 1992), Z-ratio imaging of thin sections to create contrast in unstained specimens (Garavito *et al.*, 1982; Reimer *et al.*, 1992), mapping of chromophores using low-loss electron energy loss spectroscopy (EELS) (Barfels *et al.*, 1998; Crewe *et al.*, 1971), and mapping specific elements in biological samples (Daulton *et al.*, 2002; Leapman and Rizzo, 1999; Middleton *et al.*, 2003; Radcke *et al.*, 2002). Energy filters can be obtained in a variety of configurations: the prism mirror system (Castaing and Henery, 1962), the omega type in-column filter (Rose and Pelies, 1974), the postcolumn magnetic prism spectrometer (Krivanek *et al.*, 1995), and others. It is not our purpose in this chapter to discuss the wide variety of energy-filtering devices. A complete reference with theoretical considerations can be found in Egerton (1996). In this chapter, we discuss the technique of most-probable loss (MPL) imaging for tomography of thick, selectively stained samples.

We begin with a discussion of energy-filtered electron microscopy (EM). An important method that uses this technology is EELS, which is a study of the relationship between physical material quantities and their energy spectra. This relationship is usually measured through the absorbance or emission of energy by the sample while it is probed with a beam of either light or electrons. For EELS, the electrons that have interacted with the sample and lost some energy to it are disbursed according to their energy loss across a digital-recording device, like a CCD camera. Electrons can interact with the sample in a wide variety of ways, but in the most general terms, electrons interact either through *elastic* or *inelastic* collisions. Elastic interactions include all interactions between the sample and the electrons of the primary beam where no energy is exchanged. Inelastic interactions include all such interactions where some energy is exchanged. This is usually seen in the form of energy loss from the primary electron beam, where energy is given up to the specimen through the excitation of modes specific to the sample.

Inelastic processes are the most important scattering processes for analysis with EELS. They include energy losses due to the creation of Auger electrons, x-rays, Bremsstrahlung radiation (due to electron deceleration), plasmon losses (due to collective excitation of bulk vibrational modes of the sample), visible light photons (sometimes referred to as cathodoluminescence), and losses due to excitation or ionization of atomic electrons in specific elements of the sample. For MPL imaging, we are generally interested in multiple scattering processes in the low-loss regions; these are usually associated with plasmon losses.

The main energy-filtering element in the electron optical system is a magnetic-prism spectrometer (Fig. 1). In the magnetic-prism spectrometer, electrons traveling down the column of the microscope in the z-direction, enter the prism between the poles of an electromagnet whose magnetic field points in the y-direction. Once a charged particle enters a magnetic field, it is forced to move in a circular path. For relativistic electrons, such as those used to image in an electron microscope, the radius of curvature R of electrons moving through a magnetic field B is given by

$$R = \left(\frac{\gamma m_0}{eB}\right)\upsilon \tag{1}$$

where

$$\gamma = \frac{1}{\sqrt{1 - \frac{\upsilon^2}{c^2}}} \tag{2}$$

is the relativistic correction factor, υ is the velocity of the electron, c is the speed of light in a vacuum, and m_0 is the rest mass of the electron. As illustrated in Fig. 1,

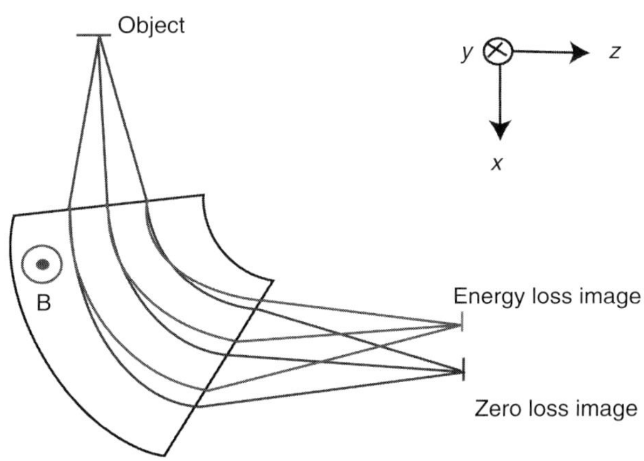

Fig. 1 A diagram depicting the prism type spatial separation of electrons according to their energies through the use of a magnetic field spectrometer.

electrons that have lost some energy in the sample will have a smaller radius of curvature than those that have not; thus, they exit the spectrometer with a slightly larger deflection angle than those that either interacted elastically or not at all. It is therefore possible to separate the electrons according to their respective energy losses as they traverse the sample.

Within the image plane of the spectrometer, there are multiple images, where each image is a convolution of the true image with the distortion introduced by the effect of chromatic aberration of the lenses on that particular electron energy. These polyenergetic images can be filtered by placing a slit in the energy selection plane of the energy filter in order to select a single energy band for imaging. By making small adjustments to the accelerating voltage of the microscope's electron gun, the beam can be steered across the slit to select the specific electron energy loss band to be used for imaging. Images can, therefore, be collected at various amounts of electron energy losses. With the use of additional lenses, the images can be magnified and focused onto a scintillator-based CCD-imaging system. Moreover, with the use of these additional lens elements, the spectrometer can be used in either an imaging mode, to collect energy-selected images, or in a spectrum mode, to collect the energy loss spectrum. Since "end-of-column" magnetic prism spectrometers, such as GIF (Gatan, Inc., Pleasanton, CA), are in wide use, we have used them to illustrate the mechanics of filtering electrons. However, the same principles apply in the omega type configuration (Egerton, 1996). With a basic understanding of the energy filter, we are now in a position to understand the application of the energy filter to the technique of MPL tomography.

II. Introduction to MPL Imaging

A. Motivation

A key goal of the modern imaging community is to obtain a multiscale understanding of structural biology, from the complete organism through the level of organs, cells, and subcellular organelles to the molecular organization of proteins and protein complexes (Martone *et al.*, 2002, 2003). In traversing these scales, an important area of research lies just beyond the light microscopic resolution of about 0.3 μm. We call this region the "mesoscale." For visualization at the spatial domain just below 0.3 μm, much successful work has been completed in specific labeling using fluorescence photooxidation and heavy metal staining for EM (Capani *et al.*, 2001; Deerinck *et al.*, 1994; Gaietta *et al.*, 2002; Maranto, 1982). With these specialized biochemical techniques, three-dimensional (3D) light and 3D EM data can be correlated, thereby bridging the spatial domains between light and EM data.

Electron contrast in tissues, cells, or subcellular organelles is often enhanced with metals. For this approach, cells are fixed, stained with osmium, lead, copper, or uranium, then, and embedded in Durcupan (Electron Microscopy Sciences, Hatfield, PA) or some other plastic resin. Nonspecific stains such as uranyl acetate and some lead salts, which react with many tissues and cellular components, are used to create contrast. Selective stains, such as copper-lead, create contrast by reacting with specific chemical groups in tissue or cellular compartments (Bozzola and Russell, 1999; Lewis and Knight, 1992; Thiery and Rambourg, 1976). These methods include photooxidation-assisted polymerization of diaminobenzadine (DAB) followed by osmium impregnation. Stained samples are then sectioned using an ultramicrotome to produce slices whose thickness is appropriate for imaging in the transmission electron microscope (TEM). Section thicknesses for use in a 100 or a 400 keV instrument are usually 100 or 500 nm, respectively. For specimens with a greater section thickness, chromatic aberration produces a drastic degradation in resolution, particularly at high-tilt angles, where the sample thickness effectively doubles at 60° and almost triples at 70°, rendering the resulting images essentially useless for high-resolution work.

The study of subcellular structures often requires the resolution afforded by the EM. Many of these cellular structures, however, can extend for tens to hundreds of micrometers and are therefore too extensive to be imaged in a single thin section. In such cases, serial section electron tomography must be performed (Soto *et al.*, 1994). In thick section studies, which are the focus of this chapter, essentially all beam electrons interact with the sample, as they pass through it. As discussed earlier, these interactions are of two types: elastic and inelastic. For thick specimens, a high percentage of the electrons will interact inelastically, producing a polyenergetic image with a large degree of chromatic aberration. The distribution of energy loss can be measured to produce an energy loss spectrum.

B. Description of the Method

Currently available electron lenses are not well corrected for "chromatic aberration," that is, they lack the ability to image all the electrons from one object point into one image point, regardless of the electron's energy. The high probability that beam electrons passing through a thick sample will experience inelastic scattering means that every point in a thick object is imaged as a spread of points, causing the image to smear. An alternative to developing chromatic-corrected lenses is to filter the electrons after they have passed through the specimen, so only a narrow range of energies is included in the image; this means that the chromatic aberration of the lenses becomes unimportant.

An energy filter can, in principle, be used to select electrons with any degree of energy loss. For example, one could select those that lost essentially no energy to the specimen (unscattered and elastically scattered electrons). However, with thick specimens these electrons are so few in number that large doses to the specimen are required to get enough electrons onto the detector to obtain a good representation of the electron scattering by the specimen. It follows that setting an electron energy loss filter so that it selects the electrons that are from the most highly populated region of the energy loss spectrum is a way to get good image "counting statistics" while keeping the range of energies in the image sufficiently narrow that chromatic aberration does not limit image quality. This approach is called "most probable loss" imaging because it is based on a selection of the region in the energy loss spectrum that contains the largest electron flux. A demonstration of the feasibility of this technique was first presented by Olins *et al.* (1989) using one of the first energy-filtering instruments available at 80 keV. In general, the MPL position will shift as a function of slit width, accelerating voltage, and mass thickness. For thick specimens and practical energy-selecting slit widths, the maximum of the plural-scattering peak will be the MPL position.

It is important to point out that that the fundamental contrast mechanism for MPL imaging of thick selectively stained samples remains scattering contrast, where contrast is created by the removal of electrons scattered into large angles beyond the acceptance angle determined by the objective aperture. This contrast results from a mixture of inelastic- and elastic-scattering events. For thick specimens, the angular distribution is broadened as a result of multiple elastic- and inelastic-scattering events (Reimer, 1989, 1993). For thin specimens, the dominant contribution to contrast is from elastic-scattering events and imaging in the low energy loss regions of the spectrum may actually result in contrast reversal, making MPL imaging inappropriate (Grimm *et al.*, 1997). In the case of low contrast or unstained specimens, contrast is created with a large defocus condition and is improved with zero-loss imaging.

III. Detailed Description of MPL Tomography

Resolution in an EM is most significantly affected by three critical factors: diffraction, spherical aberration, and chromatic aberration. Of these, diffraction and spherical aberration typically account for what is termed instrument resolution. Instrument resolution, however, is generally only achievable for very thin samples, such as graphitized carbon and amorphous germanium, where the electron mean-free-path, the mean distance traveled by an electron between collisions, is much longer than the sample thickness. In most cases, these resolutions are not attainable with thicker specimens. This is a result of energy losses and plural scattering in the sample occurring as a result of electron–specimen interaction. In the case of thick specimens, chromatic aberration can become a significant resolution-limiting factor. The shape of the energy distribution curve for multiply scattered electrons is broad and follows the Landau distribution (Reimer, 1989). As an example of the effect of sample thickness on the energy loss spectrum, Fig. 2 shows the electron energy loss spectra (EELS) as a function of tilt angle from a 1-μm thick, Durcupan-embedded section. Since the effective thickness t_{eff} of the section can be calculated geometrically to give

$$t_{\text{eff}} = \frac{t_0}{\cos\theta} \tag{3}$$

where t_0 is the sample's untilted thickness and θ is the tilt angle, we can observe how the EELS changes as a function of specimen thickness by tilting the sample.

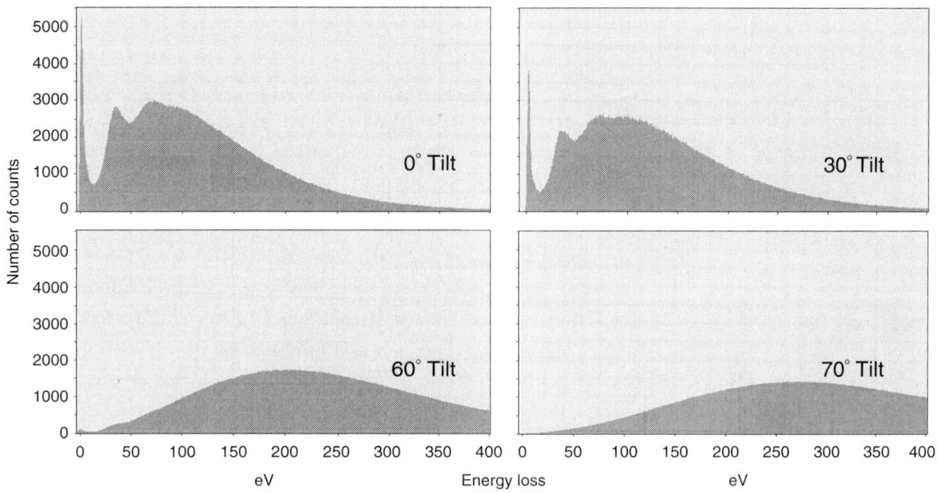

Fig. 2 Electron energy loss spectra (EELS) for a 1-μm-thick section of Durcupan-embedded specimen as a function of tilt angle. The effective thickness increases as the sample tilt angle increases. Note that the zero-loss peak decreases and eventually disappears at 50°, while the maximum of the multiple-scattering-loss peak shifts toward higher energy and broadens as the tilt angle is increased.

It is important to note that as the sample's thickness increases, the zero-loss peak decreases and eventually disappears completely, whereas the inelastic peak increases in relation to the zero-loss peak and shifts toward larger energy losses. Furthermore, the inelastic peak tends to broaden at higher specimen thickness. This shift and broadening of the inelastic peak causes a large loss of resolution due to lens chromatic aberration where electrons of different energies are focused differently by the magnetic lenses.

In Fig. 2, the shifting and broadening effects as the sample is tilted demonstrate clearly the effects of thickness on EELS. At 30°, 60°, and 70°, the 1-μm-thick sample has an effective thickness of about 1.2, 2.0, and 2.9 μm, respectively. As the thickness increases, the zero-loss peak decreases, whereas the inelastic peak increases in relation to the zero-loss peak and shifts to greater energy loss. Most importantly, the energy loss spectra broaden causing increased chromatic aberration.

The three issues that determine resolution in the TEM are characterized by considering the disks of least confusion given by the following equations (Agar, 1974).

$$\delta_{\mathrm{d}} = \frac{0.61\lambda}{\alpha} \tag{4}$$

$$\delta_{\mathrm{s}} = \frac{1}{2} C_{\mathrm{s}} \alpha^3 \tag{5}$$

$$\delta_c = C_c\left(\frac{\Delta E}{E}\right)\alpha \tag{6}$$

where δ_d, δ_s, and δ_c are the disks of least confusion due to diffraction, spherical aberration, and chromatic aberration, respectively. The semiangle objective aperture is given by α, with units of milliradians and λ is the relativistic wavelength of the electrons. E is the accelerating voltage and ΔE is the width of the energy loss dispersion of electrons forming the image. These resolution-determining factors can be plotted as a function of the objective semiangle α using the specifications provided by the manufacturer and a specified acceleration voltage.

As an example, C_s and C_c for the JEM-3200 in our laboratory are specified at 3.2 and 3.0 mm, respectively. Figure 3 shows plots of the three resolution-limiting factors versus the objective semiangle at an accelerating voltage of 300 keV. For the chromatic aberration calculations, a specimen thickness of 2 μm was used, corresponding to an untilted 2-μm section or a 1-μm section tilted to 60°. At this thickness, the dispersion in energy ΔE is determined experimentally to be ~250 eV, which is the full width of the energy distribution at half its maximal value (FWHM) (Fig. 2). For energy-filtered MPL images, an energy slit width of about 30 eV is used

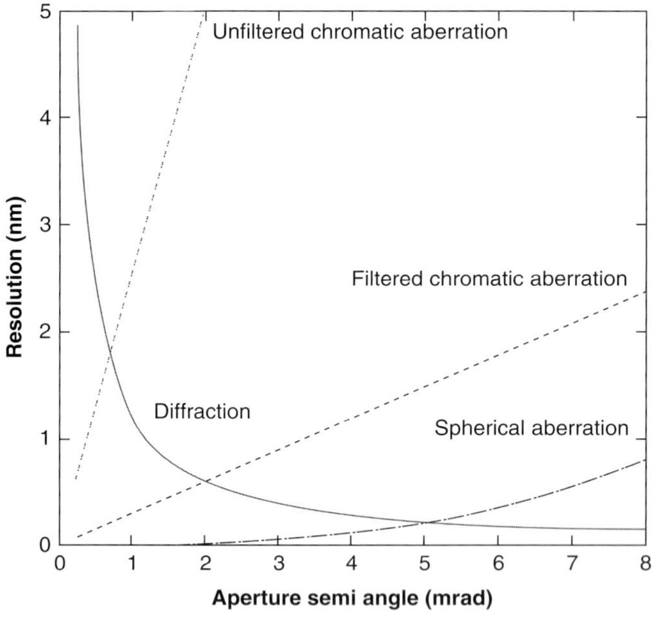

Fig. 3 JEM-3200EF electron microscope predicted resolution as a function of scattering angle due to diffraction, spherical aberration, and chromatic aberration. The relevant parameters are $E = 300$ keV, $\Delta E_{\text{unfiltered}} = 250$ eV, $\Delta E_{\text{filtered}} = 30$ eV, $C_s = 3.2$ mm, $C_c = 3.0$ mm, and $f_{\text{obj}} = 3.9$ mm. For reference, the scattering semiangle for a 30-μm objective aperture is 3.85 mrad.

as a practical value for the calculation of chromatic aberration. This degree of filtering should provide approximately one order of magnitude improvement in resolution for thick specimens where chromatic aberration is a severe resolution-limiting factor. This trend can be observed in the curves for filtered and unfiltered chromatic aberration in Fig. 3. Here the slope for chromatic aberration is reduced dramatically through the use of the energy filter. On the basis of these simple calculations, we predict that resolution in the range of 2 nm can be achieved in the raw tilt-series images by using the MPL-imaging strategy on specimens of 1-μm thickness at 300 keV. This is similar to the expected chromatic aberration in zero-loss imaging for the same slit width. However, we expect the multiple scattering processes in MPL imaging to degrade resolution slightly versus zero-loss imaging. Using unfiltered TEM on the same 1-μm section, we predict a resolution in the tilt series images on the order of 15 to 20 nm, based on Eq. (4). This is a severe problem for unfiltered TEM imaging of thick specimens with intermediate voltage electrons and precludes the use of specimens much greater than 0.5-μm thickness for high-resolution tomography work on an intermediate voltage microscope.

On the basis of the definition of MPL presented above, the MPL position can be mathematically determined by defining a new function $N_{\mathrm{Window}}(E)$, which provides the number of electrons collected into the slit window at an energy loss E. We defined this function by:

$$N_{\mathrm{Window}}(E) = \frac{1}{\Delta E_{\mathrm{slit}}} \int_{E-\Delta E_{\mathrm{slit}}/2}^{E+\Delta E_{\mathrm{slit}}/2} N(E')\mathrm{d}E' \tag{7}$$

where ΔE_{slit} is the width of the energy-selecting slit and $N(E)$ is the number of electrons collected at the energy loss value E. This is simply a moving sum across the slit width at each energy loss value E. The maximum of this function then determines the MPL position for a given slit width and energy loss distribution.

IV. Techniques

The procedure used in performing MPL tomography is simple but powerful. Once the sample has been prepared, the selected area (SA) aperture is used to select a background area that has not been stained with metal. EELS are then collected from the SA region at every 10° of tilt from 0° to 70°. These spectra are analyzed to determine the location of the MPL position according to the definition of MPL presented above. Once the locations of the MPL peaks have been determined as a function of tilt angle, the data are plotted and fit. The location of the MPL peaks have been determined empirically to be a linear function of specimen thickness for specimens between 0.5- and 3-μm thick; therefore, the data are fit to a function of the form:

$$E_{\mathrm{MPL}} = A + \frac{B}{\cos\theta} \tag{8}$$

where θ is the tilt angle, E_{MPL} is the MPL position, and A and B are the fit parameters. A least-squares fit to this cosine function for a 1-μm section of spiny dendrite is presented in Fig. 4. The parameters A and B are extracted from the fits and entered into an automated MPL tomography routine as input parameters. The autotomography routine is then set with appropriate defocus, exposure, tilt scheme, and search and track parameters. A detailed description of autotomography is presented by Koster *et al.* (1997).

Once an initial reference image is taken, MPL autotomography is performed using the following cycle of steps.

1. First, the tilt angle is read from the JEM-3200EF.
2. The MPL position is calculated according to the tilt angle and fit parameters obtained using Eq. (8). Then the energy-selecting slit is positioned about the MPL center.
3. Focusing is performed by collecting two images at a small condenser beam tilts. The amount of defocus-induced image translation is calculated by cross-correlating the two focus images. Since the image shift depends linearly on defocus and beam tilt, to first order and for small beam tilt angles, the cross-correlation provides the defocus position and the required focus correction.

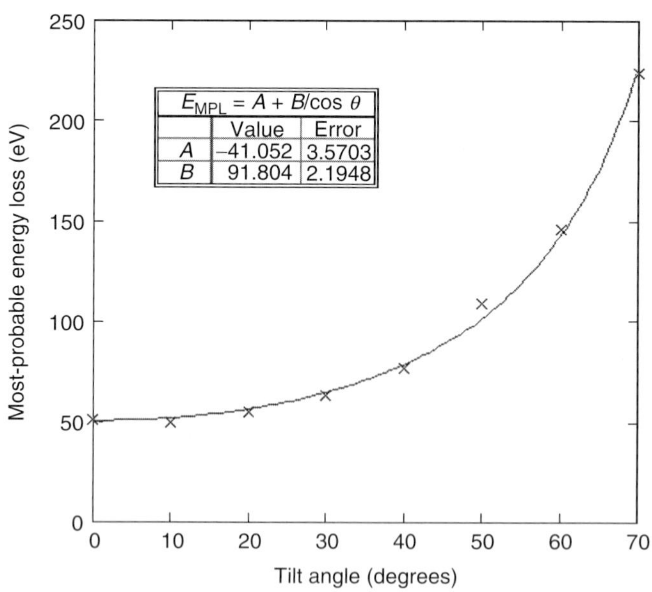

Fig. 4 Plot of MPL positions for a 35-eV-wide slit as a function of tilt angle at 300 keV. The curves are fit to the $A + B/\cos\theta$ function using the method of least-squares. The fit parameters are fed to the omega filter to control the energy filtering during automated tilt image acquisition.

4. A tracking image is collected at the current goniometer tilt angle. This image is compared to the previous reference image at the previous tilt angle.
5. Cross-correlation is performed between the reference and tracking images and a drift/shift vector is calculated. This vector is applied to the image shift coils to keep the specimen in rough alignment throughout the tilt series.
6. A tomographic image from the region of interest is acquired and stored. This image then becomes the next reference image.
7. The goniometer is rotated to the next tilt angle in the series, and the cycle is repeated beginning with step 1.

Using the steps outlined above, the entire MPL tilt series is completed automatically in 20–30 min. Variations in the method can be used, but the essential steps listed above should be included.

V. Results

To illustrate the utility of the MPL technique, two tomographic tilt series were acquired on a 1-μm section of a copper-lead stained spiny dendrite from the mouse hippocampus. The automated MPL tilt series was taken first. Then, the series was repeated in the JEM-3200EF's unfiltered mode with the slit removed. Tests were performed to ensure that sample degradation due to beam damage did not affect the unfiltered series. In Fig. 5, two representative images at 26°-tilt angle show

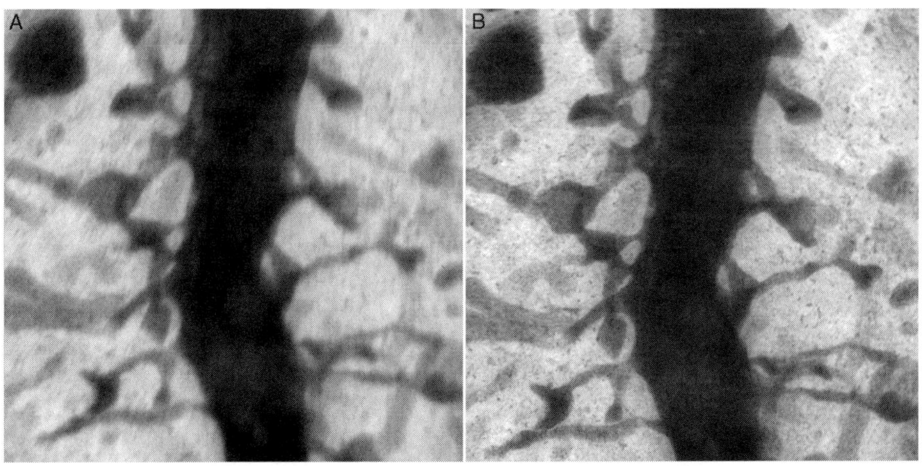

Fig. 5 Comparison between two raw images from a 4k magnification tilt series of a 1-μm-section cut from a spiny dendrite embedded in Durcupan. Images were taken at 26°. Image (A) shows an unfiltered image in imaging mode with the slit removed. Image (B) shows an image of the same region imaged using MPL tomography. For the MPL image, a slit width of 35 eV, centered at 61-eV energy loss was used. The reduction in chromatic aberration is especially evident when observing the 20-nm-diameter gold fiducial markers used for alignment.

MPL imaging versus unfiltered imaging. It is clear that the MPL image shows a dramatic improvement in resolution. In the unfiltered images, sharp edges have become blurred, and small details have been obscured as a result of chromatic aberration. Figure 6 shows images of an untilted, 2-μm section of copper-lead stained spiny dendrite for unfiltered versus MPL imaging. The improvement is once again clearly observed.

To illustrate further the utility of MPL tomography, a comparison was made between zero-loss images and MPL images for thick selectively stained samples. Images were collected from a 2-μm-thick section of a selectively stained spiny dendrite specimen. Each image was recorded using identical illumination conditions. Figure 7 shows that as a result of filtering, very few zero-loss electrons reach the detector, causing a large decrease in signal for the zero-loss images. Exposure times for the zero-loss images had to be increased to nearly 20 min to provide the same intensity values observed for 10 sec exposure times in the MPL images. The resulting MPL and zero-loss images show comparable resolution; however, far more x-ray events are observed in the zero-loss images. MPL imaging drastically reduces acquisition times and sample dose for thick specimen tomography. At zero tilt, the required dose to the specimen increases by $\sim$120; at large tilt angles this factor is greater still.

The complete tomographic reconstruction of a spiny dendrite was performed using back-projection algorithms (Sandberg *et al.*, 2003). Slices from the resulting MPL tomogram, acquired at 4,000×, are presented in Fig. 8. Figure 8A shows a central *x–y* section of the tomogram. The enlarged region in Fig. 8B shows

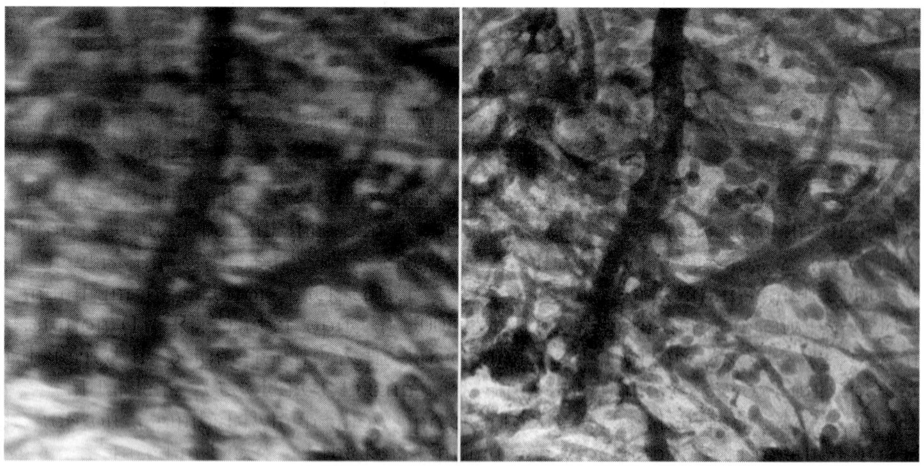

Fig. 6 Comparison between unfiltered and MPL images of an untilted 2-μm-thick spiny dendrite sample at 5k magnification with a 30-μm objective aperture. For the MPL image, the filter position was at 150-eV loss with a 40-eV-wide slit. The marked improvement in chromatic aberration is clear.

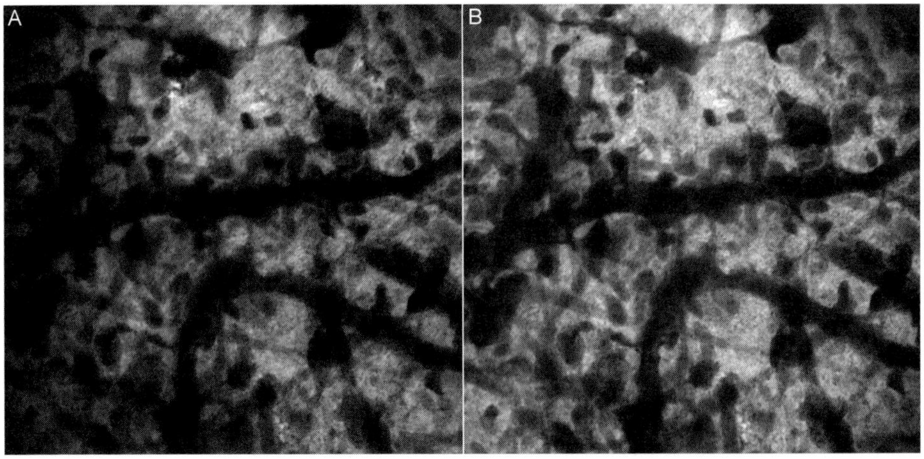

Fig. 7 Comparison of a 2-μm section of spiny dendrite imaged at 300 keV with (A) zero-loss imaging and (B) MPL imaging at 4k magnification and under identical illumination and tilt conditions. Exposure times were adjusted to produce the same mean image intensity. (A) is a zero-loss image exposed for 1200 sec with a 15-eV-wide slit. (B) is a MPL image exposed for 10 sec with a 35-eV-wide slit. Under these imaging conditions, the MPL signal is greater than the zero-loss signal by a factor of nearly 120, while the resolution in the two images is comparable.

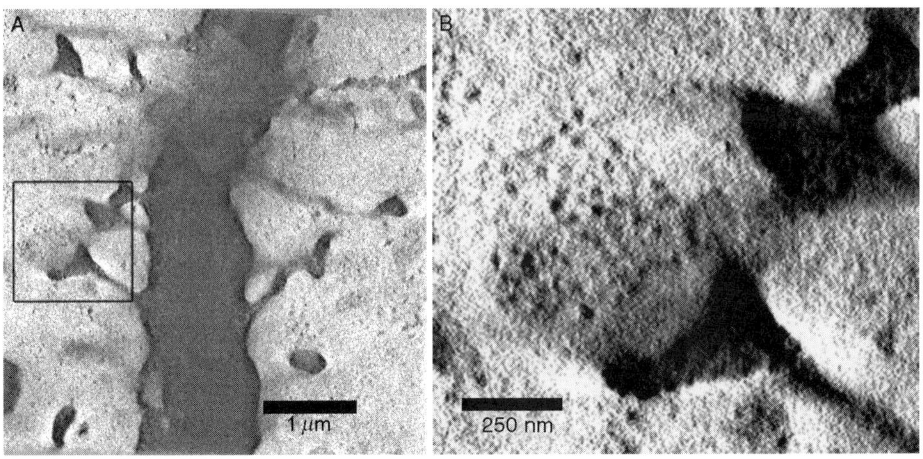

Fig. 8 (A) Slice 310 of 395 through the volume of spiny dendrite from an MPL tomographic reconstruction. Presynaptic vesicle substructure far beyond 20 nm is clearly visible in the tomogram slice. (B) A higher magnification slice from the tomogram showing individual presynaptic vesicles and staining details on the order of 4–6 nm.

structure at the subvesicle level. Operating in the lower magnification ranges allows reconstruction of a large section of the dendrite; however, higher magnification is required to determine the level of contrast provided by the MPL method at higher spatial frequencies.

At this low magnification, our ability to determine the resolution of the MPL method is limited by the response of the CCD camera system, where contrast is limited by the frequency response function or modular transfer function (MTF) of the imaging system (Fan and Ellisman, 2000). To produce a better determination of resolution in a biologically relevant sample, an MPL tomogram was produced with an instrument magnification of 20,000×, where the effects of CCD MTF contrast attenuation were less significant in determining experimental resolution. Figure 9 shows a central slice (316 of 705) from a 4k × 4k MPL tomogram at 20k magnification. At a magnification of 20k, we sampled at a quarter of the Nyquist frequency, thereby reducing detector MTF effects. At this magnification, we clearly resolve structures within the presynaptic vesicles at a 4- to 6-nm radial resolution (Bouwer *et al.*, 2004). This encouraging result supports the theoretical predictions outlined in Section III.

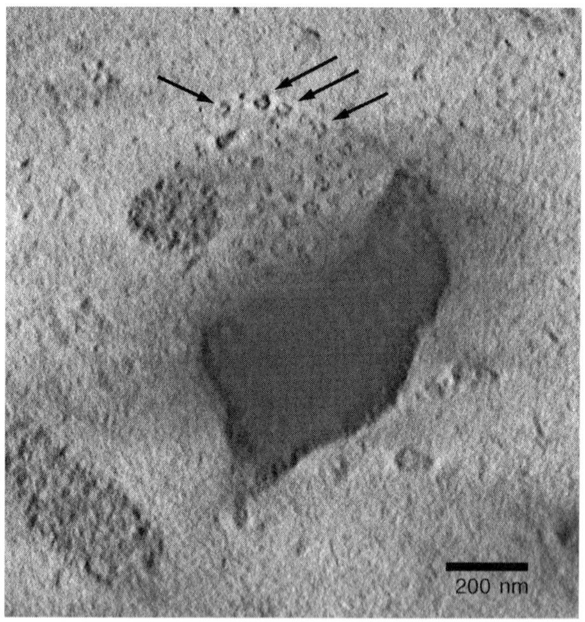

Fig. 9 A slice from a 20k magnification MPL tomogram recorded at 300 keV. The sample is a 1-μm-thick specimen containing a spine head. Details in the presynaptic vesicles (indicated with arrows) on the order of 10–14 pixels (4.4–6.2 nm) are clearly resolved.

VI. Conclusions

We have demonstrated that for thick, selectively stained biological specimens, MPL tomography can image volumes with a resolution that is far superior to those of unfiltered TEM tomography on the JEM-3200EF energy-filtered microscope. We have demonstrated the effective use of automated MPL tomography and have shown in detail how this strategy can be implemented on most energy-filtering microscopes. For a 1-μm thick selectively stained spiny dendrite specimen imaged on the JEM-3200EF without energy filtering on at 300 keV, contrast obtained at the level of 20 nm is too low to resolve even the fiducial markers. Therefore, alignment of the tilt images cannot be performed and a tomographic volume cannot be computed. If MPL tomography is implemented, however, fiducial markers are easily resolved and contrast in the tomographic volumes is improved to the level where copper-lead precipitates on the order of 4 nm can be resolved within the presynaptic vesicles. This strategy is useful for extending the range of specimen thicknesses that can be imaged on an intermediate voltage EM to more than twice that of standard unfiltered TEM on the JEM-3200EF.

The use of MPL imaging has been shown to produce a loss of resolution in thin specimens compared to the resolution attainable with zero-loss imaging primarily due to the small, natural energy width of the zero-loss peak for such samples (Grimm *et al.*, 1996); however, zero-loss imaging is impractical for thick specimens, as a result of the lack of elastically scattered electrons (Fig. 2). Moreover, the improvement in resolution resulting from imaging with the monochromatic electrons used in MPL imaging combined with the large increase in filtered image brightness exceeds the gain in resolution from using zero-loss images for thick specimens.

It has also been suggested that MPL imaging will produce problems with contrast inversion as the MPL peak shifts across the plasmon peaks and into the carbon edges (Grimm *et al.*, 1997; Olins *et al.*, 1989). Contrast inversion is clearly a problem for unstained or very thin samples. However, this phenomenon is simply not observed in thick, selectively stained samples. For these samples, the contrast is created at the objective aperture long before the image reaches the energy filter. The electrons striking the heavy metal stains, where electron mean free paths are very short, are mostly scattered into large angles at the sample and are subsequently removed by the objective aperture. Moreover, for thick specimens, where the sample is thicker than the electron mean free path, the loss edges become obscured by plural inelastic-scattering events, swamping any contrast inversion in the images (Egerton, 1996).

All of the data acquired for this study was acquired on a JEOL JEM-3200EF omega-filtered microscope at 300 keV. Unfiltered images were acquired with the slit removed from the energy selection plane. As a result, the use of an in-column energy filter may itself introduce additional chromatic aberration effects due to the spatial dispersion of polychromatic electrons in the image plane. The use of MPL imaging has been shown to dramatically reduce these effects for selectively stained thick specimen imaging. However, there remains a trade-off between MPL imaging and unfiltered imaging on a conventional column without a filter. In the case of an energy-filtered image, the added chromatic aberration of the energy filter will only yield a useful reduction in chromatic aberration versus a conventional unfiltered column for tomography of sections of thickness 1 μm or greater.

A problem inherent to thick specimen imaging that cannot be solved by MPL is the top–bottom effect. This effect arises because beam electron that experience a scattering event near the top of the specimen having a higher probability of being scattered again than electrons whose initial scattering occurred near the bottom of the sample (Fan *et al.*, 1996; Reimer and Ross-Messemer, 1987). As a result, both the resolution and the contrast at the top surface of the sample are lower than they are at the bottom. Although MPL tomography does not eliminate this effect, the quality of the images obtained for 1-μm sections show that top–bottom distortions for these selectively stained specimens are at least at or below the 4- to 6-nm chromatic aberration limit described in this report.

Acknowledgments

We thank Michael S. Isaacson from the University of Santa Cruz for thoughtful comments and suggestions and Mike Kersker at JEOL, USA for suggestions and feedback. This work is funded by NIH grants 5P41-RR004050, 1R01-NS46068, and 9R01-DA16602.

References

Agar, A. W. (1974). "Principles and Practice of Electron Microscope Operation." Elsevier, New York.

Barfels, M. M., Jiang, X., Heng, Y. M., Arsenault, A. L., and Ottensmeyer, F. P. (1998). Low energy loss electron microscopy of chromophores. *Micron* **29**(2–3), 97–104.

Bouwer, J. C., Mackey, M. R., Lawrence, A., Deerinck, T. J., Jones, Y. Z., Terada, M., Martone, M. E., Peltier, S., and Ellisman, M. H. (2004). Automated most-probable loss tomography of thick selectively stained biological specimens with quantitative measurement of resolution improvement. *J. Struct. Biol.* **148**(3), 297–306.

Bozzola, J. J., and Russell, L. D. (1999). "Electron Microscopy: Principles and Techniques for Biologists." Jones and Bartlett Publishers, Boston.

Capani, F., Deerinck, T. J., Ellisman, M. H., Bushong, E., Bobik, M., and Martone, M. E. (2001). Phalloidin-eosin followed by photo-oxidation: A novel method for localizing F-actin at the light and electron microscopic levels. *J. Histochem. Cytochem.* **49**(11), 1351–1361.

Castaing, R., and Henery, L. (1962). Filtrage magnetique des viesses en microscopie electronique. *CR Acad. Sci. Paris B* **255,** 76–78.

Crewe, A. V., Isaacson, M., and Johnson, D. (1971). Electron energy loss spectra of the nucleic acid bases. *Nature* **231**(5300), 262–263.

Daulton, T. L., Little, B. J., Lowe, K., and Jones-Meehan, J. (2002). Electron energy loss spectroscopy techniques for the study of microbial chromium(VI) reduction. *J. Microbiol. Methods* **50**(1), 39–54.

Deerinck, T. J., Martone, M. E., Lev-Ram, V., Green, D. P., Tsien, R. Y., Spector, D. L., Huang, S., and Ellisman, M. H. (1994). Fluorescence photooxidation with eosin: A method for high resolution immunolocalization and *in situ* hybridization detection for light and electron microscopy. *J. Cell Biol.* **126**(4), 901–910.

Egerton, R. F. (1996). "Electron Energy-Loss Spectroscopy in the Electron Microscope." Plenum Press, New York, London.
Fan, G. Y., and Ellisman, M. H. (2000). Digital imaging in transmission electron microscopy. *J. Microsc.* **200**(Pt. 1), 1–13.
Fan, G. Y., Young, S. J., Deerinck, T. J., and Ellisman, M. H. (1996). A new electron-optical mode for high contrast imaging and online stereo observation in TEM. *J. Microsc. Soc. Am.* **2,** 137–146.
Gaietta, G., Deerinck, T. J., Adams, S. R., Bouwer, J., Tour, O., Laird, D. W., Sosinsky, G. E., Tsien, R. Y., and Ellisman, M. H. (2002). Multicolor and electron microscopic imaging of connexin trafficking. *Science* **296**(5567), 503–507.
Garavito, R. M., Carlemalm, E., Colliex, C., and Villiger, W. (1982). Septate junction ultrastructure as visualized in unstained and stained preparations. *J. Ultrastruct. Res.* **80**(3), 344–353.
Grimm, R., Barmann, M., Hackl, W., Typke, D., Sackmann, E., and Baumeister, W. (1997). Energy filtered electron tomography of ice-embedded actin and vesicles. *Biophys. J.* **72**(1), 482–489.
Grimm, R., Typke, D., Barmann, M., and Baumeister, W. (1996). Determination of the inelastic mean free path in ice by examination of tilted vesicles and automated most probable loss imaging. *Ultramicroscopy* **63**(3–4), 169–179.
Kortje, K. H., Paulus, U., Ibsch, M., and Rahmann, H. (1996). Imaging of thick sections of nervous tissue with energy-filtering transmission electron microscopy. *J. Microsc.* **183**(Pt. 1), 89–101.
Koster, A. J., Grimm, R., Typke, D., Hegerl, R., Stoschek, A., Walz, J., and Baumeister, W. (1997). Perspectives of molecular and cellular electron tomography. *J. Struct. Biol.* **120**(3), 276–308.
Krivanek, O. L., Friedman, S. L., Gubbens, A. J., and Kraus, B. (1995). An imaging filter for biological applications. *Ultramicroscopy* **59**(1–4), 267–282.
Leapman, R. D., and Rizzo, N. W. (1999). Towards single atom analysis of biological structures. *Ultramicroscopy* **78**(1–4), 251–268.
Lewis, P. R., and Knight, D. P. (1992). "Practical Methods in Electron Microscopy: Cytochemical Staining Methods for Electron Microscopy." Elsevier, Amsterdam.
Maranto, A. R. (1982). Neuronal mapping: A photooxidation reaction makes Lucifer yellow useful for electron microscopy. *Science* **217**(4563), 953–955.
Martone, M. E., Gupta, A., Wong, M., Qian, X., Sosinsky, G., Ludascher, B., and Ellisman, M. H. (2002). A cell-centered database for electron tomographic data. *J. Struct. Biol.* **138**(1–2), 145–155.
Martone, M. E., Zhang, Y., Gupta, A., Qian, X., He, H., Price, D. L., Wong, M., Santini, S., and Ellisman, M. H. (2003). The cell-centered database: A database for multiscale structural and protein localization data from light and electron microscopy. *Neuroinformatics* **1,** 379–395.
Middleton, S. S., Latmani, R. B., Mackey, M. R., Ellisman, M. H., Tebo, B. M., and Criddle, C. S. (2003). Cometabolism of Cr(VI) by Shewanella oneidensis MR-1 produces cell-associated reduced chromium and inhibits growth. *Biotechnol. Bioeng.* **83**(6), 627–637.
Olins, A. L., Olins, D. E., Levy, H. A., Margle, S. M., Tinnel, E. P., and Durfee, R. C. (1989). Tomographic reconstruction from energy-filtered images of thick biological sections. *J. Microsc.* **154**(Pt. 3), 257–265.
Radcke, C., Stroh, T., Dworkowski, F., and Veh, R. W. (2002). Specific visualization of precipitated cerium by energy-filtered transmission electron microscopy for detection of alkaline phosphatase in immunoenzymatic double labeling of tyrosine hydroxylase and serotonin in the rat olfactory bulb. *Histochem. Cell Biol.* **118**(6), 459–472.
Reimer, L. (1989). Calculations angular and energy distribution of multiple scattered electrons using Fourier transforms. *Ultramicroscopy* **31,** 169–176.
Reimer, L. (1993). "Transmission Electron Microscopy: Physics of Image Formation and Microanalysis." Springer, Berlin.
Reimer, L., Fromm, I., Hirsch, P., Plate, U., and Rennekamp, R. (1992). Combination of EELS modes and electron spectroscopic imaging and diffraction in an energy-giltering electron microscope. *Ultramicroscopy* **46,** 335–347.
Reimer, L., and Ross-Messemer, M. (1987). Top-bottom effect in energy-selecting transmission electron microscopy. *Ultramicroscopy* **21,** 385–388.

Rose, H., and Pelies, E. (1974). Entwurf eines fehlearmen magnetishen energie-analysators. *Optik* **40,** 336–341.

Sandberg, K., Mastronarde, D. N., and Beylkin, G. (2003). A fast reconstruction algorithm for electron microscope tomography. *J. Struct. Biol.* **144**(1–2), 61–72.

Soto, G. E., Young, S. J., Martone, M. E., Deerinck, T. J., Lamont, S., Carragher, B. O., Hama, K., and Ellisman, M. H. (1994). Serial section electron tomography: A method for three-dimensional reconstruction of large structures. *Neuroimage* **1**(3), 230–243.

Thiery, G., and Rambourg, A. (1976). A new staining technique for studying thick sections in the electron microscope. *J. Microsc. Biol. Cell.* **26,** 103–106.

CHAPTER 27

Optimization of Image Collection for Cellular Electron Microscopy

Paul Mooney

Gatan, Inc., Pleasanton, California 94588

METHODS IN CELL BIOLOGY, VOL. 79

0091-679X/07 $35.00
DOI: 10.1016/S0091-679X(06)79027-6

Image collection is a key aspect of all kinds of electron microscopy (EM), including microscopy of the cell. Optimal use of existing detectors and continuing improvements to detector technology are crucial to further advancements in cellular EM. In order to pursue optimization of use and development of detectors, it is critical to understand central performance limitations and trade-offs. Spectral signal-to-noise ratio (SSNR) is presented as a means to quantify image quality and detective quantum efficiency (DQE) as a means to quantify detector performance relative to the generation of good SSNR. These quantities are used to describe a number of areas of cellular EM, and various improvements both to technique and to detector technology are described. Further work is needed to understand the SSNR implications of postacquisition processing in order to better understand the correct trade-offs to make in instrumentation development. Expected developments could help to bridge the resolution gap between cellular and molecular structural biology.

I. Introduction: Assessing the Role of Image Collection in Cellular Electron Microscopy

A. The Electron Microscope Image Collection Chain and the Role of the Detector

Cellular electron microscopy (EM), as covered in this volume, spans a diverse set of techniques used for attempting to understand cells in the range of scales between molecular structural biology and light microscopy. These techniques include stained or labeled thin-section EM, freeze-substituted thin-section microscopy and tomography, serial section techniques, and vitreous ice-embedded thin- and thick-section and whole-cell tomography. Image collection is a key component of all these varied forms of EM. In recent years, the understanding has emerged that signal-to-noise ratio is just as important or even more important than optical resolution for imaging biological material. In this context, it has become clear that efforts to improve visibility of biological structure must not neglect this very important component of the EM process.

Image collection is the fourth of four to five steps necessary for cellular EM, though all are, in a sense, imaging steps. The first, freezing and/or fixation, which is often followed by staining and/or labeling, either immobilizes the original biological system or creates a microscopic cast of it. The goal in both cases is to retain the information that is needed to study the biological structure of interest. The second and third steps are the interaction of a beam of partially coherent electrons with the fixed biological system and the formation of an electron-optical image of the prepared specimen. Electron image collection, the fourth step and the focal point of this chapter, renders the electron image into digital form—allowing it to be saved, transmitted, and processed with no risk of further loss in signal integrity. The final step, where applicable, is the processing of the acquired images, often yielding a three-dimensional (3D) map. The overall goal of cellular EM is to maximize the fidelity of this 3D map to the original object, the cell or cell section from which it came. Image acquisition is one of the most critical steps in this chain, and as such, it provides one of the biggest opportunities for further advancements of cellular EM.

Unfortunately, the goals of the various steps in cellular EM are at odds with each other in fundamental ways, creating a need for difficult trade-offs. For example, since electron image formation is done with quanta, the central disturbance in the fidelity of the system is shot noise. Thus, the electron-imaging step would be optimized with a dose high enough that variations in electron density at the electron optical image plane are less than other perturbations of image fidelity. Biological specimens are, however, sensitive to damage by the electron beam, and high-resolution structures are the first to be lost. At doses that retain high-resolution information, shot noise is a dominant disturbance. Another example is the fact that while high-resolution contrast formation and specimen penetrating power is increased by accelerating the beam electrons to high voltages, high-resolution image detection is much more difficult at higher voltages due to the long distances traveled by the high-voltage electrons inside the detector. In general, the optimization of system performance ends up being a tightly constrained trade-off among the requirements of all stages. One can hope, however, that improvements made to any one of the steps, by relieving constraints on the other steps, can lead to an improvement in overall system performance that is better than just the improvement in the component itself. One illustration of this point is provided by the way in which the limited dynamic range of film has been circumvented through the use of heavy atom stains. The higher dynamic range afforded by modern digital cameras could open up the possibility for the development and use of lighter, more structure-preserving staining protocols.

B. Quantitative Assessment of Imaging Performance Using Spectral Signal-to-Noise Ratio

Imaging fidelity is best described using the metric of spectral signal-to-noise ratio (SSNR). The "signal" in structural terms, for each of the steps listed above, would be the staining or label pattern (or molecular density in the case of frozen-hydrated specimen preparation), the distribution of electron flux, the strength in the digital record, and the densities in the 3D map, respectively. The noise is given by the disturbance—material above or below the structure of interest, fixation artifact, knife marks, shot noise, scintillation noise, and so on. In some cases, however, factors involved in the imaging process affect both signal and noise. High-voltage electrons impinging on the detector are scattered by collisions with atoms in the scintillator and its substrate. These collisions serve both to reduce signal through cross talk and also to increase noise by their random nature.

Signal-to-noise ratio should be evaluated in spatial frequency space as a way to avoid problems with analysis caused by signal and noise mixing between closely spaced pixels. Signal and noise "spectra" derived from Fourier transforms can be considered to be independent from one spatial frequency to another when the imaging system being studied is linear and the imaging performance is the same at all positions (Cunningham and Shaw, 1999). Since squared moduli of Fourier transforms, or "power spectra," add simply when two signals are independent, the preferred convention is to define the noise power spectrum as $\mathrm{NPS} = |\mathrm{FT(noise)}|^2$. The signal power spectrum is $\mathrm{SPS} = |\mathrm{FT(signal)}|^2$, and using these conventions, SSNR is

$$\mathrm{SSNR} = \frac{\mathrm{SPS}}{\mathrm{NPS}} = \frac{(|\mathrm{FT(signal)}|)^2}{(|\mathrm{FT(noise)}|)^2}$$

(Here, signal and noise are taken to be calibrated in terms of primary electrons, and the Fourier transforms are normalized to conserve the squared modulus.) A practical method for measuring SSNR will be presented in Section II.A.1. This convention, also referred to as quadratic spectral signal-to-noise, was used by Unser *et al.* (1987). This definition of the noise part of the SSNR is derived from variations in the data set itself and gives a good representation of SSNR for the entire imaging chain. In this chapter, the image acquisition device is the main focus and an instrument-based definition of the noise is needed. Specimen-free images of uniform beam will be shown to serve the purpose in Section II.A.1.

A comment on the choice of the quadratic convention is called for, since it differs from the "street" definition of SNR in terms of rms noise and linear signal. The choice is made here for a number of reasons. First, quadratic noise, also known as variance or power, has the nice property that it adds linearly for uncorrelated signals. The statistical discipline called analysis of variance (ANOVA) relies on this fact. Second, detective quantum efficiency (DQE), which will be defined in Section III.C, is more readily defined and interpreted in terms of quadratic power. Third, uncorrelated noise power is proportional to dose, leading to certain simplifications when talking about SNR or noise power as a function of dose. Finally, a small body of literature is accumulating in the field of molecular structural biology relating SSNR to other measures of resolution, such as Fourier shell correlation and differential phase residual (Unser *et al.*, 1987), and generalizing it to cover 3D data sets (Penczek, 2002). At a time when many are forming the intention to bridge the gap between molecular and cellular biology, it is appropriate to maintain consistency of terminology between these two disciplines. The term "power" spectrum will be used, despite the fact that it derives from electrical signal theory, because it has become so ingrained. An early use of two-dimensional (2D) power spectrum analysis for the description of an imaging medium was by R. Clark Jones of Polaroid Corporation (Jones, 1955), who applied Norbert Wiener's theory of harmonic analysis, first applied to one-dimensional (1D) time series (Wiener, 1930).

In the sections that follow, SSNR and some other closely related metrics will be used to describe image capture performance relevant to many areas of cellular

EM and to guide a discussion of ways to improve image capture and system performance—both in the use of existing detectors and in the design of new ones.

II. Applications: The Importance and Meaning of SSNR in Cell Biology Applications

A. Tomography

1. Tomography of Freeze-Substituted Plastic-Embedded Specimens

Tomography places strong demands on the imaging system. The need for many views increases the demands on the dose-tolerance of the specimen, on coherent beam generation by the microscope, on the stability and repeatability of the stage, and on the sensitivity of the detector. Staining and embedding in a 3D plastic matrix, while removing the imaging process one step further from the biological system, substantially relieves the dose-tolerance issue and in addition boosts signal contrast, making it the method of choice when permitted by the problem to be solved. Figure 1A shows a typical zero-tilt electron micrograph from a tilt series of a 300-nm-thick section of a *Chlamydomonas* cell in mitosis (images provided by Eileen O'Toole). The goal of 3D reconstruction encourages the inclusion of as much volume as possible, consistent with the performance of the imaging system, in this case a 300-kV FEG TEM (Tecnai TF30) and a fiber-optically coupled 2k $\times$ 2k image detector (Gatan 795 MegaScan). The use of a comparatively thick section, however, yields an image with less clarity than is typical of thin-section electron micrographs of stained, plastic-embedded specimens, but one which contains information in projection that, together with the other $\sim$100 tilted images of the same sample over a wide range of orientations, yields a 3D data set, detail from one slice of which (Fig. 1B) clearly shows high-resolution structures such as phospholipid bilayers. The features in this image demonstrate resolution at or above 5–6 nm.

2. Calculating SSNR of Detected (Unprocessed) Data

The SSNR of the tilt-series example can be calculated from azimuthally averaged squared moduli of Fourier transforms, or "power spectra," which have been normalized and calibrated such that an image of computer-generated shot noise with an average of one quantum per pixel would yield a power spectrum with a uniform value of 1. The total power spectrum (TPS) of the acquired image contains both the signal and the noise. The NPS of a uniform electron beam, unmodulated by a specimen, is produced in the same manner, either from a uniformly illuminated image at the same dose or from one at a different dose, followed by scaling the power spectrum by the ratio of the average electron doses (electrons per pixel) of specimen and uniform image (van Zwet and Zandbergen, 1996). In this case (Fig. 1C), scaling was done using the average power in the high-spatial-frequency

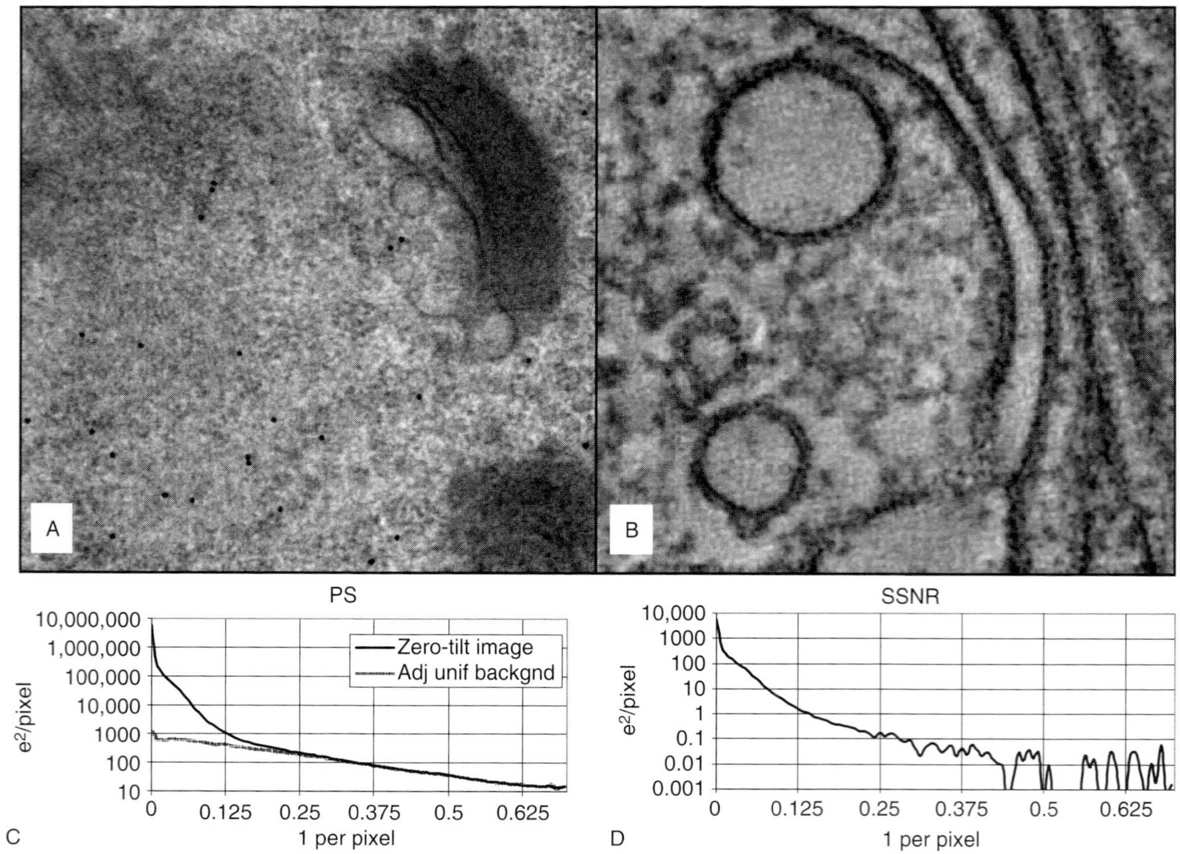

Fig. 1 (A) Zero-tilt electron micrograph from a tilt series of a 300-nm-thick section of a *Chlamydomonas* cell in mitosis (images provided by Eileen O'Toole). The specimen dose was 4000 e/nm^2 with a net 1450 e/nm (also per pixel) impinging on the detector. The dimensions of this selected subarea from the 2k × 2k raw image is 1340 pixels × 1340 pixels or 1.34 μm × 1.34 μm; (B) A subarea of a slice of the tomogram derived from the tilt series showing phospholipid bilayers in the golgi apparatus. Note that the phospholipid bilayer observed arises partially from the structure of the stain itself, which accumulates outside the membrane and thus leads to a slightly larger apparent spacing. Still, the contrast visible demonstrates tomographic resolution on the order of 6 nm. The graphical overlay illustrates the use of the data, through the use of image segmentation algorithms, for the study of 3D morphology, here mitotic spindles, only vaguely visible in the zero-tilt image; (C) Power spectra of the zero-tilt image (black) and of a uniform illumination image (gray) from the same detector which has been scaled to the same detector dose as the zero-tilt image. The latter curve is interpreted

range of the power spectrum, where it can be presumed that there is no power from the specimen signal. The SSNR is then computed with the straightforward formula:

$$\mathrm{SSNR} = \frac{\mathrm{TPS} - \mathrm{NPS}}{\mathrm{NPS}}$$

and is shown in Fig. 1D. SSNR, here at a resolution consistent with visibility of the phospholipid bilayer, is between 0.1 and 0.4. Unser *et al.* (1987) suggest a criterion of system SSNR = 4.0 to define the resolution of a two-dimensionally averaged image data set. While it might be naïve to directly apply that criterion, nothing else is currently available. Taking the visibility of bilayers spaced at ~6 nm as an indication of the resolution in the tomogram, the SSNR of the reconstruction at that spatial frequency should be at least 4, raising the question, how can the SSNR of the tomogram could be so much better than that of the input data set? We might expect some SSNR improvement from the processing step which combines ~100 images (McEwen *et al.*, 1995). Note that because of the quadratic definition of SSNR, in terms of signal power, the simplistic expectation from signal averaging of 100 frames would be a 100× improvement in SSNR. This expectation does not necessarily hold for tomographic reconstruction, though, because the data set is not reduced. In effect, the procedure moves the dose from projections into slices, leaving the total number of pixels approximately the same. It is likely that the explanation for the improvement in SNR in the final data set over the input comes from the computational removal of distracting structures from above and below the structure/slice of interest. It should also be noted that the SSNR of a reconstructed molecule (Penczek, 2002) or of a correlation averaged 2D image (Unser *et al.*, 1987) are of objects that fill the region for which the SSNR is being calculated. In the example above, the image has large regions that are not populated with any significant structure. It is for this reason, primarily, that power spectrum and SSNR studies of input data are fraught with difficulties. Still, it may be possible meaningfully to relate input to output SSNR in the imaging processing step. A proper treatment of SNR in tomogram processing is beyond the scope of this chapter, but something akin to the analyses done for molecular reconstructions would provide a useful reference point (Penczek, 2002; van Heel and Schatz, 2005). It would be very useful to be able to relate acquired input data SSNR to a 3D tomogram SSNR in such a way that aspects of detector performance could be identified for special focus and improvement.

Electron dose in plastic-section tomography is not in principal limited by specimen fragility but rather by a conspiracy of numerous practical factors that

---

as equivalent to the net camera noise background of the zero-tilt image; (D) SSNR calculated from the tilt-series image and scaled uniform background power spectra. SSNR at 6 nm (0.17 nm^{-1} or 0.17 per pixel) is about 0.4 or about 10^{-4} per specimen dose electron per pixel. No attempt is made in this chapter to either define or measure the SSNR of the slice (B) of the 3D tomogram data set.

act, nonetheless, to create a dose limitation. The trade-offs between beam intensity and coherence and between exposure time and drift, the fact that preexposure is needed to minimize specimen drift and collapse during the actual exposure (Luther *et al.*, 1988), and the use of a modest fraction of the total dose for focusing and alignment all contribute to a dose limit. For practical purposes this is ~10,000 electrons (e)/Å^2 for a tomogram or about 40–60 e/Å^2 per tilt image. At the detector, taking into consideration the elastically scattered electrons that are removed by the objective aperture and a pixel size of ~1 nm, which is appropriate to a resolution target of 4 nm, this specimen dose translates into a detector dose of 1000–1500 electrons (e)/pixel (David Mastronarde, University of Colorado, Boulder, private communication). While this dose would seem to remove shot noise from consideration, in practice the low contrast of the structures of interest, when viewed in projection through a thick 3D volume that contains other structures as well as heavy metal stain, places extreme requirements on the efficiency of the image collection device for making good use of the dose administered.

Image alignment is a less critical issue in plastic-section tomography than in tomography of frozen-hydrated specimens, because the SNR of the colloidal gold markers that are commonly used remains sufficient for the alignment process down to low electron doses. For instance, it has been stated that for alignment of 10-nm gold markers with 4-nm pixels, a minimum of 5 e/pixel is needed (Grimm *et al.*, 1998; Grimm, private communication, 2006)—well below doses typical of plastic-section tomography. However, additional demands are placed on the dose efficiency of cameras when it is necessary to use cross-correlation alignment from specimen detail alone. This happens occasionally, due simply to random fluctuations in the distribution of gold markers, and it is also called for if closer alignment is desired in a region of particular interest, when specimen distortions limit the precision of the alignment over the field of view (Richard McIntosh, private communication; Grünewald *et al.*, 2003). Further, any improvements possible in the SNR of the final tomogram will aid the success of image segmentation algorithms (see Chapter 30 by Sandberg, this volume).

3. SSNR of Detected Image Data in Vitreous Ice-Embedded Cryotomography

In cryotomography, typical doses are as much as 40–80 times lower than the ones discussed above. In this case, the specimen dose-tolerance sets the limit at 50–200 e/Å^2 (Iancu *et al.*, 2006). Figure 2 shows a montage similar to that in Fig. 1 but of a section cut from a frozen-hydrated yeast cell mitochondrion taken at 100 e/Å^2 total dose (images provided by Daniela Nicastro). With 1-nm pixels, the camera dose is 28 e/pixel after removal of some elastically scattered electrons by the objective aperture and of inelastically scattered electrons by an energy filter (Gatan GIF2002) (Krivanek *et al.*, 1992) mounted on the same microscope as the camera used in Fig. 1. While no quantitative comparison can be made between two completely different specimens, fixation methods and imaging techniques (and energy-filtered vs not), it is interesting to note that the (quadratic) SSNR curves differ by less than a factor of the ratio between the doses, despite the fact that the cryospecimen contains

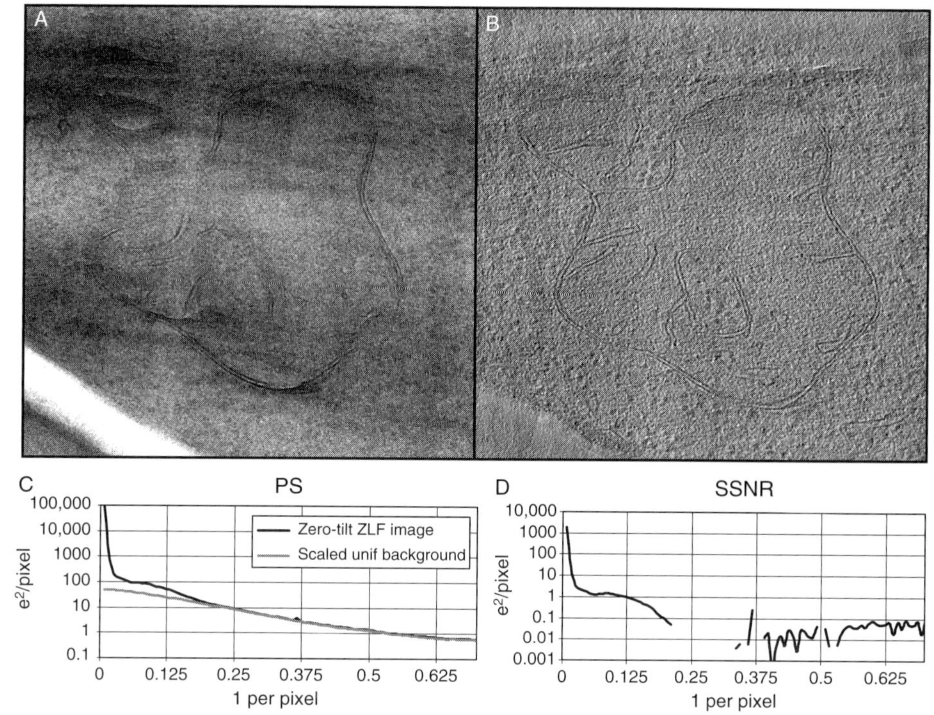

Fig. 2 (A) A 0° tilt image of a frozen-hydrated section of a fission yeast cell showing a mitochondrion (images provided by Daniela Nicastro). A 1340 × 1340 pixel subarea is shown with 1-nm specimen pixels having approximately the same field of view as in Fig. 1. The specimen dose was 100 e/nm^2 or 40 times lower than the dose in Fig. 1 with a net detector dose of 28 e/nm^2; (B) A slice of the tomogram derived from the full tilt series. The phospholipid bilayer is not visible in the tomogram of this unstained specimen; (C) Power spectra of the zero-tilt image (black) and of a uniform illumination image (gray) from the same detector scaled to the same detector dose as the zero-tilt image, as was done for Fig. 1; (D) The SSNR curve calculated from the tilt-series image and scaled uniform background power spectra. SSNR at 6 nm (0.17 nm^{-1} or 0.17 per pixel) is about 0.4 or about 4×10^{-3} per specimen dose electron per pixel—notably the same as the higher dose plastic embedded SSNR of Fig. 1D. The breakup of the curves comes from subtle variations between signal and power spectrum, possibly caused by improper matching of high-frequency power in the presence of high-frequency phase contrast.

no contrast-enhancing stain. This may be due to reduced elastic scatter in the low-Z specimen or to the removal of the energy-loss signal by the filter. The difference between the SNRs closes completely near 6 nm (0.17nm^{-1}). Since image alignment in cryotomography can also use gold markers, there are no new issues for the detector, except that the SSNR is so low that the need for SSNR improvement is all the more critical. At 100 e/Å^2 total dose, the 28 e/pixel is still well above the effective readout noise, which is typically 0.1–0.2 e/pixel for a fiber-optically coupled camera with a thick scintillator. Thus, the detector would allow acquisition under significantly lower dose conditions, for instance in the hypothetical case where dose per image was halved and the number of tilts was doubled, yielding a net dose reduction by a factor of 4. The relationship of readout noise (known as fog level for film) to the total noise power spectrum and to SSNR will be revisited in Section III.C.3.

Expected improvements to resolution in cryotomography that will be brought about by improvements in detectors can be estimated from dose-resolution curves in the published literature. McEwen reports an inverse fourth power law (McEwen *et al.*, 2002) and Grimm a third power law (Grimm *et al.*, 1998). Taking Grimm's more optimistic relation, an eightfold increase in dose would be stated to yield a resolution improvement by about a factor of 2. Using McEwen's more pessimistic relation, a resolution improvement of 1.7× would be expected. While a factor of 8 increase in dose is unrealistic under the constraints imposed by current technique in both plastic- and ice-embedded tomography, it will be shown in Section V that increases in SSNR by a factor of 8 at a constant dose are not unrealistic. These increases in SSNR will come from improvements to detector efficiency at rendering SSNR from the impinging electron images.

B. Localization of Functional Units Within the Cell

Neither conventional, thin-section EM nor tomography yield information about the components of the specimen, aside from those revealed by structure alone at the resolution obtained. A number of methods are available for rendering specific structures visible in the electron microscope. The most developed and widely used is immunolabeling with nanoparticles of heavy metal, usually gold (see Chapter 19 by Morphew, Chapter 22 by Cheutin *et al.*, and Chapter 23 by Sosinky *et al.*, this volume). Another, correlation-based identification, attempts to bridge the resolution gap between cryotomograms of cells and protein structures known at higher resolution from X-ray crystallography, electron diffraction, or averaged particle methods (Chapter 25 by Best *et al.* and Chapter 29 by Förster and Hegerl, this volume). Atomic species detection and concentration measurement (Chapter 24 by Leapman and Aronova, this volume) involve the use of electron energy-loss mapping using the signal normally discarded for contrast enhancement in zero-loss filtered tomography. Atoms specific to a macromolecule, such as the iron in ferritin or hemoglobin and phosphorus in DNA, can be used as identifiers (Chapter 24 by Leapman and Aronova, this volume) and ion concentrations can be studied directly (Leapman *et al.*, 1993). Each of these methods has its own signal-to-noise ratio issues.

1. Immunogold Labeling Signal-to-Noise Ratio Requirements

Unfortunately, heavy particles can disturb and distort the arrangement of the molecules to which they attach, affecting both directly labeled protein studies and immunolabeling. Efforts to make the labels less obtrusive also tend to make them less visible, requiring special techniques to make them visible (see Chapter 22 by Cheutin *et al.*, this volume). Immunolabeling is more reliable with smaller (5 nm) gold than with 15- to 20-nm gold but the visibility of 5-nm gold is compromised, especially at lower magnifications and in the presence of heavier stain (Chapter 20 by Pelchen-Matthews and Marsh, this volume). In addition, it is hard to label more than just a few molecular species, using size of the gold label as the discriminator. The same SNR issues arise in an attempt to discriminate different sizes of particle that arise in attempting to distinguish one class of particle from a background. In this case, it is instructive to temporarily step away from the *spectral* SNR approach and view the system in real space from the perspective proposed by Albert Rose (1946, 1948a,b). Assuming a perfect detector with no addition of noise and no pixel–pixel mixing, the threshold for visibility was found empirically to be:

$$\mathrm{SNR}[\mathrm{object}] = \mathrm{contrast}[\mathrm{object}] * \mathrm{sqrt}(\mathrm{Area}[\mathrm{object}] * [\mathrm{dose\ per\ areal}]) \geq 5$$

No attempt will be made in this chapter to relate this real-space threshold to the spectral SSNR thresholds, so Rose's rms (nonquadratic) definition will be adhered to for this instance. This criterion was developed to describe human visual interpretation of radiographs. For our purposes, it does not matter precisely which threshold is chosen, since the important task is to understand the relationship among dose per area, object size, and object contrast for the determination of visibility in a relative sense. The most interesting point made by Rose is that visibility is related to the total dose going through an object, independent of the object's size. In the case of nanogold, since the thickness of the particle determines the amount of scattering and therefore the contrast for any given section thickness, both contrast and area are related to the size of the particle. Contrast is expected to be proportional to the diameter, as long as the elastic scattering length is greater than the diameter. Area is proportional to diameter squared. It can be shown, then, from the equation above, that for a given SNR threshold and for a given section thickness and composition, the dose needed to achieve a given level of SNR ($N_{\mathrm{threshold}}$) is proportional to the diameter of the object to be detected (d_{object}) raised to the inverse fourth power. That is,

$$N_{\mathrm{threshold}} \propto \frac{1}{{d_{\mathrm{object}}}^4}$$

Thus, if under some set of conditions, a 5-nm gold particle is just distinguishable, then a 3-nm particle will require at least 7.7 times more dose per square nanometer to maintain visibility. Recall that this relationship is for a perfect detector. When the detector has blurring between pixels which reduces contrast for finite sized objects, the impact of reducing nanoparticle size will be even greater. Figure 3 shows an image of rat hippocampus labeled with 5-nm immunogold and imaged at 120 kV with 7k× nominal magnification on a fiber-optically coupled 4k × 4k charge-coupled device (CCD) camera (Gatan US4000), using a Tecnai T20 with a LaB6 filament at a dose of 840 e/pixel. The sample was provided by Kristen Harris, Georgia School of Medicine, from a study of endosomal compartments in dendritic spines (Cooney *et al.*, 2002). The pixel size, with a

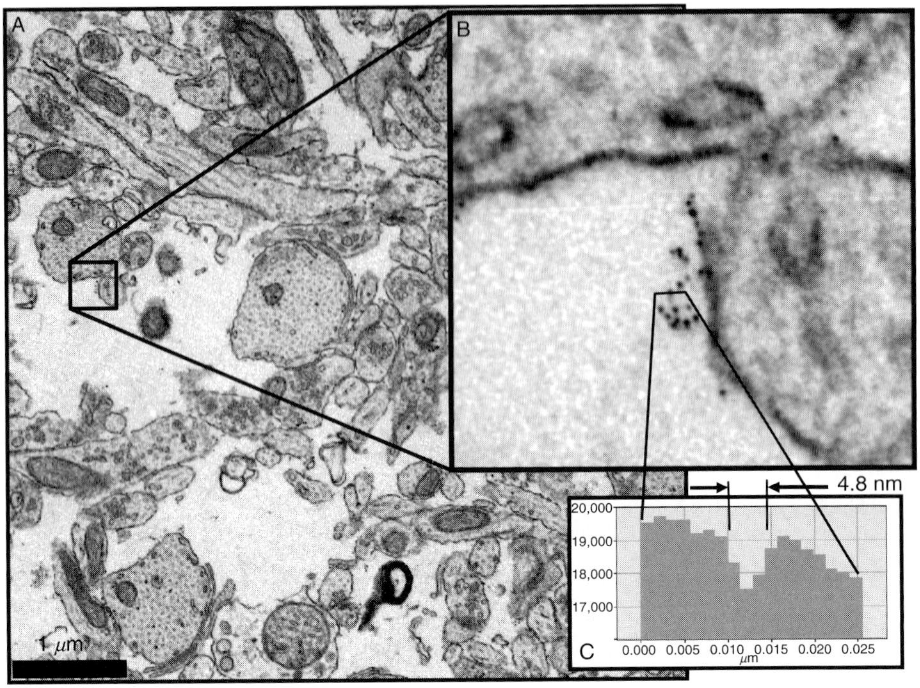

Fig. 3 (A) A 5-nm immunogold-labeled rat hippocampus imaged on a LaB_6 Tecnai T20 at 120 kV and 7k×, using a Gatan 4k × 4k with a thick scintillator (optimized for cryotomography) and a dose of 840 e/pixel. The net pixel size at the specimen was 1.6 nm. For reference, the same camera was used as for Figs. 7 and 11. The specimen was osmicated but otherwise unstained. The sample was provided by Kristen Harris, Georgia School of Medicine, as an example of nanogold-labeling and was imaged by Ken Tiekotter. Nanogold 3–4 nm was used to trace endosomal compartments in dendritic spines (Cooney *et al.*, 2002); (B) Inset showing 5-nm gold particles with sufficient contrast to be easily identified; (C) Intensity profile of one of the gold nanoparticles showing about 15% contrast and a clear 3-pixel width.

camera postmagnification of 1.35 was 1.6 nm, giving three pixels across one of the 5-nm gold particles in the example specimen with a field of view (FOV) of 6.5 μm. The specimen used in the study was labeled with 3- to 4-nm immunogold and would therefore have needed both higher dose and higher magnification to maintain the same level of visibility. Given the high contrast of the 5-nm gold images, however, some decrease in particle size without increase in magnification would have been possible.

The Rose score for the 5-nm nanogold particle in the image of Fig. 3, assuming an approximate area of 5 pixels and constant contrast of 15%, is about 10 or only double the visibility requirement. The fact that the nanoparticle images do not appear close to the visibility limit is an indication that nonindependence of neighboring pixels is altering the effective threshold. For an absolute treatment not requiring the change of magnification used above an analysis in frequency space is needed. A treatment of the relationship of the Rose criterion in spectral terms can be found in Cunningham (Cunningham and Shaw, 1999). For our purposes, however, it suffices to note the very strong relationship between visibility, dose, and size. Improvements to sensitivity and resolution in electron detectors will lead to meaningful improvements in the ability to use smaller probes to do more accurate and less disruptive labeling and to maintain the lowest magnifications possible for achieving a large FOV.

2. Correlative Techniques and Correspondence Analysis

In situ correlative techniques (Chapter 25 by Best *et al.*, and Chapter 29 by Förster and Hegerl, this volume) have the advantage that there is no limit to the number of different types of particles that can be studied. However, they are subject to the very significant challenge of cross-correlation alignment in a context that is not as well suited for high-resolution work as that of isolated, *ex situ* molecules in a thin film of vitreous ice. In any correlative technique, there is a potentiated role for SSNR in image detection. First, any improvement in raw data SSNR yields a direct improvement that is proportional to the SNR of the data set as a whole. Second, improvements in SSNR at constant dose lead to better cross-correlation alignments and consequently more rapid convergence to a high-resolution result or more accurate discrimination and categorization against reference templates from protein structure databases. The principal challenges for *in situ* protein structure determination or identification are the reduced contrast created by the thicker section and the crowding of molecules (Grünewald, 2003; Nickell *et al.*, 2006). Improvements in image capture SSNR would help to overcome these obstacles.

3. Spectrum Imaging Techniques

SNR concerns are usually thought to be raised by situations in which the dose is limited for one reason or another. However, extremely low contrast, as in the case when one is attempting to detect single atoms in a plastic-embedded biological specimen, serves as a pointed example of another important aspect of SNR and in turn, the importance of the efficient use of dose by the detector. The use of electron energy-loss spectral imaging for the detection and measurement of atoms and atom concentrations is a technique that operates in a completely different dose regime from transmission electron microscopy (TEM). It has been shown (Leapman, 2003) that using a STEM probe with 1 nA of current, 1-nm diameter and a dwell time of 0.2 sec (for a dose of 10^9 e/nm^2), single calcium atoms can be detected on a 4-nm-thick carbon film, containing on the order of 1000 atoms in each vertically projected pixel. By scanning, a 3D spectral image can be assembled that shows the locations of atoms within the matrix (Fig. 4) and, by extension, assuming the challenges of delivering sufficient signal-to-noise ratio in a considerably thicker biological specimen could be met, of proteins or nucleic acids containing those

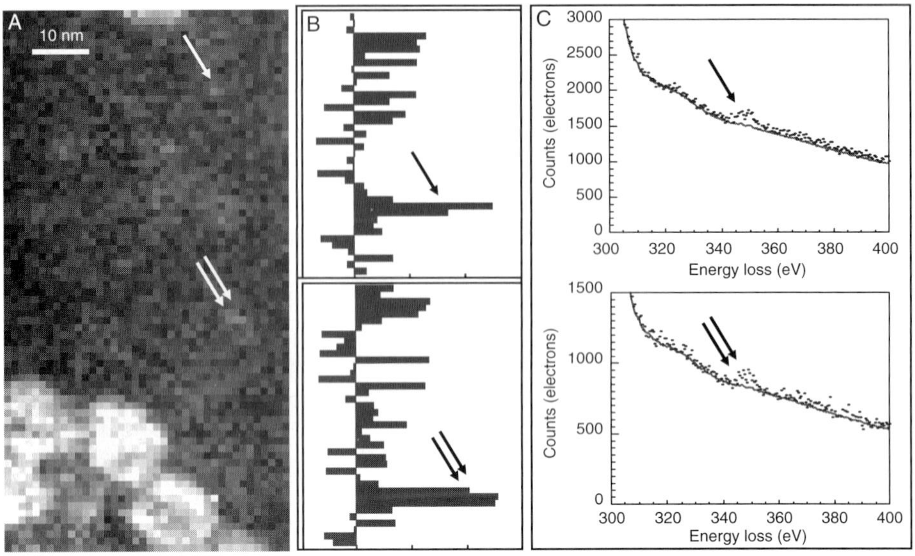

Fig. 4 (A) Calcium concentration image derived from electron energy loss spectra acquired at each point of a scanning transmission electron micrograph. The amorphous carbon substrate is 4-nm thick. The STEM probe had a 1-nm diameter and a 1-nA beam current; a dwell time of 0.2 sec/pixel was used for net dose of 10^9 e/nm^2. It was shown (Leapman, 2003) that the weak bright spots indicated by the single and double arrows correspond to one and two individual atoms of calcium, respectively; (B) Profiles through the concentration images, with single and double arrows marking one and two calcium atoms; (C) The electron energy loss spectra (EELS) of the two pixels associated with the one and two calcium atoms shown, superimposed on spectra of nearby pixels without calcium atoms. Image and graphs provided by Richard Leapman.

atoms. Measurement of concentrations to lower precision than single-atom counting allows operation at lower doses (Chapter 24 by Leapman and Aronova, this volume). Energy-loss imaging suffers in SNR from three sources. First, the energy-loss signal from core electrons, which bears the essential, elemental information, is a factor of 10^3–10^4 lower than the total signal passing through the specimen, though use of low energy-loss absorption edges offers the possibility of ratios closer to 10^2. Second, the background signal, over which the core-loss scattering signal must be detected, is proportional to the number of atoms in the matrix, not just those atoms that are being detected. Third, as specimen thickness grows, the zero-loss signal (which is used to normalize the signal to account for variations in specimen thickness) goes down, thereby increasing shot noise in the derived concentration signal. The 2D detectors currently used for spectrum imaging have very good dose efficiency for the 100-kV electrons used, as well as a slowly changing spectrum with minimal spatial variation, so there is little room for improvement of the detector in this case. Efforts in STEM-mode mapping are more focused on reducing the number of extraneous atoms by reducing the probe size through aberration correction. However, the same procedure could be carried out in an energy-filtered TEM (EFTEM) operating in an imaging mode, which would then place more extreme requirements on the detector by putting neighboring specimen pixels in close juxtaposition. In this case, intermediate to high-spatial-frequency SSNR would become important—and in this area, there is room for significant improvement in detectors. Still, it is conceivable that developments in this area, coupled with the use of low energy-loss peaks, could make molecular component concentration imaging or marker discrimination feasible in EFTEM, albeit at increased dose and reduced resolution. Two-step imaging processes can be envisioned in which high-resolution structural information is collected first with zero-loss filtering (ZLF) and then correlated with lower resolution compositional information acquired on the same instrument. The signals could complement each other in much the same way as the high-resolution brightness and the low-resolution chroma signals do in analog color TV standards.

III. Optimization of Detectors and of the Detection Process for Maximization of SSNR

A. Background: The Transition from Film to Electronic Detectors and the Value of Gain Normalization

1. Integration of the Detection Medium into the Electron Microscope

SNR is not the only issue of importance when considering image detection in cellular EM. Integration of the detector into the EM and the automation that it allows have been key to making tomography available to a broader range of users. Feedback about microscope performance derived from images analyzed online allows microscope control and optimization, which in turn makes possible new types of experiments, such as automated specimen finding and high-resolution acquisition over many hours or days, as is possible with Leginon (Suloway *et al.*, 2005) and AutoEM (Zhang *et al.*, 2003), as well as complete tomogram data-set acquisition with stage eucentricity, drift, and focus correction (Koster *et al.*, 1992; Mastronarde, 1997). It was for these reasons, rather than simply SNR, that tomography has driven the commercial introduction of CCD cameras into EM. Now that the benefit of integration has been realized and implemented across many areas, it is important to focus on the principal opportunities for further improvement.

2. Integration and Imaging Performance

In the area of image quality, it is interesting to look at the differences between the new recording modality of electronic cameras, and the film/scanner combination it is replacing. Both have detection nonuniformity or "grain" of one kind or another (silver halide particles in film, compared with phosphor particles, fiber-optic stacking patterns, and pixel sensitivity variations in CCD cameras). However, the pixel-by-pixel sensitivity variations of electronic cameras, by virtue of their fixed nature and repeatability, can be calibrated, allowing their affect on acquired images to be removed. Calibration is accomplished through the acquisition of an image of uniform illumination, usually averaged to reduce shot noise, normalized to an average value of 1, and inverted to reduce processing overhead at the time of use. The correction of newly acquired images by multiplication of the sensitivity calibration (aka gain reference) is called gain normalization (Aikens *et al.*, 1989). The improvement brought about by gain normalization can be seen in the 120-kV amorphous carbon (AmC) power-spectrum curves in Fig. 5A (data recalculated for quadratic SSNR from images taken by Dan Shi; Zhang *et al.*, 2003). At lower doses, both the gain-normalized 4k $\times$ 4k CCD camera (a US4000 with a thin scintillator) and film (scanned on a Zeiss scanner with 7-μm pixels) show a rise in their noise power spectra that is proportional to dose (data derived from same-dose images from a hole in the AmC). The signal, equal to the difference between the total and the noise power spectra, shows a rise that is proportional to the square of dose, consistent with the quadratic definition of both the signal and noise components of SSNR, as introduced above. However, at higher dose, (170 e/pixel, 1-Å pixel at the specimen, 7 μm at the film), the noise power spectrum of the film uniform image starts to rise as dose squared, just as for the signal power spectrum. This means that at doses above 170 e/Å^2 there will be no further improvement in SSNR from increased dose. This can be seen in Fig. 5C. The SSNR of the AmC images taken on the gain-normalized CCD camera on the other hand, shown in Fig. 5D, continues to rise proportional to dose. The ability to measure and factor out gain variations constitutes an important aspect of dynamic range, since there is no value to the ability to accumulate high signal if the contrast added is coming primarily from the detection medium itself.

3. Proper Gain Normalization Technique and Apparatus

It is important to note that in order to confer its intended benefit, gain normalization must be done properly. A gain reference that contains the same dose as the exposure will, when combined with the exposure to remove fixed gain artifacts, introduce the same noise variance as that present in the images of the specimen, doubling the noise component of the SSNR and halving the SSNR itself. Regular use of a gain reference underexposed in this way will effectively halve the dose efficiency of the camera. Even the standard recommendation, which is to use a gain-reference with 10 times the use-dose, will still lower the dose efficiency by 10%—not an issue for many routine applications, but certainly an issue when striving to push the limits of resolution. Indeed many efforts to optimize fiber-optically coupled cameras have resulted in dose-efficiency improvement of that same magnitude. So it is important to make sure that the gain reference is well dosed, as this can amount to a cost-free camera improvement.

Automation has raised another issue with gain-normalization, namely that subtle changes in gain-non-uniformity over the hours of an experiment can

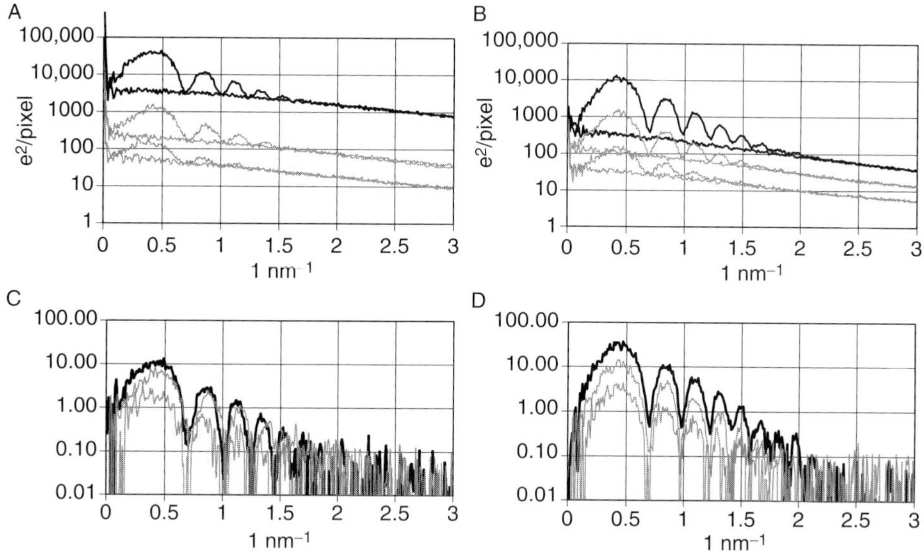

Fig. 5 (A, B) Noise power spectra of electron micrographs of a 30-nm AmC film and of uniform illumination through a nearby hole under identical conditions. The micrographs were acquired by Dan Shi at NCI/NIH using a Tecnai 12 (FEI Corporation) with LaB_6 filament at 120 kV and 700 nm underfocus both on (A) SO163 film scanned on a Zeiss SCAI scanner with 7-μm pitch and on (B) a 4k × 4k CCD Camera (Gatan, US4000P) with thin (P) scintillator. The microscope magnification was adjusted so that the pixel size at the specimen was close to 1 Å in both cases. The same three specimen dose levels were used for both film and CCD images: 17, 56, and 170 e/Å^2, respectively, displayed as light gray, dark gray, and black, respectively. The CCD images were gain-normalized using a gain reference taken with at least 100 times the experimental dose level. The noise power spectra of the AmC images were about 3% lower than the hole-image power spectra due to elastic scattering signal loss (nonphase contrast). Prior to calculation of SSNR curves, the noise power spectra were rescaled to be equal to the AmC power spectra at high spatial frequency to better approximate the true noise background of the AmC images (as shown). Note that in both (A) and (B), for doses of 17 and 56 e$^-$/Å^2, the noise power grows in proportion to dose, and the signal power (the humps riding on the noise background) grows in proportion to dose squared. However, for the 170 e/Å^2 image, the noise background power of the digitized film images grows faster than proportional to dose, indicating the rise of grain noise. No such change occurs in the gain-normalized CCD background power spectrum at 170 e$^-$/Å^2; (C) and (D) show the SSNR curves calculated from the power spectra of A and B. The film SSNR stops rising above 56 e$^-$/Å^2 while the gain-normalized CCD SSNR continues to rise in proportion to dose. For any given dose, CCD SSNR is higher at spatial frequencies below 0.4 of the Nyquist frequency, and the film SSNR is higher at spatial frequencies above that spatial frequency.

cause serious failures in the cross-correlations used to maintain system alignment during image acquisition. For this reason, recent camera developments have included measures to increase the stability of the gain-reference to the point where cross-correlated alignments can operate unperturbed by fixed structures in the gain over many hours.

4. Comparison of Film and CCD

The curves in Fig. 5A and B also allow a performance comparison between film and CCD. The AmC electron images were projected onto film and CCD with projector lens magnification adjusted to maintain close to 1-Å pixels on both detectors, so that the signal power density, referred back to the specimen, also remained the same between the two detectors prior to detection. The SSNR curves show a higher SSNR on the CCD camera than on the film/scanner combination with a shallow descent to an intersection with the film SNR at 5 Å, which corresponds to 0.4 of the Nyquist frequency in both cases. Zhang *et al.*, in the same study in which the SSNR data were taken (Zhang *et al.*, 2003), reported a reconstruction of Bacillus stearothermophilus E2CD protein to 15-Å resolution (corresponding, at 67k×, to 0.22 of the CCD camera Nyquist frequency), which was comparable to a reconstruction of the same specimen done using film digitized to 7 μm (Milne *et al.*, 2002). Aside from this example, the higher SSNR level of the CCD camera relative to film at low spatial frequency matches the experience that applications with greater requirements for sensitivity than resolution, like tomography and electron crystallography, adopted the CCD camera very quickly. The SSNR levels lower than film at higher spatial frequencies help to explain why single-particle analysis for protein structure determination, with its requirement for vast data sets with many pixels, has moved toward CCD cameras more slowly. Figure 6 shows an SSNR measurement at 200 kV (Booth *et al.*, 2004) with behavior that is similar to the 120-kV results. The curves show a crossover between film and CCD at a somewhat lower spatial frequency (0.25 of the Nyquist frequency), as would be expected for the higher microscope accelerating voltage. Reconstruction of a cytoplasmic polyhedrosis virus specimen was done on this camera to a resolution of 9 Å at a nominal magnification of 60k× or 83k× at the CCD so that the resolution limit of the specimen, referred back to the detector occurred at 0.4 of the Nyquist frequency, somewhat higher than the film/CCD crossover (Booth *et al.*, 2004). A reconstruction of the same specimen was done to 8-Å resolution on the same microscope using film (Zhou *et al.*, 2003). Encouraging results have been obtained at 300 kV as well, showing a turnaround of the film/CCD crossover and a flattening of the falloff in the CCD SSNR curve (Booth *et al.*, 2006). Other similar results by Sander *et al.* (2005) at 200 kV and by Downing and Hendrickson (1999) at 400 kV support the same conclusion that CCDs outperform film at low spatial frequencies, and that film outperforms CCDs at high spatial frequencies, given equal dose densities at the specimen. It is interesting to note the similarity between this method and the method of evaluating diffractogram spots of a metal test grid placed over a detector, in this case only for film (Downing and Grano, 1982). The earlier method is still effectively a SSNR method by virtue of the use of visibility of the diffractogram spots over the background signal as the measure of quality.

One possible explanation for the lower SSNR of the film images at low spatial frequency is its sparser distribution of grain. Film emulsions are roughly 50%

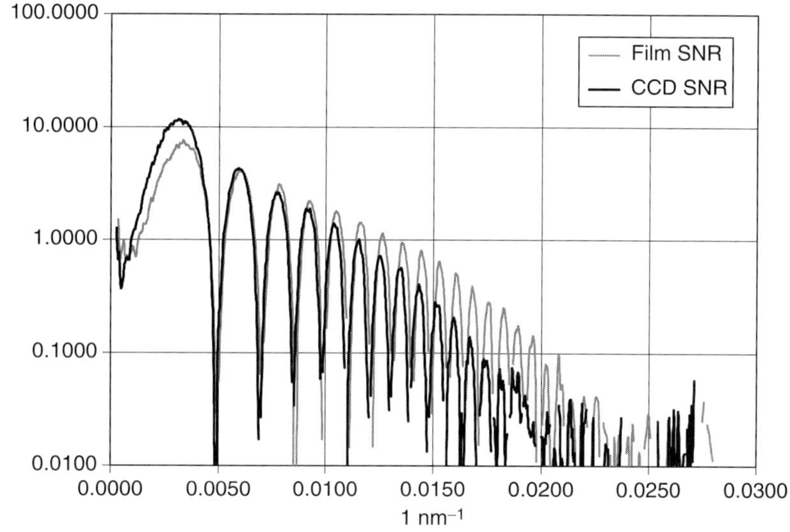

Fig. 6 Film/CCD comparison using the SSNR of AmC imaged at 200 kV (quadratic SSNR data provided by Chris Booth). The CCD Nyquist frequency is at the termination of the CCD plot. The crossover of SSNR performance (for a 7-μ film scan) is lower at 200 kV than at 120 kV, moving to approximately one-quarter of the Nyquist frequency of the CCD camera. Note that no absolute comparison between this 200 kV and the previous 120 kV result is possible, since the specimens, the microscope, and the voltage-related specimen interaction, all contribute to differences in image signal contrast power, which cannot be factored out of the comparison.

silver halide by weight, the rest gelatin (Zeitler, 1992), meaning that the volume fraction of the detective medium in film is smaller than in phosphor scintillators, as deposited in scintillator-based electronic cameras. This lower fraction could result in a higher degree of intrapixel gain variation and a drop in acquired-signal SSNR. The higher SSNR at high spatial frequency can be understood to be due to the at least partially transmissive nature of the film, which allows electrons that have traversed the detection medium to exit with a low probability of return, due to backscatter, compared with a phosphor scintillator on a glass substrate. The physical basis for the experimental differences seen between film and CCD remains an open issue.

5. Amorphous Carbon Thon Rings and Generalized SSNR Measurement

The power spectra of Fig. 5 also show that the uniform illumination power spectra track the zeros in the AmC power spectra created by the oscillations in the microscope contrast transfer function (CTF). For this to happen, it is necessary both for the specimen to be thin (30 nm in this case) and for the microscope to be stigmated very well, so the azimuthal average of the Thon ring (Thon, 1971) contrast does not wash out the zeros. This illustrates the fact that the noise background for a weak phase object is simply the random variation in a uniform illumination image at the same dose. Note that the uniform illumination noise power spectrum does not track the zeros (or lows) in the examples of Figs. 1 and 2 because those thick specimens had strong admixtures of contrast generated by incoherent processes, that is, amplitude contrast, for which there is no "zero" at a CTF crossover. These specimens will not allow the use of "zero" contrast fitting for the determination of the noise background level. In these cases, use of a uniform illumination power spectrum is the only way to provide an accurate, quantitative measure of the noise. This uniform illumination power spectrum must be scaled, however, since in general, there is always at least a small amount of elastic scattering, which will create a detector dose discrepancy between a specimen image and a "hole" image (van Zwet and Zandbergen, 1996). Rescaling a hole image taken under similar conditions to match the average dose of the specimen image will provide the correct noise power spectrum. Alternatively, the noise power spectrum can be rescaled to the level of the specimen image power spectrum if an interval of spatial frequency is available at high enough frequency that no signal power is present. In Section III.D.2, it will be shown that a uniform illumination noise power spectrum at any dose can be used in conjunction with the fog level to generate the noise background power spectrum for any particular specimen image dose.

B. The Physical Causes of SSNR Performance Reduction in Detectors

Currently, the dominant approach to the design of electronic image collection devices for cellular EM is the fiber-optically coupled scintillator CCD camera. By now, it is well understood that backscatter from the fiber optic plate into the scintillator is the cause of the shortfall in CCD cameras vis-à-vis the earliest predictions of their benefit, which were based on an understanding of single-channel scintillators and of the impressive properties of the scientific CCDs developed initially for astronomy. Published characterizations and comparisons with film (Downing and Hendrickson, 1999; Brink and Chiu, 1994) and an unpublished experiment introducing similar electron backscatter onto film by mounting a glass block directly underneath (David Mastronarde, private communication, 1996) served to delineate the circumstances under which CCD or film would be best. In addition, an earlier formal study of the effect of backscatter at high voltage on film served to elucidate the effects of backscatter in general (Valle *et al.*, 1989). Altogether, this work motivated renewed efforts to achieve the ideal performance hoped for at the initiation of the use of CCDs. Before discussing possible improvements, however, it is helpful to examine the physical factors that reduce SSNR performance. Figure 7 shows a series of line-spread functions, which illustrate the increasing severity of the backscatter problem at higher microscope accelerating voltage. These plots were generated from shifted difference images of an electron-opaque edge and are displayed in semilog coordinates to bring out the

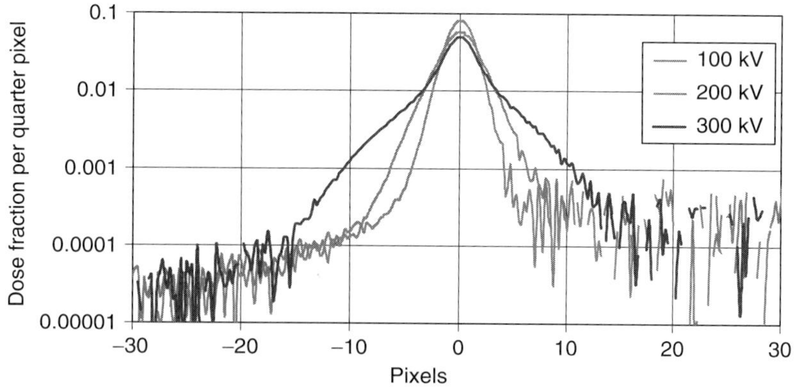

Fig. 7 Line-spread functions generated from shifted difference images from one side of a microscope beam stop. Effectively oversampled images are generated by processing images with a slight tilt according to a method proposed by Samei *et al.* (1998; ISO12223) but adapted for "found object" edges in EM, which might be slightly warped or bent through the use of slightly warped edges (Paul Mooney, unpublished work). The profiles show "shoulders" of increasing breadth as the microscope beam voltage rises from 100 to 300 kV, showing the average effect of electron scatter from the scintillator substrate (in this case a fiber-optic plate) back into the scintillator. This camera had a thick scintillator optimized for cryotomography, but it was also used for the nanogold-labeled work of Fig. 3. The CCD pixel size was 15 μm, so the extent of the backscatter plume can be seen to be over 200 μm in radius at its weak, outer extent. A kink evident in the 300-kV line-spread function shows that at higher accelerating voltage, the beam profile splits into two distinct parts, a sharp ingoing component and a long-range backscatter tail. The noisier trace on the right side of the curves corresponds to the illuminated side of the edge, the less-noisy left side to the dark shadow side of the edge.

detail of the long-range "tail" in the line response. The "shoulders" on either side of the center arise from electron backscatter, which gets stronger and wider as the incident beam energy grows from 100 to 300 kV. These "shoulders" cause a reduction in resolution. These curves, however, by virtue of showing only the average behavior of many electrons, do not show the most damaging aspect of the backscatter problem, which is the noise it generates.

Figure 8 shows two sequences of simulations that illustrate the detection process at different stages. Figure 8A and A′ show two repeats of a Monte Carlo simulation of the tracks made by 100 electrons impinging with 300 keV of energy as they pass through a typical scintillator (Gatan UltraScan "U"); they make the bright spot in the center of the field and are then backscattered by a fiber-optic plate (FOP) so they pass the scintillator a second time, hitting it in a wide range of positions and traversing it in a wide range of angles. The Monte Carlo simulation program was developed by Raynald Gauvin, McGill University, in close coordination with Gatan to assure accurate representation of proprietary material properties. Only the portions of the electron paths that are inside the scintillator are shown. Figure 8A and A′ represent the projected intensity in pixels that are 1/16th the size of the CCD pixel. About a quarter to a third of the beam electrons

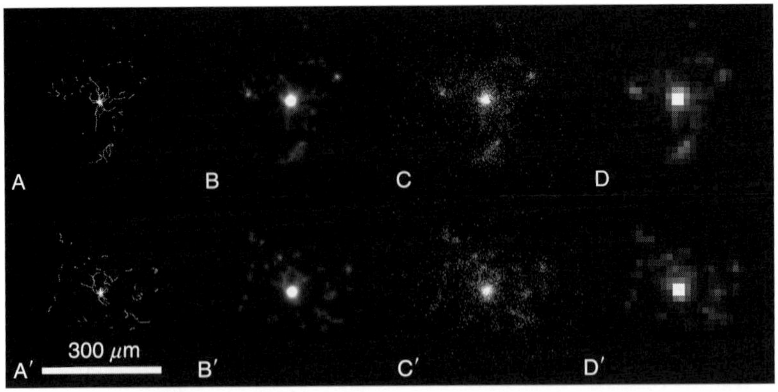

Fig. 8 Monte Carlo simulation (customized program provided by Raynald Gauvin, McGill University) of 300 kV electrons impinging on a model of a commercial fiber-optically coupled CCD camera (a Gatan US4000U) with an intermediate thickness scintillator. (A) and (A′) Two identical simulations of 100 particles each are shown to make variability apparent. Only energy deposited in the scintillator is shown, projected vertically into pixels 1/16th the size of the CCD pixel. Note that the tracks clustered near the point of entry are short compared to those created at a distance by backscatter electrons. Note also that electrons emerge up to 150 μm from the point of entry; (B) and (B′) Application of an approximate model for the probability density of the photon transfer optical process, including the effect of phosphor grain, to the energy deposited by the primary beam electrons; (C) and (C′). Poisson-model CCD electron generation based on the probability density of (B) and (B′); (D) and (D′). CCD electron collection in 15-μm pixels. It can be seen that the variation between (D) and (D′) is primarily due to the variation between the electron tracks and not to the randomness in the energy transfer process. It can also be seen that many more than the 100 pe shown would be needed to accumulate the smooth average curve of a PSF.

return to the scintillator at some distance from their initial entry point, often at reduced energy and at a higher angle, and therefore with a longer track, both factors meaning that they deposit more energy. Thus, while the backscatter coefficient can be between 25 and 30%, the energy fraction contributed by the backscattered electrons can be 30–40%. The two repeats in the figure serve to show the noise in the backscatter tail and illustrate why it is not conceptually helpful to think of the backscattered electrons as contributing to a point-spread function (PSF). Rather, the appropriate mental picture is that of electrons making tracks. Two repeats are shown to emphasize the random nature of the backscatter within the confines of the electron envelope, which at 300 kV is roughly 300 μm or 20 pixels in diameter.

The three images B, C, and D that follow each of the Monte Carlo simulation images illustrate the effects of photon coupling between the scintillator and the CCD. Figure 8B and B′ show the net effect of blurring from optical scattering by the phosphor grains and by the optical coupling from phosphor to CCD. Figure 8C and C′ show photons generated from the blurred energy deposition maps of Fig. 8B and B′ using a Poisson-noise generation function. Unlike in the case of the electrons, the net effect of the photon transfer probability distribution can be modeled with a single Poisson process, since there is no amplification, only attenuation after photon generation in the scintillator. The final images 8D and D′ show the photons binned together into the 15-μm pixels of the CCD. Readout noise is neglected for the purposes of this illustration. Clearly, the dominant source of noise in the last picture is still the electron tracks and not the random photon distribution, though it should be noted that when the photon collection efficiency of the optical coupling is low, as in the case of a lens-coupled camera, photon noise *can* be important (Section V.A).

C. Specimen-Free Detector Characterization: DQE

1. Dealing with Nonindependence of Neighboring Pixels

Because the tracks of backscattered electrons act in ways both to create crosstalk between pixels and to introduce noise above the inherent shot noise of the beam, traditional methods of characterizing detector resolution, such as point spread function (PSF), modulation transfer function (MTF), and (single-channel) detective quantum efficiency (DQE) are inadequate. The PSF, which is the average of images of many electrons entering the scintillator at the same point, fails to capture the variation of the electron images from one electron to the next and creates a highly inaccurate picture of what is happening in a cryotomography image in which only about 28 primary electrons hit each pixel. The MTF, the average amplitude attenuation of a set of idealized sine wave specimens with a range of spatial frequencies, while addressing the pixel-to-pixel mixing issue, does not describe the generation and attenuation of noise along the signal path. Single-channel DQE cannot be used in conjunction with MTF as a complement because efforts to compensate mixing-induced noise reduction with a single parameter mixing factor fail to properly handle either the shot noise in the electrons produced in the CCD or the noise of the electron tracks. Besides this, the mixing factor is very difficult to determine experimentally. To properly handle signal and noise transfer in a detector with nonindependent pixels, both signal and noise must be handled fully in frequency space.

The SSNR approach already introduced bridges the gap by merging the two issues of resolution and noise into a single concept: SNR as a function of spatial frequency. For characterization of detectors, however, it is not practical to generate a high-resolution electron image of an isotropic amorphous specimen, since changing either specimen or microscope will change the image signal due to differences in thickness, contamination, microscope CTF, beam coherence, and any number of other subtle effects. A specimen-independent method is needed and is provided by the concept of DQE generalized to the frequency domain. Here, no special nomenclature is used to distinguish "spectral DQE" from spatial-domain, single-channel DQE as defined 60 years ago (Jones, 1958; Rose, 1946; Zweig, 1965). The term and measurement techniques of spatial frequency dependent

DQE have now been in use within the medical radiography community for more than two decades (Cunningham and Shaw, 1999; Rabbani *et al.*, 1987) and have coalesced into an IEC standard (IEC 62220) for characterizing medical imagers. While these techniques have been applied primarily to X-ray imaging systems, the concepts are directly applicable to EM image acquisition (de Ruijter, 1995; de Ruijter *et al.*, 1993, 1994; Meyer and Kirkland, 1998; Meyer *et al.*, 2000).

The definition of spectral DQE (DQE from here on in this chapter) derives simply from SSNR as follows:

$$\mathrm{DQE}(N,s) = \frac{\mathrm{SSNR}_{\mathrm{out}}(N,s)}{\mathrm{SSNR}_{\mathrm{in}}(N,s)}$$

where N is dose per area and s is spatial frequency. The above equation can be rewritten in terms of measurable quantities as follows:

$$\mathrm{DQE}(N,s) = \frac{\mathrm{MTF}^2(s)}{[\mathrm{NPS}_{\mathrm{out}}(N,s)/N]}$$

where MTF, which is not a function of dose, can be measured at high dose with a low-magnification test object, for example, the edge of a pointer. $\mathrm{NPS}_{\mathrm{out}}$ is the noise power spectrum of the uniform illumination used in the power spectrum and SSNR analyses seen in Section II. Note that this formalism calibrates images and power spectra in units of beam electrons, while that of Meyer (Meyer and Kirkland, 1998) calibrates in terms of CCD electrons. Otherwise the formalisms are equivalent. The noise power spectrum of the input can be shown to be equal to the dose.

These measurements are difficult to perform accurately, and it is important to note certain cautions. These will be listed as a brief set of rules of thumb, with detail following:

Rules of thumb for detector characterization:

1. Use a slanted edge to measure MTF which renders the pre-sampled MTF accurately. The correct post-sampled MTF is then $\mathrm{MTF}_{\text{pre-sampled}}$ times the sinc function.
2. MTF must be normalized with a retracted edge image.
3. The "noise-method" for measuring MTF is incorrect.
4. MTF measurement must be done using a linear range of the detector.
5. Noise in the MTF measurement must be understood and controlled.
6. Dose must be measured using a Faraday cup.

Background of the rules of thumb:

1. The MTF measurement must be based on a method that does not allow artificial inflation from factors such as aliasing of signal from above the Nyquist frequency or from neglect of low-frequency light scatter. The slanted-edge method (Samei *et al.*, 1998) ensures that no aliased signal is injected. Figure 9 shows the

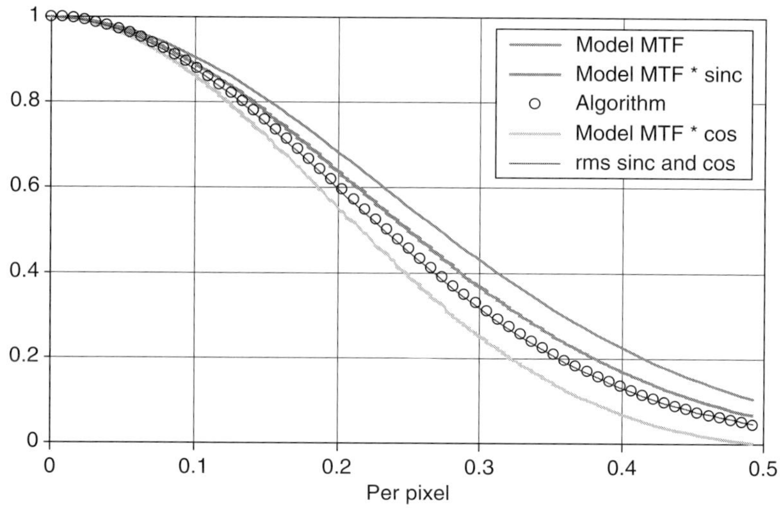

Fig. 9 Graphs showing the relationship between the adapted Samei (Samei *et al.*, 1998) MTF algorithm and presampled MTF (the MTF prior to sampling into pixels by the CCD). The top, black curve is a model Gaussian MTF defined on a grid eight times finer than the CCD pixel spacing. The darker gray curve, second to the top, is the model MTF multiplied by the sinc function, usually considered to represent the effect of sampling, since it is the Fourier transform of the square aperture of the pixel. The light gray curve at the bottom is the model MTF multiplied by the cosine function. The curve labeled "algorithm" was the result of applying the adapted Samei method to an edge slanted by 10°, which was softened by the model MTF and then "binned" into the CCD pixels. The result matches empirically with the rms average of the cos and sinc functions and is interpreted to correspond to an average over edge-pixel boundary alignments along the slanted edge. The algorithm yields consistent results over slant angles between 5° and 25°.

empirical relationship of the slanted-edge algorithm, as implemented by this author, to the presampled MTF (the MTF prior to binning of the image into the CCD pixels). The result was obtained using a simulated blurred edge image. Stated analytically, this empirically determined relationship is as follows:

$$\mathrm{MTF}_{\mathrm{algorithm}} = \mathrm{MTF}_{\text{pre-sampled}} * \mathrm{sqrt}\left[\mathrm{sinc}^2(\pi s/s_{\mathrm{Nyquist}/2}) + \cos^2(\pi s/s_{\mathrm{Nyquist}/2})\right]/\mathrm{sqrt}(2)$$

where $\mathrm{sinc}(x) = \sin(x)/x$ is the sinc function, the aperture function for a full-area pixel and s is the spatial frequency. Then

$$\mathrm{MTF}_{\text{post-sampled}} = \mathrm{MTF}_{\text{pre-sampled}} * \mathrm{sinc}(\pi s/s_{\mathrm{Nyquist}/2}).$$

2. Normalization must be to a real zero-spatial-frequency reference, such as a uniform beam with test edge removed, and not to the lowest frequency in a derived MTF curve (Cunningham, private communication, 2005). No arbitrary normalizations to one at zero spatial frequency should be done either to MTF or to NTF

(Meyer and Kirkland, 1998) since near-singularities in low-spatial-frequency performance (like the effect of lens scatter) can be missed.

3. Measured softening of the noise power spectrum itself cannot under any circumstances be used as a probe of resolution, since the detector is generating noise in the mapping between input and output (Meyer and Kirkland, 1998). The preferred, slanted-edge method for MTF measurement is codified in the IEC 62220 standard and also in the ISO 12223 standard for characterization of digital cameras (ISO 12223).

4. It may be necessary to treat nonlinear detectors such as film in a marginal-dose sense, that is, rather than using a high-dynamic range image of an edge, acquiring a double exposure of a uniform intensity and of an edge-image that has low-enough contrast to ensure linear behavior. Gain would be measured correspondingly as marginal gain:

$$\mathrm{Gain}_{\mathrm{marginal}} = \frac{\Delta\ \mathrm{counts}}{\Delta\ \mathrm{dose}}$$

though this extra effort should not be necessary if total doses are kept low.

5. Noise in images used to derive the MTF will artificially boost the MTF measurement, especially at high spatial frequency. To ensure that noise does not artificially raise high-spatial-frequency MTF, results from images at two different doses should be compared to see whether an increase in dose yields an insignificant reduction in apparent MTF. The effect of edge-length should be checked in the same manner, especially when the detection medium is not gain-normalizable so that grain-induced gain nonuniformity must be averaged out, as is the case for digitized film.

6. Errors in gain measurement translate proportionally into errors in DQE. that is, a gain measurement that is 10% high will result in a DQE measurement which is 10% high. Poor current measurement, usually caused by attempting the measurement without a Faraday cup, is the most common source of error. Microscope viewing screen calibrations are often only accurate at one microscope accelerating voltage, sometimes at none. Viewing screens, even when properly calibrated, will give different current readings from the same spot for different beam spread diameters. Beam current drift is another possible source of error.

Another rewriting of the definition of DQE makes its role as a measure of dose-efficiency clear:

$$\mathrm{SSNR}_{\mathrm{out}}(N, s) = \mathrm{DQE}(N, s) * N * C(s)$$

where $C(s)$ is the contrast power spectrum (the signal power spectrum normalized to one at zero spatial frequency). From this, it is evident that a doubling of DQE at a given spatial frequency will have the same effect on the specimen image SSNR as a doubling of dose. Yet another rewrite of the DQE definition equation:

$$\mathrm{SSNR_{in}}(N,s) = \mathrm{SSNR_{out}}(N,s)/\mathrm{DQE}(N,s)$$

suggests an approach to removing detector performance issues from comparisons of other system aspects. This would be useful not only for comparing situations with different detectors at different points, as would be true when comparing image contrast above and below an image filter, but also for comparing configurations with different system magnification onto the same detector.

2. Confirmation of the DQE Method by Comparison with a Specimen-Based SSNR Measurement

Figure 10 shows DQE measurements of film and of a CCD obtained as part of the same experiment as the SSNR curves of Fig. 5C and D. The DQE comparison agrees with the SSNR comparison. The DQE and SSNR of a CCD are both roughly double that of film at zero spatial frequency, and they both become roughly equal at 0.5 of the Nyquist frequency. It is interesting to note that the DQE at 0.22 times the Nyquist frequency, the sensor spatial frequency of the resolution obtained in the reconstruction mentioned in Section III.A.4 (Zhang *et al.*, 2003) is just over 0.3. However, as can be determined from higher voltage measurements, the resolution of the reconstruction is of course not determined by the DQE alone; it also depends on the nature of the beam-specimen

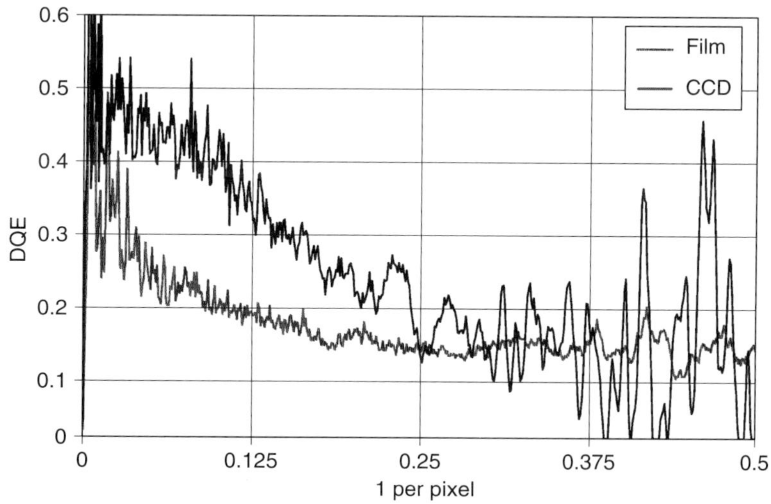

Fig. 10 Graphs of the DQE of CCD and film at 120 kV and a dose of 17 pe/pixel. The measurement was done with the same equipment used to acquire the AmC power spectra of Fig. 5. The CCD images were gain-normalized; the film images were digitized at 7 μm on a Zeiss SCAI scanner. These results are consistent with the ratio between CCD and film AmC SSNR curves, showing that this method of measuring DQE is predictive of the SSNR-generating capabilities of a detection medium. The curves were calculated from images provided by Dan Shi, NCI/NIH.

interaction. Further study is needed to better understand the interrelationship between detector DQE, microscope parameters, and reconstruction resolution.

3. DQE Assessment of FO-Coupled CCD Cameras

The DQE curves in Fig. 11 show the performance typical of a fiber-optically coupled CCD across a range of voltages and doses. They illustrate in a directly quantitative way the problem that was studied on an application basis in the work mentioned earlier (Brink and Chiu, 1994; Downing and Hendrickson, 1999). Comparing the 300 kV curves in this plot to the line-spread function in Fig. 7A, a similar inflection point can be seen at intermediate spatial frequency. The performance can be seen from both curves to be a superposition of two effects: that of the

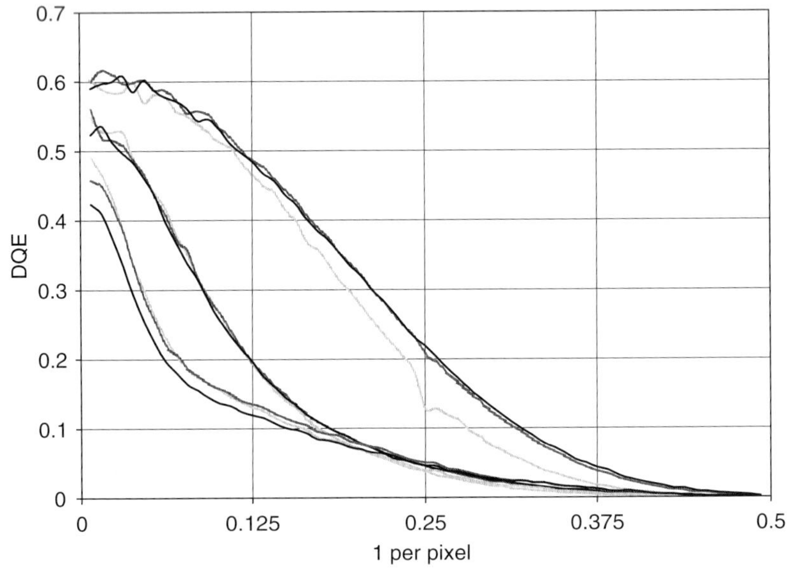

Fig. 11 DQE of a thick-scintillator, fiber-optically coupled CCD camera for electron beam voltages of 100, 200, and 300 kV. The three groups of curves each show DQE for three doses: 100 pe/pixel—black, 10 pe/pix—dark gray, 1 pe/pix—light gray. The top group of curves shows DQE for 100 kV, the middle for 200 kV and the bottom group for 300 kV. It can be seen that the DQE appears to separate out into an ingoing and a backscatter component similar to the manner noted for Fig. 6, showing a kink at low spatial frequency where the two curves meet. An interpretation of this effect in terms of backscatter noise background is that the voltage determines the size of the backscatter tail, but to first order, has little effect on the magnitude of the noise added. This can be seen in the fact that the DQE at half the Nyquist frequency ($s = 0.25$/pix) does not drop between 200 and 300 kV for this thick scintillator. These curves also illustrate that readout noise has very little effect on camera performance, even at very low doses of 1 pe/pixel. At 100 kV, where scintillation noise is lower, readout noise does start to show a slight effect, but only at higher spatial frequencies. These curves were generated with the same camera as the images and curves of Figs. 3 and 7, respectively.

incoming beam and that of the backscattered beam. No inflection is visible at 200 kV because the scales of the ingoing and outgoing beam envelopes are more similar. This camera (a US4000 "T" used in tomography applications and as the detector at the end of the GIF2002 energy filter in a binned-by-2 configuration) has had a thick scintillator applied to boost low-spatial-frequency DQE. Counterintuitively, at higher spatial frequency, the DQE does not drop when the voltage is raised from 200 to 300 kV. Three-hundred-kilovolt electrons may punch through the thick scintillator, making a higher "incoming" MTF than 200 kV electrons, compensating for the much wider range of the noisy backscatter tail. This is consistent with recent encouraging results from AmC film comparisons at 300 kV (Booth *et al.*, 2006). The DQE of this high-sensitivity camera is almost independent of dose down to 1 e/pixel. At 100 kV, where there is less noise from backscatter, the readout noise does have a small effect, but it is first seen only at higher spatial frequencies. These curves show that efforts to improve SSNR of detected images should be oriented more toward the scintillation/electron energy deposition process than toward lowering the readout noise of the detector.

D. Performance Trade-Off Example: Scintillator Thickness

1. DQE

A good use of SSNR and DQE is in weighing the relative merits of different options for scintillator thickness. Figure 12 shows the 300 kV MTF and DQE of two sensors with two standard thicknesses of scintillator currently produced by Gatan. The thick scintillator "T" is 4–5 times thicker than the thin scintillator "P," which is usually recommended for higher dose applications, like materials science. The MTF curves in Fig. 12A show a clear trade-off between thickness and resolution, as measured in terms of contrast only (without regard to noise). Low-spatial-frequency MTF is near 1 by definition, so little change is evident at low spatial frequency. At high spatial frequency, the MTF changes inversely with the thickness, as might be expected. For purposes of maximizing image quality when dose is *not* an issue, as is typically the case in materials science applications, the thin "P" would be the reasonable choice. Using DQE to compare the performance of the sensors under low-dose, noise-limited conditions (Fig. 12B) shows that the thick-scintillator performance is better at low spatial frequency and worse at high spatial frequency, though the difference at high spatial frequency is modest. The only modest reduction in DQE caused by thickening of the scintillator can most likely be attributed to the ability of a 300 kV beam to punch through the greater thickness with minimal spread, as was noted in Section III.C.3. Thus, when dose *is* limited for whatever reason, so the SSNR of the data will be leveraged by the DQE, the better choice will be the thicker scintillator.

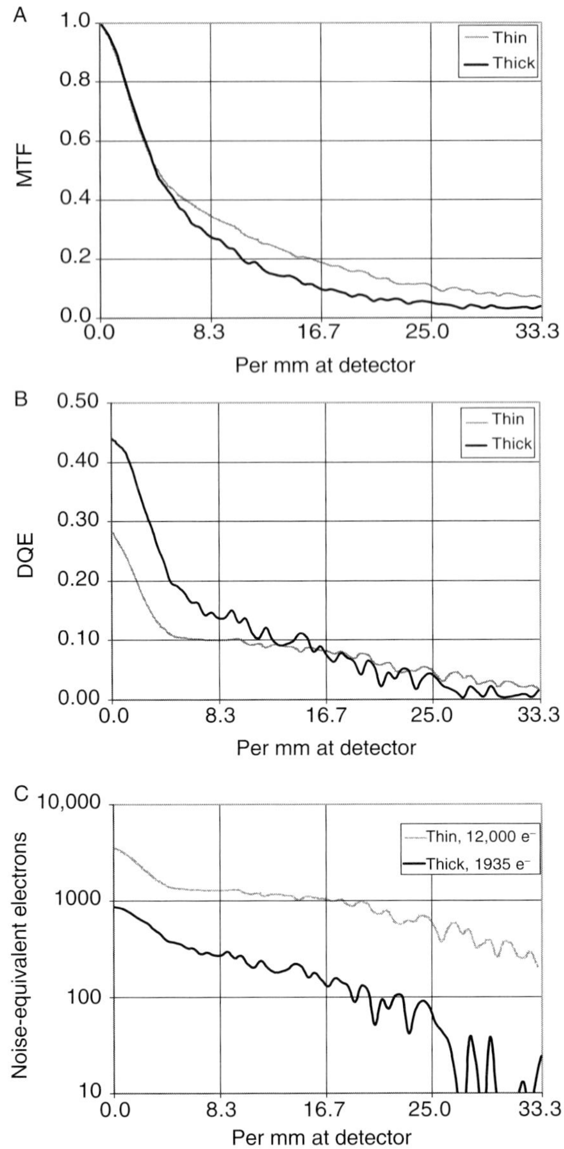

Fig. 12 Comparison of two scintillator thickness options at 300 kV using three criteria. The thick scintillator is three to five times thicker than the thin scintillator. (A) MTF curves showing higher MTF for a thin scintillator; (B) DQE curves showing higher DQE at low spatial frequency for a thick scintillator and only slightly lower DQE at high spatial frequency, a smaller difference than for the MTF due to the offsetting noise-reduction by the thicker scintillator; (C) Noise-equivalent quanta curves at saturation (full CCD well) for the two scintillators, showing an ability to acquire greater SSNR from a robust specimen on the part of the thinner scintillator when the dose is limited by the detector rather than the specimen.

2. Uniform Beam Noise Power Spectra

Power spectra calculated for a range of doses for the same thick-scintillator camera discussed in Section III.C.3 are shown in Fig. 13. The series of curves, which was calculated from a few power spectrum measurements and a readout noise measurement, shows how different noise sources contribute to the net performance. The broad slope in the high-dose curves is the primary shot noise pushed above the Poisson-noise level (equal to the dose) by the scintillation noise and then reduced at higher spatial frequencies both by correlations along the electron tracks and by blurring from the optical transfer process. The baseline at the bottom is the power spectrum of the readout noise, aka the fog level. The horizontal asymptotes on the right of each of the higher dose curves are caused by the shot noise in the CCD electrons (Fig. 8). These are proportional to dose like the scintillation power spectra but dropped in power level (measured in terms of primary electrons squared) from the incoming dose by the factor of the conversion efficiency of CCD electrons/primary electron. The curves show that the noise power spectrum is dominated by scintillation processes, not readout noise, even

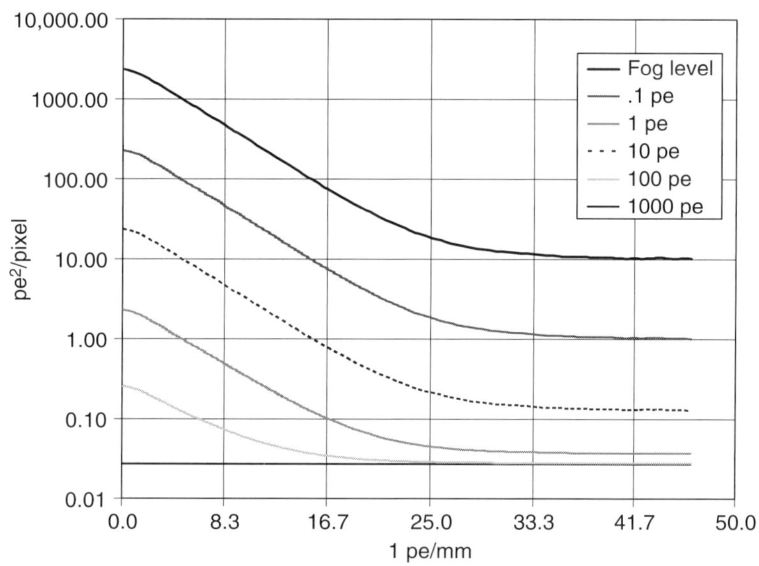

Fig. 13 Uniform-illumination background power spectrum curves at a wide range of dose. The same thick-scintillator camera was used as in Fig. 12. The higher conversion gain of the thick scintillator confers a greater immunity to the effects of readout noise, as can been seen from the fact that low-spatial-frequency noise power is uninfluenced by readout noise down to a dose below 0.1 e/pixel. The zero dose power spectrum, or fog level, is shown forming the low limit for noise power for all doses. A second high-spatial-frequency noise floor, that of the shot noise in the electrons generated in the CCD, is evident in the approach at high spatial frequency to an asymptote which is proportional to the dose level. The spatial frequency at which this noise floor is reached depends on both the conversion efficiency and on the resolution of the optical transfer. Note that these curves extend out to the corner of the square Fourier space, that is, to 1.4 of the Nyquist frequency.

at doses as low as 1 primary electron (pe)/pixel. At low spatial frequency, scintillation processes dominate the noise even when the dose is as low as 0.1 pe/pixel. The curves also show that CCD electron shot noise, and therefore photon transfer and collection processes, are unimportant in this camera. The curves also illustrate the proportionality of uncorrelated noise power to dose level, though these curves can't be taken as a proof of this fact since this fact was used to generate them. The same concept is used when rescaling background noise power spectra when calculating SSNR curves from image signal power spectra. In principle, it is possible to calculate the background noise of a particular detector at any dose from just two images: the fog level power spectrum NPS(0) and the "asymptotic variance ratio" (AVR):

$$\mathrm{AVR} = \frac{[\mathrm{NPS}(N_{\mathrm{test}}) - \mathrm{NPS}(0)]}{N_{\mathrm{test}}}$$

The noise power spectrum at any dose N can then be calculated as:

$$\mathrm{NPS}(N) = \mathrm{AVR} * N + \mathrm{NPS}(0).$$

3. Read-Noise Floor

In the following, the term "read-noise floor" will be used to mean the dose-equivalent of the readout noise in a dark-subtracted image. A side note is that since the noise variance in a dark-subtracted image is twice that of a single exposure, the noise floor can be lowered twofold by using an averaged dark reference that is based on a sufficient number of images. This is not normally stressed, due to the relative unimportance of read noise. However, it becomes an issue in lens-coupled cameras (Section V.A).

The conversion efficiency (digital counts/primary electron) determines both the noise floor and the saturation level, in terms of primary electrons. Table I gives a summary of the measurements from the three cameras tested. Higher gain pushes both the noise floor and the saturation level down. Read noise should be taken into consideration in the discussion of how fine a tilt angle to use while collecting images for tomography. Dose can be fractionated without deleterious effect until dose levels approach the level at which their shot noise equals the readout noise or fog level, in the case of film (McEwen *et al.*, 1995).

Saturation level can influence both tilt-sampling and dose-fractionation, when data collection is not dose-limited. The decision to go with a more sensitive camera for a particular sample should be made with caution, since SSNR is proportional to both DQE and dose. The product of DQE and dose is called "noise equivalent quanta" (Cunningham and Shaw, 1999) and represents the effective dose in a real exposure. It reflects the ability of the camera to render a good signal-to-noise ratio from a specimen in an exposure of that level. If a high DQE comes at the expense of a low-saturation level, by virtue of a very sensitive scintillator, the net noise-equivalent quanta at saturation will be lower. Figure 12C shows a graph of

Table I
Relationship of Conversion Gain Differences Among Various Scintillator Thickness Options to Effective Noise Floor and Saturation Level

Scintillator	Conversion gain (cnts/pe)	Readout noise (cnts)	Readout noise (pe)	Shot noise equivalent dose in dark-subtracted image (pe)	Unbinned saturation (cnts)	Binned saturation (cnts)	Unbinned saturation (pe)	Binned saturation (pe)
P	5.0	3.7	0.74	1.10	30,000	60,000	6000	12,000
U	20	3.7	0.19	0.068	30,000	60,000	1500	3000
T	31	3.7	0.12	0.028	30,000	60,000	968	1935

the noise equivalent quanta for the two phosphor thicknesses at the saturation dose of each camera at 300 kV. These indicate that the camera that would yield the higher SSNR from a given EM image at its saturation level would be the one with the thinner scintillator. An example of where this idea might be applied is in tomography tilt-series acquisition. Assuming that other aspects of the tomography acquisition protocol could be changed to provide a higher dose—and this would not be a simple matter—some improvement to acquired SSNR could be realized through the use of the lower sensitivity, thin-scintillator detector. This is true because for a given decrease in scintillator thickness, the reduction in DQE is smaller than the increase in saturation charge level. Thus the product of these two, the noise equivalent quanta, would increase. In general, while higher sensitivity cameras can achieve higher effective doses through multiple exposures, there are practical reasons in many cases that make this difficult, so a compelling benefit would have to be demonstrated (David Mastronarde, private communication). Note that this doesn't mean that any lower conversion efficiency camera is in general better, since a high DQE is still needed to make best use of the dose.

E. Performance Trade-Off Example: Microscope Magnification and Sampling Choices

Choice of the best magnification in a spatially sampled imaging system always involves a balance between FOV and resolution, as limited by sampling. However, much biological imaging has been done on film, a medium that, if sampled at 7-μm pitch, has over 150-million pixels. This situation has provided flexibility, which has allowed some trade-offs to be postponed until the development or scanning stage. With electronic detectors, the number of pixels is not as ample, so more care must be taken in the design of an experiment to make sure that the FOV provides enough context or covers enough area, while the sampling still renders the required detail.

1. Microscope Magnification (Specimen Pixel Size)

Figure 14A shows the SSNR of AmC on a 4k × 4k fiber-optically coupled CCD camera at two widely different magnifications (42k× and 150k×, data from images taken by Dan Shi at NIH). Dose per area was kept constant and only the postspecimen optics were changed. There is only a modest change in SSNR over a large range of magnification. This modest change can be attributed to differences in the DQE at the spatial frequency on the CCD at which a particular specimen spectral contrast is imaged. Division of the SSNR by the DQE, as suggested at the end of Section III.C.1 and shown in Fig. 14B, confirms the idea that the incoming SSNR of the image prior to detection is independent of magnification, and that image SSNR is a function of the detector performance issues alone. These results contradict the commonly held belief that higher magnification imaging requires higher dose. In fact it goes beyond that in indicating that at higher magnification, imaging performance is better because specimen contrast is imaged at lower spatial frequency on the camera, where the imaging performance is better by virtue of a higher DQE (Figs. 10–12). This improvement is gained, of course, at the cost of FOV. Table II gives effective pixel sizes, fields of view, and noise floors in terms of specimen coordinates. The noise floor across the entire range of magnifications shown can be seen to be below the dose levels chosen for cryotomography with 1-nm pixels. This means that if a sacrifice in FOV is acceptable, improvement in SSNR can, in principle, be made by working at higher magnification. It also shows that one means of improving DQE would be to image at higher magnification and make the detector larger to offset the loss in FOV. A possible explanation of the cultural wisdom within the biological EM community that higher magnifications require higher doses could be that the noise floor of film, also known as the fog level, is higher than that of typical electronic cameras, meaning that as a given specimen area and dose is mapped to a larger and larger area (as magnification is increased), the sum of all the fog noise grows to become a limiting factor, thus driving up the dose needed to maintain SSNR. This same issue would also emerge for fiber-optically coupled CCD cameras if magnifications were pushed to the 1 million times range.

2. Binning

Another sampling trade-off is found in the choice of the "binning" parameter. CCDs allow charge in rectangular regions to be summed before readout, increasing the device read out speed and decreasing the readout noise power contribution per sensor area. Binning, which reduces the fineness of sampling, also involves a sacrifice in resolution. Figure 15A–C show the effect of binning on the MTF, AVR, and DQE, respectively, of a typical thick-scintillator sensor. The graphs in Fig. 15A–C show the abscissa in detector coordinates of mm^{-1} to illustrate the effect relative to sensor dimensions. The reason for the commonly held understanding

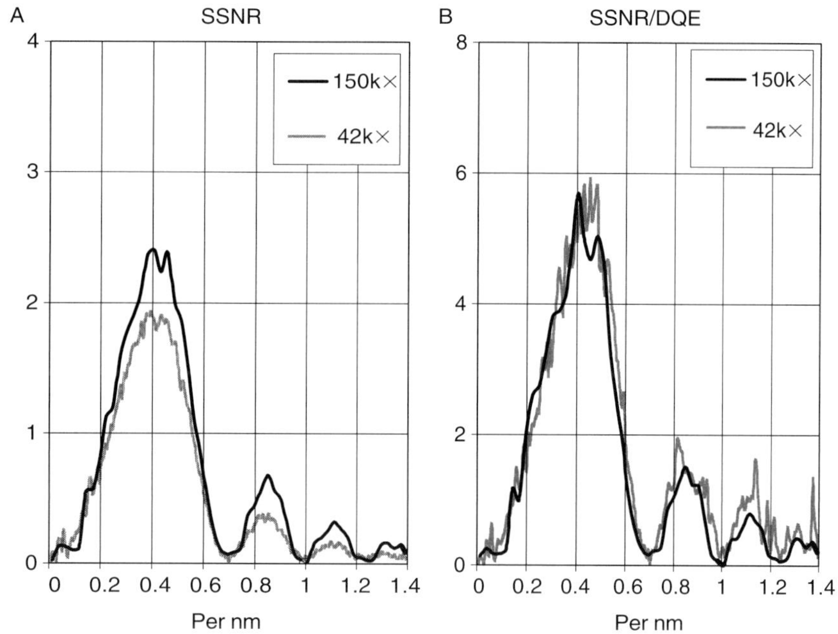

Fig. 14 (A) SSNR of a 30-nm AmC film acquired on a gain-normalized 4k × 4k CCD camera on a T12 at 120 kV at nominal magnifications of 42k× and 150k× (images courtesy Dan Shi, NCI/NIH). The SSNR curves are of the same order of magnitude showing that signal power and noise scale together with magnification when dose *per specimen area* is held constant (meaning that the dose *per detector area* is 13× lower at 150k× than at 42k×). However, the SSNR is somewhat lower at the lower mag, presumably due to the fact that specimen contrast at any given spatial frequency is projected onto a higher spatial frequency of the detector with corresponding reduction in DQE and therefore detected SSNR; (B) SSNR/DQE for the two magnifications showing that, to reasonable approximation, compensation for the DQE of the detector removes the difference between the two curves. The errors at high spatial frequency may indicate that this early DQE measurement (the same shown for the CCD camera in Fig. 10) could be a little low at high spatial frequency. This division must be done with both SSNR and DQE referenced to detector coordinates.

that binning boosts performance, aside from the reduction in readout noise per area, is illustrated by Fig. 15D. Graphed versus pixel coordinates, it can be seen that binning does confer a DQE benefit *per pixel*. However, when viewed relative to the sensor dimensions, as is shown in Fig. 15C, it can be seen that at any given spatial frequency, binning reduces signal contrast without a compensating reduction in noise power spectrum; by virtue of these effects, it has a negative effect on DQE and by extension SSNR of the acquired sample image *per sensor area*. From the standpoint of SSNR, then, the only benefit of binning is to lower

Table II

Relationship of CCD Pixel Size, Field of View and Noise Floor to *Specimen* Pixel Size, Field of View, and Noise Floor over a Range of Magnifications

Film magnification (k×)	Actual magnification at CCD (k×)	CCD pixel size at specimen (Å)	CCD noise floor referred to the specimen for various scintillator thicknesses (e/Å^2)			CCD FOV (μ)	Film pixel size at specimen (Å)	Film FOV (μ)
			Thin	Medium	Thick			
3.5	4.7	32	0.0010	0.00007	0.000030	13	20	25
5.1	7.0	22	0.0022	0.00015	0.00006	8.8	14	17
7.5	10	15	0.0046	0.00032	0.00014	6.0	9.3	12
11	15	10	0.010	0.00070	0.00030	4.1	6.4	7.9
16	22	6.8	0.022	0.0015	0.00065	2.8	4.4	5.4
23′	32	4.6	0.046	0.0032	0.0014	1.9	3.0	3.8
35	47	3.2	0.10	0.0070	0.0030	1.3	2.0	2.5
51	70	2.2	0.22	0.015	0.0065	0.88	1.4	1.7
75	102	1.5	0.46	0.032	0.014	0.60	0.93	1.2
110	150	1.0	1.0	0.070	0.030	0.41	0.64	0.79
160	220	0.68	2.2	0.15	0.065	0.28	0.44	0.54
230	323	0.46	4.6	0.32	0.14	0.19	0.30	0.38
350	474	0.32	10	0.70	0.30	0.13	0.20	0.25
510	696	0.22	22	1.51	0.65	0.09	0.14	0.17
750	1022	0.15	46	3.25	1.39	0.06	0.093	0.12
1110	1500	0.10	100	7.00	3.00	0.04	0.063	0.078
		at CCD (μ)	(e/pix)	(e/pix)	(e/pix)	at CCD (mm)	at film (μ)	at film (mm)
		15	1	0.07	0.03	61.2	7	87

Film sizes are shown for comparison but no noise floor information was available.

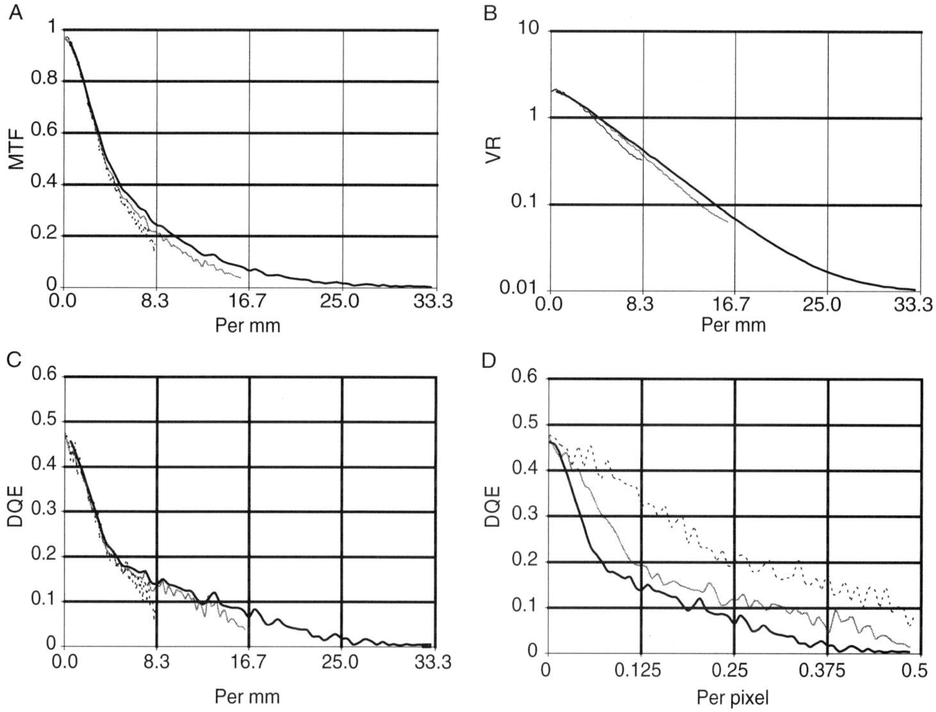

Fig. 15 Effect of binning on MTF, AVR (asymptotic variance ratio = NPS/dose) and DQE of a thick scintillator CCD camera at 300 kV. (A) MTF curves for an unbinned detector and for a detector binned by 2 and by 4 showing that relative to physical dimensions on the detector, binning reduces resolution; (B) AVR (asymptotic variance ratio) curves showing the effect of binning on noise. Binning does confer a slight noise reduction by virtue of the signal averaging effect of gathering charge into larger pixels, but aliasing from the no-longer-sampled higher spatial frequencies almost completely cancels that benefit; (C) DQE curves showing the net reduction in DQE at any given detector spatial frequency. The reduction of MTF has a strong effect on the DQE, which is proportional to MTF^2; (D) Plotting the DQE versus the spatial frequency relative to the pixels (in units of 1 per pixel, which are different for each binning state) shows that, viewed per pixel, binning does confer a benefit. The net loss per detector can also be seen to come from the fact that the total number of pixels is reduced.

the readout noise power per area. Since readout noise power is in most cases not an issue, binning should be done only when the lower readout time and/or smaller data set are of significant value. Another drawback of binning is that it reduces oversampling. This is often a real issue, because oversampling can increase the quality of the various kinds of resampling (rotations and back-projection), which are used in tomography, montaging, and general rotations for presentation. On the other hand, the choice not to bin leads not only to a slower readout speed, but it also increases the storage and processing needed to carry the extra information through all the processing steps. These changes are based on the binning reduction factor squared. The real value of the trade-offs can only be determined in practice.

3. Detector Pixel Size

An extension of the binning question is the issue of pixel size choice in the design of cameras. While in many cases the exact pixel size of a sensor in a certain size class cannot be chosen, awareness of this issue can prevent choices that are inappropriate. Commercial CCDs are now available with very large pixel counts, so it is no longer common to find that pixel sizes are too large (undersampled). Rather the issue tends to be that the data are oversampled. Summarizing the above, the only risks from oversampling are: (1) raising of the effective noise floor through more readouts per area, (2) decreasing readout speed, and (3) increasing demands on data storage and processing. The readout noise issue, while not an issue for fiber-optically coupled cameras, can be an issue for lens-coupled cameras. When a camera is oversampled by more than a factor of two, selecting a binning mode of 2× will address all three of these issues, with the consequence that the real pixel count of the sensor is reduced by four.

IV. Detectors: Proven Technologies

Early work established two basic architectures for electron microscope cameras. Fiber-optic coupling between a scintillator and the CCD chip, already in use for intensified video systems and spectrometers, has become the *de facto* standard for applications requiring high-sensitivity and resolution (Daberkow *et al.*, 1991; Krivanek and Mooney, 1993; Kujawa and Krahl, 1992; Mochel and Mochel, 1986; Spence and Zuo, 1988). Lens coupling, proposed much earlier for high-voltage electron microscopy (HVEM) video (Pawley, 1980), was adapted to the purpose of routine TEM imaging (Mancuso *et al.*, 1992) and HVEM image acquisition (Gubbens *et al.*, 1995; Mooney, 1994). Application of the lens-coupled transmission scintillator in high-resolution IVEM, first attempted for a 400-kV system (Fan and Ellisman, 1993; Fan *et al.*, 1994) and later for 300 kV (Bailey *et al.*, 2004) remain experimental for a variety of reasons and will be discussed in Section V on experimental technologies.

A third architecture, direct detection by a CCD itself without intervening scintillator and coupling optic was eliminated early on, both because the device would be too sensitive and readily damaged (Krivanek and Mooney, 1993). The idea of direct detection is currently getting a second chance (see Chapter 28 by Xuong *et al.*, this volume), but CCDs, with their reliance on perfect charge transfer and their consequential susceptibility to beam damage, remain out of consideration for this role. Some SSNR considerations for the new type of direct detection device will be addressed in Section V on experimental technologies.

A. Scintillator Fiber-Optically Coupled to a CCD

Figure 16 shows a photograph of an FOP that demonstrates the two key features of interest for its use in EM. The first is its ability to transfer an image from one position to another in a compact space. The second is its very large collection angle (here seen in reverse—by the light coming out of the fiber at an oblique angle) in comparison to even the fastest 1:1 lens. In addition to these virtues, fused fiber optic plates also have very high spatial resolution, limited primarily by the size of its component fibers, which can routinely be made with a diameter as small as 3 μm (corresponding to a hexagonal fiber sampling Nyquist frequency of 193 mm^{-1}). FOP image coupling is also extensible. The area of an optic for use in larger-format cameras can be increased with cost proportional simply to the area but with no real increase in complexity of the assembly. The disadvantages of an FOP include cosmetic defects, which can be eliminated to very good approximation through gain normalization and the contact coupling to the CCD. The latter makes assembly a critical process, requiring high levels of mechanical integration, because the sensor and electronics must reach into the microscope column. Contact coupling to the scintillator at the other end of the fiber optic leads to the problem of electron backscatter from the FOP into the scintillator (Figs. 7 and 8).

The performance of FOP-CCD cameras has matured to some extent, but it still leaves room for incremental improvements. Scintillator area can be increased, either with larger sensors or fiber-optic tapers or both. The improvement from this approach is either an increase in the effective number of resolution elements (in proportion to increased area), or an increase in DQE, through operation at higher microscope magnifications. Use of fiber-optic tapers must, however, be

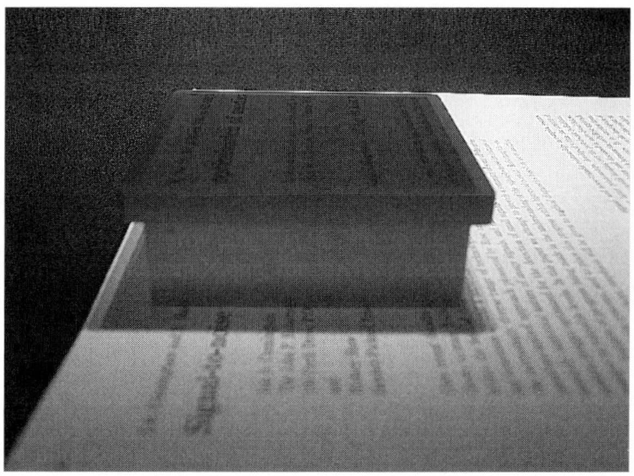

Fig. 16 An image of a fused FOP illustrating its two main benefits: compact image transfer and wide numerical aperture. Given the reciprocity of light paths, the high view angle here corresponds to the high collection angle for scintillator light.

undertaken subject to an understanding of the role of the statistics of the optical image transfer process (Cunningham and Shaw, 1999). It also incurs a requirement on distortion mapping and correction (Faruqi *et al.*, 1999).

Tiling of sensors is now widespread in detectors for synchrotron X-ray diffraction (Sweet, 1998) but in EM, where most applications are in imaging, the seams between tiles constitute a significant drawback. If seams are not allowable in the area of interest, then the effective area of the tiled system is reduced by the detector perimeter multiplied by the diameter of the typical object of interest. In applications for which the object of interest is small (nanogold, for instance) and large area throughput is desired, this option may still be useful (see Chapter 28 by Xuong *et al.*, this volume). However, significant issues associated with size and cost will have to be overcome.

Further incremental improvements to the DQE may still be available through optimization of design parameters of the scintillator and the FOP. But the curves in Fig. 12 show that the physics of the scattering processes places serious limitations on radical increases in DQE. It has become clear, however, that modest improvements can have real biological significance when an increase in resolution raises the threshold of visibility enough to allow visual access to a new structural feature. Beyond this, a few incremental improvements applied together can leverage each other for a significant overall improvement. Since different applications operate image detectors under different conditions of dose and microscope accelerating potential and have different requirements for SSNR, a further avenue for incremental improvement would be to attempt more application-tuned optimization of cameras. For this to happen, the link between camera SSNR and that of the derived 3D result needs to be better understood. Thus far, this area of study seems to be almost completely neglected.

B. Scintillator with Low-Sensitivity Lens Coupling to a CCD

Lens coupling, often with demagnification by a factor of 2–3, provides a way to interface a scintillator to a camera while allowing the sensor to be removed from the center of the column. The demagnification allows the use of smaller physical sensors for the same FOV at the scintillator. The noncontact coupling reduces the difficulty of design and assembly. These cameras, especially when mounted above the viewing chamber of the microscope, where the beam intensity is higher than at the viewing screen by about a factor of 10, are useful for acquisition setup and for applications not requiring high sensitivity. In addition, due to the use of interline CCDs, they are less susceptible to the blooming that occurs at saturation in "astronomy-style," high-sensitivity CCDs, which use the entire pixel area for exposure. This feature provides greater flexibility when working with high-dynamic range specimens and when setting up microscope conditions.

Despite the attractive functionality of lens-coupled cameras for higher dose work, it would be a misstep to use them for applications on the edge of currently feasible resolution or detectability limits for two reasons. First, lens coupling leads to sensitivities between 10 and 100 times lower than those of fiber-optically coupled cameras. Conventional, lens-coupled cameras are therefore not ideal for low-dose work. The reduction in gain by a given factor increases the effective readout noise, in terms of primary electrons, by the same factor. Using the definition of "noise floor" to be the effective dose at which read noise equals shot noise, the 10–100× reduction in sensitivity blows up to a 100–10,000 × factor increase in the effective noise floor. In addition, by reducing the number of CCD electrons per primary electron, CCD electron shot noise becomes important at all doses. Thus, even in the dose range of plastic-section tomography, which is over 1000 e/pixel, there is a reduction in DQE caused by the lower number of CCD electrons that are produced by each beam electron. For instance, a lens-coupled camera generating 1 CCD electron for each incoming primary (instead of the 60–120 CCD electrons typical of a fiber-optically coupled CCD camera) will experience a factor of 2 reduction in DQE at low spatial frequency, where the MTF is close to 1, and a significantly higher reduction in DQE at higher spatial frequencies, as the MTF rolls off the strength of the signal in comparison to the CCD electron shot noise.

The second limitation of lens-coupled cameras is their lack of extensibility. Fiber optics can be made larger more easily than lenses. Hence, to produce a lens-coupled camera with 4k × 4k resolved pixels and high-sensitivity is considerably more difficult, expensive and bulky than to make the coupling with a fiber-optic. Nonetheless, some projects have attempted to grapple with the challenges of doing just this, since lens coupling can also provide a way to use a scintillator in a transmission mode with no substrate, thus reducing backscatter. Such a camera is described in the next section.

V. Detectors: Experimental Technologies

Referring back to Figs. 11 and 12, it can be seen that there is a lot of room for improvement in the DQE and SSNR-generation capability in fiber-optically coupled EM detectors. At the 300-kV accelerating voltage that is commonly used for EM tomography of thick specimens, fiber-optically coupled detectors have a DQE at half the Nyquist frequency on the order of 0.07. An approximately threefold increase in half-Nyquist DQE from 0.07 to 0.21 would yield the same increase in half-Nyquist SSNR as a tripling of dose, but without additional beam damage. The sections that follow show some concrete examples of progress toward more-than-incremental increases in DQE. The first example, the high-sensitivity, substrate-less, lens-coupled camera, promises to deliver a threefold increase in DQE at the intermediate spatial frequency range. Other technologies, somewhat further from practical realization, promise to deliver as much as an eightfold increase.

A. High-Sensitivity Substrate-Less Transmission Scintillator Camera

1. Background of the Substrate-Less Scintillator Idea

Film serves as an example of one way to reduce electron backscatter, namely by allowing the beam to pass right out of the detection medium. While the film emulsion does have a substrate, it is light and thin enough that at 300 kV, a significant fraction of electrons can pass through it and not return to the emulsion layer. This feature can be implemented in an electronic detector by mounting the scintillator on a stretched, electron-transparent membrane, and by using a lens to form a noncontact image. If tension is applied to keep it flat, this membrane can be considerably thinner than even the backing used in film such that it doesn't contribute significantly to electron spreading. Gatan, Inc. implemented this concept in a series of cameras for HVEM ($\sim$1 MeV), both at the end of imaging energy-loss filters (Gubbens *et al.*, 1995) and directly on a microscope (Boulder Lab/Gatan 1 MeV Camera project). These applications were mentioned in Section IV on proven technologies because at microscope accelerating voltages near 1 MeV, the relative benefits of a transmission scintillator with lens coupling are completely unambiguous, due to the extreme backscatter and damage that occur at those energies in a fiber-optic. At intermediate voltages (300–400 kV), the benefits are not obvious up front, and details of implementation can have a significant impact on success. More recently, Gatan undertook to implement the concept on a 300-kV system at the end of a post-column, imaging energy filter (GIF) to combine these features with high-performance imaging on a field emission gun (FEG) system—working to improve imaging for cryotomography. Early results with a problematic initial build of the large lens have already been reported (Bailey *et al.*, 2004). Now the problems with the lens have been solved and a new system with the improved lens has been installed (UltraCam, Fig. 17A).

2. Substrate-Less Scintillator Architecture and Preliminary Implementation on a 300-kV Microscope

Figure 17B illustrates the UltraCam system optics architecture. A 45° beryllium mirror is located behind the scintillator and folds the optical path through 90°, taking glass out of the path of the electrons, and, by virtue of its low Z, generating a minimum of electron backscatter [2–3% vs 21% for Gatan's high-contrast-resolution (HCR^{TM}) fiber optic at 300 kV]. The 88-mm-diameter image field is transferred by a long focal-length, large numerical aperture (NA = 0.25), symmetrical 1:1 lens. A cooled CCD camera with a back-thinned sensor is mounted at the top on a gimbal, which allows precise focus across the entire FOV. In the system pictured in Fig. 17, a thicker scintillator has been mounted for the purpose of boosting the gain of electron/photon conversion at some expense to resolution. It has been found, however, that sensitivity issues have been adequately addressed by the backside CCD and high-numerical aperture lens, so conversion to a thinner scintillator, which will deliver higher resolution, is planned. The CCD is also a

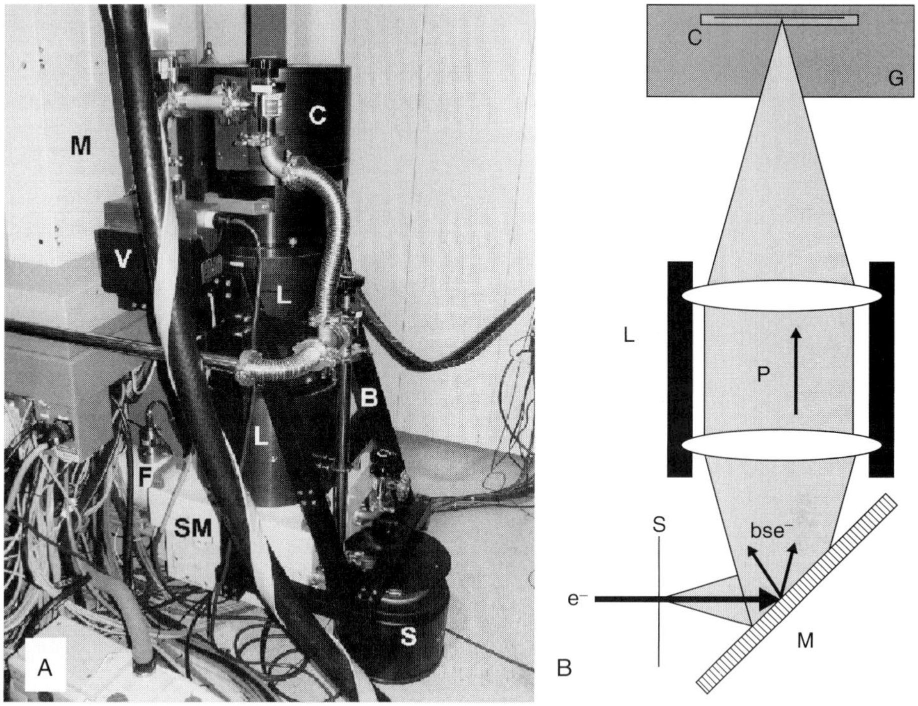

Fig. 17 (A) High-numerical aperture lens-coupled camera (Gatan UltraCam) installed at the University of Colorado, Boulder (photo courtesy of David Mastronarde) showing scintillator/mirror assembly (SM), air spring (S) and bracing (B), lens (L), camera (C) and camera vacuum pump (V) mounted on a Tecnai TF30 microscope (M) with a post-column image filter (F); (B) Schematic diagram showing incoming electrons (e^-) (bent horizontal by the bending magnet in the image filter, scintillator (S), beryllium fold-mirror and beam dump (M), backscattered electrons (bse^-), 0.25 NA image transfer lens (L) with photon beam envelope (P and light gray shape) from a single scintillator position drawn to illustrate the collection angle and the high-quantum efficiency backside CCD (C) mounted on a positioning goniometer (G).

preliminary version with an epitaxial layer tuned for red sensitivity, which sacrifices some of the resolution. Replacement of this too is planned with a CCD tuned for higher resolution at shorter wavelengths. For the present, the system serves to demonstrate the architectural benefit and viability of the transmission scintillator concept. Preliminary tests have now been performed to assess the improvement to the raw data.

3. Preliminary Results

The images in Fig. 18 were taken to evaluate the preliminary UltraCam system at 300 kv and to compare it with a 4k × 4k, fiber-optically coupled camera positioned above the GIF. Figure 18A and B show images of a 300-nm-thick

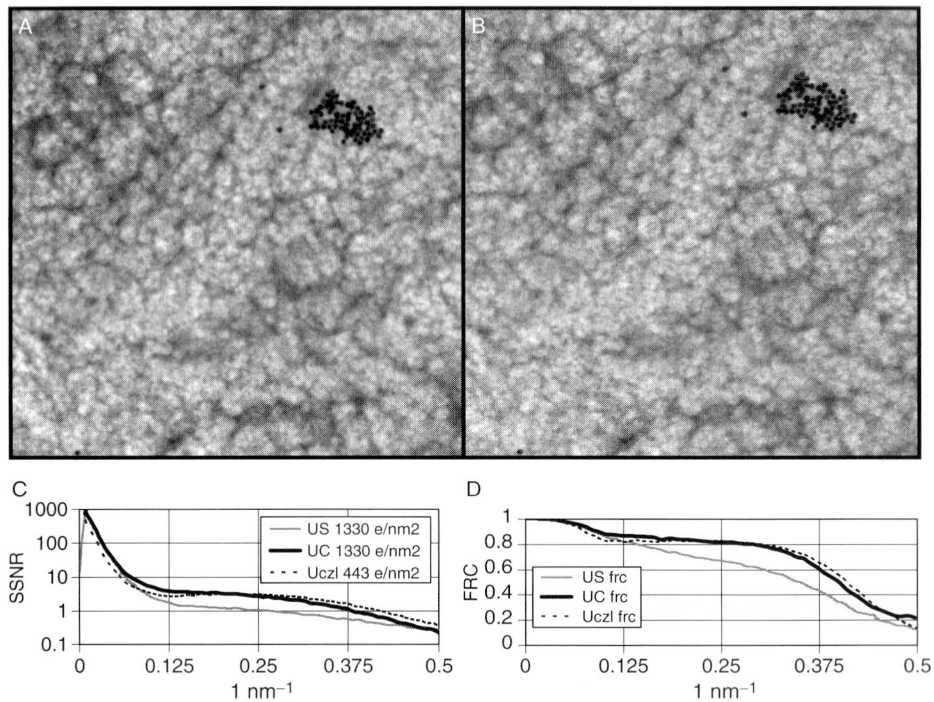

Fig. 18 (A) A selected subarea (1024 × 1024) of one of a pair of 0° tilt projection images of a 300 nm section cut from a plastic-embedded T cell and acquired with a 2k × 2k fiber-optically coupled CCD camera, which is placed above a post-column filter/UltraCam system. The pixel size is 1 nm at the specimen, 30 μm at the detector. The specimen dose is 4000 e^-/nm^2; (B) The same specimen area imaged on a lens-coupled camera with the energy-selecting slit removed to exclude the effect of energy filtering. Images provided by David Mastronarde; (C) SSNR curves derived from the images of (A) (US) and (B) (UC) and also from the same specimen area with the energy selecting slit inserted to obtain zero-loss filtering (Uczl). Significant (approximate doubling) of SSNR is evident from the use of the lens-coupled camera, despite the absence of the energy-selecting slit. The slight oscillation in the zero-loss filtered UltraCam SSNR around the unfiltered curve (UC) shows a slightly higher fraction of phase contrast in the image signal from zero loss filtering; (D) FRC curves derived from aligned image pairs showing corresponding improvement in image-to-image correlations.

plastic section cut from a fixed, stained, and embedded T cell taken with the fiber-optically coupled camera above the GIF and with the lens-coupled (UltraCam) camera at the end of the GIF, respectively. The lens-coupled image (18B) was taken with the GIF slit removed, so any benefits from image filtering were kept out of the comparison. It should be noted, however, that energy filters are not designed to image without an energy-selecting slit so that this particular comparison places the camera at the end of the image filter at a disadvantage. Separating out the issue of camera performance from imaging filter performance is beyond the scope of this chapter, though, so this rough comparison will have to suffice for our purposes here. The dose to the specimen was typical for imaging of this kind: 40 e/Å^2/exposure at the specimen, which corresponds to 13 e/Å^2 or 1300 e/pixel on both detectors. This relatively featureless area of the specimen was chosen intentionally to facilitate camera comparison by statistical means. The difference between the two images is evident to the eye. Finer details are visible in the lens-coupled image than can be seen in the fiber-optically coupled image. Statistical analysis shows the same difference in a quantitative way. Figure 18C shows the SSNRs calculated from these images and from uniform illumination images scaled to the same average dose in the manner introduced in Section II. The slit-inserted zero-loss-filtered (ZLF) SSNR curve has been included to show that for these plastic-embedded and stained specimens, the benefit of ZLF is modest, though an increase in phase contrast is evident in the ZLF SSNR curve. Again, because it is not possible in this case to separate out effects of camera improvements from ZLF improvements, this comparison is inconclusive and should not be used to assess the benefits of ZLF. For additional perspective on the system performance, it is interesting to apply the technique of Fourier ring correlation (FRC) to pairs of images taken of this featureless specimen. Figure 18D shows that these results are consistent with the SSNR curves; the lens-coupled camera shows improved performance at intermediate spatial frequencies for both ZL-filtered and for unfiltered images, though it should be noted that the resampling associated with the alignment done prior to calculation of the FRC reduces the power spectra of the images slightly at high spatial frequency. This suggests that SSNR analysis of electron microscope imaging systems should extend into the processing stage and include the effects of processing steps, such as alignments and rotations.

4. Photon Image Coupling on a Substrate-Less Scintillator Detector

Difficulties have been encountered in the development of this system, which serve to highlight the importance of optical resolution when conversion gain (the number of CCD electrons produced per beam electron) is low. Even with a numerical aperture of 0.25 and the use of a back-thinned CCD, whose sensitivity is 2.5× higher than a front-side CCD, conversion gain is on the order of 10–25 CCD electrons per primary electron, depending on scintillator thickness. The noise background created by the CCD electrons dominates over scintillation noise at spatial frequencies above the binned Nyquist frequency (16.7/mm). This situation makes the MTF of the transfer optic particularly critical, since it determines the signal strength relative to the CCD electron noise floor. The low MTF associated with problems in early builds of the lens led to a reduction in system DQE in a manner that would not have happened in a fiber-optic based system in the presence of a similar loss of MTF. On the positive side, however, two- to threefold improvements in the DQE above half the Nyquist frequency are anticipated from changes to the CCD.

B. Related Technologies for Contrast Enhancement: Zero-Loss Filtering, Higher Microscope Accelerating Voltage, and the Boersch Phase Plate

1. Zero-Loss Filtering

Zero-loss filtering is the removal of components of the beam that have lost energy in the specimen. It is not really accurate to call ZLF an experimental technology, as its value has already been demonstrated for ice-embedded specimens (Grimm *et al.*, 1998). However, there is still ongoing exploration of the parameters for optimizing both specimen thickness and resolution. In the case of plastic-embedded, stained and/or labeled specimens, the smaller ratio of inelastic to elastic scattering cross sections, due to the presence of higher *Z* staining and labeling materials, may reduce the contrast improvement value of ZLF. Higher electron beam voltages are one way to retain a larger zero-loss signal component in EM of thicker specimens and will be addressed in Section V.B.2. For thicker specimens there is also the issue of increasing attenuation of the zero-loss component of the signal; at some thickness there ceases to be an SNR advantage from removal of the energy-loss signal, since little is left after doing so. Most probable loss (MPL) imaging (Bouwer *et al.*, this volume; Bouwer *et al.*, 2004) is one way of dealing with zero-loss attenuation; one moves the energy-selecting slit to the place in the energy-loss spectrum, where most of the electrons can still be found.

Phase contrast is the main source of high-resolution contrast in specimens as thick as 0.5 μm (Agard, 2004). It is important to note that this important aspect of the electron optical image is being attacked on three fronts: (1) Portions of the beam that are scattered through large angles by the specimen are no longer available for generating phase contrast. (2) The inelastically scattered component of the beam loses coherence, for all practical purposes, due to its polychromaticity. It is therefore removed from phase contrast generation, leaving only a lower resolution background image, which is affected both by quantum delocalization and by objective lens chromatic aberration. (3) The specimen itself acts like a shower-stall window, randomizing phase relationships in the wave transmitted from the upper part of the specimen. This phenomenon is referred to as the top-bottom effect (Reimer and Ross-Messemer, 1987). As Bouwer *et al.* (2004) shows, there is still a large range of thickness and resolution for which one can abandon phase contrast and switch to elastically scattered amplitude contrast by using a moderately narrow range of energy loss, tuned to maximize signal while minimizing chromatic aberration. While this strategy does not circumvent problems like the top-bottom effect, it is a compromise that confers a large resolution benefit over light microscopy in a thickness range that is not accessible by unfiltered or zero-loss-filtered EM.

2. High-Voltage Electron Microscopy

HVEM appears to have run its course for the exploration of thicker biological specimens. Factors such as knockon damage and instrument costs become worse as accelerating voltage increases. The fact that high-voltage does not reduce the top-bottom effect, coupled with the possible benefits from the newer technology of energy filtering, have pulled research efforts away from HVEM. Nonetheless, the lengthening of scattering cross-section with increasing voltage, the fact that ZLF and field-emission gun are not incompatible with HVEM, and the fact that the resolution of tomography still falls short of the performance necessary for routine identification of densely packed proteins *in situ* (Grünewald, 2003) together imply that the overall system for EM imaging of biological material may not yet be mature. A modest increase in voltage (to 400–500 keV, for instance) coupled with an FEG electron source and zero-loss or MPL filtering could still lead to increases in the image resolution or the thickness of specimen that is currently possible (Koster *et al.*, 1997).

3. The Relationship Between Contrast Enhancement Technologies and Detector Development

This digression from the issue of detectors themselves has a purpose. It is important to point out that the exploration of the complex parameter space of 3D imaging and localization, particularly for varied biological structures in their natural contexts, has implications in how image acquisition should be optimized. For instance, to what extent has the poor SNR of film at high-voltage contributed to the retreat from high voltage? Lens-coupled transmission scintillator cameras are well suited for high-voltage work; they can actually be made with a higher DQE than is possible at 300 kV through the use of thicker scintillators that take advantage of the higher penetrating power of the HV beam. The right camera could potentially change the balance of factors that have led away from HV. Most of the design effort that has gone into the 4k $\times$ 4k lens-coupled camera described above could be transferred unchanged to an HVEM. The same logic applies equally well to ZL and MPL filtering. In the former case, cameras with higher DQE would allow a reduction in dose, which could in turn be used to allow a smaller objective lens aperture. In the latter, higher DQE could allow a narrower window on the most-probable loss peak, reducing the effect of chromatic aberration. In both cases, the increase in camera sensitivity would lead indirectly to an increase of electron-optical resolution. Further, the extent to which the system-level optimization defines the choice of microscope accelerating voltage (200–1000 keV) has a strong influence on the type of detector chosen, due to the strong influence of electron energy on detector performance (DQE curves in Fig. 11 and Downing and Hendrickson, 1999).

4. Zero-Loss Filtering in Cryotomography

Returning to ZLF, its effect on the imaging of a frozen-hydrated PtK cell is illustrated in Fig. 19. This sample is a little thicker than 300 nm, as evidenced by the difference between dose to the specimen and to the detector (Fig. 19A and caption). A dose four times higher than is normal for a single tomography tilt-series image was used to provide sufficient contrast to visualize specimen

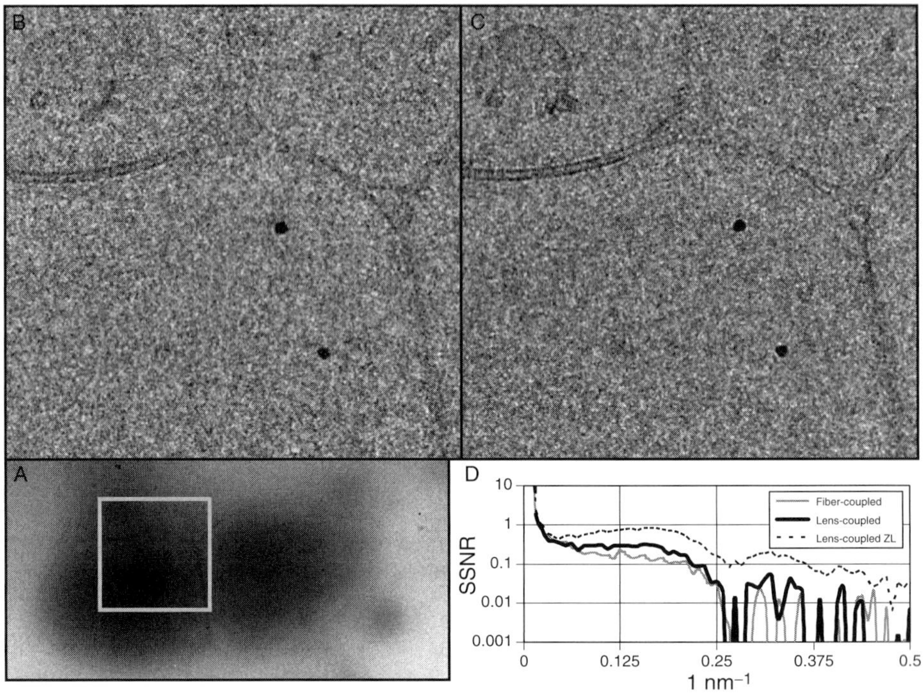

Fig. 19 (A) A subarea (2048 × 1024) of a 2k × 2k ZLF image of whole-cell-mounted, frozen-hydrated PtK cells acquired on a lens coupled camera at the end of a post-column imaging energy filter (Gatan UltraCam, as in Fig. 18). The high contrast demonstrates the strong attenuation by scattering through the thick parts of the cell. The average doses for the images were 400 e/A^2 at the specimen, 190 e/A^2 at the detector without ZL filtering, 55 e/A^2 at the detector with ZL filtering. The specimen dose is 4 times higher than normally used for a single cryotomography tilt-series image in order to give sufficient contrast in a single exposure to judge contrast improvement by eye. The binned 30-μm pixel is 1 nm at the specimen. Micrographs were provided by David Mastronarde. The square shows the area presented at full resolution in 19B and 19C; (B) A selected area (512 × 512) with the energy-selecting slit removed to show the image contrast without ZLF. (C) The same selected area with the energy-selecting slit inserted to show the effect of ZLF. Images (B) and (C) were flattened with a high-pass filter (0.008 nm^{-1} cutoff) to facilitate visual comparison; (D) SSNR of the full images of (B) and (C) and also of an image acquired on the fiber-optically coupled camera above the energy filter. No high-pass filtering was used in generating the SSNR curves. The curves show the large benefit from ZLF for a (low Z) frozen-hydrated specimen compared with Fig. 18. This preliminary version of the lens-coupled camera boosts SSNR by about 2×, ZL filtering by 3× for a net gain of 6×.

details. Figure 19B and C show the improvement in image detail brought about by the elimination of the low-contrast and noise-bearing, inelastically scattered portion of the electron beam. The significant improvement in SSNR is shown in Fig. 19D: it is about 3× for this frozen-hydrated specimen. In conjunction with the UltraCam, which in this preliminary configuration confers an additional factor of 2, the net improvement is roughly a factor of 6 in SSNR. Considered as an increase in effective dose, and raised to the 1/3rd power according to one dose/resolution relationship (Grimm *et al.*, 1998), in improved SSNR at constant real dose predicts a roughly 1.8 times improvement in resolution in the resulting tomogram.

5. The Relationship Between Embedding Material and ZLF Effectiveness

The differences between Figs. 19 and 20 demonstrate the influence of inelastic to elastic scattering cross section ratio on the benefit conferred by ZLF. Both elastic and inelastic scattering increase for higher Z, reducing the maximum thickness that still leaves sufficient unscattered beam to generate phase contrast. However, their relative strength reverses according to the following relation:

$$\frac{\sigma_i}{\sigma_e} \approx \frac{18}{Z}$$

(Egerton, 1986, p. 142) with inelastic scattering dominating for Z below about 18 (Argon) and elastic dominating for Z above 18. Thus the unscattered and low-angle, elastically scattered beam, which is the source of most high-resolution phase contrast, is "muddied" in low-Z specimens, primarily by inelastically scattered electrons, which have lost energy and can therefore be removed by ZLF. In high-Z specimens, however, scattering by out-of-plane specimen material is primarily elastic and cannot be removed by ZLF. If removed with an objective lens aperture, it is removed from phase contrast generation, if not, phase randomization produces the top-bottom effect.

This discussion suggests that there would be a benefit in making a quantitative measurement of the fraction of heavy metals used in a typically stained, plastic-embedded specimen, using the position-dependent quantitative analysis of STEM spectrum imaging, as was done for ion concentration (Leapman, this volume, Leapman *et al.*, 1993). Such a result could be used to model the trade-off between local contrast enhancement and global contrast reduction from stain through increased elastic scatter. It also suggests the possibility that lower Z contrast agents (such as vanadium) for plastic-embedded tomography or no metal contrast agents at all could yield a net contrast or resolution improvement—using plastic embedding to confer dose-tolerance vis-à-vis vitreous ice sections and using only ZLF to produce contrast as has proven effective for vitreous ice. There is no reason that plastic embedding per se, with an average Z under 18 should not benefit from ZLF, though the naïve suggestion should be taken with caution since, as illustrated in the juxtaposed images of hydrated and dehydrated specimens (Shi *et al.*, 1996); the embedding medium itself reduces contrast, since plastic is closer in density to biological material than water. The important point is that the SSNR approach provides a framework for making these comparative evaluations of embedding and

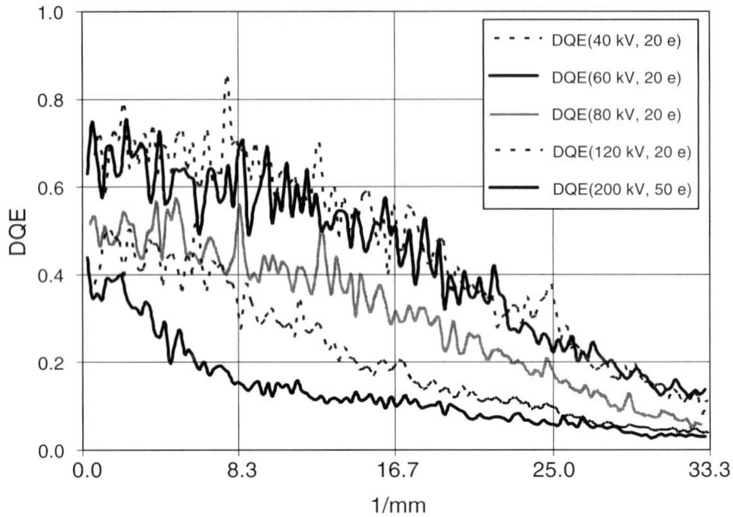

Fig. 20 DQE measurements of a thin scintillator (the same sensor used in the NCI/NIH studies shown in Figs. 5, 10, and 14) on a T20 whose voltage was varied from 40 to 200 kV. The thin scintillator shows evidence at 200 kV of the same kink seen with a thick scintillator at 300 kV; this is caused by a clear difference in length-scale between the ingoing and backscattering electron spreads. In this sequence, as the voltage is gradually decreased from 200 to 80 kV, the zero-spatial-frequency DQE stays about the same as the total amount of backscatter remains approximately constant. However, the spatial extent of the electrons gradually decreases, leading to an increase in the higher spatial frequency DQE. As the voltage is brought below 80 kV, the electrons no longer traverse and exit the scintillator so that up-and-down random walk in and out of the scintillator no longer adds noise, and the DQE of the whole spatial frequency range rises. From 60 to 40 kV, no additional benefit is gained. Backscatter from the scintillator is the only significant source of noise here, and that remains constant with further reductions in voltage. While these measurements were taken with a conventional camera by reducing microscope accelerating potential, they are indicative of what can be obtained in a camera imaging a decelerated electron image from an intermediate voltage EM.

staining options and could facilitate rapid convergence toward staining protocols better optimized for thickness, material, and resolution targets.

6. Additional Techniques Under Investigation

Other efforts to improve SSNR have focused on the contrast side of the ratio. These should be mentioned if only to emphasize that the cellular EM community has long realized that SNR is the central issue when dealing with biological specimens. Two important examples of these efforts include the Boersch phase plate and the correction of spherical and chromatic aberrations. A phase plate, while not raising the envelope of the CTF, should provide an effective factor of two boost in image contrast by allowing the elimination of the zero-crossings that are characteristic of imaging at high levels of defocus. It should also allow the simplification of postacquisition correction for CTF (Danev and Nagayama, 2001; Schultheiss *et al.*, 2006). Practical issues in the production of an electrostatic (Boersch) phase plate may make the attainment of a flat CTF at low spatial frequency difficult until spherical aberration correction makes it possible to extend the objective lens focal length, thereby reducing the low-spatial-frequency cutoff of the central phase plate structure. Spherical aberration correction, within the context of the specimen-limited resolution of biological imaging, has the goal of allowing larger objective apertures for higher signal collection at the same resolution. However, spherical aberration correction reduces low-spatial-frequency contrast (Plitzko *et al.*, 2005), and may therefore require a phase plate, just as a phase plate may require spherical aberration correction. The two thus constitute obligate symbionts, in a manner of speaking. Chromatic aberration correction may improve system resolution by allowing larger MPL windows for viewing very thick specimens with reasonable SNR. While these improvements are not part of the detector per se, improvements to the detector are leveraged by them and vice versa.

C. Direct Detection

After the initial experiments and departure from direct detection in the 1980s, the scintillator/fiber-optic/CCD chip architecture has been widely applied over a broad range of applications. However, as ambitions have grown for the application of electronic detectors in fields at the frontier of resolution and detectability, ideas that once were discarded are now being reexamined in light of recent developments in comparatively radiation-hard detectors for particle physics and medical imaging. Due to its low Z, and therefore lower backscatter, silicon is nicely suited to play the role of "scintillator," and by virtue of its electronic properties, it can be the detector of the converted energy as well. New advancements in this area will be treated fully in another chapter (Chapter 28 by Xuong *et al.*, this volume). A few things will be noted here from the SSNR perspective of this chapter.

A silicon electron detector has a high-quantum yield, which leads to effectively zero readout noise. This is not, however, the deciding factor in detector choice, as is shown in the power spectrum graphs of Fig. 13. Scintillator/fiber optic/CCD cameras already have a readout noise that is well below the level that would influence DQE for cryotomography, even with the dose per tilt fractionated to the limit of 5 e/pixel (Grimm *et al.*, 1998). Readout noise is also not important for structural biology imaging (Zhang *et al.*, 2003) with the possible exception of 2D electron crystallography. In all biological EM, it is important to understand that the object of interest is the specimen, not the electron or the electron track, so the number of electrons accumulated must be the minimum for developing *specimen* SSNR. Specimen contrast is needed in the raw data for a board range of cellular EM, both to accomplish detailed alignment for the highest resolution work, and because larger area correlations cannot usually be relied on to confer detailed alignment with subelectron per pixel doses, due to continual conformational changes during imaging. Therefore, the value of extremely low readout noise is questionable. As for other image acquisition technologies, the "high-dose" or nonreadout noise limited DQE improvement is the factor of greatest importance for image acquisition performance.

Electron dose tolerance is the key issue with direct detection devices. EM doses are significantly higher than those in high-energy physics, and at the energies of biological EM, the energy deposited in the detector is higher. Plastic section tomograms use doses to the specimen of ~10,000 e/Å^2. After aperturing and filtering this becomes ~250 ke/pixel at the detector, assuming 1-nm sampling. Even cryotomograms involve up to 10 ke/pixel/series. Imaging of labeled sections and serial sections, which requires 1–10 ke per exposure, could conceivably accumulate doses on the order of 10 Me/pixel/day. Diffraction patterns expose a detector to highly nonuniform illumination, with thousands of electrons in some spots. Addressing these dose requirements is the most significant challenge for direct detection.

D. Electron Image Deceleration

DQE curves of fiber-optically coupled scintillators (Figs. 11 and 20) suggest one possible route to detector improvement, namely reducing the energy of the electrons that hit the scintillator. Lower energy electrons will produce shorter tracks, which will simultaneously improve detector MTF and reduce the noise created by the random walk of the electrons. Since transmission microscope accelerating voltages are strongly constrained by the need to keep the beam electron scattering cross-sections in a range that is compatible with useful specimen thicknesses, the microscope accelerating voltage cannot simply be turned down, as it was when acquiring the DQE measurements above. The implementation of this principle requires that a retarding electric field be created in front of the scintillator. This in turn requires that the camera be floated at a high-negative potential. One such system has been constructed and has been used to demonstrate operational feasibility (Fig. 21; Downing and Mooney, 2004; Downing *et al.*, 2000). Further work, including use of the current system on a real biological specimen and optimization of a scintillator specifically for use at low voltage is underway as of this writing.

Figure 20 shows the behavior of DQE across a range of voltages from 200 down to 40 kV for one of Gatan's standard scintillators, the "P" or standard phosphor, the thinnest of three commercially available options and the one normally recommended for high MTF but low DQE materials science applications. Two transitions are apparent as the voltage is reduced from 200 kV. The first is the retreat of the interaction volume into the pixel of entry, which brings about an improvement of high-spatial-frequency DQE without a significant change to the low-frequency DQE. The second transition occurs between 120 and 60 kV as the interaction volume retreats vertically as well reducing the random walk noise created from

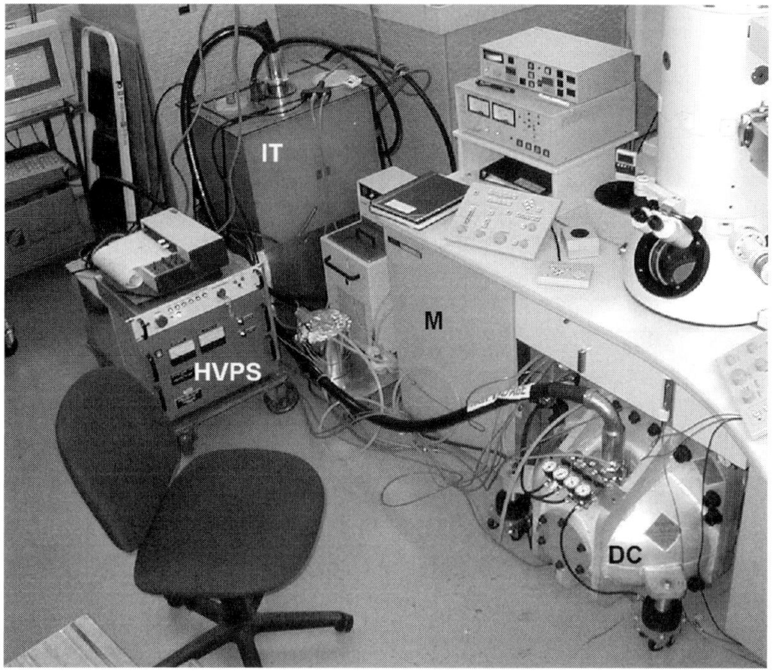

Fig. 21 Experimental implementation of a 4k × 4k decelerator camera shown on a JEOL 3100 (photograph courtesy of Ken Downing, LBNL). The decelerator camera (DC) mounts in an insulating gas pressure vessel inside the knee-well of the microscope (M). High-voltage power for the decelerating column is provided by a 200 kV power supply (HVPS) and 120 volts AC is provided to the CCD camera via an isolation transformer (IT).

electrons going up and down, in and out of the scintillator. The low-spatial-frequency DQE at 60 kV, about 0.7, is higher than the typical DQEs at 200 and 300 kV, even for a very thick scintillator. Performance descends only to 0.5 at half the Nyquist frequency. The deceleration renders the materials science "P" scintillator into a DQE-optimized scintillator that would be better than anything currently available at 300 kV for both tomography and molecular structural biology. The DQE of 0.5 at half the Nyquist frequency is roughly 7 times better than the best that can be currently expected from conventional, fiber-optically coupled cameras—not far from the factor of 8 that is sought in Section II.A.3. Aside from choosing the thinnest option of standard scintillator, no further optimization for low-voltage operation was attempted, and the customary 15 μm sampling of the standard CCD camera was used. Further optimization is planned, specifically oriented to the lower beam voltages, which should lead both to higher DQE and to the ability and need for finer sampling. This technology has the potential to provide performance significantly beyond what is currently available. Moreover, the camera will not have the beam damage issues of the direct-detection devices.

Also, as with the transmission scintillator camera, the optimization of beam-specimen and beam-detector interaction will no longer be a compromise.

VI. Discussion

The goal of instrumentation development for cellular EM is the visualization of structure and function across the gamut of length scales that are inaccessible by other means. Methods for molecular structural biology, both X-ray and electron microscopical, operate *ex situ* and thus do not allow the study of interrelationships among cellular components. Light microscopy, with its rich set of probes and its ability to reveal structure *in vivo*, is limited in resolution to ~0.3 μm. Neither X-ray microscopy nor microscopical NMR has yet developed the spatial resolution to study this regime. Thus, for the study of unique, nonaverageable cellular structures *in situ*, at scales from 1 to 300 nm, EM is the only modality available. The current practical resolution limits of tomography at about 4 nm, coupled with the issue of crowding, leave a gap between the frontier of cellular EM and the knowledge already obtained by molecular structural techniques. Bridging this gap by means of higher resolution tomography, more varied labeling, protein identification, and *in situ* structure determination is the goal for future advancements in cellular EM instrumentation. As has been outlined in the chapter, most of the challenges to bridging this gap can be understood in terms of SSNR. Improving SSNR through concerted efforts, both to increase contrast and to reduce noise, could have a significant impact on our ability to bridge this gap.

In image collection, the requirements for meeting this challenge can be summarized as follows:

1. Higher DQE at all spatial frequencies to increase the resolution limit of EM tomography under existing dose constraints. This will improve SSNR at every dose: low for cryotomography, high for plastic-section tomography, and for thin section label detection and elemental mapping.
2. More real pixels—same DQE with more sensor area or both higher DQE and larger area—to improve the span of the visualizable length scales within a single image. This could further the goal of the identification, discrimination, and structure determination of proteins *in situ*. On a larger length scale, it could facilitate the identification and relating of small organelles within the larger 3D cellular or intercellular context (e.g., Cooney *et al.*, 2002).
3. Higher data throughput (pixel rate), to cope with higher pixel counts and/or fractionation of dose into multiple readouts without significant slowdown. This will decrease overall data-set acquisition time.
4. Decoupling of image collection performance from microscope accelerating potential to allow greater flexibility in explorations of higher voltages as a way to look at thicker specimens or improve resolution in thinner specimens.

5. Study of the relationship between input SSNR and reconstruction resolution to allow application-specific design trade-offs and optimization.

Much of the above is already underway, with the possible exception of item 5, which must come from the community of instrumentation users. Both decelerated and transmission approaches provide higher DQE and decoupling from the microscope's accelerating voltage. Thinning of direct detection devices could provide the same ability for those devices to work across a range of accelerating voltages.

VII. Conclusions

SSNR is a powerful metric, which can be extended to describe input data sets through the use of uniform illumination power spectra as the noise background. Examining the SSNR of input data sets and relating those to resultant reconstructions could provide insights about the domains of operation that should be optimize in image collection devices. Specimen-free characterization of detectors allows unbiased comparison of detection technologies, subject to certain precautions regarding technique. DQE serves this role well, linking collected data SSNR to specimen/microscope contrast, and dose. DQE optimization is important both for low-dose work and for any dose level where SSNR maximization is key.

Current image collection techniques have helped to take cellular EM to the presently achievable levels of resolution, thanks to the improvements they have empowered in integration, repeatability, and low-spatial-frequency DQE. Experimental methods that are currently being developed and evaluated promise to provide dramatic further improvements to SSNR, especially at high spatial frequency. They will also improve throughput and real pixel count. Altogether, there is reason to expect great advancements in the near future in access to structure and function in the gap between cellular and molecular structural biology. These advances will come when all the contrast improvements and noise reductions currently being worked on come together, each multiplying the benefits of the other. Improvements to image collection will play a lead role in the advance of this important frontier.

Acknowledgments

I would like to thank David Mastronarde of the Boulder Laboratory for 3D Fine Structure for extensive discussions on tomography technique as it relates to image collection and for sharing data generated at that laboratory by him and others (Eileen O'Toole and Daniela Nicastro). I would also like to thank David Joyce and Brent Bailey of Gatan R&D for their contributions toward the design and characterization of the UltraScan and UltraCam cameras with which all of the CCD data presented here was obtained. It should be stated, however, that while only Gatan cameras were used for the illustrations, the discussion of this chapter applies generally to EM digital cameras from any manufacturer. I want to acknowledge the contribution of Raynald Gauvin to an enlarged understanding of beam-scintillator interactions through Monte Carlo simulation well beyond that shown briefly in Fig. 8. I would also like to thank Richard Leapman, Chris Booth, Brian Lee, and Robert Glaeser for helpful readings and criticisms of the manuscript or parts thereof.

References

Aikens, R., Agard, D. A., and Sedat, J. W. (1989). Solid-state imagers for microscopy. *Methods Cell Biol.* **29,** 291–313.

Agard, D. A. (2004). Searl Scholars Program: Scholar Profile. http://www.searlescholars.net/people/1982/agard.html

Bailey, B., Agard, D. A., Sedat, J. W., Barfels, M., Chao, H., and Mooney, P. E. (2004). Performance evaluation of a transmission scintillator-based lens-coupled 4K CCD camera for use in low-dose electron cryo-microscopy. *Microsc. Microanal.* **10**(Suppl. 2), 1204–1205.

Booth, C. R., Jakana, J., and Chiu, W. (2006). Assessing the capabilities of a 4k × 4k CCD camera for electron cryo-microscopy at 300 kV. *J. Struct. Biol.* **156,** 556–563.

Booth, C. R., Jiang, W., Baker, M. L., Zhou, Z. H., Ludtke, S. J., and Chiu, W. (2004). A 9 Å single particle reconstruction from CCD captured images on a 200 kV electron cryomicroscope. *J. Struct. Biol.* **147,** 116–127.

Bouwer, J. C., Mackey, M. R., Lawrence, A., Deerinck, T. J., Jones, Y. Z., Terada, M., Martone, M. E., Peltier, S., and Ellisman, M. H. (2004). Automated most-probable loss tomography of thick selectively stained biological specimens with quantitative measurement of resolution improvement. *J. Struct. Biol.* **148,** 297–306.

Brink, J., and Chiu, W. (1994). Applications of a slow-scan CCD camera in protein electron crystallography. *J. Struct. Biol.* **113,** 23–34.

Cooney, J. R., Hurlburt, J. L., Selig, D. K., Harris, K. M., and Fiala, J. C. (2002). Endosomal compartments serve multiple hippocampal dendritic spines from a widespread rather than a local store of recycling membrane. *J. Neurosci.* **22,** 2215–2224.

Cunningham, I. A., and Shaw, R. (1999). Signal-to-noise optimization of medical imaging systems. *J. Opt. Soc. Am. A* **16,** 621–632.

Daberkow, I., Herrmann, K. H., Liu, L., and Rau, W. D. (1991). Performance of electron image converters with YAG single-crystal screen and CCD sensor. *Ultramicroscopy* **38,** 215.

Danev, R., and Nagayama, K. (2001). Transmission electron microscopy with Zernike phase plate. *Ultramicroscopy* **88,** 243–252.

de Ruijter, W. J. (1995). Imaging properties and application of slow-scan charge coupled device cameras suitable for electron microscopy. *Micron* **26,** 247–275.

de Ruijter, W. J., Mooney, P. E., and Krivanek, O. L. (1993). Signal transfer efficiency of slow-scan CCD cameras. *In* "Proceedings of the 51st Annual Meeting of Microscopy Society of America," (G. W. Bailey and C. L. Rieder, eds.), pp. 1062–1063. San Francisco Press, San Francisco, CA.

de Ruijter, W. J., Mooney, P. E., and Völkl, E. (1994). Utilization of slow-scan CCD cameras for quantitative electron holography. *MSA Bull.* **24,** 451–458.

Downing, K. H., and Grano, D. A. (1982). Analysis of photographic emulsions for electron microscopy of two-dimensional crystalline specimens. *Ultramicroscopy* **7,** 381–404.

Downing, K. H., and Hendrickson, F. M. (1999). Performance of a 2K CCD camera designed for electron crystallography at 400 kV. *Ultramicroscopy* **75,** 215–233.

Downing, K. H., and Mooney, P. E. (2004). CCD camera with electron decelerator for intermediate voltage electron microscopy. *Microsc. Microanal.* **10,** 1378–1379.

Downing, K. H., Favia, P., and Mooney, P. E. (2000). Electron decelerator for improved CCD performance in intermediate voltage electron microscopy. *In* "Proceedings of the 58th Annual Meeting of the Microscopy Society of America," (G. W. Bailey, S. McKennan, and R. L. Price, eds.), pp. 734–735. Springer-Verlag, New York, NY.

Egerton, R. F. (1986). "Electron Energy Loss Spectroscopy in the Electron Microscope," 1st edn. Plenum Press, New York.
Fan, G. Y., and Ellisman, M. H. (1993). High-sensitivity lens-coupled slow-scan CCD camera for transmission electron microscopy. *Ultramicroscopy* **52,** 21.
Fan, G. Y., Dunkelberger, D. G., and Ellisman, M. H. (1994). Performance of thin foil scintillating screen for transmission electron microscopy. *Ultramicroscopy* **55,** 7–14.
Faruqi, A. R., Henderson, R., and Subramaniam, S. (1999). Cooled CCD detector with tapered fibre optics for recording electron diffraction patterns. *Ultramicroscopy* **75,** 235–250.
Grimm, R., Singh, H., Rachel, R., Typke, D., Zillig, W., and Baumeister, W. (1998). Electron tomography of ice-embedded prokaryotic cells. *Biophys. J.* **74,** 1031–1042.
Grünewald, K., Medalia, O., Gross, A., Steven, A. C., and Baumeister, W. (2003). Prospects of electron cryotomography to visualize macromolecular complexes inside cellular compartments: Implications of crowding. *Biophys. Chem.* **100,** 577–591.
Gubbens, A. J., Kraus, B., Krivanek, O. L., and Mooney, P. E. (1995). An imaging filter for high voltage electron microscopy. *Ultramicroscopy* **59,** 255–265.
Iancu, C. V., Wright, E. R., Heymann, J. B., and Jensen, G. J. (2006). A comparison of liquid nitrogen and liquid helium as cryogens for electron cryotomography. *J. Struct. Biol.* **153,** 231–240.
Jones, R. C. (1955). New method of describing and measuring the granularity of photographic materials. *J. Opt. Soc. Am.* **45,** 799–808.
Jones, R. C. (1958). On the quantum efficiency of photographic negatives. *Photogr. Sci. Eng.* **2,** 57–65.
Koster, A. J., Chen, H., Sedat, J. W., and Agard, D. A. (1992). Automated microscopy for electron tomography. *Ultramicroscopy* **46,** 207–227.
Koster, A. J., Grimm, R., Typke, D., Hegerl, R., Stoschek, A., Walz, J., and Baumeister, W. (1997). Perspectives of molecular and cellular electron tomography. *J. Struct. Biol.* **120,** 276–308.
Krivanek, O. L., Gubbens, A. J., Dellby, N., and Meyer, C. E. (1992). Design and 1st applications of a post-column imaging filter. *Microsc. Microanal. Microstruct.* **3,** 187–199.
Krivanek, O. L., and Mooney, P. E. (1993). Applications of slow-scan CCD cameras in transmission electron microscopy. *Ultramicroscopy* **49,** 95–108.
Kujawa, S., and Krahl, D. (1992). Performance of a low-noise CCD camera adapted to a transmission electron microscope. *Ultramicroscopy* **46,** 395–403.
Leapman, R. D. (2003). Detecting single atoms of calcium and iron in biological structures by electron energy-loss spectrum-imaging. *J. Microsc.* **210,** 5–15.
Leapman, R. D., Hunt, J. A., Buchanan, R. A., and Andrews, S. B. (1993). Measurement of low calcium concentrations in cryo-sectioned cells by parallel-EELS mapping. *Ultramicroscopy* **49,** 225–234.
Luther, P. K., Lawrence, M. C., and Crowther, R. A. (1988). A method for monitoring the collapse of plastic sections as a function of electron dose. *Ultramicroscopy* **24,** 7–18.
Mancuso, J. R., Maxwell, W. B., Camp, R. E., and Ellisman, M. H. (1992). Designing high-resolution slow-scan CCD-based TEM camera systems. *In* "Proceedings of 50th Annual Meeting of the Electron Microscopy Society of America," (G. W. Bailey, J. Bentley, and J. A. Small, eds.), pp. 946–947. San Francisco Press, San Francisco, CA.
Mastronarde, D. N. (1997). Dual-axis tomography: An approach with alignment methods that preserve resolution. *J. Struct. Biol.* **120,** 343–352.
Mastronarde, D. (1996). Private communication.
McEwen, B. F., Downing, K. H., and Glaeser, R. M. (1995). The relevance of dose-fractionation in tomography of radiation-sensitive specimens. *Ultramicroscopy* **60,** 357–373.
McEwen, B. F., Marko, M., Hsieh, C., and Mannella, C. (2002). Use of frozen-hydrated axonemes to assess imaging parameters and resolution limits in cryoelectron tomography. *J. Struct. Biol.* **138,** 47–57.
Meyer, R. R., and Kirkland, A. I. (1998). The effects of electron and photon scattering on signal and noise transfer properties of scintillators in CCD cameras used for electron detection. *Ultramicroscopy* **75,** 23–33.
Meyer, R. R., Kirkland, A. I., Dunin-Borkowski, R. E., and Hutchison, J. L. (2000). Experimental characterization of CCD cameras for HREM at 300 kV. *Ultramicroscopy* **85,** 9–13.

Milne, J. L., Shi, D., Rosenthal, P. B., Sunshine, J. S., Domingo, G. J., Wu, X., Brooks, B. R., Perham, R. N., Henderson, R., and Subramaniam, S. (2002). Molecular architecture and mechanism of an icosahedral pyruvate dehydrogenase complex: A multifunctional catalytic machine. *EMBO J.* **21,** 5587–5598.

Mochel, M. E., and Mochel, J. M. (1986). A CCD imaging and analysis system for the VG HB5 STEM. *In* "Proceedings of 44th Annual EMSA Meeting," (G. W. Bailey, ed.), Vol. 44, pp. 616–617. San Francisco Press, San Francisco, CA.

Mooney, P. E. (1994). Image coupling methods in CCD cameras for electron microscopy. *In* "Proceedings of 52nd Meeting of the Microscopy Society of America," (G. W. Bailey and A. J. Garratt-Reed, eds.), pp 406–407. San Francisco Press, San Francisco, CA.

Nickell, S., Kofler, C., Leis, A. P., and Baumeister, W. (2006). A visual approach to proteomics. *Nat. Rev. Mol. Cell. Biol.* **7,** 225–230.

Pawley, J. (1980). Recent improvements to the EM-7 at the Madison Biotechnology Resource. *In* "Proceedings of 38th Annual EMSA Meeting," (G. W. Bailey, ed.), Vol. 38, pp. 8–11. Claitor's Publishing Division, Baton Rouge, LA.

Penczek, P. A. (2002). Three-dimensional spectral signal-to-noise ratio for a class of reconstruction algorithms. *J. Struct. Biol.* **138,** 34–46.

Plitzko, J. M., Freitag, B., Hegerl, R., and Lücken, U. (2005). Aberration free microscopy for life science applications. *Microsc. Microanal.* **11**(Suppl. 2), 2140–2141.

Rabbani, M., Shaw, R., and van Metter, R.J (1987). Detective quantum efficiency of imaging systems with amplifying and scattering mechanisms. *J. Opt. Soc. Am. A* **4,** 895–901.

Reimer, L., and Ross-Messemer, M. (1987). Top-bottom effect in energy-selecting transmission electron microscopy. *Ultramicroscopy* **21,** 385–388.

Rose, A. (1946). A unified approach to the performance of photographic film, television pickup tubes, and the human eye. *J. Soc. Motion Pict. Eng.* **47,** 273–294.

Rose, A. (1948a). Sensitivity performance of the human eye on an absolute scale. *J. Opt. Soc. Am.* **38,** 196–208.

Rose, A. (1948b). Television pickup tubes and the problem of vision. *In* "Advances in Electronics and Electron Physics" (L. Marston, ed.), pp. 131–166. Academic Press, New York.

Samei, E., Flynn, M. J., and Reimann, D. A. (1998). A method for measuring the presampled MTF of digital radiographic systems using an edge test device. *Med. Phys.* **25,** 102–112.

Sander, B., Golas, M. M., and Stark, H. (2005). Advantages of CCD detectors for de novo three-dimensional structure determination in single-particle electron microscopy. *J. Struct. Biol.* **151,** 92–105.

Schultheiss, K., Perez-Willard, F., Barton, B., Gerthsen, D., and Schröder, R. R. (2006). Fabrication of a boersch phase plate for phase contrast imaging in a transmission electron microscope. *Rev. Sci. Instrum.* **77,** 033701.

Shi, S., Sun, S., Andrews, S. B., and Leapman, R. (1996). Thickness measurement of hydrated and dehydrated cryosections by EELS. *Microsc. Res. Tech.* **33,** 241–250.

Spence, J. C. H., and Zuo, J. M. (1988). Large dynamic range, parallel detection system for electron diffraction and imaging. *Rev. Sci. Instrum.* **59,** 2102–2105.

Suloway, C., Pulokas, J., Fellmann, D., Cheng, A., Guerra, F., Quispe, J., Stagg, S., Potter, C. S., and Carragher, B. (2005). Automated molecular microscopy: The new Leginon system. *J. Struct. Biol.* **151,** 41–60.

Sweet, R. M. (1998). The technology that enables synchrotron structural biology. *Nat. Struct. Biol.* **5**(Suppl.), 654–656.

Thon, F. (1971). Phase contrast electron microscopy. *In* "Electron Microscopy in Materials Science" (U. Valdre, ed.), pp. 570–625. Academic Press, New York.

Unser, M., Trus, B. L., and Steven, A. C. (1987). A new resolution criterion based on spectral signal-to-noise ratios. *Ultramicroscopy* **23,** 39–51.

Valle, R., Genty, B., and Marraud, A. (1989). Scattering and back-scattering of 1 MeV electrons. *Ultramicroscopy* **27,** 67–78.

van Heel, M., and Schatz, M. (2005). Fourier shell correlation threshold criteria. *J. Struct. Biol.* **151,** 250–262.

van Zwet, E. J., and Zandbergen, H. W. (1996). Measurement of the modulation transfer function of a slow-scan CCD camera on a TEM using a thin amorphous film as test signal. *Ultramicroscopy* **64,** 49–55.

Wiener, N. (1930). Generalized harmonic analysis. *Acta Mathematica* **55,** 117–258.

Working group FD (DR)IEC standard 62220–1.

Zeitler, E. (1992). The photographic emulsion as analog recorder for electrons. *Ultramicroscopy* **46,** 405–416.

Zhang, P., Borgnia, M. J., Mooney, P. E., Shi, D., Pan, M., Herron, P. O., Mao, A., Brogan, D., Milne, J. L. S., and Subramaniam, S. (2003). Automated image acquisition and processing using a new generation of 4 K $\times$ 4 K CCD cameras for cryo electron microscopic studies of macromolecular assemblies. *J. Struct. Biol.* **143,** 135–144.

Zhou, Z. H., Zhang, H., Jakana, J., Lu, X. Y., and Zhang, J. Q. (2003). Cytoplasmic polyhedrosis virus structure at 8A by electron cryomicroscopy: Structural basis of capsid stability and mRNA processing regulation. *Structure (Camb)* **11,** 651–663.

Zweig, H. J. (1965). Detective quantum efficiency of photodetectors with some amplifying mechanism. *J. Opt. Soc. Am.* **55,** 525–528.

CHAPTER 28

Future Directions for Camera Systems in Electron Microscopy

Nguyen-Huu Xuong,* Liang Jin,* Stuart Kleinfelder,† Shengdong Li,† Philippe Leblanc,* Fred Duttweiler,* James C. Bouwer,* Steven T. Peltier,* Anna-Clare Milazzo,* and Mark Ellisman*

*University of California, San Diego, La Jolla, California 92093

†University of California, Irvine, Irvine, California 92697

We are working to develop an optimal, direct digital replacement for electron microscopy film. Our efforts are based on a longstanding effort to push the limits of charge-coupled devices (CCDs), which has motivated us to pioneer technologies for direct electron detection. Here we review our current efforts to scale up our own lens-coupled CCD camera and make a system capable of exceeding the spatial resolution of film, while maintaining single-electron sensitivity. This 8k × 8k lens-coupled CCD system represents the current state-of-the-art in CCD-based systems, but it also demonstrates the great engineering effort required to achieve these key performance benchmarks when the detector is based on a resolution-limiting scintillation screen. We have therefore developed a parallel

0091-679X/07 $35.00
DOI: 10.1016/S0091-679X(06)79028-8

effort to produce a radiation-tolerant system that can withstand direct electron bombardment. We briefly describe our early efforts to adapt the pixel array detector (PAD) that is commonly used in X-ray diffraction. We then describe our subsequent development of a groundbreaking prototype system based on an active pixel sensor (APS). This early implementation of an APS-based direct detection detector (DDD) has already delivered unprecedented performance—in many areas exceeding the fundamental capabilities of CCD-based systems. This implementation features a 512 × 550 pixel format of 5 μm pitch with an excellent signal-to-noise ratio (~13/1 for a single incident electron in the range of 200–400 keV) and a very high spatial resolution (2–4 μm). Radiation tolerance with high-energy exposure is also impressive, especially with cooling to −15 °C. Stable performance has been demonstrated over a period of several months of continuous use before a thermal annealing process is required to restore the device. The characteristics of this new detector have exciting implications for transmission electron microscopy, especially for cryo-electron microscopy (cryo-EM) as applied to biomolecular structures. Due to its excellent spatial resolution, this DDD is capable of distinguishing macromolecule structures at very high resolution using a magnification of only 50,000×. As such, this system has the potential to deliver cryo-EM three-dimensional (3D) reconstructions with 3 Å resolution.

I. Introduction

Film has long been regarded as the gold standard for image recording for transmission electron microscopy (TEM). With a dynamic range of three orders of magnitude and a data format that is roughly equivalent to an 8k × 6.5k detector in digital terms, film has exceeded the performance of all the commercial, scintillator-based digital detection systems produced to date. Certainly, film provides excellent modulation transfer functions (MTFs), especially as compared with commercial charge-coupled device (CCD) cameras, but its use requires several post-acquisition steps, like chemical development and digitizing; these are cumbersome and time consuming. Moreover, the necessity of such processing eliminates film as a viable solution for automated, high-throughput data acquisition, especially for remote applications or highly automated processes, where the initial data are utilized to steer or refine the data acquisition process in real time. Such applications require a direct digital solution. The CCD, which does provide direct digital readout, has therefore become the standard detector option for automation. Beyond speed and flexibility, CCD imaging systems surpass film recording in many respects, including sensitivity, linearity, and dynamic range (Fan and Ellisman, 2000).

CCDs were invented in 1970 (Boyle and Smith, 1970). They soon became the sensor of choice in many imaging applications, particularly for video cameras and camcorders. A driving force behind the development of large format CCDs, with array sizes greater than 1k × 1k, came from astronomy and the establishment of space-borne observatories, where high-resolution images were needed and photographic film was impractical or impossible to use. Experimental CCD imaging systems for telescopes were built as early as 1976. However, the first experimental digital imaging system using CCDs for TEM was not described until 1982 (Roberts *et al.*, 1982). Since then, many other systems have been reported (Aikens *et al.*, 1989; Chapman *et al.*, 1989; Daberkow *et al.*, 1991, 1996; Downing and Hendrickson, 1999; Fan and Ellisman, 1993, 2000; Faruqi, 1997, 1999; Herrmann and Liu, 1992; Herrmann and Sikeler, 1995; Krivanek and Mooney, 1993; Krivanek *et al.*, 1991; Kujawa and Krahl, 1992; Spence and Zuo, 1988). Complete digital imaging systems for TEM are now produced by several companies, including Advanced Microscopy Techniques (Rowley, MA), Gatan (Pleasanton, CA), JEOL (Peabody, MA), Photometrics (Tucson, AZ), and Tietz-Video and Image Processing (Herbststrasse, Gauting, Germany). These systems have become invaluable for applications such as microscope automation, automated electron tomography, electron holography, protein crystallography, and telemicroscopy.

However, for intermediate and high-voltage electron microscopy (EM) applications, CCD devices are still far from ideal. Due to radiation damage and signal saturation of CCDs (Roberts *et al.*, 1982), direct detection of electrons with CCD technology is not practical. Instead, current CCD-based systems rely on a phosphorescent scintillation screen to convert the electron image to a photonic image within a spectral range of maximal quantum efficiency for the detector. Unfortunately, scintillation screens generally limit the resolution achievable with these systems. Specifically, with the arrival of each incident electron, a footprint or scattering plume is generated within the scintillation screen; this plume of photons is the signal that is relayed to the detector. If the extent of the plume exceeds the pixel pitch of the CCD device, spatial resolution is compromised. At 300 keV, the full width at half maximum (FWHM) of the footprint is around 30 μm for a reasonably bright scintillator; however, the full width considering 99% of the signal spread is 200 μm (Faruqi and Andrews, 1997).

The impact of this point-spread function (PSF) within the scintillator depends heavily on the relay strategy employed. Two types of CCD imaging relay systems can be classified: lens-coupled and fiber-optically coupled systems (Fig. 1). Both types of systems have been discussed extensively in the literature (Chapman, 1989; Daberkow *et al.,* 1991; Fan and Ellisman, 1993, 2000; Krivanek *et al.*, 1991; Spence and Zuo, 1988). The major issues defining the performance of each option include resolution, sensitivity, dynamic range, detection quantum efficiency (DQE), compatibility with the microscope vacuum, X-ray contamination to images, modularity, expandability, fixed pattern noise, physical dimension/packaging, shutter implementation, readout rate, and cost.

Unfortunately, there is no single system that excels in all performance categories or is optimized for all modes of imaging. Contrary to common intuition, sensitivity and DQE decrease at higher accelerating voltages, making it more difficult to implement a system that delivers satisfactory performance for intermediate and high-voltage applications. Despite the voltage dependence, some general characterization can

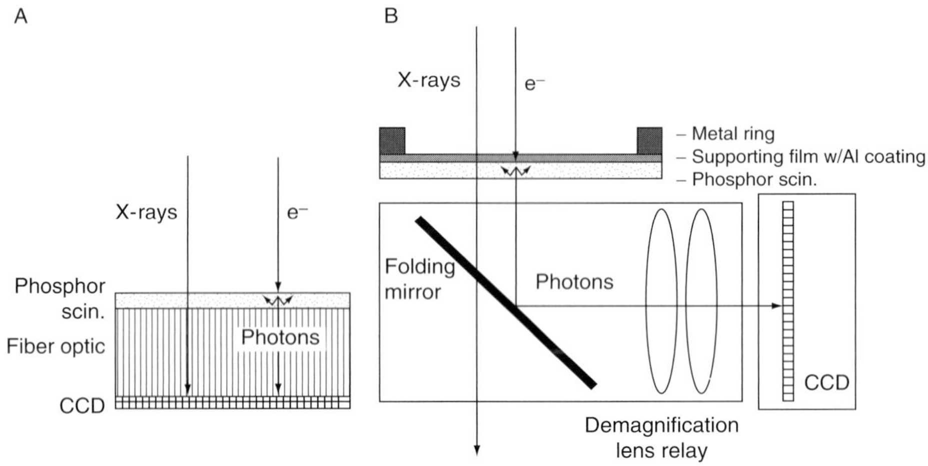

Fig. 1 CCD imaging systems for TEM. (A) Fiber-optically coupled systems provide better sensitivity and compact size, but have X-ray problems and lower resolution at voltages >200 keV due to the backscattered electrons from the fiber-optic (FO) plate. (B) A lens-coupled system is generally less sensitive but resolution can be better using a demagnification lens and a self-supporting screen. X-rays, which cause bright spots on an image, are avoided by a 90° bend in the optical path.

be made: fiber-optically coupled systems are typically more sensitive and compact, whereas the lens-coupled systems offer improved resolution, a high degree of modularity, and ease of maintenance.

At higher voltages, fiber-optically coupled systems suffer from electrons backscattering from the fiber-optic plate and reentering the scintillator; this causes an even larger PSF and hence a lower resolution (Fan and Ellisman, 1996). Demagnification of the PSF cannot be dealt with properly in a fiber-coupled system, as fiber-optic tapers introduce significant point dislocations and discontinuous spatial distortions (Faruqi and Andrews, 1997). Also, at voltages of 200 kV or higher, X-rays in the microscope, mostly from the gun area and some from the scintillator screen, penetrate the fiber-optic plate and reach the CCD. Each X-ray event may, depending on its energy, cause signal saturation in a group of 2×2 pixels, and thereby shows up as white dots on images. With a lens-coupled design, proper demagnification can be accomplished without distortion. The use of a self-supporting scintillator, positioned roughly 1 cm above the first lens element, acts to mitigate local backscattering into the scintillator, which can increase the electron footprint and lower resolution. This backscattering effect is much more prevalent at accelerating voltages of 400 keV and above, making fiber-coupled cameras unsuitable for high-voltage imaging. Furthermore, a folding mirror can also be used within the lens relay to optically redirect the photonic image 90° out of the path of the impending electrons and X-rays (Figs. 1 and 2). As previously stated, however, lens-coupled systems are characteristically less efficient (~3% total efficiency) than fiber-coupled systems (Figs. 1 and 2 below).

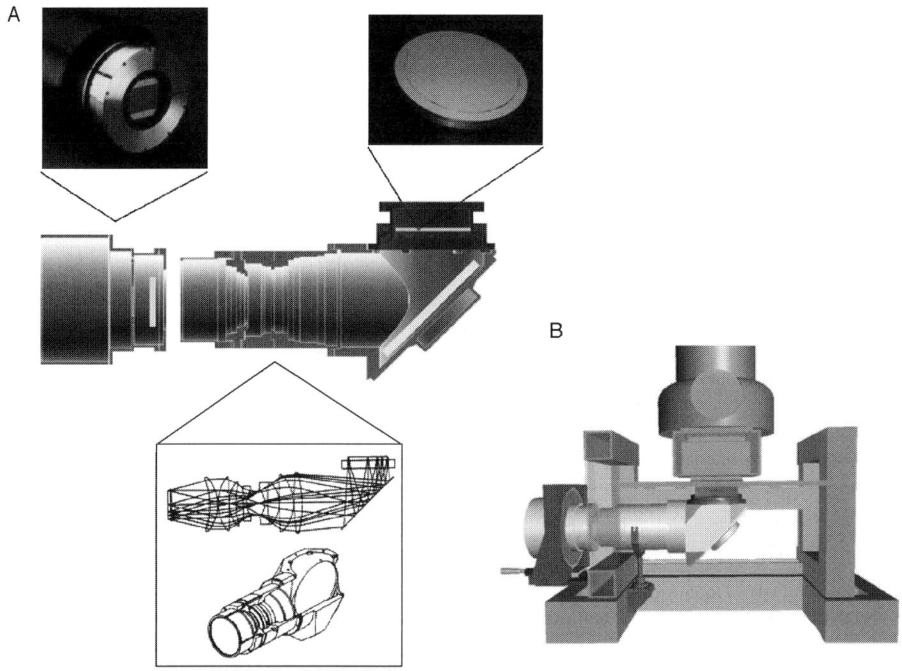

Fig. 2 Design scheme of the 2k × 2.5k CCD imaging system for the primary JEOL-4000EX microscope. This design was the highest resolution, largest format system at the time it was built. It still remains the highest resolution system at 400 keV, although it lacks the large format of more modern systems. (A) A cross-sectional schematic of the system highlighting the self-supporting scintillator, lens, and CCD camera. (B) A 3D model of the camera system as it is implemented within the lower frame (knee space) of the JEM-4000 EX IVEM.

Recognizing the problems with fiber-optically coupled systems, some of us have pioneered the development of lens-coupled CCD systems for high-voltage TEM. Starting with a proof-of-concept lens-relay system built from spare parts, we demonstrated single-electron sensitivity at 200 keV (Fan and Ellisman, 1993). From this work, it became clear that further improvements could be made at the level of the scintillation screen to improve brightness and resolution, especially for higher voltage applications. Partnering with experts from Grant Scientific and Advanced Microscopy Technique (AMT), we subsequently investigated several types of scintillating materials: yttrium aluminum garnet (YAG), phosphor (red and green), and other types of scintillators suitable for lens-coupled systems (Fan and Ellisman, 1996; Fan *et al.*, 1993, 1994). These efforts resulted in the development of a self-supporting scintillator screen design (US Patent No. 5,401,964), which demonstrated superior performance in both resolution and brightness as compared to traditional P20/glass-type designs. Encouraged by these results and the availability of an advanced astronomy grade CCD detector from MIT Lincoln Labs (CCID-16), we designed and built a 5-megapixel, lens-coupled CCD imaging system, capable of single-electron sensitivity, for our JEOL-4000EX IVEM (Fan and Ellisman, 2000). Since its commissioning in 1999, this system has operated in production use for automated electron tomography and telemicroscopy. With its modular design and convenient supporting flange (with front port for swapping scintillators), this system has also served as a platform for testing new, self-supporting scintillator screens and materials.

These activities have recently culminated in the development of an ambitious 8k $\times$ 8k detector, the Supracam, that retains the performance (single-electron sensitivity and ~15–20% contrast at Nyquist) of the MIT detector but scales its format to that of film. It incorporates a pyramid beam splitter and four high-performance demagnification lenses to split the image from a large, self-supported scintillation screen onto four custom designed, high dynamic range 4k $\times$ 4k CCD cameras (Fig. 3). Custom readout electronics have been engineered to stitch each 4k $\times$ 4k quadrant into the final image (Peltier *et al.*, 2005).

While the Supracam demonstrates an alternative means for scaling the lens-coupled CCD technology, ultimately it is cumbersome and expensive, and it demonstrates the complexity and extreme engineering required to overcome the limitations of using a scintillation screen for indirect imaging. Clearly, the future of digital imaging for TEM must overcome this fundamental limitation and provide direct electron detection at intermediate and high keV.

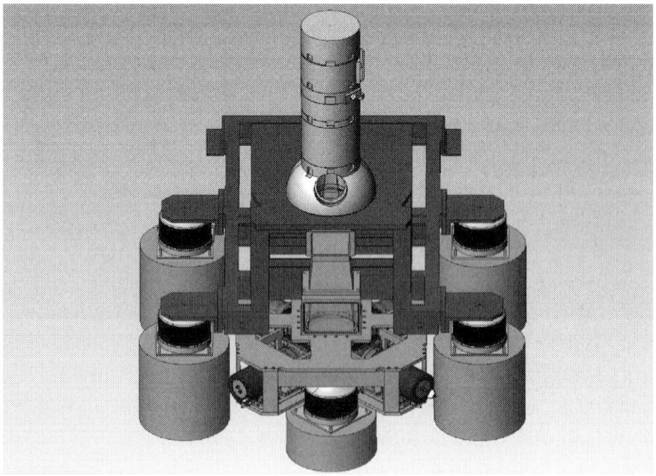

Fig. 3 Solid model of the 8k $\times$ 8k Supracam lens-coupled CCD system mounted below a JEOL 4000EX Intermediate Voltage Transmission Electron Microscope.

II. Background and Rationale

As previously mentioned, radiation damage and signal saturation prevent the practical use of CCD technology under direct electron bombardment (Roberts *et al.*, 1982). As such, all CCD-based systems for high-voltage EM are now scintillator dependent. Consequently, two significant issues continue to dominate the discussion of detector engineering. The first is the limitation imposed by the relatively large PSF of an electron striking the scintillator. This electron footprint limits the maximum attainable resolution of any scintillator system. Solutions to this issue include demagnifying the PSF to match the CCD pixel pitch or fabricating CCD chips with larger pixel sizes (which is extremely challenging and expensive at high format). The second limitation involves signal measurement, specifically single-electron sensitivity, readout noise, and dynamic range. Once again, the scintillator is a main limiting factor in these discussions. Ultimately, the amount of light output by the scintillator determines the signal-to-noise relationship and whether the detector will be single-electron sensitive. Furthermore, PSF resolution is inversely proportional to signal, resulting in a trade-off between resolution and signal. With current phosphor-based scintillators, the achievable PSF is on the order of ~20–30 μm, while the actual electron wavelength is less than an angstrom. This fact highlights the need to develop technology to replace the scintillator in the long term.

Since 1998, our group and others have pursued direct detection solutions to surmount the limitations of CCD-based systems. At first we experimented with the pixel array detector (PAD) developed for X-ray detection (Fan *et al.*, 1998). A PAD is composed of a two-dimensional (2D) pixel array of silicon sensors (300 μm thick) and a 2D application-specific integrated circuit (ASIC) device. Each individual pixel in the high-sensitivity silicon array is directly connected to its dedicated readout electronics on the ASIC chip by a solder bump bond. We tested an 8 × 8 pixel array with a 150 × 150 μm^2 pixel pitch (Fan *et al.*, 1998). The Cambridge group tested a 64 × 64 array with a 170 × 170 μm^2 pixel pitch (Faruqi *et al.*, 2003), and a 256 × 256 array with a 55 × 55 μm^2 pixel pitch (Faruqi *et al.*, 2005b).

A. Results of PAD Tests

With electron energies less than 60 keV it was possible to design a direct detection electron counting system with good uniformity and minimal cross talk between adjacent cells. The overall counting rate of this device is relatively high, since each pixel display operates independently. With electron energies in the range of 60–120 keV, the cross talk between pixel cells was small. At higher energies, however, electron scattering in the 300 μ silicon layer resulted in appreciable cross talk among the neighboring pixels. We also noted several limitations using PADs at 200–300 keV electron energies.

a. Large lateral scattering of incident electrons. Due to the thickness of the sensor (300 μm), an incident electron of 200 or 300 keV could have a large lateral scattering, up to 100 μm, making a precise measurement of its position very difficult.

b. Double counting. Even at lower incident electron energies, an electron falling near the edge of one pixel will produce two counts (due to the scattering of the electron inside the silicon), one on the original pixel and the other on the adjacent pixel. Since the readout electronics are independent between pixels, it will be extremely hard to eliminate double counting situations.

c. Low intrinsic detector size. Usually the pixel pitch of a PAD is quite large. With new *c*omplementary *m*etal *o*xide *s*emiconductor (CMOS) technology, it is possible to design a bump-bonded PAD with a pitch of 50×50 μm^2. However, it is difficult if not impossible to bump-bond a hybrid of an array with more than 250×250 pixels. Unfortunately, an array of this size is too small to properly image most protein complexes. A montage of module detectors will also be problematic, as dead space at the seams of abutting arrays will necessitate complex image reconstruction.

In consideration of these significant limitations, we discontinued the testing of bump-bonded PADs for cryo-EM in favor of the direct detection detector (DDD). The DDD is based on the active pixel sensor (APS), recently developed to detect charged particles (Kleinfelder *et al.*, 2002, 2003, 2004; Matis *et al.*, 2002, 2003; Turchetta *et al.*, 2001). Our group pioneered the use of the APS for EM (Faruqi *et al.*, 2005a; Milazzo *et al.*, 2005; Xuong *et al.*, 2004). This device detects the incident electrons directly, avoiding the requirement of a scintillation screen. The DDD's sensitive region can be very thin (as thin as 8 μm), yielding good spatial resolution with little noise by minimizing lateral scattering (thus maximizing the spatial resolution) of these incident electrons to as little as 2–3 μm (FWHM). Due to its simple design and its monolithic nature, the APS can be designed with a large number ($3.5k \times 4k$) of small pixels (5×5 μm^2), in order to match the electron footprint in the detector. For example, to resolve a 3 Å structure, we need to acquire images with an EM magnification that yields 1 pixel per $Å^2$. With the DDD's small pixel pitch, it is possible to achieve this resolution target within a reasonable EM magnification.

From a preliminary study of a DDD with 20-μm pixel pitch and beam energies in the range of 120–400 keV (Milazzo *et al.*, 2005; Xuong *et al.*, 2004), we found that the signal-to-noise ratio (SNR) for detecting a single incident electron was very high (at least 10/1), and the spatial resolution was in the range of 10 μm. However, this previous spatial resolution measurement was limited in accuracy by the 20 μm pixel pitch. We have therefore concentrated our effort into the design, fabrication, and testing of a DDD with a much smaller pixel pitch (5 μm), which will be reported below (Li *et al.*, 2006). While this technology will have obvious benefits for cryo-EM applications, it could also be useful at energies up to 3 MeV and for other applications, such as tomography of selectively stained specimens (Sosinsky, 2002) and ultrafast electron microscope imaging (Bradley *et al.*, 2003).

III. Description of the Direct Detection Detector

The APS utilized for the DDD has been developed specially for EM. It comprises a sensitive region consisting of a particularly pure and lightly doped layer of silicon that is epitaxially grown on top of an ordinary, more heavily doped silicon wafer. "Doping" refers to the amount of impurities deliberately introduced into the silicon to provide semiconductor behavior. Into this epitaxial silicon region, shallow, p-type[1] wells are implanted, wherein n-type transistors can be fabricated. Inside each pixel is a diode formed by the interface between an n-well implantation and the p-epitaxial region, as well as three transistors that read out the charge collected on the diode.

High-energy electrons from the electron microscope leave trails of ionization within the silicon. These ionization electrons quickly recombine in the highly doped bulk silicon, but have a reasonably long lifetime in the highly pure epitaxial region. Furthermore, the epitaxial silicon is sandwiched between two more heavily doped silicon regions (the bulk silicon wafer and the p-wells at the surface of the epitaxial silicon). This creates potential barriers to the electrons, causing liberated charge to be confined within the epitaxial silicon.

With an 8 μm-thick epitaxial region, an incident electron of 200 or 300 keV will liberate about 1000 electrons in the epitaxial layer, which is significantly higher than the noise of the pixel's readout circuitry (typically less than 45 electrons). The electrons are confined within the epitaxial region and diffuse randomly through it until they are collected by one of the pixel diodes (Fig. 4). Each pixel effectively integrates the collected ionization electrons. At the end of the integration period, the contents of the sensor array are read out, digitized, displayed, and ultimately stored.

Detector chips, including the prototypes, are fabricated in a standard digital CMOS process, as used for commercial microcomputers, and so on, making them convenient and inexpensive to fabricate. Unlike conventional digital cameras, in which pixels are sensitive to only a fraction of the light that strikes them, the "fill factor" (i.e., the proportion of each pixel area that is sensitive to incident electrons) is 100%, since the sensitive epitaxial layer is continuous beneath the readout circuitry.

Since the pixels are small, a major source of noise in the sensor is the reset noise. To reduce this noise, we use a correlated double sampling (CDS) method, which

[1] n-type silicon is pure silicon that is sparsely implanted ("doped") with atoms, such as phosphorus or arsenic, that have one extra electron in their valence bands when compared silicon. This electron is weakly bound to the silicon lattice and can be easily dislodged, thereby "donating" a free electron. p-type silicon is doped with atoms such as boron, with one fewer electron than silicon, and hence an atom of this type of dopant can easily accept a free electron.

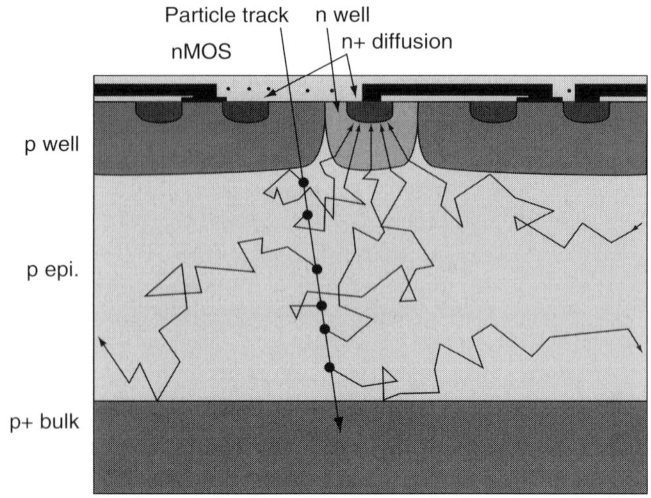

Fig. 4 Cross section of an APS pixel showing the small collection diode flanked by n-channel transistors formed in p-wells. Adapted from Kleinfelder *et al.* (2002).

employs multiple, nondestructive reads between resets. A CDS frame is then formed by subtracting subsequent nondestructive reads. Since the chip is not reset between these reads, the difference is simply the charge integrated in the diode during the time elapsed between the reads, and the reset noise is substantially cancelled. The overall exposure time, defined as the time between the two reads used for CDS subtraction, can be varied between a few milliseconds to as long as a few seconds. For the results reported here, the exposure time was about 94 msec. For faster readout, the prototype sensor system utilized four parallel analog outputs with four parallel 12-bit analog-to-digital converters (ADC), each sampling at 1.25 MHz.

All data were collected at room temperature with the exception of the radiation damage tests, which were performed both at room temperature and at $-15\,^{\circ}$C. The DDD was mounted in a modified JEOL TEM film camera drawer (Fig. 5), making it possible to install the system quickly in any JEOL electron microscope (Milazzo *et al.*, 2005). We used a JEM-2000 FX II for experiments at 200 keV and a JEM-4000 EX for experiments at 300 and 400 keV.

IV. Detector Characterization

A. Noisy Pixels

Both DDD and CCD cameras contain some pixels with excessive noise. For the DDD, we minimized the impact of these pixels through a systematic process of identification and value replacement. Noisy pixels typically yield readings that

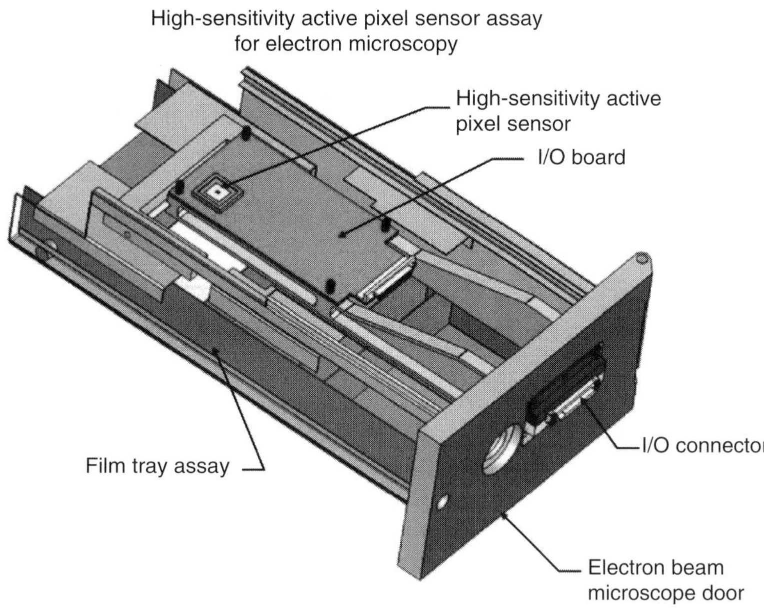

Fig. 5 Schematic diagram of the mounting of the DDD into a modified film cassette of the JOEL microscopes.

vary widely from frame to frame. Having located these pixels, we replaced their readings with the average of the neighboring pixel readings. To detect and isolate noisy pixels, we collected more than 5000 background frames (with no electron beam) and determined the standard deviation (σ) of the readings of all pixels. These readings are dark current corrected, using the average of a preliminary acquisition of 500 frames to establish the pedestal. Noisy pixels were defined as ones that gave an average reading greater than 10σ in at least 0.5% of the total number of frames. The noisy pixel positions (about 1.5% of the whole array) are stored in a file and their readings are replaced by the average of the neighboring pixel readings. In addition, we ignored all the readings from the outermost two pixels along the perimeter of the chip.

B. Signal to Noise

A strength of the DDD is its high SNR. We measured the SNR of a single 300 keV incident electron to be 13/1. For these measurements, we first collected 1000 frames without an incident beam. A histogram of the noise readings of all pixels in these frames yielded an average standard deviation of 3.2 ADC units (Fig. 6). There were very few readings equal to or above 40 ADC units. Therefore, a noise floor of 40 ADC units was used as the threshold value for the detection of an incident electron.

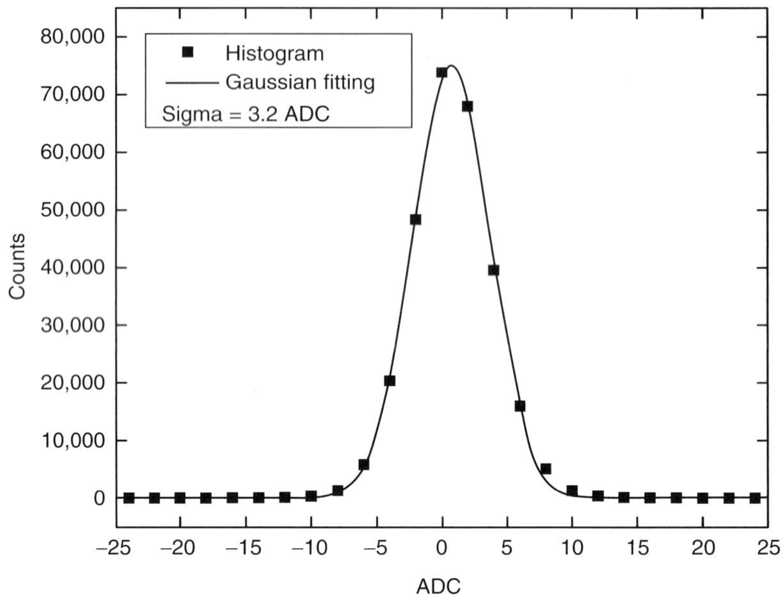

Fig. 6 Background histogram of pixel readings in a single frame after correlated double sampling. The DDD was kept in darkness.

To determine the signal level, the beam intensity was reduced so that each frame contained about 1 incident electron. Figure 7A shows a typical reading for a single incident electron. In these single-electron events, the significant pixel values are always grouped inside a 5×5 array centered on the pixel with the maximum reading. The center pixel values are typically quite high, with SNR up to 40 or 50. By averaging all the 7×7 arrays selected around the center pixel for each detected electron event, we were able to calculate the average reading distribution. Figure 7B shows such a distribution for 300 keV incident electrons. The average distribution had significant values only inside a 5×5 array centered on the maximum. By summing over this 5×5 array, we determined an average signal value of 214 ADC units for a single-electron event. Given an average noise approximation of $\sqrt{25} \times 3.2 = 16$ ADC units, so the average SNR of each detected electron is ~13/1. The same SNR was measured with 200 and 400 keV electrons.

C. Spatial Resolution

Another strength of the DDD is its high spatial resolution. A fit of a projection of the average distribution of the pixel readings from a single-electron event (shown in Fig. 7B) on either the x or y coordinate to a Gaussian curve results in an FWHM of only 5.5 μm. This FWHM is due mostly to the fact that the

A

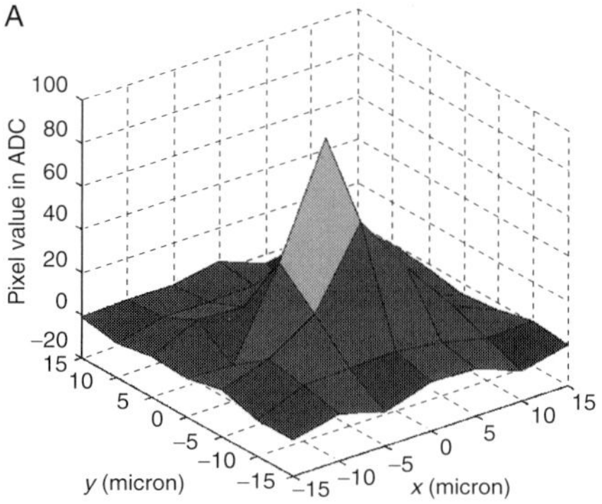

	y_1	y_2	y_3	y_4	y_5	y_6	y_7
x_1	−3	2	−4	3	2	−6	−1
x_2	−3	6	7	8	5	0	1
x_3	2	9	23	59	31	6	−3
x_4	−2	−3	38	89	29	1	−5
x_5	−1	9	8	25	6	5	−2
x_6	−4	1	1	2	−10	5	0
x_7	−1	0	0	4	−4	0	−2

B

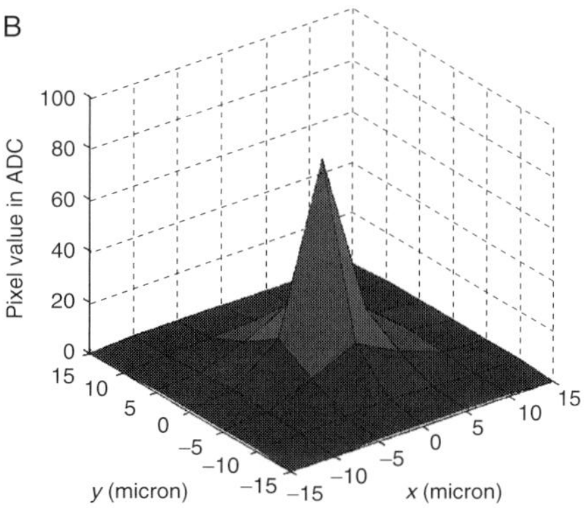

	y_1	y_2	y_3	y_4	y_5	y_6	y_7
x_1	0	0	1	1	1	1	0
x_2	1	1	2	3	2	1	1
x_3	1	2	9	17	8	2	1
x_4	1	3	17	81	17	3	1
x_5	1	2	8	17	8	2	1
x_6	0	1	2	3	2	1	1
x_7	0	1	1	1	1	1	0

$$\text{Signal-to-noise ratio} = 214/(3.2\sqrt{25}) = 13/1$$

$$\text{Spatial resolution} = \sqrt{5.5^2 - 5.0^2} = 2.3\ \mu\text{m}$$

Fig. 7 Distribution of pixel readings for single incident electron events at 300 keV. (A) Pixel readings for a typical event. (B) Average distribution.

entry points of incident electrons are distributed uniformly over the center pixel. By supposing this distribution to be Gaussian with an FWHM of 5 μm (the size of a pixel), we can deconvolute the result and obtain a spatial resolution of about 2.3 μm. Like the SNR, the spatial resolution is nearly unchanged with 200 and 400 keV incident electrons.

In order to reduce incident electrons backscattering from the 250 μm silicon substrate into the 8 μm active epitaxial layer, we trimmed the thickness of the DDD to 30 μm (Milazzo *et al.*, 2005). In this configuration, the thinner chip is capable of delivering an SNR (11/1) and the spatial resolution (4 μm) that approximates the results from the standard chips.

As a real-world example of the imaging advantages afforded by this detector, we present an electron micrograph of myelin at a magnification of 2500×. The myelin layer spacing is about 10 nm, which corresponds to about 25 μm on the image, that is, 5 pixels in the new DDD. Figure 8A shows the myelin imaged with this device, and details of the myelin band are clearly visible. These details are lost on an image (Fig. 8B) of the same myelin sample at the same EM magnification taken with a high-quality cooled, fiber-coupled CCD camera (2k $\times$ 2k Tietz TemCam-F224HD slow scan, 24 μm pixel, 16 bit). The poor spatial resolution of the scintillation screen (30 μm), together with the large pixel size (24 μm) of the CCD, makes the image very grainy. It is remarkable that this DDD image was taken at room temperature and is shown without any software correction, for example, flat fielding or gain normalization. The image was simply constructed from a sum of 50 frames taken at a very low dose (1.5 electrons/pixel/frame). In total, there was an average of only 75 incident electrons/pixel. Since counting statistics [$1/\sqrt{75}$] by itself would be about 10%, this image demonstrates the high SNR and the excellent spatial resolution of the new detector. With the DDD, microscopists will need only a reasonable EM magnification factor (50,000) in order to yield cryo-EM images that allow very high-resolution (3 Å) 3D reconstructions. MTF comparisons between CCD and DDD are currently underway and will be reported in a future manuscript.

V. Radiation Damage

A significant characteristic of the new DDD is its ability to withstand radiation damage from the microscope's high-energy electrons. At room temperature, we observed that the DDD stopped working at a fluence of 3.6×10^5 electrons/pixel.

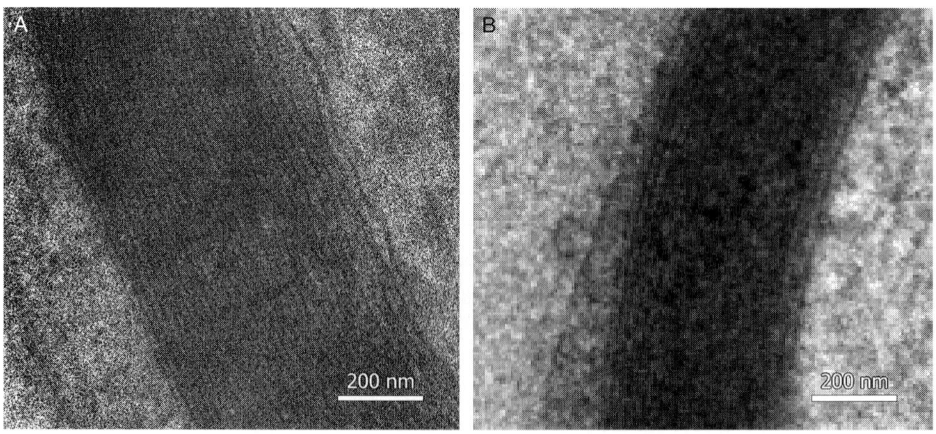

Fig. 8 Comparison of myelin images using the DDD and a CCD camera. Both images were taken from the same nerve section poststained with uranyl acetate and sato lead, at the same EM magnification of 2500 with the 200-keV electron beam. More details were captured in the DDD image (A), compared to the one from the CCD camera (B).

However, with the chip cooled down to $-15\,^{\circ}$C, it still works at a fluence of 3.6×10^6 electrons/pixel, that is, an order of magnitude more exposure.

These results were obtained from our efforts to quantify the radiation damage of the 512×550 pixel (with 5 μm pitch size) detector. The detector was submitted to a continuous 300 keV electron beam with an average intensity of 5 electrons/pixel/sec for a period of 1 sec. Figure 9A illustrates the variation of dark current in relation to the accumulated dose of electrons. Dark current, measured by the voltage across the photodiode, registered a tenfold increase after the fluence of 3.6×10^5 electrons/pixel.

At a fluence of 3.6×10^5 electrons/pixel, the corresponding signal was reduced by 10%, which was still good enough to provide a signal well above the noise level. However, when the detector was operated at near its saturation level, it became unusable after a fluence of 6.5×10^5 electrons/pixel. This is in agreement with the findings of Faruqi *et al.* (2005a). If the dark current could be reduced to a reasonable level, we would be able to continue to use the detector at higher radiation doses. One straightforward way to reduce the dark current is to cool the detector. This experiment was run in a relatively warm vacuum (33 °C). Theoretically, cooling our detector to $-15\,^{\circ}$C should reduce the dark current by a factor of 28, assuming that dark current is halved for every 10 °C reduction in temperature, and one can imagine that chilling the detector would keep the dark current low after high radiation doses.

To check this theoretical relationship, we cooled the detector to $-15\,^{\circ}$C and measured the dark current as a function of the total radiation dose. As expected, the dark current was reduced to a very small value (0.013 V/sec) and remained at low levels for a large range of radiation doses (up to 3.3×10^6 electrons/pixel) (Fig. 9B).

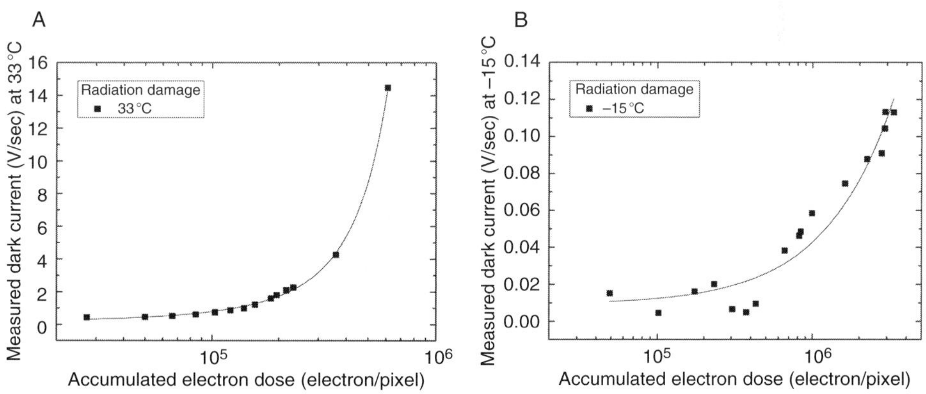

Fig. 9 Influence of DDD temperature on dark current variations with respect to the accumulated total electron dose. The dark current per second was measured by the voltage (V) across the photodiode. (A) At 33 °C the dark current increased by tenfold at a fluence of 3.6×10^5 electrons/pixel. (B) After cooling the detector to $-15\,^{\circ}$C, the dark current stayed at low levels after a total dose of 3.3×10^6 electrons/pixel.

In addition, the signal remained at 100% and the SNR stayed constant during the test (Fig. 10).

The above experiments showed that the DDD functioned well at −15 °C, even with a relatively high radiation dose of 3.3 × 10^6 electrons/pixel. Notably, however, a favorable characteristic of the DDD is the ease of repairing it from radiation damage by annealing, that is, by baking the chip in a 100 °C oven for at least 5 days (Matis *et al.*, 2005).

VI. Discussion

While protein crystallography is still the leading method to determine 3D protein structures, cryo-EM is emerging as a powerful method to determine the 3D structures of large protein complexes and viruses at 4 to 10 Å resolution (Bottcher *et al.*, 1997; Gabashvili *et al.*, 2000; Henderson, 2004; Miyazawa *et al.*, 2003; Thuman-Commike and Chiu, 2000; Valle *et al.*, 2003; Zhang *et al.*, 2003a,b). The technique could become much more important if near-atomic resolution (~3 Å) was easier to obtain. The difficulty in collecting and processing very large data sets (estimated to be up to one million images to achieve 3 Å resolution) required for statistical analysis (Frank *et al.*, 2000) has been a limitation in resolving of the structures obtained with cryo-EM.

The DDD with 5 μm pixel size has several attributes that make it highly desirable for cryo-EM. It has a large SNR (~13/1), coupled with an excellent spatial resolution (2–4 μm) at 200–400 keV. In contrast, a CCD has an SNR of ~1/1 and a spatial resolution of 30 μm. Our DDD permits us to collect near-atomic resolution (better than 3 Å) images with sufficient precision to solve a 3D macromolecule structure at 50,000× EM magnification.

A

	y_1	y_2	y_3	y_4	y_5	y_6	y_7
x_1	0	0	1	1	1	1	0
x_2	1	1	2	3	2	1	1
x_3	1	2	9	17	8	2	1
x_4	1	3	17	81	17	3	1
x_5	1	2	8	17	8	2	1
x_6	0	1	2	3	2	1	0
x_7	0	1	1	1	1	1	0

B

	y_1	y_2	y_3	y_4	y_5	y_6	y_7
x_1	0	1	1	1	1	1	1
x_2	1	1	2	2	2	1	1
x_3	1	3	8	17	7	3	1
x_4	1	3	16	85	18	4	1
x_5	0	2	6	16	9	3	1
x_6	0	1	2	3	3	1	1
x_7	0	1	1	1	1	0	0

Fig. 10 Average distribution of pixel readings for single incident electron events at 300 keV before and after the radiation damage test. The chip was maintained at −15 °C. (A) Before radiation damage, SNR = $214/(3.2*\sqrt{25}) = 13/1$. (B) After a total dose of 2.9 × 10^6 electrons/pixel, SNR = $218/(3.2*\sqrt{25}) = 14/1$.

We are in the process of designing a large chip (3.5k × 4k pixels) with a massive array of parallel, on-chip ADCs (4000 ADCs). This new feature will allow a very fast readout rate (10 frames/sec), opening up new opportunities for enhanced automation and near real-time data collection, which are not possible with slow scan CCDs.

The DDD functioned well at −15 °C, even with a relatively high radiation dose of 3.3×10^6 electrons/pixel. This dose is equivalent to 600,000 cryo-EM frames. Applied to a large (3.5k × 4k) pixel array this is roughly equivalent to 6 million single-particle images, enough to solve six structures at very high resolution (3 Å). With a mean exposure time of 1 sec/frame and the associated time requirements for collecting data we estimate a detector lifetime to be 3–6 months. Therefore, the DDD is expected to function with stable performance for about this duration in a typical high-throughput laboratory setting.

We must point out that while the cost of designing new chips is high, the cost of reproducing such chips, even large ones (3.5k × 4k pixels), is very low compared to the cost of large, sensitive CCD devices (less than $2000 per chip compared to 10–20 times more for CCDs). Therefore, the replacement cost is similar to the cost of consumable materials like film. The annealing process mentioned above holds the promise to allow continuous use of the DDD with as few as two chips in stock (one is in use while the other is being recovered).

Acknowledgments

We thank Mr. John Crum and Mr. Mason Mackey for the electron microscope technical support. We also acknowledge the help of Dr. James Matteson, Dr. Wayne Vernon, Dr. Skip Cynar, Mr. George A. White, Mr. Thomas Gasaway, and Mr. Ron Quillin. This work is supported by two grants from the Research Resource Division of the NIH (RR018841 and RR04050) and a grant from the Director Office of Science of the US Department of Energy under Contract No. DE-AC03–76SF00098. A.-C.M. was supported by NIH training grant PO1 DK54441.

References

Aikens, R. S., Agard, D. A., and Sedat, J. W. (1989). Solid-state imagers for microscopy. *Methods Cell Biol.* **29,** 291–313.

Bottcher, B., Wynne, S. A., and Crowther, R. A. (1997). Determination of the fold of the core protein of hepatitis B virus by electron cryomicroscopy. *Nature* **386,** 88–94.

Boyle, W. S., and Smith, G. E. (1970). Charge coupled semiconductor devices. *Bell Syst. Tech. J.* **49,** 587–593.

Bradley, J. S., Dwyer, J. R., Jordan, R. E., and Miller, R. J. D. (2003). An atomic-level view of melting using femtosecond electron diffraction. *Science* **302,** 1382–1385.

Chapman, J. N., Craven, A. J., and Scott, C. P. (1989). Electron detection in the analytical electron microscope. *Ultramicroscopy* **28,** 108–117.

Daberkow, I., Herrmann, K. H., Liu, L. B., and Rau, W. D. (1991). Performance of electron image converters with YAG single-crystal screen and CCD sensor. *Ultramicroscopy* **38,** 215–223.

Daberkow, I., Herrmann, K. H., Liu, L., Rau, W. D., and Tietz, H. (1996). Development and performance of a fast fibre-plate coupled CCD camera at medium energy and image processing system for electron holography. *Ultramicroscopy* **64,** 35–48.

Downing, K. H., and Hendrickson, F. M. (1999). Performance of a 2k CCD camera designed for electron crystallography at 400 kV. *Ultramicroscopy* **75,** 215–233.

Fan, G. Y., Bueno, C., Dunkelberger, D., and Ellisman, M. H. (1993). Performance characteristics of radioluminescent fiber optics as electron scintillators. *J. Electron Microsc. (Tokyo)* **42,** 419–423.

Fan, G. Y., Datte, P., Beuville, E., Beche, J.-F., Millaud, J., Downing, K. H., Burkard, F. T., Ellisman, M. H., and Xuong, N.-H. (1998). ASIC-based event-driven 2D digital electron counter for TEM imaging. *Ultramicroscopy* **70,** 107–113.

Fan, G. Y., Dunkelberger, D. G., and Ellisman, M. H. (1994). Performance of thin foil scintillating screen for transmission electron microscopy. *Ultramicroscopy* **55,** 7–14.

Fan, G. Y., and Ellisman, M. H. (1993). High-sensitivity lens-coupled slow-scan CCD camera for transmission electron microscopy. *Ultramicroscopy* **52,** 21–29.

Fan, G. Y., and Ellisman, M. H. (1996). Optimization of thin-foil based phosphor screens for CCD imaging in TEM in the voltage range of 80–400 kV. *Ultramicroscopy* **66,** 11–19.

Fan, G. Y., and Ellisman, M. H. (2000). Digital imaging in transmission electron microscopy. *J. Microsc.* **200,** 1–13.

Faruqi, A. R., and Andrews, H. N. (1997). Cooled CCD camera with tapered fibre optics for electron microscopy. *Nucl. Instrum. Methods Phys. Res. A* **392,** 233–236.

Faruqi, A. R., Cattermole, D. M., Henderson, R., Mikulec, B., and Raeburn, C. (2003). Evaluation of a hybrid pixel detector for electron microscopy. *Nucl. Instrum. Methods Phys. Res. A* **546,** 160–163.

Faruqi, A. R., Henderson, R., and Subramaniam, S. (1999). Cooled CCD detector with tapered fibre optics for recording electron diffraction patterns. *Ultramicroscopy* **75,** 235–250.

Faruqi, A. R., Henderson, R., Pryddetch, M., Allport, P., and Evans, A. (2005a). Direct single electron detection with a CMOS detector for electron microscopy. *Nucl. Instrum. Methods Phys. Res. A* **546,** 170–175.

Faruqi, A. R., Henderson, R., and Tlustos, L. (2005b). Noiseless direct detection of electrons in Medipix2 for electron microscopy. *Ultramicroscopy* **94,** 263–276.

Frank, J., Penczek, P., Agrawal, R. K., Grassucci, R. A., and Heagle, A. B. (2000). Three-dimensional cryoelectron microscopy of ribosomes. *Methods Enzymol.* **317,** 276–291.

Gabashvili, I. S., Agrawal, R. K., Spahn, C. M., Grassucci, R. A., Svergun, D. I., Frank, J., and Penczek, P. (2000). Solution structure of the *E. coli* 70S ribosome at 11.5 Å resolution. *Cell* **100,** 537–549.

Henderson, R. (2004). Realizing the potential of electron cryo-microscopy. *Q. Rev. Biophys.* **37**(1), 3–13.

Herrmann, K.-H., and Liu, L. (1992). Performance of image converters using slow-scan CCDs in MeV electron microscopy. *Optik* **92**(2), 48–50.

Herrmann, K.-H., and Sikeler, R. (1995). Low-energy electron image converters for high resolution slow-scan CCD sensor. *Optik* **98,** 119–124.

Kleinfelder, S., Bichsel, H., Bieser, F., Matis, H. S., Gulshan, R., Retiere, F., Wieman, H., and Yamamoto, E. (2002). Integrated X-ray and charged particle active pixel CMOS sensor arrays using an epitaxial silicon-sensitive region. *Proc. SPIE* **4784,** 208–217.

Kleinfelder, S., Bieser, F., Chen, Y. D., Gareus, R., Matis, H. S., Oldenburg, M., Retierc, F., Ritter, H. G., Wieman, H., and Yamamoto, E. (2003). Novel integrated CMOS pixel structures for vertex detectors. *Nuclear Science Symposium Conference Record 2003 IEEE* **1,** 335–339.

Kleinfelder, S., Bieser, F., Chen, Y. D., Gareus, R., Matis, H. S., Oldenburg, M., Retiere, F., Ritter, H. G., Wieman, H., and Yamamoto, E. (2004). Novel integrated CMOS sensor circuits. *IEEE Trans. Nucl. Sci.* **51,** 2328–2336.

Krivanek, O. L., and Mooney, P. E. (1993). Applications of slow-scan CCD cameras in transmission electron microscopy. *Ultramicroscopy* **49,** 95–108.

Krivanek, O. L., Mooney, P. E., and Fan, G. Y. (1991). Slow-scan CCD camera for transmission electron microscopy. *Inst. Phys. Conf. Ser.* **119**(119), 523–526.

Kujawa, S., and Krahl, D. (1992). Performance of a low-noise CCD camera adapted to a transmission electron microscope. *Ultramicroscopy* **45**(1), 395–403.

Li, S., Bouwer, J. C., Duttweiler, F., Ellisman, M. H., Jin, L., Leblanc, P., Milazzo, A. C., Peltier, S. T., Xuong, N. H., and Kleinfelder, S. (2006). A new direct detection camera system for electron microscopy. *Proc. SPIE* **6068,** 218–227.

Matis, H. S., Bieser, F., Kleinfelder, S., Rai, G., Retiere, F., Hans Titter, G., Samuel, K. S., Wurzel, E., Wieman, H., and Yamamoto, E. (2002). A CMOS active pixel sensor for charged particle detection. *Nuclear Science Symposium Conference Record 2002 IEEE* **1,** 259–263.

Matis, H. S., Bieser, F., Kleinfelder, S., Rai, G., Retiere, F., Ritter, H. G., Singh, K., Wieman, H., and Yamamoto, E. (2003). Charged particle detection using a CMOS active pixel sensor. *IEEE Trans. Nucl. Sci.* **50,** 1020–1025.

Matis, H. S., Bieser, F., Chen, Y. D., Gareus, R., Kleinfelder, S., Oldenburg, M., Retiere, F., Ritter, H. G., Wieman, H., Wurzel, S. E., and Yamamoto, E. (2005). Using an active pixel sensor in a vertex detector. *Nucl. Instrum. Methods Phys. Res. A* **549,** 130–136.

Milazzo, A. C., Leblanc, P., Duttweiler, F., Jin, L., Bouwer, J. C., Peltier, S. T., Ellisman, M. H., Matis, H. S., Wieman, H., Denes, P., Kleinfelder, S., Xuong, N.-H., *et al.* (2005). Active pixel sensor array as a detector for electron microscopy. *Ultramicroscopy* **104,** 152–159.

Miyazawa, A., Fujiyoshi, Y., and Unwin, N. (2003). Structure and gating mechanism of the acetylcholine receptor pore. *Nature* **423**(6943), 949–955.

Peltier, S. T., Bouwer, J. C., Jin, L., Khodjasaryan, K., Geist, S., Xuong, N.-H., and Ellisman, M. H. (2005). Design of a new 8k × 8k lens coupled detector for wide-field, high-resolution transmission electron microscopy. Proceedings of microscopy and microanalysis, digital imaging session, August 2005.

Roberts, P. T. E., Chapman, J. N., and MacLeod, A. M. (1982). A CCD-based image recording system for the CTEM. *Ultramicroscopy* **8,** 385–396.

Sosinsky, G. (2002). The nuts and bolts of electron tomography: Imaging of big and messy biological structures. *Microsc. Microanal.* **8**(S02), 876–877.

Spence, J. C. H., and Zuo, J. M. (1988). Large dynamic range, parallel detection system for electron diffraction and imaging. *Rev. Sci. Instrum.* **59,** 2102.

Thuman-Commike, P. A., and Chiu, W. (2000). Reconstruction principles of icosahedral virus structure determination using Electron Microscopy. *Micron* **31,** 687–711.

Turchetta, R., Berst, J. D., Casadei, B., Claus, G., Colledani, C., Dulinski, W., Hu, Y., Husson, D., Le Normand, J. P., Riester, J. L., Deptuch, G., Goerlach, U., *et al.* (2001). A monolithic active pixel sensor for charged particle tracking and imaging using standard VLSI CMOS technology. *Nucl. Instrum. Methods Phys. Res. A* **458,** 677–689.

Valle, M., Zavialov, A., Li, W., Stagg, S. M., Sengupta, J., Nielsen, R. C., Nissen, P., Harvey, S. C., Ehrenberg, M., and Frank, J. (2003). Incorporation of aminoacyl-tRNA into the ribosome as seen by cryo-electron microscopy. *Nat. Struct. Biol.* **10**(11), 899–906.

Xuong, N.-H., Milazzo, A. C., LeBlanc, P., Duttweiler, F., Bouwer, J. C., Peltier, S. T., Ellisman, M. H., Denes, P., Bieser, F., Matis, H. S., Wieman, H. H., and Kleinfelder, S. (2004). First use of a high-sensitivity active pixel sensor array as a detector for electron microscopy. *Proc. SPIE* **5301,** 242–249.

Zhang, W., Chipman, P. R., Corver, J., Johnson, P. R., Zhang, Y., Mukhopadhyay, S., Baker, T. S., Strauss, J. H., Rossmann, M. G., and Kuhn, R. J. (2003a). Visualization of membrane protein domains by cryo-electron microscopy of dengue virus. *Nat. Struct. Biol.* **10**(11), 907–912.

Zhang, X., Walker, S. B., Chipman, P. R., Nibert, M. L., and Baker, T. S. (2003b). Reovirus polymerase lambda 3 localized by cryo-electron microscopy of virions at a resolution of 7.6 Å. *Nat. Struct. Biol.* **10**(12), 1011–1018.

CHAPTER 29

Structure Determination *In Situ* by Averaging of Tomograms

Friedrich Förster* and Reiner Hegerl†

*Department of Pharmaceutical Sciences
University of California
San Francisco, California 94143

†Max-Planck Institute of Biochemistry
Am Klopferspitz 18, D-82152 Martinsried, Germany

Averaging single particles that have been extracted from cryoelectron tomograms is a powerful method for visualizing macromolecular complexes *in situ*. Cryoelectron tomography (CET) is uniquely suited to obtaining three-dimensional (3D) insights into pleiomorphic objects such as cells or organelles. Frozen-hydrated specimens provide excellent preservation of biological material, and thus a cryoelectron tomogram depicts a faithful image of complexes *in situ*. However, the resolution of CET is limited by the applicable electron

0091-679X/07 $35.00
DOI: 10.1016/S0091-679X(06)79029-X

dose; this limitation can be overcome for macromolecules that occur in identical copies within one or many tomograms by coherently averaging them. If the differently oriented and positioned particles are brought into precise register, their weak individual signals are amplified, resulting in an average that is of higher resolution. Here we report on the image processing that is needed to align the single particles to a common coordinate system, both in theory and in practice. Our method iteratively optimizes a scoring function that adopts its maximum value when all particles are in register; the scoring function is a cross-correlation function that accounts for the "missing-wedge effect," currently the most prominent imaging artifact in CET. Structural characterization by CET and subsequent averaging can be particularly fruitful for unraveling the architecture of membrane-bound complexes.

I. Introduction

Electron tomography (ET) has emerged as a versatile imaging method in cell biology (Lucic *et al.*, 2005; McIntosh *et al.*, 2005; Subramaniam, 2005). It is uniquely suited to obtain three-dimensional (3D) images of pleiomorphic objects, such as cells or organelles. The technical advances in electron microscopes, automation of the data acquisition, and methods for specimen preparation have enabled its use in cell biology. Depending on the type of specimen preparation, ET can be distinguished into two different types: tomography of plastic embedded specimens and tomography of frozen-hydrated samples (cryoelectron tomography, CET).

ET of plastic embedded samples is particularly suitable for characterizing the ultrastructure of cells. Prior to the embedding process, the biological sample must be fixed. The best preservation of the specimen can be achieved by rapid freezing of the aqueous sample. Rapid freezing prevents the formation of ice crystals in the aqueous sample, which would damage the architecture of biological macromolecules (Dubochet *et al.*, 1988). The specimen undergoes a phase transition into a vitreous phase at a high cooling rate. The vitrification of samples that measure several micrometer in diameter, such as whole cells, some embryos, or small invertebrates, can be achieved by means of high-pressure freezing (Hunziker *et al.*, 1984). To embed these samples into plastic, the vitrified water is replaced by organic solvent, heavy metal staining is applied subsequently, and eventually the solvent is replaced by resin (Steinbrecht and Muller, 1987). This preparation technique is commonly referred to as rapid-freezing, freeze-substitution fixation (RF-FSF) (McIntosh *et al.*, 2005). Various examples have shown that this technique can preserve cytoskeletal fibers as well as membrane structures in an excellent fashion (McDonald and Morphew, 1993; Murk *et al.*, 2003a,b). The tomography of RF-FSF samples has thus led to important insights into the ultrastructure of cells, for example, the organelles of the secretory pathway (Ladinsky *et al.*, 1999; Marsh *et al.*, 2001, 2004; Murk *et al.*, 2003a) or the organization of microtubules (Moritz *et al.*, 2000; O'Toole *et al.*, 2003a,b).

The resolution of RF-FSF tomograms is, however, limited by the specimen preparation; the dehydration of the specimen leads to alteration of the biological material and the observed features need to be judged carefully to distinguish those that depict the native structure from those that are likely to be artifacts. Moreover, the heavy metal stain is the physical quantity that is imaged and the staining process is almost impossible to quantify. It is commonly accepted that the RF-FSF preparation technique is extremely fruitful for the study of membrane structures or filaments, but in most cases it is not suitable for studying the machinery of a cell on the molecular level (McIntosh *et al.*, 2005).

A tomogram of a frozen-hydrated specimen, a cryoelectron tomogram, essentially depicts a faithful image of the electrostatic potential (for biological molecules approximately proportional to the electron density) of a cell in its close-to-native state. Therefore, a cryoelectron tomogram is in principle interpretable at the molecular level. However, omitting fixation and staining has its price: the resolution of cryoelectron tomograms is fundamentally limited by the damage that the imaging electrons cause to the frozen-hydrated specimen. Biological material is extremely sensitive to radiation, which forces us not to exceed a certain dose of electrons for imaging. In order to image biological macromolecules faithfully, that is, without alterations by the electron radiation on the nanometer range, they must not be exposed to doses beyond 20–100 e/Å^2 depending on the specimen and anticipated resolution range. This low electron dose results in a low signal-to-noise ratio (SNR) in the cryo-tomogram. Nonetheless, it has been a major achievement to demonstrate that large macromolecular complexes could be resolved in their close-to-native environment using such extremely low electron doses (Medalia *et al.*, 2002). Subsequently, many other applications have indicated that CET is able to resolve large macromolecular assemblies in cells (Hsieh *et al.*, 2006; Kurner *et al.*, 2005), viruses (Benjamin *et al.*, 2005; Briggs *et al.*, 2006; Cyrklaff *et al.*, 2005; Förster *et al.*, 2005; Grünewald *et al.*, 2003), and organelles (Beck *et al.*, 2004; Nicastro *et al.*, 2005). Moreover, it has been shown that the signal from large assemblies (>500 kDa) is sufficiently strong to distinguish different types of macromolecules automatically in cell-resembling model systems (Frangakis *et al.*, 2002).

However, the signal from an individual macromolecule does not reveal much of its substructure, due to the extremely low SNR of a tomogram: the information is typically not significant beyond 5–10 nm resolution. Computational procedures exist to enhance the SNR within this resolution range; these "denoising programs" are primarily useful for tomogram interpretation on the ultrastructure level, but they are not able to significantly increase the resolution of a single macromolecule (reviewed in Frangakis and Förster, 2004). However, substantially higher resolution information can be extracted from a macromolecule if it is present in multiple copies: by coherently averaging these copies, the electron dose can be increased *in silico*, and the average reveals structural information beyond the significance limit of an individual particle.

Any attempt to achieve a medium or high-resolution 3D structure determination of biological macromolecules by cryoelectron microscopy (cryo-EM) relies on the principle of increasing the resolution by averaging identical copies of macromolecules. The initial approaches to obtain 3D structures of biological macromolecules exploited the geometrical order of macromolecules in highly symmetric assemblies; DeRosier and Klug (1968) made use of the helical symmetry of the bacteriophage T4 tail to obtain a coherent average of its constituent macromolecules. The highest resolution cryo-EM structures published to date have exploited translational symmetries of two-dimensional (2D) crystals [electron crystallography (Gonen *et al.*, 2005)] or helical symmetries of tubular assemblies [helical reconstruction (Miyazawa *et al.*, 2003)]. If macromolecules do not assemble into symmetrical higher order structures, one can apply single-particle analysis (Frank, 2002; van Heel *et al.*, 2000): many images of arbitrarily oriented single copies of the macromolecule under scrutiny are computationally aligned to the same coordinate system and subsequently averaged. This methodology has become increasingly powerful in recent years, and nowadays it is routinely possible to resolve large macromolecules (>250–400 kDa) to 1–3 nm; in some works even subnanometer resolution has been reported.

The common approaches to single-particle analysis are either to reconstruct the macromolecule under scrutiny from single projections ["angular reconstitution" (van Heel, 1987) and "projection matching" (Penczek *et al.*, 1994)] or from projection pairs originating from different projection directions ["random conical tilting" (Radermacher *et al.*, 1986)]. A prerequisite for applying these methods is that the single particles must be well separated from each other; if the locations of single particles differ solely along the projection direction within the specimen, they overlap in the resulting projections, which is prohibitive for these protocols. Consequently, these methods are only applicable to purified macromolecules and are not suitable for characterizing complexes in their native environment, where a plethora of particles would overlap in projections. However, many biologically significant complexes are extremely hard to purify; assemblies can be fragile or even transient, which is an impediment to purification. Membrane-associated complexes are particularly difficult to deal with, since they must be solubilized in addition to being purified in order to apply these methods. Both solubilization and purification are not only labor intensive, they can also cause artifacts.

In contrast, in a tomogram particles are resolved in all 3D; particles that would overlap in projections are in distinct locations in a tomogram. Subtomograms that represent the same type of macromolecule can be averaged coherently using procedures that are analogous to those mentioned above. Thus, CET in conjunction with subsequent single-particle averaging holds the potential to provide structural insights into fragile or membrane-associated macromolecular assemblies *in situ* that are not amenable to conventional single-particle analysis protocols.

This chapter is organized into four sections. First, we briefly summarize the basics of CET to formulate the problem of tomogram averaging quantitatively. Next, we describe a specific algorithm to obtain coherent averages and discuss the details of data acquisition. Subsequently, we review selected recent applications of tomogram averaging to illustrate the current capabilities of the methodology. We conclude with a discussion of the capabilities of single tomogram averaging, its current limitations, and the future developments that will be necessary to extend the power of CET in conjunction with averaging techniques.

II. Tomography and Its Increase in Resolution by Averaging

In this section, we lay the foundations from which to derive an appropriate algorithm for averaging of tomograms. This requires recapitulating the physics that underlie tomographic reconstruction to assess the information content of tomograms prior to deriving a mathematical framework for the averaging problem.

A. The Fundamentals of Tomography

Owing to the large depth of focus of a transmission electron microscope (TEM), an electron micrograph is in good approximation with a 2D projection of the object under scrutiny. In more detail, the micrograph of a vitrified specimen is a projection of the specimen's electrostatic potential, convoluted with the inverse Fourier transform of the microscope's contrast-transfer function (CTF), which describes the imaging conditions and the properties of the TEM (Frank, 1996). The famous "projection theorem" states that a 3D object can be retrieved from its projections along different directions (Radon, 1917). Thus, in order to obtain a 3D description of an object, it must be projected along different directions. In the case of ET, this is achieved by incremental tilting of the specimen in the TEM. The fact that the data for a complete 3D description of the sample is contained in the projections of the incrementally tilted specimen, the "tilt series," is most apparent in Fourier space: the Fourier transform of a parallel projection corresponds to a central slice through the Fourier transform of the 3D object, which is also referred to as the "Projection-Slice Theorem" (Fig. 1). Due to the slab geometry of most samples studied by TEM, the specimen can only be tilted to ± 60–$70°$ and not to $\pm 90°$, as would be necessary to retrieve all the 3D information about the specimen. This tilt range leaves a wedge-shaped region in Fourier space unsampled (Fig. 1). Consequently, the final 3D reconstruction will lack spatial information, and the arising artifacts are referred to as the missing-wedge effect. There are several different algorithms to reconstruct the final 3D volume from its projections. Of these, weighted back-projection is still the most common choice (Radermacher, 1992). More elaborate reconstruction algorithms that impose *a priori* knowledge can be beneficial in retrieving some of the information in the missing wedge (reviewed in Carazo, 1992), but the extremely low SNR of cryoelectron micrographs renders these improvement rather incremental.

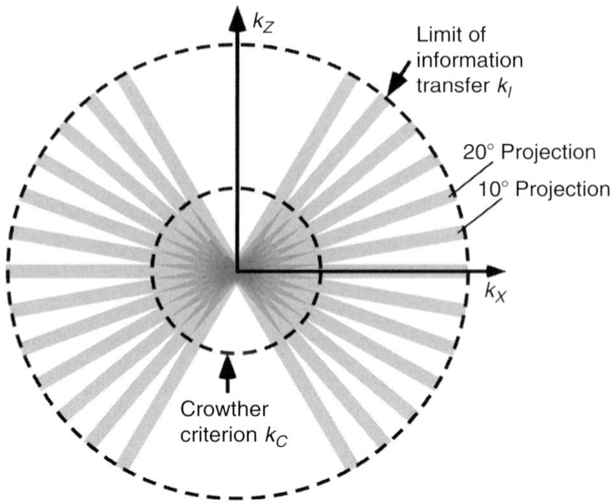

Fig. 1 Two-dimensional (2D) data sampling of a tilt series in Fourier space in 2D. In Fourier space, a projection samples a slice of the 3D object (a line in 2D). Projections along different directions sample the complete data in Fourier space as depicted. The thickness of each slice (line in 2D) is inversely proportional to the specimen thickness D (Crowther *et al.*, 1970). Thus, the information is sampled to a frequency k_C. The maximum tilt range is typically about $\pm 60^\circ$, which leaves a section of the Fourier space unsampled. Information beyond k_C can be presented in a cryoelectron tomogram: the projections can contain information to a frequency k_I that will be ultimately limited by the resolving power of the TEM or the alignment of the projections to a common coordinate system. However, the SNR is extremely low at high frequencies, due to the limited electron dose. Coherent averaging of data is a means to enhance the signal at high resolution.

B. Resolution of Cryoelectron Tomograms

To estimate the attainable resolution, one must consider two distinct issues: the angular sampling and the dose limitation. It is intuitively clear that the 3D structure of an object will be resolved to finer detail if it is reconstructed from many projections taken at small angular increments rather than from few projections at widely different orientations. Furthermore, changes in orientation will alter the appearance of the projections from a small object (one with a small diameter) less than the projections of a larger object. The relationship between a 3D object and the achievable resolution k_C as a function of the angular increment $\Delta\alpha$ (measured in radians) and the object's diameter D can be derived from Fig. 1:

$$k_C = (\Delta\alpha D)^{-1} \tag{1}$$

This relation is sometimes referred to as the Crowther criterion (Crowther *et al.*, 1970). Tomographic tilt series are commonly acquired with angular increments between 1° and 5°, a sampling that would allow the resolution of a large macromolecule ($D = 20$ nm) to 0.35–1.75 nm.

In CET, however, only a limited number of electrons can be used to image a biological specimen before the structure is significantly altered due to beam damage. Consequently, the SNR of a single projection is very low, and the information content of a projection is typically significant to only a certain resolution. In CET, the resolution limit for a macromolecule that is set by the electron dose is typically only a fraction of the resolution that the sampling would allow, and thus the resolution-limiting factor is the electron dose rather than the sampling. Assuming a perfect electron microscope and a perfect detection device for electrons, it has been estimated that the ultimate resolution of a cryoelectron tomogram is in the range of 2 nm (Henderson, 2004). In practice, however, the detection device that is currently used for ET, a CCD camera, deviates significantly from the ideal, so the resolution of cryoelectron tomograms remains far from this physical resolution limit (see Chapter 27 by Mooney and Chapter 28 by Xuong *et al.*, this volume). Various experimental studies indicated that the signal of cryoelectron tomograms is significant to a resolution that allows the location of large macromolecular assemblies such as actin filaments (Medalia *et al.*, 2002), ATP synthases (Nicastro *et al.*, 2000), or calcium release channels (Wagenknecht *et al.*, 2002) in the unprocessed tomograms. A more detailed investigation has revealed that the structural information of an individual particle is not significant beyond 6–10-nm resolution (Grünewald *et al.*, 2003).

C. Increasing the Resolution of Tomograms by Averaging

If identical macromolecules occur multiple times in a tomogram, the resolution limitation due to the low electron dose can be overcome: if the particles are brought into register they can be averaged, and the resulting volume will contain information beyond the resolution limit of a single tomogram (Fig. 2). It is well known that the dose-limited resolution theoretically grows as a function of the number N of averaged particles, proportional to $\sqrt{N}$ (Saxberg and Saxton, 1981). However, this equation holds only if the individual particles are brought into precise register and if they are identical. Furthermore, there must be no missing wedge. Technically, two procedures are needed: an algorithm that aligns the tomograms of the single particles to a common origin, and an algorithm that classifies different tomograms into different bins if there is structural heterogeneity among the individual particles. These problems are difficult to decouple, which makes it necessary to combine the operations if structural heterogeneity is present. In this chapter, however, we want to focus on the easier problem of aligning identical particles. For a discussion of the classification problem, see Chapter 13 by Taylor *et al.*, this volume.

Let us assume that we have N single tomograms P_i of the macromolecule under scrutiny. These particles are variously oriented and only roughly aligned at their centers, either by manual location in a tomogram or automated procedures, for example, template-matching algorithms (see also Chapter 25 by Best *et al.*, this volume) that search for an initially known, low-resolution template of the particle (Frangakis

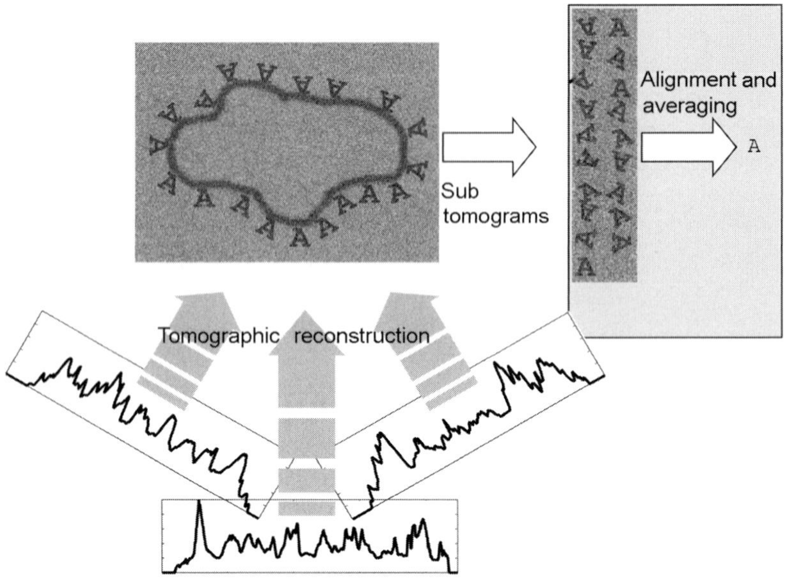

Fig. 2 2D sketch of single tomogram particle averaging. Projections of compound samples, such as organelles, are gathered using a TEM and merged into a tomogram of extremely low SNR (left). The subtomograms containing the complex of interest are iteratively aligned and averaged, resulting in an average of significantly increased SNR and resolution (box).

et al., 2002; Rath *et al.*, 2003). Each single particle is rotated (specified by the Euler angles φ_i, θ_i, and ψ_i) and subsequently translated (specified by the vectors Δx_i, Δy_i, and Δz_i) with respect to the coordinate system of the coherent average. The task of the alignment algorithm is to find the translations (Δx_i, Δy_i, and Δz_i) and rotations (φ_i, θ_i, and ψ_i) that align the ideally coherent average A with respect to the particle P_i. Given the determination of these parameters, A can be obtained by applying the inverse rotations and translations and is thus defined as:

$$A = \sum_{i=1}^{N} R_{-\psi_i,-\vartheta_i,-\varphi_i}\left[T_{-\Delta x_i,-\Delta y_i,-\Delta z_i}(P_i)\right] \tag{2}$$

Here $T_{\Delta x_i \Delta y_i \Delta z_i}$ and $R_{\varphi_i \theta_i \psi_i}$ denote translations and rotations by the specified parameters; the inverse of a rotation by the angles φ_i, θ_i, and ψ_i is a rotation by $-\psi_i$, $-\theta_i$, and $-\varphi_i$ and the inverse of a translation by the vectors Δx_i, Δy_i, and Δz_i is one by $-\Delta x_i$, $-\Delta y_i$, and $-\Delta z_i$. Moreover, the order of translation and rotation must be exchanged to invert the rotations and translations. To derive a mathematical rule for obtaining these parameters, we use the properties of the coherent average. The similarity between an individual particle and the coherent average after application of the translation vector (Δx_i, Δy_i, and Δz_i) and the Euler angles

(φ_i, θ_i, and ψ_i), which is quantified by an appropriate similarity measure, should be very high compared to other translations or rotations. In fact, the best possible alignment is achieved when the sum of similarity measures between each particle and the average that is rotated and translated by the optimal parameters is maximized. One possible measure of the similarity between two features is the cross-correlation coefficient (CCC, see also Chapter 25 by Best *et al.*, this volume). The CCC of a single tomogram P_i and the average A that is rotated by angles (φ_i, θ_i, and ψ_i) and subsequently translated by the vectors (Δx_i, Δy_i, and Δz_i) is defined as:

$$\mathrm{CCC}_i = \frac{\sum_{v=1}^{N_V} \left\{ \left[T_{\Delta x_i, \Delta y_i, \Delta z_i}(R_{\varphi_i, \vartheta_i, \psi_i}(A)) \right]_v - \langle A \rangle \right\} \cdot [P_{i_v} - \langle P_i \rangle]}{\sqrt{\sum_{v=1}^{N_V} \left(R_{\varphi_i, \vartheta_i, \psi_i}(A)_v - \langle R_{\varphi_i, \vartheta_i, \psi_i}(A) \rangle \right)^2 \cdot \sum_{v=1}^{N_V} \left(P_{i_v} - \langle P_i \rangle \right)^2}} \tag{3}$$

Here the index v denotes the voxel and N_v is the number of voxels in the volumes. The quantities $\langle P_i \rangle$ and $\langle A \rangle$ denote the mean values of the particles and the average, respectively. As a similarity measure, the CCC has the advantage of being independent of the magnitude of the signal and the average value, which typically vary throughout tomograms. The denominator of Eq. (3) essentially consists of the standard deviations of P_i and A, whereby the standard deviation of A will vary slightly with rotations, due to inevitable interpolation errors that occur when performing rotations numerically. Given that the coherent average maximizes the sum of similarities of tomograms P_i and the average A, and using Eq. (3) as a similarity measure, the mathematical task of the alignment is to maximize the functional:

$$F = \sum_{i=1}^{N} \mathrm{CCC}_i(\Delta x_i, \Delta y_i, \Delta z_i, \varphi_i, \vartheta_i, \psi_i) \tag{4}$$

Thus, a solution to the alignment problem would be to evaluate the six-dimensional (6D) function F exhaustively; the parameters that maximize F are the best choices to obtain a coherent average. In the absence of noise, the maximum value of F would be equal to the number of particles N. However, two issues complicate the alignment problem. First, the CCC as defined in Eq. (3) is only one possible measure of the similarity between a given particle and the average. It does not include many of the imaging properties and artifacts of ET, such as the missing wedge of tomographic data in Fourier space. Thus, Eq. (4) shows one possible choice of a scoring function, not necessarily the best choice. In the next section, we will introduce a more elaborate scoring function for CET. Second, the optimization of F requires an efficient algorithm, because an exhaustive evaluation of this functional is currently impossible within reasonable timescales for datasets comprising hundreds or thousands of particles.

III. Coherent Averaging of Macromolecules in Practice

Averaging single particles from tomograms is still a relatively unconventional method for characterizing the structure of macromolecules, although it was first applied in the early 1980s (Knauer *et al.*, 1983; Oettl *et al.*, 1983). In this section, we give a brief summary of the process for data collection and introduce a specific algorithm for the subsequent alignment and averaging.

A. Acquisition of Cryoelectron Tomograms

For CET, the biological samples are frozen extremely rapidly, which prevents the formation of crystals and instead yields a vitrified state (Dubochet *et al.*, 1988). Soluble samples can be vitrified by putting droplets of a few microliters onto an EM grid coated with an electron-transparent film, soaking off excess liquid to obtain thin films (~100–500 nm) using filter paper, and finally plunging the grid into a cryogen, typically liquid ethane. To facilitate the alignment of the resulting tilt series, colloidal gold is added to the solution. The EM grid is then transferred to a TEM, taking care not to allow the sample to warm enough for the vitreous water to crystallize ($T < -130\,^{\circ}C$).

The automation of data acquisition has been crucial for the recent popularity of CET (reviewed in Koster *et al.*, 1997). To obtain a useable tilt series, it is necessary to keep the object of interest in the field of view throughout the entire tilt range. However, the position of the specimen with respect to the electron beam changes on tilting the specimen due to mechanical inaccuracies of the specimen stage of the TEM. These movements need to be compensated by the acquisition software, and currently there are different programs available for this task (e.g., Mastronarde, 2005; Nickell *et al.*, 2005; Zheng *et al.*, 2004). Moreover, the specimen stages of modern TEMs that are explicitly designed for cryo-EM are mechanically very stable, which allows rapid data acquisition (Zheng *et al.*, 2004).

Data acquisition requires parameters that should be adjusted to the final aim, subtomogram averaging. A crucial parameter is the choice of the defocus value; whereas choosing a small defocus value offers the potential to extract higher resolution information, a large defocus value enhances the contrast of the tomograms, which eases coherent alignment of the single particles. Thus, a compromise must be found between contrast and resolution. Ideally, one would combine tilt series taken at different values of defocus for the final reconstruction, which is a challenge for future research (Section V). In practice, we recommend initially acquiring tilt series with large defocus values, and then recording tilt series with gradually lowered defocus values, until improvement in the quality of the resulting averages is no longer observed. Given the beam sensitivity of the sample, this requires multiple specimens. For each tilt series, the ultimately achievable resolution of the average will correspond approximately to the first zero crossing of the CTF. Moreover, the angular tilting scheme must be specified. In ET, a constant angular increment is preferred (Mastronarde, 1997). While a relatively small increment of 1–2° is often chosen for studying the ultrastructure of cells, it is advisable to choose a larger increment (3–5°) for single-particle CET; the advantage is that a higher electron dose can then be spent on each micrograph, leading to a higher SNR. Finally, the tilt range should be considered: projections taken at high tilt angles tend to have a low SNR, and we have found that a tilt range of ±60° is sufficient in many cases. However, all parameters should be carefully adjusted for the specimen under scrutiny.

The next step is to reconstruct the final 3D tomogram from the projections. Prior to the actual reconstruction process, the individual micrographs must be aligned to a common coordinate system. The most accurate alignment is typically achieved by using fiducial markers, such as gold beads with a diameter of 5–10 nm that were added to the specimen solution (Lawrence, 1992). Subsequently, the aligned micrographs are used to obtain the 3D tomogram, whereby weighted back-projection is still the most common algorithm for the very noisy CET data. Various software packages are available at no cost to perform alignment and reconstruction, for example, IMOD (Kremer *et al.*, 1996), Priism (Chen *et al.*, 1996), or the TOM toolbox (Nickell *et al.*, 2005).

To align and average macromolecules, the single particles must be located within the tomograms. This can be achieved manually using a graphical user interface (Förster *et al.*, 2005) or in an automated fashion if a model of the macromolecule exists (Frangakis *et al.*, 2002; Rath *et al.*, 2003). Finally, subtomograms that encapsulate the single particles must be extracted for use in the subsequent averaging process.

B. An Algorithm for Averaging Membrane-Bound Assemblies

The specific algorithm described here has previously been described in less detail (Förster *et al.*, 2005). It is influenced by the work of Walz *et al.* (1997b), but in particular the scoring function is developed further to account for the imaging conditions in CET. The algorithm described below has been implemented in the MATLAB scripting language (The Mathworks Inc., MA) in different programs that are assembled in the toolbox AV3, which is available from the authors. Moreover, it requires some essential routines from the TOM toolbox (Nickell *et al.*, 2005).

We divide the alignment problem into two parts: (1) an initial step to find the approximate orientations of the single particles relative to an initial, low-resolution model and (2) a refinement procedure to determine the single-particle positions and orientations with high accuracy to finally obtain an average of the highest possible resolution. This distinction is also common in the projection-based, single-particle analysis (Penczek *et al.*, 1994; Schatz *et al.*, 1995; Stewart and Grigorieff, 2004). We start with the description of the refinement because this is universally applicable, whereas the generation of the initial model is specifically designed for resolving membrane-associated complexes.

1. Refinement of Tomographic Averages

As mentioned in the previous section, the task of the refinement procedure is to optimize a scoring function, for example, Eq. (3). Thus, we require a suitable scoring function, that reaches its maximum value if all particles are coherently aligned, and an algorithm to find the translations and rotations that maximize the scoring function.

a. Scoring Function

In designing a scoring function for CET, we need to consider the missing-wedge effect. The missing-wedge effect is a notorious problem in processing electron tomograms reconstructed from single axis tilt series and hence must be considered in this process. It is particularly severe for membrane-bound particles where the lipid membrane tends to be the dominant signal of the completely sampled structure. However, in CET the membrane can be virtually invisible if the single particles adopt certain orientations; the signal from the membrane is concentrated along one direction in Fourier space, and for some orientations the missing wedge prevents resolution of the membrane. In general, for any structures whose signal is anisotropically distributed in Fourier space, it is important to account for the missing-wedge effect in the scoring function. One way to account for this effect in a correlation function is to constrain the correlation to the sampled area in Fourier space, as we first proposed in Frangakis *et al.* (2002).

Moreover, one is usually interested only in a specific macromolecule, so we want to minimize the—generally negative—effect of any random features adjacent to the individual particles, such as membranes or other proteins. This can be achieved by multiplying the particle of interest by a function Θ, which tightly masks the particle of interest. In addition, it is often beneficial to "band-limit" the correlation function (Van Heel *et al.*, 1992). Very low frequencies tend to dominate the signal of cryoelectron tomograms, in particular if an energy filter was used during imaging (Grimm *et al.*, 1996). The low frequency bands can vary significantly, for example, due to changes of specimen thickness, whereas the very high frequency bands suffer from the low SNR. If we incorporate the missing-wedge effect, the masking, and the band-pass filtering into the correlation, we should apply these operations prior to performing the correlation. Thus, the average in orientation (φ_i, θ_i, and ψ_i) should be replaced by the expression:

$$A_{\varphi_i,\vartheta_i,\psi_i} = B_{\text{low}}^{\text{high}}\{F^{-1}[F(R_{\varphi_i,\vartheta_i,\psi_i}(A))\cdot W]\cdot\Theta\} \quad (5)$$

Here F denotes a Fourier transformation and F^{-1} its inverse, $B_{\text{low}}^{\text{high}}$ is the band-pass operation in Fourier space, which filters out the signal within the lower bound "low" and the upper limit "high," and Θ is the mask. W is a function describing the tomographic sampling in Fourier space; it is specific for the acquired tomogram, and different Ws must be introduced if particles that originate from different tomograms with different tilt ranges are merged. The same operations must be applied to the particles, and hence the correlation function Eq. (3) becomes modified to:

$$\mathrm{CCC} = \frac{\sum_{v=1}^{N_V}\left[\left(T_{\Delta x_i,\Delta y_i,\Delta z_i}\left(A_{\varphi_i,\vartheta_i,\psi_i}\right)\right)_v - \langle A_{\varphi_i,\vartheta_i,\psi_i}\rangle\right]\cdot[P_{i_v} - \langle P_i\rangle]}{\sqrt{\sum_{v=1}^{N_V}\left(A_{\varphi_i,\vartheta_i,\psi_{iv}} - \langle A_{\varphi_i,\vartheta_i,\psi_i}\rangle\right)^2\cdot\sum_{v=1}^{N_V}\left(P_{i_v} - \langle P_i\rangle\right)^2}} \tag{6}$$

The different data processing steps that are needed to calculate the above correlation function are depicted in Fig. 3 (A–F). Again, the overall scoring function is the sum of all individual correlation functions. The missing-wedge effect, that is, the inhomogeneous sampling of data in Fourier space, must be considered in the actual averaging as well. This can be done analogously to the single-particle averaging from projections, where the different abundances of the various projections must be considered as well (Harauz and van Heel, 1986): first we need to evaluate how often each point **k** in Fourier space is sampled by the inversely rotated particles:

$$S_k = \sum_i R_{-\psi,-\vartheta_i,-\varphi_i} W_k \tag{7}$$

Here W denotes the sampling function of each particle in Fourier space, as also used in Eq. (5). To correct for the different sampling in Fourier space, we must downweigh intensively sampled structure factors in the coherent average of the particles by a weighting function w:

$$w_k = \begin{cases} \dfrac{1}{S_k} & \text{for } S_k > 0 \\ 0 & \text{for } S_k = 0 \end{cases} \tag{8}$$

Thus, the final average with the correctly weighted structure factors is:

$$A = \sum_{i=1}^{N} F^{-1}\left\{F\left[R_{-\phi_i,-\vartheta_i,-\psi_{i_i}}\left(T_{\Delta x_i,-\Delta y_i,-\Delta z_i}(P_i)\right)\right]\cdot w\right\} \tag{9}$$

b. Optimization

In addition to an appropriate scoring function, we need an algorithm to determine the translation vectors (Δx_i, Δy_i, and Δz_i) and Euler angles (φ_i, θ_i, and ψ_i) for those individual particles that render the maximum value of the scoring function. For this task, we choose a relatively simple algorithm that is often employed in the single-particle analysis of projections (Ludtke *et al.*, 1999;

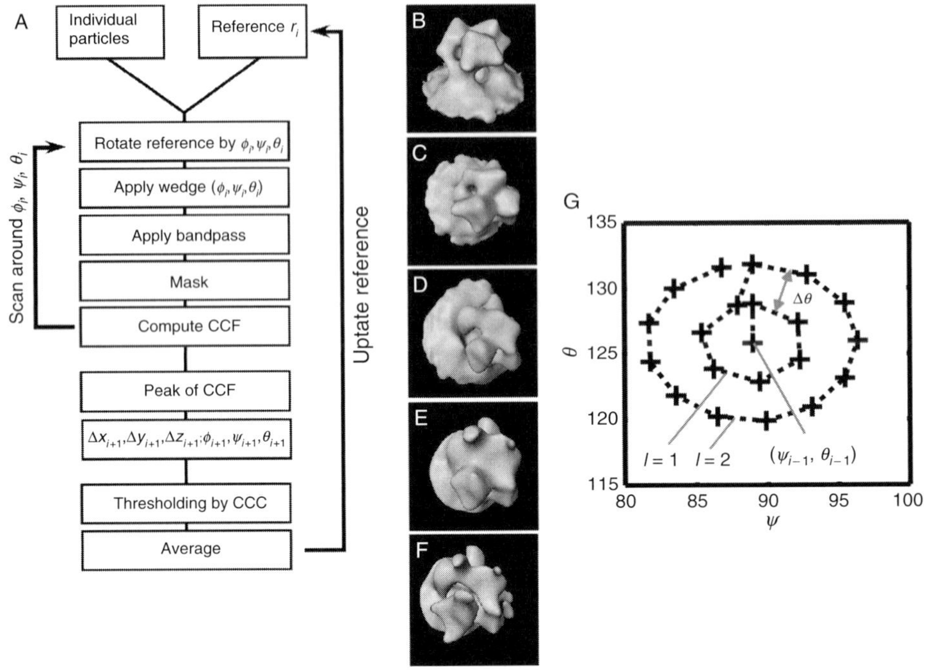

Fig. 3 Refinement procedure (A–F) and angular sampling (G). (A–F) The reference (B), which is the average resulting from the previous iteration, is rotated into an orientation (φ_i, θ_i, ψ_i), whereby orientations in proximity to the optimum orientation that was determined in the previous iteration are explored for each particle (C). Subsequently, the missing wedge is applied in Fourier space, leading to a characteristic vertical elongation of the density (D). Then the structure is masked to confine the alignment to the area of interest (E); for example, lipid membranes tend to be flexible and they should be excluded from the alignment. Finally, a band-pass filter is applied to reduce the effect of the strong, low-resolution signal and the noise in the very high frequency range (F). (G) The Euler angles ψ and θ are scanned around the angles from the previous iteration (ψ_{i-1}, θ_{i-1}), which act as "poles." The angles are scanned along L lines of equal latitude, which increases by $\Delta\theta$ between adjacent lines l. In the depicted example, $\Delta\theta$ is 3°, and L amounts to 2. To account for the increasing density of ψ near the poles, the sampling of ψ is chosen as $\Delta\psi = \Delta\theta/\sin(\Delta\theta \cdot l)$. The remaining angle φ is scanned with a constant increment $\Delta\varphi = \Delta\theta$ starting from $\Delta\varphi = -\Delta\theta \cdot L$ and ending at $\Delta\varphi = \Delta\theta \cdot L$.

Penczek *et al.*, 1994)]. Briefly, the function is optimized using an iterative process where the correlation function in Eq. (6) is computed for each particle and the average of the preceding run (the reference A') in different orientations. The reference for the next iteration is the average of the single particles that are inversely rotated and translated, using the parameters that yielded the maximum correlation values. Ideally, the reference A' will be identical to the coherent average A after convergence of this process.

In detail, we evaluate the correlation function for each particle in the following way: the correlation function Eq. (6) is calculated exhaustively for the translational parameters after application of a rotation. This exhaustive evaluation can be computed very efficiently by using Fourier transforms, which is a common trick in image processing (Frank, 1996); both the reference A' that is rotated into an orientation (φ, θ, and ψ) and the particle P_i are Fourier transformed, multiplied, and the product is transformed back to real space, which yields the correlation function as a function of (Δx, Δy, and Δz). An analogue to the Fourier transform for the rotational degrees of freedom is an expansion to rotational harmonics, which has been successfully used for an efficient reconstruction of particles from projections (Navaza, 2003; Provencher and Vogel, 1988; Vogel and Provencher, 1988). However, this transformation cannot be used to accelerate the computation of Eq. (6) for the angular parameters due to the missing-wedge effect. Consequently, the rotational degrees of freedom must be sampled by explicitly rotating A' into different orientations. The computational costs of an exhaustive sampling would be enormous, which forces us to restrict its evaluation within the vicinity of the orientations of the previous iteration. The specific sampling we have chosen is depicted in Fig. 3G. The entire refinement procedure must be iterated several times, typically 10–20 iterations. The angular sampling for the initial iterations should be relatively coarse (e.g., $\Delta\theta = 5°$ for the sampling described in Fig. 3G) and then made gradually finer. Simultaneously, a strong low-pass filter can be beneficial in the first iterations (Stewart and Grigorieff, 2004) and should then be gradually released to finally recover the maximum information content from the individual particles.

The iterative refinement described here is carried out until improvement in resolution is no longer observed. The resolution is commonly judged by Fourier shell correlation (FSC): the single particles are divided into two sets that are averaged independently. The correlation coefficient of the two averages is determined as a function of spatial frequency (Saxton and Baumeister, 1982), which typically renders a slowly decaying curve. The frequency at which the correlation coefficient drops below 0.5 is typically taken as the resolution criterion that is used to low-pass filter the final average. The shortcomings of this approach are known (Grigorieff, 2000), but it still represents the only widely accepted resolution criterion.

2. Initial Model Generation for Membrane-Associated Complexes

A prerequisite for applying the above refinement procedure is knowledge of the approximate orientations of the single particles. One option to obtain initial approximations of the orientations is to use an initial model to align the single particles with respect to this template using parallelized programs for "template matching" (Frangakis *et al.*, 2002; Rath *et al.*, 2003). However, the downside of using an initial model for the alignment cycle is that it may bias the result; optimization processes can easily be trapped in local maxima, and if a feature-rich initial template is used, the final average can correspond to a local maximum in the vicinity of the initial template.

To determine the structure of membrane-associated complexes, we have employed an approach that avoids this shortcoming (Fig. 4). We exploit the fact that the single complexes are all oriented in the same way relative to the lipid membrane, which is easily discernible in cryoelectron tomograms. Hence, the approximate manual determination of the membrane contour provides excellent initial values for two of the three Euler angles (Fig. 4). Solely the rotation within the membrane plane is initially completely unknown. To build a first reference, we choose this remaining angle φ randomly for each particle, compute this average, and use it as the initial reference. To estimate the membrane contour of many viruses, it is often sufficient to estimate their shape as a sphere (Förster *et al.*, 2005); for nuclei, ellipsoids turned out to be an excellent approximation in many cases (Beck *et al.*, 2004). Both of these models allow an easy, analytical derivation of two Euler angles for each particle. Then the above refinement procedure is used for a few (2–3) iterations, primarily to refine the translational parameters.

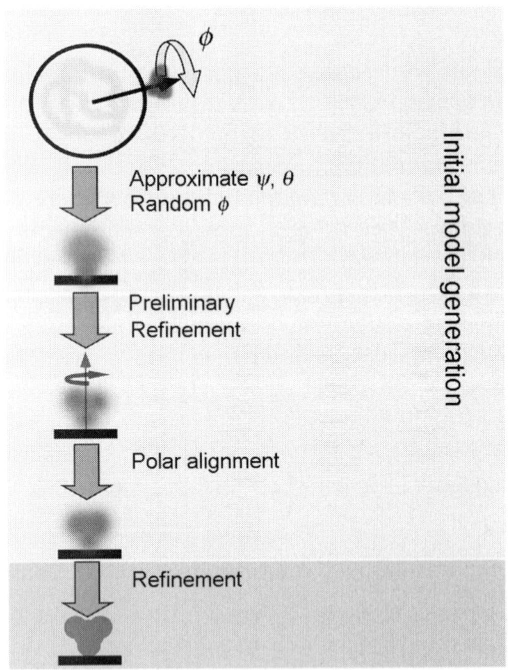

Fig. 4 Summary of the described alignment procedure for a membrane-bound complex. The generation of an initial model consists of several steps: a first model is derived from the average of the individual particles, whereby two of three Euler angles are approximated from the orientation of the membrane and the third angle is chosen randomly for each particle. Subsequently, a few refinement iterations are carried out, primarily to properly center the particles for the next step. Then a determination of the polar angle is performed by aligning the particles to a growing average as described in the text. The resulting average and orientations are then used for the ultimate refinement of the structure.

To achieve a better sampling of the polar angle φ, a further alignment step is then carried out that also allows the determination of rotational symmetries along the membrane normal. We choose a strategy similar to "reference-free alignment procedure" introduced by Penczek *et al.* (1992), which involves aligning the individual particles with respect to a growing average: We randomly choose one particle as a first reference and then correlate it to another particle, whereby φ is exhaustively sampled. The second particle is inversely rotated, as indicated by the maximum correlation value, then the two particles are added. A third particle is correlated to this average and so forth. To accelerate the computations, we perform the alignment with respect to the polar angle using the rotational correlation function as it is often used in 2D (Penczek *et al.*, 1992). The 3D formulation is:

$$C_i = (\Delta\varphi) = \sum_z \sum_\rho \rho F_\varphi^{-1}\left(F_\varphi A'_{z\rho\varphi} \cdot \left(F_\varphi P_{z\rho\varphi_i}\right)^*\right) \quad (10)$$

The function C_i measures the correlation of the particle P_i [after inversely applying the previously determined (Δx_i, Δy_i, and Δz_i) and (θ_i, ψ_i)] and the reference A' on rotation of P_i by $\Delta\varphi$. A' and P_i are transformed to cylindrical coordinates z, ρ, and φ prior to computing their Fourier transformed F_φ with respect to φ. The coordinate transformation is done by means of trilinear interpolation. By applying the Fourier transform in φ, exhaustive scanning can be achieved at moderate computational expense in Eq. (10). Although, the scoring function in Eq. (10) does not account for the missing wedge, it should be sufficiently precise to achieve an approximate determination of φ_i at moderate computational expense. For the subsequent fine alignment, the correlation coefficient defined in Eq. (6) is more appropriate. The result of the φ-alignment step can be observed for the structure of the nuclear pore complex (NPC) determined from single pores in intact nuclei (Beck *et al.*, 2004): the eightfold rotational symmetry is not recognizable prior to this alignment step, but it becomes apparent in the resulting average (Fig. 5). Moreover, rotational symmetries along the z-axis can also be verified quantitatively using the rotational autocorrelation function, which can be obtained by replacing A' in Eq. (10) by the particle P_i itself. Whereas the low SNR typically does not allow the identification of any significant symmetry for a single particle, the sum of the individual autocorrelation functions gives a clear indication of the rotational symmetry (Fig. 5E).

IV. Applications of Tomogram Averaging

A. Purified Macromolecules

The first applications of tomographic averaging employed this method for the determination of structures from biochemically purified macromolecules. The first 3D EM structures of the ribosome based on ET were obtained by averaging

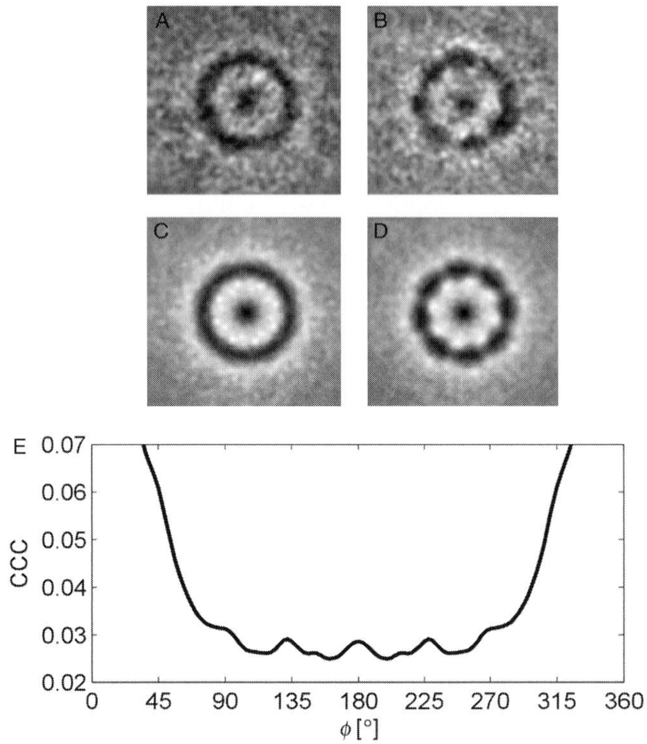

Fig. 5 (A–D) x–y slices of the NPC complex (Beck *et al.*, 2004). (A) Slice of the average without imposing eightfold rotational symmetry. (B) Slice after polar alignment as described in the text. (C) Same slice as A after eightfold symmetrization. (D) Same slice as B after eightfold symmetrization. (E) Averaged rotational correlation function of all particles (aligned without imposing symmetry). The local maxima at 90°, 135°, and 180° indicate an eightfold rotational symmetry; the expected maximum at 45° is overwhelmed by the broad peak around 0°.

tomograms of purified and negatively stained ribosomes (Knauer *et al.*, 1983; Oettl *et al.*, 1983). Nowadays, the "random-conical tilt method" (Radermacher *et al.*, 1987), which was developed later, is often preferred to assess the structures of macromolecules completely in the absence of any prior information, since it requires the acquisition of only two micrographs rather than a complete tilt series. However, the automation of ET has reestablished tomogram averaging as an alternative for obtaining initial models of macromolecules. Thus, the first insights into the structure of the fully assembled 26S proteasome were obtained by ET of negatively stained particles (Walz *et al.*, 1998). Moreover, low resolution (2–3 nm) structures of the Tricorn protease (Walz *et al.*, 1997a) and the chaperone thermosome in its open conformation (Nitsch *et al.*, 1998) have been obtained using averaged tomograms of purified and vitrified particles.

B. Highly Symmetric Assemblies

Higher resolutions could be obtained for the analysis of highly ordered arrays of actin and myosin fibers found in insect flight muscle. The structure of these assemblies has been shown to be well preserved after RF-FSF; tomographic analysis involving averaging and classification of structural inhomogeneities has allowed the fitting of atomic structures (Taylor *et al.*, 1999), and pseudo-atomic models of these assemblies have been obtained (see Chapter 13 by Taylor *et al.*, this volume). Virus capsids typically possess a high point-symmetry, for example, the capsid proteins often assemble to icosahedrons. *Herpes* viruses possess icosahedral capsids surrounded by pleiomorphic envelopes. Cryoelectron tomographic imaging of isolated *Herpes* virions has provided novel insights into their morphology, and averaging the capsids of single virions has yielded a detailed map of the capsid in the fully assembled virus to approximately 5 nm resolution (Grünewald *et al.*, 2003).

C. Assembly Structures Determined *In Situ*

It is clearly advantageous to study macromolecular assemblies in their native setting, because this rules out the potential artifacts due to *in vitro* processes such as purification or reconstitution. Moreover, many assemblies, particularly membrane-bound complexes, are difficult to prepare *in vitro*.

The NPC is a prominent membrane-bound complex; this giant assembly (mass weight ~125 MDa) is a gateway for nucleoplasmic exchange. Two recent studies employed CET and subsequent averaging to gain insights into its structure in the native membrane. In the first work, *Xenopus* nuclei were opened manually and the nuclear envelopes (NEs) were spread on an EM grid (Stoffler *et al.*, 2003). The resulting planar envelope was imaged by CET and the individual pores were subsequently aligned and averaged. This study revealed the first 3D map of the NPC in the NE to a resolution of about 12 nm. However, because all individual pores were oriented within the plane of the flat NE, the missing wedge of data that is characteristic of tomography could not be filled in completely; the random orientations of the NPCs within the membrane resulted in a cone-shaped region of missing data. This led to an extension of the resulting structure normal to the membrane.

A more recent work (Beck *et al.*, 2004) imaged the NPC even closer to its native settings: CET of purified, import-competent nuclei from *Dictyostelium discoideum* provided the raw material for images of single NPCs that were subsequently aligned and averaged using the algorithm described in the Section III. This study resolved the NPC to 8–9 nm. In addition, this study achieved isotropic sampling of the structure factors because the NE was not flat, that is, the data used to compute the structure did not miss a cone of information. Moreover, many details of the NPC, such as the nuclear basket, were recognizable in the resulting structure. These were hardly resolved in previous cryo-EM studies, presumably due to the specimen preparation. Moreover, in this study an effort was made to capture the structural dynamics of the NPC: the individual NPC particles were classified into two groups based on structural variations among them, and these two groups were hypothesized to correspond to a cargo-bound and a cargo-free state.

Eukaryotic flagella are highly ordered organelles that are used to generate motion. Nicastro *et al.* (2005) used CET to gain insights into the molecular architecture of sea urchin sperm tails. Averaging has been employed to resolve different components of the flagella in more detail, such as the outer dynein arms and the complex surrounding the two central microtubules (central pair complex, CPC). The resulting average could then reveal details of these macromolecular complexes that were not discernable in the raw data, such as the ring-shaped structure of the dynein motor domain and the periodic projections of the CPC (Fig. 6).

The aim to determine the structure of the retroviral envelope complex (Env) bound to the viral membrane initiated the development of the algorithms described in Section III (Fig. 7). Specifically, we employed the methodology to determine the Env structure of Moloney Murine Leukemia Virus (MoMuLV). The final structure that was obtained by averaging roughly 400 individual Env particles yielded a resolution of approximately 3 nm, which allowed a tentative fitting of an atomic structure of the receptor-binding domain. Some features of the MoMuLV Env morphology differed significantly from the atomic structure of the hemagglutinin complex from the influenza virus. Such differences are indicative of distinct mechanistic details in the infection process for individual retroviruses.

V. Conclusion and Outlook

CET has evolved into a powerful method to characterize macromolecular complexes in their native settings. A rapidly growing number of applications indicates that this technique holds the potential to bridge the resolution gap that exists between light microscopic techniques that have proven to be fruitful for cell biologist and high-resolution techniques employed in structural molecular biology. However, the resolution of cryoelectron tomograms is limited by the applicable electron dose, which favors the use of averaging techniques to retrieve the signals of macromolecules from tomograms at a higher resolution. Previously, averaging techniques have been applied primarily to determine the structures of purified macromolecules from projections. Methodological advances in conjunction with the ever-increasing power of computers have led to the increased popularity of this approach for the characterization of complex structures (Frank, 2002; van Heel *et al.*, 2000). Tomograms of single particles can be averaged using an analogous methodology as we have shown here. Averaging tomograms of single particles can be used to obtain the structures of biochemically purified particles; in this case its main field of application is probably the determination of medium to low-resolution structures, whereas the comparably mature methodology of single-particle averaging from projections is likely to remain more suitable for high-resolution insights. However, the major advantage of averaging

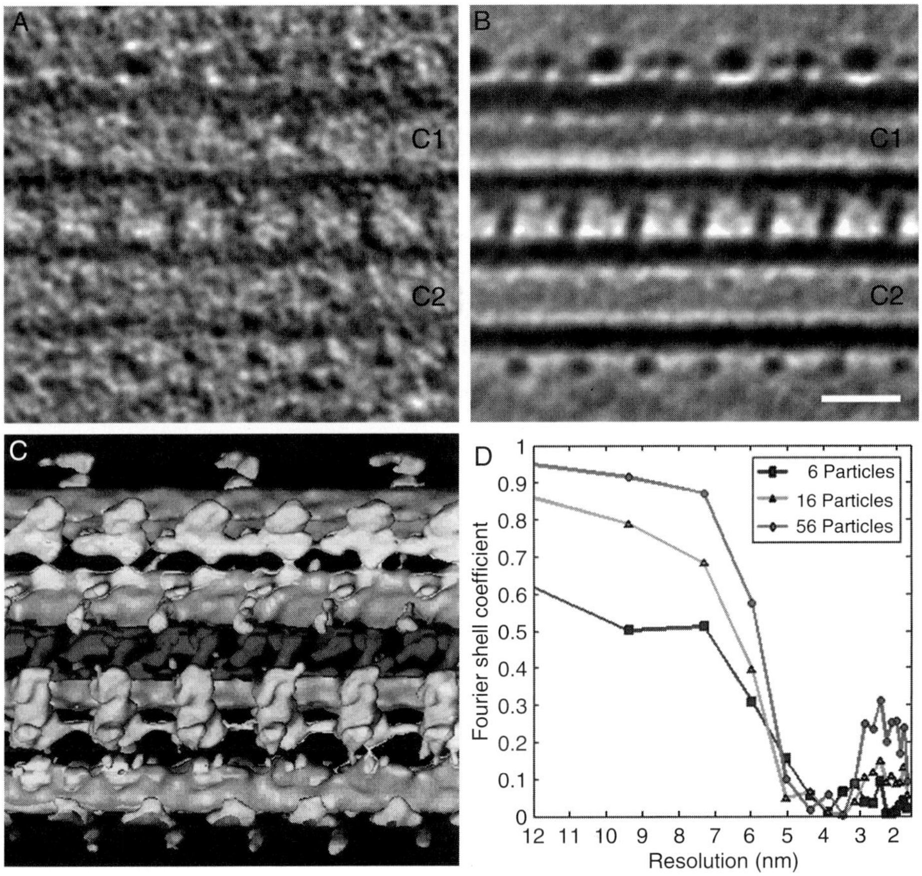

Fig. 6 Supramolecular structure of the flagellar CPC, as determined by averaging CETs (adapted from Nicastro *et al.*, 2005). (A) A 10-nm-thick tomographic slice though a raw tomogram of a CPC. C1 and C2 denote the central microtubules of the CPC. (B) Tomographic slice from an average of 56 CPC repeats. The tremendous increase in SNR allows the resolution of details such as the cross-bridges of C1 and C2. (C) An isosurface representation of the CPC. The dark central features correspond to intermicrotubule bridges; the central microtubules are light grey and the central pair projections appear white. (D) Resolution evaluation for the CPC average by FSC. The resolution curves for increasing numbers of averaged particles (6, 16, and 56) clearly demonstrate the resolution improvement by attained correlation averaging. An FSC threshold of 0.5 provided a resolution estimate of ~5.8 nm for the central region of the best average. The secondary peak at 2–3 nm suggests that the averaged data contain enough high-resolution information that other factors, like the lack of a correction for the CTF, are limiting image quality (Scale bar = 20 nm.)

from tomograms, as opposed to projections, is that it can also be applied to non-purified samples. This can be particularly valuable in the structural characterization of membrane-bound complexes.

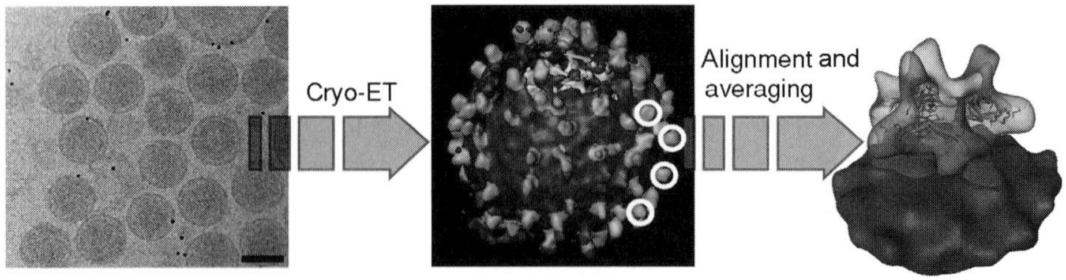

Fig. 7 *In situ* structure determination of the Env complex of MoMuLV (adapted from Förster *et al.*, 2005). Tilt series of vitrified MoMuLV virions were recorded (left: 0° projection, scale bar = 100 nm) and subsequently merged into a 3D volume [center: isosurface representation of a single virion; the density was manually segmented into lipid (dark) and Env complexes (light)]. Single Env complexes (circled in white) from various virions are subsequently aligned and averaged to yield a 3D map of Env (right). Within the extraviral portion of Env, we performed a tentative fitting of an atomic structure of a fragment of the Env complex.

In this chapter, we presented a specific algorithm to align and average single membrane-bound particles. This algorithm has been used to reconstruct the NPC and the MoMuLV Env *in situ*. In the latter, a resolution beyond 3 nm was achieved, indicating the tremendous potential of this approach for *in situ* structure determination. Despite the success of our algorithm in these studies, it is by no means perfect, and there can be no doubt that more sophisticated algorithms will be developed in the future.

Numerous innovations have been introduced to the field of single-particle averaging from projections, and many of these can be generalized to tomograms. A good scoring function is essential for the refinement algorithm, and substitution of Eq. (6) by other functions might result in less amplification of noise: the program "freealign" uses elaborate filtering in Fourier space for this aim (Stewart and Grigorieff, 2004), and maximum likelihood functions are reported to be advantageous as well (Scheres *et al.*, 2005; Sigworth, 1998). Moreover, different approaches have been reported to improve the optimizer: freealign uses a different scheme for updating the reference for each iteration compared to our approach (Stewart and Grigorieff, 2004); and Penczek and coworkers employ a quasi-Newton algorithm to achieve a global optimization (Yang *et al.*, 2005). In order to achieve averages of higher resolution from tomograms, it will be necessary to correct for the effect of the CTF. CTF correction has been a significant technological achievement in single-particle averaging from projections, and CTF corrections for tomography such as proposed by Winkler and Taylor (2003) will be important for high-resolution tomography, as well.

The alignment of tomographic tilt series can be another resolution-limiting factor. The alignment of micrographs prior to merging them into a common 3D volume can limit the resolution of the raw data; such errors cannot be overcome by averaging. Common alignment algorithms, such as discussed in Lawrence (1992), have limited accuracy. Recent work (Diez *et al.*, 2006; Lawrence *et al.*, 2006) has reported progress in reducing the alignment error, and these more elaborate procedures will be needed to obtain high-resolution averages. Moreover, the quality of averages derived from tomograms could further benefit from procedures that optimize the micrograph alignment locally in the vicinity of single particles.

Moreover, technical progress in the field of CCD cameras will be extremely beneficial for CET and subsequent averaging. Current CCD cameras for TEMs only register a fraction of the imaging electrons, in particular for high spatial frequencies. Substantial effort is currently under way to improve the efficiency of CCD cameras (see Chapter 27 by Mooney and Chapter 28 by Xuong *et al.*, this volume).

The biggest challenge for the future will likely be the classification of structural inhomogeneities. Many assemblies can adopt different conformations in their native settings, and it will be necessary to group single particles into different classes of similar conformations to obtain high-resolution structures of these assemblies (Fig. 2). The technical problem is that alignment and classification are interdependent, that is, both problems must be solved simultaneously. Moreover, classification is hampered by the missing-wedge effect. Technical advances may help tackle the latter challenge: by acquiring tilt series along two orthogonal tilt axes, the missing data can be reduced to a pyramid-shaped region (Iancu *et al.*, 2005; Mastronarde, 1997; Nickell *et al.*, 2003; Penczek *et al.*, 1995). However, the alignment of these dual tilt series to a common coordinate system can currently only be achieved to low accuracy (Iancu *et al.*, 2005), and these problems will need to be overcome to utilize dual tilt series for averaging.

In summary, averaging of tomograms is a powerful method for the structural characterization of macromolecular complexes. The current methodology readily enables the investigation of biological questions that require resolutions up to ~3 nm. There is tremendous potential to obtain even more detailed insights from future cryoelectron tomograms of complexes in their native settings. This will require numerous further technical developments, but the challenges are largely foreseeable.

Acknowledgments

We thank Daniela Nicastro for providing Fig. 6. Moreover, we are indebted to Maya Topf, Fred Davis, Daniela Nicastro, and Quanren Xiong, for carefully reading the manuscript.

References

Beck, M., Förster, F., Ecke, M., Plitzko, J. M., Melchior, F., Gerisch, G., Baumeister, W., and Medalia, O. (2004). Nuclear pore complex structure and dynamics revealed by cryoelectron tomography. *Science* **306,** 1387–1390.

Benjamin, J., Ganser-Pornillos, B. K., Tivol, W. F., Sundquist, W. I., and Jensen, G. J. (2005). Three-dimensional structure of HIV-1 virus-like particles by electron cryotomography. *J. Mol. Biol.* **346,** 577–588.

Briggs, J. A., Grunewald, K., Glass, B., Forster, F., Krausslich, H. G., and Fuller, S. D. (2006). The Mechanism of HIV-1 core assembly: Insights from three-dimensional reconstructions of authentic virions. *Structure* **14,** 15–20.

Carazo, J. M. (1992). The fidelity of 3D reconstructions from incomplete data and the use of restoration methods. *In* "Electron Tomography" (J. Frank, ed.). Plenum, New York.

Chen, H., Hughes, D. D., Chan, T. A., Sedat, J. W., and Agard, D. A. (1996). IVE (Image Visualization Environment): A software platform for all three-dimensional microscopy applications. *J. Struct. Biol.* **116,** 56–60.

Crowther, R. A., DeRosier, D. J., and Klug, A. (1970). The reconstruction of a three-dimensional structure from its projections and its applications to electron microscopy. *Proc. R. Soc. Lond.* **317,** 319–340.

Cyrklaff, M., Risco, C., Fernandez, J. J., Jimenez, M. V., Esteban, M., Baumeister, W., and Carrascosa, J. L. (2005). Cryo-electron tomography of vaccinia virus. *Proc. Natl. Acad. Sci. USA* **102,** 2772–2777.

DeRosier, D. J., and Klug, A. (1968). Reconstruction of three-dimensional structures from electron micrographs. *Nature* **217,** 130–134.

Diez, D. C., Seybert, A., and Frangakis, A. S. (2006). Tilt-series and electron microscope alignment for the correction of the non-perpendicularity of beam and tilt-axis. *J. Struct. Biol.* **154,** 195–205.

Dubochet, J., Adrian, M., Chang, J., Homo, J.-C., Lepault, J., McDowall, A. W., and Schultz, P. (1988). Cryo-electron microscopy of vitrified specimens. *Q. Rev. Biophys.* **21,** 129–228.

Förster, F., Medalia, O., Zauberman, N., Baumeister, W., and Fass, D. (2005). Retrovirus envelope protein complex structure *in situ* determined by cryo-electron tomography. *Proc. Natl. Acad. Sci. USA* **102,** 4729–4734.

Frangakis, A. S., Böhm, J., Förster, F., Nickell, S., Nicastro, D., Typke, D., Hegerl, R., and Baumeister, W. (2002). Identification of macromolecular complexes in cryoelectron tomograms of phantom cells. *Proc. Natl. Acad. Sci. USA* **99,** 14153–14158.

Frangakis, A. S., and Förster, F. (2004). Computational exploration of structural information from cryo-electron tomograms. *Curr. Opin. Struct. Biol.* **14,** 325–331.

Frank, J. (1996). "Three-Dimensional Electron Microscopy of Macromolecular Assemblies." Academic Press, San Diego.

Frank, J. (2002). Single-particle imaging of macromolecules by cryo-electron microscopy. *Annu. Rev. Biophys. Biomol. Struct.* **31,** 303–319.

Gonen, T., Cheng, Y., Sliz, P., Hiroaki, Y., Fujiyoshi, Y., Harrison, S. C., and Walz, T. (2005). Lipid–protein interactions in double-layered two-dimensional AQP0 crystals. *Nature* **438,** 633–638.

Grigorieff, N. (2000). Resolution measurement in structures derived from single particles. *Acta Crystallogr. D Biol. Crystallogr.* **56**(Pt. 10), 1270–1277.

Grimm, R., Koster, A. J., Ziese, U., Typke, D., and Baumeister, W. (1996). Zero-loss energy-filtering under low-dose conditions using a postcolumn energy filter. *J. Microsc.* **183,** 60–68.

Grünewald, K., Desai, P., Winkler, D. C., Heymann, J. B., Belnap, D. M., Baumeister, W., and Steven, A. C. (2003). Three-dimensional structure of herpes simplex virus from cryo-electron tomography. *Science* **302,** 1396–1398.

Harauz, G., and van Heel, M. (1986). Exact filter for general geometry three-dimensional reconstruction. *Optik* **73,** 146–156.

Henderson, R. (2004). Realizing the potential of electron cryo-microscopy. *Q. Rev. Biophys.* **37,** 3–13.

Hsieh, C. E., Leith, A., Mannella, C. A., Frank, J., and Marko, M. (2006). Towards high-resolution three-dimensional imaging of native mammalian tissue: Electron tomography of frozen-hydrated rat liver sections. *J. Struct. Biol.* **153,** 1–13.

Hunziker, E. B., Herrmann, W., Schenk, R. K., Mueller, M., and Moor, H. (1984). Cartilage ultrastructure after high pressure freezing, freeze substitution, and low temperature embedding. I. Chondrocyte ultrastructure–implications for the theories of mineralization and vascular invasion. *J. Cell Biol.* **98,** 267–276.

Iancu, C. V., Wright, E. R., Benjamin, J., Tivol, W. F., Dias, D. P., Murphy, G. E., Morrison, R. C., Heymann, J. B., and Jensen, G. J. (2005). A "flip-flop" rotation stage for routine dual-axis electron cryotomography. *J. Struct. Biol.* **151,** 288–297.

Knauer, V., Hegerl, R., and Hoppe, W. (1983). Three-dimensional reconstruction and averaging of 30 S ribosomal subunits of *Escherichia coli* from electron micrographs. *J. Mol. Biol.* **163,** 409–430.

Koster, A. J., Grimm, R., Typke, D., Hegerl, R., Stoschek, A., Walz, J., and Baumeister, W. (1997). Perspectives of molecular and cellular electron tomography. *J. Struct. Biol.* **120,** 276–308.

Kremer, J. R., Mastronarde, D. N., and McIntosh, J. R. (1996). Computer visualization of three-dimensional image data using IMOD. *J. Struct. Biol.* **116,** 71–76.

Kurner, J., Frangakis, A. S., and Baumeister, W. (2005). Cryo-electron tomography reveals the cytoskeletal structure of *Spiroplasma melliferum*. *Science* **307,** 436–438.

Ladinsky, M. S., Mastronarde, D. N., McIntosh, J. R., Howell, K. E., and Staehelin, L. A. (1999). Golgi structure in three dimensions: Functional insights from the normal rat kidney cell. *J. Cell Biol.* **144,** 1135–1149.

Lawrence, A., Bouwer, J. C., Perkins, G., and Ellisman, M. H. (2006). Transform-based backprojection for volume reconstruction of large format electron microscope tilt series. *J. Struct. Biol.* **154,** 144–167.

Lawrence, M. C. (1992). Least-squares method of alignment using markers. *In* "Electron Tomography" (J. Frank, ed.). Plenum, New York.

Lucic, V., Förster, F., and Baumeister, W. (2005). Structural studies by electron tomography: From cells to molecules. *Annu. Rev. Biochem.* **74,** 833–865.

Ludtke, S. J., Baldwin, P. R., and Chiu, W. (1999). EMAN: Semiautomated software for high-resolution single-particle reconstructions. *J. Struct. Biol.* **128,** 82–97.

Marsh, B. J., Mastronarde, D. N., Buttle, K. F., Howell, K. E., and McIntosh, J. R. (2001). Organellar relationships in the Golgi region of the pancreatic beta cell line, HIT-T15, visualized by high resolution electron tomography. *Proc. Natl. Acad. Sci. USA* **98,** 2399–2406.

Marsh, B. J., Volkmann, N., McIntosh, J. R., and Howell, K. E. (2004). Direct continuities between cisternae at different levels of the Golgi complex in glucose-stimulated mouse islet beta cells. *Proc. Natl. Acad. Sci. USA* **101,** 5565–5570.

Mastronarde, D. N. (1997). Dual-axis tomography: An approach with alignment methods that preserve resolution. *J. Struct. Biol.* **120,** 343–352.

Mastronarde, D. N. (2005). Automated electron microscope tomography using robust prediction of specimen movements. *J. Struct. Biol.* **152,** 36–51.

McDonald, K., and Morphew, M. K. (1993). Improved preservation of ultrastructure in difficult-to-fix organisms by high pressure freezing and freeze substitution: I. *Drosophila melanogaster* and *Strongylocentrotus purpuratus* embryos. *Microsc. Res. Tech.* **24,** 465–473.

McIntosh, R., Nicastro, D., and Mastronarde, D. (2005). New views of cells in 3D: An introduction to electron tomography. *Trends Cell Biol.* **15,** 43–51.

Medalia, O., Weber, I., Frangakis, A. S., Nicastro, D., Gerisch, G., and Baumeister, W. (2002). Macromolecular architecture in eukaryotic cells visualized by cryoelectron tomography. *Science* **298,** 1209–1213.

Miyazawa, A., Fujiyoshi, Y., and Unwin, N. (2003). Structure and gating mechanism of the acetylcholine receptor pore. *Nature* **423,** 949–955.

Moritz, M., Braunfeld, M. B., Guenebaut, V., Heuser, J., and Agard, D. A. (2000). Structure of the gamma-tubulin ring complex: A template for microtubule nucleation. *Nat. Cell Biol.* **2,** 365–370.

Murk, J. L., Humbel, B. M., Ziese, U., Griffith, J. M., Posthuma, G., Slot, J. W., Koster, A. J., Verkleij, A. J., Geuze, H. J., and Kleijmeer, M. J. (2003a). Endosomal compartmentalization in three dimensions: Implications for membrane fusion. *Proc. Natl. Acad. Sci. USA* **100,** 13332–13337.

Murk, J. L., Posthuma, G., Koster, A. J., Geuze, H. J., Verkleij, A. J., Kleijmeer, M. J., and Humbel, B. M. (2003b). Influence of aldehyde fixation on the morphology of endosomes and lysosomes: Quantitative analysis and electron tomography. *J. Microsc.* **212,** 81–90.

Navaza, J. (2003). On the three-dimensional reconstruction of icosahedral particles. *J. Struct. Biol.* **144,** 13–23.

Nicastro, D., Frangakis, A. S., Typke, D., and Baumeister, W. (2000). Cryo-electron tomography of *neurospora* mitochondria. *J. Struct. Biol.* **129**(1), 48–56.

Nicastro, D., McIntosh, J. R., and Baumeister, W. (2005). 3D structure of eukaryotic flagella in a quiescent state revealed by cryo-electron tomography. *Proc. Natl. Acad. Sci. USA* **102,** 15889–15894.

Nickell, S., Förster, F., Linaroudis, A., Del Net, W., Beck, F., Hegerl, R., Baumeister, W., and Plitzko, J. M. (2005). TOM toolbox acquisition and analysis for electron tomography. *J. Struct. Biol.* **149,** 227–234.

Nickell, S., Hegerl, R., Baumeister, W., and Rachel, R. (2003). *Pyrodictium cannulae* enter the periplasmic space but do not enter the cytoplasm, as revealed by cryo-electron tomography. *J. Struct. Biol.* **141,** 34–42.

Nitsch, M., Walz, J., Typke, D., Klumpp, M., Essen, L. O., and Baumeister, W. (1998). Group II chaperonin in an open conformation examined by electron tomography. *Nat. Struct. Biol.* **5,** 855–857.

Oettl, H., Hegerl, R., and Hoppe, W. (1983). Three-dimensional reconstruction and averaging of 50 S ribosomal subunits of *Escherichia coli* from electron micrographs. *J. Mol. Biol.* **163,** 431–450.

O'Toole, E. T., Giddings, T. H., McIntosh, J. R., and Dutcher, S. K. (2003a). Three-dimensional organization of basal bodies from wild-type and delta-tubulin deletion strains of *Chlamydomonas reinhardtii*. *Mol. Biol. Cell* **14,** 2999–3012.

O'Toole, E. T., McDonald, K. L., Mantler, J., McIntosh, J. R., Hyman, A. A., and Müller-Reichert, T. (2003b). Morphologically distinct microtubule ends in the mitotic centrosome of *Caenorhabditis elegans*. *J. Cell Biol.* **163,** 451–456.

Penczek, P. A., Grassucci, R. A., and Frank, J. (1994). The ribosome at improved resolution: New techniques for merging and orientation refinement in 3D cryo-electron microscopy of biological particles. *Ultramicroscopy* **53,** 251–270.

Penczek, P., Marko, M., Buttle, K., and Frank, J. (1995). Double-tilt electron tomography. *Ultramicroscopy* **60,** 393–410.

Penczek, P., Radermacher, M., and Frank, J. (1992). Three-dimensional reconstruction of single particles embedded in ice. *Ultramicroscopy* **40,** 33–53.

Provencher, S. W., and Vogel, R. H. (1988). Three-dimensional reconstruction from electron micrographs of disordered specimens. I. Method. *Ultramicroscopy* **25,** 209–221.

Radermacher, M. (1992). Weighted back-projection methods. *In* "Electron Tomography" (J. Frank, ed.). Plenum, New York.

Radermacher, M., Wagenknecht, T., Verschoor, A., and Frank, J. (1986). A new 3D reconstruction scheme applied to the 50S ribosomal subunit of *Escherichia coli*. *J. Microsc.* **141,** RP1–2.

Radermacher, M., Wagenknecht, T., Verschoor, A., and Frank, J. (1987). Three-dimensional reconstruction from single-exposure random conical tilt series applied to the 50S ribosomal subunit of *Eschericia coli*. *J. Microsc.* **146,** 113–136.

Radon, J. (1917). Über die Bestimmung von Funktionen durch ihre Integralwerte längs gewisser Mannigfaltigkeiten. Berichte Sächsische Akadamie der Wissenschaften, Leipzig. *Math.-Phys. Kl.* **69,** 262–277.

Rath, B. K., Hegerl, R., Leith, A., Shaikh, T. R., Wagenknecht, T., and Frank, J. (2003). Fast 3D motif search of EM density maps using a locally normalized cross-correlation function. *J. Struct. Biol.* **144,** 95–103.

Saxberg, B. E. H., and Saxton, W. O. (1981). Quantum noise in 2D projections and 3D reconstructions. *Ultramicroscopy* **6,** 85–90.

Saxton, W. O., and Baumeister, W. (1982). The correlation averaging of a regularly arranged bacterial cell envelope protein. *J. Microsc.* **127,** 127–138.

Schatz, M., Orlova, E. V., Dube, P., Jager, J., and van Heel, M. (1995). Structure of *Lumbricus terrestris* hemoglobin at 30 Å resolution determined using angular reconstitution. *J. Struct. Biol.* **114,** 28–40.

Scheres, S. H., Valle, M., Nunez, R., Sorzano, C. O., Marabini, R., Herman, G. T., and Carazo, J. M. (2005). Maximum-likelihood multi-reference refinement for electron microscopy images. *J. Mol. Biol.* **348,** 139–149.

Sigworth, F. J. (1998). A maximum-likelihood approach to single-particle image refinement. *J. Struct. Biol.* **122,** 328–339.

Steinbrecht, R. A., and Muller, M. (1987). Freeze-substitution and freeze drying. *In* "Cryotechniques in Biological Electron Microscopy" (R. A. Steinnbrecht, and K. Zierold, eds.) Springer-Verlag, Berlin.

Stewart, A., and Grigorieff, N. (2004). Noise bias in the refinement of structures derived from single particles. *Ultramicroscopy* **102,** 67–84.

Stoffler, D., Feja, B., Fahrenkrog, B., Walz, J., Typke, D., and Aebi, U. (2003). Cryo-electron tomography provides novel insights into nuclear pore architecture: Implications for nucleocytoplasmic transport. *J. Mol. Biol.* **328,** 119–130.

Subramaniam, S. (2005). Bridging the imaging gap: Visualizing subcellular architecture with electron tomography. *Curr. Opin. Microbiol.* **8,** 316–322.

Taylor, K. A., Schmitz, H., Reedy, M. C., Goldman, Y. E., Franzini-Armstrong, C., Sasaki, H., Tregear, R. T., Poole, K., Lucaveche, C., Edwards, R. J., Chen, L. F., Winkler, H., *et al.* (1999). Tomographic 3D reconstruction of quick-frozen, Ca^{2+}-activated contracting insect flight muscle. *Cell* **99,** 421–431.

van Heel, M. (1987). Angular reconstitution: A posteriori assignment of projection directions for 3D reconstructions. *Ultramicroscopy* **21,** 111–124.

van Heel, M., Gowen, B., Matadeen, R., Orlova, E. V., Finn, R., Pape, T., Cohen, D., Stark, H., Schmidt, R., Schatz, M., and Patwardhan, A. (2000). Single-particle electron cryo-microscopy: Towards atomic resolution. *Q. Rev. Biophys.* **33,** 307–369.

van Heel, M., Schatz, M., and Orlova, E. V. (1992). Correlation function revisited. *Ultramicroscopy* **46,** 307–316.

Vogel, R. H., and Provencher, S. W. (1988). Three-dimensional reconstruction from electron micrographs of disordered specimens. II. Implementation and results. *Ultramicroscopy* **25,** 223–239.

Wagenknecht, T., Hsieh, C. E., Rath, B. K., Fleischer, S., and Marko, M. (2002). Electron tomography of frozen-hydrated isolated triad junctions. *Biophys J.* **83,** 2491–2501.

Walz, J., Erdmann, A., Kania, M., Typke, D., Koster, A. J., and Baumeister, W. (1998). 26S proteasome structure revealed by three-dimensional electron microscopy. *J. Struct. Biol.* **121,** 19–29.

Walz, J., Tamura, T., Tamura, N., Grimm, R., Baumeister, W., and Koster, A. J. (1997a). Tricorn protease exists as an icosahedral supermolecule *in vivo*. *Mol. Cell* **1,** 59–65.

Walz, J., Typke, D., Nitsch, M., Koster, A. J., Hegerl, R., and Baumeister, W. (1997b). Electron tomography of single ice-embedded macromolecules: Three-dimensional alignment and classification. *J. Struct. Biol.* **120,** 387–395.

Winkler, H., and Taylor, K. A. (2003). Focus gradient correction applied to tilt series image data used in electron tomography. *J. Struct. Biol.* **143,** 24–32.

Yang, C., Ng, E. G., and Penczek, P. A. (2005). Unified 3-D structure and projection orientation refinement using quasi-Newton algorithm. *J. Struct. Biol.* **149,** 53–64.

Zheng, Q. S., Braunfeld, M. B., Sedat, J. W., and Agard, D. A. (2004). An improved strategy for automated electron microscopic tomography. *J. Struct. Biol.* **147,** 91–101.

CHAPTER 30

Methods for Image Segmentation in Cellular Tomography

Kristian Sandberg

Boulder Laboratory for 3D Electron Microscopy of Cells
Department of Molecular, Cellular, and Developmental Biology
University of Colorado, Boulder
Colorado 80309

0091-679X/07 $35.00
DOI: 10.1016/S0091-679X(06)79030-6

In this chapter, we discuss and illustrate some of the challenges for constructing robust and efficient segmentation methods for cellular tomography. We also discuss some approaches for evaluating the usability of segmentation methods and suggest a possible approach for quantifying the accuracy of segmentation algorithms. The chapter includes a review of several existing methods for denoising, contrast enhancement, and segmentation. We point out that some frequently suggested methods are often impractical for real data sets due to the computational complexity and difficulty in finding appropriate algorithm parameters. Furthermore, we claim that segmentation methods based only on intensity and contrast have problems in detecting thin, elongated structures. Instead, we propose using geometrical information, such as correlation among orientations, in order to detect membranes and microtubules. Finally, we outline a new orientation-based segmentation method and demonstrate this method on real tomograms.

I. Introduction

Three-dimensional (3D) visualization software is an important tool for studying cellular structures in electron tomograms. Tomograms with moderately complex structures can be visualized directly in 3D using software packages such as the commercial program Amira and the freely available MayaVi (Ramachandran, 2006). The result of such renderings can be improved by preprocessing the data set with thresholding or contrast adjustments. For tomograms containing more complex structures, a direct 3D rendering of the tomogram is often difficult to analyze, and the interpretation of 3D tomograms is therefore simplified by creating models where membranes are separated from the surrounding cytoplasm. Such models can be created by hand using software packages such as IMOD (Kremer *et al.*, 1996), where the user manually traces out membranes on each slice in the tomogram. The resulting collection of contours are rendered into surfaces by the program, and surfaces representing different cellular structures are given different colors. By using 3D graphics, the resulting model can be visualized and studied interactively by adjusting the viewing angle, zoom, transparency, and so on.

Creating simplified models of large data sets is of great importance in many areas of science including biological, medical, and seismic imaging. Since manual modeling of large data sets can be both time consuming and tedious, there has been active research in developing modeling software that automatically extracts features from images, a process commonly referred to as *image segmentation*.

In some applications, such as detection of calcifications in mammograms, segmentation methods have become an important and widely used tool for diagnosis of breast cancer (Nass *et al.*, 2001). Unfortunately, segmentation software for cellular tomograms has not yet reached the same level of robustness or acceptance.

The segmentation problem for cellular tomograms presents many challenges. Some of these difficulties are unique to cellular tomograms and do not manifest themselves in other segmentation applications, such as segmentation of Magnetic Resonance Imaging (MRI) data. In this chapter, we will focus primarily on the modeling of membranes in cellular tomograms by discussing some practical aspects and difficulties associated with this problem. In particular, we address the following questions:

- What particular challenges do cellular electron tomograms present for segmentation?
- What are the requirements for a useful segmentation algorithm?
- How do we objectively define the "correct" segmentation of a cell?
- What methods are currently available for cellular tomography segmentation?

Finally, we discuss and demonstrate a recently proposed method for image segmentation utilizing orientation information in the data set.

II. Problem Formulation

The goal for image segmentation can broadly speaking be described as decomposing an image into its constituent parts (Aubert and Kornprobst, 2002). We can also give image segmentation a more technical description by considering the problem of extracting features that share some characteristics such as intensity or texture (Morel and Solimini, 1995). However, extracting features based on intensity or texture alone may lead to unpredictable result for cellular tomograms that often suffer from nonuniform contrast and imaging artifacts. For segmenting cellular tomograms, a definition based on intensity or texture characteristics is therefore too restrictive. For this chapter we choose to define the segmentation problem in terms of biological structures such as membranes, specific compartments, ribosomes, and fibers (Fig. 1). Although it would be desirable to have a method that works well for all four structure types, we note that an efficient method for extracting pointlike objects, such as the ribosomes in Fig. 1C, does not necessarily work well for extracting the membranes in Fig. 1A. Therefore, it is

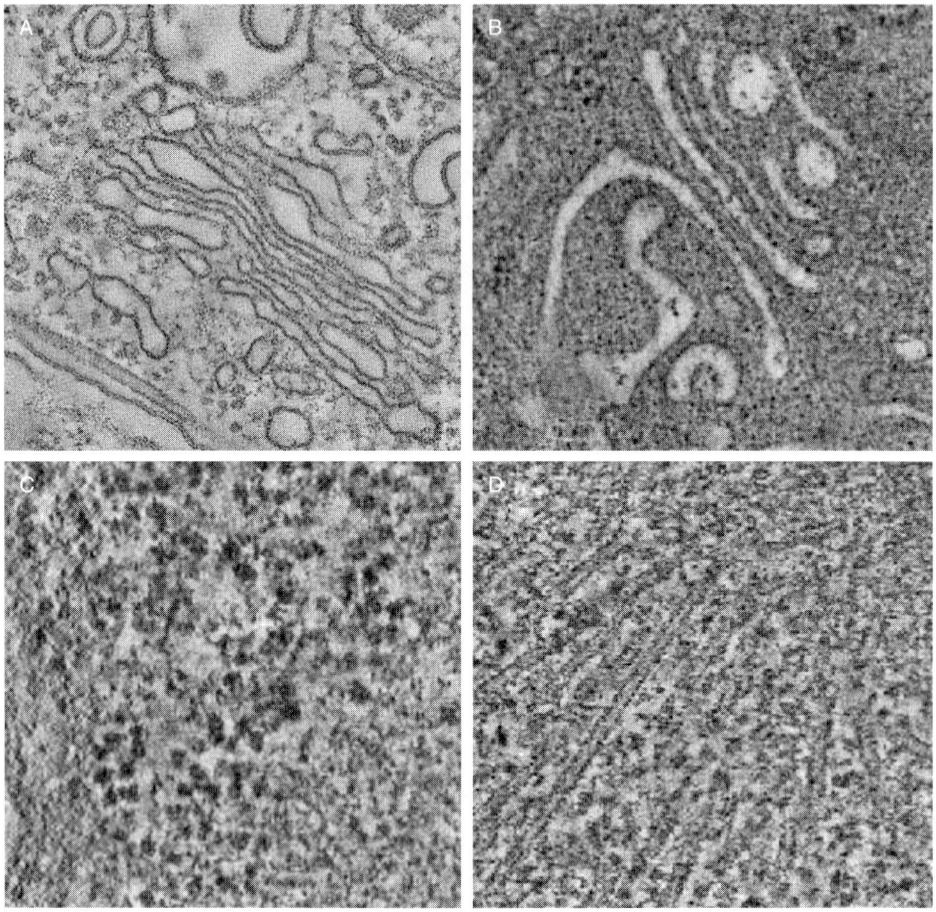

Fig. 1 Four types of data sets typically requiring different segmentation techniques. We will refer to these types as membranes (A, courtesy Mary Morphew and Richard McIntosh), compartments (B, courtesy Brad Marsh), ribosomes (C, courtesy Eileen O'Toole), and fibers (D, courtesy Eileen O'Toole).

often easier to consider these as four separate problems. In addition to the separation of the segmentation problem into these four types, it is sometimes useful to add scale as a parameter to the segmentation problem. For example, we may only want to extract significant membrane structures with a circumference exceeding a user-specified minimum length.

Besides extracting meaningful structures from a cellular tomogram, it is often desirable to label these structures by assigning a unique label to a given connected structure. This allows the user to display only selected structures, which simplifies the analysis of certain structures by enabling the user to remove occluding structures. The labeling procedure is not always associated with the image segmentation problem, but for studies of cellular tomograms the labeling is often of great importance. For the purpose of this chapter, we will therefore view the labeling process as being part of the segmentation problem.

III. Challenges in Cellular Tomography Segmentation

In this section, we discuss some of the challenges facing the segmentation methods for cellular tomograms. We note that some (but not all) of these challenges are relatively unique to these types of images, which makes many textbook examples of image segmentation methods unsuitable for cellular tomography.

A. Data Set Variety

Even if we restrict the segmentation problem to detecting membranes, there is still a great variety of data sets that a robust segmentation method is expected to handle. The variety comes not only from the many different biological structures, but also from different methods of specimen preparation. In Fig. 2, we illustrate some representative examples of membrane structures that a versatile membrane segmentation algorithm is expected to handle.

B. Low and Nonuniform Contrast

The contrast in cellular tomogram is often low and nonuniform along membranes and fibers. In some cases when the contrast is poor, the structure may only be visible at lower magnification, as illustrated in Fig. 3. This causes a problem for segmentation algorithms that focus on local intensity properties in images. In some cases, such as in Figs. 2C and 3, the intensity of a membrane or microtubule can be significantly lower than the contrast of surrounding ribosomes and other interfering structures in the cytoplasm.

C. Interfering Structures

A common problem when segmenting membranes is the presence of well-defined pointlike objects such as ribosomes in the tomogram. These objects typically have a strong uniform contrast and extend over distances larger than the width of a typical membrane or microtubule, see for example Figs. 2C and 3. Such structures are therefore not removed from consideration by many segmentation methods. In fact, some denoising methods may even emphasize such objects, as we shall see in Section VI.E. Therefore, a segmentation method for membrane extraction should use some feature that can distinguish between pointlike and linelike objects. We explore this important topic in Section VII where we discuss the benefits of measuring orientations in images.

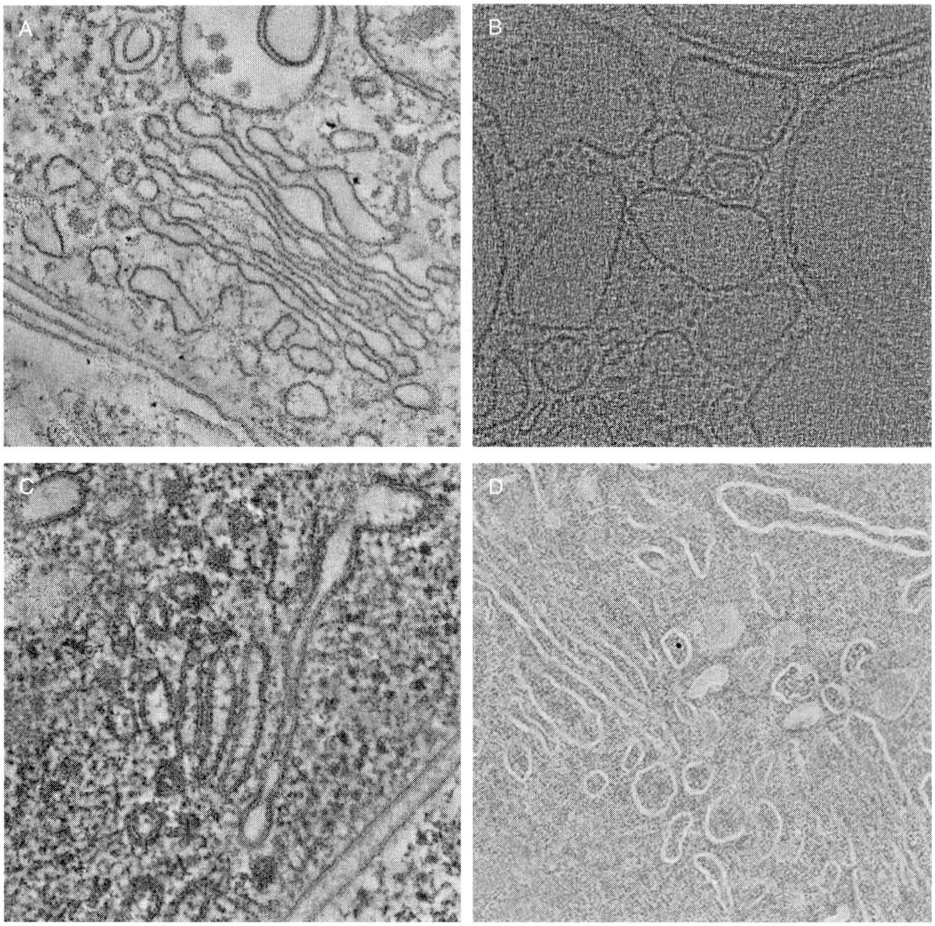

Fig. 2 Four examples of data sets that a membrane segmentation algorithm is expected to segment. (A) Trypanosome, courtesy Mary Morphew and Richard McIntosh. (B) Thylakoid membrane, courtesy Daniela Nicastro, Jason Pierson, and Richard McIntosh. (C) T-lymphocyte, courtesy Mark Davis and Hironori Ueda. (D) Golgi region labeled with antibodies against GFP–GRASP55, courtesy Mark S. Ladinsky.

D. Low Signal-to-Noise Ratio

Noise is a problem in some cellular tomograms, see Figs. 2B and 4. In some cases, the segmentation algorithm should therefore be combined with some denoising algorithm. There are several powerful denoising algorithms available for cellular tomography, and we discuss some of them in Section VI. However, it is worth pointing out that one of the simplest algorithms, median filtering, works remarkably well for cellular tomograms.

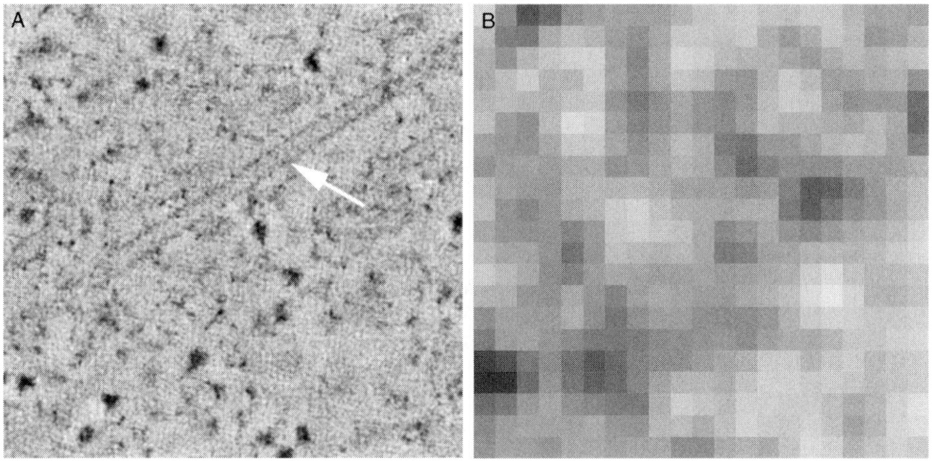

Fig. 3 Example of a microtubule, courtesy Eileen O'Toole with low and nonuniform contrast. In the right image, we show a close-up of the microtubule. Note how the microtubule is almost invisible at this scale.

Fig. 4 Cryo-electron tomography of frozen-hydrated Pelagibacter ubique. Notice the extremely low SNR. (Tomogram, courtesy Daniela Nicastro and Richard McIntosh; cell provided by Stephen Giovannoni.)

E. Anisotropic Resolution

Due to the missing wedge problem inherent to uniaxial electron tomography, some parts may have significantly lower contrast than others. This problem can be partially remedied by using dual-axis tilting (Mastronarde, 1999; Penczek *et al.*, 1995).

Recently, particle averaging has been applied to certain types of tomograms in order to increase the signal-to-noise ratio (SNR), and also to compensate for the missing wedge (Förster and Hegerl, 2007).

Even with dual-axis tilting, we can still not get isotropic resolution in a 3D tomographic reconstruction, resulting in regions of low contrast along membranes. It is therefore important for a segmentation algorithm to be able to compensate for such effects.

F. Specimen Preparation Artifacts

The specimen preparation method may introduce a number of different artifacts. The most common problem is nonuniform contrast as shown in Fig. 5A. For cryosections, cracks may appear in the specimen as in Fig. 5B. Sometimes such artifacts can make it difficult to establish the connectivity of membranes. This problem can in some cases be remedied by studying nearby sections, and the results from segmentation algorithms often improve when one uses information in all three dimensions, as opposed to segmenting each two-dimensional (2D) slice independently.

IV. Important Considerations for Segmentation Methods

Despite the rich literature on image segmentation, many practical issues are unfortunately often overlooked. As we shall see in Section VI, some contrast enhancement methods have a beautiful mathematical formulation, but they are not always practical, due to computational complexity and the difficulty in setting optimal

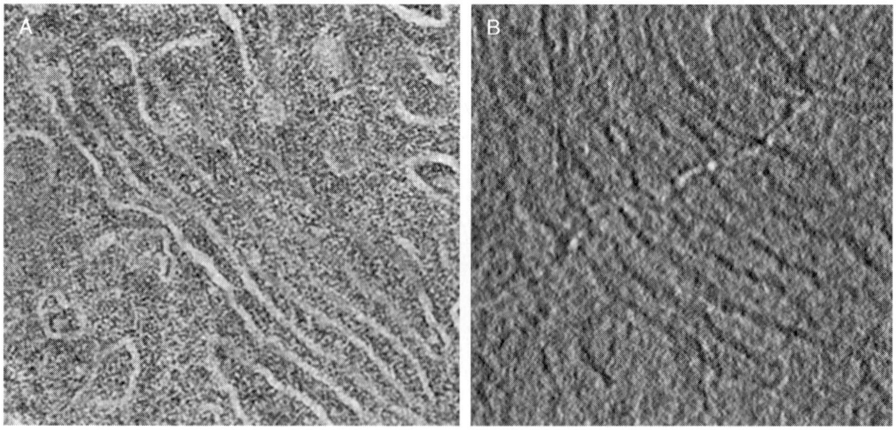

Fig. 5 (A) Example of a membrane with nonuniform contrast due to staining. (Courtesy Mark S. Ladinsky.) (B) Example of a cryosection with a preparation artifact. (Courtesy Jason Pierson.)

input parameters. In this section, we discuss some important considerations to keep in mind when evaluating the usability of segmentation methods.

A. Automation and Efficiency

In order for a segmentation algorithm to be useful, it should be as automatic as possible. Although the goal of segmentation research is an automatic algorithm, it is more realistic to construct a semiautomatic method that extracts and labels a significant fraction of the structures correctly, but requires a manual component as well. The work flow of such a semiautomatic segmentation program may be described as follows:

1. The user loads the data set and sets parameters for the segmentation algorithm.
2. The program extracts and labels features without any interaction from the user.
3. The user manually edits the automatically segmented tomogram.

In order for the segmentation program to save time for the user, the first and last steps should take a significantly shorter time than manually segmenting the full data set.

The noninteractive component of a segmentation program does not necessarily have to be fast. For example, if the interactive component takes a few minutes, running the noninteractive component overnight may be acceptable.

B. Input Parameters

Most segmentation algorithms require input in the form of one or more user-specified parameters. Such parameters may be related to scale, noise level, or numerical properties of the algorithm. Although more input parameters usually make the segmentation algorithm more precise, it is essential from a practical viewpoint that selecting the correct input parameters is easy. Searching for good parameters should not offset the benefits of the automatic component of the segmentation algorithm, and an efficient segmentation program should therefore use as few parameters as possible. In addition, if more than one parameter is used, the parameters should be independent such that each parameter can be optimized independently of the others. Exploring the parameter space of three parameters that can be optimized independently of one another is significantly easier than finding the best combination of three dependent parameters. In order to efficiently explore the parameter space, the program should quickly show the response of a selected parameter value. Furthermore, the outcome should depend in a predictable way on the value of the selected input parameter.

Finally, it is highly advantageous if the parameters are related to intuitive, physical properties such as the width of the membrane, the average size of a ribosome, or the diameter of a microtubule. Such physical or geometrical parameters are usually easier to estimate compared to more abstract parameters such as statistical noise properties.

C. Computer Requirements

Due to the large sizes of 3D tomograms, any 3D segmentation algorithm will be computationally intensive. In order for a segmentation algorithm not to require significant costs for hardware, it is important for software developers to address the processor and memory requirements for the algorithm. An algorithm that must simultaneously access the full 3D volume usually requires significant memory, and it may also be difficult to parallelize. On the other hand, if the algorithm only has to work with a few 2D slices at a time, memory requirements are modest and it may be possible to distribute the segmentation algorithm to a cluster of computers.

D. Objectivity

Although the primary goal of a segmentation algorithm may be to save time, it is also valuable if the algorithm gives an objective result with a limited dependence on user-defined parameters. This difficult but important aspect of the segmentation problem will be discussed in more depth in the following section.

V. How Do We Know if the Segmentation Method Is Reliable?

Although the results of segmentation algorithms may "look good," how do we know if the automatically extracted and labeled structures are correct? In what sense can we talk about the "accuracy" of a segmentation algorithm?

In order to illuminate some of the issues inherent to defining accuracy for segmentation algorithms, let us compare the segmentation problem to the problem of image restoration. The image restoration problem can be described as generating an improved approximation of the original image from the recorded image. In other words, we suppose the existence of a true image that is distorted during the image acquisition process, and we try to recover the true image from the recorded image. Mathematically, we construct a model of the distortion process, sometimes in terms of the statistical properties of added noise, and then solve the inverse problem by "inverting" the distortion process. Given a distortion model it is sometimes possible to define the accuracy of the reconstruction, based on the inversion algorithm. For the segmentation problem the situation is quite different. The goal is not to construct the true image. Instead we construct a significantly simplified representation of the image. Even if we knew the true image, we do not necessarily want the segmented image to be close to the true image.

Due to the nature of the cellular tomograms considered in this chapter, the correct segmentation can be ambiguous even for an experienced biologist, as illustrated in Fig. 6. Although it is outside the scope of this chapter to construct a formal accuracy measure for the segmentation of cellular tomograms, we can point out some possible approaches.

A. Measuring True/False Positives and Negatives

For Computer Aided Detection (CAD) of breast cancer from mammograms, algorithms are often evaluated by measuring ratios based on true/false positives, and true/false negatives. Using these four quantities one defines accuracy measures in terms of sensitivity, specificity, and positive prediction value (Nass *et al.*, 2001). Due to the variety and complexity of membrane structures studied in cellular tomograms, it is significantly more difficult to assign such precise notions of accuracy for segmentation algorithms of cellular tomograms. Also, for cellular tomography of dead cells there is no natural question to answer, such as "does the patient get cancer" that eventually would reveal the true answer.

The use of true/false positives and negatives can be used for measuring the accuracy of a segmentation algorithm for extracting ribosomes from cellular tomograms, but is more ambiguous for membrane extraction. For membrane extraction, we could evaluate the algorithm by measuring whether membranes are extracted or not, and if nonmembrane structures are incorrectly labeled as membranes. However, in some situations it is difficult to determine whether a structure in the data set should be interpreted as a membrane to be labeled or not

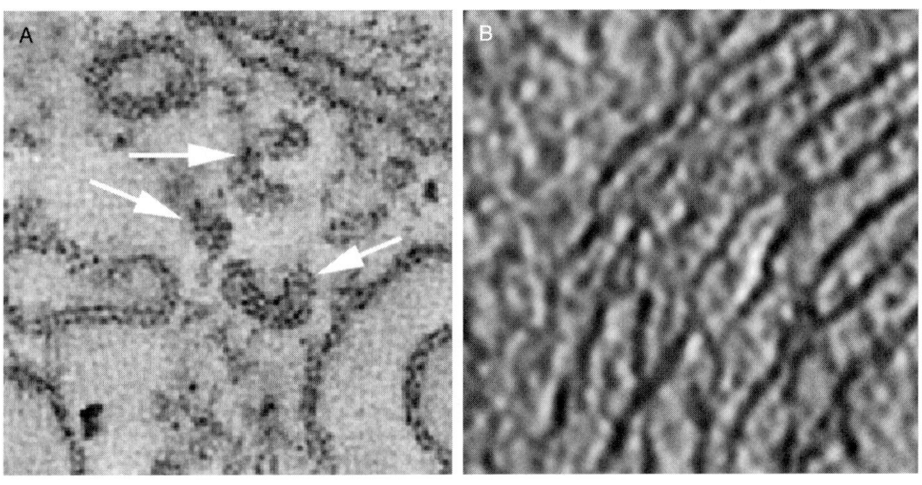

Fig. 6 (A) In this image (which is a close-up of Fig. 1A), it is not obvious whether the marked structures should be extracted as membrane structures. (B) In this cryotomogram of the Golgi apparatus, determining the "correct connectivity" is difficult. (Courtesy Jason Pierson.)

(Fig. 6A). In addition to determining the accuracy of an algorithm for correctly identifying membranes, it is also important for a segmentation algorithm to give correct descriptions of connectivity. Measuring the accuracy of connections is considerably more difficult than determining the accuracy for detecting or not detecting membranes, as illustrated in Fig. 6B.

B. Invariance

Another possibility to validate a segmentation method is to first segment a number of representative cellular tomograms manually, and then let the segmentation method segment the manually segmented result. In other words, we give a manually constructed segmentation as input (which we consider to be the true answer), and then compare the result of the segmentation algorithm to the input. We expect the input and the output to be nearly identical. If the input and output are identical, we say that the segmentation method is invariant.

Although invariance gives an indication of the reliability of segmentation methods, there are problems with this approach. First, the result is not objective. The accuracy depends on the interpretation of the person constructing the reference data set and the choice of data sets. Second, invariance is not always desirable.

To illustrate the last point, let us consider the segmentation method of thresholding the input combined with some algorithm that converts the thresholded image into labeled structures. Assuming that the input can be converted into a binary image (where structure pixels have value 1 and the background pixels have value 0), thresholding the manually segmented data using a threshold value of 0.5 will copy the input image. If we assume that the labeling part of the segmentation algorithm is perfect, we have constructed a perfectly invariant segmentation algorithm. However, if we apply this algorithm to a "real" data set that has not been manually segmented, thresholding will in many cases give useless results. This example shows that although invariance is a necessary condition, it is not a sufficient condition. In the following section, we discuss ways of overcoming this drawback.

C. Consistency

The drawbacks of the invariance measure suggested in the previous section can be partially compensated for by supplementing this measure with some additional consistency measures. Although it is outside the scope of this chapter to propose precise consistency measures, we note that such measures could be based on the sensitivity to added noise and the sensitivity to variations of input parameters. A combination of these sensitivity measures with the invariance measure suggested in the previous section could be a guiding tool in evaluating segmentation methods. Ideally, a good segmentation method would be close to invariant and relatively insensitive to added noise. Furthermore, the sensitivity to variations of user-supplied parameters would give an indication on the robustness and usability of the algorithm.

VI. Segmentation Methods for Cellular Tomography

In this chapter, we will primarily consider segmentation methods for membranes and compartments. Particle finding algorithms suitable for detecting ribosomes often use different approaches that rely on matched filters and statistical properties of the particles to be extracted (Zhu *et al.*, 2004).

A. 2D Versus 3D Methods

Segmentation algorithms for cellular tomography are mostly applied to 3D data sets. For certain types of segmentation problems, such as finding microtubules, it is essential to use an approach that uses information from all three dimensions. In other cases, such as the extraction of the membranes in Fig. 2A, using information from all three dimensions is helpful, but not as critical as for the microtubule case. In such cases, a 2D segmentation algorithm acting on each slice individually may be sufficient. Such algorithms are preferable speed wise since they usually have modest memory requirements and can easily be parallelized. However, as illustrated in Section V.A, some membrane connections can be subtle and therefore ambiguous when studying only one slice at a time. Therefore, a hybrid approach is often needed where the algorithm segments one 2D slice at a time but is able to use some information from nearby slices.

B. Variational Methods

Many segmentation methods are based on optimizing a functional chosen to have a (local) minimum for some image that simplifies the description of the underlying image. Since minimization problems for functionals are often solved by the calculus of variations, we will refer to such methods as variational methods. For a general mathematical discussion of such methods, see Morel and Solimini (1995).

Variational segmentation methods are often derived from minimizing the Mumford-Shah energy functional defined by:

$$F(u, B) = \int_{\Omega - B} [u(x, y) - u_0(x, y)]^2 \, \mathrm{d}x\mathrm{d}y + \alpha \int_{\Omega - B} |\nabla u(x, y)|^2 \, \mathrm{d}x\mathrm{d}y + \beta \int_B \mathrm{d}\sigma \quad (1)$$

where u_0 is the original image, Ω the image domain, u a simplified image, and B a set of discontinuities (boundaries). The goal is to find the simplified image u and the boundary set B. When finding compartments in cellular tomograms, u would ideally be a piecewise constant image that attains a unique value for each compartment of the image, while B would give the boundaries of the compartments.

In order to understand the basic idea of variational methods, let us interpret the different terms used in Eq. (1). The first term is sometimes referred to as a fidelity term. It attains a small value when there is a good agreement between the simplified image and the original. The second term attains a low value when the simplified image is smooth in areas not containing any boundaries. If u is piecewise constant outside boundaries, this term will be identically zero. The final term measures the length of all boundaries in the image and is small if the boundary set is small and not too "wiggly". This boundary term is sometimes supplemented by a regularity term that measures the derivative (smoothness) of the boundary set. The constants α and β are nonnegative weights that the user can adjust in order to favor increasingly smooth functions u by increasing α, or to favor small boundary sets by increasing β.

In order to minimize these types of functionals, the minimization problem is often converted into a nonlinear, time-dependent, Partial Differential Equation (PDE). The time variable is introduced such that the PDE uses the image to be segmented (u_0) as the initial condition. The PDE then evolves the segmented image $u(t)$ until an equilibrium is reached, corresponding to a local minimum of the functional. A common problem with variational methods is that there are many local minima, and it can require significant tweaking of the parameters in order for the algorithm to converge to the correct minimum.

1. Active Contours (Snakes)

One of the first variational methods to gain popularity for image segmentation was the Snake algorithm, which is a special case of so-called active contours algorithms. In a typical Snake algorithm, the user manually encloses the structure of interest with a rough contour, for example a rectangle. The contour then moves in a way determined by an underlying functional until the functional is minimized. The original Snake algorithm (Kass *et al.*, 1987) uses the functional

$$F(c) = \int_a^b |c'(s)|\, \mathrm{d}s - \alpha \int_a^b |\nabla u_0(c(s))|^2\, \mathrm{d}s$$

where $c(s) = (x(s), y(s))$ is a differentiable contour with either fixed or periodic endpoints, and $u_0(x, y)$ is the image to be segmented. Minimizing this functional numerically results in a set of time-dependent, ordinary differential equations, which can be solved rapidly. The major drawback of this method is that it can only find one contour for each initial guess and cannot handle changes in topologies, such as the splitting of one contour into two. Another drawback with the Snake algorithm in its original form is its inability to detect boundaries with sharp curvatures and nonconvex shapes. It also requires the user to specify an initial guess for each structure. However, when segmenting a 3D structure, the user can first find all contours in one slice, and then use these contours as initial guesses for nearby slices. This technique works well if the structures are well separated and do not change topology between slices. Such topology changes can occur if a structure separates into two branches when moving through the stack of slices in the tomogram.

2. Level Set Methods

Many of the drawbacks with traditional active contour methods are overcome by so-called level set methods (Chan and Vese, 2001; Osher and Fedkiw, 2003). The idea with these methods is to view contours as level sets of a function. For example, a contour in 2D may be viewed as all points (x, y) satisfying $f(x, y) = C$ for some function f and a given constant C.

Let us consider the Mumford–Shah functional in Eq. (1) but now require the simplified function u to be piecewise constant such that it can only attain two values, c_1 and c_2. Let Ω denote the image domain and let Ω_1 and Ω_2 be two subsets of Ω such that $\Omega = \Omega_1 \cup \Omega_2$. Define B as the boundary separating Ω_1 and Ω_2. The goal is to minimize the functional:

$$F(c_1, c_2, B) = \int_{\Omega_1} [u_0(x,y) - c_1]^2 \, \mathrm{d}x\mathrm{d}y + \int_{\Omega_2} [u_0(x,y) - c_2]^2 \, \mathrm{d}x\mathrm{d}y + \beta|B| \tag{2}$$

where $|B|$ denotes the length of the boundary set B. We now introduce a level set function $f(x, y)$ which has the property

$$f(x,y)\begin{cases} > 0 & \text{if} \quad (x,y) \in \Omega_1 \\ < 0 & \text{if} \quad (x,y) \in \Omega_2 \\ = 0 & \text{if} \quad (x,y) \in B \end{cases}$$

By using the Heaviside's step function

$$H(s) = \begin{cases} 1 & \text{if} \quad s \geq 0 \\ 0 & \text{if} \quad s < 0 \end{cases}$$

we can rewrite (2) using the level set function f as

$$\begin{aligned} F(c_1, c_2, f) = \textstyle\int_\Omega [u_0(x,y) - c_1]^2 H(f) \\ + [u_0(x,y) - c_2]^2[1 - H(f)] + \beta|\nabla H(f)| \, \mathrm{d}x\mathrm{d}y \end{aligned} \tag{3}$$

In practice, the functional is regularized by approximating $H(s)$ by a smooth function. This results in an algorithm that finds a level set function f and constants c_1 and c_2 that divide the image into two sets. The resulting subdivision is in a sense the "optimal" way of describing the image using only two sets. The boundary between the two sets can be represented as a contour. This contour is significantly more flexible than the contour found in the Snake algorithm, and it does not have problems following sharp curvatures and nonconvex shapes. The contour set can also split up into many contours, whereas the Snake algorithm can find only one contour at a time. The initial guess of the contour set can be fairly random and does not have to approximate the final contours set.

This algorithm can be generalized by considering M level set functions $f_1,\ldots,f_M$ each one decomposing the underlying domain into two domains, each domain associated with one constant. By overlapping the domains associated with the level set functions, we can generate a subdivision of the image consisting of 2^M subdomains (phases), each associated with a constant value. This multiphase level set method gives a significantly more flexible segmentation, but with increased computational cost. By considering another generalization, we can allow the simplified function to be piecewise bilinear instead of piecewise constant. The details of this generalized algorithm can be found in Vese and Chan (2002). We illustrate the result of this algorithm in Fig. 7.

This approach works fairly well for detecting compartment-like structures such as those in Fig. 1B, but has difficulties in segmenting the thin membrane structures in Fig. 1A. The algorithm also requires tuning of several parameters that lack obvious physical interpretations. Furthermore, as with most variational methods, finding the correct local minimum can be difficult.

C. The Watershed Transform

The Watershed Transform (Roerdink and Meijster, 2001) is based on a simple yet powerful idea inspired by topographic reliefs. We can think of a 2D image as a surface embedded in 3D where the elevation of the surface corresponds to color

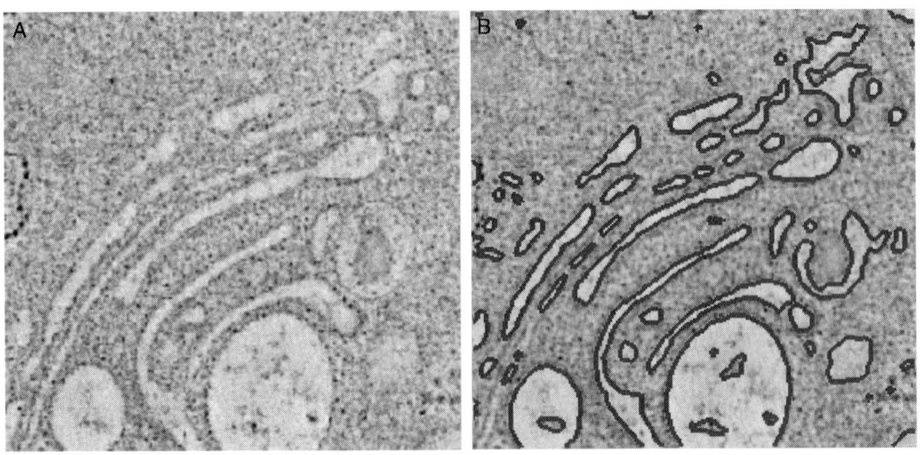

Fig. 7 Example of a level set based segmentation of compartments. This segmentation was produced using Eq. (3). The result is likely to improve using more phases and allowing piecewise bilinear level set functions. (Tomogram, courtesy Brad Marsh; Segmentation, courtesy Alejandro Cantarero.)

intensity. Let us consider an image where the intensity of the features to be extracted are darker (lower elevation) than the background (higher elevation). We punch a hole in each local minimum of the topographic surface and lower the surface into water such that water begins to fill up the reliefs through the holes. When the water level exceeds the walls of the objects, we construct a dam along the ridge of the walls to prevent the merging of water pools from different reliefs. These dams can be drawn as boundaries on the 2D image separating regions associated with different structures.

This reasoning can easily be extended to 3D data sets resulting in a simple but powerful segmentation algorithm for cellular tomograms. The main difficulty with the Watershed Transform is oversegmentation, meaning that the resulting segmentation is too sensitive, especially for noisy data sets. This problem can be partially overcome by combining the Watershed Transform with denoising and introducing more parameters. The Watershed Transform has been demonstrated to work well for certain cellular tomograms (Volkmann, 2002), but it has the drawback of requiring tuning a number of parameters lacking intuitive physical interpretations.

D. Wavelets and Brushlets

In recent years multiresolution bases, such as wavelets and brushlets, have proved useful for compression and denoising of images. Wavelet-based denoising techniques have already been used for denoising of tomograms (Stoschek and Hegerl, 1997), and can be a valuable tool for preprocessing data sets to offset some of the difficulties with other segmentation techniques.

The advantage of using brushlets (Meyer and Coifman, 2003) compared to more traditional wavelets for membranelike structures is that they have a strong notion of orientations. This makes them particularly appropriate for detection of membranelike structures, such as the membranes in Fig. 2A.

The idea with using brushlet-like bases for image segmentation is to express the image in terms of basis functions that separate elongated features from a more isotropic background. Although this procedure does not, by itself, result in a segmented image, the method can be combined with carefully chosen thresholding to highlight membranelike structures.

E. Nonlinear Diffusion Methods

Although denoising methods do not produce segmented images, certain denoising schemes can enhance the contrast of membranes such that they can be extracted using hysteresis thresholding, which we discuss in Section VI.G. A common problem with traditional linear denoising methods is that they do not preserve edges. To overcome this problem, it is helpful to view the denoising process as a diffusion process where the gray level is diffused by modeling heat diffusion. The idea with nonlinear diffusion methods is to let the rates and possibly directions of diffusion adjust based on the local geometry. Ideally, we diffuse the color intensity along membranes but not across them.

In this section, we will briefly review some of the most popular methods for such nonlinear diffusion. We note that many diffusion techniques can be derived from a variational framework, which sometimes provide useful insight to the algorithm.

For the next three methods, we introduce an additional parameter, t, which we refer to as "time" since it will play the role of time when modeled as a diffusion process. We therefore consider a parameterized family of images $u(x, y, t)$ where $u(x, y, 0)$ corresponds to the original image u_0, and $u(x, y, t)$, $t > 0$, is a diffused image. In this section, we will consider 2D images, that is, (x, y) is defined on some 2D image domain Ω. The methods can be extended to 3D data sets, but with a significant increase in computational cost.

1. The Perona-Malik Model

Let us consider the PDE

$$\frac{\partial u}{\partial t} = \text{div}[g(|\nabla u|)\nabla u] \tag{4}$$

with the initial condition $u(x, y, 0) = u_0$ (the original image) and a vanishing normal derivative at the boundary. Here $g(s)$ is a constant or decreasing function.

We first consider the linear case when $g(s) \equiv 1$. Equation (4) then reduces to the (isotropic) heat equation. By introducing the Gaussian blurring kernel

$$K(x, y, t) = \exp\frac{[-(x^2 + y^2/4t)]}{(4\pi t)} \tag{5}$$

we can write the solution to Eq. (4) as the convolution

$$u(x, y, t) = K(x, y, t) * u_0(x, y) = \iint_\Omega K(x - x', y - y', t) u_0(x', y')\, \mathrm{d}x' \mathrm{d}y',$$

resulting in an isotropically blurred image.

In order to introduce varying diffusion rates, we now select a decreasing function g such that $g(0) = 1$ and $\lim_{s\to\infty} g(s) = 0$. A common choice is the function $g(s) = 1/\sqrt{1 + s}$. With such a function we note that where the image is smooth, $|\nabla u|$ is small and $g(|\nabla u|) \approx 1$, resulting in isotropic diffusion. On the other hand, where the image has an edge, $|\nabla u|$ is large and $g(|\nabla u|) \approx 0$, that is, there is no diffusion. Hence, this results in a diffusion scheme with more diffusion in edge-free regions, and less diffusion near edges.

One drawback of this method is that it requires the final time parameter t to be set, and there is no robust way of finding the optimal choice of this parameter. Finding the right parameters is also complicated by the fact that solving this equation for a large 3D data set is computationally intensive, and finding the right parameters by trial and error can therefore result in a prohibitively long process. One should also note that establishing the well-posedness of this type of PDE can be nontrivial. The well-posedness can even depend on the numerical scheme such that the analytical problem is ill-posed, but the discretized problem is well-posed.

2. Total Variation

Total Variation (Osher and Fedkiw, 2003; Rudin *et al.*, 1992) is another useful technique for image enhancement. This method assumes that the recorded image u_0 can be decomposed as $u_0 = u + v$ where u contains homogeneous regions with sharp prominent edges, and v contains additional texture and noise. The goal is to find u by minimizing the functional

$$\int_\Omega |\nabla u| \, \mathrm{d}x\mathrm{d}y + \alpha \|u - u_0\|^2.$$

This problem is solved similarly to the Perona-Malik method, that is, by introducing a "time" parameter and then evolving the segmented image in time by solving the associated Euler-Lagrange equation

$$u = u_0 + \frac{1}{2\alpha} \nabla \cdot \left(\frac{\nabla u}{\|\nabla u\|} \right).$$

3. Anisotropic Diffusion

The idea with anisotropic diffusion is to diffuse gray level intensity along membranes, but not across membranes in order to preserve their sharpness. Anisotropic diffusion algorithms (Weickert, 1998) are based on the observation that the structure tensor of an image $u(x, y)$,

$$J = \nabla u \nabla u^T = \begin{pmatrix} \left(\frac{\partial u}{\partial x}\right)^2 & \frac{\partial u}{\partial x}\frac{\partial u}{\partial y} \\ \frac{\partial u}{\partial x}\frac{\partial u}{\partial y} & \left(\frac{\partial u}{\partial y}\right)^2 \end{pmatrix},$$

has eigenvectors parallel and perpendicular to ∇u. Since ∇u points along the direction of most rapid change (perpendicular to edges), the eigenvectors of this tensor provide local information about the directions in which to diffuse in order to blur along but not across membranes. However, in the presence of noise or other perturbations, the structure tensor J does not give a reliable estimate of edge orientation. Therefore, we first blur the image by introducing $u_\sigma \equiv K(x, y, \sigma) * \mathrm{u}(x, y)$, where K is defined in Eq. (5), and construct the modified structure tensor $\nabla u_\sigma \nabla u_\sigma^T$. Finally, we introduce another blurring parameter, ρ, by defining

$$J_{\rho,\sigma} = K(x, y, \sigma) * \nabla u_\sigma \nabla u_\sigma^T \tag{6}$$

Let us now define the matrix valued function

$$G(\lambda_1, \lambda_2, \nabla u) = [v_1, v_2]\begin{bmatrix} \lambda_1 & 0 \\ 0 & \lambda_2 \end{bmatrix}[v_1, v_2]^T$$

where v_1 and v_2 are the eigenvectors of $J_{\rho,\sigma}$, and λ_1 and λ_2 are parameters. Using this function, we define an anisotropic diffusion algorithm by solving the PDE

$$u_t = \text{div}(G\nabla u).$$

Different choices of the parameters λ_1 and λ_2 lead to different diffusion properties. In Frangakis and Hegerl (2001), the authors give a careful examination of different schemes based on this approach applied to cellular tomograms. We give an example of the result from this algorithm in Fig. 8.

This algorithm is capable of extracting membrane features even for severely noisy tomograms. However, the result is highly dependent on the choice of parameters, and finding the right parameters can be a tedious task, considering that each new set of parameters requires the numerical solution of a nonlinear PDE in 2D or 3D. In addition to the denoising parameters introduced above, the numerical scheme usually introduces one or more parameters that have to be tuned to allow for large time steps without introducing instabilities in the scheme.

Although the end result of this algorithm is anisotropic diffusion, it is important to note that the direction of diffusion is still determined by using an isotropic kernel. We will discuss drawbacks of this approach in Section VII.

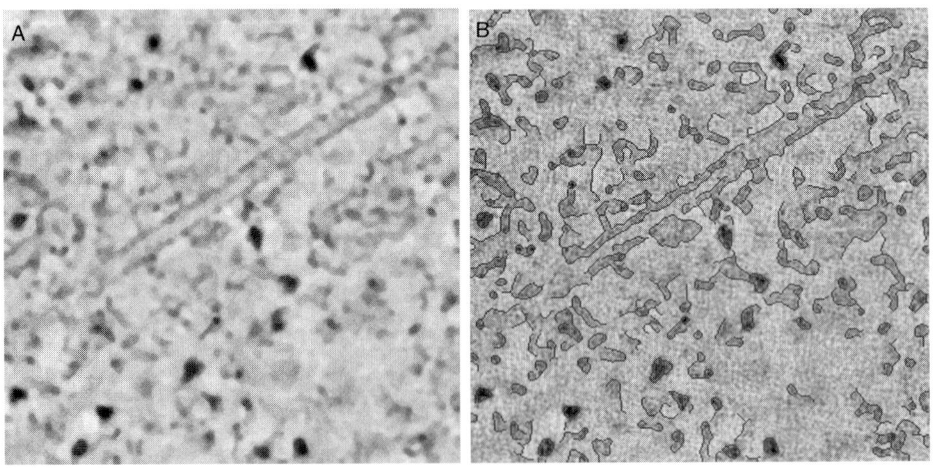

Fig. 8 Example of a segmentation of the microtubule in Fig. 3 using anisotropic diffusion combined with thresholding and Canny edge detection. These methods have difficulties extracting the microtubules due to the low and nonuniform contrast. (Tomogram, courtesy Eileen O'Toole; denoised image and segmentation, courtesy Moorea Brega.)

F. Geometric Diffusions

The next method we will study is known under many names, but we will view it as a special case of the general framework of geometric diffusions (Coifman *et al.*, 2005; Szlam, 2006). To illustrate the idea, consider an $N \times N$ image represented as the $N^2 \times 1$ vector $[u_1, \ldots u_i, \ldots u_{N^2}]$. We define the interaction matrix $\boldsymbol{W}_{ij}$, $i,j = 1, \ldots N^2$ by the rule

$$\boldsymbol{W}_{ij} = \begin{cases} 1, & \text{if } u_i = u_j \\ 0, & \text{if } u_i \neq u_j \end{cases}.$$

The goal is to subdivide the domain into two disjoint subsets A and B that minimize the cut defined by:

$$\text{cut}(A, B) = \sum_{i \in A,\, j \in B} \boldsymbol{W}_{ij}.$$

For example, if the image contains only two colors, the cut is zero if A and B are chosen as disjoint sets, each containing all indices for one of the colors. Hence, minimizing the cut gives a segmentation of the image.

Instead of considering the discrete problem of minimizing the cut, it can be shown that minimizing the cut is related to finding the eigenvector corresponding to the dominating eigenvalue of the interaction matrix $\boldsymbol{W}_{ij}$ (Weiss, 1999). Since finding the eigenvectors of the dominating eigenvalues of a sparse matrix is relatively fast, this leads to a numerical algorithm for finding approximations to the solution of the minimization problem.

The method can be generalized by considering more sophisticated interaction matrices. For example, the interaction matrix

$$\boldsymbol{W}_{ij} = \exp\left(-\|u_i - u_j\|^2 / 2\sigma\right)$$

is a better choice for noisy images. We can also add other measures in the definition of the interaction matrix depending on the properties of the features we want to extract.

The approximation of the minimization problem for the cut does not give a segmented image as a result. Instead, the resulting eigenvector corresponds to an image with enhanced contrast of the significant features in the original image. This approach was applied to cellular tomograms in Frangakis and Hegerl (2002). In Fig. 9, we demonstrate another example using geometric diffusions to enhance the contrast of weak membranes in cryotomograms.

G. Hysteresis Thresholding

Finally, we review a simple enhancement of standard thresholding that can drastically improve the thresholded image. This technique can come in use when combined with one of the denoising methods discussed earlier. For hysteresis

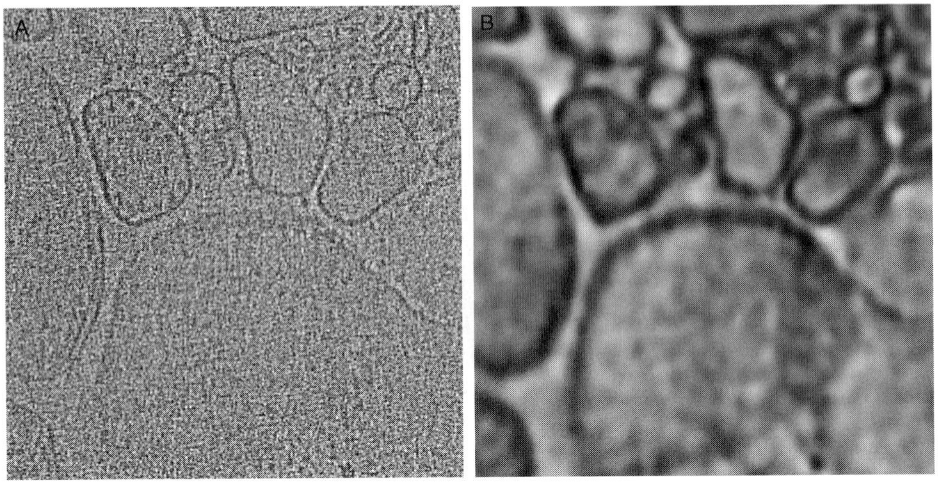

Fig. 9 Example of using diffusion geometry to enhance the contrast of membranes in noisy image. Note how the lower membrane is significantly enhanced in the right image. (Tomogram, courtesy Daniela Nicastro, Jason Pierson, and Richard McIntosh; contrast enhanced image, courtesy Nicholas Coult.)

thresholding, we set two threshold levels as opposed to one in standard thresholding. We set one large value that will "undersegment" the image (Fig. 10A), and one smaller value that will "oversegment" the image (Fig. 10B). Starting with the undersegmented image, we now (automatically) decrease the threshold level in small steps. For each new step, new pixels will be added to the thresholded image. The trick is to add only pixels that are adjacent to a pixel from the previously thresholded image. We continue this process until we have reached the threshold level corresponding to the oversegmented image. By selectively adding new pixels, we see that the final image (Fig. 10C) is significantly cleaner than the oversegmented image.

VII. Orientation-Based Segmentation

A. The Importance of Scale and Orientations

Many segmentation methods rely on the identification of regions with homogeneous intensity or texture. Although this approach makes sense for certain types of images, such as extracting the compartments in Fig. 1B, the intensity and texture information in membranes and fibers is sometimes insufficient. This is illustrated in Fig. 3, where the intensity along the microtubule is both weak and nonuniform. Nevertheless, we can easily identify the structure by eye. A likely reason for this is that on a larger scale, for example, 50×50 pixels, our eye sees a strong correlation

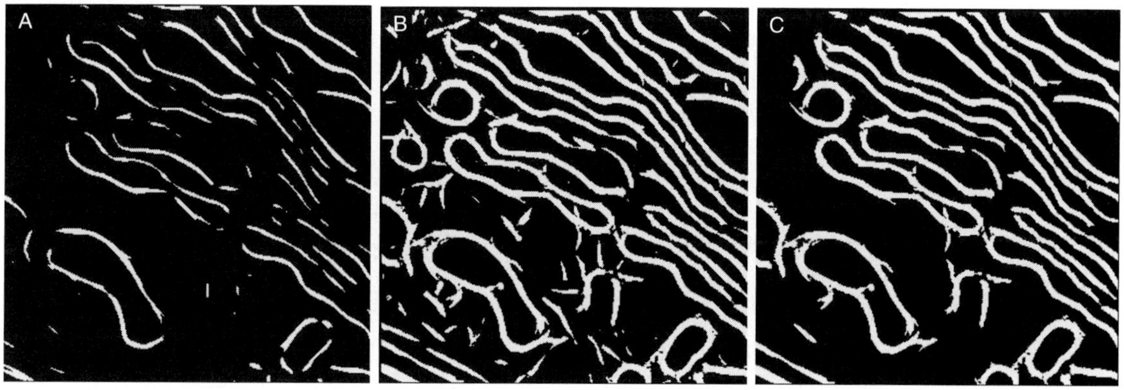

Fig. 10 Example over hysteresis thresholding. (A) Using too high threshold. (B) Using too low threshold. (C) Using hysteresis threshold with the left image as input, and the center image as the target image.

that runs in a consistent orientation along the microtubule. We therefore claim the structure is identified by looking at the image on a large scale, and by noting correlation in orientation.

In order to detect the microtubules in this type of image, it is therefore not sufficient to analyze local intensity variations. Instead, we must seek another approach that (i) analyzes the image on a large scale and (ii) measures correlation among orientations.

B. Orientation Estimation

1. Using the Structure Tensor

The idea with the anisotropic diffusion algorithm in Section VI.E.3 is to find orientations and then blur the image based on the orientation information. Although this algorithm utilizes the importance of orientations, the method has two shortcomings. First, it uses a local isotropic kernel to find the orientation. The extent of the kernel can be increased by adjusting the parameters σ and ρ in Eq. (6). Let us consider a kernel covering a circular region with a diameter of $\sim$50 pixels. Since the kernel is circular and isotropic, it is likely to contain other objects (such as ribosomes), which by their high contrast will offset the orientation estimation. Second, once the structure tensor has estimated the orientation, the anisotropic diffusion algorithm does not measure correlations in orientation, but uses the orientations only to determine directions for diffusion. If the contrast of the membrane is low, diffusion will do little to emphasize the contrast; it will merely make it more uniform (Fig. 8).

We note that a method suggested by Karssemeijer and te Brake (1996) for segmentation of mammograms does use correlations in orientations to identify structures. However, this method also uses a local isotropic kernel for orientation estimation, which will fail for extreme cases such as Fig. 3.

2. The Line Filter Transform

The idea with the Line Filter Transform (LFT) is to blur the image using an anisotropic kernel, and to generate orientations. These orientations can be used to detect membranes when combined with the orientation filtering technique discussed in the next section.

For each pixel, consider a line of a given angle and length through the center of the pixel. The LFT measures the total intensity along the line in the image and repeats this measurement for a family of lines of various orientations. For each pixel, the LFT records both the response (total intensity) and the angle for the line measurement that produced the strongest response at that pixel. Formally, the LFT of the image $u(x, y)$ is defined as

$$L(u)[x,y] \equiv \max_{\theta\in[0,\pi]} \int_{C_\theta(x,y)} u \,\mathrm{d}s$$

where $C_\theta(x,y)$ represents a line of length $2r$ and orientation θ centered at (x, y); it is formally defined by

$$C_\theta(x,y) \equiv \left\{(u,v) \in \mathbb{R}^2 | u = x - t\cos\theta \wedge v = y - t\sin\theta) \;\forall\, t \in [-r,r]\right\}.$$

There are several advantages with this nonlinear approach to blurring and finding orientations. First, we can choose the scale parameter r to be large but without requiring a large isotropic circular kernel, which may incorporate unwanted structures. Second, the method only requires one user-defined parameter, the length of the line, which can be directly related to the scale of features we are interested in. For example, for curved membranes such as in Fig. 2A, we let the algorithm choose r to be approximately twice the average width of the membranes to be extracted. Hence, the user only has to measure the width of a typical membrane to find the proper parameter for the LFT. For microtubules with low curvature, the parameter can be chosen based on the expected curvature of the microtubule, again related to a "physical" property of the data set rather than a less-intuitive parameter such as the statistical variance of estimated noise level of the data set. However, the LFT has problems in finding membranes with sharp curvatures. Further details of using this transform are given in Sandberg and Brega (2006).

C. Orientation Filtering

By recording the angle θ that maximizes the LFT for each pixel, we obtain what we refer to as an *orientation field* (Fig. 11). Once we have generated the orientation field, we look for correlations among those orientations. This approach is particularly powerful for finding membrane structures and microtubules.

Fig. 11 Orientation field of the region in Fig. 3B.

In order to measure correlations among orientations, we first note that orientations can be thought of as directed line segments. In contrast to vectors, the orientations repeat themselves every 180° in 2D rather than every 360° for vectors. In order to compute the average of orientations, addition therefore becomes less straightforward than for vector addition. We may also assign a weight to each orientation, such as the line filter response for the line that produced the best orientation in the LFT. When we add or subtract orientations with weights, we need to make sure that the addition operation is associative, such that the order in which we add orientations does not matter. One such algorithm is given in Brega (2005). Once we have an associative addition algorithm, we can construct filtering schemes for orientations that highlight curves along which we have correlated orientations. This approach is discussed in more depth in Sandberg and Brega (2006). In the next section, we provide examples where we combine orientation filtering with thresholding and labeling to produce segmentations of cellular tomograms.

VIII. Results Using Orientation-Based Segmentation

In this section, we present a few results obtained using the orientation-based segmentation technique outlined in the previous section. We illustrate the merits and problems with this approach by constructing a model of a membrane structure and extracting the microtubule in Fig. 3.

A. Method

The segmentations in this section were created using the following procedure applied to 2D slices of the 3D volume:

1. Apply the LFT.
   In order to enhance and make the contrast more uniform along membranes or fibers, we first apply the LFT. This also generates the orientations needed for the next step.
2. Orientation filter the image.
   The second step of the algorithm uses the line filtered image and the orientations found using the LFT to look for correlation in orientations. This is done using the Orientation Field Transform introduced in Sandberg and Brega (2006).
3. Apply hysteresis thresholding.
   The orientation filtered image is thresholded using the technique described in Section VI.G.

After applying these steps to one of the slices in the stack, our program automatically applies step 1–3 to the remaining slices, reusing the parameters from the first slice. Once the volume has been orientation filtered and thresholded, the contours are extracted from each slice. The contours are extracted as one pixel wide curves along the center of the thresholded membranes. The labeling algorithm then looks for neighboring contours above and below the current slice in a recursive manner, until all contours belong to some object. These objects can finally be visualized using the IMOD software (Kremer *et al.*, 1996).

B. Results

1. Trypanosome

Our first data set is an electron tomogram prepared from a chemically fixed trypanosome, courtesy Mary Morphew and Richard McIntosh. A slice from this tomograms is shown in Fig. 2A. In Fig. 12A, we show a close-up of the data set along with the same region after orientation filtering in Fig. 12B.

The program then searches for connected membranes throughout the volume, giving a unique label for each connected structure. IMOD is used to translate each label into a color that enables us to distinguish separate structures in the model view shown in Fig. 13. In Fig. 14, we show one slice where the extracted membranes have been overlaid on the original image. Note that the final model is fairly accurate, but it needs manual editing in region of low contrast, where membranes are close to each other, and along membrane sections with sharp curvature.

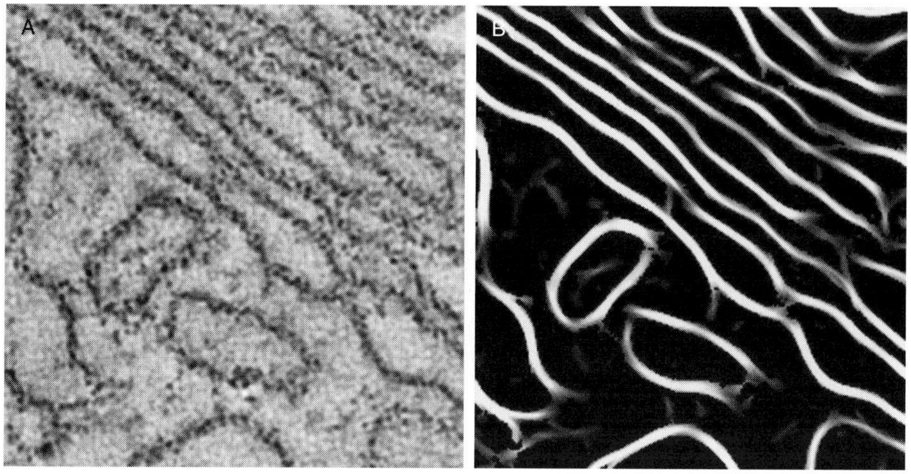

Fig. 12 (A) Close-up of Fig. 2A. (B) The result after applying the LFT and orientation filtering.

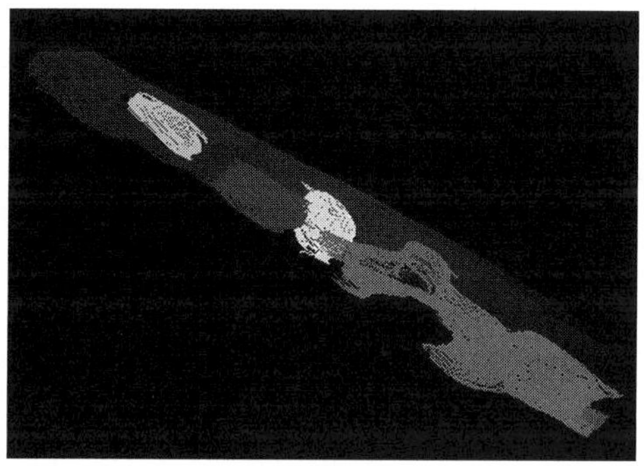

Fig. 13 Model created by the orientation-based method outlined in Section VII visualized in IMOD. The model shows a side view of 5 selected structures from a subsection of 40 slices of the tomogram. The model has been converted from color to gray scale.

2. Microtubules

We now apply the methods from Section VII to extract the microtubule in Fig. 3. This data set is an electron tomogram from a high-pressure frozen freeze-substituted mitotic centrosome from *Caenorhabditis elegans* (for complete descriptions see O'Toole *et al.*, 2003). We demonstrate the LFT and orientation filtering in 2D by studying a slice partially aligned with the microtubule such that part of the microtubule lies in the plane of the image. However, for segmentations of full 3D data sets of microtubules, it is necessary to apply the LFT and

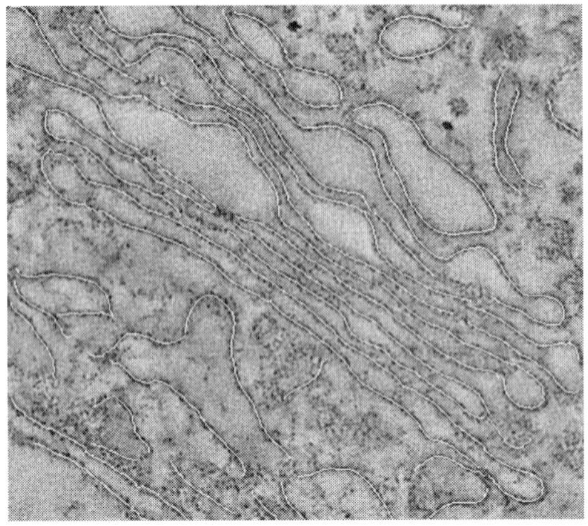

Fig. 14 The contours extracted from one of the slices overlaid the original image.

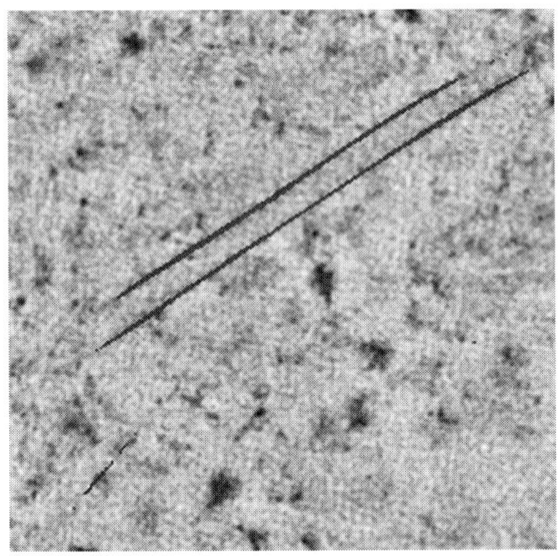

Fig. 15 *C. elegans* microtubule extracted using the methods in Section VII. (Tomogram, courtesy Eileen O'Toole; superimposed orientation filtered image, courtesy Moorea Brega.)

orientation filtering as pure 3D transforms, since there is no preferred plane for the microtubules.

We show the result in Fig. 15 and note that most of the microtubule that lies in the image plane has been extracted. Although there are a few false positives, those can be removed by requiring that any detected structure must be longer than an appropriately selected minimum length. We note that this segmentation is significantly more accurate than the edges found using anisotropic diffusion in Section VI.E.3, illustrating the importance of using correlation in orientation rather than correlation in intensity.

IX. Summary

In this chapter, we have discussed some of the methods that can be used to extract features in electron tomograms and some of the associated problems. We have pointed out the importance of using intuitive and "user friendly" input parameters to segmentation algorithms, in the sense that optimizing the parameters should be fast and easy in order not to offset any benefits of the algorithm.

Segmentation of cellular tomograms is challenging due to the great variety in the data sets encountered, and also as a result of the low and nonuniform contrast of features to be extracted. The problem is further complicated by the presence of noise and image acquisition artifacts. We also noted the importance of using orientation information in addition to contrast information in images. This is particularly important when separating low contrast membranes or microtubules from high contrast ribosomes.

Although there is currently no universally accepted segmentation method for cellular tomograms, there are several interesting methods worth further research. Some particularly interesting methods include the multiphase level set method, the Watershed Transform, and geometric diffusions. We finally demonstrated the use of a recently suggested orientation-based segmentation method and demonstrated its use for several cellular tomograms.

Acknowledgments

The author would like to thank J. Richard McIntosh and David Mastronarde for their many valuable comments on this chapter. I would also like to thank the following people for their contributions of data sets: Mark S. Ladinsky, J. Richard McIntosh, Mary Morphew, Daniela Nicastro, Jason Pierson, and Eileen O'Toole at the Boulder Laboratory for 3-D Electron Microscopy of Cells (University of Colorado at Boulder), Mark Davis and Hironori Ueda at Stanford University School of Medicine, Moorea Brega at Department of Statistics at Berkeley (University of California), Alejandro Cantarero at Department of Mathematics (UCLA), Nicholas Coult at North Country Numerics, Stephen Giovannoni at Oregon State University, and Brad Marsh at the Institute for Molecular Bioscience (University of Queensland). This work was supported by the International Human Frontiers Science Program grant, Award RGP58/2004 to J. Richard McIntosh.

References

Aubert, G., and Kornprobst, P. (2002). "Mathematical Problems in Image Processing." Springer, New York.

Brega, M. (2005). Orientation Fields and Their Application to Image Processing. Master Thesis, Department of Applied Mathematics, University of Colorado at Boulder.

Chan, T. F., and Vese, L. A. (2001). Active contours without edges. *IEEE Trans. Image Process* **10**(2), 266–277.

Coifman, R. R., Lafon, S., Lee, A. B., Maggioni, M., Nadler, B., Warner, F., and Zucker, S. W. (2005). Geometric diffusions as a tool for harmonic analysis and structure definition of data, Part I: Diffusion maps. *Proc. Natl. Acad. Sci. USA* **102**(21), 7426–7431.

Förster, F., and Hegerl, R. (2007). Structure determination *in situ* by averaging of tomograms. *In* "Methods in Cell Biology: Cellular Electron Microscopy" (J. R. McIntosh, ed.), Vol. 79.

Frangakis, A. S., and Hegerl, R. (2001). Noise reduction in electron tomographic reconstructions using nonlinear anisotropic diffusion. *J. Struct. Biol.* **135,** 239–250.

Frangakis, A. S., and Hegerl, R. (2002). Segmentation of two- and three-dimensional data from electron microscopy using eigenvector analysis. *J. Struct. Biol.* **138,** 105–113.

Karssemeijer, N., and te Brake, G. M. (1996). Detection of stellate distorsions in mammograms. *IEEE Trans. Med. Imaging* **15**(5), 611–619.

Kass, M., Witkin, A., and Terzopoulus, D. (1987). Snakes: Active contour models. *Int. J. Comp. Vision.* **1,** 321–331.

Kremer, J. R., Mastronarde, D. N., and McIntosh, J. R. (1996). Computer visualization of three-dimensional image data using IMOD. *J. Struct. Biol.* **116,** 110–118.

Mastronarde, D. N. (1999). Dual-axis tomography: An approach with alignment methods that preserve resolution. *J. Struct. Biol.* **120,** 343–352.

Meyer, F. G., and Coifman, R. R. (2003). Brushlets: Steerable wavelet packets. *In* "Beyond Wavelets" (G. V. Welland, ed.), pp. 61–82. Academic Press, San Diego.

Morel, J.-M., and Solimini, S. (1995). "Variational Methods for Image Segmentation." Birkhäuser, Boston.

Nass, S. J., Henderson, I. C., and Lashof, J. C. (eds.) (2001). "Mammography and Beyond: Developing Technologies for the Early Detection of Breast Cancer." National Academy Press, Washington, DC.

Osher, S., and Fedkiw, R. (2003). "Level Set Methods and Dynamic Implicit Surfaces." Springer, New York.

O'Toole, E. T., McDonald, K. L., Mantler, J., McIntosh, J. R., Hyman, A.A, and Müller-Reichert, T. (2003). Morphologically distinct microtubule ends in the mitotic centrosome of *Caenorhabditis elegans*. *J. Cell Biol.* **163,** 451–456.

Penczek, P., Marko, M., Buttle, K., and Frank, J. (1995). Double-tilt electron tomography. *Ultramicroscopy* **60,** 393–410.

Ramachandran, P. (2006). The MayaVi Data Visualizer. http://mayavi.sourceforge.net

Roerdink, J. B. T. M., and Meijster, A. (2001). The Watershed Transform: Definitions, algorithms, and parallelization strategies. *Fundamenta Informaticae* **41,** 187–228.

Rudin, L. I., Osher, S., and Fatemi, E. (1992). Non-linear total variation based noise removal. *Physica D.* **60,** 259–268.

Sandberg, K., and Brega, M. (2006). Segmentation of thin structures in electron micrographs using orientation fields. *J. Struct. Biol.* In press.

Stoschek, A., and Hegerl, A. (1997). Denoising of electron tomographic reconstructions using multiscale transformations. *J. Struct. Biol.* **120,** 257–265.

Szlam, A. (2006). Multiscale Analysis on Graphs and Applications. Ph.D. Thesis, Dept. of Mathematics, Yale University.

Vese, L. A., and Chan, T. F. (2002). A multiphase level set framework for image segmentation using the Mumford and Shah Model. *Int. J. Comput. Vis.* **50**(3), 271–293.

Volkmann, N. (2002). A novel three-dimensional variant of the watershed transform for segmentation of electron density maps. *J. Struct. Biol.* **138,** 123–129.

Weickert, J. (1998). "Anisotropic Diffusion in Image Processing." Teubner-Verlag, Stuttgart.

Weiss, Y. (1999). Segmentation using eigenvectors: A unifying view. *In* "Proceedings IEEE International Conference on Computer Vision," pp. 975–982.

Zhu, Y., Carragher, B., Glaeser, R. M., Fellmann, D., Bajaj, C., Bern, M., Mouche, F., de Haas, F., Hall, R. J., Kriegman, D. J., Ludtke, S. J., Mallick, S. P., *et al.* (2004). Automatic particle selection: Results of a comparative study. *J. Struct. Biol.* **145,** 3–14.

CHAPTER 31

Database Resources for Cellular Electron Microscopy

Maryann E. Martone, Joy Sargis, Joshua Tran, Willy Waiho Wong, Heather Jiles, and Cem Mangir

National Center for Microscopy and Imaging Research
Center for Research in Biological Systems
University of California, San Diego
La Jolla, California 92093

Databases have become integral parts of data management and data dissemination for biological research. In this chapter, we provide an overview of database technology and its application to both imaging in general and electron microscopy in particular. Issues involved in the creation of online community databases for electron microscopy data are discussed in the context of general issues of data sharing. We provide a detailed overview of the Cell Centered Database, a Web-accessible database for 2D, 3D, and 4D imaging data from light and electron microscopy, and discuss the lessons learned from its construction.

0091-679X/07 $35.00
DOI: 10.1016/S0091-679X(06)79031-8

I. Introduction

As the twentieth century continues to fade into memory, the time has come to move electron microscopy (EM) more firmly into the twenty-first century. Certainly, EM has experienced stunning advances in instrumentation, imaging modes, reconstruction techniques, specimen preparation, and staining (Giepmans *et al.*, 2006; Lucic *et al.*, 2005). These advances have catapulted EM once again into the forefront of tools for understanding biological complexity in the post-genomic era. EM is still the pivotal technology for solving structures in what has been termed the "mesoscale," the size range between supramolecular complexes and cells or cellular networks; it can reveal macromolecular structures and distributions *in situ*. In one area, however, EM has lagged behind, particularly compared to its genomic counterparts: complex data management. As the output of the EM community continues to increase, the utilization of databases for standard EM practice and large-scale data sharing is becoming more critical. Fortunately, segments of the community are coming together to establish the informatics infrastructure that is necessary to integrate EM images into the larger life sciences fold.

Among modern informatics tools, databases play a central role in organizing and structuring information, so it can be effectively stored, accessed, queried, and used. Databases help us to manage and archive information, to keep all the information that is associated with the data in a logical and accessible structure, to provide access control, to update and modify records, to locate data, to share data among colleagues and a wider community, and to mine data for additional information. Of these functions, it is the ability to perform data mining that has propelled publicly accessible databases, such as GenBank and SwissProt, into the collective consciousness of the biological community. Data mining can be broadly defined as the harvesting of nontrivial and interesting information by analyzing unusual patterns or trends in the raw data.

In this chapter, we will consider the design and availability of databases for EM and discuss issues involved in the creation of shared, online repositories and data management systems. Along the way, we will provide examples from our own work in creating the Cell Centered Database (CCDB), an online repository for three-dimensional (3D) information from light microscopy (LM) and EM (Martone *et al.*, 2002, 2003).

A. What Is a Database?

The simplest definition of a database is a collection of data. However, the term as generally used now refers to a computerized system for storing and retrieving data. A computerized database actually consists of both the database itself (i.e., the collection of information) and the database management system (DBMS), which provides the management of and access to the data. DBMSs are powerful programs designed for the efficient storage, query, and manipulation of information, as well as for providing concurrent access to data (transaction management). As reviewed in Bug and Nissanov (2003), databases can range in complexity from simple desktop applications, like Microsoft Access and FileMaker Pro, to enterprise-grade DBMSs, like Oracle and Informix. Enterprise-grade systems are generally necessary to support multiple hardware platforms, operating systems, and networking and storage configurations, which are often situated in different geographic locations. As the database systems become more powerful, the need for expert programmatic and administrative support increases. Most enterprise-level systems, for example, require a dedicated database administrator (DBA), an expert system administrator for installations and maintenance upgrades, and they usually require the creation of customized interfaces for data entry, display, query, and manipulation.

Multiple types of DBMSs are available, each with its own advantages and disadvantages. The most common type is the relational database management system (RDBMS), where information is organized into tables made up of rows and columns, similar to those in a spreadsheet. Each row in a table is one record (the data itself) and each column corresponds to what is often termed a database field. In general terms, the tables define the entities in the database, while the fields describe the attributes of these entities and define the relationships among them. Figure 1 shows a table from the CCDB containing information about animal subjects. Each record in the database corresponds to a single animal and the fields describe the attributes belonging to the animal, for example, age and weight. Tables may be linked together through the use of "keys," essentially a cross-reference to a record in another table.

RDBMSs are a very mature technology with a powerful query language known as the standard query language (SQL). Many commercial and open source RDBMSs are available, for example, Microsoft Access, FileMaker Pro, Oracle, Informix, MySQL, and Postgres. The well-known disadvantage of these systems is their lack of flexibility; it is very difficult to change the structure of the database

Animal Table

Column Name	ID	Pk	Null?	Data Type	Default
ANIMAL_ID	1	1	N	INTEGER	
AGECLASS	2		Y	VARCHAR2 (32 Byte)	
AGE	3		Y	MEASUREMENT	
SEX	4		Y	VARCHAR2 (16 Byte)	
WEIGHT	5		Y	MEASUREMENT	
SUBJECT_ID	6		N	INTEGER	
USER_ID	7		Y	INTEGER	
SECURITY_LEVEL	8		Y	INTEGER	
VENDOR	9		Y	VARCHAR2 (100 Byte)	

Fig. 1 Animal table from the CCDB. The table describes attributes belonging to an animal used as an experimental subject. Each field corresponds to a column in the actual database. The schema specifies additional information, like the data type for each field and which fields are required (Null = N).

once it has been established. Depending on the type of RDBMS employed, it may be possible to add additional fields, but it is difficult to change the relationships among the tables. Because of the strict requirements of database integrity, RDBMS can be costly and complicated to administer. Other types of databases exist, which are less mature but potentially more flexible, for example, object-oriented databases, object-relational databases, and XML-based databases. There are many excellent web resources that explain the differences among database technologies, for example, http://www.intersystems.com/cache/technology/white papers/hybrid.html.

Databases achieve their power by imposing a standard structure on the data. Each database requires that the structure be formalized as a database "schema." The schema describes the entities in the database, the attributes belonging to an entity, and the relationships among entities. The schema defines any constraints that may be placed on the data, for example, every animal in the database must have a species filled in; every image in the database must belong to a project. The database schema also provides the type of value that is expected for each field (Fig. 1). The declaration of a value type (or data type) for each attribute is important, because it will specify the types of operations that can be performed on the value. For instance, in the animal table shown in Fig. 1, the field recording the gender of the animal is coded as a string (VARCHAR = variable character field) and may have the value male, female, or null, while the data type expected for age is a measurement consisting of a number and a unit. Defining age as a number allows mathematical operations to be performed on it, for example, find all animals with age <4 months. As will be discussed later, defining data types for complex biological data, such as 3D reconstructions and binary images, is one of the key challenges in designing scientific imaging databases.

B. Why Are Images Different?

Any treatment of databases for scientific research should begin with the highly successful (although still far from perfect) genomics databases like GenBank and SwissProt. These databases have become highly successful because they are useful. They are useful because they are: (1) comprehensive, (2) open access, and (3) supportive of computation. The databases are comprehensive when they are contributed to by the community at large. Initially, these contributions occurred through voluntary efforts; now they occur through duress. Most journals require that all sequences be deposited at time of publication. The data are readable not only by humans but also by machines. For humans, readable means that a user can retrieve information through a search function and understand the content of the information returned. For machines, readable means that the information is stored in a way that supports computation on the information content. In the case of sequences, the information content of a gene or peptide can be symbolically represented as a simple string. Algorithms like BLAST that operate on strings can perform sequence alignment can look for interesting or significant patterns in the data stored in the database. Thus, the database becomes more than just an efficient indexing system for data retrieval but becomes a tool for *in silico* data mining. As we will see below, the content of images is much more difficult to model in a way that supports computation.

The roadblocks to the creation of similar shared repositories for complex data types, like the two-dimensional (2D) and 3D images generated by cellular EM, have been often noted but not yet solved (Kotter *et al.*, 2001; Martone *et al.*, 2004). The first obstacle is the nature of the data type. Unlike a sequence, which can be represented as a simple linear string, or protein structures, which can be represented as a set of coordinates, images and volumes are very large, often binary files. Automated image analysis of complex binary objects to extract meaningful information is still not routine. Binary files, like images, are also very large, and their storage directly in a database requires a lot of space. Because they are so large and fairly opaque to database operations, image files are generally not stored directly in a database. Rather, the database stores information *about* the image (metadata) and perhaps some information *derived* from the image, including analysis data. The image files themselves are typically stored on a separate server, and the database includes a pointer to the location where the image is stored. Most databases are set up, then, to search for images based on information *about* the image, rather than information *in* the image.

Because the contents of images are not generally stored directly in the database, most mining of image data is still usually performed by humans rather than machines. This approach may be practical and even desirable for relatively small amounts of data, but as the volume of data grows, manual analysis becomes less and less feasible. Several strategies have been implemented to query and compute on images, based on their content, through a database. The simplest is to store annotations and analyses of the data along with the image. These data are then available for query. For example, in the CCDB, measurements derived from 3D objects segmented from electron tomographic volumes are stored, allowing queries such as "Find all dendritic spines with a surface area/volume ratio of >1.5". A second strategy is to create an abstract model of the image content and store the model in the database. This approach has been used by Murphy and colleagues (Chen and Murphy, 2006; Murphy, 2005) to create an image database from fluorescence microscopy that can perform algorithmic similarity searches for subcellular patterns of gene expression. In our own work, we have developed complex data types for representing the branching patterns of neurons and the distribution of signals across a 2D image (Martone *et al.*, 2003).

Beyond the fundamental differences in the size and information content of the physical data is the nature of the data themselves. A sequence is in many respects context independent, that is, understanding the meaning of a sequence does not require a consideration of the methods used to acquire it (Goldberg *et al.*, 2005). A similar argument can be made for determination of macromolecular structure. However, for mesoscale structures such as those resolved by cellular EM, and on up the scale to whole organs, the idea of a single "average" structure has only heuristic utility. When we produce a 3D reconstruction of a subcellular structure, like a dendritic spine, we do not solve the structure of "THE dendritic spine" but rather "A dendritic spine" at one particular time, in one animal, under certain conditions. Understanding the structure of dendritic spines thus requires the solving of spine structure across many examples of spines visualized under various well-controlled experimental conditions. It also requires the careful parsing of which features are due to specimen preparation and imaging methods versus which may be accepted as universally true. Each individual data set has value as a means of assessing natural and experimental variation in structure, providing that we have the means to compare that data set to all other spines. Thus, in the realm of light and cellular EM, the need to accumulate large numbers of instances for each structure and to keep specimen preparation and imaging details together with the image data produced is particularly critical (Fellmann *et al.*, 2002). Mining information about the effect of specimen preparation on subcellular structure requires that standards be developed on reporting experimental details and that public databases be specifically designed to record this information.

In order for a Web-accessible database to have significant value, it must also contain a comprehensive sample of data. Community databases like GenBank, SwissProt, and the Protein Databank (PDB) attain their comprehensiveness through contributions from the scientific community. Efforts to promote community contributions to imaging databases to date have had mixed success. Some journals are requiring deposition into databases for selected types of image data. For example, those submitting functional brain-imaging studies to the *Journal of Cognitive Neuroscience* must agree to submit their data to The fMRI Data Center (Van Horn *et al.*, 2004). This requirement was largely achieved because the editor of the journal and the head of The fMRI Data Center were one and the same. Other imaging journals have not yet followed suit. Many researchers do voluntarily post data to Web sites and contribute additional movies and micrographs as supplemental materials to journals that will accept them. However, these modes of distribution still tend to lock data into "islands" that may have restricted accessibility to those without a subscription, and their posting on the site generally does not support cross-query with other data.

The reluctance to require deposition of imaging data, like 3D EM, in community databases likely arises for multiple reasons (Ascoli, 2006). Image-based data are often labor and resource intensive to acquire, particularly when they are segmented to extract individual objects (Marsh *et al.*, 2004). Individual 3D data sets are also information-rich and typically contain more information than is analyzed for a single study. Researchers may therefore want to hold on to their primary data, rather than releasing it to the public, so proper attribution is assured and so they can reap the full benefits of their investments. Ironically, one could argue that it is precisely because of these aspects of cellular EM data that the sharing of data is so important. The labor- and resource-intensive nature of the data make them very precious; the number of examples that can be acquired by a given researcher is very small. The only way that we will gain an understanding of biological variation in this mesoscale range will be to aggregate data across studies. The rich information content of image data also means that each reconstruction has value beyond its original intent. In our own work, for example, a 3D reconstruction that was made for a structural study of the chick ciliary ganglion cell (Shoop *et al.*, 2002), was later used for a computational modeling study to characterize neurotransmitter diffusion (Coggan *et al.*, 2005).

A second impediment to the sharing of data is the often laborious process of preparing and submitting data to a database. Because the database gains its advantage over other methods of data sharing from consistency in metadata and data structure, users must adhere to the database requirements. Thus, even if an altruistic researcher is inclined to contribute data to an online repository, gathering all of the supporting material and converting the data into the structure expected by the database can be a time-consuming task. Currently, there are few incentives for a bench researcher to spend additional time and energy depositing data into a community database when the effort often does not count toward academic advancement. Databases like the CCDB do offer at least one incentive to researchers by providing a stable archive of valuable data and by storing additional material in support of published articles. For example, Sosinsky *et al.* (2005) deposited their data into the CCDB because, unlike the journal, the database placed no restrictions on the size or amount of data to be stored.

C. Web-Accessible Databases for EM

Despite the potential challenges, several projects are underway that aim to establish a set of Web-accessible resources for microscopy data (Fig. 2). The Electron Microscopy Database (EMDB) (Henrick *et al.*, 2003; Tagari *et al.*, 2002, 2006), hosted by the European Bioinformatics Institute (EBI), contains molecular structures derived from cryo-EM, including tomography. With the advent of the EMDB, structural biology has a robust set of online resources for molecular structures derived from X-ray crystallography, NMR spectroscopy, and cryo-EM (Tagari *et al.*, 2006). The structural biology community has made great strides through these initiatives in establishing standards for data exchange (Heymann *et al.*, 2005; Tagari *et al.*, 2006). Image repositories for microscopy data at the cellular level have been established, for example BioImage (Carazo *et al.*, 1999; Catton and Shotton, 2003), although these are primarily geared toward 2D images. Our own project, the CCDB, was created to address the need for additional databases for cell level light and EM data, particularly geared toward large, 3D data sets.

D. Data Management for EM

The process of sharing EM data is considerably enhanced by the incorporation of electronic data management systems into imaging laboratories. Most of the data generated by an individual in a laboratory are locked into an individual

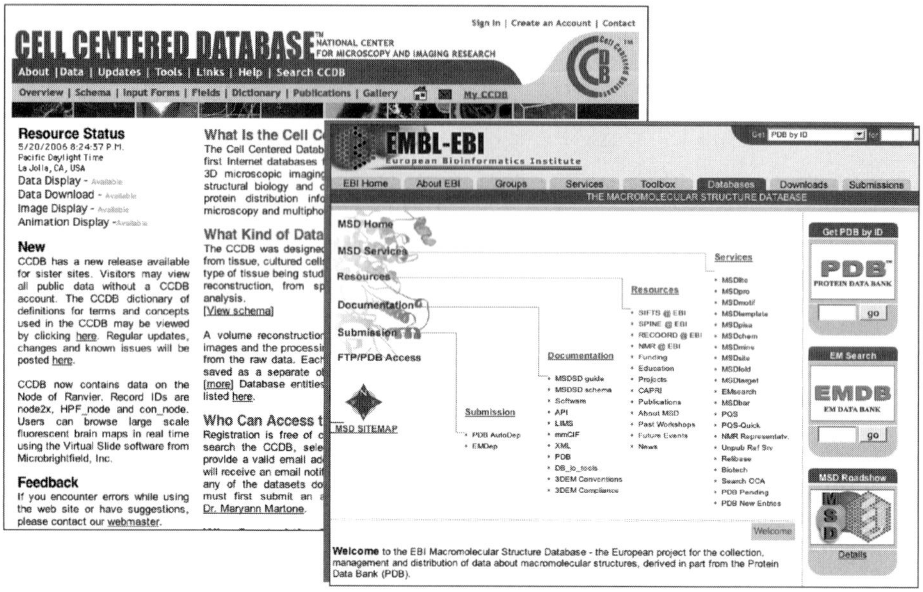

Fig. 2 Two current Web-accessible databases for sharing EM data: upper left: the CCDB (http://ccdb.ucsd.edu); lower right: EMDB (http://www.ebi.ac.uk/msd/) for macromolecular structure, hosted by the EBI.

repository, where they are neither accessible nor in many cases interpretable by another individual. Increasingly, the original image data are in a digital form, but the accompanying descriptive data for each data set are still too often stored away in a private paper notebook. Even if it is in an electronic file such as a spreadsheet or a text document, the information within the file is opaque to searching unless each document is opened separately, and the file itself may be present on a local desktop system or laptop computer. If notation is not standardized, the information in the document may also be difficult to interpret or important data may be missing.

Electronic laboratory notebooks (e-notebooks) have started to gain in popularity and several commercial and open source products are available (Butler, 2005; Kihlen, 2005). Often, these e-notebooks have a database back end and can potentially operate with a client-server architecture, so records can be cross-accessed by multiple investigators (Ludtke *et al.*, 2003). Some products are also specialized for storing image data. For example, the Axiope system (www.axiope.com), a commercial product, provides an efficient means of cataloguing and archiving images. Neurosys (http://neurosys.cns.montana.edu/), an open source project from Montana State University, allows the user to create a customized database that also works with a client-server architecture. Neurosys has no inherent facility to handle images, but it allows the user to reference a URL where the data may be available.

Some database systems have been developed specifically for microscopy data. The Open Microscopy Environment is an open source database for light microscopy that also provides a standard set of XML descriptors for imaging data (Goldberg *et al.*, 2005; Swedlow *et al.*, 2003). Image databases are also available from commercial vendors, for example, IQBase (MediaCybernetics, Inc., Silver Spring, MD, http://www.mediacy.com).

Several research groups have also produced data management systems specifically for EM, particularly in the macromolecular structure domain. These systems were developed partly out of convenience, but mostly out of necessity, given the large number of data files generated for each reconstruction (Fellmann *et al.*, 2002; Ludtke *et al.*, 2003). As will be described in the next section, the CCDB was designed to provide data management functions for reconstruction from light and electron micrographs.

II. The Cell Centered Database Project

The CCDB project was launched in 2002 as an online repository of high resolution 3D light and electron microscopic reconstructions of cells and subcellular structures (Martone *et al.*, 2002, 2004). It contains 2D, 3D, and four-dimensional (4D) structural and protein distribution information from confocal, multiphoton, and EM, including correlated light and EM. Many of the data sets are derived from electron tomography of cells and tissues. From its inception, the CCDB provided not only the final data product, usually a 3D reconstruction, but also the raw data, specimen preparation details, the imaging parameters, and any derived data products created as a result of analysis of the reconstruction. Thus, the schema of the CCDB ensures that researchers can trace the provenance of a piece of data and understand the specimen preparation and imaging conditions that led to it, while making raw and derived data available for reanalysis.

The basic organization of the CCDB is illustrated in Fig. 3. The CCDB was designed around the process of reconstruction from 2D micrographs, capturing key steps in the process from experiment to analysis. The CCDB is implemented using Oracle 9i and will soon move to Oracle 10g. The core tables shown in Fig. 3 represent the backbone of the CCDB, each of which may be further elaborated in additional tables specialized for a particular technique or data product. By designing the CCDB in this fashion, we can easily add new techniques or features without breaking the structure of the database. While many of the data sets in the current CCDB derive from the nervous system, reflecting our own area of expertise, the schema of the CCDB supports cellular data from any tissue. For example, we have made available a set of tomographic reconstruction of blue green algae (Moisan *et al.*, 2006).

The types of imaging data stored in the CCDB are quite heterogeneous, ranging from large-scale maps of protein distributions, taken by confocal microscopy, to 3D reconstructions of individual cells, subcellular structures, and organelles,

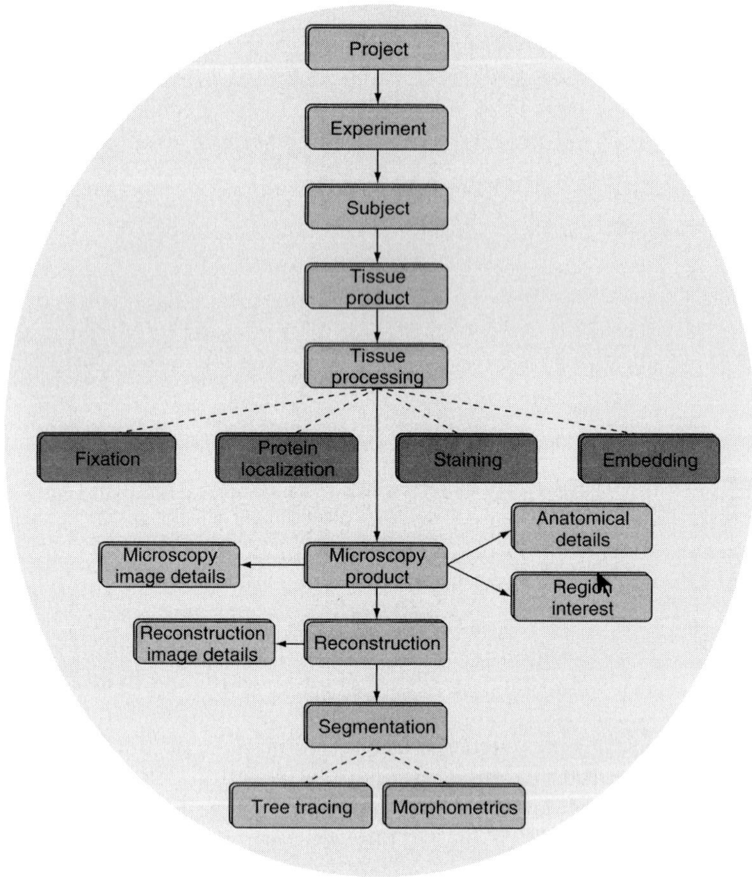

Fig. 3 Simplified view of the CCDB schema, showing the main tables that track the process of reconstruction from 2D micrographs through segmentation and analysis. Each of the tables represented may be elaborated in one or more additional tables. The full schema is available online at http://ccdb.ucsd.edu/er.html.

reconstructed using electron tomography (Fig. 4). The CCDB stores not only the original images and 3D reconstruction but also any analysis products derived from these data, including segmented objects and measurements of quantities, such as surface area, volume, length, and diameter, along with supporting image data, like animations and image maps. Thus, each data record in the CCDB consists of a set of primary images and any derived data products, for example, reconstructions and segmentations (Fig. 4). Data sets in the CCDB can be referenced to a literature citation, so that the CCDB can provide access to the full resolution data sets from the published literature (e.g., Moisan *et al.*, 2006; Sosinsky *et al.*, 2005). Each data set has a unique URL that can easily be referenced in a manuscipt.

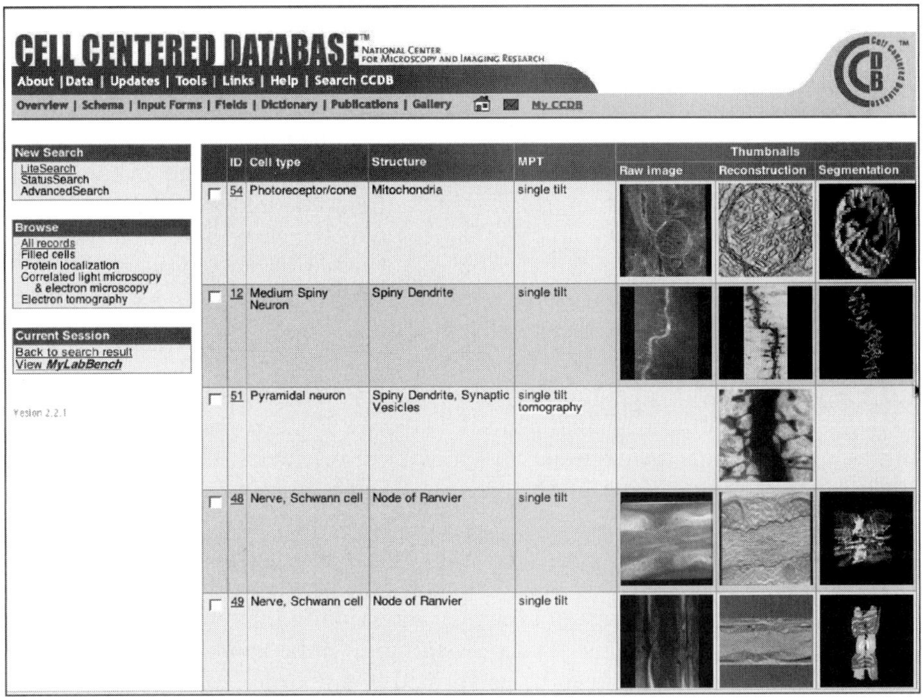

Fig. 4 Summary search result display from the CCDB. Users are presented with a list of results with the search term highlighted in yellow. Thumbnails of each of the types of image data available for the microscopy product are displayed on the right, so users can quickly get a sense of what data are available. Users may select a data set for further viewing by clicking on the microscopy product ID or by clicking one of the thumbnails. Selecting the checkbox on the left allows users to add data to their cache (MyLabBench) for future download. (See Color Insert.)

At the center of the CCDB data model is the "Microscopy Product" table. The microscopy product refers to a set of related 2D images taken by light (epifluorescence, transmitted light, confocal, or multiphoton) or EM (conventional or high-voltage transmission EM). These image sets may comprise a tilt series, optical section series, through focus series, serial sections, montages, time series, or a set of survey sections taken in a single microscopy session that are not related in any systematic way. A given set of data may be more than one product, for example, it is possible for a set of images to be both a mosaic and a tilt series. A mosaic is a type of product where the specimen is imaged systematically in x and y, and the resulting images are stitched together to create a single larger image. The microscopy product identifier serves as the accession number for the CCDB. This value was chosen because a given data set may only be taken once, so each microscopy product represents a unique data set, whereas multiple reconstructions may be derived from a single microscopy product.

All imaging data deposited in the CCDB must be accompanied by project, experimental, subject, and protocol details. Proper interpretation of imaging data requires a thorough understanding of the experimental and imaging conditions used to produce the specimen and image. The CCDB allows the user to store detailed experimental protocols in the form of text files, but it also contains fields for experimental parameters that are known to be important for interpreting structural and protein distribution data and would likely be the targets of specific queries. Examples of such details include information on specimen fixation, staining, immunolabeling, and imaging conditions.

The design of the specimen preparation tables has presented quite a challenge because of the flexible nature of specimen preparation protocols. Relational databases derive their power from the highly structured and rigid nature of the table structure; they are not very well suited to the type of fluid and changeable nature of experimental protocols. For this reason, some groups have chosen to implement databases using object-oriented principles (e.g., Qbase, EMEN; Ludtke *et al.*, 2003). In the original schema of the CCDB, users could store details about specimen processing, but the order in which the steps were performed was lost. Because the order in which steps are performed is important when comparing protocols across experiments, the new version of the CCDB has an extensively redesigned set of specimen preparation tables, whereby key specimen preparation steps such as fixation, staining, embedding, microtomy, and macromolecular localization can be performed in any order and at multiple points in a protocol.

A. CCDB Search and Display

The CCDB was created to promote the sharing of image-based reconstruction data at the tissue, cellular and subcellular level. Many of the data sets in the CCDB are freely available to the public, within the terms outlined by the usage agreement. Although earlier versions of the CCDB required a user account to view data, we recently eliminated the need for registration to view the public data. This decision was taken to relieve users of the necessity of remembering yet another user name and password, but also to promote database interoperability through remote programmatic access of the CCDB. Private data are still maintained behind a secure portal that requires an account. Both database interoperability and the CCDB secure portal will be discussed in subsequent sections.

The CCDB search and display pages were extensively redesigned in 2005. Over the past few years, conventions have grown up around Web interfaces, driven largely by the ascendancy of search engines like Google (www.google.com) and e-commerce sites like Amazon (http://www.amazon.com). Our first implementation of the CCDB provided a set of query forms whereby users could compose a query from drop down menus. With the popularization of Google, we found that it was more intuitive for users to enter a keyword, rather than having to understand anything about navigation on the Web pages or the structure of the database. To provide information on the contents of the CCDB that would guide in searching, we created the CCDB dictionary. The dictionary provides a list of current concepts in the database, along with their definitions; it is updated as data are entered into the database. On entering a keyword, users are returned a set of thumbnails that represent the three categories of image data, 2D images, reconstructions, and segmented objects, which may be stored for a given microscopy product ID (Fig. 4). In this way, users can rapidly scan through the available data and determine which data are available for each product.

Commercial sites have introduced the idea of the "shopping cart," essentially a cache of items that are selected and saved as a user peruses the site. The new CCDB search and display pages have been redesigned according to these conventions. At any point during the browsing of search results, the user can add a data set to "MyLabBench" for future download. When finished browsing, the user may "check out" by agreeing to the usage terms and then download the data. For each data set that is downloaded, the CCDB attaches a file that contains all of the descriptive data output in XML (extensible markup language). This file includes the owner of the data and any publication citation, so proper attribution can be given.

In response to a query, data shown on the CCDB display page is dynamically generated (Fig. 4). Once a data set is selected, high-level data are presented in summary form with links to additional details, such as imaging product type, reconstruction information, or project information (Figs. 5 and 6). The CCDB makes low-resolution 2D views and animations of each data set available through the Web interface. Users may also download the full resolution image data for any type of data, for example, raw data, 3D reconstruction, segmented volumes, that are available for a particular data set. Each of these types usually consists of a set of related files, for example, a set of tilt images. Thus, all files comprising a given category of information are bundled together as an archive (tar) file, so they may be downloaded as a single file. As with most image databases, the CCDB employs a separate file system for storing the images outside of the database proper. The CCDB has employed an advanced distributed file storage system, the Storage Resource Broker (SRB), an example of so-called "grid" middleware, as the main storage site for the large image files (Grethe *et al.*, 2005). The CCDB does not currently enforce a standard data format, but tools for visualizing these data sets are listed on the CCDB Web site under "Visualization and Analysis Tools." Most data can be viewed using the open source program ImageJ by using available plug-ins (http://rsb.info.nih.gov/ij/).

For interactive browsing of very large 2D images, the CCDB employs the Neuroinformatica package from Microbrightfield, Inc. Williston, VT (http://microbrightfield.org). Neuroinformatica utilizes an image server, essentially a platform optimized for serving images. Image servers allow, among other things, the efficient browsing of large images because they match the transmission of information to the resolution at which a user is browsing. In this way, the entire image does not have to be transmitted at full resolution; as the user zooms in, only the portion of image that is viewed is transmitted rather than the entire image (Fig. 6).

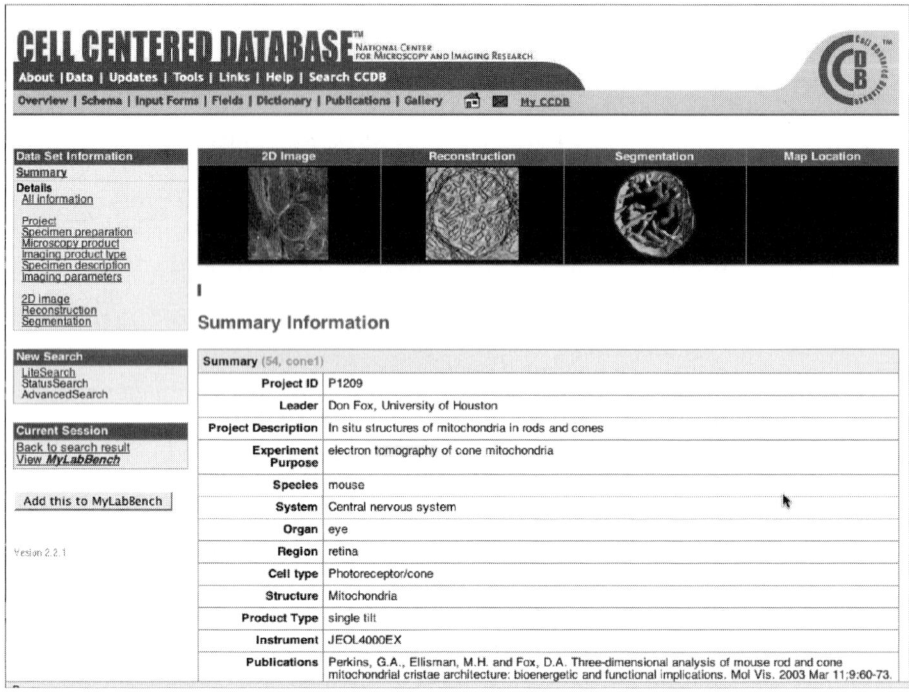

Fig. 5 Display page for a tomographic reconstruction of a mitochondrion contained in the CCDB. For this data set, 2D images, a 3D reconstruction, and a segmentation of image features were available, as indicated by the three thumbnails across the top of the display. Clicking on any of the thumbnails returns a larger image and additional details about the data. This data set was contributed by Dr. Guy Perkins (Perkins *et al.*, 2003) and may be viewed at: http://ccdb.ucsd.edu/CCDBWebSite/main?event=displaySum&mpid=54.

B. Multiscale Spatial Registration of CCDB Data

Because the ultimate goal of structural biology is to understand how complex organisms are built up from their molecular parts, the CCDB provides the ability to reference imaging data to higher levels of anatomical organization. For brain data, this reference is usually in the form of the location of the data in a standard brain atlas, for example, Paxinos and Franklin (2000). Users may also supply an "image map" that can be used to reference a higher resolution data set to one of lower resolution, for example, an EM data set may be placed within the context of a light microscopic view or within the context of a lower magnification image (Fig. 6).

To aid in the spatial registration of brain data in the CCDB, we have created the Spatial Markup and Rendering Atlas Tool ("Smart Atlas"; Martone *et al.*, 2004, 2003; Zaslavsky *et al.*, 2004). The Smart Atlas is an atlas-based query interface for brain data that is referenced to the stereotaxic coordinate system of the brain. It is

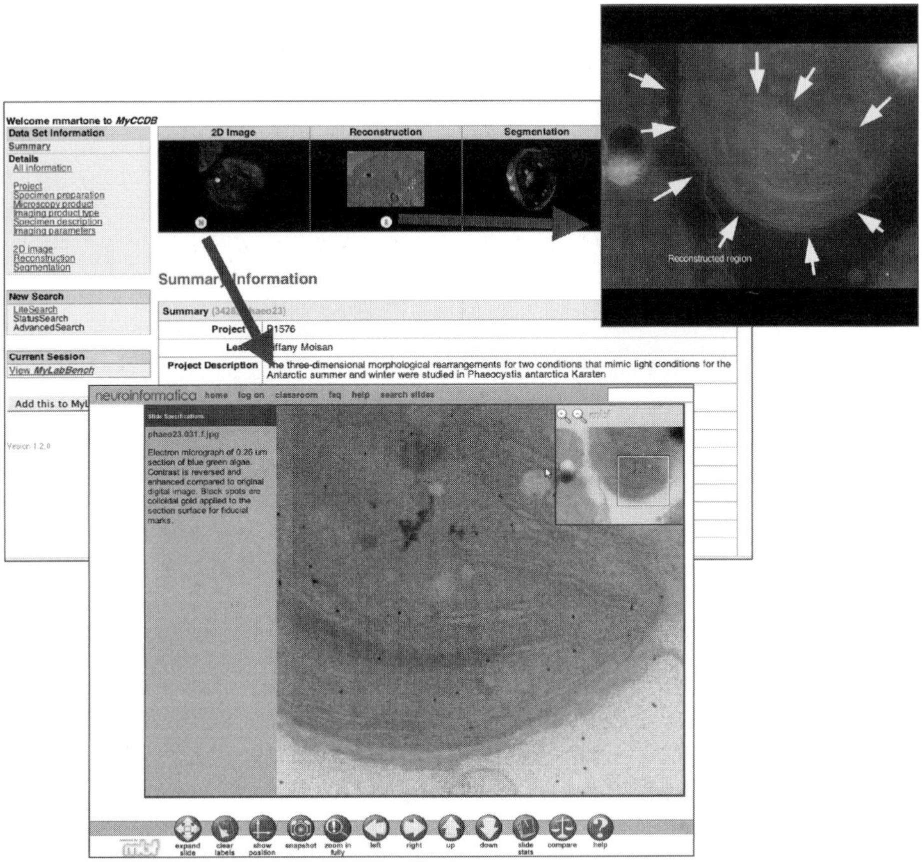

Fig. 6 The CCDB provides the ability to store supporting image files with the main record. These supporting files may include an image map (upper right), which situates the reconstruction in a broader context. For very large display images, the CCDB utilizes an image server (lower panel) that allows the user to pan and zoom on the full resolution data. Images up to 4 Gb in size may be viewed using this technology. The data set in this figure is a tomographic reconstruction of the chloroplast of a blue green alga exposed to high light conditions. Data was contributed to the CCDB by Drs. Tiffany Moisan and Gina Sosinsky (Moisan *et al.*, 2006). This data set may be viewed at: http://ccdb.ucsd.edu/CCDBWebSite/main?event=displaySum&mpid=3428.

built using geographical information system (GIS) technology, a system that has been designed specifically to deal with spatial information. Through the Smart Atlas interface, users may view light and electron microscopic data sets contained in the CCDB registered to a particular location (Fig. 7). All data are referenced to the same coordinate system, allowing data from different scales to be brought together, regardless of how the data were acquired. This lets users build up multiresolution views of the mouse nervous system.

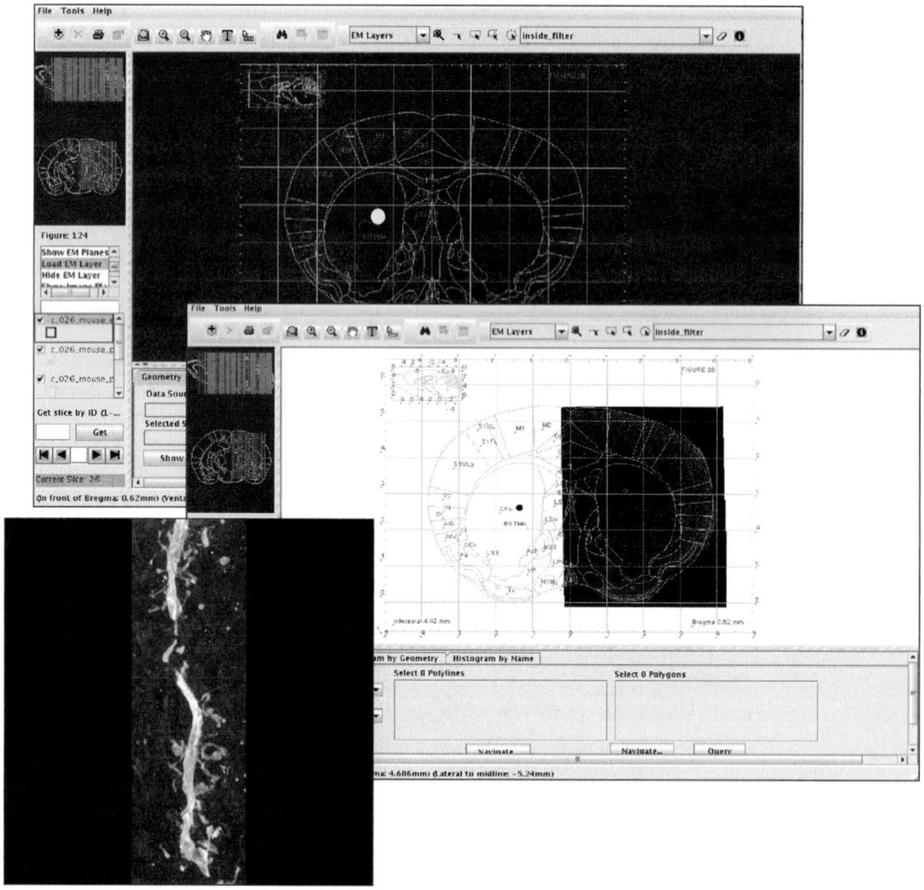

Fig. 7 Querying the CCDB through the Smart Atlas interface. The Smart Atlas displays the location of light and electron microscopic data in the CCDB registered to a standard brain atlas (yellow dot in upper panel). It also displays image data that have been normalized to the same location (middle panel). Clicking on the dot returns the data from the CCDB, in this case a tomographic reconstruction of a spiny dendrite from the mouse neostriatum. (See Color Insert.)

C. Database Interoperability

Databases like the CCDB aim to be comprehensive within their domain, but no database should expect to exist in isolation. One of the challenges facing all informatics efforts is the issue of database interoperability. Interoperability can have multiple meanings, but its essence is the idea that no database is an island, that is, no single resource can be created to house all of the information that is daily accruing in biological science. A detailed discussion of database interoperability is beyond the scope of this chapter, but several excellent treatments highlighting the promise and challenge of interoperability have been published (Grethe *et al.*, 2005; Kotter, 2001). Through the use of specialized interfaces and additional software layers, databases like the CCDB can exist as independent projects and be called remotely by other applications, including other databases. Toward this end, the CCDB has created a set of Web services (Neerincx and Leunissen, 2005), where an application can issue a remote query against the CCDB. The CCDB is also participating in several large-scale projects designed to provide software layers that promote the cross-query of databases, for example, the Biomedical Informatics Research Network (BIRN) project (Grethe *et al.*, 2005).

If the issue of database interoperability stopped at programmatic access to distributed databases, we might think that the problem was well on its way to being solved. However, because a database is about the data, any discussion of database interoperability must also consider the more thorny problem of data interoperability. While search programs are very good at finding terms that match a string of letters, merely matching a string does not say anything about the meaning of a term. To give a simple example: is a Purkinje cell in the CCDB the same as a Purkinje neuron in the SenseLab database (http://senselab.med.yale.edu/senselab/)? This problem is well known to users of online search engines and is one of the prime motivators behind efforts to create the Semantic Web (Mukherjea, 2005; Neumann, 2005).

In the CCDB, we employ several strategies to ensure that our data can be integrated with data acquired at other scales and through the use of other techniques. As discussed earlier, spatial referencing provides one important means by which disparate data can be brought together (Fig. 7). For terminological data, the CCDB employs ontologies to help disambiguate the concepts contained in the CCDB (Martone *et al.*, 2004). Ontologies are essentially formal representations of knowledge in a particular field through specification of concepts and the relationships among them (Gkoutos *et al.*, 2005). For example, a simple ontology would include statements like: cell has a nucleus; nucleus is an organelle; nucleus has a nucleolus. Several communities are working together to establish ontologies and terminologies for their domains, for example, the Gene Ontology (www.geneontology.org/). The Unified Medical Language System (UMLS; http://www.nlm.nih.gov/research/umls/), maintained by the National Library of Medicine, contains a large thesaurus that maps terms from existing terminologies to identify semantic concordances among terms used in different vocabularies.

Ontologies are used in multiple ways to support the goal of data interoperability. An ontology generally maps each concept to a unique identifier, essentially the "accession number" for that concept, to be used to distinguish it from all other concepts. In the example above, Purkinje cell and Purkinje neuron both map to the same identifier and so are assumed to be the same concept. The identifier for a Purkinje fiber from the heart would be different (Fig. 8). The CCDB has mapped each of the concepts in the database to one or more shared ontologies,

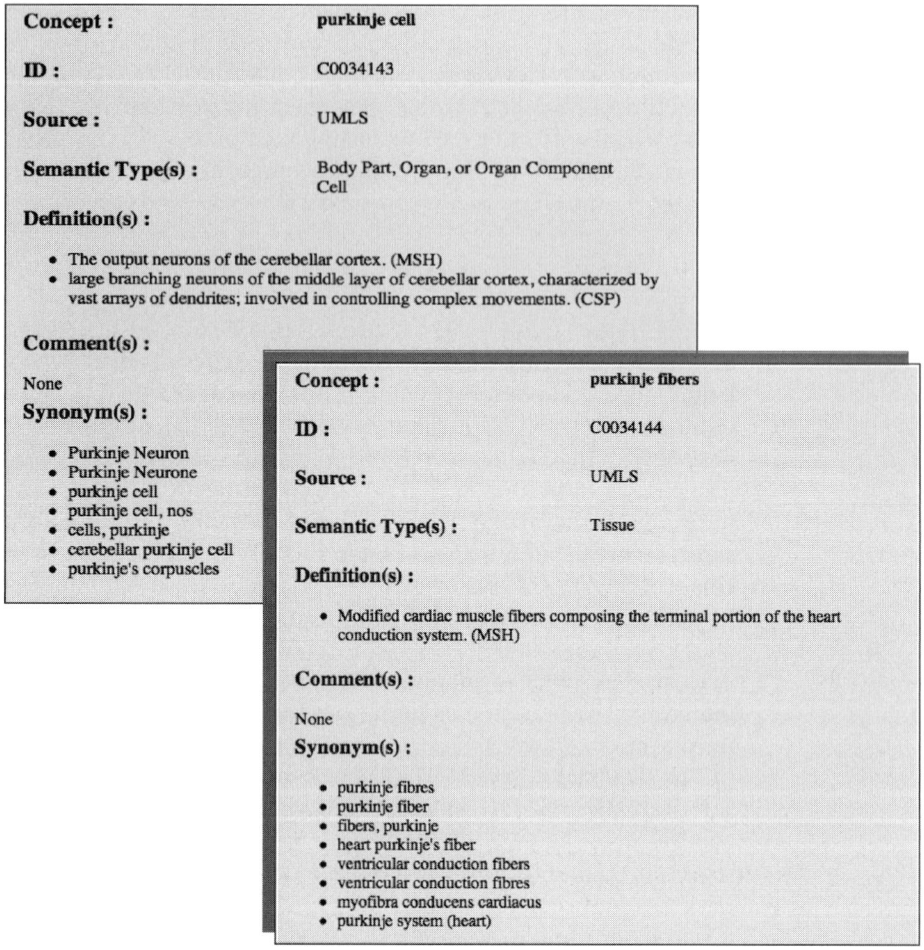

Concept : purkinje cell

ID : C0034143

Source : UMLS

Semantic Type(s) : Body Part, Organ, or Organ Component
Cell

Definition(s) :

- The output neurons of the cerebellar cortex. (MSH)
- large branching neurons of the middle layer of cerebellar cortex, characterized by vast arrays of dendrites; involved in controlling complex movements. (CSP)

Comment(s) :

None

Synonym(s) :

- Purkinje Neuron
- Purkinje Neurone
- purkinje cell
- purkinje cell, nos
- cells, purkinje
- cerebellar purkinje cell
- purkinje's corpuscles

Concept : purkinje fibers

ID : C0034144

Source : UMLS

Semantic Type(s) : Tissue

Definition(s) :

- Modified cardiac muscle fibers composing the terminal portion of the heart conduction system. (MSH)

Comment(s) :

None

Synonym(s) :

- purkinje fibres
- purkinje fiber
- fibers, purkinje
- heart purkinje's fiber
- ventricular conduction fibers
- ventricular conduction fibres
- myofibra conducens cardiacus
- purkinje system (heart)

Fig. 8 Concept properties for Purkinje cell and Purkinje fiber in the UMLS. Note that all synonyms of Purkinje cell are mapped to the same concept identifier (C0034143), while variants of Purkinje fiber are mapped to a different ID (C0034144).

for example, the UMLS. This mapping ensures that the data in the CCDB can be reliably compared to other data that map to the same sources.

The importance of ontologies and of annotating with ontologies to modern informatics goes well beyond their use as a controlled vocabulary. Their real power lies in the formalization of entities and relationships in a structure that supports computation. Ontologies that are well structured and logically sound support reasoning when coded in a language, like OWL (Web ontology language), that supports such operations. Thus, reasoning systems can be used to draw inferences like: a cell has a plasma membrane, a neuron is a cell, therefore, a neuron has a plasma membrane. Ontologies have been used in database interoperability systems to provide linkages between concepts in a database that would normally be inferred by a human (Gupta *et al.*, 2001). The recognition that the construction of ontologies provides a first step toward intelligent interrogation of biological data has led to the formation of the National Center for Biomedical Ontologies (http://www.ncbo.org), established as a national center for biomedical computing by the National Institute of Health.

D. Data Input into the CCDB

For any database to be successful, the process of data entry must be made as easy as possible (Tagari *et al.*, 2006). In an ideal system, tight integration would exist between the instruments and software tools used to produce the data and the database itself, so information is automatically entered into the database on generation. Laboratory information management systems (LIMS) are specifically designed to support this type of integration. However, these types of systems tend to be specialized around a limited set of technologies for high-throughput analysis. Experimental laboratories, such as the National Center for Microscopy and Imaging Research (NCMIR) where the CCDB is hosted, utilize many different instruments and tools in customized protocols. Some of the instruments used regularly, such as the Hitachi 3MeV ultra-high-voltage electron microscope in Osaka University (Takaoka *et al.*, 2000), are off site and under independent management. Thus, designing input strategies for the CCDB has proven to be quite a challenge.

The CCDB has recently released the first set of input forms for data upload. The CCDB input forms were built using Gridsphere, an open source project for building secure Web portals (www.gridsphere.org). Because Gridsphere components, called portlets, are built to a common specification, the CCDB input forms may be easily incorporated into any Gridsphere-compliant portal. The current input forms were incorporated into the Telescience Portal (http://.telescience.ucsd.edu), a Gridsphere-based portal developed at NCMIR for supporting the process of electron tomography using remote instruments (Peltier *et al.*, 2003). Telescience provides secure access to high-voltage electron microscopes, computational resources, and tools for electron tomography, all using a single user account. The CCDB accepts the authentication certificate issued by Telescience, so that users logged into Telescience are also authorized to enter data into the CCDB (Fig. 9).

The CCDB input forms provide the user with a data management system for tracking and organizing experimental data during the course of an experiment. Users begin the process by registering a project to the CCDB; a project is defined as a group of related experiments or studies that generally lead to a single publication. Once a project has been registered, experiments, subjects, specimens, and microscopy products may be registered to the same project in a hierarchical fashion. Each of these entities is assigned a unique identifier that can be used to

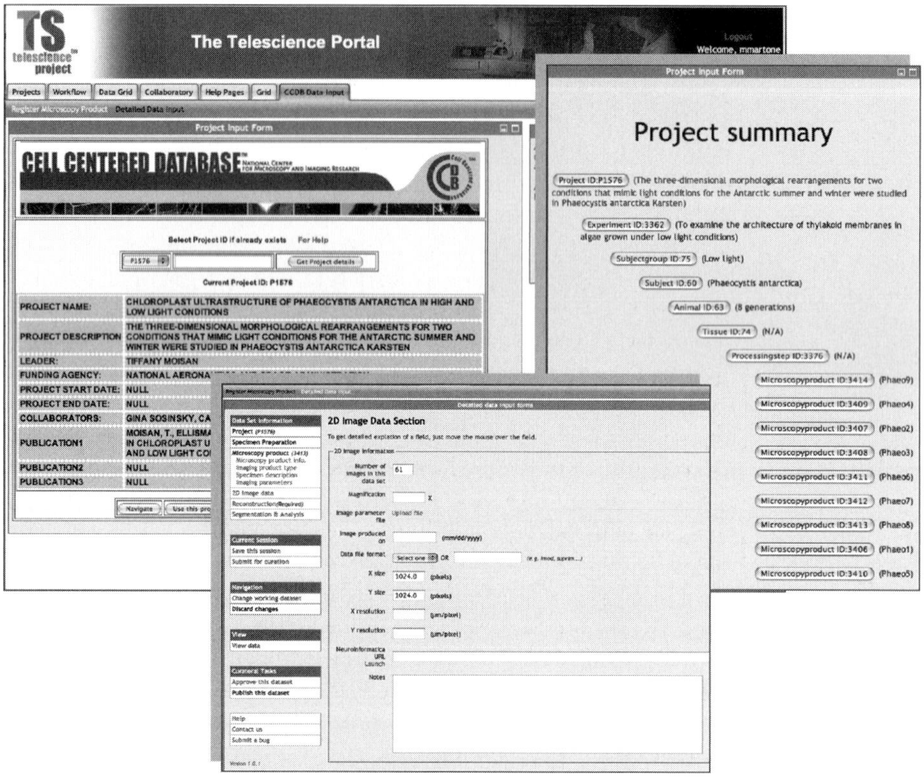

Fig. 9 Data management in CCDB through the Telescience Portal. Users register new projects, experiments, and microscopy products prior to beginning an imaging session using a hierarchical set of forms (right panel). Once a microscopy product is registered, users may add additional information on specimen preparation, imaging parameters, reconstruction, and analysis (lower panel).

tag the same entities in laboratory records, negatives, grid boxes, and so on. The Telescience Portal enforces the integrity of data in the CCDB by ensuring that no one may acquire image data from the microscope until the microscopy product has been registered within its experimental context. Once the images are acquired, users may add anatomical descriptions, microscope parameters, and image set parameters, for example, tilt range and tilt increment that are important for tomography. The CCDB is designed so that all tables do not have to be filled out in the same session. Reconstruction data and segmentation data may be added at a later time, once the images have been processed.

The CCDB actually splits the storage of information about the microscopic images into two tables: microscopy product and image 2D. The microscopy product is meant to describe the raw data collected from the microscope while the image 2D table is meant to store information about the 2D images that were actually used to create 2D, 3D, or 4D reconstructions. In the case of digital imaging, the sets of images may be the same, for example, an optical section series taken from a confocal microscope in which no further processing was performed. In the case of images taken on film, the microscopy product and the image 2D set are not the same; the 2D images are digitized from the negatives. Even in the case of digital images, the original data may be down-sampled or cropped before reconstruction.

The process of data entry into the CCDB is currently manual for many types of data. However, for data acquired and processed through the Telescience Portal, we are creating a set of services that allow communication between CCDB and Telescience resources. These resources may include instruments like our JEOL intermediate-voltage EMs, reconstruction algorithms like TxBr (Lawrence *et al.*, 2006), and tools for segmentation and analysis. Currently, once a user registers a microscopy product, they may acquire a tomographic tilt series through the remote microscopy client. As the images are acquired, they are stored in the SRB while the instrument parameters are input to the CCDB. Similar services will be created so that applications like TxBr can read from and write to the database.

All the data entered into the CCDB through the portal are considered private data. Users may assign group privileges to allow their collaborators to view the data, but they are not displayed in the public CCDB. While the data are private, users may edit the data and delete records as necessary. When data are made public, the CCDB replicates the data, performs the necessary curatorial editing to ensure that the forms are filled out correctly, and changes the security access in Oracle. Contributors still own the copyright to their data, but once the data are made public, contributors are no longer able to edit or delete their data. If corrections are necessary, the CCDB curators will add an erratum or addendum to the record, similar to the procedures used by journals. By utilizing the CCDB for data management, the process of making data available to the public is greatly simplified; there is no need to reformat the data for public display and all necessary data are already in the database.

III. Looking Ahead

Database resources for EM have matured significantly in the past few years, providing venues for online sharing of data and platforms for managing data in the laboratory. Whether the community will adopt these as forums for large-scale data sharing on a par with the genomics and structural biology communities remains to be seen, but there are signs that the EM community is taking steps to ensure that its unique and important data are not kept in a lockbox (Tagari *et al.*, 2006). We believe that as data management tools are incorporated into the daily workflow of the modern EM laboratory, sharing these data with broader communities will be facilitated. Toward this end, we make the CCDB freely available to anyone that would like to use it: for data management, collaborative science among smaller communities, or public sharing for the community at large. Users may employ the CCDB infrastructure for secure storage of data through "MyCCDB", due to be launched in 2006. Or, groups are free to establish instances of the CCDB at their own sites. For example, we have been working with the tomography groups of Dr. Brad Marsh in Australia and Dr. Abraham Koster in the Netherlands to establish the CCDB in these venues, both for data management and data dissemination. From these interactions, it is likely that great improvements will be made to the CCDB, and a set of standards established for sharing cell level tomography data will emerge.

Many countries have made it a priority to modernize the infrastructure for science to take advantage of the power of networks, distributed computers, and remote instruments to advance discovery. Projects like the BIRN in the United States, the Virtual Laboratory project from the Netherlands, and the myGrid project from the United Kingdom (Grethe *et al.*, 2005) all seek to provide the substrate for new modes of collaboration and more powerful data analysis. Databases like the CCDB form one of the core components of these new e-science platforms. As their usage is embraced by microscopists, EM will firmly establish itself as a twenty-first century technology.

Acknowledgments

Supported by NIH grants NIDA DA016602 (CCDB), NCRR RR04050, and RR08605. The Bioinformatics Research Network is supported by NIH grants RR08605–08S1 (BIRN-CC) and RR021760 (Mouse BIRN).

References

Ascoli, G. A. (2006). Mobilizing the base of neuroscience data: The case of neuronal morphologies. *Nat. Rev. Neurosci.* **7,** 318–324.

Bug, W., and Nissanov, J. (2003). A guide to building image-centric databases. *Neuroinformatics* **1,** 359–377.

Butler, D. (2005). Electronic notebooks: A new leaf. *Nature* **436,** 20–21.

Carazo, J. M., Stelzer, E. H., Engel, A., Fita, I., Henn, C., Machtynger, J., McNeil, P., Shotton, D. M., Chagoyen, M., de Alarcon, P. A., Fritsch, R., Heymann, J. B., *et al.* (1999). Organising multidimensional biological image information: The BioImage Database. *Nucleic Acids Res.* **27,** 280–283.

Catton, C., and Shotton, D. (2003). Illustration: Database pictures tell a true story. *Nature* **422,** 374.

Chen, S. C., and Murphy, R. F. (2006). A graphical model approach to automated classification of protein subcellular location patterns in multi-cell images. *BMC Bioinformatics* **7,** 90.

Coggan, J. S., Bartol, T. M., Esquenazi, E., Stiles, J. R., Lamont, S., Martone, M. E., Berg, D. K., Ellisman, M. H., and Sejnowski, T. J. (2005). Evidence for ectopic neurotransmission at a neuronal synapse. *Science* **309,** 446–451.

Fellmann, D., Pulokas, J., Milligan, R. A., Carragher, B., and Potter, C. S. (2002). A relational database for cryoEM: Experience at one year and 50000 images. *J. Struct. Biol.* **137,** 273–282.

Giepmans, B. N., Adams, S. R., Ellisman, M. H., and Tsien, R. Y. (2006). The fluorescent toolbox for assessing protein location and function. *Science* **312,** 217–224.

Gkoutos, G. V., Green, E. C., Mallon, A. M., Hancock, J. M., and Davidson, D. (2005). Using ontologies to describe mouse phenotypes. *Genome Biol.* **6,** R8.

Goldberg, I. G., Allan, C., Burel, J. M., Creager, D., Falconi, A., Hochheiser, H., Johnston, J., Mellen, J., Sorger, P. K., and Swedlow, J. R. (2005). The open microscopy environment (OME) data model and XML file: Open tools for informatics and quantitative analysis in biological imaging. *Genome Biol.* **6,** R47.

Grethe, J. S., Baru, C., Gupta, A., James, M., Ludaescher, B., Martone, M. E., Papadopoulos, P. M., Peltier, S. T., Rajasekar, A., Santini, S., Zaslavsky, I. N., and Ellisman, M. H. (2005). Biomedical informatics research network: Building a national collaboratory to hasten the derivation of new understanding and treatment of disease. *Stud. Health Technol. Inform.* **112,** 100–109.

Gupta, A., Ludäscher, B., and Martone, M. E. (2001). An extensible model-based mediator system with domain maps. *In* "17th International Conference on Data Engineering (ICDE), Demonstration Track." Heidelberg, Germany.

Henrick, K., Newman, R., Tagari, M., and Chagoyen, M. (2003). EMDep: A web-based system for the deposition and validation of high-resolution electron microscopy macromolecular structural information. *J. Struct. Biol.* **144,** 228–237.

Heymann, J. B., Chagoyen, M., and Belnap, D. M. (2005). Common conventions for interchange and archiving of three-dimensional electron microscopy information in structural biology. *J. Struct. Biol.* **151,** 196–207.

Kihlen, M. (2005). Electronic lab notebooks—do they work in reality? *Drug Discov. Today* **10,** 1205–1207.

Kotter, R. (2001). Neuroscience databases: Tools for exploring brain structure-function relationships. *Philos. Trans. R. Soc. Lond. B Biol. Sci.* **356,** 1111–1120.

Lawrence, A., Bouwer, J. C., Perkins, G., and Ellisman, M. H. (2006). Transform-based backprojection for volume reconstruction of large format electron microscope tilt series. *J. Struct. Biol.* **154,** 144–167.

Lucic, V., Forster, F., and Baumeister, W. (2005). Structural studies by electron tomography: From cells to molecules. *Annu. Rev. Biochem.* **74,** 833–865.

Ludtke, S. J., Nason, L., Tu, H., Peng, L., and Chiu, W. (2003). Object oriented database and electronic notebook for transmission electron microscopy. *Microsc. Microanal.* **9,** 556–565.

Marsh, B. J., Volkmann, N., McIntosh, J. R., and Howell, K. E. (2004). Direct continuities between cisternae at different levels of the Golgi complex in glucose-stimulated mouse islet beta cells. *Proc. Natl. Acad. Sci. USA* **101,** 5565–5570.

Martone, M. E., Gupta, A., and Ellisman, M. H. (2004). E-neuroscience: Challenges and triumphs in integrating distributed data from molecules to brains. *Nat. Neurosci.* **7,** 467–472.

Martone, M. E., Gupta, A., Wong, M., Qian, X., Sosinsky, G., Ludäscher, B., Ellisman, M. H. (2002). A cell-centered database for electron tomographic data. *J. Struct. Biol.* **138,** 145–155.

Martone, M. E., Zhang, S., Gupta, A., Qian, X., He, H., Price, D. L., Wong, M., Santini, S., and Ellisman, M. H. (2003). The cell-centered database: A database for multiscale structural and protein localization data from light and electron microscopy. *Neuroinformatics* **1,** 379–395.

Moisan, T., Ellisman, M. H., Buitenhuys, C. W., Sosinsky, G. E. (2006). Differences in chloroplast ultrastructure of *Phaeocystis antarctica* in high and low light conditions. *Mar. Biol.* **149**(6), 1281–1290.

Mukherjea, S. (2005). Information retrieval and knowledge discovery utilising a biomedical Semantic Web. *Brief Bioinform.* **6,** 252–262.

Murphy, R. F. (2005). Location proteomics: A systems approach to subcellular location. *Biochem. Soc. Trans.* **33,** 535–538.

Neerincx, P. B., and Leunissen, J. A. (2005). Evolution of web services in bioinformatics. *Brief Bioinform.* **6**(2), 178–188.

Neumann, E. (2005). A life science Semantic Web: Are we there yet? *Sci. STKE* **2005,** pe22.

Paxinos, G., and Franklin, K. B. J. (2000). "The Mouse Brain in Stereotaxic Coordinates." Academic Press, San Diego.

Peltier, S. T., Lin, A. W., Lee, D., Mock, S., Lamont, S., Molina, T., Wong, M., Dai, L., Martone, M. E., and Ellisman, M. H. (2003). The Telescience Portal for advanced tomography applications. *J. Parallel Distrib. Comput.* **63,** 539–550.

Perkins, G. A., Ellisman, M. H., and Fox, D. A. (2003). Three-dimensional analysis of mouse rod and cone mitochondrial cristae architecture: Bioenergetic and functional implications. *Mol. Vis.* **11,** 60–73.

Shoop, R. D., Esquenazi, E., Yamada, N., Ellisman, M. H., and Berg, D. K. (2002). Ultrastructure of a somatic spine mat for nicotinic signaling in neurons. *J. Neurosci.* **22,** 748–756.

Sosinsky, G. E., Deerinck, T. J., Greco, R., Buitenhuys, C. H., Bartol, T. M., and Ellisman, M. H. (2005). Development of a model for microphysiological simulations: Small nodes of ranvier from peripheral nerves of mice reconstructed by electron tomography. *Neuroinformatics* **3,** 133–162.

Swedlow, J. R., Goldberg, I., Brauner, E., and Sorger, P. K. (2003). Informatics and quantitative analysis in biological imaging. *Science* **300,** 100–102.

Tagari, M., Newman, R., Chagoyen, M., Carazo, J. M., and Henrick, K. (2002). New electron microscopy database and deposition system. *Trends Biochem. Sci.* **27,** 589.

Tagari, M., Tate, J., Swaminathan, G. J., Newman, R., Naim, A., Vranken, W., Kapopoulou, A., Hussain, A., Fillon, J., Henrick, K., and Velankar, S. (2006). E-MSD: Improving data deposition and structure quality. *Nucleic Acids Res.* **34,** D287–D290.

Takaoka, A., Yoshida, K., Mori, H., Hayashi, S., Young, S. J., and Ellisman, M. H. (2000). International telemicroscopy with a 3 MV ultrahigh voltage electron microscope. *Ultramicroscopy* **83,** 93–101.

Van Horn, J. D., Grafton, S. T., Rockmore, D., and Gazzaniga, M. S. (2004). Sharing neuroimaging studies of human cognition. *Nat. Neurosci.* **7,** 473–481.

Zaslavsky, I., He, H., Tran, J., Martone, M. E., and Gupta, A. (2004). Integrating brain data spatially: Spatial data infrastructure and atlas environment for online federation and analysis of brain images, Second International Workshop on Biological Data Management (BIDM'04) in conjunction with 15th international workshop on data base and expert systems applications (DEXA' 04) pp. 389–393, Zaragosa, Spain, Aug/Sept 2004.

INDEX

T

VOLUMES IN SERIES

Founding Series Editor
DAVID M. PRESCOTT

Advisory Board Chairman
KEITH R. PORTER

Series Editor
LESLIE WILSON

Volume 29 (1989)
Fluorescence Microscopy of Living Cells in Culture, Part A: Fluorescent Analogs, Labeling Cells, and Basic Microscopy
Edited by Yu-Li Wang and D. Lansing Taylor

Volume 30 (1989)
Fluorescence Microscopy of Living Cells in Culture, Part B: Quantitative Fluorescence Microscopy—Imaging and Spectroscopy
Edited by D. Lansing Taylor and Yu-Li Wang

Volume 31 (1989)
Vesicular Transport, Part A
Edited by Alan M. Tartakoff

Volume 32 (1989)
Vesicular Transport, Part B
Edited by Alan M. Tartakoff

Volume 33 (1990)
Flow Cytometry
Edited by Zbigniew Darzynkiewicz and Harry A. Crissman

Volume 34 (1991)
Vectorial Transport of Proteins into and across Membranes
Edited by Alan M. Tartakoff

Selected from Volumes 31, 32, and 34 (1991)
Laboratory Methods for Vesicular and Vectorial Transport
Edited by Alan M. Tartakoff

Volume 35 (1991)
Functional Organization of the Nucleus: A Laboratory Guide
Edited by Barbara A. Hamkalo and Sarah C. R. Elgin

Volume 36 (1991)
***Xenopus laevis:* Practical Uses in Cell and Molecular Biology**
Edited by Brian K. Kay and H. Benjamin Peng

Series Editors
LESLIE WILSON AND PAUL MATSUDAIRA

Volume 37 (1993)
Antibodies in Cell Biology
Edited by David J. Asai

Volume 38 (1993)
Cell Biological Applications of Confocal Microscopy
Edited by Brian Matsumoto

Volume 39 (1993)
Motility Assays for Motor Proteins
Edited by Jonathan M. Scholey

Volume 40 (1994)
A Practical Guide to the Study of Calcium in Living Cells
Edited by Richard Nuccitelli

Volume 41 (1994)
Flow Cytometry, Second Edition, Part A
Edited by Zbigniew Darzynkiewicz, J. Paul Robinson, and Harry A. Crissman

Volume 42 (1994)
Flow Cytometry, Second Edition, Part B
Edited by Zbigniew Darzynkiewicz, J. Paul Robinson, and Harry A. Crissman

Volume 43 (1994)
Protein Expression in Animal Cells
Edited by Michael G. Roth

Volume 44 (1994)
***Drosophila melanogaster:* Practical Uses in Cell and Molecular Biology**
Edited by Lawrence S. B. Goldstein and Eric A. Fyrberg

Volume 45 (1994)
Microbes as Tools for Cell Biology
Edited by David G. Russell

Volume 46 (1995) (in preparation)
Cell Death
Edited by Lawrence M. Schwartz and Barbara A. Osborne

Volume 47 (1995)
Cilia and Flagella
Edited by William Dentler and George Witman

Volume 48 (1995)
***Caenorhabditis elegans:* Modern Biological Analysis of an Organism**
Edited by Henry F. Epstein and Diane C. Shakes

Volume 49 (1995)
Methods in Plant Cell Biology, Part A
Edited by David W. Galbraith, Hans J. Bohnert, and Don P. Bourque

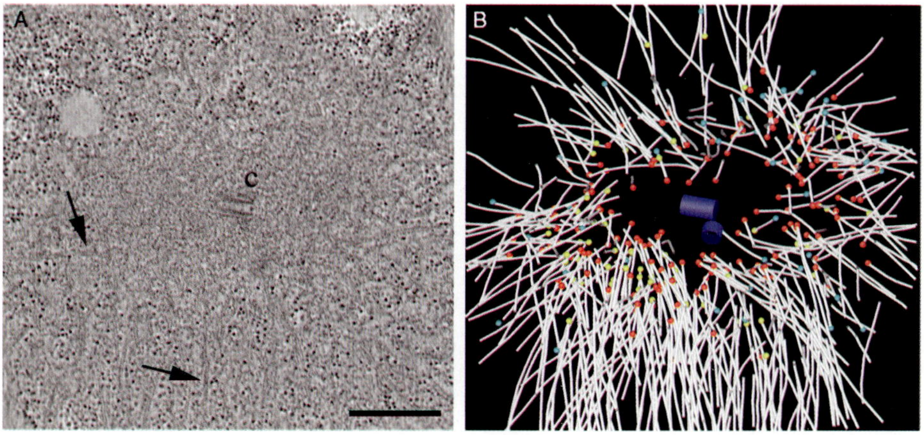

Chapter 4, Fig. 6

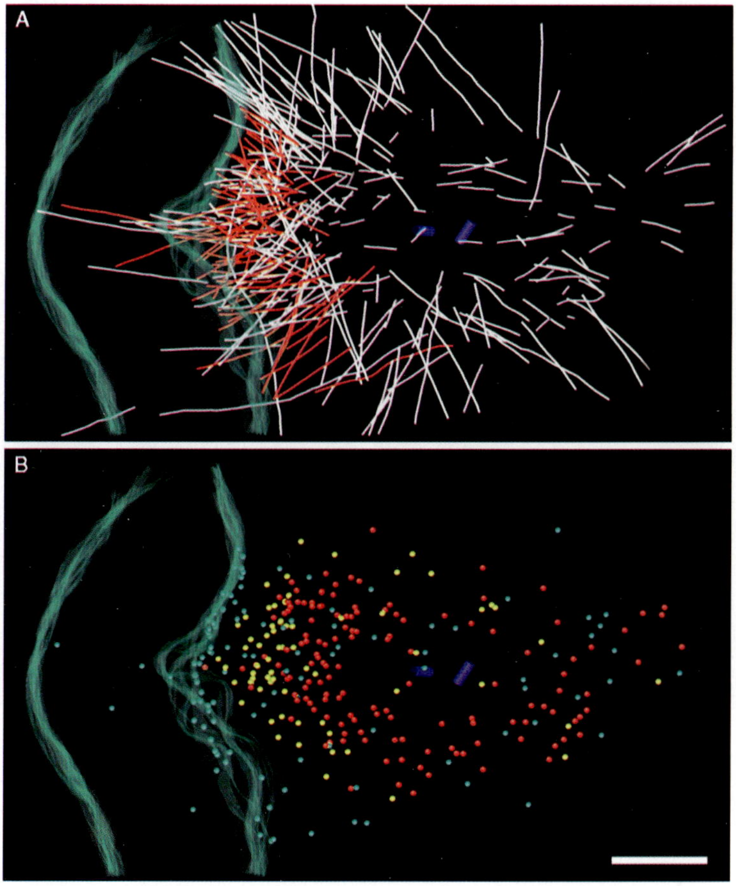

Chapter 4, Fig. 7

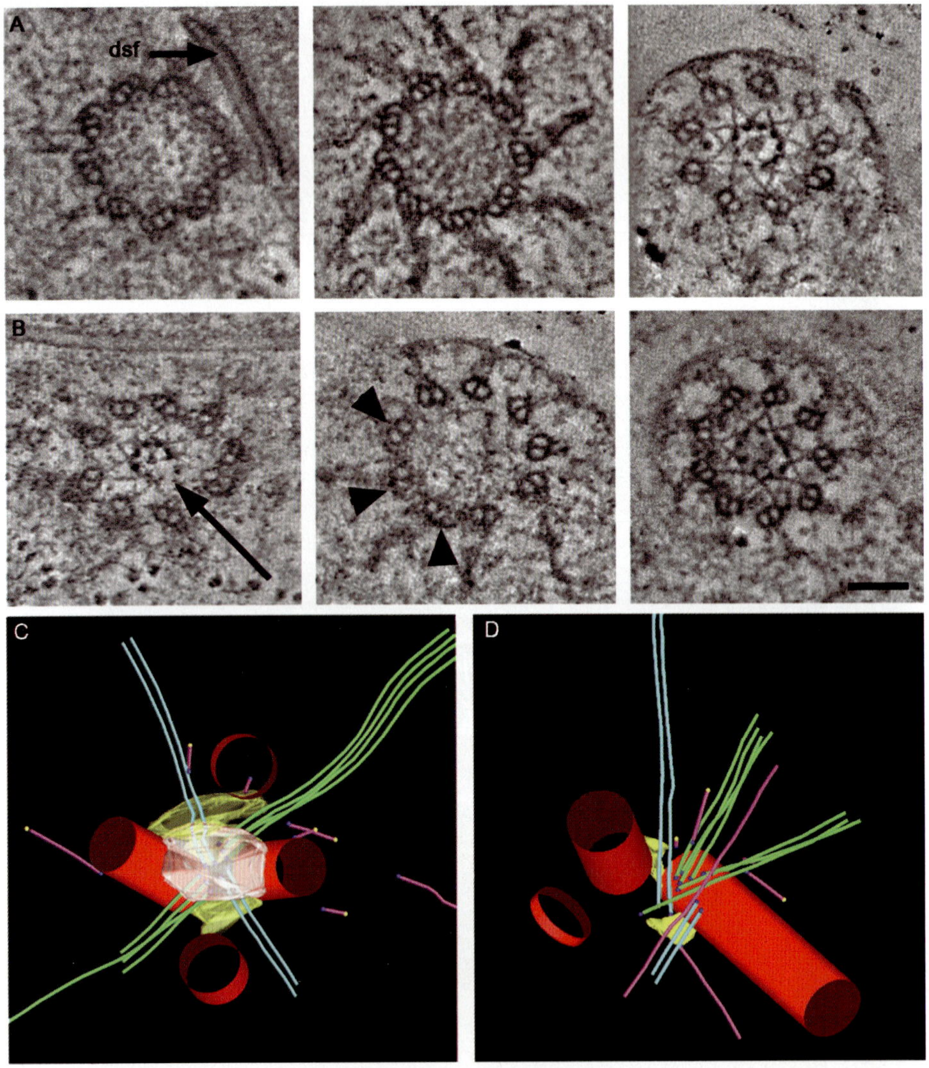

Chapter 5, Fig. 6

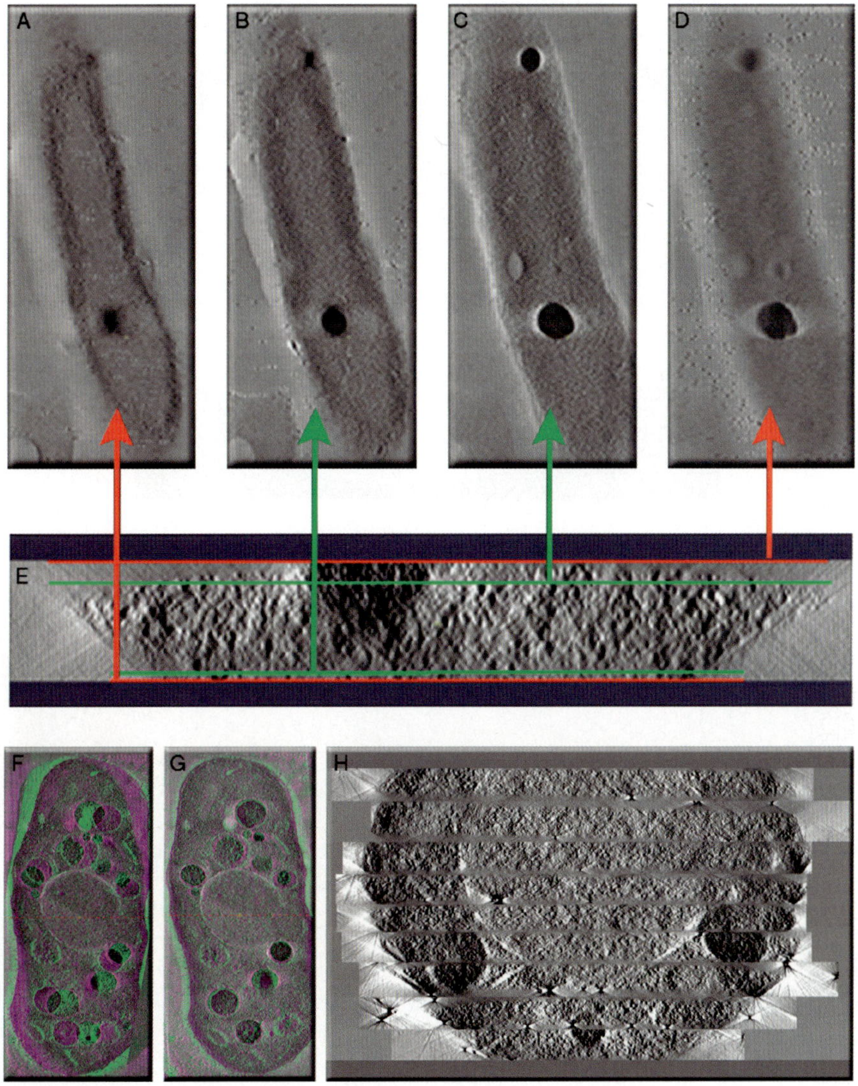

Chapter 6, Fig. 4

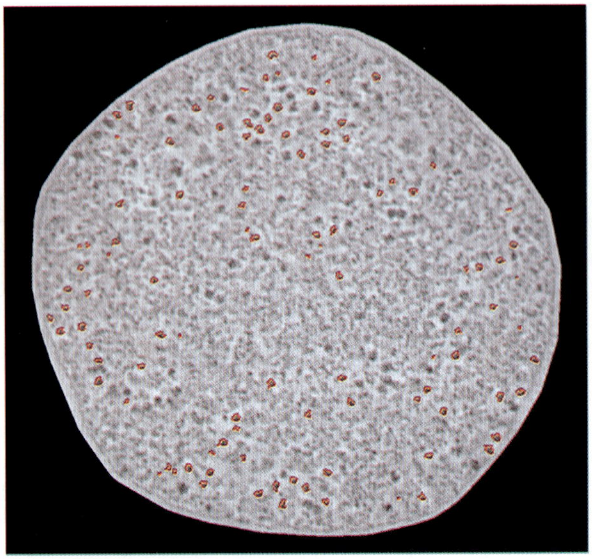

Chapter 7, Fig. 7

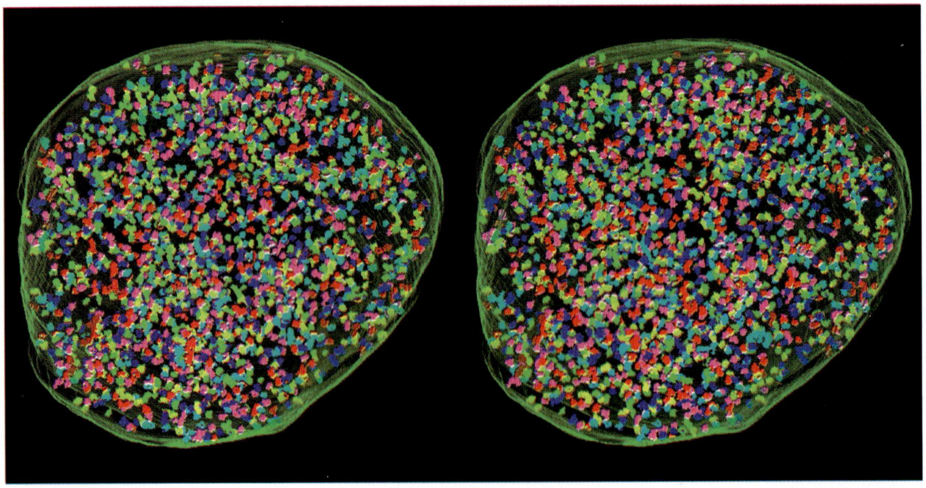

Chapter 7, Fig. 8

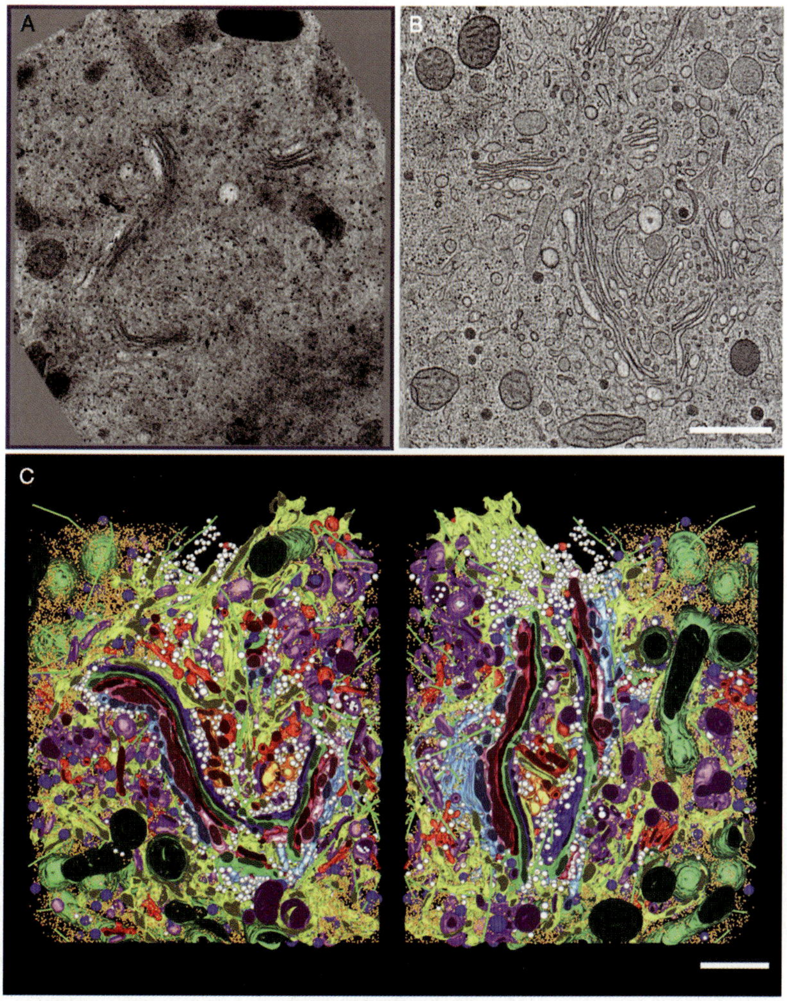

Chapter 8, Fig. 1

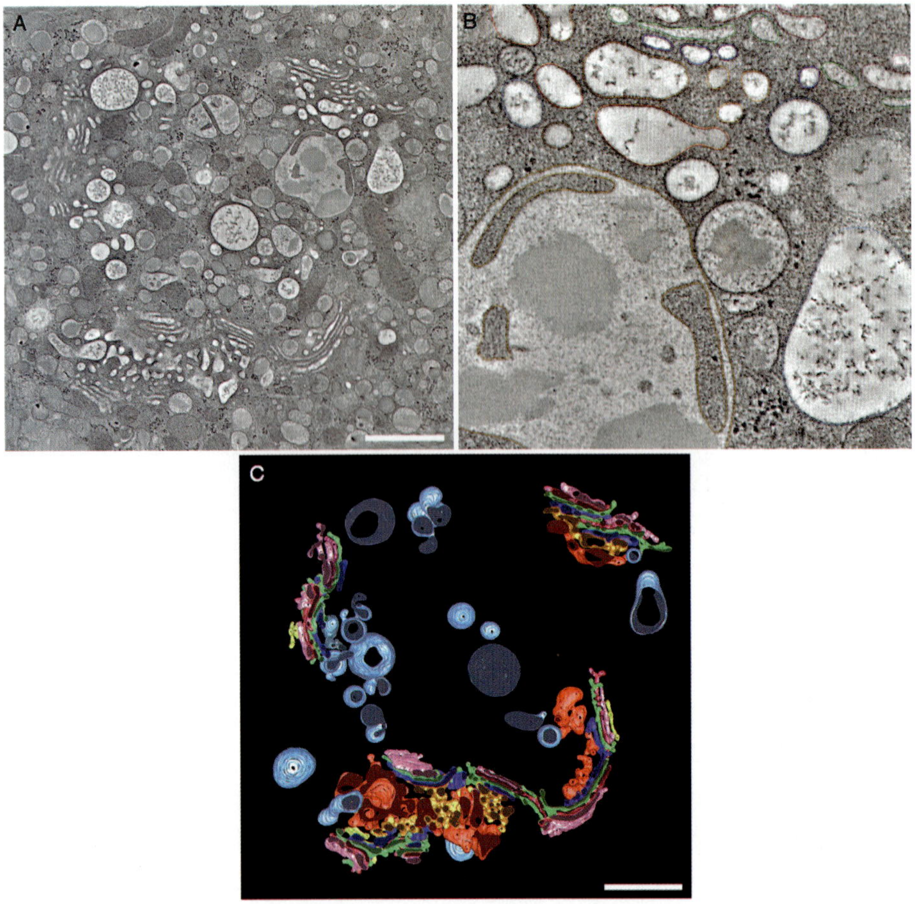

Chapter 8, Fig. 5

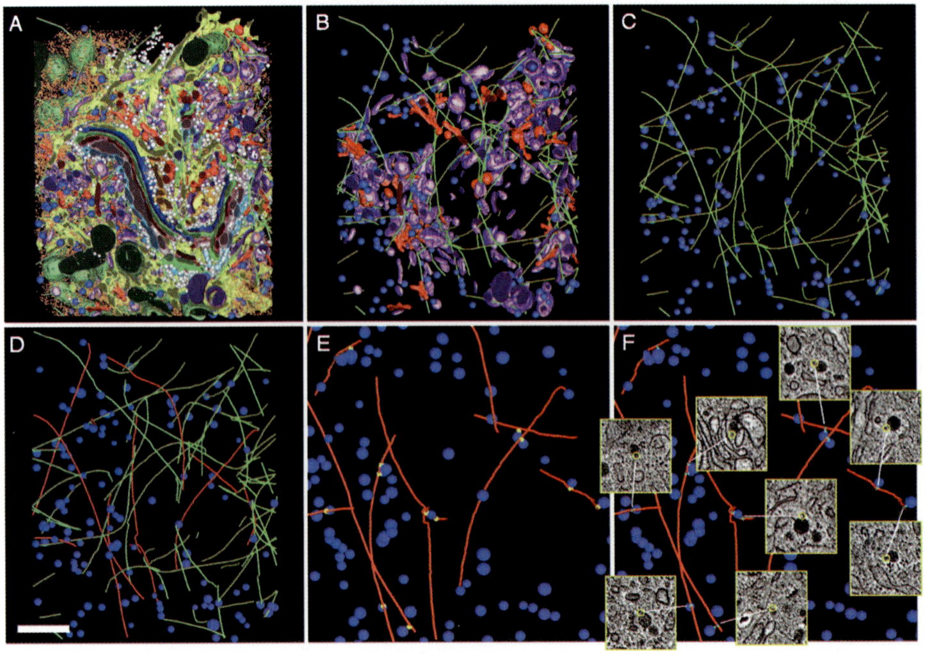

Chapter 8, Fig. 6

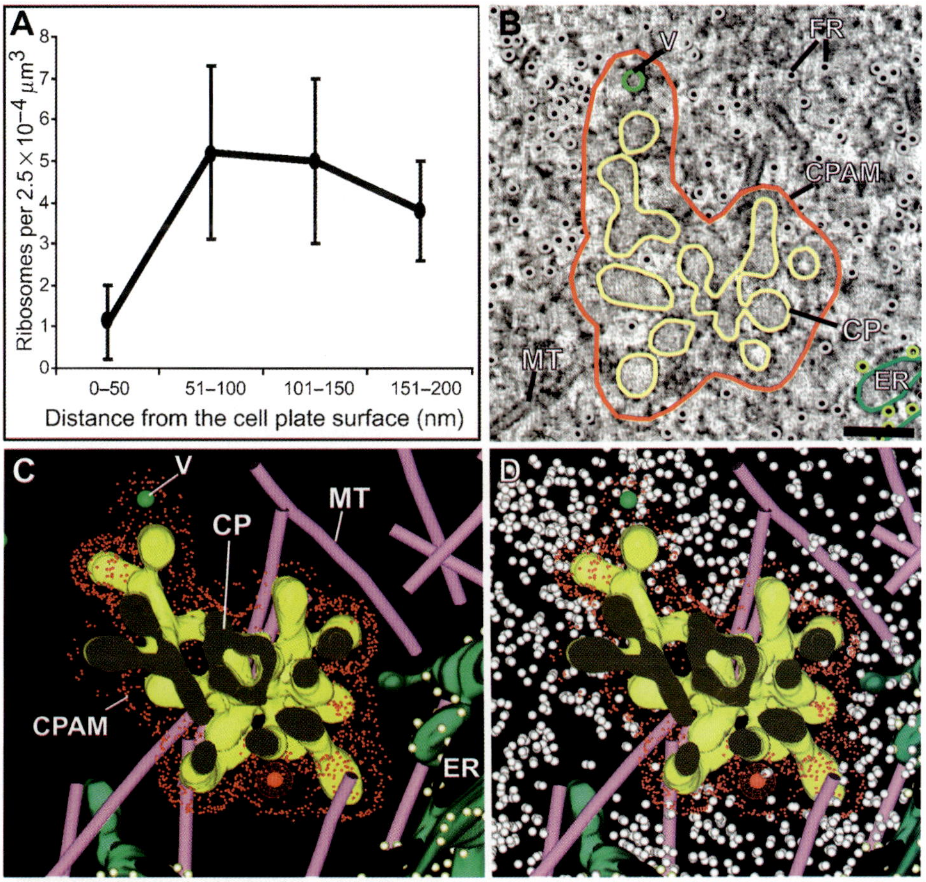

Chapter 9, Fig. 4

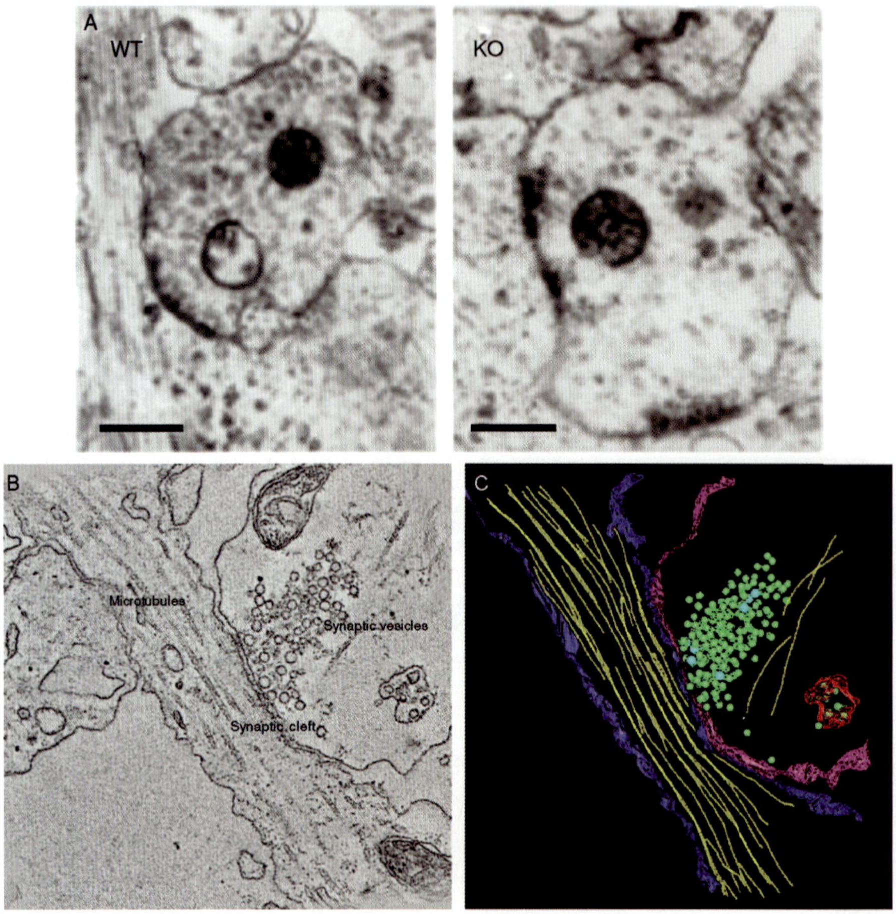

Chapter 10, Fig. 4

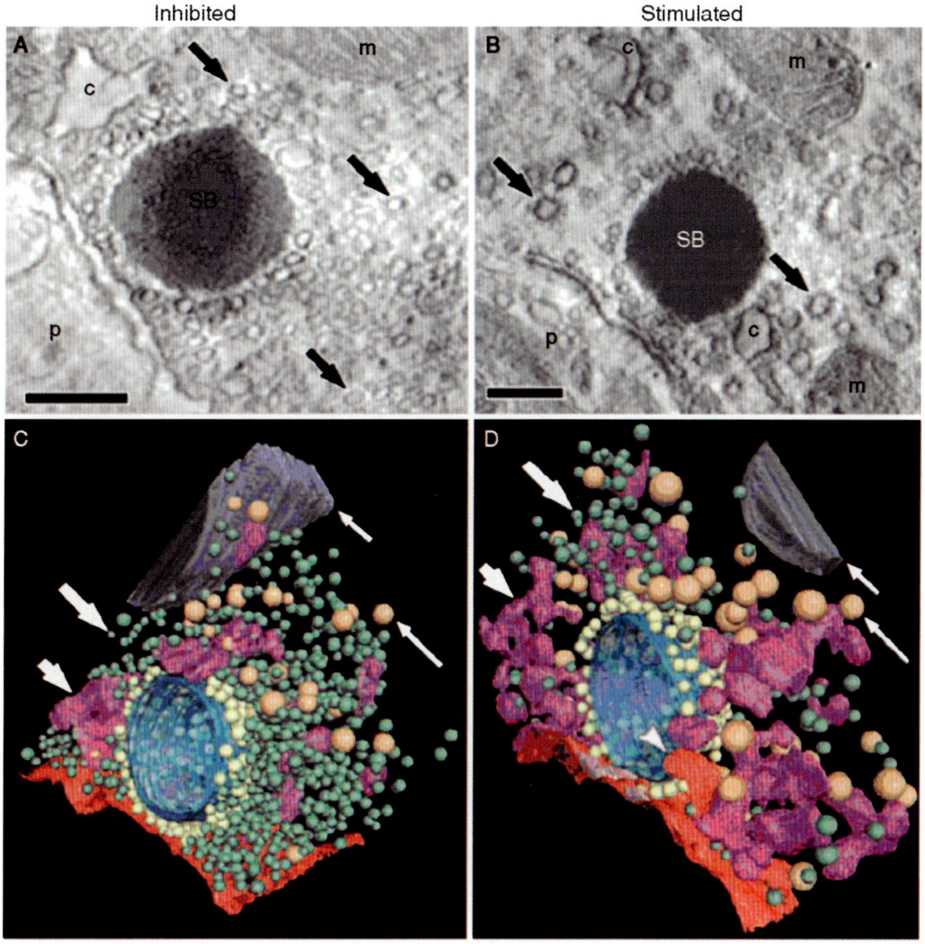

Chapter 10, Fig. 5

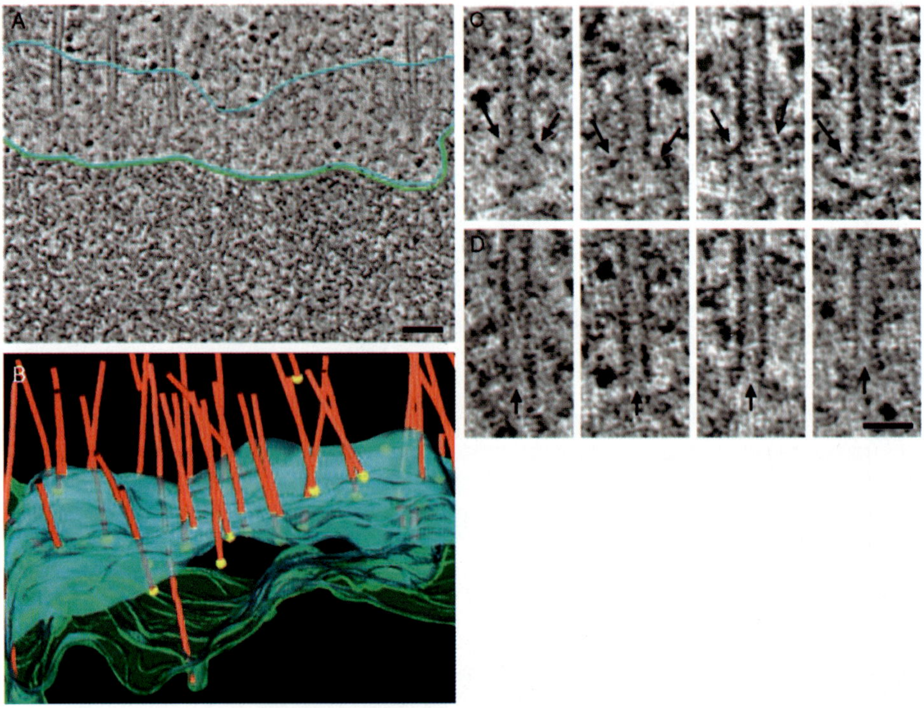

Chapter 11, Fig. 4

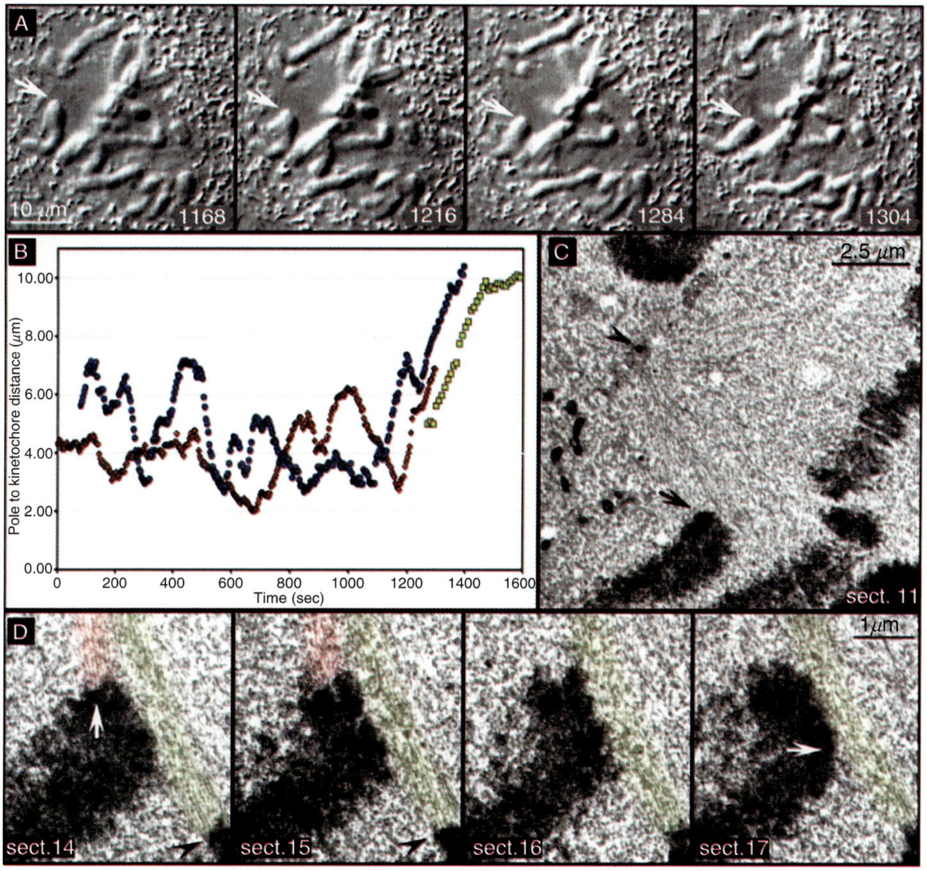

Chapter 11, Fig. 7

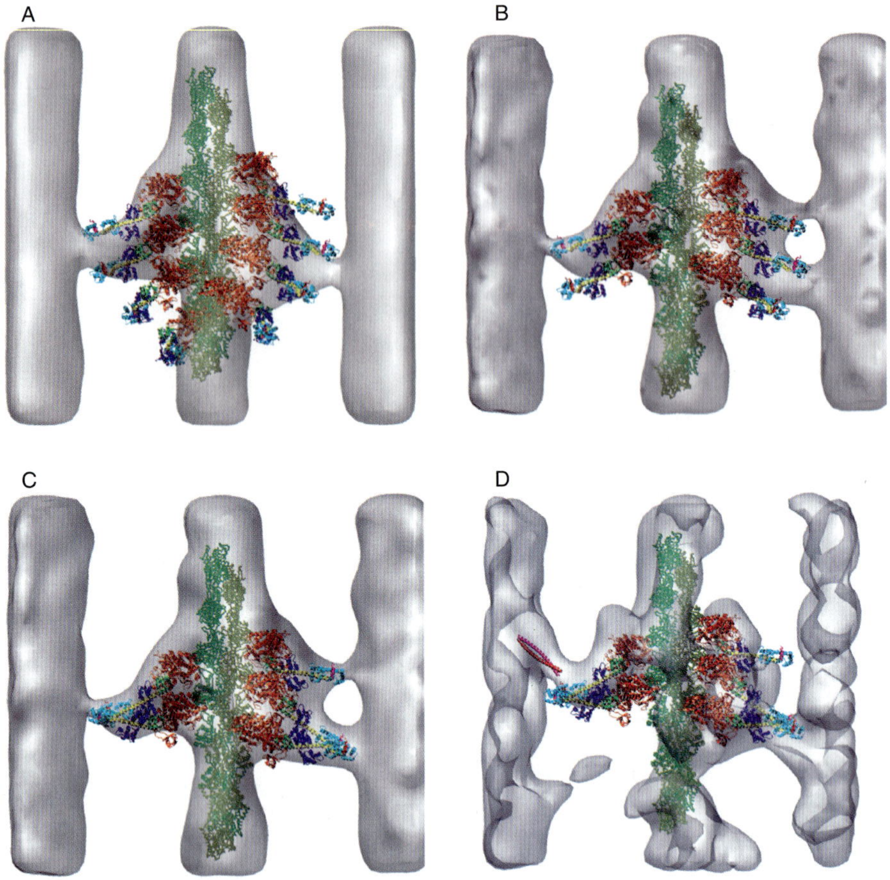

Chapter 13, Fig. 10

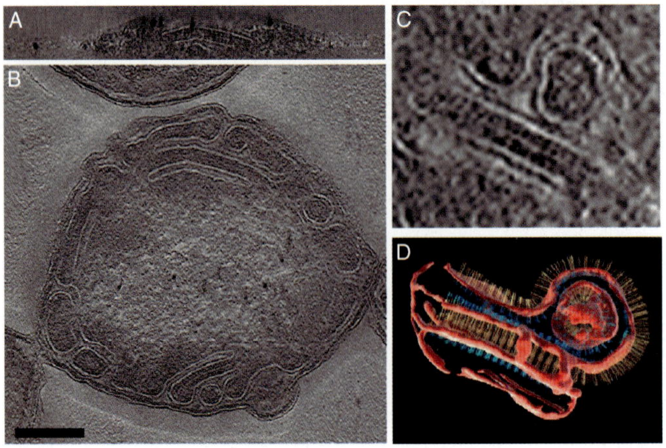

Chapter 14, Fig. 2

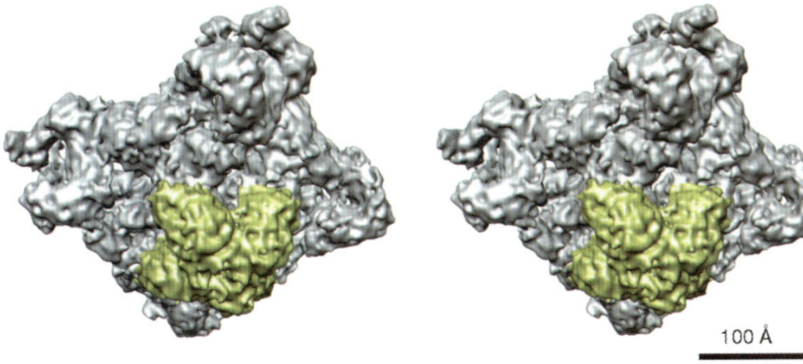

Chapter 16, Fig. 6

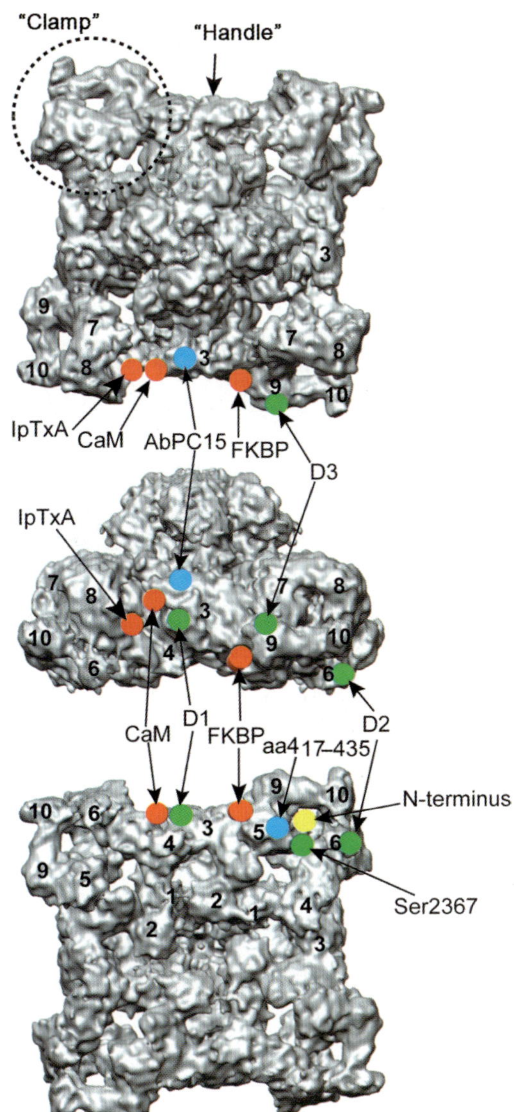

Chapter 16, Fig. 9

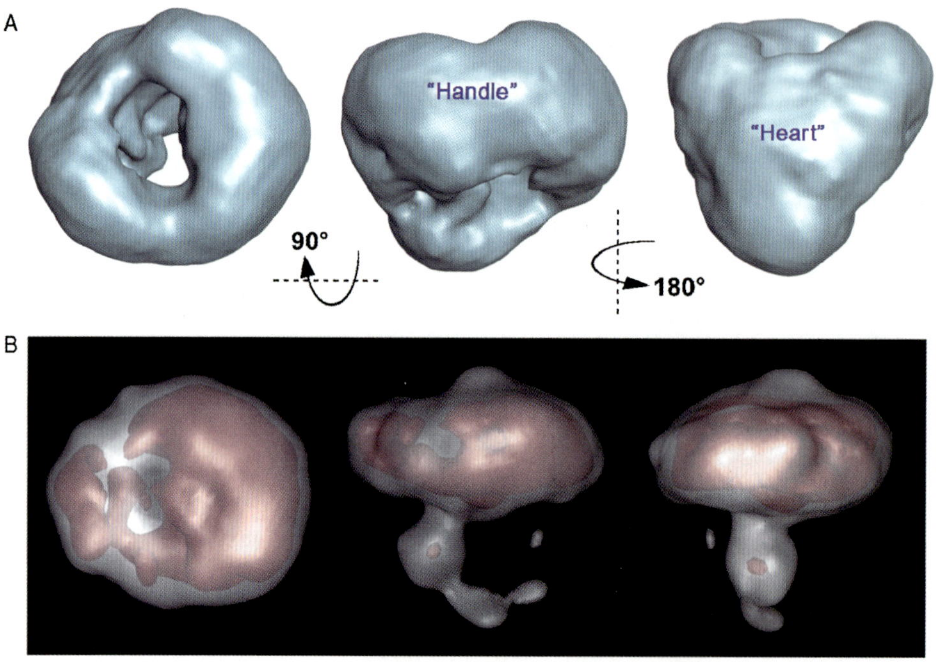

Chapter 16, Fig. 12

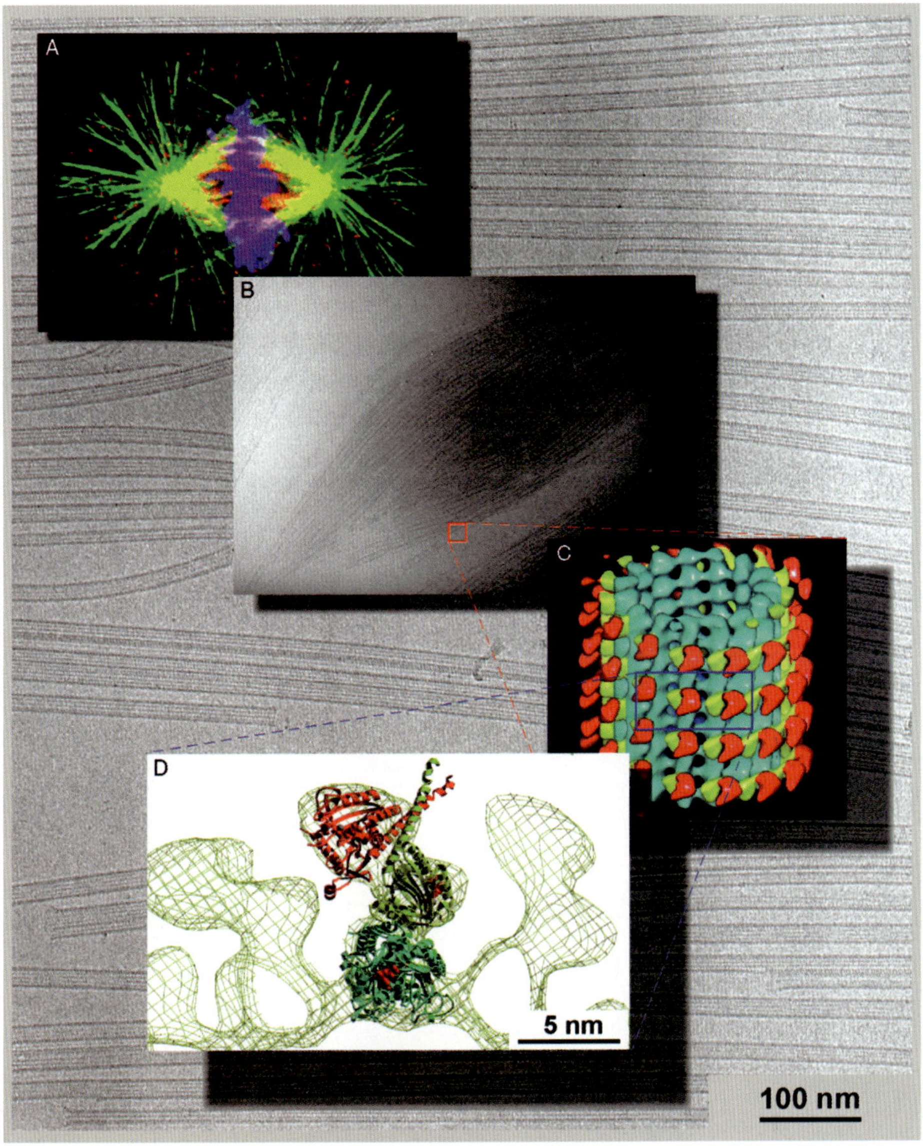

Chapter 17, Fig. 1

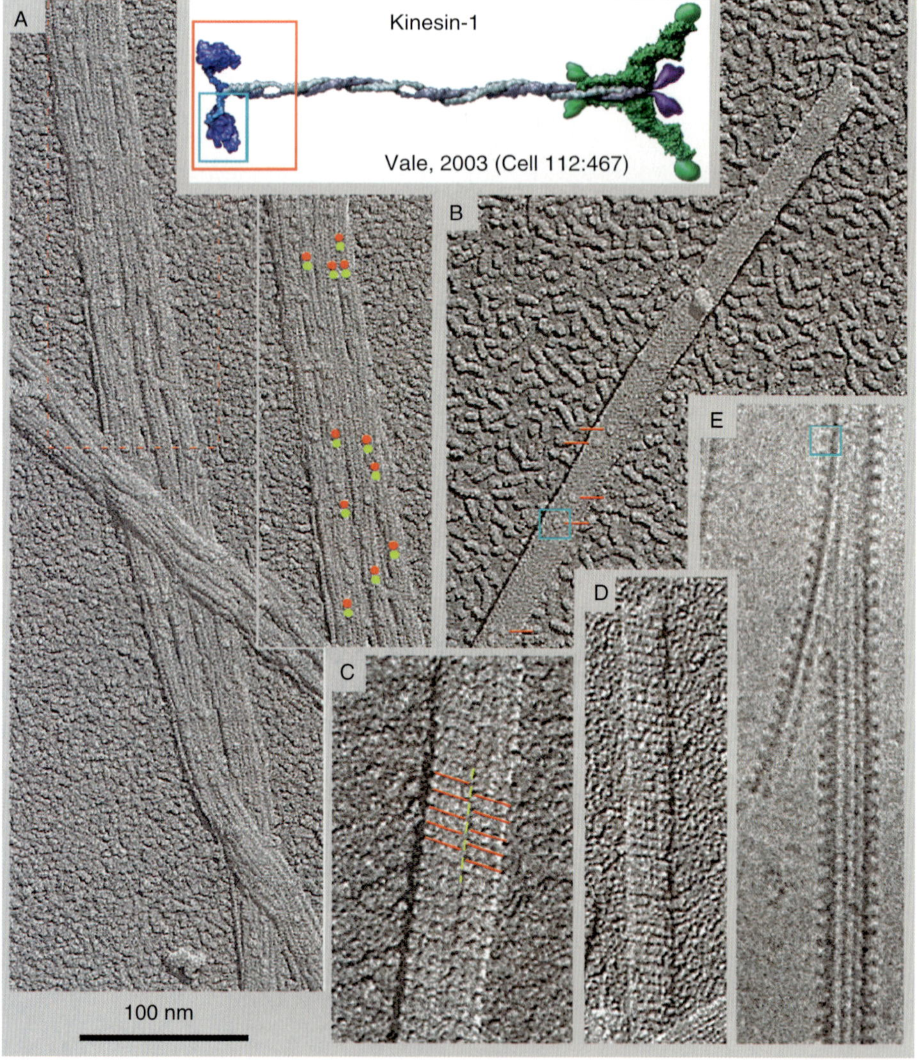

Chapter 17, Fig. 3

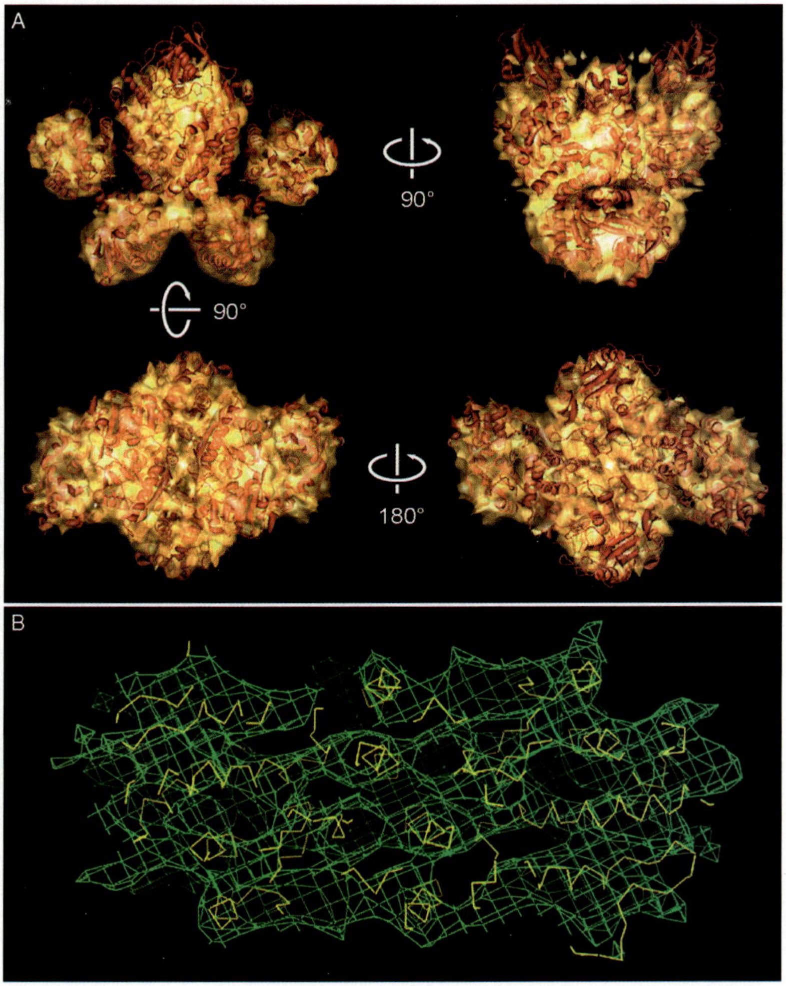

Chapter 18, Fig. 4

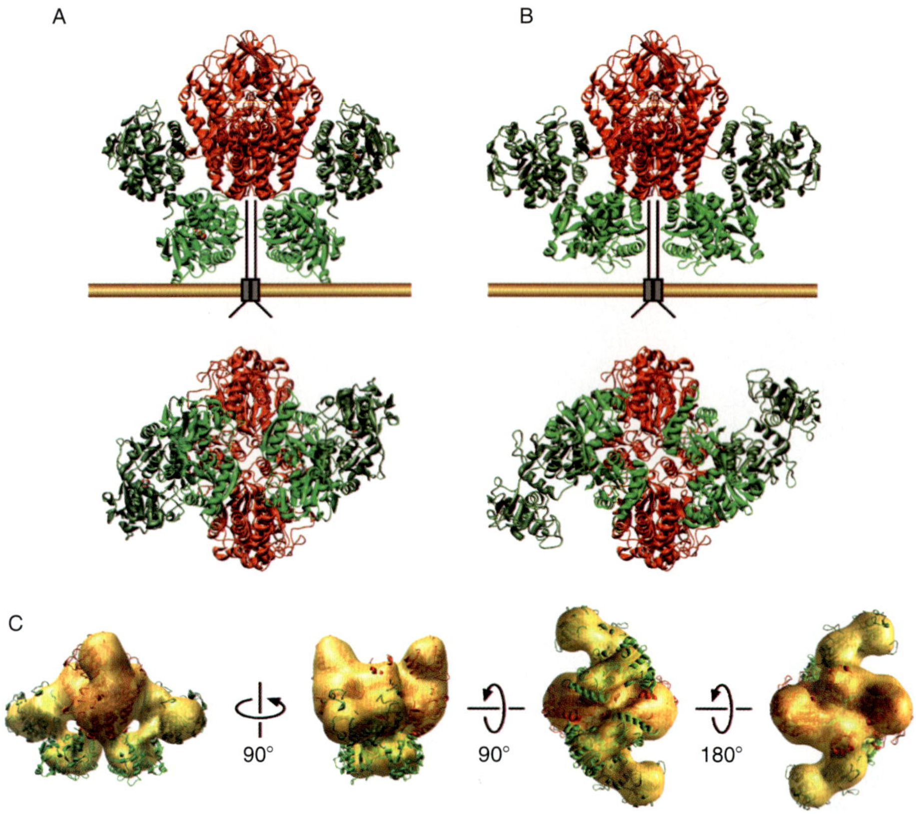

Chapter 18, Fig. 5

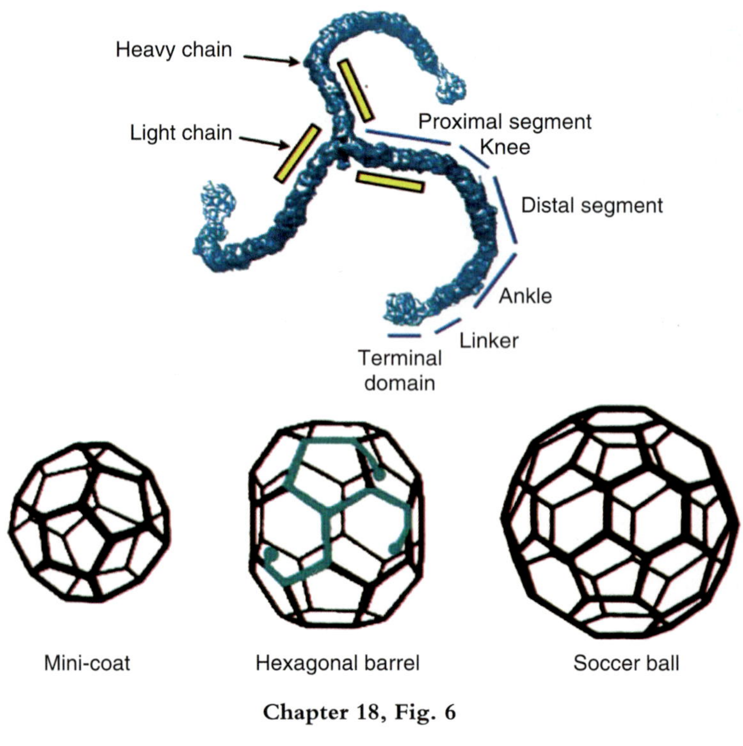

Chapter 18, Fig. 6

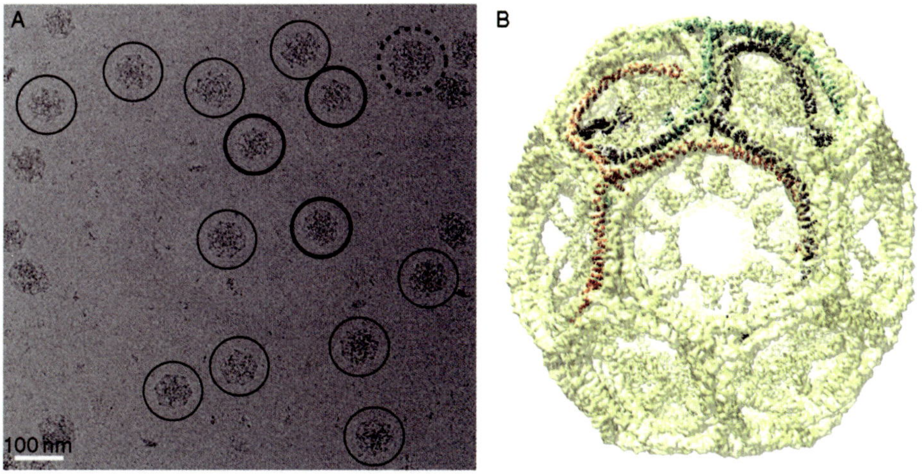

Chapter 18, Fig. 7

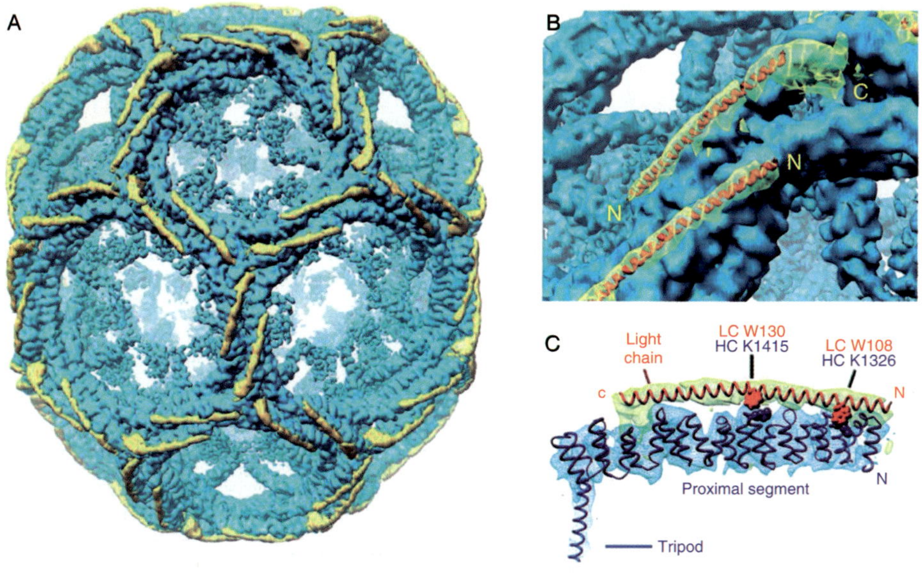

Chapter 18, Fig. 8

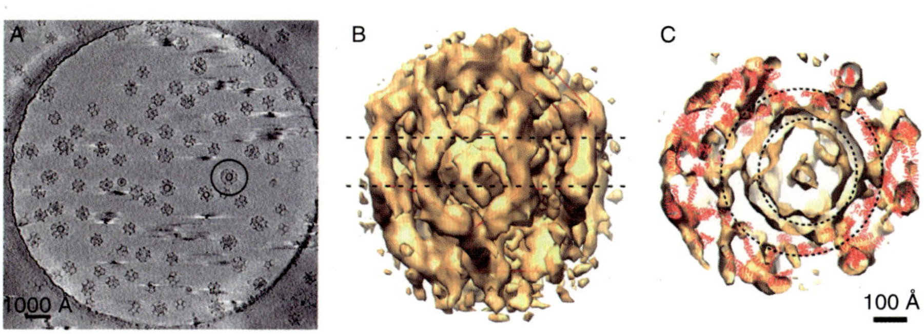

Chapter 18, Fig. 10

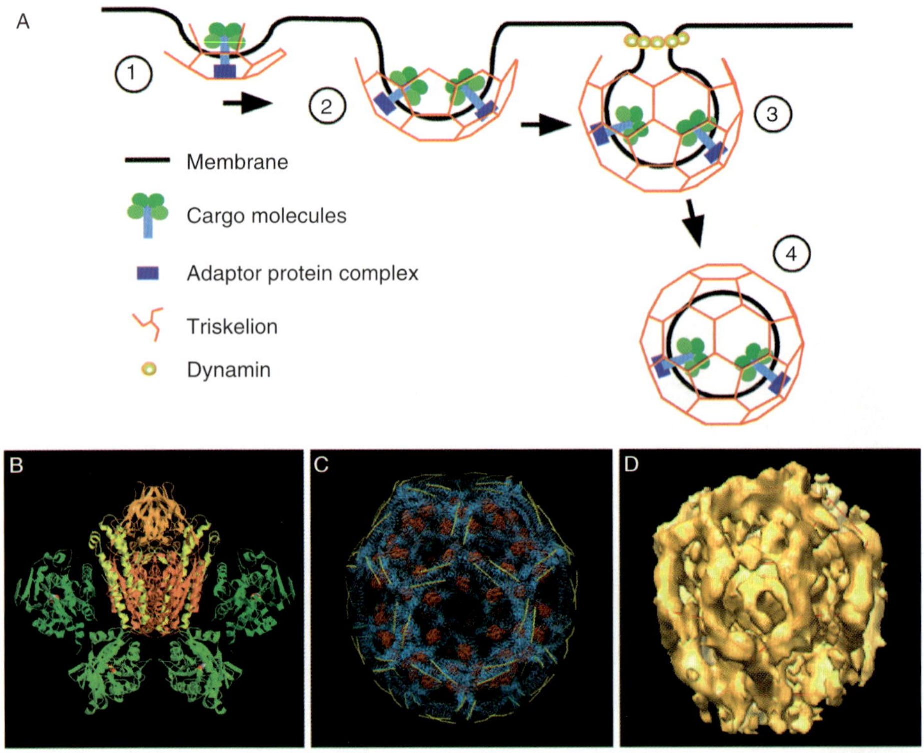

Chapter 18, Fig. 12

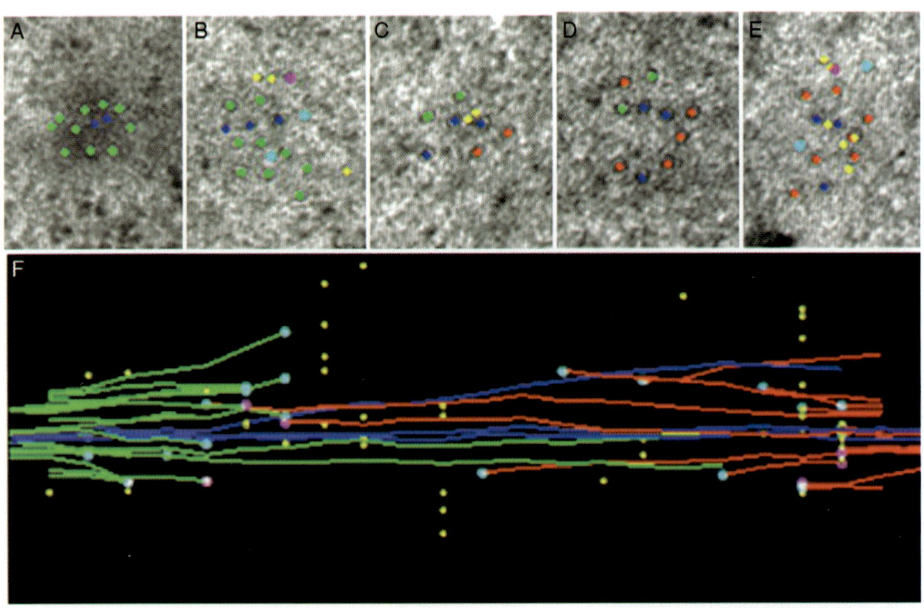

Chapter 19, Fig. 3

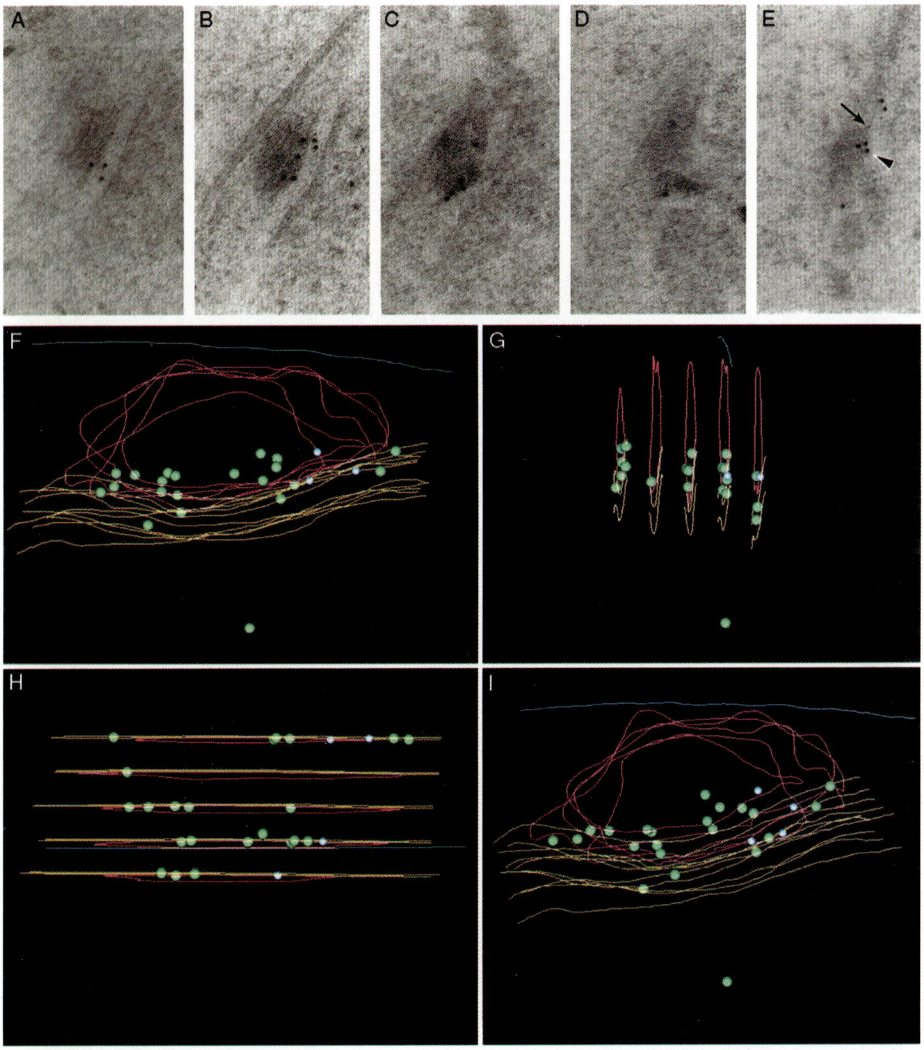

Chapter 19, Fig. 4

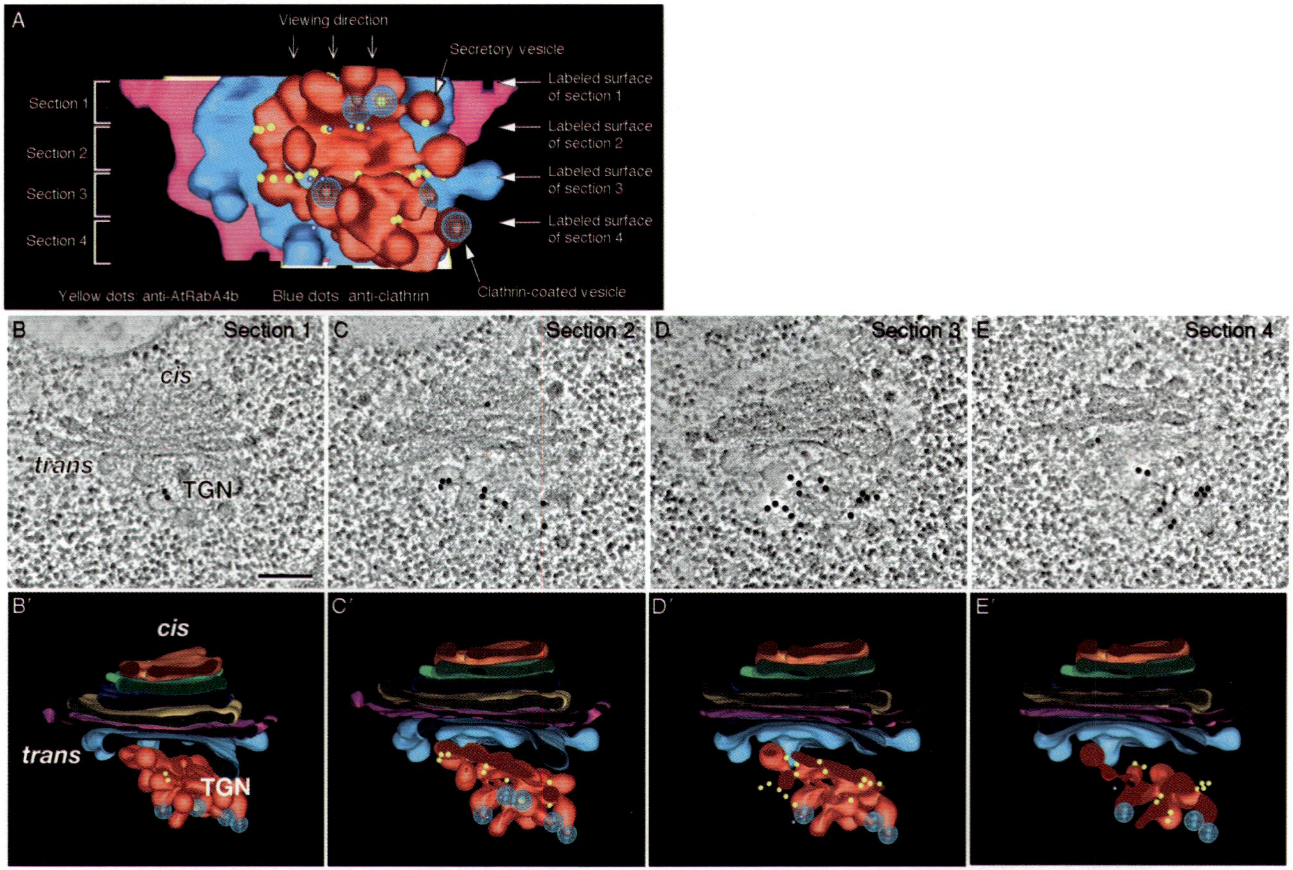

Chapter 19, Fig. 5

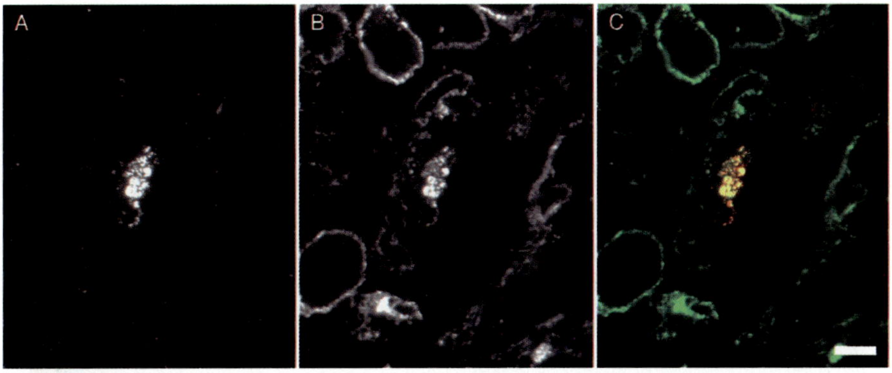

Chapter 20, Fig. 4

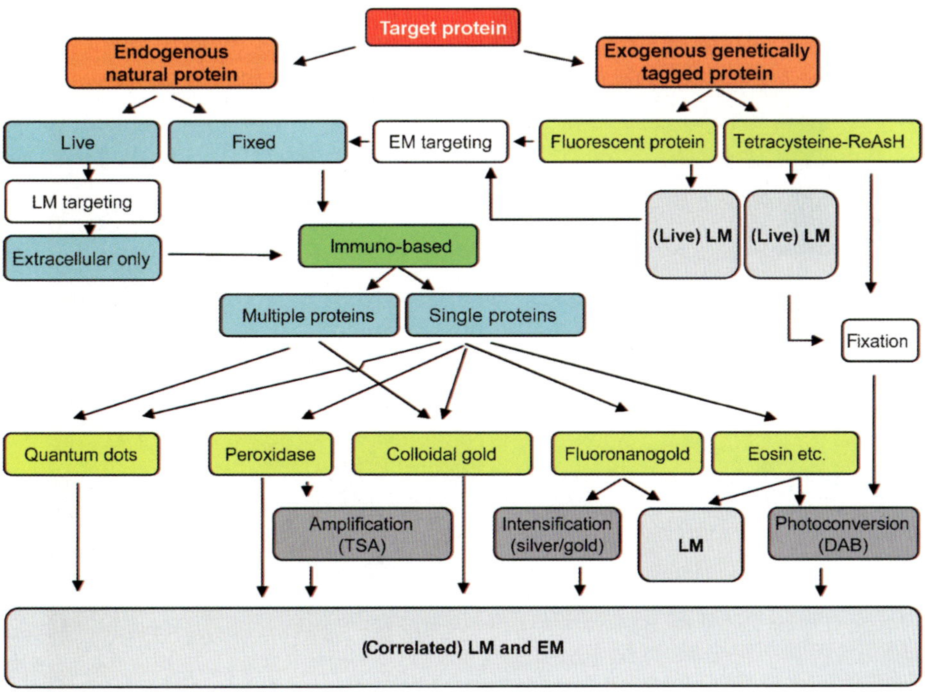

Chapter 23, Fig. 1

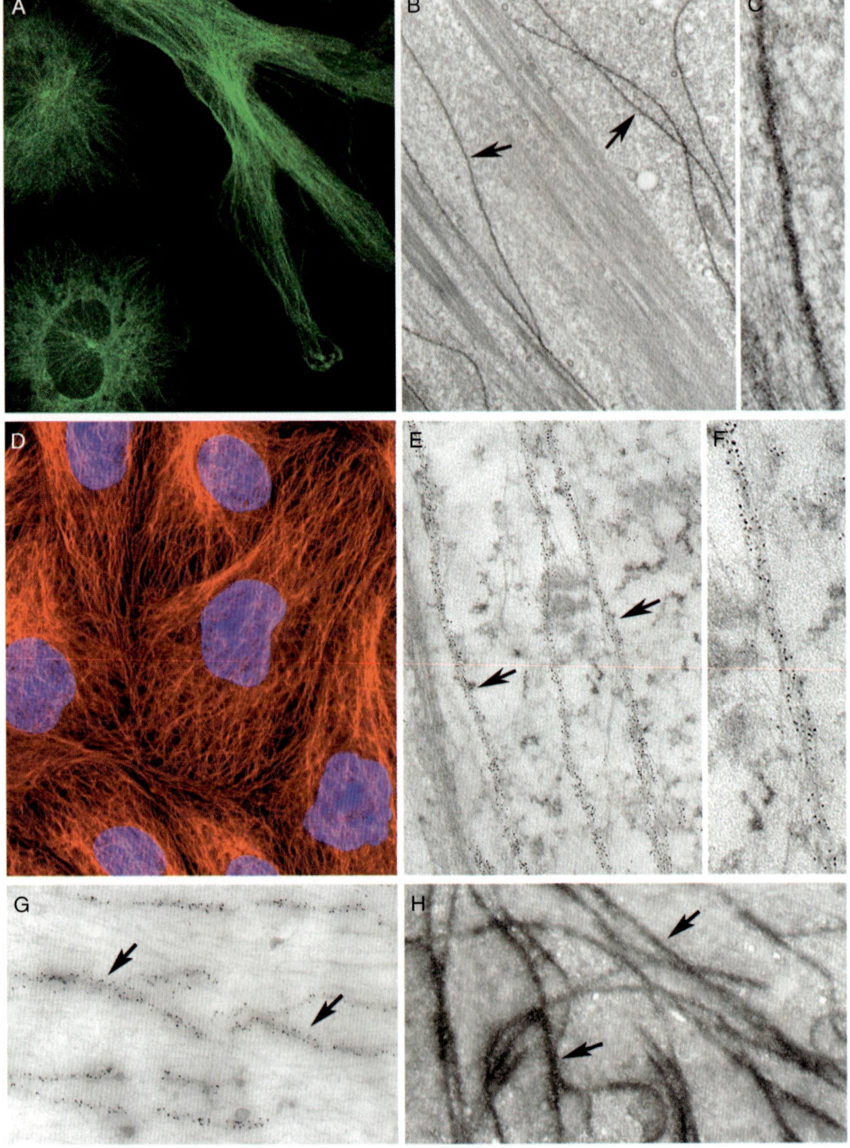

Chapter 23, Fig. 4

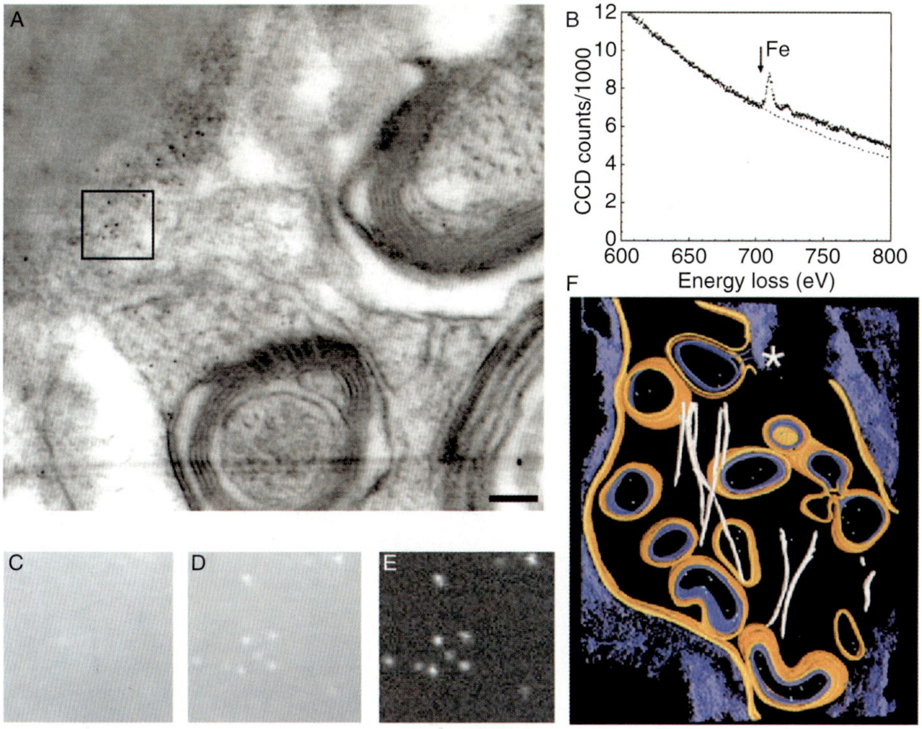

Chapter 24, Fig. 7

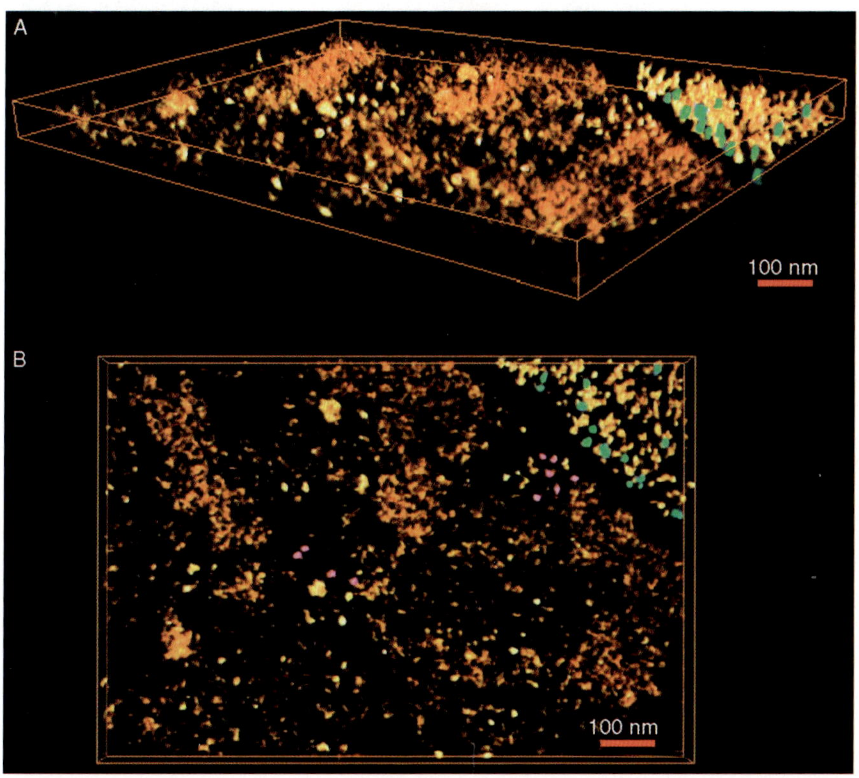

Chapter 24, Fig. 8

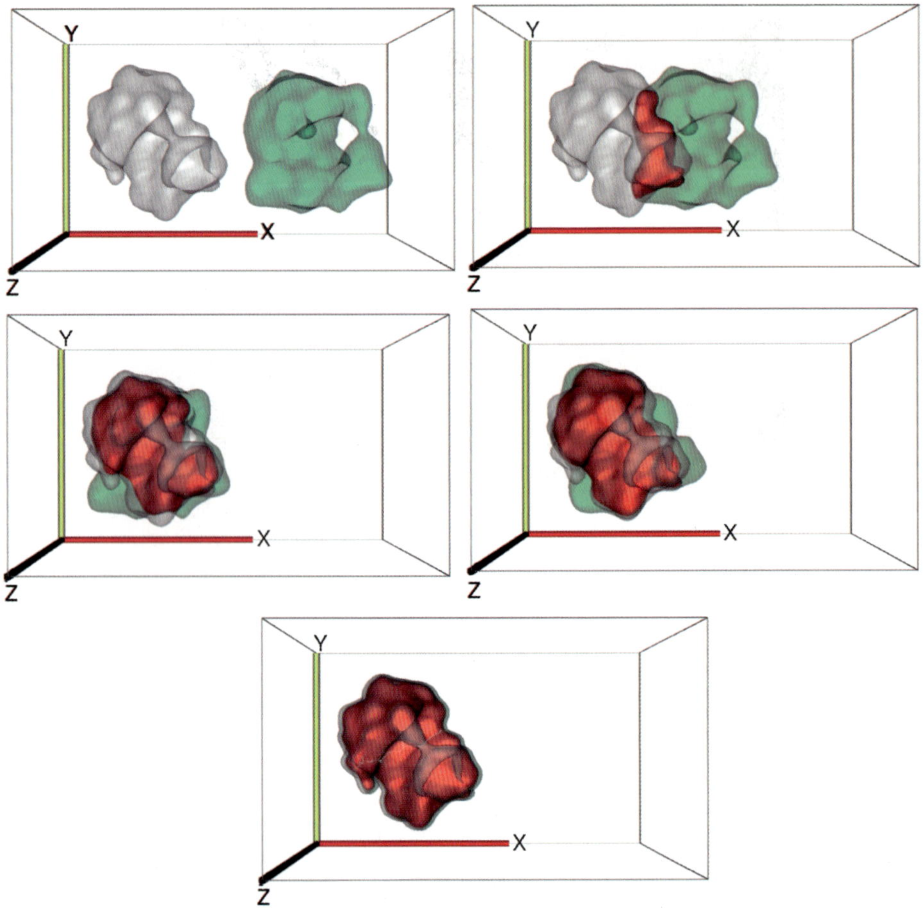

Chapter 25, Fig. 1

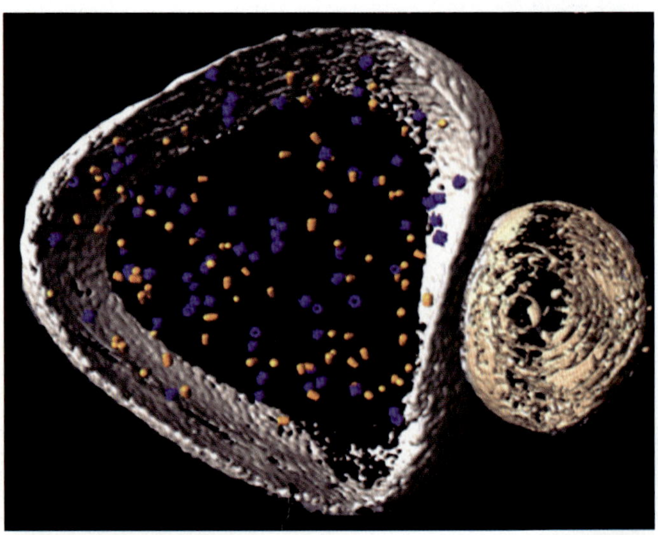

Chapter 25, Fig. 4

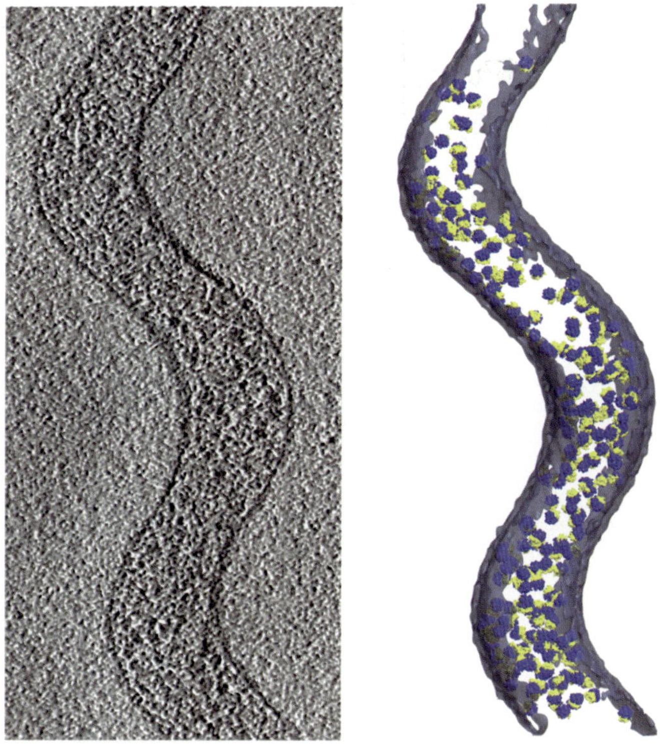

Chapter 25, Fig. 5

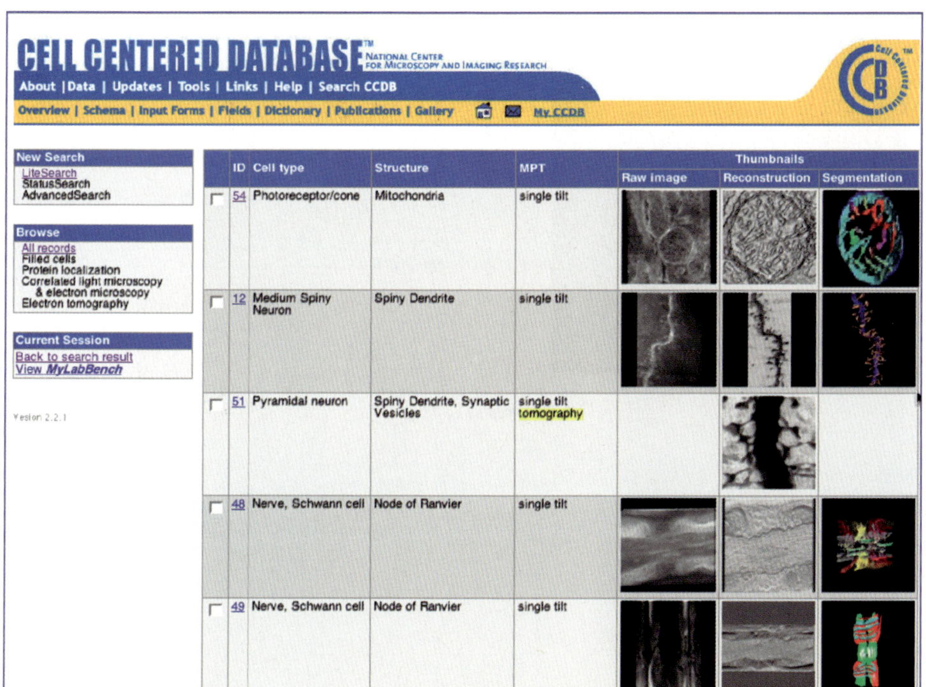

Chapter 31, Fig. 4

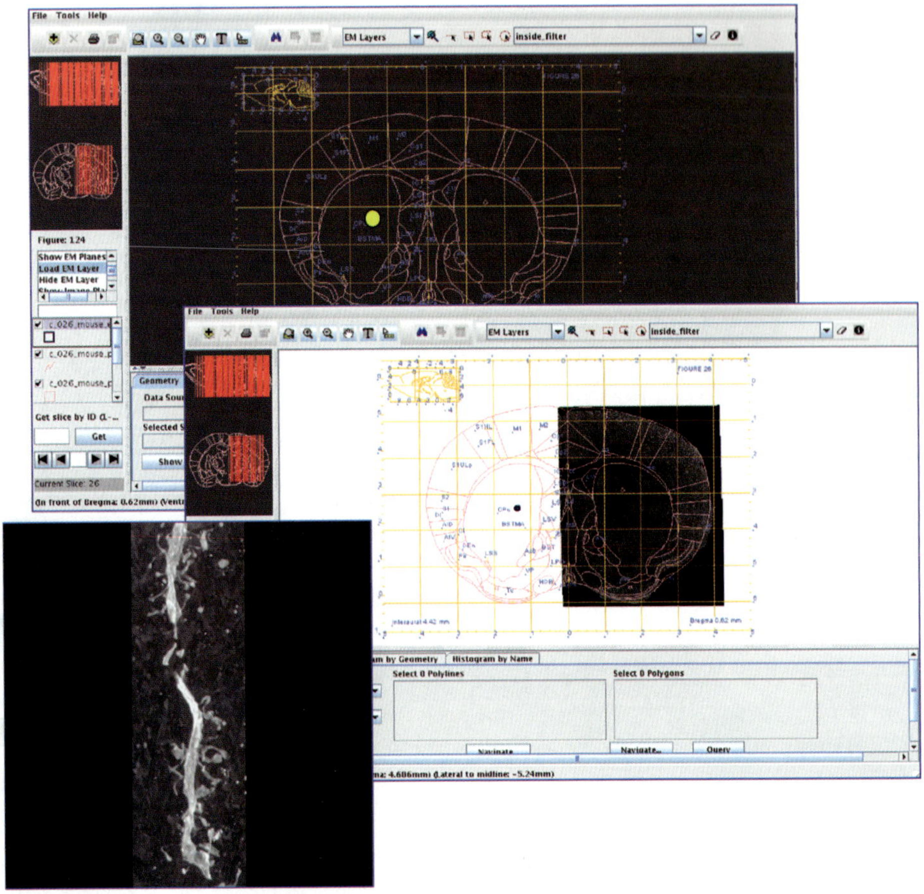

Chapter 31, Fig. 7